Chemistry of the Natural Atmosphere

Second Edition

This is Volume 71 in the
INTERNATIONAL GEOPHYSICS SERIES
A series of monographs and textbooks
Edited by RENATA DMOWSKA, JAMES R. HOLTON, AND
H. THOMAS ROSSBY

A complete list of the books in this series appears at the end of this volume.

Chemistry
of the
Natural
Atmosphere
Second Edition

Max-Planck-Institut für Chemie
Mainz, Germany

ACADEMIC PRESS
A Harcourt Science and Technology Company

San Diego San Francisco New York Boston London Sydney Tokyo

Cover: Images © 1999 PhotoDisc.

This book is printed on acid-free paper. ∞

Academic Press
A Harcourt Science and Technology Company
525 B Street, Suite 1900, San Diego, California 92101-4495, U.S.A.
http://www.apnet.com

Academic Press
24-28 Oval Road, London NW1 7DX, UK
http://www.hbuk.co.uk/ap/

Library of Congress Catalog Card Number: 99-64620

International Standard Book Number: 0-12-735632-0

PRINTED IN THE UNITED STATES OF AMERICA
99 00 01 02 03 04 MM 9 8 7 6 5 4 3 2 1

To the memory of
Christian E. Junge
(1912–1996)
a pioneer in the exploration of atmospheric trace substances

CONTENTS

7 The Atmospheric Aerosol

8 Chemistry of Clouds and Precipitation

11 Geochemistry of Carbon Dioxide

12 The Evolution of the Atmosphere

PREFACE TO THE SECOND EDITION

Our current knowledge and understanding of atmospheric chemistry are largely founded on scientific advances made in the 1970s. Work done during the past decade has consolidated many of the earlier results. In particular, reaction rate data and mechanisms have been refined, but new insight has also been obtained in several other areas. Examples are polar stratospheric chemistry and the chemistry in clouds. The preoccupation of atmospheric scientists in recent years with the problem of climate change and the role of radiatively active traces gases, such as methane and nitrous oxide, in the radiation balance of our planet has led to a reexamination and better description of their budgets. In preparing the second edition of this book I have tried to incorporate the most obvious advances that were made during the past decade. I have also noted, however, that comparatively little progress has been made in other areas that have received less attention. For example, our knowledge of the organic fraction of the atmospheric aerosol is still largely that of 10 years ago.

Although the text has been entirely rewritten, many sections have been left essentially unchanged, either because concepts based on the earlier data are still valid or because new data are scant. Some sections have been substantially revised to include new data, especially where they have changed the perception, even though final conclusions may still be out of reach. In such cases the treatment assumes the character of an abridged literature review. The ever-increasing rate of research publications makes it impossible, more so than ever before, to include a comprehensive list of literature citations. The reader is urged to consult the research papers cited to obtain additional information.

In this edition I have made an effort to use SI (Système International) units more consistently by following the recommendations of the IUPAC

Atmospheric Chemistry Commission, which appeared in *Pure and Applied Chemistry* (Vol. 67, pp. 1377–1406 (1995)). One of the recommendations was to replace the notation ppmv, ppbv, etc., for gas-phase mixing ratios with the unit mol mol^{-1} in combination with the appropriate SI prefix for submultiples; that is, ppmv should be replaced by μmol mol^{-1}, ppbv by nmol mol^{-1}, and so forth. Another suggestion was to retain molecule cm^{-3} as a unit of concentration despite the violation of the SI rule that a qualifying name should not be part of a unit, and the number of an entity such as a molecule has no dimension. (Note that in this unit *molecule* appears in the singular, although a great many molecules are present in 1 cm^3 of gas.)

As before, I have received much support from colleagues of the local science community, and I am grateful for their advice. I am also greatly indebted to I. Bambach and G. Feyerherd, who have prepared new illustrations and brought earlier ones up to date. Finally, I owe my thanks to the editor of this edition at Academic Press, David Packer, for his patience in bearing with me when, because of unforeseen circumstances, the preparation of the manuscript was delayed.

Peter Warneck

PREFACE TO THE
FIRST EDITION

Atmospheric chemistry deals with chemical compounds in the atmosphere, their distribution, origin, chemical transformation into other compounds, and finally, their removal from the atmospheric domain. These substances may occur as gases, liquids, or solids. The composition of the atmosphere is dominated by the gases nitrogen and oxygen in proportions that have been found to be invariable in time and space at altitudes up to 100 km. All other compounds are minor ones, with many of them occurring only in traces. Atmospheric chemistry thus deals primarily with trace substances.

As an interdisciplinary field of science, atmospheric chemistry has its main roots in meteorology and chemistry, with additional ties to microbiology, plant physiology, oceanography, and geology. The full range of the subject was last treated by C. E. Junge in his 1963 monograph, "Air Chemistry and Radioactivity." The extraordinarily rapid development of the field in the past two decades has added much new knowledge and insight into atmospheric processes, so that an updated account is now called for. To some extent, the new knowledge has already been incorporated into the recent secondary literature. Most of these accounts, however, specifically address the problems of local air pollution, whereas the natural atmosphere has received only a fragmentary treatment, even though it provides the yardstick for any assessment of air pollution levels. The recognition that humanity has started to perturb the atmosphere on a global scale is now shifting attention away from local toward global conditions, and this viewpoint deserves a more comprehensive treatment.

Atmospheric chemistry is now being taught in specialty courses at departments of chemistry and meteorology of many universities. The purpose of this book is to provide a reference source to graduate students and other interested persons with some background in the physical sciences. In preparing the text, therefore, I have pursued two aims: one is to assemble and review observational data on which our knowledge of atmospheric processes is founded; the second aim is to present concepts for the interpretation of the data in a manner suitable for classroom use. The major difficulty that I encountered in this ambitious approach was the condensation of an immense volume of material into a single book. As a consequence, I have had to compromise on many interesting details. Observational data and the conclusions drawn from them receive much emphasis, but measurement techniques cannot be discussed in detail. Likewise, in dealing with theoretical concepts I have kept the mathematics to a minimum. The reader is urged to work out the calculations and, if necessary, to consult other texts to which reference is made. As in any active field of research, atmospheric chemistry abounds with speculations. Repeatedly I have had to resist the temptation to discuss speculative ideas in favor of simply stating the inadequacy of our knowledge.

The first two chapters present background information on the physical behavior of the atmosphere and on photochemical reactions for the benefit of chemists and meteorologists, respectively. Chapter 3, which deals with observations and chemistry of the stratosphere, follows naturally from the discussion in Chapter 2 of the absorption of solar ultraviolet radiation in that atmospheric region. Chapter 4 develops basic concepts for treating tropospheric chemistry on a global scale. Methane, carbon monoxide, and hydrogen are then discussed. Subsequent chapters consider ozone, hydrocarbons, and halocarbons in the troposphere. Chapters 7 and 8 are devoted to the formation, chemistry, and removal of aerosol particles and to the interaction of trace substances with clouds and wet precipitation. These processes are essential for an understanding of the fate of nitrogen and sulfur compounds in the atmosphere, which are treated in Chapters 9 and 10. The last two chapters show the intimate connection of the atmosphere to other geochemical reservoirs. Chapter 11 introduces the underlying concepts in the case of carbon dioxide; Chapter 12 discusses the geochemical origin of the atmosphere and its major constituents.

A major problem confronting me, as it does many other authors, was the proper choice of units. Atmospheric scientists have not yet agreed on a standard system. Values often range over many orders of magnitude, and SI units are not always practical. I have used the SI system as far as possible, but have found it necessary to depart from it in several cases. One is the use of molecule per cubic centimeter as a measure of number density, since rate coefficients are given in these units. I have also retained moles per liter instead of moles per cubic decimeter, and mbar instead of hPa for simplicity and because of widespread usage, although the purist may disapprove of this.

Literature citations, although extensive, are by no means complete. A comprehensive coverage of the literature was neither possible nor intended. In keeping with the aim of reviewing established knowledge, the references are to document statements made in the text and to provide sources of observational data and other quantitative information. On the whole, I have considered the literature up to 1984, although more recent publications were included in some sections.

Thanks are due to many colleagues of the local science community for advice and information. S. Dötsch and C. Wurzinger compiled the list of references. I. Bambach, G. Feyerherd, G. Huster, and P. Lehmann patiently prepared the illustrations. I am grateful to all of them for essential help in bringing this volume to completion.

Peter Warneck

CHAPTER 1

Bulk Composition, Physical Structure, and Dynamics of the Atmosphere

1.1. OBSERVATIONAL DATA AND AVERAGES

Our knowledge about atmospheric phenomena derives largely from observations. In dealing with atmospheric data, one must make allowances for the mobility of the atmosphere and a tendency of all measurable quantities to undergo sizable fluctuations. The variability of meteorological observables such as temperature, wind speed, or wind direction is common knowledge from daily experience. Concentrations of atmospheric trace constituents are often found to exhibit similar fluctuations. These variations arise in part from temporal changes in the production mechanisms that we call sources and the removal processes that we call sinks. The irregularities of air motions responsible for the spreading of trace substances within the atmosphere impose additional random fluctuations on local concentrations. It is difficult to evaluate short-term, random variations except by statistical methods, so that for most purposes we are forced to work with mean values obtained by averaging over a suitable time interval. This situation is different from that in the laboratory, where it usually is possible to keep parameters influencing experimental results sufficiently well under control. In addition to random fluctuations, atmospheric data often reveal periodic variations, usually in the form of diurnal or seasonal cycles. The proper choice of averaging period obviously becomes important for bringing out suspected periodicities, and this point deserves some attention.

To illustrate the variability of observational data and the applied averaging procedures, consider the local abundance of carbon dioxide at Mauna Loa, Hawaii, as reported by Pales and Keeling (1965). The measurements were performed by infrared analysis with a response time of about 10 min. Figure 1.1 shows from left to right (a) the sequence of hourly means for the day of July 29, 1961; (b) averages of hourly means obtained by superimposition of all days of measurement during the month of July 1961; and (c) monthly means for the year 1961. The first frame indicates the variability of the CO_2 mixing ratio during a single day. The second frame shows that the afternoon dip observed on July 29 is not an isolated event but occurs regularly, although generally with smaller amplitude. In fact, the afternoon dip is a frequent year-round feature. It is interpreted as an uptake of CO_2 by island vegetation on the lower slopes of Mauna Loa, made manifest by the local circulation pattern (upslope winds during the day, subsiding air at night). The assimilation of CO_2 by plants in daylight is a well-known process, as is its release at night due to respiration. The third frame in Figure 1.1

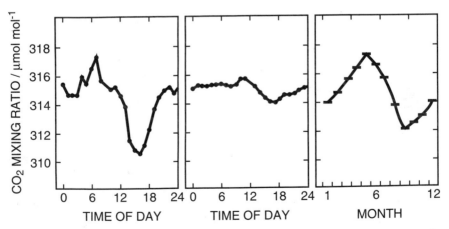

FIGURE 1.1   Volume mixing ratio of CO_2 at Mauna Loa, Hawaii (Pales and Keeling, 1965). Left: Hourly means for July 29, 1961. Center: Averages for each hour of the day for the days of July, 1961. Right: Monthly means for 1961.

demonstrates the existence of a seasonal variation of CO_2, with a maximum in May and a minimum in September. In contrast to the diurnal variation, which is a local feature, the annual cycle has global significance. The decline of CO_2 in the Northern Hemisphere during the summer months is believed to result from the incorporation of CO_2 by plants during their growth period, whereas the rise of CO_2 in winter and early spring must then be due to the decay of leaf litter and other dead plant material. Carbon dioxide is one of the few atmospheric trace gases that have been measured for many decades. Figure 1.2 shows the record of monthly means during the period 1958–1984. The seasonal oscillation is seen to occur with similar amplitude every year, but in addition there is a steady increase of the mixing ratio, averaging about 0.2% per year. This long-term variation arises largely from the combustion of fossil fuels by humans and the accumulation of CO_2 in the atmosphere–ocean system. During the last decade, the CO_2 content of the atmosphere has risen further by about 0.4% per year. In 1994 the average mixing ratio was 358 μmol mol^{-1} (Houghton et al., 1995). A more detailed discussion of CO_2 is given in Chapter 11.

Total ozone, that is, the column density of ozone in the atmosphere, provides another example of the variability inherent in observational data. This quantity is determined by optical absorption, with the sun or the moon as the background light source. Figure 1.3 presents results of measurements in Arosa, Switzerland. Shown from left to right are (a) the variation of total ozone on a single day; (b) daily means for the month of February 1973; (c) February averages for the period 1958–1971; (d) monthly means for 1962; and (e) the annual variation averaged over a greater number of years. The day-to-day variability of total ozone is impressive. It is caused by the influence of meteorological changes, mainly the alternation of high and low pressure systems. And yet, by using monthly means, it is possible to show the existence of an annual cycle, with a maximum in early spring and a minimum in the autumn. The origin of the cycle must be sought in the varying strength of the meridional circulation. A detailed discussion of this and other aspects of total ozone is given in Chapter 3. It should be noted, however, that the annual oscillation has the same magnitude as the fluctuation of monthly means (compare Figure 1.3c), and that the annual cycle itself undergoes considerable variations. Several years of observations are required to obtain a useful average of the atmosphere ozone level.

The foregoing examples of atmospheric data and their behavior suggest a hierarchy of averages, which are summarized in Table 1.1. The lengths of the averaging periods are chosen to be hour, day, month, and year, and these units are essentially predetermined by the calendar. Each higher unit increases the averaging period by about $1\frac{1}{2}$ orders of magnitude. Hourly means require a sufficiently fast instrument response time. It is clear that an

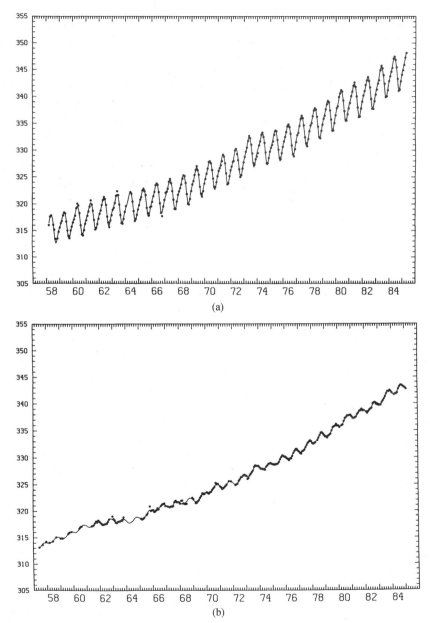

FIGURE 1.2   Secular rise of CO_2 mixing ratios in the atmosphere. Dots represent monthly mean values in μmol mol^{-1} as observed with a continuously recording nondispersive infrared analyzer. The smooth curve represents a fit of the data to a fourth harmonic annual cycle, which increases linearly with time, and a spline fit of the interannual component of variation. Top: Mauna Loa Observatory, Hawaii; bottom: South Pole. (From C. D. Keeling *et al.* (1982) and unpublished data courtesy C. D. Keeling.)

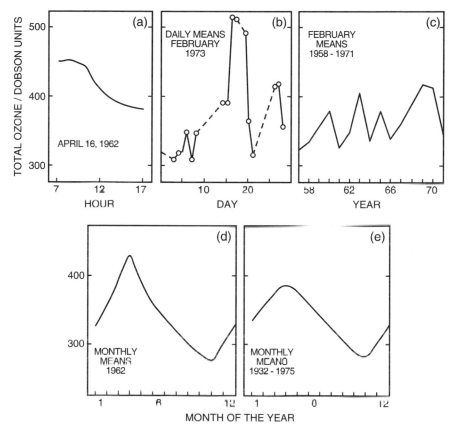

FIGURE 1.3 Total ozone at Arosa, Switzerland (in Dobson units; 1 D.U. corresponds approximately to 446.2 μmol m^{-2}). (a) Variation during the course of 1 day. (b) Day-to-day fluctuation in February 1973. (c) Variation of February means 1957–1971. (d) Monthly means for 1962. (e) Annual variation averaged over many years. (Adapted from Dütsch (1980).)

analytical technique necessitating the collection of material for 24 h cannot reveal diurnal variations, but it would still disclose seasonal changes. The fourth column of Table 1.1 indicates averages derived from the superimposition of several periods. This procedure, as we have seen, is useful for enhancing periodicities or trends. Diurnal or annual cycles ultimately result from periodic changes associated with the earth's rotation on its axis and its revolution around the sun. Also known are longer periodicities, such as the 11-year sunspot cycle.

Long-term measurements to reveal trends of trace components such as CO_2 or ozone cover a few decades, at most. For much longer periods, the supposition of a quasi-stationary atmosphere makes it likely that the trend

TABLE 1.1   Hierarchy of Averages[a]

Time scale	Time average $\bar{y}_k$	k	Superimposed epoch average, $\tilde{y}_k$	Cycle or trend
Minutes	Not considered	—	—	Short-term fluctuations
Hours	Hourly mean	24	Hour by hour for many days $(n \approx 10)$	Diurnal cycle
Day	Daily mean	30	Day by day for several months $(n \approx 3)$	Seasonal variation
Month	Monthly mean	12	Month by month for several years $(n \geq 3)$	Annual cycle
Year	Annual mean	—	—	Long-term trend

[a] Time averages are defined as $\bar{y}_k = \int_\tau y(t)\, dt/\tau$, where τ is a fixed length of time and k designates an individual mean value within a sequence of averages over the next longer period. Associated with each period is a superimposed epoch average $\tilde{y}_k = \sum_{i=1}^{i=n} \bar{y}_{ki}/n$, where n is the number of periods available for averaging. The variability in each case can be described by the standard deviation from the mean.

will not continue indefinitely, but that eventually a new steady state is reached or the trend reverses. In recent years it has become possible to obtain information on the behavior of trace substances in past epochs by the exploration of deep ice cores from the glaciers in Greenland and Antarctica. They provide a record of atmospheric conditions dating back about 100,000 years. An example is the spectacular record of atmospheric CO_2 and its covariation with temperature during this period (Barnola *et al.*, 1987; Jouzel *et al.*, 1987).

Finally, a few words must be added about spatial averages. They are important in all global considerations of atmospheric trace constituents. A continuous three-dimensional representation of data is not practical, and the data are not available in sufficient detail. Table 1.2 lists common spatial averages based on longitude, latitude, and height as space coordinates. Most important, inasmuch as they are readily justifiable because of the comparatively rapid circumpolar atmospheric circulation, are zonal means, which are defined as an average along latitude circles. Thereby one obtains a two-dimensional representation in the form of a meridional cross section. This treatment is always desirable for data that depend on both latitude and altitude. Often, however, the data base is not broad enough or the variations with either latitude or altitude are small. In this case a one-dimensional representation of the data is more convenient. Table 1.2 includes the two resulting cases. If the averaging process is carried still further, one obtains global means, which are important for budget considerations. Since the northern and southern hemispheres are to some extent decoupled, it is

TABLE 1.2   Spatial Averages Commonly Applied in Global Considerations of Data

Type of average	Averaging coordinate	Resulting data representation	Applicability
Zonal means	Longitude	Two-dimensional, meridional cross section	General
Zonal + meridional	Longitude + latitude	One-dimensional, versus height	Average altitude profile
Zonal + height	Longitude + height	One-dimensional, versus latitude	Selected altitude domains, usually lower atmosphere
Hemispheric mean	All three coordinates	Average value for each hemisphere	Mass balance
Global average	All three coordinates	Both hemispheres combined	Mass balance

sometimes necessary to consider averages for each hemisphere separately, and Table 1.2 makes allowances for this possibility.

1.2. TEMPERATURE STRUCTURE AND ATMOSPHERIC REGIONS

Meteorological sounding balloons (radiosondes) are used routinely to measure from networks of stations on all continents the vertical distribution of pressure, temperature, wind velocity, and, to some extent, relative humidity, as a function of altitude up to 30 km. Atmospheric conditions at greater heights have been explored by rocket sondes and by infrared sounding techniques from satellites.

Figure 1.4 shows zonal mean temperatures thus obtained as a function of latitude and height. In this form, the temperature distribution is introduced here mainly for illustration, and details are not discussed. Alternatively, one may set up mean annual temperature–altitude profiles at several latitudes and derive, by interpolation, an average profile valid for middle latitudes. The result for the Northern Hemisphere is shown in Figure 1.5. The solid curve represents a so-called standard profile composed of straight-line sections, which allow the calculation of standard pressures and densities. The need for a standard or model atmosphere arose originally with the advent of commercial air flight, but atmospheric chemistry also benefits from having available a model atmosphere providing average values of the principal physical parameters as a function of altitude.

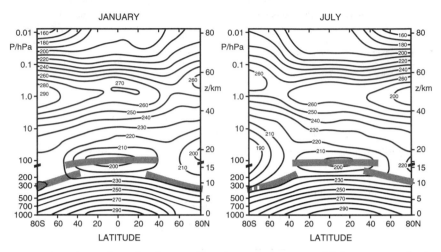

FIGURE 1.4   Atmospheric distribution of zonal mean temperatures (K) for January and July. (Data from Newell *et al.* (1972) for the lower 20 km and from Labitzke and Barnett (1979) for the upper atmosphere are combined.) The tropopause levels are indicated by shaded bars.

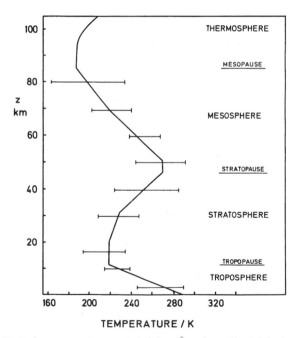

FIGURE 1.5   Vertical temperature structure of the atmosphere. The solid curve represents the U.S. Standard Atmosphere, the horizontal bars indicate the range of monthly means observed between the equator and the North Pole. (Adapted from U.S. Standard Atmosphere (1976).)

The right-hand side of Figure 1.5 shows the nomenclature adopted by international convention to distinguish different altitude regimes. The division is based on the prevailing sign of the temperature gradient in each region. The troposphere is the region nearest to the earth's surface. Here, the temperature decreases with height in a fairly linear fashion up to the tropopause. Occasional temperature inversions are restricted to narrow layers. The negative temperature gradient (or lapse rate) averages 6.5 K km^{-1}. A well-defined tropopause is marked by an abrupt and sustained change in the lapse rate to low values. Examples of radiosonde data illustrating this point are contained in Figure 3.3. The altitude level of the tropopause depends considerably on latitude, as Figure 1.4 shows, and varies with the seasons and with meteorological conditions. The variations are important for the exchange of air between the troposphere and the stratosphere, which is the next higher atmospheric region. Here, the temperature rises again, reaching a peak at the stratopause. The stratosphere is followed by the mesosphere, where the temperature declines to an absolute minimum of 185 K, on average. At still greater heights lies the thermosphere, which has no well-defined upper boundary. In the thermosphere the temperature rises to values on the order of a few thousand degrees at an altitude of 250 km but varies widely with solar activity. The outermost region of the atmosphere before it merges with the interplanetary medium is called the exosphere. Here, the density of air is so low and the mean free path of gas molecules between two collisions is so great that the molecules describe ballistic orbits in the earth's gravitational field. Altogether, the physical and chemical conditions of the thermosphere differ radically from those existing in the lower atmosphere. The differences are due, on the one hand, to the exposure of the outer atmosphere to high energy solar radiation and, on the other hand, to the low density, which promotes the separation of lighter and heavier components of air by molecular diffusion. A treatment of the outer atmosphere is outside the scope of this book, but one physical phenomenon connected with it, namely the escape of hydrogen and helium into space, will be considered in Chapter 12.

Temperature is a measure of the internal energy of air, and the temperature at a specific altitude generally reflects a quasi-stationary energy balance (input versus loss). The energy spectrum of the sun has its maximum in the visible region of the electromagnetic spectrum. Only a small portion of the incoming solar radiation is absorbed directly into the atmosphere. Most of it reaches the Earth's surface, is largely absorbed there, and thus heats the atmosphere from below. The global energy budget of the atmosphere is balanced by outgoing infrared (thermal) radiation at wavelengths greater than 3 μm. Two processes are important in the vertical transfer of heat. One is absorption and reemission of thermal radiation, and the other is local convection. The infrared-active atmospheric components are H_2O, CO_2, O_3, and aerosol particles. The main constituents of air, nitrogen and oxygen, cannot emit dipole radiation and thus do not participate in the thermal radiation budget. Local convection arises from the heating of the Earth's surface and the transfer of heat to adjacent layers of air. As a heated air parcel rises, it expands because of the pressure decrease with height and undergoes adiabatic cooling. A subsiding air parcel contracts so that its temperature is correspondingly raised. This mechanism leads to a rapid vertical exchange of air in the troposphere and causes the temperature to decrease with altitude.

Convection ceases to be important at the level of the tropopause, and the temperature in the stratosphere and mesosphere is determined strictly by radiation balance. At altitudes above 10 km, the absorption of solar ultraviolet radiation becomes increasingly important. The temperature peak at the stratopause has its origin in the absorption of near-ultraviolet radiation by stratospheric ozone. In fact, the very existence of the ozone layer is a consequence of the ultraviolet irradiation of the atmosphere. The enormous temperature increase in the thermosphere is due to the absorption of solar radiation in the extreme ultraviolet coupled with the tenuity of the atmosphere, which prevents an effective removal of heat by thermal radiation. Instead, the heat must be carried downward by conduction toward denser layers of the atmosphere, where H_2O and CO_2 are sufficiently abundant to permit the excess energy to be radiated into space.

1.3. PRESSURE, DENSITY, AND MIXING RATIO

At the pressures and temperatures prevailing in the atmosphere, air and its gaseous constituents have the properties of an ideal gas to a very good approximation. Even water vapor can be treated in this way. Absolute temperature T, total pressure p, and air density ρ thus are related by the ideal gas law,

$$p = \nu R_g T/V = \rho R_g T/M_{air} \qquad (1.1)$$

where $R_g = 8.31$ m^3 Pa mol^{-1} K^{-1} is the gas constant, $\nu/V = \Sigma \nu_i/V$ (mol m^{-3}) denotes the total chemical amount per volume V of all gaseous constituents i, and

$$M_{air} = \sum (\nu_i/\nu) M_i = \sum x_i M_i \qquad (1.2)$$

is the molar mass of air derived from the molar masses M_i of the individual constituents and their mixing ratios expressed as mole fractions, that is, the amount of substance per amount of air

$$x_i = \nu_i/\nu = p/p_i \qquad (1.3)$$

Since pressure and density are additive properties of ideal gases, one finds that Equation (1.1) applies also to each gaseous constituent separately:

$$p_i = (\nu_i/V) R_g T = \rho_i R_g T/M_i \qquad (1.4)$$

The quantities ν_i/V and ρ_i are called concentrations. Although they may be used to specify the abundance of a constituent in a parcel of air, it is customary to use mixing ratios for this purpose, because these are independent of pressure and temperature. For a constituent that mixes well in the atmosphere, the mixing ratio will be constant. Note that for ideal gases the amount fraction (ν_i/ν) is identical to the volume fraction (V_i/V), which has been widely used in atmospheric chemistry. The amount fraction is preferable because it does not require the assumption that the gases have ideal properties, and it may also be applied to species in the condensed phase if present in the same volume. Table 1.3 summarizes notations for mixing ratios currently in use. Because of inherent ambiguities (for example, whether one is dealing with volume or mass mixing ratios), it is preferable to express mole fractions by SI units such as μmol mol^{-1} for ppm. Table 1.3 includes the corresponding units.

The mass mixing ratio of a trace substance, that is, its mass fraction, is approximately related to the amount fraction x_i by

$$w_i = (M_i/M_{air}) x_i \qquad (1.5)$$

For trace gases, the mass mixing ratio is rarely used because it enjoys no particular advantage over the amount fraction, but it is appropriate when the chemical composition of the constituent, and thus the molar mass, is not

TABLE 1.3   Units of Mixing Ratios

Percent (%)	1 in 100	1×10^{-2}	cmol/mol[a]
Per mille (‰)	1 in 1000	1×10^{-3}	mmol/mol
Parts per million (ppm)	1 in 10^6	1×10^{-6}	μmol/mol
Parts per hundred million (pphm)	1 in 10^8	1×10^{-8}	—
Parts per billion (ppb)	1 in 10^9	1×10^{-9}	nmol/mol
Parts per trillion (ppt)	1 in 10^{12}	1×10^{-12}	pmol/mol

[a] SI units.

known or is not well defined, so that the amount fraction cannot be calculated. This is generally the case with aerosol particles. Accordingly, the mass fraction is used to specify the local abundance of suspended particulate matter. For comparison with the abundance of other constituents, it is then convenient to normalize the volume of air to a specific standard density, such as that existing at a standard temperature and pressure (STP). This leads to the relation

$$C_{i\,\text{STP}} = \rho_{\text{STP}}(M_i/M_{\text{air}})\,x_i \qquad (1.6)$$

If $T = 273.15$ K and $p = 1.01325 \times 10^5$ Pa are chosen as standard conditions, $\rho_{\text{STP}} = 1.293$ kg m^{-3} for dry air, but other conventions have also been used. We shall reserve this form of the mass mixing ratio to describe the abundance of particulate matter and aerosol-forming trace gases in the atmosphere.

Concentrations rather than mixing ratios determine the rate of chemical reactions. The traditional way of expressing concentration in chemistry is amount per volume, ν_i/V. However, atmospheric chemists have adopted for gaseous constituents the number concentration n_i, that is, the number of molecules of type i contained in a volume of air,

$$n_i = \nu_i N_A/V = p_i N_A/R_g T = p_i/k_B T \qquad (1.7)$$

Here, $N_A = 6.02 \times 10^{23}$ molecules mol^{-1} is Avogadro's number, and $k_B = R_g/N_A = 1.38 \times 10^{-23}$ J K^{-1} is Boltzmann's constant. The customary unit for n_i is molecule cm^{-3}. Although the two quantities are different, the terms "mixing ratio" and "concentration" are often indiscriminately used in the literature to describe local abundance, and this may give rise to confusion. Both expressions are justifiable as long as one talks loosely about the presence of a trace gas in surface air, but when units are involved the two quantities must be distinguished.

Table 1.4 gives amount fractions of the main constituents of dry air: nitrogen, oxygen, argon, and carbon dioxide. The average molar mass of air as calculated from Equation (1.2) is $M_{\text{air}} = 28.97 \approx 29$ g mol^{-1}. This value can be markedly lowered in the presence of water vapor, whose abundance in the atmosphere is highly variable. It is determined primarily by the saturation vapor pressure, which is a strong function of temperature. The highest H_2O mixing ratios occur in the surface air of the tropics. An H_2O mixing ratio of 5% lowers the molar mass of air to 28.4 g mol^{-1}. In the middle and upper atmosphere the H_2O mixing ratio is so low that the air is essentially dry. The sum of nitrogen, oxygen, and argon alone amounts to 99.96% of the total contributions from all constituents. The mixing ratios of these gases are fairly constant with time and show no variation in space up to

TABLE 1.4   Bulk Composition of Dry Air

Constituent	M (g mol^{-1})	(% amount fraction) (a)	(b)	Remarks
N_2	28.01	78.084 ± 0.004	78.094 ⎫	"permanent gases"
O_2	32.00	20.946 ± 0.002	20.936 ⎬	(small variability)
Ar	39.95	0.934 ± 0.001	0.934 ⎭	
CO_2	44.01	0.033 ± 0.001	0.036	variable

(a) From Glueckauf (1951); the CO_2 content was extrapolated from the data of Callendar (1940) and did not correspond to the actual global value in 1951.
(b) Recalculated according to the CO_2 content of 1992 and the oxygen consumption due to the combustion of fossil fuels, assuming 48% of CO_2 to have remained in the atmosphere.

an altitude of 100 km. Gas analytical procedures developed in the last century and modern instrumental techniques provide comparable accuracy. The measurements have primarily quantified the volume percentages of oxygen, argon, and CO_2 (Glueckauf, 1951). That of nitrogen was obtained by difference. Keeling and Shertz (1992) have recently shown that oxygen is declining somewhat relative to nitrogen because of consumption by the combustion of fossil fuels. Oxygen consumption and the rise of atmospheric CO_2 now are beginning to cause a slight redistribution among the principal components of air. Nitrogen, oxygen, and argon are often called "permanent gases" in the atmosphere. This should not be taken to imply that the present abundances have existed throughout the entire history of Earth's atmosphere. The evolution of these gases is discussed in Chapter 12.

According to the hydrostatic principle, the pressure decreases with altitude because of the decrease in column density of air overhead. The differential pressure decrease for an atmosphere in hydrostatic equilibrium is given by

$$dp = g\rho(-dz) = -(gM_{air}/R_g T)p\,dz \qquad (1.8)$$

where g is the acceleration due to gravity, and the expression on the right-hand side is obtained by virtue of Equation (1.1). Up to an altitude of 100 km, g decreases by less than 3%. Its variation with latitude is even smaller, so that it may be taken as constant. The standard value is $g = 9.807$ m s^{-2}. Equation (1.8) is integrated stepwise over altitude z, starting at ground level and using the linear sections of the standard temperature profile in Figure 1.5. In regions of constant temperature one obtains

$$p/p_0 = \rho/\rho_0 = \exp(-z/H) \qquad (1.9)$$

where the subscript 0 refers to the bottom of the atmospheric layer considered, and $H = R_g T / g M_{air} = 29.2\ T$ (m) is called the *scale height*. Its value ranges from 5.8 to 8.7 km for temperatures between 200 and 300 K. In regions with linearly increasing or decreasing temperature, the scale height varies with z. Integration of Equation (1.8) then yields

$$p/p_0 = (T/T_0)^{-\beta}$$

$$\rho/\rho_0 = (T/T_0)^{-(1+\beta)}$$

(1.10)

with

$$\beta = (dH/dz)^{-1} = (gM_{air}/R_g)/(dT/dz)$$

It is an interesting fact that with any prescribed vertical temperature distribution, the column density of air above a selected altitude level z is given by the product of air density and scale height at that level,

$$W = \rho_z H_z = p_z/g$$

(1.11)

The substitution $\rho_z H_z = p_z/g$ follows from Equation (1.8) and the definition of the scale height. Proof of Equation (1.11) is obtained by integrating Equations (1.9) and (1.10) for all layers from the level z up to infinite heights.

The application of Equation (1.11) makes it convenient to estimate the air masses in the atmospheric domains of principal interest here, namely the troposphere and the stratosphere. The total mass of the atmosphere is obtained as follows. The air mass over the oceans $(p_0/g)\ A_{sea}$, where the subscript 0 refers to sea level and $A_{sea} = 3.61 \times 10^{14}$ m^2 is the total ocean surface area. The average elevation of the continents was estimated by Kossina (1933) to be 874 m above sea level. From Equation (1.10) one estimates with $T_0 = 284$ K that the average column density of air over land is reduced to 0.91 of that over the ocean. The surface area of the continents is $A_{cont} = 1.49 \times 10^{14}$ m^2. The total mass of the atmosphere thus sums to

$$G = (p_0/g)(A_{sea} + 0.91\ A_{cont}) = 5.13 \times 10^{18}\ \text{kg}$$

(1.12)

The pressure at sea level varies somewhat with the seasons and with latitude, primarily because of fluctuations in the pressure of water vapor. In 1981 Trenberth comprehensively reviewed the data, including a revision of average height of the continents, and calculated the mass of the atmosphere to be 5.137×10^{18} kg, with an annual cycle of amplitude of 1×10^{15} kg. The result of recent calculations of Trenberth and Guillemot (1994), based on a 1985–1993 time series of surface pressures, was 5.1441×10^{18} kg for the total mass of the atmosphere, with a range of 1.93×10^{15} kg associated with changes in water vapor column density. The dry air mass obtained was 5.132×10^{18} kg, and the mean mass of water vapor was 1.25×10^{16} kg.

The air mass of the troposphere cannot be derived with the same confidence as the total air mass, because of the variation in tropopause height. The variation with latitude may be taken into account by adopting different average tropopause levels in the tropical and extratropical latitude belts. It is convenient to subdivide the tropospheric air space in each hemisphere along the 30° latitude circle, because equatorial and subpolar regions then cover equal surface areas. Temperatures at the average tropopause levels can be estimated from Figure 1.4, and pressures and densities are then obtained from Equation (1.10). The results are shown in Table 1.5. The residual air masses above the tropopause in the equatorial and subpolar latitude regions of both hemispheres combined are 3.12×10^{17} kg and 5.98×10^{17} kg, respectively. The sum must be deducted from the total mass of the atmosphere to give the air mass of the troposphere, 4.22×10^{18} kg. The air mass of the stratosphere can be calculated similarly from the difference in air masses above the tropopause and the stratospause. The results are summarized in Table 1.6. Only about 1% of the entire mass of the atmosphere resides at altitudes above 30 km, and less than 0.1% above the stratopause. The troposphere contains roughly 80% of the entire air mass and the lower stratosphere most of the remainder. The seasonal variation of the tropopause at middle and high latitudes has been estimated by Reiter (1975) and Danielsen (1975) to cause a variation of 10–20% in stratospheric air mass. The corresponding variation of the tropospheric air mass is 4%, which is almost negligible.

1.4. GLOBAL CIRCULATION AND TRANSPORT

An outstanding feature of atmospheric motions is that they are turbulent. Wind speed and wind direction fluctuate considerably, giving rise to small- and large-scale eddies. These are responsible for the intermingling of air

TABLE 1.5 Tropopause Temperatures, Pressures, Densities, and Number Densities

	T_0 (K)	T_{tr} (K)	p_{tr} (hPa)	ρ_{tr} (kg m^{-3})	ρ_{tr}/ρ_0	ρ_{tr}/ρ_{STP}	n (molecule cm^{-3})
Tropical	295	195	120	0.21	0.17	0.166	4.4×10^{18}
Extratropical	285	215	230	0.37	0.30	0.280	7.5×10^{18}

TABLE 1.6   Air Masses (kg) in Various Regions of the
Lower Atmosphere

Total atmosphere	5.13×10^{18}
Tropical troposphere	2.25×10^{18}
Extratropical troposphere	1.97×10^{18}
Total troposphere	4.22×10^{18}
Lower stratosphere (<30 km)	8.48×10^{17}
Upper stratosphere (30–50 km)	5.80×10^{16}
Total stratosphere	9.06×10^{17}
Remaining atmosphere	4×10^{15}

parcels and the spreading of trace substances by turbulent mixing. From the observed wind patterns it is possible to derive the mean wind field by appropriate averaging. The procedures were described, for example, by Newell *et al.* (1972). Thereby one obtains the mean zonal circulation around the globe and the mean meridional motion as a function of both latitude and height. Together, the mean air flow and turbulent mixing determine the time scale for the spreading of a tracer in the atmosphere.

1.4.1. ATMOSPHERIC MEAN MOTIONS

Figure 1.6 illustrates the mean zonal wind field for the Northern Hemisphere. The situation in the Southern Hemisphere mirrors that of the Northern Hemisphere, although not quantitatively. In mid-latitudes the dominant wind direction is from west to east, with maximum velocities in the vicinity of the subtropical jet stream near 30° latitude. The location of the polar jet stream associated with the polar front is more diffuse, and it does not show up in the averaged wind field. The westerlies encircle the globe in a wavelike pattern (Rossby waves). Perturbations are due to the buildup and decay of temporary as well as quasi-stationary cyclones and anticyclones. Toward the equator, in the trade wind region, the main wind direction is from the east to the southwest. The trade winds are caused by the uprising air motion in the Tropics, which gives rise to the meridional Hadley circulation (see Fig. 1.7). In summer, the region of the easterlies extends far into the stratosphere and overlays the domain of the westerlies at mid-latitudes. In winter, the Hadley cell is located farther to the south, and the easterlies are confined to the lower atmosphere. In the region of the westerlies the travel time of an air parcel encircling the globe is about 3 weeks. This gives an impression about the time scale involved in the transport of a tracer by the zonal circulation.

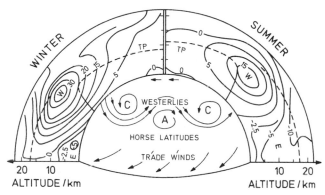

FIGURE 1.6 Mean zonal circulation in the Northern Hemisphere, 0–20 km. Distribution of wind velocities (in s^{-1}) from Labitzke (1980). W, mean winds from the west; E, mean winds from the east. Heavy lines indicate the approximate location of the polar front; broken lines indicate the tropopause. The maximum wind speed coincides approximately with the subtropical jet stream. The location of the polar jet fluctuates considerably and does not show up in the average. The center is intended to illustrate wind directions near the Earth's surface (trade winds and westerlies). Cyclones (C) and anticyclones (A) imbedded in the westerlies are only sketched. The frontal systems associated with cyclones cannot be shown in this extremely simplified diagram.

Figure 1.7 shows meridional circulation patterns and their seasonal variations. The meridional component of the mean wind field is comparatively weak, yet it is extremely important in the north–south transport of heat and angular momentum. In the Tropics the energy budget is positive; near the Poles it is negative. Here, more energy is radiated into space than the Earth receives from the Sun. The meridional circulation maintains the heat flux required to balance the global energy budget. The major features of the meridional wind field are the two tropical Hadley cells. They are separated by a region of ascending air called the interhemispheric tropical convergence zone (ITCZ), which forms a barrier to the exchange of air between the two circulation systems and thus divides the troposphere into a northern and southern part. In December–February, and again in July–August, one large thermally driven Hadley cell dominates the tropical circulation, while the second Hadley cell diminishes in strength. This causes the ITCZ to oscillate northward or southward about an average position near the equator. The exchange of air between the two hemispheres is largely due to the seasonal relocation of the ITCZ.

The descending branch of the Hadley cell gives rise to a high-pressure belt at about 30° latitude. Further poleward the meridional circulation is opposite in direction to that required for thermally driven motion. These features,

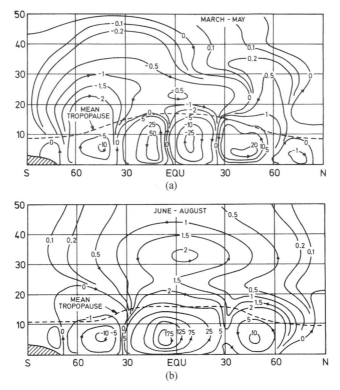

FIGURE 1.7 Mean meridional circulation. (a) March–May. (b) June–August. (c) September–November. (d) December–February. Stream lines indicate mass fluxes in Tg/s. (Adapted from Louis (1975); below 15 km from Newell *et al.* (1972).)

which are called Ferrel cells, must be indirect circulation systems. Yet another pair of cells is evident in polar regions. These are very weak systems.

   The mean meridional motion in the lower stratosphere is much less well defined than that in the troposphere, and the streamlines in Figure 1.7 are rather sketchy. The circulation is largely eddy-driven and the mean and eddy mass fluxes partly compensate each other. The net meridional circulation (often called diabatic, because it is solely driven by the stratospheric heat budget) indicates nevertheless that upward motion occurs in the tropics, where net diabatic heating is positive, whereas downward motion prevails in the extra-tropical latitudes and over the poles where the heating is negative. For details, reference is made to reviews of Brasseur and Solomon (1986) and Andrews *et al.* (1987).

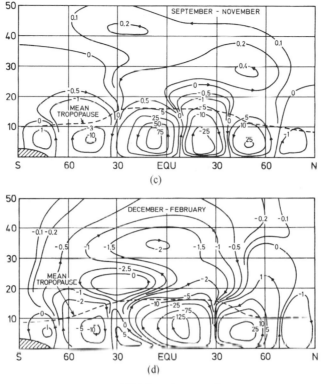

FIGURE 1.7 (Continued)

1.4.2. Eddy Diffusion

Small scale turbulence in the atmosphere results primarily from the fact that air motions usually are discontinuous, often occurring in a succession of brief bursts, so that the wind speed fluctuates considerably around a mean value. Because of shear stresses, small vortices or eddies develop that carry momentum, heat, and trace materials in the direction perpendicular to the mean air flow. The eddies can be made visible by a suitable tracer such as smoke particles. Any observer of the plume emerging from a smoke stack will be impressed by the apparently random dispersal of material by small eddies in a manner resembling the spreading of particles by Brownian motion. Although individual eddies never move very far before viscous drag dissipates them into the surroundings, new eddies are constantly generated in a statistically homogeneous turbulent flow field so that the dispersal of material continues. Although turbulent mixing may be treated mathematically in different ways (Sutton, 1955; Pasquill, 1974), the most common approach in air chemistry is by analogy to Fickian diffusion, that is the flux of a trace substance in the air is assumed to arise from the existence of a gradient of the mixing ratio established by the injection (or removal) of material into (or from) the atmosphere. Because of this analogy, turbulent mixing is frequently referred to as "eddy diffusion." This section discusses basic concepts of eddy-diffusion in the framework of global circulation and in the atmospheric boundary layer.

Let $\bar{u}$ denote the mean velocity of turbulent air motion and x_s the local mean value of the mixing ratio of a gaseous constituent s. The flux of s can then be written as the sum of two terms, the first describing the mean flux in the direction of the air flow, and the second describing the flux due to eddy diffusion:

$$F_s = \bar{u} n x_s - \mathbf{K} n \, \mathrm{grad} \; x_s \qquad (\text{molecule m}^{-2} \, \text{s}^{-1})$$

$$F_s = (M_s/M_{air})(\bar{u} \rho x_s - \mathbf{K} \rho \, \mathrm{grad} \; x_s) \quad (\text{kg m}^{-2} \, \text{s}^{-1}) \qquad (1.13)$$

As indicated, F_s may be expressed as a flux of either the chemical amount or the mass of the material. Here, ρ and n are the density and the number concentration of air, respectively, and $\mathbf{K}$ is the *eddy diffusion coefficient*. This quantity must be treated as a tensor, because eddy diffusion in the atmosphere is highly anisotropic because of gravitational constraints on the vertical motion and large-scale variations in the turbulence field. Eddy diffusivity is a property of the flowing medium and is not specific to the tracer. Contrary to molecular diffusion, the gradient is applied to the mixing ratio and not to concentration, and the eddy diffusion coefficient is independent of the type of trace substance considered. In fact, aerosol particles and trace gases are expected to undergo transport at the same rate.

Mass conservation requires that Equation (1.13) be used in conjunction with the continuity equation

$$n\left(\frac{\partial x_s}{\partial t}\right) = -\,\mathrm{div}\, F_s + q_s - s_s$$

$$= \mathrm{div}(\mathbf{K} n \, \mathrm{grad} \; x_s - \bar{u} n x_s) + q_s - s_s \qquad (1.14)$$

which describes the change with time of the local concentration of a trace substance s due to aerial transport in the presence of local chemical sources q_s and sinks s_s. This equation must be applied to each trace constituent separately. Laboratory rate coefficients for chemical reactions and photo-chemical processes furnish the data needed to define the local chemical sources and sinks. Information on the mean wind fields and the degree of turbulent mixing are taken from meteorological observations. Finally, a number of boundary conditions must be added to describe incoming and outgoing fluxes of trace substances at the Earth's surface and at the high altitude boundary of the air space, frequently the stratopause. For simplicity, the assumption of steady state is frequently made, which requires that appropriate temporal averages are applicable to all quantities on the right-hand side of Equation (1.14). The equations can describe the dispersal of material in the plume of a power plant as well as the global transport of a tracer, provided allowance is made for the difference in scales and a corresponding change in the magnitude of the diffusion coefficients. The reason for the dependence on scale is that the atmosphere harbors eddies of many sizes. In the troposphere, turbulent mixing on a global scale is effected by large cyclones, whereas local mixing is due to small-sized eddies. For a more detailed discussion of this problem see Bauer (1974). Large-scale eddies are responsible for the wave-like pattern of the circumpolar circulation in the middle latitudes. These planetary waves consequently harbor a potential for the meridional exchange of air masses, and they may also be used to describe the transport of trace components

Computer models for the transport of trace substances in the atmosphere that are based on the above equations utilize an Eulerian description of space by subdividing the atmosphere into an assembly of boxes, which exchange air and trace constituents with adjacent boxes in accordance with the overall flow field impressed on the atmosphere. Even the largest computers available today cannot simultaneously simulate the daily variations in flow patterns imposed by meteorological changes and chemical changes taken together, so that a considerable degree of averaging is required. One simplification makes use of the fact that reactive species in chemical mechanisms are short-lived and that steady state conditions are achieved quickly. For such substances the transport term can be neglected, whence $q \approx s$. Other simplifications make use of the averaging schemes of Table 1.2. As the zonal mean circulation is more rapid than meridional transport, it is possible to average over all longitudes and to study the distribution of trace substances in the plane defined by latitude and altitude. This approach is called a two-dimensional model. One-dimensional models drive spatial averaging one step further and use either latitude or altitude as space coordinates. This approach requires that the wind vector $\bar{u}$ in the flux equation be set to zero in order to satisfy the continuity equation. Transport in one-dimensional models is then assumed to occur exclusively by eddy diffusion. As the mean motion is now included in the eddy diffusion term, the associated eddy coefficients differ somewhat from those used in the two-dimensional representation. Because one-dimensional models keep the transport term simple, they allow a very detailed description of chemical processes. Such models have been most successful in describing the vertical distribution of trace gases in the strato-sphere. The approach benefits from the fact that horizontal transport in the stratosphere is much faster than vertical transport.

Three-dimensional models are the most demanding in terms of computer time. Here, one must distinguish between models based on Equation (1.14) that make use of empirical transport parameters derived from meteorological observations and so-called general circulation models, which are based on the fundamental equations of hydrodynamics and thermodynamics. The latter description does not make use of the concept of eddy diffusion, and we shall not discuss the details. Three dimensional models require a detailed specification of input parameters such as the spatial distribution of sources and sinks of atmospheric trace gases. But these models are the only ones that furnish a global coverage of the distribution of trace components for compar-ison with satellite data.

Our knowledge of eddy diffusion coefficients is obtained largely from observations of the spreading of natural or artificial tracers in the atmosphere and from the evaluation of such data by eddy diffusion models based on Equations (1.13) and (1.14). The two-dimensional representation requires

$$\mathbf{K} = \begin{pmatrix} K_{yy} & K_{yz} \\ K_{zy} & K_{zz} \end{pmatrix}$$

where K_{yy} and K_{zz} are the components in the north–south and vertical directions, respectively. In the lower stratosphere the off-diagonal elements of the $\mathbf{K}$ tensor are nonzero because mixing occurs preferentially in a direction roughly parallel to, although somewhat more inclined than, the tropopause. The term $K_{yz} = K_{zy}$ vanishes when the vertical and horizontal motions are statistically uncorrelated. Empirical values for K_{yy}, K_{yz}, K_{zz} depend on latitude and altitude, and to some extent on the season. In one-dimensional models one must set u_y or u_z arbitrarily to zero; otherwise it is difficult to satisfy the continuity equation. This is particularly necessary in the vertical one-dimensional model because of the change in air density with altitude. In the one-dimensional models the effects of transport by mean motions and eddy diffusion must be combined and expressed by $\mathbf{K}_z(z)$. It should be clear that these $\mathbf{K}$ values differ from the corresponding ones in two-dimensional models.

Order-of-magnitude estimates for K_y and K_z may be obtained from wind variance data with the help of the mixing length hypothesis, which is again made in analogy to molecular diffusion. The hypothesis is based on the assumption that an air parcel moves a distance l before it mixes suddenly and completely with the surrounding air, and that during the displacement the mixing ratio of a tracer (or another property such as heat) is conserved. The momentary flux F_s' resulting from the variance u' of the wind vector is given by

$$F_s' = nu'[x_s(x_1) - x_s(x_2)] = nu'(x_1 - x_2)[x_s(x_1) - x_s(x_2)]/(x_1 - x_2)$$

$$= -nu'(x_2 - x_1)\,\text{grad}\,x_s = -nu'l'\,\text{grad}\,x_s \qquad (1.15)$$

where x_1 and x_2 are the space coordinates connected with the displacement l'. Averaging over all such displacements and combining the average flux due to turbulent mixing with that due to the mean motion gives

$$F_s = nx_s\bar{u} - n\overline{u'l'}\,\text{grad}\,x_s \qquad (1.16)$$

Comparison with Equation (1.13) then shows that $\mathbf{K} = \overline{u'l'}$ with the components $K_{jk} = \overline{u_j'l_k'}$. In contrast to the wind variance u', it is not generally possible to observe the mixing length directly in the atmosphere. Accordingly, the expression derived for $\mathbf{K}$ has mainly a heuristic value. It is possible, however, to observe the time periods associated with the displacement of an air parcel, for example, from the motions of a small balloon. From such observations, Lettau and Schwerdtfeger (1933) obtained one of the first estimates for K_z. The value derived was 5 m^2 s^{-1} in the lower troposphere. Since l' and u' have the same direction, one may set $l' = u'\bar{\tau}$, so that the eddy diffusion coefficient becomes proportional to the mean square of the wind fluctuation:

$$\mathbf{K} = \begin{pmatrix} \overline{u_y'u_y'} & \overline{u_y'u_z'} \\ \overline{u_z'u_y'} & \overline{u_z'u_z'} \end{pmatrix}\bar{\tau} \qquad (1.17)$$

According to Möller (1950), $\bar{\tau}$ may be regarded as an average time constant for the rate at which the property transferred by individual air parcels adjusts to the mean value of the surroundings. In statistical turbulence theory $\bar{\tau}$ is identical with the time integral over a correlation coefficient (Pasquill, 1974). From 13 years of meteorological observations in Europe, Möller (1950) showed that $\bar{\tau}$ is quite constant, with a value of 12 h (4.2×10^4 s) for horizontal motions. The magnitude of the time constant confirms that mixing by large-scale horizontal transport is dominated by cyclone activity and waves. If $\bar{\tau}$ is assumed to be roughly independent of latitude and height, typical wind variances $u_y' = 7$ m s^{-1} and $u_z' = (0.5-1.5) \times 10^{-2}$ m s^{-1} yield the following order-of-magnitude values for the eddy diffusion coefficients:

$$K_{yy} = 2 \times 10^6 \quad |K_{yz}| \le 4 \times 10^3 \quad K_{zz} = 1\text{-}10 \quad (\text{m}^2\,\text{s}^{-1})$$

Below, we shall consider the variation of **K** with latitude and altitude.

Czeplak and Junge (1974), who discussed a one-dimensional horizontal diffusion model for the troposphere, derived the dependence of K_y with latitude from wind variance data of Flohn (1961) and Newell et al. (1972) on the basis of $\bar{\tau} = 4.2 \times 10^4$ s, as given by Möller (1950). The results are shown in Figure 1.8. They supersede earlier estimates reviewed by Junge and Czeplak (1968). All data show that K_y is greatest in mid-latitudes, where cyclone activity is most prominent. This results in a very efficient mixing by turbulence in the region of the westerlies. The much lower K_y values near the equator are due, in part, to the increase in tropopause height, as K_y is an average over the vertical extent of the troposphere. The low equatorial K_y

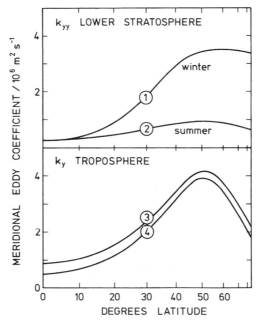

FIGURE 1.8   Eddy diffusion coefficients in the Northern Hemisphere. Upper frame: Lower stratosphere (100 hPa, about 16 km altitude), curve 1 for December–January, curve 2 for June–August, approximated from Luther (1975) and Hidalgo and Crutzen (1977). Lower frame: One dimensional values for the troposphere after Czeplak and Junge (1974). Curve 3 is based on wind variance data of Flohn (1961) and Newell et al. (1966). Curve 4 is based on data of Newell et al. (1972), annual average.

values are mainly responsible for the relatively slow transport of trace substances across the interhemispheric tropical convergence zone.

Figure 1.8 includes for comparison K_{yy} values at the 100-hPa pressure level, that is, in the lower stratosphere. These data likewise are based on the mixing-length hypothesis combined with observations of heat flux, temperature gradients, and wind variances (Luther, 1975). The first estimates of K_{yy} values in the stratosphere were obtained by Reed and German (1965) and Gudiksen et al. (1968). Both groups sought to simulate the observed spreading of tungsten 185 debris injected into the tropical stratosphere by the atomic weapons tests in the late 1950s. Observations and model calculations agreed remarkably well and showed that eddy diffusion in north-south direction was more important in spreading the tungsten-185 cloud than was the relative weak meridional circulation. Later two-dimensional models, such as that of Gidel et al. (1983) used K_{yy} values that were slightly modified from those of Luther (1975). As in the troposphere, stratospheric K_{yy} values are higher in midlatitudes compared to those near the equator. Figure 1.8 also illustrates a seasonal dependence such that (meridional) turbulent mixing is greatly enhanced during the winter months.

Values for the off-diagonal component for the **K** tensor usually are scaled to those of K_{yy} by means of heat flux data. The K_{yy} values determine the slope of the mixing surface. In the lower stratosphere of the Northern Hemisphere, the values are negative—that is, the mixing surface is slanted downward as one goes from equator to Pole. Since the K_{yy} terms do not contribute directly to the rate of transport, we shall not discuss them further.

Current formulations of two-dimensional models for the stratosphere have largely abandoned the concept of the empirical **K** tensor. Instead they incorporate comprehensive radiative transfer to determine the diabatic circulation (mean and turbulent motions combined) and calculate diffusion coefficients for transport along isentropic surfaces from the rate of planetary wave dissipation at altitudes in the stratosphere (see, for example, Garcia et al., 1992; Garcia and Solomon, 1994; Yang et al., 1992).

Values for K_z have been derived mainly from tracer studies. To illustrate the application of the one-dimensional diffusion equation in a simple case, consider radon as a tracer. The half-life of radon 222 is 3.8 days. The main source is emanation from continental soils. The oceans are not a significant source. In the middle of a continent the concentration of radon decreases with altitude, because it decays during upward transported by vertical mixing. Figure 1.9 shows a few measurements of vertical profiles of the radon mixing ratio. Individual profiles exhibit considerable variability, which is partly due to short-term changes in meteorological conditions, such as enhanced convection during the day. The heavy line is an average of six profiles. Above the boundary layer, the average declines quasi-exponentially

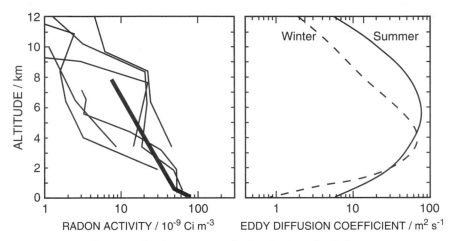

FIGURE 1.9   Left: Decay of radon activity with altitude. Six individual measurements over Colorado, Nebraska, Utah (Moore *et al.*, 1973). The heavy line is an average of six profiles over New Mexico (Wilkening, 1970); to convert Curie to SI unit Bequerel (s^{-1}): 1 Ci = 3.7×10^{10} Bq. Right: Variation of the eddy diffusion coefficient with altitude derived from averaged radon profiles (Liu *et al.*, 1984).

with a scale height $h = 3.7$ km, that is,

$$x(\mathrm{Rn}) = x_0(\mathrm{Rn}) \exp(-z/h)$$

The steady state solution of Equation (1.14) for constant K_z and $s_s = nx_s\lambda^*$ in the one-dimensional case, where $\lambda^* = 2.1 \times 10^{-6}$ s^{-1} is the radon decay constant, then yields an algebraic equation for h from which one obtains

$$K_z = \lambda^* h/(1/H + 1/h) = 20 \ \mathrm{m^2 \ s^{-1}} \qquad (1.18)$$

Here, $H = 8500$ m denotes the atmospheric scale height. Obviously radon would be a good tracer for turbulent mixing in the troposphere. Unfortunately, it is hard to measure. It should be noted that the assumption of a constant K_z in the troposphere is a reasonable approximation only for the region outside the planetary boundary layer ($z \geq 2$ km). Air motions closer to the earth surface are greatly influenced by friction, which causes eddy diffusion coefficients to decrease from their values in the free troposphere. Liu *et al.* (1984) have averaged a variety of radon measurements such as those shown in Figure 1.9 on the left to derive eddy diffusion coefficients as a function of altitude. Their results, which are shown in Figure 1.9 on the right, clearly show the lower values in the boundary layer. At higher altitudes, the values also are lower than in the middle troposphere, presumably because of the influence of the tropopause level.

Table 1.7 summarizes vertical eddy diffusion coefficients for the troposphere and lower stratosphere as derived from observations, mainly the distribution of tracers. For the troposphere K_z data of Davidson et al. (1966) are the least trustworthy because they are based on tracers originating in the stratosphere. The best values probably are those of Bolin and Bischof (1970) from the seasonal variation of CO_2. Above the tropopause the eddy coefficient K_z decreases quite abruptly, from high values in the troposphere to low values in the range 0.2–0.4 m^2 s^{-1}. Methane and nitrous oxide are useful tracers in the stratosphere. Both gases are well mixed in the troposphere, but above the tropopause their mixing ratios decrease with altitude because of losses by chemical reactions. Figure 1.10 shows the dependence of K_z on altitude as determined from measurements of vertical profiles CH_4 and N_2O and one-dimensional model evaluations. K_z values adjacent to the tropopause are low, but K_z increases with increasing altitude, and eventually

TABLE 1.7   Values for the Eddy Diffusion Coefficient K_z Derived Mainly from Tracer Observations

Authors	$K_z{}^a$	Remarks
Troposphere		
Davidson et al. (1966)	1–10	From the fall-out of bomb-produced ^{185}W and ^{90}Sr; source in the stratosphere
Bolin and Bischof (1970)	14–26	From the seasonal oscillation of $x(CO_2)$ imposed by the biosphere
Machta (1974)	40	Coarse estimate from radon decay data
Present	20	From the radon decay data of Figure 1.9
Stratosphere		
Davidson et al. (1966)	0.1–0.6	From the distribution of ^{185}W and ^{90}Sr
	0.1	Best value
Gudiksen et al. (1968)	0.15–3.6	Two-dimensional model including mean motions of ^{185}W tracer distribution
Wofsy and McElroy (1973)	0.2	From one-dimensional diffusion model and observed altitude profile of methane in the stratosphere
Luther (1975)	0.2–2.0	From heat flux, temperature, and wind variance data
Schmeltekopf et al. (1977)	0.3–0.4	From one-dimensional diffusion model and measured altitude distribution of N_2O averaged over one hemisphere

^{a} Vertical transport (m^2 s^{-1}).

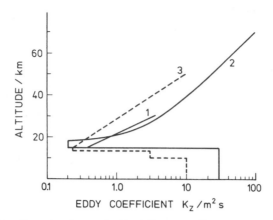

FIGURE 1.10   One-dimensional vertical eddy diffusion coefficient K_z derived from trace gas observations in the stratosphere. Curve 1: Nitrous oxide (Schmeltekopf *et al.*, 1977). Curves 2 and 3: Methane (Wofsy and McElroy, 1973), Hunten, 1975).

the values reach a magnitude similar to or greater than that in the troposphere.

Finally it is necessary to assess the importance of transport by eddy diffusion relative to that by mean meridional motions. For this purpose, we define time constants for the transport over a characteristic distance d_c as follows:

Mean motion:
$$\tau_{\text{trans}} = d_c / \bar{u}_d \qquad (1.19a)$$

Eddy diffusion:
$$\tau_{\text{trans}} = \overline{d_c^2} / 2 K_d = \pi \overline{d_c^2} / 4 K_d \qquad (1.19b)$$

where d_c equals y_c in the meridional and z_c in the vertical direction. The characteristic distance parallel to the surface of the Earth is taken as 3300 km, corresponding to one-third of the distance between equator and Pole. For z_c the atmospheric scale height H is used. The results are assembled in Table 1.8. The data indicate that a tracer is transported along a meridian about three times faster by eddy diffusion than by mean motions of the Hadley circulation. Stratosphere and troposphere behave rather similarly in this regard. The characteristic time for the spreading of a tracer in the north–south direction is about 3 months in both directions. Vertical transport in the troposphere is dominated by eddy diffusion with a characteristic time of about 4 weeks. This applies to average meteorological conditions. High-reaching convection by cyclone activity may achieve a much shorter exchange time. In the lower stratosphere, eddy diffusion and mean motions contribute about equally to vertical transport. The direction of the mean

TABLE 1.8 Time Constants for Horizontal and Vertical Transport by Atmospheric Mean Motions and Eddy Diffusion (Northern Hemisphere)[a]

Altitude level	Horizontal transport					Vertical transport				
	y_c (km)	$\bar{u}_y$ (m s^{-1})	τ_{trans} (months)	K_{yy} (10^6 m^2 s^{-1})	τ_{trans} (months)	z_c (km)	$\bar{u}_z$ (m s^{-1})	τ_{trans} (months)	K_z (m^2 s^{-1})	τ_{trans} (months)
Troposphere (800 hPa, 2 km)	3300	0.20 w / 0.05 s	6 / 25	1.8 w / 0.5 s	1.8 / 6.6	8.5	0.00015	20	15	1.2
Stratosphere (100 hPa, 16 km)	3300	0.20 w / 0.12 s	6 / 10	3.0 w / 0.6 s	1.1 / 5.5	6	0.00012	19	0.4	27

[a] Data for $\bar{u}_y$, $\bar{u}_z$, and K_{yy} were taken from Hidalgo and Crutzen (1977) for 40° northern latitude. w = winter, s = summer.

motion thus becomes important; it may either counteract or enhance transport by eddy diffusion. The time constants are such that a substance injected into the stratosphere will spend 1–2 years there before it is drained to the troposphere by downward transport, but during this time it will become meridionally well mixed. The time constants in Table 1.8 are important for all considerations of chemical reactivity in the atmosphere. If the rate of chemical consumption is greater than that of transport, the mixing ratio of the substance will decrease with distance from the point of injection. Nonreactive compounds, in contrast, will spread to fill the entire atmosphere within a time period determined solely by the rate of transport.

1.4.3. MOLECULAR DIFFUSION

The coefficient for molecular diffusion D is determined by the mean free path between collisions of molecules, which is inversely proportional to pressure. In the ground-level atmosphere the value for the diffusion coefficient typically is $D \approx 2 \times 10^{-5}$ m^2 s^{-1}, which makes molecular diffusion unsuitable for large-scale transport. Turbulent mixing is faster. The pressure decreases with altitude, however, causing the molecular diffusion coefficient to increase until it reaches a value of 100 m^2 s^{-1} at $z \approx 100$ km. Above this level molecular diffusion becomes the dominant mode of large-scale transport.

In the troposphere, molecular diffusion is important in the size range of 1 mm or less. This may be gleaned from Equation (1.19b) by substituting $D \approx 2 \times 10^{-5}$ m^2 s^{-1} for K_d, which gives a time constant for the displacement $d_c \leq 10^{-3}$ m of 0.1 s or less. Molecular diffusion is responsible for the growth of cloud drops by condensation of water vapor, and for the exchange of water-soluble trace gases between cloud drops and the surrounding air. In the case of a spherical drop there is no dependence on angular coordinates because of the isotropic nature of molecular diffusion. In the absence of volume sources or sinks the diffusion equation reduces to

$$\partial n_s / \partial t = D \operatorname{div} \operatorname{grad} n_s = \frac{D}{r^2} \frac{\partial}{\partial r} r^2 \frac{\partial n_s}{\partial r} \qquad (1.20)$$

Solutions to this equation for various boundary conditions have been assembled by Crank (1975). For a molecular flux that varies slowly with time, one may apply the steady-state approximation, which in the case of an infinite supply of molecules of substance s leads to a total flux into (or out of) a drop with radius r of

$$F_{s,\,tot} = 4\pi Dr(n_{sr} - n_{s\infty})   \quad\quad (1.21)$$

where $n_{s\infty}$ is the number concentration of s at a distance far from the drop. We shall discuss applications in Sections 7.3.1 and 8.4.2.

1.5. AIR MASS EXCHANGE BETWEEN PRINCIPAL ATMOSPHERIC DOMAINS

A simple yet very useful concept in treating material transport in geochemistry, and in atmospheric chemistry as well, is to consider a small number of reservoirs that communicate with each other across common boundaries. Ideally, the contents of the reservoirs should be well mixed, that is, the rate of internal mixing should exceed the rate of material exchange between adjoining reservoirs. In these so-called box models the exchange is treated kinetically as a first-order process.

The relatively slow rate of transport across the interhemispheric tropical convergence zone naturally subdivides the troposphere into a northern and a southern hemispheric part. In a similar manner one may consider the stratosphere and the troposphere as separate reservoirs, with the tropopause as a common boundary. Vertical and horizontal mixing in each part of the troposphere is faster than transport of air across the ITCZ or across the tropopause, so that box model conditions are met. The stratosphere, in contrast, is not an ideal reservoir, because its upper boundary is not well defined. In addition, in its lower part the rate of vertical mixing is much reduced from that in the troposphere. These shortcomings must be kept in mind when one interprets observational data from the stratosphere in terms of a box model. The rate of exchange between stratosphere and troposphere inferred from the model will reflect to some degree the rate of transport of material from greater heights downward to the tropopause rather than the true exchange rate across the tropopause. We discuss below two examples for the application of the four-reservoir atmospheric box model and provide numerical values for the exchange times as derived from tracer studies.

The model is illustrated in Figure 1.11. The exchange coefficients κ are distinguished by subscripts whose sequence designates the direction of mass flow. The flux of material across a boundary is given by the product of the corresponding exchange coefficient and the content of material in the donor box. If the mass flux is expressed in kg year^{-1}, the exchange coefficients have the unit year^{-1}. The inverse of κ represents the time constant for exponential decay of the reservoir content in the case of negligible return

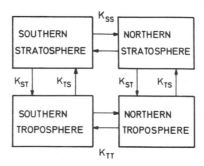

FIGURE 1.11   Four-box model of the atmosphere. Common boundaries are the tropopause and the interhemispheric tropical convergence zone. The rate of air exchange between individual reservoirs is expressed by the exchange coefficients shown and the mass content of the donor reservoir.

flow. The coefficients for exchange between the two tropospheric reservoirs are equal because the air masses are equal. The coefficients for the exchange between stratosphere and troposphere differ, because the air mass in the troposphere is greater by a factor of 4.7 than that of the stratosphere (a factor of 5 for the lower stratosphere; see Table 1.6), so that $\kappa_{ST} \approx 5\kappa_{TS}$. The model in Figure 1.11 assumes for simplicity that κ_{ST} has the same value in the Northern and Southern Hemispheres. This is an approximation and may not be true. In fact, estimates of the influx of stratospheric ozone into the troposphere suggest a smaller flux in the Southern Hemisphere when compared to the Northern Hemisphere.

Tracers are again used to obtain information on the exchange coefficients. To illustrate the approach in the case of interhemispheric air mass exchange consider the rise of CO_2 in the atmosphere due to the combustion of fossil fuels. During the period 1958–1965 the rise was nearly linear in both hemispheres. This is evident from Figure 1.2 if the seasonal variation is subtracted. A linear increase means that similar amounts of CO_2 were added to the atmosphere each year. Since over 90% of the addition occurred in the Northern Hemisphere, we may for simplicity ignore the sources in the Southern Hemisphere. The small fluxes into the stratosphere and the ocean also are neglected. Let x_N and x_S denote the mixing ratios and $G_N(CO_2) = x_N G_T/2$ and $G_S(CO_2) = x_S G_T/2$ the masses of CO_2 in the Northern and Southern Hemispheres, respectively. Furthermore, let $P_a G_T$ be the constant annual rate of CO_2 increase in the atmosphere. The behavior of CO_2 with time should then be described by the two coupled equations

$$dG_N(CO_2)/dt = -\kappa_{TT}[G_N(CO_2) - G_S(CO_2)] + P_a G_T$$
$$dG_S(CO_2)/dt = \kappa_{TT}[G_N(CO_2) - G_S(CO_2)]$$

(1.22)

After subtraction of the second equation from the first and dividing by $G_T/2$, one obtains

$$d(x_N - x_S)/dt = -2\kappa_{TT}(x_N - x_S) + 2P_a \qquad (1.23)$$

The last equation has the solution

$$\Delta x = (x_N - x_S) = \Delta_0 x \exp(-2\kappa_{TT}t) + (2P_a/\kappa_{TT})[1 - \exp(-2\kappa_{TT}t)]$$
$$(1.24)$$

which after a time of adjustment on the order of $1/\kappa_{TT}$, relaxes to steady-state conditions with

$$\Delta x = 2P_a/\kappa_{TT} \quad \text{so that} \quad \kappa_{TT} = 2P_a/\Delta x \qquad (1.25)$$

From Figure 1.2 one estimates that $\Delta x = 0.66$ μmol mol^{-1} during the period considered. P_a is given by the slope of the rise curve, or more precisely by the average slope $(\Delta x_N + \Delta x_S)/2 \Delta t = 0.64$. The exchange coefficient thus is $\kappa_{TT} = 0.64/0.66 = 0.97$ year^{-1}. The corresponding exchange time is $\tau_{TT} = (1/\kappa_{TT}) \approx 1$ year.

The inventory of strontium 90 in the stratosphere provides another set of observational data amenable to box model analysis. Strontium 90 is an artificial radioactive isotope with a half-life due to beta decay of 28.5 years. It was brought into the atmosphere by atomic weapons tests and has been monitored for at least two decades, starting in the late 1950s, with aircraft and balloons serving as sampling platforms. Figure 1.12 shows the integrated activity of ^{90}Sr for the northern and southern stratospheric reservoirs separately and for both regions combined. In the south the amount of ^{90}Sr rises initially as material enters from the north, where most of it was deposited. Eventually the activity declines in both hemispheres because of losses to the troposphere. Strontium combines with other elements in the atmosphere to form inorganic salts, and once it reaches the troposphere it is quickly removed by wet precipitation on a time scale shorter than 4 weeks. For a treatment of the observational data, the four-box model must be used, although it suffices to consider only the stratospheric reservoirs. Since the mass or atom content of ^{90}Sr in both reservoirs is directly proportional to the integrated activities plotted in Figure 1.12, we may use the activities to set up the budget equations

$$da_N/dt = -\kappa_{ST} a_N - \kappa_{SS}(a_N - a_S)$$
$$da_S/dt = -\kappa_{ST} a_S + \kappa_{SS}(a_N - a_S)$$
$$(1.26)$$

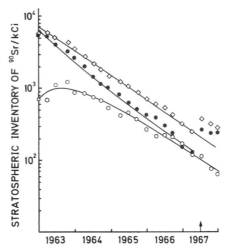

FIGURE 1.12 Inventory of strontium 90 in the stratosphere following the injection by nuclear weapons tests (1 Ci = 3.7 × 10^{10} Bq). The arrow indicates a new injection due to Chinese tests in June 1969. Filled circles, northern stratosphere; open circles, southern stratosphere; diamonds, sum of both. (Redrawn from Reiter (1975), originally Krey and Krajewsky (1970) and Krey *et al.* (1974).)

The solution to these equations is derived most conveniently from two new equations obtained from the above by addition and subtraction, respectively:

$$d(a_N + a_S)/dt = -\kappa_{ST}(a_N + a_S)$$
$$d(a_N - a_S)/dt = -(\kappa_{ST} + 2\kappa_{SS})(a_N - a_S) \tag{1.27}$$

Integration is now straightforward and yields

$$(a_N + a_S) = (a_N^0 + a_S^0)\exp(-\kappa_{ST}t) \tag{1.28a}$$

$$(a_N - a_S) = (a_N^0 - a_S^0)\exp[-(\kappa_{ST} + 2\kappa_{SS})t] \tag{1.28b}$$

where a_N^0 and a_S^0 denote the initial activities at $t = 0$. In the semi-log plot of Figure 1.12, the data for the sum of a_N and a_S follow a straight line, thereby verifying the exponential decay predicted by Equation (1.28a). From the slope of the line one obtains $\kappa_{ST} = 0.77$ year^{-1}, which corresponds to an exchange time $\tau_{ST} = 1.3$ years. The second exchange coefficient can be obtained by fitting Equation (1.28b) to the data for the ^{90}Sr activity in the southern stratosphere. Such a fit is shown in Figure 1.12 for $\kappa_{SS} = 0.28$ year^{-1}. The corresponding exchange time is $\tau_{SS} = 3.5$ years. The loss of strontium 90 by radioactive decay was neglected in the above treatment. The decay constant $\lambda^* = 0.0248$ year^{-1} is sufficiently small to permit this approximation. When the decay is taken into account one finds that κ_{ST} is slightly reduced to $\kappa_{ST} = 0.74$ year^{-1}, whereas κ_{SS} remains unaffected.

Table 1.9 summarizes exchange times derived from observations of a variety of tracers. Most investigators made use of radioactive isotopes brought into the atmosphere by atomic weapons tests in 1958 and then again in 1962. The increase in carbon 14 in particular has been studied in detail. The analysis of radiocarbon is complicated by the fact that CO_2 exchanges with the oceans and the biosphere, the latter giving rise to seasonal variations. A four-box model is inadequate to describe the temporal behavior of $^{14}CO_2$ in the atmosphere, and more complicated models must be applied. The values obtained for the exchange times nevertheless are in good agreement with those derived from other tracers. The results of Walton et al. (1970) are an exception. These measurements were made in 1967/68, at a time when the differences in activities between different reservoirs had decayed to small values. The early results based on the use of ^{85}Kr as a tracer were too high. Jacob et al. (1987) and Levin and Hesshaimer (1996) have shown that the reason for this is the use of surface observations in the Northern Hemisphere, that is in the source regions, where ground-level mixing ratios exceed the vertical tropospheric average. Simulation of the interhemispheric ^{85}Kr distribution by two- or three-dimensional models that take the decrease with altitude into account lead to an exchange time of 1.1 years. The same situation applies to SF_6 as a tracer. On the whole, the data indicate time constants of about 1 year for the air exchange between the northern and southern troposphere, 1.4 years for the exchange from the stratosphere to the troposphere, and 3.5 years for the exchange between the two stratospheric reservoirs. Two estimates based on meteorological data are in agreement with the results from tracer studies.

The exchange of air between the northern and southern troposphere is caused partly by eddy diffusion in the equatorial upper troposphere and partly by the seasonal displacement of the interhemispheric tropical convergence zone, which lies north of the equator in July and south of the equator in January. The displacement is greatest in the region of the Indian Ocean, where the ITCZ is located over the Indian subcontinent in July. By tracing the transport of fission products from Chinese and French weapons tests, Telegadas (1972) has demonstrated the importance of monsoon systems for the interhemispheric exchange process over the Indian Ocean.

The various mechanisms contributing to the air exchange across the tropopause were reviewed by Elmar R. Reiter (1975). Four processes must be distinguished: (a) the seasonal adjustment of the average tropopause level; (b) organized large-scale mean motions due to the meridional circulation; (c) turbulent exchange processes via tropopause gaps or folds that are associated

TABLE 1.9 Exchange Times between Different Atmospheric Reservoirs Derived by Various Authors[a]

Authors	Tracer	τ_{TT} (years)	τ_{ST} (years)	τ_{SS} (years)	Remarks
Czeplak and Junge (1974)	CO_2	> 0.7			From difference in the annual variation in the two hemispheres
Czeplak and Junge (1974)	CO_2	1.0			From difference in the secular increase between the hemispheres (treated here)
Münnich (1963)	$^{14}CO_2$	< 1			From difference in the increase of bomb-produced ^{14}C in both hemispheres
Lal and Rama (1966)	$^{14}CO_2$	1.2	0.8 ± 0.3		From tropospheric increase of bomb-produced ^{14}C injected from the stratosphere
Feely et al. (1966)	$^{14}CO_2$		2.2		From stratospheric inventory and its change with time
Young and Fairhall (1968)	$^{14}CO_2$		1.5		From stratospheric inventory and its change with time
Nydal (1968)	$^{14}CO_2$	1.0 ± 0.2	2.0 ± 0.5	5.0 ± 1.5	Detailed box model of ^{14}C variations in the atmosphere
Walton et al. (1970)	$^{14}CO_2$	4.4	2.1		Difference in different reservoirs from 1967/1968 data
Czeplak and Junge (1974)	$^{14}CO_2$	1.0			Using the data of Münnich (1963)

Reference	Tracer	τ_{TT}	τ_{ST}	τ_{SS}	Comments
Gudiksen et al. (1968)	[185]W		1.2 ± 0.5		From stratospheric inventory and its change with time
Peirson and Cambray (1968)	[144]Ce, ^{137}Cs, [90]Sr		1.37 ± 0.05	1.5 ± 1.0	From ratio of [144]Ce to ^{137}Cs fission products in surface air and ^{90}Sr in stratosphere and fallout
Feely et al. (1966)	[54]Mn, [90]Sr		1.2		From surface fallout data
Fabian et al. (1968)	[90]Sr		1.56 ± 0.13	3.3 ± 0.3	From stratospheric inventory and its change with time
This study	[90]Sr		1.35	3.5	From the data of Krey and Krajewski (1970) and Krey et al. (1974)
Pannetier (1970)	[85]Kr	2			From latitudinal [85]Kr profile and [85]Kr increase
Czeplak and Junge (1974)	[85]Kr	1.8			Using the data of Pannetier (1970)
Weiss et al. (1983)	[85]Kr	1.7			Two-box model, surface data
Jacob et al. (1987)	[85]Kr	1.6			Two-box model, surface data
	[85]Kr	1.1			3-D general circulation model
Maiss et al. (1996)	SF_6	1.7 ± 0.2			Two-box model, surface data, 4 year record
Levin and Hesshaimer (1996)	[85]Kr, SF_6	1.5			Two-box model, surface data
		1.1			2-D transport model
Newell et al. (1969)	Meteorological data	0.9			Calculated form the mean motion across the equator
Reiter (1975)	Meteorological data		1.4	6.6	Estimated from mean and eddy motions
Averaged values		1.0	1.4	4.0	Results of Walton et al. and Pannetier omitted.

[a] τ_{TT}, Troposphere-troposphere; τ_{ST}, stratosphere-troposphere; τ_{SS}, stratosphere-stratosphere.

with jet streams; and (d) small-scale eddy transport across the entire tropopause. The last process does not contribute to the exchange of bulk air, but it is important in the exchange of trace constituents if a vertical gradient of the mixing ratio exists. In the following discussion we consider each of the four processes separately. More recently, Holton *et al.* (1995) have discussed stratospheric-tropospheric exchange with emphasis on the global-scale transport.

The location of the tropopause is not fixed but adjusts according to meteorological conditions on a time scale of 1–2 days because of radiative balancing. A double tropopause (i.e., two breaks in the temperature lapse rate, each qualifying as a tropopause according to the meteorologist's rule book), is frequently evident in radiosonde data. Under normal conditions, one of the two may be considered a remnant, which will subsequently disappear. In the vicinity of a jet stream, the double tropopause is a regular phenomenon (see Fig. 1.13). Although in mid-latitudes the tropopause level fluctuates somewhat with high- and low-pressure systems, these fluctuations are not considered important in the exchange of air between troposphere and

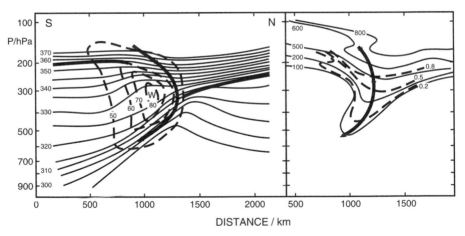

FIGURE 1.13 South–north cross section through a jet stream, simplified from Danielsen *et al.* (1970). Left: Isotachs (dashed lines, m s^{-1}) and isotherms (solid lines) of potential temperature θ/K (the temperature an air parcel would assume if it were compressed adiabatically to a pressure of 1.013×10^5 Pa). Heavy lines indicate the tropopause. Right: Extrusion of ozone-rich air from the stratosphere (dashed lines, $x(O_3)$ in μmol mol^{-1}). Also shown for comparison are isolines of potential vorticity (solid lines) ($-Q_z \, \partial\theta/\partial z$, where Q_z is the vertical component of absolute vorticity). Potential vorticity correlates well with tracer observations. Both parameters may be used to discriminate stratospheric from tropospheric air. The arrows in both parts of the figure indicate the direction of the air flow. A return flow of about half the magnitude occurs on the opposite side of the jet stream above its core.

stratosphere. The seasonal displacement is more substantial. Poleward of 30°
latitude the tropopause is relocated to a lower level in winter, and equator-
ward it is relocated to a higher level compared to summer conditions. The
relocation incorporates tropospheric air into the stratosphere when the
tropopause is lowered, and stratospheric air into the troposphere when it is
raised. Reiter (1975) estimated that 4×10^{16} kg of air is displaced in this
manner in the Northern Hemisphere, which corresponds to about 10% of
the air mass residing in the stratosphere.

The contribution of mean meridional circulation to the air-mass exchange
across the tropopause is substantial. Air enters the stratosphere primarily
with the equatorial uprising motion, gets directed toward the winter pole and
leaves the stratosphere primarily through the tropopause gaps associated
with the subtropical jet stream. From the mass stream lines shown in Figure
1.7 one estimates that about 10 Tg s^{-1} of air migrate across the equatorial
tropopause during northern hemispheric winter (December-February). About
80% of this flux is directed northward. The flux entering the northern
hemisphere in spring is 3.5 Tg s^{-1}, in the autumn it is 6.5 Tg s^{-1}, whereas
in summer the flux is negligible. The total annual air mass transfer is
obtained by summing and amounts to 1.4×10^{17} kg, or 32% of the
northern hemispheric stratospheric air mass.

The physical structure of a jet stream can be detailed from radiosonde and
aircraft observations. Figure 1.13 shows, somewhat schematically, a cross
section through a jet stream. Isotachs (i.e., regions of constant wind velocities) are shown by dashed lines. Isotherms of potential temperature are
shown by solid lines. In this representation, the stratosphere is indicated by
the crowding of isotherms. The location of the tropopause is shown by the
heavy lines. In the center of the jet stream the tropopause cannot be defined
and a gap appears in the tropopause surface. A frontal zone extends from the
tropopause section on the right into the lower troposphere. The mean
circulation relative to the position of the jet stream causes an extrusion of
stratospheric air into the frontal zone, as indicated by the curved arrow
(tropopause fold). Several tracers have been used to identify such extrusions.
Briggs and Roach (1963) used humidity and ozone; Danielsen (1968),
Danielsen et al. (1970), and Danielsen and Mohnen (1977) have considered
strontium 90, zirconium 95, and ozone as tracers. The distribution of ozone
is shown on the right-hand side of Figure 1.13 by the dashed lines. A tongue
of ozone-rich air is seen to extend from the stratosphere into the frontal zone
in the direction of the arrow. This demonstrates most directly the outflow of
stratospheric air through the tropopause gap.

Jet streams occur fairly regularly in the 30–35° latitude belt (subtropical
jet stream), and with lesser frequency near 50–60° latitude, where they are
associated with cyclone activity. From a number of case studies carried out

by aircraft, Reiter and Mahlmann (1965) estimated the mass transfer of stratospheric air for a cyclone of average intensity over North America as 6×10^{14} kg. About 22 such events were estimated to occur over North America each year. By extrapolation to all longitudes, the polar jet-stream belt would cause a total mass transfer of 4×10^{16} kg annually. If the subtropical jet stream contributed equally, the total mass transfer in the Northern Hemisphere due to jet-stream activity would be 8×10^{16} kg annually or about 19% of the stratospheric air mass. Actually, this would be an overestimate because the outflow via the subtropical jet stream is mainly fed by the Hadley circulation. A weaker return flow also exists, which is carried poleward in the stratosphere and must be discharged to the troposphere in middle and high latitudes. Recently it became possible to simulate tropopause folding events in synoptic-scale three-dimensional numerical models initialized with meteorological data (Ebel *et al.*, 1991; Lamarque and Hess, 1994). The results indicate that the intensity of downward flow is twice that in the upward direction. Elbern *et al.* (1998) performed a statistical analysis of tropopause folds based on a 10 year global data set derived from daily routine weather analyses. Because of the limited horizontal resolution, potential vorticity maxima and related criteria were used as indicators. On a global scale they identified about 10,800 events per year of varying intensity, all of them located poleward of 30° latitude, with 62% of them occurring in the Northern Hemisphere. The geographical distribution of such events is reasonable, but the critical threshold may have been set too low so that the frequency is too high, and the air mass transfer rate inferred (8×10^{17} kg year^{-1}) is twice that found in other studies.

Van Velthoven and Kelder (1996) have used a three-dimensional global tracer transport model based on observational meteorological data to derive the downward air mass transport in the region between 85 and 225 hPa ($\sim 12-16$ km altitude). They found 2.3×10^{17} kg year^{-1} in the Northern Hemisphere and 1.6×10^{17} kg year^{-1} in the Southern Hemisphere. The first value is similar in magnitude to the sum of the rates estimated above, 2.4×10^{17} kg year^{-1}. Previous efforts of Holton (Holton, 1990; Rosenlof and Holton, 1993) based on a coarser horizontal resolution had led to smaller rates.

Table 1.10 summarizes the individual exchange rates as given by Reiter (1975). The total exchange rate resulting from the three processes discussed is 0.68 year^{-1}, which corresponds to an exchange time of 1.45 years. For comparison with the strontium 90 data, one would have to add the flux of atoms due to eddy transport. List and Telegadas (1969) have sketched the distribution of ^{90}Sr with latitude and altitude as determined in 1965. The data are somewhat coarse-grained, and gradients near the tropopause were not specifically measured. The vertical gradient is nevertheless reasonably well defined as 50 disintegrations/min/1000 ft^3/2 km in the region north of 30° latitude, which translates to a gradient of mixing ratio, 6.9×10^{-22} m^{-1}. In the equatorial region the downward eddy flux is largely counterbalanced by the upward mean motion branch of the Hadley circulation, so that one needs to consider only the extratropical latitude region. According to Equation (1.13) the local eddy flux is

$$F_z(^{90}\text{Sr}) = -n_T K_z \, dx(^{90}\text{Sr})/dz$$

$$= -(7.5 \times 10^{24})(0.3)(6.9 \times 10^{-22})$$

$$= -1.6 \times 10^3 \text{ atom m}^{-2} \text{ s}^{-1}$$

where n_T, the number density of air molecules at the tropopause, is taken from Table 1.5. The global area north of 30° latitude is 1.25×10^{14} m^2, and the number of seconds per year is 3.15×10^7. This leads to an annual flux of 6.3×10^{24} atoms by eddy diffusion. The total inventory of ^{90}Sr in the northern stratosphere in March 1965 was 8×10^5 Ci, which with 1 Ci = 3.7 $\times 10^{10}$ Bq corresponds to 2.96×10^{16} Bq (disintegrations per second) or $2.96 \times 10^{16}/\lambda^* = 3.74 \times 10^{25}$ atoms. Accordingly, the eddy flux of ^{90}Sr into the troposphere accounted for 16% of the strontium 90 inventory in the stratosphere. Comparison of the individual entries in Table 1.10 shows that all processes are significant but that the meridional mean motion contributes the lion's share.

Transport of a trace substance by eddy diffusion across the intact tropopause will be significant only if a vertical gradient exists in the lower stratosphere. An assessment of the flux requires that the gradient be measured. Figure 1.14 illustrates the behavior of several trace gases in the vicinity of the tropopause. The mixing ratio of CO declines with altitude above the tropopause, indicating that the flux is directed from the tropo-

TABLE 1.10 Contributions of Individual Processes to the Air Exchange Across the Tropopause (Stratosphere-Troposphere, Expressed as a Fraction of Stratospheric Air Mass)

Type of process	Rate (per year)	
	Reiter (1975)	Text
Seasonal relocation of the tropopause	0.10	0.15
Large-scale meridional motion	0.38	0.32
Tropopause folding events	0.20	0.19
Small-scale eddy diffusion (for ^{90}Sr)	0.01	0.16
Total rate	0.69	0.82
Stratospheric residence time for ^{90}Sr (yr)	1.45	1.22

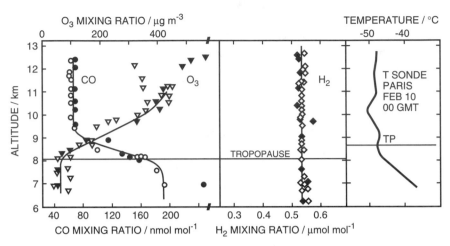

FIGURE 1.14   Vertical profiles of mixing ratios for ozone, carbon monoxide, and hydrogen in the region of the tropopause, from measurements over western France on February 9–10, 1972. Open symbols indicate ascents, filled symbols descents. The right-hand side gives the temperature profile from radiosonde measurements at Paris. Tropopause levels derived from ozone and temperature profiles differ somewhat, presumably because of the difference in location. (Compiled from data of Warneck *et al.* (1973) and Schmidt (1974).)

sphere into the stratosphere. The mixing ratio of ozone increases with altitude, so that in this case the flux is directed from the stratosphere into the troposphere. The mixing ratio of hydrogen exhibits no significant gradient, which means that the flux in either direction must be negligible. To the troposphere, the stratosphere thus represents a source region with regard to ozone, and a sink region with regard to carbon monoxide.

1.6. THE PLANETARY BOUNDARY LAYER

The importance of the Earth's surface to atmospheric chemistry lies in the fact that it may act as an emitter and/or as a receiver of atmospheric trace substances. In this section we discuss concepts that have been found useful in describing these processes. The transport of gases across the ocean surface is reasonably well understood. The release of gases from soils cannot be modeled and must be explored by field studies. The permanent absorption of gases and particles at the ground surface is called dry deposition.

   The physical state of the atmosphere in the vicinity of the Earth's surface is determined by friction and by heat transfer. Friction increasingly reduces wind velocity and turbulence as one approaches the surface, so that the rate of eddy transport declines. Heating of the surface by solar radiation imparts energy to the overlying air, causing an enhancement of vertical motions and eddy transport due to convection. An increase of temperature with altitude in the atmosphere for a certain distance instead of the normal, adiabatic decrease is called a temperature inversion layer. Temperature inversions often occur a few 100 m above the ground, because of radiative cooling of the air below or advection of warmer air above the inversion level. Such inversion layers impede vertical mixing in the atmosphere because they stabilize buoyant air parcels, and they cause an accumulation of pollutants in the atmospheric boundary layer. As a consequence of these phenomena, the conditions in the boundary layer can vary considerably, making it impossible to establish a working average that is universally applicable.

Boundary layer conditions are called labile when convection predominates in the vertical air exchange; stable when the vertical air motions are dampened, such as by a temperature inversion; and indifferent or neutral when the vertical turbulence is induced solely by the horizontal wind field. In the last case the horizontal wind speed generally is found to increase with height above the ground in accordance with a logarithmic function,

$$\bar{u} = (u^*/k^*)\ln\left(\frac{z + z_0}{z_0}\right) \tag{1.29}$$

where z_0 is an empirical parameter, called the surface roughness length, $k^* \approx 0.4$ is von Karman's constant, and u^* is the friction velocity, which is a measure of the drag exerted on the wind at the ground surface. Because the expression is simple, neutral stability conditions are frequently assumed when the vertical flux of a trace gas in the boundary layer is considered.

If a trace gas in the atmosphere undergoes irreversible absorption or chemical reaction at the ground surface, the process will set up a vertical gradient of the mixing ratio, leading to a downward flux

$$F_s = -K_z(z)n\,dx_s/dz = -K_z(z)\,dn_s/dz \tag{1.30}$$

Here, the number concentration of air molecules, n, may be taken as constant within the first 100 m above ground, whereas $K_z(z)$ must be treated as a function of height z. Under steady-state conditions the flux remains constant (unless the gas undergoes rapid chemical reactions). For a constant flux the above equation can be integrated to yield

$$-F_s = [n_s(z) - n_s(0)] \Big/ \int_0^z \frac{dz}{K_z(z)}$$

$$= [n_s(z) - n_s(0)]/r_g(z) \tag{1.31}$$

where $r_g(z)$ represents the gas-phase resistance to material transfer. It includes the resistance of turbulent transport as well as the resistance caused by molecular diffusion in the laminar layer adjacent to the surface. An additional resistance r_s occurs at the surface itself in the uptake of the trace gas. If the surface resistance is defined as $r_s = n_s(0)/F_s$, the flux takes the form

$$-F_s = n_s(z)/[r_g(z) + r_s] = v_d n_s(z) \qquad (1.32)$$

$v_d = [r_g(z) + r_s]^{-1}$ is called the *deposition velocity*. It is usually determined from the concentration measured at a height about 1 m above the ground, or at a similar height above a canopy of high-growing vegetation. As n_s generally varies slowly with z, $v_d(z)$ is not strongly dependent on the choice of reference height.

Wind tunnel measurements combined with turbulence theory for hydro-dynamic flows influenced by surface friction have led to semiempirical expressions for the vertical fluctuation of wind speed $u' = l'\, d\bar{u}/dz$ and average mixing length $|\bar{l}'| = k_* z$ (Sutton, 1953). When these expressions are used together with Equations (1.16), (1.29), and (1.31), one obtains $r_g(z)$ in terms of wind and friction velocities:

$$r_g(z) = \bar{u}(z)/u_*^2 + B/u_* \qquad (1.33)$$

Here, the first term represents the resistance due to eddy transport, and the second term that of molecular diffusion. The factor $B \approx 0.1$ depends on the type of surface, but it is only mildly dependent on u^* (Chamberlain, 1966a; Garland, 1977). In this manner, $r_g(z)$ can be estimated, and Table 1.11 presents some numerical values. The results are maximum velocities, as they do not yet include the surface resistance r_s. For the uptake of gases by soils

TABLE 1.11   Turbulent Vertical Transport under Neutral Stability Conditions[a]

Type of surface	$\bar{u}$ (m s^{-1})	u_* (m s^{-1})	$r_g\ (z = 1)$ (10^2 s m^{-1})	$v_d\ (z = 1)$ (10^{-2} m s^{-1})
Grass	3	0.13	1.25	0.8
	10	0.45	0.41	2.44
Cereal crop	3	0.18	0.87	1.15
	10	0.59	0.32	3.12
Forest	3	0.27	0.26	3.85
	10	0.89	0.11	9.09

[a] Aerial resistance r_g and corresponding maximum deposition velocity v_d for gases according to the calculations of Garland (1979) for three surfaces and two wind speeds 100 m above ground.

and vegetation, r_s must be considered a quantity to be determined by experiment.

One of the techniques applied to determination of the rate of absorption or release of a trace gas from soils is to monitor its concentration in a volume of air inside a box or tent placed tightly on the ground surface, open face down. The walls of the enclosure must be inert to the gas being studied, and the air volume must be stirred to ensure adequate mixing. For high growing vegetation, leaf enclosure techniques are similarly used. The increase or decrease of the concentration shows directly whether the soil under study acts as a source or a sink of the trace gas. Figure 1.15 illustrates this point for the trace gases CO, H_2, and N_2O. In all three cases the gases undergo consumption when the initial mixing ratios in the air are high, whereas the gases are released when the initial mixing ratios are low. The mixing ratios tend toward a steady state at longer observation times (the so-called compensation point), which indicates that the rates of production and consumption in the soil are balanced. The balance is largely controlled by microbiological processes. The soil thus acts as a sink when the normal mixing ratio in ambient air is higher than the steady-state value (CO, H_2); otherwise the soil acts as a source (N_2O). Measurements of the uptake rates under actual field

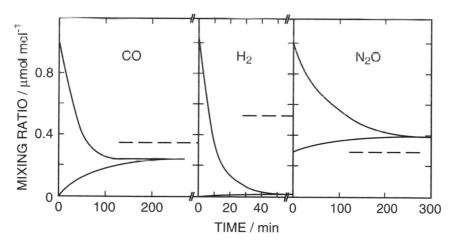

FIGURE 1.15   Temporal development of CO, H_2, and N_2O mixing ratios in a fixed volume of air in contact with natural soils. An increase in the mixing ratio indicates that the trace is released from the soil, whereas a decline indicates that it is absorbed. In all cases shown, a temperature-dependent steady-state level is reached after a certain time, regardless of the initial mixing ratio. The dashed lines indicate typical mixing ratios in ambient continental air. When it is higher than the steady-state level, the soil acts as a sink, otherwise it provides a source. (Data from Seiler (1978) and Seiler and Conrad (1981).)

conditions provide values for the surface resistance r_s. For CO, H_2, and a number of other trace gases, $r_s \gg r_g$, so that $v_d \approx 1/r_s$. However, for several important trace gases in the atmosphere, notably sulfur dioxide and ozone, the resistances r_s and r_g are comparable in magnitude. In these cases, the box method must be supplemented by measurements (or estimates) of r_g to determine the true deposition velocity.

Another technique for the determination of turbulent fluxes is the eddy correlation method, which involves the simultaneous measurement of the vertical wind component and trace gas concentration or mixing ratio on a fast time scale (1–10 Hz). Reynold's averaging splits the time dependent variable into a constant time average $\bar{a}$ and a fluctuating part $a'(t)$. The total flux due to the vertical wind component is given by $\bar{F} = \overline{wa} + \overline{w'a'}$, where the second term represents the turbulent part and the first term generally is close to zero unless convection induces an appreciable mean vertical wind component. Figure 1.16 shows results of flux measurements of ozone and nitrogen dioxide at two height levels above a coniferous forest. Whereas the O_3 fluxes are entirely negative, in agreement with the notion that ozone is deposited, the NO_2 fluxes are largely positive, indicating the existence of a source of nitrogen oxides. The primary source presumably is soil emission of NO, which is oxidized to NO_2 by ozone. Figure 1.16 also demonstrates the large daily variation of the fluxes, which is caused partly by increased turbulence during the day and partly by the higher resistance to trace gas uptake by plants at night. The uptake of trace gases by vegetation occurs primarily via the stomata (leaf breathing pores) followed by destruction within the plant tissue. Plants control the rate of water transpiration by opening or closing the stomata, and the resistance to trace gas uptake decreases or increases correspondingly. Table 1.12 presents selected values for deposition velocities determined by field experiments. McMahon and Denison (1979) have given a more comprehensive compilation.

Finally, we consider the exchange of trace gases between atmosphere and ocean. The simplest concept is that of the thin-film model, which assumes the flux across the ocean surface layer to depend on solubility of the substance in sea water and on the rate of molecular diffusion within the laminar layers adjacent to the air–sea interface. The situation is illustrated in Figure 1.17 (Danckwerts, 1970; Liss and Slater, 1974). The resistance to turbulent transport in both media is considered small compared to that exerted by molecular diffusion. Under steady state conditions, the flux through the interface is

$$F_s = \frac{D_g}{\Delta z_g}(c_{gi} - c_g) = \frac{D_L}{\Delta z_L}(c_L - c_{Li}) \qquad (1.34)$$

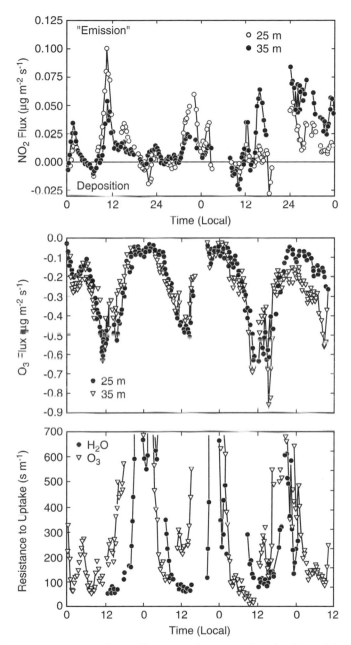

FIGURE 1.16 Upper parts: Fluxes of ozone and nitrogen dioxide observed by the eddy correlation technique at two levels (25 and 35 m) above a coniferous forest at Speulderbos, the Netherlands, during the period from June 29 to July 2, 1993. Lower parts: Resistance to the uptake of ozone compared with water vapor. (Reprinted with permission from Choularton *et al.*, © 1994 Backhuys Publishers.)

TABLE 1.12 Deposition Velocities for a Number of Atmospheric Trace Gases, as Derived from Field Measurements

Trace gas	Type of surface	v_d (m s^{-1})	Resistance (s m^{-1})	Method	Authors
H$_2$	Soil, bare and grass-covered	$(2-12) \times 10^{-4}$	$r_s = (0.8-5) \times 10^3$	Flux box	Liebl and Seiler (1976)
		7×10^{-4}		Flux box	Conrad and Seiler (1985a)
CO	Soil, bare and grass-covered	$(2-7) \times 10^{-4}$	$r_s = (1.4-5) \times 10^3$	Flux box	Liebl and Seiler (1976)
		$(0.8-5) \times 10^{-4}$	$(2-12.5) \times 10^3$	Flux box	Conrad and Seiler (1985a)
SO$_2$	Grass-covered soil	$(3-12) \times 10^{-3}$	$r_g + r_s = 83-330$	Gradient	Shepherd (1974), Garland (1977)
		8×10^{-3}			
	Wheat	7.4×10^{-3}	$r_g + r_s = 135$	Gradient	Fowler (1978)
	Pine forest	$(2-22) \times 10^{-3}$	$50-500$	Gradient, tracer	Garland (1977), Garland and Branson (1977)
	Water	4.5×10^{-3}	$r_g + r_s = 222$	Gradient	Garland (1977)
O$_3$	Soil, bare and grass-covered	$(4-8) \times 10^{-3}$	$r_g + r_s = 125-250$	Flux box	Aldaz (1969), Garland (1977), Galbally and Roy (1980)
		$(6-10) \times 10^{-3}$			
	Plants	$(0.3-3.3) \times 10^{-3}$	$r_g + r_s = 300-3300$	Flux box	Turner et al. (1974), Thorne and Hanson (1972)
HNO$_3$	Water, snow	$(2-4) \times 10^{-4}$	$r_s = (2.5-5) \times 10^3$	Flux box	Aldaz (1969), Galbally and Roy (1980)
	Grass-covered soil	$(2-3) \times 10^{-2}$	$r_g + r_s = 33-50$	Gradient, heat flux	Huebert and Robert (1985)
	Deciduous forest	$(2-5) \times 10^{-2}$		Gradient	Meyers et al. (1989)
NO$_2$	Pine needles	$(3-8) \times 10^{-3}$	$r_s = 125-250$	Flux box	Grennfelt et al. (1983)
	Soils	$(3-6) \times 10^{-3}$	$r_s = 167-360$	Flux box	Böttger et al. (1978)
	Vegetation	$(0.5-6) \times 10^{-3}$	$r_g + r_s = 167-2000$	Eddy correlation	Weseley et al. (1982)

The notation is explained in Figure 1.17, and D_g and D_L are the molecular diffusion coefficients in air and water, respectively. The flux is maintained by the concentration difference across the molecular diffusion layers. The flux is directed from the ocean to the atmosphere when the concentration in sea water is greater than that at the interface (the sign of the flux then is positive); the flux is directed from the atmosphere into the ocean when the concentration difference and the flux are negative. If the trace gas under consideration obeys Henry's law, the concentrations at the gas–liquid interface are connected by

$$c_{gi} = H^\circ c_{l,i} \qquad (1.35)$$

where H° is the Henry's law coefficient that is obtained when gas and liquid concentrations are expressed in the same units (e.g., mass per volume). This coefficient is related to that used in Section 8.4 by $H^\circ = H M_w / R_g T \rho_w$, where M_w is the molar mass, ρ_w is the density of sea water, R_g is the gas constant, and T is the absolute temperature. With the substitutions

$$\frac{D_g}{\Delta z_g} = k_g \quad \text{and} \quad \frac{D_l}{\Delta z_l} = k_L \qquad (1.36)$$

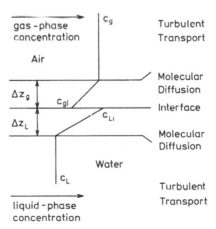

FIGURE 1.17 Two-layer thin-film model of the air–sea interface, adapted from Liss and Slater (1974). The concentrations at the interface, c_{gi} (gas phase) and c_{Li} (liquid phase), are related by Henry's law. Material transport across the interface occurs by molecular diffusion within the film layers.

one obtains the transfer coefficients k_g and k_L, and by combining the preceding equations

$$F_s = \left(c_L - c_g/H^0\right) \Big/ \left(\frac{1}{k_g H^0} + \frac{1}{k_L}\right) = \frac{1}{r}\left(c_L - c_g/H^0\right) \quad (1.37)$$

where

$$r = r_g + r_L = \frac{1}{k_g H^0} + \frac{1}{k_L} \quad (1.38)$$

represents the sum of the transfer resistances in both media. For water molecules crossing the air-sea interface, the liquid phase resistance is nil, so that in this case the resistance to transfer is that of the gas phase alone. Measurements of the transfer rate indicate a dependence on wind velocity. Liss and Slater (1974) have used the field data of Schooley (1969) to derive a mean value $k_g(H_2O) = 8.3 \times 10^{-3}$ m s^{-1}. For other gases the different molecular diffusivity must be incorporated, that is, $k_g(H_2O)$ must be multiplied by the ratio of the square roots of the molecular masses for H_2O and the other gas.

Gas exchange rates across the sea surface have been studied by various techniques, for example, by tracers such as radon or $^{14}CO_2$. The results indicate a liquid film thickness in the range of 25–60 μm, depending on wind speed and surface roughness (Liss, 1973; Broecker and Peng, 1971, 1974; Peng et al., 1979). For values such as $D_L \approx 2 \times 10^{-9}$ m^2 s^{-1} and $\Delta z_L = 36$ μm, one finds $k_L \approx 5.5 \times 10^{-5}$ m s^{-1}. In the case of gases that form ions upon dissolution in water, the transport across the liquid film is complicated by several diffusing species. For SO_2, for example, Liss and Slater (1974) estimated that the effect enhances k_L several thousandfold. Details of the problem need not be considered, however, because in the alkaline environment of sea water SO_2 is rapidly oxidized, so that its concentration is reduced to near zero.

Table 1.13 presents some examples for the exchange of trace gases between atmosphere and ocean. For SO_2 the ocean provides an efficient sink. In this case, the gas phase resistance exceeds that of the liquid phase. The deposition velocity $v_d = 4.4 \times 10^{-3}$ m s^{-1} agrees with the observational data of Table 1.12. Sea water absorbs ozone less well than SO_2, but the reaction of ozone with iodide reduces the liquid-phase resistance by an order of magnitude, as Garland et al. (1980) have shown. The deposition velocity calculated by taking this effect into account agrees reasonably with the field

TABLE 1.13 Air–Sea Exchange of Various Gases[a]

Trace gas	k_g (m s^{-1})	k_L (m s^{-1})	H^0	r_g (s m^{-1})	r_L (s m^{-1})	x_s	$(c_L - c_g/H^0)$ (μg m^{-3})	F_s (μg m^{-2} s^{-1})	Remarks
SO$_2$	4.4×10^{-3}	9.6×10^{-3}	0.02	1.1×10^4	10.4	45 pmol mol^{-1}	-6.4	-5.7×10^{-4}	$c_L = 0$, $v_d = 4.4 \times 10^{-3}$
O$_3$	5.1×10^{-3}	4.9×10^{-4}	3	6.5×10^1	2.0×10^3	25 nmol mol^{-1}	-17.9	-8.5×10^{-3}	$v_d = 5 \times 10^{-4}$
CCl$_4$	2.9×10^{-3}	3.0×10^{-5}	1.1	3.2×10^2	3.3×10^4	71 pmol mol^{-1}	-0.04	-1.2×10^{-6}	According to measurements of Lovelock et al. (1973)
CFCl$_3$	3.0×10^{-5}	3.1×10^{-5}	5	6.6×10^1	3.2×10^4	50 pmol mol^{-1}	-0.015	-4.7×10^{-7}	
N$_2$O	5.3×10^{-3}	5.5×10^{-5}	1.6	1.2×10^2	1.8×10^4	300 nmol mol^{-1}	$+68$	$+3.7 \times 10^{-3}$	For measurements see Chapters 4, 9, and 10
CO	6.7×10^{-3}	5.3×10^{-5}	50	3.0	1.8×10^4	150 nmol mol^{-1}	$+113$	$+6.2 \times 10^{-3}$	
CH4	8.8×10^{-3}	5.5×10^{-5}	42	3.7×10^{-1}	1.8×10^4	1.7 μmol mol^{-1}	$+9.5$	$+5.3 \times 10^{-4}$	
(CH3)$_2$S	4.5×10^{-3}	3.1×10^{-5}	0.3	7.4×10^2	3.2×10^4	100 pmol mol^{-1}	$+217$	$+6.5 \times 10^{-3}$	

[a] Given are the gas and liquid phase transfer coefficients, dimensionless Henry coefficients, and the corresponding resistances to transfer as defined in the text, according to Liss and Slater (1974), except for ozone (Garland et al., 1980). Approximate mixing ratios of the gases in surface air and liquid phase concentration differences resulting from the consumption (−) or production (+) in sea water determine the fluxes through the air–sea interface.

measurement value entered in Table 1.12. For the halocarbons CCl_4 and CCl_3F, the flux is also directed from the atmosphere toward the ocean because concentrations are higher in the atmosphere than in sea water. The other gases shown in Table 1.13 are released from the ocean to the atmosphere. In these cases the gases are produced by biological processes in sea water so, that a positive concentration difference is established.

While the thin-film model is useful in identifying whether r_g or r_L is the controlling resistance, it is not nearly as useful in calculating the total flux of trace gases emanating from the ocean, because of the wide variation in wind speed and its influence on the transfer coefficients. Empirical relationships are often used. Liss and Merlivat (1986), for example, distinguished three regimes of wind speed, $u \le 3.6 \text{ m s}^{-1}$(smooth surface), $3.6 < u < 13 \text{ m s}^{-1}$

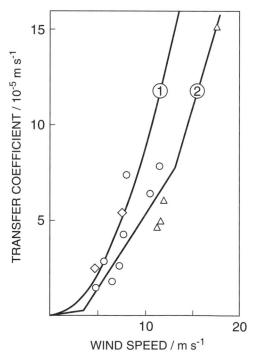

FIGURE 1.18   Liquid phase gas transfer velocity as a function of wind speed over the ocean, normalized to a Schmidt number Sc = 600 corresponding to CO_2 at 20°C. Open points are from radon measurements, diamonds from ^{14}C measurements, triangles from dual tracer measurements (for details see Wannikhof, 1992). Curve 1 is as proposed by Wannikhof (1992); curve 2 is as proposed by Liss and Merlivat (1986).

(rough surface), and $u > 13$ m s^{-1} (breaking waves), and combined results from wind tunnel studies and artificial tracer gas-exchange experiments in a lake to derive an empirical formula, in which the transfer coefficient depends linearly on wind speed in each regime. Wanninkhof (1992) and Erickson (1993) have used similar approximations. Figure 1.18 shows the empirical relationships suggested by Liss and Merlivat (1986), Tans *et al.* (1990), and Wanninkhof (1992) and compares them to transfer velocities derived from oceanic measurements, based primarily on the radon deficiency method.

CHAPTER 2

Photochemical Processes and Elementary Reactions

Chemical reactions that occur spontaneously with the admixture of two or more stable reagents are quite rare, because energy is usually required to stimulate a reaction. In the laboratory, heat is the most common source of energy; other forms of energy, such as visible or ultraviolet light, electric discharges, or ionizing radiation, are used less frequently. In the atmosphere, in contrast, most chemical processes are initiated by solar radiation, because the heat content of ambient air is insufficient for thermal activation, except in a few special cases, and lightning discharges and cosmic radiation are comparatively insignificant on a global scale.

The description of complex photochemical reactions is facilitated by subdividing the total reaction into a series of steps, which separate the primary (or initiation) process caused by the absorption of a photon from the subsequent thermal (or dark) reactions. The interaction of solar radiation with the atmosphere and its constituents and primary photochemical processes are examined in this chapter. Dark reactions will be discussed whenever appropriate. The identification of the individual reaction steps and their characterization by rate laws are the subject of chemical kinetics. Some

54

familiarity with reaction kinetics and the behavior of elementary reactions is essential to any discussion of atmospheric chemistry. Hence, a brief summary of the basic concepts is initially provided in this chapter, with emphasis on gas-phase reactions. Reactions occurring in the aqueous phase of clouds are discussed in Chapter 8. Reactions proceeding on the surface of aerosol particles may also be important in special cases.

Chemistry is basically a practical science, and knowledge of atmospheric reactions has been derived mostly from laboratory studies. Despite considerable advances in laboratory techniques it is often difficult to establish experimental conditions such that a reaction can be isolated from other concurrent ones so that its rate behavior can be unambiguously determined. Rate data generally are considered reliable only if at least two independent experimental techniques lead to comparable results. Great strides have been made in past decades in accumulating dependable information on rates of homogeneous reactions taking place in the gas phase and in the aqueous phase. Tables A.4, A.5, and A.6 contain reaction rate data as a reference for all subsequent discussions. The numbering in those tables is used throughout this book.

2.1. FUNDAMENTALS OF REACTIONS KINETICS

Chemical reactions manifest themselves by a change in the composition of a mixture of chemical substances. In the course of time, some substances disappear, while others are formed. The former are called *reactants*, and the latter *products*. Chemical reactions lead to a rearrangement of the atoms among the molecules making up the reacting mixture. Products may arise from a rearrangement of atoms within an individual molecule, from the disintegration or combination of molecules, from atom interchange among two or more reactant molecules, etc. In chemistry the integrity of the atoms is preserved. The law of mass conservation then leads to an important restriction: the number of atoms of each kind taking part in the chemical reaction remains constant, no matter how extensive a rearrangement of atoms takes place. To give a simple example, the complete oxidation of methane to carbon dioxide and water must be written as

$$CH_4 + 2O_2 \rightarrow CO_2 + 2H_2O$$

so that the same numbers of carbon, hydrogen, and oxygen atoms appear on the reactant and product sides of the equation. This requirement is known as the law of stoichiometry.

The oxidation of methane proceeds at measurable rates at temperatures between 400°C and 500°C. Although the final products are CO_2 and H_2O, one should not assume that the reaction occurs in a single step as written above. When the reaction is studied in a closed vessel, several intermediate products can be discerned, as Figure 2.1 shows. Formaldehyde and hydrogen peroxide appear first, followed by carbon monoxide, before carbon dioxide emerges. Thus, the reaction proceeds in stages. The above way of writing the reaction indicates only the overall chemical change, but it says nothing about the detailed pathways by which the final products are formed, nor does it account for the intermediate products occurring along the way.

Oxidation reactions, like other chemical reactions, are complex processes involving many individual reaction steps. A minimum set of reactions required to explain the experimental observations is called a *reaction mechanism*. The elucidation of such mechanisms is one of the principal aims of reaction kinetics. Mechanisms usually evolve from initial postulates, which are modified as a better understanding of the reaction system is obtained by further studies. Data such as those shown in Figure 2.1 do not contain enough information to support a specific reaction mechanism. Yet postulated

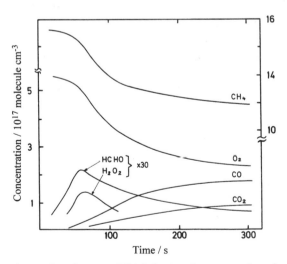

FIGURE 2.1   Oxidation of methane at 500°C: behavior of reactant and product concentrations with time. Initial number densities were $n(O_2) = 6.8 \times 10^7$, $n(CH_4) = 1.56 \times 10^{18}$, $n(N_2) = 2.63 \times 10^{18}$ (molecule cm^{-3}). The number densities for H_2O_2 were scaled from the data of Egerton *et al.* (1956). Adapted from measurements of Blundell *et al.* (1965).

reactions can be isolated for individual study in many cases, and experiments may be devised to verify or reject postulated mechanisms. In this way, many of the reactions participating in the oxidation of methane have been identified. They are listed in Table 2.1. The experimental evidence for their occurrence was reviewed by Hoare (1967).

The reactions in Table 2.1 are elementary in the sense that they cannot or need not be broken up further. It is evident that all of the reactions shown involve molecular fragments, either as reactants or as products. Species such as CH_3, CHO, OH, or HO_2 are called *radicals*. By their nature they are very reactive, so that reactions involving radicals usually are fast. The mechanism of methane oxidation contains three sequences of reactions in which a radical consumed in one reaction is regenerated in a following reaction. Such sequences are called *chain reactions*. The rate of methane consumption is almost entirely due to the first of the three reaction chains shown in Table 2.1. The direct interaction of methane with oxygen is slow. It serves to initiate the chain reaction by generating radicals in the first place. Figure 2.1 shows that the rate at which methane is consumed accelerates as the reaction progresses. This kind of behavior is known as *autocatalysis*. It is frequently observed when the production rate of one or more of the radicals increases. In the oxidation of methane the acceleration of the reaction rate is due to reaction (g), which describes the decomposition of H_2O_2 and the formation of two new OH radicals in place of one HO_2 radical consumed in reaction (f) to form H_2O_2. Ultimately, the propagation of reaction chains is stopped by

TABLE 2.1 Reaction Mechanism for the Thermal Oxidation of Methane[a]

(a)	$CH_4 + O_2 \rightarrow HO_2 + CH_3$	Initiation step
(b)	$CH_3 + O_2 \rightarrow CH_2O + OH$	First chain
(c)	$OH + CH_4 \rightarrow H_2O + CH_3$	First chain
(d)	$OH + CH_2O \rightarrow H_2O + CHO$	Second chain
(e)	$CHO + O_2 \rightarrow HO_2 + CO$	Second chain
(f)	$HO_2 + CH_2O \rightarrow H_2O_2 + CHO$	Second chain
(g)	$H_2O_2 + M \rightarrow 2OH + M$	Chain branching
(h)	$OH + CO \rightarrow CO_2 + H$	Third chain
(i)	$H + O_2 + M \rightarrow HO_2 + M$	Third chain
(j)	$HO_2 + CO \rightarrow CO_2 + OH$	Third chain
(k)	$OH + HO_2 \rightarrow H_2O + O_2$	Chain termination
(l)	$HO_2 + HO_2 \rightarrow H_2O_2 + O_2$	Chain termination
(m)	Loss of HO_2 and H_2O_2 at the walls of the vessel	Chain termination

[a] At about 400°C M indicates a chemically inert constituent, such as N_2, which acts mainly as an energy transfer agent.

the termination reactions (k)–(m). These reactions convert the chain carriers OH and HO_2 to less reactive products.

The occurrence of many of the reactions in Table 2.1 is not restricted to methane oxidation. OH and HO_2 are important chain carriers in most oxidation processes involving organic compounds. Whenever formaldehyde appears as an intermediate product, it will enter into the same reactions as those shown in Table 2.1. Methane and carbon monoxide are removed by OH radicals in the atmosphere as well. The reactions shown in Table 2.1 thus have a more universal significance than appears at first sight. In fact, by tacit agreement, universality is considered a requirement that any genuine elementary reaction must meet. This feature makes it possible to acquire knowledge about an elementary reaction from one particular experiment and then apply it to another experimental situation or to an entirely different environment, such as the atmosphere.

In the foregoing, mention has been made of reaction rates. This quantitative aspect of reaction kinetics requires additional discussion. The rate of change of any substance in a reaction mixture is defined by the derivative of its concentration with respect to time. The derivative is negative for reactants and positive for products. Explicit mention of the sign is rarely made, however. Instead it is customary to speak of rates of consumption and rates of formation, respectively. Intermediates in complex chemical reactions are first products, then reactants, so that their rates change from positive to negative during the reaction. Formaldehyde in Figure 2.1 provides an example of this behavior.

For complex reactions such as the oxidation of methane, no mathematically simple rate laws can be formulated. The rates of many of the species involved are complex mathematical functions of the concentrations of all of the other participating reactants and intermediates, temperature, pressure, sometimes also wall conditions of the vessel, and other parameters. Isolated elementary reactions, in contrast, obey comparatively simple rate laws. Table 2.2 summarizes rate expressions for several basic types of homogeneous elementary reactions. Let us single out for the purpose of illustration the bimolecular reaction $A + B \rightarrow C + D$. On the left-hand side of the rate expression one has equality between the rates of consumption of each reactant and the rates of formation of each product, in accordance with the requirement of stoichiometry. On the right-hand side, the product of the reactant concentrations expresses the notion that the rate of the reaction at any instant is proportional to the number of encounters between reactant molecules of type A and B occurring in a unit time and volume. The rate coefficient k_{bim} is still a function of temperature, but it is independent of concentration. The same is assumed to hold for the rate coefficients of other types of elementary reactions in Table 2.2. At a constant temperature the rate

TABLE 2.2 Rate Laws for Three Common Types of Elementary Chemical Reactions[a]

Reaction	Type	Rate law	Units of k
$A \rightarrow B + C$	Unimolecular decomposition	$-\dfrac{dn_A}{dt} = \dfrac{dn_B}{dt} = \dfrac{dn_C}{dt} = k_{uni}n_A$	s^{-1}
$A \rightarrow B + B$	Unimolecular decomposition	$-\dfrac{dn_A}{dt} = \dfrac{1}{2}\dfrac{dn_B}{dt} = k_{uni}n_A$	s^{-1}
$A + B \rightarrow C + D$	Bimolecular reaction	$-\dfrac{dn_A}{dt} = -\dfrac{dn_B}{dt} = \dfrac{dn_C}{dt} = \dfrac{dn_D}{dt} = k_{bim}n_A n_B$	$cm^3\ molecule^{-1}\ s^{-1}$
$A + A \rightarrow B + C$	Bimolecular reaction	$-\dfrac{1}{2}\dfrac{dn_A}{dt} = \dfrac{dn_B}{dt} = \dfrac{dn_C}{dt} = k_{uni}n_A^2$	$cm^3\ molecule^{-1}\ s^{-1}$
$A + B + M \rightarrow C + M$	Termolecular reaction	$-\dfrac{dn_A}{dt} = -\dfrac{dn_B}{dt} = \dfrac{dn_C}{dt} = k_{ter}n_A n_B n_M$	$cm^6\ molecule^{-2}\ s^{-1}$
$A + A + M \rightarrow B + M$	Termolecular reaction	$-\dfrac{1}{2}\dfrac{dn_A}{dt} = \dfrac{dn_B}{dt} = k_{ter}n_A^2 n_M$	$cm^6\ molecule^{-2}\ s^{-1}$

[a] M signifies an inert constituent that acts as a catalyst but whose concentration is not changed by the reaction.

coefficients are constants, and the equations can be integrated to yield concentrations of reactants and products as a function of time.

In laboratory studies of elementary reactions one will strive for conditions by which the system is isolated so that simple rate laws apply. In complex reaction systems a chemical species usually is involved in several reactions simultaneously. The appropriate rate expression then contains on the right-hand side as many terms as there are reactions in which the species participates. Formaldehyde in methane oxidation may again be cited as an example. According to Table 2.1, it enters into reactions (b), (d), and (f), so that the rate is given by the sum of the three individual rates

$$dn[\text{HCHO}]/dt = k_b n(\text{CH}_3)n(\text{O}_2) - k_d n(\text{OH})n(\text{HCHO})$$
$$- k_f n(\text{HO}_2)n(\text{HCHO})$$

where the $n(x)$ denote number concentrations of the various reactants. The rate equations for the other species appearing in the mechanism can be set up in a similar manner, but care must be taken to apply the correct stoichiometric factors associated with each term. There are altogether 11 species, so that one has a total of 11 equations. This result may be generalized: for any complex chemical mechanism, the number of rate equations is equal to the number of chemical species participating in the reaction. A complete set of rate equations forms a mathematically well defined nonlinear system of differential equations, which together with the initial conditions can be solved to yield the individual concentrations as a function of time. It is clear that closed solutions will exist only in special cases, but numerical solutions can always be obtained with the help of modern computers. In this way the main features of any reaction mechanism, even a hypothetical one, can be exposed. Reaction kineticists make liberal use of this technique to verify reaction mechanisms by comparing computed and experimental results. With appropriate modifications, the same procedure of computer simulation is used in atmospheric chemistry to interpret observational data in terms of specific reaction mechanisms. The success of such models depends on a sufficient knowledge of elementary reactions, the associated rate coefficients, and temperature and pressure dependence. This last aspect is considered next.

2.2. PROPERTIES OF RATE COEFFICIENTS

Temperature and pressure represent external variables that can be controlled in experimental studies of reaction rates. It has been found that the rate coefficients associated with the three basic types of elementary reactions in

Table 2.2 behave differently with regard to both variables, so that it is convenient to treat each reaction type separately.

Bimolecular reactions are basically independent of pressure. If any influence of pressure is observed, it is an indication that the reaction is not truly elementary, but involves more than one chemical process. This complication is ignored here. The rate of a bimolecular reaction increases as the temperature is raised. Laboratory experience shows that the increase in the rate coefficient over not too wide a temperature range can almost always be expressed by an equation of the form

$$k_{\text{bim}} = A \exp\left(-E_a/R_g T\right)$$

where T is absolute temperature, R_g is the gas constant (here expressed in J/mol K), and A and E_a are constants. The preexponential factor A is related to the collision frequency between reactants. In the gas phase, the upper limit is $\sim 10^{10}$ cm^3 molecule^{-1} s^{-1}, but usually it is smaller. The constant E_a has the dimension of energy/mole. This quantity is called the *activation energy*, following a proposal first made by S. Arrhenius in 1889, according to which only those molecules can enter into a reaction that are thermally activated to energies sufficient to surmount an energy barrier E_a. The exponential form of the temperature dependence follows from the Boltzmann distribution of energies among the various degrees of freedom of the molecules.

Figure 2.2 illustrates the modern concept of activation energy for the simplest bimolecular reaction, the atom exchange process A + BC → AB +

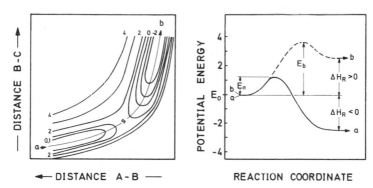

FIGURE 2.2   Left: Map of potential energy surface established in the interaction of three atoms A, B, and C. The reaction A + BC → AB + C proceeds from point a via a saddle point s to point b. Energy units are arbitrary and relative to the starting point. Right: Potential energy along the reaction coordinate for the forward reaction (solid line) and the reverse reaction (dashed line) relative to the starting points a and b, respectively. The heat of reaction is negative in the first case, and positive in the second.

C. The diagram on the left shows the contours of the potential energy surface established by the interaction of the three atoms involved in the colinear A-B-C collision complex. The preferred reaction path, shown by the solid line, starts out at point (a) in the reactant valley, passes a saddle point, then enters the product valley to end up at point (b). The diagram on the right shows the local potential energy along the reaction path relative to the energy at the starting position. The activation energy E_a corresponds to the height of the potential barrier at the saddle point. In the example chosen, the bottom of the product valley ultimately reaches an energy level lower than that at the starting point. Thus, energy is set free during the reaction. The excess energy appears temporarily in the vibrational, rotational, and translational modes of the products before it is dissipated to the surroundings. Reactions of this kind are called *exothermic*. The total energy made available, the heat of the reaction, ΔH_R, is negative, as the right hand-side of Figure 2.2 shows. The opposite case, a reaction requiring an energy input, is called *endothermic*. One can visualize this case by reversing the reaction path in Figure 2.2 (left) to proceed from point (b) to point (a). The reaction then reads $C + AB \rightarrow BC + A$. The behavior of potential energy along the reaction path relative to the energy at the starting point (b) is shown in Figure 2.2 (right) by the dashed curve. In this case, ΔH_R is positive. The activation energy now is the sum of the energies E_a and ΔH_R, that is, the activation energy is somewhat greater than the endothermicity of the reaction. Because of the exponential energy dependence of the rate coefficient, it follows that the endothermic pathway of our model reaction will be much slower at ambient temperatures than the exothermic pathway. At low temperatures, the reaction will proceed from point (a) to (b), whereas the reverse reaction can be neglected. At elevated temperatures the reverse reaction must be taken into account.

In the atmosphere, most endothermic reactions are unimportant because temperatures are not high enough. In addition, many exothermic reactions between stable atmospheric species are negligibly slow because of high activation energies. For example, the reaction $CO + NO_2 \rightarrow CO_2 + NO$ is exothermic by $\Delta H_R = -229.5$ kJ/mol, but the activation energy is $E_a = 132.3$ kJ mol^{-1} (Johnston *et al.*, 1957), leading to an exponential factor of 4×10^{-27}. Only reactions with activation energies lower than about 50 kJ mol^{-1} need to be considered in the atmosphere. Reactions involving radicals owe their importance largely to the fact that the associated activation energies lie predominantly in the low 5–40 kJ mol^{-1} range.

Thermal decomposition reactions require the breakage of a chemical bond and a corresponding energy input. These reactions are endothermic, and the rate coefficient displays an exponential temperature dependence. The activation energy is related to the endothermicity of the reaction. In the atmosphere, only a few molecules with low bond dissociation energies are capable of thermal decomposition, among them nitrogen pentoxide ($E_a = 88$ kJ mol^{-1}) and peroxyacetyl nitrate ($E_a = 104$ kJ mol^{-1}). For temperatures and pressures existing near the Earth's surface, 293 K and 1000 hPa, these substances decompose with time constants $\tau = 1/k$ of 10 and 6500 s, respectively (Cox and Roffey, 1977; Hendry and Kenley, 1977; Connell and Johnston, 1979). In the upper atmosphere and in the stratosphere, where temperatures and pressures are lower, both substances are considerably more stable. Note that admissible activation energies for unimolecular reactions in the atmosphere may be higher than those for bimolecular reaction because of more favorable preexponential factors.

In the gas phase, thermal decomposition reactions generally follow unimolecular rate behavior only at sufficiently high pressures (high pressure limit). Then the number of collisions effecting the transfer of energy from the carrier gas to the reactant molecules is high enough to establish an equilibrium of activated molecules with the surrounding heat bath; activating and deactivating collisions are in balance, and their rates exceed the decomposition rate. At reduced pressures the equilibrium may be disturbed in that the rate of decomposition of activated molecules becomes faster than that of deactivating collisions, so that the decomposition rate is collision-limited and a bimolecular rate law applies. In the intermediate pressure region between both limits the first order-rate coefficient depends on both the pressure and the nature of the carrier gas molecules. An example of the behavior of the rate coefficient as a function of pressure and temperature is shown in Figure 2.3 for the thermal decomposition of nitrogen pentoxide. Johnston (1966) and Connell and Johnston (1979) have discussed formulas by which the functional relationship $k(p, T)$ for the decomposition of N_2O_5 can be approximated.

Recombination reactions are the inverse of unimolecular dissociation processes, and the associated rate coefficients also exhibit a pressure dependence. Recombination reactions, like thermal decomposition processes, require an energy transfer by collisions. A third molecule M serves to remove the excess energy from the incipient product molecule, and the pressure dependence results from the efficiency of this process. The situation can be made clearer by writing the reaction as a sequence of two steps

$$A + B \rightleftharpoons AB^* \qquad\qquad k_q, k_r$$
$$AB^* + M \rightarrow C + M \qquad k_s$$

where AB^* is an energy-rich intermediate capable of redissociation to the original reactants. Only a collision with a chemically inert molecule stabilizes the AB^* product. If one treats the concentration of AB^* molecules as

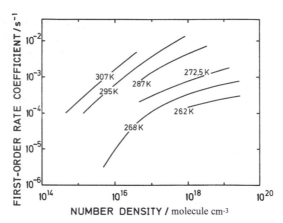

FIGURE 2.3   Empirical first-order rate coefficients for the thermal decomposition of N_2O_5 in nitrogen as a function of the effective number density and temperature. Interpolated data from Connell and Johnson (1979).

stationary so that the rates of formation and consumption are equal, one has

$$dn(AB^*)/dt = k_q n(A) n(B) - [k_r + k_s n(M)] n(AB^*) = 0 \qquad (2.1)$$

and the rate of formation of the final products becomes

$$\frac{dn(C)}{dt} = k_s n(AB^*) n(M) = \frac{k_s k_q n(A) n(B) n(M)}{k_r + k_s n(M)} \qquad (2.2)$$

It is apparent that two limits exist. At low pressures when $k_s n(M) \ll k_r$, the reaction will follow a termolecular rate law with an effective rate coefficient $k_0 = k_s k_q / k_r$, whereas at high pressures, when $k_s n(M) \gg k_r$, the $n(M)$ cancel, and the formation of the product C proceeds in accordance with a bimolecular rate law. The effective rate coefficient then is $k_\infty = k_q$. In the intermediate pressure range neither rate law applies, and the rate coefficient becomes pressure dependent. Usually, the rate of the reaction depends somewhat on the nature of the third body M as well.

   Recombination reactions differ from bimolecular and thermal decomposition reactions also by their weak and usually negative temperature dependence. Recombination reactions do not require an activation energy. Rather, the excess energy has to be dissipated. If the temperature dependence of the rate coefficient is expressed nonetheless by an exponential factor, the exponent $-E_R/R_g T$ is positive and E_R constitutes just a parameter with no physical meaning. Frequently, the temperature dependence of the rate coefficient k_R for a recombination reaction is written in the form

$$k_R = k_R^{300}(300/T)^m \qquad (2.3)$$

with k_R^{300} referring to the value of k_R at 300 K. Values for m range from 0.5 to 5. For recombination coefficients with a strong pressure dependence, Zellner (1978) has given an approximation formula based on the theoretical work of Troe (1974):

$$k_R(M, T) = \left[\frac{k_0(T)n(M)}{1 + k_0(T)n(M)/k_\infty(T)} \right] \times 0.8^a$$

$$a = \left(1 + \left\{ \log_{10}[k_0(T)n(M)/k_\infty(T)] \right\}^2 \right)^{-1} \qquad (2.4)$$

which is well applicable to atmospheric conditions. The terms $k_0(T)$ and $k_\infty(T)$, respectively, are the low- and high-pressure limiting values for the rate coefficient k_R. Their temperature dependencies are given by Equation (2.3). Figure 2.4 compares recombination rate coefficients for the reaction $OH + NO_2 \rightarrow HNO_3$ measured as a function of temperature and N_2 carrier gas pressure by Anastasi and Smith (1976) with rate coefficients interpolated using Equation (2.4). Zellner (1978) has compiled experimental data for several recombination reactions of importance to atmospheric chemistry to compute the effective rate coefficients as a function of altitude. The data are shown in Table 2.3. Further data on recombination reactions occurring in the atmosphere are summarized in Table A.5.

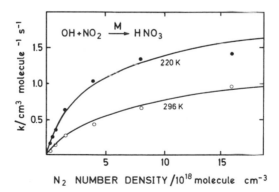

FIGURE 2.4 Bimolecular rate coefficient for the reaction $OH + NO_2 \rightarrow HNO_3$ in nitrogen as a function of number density and temperature. Points indicate measurements of Anastasi and Smith (1976); solid curves were calculated with $k_0 = 2.6 \times 10^{-30}(300/T)^{2.9}$ and $k_\infty = 1.5 \times 10^{-11}(300/T)^{1.3}$, using the approximation formula of Zellner (1978).

TABLE 2.3 Effective Bimolecular Rate Coefficients (cm^3 molecule^{-1} s^{-1}) for Several Recombination Reactions as a Function of Altitude in the Atmosphere[a] (Zellner, 1978)

z (km)	p (hPa)	T (K)	n_M (molecule cm^{-3})	Effective rate coefficient k_{bim} (cm^3 molecule^{-1} s^{-1})				
				OH + NO$_2$	OH + NO	OH + SO$_2$	ClO + NO$_2$	HO$_2$ + NO$_2$
0	1013	288	2.5 (19)	1.2 (−11)	4.9 (−12)	8.9 (−13)	2.3 (−12)	1.2 (−12)
5	487	268	1.3 (19)	1.0 (−11)	4.1 (−12)	8.2 (−13)	1.8 (−12)	1.0 (−12)
10	237	233	7.4 (18)	9.6 (−12)	3.7 (−12)	7.8 (−13)	1.6 (−12)	9.6 (−13)
15	121	220	4.0 (18)	7.8 (−12)	2.8 (−12)	6.8 (−13)	1.1 (−12)	7.7 (−13)
20	55	217	1.8 (18)	5.4 (−12)	1.7 (−12)	5.0 (−13)	5.8 (−13)	5.1 (−13)
25	26.7	222	8.5 (17)	3.1 (−12)	8.9 (−13)	3.1 (−13)	2.6 (−13)	2.8 (−13)
30	11.9	227	3.8 (17)	1.5 (−12)	4.1 (−13)	1.7 (−13)	1.1 (−13)	1.4 (−13)
35	5.7	235	1.8 (17)	7.2 (−13)	1.9 (−13)	8.2 (−14)	4.9 (−14)	6.3 (−14)
40	2.9	250	8.5 (16)	3.0 (−13)	7.9 (−14)	3.5 (−14)	1.9 (−14)	2.6 (−14)
45	1.47	260	4.0 (16)	1.3 (−13)	3.4 (−14)	1.5 (−14)	8.2 (−15)	1.1 (−14)
50	0.67	270	1.8 (16)	5.6 (−14)	1.4 (−14)	6.5 (−15)	3.4 (−15)	4.6 (−15)

[a] Powers of 10 are indicated in parentheses.

2.3. PHOTOCHEMICAL PROCESSES

As stated earlier, chemical reactions in the atmosphere are initiated mainly by photochemical processes rather than by thermal activation. This section gives a brief account of the principles of photochemistry.

Light can become photochemically effective only if it is absorbed—that is, the energy associated with radiation must be incorporated into a molecule. The mere interaction of light with matter, such as in scattering, is not sufficient to initiate a chemical reaction. Moreover, energy is quantized. A ray of light may be described as a flux of photons, each of which represents an energy quantum $E = h\nu$ proportional to the frequency ν of light and inversely proportional to the wavelength $\lambda = c/\nu$, where c is the velocity of light. In the act of optical absorption, exactly one photon is taken up by the absorbing molecule, whose internal energy is thereby raised by exactly the amount of energy supplied by the photon. According to a requirement of quantum chemistry, the energy of the photon must coincide with the energy level of one of the many higher states of the molecule relative to its initial state. If the molecular states are widely spaced, the spectrum of absorption as a function of wavelength will consist of discrete lines. A dense spacing will give the appearance of a continuous absorption spectrum. The strength of the absorption is determined by the probability of such transitions between the energy states.

It is customary to distinguish energy levels due to rotational, vibrational, and electronic motions within the molecule. Under normal conditions, the excitation of rotational and vibrational levels does not provide enough energy for photochemical action. Because of spectroscopic selection rules, high-lying vibrational levels are not directly accessible from the vibrational ground state. Thus, photochemical processes generally require transitions to higher electronic states. In the excitation process a part of the energy usually is channeled also into vibrational modes. Commonly, energies in excess of 100 kJ per mole of photons are involved, corresponding to light in the very near infrared, visible, and ultraviolet spectral regions.

The fate of the excitation energy depends on the nature of the molecule and on the amount of energy it receives. The excited molecule may give off the energy as radiation (fluorescence), dissipate it by collisions (quenching), utilize the energy for chemical transformations (isomerisation, dissociation, ionization, etc.), transfer all or a part of the energy to other molecules that then react further (sensitization), or enter into chemical reactions directly. Several of these processes are written in Table 2.4 in the form of chemical reactions. They are considered primary processes in the sense that they all involve the excited molecule formed initially by photon absorption. Any

TABLE 2.4   Types of Primary Photochemical Processes Following
Photon Absorption AB + $h\nu \rightarrow$ AB*

(1)	AB* $\rightarrow$ AB + $h\nu'$	Fluorescence
(2)	AB* $\rightarrow$ A + B	Dissociation
(3)	AB* + M $\rightarrow$ AB + M	Quenching
(4)	AB* + C $\rightarrow$ AB + C*	Energy transfer
(5)	AB* + C $\rightarrow$ A + BC	Chemical reaction

subsequent thermal chemical reactions or physical energy dissipation processes are of a secondary nature. Specific examples for both primary and
secondary reactions are given in later chapters.

The quantitative assessment of photochemical activity is facilitated by
introducing the quantum yield. In the practice of photochemistry a variety of
quantum yield definitions are in use, depending on the type of application.
There are quantum yields for fluorescence, primary processes, and final
products, among others. In atmospheric reaction models, the primary and
secondary reactions are usually written down separately, so that the primary
quantum yields become the important parameters. Only these will be considered here. Referring to Table 2.4, it is evident that an individual quantum
yield must be assigned to each of the primary reactions shown. The quantum
yield for the formation of the products P_i in the ith primary process is the
rate at which this process occurs in a given volume element divided by the
rate of photon absorption within the same volume element.

$$\phi_i = \frac{\text{product molecules } P_i \text{ formed/cm}^3 \text{ s}}{\text{photons absorbed/cm}^3 \text{ s}} \qquad (2.5)$$

The quantum yield thus represents the probability of the occurrence of a
selected process compared to the total probability for all primary processes
taken together. The sum of the individual primary quantum yields is unity.

A complete set of primary and secondary photochemical processes constitutes the mechanism of the overall photochemical reaction. The resulting
changes in concentrations with time are described by the appropriate system
of rate equations, as explained in Section 2 of this chapter. In setting up the
rate equations, the primary processes can be treated as elementary reactions
in the same way as thermal elementary reactions. The rate of chemical
change associated with an individual primary process can be quantified in
the following manner. Let light of wavelengths within a range $\Delta\lambda$ pass
through a layer, not necessarily homogeneous, containing different kinds of
absorbing molecules. Consider molecules of type a, present in number
concentration n_a. They contribute to the local rate of light absorption at each

wavelength the term

$$[-dI(\lambda)/dl]_a = \sigma_a(\lambda) n_a I(\lambda) \tag{2.6}$$

where $I(\lambda)$ is the local photon flux at the wavelength considered, l is the absorption path parameter, and σ_a is the absorption cross section. The local rate of product formation at the same wavelength λ, resulting from the ith primary process associated with the absorption of type a molecules, is the product of $[-dI(\lambda)/dl]_a$ with the appropriate quantum yield $\phi_i(\lambda)$. The total local rate for the formation of a specific product from light of all wavelengths is given by the integral over the entire wavelength range

$$dn_i/dt = \int_{\Delta\lambda} \phi_i(\lambda)[-dI(\lambda)/dl]_a \, d\lambda$$

$$= n_a \int_{\Delta\lambda} \phi_i(\lambda) \sigma_a(\lambda) n_a I(\lambda) \, d\lambda - j_i n_a \tag{2.7}$$

As the right-hand side shows, the primary photochemical process can be treated as a unimolecular reaction, where the integral, denoted by j_i, takes the place of a rate coefficient. If the primary process is a photodissociation, j_i is called the *photodissociation coefficient*. Its determination requires an evaluation of the integral and the parameters that enter into it. The terms $\sigma_a(\lambda)$ and $\psi_i(\lambda)$ are molecular properties, which can be determined by measurement. Both parameters may also depend on pressure. The photon flux $I(\lambda)$ is a local parameter, which in the laboratory can be confined to a narrow wavelength band. For optically thin conditions it equals the incident light flux. In the atmosphere, one must take into account the attenuation of solar radiation and corresponding changes in the spectral intensity distribution. Solar radiation additionally undergoes scattering that gives rise to a diffuse flux component. Both the direct radiation flux and the diffuse radiation flux contribute to the photochemical effects. In the visible and ultraviolet spectral regions, attenuation by absorption is confined to the upper atmosphere, whereas scattering of radiation becomes important in the lower atmosphere. The next section will give a brief description of the solar spectrum and its attenuation in the atmosphere.

2.4. ATTENUATION OF SOLAR RADIATION IN THE ATMOSPHERE

Figure 2.5 shows the spectral distribution of solar radiation outside the earth's atmosphere and the direct flux reaching the ground surface. Table A.3 of the Appendix lists numerical data. Fluxes in the visible and near-infrared

70

Chapter 2. Photochemical Processes and Elementary Reactions

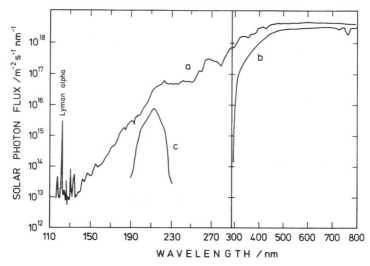

FIGURE 2.5 The solar flux spectrum in the 120–800-nm wavelength region (a) outside the atmosphere, (b) at sea level, and (c) at about 30 km altitude to show the atmospheric window in the 185–215 nm wavelength region.

spectral regions were mainly derived from ground-based measurements at mountain stations, supplemented by aircraft and balloon measurements. The ultraviolet portion of the spectrum has been explored with instruments on board rockets and satellites. These data are of special interest here. Measurements before 1980 have been reviewed by Ackerman (1971), Simon (1978), and Nicolet (1981). More recent data are available from Mount and Rottman (1981, 1983, 1985) and Mentall et al. (1985).

The solar spectrum consists of a continuum superimposed by line structure. Emission lines as well as (Fraunhofer-type) absorption lines are present. The continuum is dominant in the visible and near-ultraviolet spectral regions. At wavelengths above 400 nm the continuum flux distribution corresponds to that of a black body at 5900 K. Below 400 nm the solar flux decreases more strongly than a black body temperature of 5900 K would allow. At wavelengths between 200 And 250 nm the flux corresponds to a black body temperature of 5100 K, and at wavelengths between 130 and 170 nm to 4600 K. Below 200 nm there is an increasing contribution of solar emission lines, the strongest of which is the hydrogen Lyman alpha line at 121.6 nm. Most of the other lines in the intensity distribution cannot be separated from the continuum because the absolute intensity measurements, on which the data in Figure 2.5 are based, were made with a moderate spectral resolution, causing the line structure to be smeared out.

An adequate knowledge of the solar ultraviolet flux is essential to an understanding of stratospheric photochemistry. For the evaluation of photodissociation coefficients, a spectral resolution of 1 nm usually suffices. A higher resolution is required in some important cases involving molecules subject to predissociation, such as oxygen or nitric oxide. Predissociation occurs from a bound electronic state due to perturbation by another dissociating state. Then the fine structure of the absorption bands and their overlap with solar lines must be considered in detail. Simon (1978) has emphasized that photochemical calculations require fluxes representative of integration over the entire solar disk. Measurements made with high spatial resolution at the center of the disk provide only upper-limit fluxes due to the effect of limb darkening. Partly for this reason, the fluxes derived from the early measurements by Detwiler *et al.* (1961) are now considered too high. All of the data, however, are uncertain by about ±50% because of calibration problems. In addition, the solar flux exhibits a certain natural variability. This aspect has been discussed in some detail by Brueckner *et al.* (1976) and by Heath and Thekaekara (1977). The flux in the ultraviolet is more variable than that in the visible range of the spectrum. Solar emission lines originate mainly from the sun's chromosphere and corona and are influenced to a greater extent by sunspots and related phenomena than the continuum at wavelengths above 300 nm, which is emitted from the photosphere. Variations occur on a time scale of minutes, due to solar flares; days, caused by the rise and decay of active regions; 27 days, due to the sun's rotation; and 11 years, because of the different phases of the solar cycle. Satellite measurements have shown that at wavelengths near 120 nm the peak-to-peak variations amount to 35% during one solar rotation (Heath, 1973; Vidal-Madjar, 1975, 1977). Individual flares can change the flux of the Lyman alpha line by up to 16%. At wavelengths near 190 nm the flux varies by less than 10%, whereas at wavelengths greater than 210 nm the variation is less than 1%. These intensity variations occur in addition to the cyclic modulation by ±3.3% of the entire solar flux due to the eccentricity of the earth's orbit around the sun. All fluctuations are generally neglected when considering photochemical processes in the atmosphere.

With regard to absorption processes in the atmosphere, the spectrum of the sun may be subdivided into three wavelength regions: < 120 nm, 120–300 nm, and 300–1000 nm. Radiation within the extreme ultraviolet portion of the spectrum is absorbed at altitudes above 100 km, which will not be considered here. Note, however, that the high photon energy leads partly to photoionization and to the formation of the earth's ionosphere (for details see, for example, the treatise of Banks and Kockarts, 1973). Radiation within the second spectral region, 120–300 nm, is absorbed mainly in the mesosphere and stratosphere. The associated attenuation of solar radiation and the resulting photochemical processes are discussed below. Radiation within the third spectral region penetrates into the troposphere, although it is somewhat reduced by Rayleigh scattering. Radiation at wavelengths longer than 1000 nm does not supply enough energy for photochemical action. As was noted earlier, however, this infrared radiation is of prime importance to the heat balance of the atmosphere.

Nitrogen, although the major constituent of the atmosphere, is not an effective absorber at wavelengths greater than 120 nm, because the principal absorption system, the Lyman–Birge–Hopfield bands, covers only the range below 145 nm and the absorption is extremely weak. The main absorbers of solar radiation in the ultraviolet spectral region are oxygen and ozone. Figure 2.6 shows the absorption cross section of oxygen as a function of wavelength. Prominent features are the Herzberg continuum, the Schumann–Runge bands, and the Schumann continuum. A discussion of the O_2 absorption spectrum and the associated photochemical processes is facilitated by considering the energy level diagram of oxygen shown in Figure 2.7. Absorption is due to transitions from the ground state to electronic states at

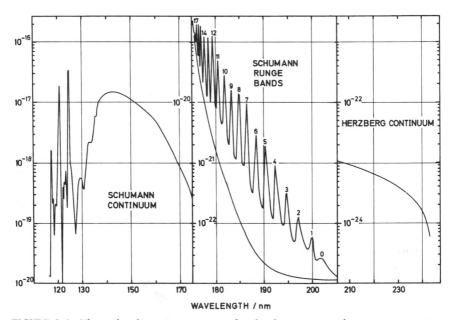

FIGURE 2.6   Ultraviolet absorption spectrum of molecular oxygen, with cross sections given in cm^2 molecule^{-1}. (Data compiled from Ackerman and Biaumé (1970), Ditchburn and Young (1962), Hudson *et al.* (1966), Ogawa (1971) Shardanand (1969), Shardanand and Rao (1977), and Watanabe *et al.* (1953).)

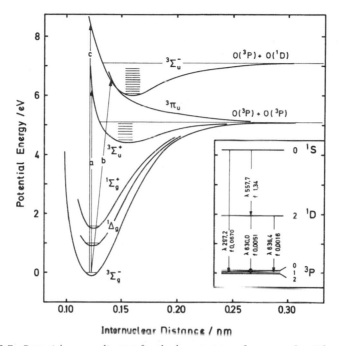

FIGURE 2.7   Potential energy diagram for the lowest states of oxygen, after Gilmore (1965). The indicated transitions produce the following absorption spectra. (a) Herzberg continuum, (b) Schumann–Runge bands, (c) Schumann continuum. The inset shows the lowest energy levels of the oxygen atom, the wavelengths of emission lines resulting from the transitions indicated, and the associated oscillator strengths. Energy units: 1 eV = 96.5 kJ mol^{-1}.

higher energies. Transitions to the low-lying excited states $^1\Delta_g$ and $^1\Sigma_g$ are forbidden by spectroscopic selection rules, so that the transition probabilities are low. The corresponding weak absorption bands lie in the near-infrared spectral region, where they are observed in the solar spectrum at long atmospheric path lengths. The transitions to the discrete levels of the $^3\Sigma_u^+$ state give rise to the weak Herzberg bands in the wavelength region 250–300 nm. These bands are shielded by the much stronger absorption of ozone in the atmosphere, which also makes them unimportant. At wavelengths below 242.2 nm (i.e., at energies exceeding 5.11 eV or 492 kJ mol^{-1}), the spectrum becomes continuous, because the transition to the $^3\Sigma_u^+$ state now leads to the dissociation of oxygen. A portion of the Herzberg continuum is apparent in Figure 2.6 at wavelengths above 190 nm.

The transition to the next higher $^3\Sigma_u^-$ state is the origin of the Schumann–Runge bands. They merge with the Schumann continuum when, at

175 nm, the second dissociation limit of O_2 is reached. Another transition in the same energy range is that to the repulsive $^3\Pi_u$ state. This absorption process probably is responsible for the continuum underlying the Schumann–Runge bands. The intensity of such transitions is governed largely by the Franck–Condon principle, which states that the positions of atomic nuclei in the molecule should change little if the transition is to be favorable. The rise in the absorption cross section with decreasing wavelength (increasing energy) can be ascribed to the increase in the Franck–Condon probability. The absorption intensity attains a maximum value near 145 nm for a nearly vertical transition. At still shorter wavelengths the O_2 absorption spectrum consists of a greater number of diffuse bands associated with transitions to higher, not well characterized dissociating states.

Ozone absorbs radiation throughout the near-infrared, visible, and ultraviolet spectral regions. The dissociation $O_3 \rightarrow O_2 + O$ has a low energy requirement ($100\ kJ\ mol^{-1}$), corresponding to the wavelength limit 1190 nm, so that photodissociation is effective in all of these absorption regions. In the visible and near-infrared, the diffuse Chappuis bands represent a broad yet weak dissociation regime. Owing to the weakness of the absorption, the attenuation of the solar flux is comparatively minor (see Fig. 2.5). Absorption cross sections for ozone in the spectral region below 350 nm are shown in Figure 2.8. At the long-wavelength limit the absorption sets in with the Huggins bands, then blends into the Hartley band, which has the appearance of a continuum topped by weak discrete structure. Here, the cross section reaches a first maximum at 255 nm. Toward shorter wavelengths follows a region of lower absorption intensity before the cross section rises again toward a maximum at 135 nm and additional peaks at still shorter wavelengths. This absorption region is of lesser interest here, because it is dominated by oxygen, and ozone contributes little to the attenuation of the solar flux.

In those spectral regions where the attenuation of incoming solar radiation is due primarily to absorption, the intensity of the solar flux having penetrated to the altitude level z_0 follows from the generalized form of Beer's law,

$$I(\lambda, z_o, \chi) = I_0(\lambda)\exp\left\{-(\cos \chi)^{-1}\int_{z_o}^{\infty}\left[\sum_k \sigma_k(\lambda)n_k(z)\right]dz\right\} \qquad (2.8)$$

where $I_0(\lambda)$ the incident solar flux outside the earth's atmosphere, χ is the solar zenith angle, and the summation

$$\sum_k \sigma_k(\lambda)n_k(z) = \sigma_{O_2}(\lambda)n_{O_2}(z) + \sigma_{O_3}(\lambda)n_{O_3}(z) \qquad (2.9)$$

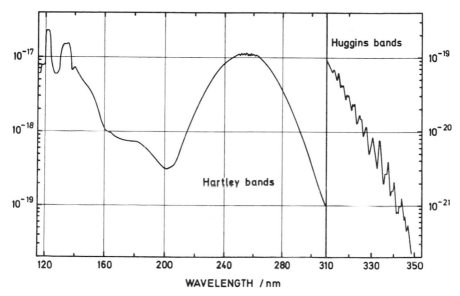

FIGURE 2.8 Absorption spectrum of ozone in the wavelength region 115–350 nm, with cross sections given in cm^2 molecule^{-1}. (Data assembled from Inn and Tanaka (1953), Tanaka *et al.* (1953), and Griggs (1960).)

involves only the concentrations and absorption cross sections of oxygen and ozone. The contributions of all other constituents of the atmosphere can be neglected.

The attenuation of solar radiation as a function of wavelength is most conveniently illustrated by plotting the altitude at which the solar flux is diminished to one-tenth of its initial value at normal incidence. Such a plot is shown in Figure 2.9. The absorption by oxygen in the Schumann region removes most radiation at wavelengths shorter than 160 nm already at altitudes above 90 km, but a number of windows exist that allow several wavelengths to penetrate to lower altitudes. Most notable is the Lyman alpha line, which falls exactly into one of the windows. In the wavelength region below 190 nm, where absorption cross sections of oxygen and ozone are comparable in magnitude, oxygen is the principal absorber because the mixing ratio of ozone is always much smaller. At wavelengths above 210 nm, the absorption cross section of oxygen falls to very low values, so that ozone becomes the principal absorber. The number concentration of ozone does not decrease quasi-exponentially with altitude like that of oxygen. Ozone arises as a photodissociation product from oxygen and attains its maximum number density at altitudes near 25 km. In this region an ozone layer is

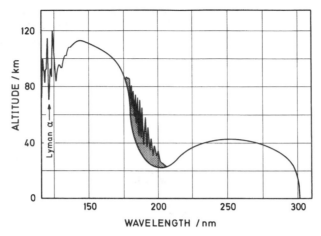

FIGURE 2.9   Altitude at which incoming solar radiation from an overhead sun is attenuated to on-tenth the initial intensity.

formed that shields the lower atmosphere from solar radiation at wavelengths between 210 and 300 nm. Between the two principal absorption regimes of oxygen and ozone, in the 190–210 nm wavelength region, there exists an optical window where solar radiation reaches deeper into the atmosphere. Radiation in this window region is responsible for the photodissociation of several minor atmospheric constituents that absorb in this wavelength range but not at wavelengths above 300 nm. Water vapor, H_2O, nitrous oxide, N_2O, and the halocarbons CF_2Cl_2 and $CFCl_3$ are thus affected. The compounds are chemically quite inert, and processes removing them from the troposphere are unknown (except for H_2O). Upward transport into the photodissociation region of the stratosphere appears to provide the only loss processes.

Beginning in the lower stratosphere, but becoming most important in the lower troposphere, is the phenomenon of radiation scattering. The principal processes involved are Rayleigh scattering of light by air molecules and Mie scattering by airborne solid and liquid particles. The attenuation of incoming solar radiation due to scattering is again expressed by Equation (2.8), where now

$$\sum_k \sigma_k(\lambda) n_k(z) = \sigma_{N_2}^R(\lambda) n_{N_2}(z) + \sigma_{O_2}^R(\lambda) n_{O_2}(\lambda) + \sigma^M(\lambda) n_p(z)$$

$$(2.10)$$

Here, the superscripts R and M signify Rayleigh and Mie scattering, respectively, and n_p is the number concentration of particles. Rayleigh cross sections vary with λ^{-4} and have values ranging from about 2×10^{-25} to less than 10^{-26} cm^2 molecule^{-1} in the wavelength region 300–700 nm. The scattering cross section for particles rises with the square of particle radius. In the absence of clouds only particles in a narrow range of the aerosol size spectrum, about 0.2–2 μm, are effective in scattering; larger particles are not sufficiently numerous, and smaller particles contribute little because of their small cross section. Number concentrations of particles maximize near ground level and decrease rapidly with altitude (see Fig. 7.28). The Mie scattering coefficient is roughly independent of wavelength in the 300–800 nm spectral region. Under urban conditions, for particle concentrations that are elevated compared to the normal background, Rayleigh scattering dominates the attenuation of solar radiation in the wavelength region 300–500 nm, whereas Mie scattering takes over at longer wavelengths.

Unlike absorption, scattering does not result in a loss of radiation. Whereas the direct flux of incoming solar radiation is attenuated, the diverted light undergoes multiple scattering and is available as a diffuse component, which must be added to the direct flux. In addition, some radiation is reflected at the ground surface (or from clouds, although clouds are rarely considered). Thus the total flux may be written

$$I_{\text{total}} = I_{\text{direct}} + I_{\text{diffuse}} + I_{\text{reflected}} \tag{2.11}$$

The reflected light also undergoes scattering. A photochemically active volume of the lower atmosphere evidently receives radiation from all directions and not just from the preferred direction pointing toward the sun. Accordingly, it is necessary to integrate over fluxes incident on the air volume from all directions to determine the photochemically effective or actinic flux. This integrated flux must then be inserted into Equation (2.7) to derive the photodissociation coefficients.

The diffusive component of the total flux and its contribution to photodissociation rates can be calculated from radiative transfer theory with the help of fast computers. (Braslau and Dave, 1973; Peterson, 1976; Luther and Gelinas, 1976). To save computing time, simplified models of radiative transfer usually are invoked. Leighton (1961), who first recognized the problem to occur as part of photochemical air pollution models, devised a simple routine for calculating the actinic flux at the earth surface. For more complex photochemical models of the lower atmosphere involving many photochemical processes, Isaksen et al. (1977) argued that because Rayleigh scattering directs light preferentially forward and backward with equal probability, it would simplify the calculations if scattering were assumed to occur primarily in the direction of the solar light beam. Although this is an oversimplification, the method gives reasonably accurate results and is widely applied (Madronich, 1987a). Improved algorithms of the two-stream method distinguish between diffuse and direct light and include reflection from the ground surface (Brühl and Crutzen, 1989). Peterson (1976) included scattering due to aerosol particles, which scatter light preferentially in the forward direction. Figure 2.10 shows some of his results to illustrate the relative contribution of upward scattered radiation to the total radiation flux. Generally, the upward component increases with altitude near the earth surface, partly because aerosol particles were assumed to absorb some of the radiation and the concentration of particles is greatest near the surface. The upward scattered flux is largest at short wavelengths and for small zenith angles. For large zenith angles the upward-scattered flux is comparatively small, because less radiation reaches the ground surface; in this case, however, downward scattering makes a fairly large contribution to the total flux.

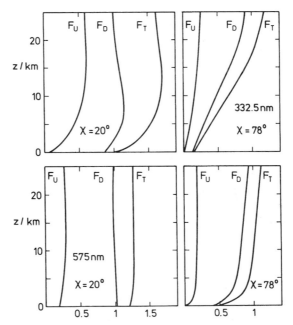

FIGURE 2.10   Upward directed flux F_U, downward directed flux F_D, and total actinic flux F_T as a function of altitude in the lower atmosphere. Values are given relative to a solar constant of unity for two wavelengths, 332.5 and 575 nm, and for two zenith angles, 20° and 78°. The calculations of Peterson (1976) included Rayleigh scattering, absorption by ozone, and scattering and absorption by aerosol particles.

A further complication of an already complicated subject is caused by the presence of clouds. Small clouds such as fair weather cumuli produce much forward scattering, so that the situation differs little from that for a clear sky. An extended cloud cover modifies the situation considerably, however. The calculations of Madronich (1987a) indicate that above the cloud and within its upper layers, the flux is elevated over that in cloudless air; with increasing penetration into the cloud, that is with increasing cloud thickness, the flux is progressively attenuated. Madronich (1987a) and Bott and Zdunkowsky (1987) also have shown that the actinic flux inside cloud drops is raised by a factor of about 1.8 above that in the surrounding air, mainly because of internal reflection. This effect is important for photochemical processes occurring in the aqueous phase of clouds.

2.5. PHOTODISSOCIATION OF OXYGEN AND OZONE

Continuing the discussion of ultraviolet photon absorption processes, we now consider the production of oxygen atoms from the photodecomposition of oxygen and ozone. Both ground-state and electronically excited oxygen atoms are involved. The lowest energy levels of the oxygen atom and the spectroscopic term symbols are shown in Figure 2.7. The ground state is a triplet, because the coupling between electron spin and angular momentum vectors permits three quantum states of total angular momentum. The two lowest excited states, singlet D and singlet S, are highly metastable, that is the transitions ^1D-^3P, ^1S-^1D, and ^1S-^3P have low transition probabilities in accordance with spectroscopic selection rules. Radiative lifetimes for $O(^1D_2)$ and $O(^1S_0)$ are 148 and 0.71 s, respectively. In the lower atmosphere the excitation is removed mainly by collisional quenching rather than by radiation, but at high altitudes the corresponding red and green emission lines can be observed. Table 2.5 summarizes data regarding the production and loss processes of excited oxygen atoms in the atmosphere. From the quenching coefficients one estimates that collisional deactivation of $O(^1D_2)$ becomes important at altitudes below 250 km, and that of $O(^1S_0)$ below 95 km.

As Figure 2.7 shows, the photoexcitation of O_2 in the wavelength region of the Herzberg and Schumann continua leads to the repulsive branches of the corresponding potential energy curves. Dissociation then takes place within the time span of one vibrational period, that is, within some 10^{-12} s, unaffected by collisions. This leads to a quantum yield of two for oxygen atom production. Excitation of O_2 by radiation in the wavelength region of the Schumann–Runge band system populates the discrete levels of the $^3\Sigma_u^-$

TABLE 2.5   Data Regarding Excited Oxygen Atoms

Excitation level	Type of process		Altitude regime
(a) Production			
$O(^1D)$ Photodissociation	$O_2 + h\nu \rightarrow O + O(^1D)$	$\lambda < 175$ nm	> 85 km
	$O_3 + h\nu \rightarrow O_2(^1\Delta_g) + O(^1D)$	$\lambda < 310$ nm	Entire atmosphere
Ion–electron recombination	$O_2^+ + e^- \rightarrow O + O(^1D)$		
Electron impact	$O + e_{fast}^- \rightarrow O(^1D) + e_{slow}^-$		
$O(^1S)$ Photodissociation	$O_2 + h\nu \rightarrow O + O(^1S)$	$\lambda < 133$ nm	
O atom recombination	$O + O + O \rightarrow O_2 + O(^1S)$		> 80 km

(b) Quenching coefficients (cm^3 molecule^{-1} s^{-1})^a

	N_2	O_2	CO_2	O_3	H_2O	$\tau_Q(s)$	$z(\tau_R = \tau_Q)$ (km)
$O(^1D)$	$2.4(-11)$	$4.0(-11)$	$1.0(-10)$	$2.4(-10)$	$2.0(-10)$	148	~ 250
$O(^1S)$	$< 5(-17)$	$2.6(-13)$	$3.7(-13)$	$5.8(-10)$	$> 1(-10)$	0.71	96

(c) Observed emissions		
$O(^1D)$	$\lambda = 630, 636.4$ nm	> 110 km (day), > 160 km (night)
$O(^1S)$	$\lambda = 557.7$ nm	85–110 km

(a) Known production processes; (b) 298 K quenching coefficients, radiative lifetimes τ_R, and altitudes for equivalent loss rates for quenching by collisions versus radiative emission; (c) emission altitudes.
^{a} Powers of 10 are given in parentheses.

state. Here, one would not expect dissociation to take place, but a crossing and perturbation by the repulsive $^3\Pi_u$ state exist, giving rise to predissociation from vibrational levels of the $^3\Sigma_u^-$ state with v $\geq$ 3. Thus, absorption within the Schumann–Runge bands can also produce oxygen atoms. Predissociation in the region of the Schumann–Runge bands is evidenced by the absence of fluorescence and by a broadening of rotation lines to an extent implying a lifetime of the $^3\Sigma_u^-$ state on the order of 10^{-11} s (Hudson et al., 1969; Ackerman and Biaume, 1970; Frederick and Hudson, 1979a). This may be compared to a radiative lifetime of 2×10^{-9} s. Since the time constant for predissociation is still shorter than the collision frequency, a quantum yield of unity is indicated. Whereas for the wavelength region of the Schumann continuum laboratory studies have confirmed quantum yields of unity for the dissociation of O_2 (Sullivan and Warneck, 1967; Lee et al., 1977), similar studies at longer wavelengths are more difficult. Washida et al. (1971) have explored ozone formation in oxygen at 184.9 and 193.1 nm and found O_3 quantum yields of 2.0 and 0.3, respectively. The first value agrees with expectation, the second is lower.

Knowledge of quantum yields, absorption cross sections, and solar intensities, all as a function of wavelength, allows the calculation of photodissociation coefficients for oxygen by means of Equation (2.7). The calculation presents no particular difficulties for those spectral regions featuring absorption continua, but in the region of the Schumann–Runge bands the calculation becomes exceedingly difficult, for an exact evaluation of the integral (Eq. 2.7) makes it necessary to take into account each rotation line, including the effects of temperature, line width, and line overlap. The calculations of the past have used measured absorption or transmission data averaged over separate bands (Hudson and Mahle, 1972; Kockarts, 1976; Blake, 1979). More recently, Yoshino et al. (1987, 1992) measured the absorption cross sections in the Schumann–Runge region at high spectral resolution, and Minschwaner et al. (1992) developed a line-by-line model of the Schumann–Runge band system to allow calculation of cross sections at stratospheric temperatures, which has greatly improved the situation. Figure 2.11 shows results from the review of Nicolet and Peetermans (1980) for an overhead sun as a function of altitude. Below 50 km, which is the main region of interest here, only the tail of the Schumann–Runge band system and the Herzberg continuum remain effective. Shorter wavelengths of the solar spectrum are much too attenuated.

The behavior of the total O_2 photodissociation coefficient in the stratosphere is shown in Figure 2.12. The strong decrease of the j value toward lower altitudes by more than five orders of magnitude is due to the

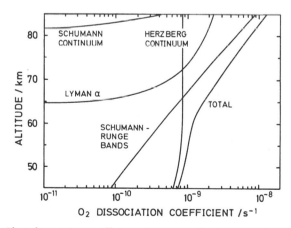

FIGURE 2.11 Photodissociation coefficients for oxygen in the mesosphere for an overhead sun according to calculations of Nicolet and Peetermans (1980).

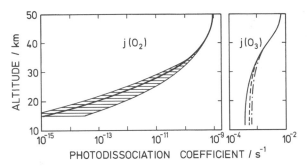

FIGURE 2.12 Photodissociation coefficients for oxygen and ozone in the stratosphere. A range of values is given for $j(O_2)$ corresponding to data from Dütsch (1968), Nicolet (1975 and private communication), Luther and Gelinas (1976), Johnston and Podolske (1978); the upper limit is for minimum ozone number densities. The dashed lines for $j(O_3)$ include back-scattering: - - -, scattering only; - · - · · -, surface albedo 0.25. $j(O_2)$ is not affected by back-scattering. The heavy lines were used to calculate ozone number concentrations in Figure 3.5.

attenuation of solar radiation by both oxygen and ozone combined. In fact, $j(O_2)$ depends sensitively on the amount and the vertical distribution of ozone assumed in the calculations. Additional uncertainties arise from computational problems at high degrees of attenuation (in the lower stratosphere). Figure 2.12 indicates a range of values.

The uneven distribution of rotational lines in the Schumann–Runge band system admits radiation that falls between the lines to leak through to lower altitudes. Cicerone and McCrumb (1980) have pointed out that a portion of this radiation is absorbed by the $^{18}O^{16}O$ isotope, which is present in the atmosphere with a mixing ratio of 0.00408 and whose rotational lines are wavelength-shifted from those of the normal $^{16}O_2$ molecule. Although $^{18}O^{16}O$ features twice as many lines and thus absorbs radiation more effectively, the mixing ratio is too small to increase the total O_2 photodissociation rate significantly.

The photochemistry of ozone has been studied in detail in the laboratory, and quantum yields for the formation of oxygen atoms are known for the most important wavelength regions. Schiff (1972), Welge (1974), and Wayne (1987) have reviewed the data. The weak Chappuis bands at wavelengths near 600 nm show a diffuse structure that suggests a predissociation process. The products are molecular oxygen and an oxygen atom in their respective ground electronic states. Castellano and Schumacher (1962) have established a quantum yield of unity for O atom formation. In pure ozone, the quantum yield for O_3 loss is two, because of the subsequent reaction $O + O_3 \rightarrow 2O_2$. The Huggins bands in the wavelength region 310–350 nm likewise are diffuse, even when sharpened by a lowered temperature, so that predissociation is likely. At 334 nm the quantum yield for O_3 removal in pure O_3 is four. This suggests a quantum yield of unity for oxygen atom production and the formation of excited O_2 ($^1\Delta_g$) or O_2 ($^1\Sigma_g$), which are both capable of dissociating ozone by energy transfer whereby another oxygen atom is formed (Jones and Wayne, 1970; Castellano and Schumacher, 1972). The main dissociation process of the Hartley band, at wavelengths between 200 and 300 nm, is $O_3 \rightarrow O_2(^1\Delta_g) + O(^1D)$. The nominal threshold for this process lies at wavelengths near 310 nm. Figure 2.13 shows primary quantum yields for $O(^1D)$ as a function of wavelength. The values are near unity throughout most of the spectral region. The associated production of $O_2(^1\Delta_g)$ has been observed at 253.7 nm to occur also with a quantum yield of unity (Gauthier and Snelling, 1971). Fairchild et al. (1978), who used molecular beam photofragment spectroscopy, have found up to 10% production of $O(^3P)$ in the wavelength region 270–300 nm. The $O(^1D)$ quantum yield accordingly must be somewhat lower than unity. Uncertainties of this magnitude are common in photochemical quantum yield determinations.

Figure 2.12 includes photodissociation coefficients for ozone as a function of altitude. In the stratosphere, because of the occurrence of the ozone layer, wavelengths in the spectral region of the Hartley bands are filtered out so that $j(O_3)$ decreases toward the lower atmosphere. However, photodissociation due to absorption by the Huggins and Chappuis bands remains effective in the troposphere.

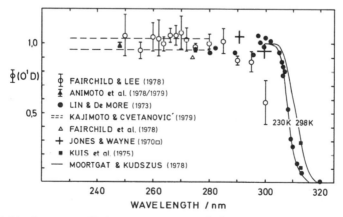

FIGURE 2.13 Quantum yields for the formation of $O(^1D)$ in the photodissociation of ozone as a function of wavelength. Note the temperature dependence of the quantum yield in the threshold region.

The threshold to $O(^1D)$ formation by photodecomposition of ozone is of considerable importance to tropospheric chemistry, because it is located above the cutoff wavelength of solar radiation penetrating to the Earth's surface. $O(^1D)$ atoms react with water vapor (whereas $O(^3P)$ atoms do not because the reaction would be endothermic), and this reaction leads to the production of OH radicals, which initiate many oxidation processes. As Figure 2.14 shows, the $O(^1D)$ quantum yield decreases at wavelengths greater than 305 nm, but the tail extends beyond the nominal energy threshold for $O(^1D)$ formation, mainly because of the thermal excitation of the ozone molecule. While this causes a temperature dependence of the quantum yield, the effect is greater than that predicted from the energy content due to thermal excitation of ground-state rotational levels (Moortgat et al., 1977), so that a contribution due to vibrational excitation appears to be present. At low temperatures, the thermal excitation and much of the tail vanish, but a small contribution from the spin-forbidden process $O_3 \rightarrow$ $O_2(^1\Sigma_g^-) + O(^1D)$ remains. Brock and Watson (1980) made use of N_2O as a scavenger of $O(^1D)$ and observed the chemiluminescence from the reaction of the product NO with ozone; Trolier and Wiesenfeld (1988) used CO_2 to scavenge $O(^1D)$ and observed the infrared fluorescence of vibrational excited CO_2; Armerding et al. (1995) used laser-induced fluorescence to detect the OH radicals generated by reaction of $O(^1D)$ with water vapor; and Talukdar et al. (1998) converted $O(^1D)$ to $O(^3P)$, which was observed by resonance fluorescence. Figure 2.14 shows that the results are in very good agreement. Ball and Hancock (1995) have additionally measured the second product, $O_2(^1\Delta_g)$, and found quantum yields equivalent to $O(^1D)$. A comparatively narrow wavelength range is responsible for $O(^1D)$ formation in the troposphere. In addition, the rate varies considerably with attenuation of solar radiation by the ozone layer overhead. The natural distribution of stratospheric ozone (see Fig. 3.4) features greater ozone densities at high latitudes, leading to comparatively higher rates of $O(^1D)$ formation in the Tropics.

The production of $O(^1S)$ by photodissociation of ozone is energetically allowed at wavelengths less than 237 nm, provided a ground-state O_2 molecule is the second product. The process does not conserve spin, so that a low probability of its occurrence is expected. The threshold for the spin-allowed process $O_3 \rightarrow O_2(1\Delta_g) + O(^1S)$ lies at 199 nm. Although the onset of a new continuum in this region coincides with the threshold wavelength, it appears not to be associated with the production of $O(^1S)$. Lee et al. (1980) reported quantum yields less than 0.1% for the 170–240-nm wavelength region. Table 2.5 indicates that rate coefficients for $O(^1S)$ quenching by oxygen and nitrogen are small compared with reactions of $O(^1S)$ with H_2O and O_3. This contrasts with the behavior of $O(^1D)$.

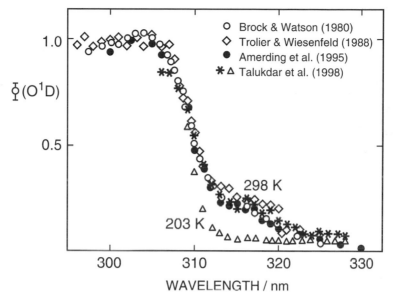

FIGURE 2.14 Threshold behavior of the $O(^1D)$ quantum yield for the photodissociation of ozone at 298 K and 203 K.

2.6. PHOTOCHEMISTRY OF MINOR ATMOSPHERIC CONSTITUENTS

The atmosphere contains many trace gases that are photochemically active. Few, however, attain the significance of oxygen or ozone in driving atmospheric chemistry. Table 2.6 gives an overview of the photochemical behavior of atmospheric constituents. The last column indicates whether photodecomposition of the substance is important to atmospheric chemistry. The species listed may be subdivided into three groups: (1) For a number of trace gases such as methane or ammonia that enter into other reactions, photodecomposition is relatively unimportant because the photochemically active wavelength region is shielded by oxygen and/or ozone. (2) A second group includes gases that are rather inert chemically in the troposphere, but undergo photodecomposition in the stratosphere, primarily in the 185–215-nm wavelength region of the atmospheric window. Nitrous oxide and the halocarbons $CFCl_3$, CF_2Cl_2, and CCl_4 belong to this group. (3) The third group comprises all gases that undergo photodecomposition by radiation reaching the troposphere. Examples are nitrogen dioxide and formaldehyde. Whether photodecomposition is important as a loss process depends on the

TABLE 2.6 Photochemical Behavior of Trace Gases in the Atmosphere[a]

Constituent	Absorption spectrum — Type and approximate long wavelength limit (nm)	Absorption spectrum — Dissociation limit (nm)	Photochemistry — Main primary photochemical process	Photochemistry — Quantum yield	Significance to the atmosphere
O_3	Hartley bands	320	310	$O_3 \rightarrow O_2(^1\Delta_g) + O(^1D)$	$\phi \approx 0.9$	Important in the stratosphere (see Figs. 2.8, 2.13).
	Huggins bands	360	410	$O_3 \rightarrow O_2(^3\Sigma_g) + O(^3P)$	$\phi \approx 1.0$	Important in the troposphere
	Chappuis bands	850	1180	$O_3 \rightarrow O_2(^3\Sigma_g^-) + O(^3P)$	$\phi \approx 1.0$	
CO_2	Continuum over-lapped by bands	200	226	(a) $CO_2 \rightarrow CO + O(^3P)$	$\phi \approx 1.0$	(a) Mesosphere and stratosphere, 175–200 nm
			167	(b) $CO_2 \rightarrow CO + O(^1D)$		(b) Shielded by O_2 Not significant
CO	Resonance bands 4th positive system	155	111	Fluorescence/quenching		Not significant; shielded by O_2
CH_4	Continuum, some band structure	160	277	Dissociation		Not significant
H_2	Lyman bands	111	84.4	Fluorescence/quenching		Not significant
H_2O	Continuum, some band structure	200	243	$H_2O \rightarrow OH + H$	$\phi \approx 1.0$	Mesosphere and stratosphere; in atmospheric absorption window (185–215 nm)
			176	$H_2O \rightarrow H_2 + O(^1D)$	Small	
H_2O_2	Continuum	350	557	$H_2O_2 \rightarrow OH + OH$	$\phi \approx 1.0$	Mainly stratosphere and mesosphere, > 185 nm
			324	$H_2O_2 \rightarrow HO_2 + H$		
N_2O	Continuum, some band structure	230	741.5	(a) $N_2O \rightarrow N_2 + O(^3P)$	Spin-forbidden	(b) Stratosphere and in the atmospheric absorption window > 185 nm
			340.7	(b) $N_2O \rightarrow N_2 + O(^1D)$	$\phi \approx 1.0$	
NH_3	Bands overlying a continuum	235	281	Dissociation		Not significant
NO	Resonance bands Continuum + bands	230	191	Fluorescence/quenching (a) Predissociation, mainly the bands 0-0 and 0-1		(a) Mesosphere and upper stratosphere in the atmospheric absorption window; (b) by Lyman α in the upper mesosphere
				(b) Photoionization		
NO_2	Bands, diffuse structure	135	134	(a) $NO_2 \rightarrow NO + O(^3P)$	See Fig. 2.17	(a) Very important throughout entire atmosphere
		700	398	(b) $NO_2 \rightarrow NO + O(^1D)$		
	Continuum	< 250 nm	244			

Species	Band type			Reaction	ϕ	Comment
NO_3	Diffuse bands	690	587	$NO_3 \rightarrow NO_2 + O$	$\phi \approx 1.0$	Important
			1195	$NO_3 \rightarrow NO + O_2$	$\phi\ 0\text{--}0.5$	$586 < \lambda < 640$ nm
HNO_2	Diffuse bands	390	475	$HNO_2 \rightarrow NO + OH$	$\phi \approx 0.9$	Important
			356	$HNO_2 \rightarrow H + NO_2$	$\phi \approx 0.1$	
HNO_3	Continuum	330	599	$HNO_3 \rightarrow NO_2 + OH$	$\phi \approx 1.0$	Mainly in the stratosphere
N_2O_5	Continuum	380	1340	$N_2O_5 \rightarrow NO_2 + NO_3$	See text	Mainly in the stratosphere
			307	$N_2O_5 \rightarrow NO_3 + NO + O$		
HO_2NO_2	Continuum	330	1340	$HO_2NO_2 \rightarrow HO_2 + NO_2$	$\phi \approx 0.56$	Stratosphere
			707	$HO_2NO_2 \rightarrow OH + NO_3$	$\phi \approx 0.33$	
HCl	Continuum	220	283	$HCl \rightarrow H + Cl$	Expected	Possibly in the mesosphere, > 175 nm
$HOCl$	Continuum	390	513	$HOCl \rightarrow OH + Cl$	$\phi = 1.0$	Entire atmosphere
CH_3Cl	Continuum	220	347	$CH_3Cl \rightarrow CH_3 + Cl$ (expected)	$\phi \approx 1.0$	Stratosphere
CH_3Br	Continuum	300	417	$CH_3Br \rightarrow CH_3 + Br$	$\phi \approx 1.0$	Not competitive
CH_3I	Continuum	380	512	$CH_3I \rightarrow CH_3 + I$	$\phi \approx 1.0$	Important
CF_4	Continuum + bands	103	220	$CF_4 \rightarrow CH_3 + F$ (presumably)		Only known loss process
$CFCl_3$	Continuum	230	375	$CFCl_3 \rightarrow CFCl_2 + Cl$; $CFCl_3 \rightarrow CFCl + 2Cl$	See Fig. 2.16	Important in the stratosphere, in the atmospheric absorption window
CF_2Cl_2	Continuum	220	354	$CF_2Cl_2 \rightarrow CF_2Cl + Cl$; $CF_2Cl_2 \rightarrow CFCl + 2Cl$	See Fig. 2.16	Important in the stratosphere in the atmospheric absorption window
CCl_4	Continuum	240	407	$CCl_4 \rightarrow CCl_3 + Cl$; $CCl_4 \rightarrow CCl_2 + 2Cl$	See Fig. 2.16	Important in the stratosphere in the atmospheric absorption window
HF	Continuum	162	211	$HF \rightarrow H + F$		Shielded by O_2; not significant
$ClONO_2$	Continuum	450	509	$ClONO_2 \rightarrow Cl + NO_3$; $ClONO_2 \rightarrow ClO + NO_2$	$0.6 < \phi < 1.0$; $0 < \phi < 0.4$	Stratosphere
$ClONO$	Continuum	400	400	$ClONO_2 \rightarrow ClO + NO$ (expected)		Not established

(Continues)

TABLE 2.6 (Continued)

Constituent	Absorption spectrum Type and approximate long wavelength limit (nm)	Dissociation limit (nm)	Photochemistry Main primary photochemical process	Quantum yield	Significance to the atmosphere
Cl_2	Continuum	470	495	$Cl_2 \to Cl + Cl$	$\phi \approx 1$	Entire atmosphere
$ClOOCl$	Continuum	450	1615	$ClOOCl \to ClOO + Cl$	$\phi \approx 1$	Polar stratosphere
PAN^b	Continuum	350	1086	$PAN \to CH_3CO_3 + NO_2$ $PAN \to CH_3CO_2 + NO_3$	$\phi \approx 0.75$ $\phi \approx 0.25$	Not established
C_2–C_5 Alkanes	Continuum	170		Photodecomposition		Not significant
C_2–C_5 Alkenes	Continuum	205		Photodecomposition		Not significant
C_2H_2	Continuum, weak band structure	237	230	$C_2H_2 \to C_2H + H$		Not established
HCHO	Bands	360	335	$HCHO \to H + HCO$ $HCHO \to H_2 + CO$	See Fig. 2.20	Important in the entire atmosphere
CH_3CHO	Quasi-continuum, some structure	340	350	$CH_3CHO \to CH_3 + CHO$	ϕ pressure dependent	Somewhat significant in the troposphere
CH_3OOH	Continuum	350	647	$CH_3OOH \to CH_3O + OH$	Expected $\phi \approx 1.0$	Possibly stratosphere
SO_2	Very weak bands Stronger bands	390 340		Excited state quenching and reaction		Not significant because of quenching
	Continuum + bands	220	220	$SO_2 \to SO + O$		Stratosphere
OCS	Continuum	255	397	$OCS \to CO + S(^3P)$ $OCS \to CO + S(^1D)$	$\phi \approx 0.27$ $\phi \approx 0.67$	Stratosphere
CH_3SH	Continuum	280	311	$CH_3SH \to CH_3S + H$	$\phi \approx 0.9$	Not significant
CS_2	Two strongly structured bands	220 350	223 281	$CS_2 \to CS + S(^3P)$ $CS_2 \to CS + S(^1D)$		Not important

[a] Shown are type and wavelength regions of spectra, dissociation limits, photochemical main processes, and an assessment of their significance to atmospheric chemistry; data taken mainly from Okabe (1978) and DeMore et al. (1997).
[b] PAN = peroxyacetyl nitrate.

extent of other competing loss reactions. Even if photolysis does not cause an appreciable loss of the trace component itself, the photodecomposition products may still initiate important chemical reactions.

This section deals with absorption spectra, quantum yields, and photodissociation coefficients for a number of such species. DeMore *et al.* (1997), Baulch *et al.* (1980, 1982, 1984), and Atkinson *et al.* (1989, 1992, 1997) have critically reviewed data relevant to atmospheric processes. Table 2.7 summarizes important photochemically active molecules and associated photodissociation coefficients. These were computed with the full, nonattenuated solar spectrum so that they represent maximum values. But an attempt has been made to separate critical wavelength regions as a discriminator for the significance of the process in the stratosphere and troposphere, respectively.

Figure 2.15 shows absorption spectra for CO_2, H_2O, N_2O, HNO_3, H_2O_2, and N_2O_5 in the 170–370-nm wavelength region. At the longer wavelengths,

TABLE 2.7   Some Photochemically Active Atmospheric Trace Constituents and the Associated Photodissociation Coefficients Calculated from the Radiation Flux Outside the Earth's Atmosphere[a]

Constituent or process	Wavelength region (nm)	j (s^{-1})	Constituent or process	Wavelength region (nm)	j (s^{-1})
O_3	< 310	8.7(−3)	HO_2NO_2	< 300	6.3(−1)
	> 310	4.6(−4)		> 300	1.5(−4)
CO_2	Ly α	2.2(0)	HCl	> 175	1.2(−6)
	> 175	1.9(−9)	$HOCl$	< 300	9.2(−5)
H_2O	Ly α	4.2(−6)		> 300	3.7(−4)
	> 175	8.6(−7)	CH_3Cl	> 175	1.0(−7)
H_2O_2	< 300	1.0(−4)	$CFCl_3$	> 175	1.5(−5)
	> 300	1.3(−5)	CF_2Cl_2	> 175	6.3(−7)
N_2O	> 175	7.3(−7)	CCl_4	> 175	3.5(−5)
NO_2	175–240	5.9(−5)	ClO	< 300	6.5(−3)
	240–307	2.3(−4)		> 300	5.5(−4)
	> 310	8.0(−3)			
$NO_3 \rightarrow NO_2 + O$	> 400	1(−1)	$ClONO_2$	< 310	2.0(−5)
$NO_3 \rightarrow NO + O_2$		4(−2)		> 310	9.5(−4)
N_2O_5	< 310	5.8(−4)	$ClONO$	< 310	2.0(−5)
	> 310	3.6(−5)		> 310	9.5(−4)
HNO_3	< 200	7.1(−5)	$HCHO \rightarrow H$	< 300	4.5(−5)
	200–307	7.2(−5)	$+ HCO$	> 300	1.1(−5)
	> 307	2.0(−6)			
			$HCHO \rightarrow H_2$	< 300	4.0(−5)
			$+ CO$	> 300	1.1(−5)
HNO_2	< 310	3.4(−4)			
	> 310	2.5(−3)	CH_2OOH	> 300	1.2(−5)

[a] According to Nicolet (1978). Powers of 10 in parentheses.

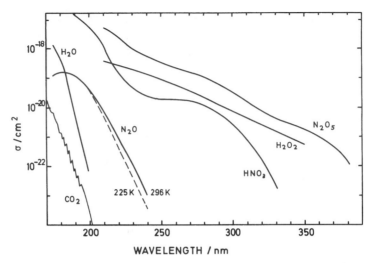

FIGURE 2.15   Absorption spectra of CO_2, H_2O, N_2O, HNO_3, H_2O_2, and N_2O_5. (From data in De More *et al.* (1994), Baulch *et al.* (1982), and Atkinson *et al.* (1989); data for CO_2 are from Ogawa (1971).)

the absorption cross sections are low, but they rise with decreasing wave-length and attain appreciable values in the spectral region of the atmospheric window near 200 nm. Except for CO_2, these spectra are continuous, so that one expects photodissociation to occur in all cases. Laboratory studies have largely confirmed the expectation. For water vapor, the major dissociation process in the wavelength region considered is $H_2O \rightarrow OH + H$, which becomes energetically feasible at wavelengths below 242 nm. A second spin-allowed process, $H_2O \rightarrow H_2 + O(^1D)$, requires more energy ($\lambda \leq 177$ nm). However, Chou *et al.* (1974), who studied it by using the HTO isotope, established a quantum yield of $\leq 0.3\%$ at 175 nm. Okabe (1978) discussed the photodissociation of water vapor in more detail.

The main photodissociation process for nitrous oxide is $N_2O \rightarrow N_2 + O(^1D)$, which occurs with a quantum yield of nearly unity (Cvetanovic, 1965; Greiner, 1967a; Paraskepopoulos and Cvetanovic, 1969). The photo-products $N_2 + O(^3P)$ and $NO + N(^4S)$, although energetically allowed, would violate the spin conservation rule and are expected to be generated with low probability. For the second of these processes, Preston and Barr (1971) have established a quantum yield of less than 2%.

The photodecomposition of HNO_3 in the 200–300-nm wavelength region has been studied by Johnston *et al.* (1974). In the presence of excess CO and O_2 to scavenge OH radicals so as to prevent their reaction with HNO_3, nitrogen dioxide is formed with a quantum yield of unity. This suggests that the principal primary photodissociation process is $HNO_3 \rightarrow OH + NO_2$, which represents a reversal of the recombination reaction between OH and NO_2 discussed earlier. By the direct observation of OH radicals, Turnipseed et al. (1992) confirmed a quantum yield of essentially unity in the wavelength region greater than 220 nm, although a small contribution of oxygen atoms was detected with a yield of 0.03 and 0.1 at 222 and 248 nm, respectively. At 193 nm, the O-atom quantum yield was 0.4, and that of OH radicals was 0.6. At this wavelength mainly $O(^1D)$ oxygen atoms are produced.

Hydrogen peroxide generally is assumed to undergo photodecomposition to form two OH radicals at wavelengths greater than 230 nm. With this assumption the experimental results of Volman (1963) indicated a quantum yield of unity for the loss of H_2O_2 and a quantum yield of two for the formation of OH radicals. Wavelengths below 365 nm provide enough energy for the formation of the products $H_2O + O(^1D)$. Greiner (1966) pointed out that the data of Volman (1963) do not entirely preclude O atom formation, but recent observations of Vaghjiani and Ravishankara (1990) indicate that at 248 nm the yield of oxygen atoms is negligible. At 193 nm hydrogen atoms are formed with a 15% contribution.

Nitrogen pentoxide undergoes photodissociation to form NO_3 with a quantum yield of nearly unity in the wavelength region 210–350 nm, as determined by laser pulse experiments at selected wavelengths with detection of NO_3 by absorption at 662 nm (Swanson et al., 1984; Barker et al., 1985; Ravishankara et al., 1986; Oh et al., 1986). In several of these studies $O(^3P)$ atoms were observed by resonance fluorescence with a quantum yield decreasing from 0.72 at 248 nm toward zero in the vicinity of the energy threshold at 307 nm. The oxygen atom results from excess energy imparted to the NO_2 molecule that is produced together with NO_3. The latter features a series of diffuse absorption bands in the visible portion of the spectrum, where the solar flux has its maximum. The dissociation products from NO_3 are mainly oxygen atoms and NO_2. Accordingly, the photolysis of N_2O_5 leads to the same products, regardless of whether the oxygen atom is ejected directly or is formed via NO_3 as an intermediate.

The photochemical behavior of methyl chloride, CH_3Cl, and the halocarbons CF_2Cl_2, $CFCl_3$, and CCl_4 is illustrated in Figure 2.16. These gases begin to absorb radiation at wavelengths below 250 nm. The absorption spectra are again quasi-continuous, and the cross sections rise with decreasing wavelength. Thus photodecomposition occurs mainly in the 185–215-nm wavelength region of the atmospheric window. The main photodecomposition process for chlorine-containing gases is the release of a chlorine atom. The halocarbons admit the release of a second chlorine atom from the

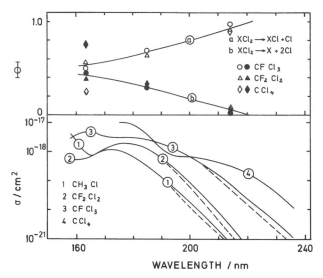

FIGURE 2.16   Absorption spectra and quantum yields for the photodecomposition of CH_3Cl, CF_2Cl_2, $CFCl_3$, and CCl_4. (Absorption cross sections are from Chou *et al.* (1977) and Hubrich *et al.* (1977); quantum yields are from Rebbert and Ausloos (1975, 1977).)

remaining fragment, provided additional energy is imparted to the molecule. The upper half of Figure 2.16 shows quantum yields for both processes. Near the long-wavelength limit the quantum yield for the process $XCl_2 \rightarrow X + 2Cl$ is nearly zero, whereas that for the process $XCl_2 \rightarrow XCl + Cl$ is close to unity. The first process increases in significance at the expense of the second as the wavelength is lowered. (i.e., the photon energy is raised), until at wavelengths near 166 nm the two processes occur with about equal probability.

Figure 2.17 shows the absorption spectrum and quantum yields for the photodissociation of nitrogen dioxide. This process is of major importance to atmospheric chemistry in addition to that of ozone. NO_2 absorbs radiation throughout the entire visible and near-ultraviolet spectra region. The spectrum is extremely complex, with little regularity in rotational and vibrational structure. Excitation of NO_2 at wavelengths above 430 nm leads to fluorescence, which is quenched by collisions with air molecules. Near the first dissociation limit at 398 nm for the process $NO_2 \rightarrow NO + O(^3P)$, the rotational structure becomes diffuse and fluorescence vanishes, indicating that predissociation takes place. In the 290–390-nm wavelength region, quantum yields for O atoms and NO are near unity throughout. At wavelengths above the dissociation threshold, the O atom quantum yields gradu-

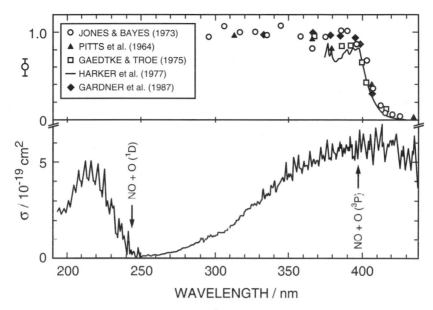

FIGURE 2.17 Absorption spectrum and O(^{3}P) quantum yields for nitrogen dioxide. The threshold values for the formation of O(^{3}P) and O(^{1}D) are indicated. (Absorption cross sections are from Bass *et al.* (1976) and Schneider *et al.* (1987); quantum yields are from the authors listed).

ally fall to zero. The fall-off curve is explained by the thermal energy content of the NO$_2$ molecule, residing mainly in rotation and some vibration, which supplements the energy provided by the absorbed photon. Toward shorter wavelengths, the absorption cross section for NO$_2$ first decreases to low values and then rises again below 244 nm. This wavelength coincides with the second dissociation limit, which leads to the production of O(^{1}D) atoms. In the wavelength range 214–242 nm, Uselman and Lee (1976) have found an O(^{1}D) quantum yield of 0.5 ± 0.1, indicating that other primary processes are also at work. Table 2.7 shows, however, that this wavelength region contributes little to the total photodissociation coefficient of NO$_2$ in the atmosphere, so that a precise knowledge of the photodissociation products is less important.

Nitrous acid, HNO$_2$, features a series of diffuse absorption bands in the spectral region 300–400 nm (Cox and Derwent, 1976; Stockwell and Calvert, 1978; Bongartz *et al.*, 1991). The photochemical behavior of HNO$_2$ is difficult to study because the substance is in chemical equilibrium with NO, NO$_2$, and H$_2$O, and when one attempts to isolate HNO$_2$ from the mixture, it decomposes. Cox (1974) and Cox and Derwent (1976) have studied the photolysis of such mixtures at 365-nm wavelength. The main photolysis products were NO and NO_2, which suggests that HNO_2 photolyzes to give NO + OH, followed by the reaction of OH + HNO_2 → H_2O + NO_2. The measured rate of product formation gave a quantum yield of 0.92 ± 0.16 for the photodissociation of HNO_2. The possibility of a second primary process HNO_2 → H + NO_2 exists at wavelengths below 361 nm and is suggested by a smaller OH radical yield relative to that at λ > 370 nm (Vasudev, 1990).

Formaldehyde, HCHO, is another important atmospheric compound subject to photolysis. The absorption spectrum of formaldehyde, shown in Figure 2.18, consists of a series of fairly sharp bands extending from about 355 nm toward shorter wavelengths. The system leads to excitation in the first electronic singlet state, which undergoes internal energy conversion to highly excited vibrational levels. Photodissociation occurs along two routes: HCHO → H_2 + CO, and HCHO → H + HCO. The first process occurs throughout the entire absorption region; the second has its threshold near 335 nm. Figure 2.18 shows that the quantum yields are a function of wavelength. The total quantum yield for the sum of both processes is unity in the region below 330 nm. Above this wavelength the quantum yield is pressure dependent because of collisional quenching of the excited states. The data in Figure 2.18 refer to 100-hPa air pressure. Lowering the pressure

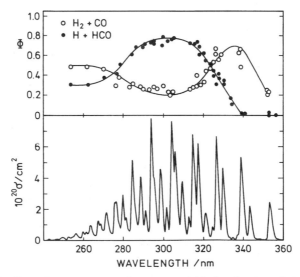

FIGURE 2.18   Absorption spectrum and quantum yields for the two photodecomposition channels of formaldehyde in air at atmospheric pressure. (From the data of Clark *et al.* (1978), Horowitz and Calvert (1978), Tang *et al.* (1979), and Moortgat *et al.* (1979, 1983).)

causes the quantum yield to increase, and at very low pressures it tends toward unity. The pressure dependence complicates calculations of the photodissociation coefficient for the two product channels of formaldehyde.

Like formaldehyde, acetaldehyde (CH_3CHO) and acetone (CH_3COCH_3), are products resulting from the oxidation of hydrocarbons. The near ultraviolet spectra feature cross sections maximizing near 270 nm for both molecules and a long wavelength limit somewhere near 340 and 330 nm, respectively, so that they are susceptible to photolysis by solar radiation reaching the troposphere. The principal photodecomposition pathways in this wavelength region are $CH_3CHO \rightarrow CH_3 + CHO$ and $CH_3COCH_3 \rightarrow CH_3CO + CH_3$. Both acetaldehyde and acetone undergo collisional deactivation after photoexcitation, so that quantum yields for photodissociation at ground-level ambient pressures are low (Horowitz and Calvert, 1982; Meyrahn et al., 1982, 1986; Gardner et al., 1984). As a consequence, the photodissociation coefficients are at least an order of magnitude smaller than that of formaldehyde.

Figure 2.19 shows the altitude dependence of the photodissociation coefficients for several important photoactive atmospheric trace constituents. These data should be compared with those in Table 2.7. For the group II of gases H_2O, N_2O and CF_2Cl_2, whose absorption spectra are located at wavelengths below 300 nm, the photochemical activity is restricted to the upper atmosphere. These gases are photolyzed mainly in the region of the atmospheric absorption window, and at altitudes below 20 km the photodissociation coefficients decrease rapidly to low values. The absorption continua

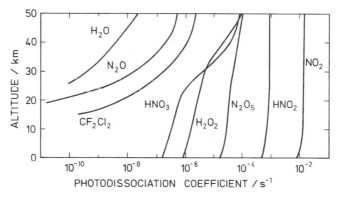

FIGURE 2.19 Photodissociation coefficients for several photochemically active atmospheric trace components as a function of altitude. (Data for H_2O are from Park (1974) for an overhead sun, for N_2O and HNO_3 from Johnston and Podolske (1978) for global average conditions, for CF_2Cl_2 from Rowland and Molina (1975) for an overhead sun, and for H_2O_2, N_2O_5, HNO_2, and NO_2 from Isaksen et al. (1977) for a 60° solar zenith angle.)

of HNO_3, H_2O_2, and N_2O_5 extend beyond the 300-nm limit, so that these molecules are also photolyzed in the troposphere. Here, however, their photochemical activity is of lesser importance because these molecules show a high affinity toward liquid water, that is, HNO_3, H_2O_2, and N_2O_5 are readily scavenged by clouds. The photodissociation coefficients for NO_2 and HNO_2 show fairly little dependence on altitude. The photodecomposition of these molecules in the atmosphere is dominated by radiation in the 300–400-nm wavelength region. Shorter wavelengths contribute in a minor way to the total photodissociation coefficients, as Table 2.7 shows.

Figure 2.20 shows the altitude dependence of the photodissociation coefficients for the two decomposition pathways of formaldehyde

$$HCHO \rightarrow H + HCO \qquad j_1$$

$$\rightarrow H_2 + CO \qquad j_2$$

The overall behavior is similar to that of N_2O_5, except that the rise in the H_2 production rate is due to the decrease in pressure and a corresponding increase in the quantum yield for the process. At altitudes above 30 km the rate of H atom formation increases by at least a factor of 2, as ozone no longer shields the wavelength region below 300 nm and additional bands within the HCHO absorption spectrum begin to absorb radiation.

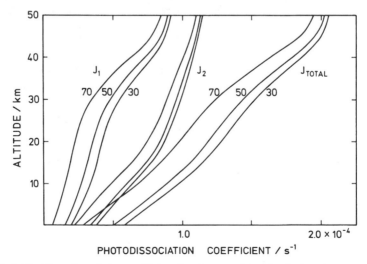

FIGURE 2.20 Photodissociation coefficient for the two channels of photodecomposition of formaldehyde as a function of altitude and solar zenith angle (30°, 50°, and 70°); calculated by the two-stream method (G. K. Moortgat and P. Warneck, unpublished data).

The calculated photodissociation coefficients shown in Figures 2.19 and 2.20 are intended to indicate orders of magnitude only. Even for a cloudless sky, the exact values depend critically on the solar zenith angle, on the thickness of the ozone layer overhead (which varies with season and latitude), on the reflective properties of the Earth's surface, and on the particle load in the lower atmosphere, which determines the scattering properties of the atmosphere. This has led to studies in which important photodissociation coefficients were derived directly from field measurements by chemical actinometry. Frequently a transparent vessel containing the substance of interest is exposed to sunlight, and the rate of decomposition is measured together with the solar actinic flux. Back-reactions must be suppressed in a suitable manner. Stedman *et al.* (1975), Harvey *et al.* (1977), Bahe *et al.* (1980), Dickerson *et al.* (1982), Madronich *et al.* (1983), and Madronich (1987b) have studied the photodissociation of NO_2 under field conditions by measuring the product NO by means of chemiluminescence detection. Bahe *et al.* (1979), Dickerson *et al.* (1979), and Blackburn *et al.* (1992) have studied the formation of $O(^1D)$ from the photodissociation of ozone in the ground-level atmosphere. N_2O was used to scavenge $O(^1D)$, followed by detection of either of two products, N_2 or NO. Marx *et al.* (1984) included

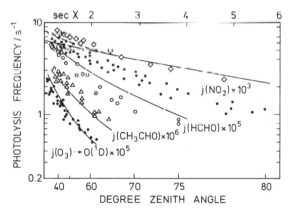

FIGURE 2.21 Photodissociation coefficients for NO_2, HCHO, CH_3CHO, and $O(^1D)$, the latter produced from ozone, for ground-level, clear sky conditions. Nitrogen dioxide: diamonds (Madronich *et al.*, 1983), solid points (Marx *et al.*, 1984); the solid line represents calculations based on the two-stream method; and the dashed line is for isotropic scattering (Madronich *et al.*, 1983). Formaldehyde: open points (Marx *et al.* (1984), the solid line calculations (Calvert, 1980). Acetaldehyde: triangles (Marx *et al.*, 1984); the solid line represents calculations of Meyrahn *et al.* (1982). Ozone: solid points (Dickerson *et al.*, 1979; Bahe *et al.*, 1989) measurements normalized to 325 Dobson units of total ozone overhead; the solid line represents calculations (Dickerson *et al.*, 1979).

formaldehyde and acetaldehyde in their studies. By and large, the results were in reasonable agreement with dissociation rates calculated from the solar radiation field. Figure 2.21 compares ground-based, clear sky measurements of the photolysis frequencies for NO_2, HCHO, and ozone to form $O(^1D)$ with calculated values as a function of solar zenith angle. Madronich et al. (1983) found that for NO_2, the calculations based on isotropic Rayleigh scattering led to a better fit to the observational data than the columnar scattering approximation of Isaksen et al. (1977) at solar zenith angles less than 60°. The data of Marx et al. (1984) for NO_2 fall below this line. They measured the loss of NO_2, whereas Madronich et al. (1983) determined the pressure increase resulting from the dissociation of NO_2. The general behavior of j values is to decline with solar zenith angle more rapidly as the onset of the photodissociation process is shifted toward the 300-nm solar radiation cut-off limit. Thus the steepest curve occurs for the formation of $O(^1D)$ from ozone. This process also depends considerably on the thickness of the ozone layer overhead.

CHAPTER 3

Chemistry of the Stratosphere

Chemical processes in the stratosphere are intimately connected with the phenomenon of the ozone layer at altitudes between 15 and 35 km. The significance of the ozone layer is twofold: on the one hand it absorbs ultraviolet radiation at wavelengths below 300 nm, so that this biologically harmful radiation is prevented from reaching the Earth's surface; on the other hand it dissipates the absorbed energy as heat, thereby giving rise to the temperature peak at the stratopause. Because of its importance to the terrestrial radiation budget, ozone has become one of the most thoroughly studied atmospheric constituents. However, other trace gases and their reactions have a decisive influence on the budget of ozone in the stratosphere. In contrast to ozone, which is formed in the stratosphere, the other trace gases have their origin mainly in the troposphere. They are cycled through the stratosphere in the sense that they are brought upward by eddy diffusion, undergo chemical modifications, and finally leave the stratosphere as a different chemical entity. In this chapter we describe first the behavior of ozone, then discuss the chemistry of other trace components that are considered important.

3.1. HISTORICAL SURVEY

Hartley (1881) appears to have been the first to point out that the presence of ozone in the atmosphere would explain the abrupt 300-nm cut-off of the solar spectrum observed at ground levels. Forty years later, Fabry and Buisson (1921) confirmed Hartley's conclusion and obtained the first reliable estimate of total ozone overhead from optical absorption measurements with the sun as the background source. At about the same time, Strutt (1918) had shown in a milestone experiment involving a 6.5-km optical path of 254-nm radiation derived from a mercury lamp that the ground-level O_3 mixing ratio was 40 nmol mol^{-1} or less, which was insufficient to account for the column density of ozone if it were present with a vertically uniform mixing ratio. The discrepancy gave rise to the idea of an ozone layer occurring at greater heights. In 1929, during an expedition to Spitsbergen, Götz studied the zenith sky radiation as a function of solar elevation and discovered that the ratio of intensities of two neighboring ultraviolet wavelengths exhibited a reversal of the trend when solar rays approached large zenith angles (Umkehr effect). He attributed the effect to the layered structure of atmospheric ozone, and after evaluating further Umkehr measurements in Arosa in collaboration with Meetham and Dobson (Götz et al., 1934), he located the maximum density of ozone at 22 km altitude. His conclusions were confirmed in the same year by the first balloon sounding of ozone by Erich and Victor Regener (1934).

In 1930, Chapman offered an explanation for the origin of stratospheric ozone and its layered structure by means of a theory involving the photodissociation of oxygen. Because of uncertainties in rate parameters, solar ultraviolet intensities, and physical conditions in the stratosphere, Chapman could draw few quantitative conclusions, yet his results as well as the more detailed calculations of Wulf and Deming (1937) revealed the formation of an ozone layer at about the correct altitude. At the same time, Schumacher (1930) studied the photodecomposition of ozone in the laboratory and interpreted his data in terms of three reactions that are part of the Chapman mechanism. While Schumacher (1930, 1932) was unaware of Chapman's work and the significance of the laboratory data to atmospheric chemistry, Eucken and Patat (1936) fully recognized the implications when they redetermined the ratio of rate coefficients associated with the two dark reactions in the mechanism. Thirty-five years later, after chemical kineticists had mastered experimental difficulties in measuring rate coefficients of individual reactions, the values obtained by Eucken and Patat (1936) were found to be in error by a factor of 6, while their temperature dependence was essentially confirmed. In 1950, Craig reviewed all important observations and laboratory data entering into the calculations and concluded that calculated and observed vertical distributions of atmospheric ozone were in reasonable agreement.

A problem that had come to light early was the latitude distribution of total ozone, which was first explored by Dobson and Harrison (1926). Total ozone, that is the ozone column density, was found to increase from the equator to higher latitudes, whereas the Chapman theory predicted an opposite trend. Equally at variance with theory was the observation of a winter maximum at high latitudes. It was then recognized that photochemical activity in the lower stratosphere is so weak that ozone would be influenced more by transport than by photochemical processes, and eventually a poleward flux of ozone due to meridional circulation was postulated. Measurements of stratospheric water vapor by Brewer (1949) and his interpretation of the low mixing ratios by a meridional circulation model were taken to provide support for the transport of ozone toward high latitudes. It now appeared that stratospheric ozone was more a problem of meteorology, and this view was to persist for the following two decades.

During this period considerable progress was made in the exploration of the upper atmosphere, which, in turn, stimulated extensive laboratory studies of reactions deemed important in the atmosphere. The installation of a global network of stations dedicated to ozone observations improved the data base on total ozone, its vertical distribution, and seasonal variations, although a better global coverage was achieved only much more recently with the advent of satellite observations. Ironically, it was not the ozone measurement program that provided the most clear-cut evidence for the existence of meridional transport, but the behavior of radioactive debris from nuclear weapons tests in the early 1960s. Finally, space exploration produced for the first time realistic information on ultraviolet solar radiation intensities. The improved data base stimulated new calculations of the ozone density profile in the altitude regime where the Chapman equations were considered applicable, namely above 25 km in the Tropics, and it was now found that theory overestimated the ozone density by at least a factor of 2. At this time, the discrepancy was considered serious, and one began to look into possible causes, especially reactions other than those contained in the Chapman mechanism.

Laboratory experience had convinced chemists earlier that the Chapman mechanism needed a supplement of additional reactions. In 1960, McGrath and Norrish discovered the formation of OH radicals when ozone was photolyzed by ultraviolet radiation in the presence of water vapor, and they proposed a chain decomposition of ozone by water radicals. Meinel (1950) had already demonstrated the existence of OH radicals in the upper atmosphere by observing the emission spectrum, and Nicolet (1954) had discussed the chemical significance of H, OH, and HO_2 in the mesosphere. Hunt (1966a, b) then showed that a modification of the Chapman mechanism by such reactions might bring it into harmony with observations. When, however, laboratory measurements of the rate coefficients were finally successful in the early 1970s, it was found that water radical reactions were unable to reduce stratospheric ozone sufficiently to achieve agreement with observations. But Hunt's study had set a new trend, and from now on stratospheric trace gases other than ozone came under closer scrutiny.

An important step forward was made by Crutzen (1970, 1971) and Johnston (1971), who drew attention to the catalytic nature of the nitrogen oxides NO and NO_2 with regard to ozone destruction. At that time, aircraft companies had planned the construction of commercial airplanes with a cruising altitude of about 20 km. Atmospheric scientists became concerned about the possibility of ozone layer depletion by nitrogen oxides contained in the engine exhaust gases, because the concomitant increase in ultraviolet radiation at the Earth's surface was predicted to lead to an increased incidence of skin cancer. The stratosphere thus became a political issue, which led the U.S. Congress to block subsidies for the development of commercial supersonic aircraft, and the Department of Transportation to initiate a research program for the study of conceivable perturbations of the stratosphere. By 1974, exploratory measurements had traced the vertical concentration profiles of N_2O, NO, NO_2, and HNO_3, in addition to methane and hydrogen, and chemical reaction models were able to explain most of these data (CIAP, 1975). The issue of high-flying jet liners had hardly been settled when, in a dramatic new development, Molina and Rowland (1974) discovered that chlorofluoromethanes have an adverse effect on stratospheric ozone. These human-made compounds were widely used as refrigerants and aerosol propellants. In the atmosphere they are degraded primarily by photodecomposition in the upper stratosphere. This process liberates chlorine atoms, which can enter into similar catalytic ozone destruction cycles as the nitrogen oxides. The worst aspect of the halocarbons was the predicted long-term effect. Continued production causes accumulation in the atmosphere, but because of slow diffusion into the stratosphere the full impact is delayed. This threat eventually led to a worldwide ban of nonessential uses of chlorofluorocarbons according to the Montreal protocol (UNEP, 1987) and its amendments.

Another problem that was brought closer to solution is the origin of the stratospheric aerosol layer discovered by Junge (1961). The layer centers at an altitude near 20 km and consists mainly of particles representing a mixture of sulfuric acid and water. High-reaching volcanic eruptions were found to contribute significantly to the strength of the aerosol layer because of the direct injection of ash particles, SO_2, and H_2O into the stratosphere. But even during extended periods of volcanic quiescence, long after the excess material has cleared the stratosphere, a layer of sulfate particles persists. The background can be explained only if a gaseous sulfur compound diffuses upward from the troposphere, undergoing oxidation and conversion to H_2SO_4, not before having penetrated the 20-km altitude level. Initially SO_2 was thought to meet these conditions. The rate of SO_2 photooxidation increases markedly as the excitation wavelength is lowered toward 200 nm, and this radiation becomes available above 20 km. In the mid-1970s, when stratospheric chemistry became better understood, the reaction of SO_2 with OH radicals came into focus, and it was realized (by Moortgat and Junge, 1977) that its rapid rate might prevent SO_2 from actually reaching the photochemically active region. At the same time carbonyl sulfide was detected in the atmosphere, leading Crutzen (1976a) to explore its behavior at high altitudes. He showed that OCS is converted to SO_2 in the 20–25-km altitude region, and that the resulting flux of sulfur into the stratosphere is sufficient to maintain the Junge layer. In addition, OH radicals were found to react with OCS much more slowly than with SO_2. Finally, Inn et al. (1981) obtained direct evidence for the presence of OCS in the stratosphere from in situ measurements.

In the Antarctic, total ozone was known to show a temporary depletion in September when the dark period ceases. The discovery by Farman et al. (1985) that the annual minimum, which subsequently was dubbed the "Antarctic ozone hole," had increased since the mid-1970s led to renewed research activities in search of an explanation. Although the phenomenon appeared to be connected with the very stable polar vortex existing in the Antarctic during the winter, it was initially not clear whether the cause was associated with the dynamics or the chemistry of the Antarctic stratosphere. However, a large scale experiment led by U.S. scientists (Tuck et al., 1989) soon provided evidence for the dramatic chemical changes that accompany the loss of ozone, especially the anticorrelation between ozone and chlorine monoxide, which suggested that an abnormal chlorine chemistry is responsible for the loss. The new data revealed that large-scale motions could not explain the vertical extent of the depletion. Other measurements indicated a deficiency in the concentration of nitrogen oxides in the presence of ice particles, which led to the postulation of heterogeneous reactions occurring on the surface of the ice particles that could convert chlorine from inactive compounds such as HCl or $ClONO_2$ to more reactive species (Solomon et al., 1986a). Since then, laboratory studies have identified the reactions and confirmed the necessity of including heterogeneous reactions in the models.

3.2. OZONE OBSERVATIONS

Measurements of stratospheric ozone fall into two classes: total ozone observed from the ground or by means of satellites, and vertical distributions derived from balloon (rarely rocket) sondes. The data are too numerous for individual discussion; only the most obvious features can be treated. Reviews of the subject by Craig (1950), Dütsch (1971, 1980), and the WMO (1986, 1990, 1995) trace the progress achieved in understanding the behavior of stratospheric ozone.

3.2.1. PRINCIPAL FEATURES AND BEHAVIOR OF THE OZONE LAYER

Ground-based measurements of total ozone are routinely made at many meteorological stations by means of the Dobson spectrometer. The instrument measures the column abundance of ozone in the atmosphere by optical absorption, with the sun or the moon as background source. The data are expressed as an equivalent column height of a layer of uniform density reduced to 1 atmosphere and 273.15 K. One Dobson unit (DU) corresponds to a layer thickness of 10^{-3} cm. This is equivalent to about $446.2\,\mu\text{mol m}^{-2}$ or 21.4 mg m^{-2}.Because of the characteristic vertical distribution, total ozone refers mainly to stratospheric ozone. The contribution of tropospheric and mesospheric ozone to the total column content is small. As discussed in Section 1.1, total ozone displays a considerable day-to-day fluctuation requiring extensive smoothing of the data by averaging. The data from Arosa, Switzerland, in Figure 1.3 showed monthly mean values to undergo an almost sinusoidal annual variation with a maximum in spring and a minimum in the autumn. A similar annual variation is observed at other stations at high latitudes, whereas in the Tropics the seasonal variation in total ozone is small.

Figure 3.1 shows the distribution of total ozone with latitude and the temporal variation for two consecutive years. The data were derived from satellite observations by averaging over narrow latitude belts. The results agree quite well with those deduced from the network of Dobson stations around the world, but the satellite data provide a more complete global coverage. The lowest values of total ozone occur in the equatorial region, and the highest values occur near the North Pole in spring. Here, the annual peak-to-peak variation approaches 40% of the annual average. In the Southern Hemisphere, the spring maximum appears first at about 60° latitude, then migrates toward the Antarctic, where it arrives 1–2 months later. The

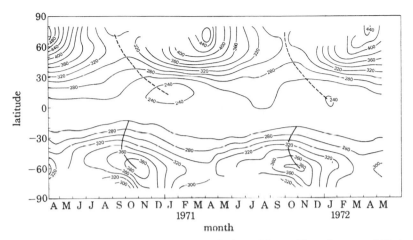

FIGURE 3.1   Total ozone versus time and latitude derived from 10° zonal means of *Nimbus 4* satellite data. Contours given in Dobson units (1 DU = 446.2 μ mol m^{-2}). (From Hilsenrath *et al.* (1979) with permission.)

maximum amount of total ozone in the Southern Hemisphere is lower than that of the Northern Hemisphere. The cyclic behaviors of total ozone in the two hemispheres are similar but evidently not identical. The differences have been known for some time from ground-based measurements, but until the advent of satellite data their reality had been in doubt because of the sparseness of Dobson stations in the Southern Hemisphere. As noted previously, the spring maximum is thought to arise from meridional transport of ozone from low to high latitudes, mainly in the winter hemisphere. The dissimilarity between the hemispheres thus suggests differences in the circulation patterns.

Satellite and ground-based observations also reveal differences with regard to the distribution of total ozone with longitude, which is shown in Figure 3.2. In the Northern Hemisphere, total ozone features three maxima, which coincide approximately with the quasi-stationary low pressure regions near 60° latitude (wave number 3), with locations over Canada, northern Europe, and eastern Russia. In the Southern Hemisphere, in contrast, only one broad maximum occurs at 60° latitude (wave number 1), located southeast of Australia. The occurrence of these maxima may be explained in the same way as the day-to-day variation due to the ever-changing meteorological situation. A correlation is known to exist between total ozone and the pressure at the tropopause level. The height of the tropopause serves only as an indicator of cyclone activity, however. Low-pressure systems are associated with a lower than average tropopause level, whereby total ozone is

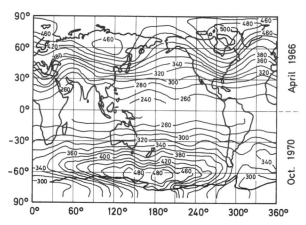

FIGURE 3.2. Total ozone averaged versus geographic latitude and longitude. (From Dütsch (1980) with permission.)

above average. Upper tropospheric troughs generally are displaced westward of the surface center of cyclones, and it is here that the highest amounts of total ozone are found. The variation in total ozone follows the local dynamics of the lower stratosphere. Low-pressure systems induce convergence and subsidence of air in the middle stratosphere, forcing ozone-rich air from the 25–20 km altitude regime downward into the lower stratosphere. The ozone column density thus increases. A reversal of the situation occurs in high pressure systems. The velocity of vertical air motions can reach 1 km/day, which is sufficient to change total ozone by 25% within a few days. As described by Dütsch (1980), the process has been confirmed by successive balloon soundings.

Figure 3.3 shows two examples for the vertical distribution of ozone. The data were selected to present typical concentration–altitude profiles for low and high latitudes. Both examples demonstrate the existence of the ozone layer, but they also reveal differences. In the Tropics, because of the high-lying tropopause, the ozone layer is fairly narrow, with a maximum at 26 km altitude. In middle and high latitudes the distribution is broader and much more ragged, and the maximum occurs at altitudes between 20 and 23 km. At altitudes above 35 km, the rocket-sonde data of Krueger (1973) indicated that the ozone concentration decreases with a scale height of about 4.6 km, regardless of latitude. The steep rise in ozone above the tropopause at high latitudes is evidence for the lower rate of vertical eddy diffusion in the stratosphere compared to that in the troposphere. The resulting vertical gradient of mixing ratios gives rise to a downward flux of ozone toward the tropopause and ultimately into the troposphere. The gradient may be en-

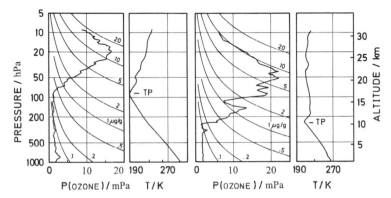

FIGURE 3.3 Vertical ozone profiles (partial pressures) from balloon soundings at low latitude (left) and high latitude (right). Mass mixing ratios are shown by the array of smooth curves. Vertical temperature profiles are included to indicate the tropopause levels. (Adapted from Hering and Borden (1967).)

hanced by large-scale mean motions, which at higher latitudes usually are directed downward. The periodic relocation of the tropopause due to heat-exchange processes causes a portion of the ozone-rich air to become part of the troposphere when the tropopause level is raised. Such an event appears to have taken place not too long before the data on the right-hand side of Figure 3.3 were obtained. This is suggested by the spike of ozone at 8 km altitude, and by the abrupt cut-off of the stratospheric ozone profile near the tropopause.

In the Tropics the situation is somewhat different. Here, the mean motion is directed upward as it follows the ascending branch of the Hadley cell, thus counteracting the downward flux of ozone by eddy diffusion. The ozone concentration rises more gradually with altitude, and the gradient at the tropopause is essentially negligible. Since in addition the tropopause level is fairly stable, there is practically no influx of ozone into the troposphere. Instead, ozone undergoes lateral, poleward transport and can then enter the troposphere at mid-latitudes, partly via the subtropical tropopause breaks. Balloon soundings at higher latitudes often reveal a remarkable pastry-like structure of the vertical ozone distribution. Layers rich in ozone are interspersed with layers containing lesser amounts. The right-hand side of Figure 3.3 may serve to demonstrate the extent of stratification. Simultaneous measurements with two instruments (Attmannspacher and Dütsch, 1970) have proved beyond doubt that the phenomenon is real and that it is not caused by instrumental errors. Dobson (1973) performed a statistical analysis of the frequency of occurrence of laminar structure, based on data from the

Northern Hemisphere. He found the effect to be most frequent in spring and in the region poleward of 30° latitude, whereas in the equatorial stratosphere the ozone–altitude profiles generally had a smooth appearance. At mid-latitudes, a particularly well-developed minimum of stratospheric ozone was often encountered at or near the 15-km altitude level. Figure 3.3 also displays this feature. The minimum appears to derive from the injection of tropospheric air into the stratosphere at the location of the subtropical tropopause break. On its poleward journey the air then is gradually enriched with ozone from adjacent layers. According to Table 1.8, horizontal transport by eddy diffusion is much faster than vertical transport. The layered structure of the ozone layer thus shows most directly, although qualitatively, the relatively slow rate of vertical mixing in the stratosphere versus a fairly rapid transport in a quasi-horizontal direction.

Figure 3.4 shows the meridional distribution of ozone derived by averaging data from many soundings at different latitudes. The left-hand side presents the distribution in units of partial pressure, which is proportional to number concentration. This figure illustrates again the gradual broadening of the layer and its displacement toward lower altitudes as one goes from the equator toward the Poles. The highest densities occur at high latitudes in the winter hemisphere. The right-hand side of Figure 3.4 shows the ozone distribution in terms of mixing ratios. Maximum ozone mixing ratios are found in the 30–40-km altitude regime over the equator, that is, in the domain of the greatest photochemical activity. Mixing ratios decline toward lower altitudes and toward the Poles. The gradient of mixing ratios gives rise to an eddy flux of ozone in the poleward direction. Transport toward higher latitudes thereby is seen to arise partly from eddy diffusion and not solely from large-scale meridional transport.

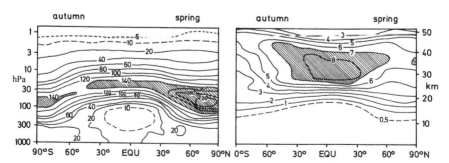

FIGURE 3.4 Meridional distribution of ozone in the atmosphere. Left frame: expressed as partial pressure (unit: 100 μPa). Right frame: expressed as mixing ratio (unit: μmol mol^{-1}). (Adapted from Dütsch (1980), with permission.)

3.2.2. GLOBAL TRENDS IN TOTAL OZONE

The altitude profile and the total depth of the stratospheric ozone layer are subject to considerable fluctuations, which are due partly to atmospheric motions and partly to the influence of chemical reactions to be discussed further below. The current interest in trends derives from the influence of certain trace gases on stratospheric chemistry and changes in their concentrations caused by anthropogenic activities. The analysis of ozone measurement series for trends is based on statistical regression models that take into account seasonal variations in mean ozone, seasonal variations in ozone trends, and the effects of cyclic phenomena such as the 11-year solar cycle or the biennial alternation between easterlies and westerlies in the equatorial stratosphere. The residuals from the model are autocorrelated, and this feature is used to ensure reliable standard errors in calculating the trend (Bojkov et al., 1990; Reinsel et al., 1994). A proper error analysis requires a weighted regression procedure, because ozone levels are more variable during the winter months than in summer.

The longest data series that may be analyzed for trends is available from the network of Dobson stations. The initial efforts of Komhyr et al. (1971), Angell and Korshover (1978), London and Oltmans (1979), and Reinsel (1981) indicated a slight increase in total ozone during the 1960s and fairly little change during the 1970s. In the meantime, many of the data have been reevaluated because changes in instruments and calibrations caused breaks and other inconsistencies in the record. The statistical analysis currently uses the ozone level prior to 1970 as a baseline and assumes a linear trend afterward. The results for the two decades 1970–1990 assembled by the Ozone Trends Panel (WMO 1995) indicated a negative trend in all seasons and year-round at essentially all stations.

Long-term trends of total ozone may also be derived from satellite ultraviolet back-scatter (SBUV) observations. Satellites provide a better global coverage than the network of surface stations, but the measurements have suffered from drift and other deficiencies. Hilsenrath and Schlesinger (1981) discussed data obtained with the instrument onboard the Nimbus 4 satellite launched in 1970. Nimbus 7 carried the Total Ozone Mapping Spectrometer (TOMS), which was operational from 1978 until 1993, and an ultraviolet solar back-scatter instrument, which failed in 1990. These instruments have since been replaced by similar ones on other satellites launched more recently. Fleig et al. (1986) and Herman et al. (1991) have discussed correction procedures for the observed downward drift of the Nimbus 7 instruments, and a reprocessed data set exists for the 1980s that is currently considered the most reliable satellite-based data set.

Figure 3.5 (left) compares trends of total ozone in various latitude bands as derived for the period 1979–1991 from the satellite and ground-based measurements to show that good agreement exists between the different data sets. In the equatorial region, the trend is statistically insignificant. In the middle and high latitudes a negative trend is observed, increasing toward the Poles, more so in the Southern compared to the Northern Hemisphere. This feature is primarily caused by large southern ozone losses during the winter half-year, that is during June–November. While the trends depicted in Figure 3.5 provide an indication of the average annual changes of total ozone during the 1980s, the trend would be 1–2.5% smaller, if the period had been extended backward in time to include the 1970s. Accordingly, it appears that losses of total ozone have increased with time.

When it was recognized in the late 1970s that chlorofluorocarbons have a detrimental effect on the ozone layer, it was noted that the greatest depletion of ozone would occur at altitudes above 30 km. At that time, ground-based routine analyses of the ozone altitude distribution were made with balloon-borne ozone sondes and the optical (Umkehr) technique pioneered by Götz (1931) based on measurements with the Dobson spectrometer. The mathematical inversion techniques required to evaluate the optical data made the method cumbersome until fast-speed computers became available. In addition, corrections are needed to account for the presence of aerosol particles. Thus, fewer data of this type existed in comparison to total ozone. The trend

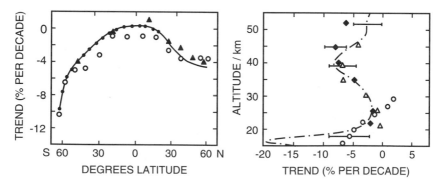

FIGURE 3.5   Left: Annual average trend of total ozone 1979–1991 as a function of latitude. ---, Total Ozone Mapping Spectrometer (TOMS), O, satellite back-scattered ultraviolet (SBUV), ▲, averages of individual trends at 59 Dobson stations within latitude zones. Right: Trends in the vertical distribution of ozone 30–50° N. ---, Stratospheric Aerosol and Gas Experiment (SAGE I & II); ◆, SBUV; △, ground-based Umkehr measurements; O ozone sonde data. (Adapted from WMO (1995).)

during the 1970s in the 40-km altitude region, which was initially explored by Bloomfield *et al.* (1982) and Reinsel *et al.* (1983, 1984), turned out to be small, less than minus 0.3% annually. More recently, Mateer and DeLuisi (1992) have developed an improved inversion algorithm based on temperature dependent ozone absorption coefficients, and the existing Umkehr data base has been reevaluated. These authors concluded that reliable ozone trends can be found for the altitude region above about 20 km. Satellite ultraviolet back-scatter measurements have also been used to determine ozone profiles, especially the Stratospheric Aerosol and Gas Experiment (SAGE) instruments, which measure ozone in absorption at 600-nm wavelength by solar occultation.

The right-hand side of Figure 3.5 shows the trend derived from such measurements during the 1980s at middle latitudes of the Northern Hemisphere. Good agreement is reached between Umkehr and satellite-based measurements, whereas the results from the ozone sondes appear to deviate markedly from the others. Strong ozone losses evidently occur at altitudes around 45 km, in agreement with expectation, and relatively independently of latitude. The magnitude of this loss is smallest near the equator and increases toward the Poles, reaching values in excess of 10% per decade poleward of 60° latitude. The trend is smallest at 20 km, near the maximum of the layer density, where it amounts to 1–2% per decade, which is almost within the error limits. Stronger losses occur also at altitudes below about 20 km, in a region where the ozone density decreases. In this region the density of aerosol particles, which attenuate the optical signal, increases, whereas the ozone sonde measurements would not be affected.

The temporal decrease in total ozone over the Antarctic region occurring during austral spring each year represents the strongest perturbation discovered so far. Observations at several Antarctic sites show that ozone declines largely during the month of September and reaches a minimum early in October. The abundance of ozone has reached levels typically 50% lower than the historical value of about 300 Dobson units, corresponding to about 134 mmol m^{-2}. This seasonal void is then filled by the rapid influx of ozone-rich air from lower latitudes as the winter polar stratospheric vortex breaks down, typically in late October or November. Figure 3.6 (left) shows monthly mean total ozone values in October, measured at the stations Halley Bay (76° S) and South Pole (90° S), indicating the decline since 1975. The data also indicate that in some years the depletion is worse than in others, but a regular pattern does not appear to exist. Figure 3.6 (right) shows that ozone losses during the event are greatest in the center of the layer in the stratosphere, whereas the upper part of the altitude profile is least affected. This feature differs from the situation at mid-latitudes, where the greatest

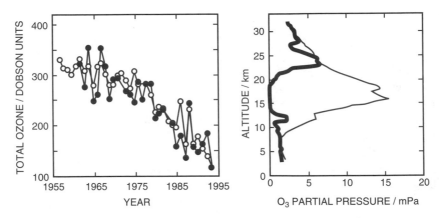

FIGURE 3.6   Left: October monthly means of total ozone since 1962 observed at the South Pole (points) and Halley Bay (76° S, circles). Right: Vertical distribution of ozone over the South Pole in austral spring 1993. Thin line, late August, column density 121 mmol m^{-2} (272 Dobson units); thick line, October 12, column density 41 mmol m^{-2} (91 Dobson units). (Adapted from WMO (1995).)

losses occurred in the upper stratosphere, as Figure 3.5 shows. The difference arises from the low temperatures in the Antarctic stratosphere. Low temperatures allow the formation of polar stratospheric clouds, giving rise to heterogeneous reactions on the surface of the cloud particles. The horizontal extent of the springtime Antarctic ozone depletion has been increasing in parallel with the loss of total ozone. The area in which total ozone is reduced to less than 100 mmol m^{-2} (220 DU) at the height of the event has now reached about 25 million km^2, corresponding to about 10% of the total area of the Southern Hemisphere.

In addition to low temperatures, the Antarctic is characterized by a very strong polar winter vortex, which extends from the ground through the entire stratosphere and essentially prevents the admixture of air from lower latitudes outside the vortex. While ozone losses occur also in the northern polar stratosphere, the winter vortex over the North Pole is much weaker, and temperatures in the stratosphere are about 10° higher (Newman *et al.*, 1993; Pawson *et al.*, 1995), so that the potential for the formation of polar stratospheric clouds is much reduced. When they occur, they do not persist for long. Accordingly, the ozone losses are temporary and less severe compared to the Antarctic. Nevertheless, larger column losses have been observed in recent years. For example, according to Bojkov *et al.* (1995) total ozone over Siberia in mid-January 1995 fell 25% below the long-term average.

3.3. THE CHAPMAN MODEL

The formation of the ozone layer can be understood most simply on the basis of a reaction model composed of a minimum set of four elementary processes: (a) the dissociation of oxygen molecules by solar radiation in the wavelength region 180–240 nm; (b) the association of oxygen atoms with molecular oxygen, causing the formation of ozone; (c) the photodissociation of ozone in the Hartley band between 200 and 300 nm; and (d) the destruction of ozone by its reaction with oxygen atoms.

$$O_2 + h\nu \rightarrow O + O \qquad j_a \qquad \text{(a)}$$

$$O + O_2 + M \rightarrow O_3 + M \qquad k_b \qquad \text{(b)}$$

$$O_3 + h\nu \rightarrow O_2 + O \qquad j_c \qquad \text{(c)}$$

$$O + O_3 \rightarrow O_2 + O_2 \qquad k_d \qquad \text{(d)}$$

The reactions and rate coefficients are here identified by letters rather than by the numbers of Table A.1 so as to keep the following discussion simple. In reaction (b), M denotes an inert third body, which in the atmosphere is supplied by either nitrogen or oxygen. In this case, the efficiencies of N_2 and O_2 as chaperones are almost equal. Therefore it is not necessary to distinguish between them. The temperature dependence of reaction (b) differs somewhat for N_2 and O_2, but the effect may be neglected to a first approximation. The excited $O(^1D)$ atoms formed in reaction (c) are quite rapidly deactivated by collisions with nitrogen or oxygen, so that the oxygen atoms may be taken to react in the ^3P ground state. Let n_1, n_2, and n_3 represent the number concentrations of the three species O, O_2, and O_3, respectively, and let n_M be the number concentration of N_2 and O_2 combined, at the selected altitude. The local variations with time of the concentrations of interest are given by the rate equations

$$dn_1/dt = 2j_a n_2 - k_b n_1 n_2 n_M + j_c n_3 - k_d n_1 n_3 \qquad (3.1)$$

$$dn_2/dt = -j_a n_2 - k_b n_1 n_2 n_M + j_c n_3 + 2k_d n_1 n_3 \qquad (3.2)$$

$$dn_3/dt = +k_b n_1 n_2 n_M - j_c n_3 - k_d n_1 n_3 \qquad (3.3)$$

As a check for the correct application of stoichiometry factors, we multiply the second equation by 2, the third by 3, and obtain after summing

$$dn_3/dt + 2\,dn_2/dt + 3\,dn_3/dt = 0 \qquad (3.4)$$

Integration then yields the mass balance equation

$$n_1 + 2n_2 + 3n_3 = \text{constant} = n_2^0 \qquad (3.5)$$

which demonstrates that mass conservation is obeyed and that the stoichiometry factors were applied correctly. The observations of ozone further establish that $n_2 \gg n_1 + n_3$, so that $n_2 = n_2^0$ may be taken as a constant. The first and third of the rate equations then suffice to determine n_1 and n_3. These equations form a system of coupled differential equations whose exact solutions can be obtained only by numerical integration. Fortunately, several simplifications are possible. First, it should be noted that the O atom concentration approaches a steady state much faster than the concentration of ozone. At 50 km altitude, the time constant for the adjustment to steady state of n_1 is $\tau = 1/k_b n_2 n_M \approx 20$ s, and the time constant decreases toward lower altitudes as n_M increases. The time constant for the approach to the steady state of ozone is much longer, and it increases with decreasing altitude (see below). Thus it is reasonable to assume that oxygen atoms are always in a steady state, that is, $dn_1/dt = 0$. In addition, it turns out that the second and third terms in Equation (3.1) are dominant compared with the other two. Accordingly, one has approximately

$$n_1 = j_c n_3 / k_b n_2 n_M \qquad (3.6)$$

The first and the last terms of Equation (3.1) cannot be neglected, of course, because they are ultimately responsible for the production and removal of oxygen atoms and ozone. This fact may be taken into account by summing Equations (3.1) and (3.3), and by using the sum equation in conjunction with Equation (3.6). These simplifications yield

$$dn_3/dt + dn_3/dt \approx dn_3/dt = 2j_a n_2 - 2k_d n_1 n_3$$

$$= 2\left(j_a n_2 - k_d j_c n_3^2 / k_b n_2 n_M\right) \qquad (3.7)$$

This equation can be integrated without difficulty. We assume for simplicity that the ozone concentration is initially zero and obtain

$$n_3 = \left(\frac{B}{A}\right)^{1/2} \frac{1 - \exp\left[-2(AB)^{1/2}t\right]}{1 + \exp\left[-2(AB)^{1/2}t\right]} \qquad (3.8)$$

where $A = 2k_d j_c / k_b n_2 n_M$ and $B = 2j_a n_2$.

The number densities n_2 and n_M are connected by the mixing ratio of oxygen, $x_2 = n_2/n_M = 0.2095$. After a sufficiently long period of time, the number density of ozone tends toward a steady state with the value

$$n_3 = \left(\frac{B}{A}\right)^{1/2} = n_2 \left(\frac{k_b n_M j_a}{k_d j_c}\right)^{1/2} = x_2 n_M \left(\frac{k_b j_a n_M}{k_d j_c}\right)^{1/2} \qquad (3.9)$$

The simplicity of Equation (3.9) is only apparent. Making use of Equation (2.10), the photodissociation coefficients are

$$j_a = \int \sigma_2 I_0 \exp(-\sec \chi)\left[\int_z^\infty (\sigma_2 n_2 + \sigma_3 n_3)\,dz\right] d\lambda \qquad (3.10a)$$

$$j_c = \int \sigma_3 I_0 \exp(-\sec \chi)\left[\int_z^\infty (\sigma_2 n_2 + \sigma_3 n_3)\,dz\right] d\lambda \qquad (3.10b)$$

where the subscripts 2 and 3 refer again to oxygen and ozone, respectively, and quantum yields of unity are assumed. Since the expressions for j_a and j_c involve the number density of ozone, it is not possible to calculate n_3 directly from Equation (3.9). A stepwise approximation procedure is required, starting from an assumed ozone density profile and improving it by iteration, working from high altitudes downward. For the purpose of illustration, the photodissociation coefficients shown in Figure 2.12 may be used to calculate ozone number densities as a function of altitude. The ratio of rate coefficients $k_b/k_d = k_1/k_2$ may be gleaned from Tables A.4 and A.5. At an altitude of 15–50 km the temperature increases from 210 to 280 K, and k_b/k_d decreases in magnitude from 3.4×10^{-18} to 1.6×10^{-19} cm^3 molecule^{-1}. The number density of oxygen decreases from 8.1×10^{17} to 4.8×10^{15} molecule cm^{-3}. Figure 3.7 shows the number densities of ozone calculated with these data. Included for comparison is the density profile measured at low latitudes. The calculation gives approximately the correct altitude distribution, but the concentrations derived are greater than those observed. This is partly due to the assumption of an overhead sun. More realistically, the photodissociation coefficients should be averaged over the day. Johnston (1972) has made such calculations and found the maximum ozone number density to reach 9.7×10^{12} molecule cm^{-3}, only slightly lower. Another part of the discrepancy arises from uncertainties in the O_2 absorption cross sections. More recent measurements in the 200–240-nm wavelength region, both in situ in the stratosphere (Herman and Mentall, 1983) and in the laboratory (Jenouvrier et al., 1986; Yoshino et al., 1988) indicate lower cross sections than were employed earlier.

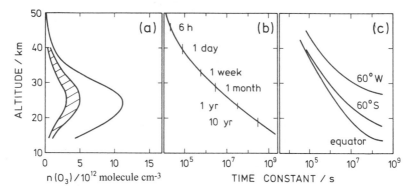

FIGURE 3.7   (a) Vertical profile of ozone number density calculated from Equation (3.9). The hatched area shows the range of observations at low latitudes from the data of Krueger (1969), Randhawa (1971), and Mauersberger *et al.*, (1981). (b) Time constant for the approach to the photostationary state of ozone calculated from Equation (3.11). (c) Ozone replacement times calculated from Equation (3.12) by Johnston and Whitten (1973), here for 60° N, summer and winter, and at the equator.

According to Equation (3.8) the characteristic time for the approach to steady state is

$$\tau_{ss} = \frac{1}{2}(AB)^{-1/2} = \frac{1}{4}\left(\frac{k_b n_M}{k_d j_a j_c}\right)^{1/2} \tag{3.11}$$

Again, the appropriate diurnal averages of the photodissociation coefficients must be inserted. For the purpose of illustration, half-values of those for an overhead sun are used here. Figure 3.7 shows the resulting time constants as a function of altitude. At altitudes above the ozone maximum, the characteristic times are sufficiently short to justify the assumption of steady-state conditions. Below the ozone maximum the times scale becomes longer than a year. Under such conditions, it is no longer possible to neglect the influence of transport by eddy diffusion and large-scale mean motions, so that photostationary conditions are not established.

Instead of the approach to steady-state conditions, one may also consider the time required to replenish the average amount of ozone at a given altitude by the photodissociation of oxygen. This replacement time is

$$\tau_{replacement} = \frac{n_{3, obs}}{2 j_a n_2} \tag{3.12}$$

Johnston and Whitten (1973, 1975) have calculated ozone replacement times from zonally averaged O_2 photodissociation rates as a function of latitude and height. The right-hand side of Figure 3.7 shows results for 60° latitude and for the equatorial region to provide a comparison with the characteristic time for the approach to steady state. The two time constants exhibit a similar behavior with regard to altitude, so that both are suitable as indicators for the stability of ozone in the stratosphere. At high latitudes the replacement time is larger in winter than in summer, in agreement with expectation because of the greater solar inclination during the winter months. The time constant for the approach to steady state shows the same behavior. An ozone replacement time of 4 months or less may be taken to characterize the main region of ozone formation. This region coincides approximately with the region of high O_3 mixing ratios shown in Figure 3.4. In the other regions, where the replacement time exceeds 1 year, ozone must be considered to be completely detached from the region of O_2 photodissociation.

Transport of ozone over large distances into those regions of the lower stratosphere where ozone is no longer replaced requires a sufficiently long lifetime against destruction by photolysis. The associated time constant differs from those discussed above. Within the framework of the Chapman theory, the photochemical lifetime of ozone is obtained from Equation (3.7) by setting $j_a = 0$. Integration then gives the decay of the ozone concentration with time

$$n_3 = n_3^0 / \left(1 + A n_3^0 t \right) \tag{3.13}$$

which leads to a half-life time

$$\tau_{1/2} = 1/A n_3^0 = \frac{k_b n_2 n_M}{2 k_d j_c n_3^0} \tag{3.14}$$

For an altitude of 15 km or less one finds half-life times greater than 100 years. This value must be considered an upper limit because it ignores the influence of other trace gases that catalyze ozone destruction. Up to 80% of ozone destruction may occur via such catalytic cycles, and the lifetime of ozone in the lower stratosphere will be shortened to about 20 years. As discussed in Chapter 1, the meridional transport of trace substances toward the Poles takes place mainly during the winter months in the respective hemisphere, and this is supported by the observed behavior of ozone (Fig. 3.4). A large portion of the ozone content at high latitudes is discharged into the troposphere in spring and early summer. Accordingly, the residence time of ozone in the lower stratosphere cannot be longer than a few years at most. This is much shorter than the photochemical lifetime derived above. Consequently, ozone can be considered to be chemically stable in those regions of the stratosphere that lie outside the principal domain of ozone formation and where the influence of transport is preponderant.

The importance of transport has led to the suggestion that the excess of ozone derived from the calculations (see Fig. 3.7) is an artifact of the steady-state assumption resulting from the neglect of losses due to transport. It must be recognized, however, that the main effect of transport is to displace ozone from one region of the stratosphere to another. The global budget of ozone should not be affected. Johnston (1975) has calculated globally integrated rates of (a) ozone formation due to the dissociation of oxygen, (b) ozone losses due to reaction with oxygen atoms, and (c) transport to the troposphere. The results, which are shown in Table 3.1, make it apparent that the global loss of ozone to the troposphere is only a small fraction of the total budget of stratospheric ozone. Table 3.1 shows also that, on a global scale, the total loss rate does not balance the production rate, so that there must be other loss processes that the Chapman model ignores. This comparison provides the most compelling argument for an expansion of the Chapman reaction scheme by additional reactions involving trace constituents other than ozone. These are discussed below.

3.4. TRACE GASES OTHER THAN OZONE

Processes causing the loss of ozone involving reactions other than those with oxygen atoms involve substances, that contain nitrogen (NO_x, odd nitrogen), hydrogen (HO_x, odd hydrogen), and chlorine (ClO_x), or other halogens. The various species derive from source gases entering the stratosphere, mainly from the troposphere. The most prominent source gases are nitrous oxide, (N_2O), water vapor and methane (H_2O, CH_4), and the chlorine-containing

TABLE 3.1   Globally Integrated Rates of Ozone Formation and Ozone Destruction[a]

Process	January 15 (Mmol s^{-1})	March 22 (Mmol s^{-1})
Gross rate of O_3 formation by O_2 photodissociation	83.3	81.0
Chemical loss from Chapman reactions	14.3	14.8
Transport to the troposphere	1.0	1.0
Imbalance between production and losses	68.0	65.2
Additional chemical losses due to water reactions	9.3	9.0
Unbalanced difference	58.7	56.2

[a] According to Johnston (1975).

compounds CH_3Cl, CCl_4, CF_2Cl_2, and $CFCl_3$. These gases do not react with ozone, and they must first be converted into the reactive species. It will be convenient to treat each class of compounds separately, although it should be noted that there is some interaction between the three groups, so that they are not entirely independent. The origin of the source gases and their behavior in the troposphere will be discussed in later chapters.

3.4.1. NITROGEN OXIDES

Figure 3.8 shows that the mixing ratio of N_2O in the stratosphere declines with altitude from about 300 nmol mol^{-1} at the tropopause to about 20 nmol mol^{-1} at 40 km. The decline is somewhat more pronounced in mid-latitudes because of the lower-lying tropopause compared with the Tropics. Evidently, N_2O is destroyed at altitudes above 25 km. The vertical gradient of the mixing ratio maintains an eddy flux of N_2O from the tropopause up to the region of destruction. The principal loss process is photodecomposition

$$N_2O + h\nu \rightarrow N_2 + O(^1D)$$

which becomes effective at wavelengths below 220 nm, but occurs mainly in the atmospheric window of absorption near 200 nm (see Fig. 2.9) A second

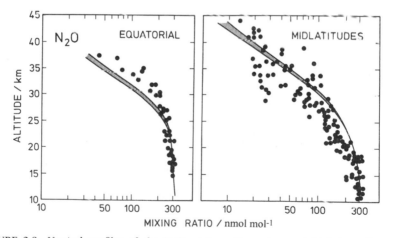

FIGURE 3.8 Vertical profiles of the N_2O mixing ratio at low and high latitudes. From measurements of Tyson et al. (1978a), Vedder et al. (1978, 1981), Fabian et al. (1979, 1981), Goldan et al. (1980, 1981). The solid lines are results of calculations by Gidel et al. (1983) based on a two-dimensional model.

process contributing to N_2O loss is the reaction

$$O(^1D) + N_2O \rightarrow N_2 + O_2$$

$$\rightarrow 2NO$$

where the $O(^1D)$ oxygen atoms arise primarily from the photodissociation of ozone in the 200–300-nm wavelength region. The branching ratio of the reaction is approximately unity. Only the second process leads to reactive nitrogen oxides. Table 3.2 shows several estimates for the global loss rate of N_2O and the associated production of NO. The latter is about 12% of the former. The principal region of this photochemical activity occurs at 25–35 km altitude in the Tropics. Other mechanisms of NO production have been considered. They include ionization of air by cosmic radiation (Warneck, 1972; Nicolet, 1975) and the injection into the atmosphere of solar protons resulting from sunspot activity (Crutzen et al., 1975; Frederick, 1976). An intercomparison of the various sources by Jackman et al. (1980) has shown that nitrous oxide is by far the largest source of NO on a global scale.

Nitric oxide is quite rapidly oxidized to NO_2 by reacting with ozone. Nitrogen dioxide, in turn, is subject to photolysis, whereby NO is regenerated. This leads to the following reaction sequence:

$$NO_2 + h\nu \rightarrow NO + O$$

$$O + O_2 + M \rightarrow O_3 + M$$

$$NO + O_3 \rightarrow NO_2 + O_2$$

which leaves the sum of NO and NO_2 unaffected. At altitudes below 40 km, the time constant for the approach to the photostationary state between NO and NO_2 is less than 100 s, and the steady state is rapidly established. After sunset, NO decays to low values, and it comes up again following sunrise.

TABLE 3.2   Estimates for the Global Loss of N_2O Due to Photodissociation in the Stratosphere, and the Production of NO Due to the Reaction $O(^1D) + N_2O \rightarrow 2NO$

Authors	Loss of N_2O nitrogen		Gain of NO nitrogen	
	(kmol s^{-1})	(Tg year^{-1})	(kmol s^{-1})	(Tg year^{-1})
Schmeltekopf et al. (1977)	16.7	15	3.6	1.6
Johnston et al. (1979)	10	9	2.3	1.0
Levy et al. (1980)	—	—	1.2	0.5
Jackman et al. (1980)	—	—	2.8	1.0
Crutzen and Schmailzl (1983)	7.5–11.7	6.8–10.3	0.9–1.5	0.4–0.7
Minschwaner et al. (1993)	13.8	12.2	1.6	0.7

The diurnal variation of NO has been documented by Ridley *et al.* (1977) and Ridley and Schiff (1981) from in situ measurements by instruments onboard a balloon floating at 26 km altitude, and more recently from satellite measurements (Russell *et al.*, 1984; Gordley *et al.*, 1996).

Figure 3.9 summarizes mixing ratios of NO, NO_2, and HNO_3 as a function of altitude. These data were obtained by means of balloons or rockets as platforms. NO_2 and HNO_3 were mainly determined by infrared spectroscopy, and NO with a chemiluminescence technique. More recently, vertical profiles for odd nitrogen components were measured from the space shuttle platform by means of limb-sounding optical techniques. Figure 3.10 shows some results from this experiment. Both data sets demonstrate that in the lower stratosphere nitric acid is the major component derived from NO_x. The vertical gradient of the mixing ratios of all odd nitrogen species is opposite that for N_2O, indicating that they are transported downward from the stratosphere toward the troposphere. The degree of NO_2 dissociation increases with altitude because the number density of ozone declines above 30 km altitude, so that the $NO + O_3$ reaction is less and less able to reoxidize NO to NO_2. In the mesosphere, NO is the dominant species. The

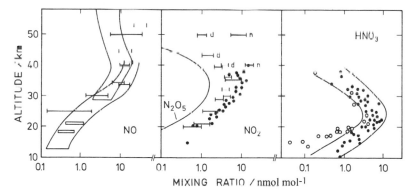

FIGURE 3.9 Vertical distribution of nitrogen oxides and nitric acid in the stratosphere. Left: Nitric oxide in the sunlit atmosphere. Fields enclose data obtained with the chemiluminescence technique (Horvath and Mason, 1978; Roy *et al.*, 1980; Ridley and Schiff, 1981; Ridley and Hastie, 1981); horizontal lines represent measurements by infrared optical techniques (Drummond and Jarnot, 1978; Roscoe *et al.* 1981; Loewenstein *et al.*, 1978a, b). Center: Nitrogen dioxide observed by infrared optical techniques, day (d) and night (n). Points are data from Goldman *et al.* (1978), Blatherwick *et al.* (1980). Horizontal bars are data from Drummond and Jarnot (1978) and Roscoe *et al.* (1981). The N_2O_5 profile was obtained by Toon *et al.* (1986a) at sunrise. Right: Nitric acid observed by in situ filter sampling (open points, Lazrus and Gandrud, 1974) and by infrared spectroscopy and mass spectrometry (solid points, Fontanella *et al.*, 1975; Harries *et al.*, 1976; Arnold *et al.*, 1980; Murcray *et al.*, as quoted by Hudson, 1982; Fischer *et al.*, 1985). The envelope indicates the error range.

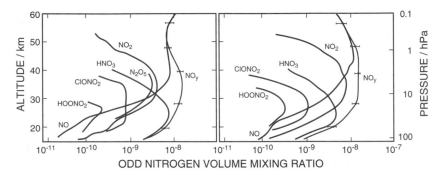

FIGURE 3.10   Vertical profiles of several important odd nitrogen species in the stratosphere, determined by infrared measurements from a spacecraft platform on May 1, 1985 (Redrawn from Russell *et al.*, 1988). Left: at sunrise, 48° S; right: at sunset, 30° N.

mixing ratio of NO at the stratospause is approximately 10 nmol mol^{-1}. With increasing altitude it rises toward 300 μmol mol^{-1} at 130 km (Barth *et al.*, 1996). This region receives NO primarily from the ionosphere, where it is produced by ion reactions. In addition, NO undergoes photolysis (predissociation) at wavelengths shorter than 200 nm, which leads to destruction due to the reactions

$$NO + h\nu \rightarrow N + O$$

$$N + O_2 \rightarrow NO + O$$

$$N + NO \rightarrow N_2 + O$$

The situation has been discussed by Strobel (1971), Brasseur and Nicolet (1973), Frederick and Hudson (1979b), and Jackman *et al.* (1980). It appears that the flux of NO at the stratopause, either into or out of the mesosphere, is essentially negligible compared with the stratospheric NO production rate.

The significance of NO$_x$ for stratospheric ozone arises from the competition between NO$_2$ and O$_3$ for oxygen atoms:

$$O + NO_2 \rightarrow NO + O_2$$

$$O + O_3 \rightarrow 2O_2$$

and the fact that the rate coefficient for the first of these processes is much greater than that for the second. Accordingly, the first reaction is favored despite the fact that NO$_2$ mixing ratios are about two orders of magnitude smaller than those of ozone. This leads to the reaction sequence

$$O_3 + h\nu \rightarrow O + O_2$$
$$O + NO_2 \rightarrow O_2 + NO$$
$$\underline{NO + O_3 \rightarrow NO_2 + O_2}$$

$$\text{net} \quad O_3 + O_3 \rightarrow 3O_2$$

NO_x is not consumed in these reactions. Rather it acts as a catalyst to ozone destruction. The catalytic cycle is fast and proceeds several times during the time interval in which the ozone loss reaction of the Chapman mechanism occurs once.

In addition to reactions generating NO, nitrogen dioxide undergoes oxidation to N_2O_5 and to nitric acid. The formation of N_2O_5 follows from the relatively slow reaction of NO_2 with ozone,

$$NO_2 + O_3 \rightarrow NO_3 + O_2$$
$$NO_3 + NO_2 \rightleftharpoons N_2O_5$$

where the NO_3 radical serves as an intermediate. The second equation shows that NO_3 and N_2O_5 are in thermal equilibrium. At the low temperatures in the stratosphere, the equilibrium favors N_2O_5 (see Table 9.8). While these reactions occur at all times, they are supplemented during the day by the photodecomposition reactions

$$N_2O_5 + h\nu \rightarrow NO_2 + NO_3$$
$$NO_3 + h\nu \rightarrow NO_2 + O$$

The photodecomposition of NO_3 is very rapid with $j(NO_3) \approx 0.1 \text{ s}^{-1}$, as it occurs mainly by radiation in the visible region of the solar spectrum. This prevents the production of N_2O_5 during the day. The absorption of light by N_2O_5 becomes appreciable at wavelengths shorter than 350 nm, although it is most effective at wavelengths in the atmospheric window near 200 nm. N_2O_5 thus builds up during the night and undergoes photodecomposition during the day (with a time constant of about 7 h; see Fig. 2.19). The consequence is a diurnal variation of NO_x; low values occur in the morning and higher ones—by a factor of 2—in the late afternoon. Such a variation was first observed by Brewer and McElroy (1973) from twilight measurements of NO_2 and subsequently confirmed by measurements made onboard balloons and from the ground (Kerr and McElroy, 1976; Evans et al., 1977; Noxon, 1979a, b). Clear-cut evidence for the presence of N_2O_5 in the stratosphere was first obtained by Toon et $al.$ (1986a) from infrared spectra. Figures 3.9 and 3.10 include altitude profiles of N_2O_5.

Nitric acid (HNO_3) arises from the interaction of NO_2 with water radicals. The relevant reactions are

$$NO_2 + OH + M \rightarrow HNO_3 + M$$

$$HNO_3 + h\nu \rightarrow NO_2 + OH$$

$$HNO_3 + OH \rightarrow H_2O + NO_3$$

The ultraviolet absorption spectrum of nitric acid has a long-wavelength tail reaching beyond 300 nm, so that HNO_3 is somewhat susceptible to photodecomposition in the lower stratosphere and in the troposphere. In these regions, $j(HNO_3) \approx 5 \times 10^{-7}$ s^{-1} (see Fig. 2.19); the corresponding time constant for photodestruction is about 25 days, so that at altitudes below 25 km, nitric acid is relatively stable and provides the major reservoir of NO_x. Nitrogen dioxide then occurs as a photodissociation product of HNO_3, but because of the long time constant for photolysis it is doubtful whether a photostationary state between HNO_3 and NO_2 is actually achieved. Above 30 km altitude, $j(HNO_3)$ increases by more than an order of magnitude and $n(OH)$ increases as well, so that the time constant for the approach to steady-state conditions decreases to less than 1 day. In the upper stratosphere, therefore, HNO_3 and NO_2 are in a photostationary state, although in this region HNO_3 becomes less important as a reservoir of NO_2. Peroxynitric acid, which arises from the recombination–dissociation reaction $HO_2 + NO_2 \rightleftharpoons HO_2NO_2$, represents another reservoir of NO_x, although it is not very significant compared to the other species, as Figure 3.10 shows. More important in this regard is chlorine nitrate, $ClONO_2$, which is formed by the association of NO_2 with ClO radicals (see Section 3.4.3). Chlorine nitrate is thermally quite stable, but it is fairly rapidly photolyzed and has a lifetime of several hours.

Similar to ozone, the nitrogen oxides and nitric acid undergo poleward transport and partly accumulate in the lower stratosphere at high latitudes before they are discharged into the troposphere. Murphy et $al.$ (1993) have measured simultaneously ozone and total reactive odd nitrogen (NO_y) in the lower stratosphere by means of high-flying aircraft and found a striking positive correlation. As Figure 3.10 indicates, the definition of NO_y includes all NO_x-derived species but not the N_2O precursor, with which NO_y is anti-correlated. At altitudes below 20 km, NO_y is composed primarily of HNO_3. As ozone in the lower stratosphere is transported from the equator to the Poles, the correlation confirms that NO_y also undergoes poleward transport to accumulate at high latitudes. At the 20-km altitude level, the aircraft measurements showed NO_y mixing ratios near 2 nmol mol^{-1} in the equatorial region, rising to 10 nmol mol^{-1} at 60° latitude in both hemispheres.

These results confirmed earlier observations of Coffey *et al.* (1981), who determined the latitudinal dependence of NO_2 and HNO_3 column densities in the Northern Hemisphere by means of infrared absorption measurements at sunrise and sunset. The data are shown in Figure 3.11. A poleward increase in both NO_2 and HNO_3 is evident, but NO_2 shows a ridge near 35° N in winter and then declines further northward. The decline is largely compensated by an increase in HNO_3, which suggests that photodissociation of HNO_3 is greatly reduced because of the low elevation of the sun. This phenomenon gives rise to a pronounced seasonal variation in the NO_2 column density, with high values in summer and lower ones in winter. The ridge was first noted by Noxon (1975, 1979b). He suggested that it occurs after the winter polar vortex is established, which then blocks meridional transport of NO_x toward the Arctic, and that the ridge disappears when the vortex breaks down in early spring. However, low dissociation rates alone would suffice to explain the data. Gordley *et al.* (1996) compared the seasonal variation of the NO_2 column density determined from zenith sky

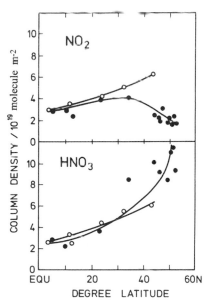

FIGURE 3.11   Total column densities of NO_2 and HNO_3 in the stratosphere as a function of latitude in the Northern Hemisphere (O, summer, ●, winter). (Adapted from Coffey *et al.* (1981).)

light measurements at Fritz Peak Observatory, Colorado, and from satellite solar occultation (sunrise and sunset) data and found good agreement. Figure 3.12 is added to show the distribution of NO_2 with latitude and altitude and its seasonal variation as derived from satellite data during the periods of winter and summer solstice. The principal maximum occurs in the Tropics at about 40 km altitude. The seasonal maximum at high latitudes occurs in summer and is located near 30 km altitude, that is near the upper fringe of the HNO_3 reservoir.

3.4.2. WATER VAPOR, METHANE, AND HYDROGEN

In the troposphere, the mixing ratio of water vapor decreases with altitude because of the lowering of the saturation vapor pressure with decreasing temperature and condensation of the excess. Minimum temperatures between troposphere and stratosphere occur at the tropical tropopause. Because most of the air entering the stratosphere passes through this region, it represents a cold trap that regulates the mixing ratio of water vapor in the stratosphere. Table 1.5 indicates that the temperature and pressure of air at the tropical tropopause are 195 K and 120 hPa, respectively. The equilibrium vapor pressure of H_2O at 195 K is roughly 5×10^{-5} Pa, corresponding to an H_2O mixing ratio of 5 μmol mol^{-1}. Harries (1976) and Elsaesser et $al.$ (1980) have reviewed a great number of stratospheric water vapor measure-

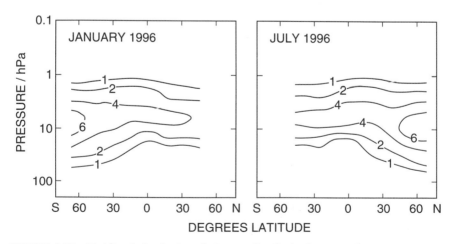

FIGURE 3.12 Meridional distribution of nitrogen dioxide in the stratosphere at sunrise, in January and July 1996. Contours show the mixing ratio in nmol mol^{-1}. Data are from the Halogen Occultation Experiment on the Upper Atmosphere Research Satellite (HALOE, 1998).

ments. Most of the data, as far as they were considered reliable, lie in the range of $2-7 \, \mu$mol mol^{-1}, with the average rising slightly with altitude up to 40 km, but the data exhibit considerable scatter. Water vapor may enter the stratosphere with the Hadley circulation, as originally proposed by Brewer (1949). High-reaching convective clouds also provide a transport mechanism. Kley et al. (1982) showed that a minimum in the H_2O mixing ratio generally occurs 2-3 km above the tropical tropopause. This feature is not well understood. Additional water is brought into the stratosphere by the oxidation of methane. This process must be the origin of the weak rise in the H_2O mixing ratio with increasing altitude up to the stratopause. Jones et al. (1984) have used satellite data to show that despite considerable seasonal variations in the mixing ratios of both constituents, the mixing ratio of hydrogen contained in H_2O and CH_4 averages to 6 μmol mol^{-1} over the stratosphere. The flux of methane into the stratosphere exceeds the planetary escape flux of hydrogen by two orders of magnitude, so that most of the water produced from methane as well as that entering the stratosphere directly must return to the troposphere to maintain a global balance. Stanford (1973) pointed out that the winter temperatures of the stratospheres above the poles fall below that of the tropical tropopause, giving rise to the formation and precipitation of ice crystals. This is especially true of the Antarctic. The winter poles, therefore, are natural sinks for stratospheric water vapor.

The reaction of water vapor with $O(^1D)$ atoms is a major source of hydroxyl radicals in the stratosphere

$$O(^1D) + H_2O \rightarrow 2OH$$

Ground-state oxygen atoms do not react with water vapor; the photodissociation of H_2O is shielded by oxygen and ozone and becomes effective not before one reaches the mesosphere. Figure 3.13 shows the altitude profile of the daytime mixing ratios of OH and HO_2 radicals as derived mainly from in situ measurements with balloons as carriers. The OH mixing ratio is controlled by a greater number of chemical reactions. Among them are the chain reactions

$$O + OH \rightarrow O_2 + H$$

$$H + O_3 \rightarrow O_2 + OH$$

$$H + O_2 + M \rightarrow HO_2 + M$$

$$O + HO_2 \rightarrow OH + O_2$$

which are of major importance in the upper stratosphere and mesosphere.

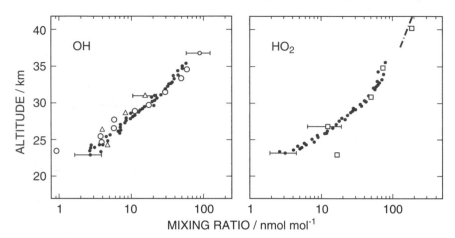

FIGURE 3.13   Vertical distribution of HO_x radicals in the stratosphere. Left: OH mixing ratio obtained on balloon flights in August 1987, July 1988, and July 1989 by the resonance fluorescence technique (Stimpfle and Anderson, 1988; Stimpfle *et al.* 1989, 1990). Right: HO_2 mixing ratio. Points, by resonance fluorescence (Stimpfle *et al.*, 1990), squares, by far infrared spectroscopy (Traub *et al.*, 1990), dotted line, by ground-based millimeter wave spectroscopy (de Zafra *et al.*, 1984, as given by Traub *et al.*, 1990).

These reactions shuffle odd hydrogen back and forth between OH and HO_2 radicals, and in the process they cause the destruction of odd oxygen, that is, the sum of oxygen atoms and ozone. The odd hydrogen cycle is terminated by the recombination reaction

$$OH + HO_2 \rightarrow H_2O + O_2$$

A similar recombination reaction of HO_2 radicals to form H_2O_2 is not very effective in removing odd hydrogen because H_2O_2 is subject to photodissociation, whereby OH radicals are regenerated:

$$HO_2 + HO_2 \rightarrow H_2O_2$$

$$H_2O_2 + h\nu \rightarrow 2OH$$

The loss of HO_2 by the formation of H_2O_2 is more important in the troposphere because of the reduced photolysis frequency and because H_2O_2 partly dissolves in clouds and is removed by rainout.

   In the lower stratosphere the interaction of OH and HO_2 with oxygen atoms becomes fairly unimportant, whereas the direct reactions with ozone gain significance. These are

$$OH + O_3 \rightarrow HO_2 + O_2$$
$$\underline{HO_2 + O_3 \rightarrow OH + 2\,O_2}$$

net $\quad O_3 + O_3 \rightarrow 3O_2$

The second of these reactions is fairly slow, and it must compete with the reaction of HO_2 with nitric oxide:

$$HO_2 + NO \rightarrow NO_2 + OH$$
$$NO_2 + h\nu \rightarrow NO + O$$
$$O + O_2 + M \rightarrow O_3 + M$$

which regenerates the ozone lost by reaction with OH so that it causes no net loss. In the lower stratosphere, odd hydrogen is removed mainly by reaction with nitric acid, as discussed previously. The reaction $OH + HO_2$ becomes comparatively unimportant.

Hydroxyl radicals also initiate the oxidation of methane in the stratosphere by virtue of the reaction

$$OH + CH_4 \rightarrow CH_3 + H_2O$$

The mechanism for the oxidation of methane will be discussed in more detail in Section 4.2. Here it suffices to note that formaldehyde is the first nonradical product arising from the attack of OH upon methane. Formaldehyde undergoes photodecomposition in the near-ultraviolet spectral region. There are two decomposition channels,

$$HCHO + h\nu \rightarrow H + HCO$$
$$\rightarrow H_2 + CO$$

one of which yields hydrogen and carbon monoxide as stable products. Both H_2 and CO enter into reactions with OH radicals:

$$OH + CO \rightarrow CO_2 + H$$
$$OH + H_2 \rightarrow H_2O + H$$

The reaction with CO is two orders of magnitude faster than that with H_2, so that the mixing ratio of CO is kept at a fairly low level, about 10 nmol mol^{-1}, according to measurements of Farmer et al. (1980) and Fabian et al. (1981b), at altitudes between 20 and 30 km. In the mesosphere, CO is produced by photodissociation of CO_2, and the CO mixing ratio rises again. In the lower stratosphere, where the main fate of the hydrogen atoms is attachment to O_2, the generation of hydrogen atoms by the reactions OH + CO and OH + H_2 represents a source of ozone, because the HO_2 radicals formed react further with NO to produce NO_2, which subsequently photolyzes to form O atoms. The process is similar to that discussed in Section 5.2, which leads to the formation of ozone by photochemical smog.

Figure 3.14 shows the altitude profiles of methane and hydrogen in the stratosphere. The loss of methane is made evident by the decrease in the CH_4 mixing ratio with increasing altitude. In the lower stratosphere, the reaction with OH radicals is the main loss process. In the upper stratosphere methane also reacts with $O(^1D)$ and Cl atoms. Crutzen and Schmailzl (1983) estimated from two-dimensional model calculations that OH and $O(^1D)$ and Cl, respectively, contribute about 50% each to the total loss of methane. The global loss rate of stratospheric CH_4 removal according to their estimate is 42 Tg year^{-1}, whereas Prather and Spivakovsky (1990) obtained 29 Tg year^{-1} from calculations based on a three-dimensional model.

The chemical reactions discussed above suggest a yield of two H_2O molecules for each molecule of methane that is oxidized. This leads to a global stratospheric production rate of $2(M_{H_2O}/M_{CH_4})F(CH_4) = 65-95$ Tg year^{-1}. In Section 1.5 the air flow into the stratosphere due to the Hadley

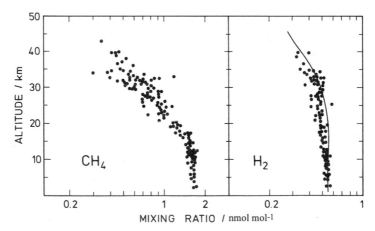

FIGURE 3.14   Vertical distribution of methane and hydrogen in the stratosphere at mid-latitudes (40–60° N). Data compiled from observations of Bush *et al.* (1978), Ehhalt and Heidt (1973a, b), Ehhalt *et al.* (1975), Pollock *et al.* (1980), Heidt and Ehhalt (1980), Fabian *et al.* (1979, 1981b), and Volz *et al.* (1981b).

circulation is estimated to be 1.4×10^{17} kg year^{-1}. If one adopts an average molar H_2O mixing ratio of 4 μmol mol^{-1}, corresponding to a mass mixing ratio of 2.5 μg g^{-1}, the flux of water vapor cycled through the stratosphere is 350 Tg year^{-1}. The oxidation of methane thus adds 20–25% to the flux of water vapor entering the stratosphere directly through the tropical tropopause.

In contrast to methane, the vertical distribution of hydrogen shows little change with altitude, although reaction rates for the interaction of H_2 and CH_4 with OH and $O(^1D)$ are approximately equivalent. Accordingly, a source of H_2 is required in the stratosphere order to explain the observed H_2 altitude profile. At altitudes below 40 km, the oxidation of methane is the only known source of molecular hydrogen. Above 40 km the reaction

$$H + HO_2 \rightarrow H_2 + O_2$$

must be added to maintain the high level of the hydrogen mixing ratio, because the concentration of methane declines and can no longer provide an adequate supply. Figure 3.14 includes a vertical H_2 profile derived from a one-dimensional eddy diffusion model. The calculated distribution shows a slight bulge near 20 km, indicating that in this region the production of H_2 exceeds the losses. The observations do not generally exhibit this feature, although measurements and calculations are in good agreement.

3.4.3. HALOCARBONS

Important source gases for chlorine-containing species in the stratosphere are methyl chloride (CH_3Cl), the chlorofluoromethanes ($CFCl_3$ (CFC-11), CF_2Cl_2 (CFC-12)), carbon tetrachloride (CCl_4), and methylchloroform (CH_3CCl_3). Of these, only methyl chloride has a natural origin, whereas the other compounds are released from anthropogenic sources. Figure 3.15 shows vertical profiles of the first four species in the stratosphere. The behavior is similar to that of nitrous oxide or methane—that is, the halocarbon diffuses upward from the tropopause and undergoes destruction somewhere in the middle and upper stratosphere. The hydrogen-bearing species are lost mainly by reactions with OH radicals and $O(^1D)$ atoms, and the others undergo photodecomposition. $CFCl_3$, CF_2Cl_2, and CCl_4 begin to absorb solar radiation at wavelengths below about 230 nm. Absorption rates and photodissociation become most effective in the wavelength region of 175–210 nm, which is partly shielded by the Schumann-Runge bands of oxygen. The photolysis of a halocarbon leads to the generation of chlorine

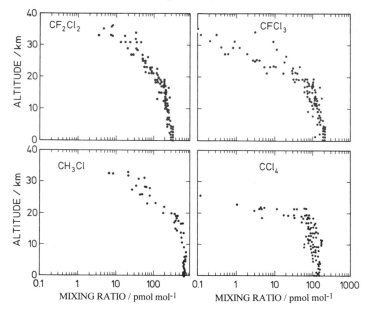

FIGURE 3.15   Vertical distribution of CF_2Cl_2, $CFCl_3$, CH_3Cl, and CCl_4 in the stratosphere. Data compiled by Fabian (1986) from Lovelock (1974a), Heidt *et al.* (1975), Schmeltekopf *et al.* (1975), Krey *et al.* (1977), Robinson *et al.* (1977), Seiler *et al.* (1978b), Tyson *et al.* (1978a), Vedder *et al.* (1978, 1981), Fabian *et al.* (1979, 1981a), Goldan *et al.* (1980), Penkett *et al.* (1980), Leifer *et al.* (1981), Rasmussen *et al.* (1980, 1982c), Schmidt *et al.* (1981), and Borchers *et al.* (1983).

atoms, for example:

$$CFCl_3 + h\nu \rightarrow Cl + CFCl_2$$

$$CFCl_2 + O_2 \rightarrow CFCl_2OO$$

$$CFCl_2OO + NO \rightarrow NO_2 + COFCl + Cl$$

The carbonyl halides thus formed are also subject to photodecomposition, whereby additional chlorine (and fluorine) atoms are released. Their interaction with ozone results in the catalytic cycle

$$Cl + O_3 \rightarrow ClO + O_2$$
$$\underline{ClO + O \rightarrow Cl + O_2}$$

$$\text{net} \quad O + O_3 \rightarrow 2\,O_2$$

which represents an important ozone loss process. The ClO_x and NO_x families are coupled by the reactions

$$ClO + NO \rightarrow NO_2 + Cl$$

$$Cl + O_3 \rightarrow ClO + O_2$$

$$NO_2 + h\nu \rightarrow NO + O$$

$$O + O_2 + M \rightarrow O_3 + M$$

In this reaction sequence ozone is regenerated so that there is no net loss. Yet another link between the ClO_x and NO_x families is the formation of chlorine nitrate and its photolysis:

$$ClO + NO_2 \rightarrow ClONO_2$$

$$ClONO_2 + h\nu \rightarrow Cl + NO_3 \quad \text{(predominantly)}$$

$$Cl + O_3 \rightarrow ClO + O_2$$

$$NO_3 + h\nu \rightarrow NO_2 + O$$

$$O + O_2 + M \rightarrow O_3 + M$$

This reaction sequence also does not lead to a net loss of ozone, but it temporarily ties up some chlorine in an inactive reservoir. Figure 3.10 shows that the mixing ratio of chlorine nitrate at 30 km altitude is about 1 nmol mol^{-1}.

An interaction also occurs between the chlorine and HO_x families in that ClO reacts with HO_2 to form hypochlorous acid. This leads to the reaction sequence

$$ClO + HO_2 \rightarrow HOCl + O_2$$

$$HOCl + h\nu \rightarrow OH + Cl$$

$$Cl + O_3 \rightarrow ClO + O_2$$

$$\underline{OH + O_3 \rightarrow HO_2 + O_2}$$

$$\text{net} \quad O_3 + O_3 \rightarrow 3O_2$$

which represents again an ozone loss process. The maximum mixing ratio of $HOCl$ is approximately 100 pmol mol^{-1} and occurs at altitudes near 35 km (Chance *et al.*, 1989).

The principal sink for active chlorine species is the conversion to hydrochloric acid. The reactions involved are

$$Cl + CH_4 \rightarrow HCl + CH_3$$

$$Cl + HO_2 \rightarrow HCl + O_2$$

$$Cl + H_2 \rightarrow HCl + H$$

with methane providing the main agent for the removal of Cl atoms and HCl formation. From HCl the chlorine atoms are regenerated, at a much slower rate, by reaction with OH radicals:

$$OH + HCl \rightarrow H_2O + Cl$$

and this reaction provides the principal source of ClO_x in the upper stratosphere. Fluorine atoms are similarly converted to hydrogen fluoride. Its reaction with OH is endothermic, however. The photolysis of HF is shielded by oxygen so that fluorine atoms do not play the same role as chlorine atoms.

Figure 3.16 shows the vertical distribution of HCl and HF in the stratosphere in addition to that of (daytime) ClO. The mixing ratio of ClO, although appreciable, generally falls below that of HCl, so that HCl is the major chlorine reservoir in the upper stratosphere. In the lower stratosphere chlorine nitrate also contains nonnegligible amounts of chlorine. The gradients of mixing ratios of HCl and HF are positive, which indicates that these substances are transported downward into the troposphere. Girard *et al.* (1978, 1983) and Mankin and Coffey (1983) have measured total column densities of HCl and HF and reported an increase from the equator toward higher latitudes by a factor of about 4. This indicates that HCl and HF are transported poleward and accumulate in the lower stratosphere to some extent before they enter the troposphere. The behavior is qualitatively similar to that of ozone and nitric acid and shows that HCl and HF indeed represent the terminal sinks of chlorine and fluorine.

At the stratopause, the mixing ratio of HCl extrapolates to about 3 nmol mol^{-1}, whereas that of HF extrapolates to about 0.9 nmol mol^{-1}. Zander *et al.* (1992) have derived more precise stratospheric inventories of chlorine and fluorine for the year 1985 by considering not only HCl and HF, but other chlorine- and fluorine-bearing species as well. The stratospheric chlorine and fluorine contents at that time were 2.58 and 1.15 nmol mol^{-1}, which they found to be in good agreement with the 1980 chlorine and fluorine contents in the troposphere, 2.58 and 1.19 nmol mol^{-1}, respectively, whereas in 1985 the total content of chlorine and fluorine in the troposphere had risen to 3.19 and 1.55 nmol mol^{-1}. The data thus indicate a 5-year time lag (troposphere

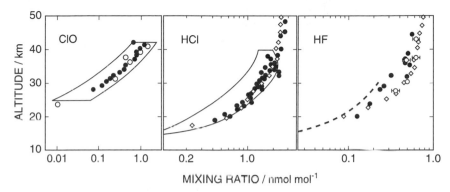

FIGURE 3.16 Vertical distribution of ClO, HCl, and HF in the stratosphere. Left: Closed points shows averages of eight altitude profiles for ClO, measured in 1976–1979 by in situ resonance fluorescence; the envelope indicates ranges of values (Weinstock et al., 1981), two additional high-mixing ratio profiles are omitted. The open points are from balloon-borne infrared remote measurements by Waters et al. (1981) and Menzies (1983). Center: The envelope encompasses observational data for HCl obtained by balloon-borne infrared measurements of Farmer et al. (1980), Buijs (1980), Raper et al. (1977), Eyre and Roscoe (1977), Williams et al. (1976), Zander (1981); closed points are from WMO (1985); diamonds indicate measurements in 1985 from a space platform (Zander et al., 1992). Right: Balloon-borne infrared measurements of HF by Buijs (1980) (dashed line), Zander (1981) (open points), WMO (1985) (closed points), and remote infrared measurements in 1985 from a space platform (diamonds) (Zander et al., 1992).

1980 versus stratosphere 1985) for transport of halocarbons to the stratosphere. Table 3.3 shows recent mixing ratios of the main chlorine- and fluorine-containing source gases. The mixing ratios of chlorine and fluorine in the troposphere now are about 3.6 and 1.8 nmol mol^{-1}, respectively, and stratospheric abundances will have increased accordingly.

Since the concentration of ClO is directly proportional to that of HCl, total chlorine determines the rate of ozone destruction in the upper stratosphere (by the reaction of ClO with O atoms). Table 3.3 thus shows that total chlorine from human-made halocarbons overrides natural chlorine from CH_3Cl by a factor of 5. The fluxes of chlorine are not apportioned in the same way, because they are determined by the global loss rates of the source gases in the stratosphere. In Table 3.3, the fluxes are estimated from the individual stratospheric lifetimes of the components. The estimates indicate that CH_3Cl contributes about 23% to the total input. $CFCl_3$, CF_2Cl_2 and CCl_4 make up most of the remainder. It should be reemphasized that for the last three components photochemical destruction in the stratosphere is the principal loss process, whereas CH_3Cl and CH_3CCl_3 react mainly with OH radicals in the troposphere.

TABLE 3.3   Halocarbons: Mixing Ratios of Total Chlorine and Fluorine in the Troposphere and Global Fluxes of Chlorine and Fluorine into the Stratosphere[a]

Halocarbon	x(HC)	x(Cl) (pmol mol^{-1})	x(F)	τ_{str}^{b} (years)	F_{Cl} F_F (Gmol year^{-1})	
CH_3Cl	610	610	—	—	1.91	—
$CFCl_3$	270	810	270	55	2.14	0.71
CF_2Cl_2	530	1060	1060	116	1.33	1.33
CCl_4	105	420	—	47	1.30	—
CH_3CCl_3	110	330	—	47	1.02	—
$CHClF_2$	115	115	230	240	0.07	0.14
CCl_2FCClF_2	85	255	255	110	0.34	0.34
Total Cl, F	—	3600	1815	—	8.11	2.52

[a] Tropospheric halocarbon abundances from Table 6.18.
[b] Stratospheric residence times from WMO (1992), except that for CH_3Cl. In this case the flux was estimated from the altitude profile in the lower stratosphere.

Finally it should be noted that bromine interacts with ozone in the same manner as chlorine (Wofsy *et al.*, 1975). The catalytic cycle involving Br atoms (and BrO) is quite efficient, because the reaction of Br with methane is slower than that of Cl atoms, and the reaction of OH with HBr is faster than that of OH with HCl. The most abundant bromine-containing source gas is methyl bromide (CH_3Br), which is produced both naturally and by anthropogenic processes. The mixing ratio of methyl bromide in the troposphere is approximately 12 pmol mol^{-1} (see Table 6.18). Although this seems small, we shall see below that bromine-catalyzed ozone destruction is important in the polar stratosphere.

3.4.4. CARBONYL SULFIDE AND SULFUR DIOXIDE

Most sulfur compounds in the troposphere are fairly rapidly oxidized and do not reach the stratosphere. A notable exception is OCS, which is rather inactive in the troposphere and occurs with a mixing ratio of about 500 pmol mol^{-1}, so that it represents a major reservoir of sulfur. In the upper troposphere, sulfur dioxide also appears to be quite stable at mixing ratios in the range of 40–90 pmol mol^{-1}. Both components can be traced into the stratosphere, as Figure 3.17 illustrates. The mixing ratio of OCS decreases with increasing altitude, whereas that of SO_2 remains approximately constant up to 25 km. Such a behavior suggests that SO_2 is a product resulting from the oxidation of OCS. Carbonyl sulfide photolyzes at wavelengths below about 250 nm, which become available at altitudes above 20 km. In

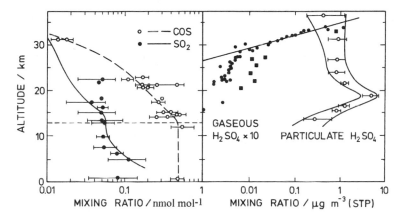

FIGURE 3.17 Left: Vertical distribution of carbonyl sulfide and sulfur dioxide in the strato sphere (from data of Maroulis *et al.* (1977), Sandalls and Penkett (1977), Torres *et al.* (1980), Mankin *et al.* (1979), and Inn *et al.* (1979, 1981) for OCS, and from Jaeschke *et al.* (1976), Maroulis *et al.* (1980), Georgii and Meixner (1980), and Inn *et al.* (1981) for SO₂) Curves represent calculations of Turco *et al.* (1980, 1981) for an assumed cut-off of OCS photodissociation at 312-nm wavelength. Right: Vertical distribution of gaseous and particulate sulfuric acid. Solid squares and circles are from mass spectrometric measurements of Arijs *et al.* (1982) and Viggiano and Arnold (1983), respectively. Open circles with error bars (one standard deviation) are from filter collections of Lazrus and Gandrud (1977). The range given by the thin lines indicates the seasonal variability of particulate sulfate. The solid line indicates the vapor pressure of sulfuric acid over a 75% H_2SO_4/25% H_2O mixture.

addition, OCS reacts with oxygen atoms. These processes lead to reactions that convert OCS to sulfur dioxide:

$$OCS + h\nu \rightarrow CO + S(^3P, {}^1D)$$

$$S + O_2 \rightarrow SO + O$$

$$O + OCS \rightarrow SO + CO$$

$$SO + O_2 \rightarrow SO_2 + O$$

$$SO + O_3 \rightarrow SO_2 + O$$

Odd oxygen is both consumed and produced. At 30 km altitude and below, the OCS loss rate due to the reaction with oxygen atoms is an order of magnitude smaller than that caused by photodissociation. At the same altitude, the reaction of SO with O_2 contributes about 70% to the total rate of SO-to-SO_2 conversion. Generally, it appears that the above reaction sequence forms more odd oxygen than it removes.

Sulfur dioxide produced from the oxidation of OCS enters into the reactions

$$SO_2 + h\nu \rightarrow SO + O$$

$$SO + O_2 \rightarrow SO_2 + O$$

which are seen to generate odd oxygen, and

$$SO_2 + OH + M \rightarrow HOSO_2 + M$$

$$HOSO_2 + O_2 \rightarrow HO_2 + SO_3$$

$$SO_3 + H_2O \rightarrow H_2SO_4$$

whereby SO_2 is permanently converted to sulfuric acid. No reactions are known that reactivate sulfur from H_2SO_4. The photodissociation of SO_2 becomes energetically possible at wavelengths below 228 nm. Thus, it is effective in the atmospheric radiation window near 200 nm. The SO_2 absorption spectrum in this wavelength region consists of a continuum overlapped by absorption bands. Only the continuum leads to dissociation, whereas excitation of SO_2 due to the discrete absorption features is removed by collisional quenching (Warneck et al., 1964; Driscoll and Warneck, 1968). At 30 km altitude, taking into account only the continuum, the photodissociation coefficient is $j(SO_2) \approx 1 \times 10^{-5}$ s^{-1}. The rate of the competing reaction of SO_2 with OH radicals at the same altitude is one order of magnitude smaller, so that SO_2 undergoes about 10 cycles of regenerating oxygen atoms before it is permanently stored in the form of H_2SO_4. For the purpose of illustrating the effects of OCS and SO_2 on stratospheric odd oxygen production, it may be assumed that the conditions at 30 km are representative of the entire OCS destruction region.

Estimates for the OCS column loss rate fall in the range $(0.3-1.0) \times 10^{11}$ molecule m^{-2} s^{-1} (Inn et al., 1981; Engel and Schmidt, 1994; Chin and Davis, 1995). The corresponding global rate of SO_2 production in the stratosphere is 25–83 mol s^{-1}. With the assumption that SO_2 undergoes 10 photolysis cycles, one obtains a global odd oxygen production rate of $(0.5-1.7)$ kmol s^{-1}. The gross rate of odd oxygen formation from the photodissociation of oxygen according to Table 3.1 is 81 Mmol s^{-1}, which is 45,000 times greater than that caused by the injection of OCS into the stratosphere. The comparison shows that the effect of carbonyl sulfide on the production (and destruction) of stratospheric ozone is negligible.

Sulfuric acid in the stratosphere exists as a vapor as well as a condensation aerosol. Figure 3.17 includes data for gaseous H_2SO_4 and for particulate sulfate. The mass mixing ratio of the sulfate layer has its peak at altitudes between 20 and 25 km (compare this with Fig. 7.28, which shows the altitude profile of aerosol particles in terms of number density). The maximum occurs markedly below the altitude regime where SO_2 is most efficiently converted to H_2SO_4. Junge *et al.* (1961) noted that particles with radii near 0.3 μm have a tendency to accumulate at the 20-km altitude level by achieving a balance between downward sedimentation and upward eddy diffusion. This prediction agrees roughly with observations. However, the time constant for the system to approach steady state is appreciable for particles smaller than 1 μm. The particle size distributions usually observed exhibit maxima at radii near 0.1 μm, according to the review of Inn *et al.* (1982), so that a steady state is unlikely to be established. At altitudes above 30 km the concentration of the H_2SO_4 vapor rises, whereas that of sulfate-containing particles declines. As Figure 3.17 shows, the data for gas-phase H_2SO_4 in this altitude region follow approximately the saturation vapor pressure curve, suggesting the existence of an equilibrium between the vapor and condensed material. Owing to the temperature rise with increasing altitude above 30 km, the saturation vapor pressure rises as well, so that with increasing height more and more particles evaporate.

Turco *et al.* (1982) published a comprehensive review of observations and knowledge of the stratospheric aerosol. Chemical analysis of particles collected in situ indicate that the stratospheric aerosol consists of a solution of 70–75 wt% sulfuric acid in water with an admixture of dissolved nitrosyl sulfates and solid granules containing silicates (Cadle *et al.*, 1973; Rosen, 1971; Farlow *et al.*, 1977, 1978). The nonsulfate constituents derive partly from the influx of micrometeorites (Ganapathy and Brownlee, 1979), and to a larger extent from the sporadic injection of solid ash particles due to high-reaching volcanic eruptions. The solid particles may assist in the formation of sulfate aerosol in that they serve as condensation nuclei. Particles containing no solid nucleus have also been observed, but model calculations of Hamill *et al.* (1982) showed that the homogeneous (that is, purely vapor phase), formation of sulfate particles is quite inefficient. Mechanisms of homogeneous and heterogeneous nucleation processes are discussed in more detail in Chapter 7.

In addition to ash particles, high-reaching volcanic eruptions also carry sulfur dioxide into the stratosphere. Cadle *et al.* (1977) estimated that the 1963 eruption of Mt. Agung (Bali), which was a major event, brought 12 Tg SO_2 into the stratosphere. Other major eruptions in recent times were that of El Chichón (Mexico) in 1982 and Mt. Pinatubo (Philippines) in 1991. The last event was estimated to have injected about 20 Tg SO_2 into the stratosphere (Bluth *et al.*, 1992; McPeters, 1992), three times as much as that delivered by El Chichón. High-reaching volcanic eruptions cause considerable perturbations of the stratospheric sulfate layer. Castleman *et al.* (1973,

1974), from a study covering the period 1963–1973, found sulfate concentration to vary from 0.1 to 40 μg m^{-3} at altitudes near the peak of the sulfate layer. The highest values occurred in the Southern Hemisphere about 1 year after the Mt. Agung eruption. Subsequently, the values decayed to a level of quiescence in 1967. Castleman *et al.* (1974) further studied the ^{34}S/^{32}S isotope ratio and its variation with time. Invariably the isotope ratio decreased after a major volcanic event and then increased again. This aspect shows perhaps most clearly that the stratospheric sulfate layer has two sources, one that is always present and has its origin in the diffusion of source gases such as OCS, and a second input due to high-reaching volcanic clouds that carry additional sulfur gases into the stratosphere.

The perturbations of the stratospheric sulfur layer by El Chichón and Mt. Pinatubo have been studied by means of optical back-scattering with ground-based techniques and from satellites (Fiocco *et al.*, 1996). At midlatitudes, the integrated signal reached a maximum within about 6 months, indicating mainly the time period required to convert SO_2 to sulfate aerosol as the dispersal from the Tropics to higher latitudes is more rapid. Subsequently the signal decayed quasiexponentially with a time constant of about 10 months. The increase in the stratospheric sulfate aerosol layer by each of the three volcanic eruptions, which was more than 10 times above background, caused a warming of the middle stratosphere that was greatest in the Tropics (about 1 K) but decreased poleward. The Mt. Pinatubo eruption also led to a temporary decrease in stratospheric ozone concentrations, observed in the total ozone column at various stations in the Northern Hemisphere (Hofmann *et al.*, 1994a). Total ozone began to decrease below the long-term average about 6 months after the eruption, the loss reached its peak of 10–15% about 1 year later, but ozone had largely recovered $2\frac{1}{2}$ years after the event. Hofmann *et al.* (1994b) showed that the altitude region above Boulder, Colorado (40° N), where most of the ozone loss occurred, coincided with the observed altitude of the Pinatubo aerosol layer as determined by lidar back-scatter. These observations confirmed suggestions by Hofmann and Solomon (1989) that heterogeneous chemical reactions occurring on sulfate aerosol particles can also lead to ozone losses. These will be discussed further below.

3.5. HETEROGENEOUS AND POLAR STRATOSPHERIC CHEMISTRY

The stratosphere is usually too dry to give rise to liquid water clouds, but aerosol particles are present, and ice particles can be formed in the cold regions above the Poles. Such particles present surfaces on which gas-phase species can undergo chemical reactions. The occurrence of heterogeneous reactions in the polar stratosphere was initially postulated by Solomon *et al.*, (1986a) because gas-phase reactions alone could not explain the high concentrations of ClO during the temporary depletion of ozone in the Antarctic in October. The importance of heterogeneous reactions has since been reaffirmed by laboratory studies in which reaction probabilities were determined. Peter (1997) has reviewed the current knowledge of the subject.

Satellite observations evaluated by McCormick *et al.* (1982, 1989) have shown that polar stratospheric clouds (PSCs) are common over the Antarctic continent during the winter season at altitudes below 25 km and that they are strongly correlated with temperatures of less than 195 K. Two broad categories of such clouds exist: (a) Type I PSCs appear as a micrometer-sized haze of particles at temperatures about 5 K above the frost point of pure water ice. They consist most likely of crystals of nitric acid trihydrate, because nitric acid and water vapor both condense in this temperature regime, and nitric acid trihydrate is the thermodynamically stable entity. (b) Type II PSCs resemble polar cirrus (Rosen *et al.*, 1988; Poole and Mc-Cormick, 1988). They are observed to form only when the temperature drops to values below the frost point of ice (below ~ 189 K), so that they must consist of pure water ice. The particles can reach a size of up to 100 μm, significantly larger than those in type I PSCs, and they are presumably formed by the condensation of water vapor on preexisting particles. (c) Type III PSCs, or mother-of-pearl clouds, are similar to type II PSC. They occur as lee wave clouds formed behind mountains, are usually lenticular in shape, and have been reported most frequently over Scandinavia (Stanford and Davis, 1974). (d) Sulfate aerosol particles must be added to the list, as they are ubiquitous in the stratosphere. Table 3.4 summarizes some properties of these particles. It is evident that the dominant constituents are H_2O, H_2SO_4,

TABLE 3.4 Properties of the Background Sulfate Aerosol and of Polar Stratospheric Clouds[a]

Property	Sulfate aerosol	Type I PSC	Type II PSC	Type III PSC
Diameter, μm	0.01–1.0	~ 1	10–100	~ 5
Surface area,[b] cm^2 cm^{-3}	$\sim 2 \times 10^{-9}$	$\sim 1 \times 10^{-8}$	$\sim 1 \times 10^{-7}$	$\sim 1 \times 10^{-5}$
Composition[c]	40–80 wt% H_2SO_4	$HNO_3 \cdot 3H_2O$	H_2O	H_2O
Formation temperature,[d] K	195–240	<195	<189	<189
Relative frequency, %	Ubiquitous	~ 90	~ 10	~ 1

[a] Compiled mainly from Turco *et al.* (1989).
[b] Total particle surface area per volume of air.
[c] Traces of other materials are present in all cases.
[d] Approximate values.

and HNO_3, whereas other species are present only in traces. For example, the solubility of HCl in sulfuric acid aerosols is lower than that of HNO_3 (Hanson and Ravishankara, 1993), and crystalline hydrates of HCl are unstable under polar stratospheric conditions (Shen et al., 1995).

The development of polar stratospheric clouds is thought to begin with an initially liquid sulfate aerosol particle. The liquid–solid phase diagram of the binary system sulfuric acid–water, which is shown in Figure 3.18, features two main eutectic points. One occurs at a mole fraction $x(H_2SO_4) = 0.3$ (~ 70 mass % H_2SO_4), midway between the stable hydrates $H_2SO_4 \cdot H_2O$ and $H_2SO_4 \cdot 4H_2O$, at a temperature near 215 K. This mixture appears to be the normal state of sulfate particles at about 16 km altitude in the mid-latitude stratosphere. The second eutectic point occurs at about 195 K. Here, the mole fraction is $x(H_2SO_4) \approx 0.1$ (~ 40 mass % H_2SO_4), and the point is located near the boundary to the ice phase. Under these conditions, H_2SO_4 is essentially nonvolatile. Accordingly, when the temperature is lowered, particles with a composition of about $70\% H_2SO_4$ will add water, while the sulfate content remains fixed. The abundance of water vapor is not affected by the condensation process, as it is present in excess. As the temperature drops, the mixture becomes further diluted until the composition of the particle reaches that of the second eutectic point. The phase diagram suggests that the particle should at least partly solidify to form the sulfuric acid tetrahydrate, but lidar observations that were able to distinguish between spherical and nonspherical particles (Toon et al., 1990; Beyerle et al., 1994), labora-

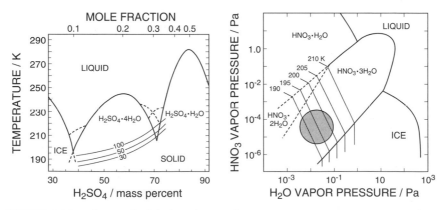

FIGURE 3.18   Left: Liquid-solid phase diagram for H_2SO_4–H_2O mixtures. Superimposed are typical sulfuric acid concentrations in the altitude regime at 30–100 hPa pressure. (Adapted from Carslaw et al. (1997).) Right: Phase diagram and isothermal vapor pressure curves for HNO_3–H_2O mixtures. The gray area indicates the pressure regime in the polar stratosphere. (Adapted from Hanson and Mauersberger (1988).)

tory freezing experiments (Koop *et al.*, 1995; Zhang *et al.* 1993a), and nucleation calculations (Luo *et al.*, 1994) all indicate that supercooling keeps the particle largely liquid until the temperature reaches about 195 K.

Although nitric acid dissolves in sulfuric acid solutions, only small amounts of HNO_3 are absorbed at high H_2SO_4 concentrations. However, the uptake capacity for HNO_3 increases with increasing water content (Reihs *et al.*, 1990; Zhang *et al.*, 1993b). When the uptake of water reduces the H_2SO_4 mole fraction to 0.1 ($\sim$ 40 mass % H_2SO_4), most of the nitric acid present in the gas phase is taken up by the sulfuric acid aerosol. Thus, at temperatures below 195 K, the particulate phase changes rapidly from one dominated by sulfuric acid to one dominated by nitric acid. This is clearly shown by the data in Figure 3.19 discussed further below.

McElroy *et al.* (1986a) first recognized from the phase diagram of the binary HNO_3-H_3O system included in Figure 3.18 that solid nitric acid trihydrate, $HNO_3 \cdot 3H_2O$, is the principal thermodynamically stable entity in this system under conditions typical of the polar stratosphere. Here, the mixing ratios are 3–5 μmol mol^{-1} H_2O and 5–10 nmol mol^{-1} HNO_3, corresponding to vapor pressures of 17–28 mPa and 27–55 μPa, respectively. Laboratory data of Hanson and Mauersberger (1988) confirmed that equilibrium vapor pressures of water and nitric acid above $HNO_3 \cdot 3H_2O$ follow the Duhem–Margules equation $p(HNO_3) \cdot p^3(H_2O) = f(T)$ and that solid nitric acid trihydrate is formed at temperatures 5–7 K higher than ice. The abundance of water vapor remains essentially constant, while gaseous nitric acid is converted to particulate $HNO_3 \cdot 3H_2O$. Molina *et al.* (1993) have explored the freezing characteristics of ternary mixtures of sulfuric acid, nitric acid, and water by cooling them down to 190 K. Only solutions containing less than 40% H_2SO_4 and at least 6.6% of HNO_3 by mass crystallized upon cooling. For this process, $HNO_3 \cdot 3H_2O$ and $H_2SO_4 \cdot 6.5H_2O$ were found to be the dominant crystalline products. These results suggest that crystallization occurs not before the temperature drops to 195 K, in agreement with the satellite observations on the formation of type I polar stratospheric clouds.

Figure 3.19 shows the total particle volume measured by Dye *et al.* (1992) in the Arctic with an optical particle counter onboard high-altitude aircraft flying at about 19 km altitude. While total particle volume increased with decreasing temperature, a steep rise was observed only when the temperature reached 193 K. The lower part of the figure shows the behavior of the ternary solution as a function of temperature according to the calculations of Carslaw *et al.* (1994, 1997). The weak dependence of particle volume on the temperature above 193 K may be identified with the uptake of water by sulfate particles, whereas the strong rise below 193 K must be associated

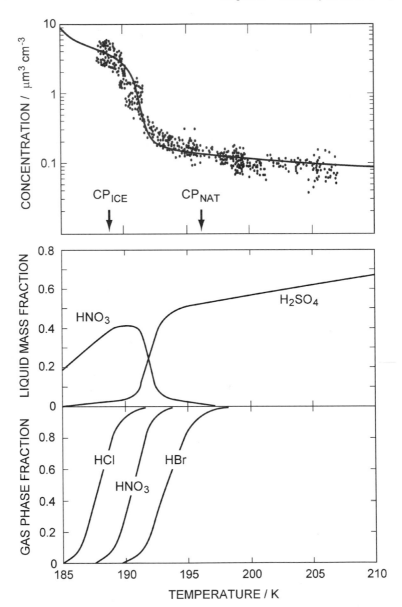

FIGURE 3.19 Top: Total particle volume concentration versus temperature at the 55 hPa ($\sim$ 19 km) altitude level observed during a flight of the high-altitude research aircraft ER-2 on January 24, 1989, north of Stavanger, Norway (Dye *et al.*, 1992). The solid line shows the calculated volume increase of the ternary H_2SO_4–HNO_3–H_2O system. Bottom: Composition of the liquid and the fraction of HNO_3 and HCl remaining in the gas phase, as calculated by Carslaw *et al.* (1994). Note that HCl is absorbed at lower temperatures than HNO_3.

with particle growth caused by the uptake of nitric acid and more water, leading ultimately to $HNO_3 \cdot 3H_2O$.

Ice crystals that grow large enough can sediment over appreciable distances in the lower stratosphere within time periods shorter than the winter season, thereby removing both water and nitric acid from the region. Crutzen and Arnold (1986) and Toon *et al.* (1986b) first recognized that the cocondensation of nitric acid and water vapor in the cold polar winter stratosphere is one of the processes depleting HNO_3 from the gas phase. As nitric acid is the principal precursor of NO_2 when sunlight returns in spring and NO_2 serves as a scavenger of ClO radicals by forming chlorine nitrate, the chemistry shifts from one in which O_3 is largely controlled by nitrogen oxides to one in which chlorine compounds dominate the control. The situation is further aggravated by the occurrence of heterogeneous reactions that convert inactive chlorine compounds into far more reactive species. These will be discussed next.

The maximum rate of a heterogeneous reaction is given by the collision frequency of gas-phase molecules with the surface of a particle (see Section 7.3.1). However, few of the molecular encounters lead to absorption, and most molecules return to the gas phase. This is accounted for by the mass accommodation coefficient α. Even fewer collisions lead to reaction. The reactive uptake coefficient γ, which defines an overall reaction probability, is the fraction of collisions that result in a chemical loss. For reactions in liquid aerosols, γ is a function of the solubility K_H of the gaseous substances in the liquid and the coefficients of diffusion D and reaction k in the solution (Schwartz, 1986; Hanson *et al.*, 1994; Hanson and Lovejoy, 1995). The probability of a reaction between X and Y under first-order conditions, that is with Y in excess, is given by

$$\gamma_X^{-1} = \alpha_X^{-1} + f(r)\bar{v}\left[4K_H^* R_g T\left(D_X k c_Y\right)^{1/2}\right]^{-1} \qquad (3.15)$$

where $\bar{v}$ is the gas-phase mean velocity of X molecules, K_H^* is the effective Henry's law coefficient of X, and c_Y is the concentration of Y molecules in the liquid. The function $f(r) = \coth(q) - 1/q$, where $q = r(k/D_X)^{1/2}$ takes into account the spherical geometry of the aerosol droplet. The uptake coefficient saturates when mass accommodation is limiting ($\gamma_X = \alpha_X$). Far below this limit, γ is proportional to the solubility of X, to the square root of Y, and to the size function $f(r)$.

The uptake coefficient for a reaction on the surface of solid particles may be derived in a similar manner by considering absorption to be replaced by

Langmuir type adsorption (Tabazadeh and Turco, 1993; Carslaw and Peter, 1997):

$$\gamma_X^{-1} = \alpha_X^{-1} + \bar{v}\sigma^2[4K_X^* k_B Tk\theta_Y]^{-1} \qquad (3.16)$$

Here, σ is the surface area per adsorption site; $K_X^* = K_X/(1 + K_X p_X)$, with p_X denoting the partial pressure of X, is an effective Langmuir term that tends to K_X for weak adsorption and takes self-competition of X into account; k_B is the Boltzmann constant; and θ_Y is the fractional coverage of the excess species Y.

Table 3.5 lists several reactions that are considered important and have been studied in the laboratory, together with the reactive uptake coefficients derived from the experimental data. We discuss first the three reactions involving chlorine species and consider nitrogen pentoxide reactions later. The most important heterogeneous reaction activating chlorine in the polar stratosphere is

$$ClONO_2 + HCl \overset{het}{\rightarrow} Cl_2 + HNO_3$$

Elemental chlorine is released to the gas phase, whereas nitric acid is largely retained by the particulate phase. Chlorine nitrate can also react directly with water in the condensed phase

$$ClONO_2 + H_2O \overset{het}{\rightarrow} HOCl + HNO_3$$

and both reactions will be in competition. Working with aqueous H_2SO_4 solutions, Hanson and Ravishankara (1994) succeeded in separating the two reactions by varying the HCl partial pressure over several orders of magnitude. The solubility of HCl in such solutions increases strongly with decreas-

TABLE 3.5   Reactive Uptake Coefficients γ for Heterogeneous Reactions Considered Important in the Stratosphere[a]

Reaction	H_2SO_4 solution	NAT	Ice
$ClONO_2 + HCl$	0.001–0.3 ($\sim$ 200 K)[b]	0.001–0.3 ($\sim$ 200 K)[c]	0.3 (180–200 K)
$ClONO_2 + H_2O$	0.0002–0.06 (200–265 K)[b]	$\leq$ 0.001 ($\sim$ 200 K)	0.3 (180–200 K)
$HOCl + HCl$	0.002–0.5 (200–215 K)[b]	0.1 (195–200 K)	0.3 (195–200 K)
$N_2O_5 + HCl$	—	0.003 ($\sim$ 200 K)	0.03 (190–220 K)
$N_2O_5 + H_2O$	0.1 (195–295 K)	0.0003 ($\sim$ 200 K)	0.01 (195–200 K)

[a] From the compilation of DeMore et al. (1997).
[b] Increasing with decreasing H_2SO_4 mass fraction (40–75 mass %).
[c] Rising with increasing relative humidity over ice.

ing H_2SO_4 content, following the decrease in temperature, and the probability of the first reaction increases accordingly. It begins overriding the reaction of $ClONO_2$ with liquid water when the mass fraction of H_2SO_4 falls below about 60%. On the surface of ice particles, however, the two reactions have similar probabilities. The reaction

$$HOCl + HCl \overset{het}{\to} Cl_2 + H_2O$$

also proceeds rapidly on water ice and nitric acid trihydrate surfaces. In sulfuric acid solutions the reaction is slower than that of chlorine nitrate. During the early stage of polar chlorine activation this reaction augments that between chlorine nitrate and HCl because HOCl is a product of the latter. Eventually, however, HCl may become locally depleted so that the concentration of HOCl builds up.

Toward the end of polar winter, heterogeneous reactions occurring during the period of darkness have converted chlorine from the reservoir species $ClONO_2$ and HCl to compounds that are easily photolyzed by sunlight. This situation is particularly striking in the Antarctic stratosphere because the winter polar vortex causes it to be largely isolated from exchange with mid-latitude air, so that the constituents become chemically preconditioned during the polar night. When solar radiation reaches the polar region in spring, Cl_2 and HOCl undergo photodissociation, whereby the chlorine-catalyzed ozone destruction cycle is activated:

$$Cl_2 + h\nu \to Cl + Cl$$

$$HOCl + h\nu \to OH + Cl$$

$$Cl + O_3 \to ClO + O_2$$

$$ClO + O \to Cl + O_2$$

At mid-latitudes, the last two reactions occur in the upper stratosphere, while in the lower stratosphere ClO associates with NO_2 to form chlorine nitrate. It is important to note that the bulk of the heterogeneous chemistry in polar regions occurs in the lower stratosphere. Because nitric acid is largely incorporated in the particulate phase, its availability as a source of NO_2 is limited. Moreover, oxygen atoms in the lower stratosphere are considerably less abundant than in the upper part, so that the concentration of ClO builds up. The destruction of ozone thus requires a different reaction to regenerate chlorine atoms. As Cox and Hayman (1988) and Molina *et al.*

(1990) have shown, two ClO radicals can combine to form the dimer ClOOCl, which undergoes photolysis

$$ClO + ClO + M \rightarrow ClOOCl + M$$
$$ClOOCl + h\nu \rightarrow Cl + ClOO$$
$$ClOO + M \rightarrow Cl + O_2 + M$$

such that the ClO radicals are reconverted to chlorine atoms and the chain can continue. Two ClO radicals may also enter into the bimolecular reactions

$$ClO + ClO \rightarrow Cl_2 + O_2$$
$$\rightarrow Cl + ClOO$$
$$\rightarrow Cl + OClO$$

but these reactions are very slow and of little importance in the stratosphere. The formation of ClOOCl is facilitated by the low temperatures within the polar vortex. At temperatures above 220 K, the thermal dissociation of the molecule begins to compete with photodissociation, but this leads to ClO radicals and would not support the chain destruction of ozone.

While the bimolecular reactions between two ClO radicals are slow, the corresponding reaction between BrO and ClO radicals is considerably faster, so that they contribute to polar stratospheric chemistry:

$$BrO + ClO \rightarrow Cl + Br + O_2$$
$$\rightarrow Br + OClO$$

Both Br and Cl atoms react rapidly with ozone, so the destruction chain can continue. The catalytic cycle involving bromine was initially proposed by McElroy et al. (1986b) to be important in the polar stratosphere. Subsequent laboratory studies have shown that the two channels indicated above are equally fast, whereas the formation of BrCl, which is also possible, is a minor pathway (Friedl and Sander, 1989; Turnipseed et al., 1991). The bromine cycle contributes about 25% to the total destruction of ozone in austral spring (Anderson et al., 1989a).

The activation of chlorine over the winter poles has clearly been demonstrated by in situ and remote measurements of ClO and its anticorrelation with ozone (DeZafra et al., 1987; Anderson et al., 1989b, 1991; Toohey et al. 1993) and by satellite observations (Waters et al., 1993). Figure 3.20 shows results of measurements made onboard the high-altitude ER-2 aircraft on a southbound flight from Puntas Arenas, Chile, at a time during ozone depletion in the Antarctic. The rim of the polar winter vortex occurred at about 64° southern latitude. Further south, the mixing ratios of ozone

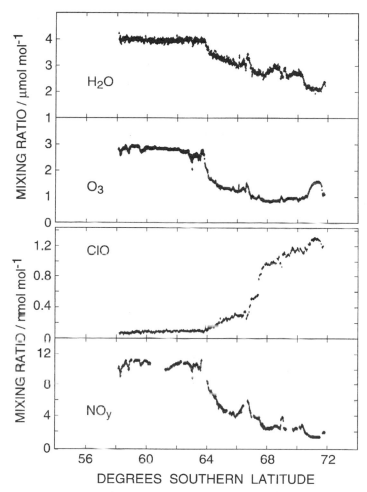

FIGURE 3.20 Mixing ratios of water vapor, ozone, ClO, and NO_y (primarily HNO_3) measured onboard high-altitude ER-2 aircraft during the southbound flight of September 22, 1987, starting from Puntas Arenas, Chile. (Adapted from Fahey *et al.* (1989a).) The boundary of the polar vortex is indicated by the rise in ClO and the concomitant decrease in ozone. Data were discussed in detail by Kelly *et al.* (1989), Anderson *et al.* (1989), Starr and Vedder (1989), and Fahey *et al.* (1989b).

declined, whereas that of chlorine monoxide increased as expected if chlorine activation was responsible for the depletion of ozone. In addition, the mixing ratios of both nitric acid and water vapor decreased, confirming the loss of both constituents by condensation to form nitric acid-rich ice particles. The involvement of bromine in the activation of chlorine is made evident by the increase in OClO (Solomon *et al.*, 1989; Sanders *et al.*, 1993), for which the reaction between BrO and ClO is the only known source, and the direct detection of BrO by Anderson *et al.* (1989a) and Carroll *et al.* (1989) within the Antarctic vortex. Schiller *et al.* (1990) and Brandtjen *et al.* (1994) have also observed OClO in the Arctic stratosphere, but the largest column abundances were found in the Antarctic.

The situation during winter in the Arctic is basically similar to that in the Antarctic, but temperatures are higher and polar stratospheric clouds are less frequent. Simultaneous measurements of ClO, HCl, and total inorganic chlorine in the Arctic winter vortex by Webster *et al.* (1993) show that the concentration of ClO increases, while HCl and chlorine nitrate decrease appreciably. Losses of HNO_3 have also been observed in the Arctic (Schlager and Arnold, 1990; Kondo *et al.*, 1992), but the incorporation of HNO_3 into the particles appears to be reversible, whereas in the Antarctic HNO_3 is more or less permanently removed as the particles precipitate (Fahey *et al.*, 1990). Satellite observations of HNO_3 at high latitudes have confirmed the contrast between the two polar regions (Santee *et al.*, 1994). As a consequence, winter losses of ozone in the Arctic are smaller. In both regions the recovery of stratospheric ozone to prewinter values occurs when, following the temperature rise in spring, PSC formation ceases and HNO_3 becomes available again as a photochemical source of NO_x. The reservoir of chlorine nitrate can then be reestablished. In addition, chlorine atoms react with methane to convert reactive chlorine back into hydrochloric acid. The rate of recovery depends on the extent of denitrification (HNO_3 loss). Full recovery requires the breakup of the vortex, which allows the admixture of air from mid-latitudes.

The preceding considerations did not yet include the heterogeneous reactions of nitric pentoxide listed in Table 3.5:

$$N_2O_5 + HCl \overset{het}{\to} ClNO_2 + HNO_3$$

$$N_2O_5 + H_2O \overset{het}{\to} 2HNO_3$$

which require additional comments. The first reaction is not very important in the polar stratosphere, because it must compete with the second, which occurs with a high probability on liquid sulfuric acid particles, so that N_2O_5 is removed and converted to HNO_3 before temperatures are reached that make the reaction with HCl effective. The probability for the reaction of the N_2O_5 with liquid water is essentially independent of the sulfuric acid concentration, which contrasts markedly with the reactions of chlorine nitrate on sulfuric acid solutions. The reaction will therefore occur also in the mid-latitude stratosphere and not only under polar winter conditions. Normally, the sulfate aerosol layer is not dense enough for this reaction to cause a serious impact upon the odd nitrogen chemistry in the stratosphere. However, after large volcanic eruptions the surface area available for heterogeneous reactions is increased 10–100 times, and this factor can lead to a more substantial perturbation. The increased loss of N_2O_5 by reactive uptake on sulfate aerosol converts nitrogen oxides to less reactive HNO_3, which makes mid-latitude stratospheric ozone less vulnerable to odd nitrogen but more vulnerable to chlorine species. Model calculations (Brasseur and Granier, 1992; Solomon et al., 1996) indicate that the concentrations of OH and HO_2 radicals should also increase. The corresponding shift in stratospheric chemistry, which controls the loss of ozone, can lead to temporary but not catastrophic losses. Fahey et al. (1993) have studied the effect of the increased particle load due to the Mt. Pinatubo cloud on stratospheric chemistry by comparing the results of measurements of nitrogen and chlorine species made onboard high-flying aircraft with model calculations. They showed that in accord with expectation, the NO_x/NO_y ratio is reduced and the ClO/Cl_y ratio enhanced, where the synonyms NO_y and Cl_y are again used to indicate the total reservoirs of odd nitrogen and of chlorine. The changes in ClO/Cl_y were consistent with the role of $ClONO_2$ and HCl in the odd nitrogen and chlorine reservoirs. The lifetime of N_2O_5 due to heterogeneous reactions was reduced from several days under normal conditions to about 1 day. The heterogeneous loss of $ClONO_2$ by reaction with liquid water was negligible in comparison.

3.6. THE BUDGET OF OZONE IN THE STRATOSPHERE

Having discussed the chemistry associated with various trace gases in the stratosphere, we return to the problem of balancing the ozone budget. The impact of the additional chemistry upon stratospheric ozone is most conveniently assessed by considering the results of model calculations. Both one- and two-dimensional transport models incorporating the reactions introduced in the preceding sections are able to reproduce the vertical distributions of trace components that have been observed in the stratosphere. Although these models achieve a reasonably quantitative simulation of stratospheric chemistry, the built-in uncertainties are such that one should not expect models to describe the complex behavior of the stratosphere in every detail. As discussed previously, all transport models are based on the continuity equation (Eq. 1.14), which must be supplemented by appropriate boundary conditions, for example fixed mixing ratios at the tropopause and vanishing fluxes at high altitudes. Flow and temperature fields and rate coefficients of elementary reactions are specified parameters. Solar fluxes are calculated as needed from the extraterrestrial radiation field.

Figure 3.21 shows vertical ozone distributions at several latitudes calculated with a two-dimensional model and compares the results with observations. Although the agreement is reasonable, the critical test of the applicability of models lies in their ability to reproduce the distributions of other trace gases and radicals, especially those that participate in the ozone destruction reactions. Satellite observations probing the middle and upper stratosphere for a variety of important constituents have increasingly been used to test models and chemical mechanisms. For example Crutzen and Schmailzl (1983) based their evaluation on LIMS (Limb Infrared Monitor of the Stratosphere) in 1978–1979 (Gille and Russell, 1984; Remsberg et al., 1984), and McElroy and Salawitch (1989) made use of space platform observations ATMOS (Atmospheric Trace Molecules Observed by Spectroscopy) during April and May 1985 (Raper et al., 1987; Russell et al., 1988). The latter study in particular showed excellent agreement for many trace components between observed and calculated altitude profiles.

It should be noted that when the Chapman mechanism is extended to include other trace species, a redefinition of odd oxygen is necessary. Whereas the Chapman mechanism defined odd oxygen as the sum of free oxygen atoms and ozone only, it is now necessary to account also for odd

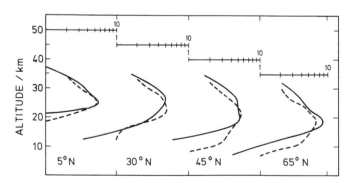

FIGURE 3.21 Comparison of calculated with observed vertical ozone distributions at four latitudes (unit: 10^{12} molecule cm^{-3}). Solid lines: Calculated by Miller et al. (1981) with a two-dimensional model. Broken lines: Observed by Wilcox et al. (1977). Note that the agreement of ozone profiles by itself does not provide validation of the model; agreement with other trace components must also be accomplished. Miller et al. (1981) reported this to have been achieved.

oxygen contained in a greater number of trace species. Thus,

$$[O_x] = [O] + [O(^1D)] + [O_3]$$
$$+ [NO_2] + 2[NO_3] + 3[N_2O_5] + [HNO_3] + [HO_2NO_2]$$
$$+ [ClO] + [HOCl] + 2[Cl_2O_2] + 2[ClONO_2] \tag{3.17}$$

even if several of the trace species contribute little to the overall budget. The budget is determined by production and loss terms in the kinetic equation

$$d[O_x]/dt = P_O + P_N - D_O - D_N - D_{Cl} - D_H \tag{3.18}$$

where P_O is the production by O_2 photolysis, D_O is the destruction by the Chapman reaction, and the remaining terms represent destruction by catalytic cycles involving NO_x, ClO_x, and HO_x radicals, respectively. P_N is the production by NO_x reactions, but this term becomes important only in the lower stratosphere.

The principal reactions that produce and destroy odd oxygen are

P_O	$O_2 + h\nu \rightarrow 2O$	$(+2)$	
P_N	$HO_2 + NO \rightarrow OH + NO_2$	$(+1)$	
D_O	$O + O_3 \rightarrow 2O_2$	(-2)	upper stratosphere
D_{Cl}	$O + ClO \rightarrow Cl + O_2$	(-1)	middle stratosphere
	$Cl + O_3 \rightarrow ClO + O_2$	(-1)	
D_N	$O + NO_2 \rightarrow NO + O_2$	(-1)	middle stratosphere
	$NO + O_3 \rightarrow NO_2 + O_2$	(-1)	
D_H	$H + O_3 \rightarrow OH + O_2$	(-1)	upper stratosphere
	$O + OH \rightarrow H + O_2$	(-1)	
	$O + HO_2 \rightarrow OH + O_2$	(-1)	
	$OH + O_3 \rightarrow HO_2 + O_2$	(-1)	lower stratosphere
	$HO_2 + O_3 \rightarrow OH + 2O_2$	(-1)	

Not all of the reactions are equally effective, however, and their relative significance also depends on altitude. Figure 3.22 shows the altitude dependence of the individual loss reactions determined by Logan *et al.* (Logan *et al.*, 1978; Wofsy, 1978) from calculations based on a one-dimensional model. The reaction $H + O_3$ is much less important than the subsequent two reactions in the sequence shown above. These reactions again are more important in the upper stratosphere, whereas the last two reactions are significant only in the lower stratosphere. Above 45 km the Chapman reaction $O + O_3$ remains an important loss process for odd oxygen, even though the reactions $O + OH$ and $O + HO_2$ contribute appreciably. In the

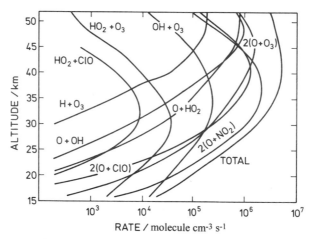

FIGURE 3.22   Rates for the loss of odd oxygen in the stratosphere due to various reactions, averaged over a 24-h period. (Adapted from Wofsy and Logan (1982).)

40-km altitude region, the Chapman process and the loss cycles involving NO_x, HO_x, and ClO_x are about equally important. In the lower stratosphere odd oxygen is predominantly lost by catalytic reactions with NO_x. The reaction $OH + O_3$ comes to prominence at altitudes below 20 km, whereas the reaction $HO_2 + O_3$ contributes less. Since the time when the data in Figure 3.22 were generated, rate coefficients for some of the reactions have undergone changes, but the general behavior exhibited in the figure has remained valid. This is demonstrated in Figure 3.23, which shows results of box-model calculations of Crutzen *et al.* (1995) initialized by sunset and sunrise observations of HALOE (Halogen Occultation Experiment) on the Upper Atmosphere Research Satellite (Russell *et al.*, 1993).

For comparison with the data in Table 3.1, the column loss rates must be calculated and extrapolated to a global scale. Table 3.6 shows two sets of results. The first was obtained by Wofsy and Logan (1982), who applied a one-dimensional model, corresponding to the vertical distributions of individual rates shown in Figure 3.22. The other set represents results of Crutzen and Schmailzl (1983), who used the two-dimensional model of Gidel *et al.* (1983). In the first case, the loss rate associated with the Chapman reaction is slightly higher than the value derived from the observed distribution of ozone (Table 3.1). The difference may be caused by uncertainties in the model or in the measurements, but it is not a serious discrepancy. The total rate of ozone destruction, 81.4 Mmol s^{-1}, compares well with the global rate of odd oxygen production due to photodissociation of O_2 calculated by Johnston (1975). The model indicates that odd oxygen is consumed to 25%

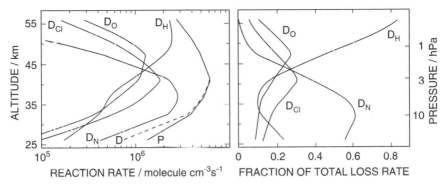

FIGURE 3.23 Left: Total odd oxygen production and destruction rates as a function of altitude on January 12, 1994, at 23° S, derived from HALOE observations and model calculations. The individual contributions to the destruction rates due to O_y, NO_x, ClO_x, and HO_x reactions are shown. Right: Fractional contributions of the individual loss terms. (Adapted from Crutzen *et al.* (1995).)

by the $O + O_3$ reaction, 33.8% by the $O + NO_2$ reaction, 21.7% by HO_x radicals, and 18.7% by chlorine radicals, a fairly even apportionment. The results from the two dimensional model of Crutzen and Schmailzl (1983) provide a better agreement between calculated and observed losses from the Chapman reaction (Table 3.1). The total rate of ozone destruction, 78.0 Mmol s^{-1}, also agrees well with that calculated by Johnston (1975). With regard to the other destruction pathways, the two sets of calculations are in good agreement, even though the two-dimensional model shows a more

TABLE 3.6   Globally Integrated Stratospheric Ozone Loss Rates Due to the Chapman Reaction and Catalytic Destruction Cycles Involving ClO_x, NO_x, and HO_x[a]

Loss process	Column loss rate (10^{14} molecule m^{-2} s^{-1})		Global loss rate (Mmol s^{-1})		Percentage of total loss	
	L & W	C & S	L & W	C & S	L & W	C & S
$O + O_3$	253	191	21.1	15.9	25.9	20.4
$2 \times (O + ClO)$	182	187	15.2	15.6	18.7	20.0
$2 \times (O + NO_2)$	330	289	27.5	24.1	33.8	30.9
$\Sigma (HO_x + O_3)$	51	[b]	4.2	[b]	5.2	[b]
$\Sigma (HO_x + O)$	161	269	143.4	22.4	16.5	28.7
Total loss rate	977	936	81.4	78.0	100	100

[a] According to the one-dimensional model of Logan *et al.* (Logan *et al.*, 1978; Wofsy and Logan, 1982) (L & W) as shown in Figure 3.22, and the two-dimensional model of Crutzen and Schmailzl (1983) (C & S).
[b] Included in the next line.

pronounced effect of the HO_x cycle at high altitudes. Nevertheless, both sets of results clearly demonstrate that the inclusion of reactions other than the Chapman reaction greatly improves the ozone balance in the stratosphere. McElroy et al. (1992) further explored the impact of bromine radicals and the hydrolysis of N_2O_5 on sulfuric acid aerosol particles on the fractional loss of odd oxygen. The effect is slight and occurs primarily in the lower stratosphere below 20 km altitude.

Several model studies based on experimental data simulating stratospheric chemistry have concluded that odd oxygen destruction exceeds its production by O_2 photodissociation in the upper stratosphere (e.g., Crutzen and Schmailzl, 1983; McElroy and Salawitch, 1989; Eluszkiewicz and Allen, 1993; Patten et al., 1994). To account for the deficit, additional autocatalytic O_3 production mechanisms involving vibrationally or electronically excited O_2 molecules have been proposed (Slanger et al., 1988; Miller et al., 1994; Toumi et al., 1991). However, Crutzen et al. (1995) have used the HALOE measurements of ozone and other compounds (HCl, H_2O, CH_4, NO, NO_2) to show that a good balance exists. The results are shown in Figure 3.23. Discrepancies in the earlier studies appear to have been caused mainly by uncertainties in the satellite data. For example, LIMS ozone values at high altitudes were systematically too high because of chemical excitation effects (Solomon et al., 1986b; Connor et al., 1994), whereas the HALOE ozone measurements have been successfully validated against a large set of supportive measurements such ozone sondes, lidar, ground-based microwave sounders, and different balloon-borne optical instruments (Brühl et al., 1996). The preliminary ATMOS data used by McElroy and Salawitch (1989) have since been revised (Gunson, 1990) and were shown to agree with HALOE within the common error margin. Thus, experimental data and model simulations are in good accord within common error limits, and no compelling need for significant ozone production by unconventional mechanisms exists.

Chemistry of the Troposphere: The Methane Oxidation Cycle

A fairly general treatment of trace gases in the troposphere is based on the concept of the tropospheric reservoir introduced in Section 1.5. The abundance of a trace gas in this reservoir is determined by the supply of material to the atmosphere (sources) and the removal by chemical transformation processes and loss processes at the boundaries (sinks). A steady state is established when sources and sinks are balanced. To quantify sources and sinks for any specific substance, it is necessary to identify the most important production and removal processes, to determine the associated yields, and to set up a detailed account of the budget. In the present chapter these concepts are applied to the trace gases methane, carbon monoxide, and hydrogen. Initially it will be useful to discuss the model of a reservoir at steady state and the importance of tropospheric OH radicals in the oxidation of methane and many other trace gases.

4.1. THE TROPOSPHERIC RESERVOIR

The lower bound to the troposphere is the Earth's surface, and its upper bound is the tropopause. Exchange of material occurs via these boundaries with the ocean surface, the terrestrial biosphere, and the stratosphere. Justification for treating the tropopause as a boundary layer derives from the fact that the rate of vertical transport by turbulent mixing changes abruptly over a short distance, from high values in the troposphere to low values in the stratosphere. This limits the exchange of trace species with the strato-sphere while at the same time keeping the troposphere reasonably well mixed vertically. The behavior of trace gases in the tropospheric reservoir is determined to some extent by the competition between losses due to chemical reactions and the rate of vertical and horizontal mixing. The assumption of a well-mixed reservoir is frequently made, but it is not a necessary precondition for the application of reservoir theory.

For a trace gas whose mixing ratio varies within the troposphere, the local number concentration, as always, is determined by the continuity equation

$$\frac{dn_s}{dt} = \text{div} f + q - s$$

where q denotes the local production rate and s is the local consumption rate due to chemical reactions, and the flux f is given in the usual way by the gradient of the mixing ratio augmented by an advection term. For the present purpose we shall not seek possible solutions to this equation, but integrate it over the entire tropospheric air space. After multiplying with the molar mass M_s of the trace species considered and dividing by Avogadro's number N_A (6.02×10^{23} molecule mol^{-1}), we obtain the time derivative of the total mass of the trace substance in the troposphere:

$$\frac{M_s}{N_A} \int \frac{\partial n_s}{\partial t} \, dV = \frac{dG_s}{dt} = \frac{M_s}{N_A} \left(\int (\text{div} f) \, dV + \int q \, dV - \int s \, dV \right) \quad (4.1)$$

which is equivalent to the budget equation

$$\frac{dG_s}{dt} = F_{in} - F_{out} + Q - S \quad (4.2)$$

Capital letters are now used to denote the integrated terms. Integration of the divergence term yields the difference in mass fluxes entering and leaving the tropospheric reservoir through the boundaries. The terms Q and S denote the sums of the internal sources and sinks, respectively. The global source and sink strengths are given by the sums

$$Q_T = F_{in} + Q \qquad S_T = F_{out} + S \qquad (4.3)$$

The mass content G_s of the trace constituent in the tropospheric reservoir may be derived from measurements of the mole fraction at various locations. Let $\bar{x}_s$ denote the spatial average of the mixing ratio; G_s then is obtained from

$$G_s = \frac{M_s}{M_{air}} \bar{x}_s G_T \qquad (4.4)$$

where $M_{air} \approx 29$ g mol^{-1} is the molar mass of air and $G_T = 4.22 \times 10^{18}$ kg is the tropospheric air mass. Of the four quantities on the right hand side of Equation (4.2), the outgoing flux and the volume sink are functions of G_s, whereas the input flux and the volume source are independent of G_s. The flux leaving a well-mixed reservoir is proportional to G_s. This relation is expected to hold also when the condition of complete mixing is not achieved. Accordingly, we set

$$F_{out} = k_F G_s \qquad (4.5)$$

where k_F must be treated as an empirical parameter. The functional dependence of S on G_s is less transparent. The local consumption rate of a trace gas depends on the type of reaction into which it enters. For photodecomposition the loss rate is given by $j_s n_s$. For bimolecular reactions the loss rate is proportional to the concentration of the second reactant n_r and the associated rate coefficient k_r. The local volume sink including all such processes thus is

$$s = n_s \left(\sum_r k_r n_r + j_s \right) \qquad (4.6)$$

which is proportional to n_s. A rate proportional to n_s^2 would be obtained if the trace gas reacted mainly with itself, but no such case is known. In the process of integrating s over the volume of the troposphere, one may express the summation over individual reaction rates by a suitable tropospheric average rate:

$$S = \frac{M_s}{N_A} \int n_s \left(\sum_r k_r n_r + j_s \right) dV = \left(\overline{\sum_r (k_r n_r + j_s)} \right) \frac{M_s}{N_A} \int n_s \, dV$$

$$= k_v \frac{M_s}{M_{air}} \bar{x}_s \frac{M_{air}}{N_A} \int n_T \, dV = k_v G_s \qquad (4.7)$$

Thus, if s is proportional to n_s, the integrated sink strength is also proportional to G_s. It may happen that the concentrations n_r depend in part on n_s because of chemical feedback processes, and when this occurs k_v will depend to some extent on G_s. We cannot pursue this problem here and must assume that the influence of G_s on k_v is negligible to a first approximation. If this is so, one obtains the rather general budget equation

$$\frac{dG_s}{dt} = F_{in} + Q - (k_F + k_V)G_s \qquad (4.8)$$

in which all quantities are functions of time. It is useful to distinguish two time scales: a short-term component such as diurnal or seasonal variations, and a long-term component on a time scale greater than 1 year. The short-term variations may be smoothed by averaging over a suitable time period. If the long-term variation is small during this period, $\overline{dG_s/dt} \approx 0$; that is, G_s will be nearly constant. This leads to the steady-state condition

$$\overline{F}_{in} + \overline{Q} - \left(\overline{k}_F + \overline{k}_V\right)G_s = \overline{Q}_T - \overline{S}_T \approx 0 \qquad (4.9)$$

For such conditions the budget is balanced, global source and sink strengths have the same magnitude, and G_s is constant:

$$G_s = \frac{\overline{F}_{in} + \overline{Q}}{\overline{k}_F + \overline{k}_V} \qquad (4.10)$$

The mean residence time of the trace substance in the tropospheric reservoir is

$$\tau = \frac{G_s}{\overline{F}_{in} + \overline{Q}} = \frac{G_s}{\overline{F}_{out} + \overline{S}} = \frac{1}{\overline{k}_F + \overline{k}_V} \qquad (4.11)$$

The two expressions on the left are independent of any assumptions concerning the functional relationship between $\overline{S}$ and G_s, but if there exists a first-order dependence, τ is the inverse of $\overline{k}_F + \overline{k}_V$. In the literature τ is frequently designated as the "lifetime" of the substance in the troposphere. This term implies that losses are due exclusively to chemical reactions within the tropospheric reservoir, that is, it does not account for transport of material across the boundaries. The term "residence time" is preferable because it encompasses both types of losses. However, we retain the expression "lifetime" when gas-phase reactions are the only loss processes.

    If the residence time is longer than the short-term averaging period and long-term variations are negligible, the system will always tend toward a steady state. After any perturbation, that is, when any of the fluxes, sources, or sinks take on a new value sufficiently rapidly, G_s gradually adjusts to the new conditions. The change of G_s with time after the event is obtained by integration of Equation (4.8), taking the new values of F_{in}, F_{out}, Q, and S as constant:

$$G_s = G_s^\circ \exp\left[-\left(\bar{k}_F + \bar{k}_V\right)t\right] + \frac{\bar{F}_{in} + \bar{Q}}{\bar{k}_F + \bar{k}_V}\left\{1 - \exp\left[-\left(\bar{k}_F + \bar{k}_V\right)t\right]\right\}$$

(4.12)

The first term on the right describes the decay of the initial mass of G_s with time due to the consumption processes when the sources are arbitrarily set to zero. The second term describes the rise of G_s with time due to the input of new material. A new steady state is reached after sufficient time has passed and the exponential has decreased to small values. The time constant for the approach to steady state is the same as the residence time of the substance in the tropospheric reservoir. This equivalence holds only in the case of proportionality of S_T with G_s and not in other cases.

Gradients of mixing ratios arise whenever the residence time of a trace gas in the troposphere is shorter than the time scale of transport by turbulent mixing. The longest mixing time in the troposphere is that associated with the exchange of air between the two hemispheres, which requires about 1 year. A nearly uniform mixing ratio of a trace gas in both hemispheres thus implies a residence time much greater than 1 year unless sources and sinks are extremely evenly distributed. Conversely, the existence of a latitudinal gradient of the mixing ratio within one hemisphere indicates a residence time of less than 1 year. In that case, the Northern and Southern Hemispheres should be treated as two separate, coupled reservoirs, with the interhemispheric tropical convergence zone as a common boundary.

Similar arguments lead one to expect that, in general, an inverse relation exists between residence time and spatial variation of the mixing ratio for any tropospheric constituent. A suitable measure for the variability of the mixing ratio $x' = (x_i, t)$ is the relative standard deviation

$$\sigma_s = \sigma_s^*(x')/\bar{x}$$

(4.13)

where $\sigma_s^*(x')$ is the absolute standard deviation, and $\bar{x}$ is the mixing ratio averaged over space and time. Junge (1974) has used information on a number of atmospheric trace gases to estimate σ_s and τ. The results are shown in Figure 4.1. The sizes of the boxes in the figure indicate the estimated uncertainties. The solid line is an interpolation, which approximates to $\sigma_s^* \tau = 0.14$. The fact that helium does not fall on the solid line is almost certainly due to the limitations of the measurement techniques required to establish the natural variability of helium. The scatter of the other data must be accepted as real, because the residence time is only one of the factors having an influence on σ_s. Slinn (1988) provided an illuminating analysis of Junge's (1974) empirical relation in terms of a simple model in which the mixing ratio of a trace gas decreases exponentially with distance from a source because of chemical consumption in a uniform wind field. Hamrud (1983) used a two-dimensional steady-state model with various source functions to test the empirical relationship. He found the empirical relation to be largely confirmed, even though the results depended greatly on the source distribution.

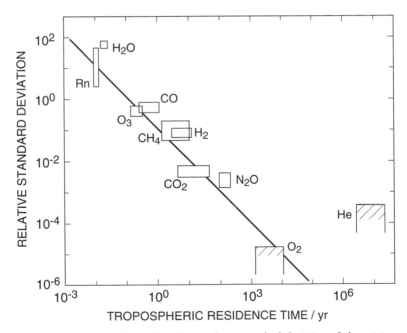

FIGURE 4.1 Double logarithmic plot of the relative standard deviation of the mixing ratio versus the residence time in the troposphere for several atmospheric constituents. Data are mostly from Junge (1974), except for N_2O from Hamrud (1983). The solid line shows the relation $\sigma \times \tau = 0.14$.

4.2. HYDROXYL RADICALS IN THE TROPOSPHERE

The recognition that OH radicals might be important in the troposphere has come comparatively late and then by indirect reasoning. Early work directed at understanding the origin of photochemical smog by simulation in large photochemical reactors (smog chambers) showed that hydrocarbon oxidation in the presence of nitric oxide and its concomitant conversion to NO_2 are rapid, and this phenomenon had baffled researchers since the 1950s. In 1969, Weinstock (Weinstock, 1971, see also Niki et al., 1972) and Heicklen et al. (1971) independently suggested that the unexplained high rate of hydrocarbon consumption is due to reaction with OH radicals and that these are regenerated by a chain mechanism of the type

$$OH + CO \rightarrow CO_2 + H$$

$$H + O_2 + M \rightarrow HO_2 + M$$

$$HO_2 + NO \rightarrow NO_2 + OH$$

where M is either nitrogen or oxygen, which serve as abundant third-body molecules. The last reaction, which continues the chain, is now known to be fast, and the reaction scheme is generally accepted.

A concurrent development concerned the fate of carbon monoxide in the atmosphere. A survey of measurements (Pressman and Warneck, 1970) had shown that the CO mixing ratio in the air of the Northern Hemisphere had stayed fairly constant at about 0.1 μmol mol^{-1} during the period 1953–1969, despite a sizable increase in the global rate of anthropogenic emissions, mainly from automobiles. Weinstock (1969) considered the formation of ^{14}CO by cosmic rays and utilized measurements of the $^{14}CO/^{12}CO$ ratio in air reported by McKay et al. (1963) to derive an order-of-magnitude estimate of 0.1 year for the residence time of CO in the troposphere. Losses of CO to the stratosphere turned out to be insufficient as a sink for tropospheric CO. The possibility of OH radicals as a reagent for the oxidation of atmospheric CO gained ground as other conceivable oxidation reactions such as that of CO with ozone (Arin and Warneck, 1972) were found to be slow. The breakthrough came when Levy (1971) identified a route to the formation of OH radicals in the troposphere, namely, photolysis of ozone by ultraviolet radiation in the vicinity of 300 nm to produce $O(^1D)$ atoms, which subsequently react with water vapor

$$O_3 + h\nu \rightarrow O(^1D) + O_2(^1\Delta_g)$$

$$O(^1D) + N_2(\text{or } O_2) \rightarrow O(^3P) + N_2(\text{or } O_2)$$

$$O(^1D) + H_2O \rightarrow 2OH$$

A large fraction of the $O(^1D)$ atoms are deactivated by collisions with N_2 or O_2 molecules, but the remainder generates enough OH to provide a sink for CO of the required magnitude. In addition, OH oxidizes methane to formaldehyde and, by reaction with and photolysis of the aldehyde, provides a source of CO. The actual presence of OH in ground-level, continental air during daylight was demonstrated by its characteristic ultraviolet absorption feature with the help of a long-path laser beam as the background light source (Perner et al., 1976). While the early work was plagued by high noise levels, the technique has been gradually improved (Hübler et al., 1984; Perner et al., 1987; Platt et al., 1988). Recent differential absorption data obtained with high spectral resolution have unequivocally confirmed the existence of OH and have traced its diurnal variation (Dorn et al., 1996). Figure 4.2a compares the OH signal with a laboratory reference spectrum; Figure 4.2b shows that the OH concentration follows quite closely the diurnal variation of ozone photodissociation to produce $O(^1D)$, which was measured simultaneously. Early attempts to determine the OH concentration by laser-induced fluorescence (Wang and Davis, 1974; Wang et al., 1981; Davis et al., 1976) have met with difficulties because the laser pulse generates spurious OH by photodissociation of ozone (Ortgies et al., 1980; Davis et al., 1981). The problem has been largely overcome by operating at lower pressures, and a recent comparison of the two techniques has led to remarkably good agreement (Brauers et al., 1996). A third method for determining OH concentrations is based on the reaction of OH with $^{34}SO_2$ to form $H_2{}^{34}SO_4$, which is monitored by mass spectrometry (Eisele and Tanner, 1991). Here again, good agreement with long-path differential optical absorption has been achieved (Eisele et al., 1994). Despite the progress made in OH measurement techniques, however, our knowledge of average OH concentrations in the troposphere still relies mainly on model calculations and indirect determinations rather than direct observations.

The local primary OH production rate according to the above mechanism is

$$\bar{P}_0(OH) = \frac{2k_{40}\,x(H_2O)\,j(O^1D)\,n(O_3)}{k_{37}\,x(N_2) + k_{38}\,x(O_2) + k_{40}\,x(H_2O)} \qquad (4.14)$$

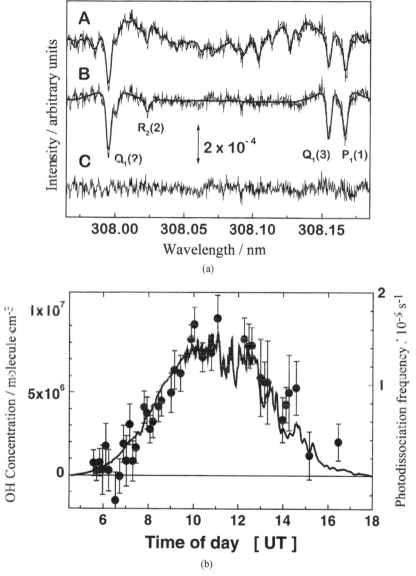

FIGURE 4.2 (a) Folded long-path differential absorption spectra of the OH radical in ground-level air. (A) original spectrum. (B) After subtraction of absorption spectra of interfering compounds, mainly HCHO, SO_2, $C_{10}H_8$. The thick line is a superimposed OH reference spectrum. (C) Residual after subtraction of all known trace species. (b) Diurnal variation of OH concentration (points with 1 sigma error bars) superimposed on the ozone photolysis frequency (line) to produce $O(^1D)$. (Adapted from and courtesy of Dorn *et al.* (1996).)

Diurnally averaged values $\bar{P}_0(\text{OH})$ are given in Table 4.1 as a function of latitude and height in the Northern Hemisphere. Production rates at noon are about five times higher. The rates are greatest in the lower troposphere, because of the highest water vapor mixing ratios there, and in the equatorial region, where the solar ultraviolet flux is least attenuated by stratospheric ozone. These production rates are augmented by chain reactions in which OH is regenerated from HO_2 (see below).

Hydroxyl does not react with the main atmospheric constituents N_2, O_2, H_2O, and CO_2, but it reacts readily with many trace gases. Table 4.2 lists several trace gases, approximate mixing ratios, and rate coefficients for reaction with OH for several tropospheric conditions. The reactions with methane and carbon monoxide are dominant in the entire troposphere; nitrogen dioxide and formaldehyde are important in the continental boundary layer. The other trace gases are less significant with regard to OH losses. The inverse of the sum of the individual reaction rates gives the lifetime of OH, which falls into the range 0.3–2.5 s. Before the concept of OH reactions gained ground, researchers thought that radicals would be quickly scavenged by aerosol particles so that gas-phase reactions would be less important. Warneck (1974) investigated this possibility and found that OH radicals are not thus affected. Collision frequencies of molecules with particles of all

TABLE 4.1   Diurnally Averaged Primary OH Production Rate (Not Including Amplification by Secondary Chain Reactions) as a Function of Latitude and Height in the Troposphere (Northern Hemisphere)[a]

	Degrees latitude									
z (km)	0	10	20	30	40	50	60	70	80	90
					July					
0	716	677	1080	865	530	386	186	75.2	34.9	10.9
2	515	543	847	589	379	316	154	66.7	30.9	8.21
4	311	360	541	348	209	187	89.1	41.9	19.6	5.45
6	137	190	291	156	87.5	77.6	35.9	17.1	8.39	2.24
8	53.2	68.8	113	60.7	34.3	39.2	15.5	9.41	6.81	1.84
					January					
0	504	444	318	110	18.6	1.34	0.105	—	—	—
2	357	259	171	52.4	10.7	0.72	0.054	—	—	—
4	218	115	73.3	23.3	5.20	0.32	0.020	—	—	—
6	105	43.7	27.4	7.83	1.99	0.11	0.006	—	—	—
8	41.2	15.9	8.92	3.26	0.83	0.085	0.005	—	—	—

[a] Units: 10^3 molecule cm^{-3} s^{-1}; the values were calculated with Eq. (4.14) from observational data for $n(O_3)$, $x(H_2O)$, and solar radiation fluxes, as well as laboratory data for $\sigma(O_3)$, $q(O^1D)$, and $O(^1D)$ rate coefficients (Warneck, 1975). The data must be considered approximate because of recent improvements in quantum yields.

TABLE 4.2. Removal of OH Radicals by Reactions with Trace Gases: Mixing ratios x, Rate Coefficients k, and Relative Efficiencies for Various Tropospheric Constituents[a]

Trace gas	x (nmol mol^{-1})			k (cm^3 molecule^{-1} s^{-1})		Rate factor $kn_M x$			Percentage of total		
	A	B	C	A,B	C	A	B	C	A	B	C
CH$_4$	1.7 (3)	1.7 (3)	1.7 (3)	4.9 (−15)	1.5 (−15)	2.1 (−1)	2.1 (−1)	3.0 (−2)	7.3	21.9	10.2
CO	2.5 (2)	1 (2)	1 (2)	2.4 (−13)	1.9 (−13)	1.5	6.0 (−1)	2.3 (−1)	52.1	62.6	78.2
H$_2$	6 (2)	5 (2)	5 (2)	5.2 (−15)	1.2 (−15)	7.8 (−2)	6.6 (−2)	7.3 (−3)	2.7	6.9	2.5
O$_3$	2.5 (1)	2.5 (1)	3 (1)	6.1 (−14)	3.2 (−14)	3.8 (−2)	3.8 (−2)	1.2 (−2)	1.3	4.0	4.1
NO$_2$	2	3 (−2)	3 (−2)	9.7 (−12)	7.2 (−12)	4.9 (−1)	7.3 (−3)	2.6 (−3)	17.0	0.8	0.9
SO$_2$	2	5 (−2)	5 (−2)	8.5 (−13)	4.8 (−13)	4.2 (−2)	1.0 (−3)	2.9 (−4)	1.4	0.1	0.1
NH$_3$	4	1 (−1)	1 (−1)	1.4 (−13)	8.8 (−14)	1.4 (−2)	3.6 (−4)	1.0 (−4)	0.5	—	—
HCHO	2	3 (−1)	1 (−1)	1.0 (−11)	1.0 (−11)	5.1 (−1)	3.6 (−2)	1.2 (−2)	17.7	3.7	4.1

[a](A) Continental boundary layer ($n_M = 2.5 \times 10^{19}$ molecule cm^{-3}, $T = 288$ K); (B) marine boundary layer; and (C) upper troposphere, $z = 7$ km ($n_M = 1.2 \times 10^{19}$ molecule cm^{-3}, $T = 240$ K) (numbers in parentheses indicate powers of 10).

sizes fall into a range of 9×10^{-4} to 6×10^{-2} s^{-1}. The corresponding time constants (see Table 7.5) are 16 and 1150 s, respectively. Although the aerosol is not a good scavenger for OH in comparison with gas-phase reactions, other radicals with lesser gas-phase reactivities may suffer such losses. Thus it is necessary to treat this problem separately for each type of radical.

The reactions following the attack of OH on methane were first discussed by Levy (1971) and McConnell *et al.* (1971). The reactions include

$$OH + CH_4 \rightarrow H_2O + CH_3$$
$$CH_3 + O_2 + M \rightarrow CH_3O_2 + M$$
$$CH_3O_2 + NO \rightarrow CH_3O + NO_2$$
$$CH_3O + O_2 \rightarrow HCHO + HO_2$$

Formaldehyde is the immediate stable product resulting from this reaction sequence. In addition, an HO_2 radical is formed for each OH radical consumed. This is similar to the reaction of OH with CO discussed earlier. The reaction of OH with nitrogen dioxide

$$OH + NO_2 \overset{M}{\rightarrow} HNO_3$$

in contrast, does not produce HO_2 because it is an addition reaction leading to a stable product. The photodecomposition rate of HNO_3 in the troposphere is small, and according to present knowledge HNO_3 is mainly precipitated.

Photodissociation of formaldehyde or its reaction with OH constitutes another source of HO_2 radicals:

$$HCHO + h\nu \overset{a}{\rightarrow} H_2 + CO$$
$$\overset{b}{\rightarrow} HCO + H$$
$$HCHO + OH \rightarrow HCO + H_2O$$
$$HCO + O_2 \rightarrow HO_2 + CO$$
$$H + O_2 + M \rightarrow HO_2 + M$$

The relative quantum yield of HO_2 from formaldehyde, $2j_b(HCHO)/j(HCHO)$, is about 0.8 at ground level. Since photodecomposition is the dominant loss process for formaldehyde in the atmosphere, the oxidation of methane ultimately produces more than one HO_2 for each OH radical entering into a reaction with methane, provided enough NO is present to convert all of the CH_3O_2 radicals to formaldehyde. Losses of formaldehyde from the atmosphere due to in-cloud scavenging and wet precipitation amount to less than 15% of those caused by photolysis and reaction with OH (Warneck *et al.*, 1978; Thompson, 1980).

The main reactions of HO_2 radicals that must be considered are

$$HO_2 + NO \rightarrow NO_2 + OH$$

$$HO_2 + O_3 \rightarrow 2O_2 + OH$$

$$HO_2 + HO_2 \rightarrow H_2O_2 + O_2$$

$$HO_2 + CH_3O_2 \rightarrow CH_3OOH + O_2$$

$$HO_2 + OH \rightarrow H_2O + O_2$$

Only the first two of these reactions convert HO_2 back to OH. The other reactions terminate the reaction chain. H_2O_2 and CH_3OOH in the background atmosphere may react with OH radicals or undergo photodissociation (thereby creating new OH radicals). Compared with formaldehyde, however, the rates of these processes are smaller, whereas aqueous solubilities are greater (cf. Table 8.2). This suggests that hydroperoxides are preferentially lost by reactions in clouds or by precipitation. Methyl hydroperoxide is less well soluble than H_2O_2, so that gas-phase destruction may be more important.

The source of nitric oxide in the unpolluted troposphere is the photodissociation of nitrogen dioxide, which leads to the reactions

$$NO_2 + h\nu \rightarrow NO + O$$

$$O + O_2 + M \rightarrow O_3 + M$$

$$O_3 + NO \rightarrow NO_2 + O_2$$

$$HO_2 + NO \rightarrow NO_2 + OH$$

$$RO_2 + NO \rightarrow NO_2 + RO$$

where RO_2 and RO stand for alkylperoxy and alkoxyl radicals derived from the oxidation of hydrocarbons (including methane). The reactions with ozone, HO_2, and RO_2 radicals cause NO to rapidly revert back to NO_2, so that steady-state conditions are set up for NO. The lifetimes of HO_2 and CH_3O_2 are quite sensitive to this stationary NO concentration. The tropospheric background level of NO_2 in marine air is on the order of 30 pmol mol^{-1} and higher over the continents. Measurements of NO in the free troposphere indicate daytime mixing ratios of about 10 pmol mol^{-1} or less (see Table 9.12). In rural continental air, values of up to a few nmol mol^{-1}

are encountered. The assumption of an NO mixing ratio of 10 pmol mol^{-1} leads to lifetime estimates for HO_2 and CH_3O_2 of 500 s; increasing the NO mixing ratio to 500 pmol mol^{-1} reduces the lifetime to about 10 s. The values are high enough to make losses of radicals by collisions with aerosol particles likely, but this process is usually ignored.

The reactions presented above form the basis of a tropospheric chemistry model in which radicals adjust to the photostationary state within seconds, formaldehyde within hours, and hydroperoxides within days. Input parameters are the solar radiation flux and the concentrations of ozone, water vapor, methane, carbon monoxide, and nitrogen oxides. A two-dimensional analysis is called for, as most of the input data vary with latitude and altitude. The model yields the distribution of diurnal averages of OH concentrations in the troposphere, the seasonal variation, and the annual and spatial averages. Figure 4.3 shows as an example the seasonally averaged OH distribution. It again makes evident that maximum OH concentrations occur near the equator.

Table 4.3 summarizes estimates for the globally and seasonally averaged OH concentration. The early estimates of Levy (1972), McConnell *et al.* (1971) and Wofsy *et al.* (1972), which were based on a one-dimensional analysis at mid-latitudes of the Northern Hemisphere, are not included. These values were too high, partly because the latitudinal dependence of the radiation field was ignored, but also because the adopted NO_2 concentrations were too high. Even today the NO/NO_2 concentration field is poorly known, and it is calculated from the distribution of NO_x sources.

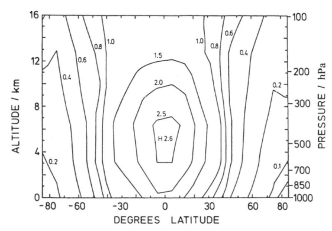

FIGURE 4.3   Meridional cross section of annual average OH number concentrations (10^6 molecule cm^{-3} s^{-1}) in the troposphere calculated with a two-dimensional model. (Adapted from Crutzen (1982).)

TABLE 4.3 Annual Average OH Number Concentration (10^5 Molecule cm^{-3}) in the Troposphere, Derived Mainly from Two-Dimensional and Box Models

Authors	Northern Hemisphere	Southern Hemisphere	Global	Remarks
Warneck (1975)	7	—	—	Based on observational data, primary OH production rate emphasized; major uncertainty: NO_x distribution
Crutzen and Fishman (1977)	2.6	—	4.8	Mainly based on observational data
Singh (1977b)	2.6	6.5	5.2	From production and distribution of CH_3CCl_3 and $CHCl_3$
Neely and Ploncka (1978)	4.8	18	11	From production and distribution of CH_3CCl_3
Chameides and Tan (1981)	2.9–7.6	2.1–8.1	3–8	Continental and marine regions considered separately
Volz et al. (1981)	5.5	6.5	6.5	Based on ^{14}C and ^{12}C budgets
Derwent (1982)	6.3	5.7	6	Constrained by ^{14}CO and ^{12}CO and halocarbon data
Crutzen and Gidel (1983)	4.9	6.3	5.5	NO_x distribution derived from budget considerations
Singh et al. (1983a)	—	—	5.2 ± 1.7	From production and distribution of CH_2Cl_2, CH_3CCl_3, and C_2Cl_4
Prinn et al. (1983b)	—	—	5 ± 2	From production rates and trend of CH_3CCl_3
Prinn et al. (1987)	—	—	7.7 ± 1.4	From production rates and 7-year trend of CH_3CCl_3
Spivakovsky et al. (1990)	—	—	8	From three-dimensional chemical tracer model
Prinn et al. (1992)	8.9	3.5	8.1 ± 0.9	From 1978–1990 production rates and trend of CH_3CCl_3; 1% annual increase in n(OH) deduced. Global value corrected for loss of CH_3CCl_3 to the ocean
Prinn et al. (1995)	—	—	9.7 ± 0.6	From 1978–1994 production rates and trend of CH_3CCl_3 after instrument recalibration, no trend of n(OH) detected

In addition to model computations, the average OH concentration can also be derived from tracer measurements if the source strength of the tracer is known and reaction with OH is the dominant sink. The associated rate coefficient must be known, and the abundance of the tracer should be small enough to preclude any interference with major routes of tropospheric OH chemistry. Singh (1977a) has pointed out that the chlorinated hydrocarbons CH_3CCl_3 and $CHCl_3$ are useful tracers because they are entirely human-made and their source strengths can be obtained from industrial production data. Singh (1977a) approximated the source data by an exponential growth function $Q_s(t) = a \exp(bt)$. It is then simple to integrate the budget equation (Eq. (4.8)) with the result

$$G_s(t) = \frac{a\tau}{b\tau + 1} \exp(bt)\{1 - \exp[-(b + 1/\tau)t]\} \qquad (4.15)$$

where τ is the residence time of the tracer in the troposphere. It is convenient to introduce the ratio R_s of the total mass of the tracer in the troposphere to the cumulative mass emitted by the sources:

$$\Delta G_s(t) = \int_0^t Q_s(t)\, dt = (a/b)[\exp(bt) - 1]$$

For sufficiently long times t the ratio approximates to

$$R_s = G_s(t)/\Delta G_s(t) = b\tau/(b\tau + 1) \qquad (4.16)$$

and one obtains for the residence time

$$\tau = [\gamma \bar{k}\bar{n}(OH)]^{-1} = R_s/b(1 - R_s) \qquad (4.17)$$

where the factor $\gamma = 3.15 \times 10^7$ is the number of seconds per year. From measurements of the mixing ratios of CH_3CCl_3 and $CHCl_3$ performed in 1976 in both hemispheres, and from the emission estimates available at that time, Singh (1977a) derived the data assembled in Table 4.4. The calculation of the OH concentration requires a suitable choice of temperature to determine the effective rate coefficient $\bar{k}$. Trace gas concentration and temperature decline with altitude, so that the reaction rate is weighted in favor of the lower troposphere. Prather and Spivakovsky (1990) have used a detailed analysis to derive 277 K. The corresponding rate coefficients and the average OH concentrations are given in Table 4.4. Singh (1977a) used an average tropospheric temperature of 265 K, which is too low, but with the preliminary rate coefficient data available to him he obtained similar $\bar{n}(OH)$ values.

TABLE 4.4 Emission Growth Factor (b), Cumulative Emissions into the Atmosphere (ΔG_s), Total Mass of Trace Constituent Still Present in the Troposphere in the Year Given (G_s), Ratio $(G_s / \Delta G_s)$, Residence Times, and Average OH Number Density, Deduced from Budgets of CH_3CCl_3, $CHCl_3$, and $HClF_2$

Trace gas		b (year^{-1})	ΔG_s (Tg)	G_s (Tg)	R_s	τ (yr)	$\bar{n}$(OH) (10^5 molecule cm^{-3})	
CH_3CCl_3	1976[a]	0.166	3.0	1.76	0.58	8.3[c]	5.8[d]	5.1[a]
	1979[b]	0.159	4.3	1.92	0.45	5.1[c]	9.5[d]	—
$CHCl_3$	1976[a]	0.121	1.1	0.20	0.15	1.7	3.4[d]	3.0[a]
$HClF_2$	1990[b]	0.075	2.7	1.33	0.49	12.9	7.9[d]	—

[a] Singh (1977a).
[b] Singh (1995).
[c] Some methyl chloroform is lost to the ocean and the stratosphere, so that the calculated OH concentrations are somewhat too high.
[d] $k(CH_3CCl_3) = 6.5 \times 10^{-15}$, $k(CHCl_3) = 5.6 \times 10^{-14}$, $k(HClF_2) = 3.1 \times 10^{-15}$ (cm^{-3} molecule^{-1} s^{-1}), $T = 277$ K.

Recently Singh (1995) applied his method to $HClF_2$; Table 4.4 includes the results.

Table 4.3 shows that for a number of years the global average OH concentrations derived from both models and tracer data converged on a value of about 3×10^5 molecules cm^{-3} with an uncertainty of less than a factor of 2. The more recent data, in particular those derived from the trend analysis of CH_3CCl_3, indicate a higher value, $(8 \pm 2) \times 10^5$ molecule cm^{-3}. The difference is partly due to new laboratory data that changed the temperature dependence of the rate coefficient, and partly due to changes in instrument calibration. A real increase in OH concentrations may have occurred as a consequence of losses of stratospheric ozone and the associated increase in ultraviolet radiation in the troposphere. Some of the results have indicated a slightly higher average OH concentration in the Southern Hemisphere compared to the Northern Hemisphere. According to the more recent data the difference is marginal. Spivakovsky et al. (1990) noted, for example, that the effect of higher CO concentrations in the Northern Hemisphere, which would reduce OH, is more than offset by the higher levels of O_3 and NO_x. Lower mixing ratios of CH_3CCl_3 in the Southern Hemisphere compared to the Northern Hemisphere, are mainly caused by the sluggishness of transport across the interhemispheric tropical convergence zone, combined with the fact that more than 90% of the sources of CH_3CCl_3 are located in the Northern Hemisphere. The influence of chemistry on the latitudinal distribution of CH_3CCl_3 is smaller than the interannual variations.

4.3. THE BUDGET OF METHANE

The discovery of methane (CH_4) in the atmosphere is due to Migeotte (1948), who observed the characteristic infrared absorption features in the solar spectrum and estimated a mixing ratio of 1.5 μmol mol^{-1}. The majority of recent measurements have employed gas chromatographic techniques. Wilkness (1973) first explored the latitudinal distribution of methane in the marine atmosphere of the Pacific Ocean. He found an almost uniform distribution of 1.4 μmol mol^{-1}, with a small excess of about 6% in the Northern Hemisphere. The difference has been confirmed by several subsequent studies (e.g., Ehhalt and Schmidt, 1978; Steele *et al.*, 1987). Figure 4.4 shows a recent distribution. The increase toward middle and high latitudes of the Northern Hemisphere is due to the uneven distribution of CH_4 sources, which favors northern regions. Methane has also been shown to undergo a weak seasonal cycle with amplitude depending on location (Khalil and Rasmussen, 1983; Khalil *et al.*, 1993a).

When Ehhalt (1974) reviewed the early measurements, he concluded that the mixing ratio was constant with time. However, improvements in the

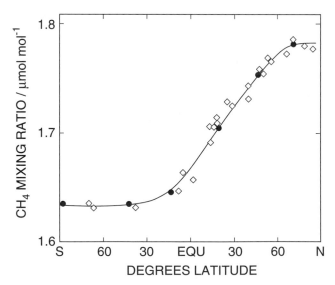

FIGURE 4.4   Latitudinal distribution of methane in the troposphere in 1988. Solid points indicate stations at Point Barrow, Alaska; Cape Meares, Oregon; Mauna Loa, Hawaii; Cape Matatula, Samoa; Cape Grim, Tasmania; and the South Pole. The open points are from the National Oceanic and Atmospheric Administration flask sampling program at various marine background locations. (Adapted from Khalil *et al.* (1993a).)

precision of measuring techniques and a more systematic monitoring since 1978 have revealed global changes. In the period 1980–1984 the average mixing ratio increased by about 20 nmol mol^{-1}/year. Subsequently the rate declined, and during 1989–1992 it was about 11 nmol mol^{-1}/year (Rasmussen and Khalil, 1981; Blake *et al.* 1982; Blake and Rowland, 1988; Khalil *et al.*, 1993a; Houghton *et al.*, 1996). In 1994 the average global mixing ratio was close to 1.7 μmol mol^{-1}. The long-term trend in atmospheric methane has been explored by the analysis of air bubbles incorporated in the great ice sheets of Greenland and Antarctica. Figure 4.5 shows results of several studies that indicate a CH$_4$ mixing ratio of 700 nmol mol^{-1} in the period before 1700 A.D. and an almost exponential increase during the past 300 years. Khalil and Rasmussen (1985a) have pointed out that the increase parallels the growth in human population, which suggests that anthropogenic activities are responsible for the trend. Going further back in time, samples of ice from both Greenland (Stauffer *et al.*, 1988) and Antarctica (Raynaud *et al.*, 1988) show that during the past glacial–interglacial climate transition the atmospheric CH$_4$ mixing ratio nearly doubled from 0.35 to 0.65 μmol mol^{-1}. The full record, which can be traced back 120 thousand

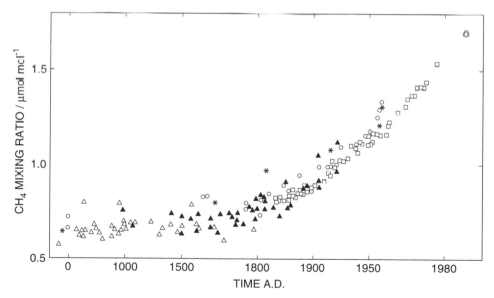

FIGURE 4.5 Trend of methane for the past 2000 years, derived from air bubbles in ice cores. Greenland: crosses, Craig and Chou (1982); filled triangles, Rasmussen and Khalil (1984); Antarctica: open triangles, Rasmussen and Khalil (1984); squares, Etheridge *et al.* (1992); circles, Stauffer *et al.* (1985). The double circle indicates the approximate present value.

years (Chappellaz *et al.*, 1990), shows oscillations that—similar to CO_2—correlate well with temperature. The main natural source of atmospheric methane is bacterial decomposition of organic matter in the soils of wetlands. Although bacterial processes depend strongly on temperature, it is likely that a redistribution of wetlands and vegetation is mainly responsible for the variation of methane during glacial periods.

Figure 4.6 illustrates the production of methane at the bottom of a lake. Methanogenic bacteria belong to a small group of strict anaerobes living in symbiosis with other bacteria, which derive their energy needs from the fermentation of cellulose and other organic materials that accumulate in the lake sediments and wetland soils. Disproportionation brakes the material down toward carbon dioxide, alcohols, fatty acids, and hydrogen. A small group of methanogens produces methane from CO_2 and H_2. Other bacteria utilize methanol, acetic acid, and butyric acid. As methane migrates from the anaerobic zone to the overlying lake waters, where the environment turns aerobic, CH_4 becomes subject to oxidation by over 100 different kind of bacteria, most of which are obligately bound to methane as substrate. Methane obviously can rise to the lake surface and enter the atmosphere only if the lake is sufficiently shallow. Otherwise it is completely consumed. In freshwater lakes more than 10 m deep the release of methane becomes negligible. The most important sources of atmospheric methane therefore are the shallow waters of swamps, marshes, and rice-paddy fields.

The early estimates of CH_4 production from such sources were based on a laboratory study of Koyama (1963), who measured release rates from nine Japanese paddy soils mixed with water and incubated at different temperatures. From statistical data for the area of rice paddy fields in Japan (2.9×10^4 km^2), the average depth of paddy soils (15 cm), the annual average of the temperature (16°C), and the fact that Japanese rice paddies are water-logged 4 months of the year, Koyama (1963) estimated an annual methane production rate of 80 g m^{-2} year^{-1}. In a similar manner he derived

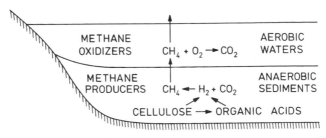

FIGURE 4.6   Production and consumption of methane at the bottom of a lake. (Adapted from Schlegel (1974).)

a rate of 0.44 g m^{-2} year^{-1} for upland fields and grasslands, and 8×10^{-3} g m^{-2} year^{-1} for forest soils. These data then led him to estimate a world production of 190 Tg year^{-1} from rice paddies (0.9×10^6 km^2), taking into account multiple crops, 10 Tg year^{-1} from other cultivated soils (30×10^6 km^2), and 0.4 Tg year^{-1} from forest areas (44×10^6 km^2).

In reviewing these data Ehhalt (1974) noted an increase in the rice growing area to 1.35×10^6 km^2, based on 1970 United Nations statistics. In addition, he adopted a global area of swamps and marshes of 2.6×10^6 km^2 from Twenhofel (1951) to estimate a methane flux of 130–260 Tg year^{-1} from this source. He also pointed out that water-logged tundra may be a source of methane and estimated a production rate of 1.3–13 Tg year^{-1}. This set the stage for numerous field studies of methane emissions from wetlands and rice paddy fields followed by new assessments.

Rather detailed information on the geographic distribution of wetlands is now available. Matthews and Fung (1987) used UNESCO vegetation and soil maps to derive a $1° \times 1°$ resolution global distribution of wetlands, classified to distinguish forested and nonforested bogs, forested and nonforested swamps, and alluvial formations. Aselmann and Crutzen (1989) compiled similar data from many other sources to obtain a 2.5° latitude by 5° longitude global coverage of wetlands that distinguished six major categories: bogs, fens, swamps, marshes, flood plains, and lakes. *Bogs* are peat-producing wetlands, largely boreal, that receive water and nutrient input entirely through precipitation. *Fens* are peat-producing wetlands influenced by soil nutrients from water flowing through the system. In comparison to bogs, because of better nutrition, fens commonly are more prolific under similar climatic conditions. *Swamps* are forested freshwater wetlands on waterlogged or inundated soils with little peat accumulation. *Marshes* are herbaceous mires with vegetation dominated by grasses, either permanently or seasonally subjected to gravitational water levels. Salt marshes are not included because they do not produce significant methane emissions. *Flood plains* are periodically flooded areas along rivers or lakes. *Shallow lakes* are open water bodies of a few meters depth, which are likely to emit methane through the water column. Both groups of authors conclude that the total area of wetlands is about 5.5×10^6 km^2. About half the area is located at high latitudes north of 50° N, and about a third are tropical or subtropical swamps and alluvial formations.

Measured fluxes of methane from wetlands have been summarized by Aselmann and Crutzen (1989) and by Bartlett and Harriss (1993). Fluxes can vary widely because of a combination of environmental factors such as temperature, moisture or depth of water level, nutrient input, accumulation of organic matter, and type of vegetation. A clear picture of what governs the flux rates has not yet emerged, and emission rates may range over several orders of magnitude, even within the same habitat. Matthews and Fung (1987) adopted CH_4 emission rates of 200 mg m^{-2} day^{-1} for both forested and nonforested bogs, 70 for forested swamps, 120 for nonforested swamps, and 30 for alluvial formations. The productive periods during the year were assumed to be 100 days poleward of 60°, 150 days from 30° to 60°, and 180 days between 30° and the equator. In this manner they obtained a global CH_4 emission of 111 Tg year^{-1}. Aselmann and Crutzen (1989) applied the following geometric mean CH_4 emission rates (in mg m^{-2} day^{-1}, ranges in parentheses): bogs 15 (1–50), fens 80 (28–216), swamps 84 (57–112), marshes 253 (137–399), flood plains 100 (50–200), and lakes 43 (17–89), combined with productive periods (in days) of 178, 169, 274, 249, 122, and 365, respectively. The global CH_4 emission thus obtained is 80 Tg year^{-1} with a range of 40–160. Bartlett and Harriss (1993) calculated 105 Tg year^{-1}, using a more extensive flux data base. These estimates agree reasonably well.

Rice production has risen by about 30% since the mid-1970s. Although the harvest area increased only by 4%, higher yields (40%) were achieved by multiple cropping. Globally, 60% of the area is managed under a triple-crop system, 15% is double-cropped, and 25% is planted to rice once a year. The total rice-growing area is $(1.3–1.5) \times 10^6$ km^2, about half of it located in a narrow subtropical latitude zone (10°–30°), over 90% in Asia (Aselmann and Crutzen, 1989; Matthews et al., 1991). Rice growing includes three types: dry land rice, rain-fed rice, and irrigated rice cultivation. Only the latter two are expected to release methane. In India, according to Singh (1994), dry land rice is grown on about 15% of the total area, whereas in surrounding countries the fraction is less. Numerous measurements of methane emissions from irrigated rice fields are now available, and Shearer and Khalil (1993) have recently summarized the data. All of the measurements are based on the flux chamber technique (see Section 1.6), in which a closed-off collector box is placed over a part of the field, and the increase in methane in the box is determined as a function of time. Already the early studies of rice fields in California (Cicerone et al., 1983) and Spain (Seiler et al., 1984a) had shown that the release of methane proceeds mostly (to more than 60%) via the stems and leaves of the rice plant, and to a lesser extent by gas bubbles rising directly through the water column. As in the case of natural wetlands, methane emission fluxes vary with soil type, growing season, and temperature, but in addition the flux depends on the type of fertilizer applied. Rice straw, manure, and other organic fertilizers appear to enhance methane production, whereas nitrogen fertilizers such as ammonium sulfate may inhibit it (Schütz et al., 1989; Chen et al., 1993). Because of the daily and seasonal variations in the CH_4 emission rate, measurements over a whole season are required to calculate an average. Fluxes ranged from about 25 to

1175 mg m^{-2} day^{-1}, depending on the type and extent of fertilization. While the growing season is 120–150 days, the methane emission period is shorter, typically 85–115 days. Aselmann and Crutzen (1989) used an average flux of 310 mg m^{-2} day^{-1} from the work of Holzapfel-Pschorn and Seiler (1986) in Italy and a 130-day season to calculate a global emission rate of 53 (range 30–75) Tg year^{-1}. As the flux refers to temperate latitudes, and the emission rate was observed to increase with increasing soil temperature in the range 17–30°C, they also used an expression for the temperature dependence to calculate a global emission of 90 Tg year^{-1} based on local mean monthly temperatures. Shearer and Khalil (1993) used averaged fluxes from all of the available measurements in China, India, Japan, Italy, Spain, and the United States and extrapolated the data to other countries, where measurements are lacking. For areas of rain-fed rice the flux rates were reduced by 40%, similar to the reduction found by Chen (1993) for intermittent irrigation schemes. The global rate thus obtained for the year 1990 was 66 Tg year^{-1}. In a complementary study, Bachelet and Neue (1993) worked with rice yield data assembled by the International Rice Research Institute (Manila), regional estimates of organic matter incorporated in the soils, and associated CH_4 emission factors to derive a range of 40–80 Tg year^{-1} for CH_4 emissions in the major rice-growing countries in South and East Asia, in agreement with the other estimates. The global flux of methane from rice paddy fields still is rather uncertain, but with about 60 Tg year^{-1} it clearly is much lower than the early estimates.

The microbial breakdown of carbohydrates in the rumen and lower gut of domestic animals was identified early as a source of atmospheric methane. Daily production yields reported by Hutchinson (1948, 1954) per head of cattle, horses, sheep, and goats were 200, 106, 15, and 15 (in g day^{-1}), respectively. From these data he derived a global rate of 45 Tg year^{-1} for 1940. Ehhalt (1974) used United Nations statistics that had indicated an increase in the number of heads to, in millions, 1198 cattle, 125 horses, 1026 sheep, and 384 goats to derive a global CH_4 emission rate of 100 Tg year^{-1} in 1970.

Crutzen et al. (1986) have pointed out that the release of methane from domestic animals depends on the type of animal husbandry and feeding schemes. Protein-rich food, for example, would lower the methane production rate. Instead of fixed CH_4 emission factors, they used a methane yield expressed as a fraction of the gross energy intake based on energy needs at maintenance levels. In the United States, three types of cattle were distinguished: dairy cows (10%), beef cattle on feed lots (12.5%), and cattle on range (77.5%). The daily energy intake is 230, 150, and 110 MJ per head, corresponding to 2–3, 1.75, and 1.3 times the subsistence level. With Blaxter and Clapperton's (1965) methane yields of 5.5%, 6.5%, and 7.5% per gross energy intake for these categories, and an energy content of methane of 55.65 kJ kg^{-1}, the daily production rates (in grams per head) are 230, 178, and 148, respectively. This corresponds to 84, 65, and 54 kg annually, averaging to 58 kg. A similar detailed accounting for West Germany gave 57 kg per animal per year, whereas in India, which was considered exemplary for the developing world with low-quality feed, a mean annual CH_4 production rate of 35 kg per head was calculated. According to the Food and Agricultural Organization yearbook (FAO, 1984), the world cattle population in 1983 was 1.2×10^9, with 47% in the developed and 53% in the developing countries. This leads to 54 Tg $year^{-1}$ as global methane production rate from cattle. Crutzen et al. (1986) also reported global production rates from buffaloes, sheep, goats, camels, pigs, horses, mules and asses, and humans of (in Tg $year^{-1}$) 6.2, 6.9, 2.4, 1.0, 0.9, 1.2, 0.5, and 0.3, respectively. The total sum is 74 Tg $year^{-1}$ from all these sources. The uncertainty was estimated to be $\pm 15\%$. Cattle obviously contributes the major share. Wild ruminants were also considered, but their contribution is minor. A recent review of CH_4 emissions from ruminants by Johnson et al. (1993) confirmed this assessment.

Methane is also produced in the digestive tract of herbivorous insects, of which termites have been studied most extensively. The early work of Zimmerman et al. (1982), Rasmussen and Khalil (1983a), and Seiler et al. (1984b) had led to diverging conclusions on global methane emission rates from termites (range 2–150 Tg $year^{-1}$), because the few species studied were not globally representative. This was first pointed out by Fraser et al. (1986), who summarized the rate data available at that time. Wood-eating (xylophageous) termites produce fairly little methane; soil-feeding termites consume organic matter mixed with mineral particles; fungus-growing termites develop a symbiotic relationship with a fungus, which they maintain in their nests with collected plant materials. Rouland et al. (1993) have measured methane yields from four wood-eating, five soil-feeding, and six fungus-growing African termite species and found ranges of 0.34–2.4, 8.5–17, and 5.6–14 $\mu g\ h^{-1}$, respectively, per gram of termite. Fraser et al. (1986) have reported similar values for several other species. The greatest uncertainty is the density of termites in the areas affected, mainly tropical rain forests and savannas, with areas of about 18.5×10^6 km^2 each, according to Fraser et al. (1986). They adopted termite densities from Wood and Sands (1978) of 5.6 and 4.5 g m^{-2}, respectively, to calculate a global emission rate of 14 Tg $year^{-1}$ (range 6–42). A recent estimate by Judd et al. (1993) of 27 Tg $year^{-1}$ falls into this range. If all of the methane produced by soil-eating termites were oxidized during diffusion through the soil, the emission would decrease to 21 Tg $year^{-1}$.

Landfills of municipal solid wastes are subject to bacterial degradation of organic matter and lead to emissions of methane. Bingemer and Crutzen (1987) first derived an estimate of methane from the landfill of municipal refuse by considering the amount of waste generated per capita, which (in kg day^{-1}) is 1.8 in North America and Australia, 0.8 in other OECD countries, 0.6 in the former Eastern Bloc countries, and 0.5 in the developing world. The fraction buried was estimated as 91%, 71%, 85%, and 80%, respectively. Consideration of the detailed composition and carbon content of paper (40% C by weight), textiles (40%), wood and straw (30%), garden and park waste (17%), and food waste (15%) resulted in a range of degradable organic carbon of 14–31% by weight in the OECD, 17% in former Eastern Bloc countries, and 8–17% in developing countries. Finally, urban population statistics (United Nations Demographic Year Book, 1984) listed the number of people living in cities, in millions, as 272 in North America and Australia, 471 in other OECD countries, 400 in the former Eastern Bloc countries, and 736 in the rest of the world. This leads to a total of 85 Tg of biodegradable carbon deposited annually in landfills worldwide. In a similar manner, by extrapolating from the situation in Germany to the whole world, Bingemer and Crutzen (1987) estimated that about 28 Tg year^{-1} biodegradable carbon results from industrial and commercial wastes. In the degradation process, fermentation raises the temperature to about 35°C, causing about 80% of organic matter to be dissimilated. It was assumed that 50% of it yields methane, and that all of the CH_4 produced enters the atmosphere during the same year that refuse is placed in the landfill. The amount thus calculated was 56 Tg year^{-1}, with a range of 30–75 Tg year^{-1}. This corresponds to an average yield of methane from refuse of about 100 g kg^{-1}. Landfill gas recovery measurements indicate, however, that the generation time for gas production from refuse is 10–30 years, and that an average recovery rate of 42 g CH_4 per kg of wet refuse gives a better fit to the observations (Peer et al., 1993; Thorneloe et al., 1993). A more recent estimate for the global rate of CH_4 emissions from landfills thus is 22 Tg year^{-1} (range 11–32).

The surface waters of the ocean are slightly supersaturated with dissolved methane compared with concentrations expected from the CH_4 mixing ratio in air (Swinnerton and Linnenboom, 1967; Swinnerton et al., 1969; Lamontagne et al., 1974). The concentration gradient gives rise to a flux of methane from the ocean to the atmosphere. The flux can be estimated from the stagnant film model discussed in Section 1.6. The data in Table 1.13 suggest a typical flux of 5.3×10^{-10} g m^{-2} s^{-1}, which must be extrapolated to the entire sea surface (3.61×10^{14} m^2) and the whole year (3.15×10^7 s) to give 6 Tg year^{-1}. Ehhalt (1974) derived a flux from this source of 4–7 Tg year^{-1}, whereas Seiler and Schmidt (1974) obtained 16 Tg year^{-1}. The difference arises from uncertainties regarding the film thickness of the laminar layer at the sea surface, which varies somewhat with agitation by wind force and therefore depends on the wind speed. In view of the fact that methanogens are strict anaerobes, it is surprising to find an excess of methane in the aerobic surface waters of the ocean. Pockets of anaerobic regions must exist in or around organic particles undergoing bacterial degradation so that methane can be produced in this environment. The marine source of methane is minor compared with other sources.

The amount of methane released in the retrieval and use of fossil fuels has still not been fully quantified. Table 4.5 shows several estimates. The amount of methane emitted from coal mining operations, including both hard coal and lignite, was first estimated by Hitchcock and Wechsler (1972). Recently, two country-specific reassessments by Boyer et al. (1990) and Kirchgessner et al. (1993), based on coal production, coal properties, and mine emission rates, gave almost identical results, 47 and 46 Tg year^{-1}, with a range of 35–65 Tg year^{-1}. Methane emissions associated with natural gas production and consumption include venting and flaring at oil and gas well sites, and losses during processing, transmission, and distribution. Pipe leakage rates are usually calculated from losses that the gas industry refers to as unaccounted for, that is, the difference between the volume of gas reported as purchased versus sold, less any company use or interchange. This statistical figure includes numerous diverse components such as meter inaccuracies, variations in temperature and pressure, and billing cycle differences, in addition to actual gas leakage or losses, so that it represents an overestimate as, Beck et al. (1993) have pointed out. The rates fall in the range of 1–5% and vary considerably by country. An average of 2–3% is usually taken. The

TABLE 4.5   Estimates of Global Emission of Methane from Fossil Fuel Production and Use and from Volcanoes (Tg year^{-1})

Coal mining (including lignite)	20,[a] 7.9–27.7,[b] 47.1,[c] 45.6,[d]
Automobile exhaust	0.5,[b] 35,[e] 25,[f]
Natural gas losses	7–21,[b] 14,[e] 40,[g] 10–35,[h]
Oil wells	10,[g] 14,[h]
Volcanoes	0.2[e]

[a] Koyama (1963).
[b] Hitchcock and Wechsler (1972).
[c] Boyer et al. (1990).
[d] Kirchgessner et al. (1993).
[e] Robinson and Robbins (1968b).
[f] Junge and Warneck (1979).
[g] Sheppard et al. (1982).
[h] Cicerone and Oremland (1988).

global consumption of natural gas is on the order of 1 Pg year^{-1}. This leads to a potential global emission rate of about 25 Tg year^{-1}, with a range of 15–50 Tg year^{-1}. Emissions from venting and flaring at oil well heads may amount to about 15 Tg year^{-1}, but this value is highly uncertain. Automobile exhaust emissions contain appreciable amounts of methane; this source is usually ignored, however.

Fossil sources of methane are expected to be free of ^{14}C, because its half-life due to beta decay is about 5700 years. Carbon 14 is produced naturally by the interaction of cosmic ray-produced neutrons with nitrogen in the atmosphere (Lal and Peters, 1967), and it enters the biosphere by assimilation of ^{14}CO$_2$. The ^{14}C content of atmospheric methane therefore should provide an indirect estimate for the contribution of methane from fossil sources. Ehhalt (1974) and Ehhalt and Schmidt (1978) have discussed data for the ^{14}C content of atmospheric methane collected from cylinders of liquefied air. The data suggested that prior to 1960 the atmospheric ^{14}C methane content was 83 ± 13% of a recent wood standard. The possibility of contamination was recognized because air liquefaction plants are frequently located in urban areas. These samples were taken before the atmosphere became contaminated with ^{14}C from large-scale nuclear weapons tests. While some of the excess has been transferred to the ocean since the test moratorium took effect in 1963, another part has entered the biosphere and reappears as methane. The ^{14}C content of atmospheric methane is currently rising due to the release of ^{14}CH$_4$ from nuclear power plants. Quay et al. (1991) have estimated that in 1988 this source contributed about 26% to ^{14}CH$_4$ from all sources. These authors used the observed rise rate (1.4% per year), the average ^{14}C content of atmospheric methane (122%), and the mass balance equation to calculate the ^{14}C content of the total CH$_4$ source, including nuclear, as 136% (all values relative to a standard). If nuclear reactors contribute 26%, the nonnuclear CH$_4$ source has a ^{14}C content of about 100%. Numerous measurements of bacterial sources (wetlands, rice, rumen gas) have given values ranging from 111% to 124%, reasonably close to the atmospheric ^{14}CO$_2$ content of about 120% (relative to the same standard). The difference between these nonfossil sources and the overall source (120% versus 100%) suggests that 84% of the total is modern, and 16% is due to the contribution of fossil sources The uncertainty is 12%. Wahlen et al. (1989) have used similar arguments to estimate that 21% of atmospheric CH$_4$ was derived from fossil carbon in 1988. The percentage corresponds to a CH$_4$ source strength of 80–105 Tg year^{-1} (see further below).

Another important anthropogenic source of methane is the incomplete combustion of living and dead plant organic matter (biomass) in forest fires, the burning of agricultural wastes, and the clearance of forest and brush lands by fire for agricultural purposes. The burning of biomass occurs in three stages: the first involves pyrolysis of the fuel and the emission of volatile organic compounds; in the second stage the organic volatiles are burned in a turbulent, high-temperature flame; and the last stage consists of smoldering combustion, which sets in when the rate of volatile formation decreases so much that it can no longer sustain an open flame. The combustion zone then retreats to the surface of the residual, highly charred fuel, to which oxygen is carried by molecular diffusion rather than turbulent mixing. Figure 4.7 shows the general behavior of several combustion products with time in open scale laboratory fires fueled by savanna grasses, deciduous wood, pine needles, etc. The formation of CO_2, NO_x, and N_2O is

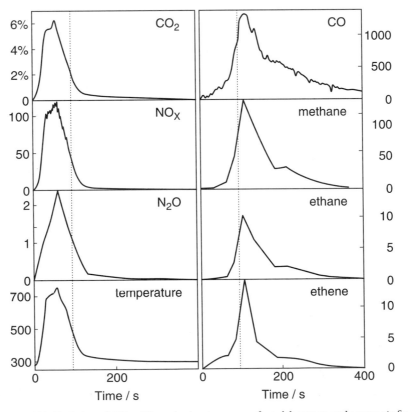

FIGURE 4.7   Emission of CO_2, CO, and other trace gases from laboratory-scale open-air fires. The plots show mixing ratios above background in volume percentage for CO_2 and in μmol mol^{-1} for the other gases, stack-gas temperature in K. The vertical dotted lines indicate the transition from flaming to smoldering conditions. (Adapted from Lobert et al. (1991).)

associated with the flaming stage, whereas CO, CH_4, and other hydrocarbons are formed primarily during smoldering combustion. Average emission factors derived from laboratory studies (Lobert et al., 1991) are, in percentage of fuel carbon, 82.58 for CO_2, 5.73 for CO, 0.42 for CH_4, 1.18 for nonmethane hydrocarbons, and 5.00 for ash, summing to a total carbon of about 95%. Individual values are subject to larger variations, for example, 2.83–11.19 for CO and 0.14–0.94 for CH_4. Emission rates derived from field observations are normalized to CO_2 for convenience. Summaries of emission ratios by Levine et al. (1993) and by Andreae and Warneck (1994) also indicate larger variations. The range of CH_4/CO_2 mole ratios, for example, is 0.26–1.5%, similar to those in the laboratory (0.11–1.9%). Delmas et al. (1991) have compared CH_4/CO_2 mole ratios in plumes of several fires in the African savanna and reported the following: bush fires 0.28 ± 0.4, forest fire 1.23 ± 0.6, firewood 1.79 ± 0.81, burning of charcoal 0.14, and emissions from charcoal kilns 12.06 ± 2.86.

Global rates of biomass burning were first estimated by Seiler and Crutzen (1980) on the basis of land use and population statistics. More recent estimates are due to Hao et al. (1990) and Andreae (1991). Table 4.6 summarizes the amounts of biomass exposed to fire, the amounts actually burned, and the mass of carbon released. In shifting agriculture fallow forests are cleared and the land is cultivated for about 2 years. Thereafter the land is abandoned and left to grow a secondary forest for 10–20 years. This type of operation affects an area of about 3×10^5 km^2 annually, with biomass densities of 1–12 kg m^{-2}, 70–80% of it above ground. An estimated 40% is burned and released promptly, mainly as CO_2; the rest is left to rot. Permanent removal of forests now occurs at an increasing rate, most of it in South America, which contributes as much as Africa and Asia taken together. The area cleared annually is about 1×10^5 km^2, containing a total biomass of 2.4–8 Pg. About 60% of the biomass is below ground, including soil organic matter, so that 1.0–3.2 Pg dry biomass is exposed to fire annually. Forty to fifty percent of the material is consumed by fire; the rest undergoes decomposition by microorganisms. Tropical savannas and brush lands are estimated to cover an area of about 13×10^6 km^2 worldwide. Savannas are burned every 1–4 years during the dry season, with the highest frequency in the humid Tropics. The amount of above-ground biomass is 200–670 g m^{-2}, depending on location. Hao et al. (1990) estimated total biomass exposed to fire in savannas as 4.4 Pg year^{-1}, which he considered uncertain by a factor of 2. Fuel wood and agricultural waste are difficult to separate because much of the waste is consumed domestically. The estimated per capita fuel need in developing countries ranges from 350 to 2500 kg year^{-1}. The number of people living in the Tropics is about 3800 million. Charcoal for domestic use finds increasing application but still represents a relatively minor fraction of

TABLE 4.6 Global Estimates of Amounts of Biomass Burned Annually and the Resulting Release of CO_2, CO, and CH_4 into the Atmosphere[a]

Type of activity	Biomass exposed to fire (Pg dry matter year^{-1})			Carbon released (%)	Carbon released[b] (Tg year^{-1})	CH_4/CO_2[c] (vol. %)	CH_4 released (Tg year^{-1})	CO released[d] (Tg year^{-1})
	Tropical	Extratropical	Total					
Shifting agriculture[e]	0.8–2.3	—	0.8–2.3	45	170–470	1.2	5.2 (2.7–7.5)	54.5 (29–80)
Permanent deforestation[f]	1.0–3.2	—	1.0–3.2	45	200–650	1.2	6.7 (3.2–10)	72.4 (34–110)
Forest fires (extratropical)[g]	—	0.6–1.8	0.6–1.8	40	110–320	1.2	4.6 (2.3–6.9)	36.6 (19–54)
Savanna burning[e]	2.2–6.6	—	2.2–6.6	80	790–2410	0.4	8.6 (4.2–13)	272.5 (135–410)
Biomass used as fuel[g]	0.9–3.5	0.1–0.4	1.0–3.9	90	400–1590	1.4	14.8 (7.4–30)	169.5 (68–271)
Agricultural waste in the fields[g]	0.2–0.8	0.2–0.8	0.4–1.7	90	190–680	0.8	3.6 (1.8–7.2)	74.1 (32–116)
Totals	6.9–14.1	0.9–3.1	6.0–17.2		1860–6120		43 (22–75)	679.6 (347–1042)

[a] Biomass dry matter contains about 45% carbon by mass.
[b] Mass of carbon released as CO_2 and CO.
[c] Emission factors taken from Andreae and Warneck (1994).
[d] The emission factor applied throughout is 7.3% by volume (Lobert *et al.*, 1991).
[e] Biomass exposed to fire according to estimates of Hao *et al.* (1990).
[f] Estimated by Crutzen and Andreae (1990).
[g] Estimated by Andreae (1991).

the total, so that it can be ignored, even though its production leads to high methane emissions. Agricultural wastes (that is, crop residues) have been estimated to amount to 3.4 Pg dry matter per annum, with about 1.5 Pg produced in the developed and 1.9 Pg in the developing countries. However, only a fraction of it is actually burned. The total rate of methane released by biomass burning averages to 45 Tg year^{-1}, with a range of 22–75. It is interesting to note that because of the small emission factor, savannas contribute only about 20% to the total, despite the large area involved, whereas with regard to CO_2 emissions, savanna fires are most productive.

Table 4.7 summarizes methane emission rates estimated by several authors. The earliest estimate by Ehhalt (1974) is of historical interest. His major sources were domestic ruminants, rice fields, and wetlands. While the strengths of these sources have been reduced, subsequent studies have identified many other sources with smaller contributions. This is exemplified by the distribution of Bolle et al. (1986). Cicerone and Oremland (1988) have additionally used the measured isotope ratios to provide constraints for the total budget. It is interesting to note that the budget has remained constant at about 500 Tg year^{-1}. Anthropogenic activities account for about two thirds of the emissions. Landfills, coal mining and natural gas leakage would contribute to methane depleted in ^{14}C. These sources amount to 66–112 Tg year^{-1}, which covers the same range as that estimated from the ^{11}C measurements.

The major sink for methane in the troposphere is reaction with OH radicals, as discussed in Section 4.2. Losses to the stratosphere provide an

TABLE 4.7   Estimates of CH_4 Emission Rates (Tg year^{-1}) from Individual Sources

Type of source	Ehhalt (1974)	Bolle et al. (1986)	Cicerone and Oremland (1988)	Khalil and Shearer (1993)
Ruminants etc.	101–220	70–100	80 (65–100)	55–90
Termites	—	2–5	40 (10–100)	15–35
Rice paddy fields	280	70–170	110 (60–170)	55–90
Natural wetlands	130–260	25–70	115 (100–200) ⎫	110
Tundra	—	2–15	5 ⎬	
Ocean	6–45	1–7	15 (2–20)	4
Domestic sewage	—	—	–	27–80
Landfills	—	10	40 (30–70)	11–32
Animal waste	—	—	—	20–30
Coal mining	15–50	35	35 (25–45)	25–50
Natural gas leakage	—	30–40	45 (25–45)	30
Biomass burning	—	55–100	55 (50–100)	50
Total	533–854	300–552	540 (370–855)	402–601

additional sink. According to Equation (4.7), the global sink for destruction by OH is

$$S_{OH} = \gamma \bar{k} \bar{n}(OH) G(CH_4)$$

where $\gamma = 3.15 \times 10^7$ s year^{-1}, $\bar{k}$ is the rate coefficient for the reaction, suitably averaged over tropospheric temperatures, and $\bar{n}(OH) \approx 8 \times 10^5$ molecules cm^{-3} is the average OH number concentration in the troposphere. With an average CH_4 mixing ratio $\bar{x} = 1.7 \times 10^{-6}$, the total content of methane in the troposphere is calculated as

$$G(CH_4) = \bar{x}(M_{CH_4}/M_{air}) G_T \approx 4000 \text{ Tg}$$

The temperature dependence of the rate coefficient is $k_{78} = 2.65 \times 10^{-12}$ exp$(-1800/T)$. The average tropospheric temperature for OH reactions is about 277 K, so that $\bar{k} = 4 \times 10^{-15}$ cm^3 molecule^{-1} s^{-1}. With these values one obtains $S_{OH} = 403$ Tg year^{-1}.

The flux of methane into the stratosphere is determined by eddy diffusion in the lower stratosphere. Here, the vertical profile of CH_4 mixing ratios may be approximated by an exponential decrease with altitude:

$$x(z) = x_{tr} \exp(-z/h)$$

where x_{tr} denotes the mixing ratio at the tropopause and h is an empirical scale height. At middle and high latitudes $h \approx 25$ km, whereas at the tropical tropopause $h \approx 50$ km. The local upward flux is given by

$$F_{CH_4} = -K_z n_{tr}(dm/dz)|_{tropopause}$$
$$= -K_z \rho_{tr}(M_{CH_4}/M_{air}) x_{tr}/h$$

Here, n_{tr} is the number density of air molecules at the tropopause and ρ_{tr} is the corresponding air density. These data may be taken from Table 1.5. Values of the eddy diffusion coefficients in the lower stratosphere are 1 m^2 s^{-1} in the equatorial updraft region and 0.3 m^2 s^{-1} at higher latitudes. The second of the above equations gives the flux in kg m^{-2} s^{-1}. The global flux is obtained after converting from seconds to years and integrating over the Earth's surface area. This yields loss rates of 29 and 31 Tg year^{-1} for the tropical and extratropical latitudes, respectively, or a total of 60 Tg year^{-1}. Modeling studies have indicated slightly smaller stratospheric loss rates. Crutzen and Schmailzl (1983), for example, found 42 Tg year^{-1}; Prather and Spivakovsky (1990) obtained 29 Tg year^{-1}.

Bacterial degradation of methane in aerobic soils may represent an additional sink. Bartlett and Harris (1993) have listed a variety of measurements of CH_4 uptake by soils. The uptake efficiency is primarily controlled by the rate of diffusion in soils, that is by soil porosity, and upland forest ecosystems appear to provide the most conducive conditions for the uptake of methane. The global uptake rate is rather uncertain, however. Keller et al. (1983) suggested less than 5 Tg year^{-1}, whereas Seiler et al. (1984b) and more recently Khalil et al. (1993a) derived 30 Tg year^{-1}.

The sum of loss rates (reaction with OH in troposphere and stratosphere and uptake by soils) is 475–495 Tg year^{-1}, which agrees quite well with the global emission estimates in Table 4.7, that is, the budget is approximately balanced. A rise rate of 1% per year corresponds to an excess in emissions of about 40 Tg year^{-1}. This is smaller than the range of uncertainty in the global source strength. If sources and sinks are assumed to be approximately in balance, the tropospheric residence time for methane is

$$\tau_{CH_4} = G(CH_4)/S(CH_4) = 4000/485 = 8.3 \text{ years}$$

The time is long compared to interhemispheric transport and guarantees an essentially even distribution of methane in the troposphere.

4.4. FORMALDEHYDE

This compound is the first relatively stable product resulting from the oxidation of methane. In the marine atmosphere methane is expected to be the major precursor of formaldehyde (HCHO), whereas over the continents multiple other sources exist. Among them are direct emissions from industries, automotive exhaust, stationary combustion, biomass burning, and secondary formation of HCHO by the photooxidation of nonmethane hydrocarbons, both natural and human-made. Carlier et al. (1986) have reviewed formaldehyde measurements as well as primary and secondary sources. In cities such as Denver, Colorado, the dominant source is emission from motor vehicles (Anderson et al., 1996). In the continental background atmosphere, most of the formaldehyde is produced by the photooxidation of isoprene and terpenes emitted from plants. In this environment, diurnal variations with maxima at noontime are often observed. In the atmosphere, formaldehyde undergoes photodecomposition and reaction with OH to form CO, so that the two budgets are coupled. Uncertainties in the continental budget of formaldehyde thus carry over to the budget of CO. The following brief discussion deals primarily with formaldehyde in the marine atmosphere.

Wet-chemical and optical methods have been used to determine formaldehyde in air. The preferred analytical technique currently is reaction with 1,4 dinitro-phenyl-hydrazine, followed by liquid chromatography to separate the product from other compounds, and optical detection of the product. Table 4.8 shows mixing ratios observed at various locations. The highest values are associated with urban air, especially under conditions of photochemical smog. Grosjean (1991) has summarized measurements made in the greater Los Angeles area, where the mixing ratio may reach 70–100 nmol mol^{-1}.

TABLE 4.8   Molar Mixing Ratios (nmol mol^{-1}) of Formaldehyde in Urban, Rural Continental, and Marine Locations

Authors	Location	Mixing ratio
	Urban air	
Cleveland et al. (1977)	New York City	2 (av.), 10 (max.)
	Bayonne, New Jersey	3 (av.), 20 (max.)
Fushimi and Miyake (1980)	Tokyo, Japan	1.24
Tuazon et al. (1981)	Claremont, California	16–49 (daily average)
		23–71 (daily max.)
Kuwata et al. (1983)	Osaka, Japan	1.6–8.5 (34 high. max.)
Grosjean et al. (1983)	Los Angeles, California	2–40
	Claremont, California	3–48
Lipari et al. (1984)	Warren, Michigan	1.3–6.5
Kalabokas et al. (1988)	Paris, France	4–32
Haszpra et al. (1991)	Budapest, Hungary	7–176
Possanzini et al. (1996)	Rome, Italy	8–17
	Rural air	
Platt et al. (1979), Platt and Perner (1980)	Jülich, Germany	0.1–6.5
Fushimi and Miyake (1980)	Mt. Norikura, Japan	1–4
Lowe et al. (1980)	Eifel region, Germany	0.03–5
Neitzert and Seiler (1981)	Suburban Mainz, Germany	0.7–5.1
Puxbaum et al. (1988)	Suburban Vienna, Austria	6.4–13.4
Harris et al. (1989)	Sarnia, Ontario	0.5–5.7 (diurnal variation)
	Lewes, Delaware	0.2–0.7 (no diel variation)
Shepson et al. (1991)	Egbert, Ontario	0.7–4.3 (1.8 av.)
	Dorset, Ontario	0.6–4.4 (diurnal variation)
Trapp and Serves (1995)	Calabozo, Venezuela	0.3–1.5 (diurnal variation)
Slemr et al. (1996)	Mt. Schauinsland, Germany	0.4–2.3 (diurnal variation)
	Marine air	
Platt and Perner (1980)	Dagebüll, Germany	0.2–0.5
	Loop Head, Ireland	0.3
Fushimi and Miyake (1980)	Pacific Ocean	0.2–0.8
	Indian Ocean	0.8–11
Zafiriou et al. (1980)	Enewak Atoll, Pacific	0.3–0.6
Lowe et al. (1980)	Irish west coast	0.1–0.42
Neitzert and Seiler (1981)	Cape Point, South Africa	0.2–1.0
Lowe and Schmidt (1983)	Central Atlantic Ocean	0.12–0.33
Harris et al. (1992)	Central Atlantic Ocean	0.47 ± 0.2

Mixing ratios in rural areas are on the order of a few nmol mol^{-1}; still lower values are found in marine air masses.

Figure 4.8 shows mixing ratios of formaldehyde over the Atlantic Ocean as a function of latitude as measured by Lowe and Schmidt (1983). In the region 33° S–40° N the values scatter around a mean value of 0.22 nmol mol^{-1}. Further northward they decline. At mid-latitudes, the lifetime of formaldehyde due to photodecomposition by sunlight is about 5 h, and 15 h when averaged over a full day (Warneck et al., 1978). The rate coefficient for the reaction with OH is $k_{86} = 1.5 \times 10^{-11}$ cm^3 molecule^{-1} s^{-1}. Where the oxidation of methane is the sole source, the formaldehyde mixing ratio is expected to adjust to a steady state with methane:

$$x(\mathrm{HCHO}) = \frac{k_{78}\,\bar{n}(\mathrm{OH})\,x(\mathrm{CH_4})}{j_{\mathrm{HCHO}} + k_{86}\,\bar{n}(\mathrm{OH})}$$

$$= \frac{(6 \times 10^{-15})(8 \times 10^5)(1.7 \times 10^{-6})}{(1.8 \times 10^{-5}) + (1.5 \times 10^{-11})(8 \times 10^5)}$$

$$= 0.27 \text{ nmol mol}^{-1}.$$

Comparison with Figure 4.8 shows that the steady-state prediction is in good agreement with the observational data. This justifies the assumption that methane is the dominant source of formaldehyde in the marine atmosphere.

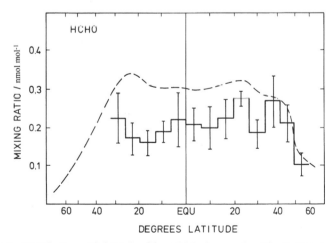

FIGURE 4.8   Distribution with latitude of formaldehyde over the Atlantic Ocean; vertical bars indicate two-sigma variances. (Adapted from Lowe and Schmidt (1983).) The dashed curve shows the results of a two-dimensional model calculation (Derwent, 1982).

Figure 4.8 includes results from a two-dimensional model calculation (Derwent, 1982). The values are somewhat higher than the measurements indicate, but they confirm the uniformity with latitude in the equatorial region. The decline of x(HCHO) at high latitudes, which is evident in both model and measurements, arises from the fact that the average OH concentration decreases faster toward the Poles than the HCHO photodissociation coefficient. The reason is the increasing optical density of the ozone layer.

The photochemical lifetime of formaldehyde is short enough to allow some diurnal variation of its mixing ratio. Lowe and Schmidt (1983) have found that diurnal variations do occur, but not as much as predicted. The model on which the prediction was based included losses of HCHO due to dry deposition on the ocean surface. This process was not considered in the above equation because it is less important than daytime photochemistry, but it would continue during the night, thereby causing the HCHO mixing ratio to decrease below the steady state value. In seawater formaldehyde is biologically consumed, so that the ocean is not a source (Thompson, 1980).

In the free troposphere the steady-state HCHO mixing ratio is expected to decline with increasing altitude. On the one hand, the OH production rate decreases with altitude; on the other hand, the rate coefficient for the reaction with methane decreases because of the decrease in temperature. In addition, the HCHO photodissociation coefficient increases with altitude because of the pressure dependence of the quantum yield (Fig. 2.20). Exploratory measurements by Lowe *et al.* (1980) over western Europe have confirmed the expected trend. The HCHO mixing ratio was found to decline from values near 1 nmol mol^{-1} at ground level to less than 0.1 nmol mol^{-1} at 5 km altitude. Still lower values are expected in the upper troposphere.

4.5. CARBON MONOXIDE

Like many other gases in the atmosphere, carbon monoxide (CO) was discovered as a terrestrial absorption feature in the infrared solar spectrum (Migeotte, 1949). A variety of sporadic measurements until 1968, reviewed by Pressman and Warneck (1970), established a CO mixing ratio on the order of 0.1 μmol mol^{-1}. Robinson and Robbins (1968a) and Seiler and Junge (1970) then developed a continuous registration technique for CO in air based on the reduction of hot mercury oxide to Hg and its detection by atomic absorption spectrometry. In addition, gas chromatographic detection techniques have been utilized.

In the cities, the CO mixing ratio is in the range of 1–10 μmol mol^{-1} because of nearby sources, mainly emissions from automobiles. In clean-air regions, the mixing ratio decreases to values of about 150 nmol mol^{-1} at mid-latitudes of the Northern Hemisphere and 50 nmol mol^{-1} in the Southern Hemisphere. The existence of a latitudinal gradient was initially reported by Robinson and Robbins (1970a) for the Pacific Ocean, and by Seiler and Junge (1970) for the Atlantic Ocean. The hemispheric imbalance was subsequently confirmed by measurements onboard ships and aircraft (Swinnerton and Lamontagne, 1974a; Seiler, 1974; Heidt *et al.*, 1980; Seiler and Fishman, 1981; Robinson *et al.*, 1984; Marenco *et al.*, 1989). Long-term measurements have been conducted at a number of stations in the remote atmosphere. Figure 4.9 shows the distribution of CO with latitude obtained from measurements at background stations. Figure 4.10 shows the meridional distribution of CO above the atmospheric boundary layer, derived from aircraft flights. Although only marine air masses were considered, the CO mixing ratio shows a maximum in the lower troposphere at mid-latitudes of the Northern Hemisphere, indicating a source of considerable magnitude in this region. By following the gradient one may trace the direction of transport from the surface toward the middle troposphere and then further on into the Southern Hemisphere. This is indicated by the arrow. The principal pathway of air exchange between the two hemispheres is located in the middle and upper troposphere. This agrees with conclusions of Newell *et al.* (1974) from air motions driven by the Hadley cells. One should also note, however, that reaction with OH is the principal sink for tropospheric CO, and this sink is strongest in the equatorial region. Thus, only a fraction of

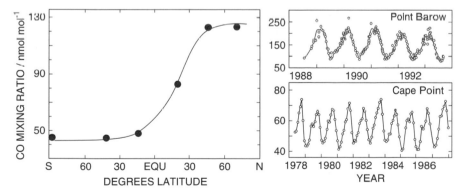

FIGURE 4.9 Left: Latitudinal distribution of carbon monoxide in 1988, as determined from measurements at six background stations (Point Barrow, Alaska; Cape Meares, Oregon; Cape Kumukahi and Mauna Loa Observatory, Hawaii; Cape Matatula, Samoa; Cape Grim, Tasmania; and South Pole, Antarctica). Right: Seasonal variation of the CO mixing ratio at Point Barrow, Alaska, and Cape Point, South Africa. Data are from Khalil and Rasmussen (1994), Brunke *et al.* (1990), and Novelli *et al.* (1994).

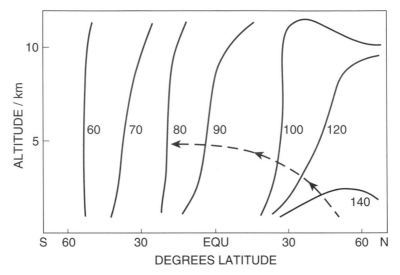

FIGURE 4.10   Meridional cross section of carbon monoxide in the troposphere. Contours give CO mixing ratios in nmol mol^{-1}. The broken arrow indicates the approximate direction of meridional transport. Greatly simplified from Marenco *et al.* (1989).

CO generated in the Northern Hemisphere enters the Southern Hemisphere. The fact that a gradient exists points to a CO residence time in the troposphere of less than about 1 year, according to the discussion in Section 4.1. The prediction is confirmed by the CO budget to be discussed later.

In the Southern Hemisphere, south of 50° latitude, the CO mixing ratio assumes a fairly constant value of about 50 nmol mol^{-1}. McConnel *et al.* (1971) first pointed out that a photostationary state will be set up for CO, if the oxidation of CH_4 is the sole source of CO and its reaction with OH radicals is the sole sink. In this case one has

$$dx(CO)/dt = k_{78}\,n(OH)\,x(CH_4) - k_{132}\,n(OH)\,x(CO) = 0$$

$$x(CO) = (k_{78}/k_{132})\,x(CH_4)$$

$$= \left[(6.3 \times 10^{-15})/(2.4 \times 10^{-13})\right]1.7 \times 10^{-6}$$

$$= 45 \text{ nmol mol}^{-1}$$

Note that the OH concentration cancels. The comparison of calculated with observed mixing ratios shows that methane is indeed the major source of CO under these conditions.

The CO mixing ratio exhibits an annual cycle, which appears to have been first noted by Seiler *et al.* (1984c) from measurements at Cape Point, South Africa (34° S), over a 3-year period. The seasonal variation is more pronounced in the Northern Hemisphere than in the Southern Hemisphere. Figure 4.9 includes data obtained at Point Barrow, Alaska, and at Cape Point, South Africa, to demonstrate the observations. Figure 4.11 shows for the data from Cape Point that the oscillations of CO and CH_4 occur in phase. The annual cycle of methane is largely caused by the seasonal variation of the average OH concentration. If the oxidation of methane were the only source of CO in the Southern Hemisphere, the relative amplitude of CO should be the same as that of CH_4. This is not the case, because $\Delta[CO]/[CO] \approx \pm 0.24$ compared to $\Delta[CH_4]/[CH_4] \approx \pm 0.01$, where $[CO] \approx 55$ and $[CH_4] \approx 1600$ nmol mol^{-1} are the annual averages for 1986. Thus, the oxidation of

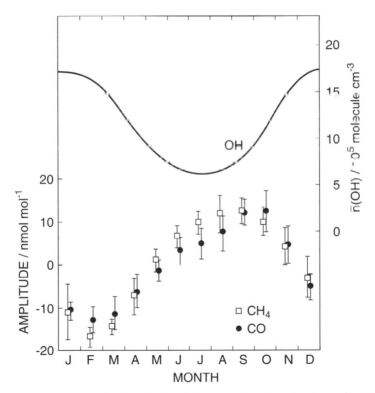

FIGURE 4.11 Average residual amplitude of the seasonal cycles of carbon monoxide and methane for the years 1983–1987 at Cape Point, South Africa (Brunke *et al.*, 1990). Added for comparison is the seasonal variation of OH number density according to calculations of Logan *et al.* (1981) and Crutzen (1982).

methane cannot be the only source of CO in the Southern Hemisphere, and the annual oscillation of CO must be caused by another mechanism. Seiler *et al.* (1984c) suggested that the seasonal displacement of the intertropical convergence zone (ITCZ) would modulate the influx of CO from the Northern to the Southern Hemisphere and thereby may explain the seasonal variation of CO at Cape Point. Biomass burning in South America and Africa represents another seasonal activity maximizing in August–September. Allen *et al.* (1996) have used a global transport model to show that much of the interannual variability of CO is indeed related to transport within and between the two hemispheres. They were able to reproduce seasonal variations at many stations where long-term measurements have been made.

The rate of CO production from the oxidation of methane in the troposphere is $Q(CO) = (M_{CO}/M_{CH_4})S_{OH}(CH_4) = (28/16) \times 403 = 700$ Tg year^{-1}. About 50%, that is 350 Tg year^{-1}, is produced in the Southern Hemisphere. From the distribution of CO with latitude as shown in Figure 4.9 and the air masses given in Table 1.6 for tropical and extratropical latitudes, one estimates average mixing ratios of 128 nmol mol^{-1} in the Northern and 52 nmol mol^{-1} in the Southern Hemisphere. The average is about 90 nmol mol^{-1}. The mass contents of CO in the Northern and Southern Hemispheres would be 265 and 105 Tg, respectively. The difference causes an influx of CO from the Northern into the Southern Hemisphere of $(1/\tau_{exch})(265 - 105) = 160$ Tg year^{-1}, on average. The seasonal variation in flux needed to explain the annual oscillation of the CO mixing ratio in the Southern Hemisphere is approximately 50 Tg year^{-1}, so that this mechanism can contribute significantly to the phenomenon.

Anthropogenic emissions are mainly responsible for the excess of CO in the Northern Hemisphere. A large share of it is associated with the combustion of fossil fuels and industrial activities. The usual procedure for estimating the anthropogenic contribution to the total budget is to combine statistical data for the consumption or production of a commodity with the observed emission factors, which are defined as the amount of CO produced per unit of fuel consumed or material produced. Typical emission factors are summarized in Table 4.9, and fuel use patterns are shown in Table 4.10. The data are combined in Table 4.11 to obtain the global rate of CO production due to energy- and industry-related activities. Automobiles account for 80% of CO from the combustion of fossil fuels, and for 50% of all emissions from these sources. The other large source, which contributes about 30% to the total, is due to industrial processes, mainly the production of steel and the catalytic cracking of crude oil. The individual emission estimates in Table 4.11 combine to a total of 445 Tg year^{-1}. Seiler (1974) had derived a higher estimate of 640 Tg year^{-1}. He considered CO emissions from home heating to be underestimated, because Baum (1972) had shown for West Germany

TABLE 4.9 Emission Factors for CO from Combustion and Industrial Processes[a]

Combustion of hard coal (g/kg)		Natural gas (g/m^3)	
Utility boilers	0.5–1.0	Utility boilers	0.04–0.27
Coke ovens	0.6	Industrial boilers	0.08–0.27
Industrial boilers	0.5–1.0	Residential units	0.32–2.73
Residential units	5–290	Industrial processes	
		(g/kg of product)	
Combustion of lignite (g/kg)		Pig iron production	80
Utility boilers	1.0	Steel production	70
Coke ovens	1.0	Cracking of crude oil	39
Industrial boilers	1.0	Carbon black production	2000
Residential units	45–137	Ammonia production	100
Combustion of oil (g/dm^3)		Pig iron foundry	72
Utility boilers	0.1–63	Biomass burning (volume	
		% CO/CO$_2$)	
Industrial boilers	0.4–63	Forest wild fires	11–25
Industrial gasoline engines	472	Grass, stubble, straw	3–15
Industrial diesel engines	12	Agricultural wastes	3–16
Residential boilers	0.1–63	Burning of wood for fuel	6–18
Transportation (g/dm^3)			
Gasoline engines	180–360		
Diesel engines	15–30		

[a] Compiled from Logan et al. (1981) and Crutzen et al. (1979).

that CO emitted from residential heating amounted to 30% of the annual emission from motor vehicles. The emission factors in Table 4.9 for this source indeed admit a large range of uncertainty. In this regard it is necessary to keep in mind that the data in Table 4.11 are based on 1976 statistics, and changes have occurred since then. One important development is the catalytic converter for automobiles, which has led to reduced exhaust emissions from vehicles, at least in the United States, where during the period 1970–1984 emissions of CO from transportation have decreased by 30% (Müller, 1992). The reduction is partly off-set by the wordwide increase in the population of automobiles. Müller (1992) estimated for the year 1986 that the worldwide anthropogenic CO emission amounted to 383 Tg year^{-1}, with 52% derived from transportation, which is similar to the total obtained by Logan et al. (1981) for 1976.

Khalil and Rasmussen (1994) noted that during the first half of the 1980s the atmospheric level of CO increased at a rate of about 1.2% year^{-1}, whereas from 1988 to 1992 the global CO concentration declined at a rate of about −2.6% year^{-1}. Novelli et al. (1994) also found the CO mixing ratio to decrease markedly during the period 1990–1993, by −6.1% in the Northern Hemisphere and −7% in the Southern. These authors consider a reduction

TABLE 4.10   Fossil Fuel Consumption Patterns, 1976, in Tg Carbon[a]

	United States	Europe	Rest of the world
Energy conversion			
Hard coal	275.3	152.6	360.4
Lignite	—	390.5	43.5
Oil	93.8	125.8	141.3
Natural and manufactured gas	108.2	85.3	30.5
Industrial consumption			
Hard coal	115.7	156.0	237.1
Lignite	—	65.1	7.2
Oil	53.6	188.7	135.8
Natural and manufactured gas	127.9	100.4	35.9
Residential uses			
Hard coal	8.0	30.5	34.8
Lignite	—	9.3	1.0
Oil	134.0	207.2	214.8
Natural and manufactured gas	91.8	65.3	24.6
Transportation (oil-derived)			
Gasoline and diesel oil	388.6	218.3	108.0
Total consumption rates	1397.0	1795.0	1375.0

[a] Data compiled by Logan et al. (1981) from United Nations statistics (1974–1978), taking into account the percentages of carbon in different grades of coal and lignite. The carbon content of solid fuel is 70%, that of liquid fuel is 84%, and that of natural gas is 0.536 g/l, according to Keeling (1973a).

of anthropogenic emissions insufficient to account for the observed changes in the CO mixing ratio, and they suggest that an increase in the average OH concentration should be responsible.

Biomass burning is another human activity causing the release of CO to the atmosphere. The annual rate of biomass destruction was discussed in the previous section (see Table 4.6). Open fires emit about 80% of CO during the smoldering stage (see Fig. 4.6). The CO emission factor depends greatly on the extent of moisture, which enhances smoldering. Numerous field measurements summarized by Cofer et al. (1991) and Radke et al. (1991) have confirmed the range of emission factors shown in Table 4.9 established in earlier studies. Similar results have been obtained in laboratory experiments (Lobert et al., 1991). The CO/CO_2 volume ratio usually is in the range 3–16%, with an average of about 7.3%, fairly independent of the type of material being burned. In Table 4.6 this average factor was applied to derive the global amount of CO from biomass burning, 680 Tg year^{-1} (range 350–1040). Most of the emissions are associated with agricultural practices

TABLE 4.11   Anthropogenic CO Emissions from Fossil Fuel Combustion and Industrial
Activities in Various Parts of the World (in Tg year $^{-1}$ of CO)[a]

Source type	North America	Europe	Rest of the world	Total
Combustion				
Coal	1	24	23	48
Lignite	—	3	—	3
Gas	0.2	0.3	0.1	0.6
Oil	2.2	4.6	3.7	11
Total combustion	3.4	32	27	62
Transportation	94	71	66	233
Industries[b]				
Pig iron production	7.8	16.2	18.0	42
Steel production	9.5	19.8	17.7	47
Cracking of crude oil	4.5	5.0	4.1	13.6
Miscellaneous	10.5	9.6	7.3	27.4
Total from industries	32.3	50.6	47.1	130
Waste disposal	3	6	11	20
Sum total	137	155	152	445

[a] Compiled by Logan et al. (1981) from United Nations statistics (1974–1978).
[b] Distribution from the Yearbook of Industrial Statistics, United Nations (1978).

of the Tropics, namely the burning of savanna grass at the end of the dry period, and the use of agricultural wastes as fuel. Biomass burning represents a source of CO similar in magnitude to fossil fuel combustion, but it has a greater range of uncertainty because of the indirect method of assessment. In addition, the global distribution is different.

Yet another anthropogenic CO source results from the oxidation of hydrocarbons that are released into the atmosphere, mostly from automobiles and by the use of organic solvents by industry, dry cleaning, etc. Logan et al. (1981) combined emission data for the United States (EPA, 1976) with fuel use patterns to estimate a global source strength of 90 Tg year^{-1}. It was assumed that 50% of carbon in volatile hydrocarbons is converted to CO. Even if the conversion efficiency were higher, it is clear that hydrocarbons from these sources contribute fairly little to the total anthropogenic CO source strength. Hydrocarbons, however, are released also from plants and soils, and this contribution is more substantial.

The discussion in Chapter 6 will show that isoprene and terpenes are the main hydrocarbons emitted from vegetation. The former is the dominant compound arising from deciduous plants, notably oak, and it is emitted only during daylight. The latter compounds derive primarily from conifers, and emission rates vary with temperature, so that they depend on season and latitude. Numerous field measurements have established leaf emission rates and their dependence on various parameters. Plant isoprene production is related to photosynthesis by green plants and may be scaled to net primary productivity. This quantity is the amount of carbon from atmospheric CO_2 fixed as plant matter and not immediately respired. Globally extrapolated emission rates of isoprene range from 175 to 500 Tg carbon year^{-1} (Zimmerman, 1978; Müller, 1992; Guenther et al., 1995), with roughly 60% estimated to occur in the Tropics. Estimates for the global emission rate of terpenes have a similar range, 125–480 Tg year^{-1} (as carbon). To illustrate the procedure for estimating the CO source from hydrocarbon oxidation, we use the recent estimates of Guenther et al. (1995), 500 Tg year^{-1} for isoprene and 127 Tg year^{-1} for terpenes.

In the atmosphere, hydrocarbons are subject to attack by OH radicals and ozone, which initiate an oxidation mechanism converting the materials first to oxygenated compounds, and then partly to CO. Hydrocarbon oxidation mechanisms are discussed in Section 6.3. Here it should be noted that not every carbon atom is converted to CO. Accordingly a yield estimate is required, if one wishes to use the above data in estimating the production of CO from the oxidation of hydrocarbons. For isoprene the oxidation mechanism has been staked out, and one expects a yield of 80% CO and 20% CO_2 (mole fraction). A laboratory study by Hanst et al. (1980) essentially confirmed these yields. A reinvestigation by Myoshi et al. (1994) gave 74% CO under NO_x-rich conditions, but only 35% in NO_x-free air. The first condition applies to air over the continents. Terpenes pose much larger uncertainties, because a substantial portion of the material is converted to low-volatility products, which condense to form aerosol particles. (see Section 7.4.3). The experiments of Hanst et al. (1980) on α-pinene indicated a total yield of CO + CO_2 of 30% and a CO/CO_2 ratio of 0.7. Thus, about 20% of carbon in α-pinene was converted to CO. If the conversion efficiencies for other terpenes were similar, one would obtain the following CO production rates from the oxidation of natural hydrocarbons:

From isoprene: $0.75(M_{CO}/M_C)500 = 875$

From terpenes: $0.2(M_{CO}/M_C)127 = 59$

The combined rate of CO production, 935 Tg year^{-1}, is similar in magnitude to that arising from the oxidation of methane, but it is confined to the continents. Guenther et al. (1995) also discussed possible emissions of other reactive organic compounds from vegetation, which would further increase the production of CO from this source. The world ocean is a source of light (C_2-C_8) and heavier (C_9-C_{30}) hydrocarbons, but the emission rate is not competitive on a global scale with that from continental vegetation. Rudolph and Ehhalt (1981) and Eichmann et al. (1980) gave emission estimates for marine hydrocarbons on the order of 30–50 Tg year^{-1}, whereas Guenther et al. (1995) obtained 5 Tg year^{-1}. Nonmethane hydrocarbons are also emitted from open fires, but the amounts are only a few percent of the direct emission of CO from this source (Darley et al., 1966; Gerstle and Kemnitz, 1967; Andreae, 1991). Accordingly, hydrocarbons from biomass burning can be neglected to a first approximation.

In addition to the secondary formation of CO from the oxidation of hydrocarbons, carbon monoxide is emitted directly from the oceans and from land plants. Swinnerton et al. (1970) discovered that the surface waters of the oceans are supersaturated with CO compared with CO in the atmosphere. Subsequent measurements of Swinnerton and Lamontagne (1974a), Seiler (1974), Seiler and Schmidt (1974), and Conrad et al. (1982) confirmed the excess CO concentration in the surface waters of the Pacific and the Atlantic Oceans, which are fairly independent of latitude. Strong diurnal variations of aqueous CO have been observed with maxima in the early afternoon. This indicates a photoactive source mechanism. Swinnerton and Lamontagne (1974a) assumed that CO production is due to algae, but Conrad et al (Conrad et al., 1982; Conrad and Seiler, 1980a) applied filtration to remove algae and bacteria from the water samples. The production of CO continued in sunlit vessels but not in the dark. Thus, the source of CO must be organic compounds present in seawater. The rapid diurnal changes, however, can be understood only if CO is consumed as well as produced. Losses by molecular diffusion are slow and may be ignored. Conrad et al. (Conrad et al., 1982; Conrad and Seiler, 1980a) conducted laboratory and field experiments to show that the consumption is caused by bacteria. The combined action of photoproduction and bacterial consumption leads to a steady-state CO concentration that depends on light intensity and dissolved organic matter. The observations suggest seawater to be supersaturated with CO by a factor of 30 ± 20, on average. The associated flux of CO into the atmosphere may be derived by the method outlined in Section 4.3 for methane. In this manner, Linnenboom et al. (1973), Seiler (1974), Seiler and Schmidt (1974), and Conrad et al. (1982) have obtained global CO flux values of 220, 100, 70, and 75 Tg year^{-1}. The possible range of values given by the last authors is 10–180 Tg year^{-1}.

Green plants may both absorb carbon monoxide and release it. Wilks (1959) found alfalfa, sage plants, and cedar, among other species, to produce CO, whereas Krall and Tolbert (1957), Bidwell and Fraser (1972), and Peiser et al. (1982) reported an uptake of CO by barley, bean, and lettuce leaves. Some of the results were obtained by the ^{14}C method, which is very sensitive. Seiler et al. (1978a) and Bauer et al. (1979) have studied the behavior of CO in enclosures surrounding either parts of or whole plants. The species studied were *Fagus silvatica, Pinus silvestris, Vicia faba,* and *Platanus acerifolia.* A net production of CO was observed in all cases. The production rate was proportional to leaf area, and it rose almost linearly with radiation intensity. There were no signs of saturation, even at 800 W m^{-2}, an intensity at which net photosynthesis was already inhibited. Except for the dependence on light intensity, the CO emission was essentially independent of season, even in the autumn, when the leaves showed discoloration. The net CO production rate for a 50 W m^{-2} radiation flux was 3×10^{-9} g m^{-2} s^{-1}. Bauer *et al.* (1979) used their results to estimate a global CO emission rate from plants of 75 ± 25 Tg year^{-1}. The mechanism for the production of CO in plant leaves has not been clearly identified. CO may be produced as a metabolite during the degradation of porphyrins such as chlorophyll (Troxler and Dokos, 1973); it may be formed by the degradation of glycolate, a product of photorespiration (Fischer and Lüttge, 1978); or it may arise from the photooxidation of plant cellular material. The last pathway occurs in phototrophic microorganisms (Bauer *et al.,* 1980).

Table 4.12 summarizes global CO source estimates derived by several research groups. The total source strength is about 3000 Tg year^{-1} of CO. Anthropogenic activities consisting of fossil fuel combustion, industries, and biomass burning contribute about 50% to the total. The oxidation of hydrocarbons, both methane and others, contributes about 45%. Emissions from the ocean and from plants are relatively minor. Table 4.12 includes estimates for global CO removal rates. The dominant sink, with a consumption rate of about 2000 Tg year^{-1}, is the oxidation of CO by OH radicals. This process does not quite suffice to balance the sources, however. Two other sinks are losses to the stratosphere and consumption of CO by soils.

The stratosphere as a sink was first identified by Seiler and Junge (1969). The flux of CO into the stratosphere is due to a decrease in the CO mixing ratio above the tropopause toward a steady-state level lower than that normally found in the upper troposphere (see Fig. 1.14). In the lower stratosphere, CO is produced from methane and other long-lived hydrocarbons, and it is consumed by reaction with OH as in the troposphere, but the rate of vertical mixing is much lower (Seiler and Warneck, 1972; Warneck *et al.,* 1973). The flux of CO into the stratosphere can be derived from the observed gradient of the CO mixing ratio above the tropopause in the manner described in Section 4.3 for methane. The loss rate obtained, 110 Tg year^{-1}, is small compared with that for reaction of CO with OH radicals in the troposphere.

The importance of soil surfaces as a sink for atmospheric CO was first pointed out by Inman *et al.* (Inman *et al.,* 1971; see also Ingersoll *et al.,* 1974). Their experiments were carried out with rather high CO mixing

TABLE 4.12 The Global Budget of Carbon Monoxide in the Troposphere (Tg year^{-1} of CO)

Type of source or sink	Seiler (1974)	Logan et al. (1981) Global	Logan et al. (1981) NH	Logan et al. (1981) SH	Seiler and Conrad (1987)	Khalil and Rasmussen (1990a)	Pacyna and Graedel (1995)
Sources							
Fossil fuel combustion and related activities	640	450	425	25	640 ± 200	500	440 ± 150
Biomass burning	60	655	415	240	1000 ± 600	680	700 ± 200
Oxidation of human-made hydrocarbons	—	90	85	5	—	90	—
Oxidation of natural hydrocarbons	60	560	380	180	900 ± 500	600	800 ± 400
Ocean emissions	100	40	13	27	100 ± 90	40	50 ± 40
Emissions from vegetation	—	130	90	40	75 ± 25	100	75 ± 25
Oxidation of methane	1500–4000	810	405	405	600 ± 300	600	600 ± 20
Total source strength	2360–4860	2735	1813	922	3315 ± 1700	2600	2700 ± 1000
Sinks							
Reaction with OH radicals	1940–5000	3170	1890	1280	2000 ± 600	2200	2000 ± 600
Consumption by soils	450	250	210	40	390 ± 140	250	250 ± 100
Flux into the stratosphere	110	—	—	—	110 ± 30	100	110 ± 30
Total sink strength	2500–5560	3420	2100	1320	2500 ± 770	2550	2400 ± 750

ratios, however, which led them to overestimate the sink strength. Liebl and Seiler (1976) then showed that soils may represent either a source of CO or a sink, because the mixing ratio of CO in the air above soils confined by an enclosure adjusts to a steady-state value determined by conditions within the soils (see Fig. 1.15). The steady-state value is strongly dependent on temperature, moisture content, population of bacteria, and other soil parameters. A soil will act as a source of CO when the steady-state CO mixing ratio is higher than that of ambient air, and it will act as a sink when this condition is reversed. Usually, atmospheric CO is lost at the soil surface; only arid soils release CO to the atmosphere (Conrad and Seiler, 1982, 1985a).

Field and laboratory studies leave no doubt that the consumption of CO in soils is a biological process (Ingersoll *et al.*, 1974; Bartholomew and Alexander, 1979, 1982; Conrad and Seiler, 1980b, 1982), probably involving microorganisms. For example, Conrad and Seiler have shown that the activity for CO consumption follows Michaelis–Menten kinetics, that it is removed by filtration through a 0.2-μm filter but not a 0.3-μm filter, and that it is inhibited by antibiotics. Uncertainties exist about the types of microorganisms responsible for the CO loss. Although CO may serve as a growth substrate for a variety of aerobic bacteria, their affinity for CO utilization is low, and they cannot account for the observed CO consumption (Conrad, 1996). The production of CO in soils, in contrast to its consumption, is not related to microorganisms, even though a number of bacteria and fungi are known to produce CO from certain substrates. Instead, CO is produced from the oxidation of organic compounds present in soil humus. Evidence for the abiotic nature of the process was obtained by Conrad and Seiler (1980b, 1985b) by autoclaving the soil. The rate of CO production was stimulated when soil pH was raised, or when the air was replaced by pure oxygen, whereas replacement by nitrogen decreased the rate. Sterile mixtures of humic acids in quartz sand produced CO, while pure quartz sand did not. Field measurements of the steady-state CO mixing ratio and its variation with temperature above arid soils enabled Conrad and Seiler (1985a) to separate the effects of CO consumption and CO production. The consumption may be expressed in terms of a deposition velocity, which was found to vary little with temperature. The CO production rate, in contrast, was strongly temperature-dependent, following an Arrhenius law with an apparent heat of activation in the range of 57–110 kJ mol^{-1}. The rate was proportional to the organic carbon content of the soil. From these and other observations, Conrad and Seiler (1985a) deduced a globally averaged deposition velocity of $(2$–$4) \times 10^{-4}$ m s^{-1}, which is equivalent to a global CO loss rate of 190–580 Tg year^{-1}, with an average of 390 Tg year^{-1}. The global emission from soils was estimated as 17 ± 15 Tg year^{-1}. Because of the temperature dependence, these emissions are mainly from tropical and desert soils. The global emission rate is much smaller than the consumption rate, so it may be neglected.

The various estimates in Table 4.12 for the total production and loss rates show that the CO budget is approximately balanced. As Table 4.12 shows, the individual estimates for sources and sinks have not changed much in the recent decade, but they still carry large uncertainties. The average residence time for CO in the troposphere is

$$\tau = G(CO)/S(CO) = 370/2500 = 0.15 \text{ year}$$

or about 2 months. This is short enough to cause the spatial variation within the troposphere displayed in Figures 4.9 and 4.10.

Volz et al. (1981a) have measured the concentration of ^{14}CO as a function of season and latitude. Radiocarbon is produced by cosmic ray neutrons in the atmosphere and generates ^{14}CO by recoil reaction with oxygen. The global production rate is about 330 mol year^{-1}, and a large part of it is generated in the troposphere, where it undergoes oxidation to CO_2 by reaction with OH radicals. From their measurements Volz et al. (1981a) established the tropospheric ^{14}CO content, and with the known production rate and a two-dimensional model they calculated a residence time for ^{14}CO of 2 months, in agreement with the above value.

4.6. HYDROGEN

The presence of hydrogen (H_2) in the atmosphere appears to have been discovered as a byproduct in the liquefaction of air. An early report by Paneth (1937) indicated an H_2 mixing ratio of about 0.5 μmol mol^{-1} derived in 1923. Schmidt (1974) has reviewed a variety of measurements made during the period 1950–1970 that led to a similar value. During this time the main interest was the tritium content of atmospheric hydrogen and its rise due to atomic weapons tests and emissions from the nuclear industry (Ehhalt et al., 1963; Begemann and Friedman, 1968). According to Suess (1966), it was generally assumed that hydrogen in the air originated from the photodissociation of water vapor in the upper stratosphere and mesosphere. This idea had to be abandoned when Ehhalt and Heidt (1973b) and Schmidt (1974) showed that the H_2 mixing ratio above the tropopause lacked a vertical gradient (see Fig. 1.14), indicating that the flux of H_2 between troposphere and stratosphere is negligible. In the early 1970s, when the methane oxidation cycle was explored, it was recognized that hydrogen is a product of the photodissociation of formaldehyde (Calvert et al., 1972). A search for other H_2 sources indicated emissions from the oceans and from anthropogenic activities. In addition, the reaction of hydrogen with OH radicals and its uptake by soils were identified as sinks. These studies led to a budget of hydrogen in the troposphere that is largely controlled by the biosphere. The budget of hydrogen in the stratosphere is detached from that in the troposphere.

The spatial distribution of hydrogen in the atmosphere has been studied by Schmidt (1974, 1978), Ehhalt *et al.* (1977), and Khalil and Rasmussen (1990b). A summary of the data shown in Table 4.13 indicates that hydrogen is fairly evenly distributed in the troposphere. The good precision of the individual data contrasts with greater differences in the absolute values reported by each research group, which may reflect instrument calibration problems. The measurements of Schmidt (1978) indicated a slight excess of about 2% in the Northern Hemisphere, suggesting that anthropogenic sources are important, but Khalil and Rasmussen (1990b), who made measurements at six remote surface sites, found H_2 mixing ratios to be about 3% higher in the Southern Hemisphere. These authors observed also seasonal variations, which were unusual in that they did not display the normal 6 month phase lag between the Northern and Southern Hemisphere, and they found an increase in the hydrogen mixing ratio of about 0.6% $year^{-1}$. It is not known whether this is a temporary or a longer lasting phenomenon.

Higher values than those in background air occur in continental surface air, particularly in the cities. Schmidt (1974), for example, reported 800 ± 9 nmol mol^{-1} in the city of Mainz, Germany. Scranton *et al.* (1980) have studied the diurnal variation of hydrogen in Washington, DC. In winter, the mixing ratio attained maximum values up to 1.2 μmol mol^{-1} twice daily. The maxima coincided with rush-hour traffic, indicating that motor vehicle exhaust is a major source of H_2. Indeed, hydrogen correlated well with CO

TABLE 4.13   Mixing Ratios (nmol mol^{-1}) of Hydrogen in the Background Troposphere

	Northern Hemisphere	Southern Hemisphere	Remarks
Ehhalt *et al.* (1977)	503 ± 10	—	From altitude profiles[a]
Schmidt (1978)	559 ± 17	551 ± 13	Upper troposphere[b]
Schmidt (1978)	584 ± 27	552 ± 10	Atlantic Ocean surface[c]
Schmidt (1978)	576 ± 1	552 ± 1	Averages
Khalil and Rasmussen (1990b)	498 ± 3	513 ± 3	Three-year averages from six remote sites[d]

[a] Aircraft grab samples, 26–42° N latitude.
[b] Aircraft continuous monitoring, 67° N–53° S.
[c] Shipboard measurements, 54° N–23° S.
[d] Alaska, 71.5° N; Oregon, 45.5° N; Hawaii, 19.3° N; Samoa, 14.1° S; Tasmania, 42° S; Antarctica, 71.4° S. The values were calculated from data presented in graphs.

and with the nitrogen oxides. From aircraft measurements in the plume of the city of Munich, Germany, Seiler and Zankl (1975) found that the H_2/CO volume ratio resulting from anthropogenic emissions is close to unity. This result, when combined with the rate of CO emissions from transportation (Table 4.12), leads to a source strength for anthropogenic hydrogen of 17 Tg year^{-1}. Liebl and Seiler (1976) derived a higher value of 20 Tg year^{-1}, whereas Schmidt (1974), who did not yet have this information, suggested 13 Tg year^{-1}. Since there may be other anthropogenic sources in addition to motor vehicle exhaust, a source strength of 20 Tg year^{-1} appears reasonable. The global scale emission of hydrogen from the burning of biomass has been estimated initially by Crutzen *et al.* (1979) in a manner similar to that discussed for CO. They obtained 15 Tg year^{-1}, whereas Andreae (1991) calculated 19 Tg year^{-1}, with an uncertainty of about 50%.

Hydrogen is produced during the oxidation of methane in the atmosphere by photolysis of formaldehyde. From the photodissociation coefficients of formaldehyde (see Fig. 2.20) one finds that the channel leading to $H_2 + CO$ as dissociation products contributes roughly 68% to the overall process. Not all of the formaldehyde resulting from the oxidation of methane is photolyzed, however, because a certain fraction undergoes reaction with OH radicals, and this reaction does not produce molecular hydrogen. In ground-level air, the fraction is about 40%, but it decreases with altitude because of the decrease in OH concentration and the increase in the HCHO photolysis rate. If 60% of formaldehyde derived from methane is photolyzed, the rate of H_2 formation will be

$$0.68\,\alpha S_{OH}(CH_4)\left(M_{H_2}/M_{CH_4}\right) \approx 20 \text{ Tg year}^{-1}$$

where $\alpha = 0.6$ and $S_{OH} = 403$ Tg year^{-1}. More difficult to estimate is the rate of hydrogen production from the oxidation of nonmethane hydrocarbons. The principal hydrocarbons emitted from plants are terpenes and isoprene. It should be recognized that formaldehyde is the only aldehyde whose decomposition yields H_2, although all of them generate CO. Of the naturally emitted hydrocarbons, the terpenes are not expected to produce much formaldehyde when oxidized, so that these compounds may be disregarded to a first approximation. The oxidation of isoprene is known to produce approximately three molecules of formaldehyde per molecule of isoprene that is oxidized. With an H_2 yield of 0.68 from formaldehyde and $\alpha = 0.6$ as above, the amount of hydrogen derived from the oxidation of isoprene is

$$3 \times 0.68 \times \alpha \times Q(C_5H_8)\left(M_{H_2}/M_{C_5H_8}\right) \approx 18 \text{ Tg year}^{-1}$$

where $Q(C_5H_8) = 500$ Tg year^{-1}, as discussed in the previous section.

The oceans represent a source of atmospheric hydrogen because the surface waters are supersaturated with H_2—that is, the concentrations are higher than expected if dissolved hydrogen were in equilibrium with that in the air above the ocean. Saturation factors observed by Schmidt (1974) in the northern and southern Atlantic ranged from 0.8 to 5.4, with an average of 2.5. From these data, extrapolated to other oceans, Schmidt (1974) and Seiler and Schmidt (1974) estimated a global emission rate of 4 Tg year^{-1}. In contrast to CO, the aqueous concentration of H_2 was not found to undergo diurnal variations. This does not preclude a photochemical production as in the case of carbon monoxide, but another production mechanism, such as bacteria or the digestive tracts of zooplankton, may be more effective.

Finally, the behavior of hydrogen in soils will be discussed. In water-logged, anaerobic soils, hydrogen is generated in large amounts because of the bacterial fermentation of organic matter. Hydrogen emissions from such soils nevertheless are marginal because nearly the entire production is consumed by other bacteria, mainly denitrifiers, sulfate reducers, and methanogens, which, as oxygen becomes scarce, successively utilize nitrate, sulfate, and CO_2 as an oxygen supply. After the onset of methogenesis, hydrogen occurs only in traces in such soils, although the turnover rate remains high. Thus methane is emitted rather than hydrogen.

The only occurrence of significant net emission of H_2 into the atmosphere was reported for soils covered with legumes (Conrad, 1988). Members of this plant family live in symbiosis with *Rhizobium* bacteria. Contained in root nodules, these bacteria are able to fix atmospheric N_2 (see Section 9.1) by means of the enzyme nitrogenase, which catalyzes the reduction of N_2 to NH_3. In addition, nitrogenase is able to reduce protons to hydrogen, and it appears that this process always occurs parallel to nitrogen reduction. About 30% of the energy flow needed for N_2 fixation is diverted to reduce protons to H_2 (Schubert and Evans, 1976). Most of them have developed enzymes that recycle the H_2 so formed internally (Robson and Postgate, 1980), so that hydrogen does not leave the cell. *Rhizobium* species lack these enzymes, causing hydrogen to be released to the environment. Conrad and Seiler (1980c) have studied the release of hydrogen from soils planted with clover, in comparison to soils overgrown exclusively with grass. Similar to CO, hydrogen was observed to approach a steady-state level within 60 min after the plot was covered with an airtight box. Hydrogen as well as CO thus undergo simultaneous production and loss. From the rate of change of the H_2 mixing ratio inside the box, Conrad and Seiler (1980c) were able to separate the contributions of the two processes. The results indicated consumption rates that were approximately equal for all soil plots and similar to those derived by Liebl and Seiler (1976). The corresponding H_2 deposition velocities fell into the range $(2-10) \times 10^{-4}$ m s^{-1}, with a minimum in winter and a maximum in early autumn. Hydrogen emissions occurred only from plots planted with clover, indicating that some of the hydrogen resulting from nitrogen fixation is released to the atmosphere. A strong seasonal variation of the release rate was observed, with a maximum during the main growth period, coinciding with the maximum of nitrogen fixation activity. Burns and Hardy (1975) have estimated the global rate of N_2 fixation by agricultural legumes as 35 Tg year^{-1}, or about 25% of the global rate of nitrogen fixation of 139 Tg year^{-1}(N). The remainder is mainly due to grasslands (45 Tg year^{-1}) and forest and woodlands (40 Tg year^{-1}). The volume ratio of hydrogen produced to nitrogen consumed is estimated to be about 30%, so that the total rate of hydrogen production is

$$0.3 \times 139\left(M_{H_2}/M_N \right) = 6 \text{ Tg year}^{-1}$$

Not all of it is released, however. Conrad and Seiler (1980c) have considered the individual contributions more carefully. From their observations they derived a global H_2 release rate of 3 Tg year^{-1}, with a possible range 2.4–4.9 Tg year^{-1}.

Liebl and Seiler (1976) showed that losses of hydrogen in soils increase with temperature toward an optimum at 30–40°C. Soil moisture also influenced the deposition rate. Conrad and Seiler (1980c, 1985a), in contrast, found comparable H_2 deposition velocities for soils in subtropical and temperate latitude regions, with an average of 7×10^{-4} m s^{-1}. In estimating the effective land area, one must take into account winter seasons at high latitudes and subtract urban areas, deserts, and regions subject to permafrost. The effective land area estimated by Conrad and Seiler (1980c, 1985a), 90×10^6 km^2, led to 90 Tg year^{-1} for the global rate of H_2 uptake by soils, with a range of 70–110 Tg year^{-1}. This is much larger than the potential rate of hydrogen emission caused by nitrogen fixation.

Although aerobic soils favor the consumption of H_2 by a large variety of oxidizing bacteria, they generally require H_2 concentrations higher than those offered by ambient air (Conrad, 1996), so that they cannot be responsible for losses of H_2 by soils. Conrad and Seiler (1981) have shown that extracellular enzymes are mainly responsible for the oxidation of hydrogen in soils. This activity appears to represent the largest sink of hydrogen in the atmosphere.

Table 4.14 summarizes the known sources and sinks for atmospheric hydrogen. The total budget amounts to about 85 Tg year^{-1}. Roughly 40% of it is due to anthropogenic activities, that is, automotive exhaust and biomass burning. The major natural sources are the oxidation of methane and isoprene. Emissions from the oceans and from legumes are relatively minor. The fraction of hydrogen consumed by the reaction $OH + H_2 \rightarrow H_2O + H$

TABLE 4.14   Estimates for the Budget of Hydrogen (Tg year^{-1}) in the Troposphere

Type of source or sink	Schmidt (1974)	Conrad and Seiler (1980c)	Present
Sources			
Anthropogenic emissions	13–25.5	20 ± 10	17[a]
Biomass burning	—	10 ± 10	19[b]
Oceans	4	4 ± 2	4[c]
Methane oxidation	4.6–9.2	15 ± 5	20[a]
Oxidation of nonmethane hydrocarbons	—	25 ± 10	18[a]
Biological N$_2$ fixation	—	3 ± 2	3[d]
Volcanoes	0.1	—	0.2[e]
Total sources	21.6–38.7	87 ± 38	81
Sinks			
Oxidation by OH radicals	3.7–7.3	8 ± 3	16[f]
Uptake by soils	12–31	90 ± 20	70[a]

[a] See text.
[b] Crutzen and Andreae (1990).
[c] Seiler and Schmidt (1974).
[d] Identical to the average of Conrad and Seiler (1980c).
[e] From Section 12.4.1.
[f] The rate coefficient $k_{119} = 7.7 \times 10^{-12}$ exp $(2100/T)$ and $T = 277$ K were employed.

is smaller than that consumed by soils. The global distribution of the sources favors the continents. The residence time of hydrogen in the troposphere is

$$\tau = G(\mathrm{H}_2)/Q(\mathrm{H}_2) \approx 160/85 = 1.9 \text{ years}$$

The time is sufficiently long to ensure a fairly uniform distribution of the H$_2$ mixing ratio within each hemisphere, but it is not long enough to guarantee *a priori* a uniform distribution between the Northern and the Southern Hemispheres. The fact that the average H$_2$ mixing ratios in the two hemispheres are nevertheless almost identical suggests that sources and sinks in each hemisphere are closely balanced. As the major natural sources are concentrated in the Tropics, each hemisphere presumably receives a similar share of hydrogen, so that the seasonal variation should be small. The observed seasonal variation at high latitudes remains unexplained.

CHAPTER 5

Ozone in the Troposphere

5.1. INTRODUCTION

Schönbein (1840, 1854), who discovered ozone as a byproduct of electric discharges in air, found that ozone liberates iodine from iodide. He exposed to the atmosphere paper strips impregnated with starch and potassium iodide to demonstrate the presence of ozone in the air by the characteristic blue color of the starch—iodine complex. Paper ozonometry became very popular in central Europe in the middle of the nineteen century because of the then widespread belief that ozone-rich air is beneficial to health because ozone is a disinfectant. However, the technique does not lend itself to quantitative measurement (Kley *et al.*, 1988). The first quantitative method for the determination of atmospheric ozone, which was based on the use of arsenite to scavenge iodine (Albert-Levy, 1878), was applied between 1876 and 1911 at the Observatoire Montsouris near Paris. Volz and Kley (1988) reevaluated this historic record. More recently, Cauer (1935) determined the loss of iodine from a potassium iodide solution after ozone-containing air had passed through the solution. Regener (1938) used thiosulfate as a reagent for iodine followed by back-titration. Ehmert (1949, 1951) then perfected the system by introducing a coulometric titration technique. The current derived can be amplified and a continuous record obtained. The principle was adopted by Brewer and Milford (1960) in the construction of a simple balloon ozonesonde, which is still in use today for stratospheric ozone measurements. The potassium iodide reaction is subject to some interference by other oxidants in polluted air, but the effect is considered minor in clean air regions and in the free troposphere. In the meantime, commercial instruments have been developed that determine ozone either by ultraviolet optical absorption or by gas phase chemiluminescence. These instruments have a greater dynamic range, and they are relatively free of interference effects.

Ozone in the troposphere occurs with mixing ratios in the range of, roughly, about $10-100$ nmol mol^{-1}. Ultraviolet radiation capable of photodissociating oxygen does not reach the troposphere, so that in this atmospheric domain ozone must have a different origin. The classic concept described by Junge (1963) considered the stratosphere to be the main source, from which ozone enters the troposphere via tropopause exchange processes (see Section 1.5). In the troposphere, ozone is carried downward by turbulent mixing until it reaches the ground, where it is destroyed.

While this concept was developed, a phenomenon called *photochemical smog* made its appearance, first in Los Angeles, and later in many other regions of the world. Here, ozone is a byproduct of the photooxidation of hydrocarbons, which takes place as the urban plume of polluted air spreads over the countryside. For many years, photochemical air pollution was considered a problem of mainly local or regional significance, somewhat affecting clean air sites by the advection of polluted air. Crutzen (1973, 1974) first made the farther-reaching suggestion that smog-like reactions associated with the oxidation of methane and other hydrocarbons might induce the photochemical production of ozone also in the unpolluted troposphere. Chameides and Walker (1973, 1976) went even further by suggesting that ozone in the lower troposphere is controlled entirely by photochemistry. Their proposal has met with considerable skepticism among proponents of an ozone budget dominated by downward transport (Fabian, 1974; Chatfield and Harrison, 1976). Much of the controversy arose from observations such as the seasonal variation of surface ozone showing a maximum in the summer that can be interpreted to support either transport from above or local photochemical production of ozone. More recently, records of longer-lasting ozone measurement series in Europe have been examined for trends, and it was found that the concentration of ozone has increased not only in the air near the Earth's surface (Feister and Warmbt

(1987), but also in the free troposphere (Wege *et al.*, 1989; Staehelin and Schmid, 1991; Staehelin *et al.*, 1994). This result has been taken as evidence for an increase in the photochemical production of ozone in the atmosphere due to the growing emissions of ozone precursors.

5.2. PHOTOCHEMICAL AIR POLLUTION

Los Angeles was the first city where, starting in the mid-1940s, the phenomenon of photochemical smog became evident. The frequency and severity of its occurrence soon made it unbearable and called for study and abatement (Hagen-Smit, 1952). Although many other cities of the world now are burdened with photochemical smog, Los Angeles is still the prime example of a region thus affected.

The term *smog* combines the words *fog* and *smoke*. It was originally applied to episodes of London-type fogs containing abnormally high concentrations of smoke particles and sulfur dioxide. This type of smog occurred in the winter and was caused by the accumulation of products resulting from the combustion of fossil fuels in home heating, industry, and power stations under adverse meteorological conditions. It was severely taxing the bronchial and pulmonary system. The Los Angeles type of smog is more a haze than a fog, characteristic of a low-humidity aerosol. It arises from photochemical processes involving nitrogen oxides and hydrocarbons emitted from automobiles, petroleum industries, dry cleaning, etc. A major feature is the high level of oxidant concentration, mainly ozone and peroxidic compounds such as peroxyacetyl nitrate, produced by the photochemical reactions.

The favored season for the occurrence of photochemical smog is summer, when solar intensities are high, but meteorological conditions also are important. Specifically, the formation of a temperature inversion layer appears to be a prerequisite for the accumulation of smog precursors. Temperature inversions obstruct the vertical exchange of air by convective mixing and thereby prevent the dispersal of polluted air toward the free troposphere. In Los Angeles, temperature inversions are frequent, occurring on about 300 days of the year. They are caused by an inflow of cool marine air during the daytime. The air moves inland, close to the ground beneath warmer continental air. The stratification is not easily broken up because mountains surrounding the Los Angeles basin impede a horizontal advection of continental air. As a result, polluted air is pushed inland by the sea breeze until in the late afternoon solar heating of the continental surface subsides and the local circulation reverses. By this time the pollution has been carried toward many neighboring communities and fills almost the entire basin.

Photochemical air pollution arises largely from gas-phase chemistry. Reaction mechanisms are complex, involving many different chemical species, and have still not been fully elucidated. Mechanisms of hydrocarbon oxidation will be treated in Section 6.2. Here, the main focus is the photochemical production of ozone. The subject was first reviewed by Leighton (1961), and subsequently by Altshuller and Bufalini (1965, 1971) and Finlayson and Pitts (Finlayson and Pitts, 1976; Finlayson-Pitts and Pitts, 1986). Two important facts emerged early: nitrogen oxides are catalysts of hydrocarbon oxidation in the atmosphere; and automobiles are a major source of both hydrocarbons and nitrogen oxides. It is appropriate, therefore, to begin the discussion of ozone formation with a brief description of the chemical composition of emissions derived from automobiles. In the late 1960s, catalytic converters were introduced in an effort to reduce hydrocarbons, carbon monoxide, and nitrogen oxides in automobile exhaust gases. Following the lead of the state of California, the use of such converters is increasingly being enforced by legislation in various countries around the world. The discussion below will refer to the situation existing before the introduction of the converter.

Gasoline-fueled vehicles account for most of the emissions from mobile sources, and in the present discussion the contribution from diesel and jet engines will be neglected. The bulk composition of exhaust gases from piston engines run on gasoline consists of, roughly, 78% nitrogen, 12% carbon dioxide, 5% water vapor, 1% unused oxygen, and about 2% each of carbon monoxide and hydrogen. The remainder is about 0.08% hydrocarbons, 0.06% nitric oxide, and several hundred μmol mol^{-1} partly oxidized hydrocarbons, mostly aldehydes, with formaldehyde making up the largest fraction. The exact composition is variable and depends on the type of engine and the way the engine is run.

Nitric oxide is formed in combustion engines by the interaction of oxygen and nitrogen in air at the high temperatures reached during the combustion cycle. The percentage of NO found in the exhaust gas is close to that calculated from the chemical equilibrium $N_2 + O_2 = 2NO$ at peak temperatures near 2500 K. Once NO is formed, its concentration appears to be effectively frozen in. Most authors have adopted the reaction chain first proposed by Zeldovich (1946):

$$O + N_2 \rightarrow NO + N$$

$$N + O_2 \rightarrow NO + O$$

The first reaction is quite endothermic (316 kJ mol^{-1}), and it is questionable whether it can proceed sufficiently rapidly, even at 2500 K, to bring the NO level up to the equilibrium values within the millisecond time scale available during the combustion cycle. Fenimore (1971) noted that NO is formed in fuel-rich combustion flames, and he argued that NO formed by reactions involving hydrocarbon radicals cannot be neglected, for example,

$$CH + N_2 \rightarrow HCN + N$$

$$N + O_2 \rightarrow NO + O$$

and further reactions oxidizing HCN to NO. Both mechanisms appear to contribute to NO formation in the internal combustion engine. Glassman (1996) discussed the state of knowledge regarding this problem. In the combustion of coal and heavy fuel oils, organic nitrogen compounds already present in the fuel contribute appreciably to nitric oxide formation. Gasoline contains no nitrogen compounds.

Table 5.1 gives the approximate composition of hydrocarbons present in the exhaust of gasoline-powered engines. Also listed for comparison are typical compositions of gasoline, of fuel evaporates, and of Los Angeles city air sampled in the morning. Hydrocarbons are grouped as alkanes, alkenes, alkynes, and aromatics, but a differentiation between the various isomers is not made. The data were collected primarily in the 1960s. Gasoline contains mostly saturated hydrocarbons in the C_4–C_9 range of carbon numbers plus varying fractions of aromatics and unsaturated hydrocarbons. The exact composition depends greatly on the supplier and is rarely known, as chromatographic separation of all isomeric components in gasoline appears to have been achieved relatively recently (Maynard and Sanders, 1969). The "typical" composition shown in Table 5.1 consists of 60% alkanes, 30% aromatics, and 10% alkenes. Earlier, the percentage of alkenes was greater, and it has subsequently been reduced further because of their high reactivity in the atmosphere.

The lower hydrocarbons have a tendency to evaporate, and column 2 of Table 5.1 gives the composition of vapor in the head space of a fuel tank. Starting in 1971, cars in the United States were equipped with evaporation control systems whereby the direct emission of fuel into the atmosphere from tanks and carburetors was greatly reduced. Fuel evaporation remains important, however, in the handling of fuel at filling stations.

The composition of hydrocarbons in the engine exhaust gases is partly correlated with fuel composition inasmuch as a small portion of the fuel leaves the combustion chamber unburned. The lighter hydrocarbons, such as methane, ethane, ethene, and acetylene, are formed during the combustion process. Certain other species—for example, benzene—also are present at higher relative concentrations than in the fuel, indicating that they must be formed by combustion reactions. A complete assay of motor vehicle emis-

TABLE 5.1   Hydrocarbons Derived from Automobiles[a]

Hydrocarbon	Fuel (%)	Evaporate (%)	Exhaust		Los Angeles air	
			μmol mol^{-1}	%[b]	nmol mol^{-1}	%[b]
Alkanes						
C_1	—	—	86	15.0	3220	70[c]
C_2	—	—	14	2.4	100	4.1
C_3	—	—	1	0.1	47	1.9
C_4	7.0	33	23	4.0	74	3.0
C_5	19.0	35	23	4.0	76	3.1
C_6	11.0	7.8	14	2.4	42	1.7
C_7	12.0	—	13	2.3	35	1.4
C_8	8.0	—	48	8.4	26	1.1
$\geq C_9$	3.5	—	3	0.5	—	–
Alkenes						
C_2	—	—	97	17.0	67	2.7
C_3	—	—	40	7.0	19	0.8
C_4	0.5	3.2	17	3.0	15	0.5
C_5	4.5	10.8	16	2.8	12	0.5
$\geq C_6$	5.0	—	9	1.6	2.5	0.1
Alkynes						
C_2	—	—	55	9.7	55	2.2
C_3	—	—	7	1.2	10	0.4
Aromatics						
C_6	1.0	0.4	14	2.4	23	0.9
C_7	16.5	0.9	43	7.5	39	1.6
C_8	8.5	—	18	3.1	57	2.3
$\geq C_9$	3.5	—	29	5.1	28	1.1

[a] Comparison of hydrocarbon composition of gasoline, fuel evaporate, exhaust, and ambient air sampled in Los Angeles. Adapted from data presented by Glasson and Tuesday (1970a, b), Heaton and Wentworth (1959), Kopczynski et al. (1972), Leach et al. (1964), Lonneman et al. (1974), Maynard and Sanders (1969), McEwen (1966), Stephens et al. (1958), Stephens and Burleson (1967), and Tuesday (1976).
[b] Percent of total hydrocarbons.
[c] After deducting the contribution of natural methane.

sions of C_2–C_{20} hydrocarbons has recently been made by analyzing air sampled in highway tunnels (Sagebiel et al., 1996). The results show that automobiles emit mainly hydrocarbons smaller than C_{10}, whereas diesel-powered trucks emit about 50% of hydrocarbons above C_{10}. The emission consists of roughly 40% alkanes, 18% alkenes, and 42% aromatic compounds.

Essentially all of the hydrocarbons, including the isomers found in the exhaust gases, show up again in the urban air. Relative concentrations are different, however. A striking example is methane. After correcting for the approximately 1.5 μmol mol^{-1} contribution by natural methane, one finds that CH$_4$ accounts for 70% of all hydrocarbons in urban air, although its relative abundance in the exhaust gases is only 15%. In fact, all of the alkanes appear to be enriched compared with alkenes and aromatic compounds. One possible reason is that because of their lower reactivity and hence longer lifetime in the atmosphere, the alkanes carry over from previous days. Note also that propane was present at 46 nmol mol^{-1} in Los Angeles air, although automobile exhaust gases contain negligible amounts of propane. It is evident that propane must have another source. Neligan (1962) and Stephens and Burleson (1967) have considered natural gas, which contains 1.3% propane in addition to 6% ethane, as a source of propane in urban air. Propane is an extremely minor constituent of gasoline, on the order of 0.05 mg kg^{-1}, but its vapor pressure is high. Mayrson and Crabtree (1976) have given a reasonable assignment to various sources of hydrocarbons in Los Angeles air (see Table 6.5).

Consider now the interaction of nitrogen oxides with hydrocarbons in sunlit urban air and the generation of ozone from this mixture. Figure 5.1 shows the variation with time of the nitrogen oxides, hydrocarbons, and ozone in Los Angeles during the day. In the city, a great deal of NO and hydrocarbons is deposited in the early morning during rush hour traffic. In the course of time, NO is converted to NO$_2$, hydrocarbons are oxidized to aldehydes, and ozone rises as a product, peaking in midday before decaying again because of reactions of its own and/or dispersal processes. Figure 5.1, center, shows results of experiments with automobile exhaust, diluted with air, filled into transparent bags or vessels, and irradiated with either natural or artificial sunlight. The variation of nitrogen oxides, hydrocarbons and aldehydes, and ozone resembles rather closely the behavior observed in the urban atmosphere, although it occurs on a shorter time scale. Figure 5.1, bottom, is added to show that equivalent results are obtained when mixtures of individual hydrocarbons with air containing also nitric oxide are irradiated in smog chambers. The example shown, with propene as the hydrocarbon, has been selected from many similar results described in the literature because it clearly demonstrates the autocatalytic nature of the process. The rate of hydrocarbon consumption accelerates with time, indicating the occurrence of a chain reaction accompanied by a multiplication of chain carriers (compare with Fig. 2.1). The common features of the data assembled in Figure 5.1 are the formation of aldehydes as intermediate products and, more importantly, the rapid conversion of NO to NO$_2$, preceding the formation of ozone. These results lead to the conclusion that the reaction mechanisms at work follow a common scheme, even though the individual hydrocarbon species involved may differ.

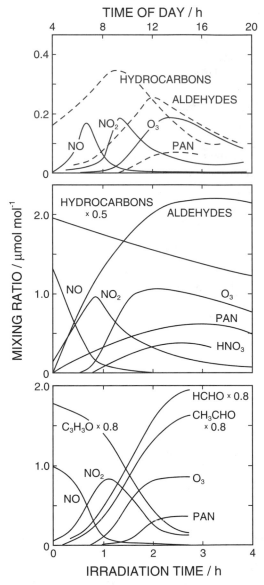

FIGURE 5.1 Variation with time of hydrocarbons, aldehydes, nitrogen oxides, ozone, and peroxyacetyl nitrate (PAN) for three conditions. Top: Downtown Los Angeles during the course of a day with eye irritation. From data presented by Leighton (1961) and Air Quality Criteria for Photochemical Oxidants (1970). Center: Irradiation of automobile exhaust diluted with air in a smog chamber of a plastic bag exposed to sunlight. From data of Leighton (1961), Kopzcynski *et al.* (1972), Wilson *et al.* (1973), Miller and Spicer (1975), Jeffries *et al.* (1976), and Wayne and Romanofsky (1961). Bottom: Smog-chamber irradiation of a mixture of propene, nitric oxide, and air. (Adapted from data presented by Altshuller *et al.* (1967) and Pitts *et al.* (1975).) Note differences in time scales.

Figure 5.1 suggests that the formation of ozone is the result of two consecutive steps. The first is the oxidation of NO to NO_2, and the second is a reaction that converts NO_2 to ozone. The latter process was recognized early to result from the photochemical activity of NO_2 (Leighton, 1961). As discussed in Chapter 2, nitrogen dioxide is an effective absorber of visible and near-ultraviolet solar radiation. At wavelengths below about 400 nm, NO_2 undergoes photodissociation, giving rise to the following reaction sequence:

$$NO_2 + h\nu \rightarrow NO + O \qquad j(NO_2) \approx 5 \times 10^{-3} \text{ s}^{-1}$$

$$O + O_2 + M \rightarrow O_3 + M \qquad k_1 n(O_2) n(M) \approx 1 \times 10^5 \text{ s}^{-1}$$

$$O_3 + NO \rightarrow NO_2 + O_2 \qquad k_{66} n(O_3) \approx 1 \times 10^{-2} \text{ s}^{-1}$$

where the rates, given on the right, involve the number densities of the species. The second reaction is much faster than the first, so that the first and the last reactions are rate-determining.

The photodissociation of NO_2 is the only definitely established process for the formation of ozone under such conditions. The reaction of alkylperoxy radicals with oxygen, $ROO + O_2 + RO + O_3$, has occasionally been invoked—for example, by Cadle and Allen (1970)—but according to current knowledge of bond dissociation energies, the RO–O bond is stronger than the O_2–O bond in ozone, so that the reaction would be endothermic and thus slow.

Without knowledge of the individual rate coefficients associated with the above reactions, one might interpret the rise of ozone at the expense of NO_2, apparent in all data in Figure 5.1, to result from a slow approach to photochemical steady state. This conclusion is unjustified, however, because all three reactions are rapid. Approximate rates are indicated on the right-hand side of the above reactions for ground-level conditions, 1000 hPa atmospheric pressure, and $x(O_3) = 25$ nmol mol^{-1}. With regard to establishing the steady state, the O_3 + NO reaction is rate-determining. The numbers show that this steady state is reached within about 200 s. It is important to note the stoichiometric requirements:

$$n(NO) + n(NO_2) = n(NO_x)$$

$$n(O_3) + n(NO_2) = n(NO_x)$$

$$n(O_3) = n(NO_x) - n(NO_2) = n(NO)$$

Even a cursory inspection of Figure 5.1 shows that the required stoichiometric relations are not obeyed. The amount of ozone rises while that of NO_x

decays. If one ascribes the loss of NO_x to the formation of nitric acid and peroxyacetyl nitrate (PAN), which are indeed seen to occur as long-lived products, not only NO_2 should decline but ozone should follow suit, whereas, in fact, it continues to rise. It will be clear, therefore, that the three reactions coupling NO, NO_2, and O_3 cannot by themselves explain the formation of ozone in photochemical smog. Additional reactions must be involved, especially a reaction that converts NO to NO_2 without affecting ozone. The same reaction would then explain the rapid oxidation of NO to NO_2 that initially precedes the formation of ozone.

It is also important to note that the reaction $NO + O_3$ cannot be assumed to be the mechanism for the conversion of NO to NO_2 during the initial stage of smog formation, because the amount of ozone present initially is too small. In urban air, any ozone left over from the preceding day is rapidly consumed by reaction with NO emitted early in the morning. Chemists generally are familiar with the Bodenstein (1918) reaction,

$$2NO + O_2 \rightarrow 2NO_2$$

which oxidizes nitric oxide directly. In the atmosphere, however, this process is extremely slow at the usual low concentration of NO, because the rate of the reaction involves the square of the NO concentration,

$$-dn(NO)/dt = 2k_{73}n(O_2)n^2(NO)$$

where $n(O_2)$ and $n(NO)$ are the number densities of oxygen and nitric oxide, respectively. For this second-order process, the time $t_{1/2}$ after which the NO number density is reduced to one-half the initial value $n_0(NO)$ is given by

$$t_{\frac{1}{2}} = 1/[2k_{73}n(O_2)n_0(NO)]$$

$$= 1/[2k_{73}x(O_2)x_0(NO)n_T^2]$$

where for convenience mixing ratios are introduced in the second equation, with n_T denoting the total number density of the $NO-NO_2-$air mixture. The rate coefficient, taken from Table A.5 in the Appendix, is $k_{73} = 2 \times 10^{-38}$ cm^6 molecule^{-2} s^{-1} for $T = 300$ K. Taking $x(O_2) = 0.21$ and $n_T = 2.5 \times 10^{19}$ molecule cm^{-3}, one calculates the following NO half-lifetimes for various initial NO mixing ratios:

$x_0(NO)$ (μmol mol^{-1})	0.1	1.0	10	100
$t_{1/2}$ (h)	530	53	5.3	0.53

Mixing ratios in excess of 10 μmol mol^{-1} are required to achieve a reasonably rapid conversion of NO to NO_2 by the Bodenstein reaction. Such high values may occur in the immediate vicinity of emission sources, but they would be unusual in urban air, where they are in the range of nmol mol^{-1}. This result makes the Bodenstein reaction essentially negligible in the atmosphere.

The high rate at which NO is converted to NO_2 in photochemical smog has baffled researchers for many years, but it is now established to result from reactions of peroxy radicals of the general type ROO· formed by the associated of alkyl or other hydrocarbon radicals with molecular oxygen. The reaction sequence can be written:

$$R + O_2 + M \rightarrow ROO + M$$

$$ROO + NO \rightarrow NO_2 + RO$$

At atmospheric pressure, the rate coefficients for the $R + O_2$ association reaction are already close to the second-order pressure limit. The high concentration of oxygen causes these reactions to occur almost instantaneously, so that other conceivable reactions of R radicals are negligible. Laboratory data show that reactions of ROO with NO, including HO_2, are rapid (see Table A.4). In addition to converting NO to NO_2, these reactions also convert the ROO radical to the corresponding RO radical, which reacts further to form aldehydes or ketones. This will be discussed further below.

The hydrocarbon radicals ROO and RO occur as a consequence of hydrocarbon breakup during their oxidation and act as chain carriers in the reaction mechanism. The above reactions contain no information on the initiation of hydrocarbon oxidation. This aspect will now be discussed. Historically there was a period when the consumption of hydrocarbons in photochemical smog was treated in terms of reactions of oxygen atoms and ozone, the oxygen atoms arising mainly from the photolysis of NO_2, and ozone coming in during the later stages of smog formation. The steady-state number density of oxygen atoms can be fairly reliably estimated as about 2.5×10^5 molecule cm^{-3}. When rate coefficients for their reactions became available and rates for the consumption of hydrocarbons were calculated, it was found that they fall far short of explaining the observed loss rates. Niki *et al.* (1972) were among the first to point out the inadequacy of reactions involving oxygen atoms and ozone, and they suggested that hydrocarbons were preferentially attacked by OH radicals. The realization that hydroxyl radicals are important intermediates and that they are more effective than other species in reacting with hydrocarbons came relatively late, presumably because rate coefficients for OH radicals were initially not well known. The development of the flash photolysis technique for the generation of OH radicals from H_2O greatly facilitated the determination of rate coefficients. Greiner (1967b, 1970) furnished the first reliable data for reaction of OH with CO and several alkanes. Stuhl and Niki (1972) then perfected the detection of OH by means of resonance fluorescence, and the combination of these methods has provided a wealth of data during the subsequent years. Atkinson et al. (Atkinson et al., 1979; Atkinson 1986, 1994) has reviewed these data.

Table 5.2 compares rate coefficients for reactions of oxygen atoms, ozone, and hydroxyl radicals with selected hydrocarbons and the corresponding rates for the loss of the same hydrocarbons calculated with appropriate steady-state values for the number densities of O, O_3, and OH. The rates for reactions of OH radicals exceed those of the other reactants in almost every case. The OH number density adopted, 2×10^6 molecule cm^{-3}, corresponds to that expected at noon on a clear summer day in the natural atmosphere at 30° northern latitude.

By itself, the high reactivity of OH radicals does not constitute sufficient evidence for the hypothesis that they are the principal species responsible for the degradation of hydrocarbons in photochemical smog. A more tangible support is the finding that hydrocarbon reactivities in smog chambers (reaction vessels irradiated with natural or artificial sunlight) correlate closely

TABLE 5.2   Selected Rate Coefficients and Hydrocarbon Consumption Rates for Reactions with Oxygen Atoms, Ozone, and OH Radicals at 298 K[a]

Hydrocarbon	k_O	k_{O_3}	k_{OH}	R_O	R_{O_3}	R_{OH}
Methane, CH_4	5.0 (−18)	< 7 (−24)	6.2 (−15)	8.0 (−13)	< 1.7 (−11)	1.3 (−8)
Ethane, C_2H_6	4.7 (−16)	< 1 (−20)	2.5 (−13)	7.5 (−11)	< 2.5 (−8)	5.4 (−7)
Propane, C_3H_8	7.9 (−15)	< 1 (−20)	1.1 (−12)	1.3 (−9)	< 2.5 (−8)	2.2 (−6)
n-Butane, C_4H_{10}	1.8 (−14)	< 1 (−20)	2.4 (−12)	2.9 (−9)	< 2.5 (−8)	5.0 (−6)
Ethene, C_2H_4	8.1 (−13)	1.6 (−18)	8.5 (−12)	1.3 (−7)	4.0 (−6)	1.7 (−5)
Propene, C_3H_6	4.8 (−12)	1.0 (−17)	2.6 (−11)	7.7 (−7)	2.5 (−5)	5.2 (−5)
Cis/trans-2-Butene, C_4H_8	1.9 (−11)	1.5 (−16)	6.0 (−11)	3.0 (−6)	3.9 (−4)	1.2 (−4)
Acetylene, C_2H_2	1.4 (−13)	7.8 (−21)	8.5 (−13)	2.2 (−8)	2.0 (−8)	1.7 (−6)
Benzene, C_6H_6	2.5 (−14)	1.7 (−22)	1.2 (−12)	4.0 (−9)	4.2 (−10)	1.2 (−5)
Toluene, $CH_3C_6H_5$	7.6 (−14)	4.1 (−22)	6.2 (−12)	1.2 (−8)	1.0 (−9)	1.2 (−5)

[a] Powers of 10 are shown in parentheses. Units for rate coefficients are cm^3 molecule^{-1} s^{-1}. Hydrocarbon consumption rates, $R = dn(HC)/n(HC)\, dt$, assume steady-state concentrations of 1.6×10^5, 2.5×10^{12}, and 2.0×10^6 molecule cm^{-3} for oxygen atoms, ozone, and OH radicals, respectively.

with the rate coefficients associated with OH reactions. This aspect seems to have been recognized first by Morris and Niki (1971) for a number of alkenes and aldehydes. Hydrocarbon reactivity, or the effectiveness of an individual hydrocarbon in producing photochemical smog, may be defined in various ways. Examples are the rates of hydrocarbon consumption, of NO to NO_2 conversion, or of oxidant formation, all measured under controlled experimental conditions. Biological effects may provide another, quite different standard, which will not be considered here, however. Of the first category, only hydrocarbon consumption will be examined as a measure of reactivity. The existence of a coarse relationship between hydrocarbon consumption and oxidant formation has been established by Farley (1978) and by Winer et al. (1979). The fact that hydrocarbon reactivities differ has been known since the earliest days of smog investigation and can be understood to derive from the differences in reaction rates. Table 5.2 may serve to indicate the range. The lower alkenes react faster than benzene and toluene, and these react faster again than the light alkanes. The differences in reactivities had given rise to the expectation that one might curb oxidant formation through a stringent emission control of the more reactive hydrocarbons, and for this purpose much effort has gone into setting up relative reactivity scales by means of smog chamber experiments. A review of such studies by Altshuller and Bufalini (1971) showed that the data unfortunately scattered widely because of different and sometimes poorly controlled experimental conditions, in addition to influences of wall and memory effects. Subsequently, Pitts et al. (1978) were able to reduce the uncertainties to a tolerable level by studying hydrocarbon disappearance rates relative to a compound for which the OH rate constant was well known and which thus served as an internal standard. By scaling the results it was found that the hydrocarbon loss rates agreed favorably with OH rate coefficients obtained in independent studies using the flash photolysis-resonance method. Figure 5.2 compares results presented by Pitts et al. (1978), who used n-butane as an internal standard, with similar data obtained by Wu et al. (1976), who used cis-2-butene as a standard. The correlation is so good that the smog chamber has been proposed as a device for the determination of OH rate coefficients for those hydrocarbons that cannot be investigated by any other experimental technique (Darnall et al., 1976). It should be emphasized that hydrocarbon loss rates correspond to OH reaction rates only during the initial stage of the reaction in smog chambers, that is, during the NO to NO_2 conversion stage. Once the concentration of ozone reaches an appreciable level, reactions of ozone with hydrocarbons must be included. This qualification applies specifically to the alkenes, which are most reactive with ozone. Table 5.2 may

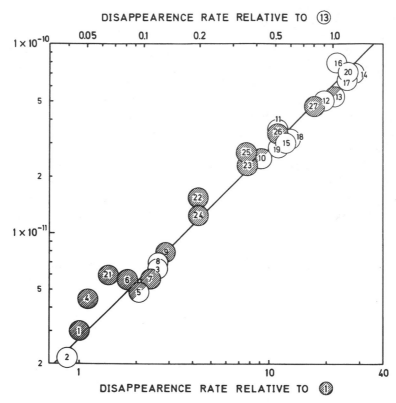

FIGURE 5.2 Plot of relative disappearance rates for hydrocarbons in smog chambers versus OH rate coefficients (units: cm^3 $molecule^{-1}$ s^{-1}) determined by independent measurements. Key: (1) n-butane, (2) isobutane, (3) n-pentane, (4) isopentane, (5) n-hexane, (6) 2-methylpentane, (7) 3-methylpentane, (8) cyclohexane, (9) ethane, (10) propene, (11) 1-butene, (12) isobutene, (13) cis-2-butene, (14) trans-2-butene, (15) 1-pentene, (16) methyl 1-butene, (17) cis-2-pentene, (18) 1-hexene, (19) 3,3-dimethylbutene, (20) cyclohexene, (21) toluene, (22) o-xylene, (23) m-xylene, (24) p-xylene, (25) 1,2,3-trimethylbenzene, (26) 1,2,4-trimethylbenzene, (27) 1,3,5-trimethylbenzene. Open circles are from Wu *et al.* (1976) relative to cis-2-butene, upper scale; hatched circles are from Lloyd *et al.* (1976) and Pitts *et al.* (1978) relative to n-butane, lower scale.

again serve to demonstrate this point. Although the rate coefficients for such reactions are not particularly large on an absolute scale, the reactions are competitive with OH reactions when the concentration of ozone builds up.

The preceding considerations form the basis for a general mechanism of ozone generation in photochemical smog by the oxidation of hydrocarbons.

Early in the morning rush-hour traffic leads to a high concentration of NO. Before ozone can appear, NO must first be converted to NO_2 by the degradation of hydrocarbons. As an example of such a reaction scheme, consider the oxidation of a straight-chain alkane $R_1CH_2R_2$, where R_1 and R_2 stand for suitable alkyl groups. The reaction is started by the abstraction of a hydrogen atom from the carbon skeleton, preferentially at sites of secondary hydrogen atoms, and the chain is propagated by reactions of alkyl peroxy radicals:

$$R_1CH_2R_2 + OH \rightarrow R_1R_2\dot{C}H + H_2O$$

$$R_1R_2\dot{C}H + O_2 \rightarrow R_1R_2CHOO\cdot$$

$$R_1R_2CHOO\cdot + NO \rightarrow R_1R_2CHO\cdot + NO_2$$

$$\alpha \quad R_1R_2CHO\cdot + O_2 \rightarrow R_1R_2CO + HO_2$$

$$\beta \quad R_1R_2CHO\cdot \rightarrow R_1\cdot + R_2CHO$$

$$R_1\cdot + O_2 \rightarrow R_1OO\cdot$$

$$R_1OO\cdot + NO \rightarrow R_1O\cdot + NO_2$$

$$R_1O\cdot + O_2 \rightarrow R_1'CHO + HO_2$$

$$HO_2 + NO \rightarrow NO_2 + OH$$

net: $\quad R_1CH_2R_2 + (2 + \beta)\{NO + O_2\} \rightarrow (2 + \beta)NO_2 + \alpha R_1R_2CO$
$$+ \beta R_1'CHO + H_2O$$

$(2 + \beta) \times \quad NO_2 + h\nu + O_2 \rightarrow NO + O_3$

Here, R_1' represents an alkyl group containing one carbon atom less than R_1, so that $R_1O\cdot = R_1'CH_2O\cdot$. Note that the OH radical initiating the reaction sequence is regenerated so that it can continue the chain. The net reaction converts an n-alkane to ketones and aldehydes, and it oxidizes at least two molecules of NO to NO_2. The subsequent photodissociation of NO_2 is the source of ozone in photochemical smog. More than two molecules of ozone are formed for each molecule of hydrocarbon being oxidized. The excess depends on the degree of decomposition of $R_1R_2CHO\cdot$ radicals; this fraction is denoted β in the above scheme, but decomposition becomes important only in longer-chained hydrocarbons.

Aldehydes are considerably more reactive with OH radicals than are alkanes, so that the aldehydes produced are subject to further oxidation and additional NO is converted to NO_2. To give an example, let R_1' stand for CH_3, so that $R_1'CHO$ represents acetaldehyde. The mechanism for acetaldehyde oxidation may be written as follows:

$$CH_3CHO + OH \rightarrow CH_3\dot{C}O + H_2O$$
$$CH_3\dot{C}O + O_2 \rightarrow CH_3(CO)OO\cdot$$
$$CH_3(CO)OO\cdot + NO \rightarrow CH_3 + CO_2 + NO_2$$
$$CH_3 + O_2 \rightarrow CH_3O_2$$
$$CH_3O_2 + NO \rightarrow CH_3O + NO_2$$
$$CH_3O + O_2 \rightarrow HCHO + HO_2$$
$$HO_2 + NO \rightarrow NO_2 + OH$$

net: $CH_3CHO + 3NO + 3O_2 \rightarrow HCHO + 3NO_2 + CO_2 + H_2O$
3 × $NO_2 + h\nu + O_2 \rightarrow NO + O_3$

Here again, the OH radical initiating the reaction sequence is regenerated. In the mechanism shown, three molecules of NO are oxidized to NO_2, and an equivalent number of ozone molecules is subsequently formed. Additional NO_2 and ozone arises from the photooxidation of formaldehyde, HCHO, produced from acetaldehyde and from the photooxidation of the ketones occurring as products of alkane oxidation.

The full potential of ozone production from hydrocarbon oxidation will be utilized only if the oxidation is carried to completion. Locally, this is rarely the case. The build-up of aldehydes in the Los Angeles atmosphere and in smog chambers, displayed in Figure 5.1, shows that the time of exposure to solar radiation is too short for all of the aldehydes to react. It is useful, therefore, to distinguish two time scales: a short one covering only the initial oxidation stage leading to the formation of carbonyl compounds, and a longer time scale during which the oxidation of carbonyl compounds is also completed. The first time scale indicates a minimum, and the second a maximum potential for ozone formation. Table 5.3 shows for several alkanes and alkenes the number of NO molecules converted to NO_2 during the oxidation of one molecule of hydrocarbon. The yields of NO_2 (and ozone) are based on reaction mechanisms similar to those shown above. A more detailed discussion will be given in Section 6.2. The second column of Table 5.3 shows the intermediate aldehydes and ketones formed, the third column indicates the yield of NO_2 molecules during the first stage of hydrocarbon oxidation, and the fourth column indicates the yield of NO_2 in the subsequent oxidation of carbonyl compounds. In the first stage the $NO-NO_2$ conversion factor is 2 in most cases. This agrees with measurements. Niki et al. (1978) have studied ethene, propene, and trans-2-butene, and Hoffmann et al. (1992) and Zellner et al. (1997) have additionally studied the light alkanes. The decomposition of RO· radicals, which increases the chain

TABLE 5.3   Potential for Ozone Formation in the Oxidation of Several Hydrocarbons
Following Reaction with OH Radicals[a]

| Compound | Intermediate aldehydes and ketones | NO to NO$_2$ conversion factors | | |
		Initial	From carbonyl compounds	Total
Ethene	2HO	2	2	4
Propene	HCHO, CH$_3$CHO	2	5	7
1-Butene	CH$_3$H$_2$CHO, HCHO	2	8	10
Cis/trans-2-Butene	2CH$_3$CHO	2	8	10
Isobutene	CH$_3$COCH$_3$, HCHO	2	5	7
Ethane	CH$_3$CHO	2	4	6
Propane	CH$_3$COCH$_3$, HCHO, CH$_3$CHO	2	6	8
n-Butane	CH$_3$CH$_2$COCH$_3$, CH$_3$CHO	2	6	8
Isobutane	CH$_3$COCH$_3$, HCHO	3	6	9

[a] Decomposition of RO· radicals can be largely neglected in the oxidation schemes of the hydrocarbons shown, that is, $\beta \ll \alpha$.

length, can be neglected for the hydrocarbons listed in Table 5.3 (for n-butane, $\beta/(\alpha + \beta) \approx 0.1$). The extent of NO–NO$_2$ conversion occurring during the second stage grows roughly with the number of carbon atoms in the hydrocarbon considered, with alkanes having a slightly higher potential than the alkenes.

In the atmosphere, the maximum potential for ozone formation will not be reached for several reasons. On the one hand, aldehydes and ketones are removed not only by oxidation reactions, but also by physical processes such as wet and dry deposition. In addition, the oxidation of the higher hydrocarbons and aromatic compounds may lead to condensable products that attach to aerosol particles, thus escaping further oxidation. On the other hand, the ROO radicals from the higher hydrocarbons not only react with NO by oxygen atom transfer; they also can form alkyl nitrates:

$$ROO + NO \rightarrow NO_2 + RO$$

$$\rightarrow RONO_3$$

and this route becomes more effective with increasing hydrocarbon chain length, although it rarely exceeds 10% of the total reaction (Atkinson, 1994). On the other hand, an appreciable fraction of NO$_2$ is removed by the formation of peroxyacetyl nitrate and nitric acid, whereby the yield of ozone is further diminished.

The preceding discussion has eluded the question about the origin of OH radicals that are necessary to trigger hydrocarbon oxidation. In the natural atmosphere hydroxyl radicals originate from $O(^1D)$ atoms formed in the near-ultraviolet photolysis of ozone and their subsequent reaction with water vapor (cf. Section 4.2). Although the same process will be important in the urban atmosphere after ozone builds up to a sufficient level, it cannot serve to initiate photochemical smog in the early morning, because at that time high NO concentrations keep ozone concentrations low. Hydroxyl radicals may be formed by hydrogen abstraction from hydrocarbons due to reactions of oxygen atoms resulting from the photolysis of NO_2, but again it is first necessary to convert NO to NO_2, before this source can attain significance. Other conceivable sources of OH are photolysis of H_2O_2 or HNO_3. Their absorption cross sections at wavelength above 300 nm are very low, however, as Figure 2.14 shows, so that these processes are rather ineffective.

The most promising process for the generation of OH radicals in the morning appears to be the photolysis of nitrous acid:

$$HNO_2 + h\nu \rightarrow NO + OH \qquad \lambda = 395 \text{ nm}$$

Nitrous acid has been observed in the atmosphere by long-path optical absorption (Platt and Perner, 1980; Platt et al., 1980b; Pitts et al., 1984c). Substantial evidence for the importance of HNO_2 as an OH source was obtained at Riverside, California, a community located about 100 km east of Los Angeles. Riverside is subjected almost daily to a plume of smog-laden air carried eastward. Figure 5.3 shows the observational data for HNO_2 at this site. The mixing ratio of nitrous acid increases during the night until it reaches a level of about 2 nmol mol^{-1}. After daybreak, HNO_2 disappears at a rate in agreement with the HNO_2 photolysis rate. During the day the mixing ratio of HNO_2 is kept at a low value because of photodissociation.

Formaldehyde, which is emitted from automobiles together with NO and hydrocarbons, provides yet another, although indirect source of OH radicals. As discussed in Section 2.5, one of the photodecomposition channels of formaldehyde leads to the formation of radicals, and this causes the following reaction sequence:

$$HCHO + h\nu \rightarrow HCO + H$$
$$HCO + O_2 \rightarrow HO_2 + CO$$
$$H + O_2 + M \rightarrow HO_2 + M$$
$$HO_2 + NO \rightarrow NO_2 + OH$$

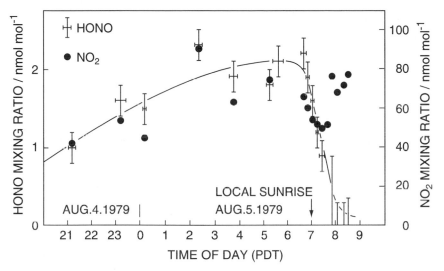

FIGURE 5.3 Behavior of nitrous acid and nitrogen dioxide during the night of August 4–5, 1979, at Riverside, California. PDT, Pacific daylight Time. (From Platt *et al.* (1980b) with permission.)

Comparison of the data in Table 2.7 (and in Figs. 2.18 and 2.19) shows that the photodissociation coefficient for the formation of radicals from formaldehyde is smaller by almost two orders of magnitude than that for the formation of OH from HNO_2. Formaldehyde is definitely more abundant in urban air than nitrous acid, but in the morning the HCHO mixing ratio is not much greater than 50 nmol mol^{-1}, if one accepts the data of Figure 5.1. At sunrise, formaldehyde will be an inferior source of OH compared with HNO_2. Later in the day the mixing ratio of formaldehyde increases, whereas that of HNO_2 declines. Formaldehyde then becomes more important as a radical source. At the same time, the mixing ratio of ozone rises as well, so that it may contribute to OH formation via $O(^1D)$ production by photolysis at wavelengths below 310 nm. These processes sustain the smog reactions during the day.

A continuous source of radicals is required, in addition to the regeneration of OH radicals in the hydrocarbon oxidation chains, to replace radicals that are lost in chain-terminating reactions. No particular attention has been paid to chain-terminating reactions in the foregoing description of smog chemistry, but that they cannot be neglected was made clear in the general discussion of chain reactions in Section 2.1. Three specific chain-terminating reactions leading to identifiable products are

$$HO_2 + HO_2 \rightarrow H_2O_2 + O_2$$

$$OH + NO_2 \rightarrow HNO_3$$

$$CH_3(CO)OO + NO_2 \rightleftharpoons CH_3(CO)OONO_2 \qquad (PAN)$$

The first two of these reactions lead to stable products, whereas peroxyacetyl nitrate (PAN) is thermally unstable and redissociates when temperatures are high. PAN is an eye irritant, which is responsible for much of the physical discomfort experienced by humans exposed to photochemical smog. Hydrogen peroxide, nitric acid, and PAN have been observed in the Los Angeles area as well as in smog chambers (Bufalini et al., 1972; Miller and Spicer, 1975; Lonneman et al., 1976; Kok et al., 1978). All three products are longer-lived, and they accumulate in the air in a way similar to that in which ozone does. Thus, it is not surprising that they correlate well with ozone and show a similar diurnal variation, with maxima occurring during the early afternoon (Tuazon et al., 1981).

5.3. DISTRIBUTION AND BEHAVIOR OF TROPOSPHERIC OZONE

Measurements of ozone in the troposphere fall into three categories: balloon soundings, aircraft observations, and surface measurements. The great variability of ozone requires longer measurement series so that average mixing ratios, seasonal trends, and other features can be determined. Although extended measurement series have been made at many surface stations, mainly within national air pollution networks, the information obtained refers largely to the local situation and not to the troposphere as a whole. Two past measurement programs have furnished information on the meridional distribution of ozone in a systematic fashion: the North American Ozonesonde Network, which was operative during the 1960s (Hering and Borden, 1967), and the network of stations for the measurement of surface ozone set up by Fabian and Pruchniewicz (1976) in western Europe and on the African continent during the 1970s. This program was supplemented by aircraft measurements at high altitudes. Satellite data provide a more complete global coverage, but the larger amount of ozone in the stratosphere masks tropospheric ozone, and the residual is difficult to retrieve. Yet Fishman et al. (1990) have shown that retrieval is possible, and Fishman et al. (1991) used the data to demonstrate the existence of elevated ozone concentrations over the South Atlantic Ocean in September and October as a consequence of widespread pollution arising from biomass burning on the African continent.

5.3.1. BALLOON SOUNDINGS

Balloon sondes for the determination of stratospheric ozone profiles often use electrochemical instruments based on the ozone-iodide reaction (Brewer and Milford, 1960; Komhyr, 1969). For a number of years the North American Ozonesonde Network used a different technique based on dry chemiluminescence from an organic dye (Regener, 1964). The sensor required calibration in the field, which was achieved most conveniently by Dobson spectrophotometry. The electrochemical sonde is potentially an absolute instrument, although in practice it requires a number of corrections, and comparison with total ozone measurements is always made. A field comparison of the two sondes (Hering and Dütsch, 1965) gave good agreement in the stratosphere. In the troposphere the concentrations obtained with the electrochemical sonde were often higher than those found with the chemiluminescence sonde. Both may be influenced by trace constituents other than ozone, particularly in the lowest layers of the atmosphere, where concentrations are high. During a certain period in 1963–64, many chemiluminescence sondes were found to experience a loss in sensitivity during ascent. The problem was traced to a modified procedure in the fabrication of the dye-covered chemiluminescent disk and was subsequently corrected, but it undermined confidence in the validity of the data. A reduction of sensitivity during ascent should affect stratospheric ozone values more than tropospheric ones, so that after normalization to total ozone, higher rather than lower ozone concentrations should have resulted in the troposphere compared to the electrochemical sonde. Chatfield and Harrison (1977a, b) have performed a critical analysis of both data sets. They confirmed that in the lower troposphere, the Regener sonde measured 35% less ozone, on average, than the Brewer sonde. They also noted that the former showed larger fluctuations than the latter, and for this reason they considered the chemiluminescence device less reproducible and hence less reliable. This ignores its very rapid response, however. The Regener sonde records natural fluctuations of ozone almost instantaneously, whereas the Brewer sonde has a response time of 20 s, which tends to dampen local variations in the O_3 mixing ratio. One should further appreciate that the sensitivities of both sondes are adjusted to be optimal in the stratosphere and that readings in the troposphere are low and correspondingly less accurate.

Figure 5.4 compares averaged altitude profiles of tropospheric ozone for the summer season. The differences between the two sensors are clearly

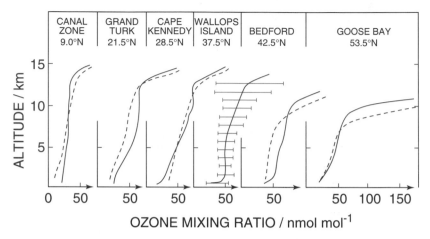

FIGURE 5.4 Averaged altitude profiles of ozone in the troposphere above six stations of the North American Ozonesonde Network. Solid curves represent data for July, obtained with the electrochemical sonde in the years 1966–1969 (Chatfield and Harrison, 1977b). The dashed curves represent data for the summer season, obtained with the chemiluminescence sonde in the years 1963–1965 (Hering and Borden, 1967). Since they do not include data for Cape Kennedy, the data for Tallahassee, Florida, at 30.4° N were used instead.

displayed. While the electrochemical instrument usually shows higher readings in the lower troposphere, a reversal of the trend occurs in the upper troposphere. Variances also are considerable. The horizontal bars superimposed on the profile at Wallops Island indicate ±1 standard deviation. Variances at other stations were similar except in the Tropics, where they were smaller. Note the increase in variability as one approaches the tropopause level. The effect is caused mainly by a part-time sampling of the stratosphere rather than by the influx of stratospheric ozone. Gradients of the ozone mixing ratio at extratropical latitudes are similar in magnitude. The increase in all gradients in the upper troposphere must again be assigned to tropopause variability. In the Tropics the gradients are smaller. If tropospheric ozone had its origin in the stratosphere, the gradient would be a consequence of the continuous eddy flux of ozone from the tropopause to the ground surface. In the Tropics the upward motion of the Hadley circulation would partly impede the injection of ozone across the tropical tropopause. The existence of a gradient as such does not, however, provide evidence for the dominance of downward transport, as the local generation and loss of ozone by photochemical processes may be fast enough to overrule the transport process.

Chatfield and Harrison (1977b) have analyzed the seasonal dependence of the averaged ozone mixing ratio in the air above six stations from which regular balloon soundings were made with the electrochemical sonde. Figure 5.5 shows time histories as a function of altitude, with the contours tracing ozone mass mixing ratios (1 μg g^{-1} = 1.29 × 10^3 μg m^{-3} at STP = 604 nmol mol^{-1}). The most striking feature is the annual wave in middle and high latitudes. Again the data in the uppermost altitude range refer partly to stratospheric air. Chatfield and Harrison did not delineate the boundary region, but one may obtain an indication of it from the variances, because they tend to maximize near the tropopause because of the natural fluctuation

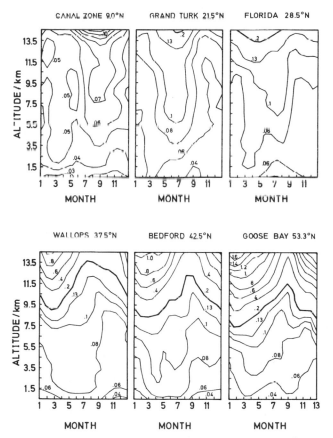

FIGURE 5.5   Seasonal variation of ozone mass mixing ratios (unit: μg g^{-1}) at six stations of the North American Ozonesonde Network. (Adapted from Chatfield and Harrison (1977b).)

of the tropopause level. The appropriate curves are shown in Figure 5.5 by the heavy lines, which may be taken to represent monthly mean tropopause levels. The seasonal variation is consistent with this interpretation; low levels occur in winter, and high levels in late summer. Ozone mixing ratios covary with $x(O_3)$ at the tropopause, indicating that the stratosphere is a source of tropospheric ozone. Ozone-rich air penetrates deeper into the troposphere in spring than at any other time of the year at all locations in north Florida. The spring maximum extends all the way down to the ground surface, although it experiences a time lag with increasing distance from the tropopause. In the surface layer, the maximum thus occurs in early summer rather than in spring. This feature provides strong support for a downward transport of ozone into the lower troposphere. It is not possible to explain the phase shift alternatively in terms of a photochemical mechanism.

It is considerably more difficult to interpret the data in the equatorial and subtropical regions. Near the equator, a semiannual wave is indicated, but the amplitude is low, the variability is high, and the data may be taken equally well to show no seasonal dependence at all. At Grand Turk, during the months May–July, a vertical tongue of ozone-rich air appears that reaches downward into the lower troposphere, apparently unrelated to the tropopause level. Chatfield and Harrison (1977b) noted that the tongue occurs roughly synchronously with the maximum of annual rainfall activity associated with the upward branch of the Hadley cell, which approaches its northernmost position in July and then retreats again to the south. It is possible that high-reaching cumulus towers coupled with cyclone activity disturb the tropopause sufficiently to trigger the injection of ozone from the stratosphere. Stallard et al. (1975) found evidence for a locally significant influx of stratospheric ozone from mesoscale squall systems. Alternatively, one must take into account a transport of ozone from higher to lower latitudes. The most likely injection regime near the station of Grand Turk is the subtropical jet stream, which in summer is located north of 40° northern latitude. This location lies a considerable distance to the north of the station, but during the summer months the meridional circulation system would favor a southward dispersal of injected ozone, as opposed to the pronounced northward transport in winter.

Fishman and Crutzen (1978) have combined data from the North American Ozonesonde Network with observations at other stations in the Northern and Southern Hemispheres to derive an approximate meridional distribution for both hemispheres. Data for the Southern Hemisphere were taken mainly from Canton Island (3° S), La Paz (16° S), Aspendale (38° S), Christchurch (43° S), and Syowa (69° S). Pittock (1977) had previously analyzed the more than 700 soundings at Aspendale. Data from the other stations were less numerous. Figure 5.6 shows the annual average meridional distribution of

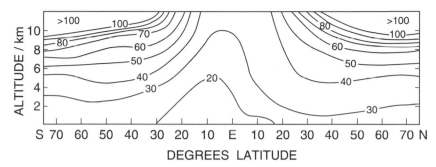

FIGURE 5.6 Seasonally averaged meridional distribution of ozone according to the analysis of Fishman and Crutzen (1978a, b). Contours indicate equal mixing ratios (unit: nmol mol^{-1}).

tropospheric ozone. Values exceeding 100 nmol mol^{-1} may be taken to represent stratospheric air. The distribution indicates an excess of ozone in the Northern Hemisphere at altitudes below about 5 km. From an analysis of average vertical ozone profiles at similar latitudes in both hemispheres, Fishman and Crutzen (1978) and Fishman et al. (1979) showed that in addition to greater absolute amounts, the Northern Hemisphere features higher ozone concentrations in summer compared to winter, in contrast to the behavior in the Southern Hemisphere. Fishman et al. (1979) argued that the excess of ozone might be a by-product of the higher level of carbon monoxide in the Northern Hemisphere, because the oxidation of CO would generate excess ozone via the OH reaction chain discussed in Section 4.2, provided NO_x concentrations are high enough. On the other hand, the increase in anthropogenic NO_x emissions, which has occurred primarily in the Northern Hemisphere, is probably in itself an important factor (Crutzen, 1995) in producing an excess of ozone.

Fishman et al. (1990) have used satellite data to derive the distribution of tropospheric ozone in terms of column densities. The results confirm the salient features of the meridional distribution of Figure 5.6, showing that the Northern Hemisphere contains 40% more ozone on average than the Southern, and that ozone concentrations in the Tropics are lower than at mid-latitudes. The data also indicate a seasonal variation such that in both hemispheres a temporary increase in ozone occurs during the spring, whereas an additional increase occurs in the Northern Hemisphere in the summer.

Marenco and Said (1989) derived the meridional cross section of the ozone distribution in the background troposphere from aircraft measurements between Europe and South America. The flights were made in June 1984 mainly along coastal routes, so that they do not provide an annual average. The results are nevertheless in reasonable agreement with the data in Figure 5.6, demonstrating in particular the greater concentration of ozone in the Northern Hemisphere.

5.3.2. Aircraft Observations

Owing to the great variability of ozone, isolated aircraft measurements add little to our knowledge of the large-scale distribution of tropospheric ozone, so that more comprehensive measurement programs are required. Commercial airliners have been used for this purpose. Most observations were made at cruising altitudes, at 10–12 km. Sampling of outside air can be done conveniently via the fresh-air system. A number of studies by Tiefenau *et al.* (1973) have shown that ozone is unaffected by the compression of air within the air ducts of the aircraft. Nastrom (1977) compared mean ozone values from measurements onboard commercial airliners with those obtained from the North American Ozonesonde Network and found them to be comparable. This established additional confidence in the aircraft data. Fabian and Pruchniewicz (1977) made a series of measurements within a flight corridor between 20° western and 35° eastern longitude, extending all the way from Norway to South Africa. A continuous record of ozone mixing ratios was obtained on all flights, and data pertaining to the stratosphere were subsequently removed as far as necessary. The top of Figure 5.7 shows a typical ozone record in the upper troposphere obtained on one such flight. Fabian and Pruchniewicz (1977) have averaged data from about 50 flights to derive seasonal averages as a function of latitude. The results are shown in the central portion of Figure 5.7. To minimize systematic errors the authors have chosen a relative scale of ozone mixing ratios that diminishes the usefulness of the data if one wants to compare them with others. Routhier *et al.* (1980) reported mid-tropospheric ozone mixing ratios at latitudes between 58° S and 70° N over the Pacific Ocean and the North American continent. Figure 5.7 includes an interpolation of this data set.

Two aspects of the data are of interest. The absence of a significant meridional gradient between 25° S and 25° N indicates an efficient horizontal equilibration of ozone mixing ratios in the equatorial upper troposphere by eddy diffusion. This agrees with the notion that the interhemispheric air mass exchange is most effective in this altitude regime. Pruchniewicz *et al.* (1974) noted a slight seasonal dependence of the ozone mixing ratio, with a maximum in May and a minimum in November–December. The variation was essentially in phase on both sides of the equator. This type of variation is not compatible with the balloon sounding data in Figure 5.5 and may be an overinterpretation of a limited data set. The second aspect of the data in

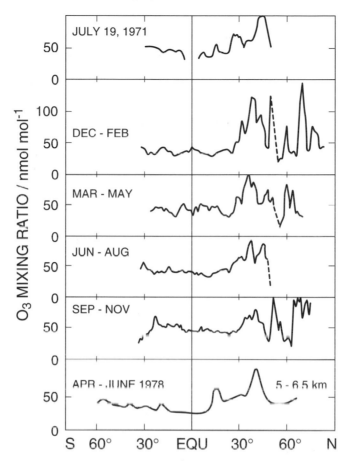

FIGURE 5.7 Ozone mixing ratios (nmol mol^{-1}) in the upper troposphere as a function of latitude, from aircraft observations. Top: Data from one flight on July 19, 1971, flight altitude 11–12 km (Fabian and Pruchniewicz, 1977). Center: Seasonally averaged data from 40 individual flights between Norway and South Africa (Fabian and Pruchniewicz, 1977) in relative units. Bottom: average ozone mixing ratios at 5–6 km altitude over the North American continent and the Pacific Ocean (Routhier et al., 1980).

Figure 5.7 is the occurrence of peaks of ozone-rich air at those latitudes where ozone intrusions are expected to occur with a high frequency. If the peak heights are taken to indicate the extent of ozone injection, the ozone flux is seen to be strongest during the first quarter of the year and weakest during the third quarter, which coincides with the build-up of stratospheric ozone during winter and its depletion during summer, as well as with the strength of subtropical jet-stream activity. Similarly intense stratospheric ozone injections are not apparent during the winter season of the Southern Hemisphere. The data of Fabian and Pruchniewicz (1977) unfortunately do not extend beyond the 35° S latitude circle. If the stratospheric injection rates through the tropopause gaps were the same in the two hemispheres, a steeper rise of the ozone mixing ratio would be expected at 30° S than is actually observed. The mid-tropospheric measurements of Routhier *et al.* (1980) confirmed the existence of stratospheric ozone intrusions in the Northern Hemisphere and their comparative weakness in the Southern Hemisphere, but it should be noted that the data in the Southern Hemisphere were obtained during the quiet season in that region. Nevertheless, both data sets indicate a smaller injection rate in the Southern Hemisphere in comparison with that observed in the Northern Hemisphere.

5.3.3. SURFACE MEASUREMENTS

Ozone mixing ratios in surface air are highly variable for a variety of reasons. Figure 5.8 shows a time series at a rural continental station demonstrating a typical diurnal variation. The advection of air with high NO_x levels, which is

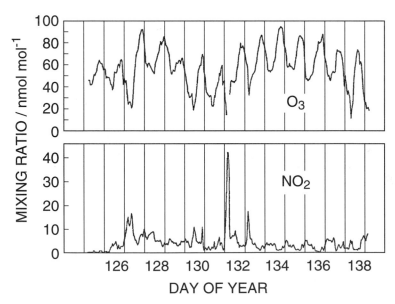

FIGURE 5.8 Two-week record of ozone and NO_2 in Deuselbach, Germany, in May 1988, during a fair-weather period, to demonstrate diurnal variation and loss of ozone during temporarily high NO_x levels.

frequent in regions affected by the vicinity of NO_x sources, causes the concentration of ozone to decline temporarily to a low value because of its reaction with NO, whereas the advection of polluted air having reacted to form photochemical smog gives rise to higher ozone concentrations. Summer smog is a regional phenomenon of the atmospheric boundary layer. As first discussed by Guicherit and van Dop (1977) for Europe and by Vukovich et al. (1977) for the United States, smog episodes usually are triggered by a high-pressure system, in which subsiding air causes not only the dissipation of clouds and long periods of sunshine, but also the slow build-up of an inversion layer that forms a barrier to the exchange of air with the free troposphere. The increase in the ozone level occurs gradually over a period of days. The O_3 mixing ratio is reduced to normal values when the meteorological situation changes and the inversion layer breaks up, so that the polluted air is diluted by fresh air moving both vertically and horizontally.

In all continental regions ozone is lost upon contact with the ground surface and the plant cover. The losses establish a vertical gradient of the ozone mixing ratio, which for a constant deposition flux decreases with height above the ground because of the increase in eddy diffusivity (see Section 1.6). The local behavior of ozone loss depends greatly on the topography and micrometeorology at the sampling site. Under fair-weather conditions, flatland stations show a characteristic diurnal ozone variation, with maximum values occurring in the early afternoon (compare Fig. 5.8), when convective turbulence due to solar heating is strongest and tropospheric ozone is rapidly carried downward. Minimum values are observed during the early morning hours, when stable conditions prevail in the continental boundary layer. Diurnal variations at mountain stations, in contrast, often show small amplitudes with maxima occurring preferentially during the night. This behavior is explained by the typical local mountain circulation pattern, with air directed upslope during the day and downslope during the night. Upslope circulation brings with it air depleted in ozone because of contact with the ground, whereas downslope circulation imports air from the free troposphere, which is richer in ozone. Mountain stations thus are favorable sites for sampling the free troposphere. A similar situation exists at coastal sites of the continents, where convective cells established during the day give rise to surface winds directed inland, whereas at night, when the land surface cools, subsidence sets in, so that the air flow is reversed. Maximum ozone mixing ratios are again found mostly in the later evening. Figure 5.9 shows frequency distributions for the occurrence of the daily ozone maximum at three stations: Hermanus, South Africa, a continental flatland station; Zugspitze, Germany, a mountain station; and Westerland, Germany, a coastal station. Although the frequency distributions are fairly

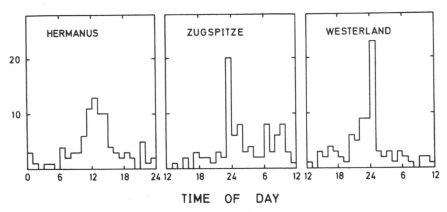

FIGURE 5.9   Histograms for the occurrence of the daily maximum ozone mixing ratio at three types of measurement sites. Left: Flatland station (Hermanus, South Africa). Center: Mountain station (Zugspitze, Germany). Right: Coastal station (Westerland, Germany). Observation periods were about 90 days each. (From Fabian and Pruchniewicz (1977).)

broad in all three cases, it is evident that in Hermanus the ozone maximum occurs preferentially at midday, whereas Zugspitze and Westerland show it shortly before midnight.

Figure 5.10 is added to show schematically ozone mixing ratios as a function of height above ground at a flatland site. In the middle of the day mixing ratios are high and the vertical gradient is small, because turbulent mixing maintaining the downward ozone flux is appreciable. Later in the afternoon the intensity of turbulence weakens, the eddy diffusivity decreases, and the gradient of ozone mixing ratios increases. The surface mixing ratios then decrease because of ground losses, while the effect at 12 m height is still

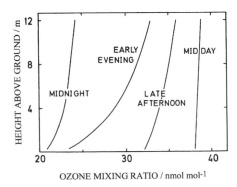

FIGURE 5.10   Vertical profiles of ozone mixing ratio at a flatland station during the course of 1 day. Schematic representation of data obtained by Regener (1957).

minimal. In the evening the situation is qualitatively still the same, but the supply of ozone from above now ceases and an accelerated decay of ozone occurs at all height levels. At midnight, the ozone profile is essentially stabilized and the further decrease in the O_3 mixing ratio is slow. After daybreak the turbulence in the boundary layer resumes, and eventually the ozone mixing ratio rises again.

The preceding discussion suggests that in unpolluted air the daily maximum of surface ozone is caused by the down flow from greater heights, when the influence of surface destruction is weakest. Fairly convincing evidence for this concept was obtained by Pruchnewicz et al. (1974) from measurements as a function of height above ground with the help of a 120-m tower at Tsumeb, South Africa. Meteorological parameters provided information on the stability of the atmospheric boundary layer. The dimensionless Richardson number was used as a characteristic turbulence indicator to differentiate between labile condition and stable stratification. In the first case, the intensity of turbulence is high and the near-surface ozone mixing ratio was found to correlate well with that 100 m above the ground. A one-to-one relationship was closely approached. In the second case of stable stratification the correlation was poor and proportionality was not observed. Pruchniewicz et al. (1974) also pointed out that during the entire two-month measurement period the daily maxima were always found to occur during labile condition. It must be emphasized, however, that the observed positive correlation shows evidence only for a downward transport of ozone. It gives no clues to the origin of ozone, which may well be photochemically produced. Mountain stations are better suited to sample air from the middle troposphere, and these data are less likely to be affected by surface air pollution events.

Occasionally, rather high ozone mixing ratios are encountered, with values exceeding 100 nmol mol^{-1}. Attmannspacher and Hartmannsgruber (1973) reported values up to 390 nmol mol^{-1} in southern Germany during an episode associated with a passing cold front. Lamb (1977) observed ozone mixing ratios well above 100 nmol mol^{-1} for several hours during a rainy night in Santa Rosa, California, with a maximum hourly value of 230 nmol mol^{-1}. Reiter et al. (1977) described an event with peak O_3 mixing ratios exceeding 145 nmol mol^{-1} at the Zugspitze Mountain station [2962 m above sea level, (a. s. l.)]. None of these events can be related to air pollution. They are attributed to local intrusions of stratospheric ozone via tropopause folds connected with frontal jet streams (see Fig. 1.13). Under favorable conditions the ozone-rich air parcels are brought straight down to the surface layer with the frontal air flow. The meteorology associated with jet streams has been discussed, for example, by Danielsen et al. (1970) and Danielsen and Mohnen (1977). In the case of the Zugspitze event the stratospheric origin of the air was demonstrated by the simultaneous observation of beryllium 7, a spallation product of cosmic ray interaction with the air in the stratosphere. The observation of a stratospheric intrusion by an increase in ground-level ozone is a fairly rare event. Dutkiewicz and Husain (1979) examined data taken at Whiteface Mountain in upper New York state and found coincidences of high ozone and beryllium 7 on 11 days out of 61 in June and July 1977. On these 11 days fresh stratospheric ozone contributed between 26% and 94% to the total 24-h ozone mixing ratio. Elbern *et al.* (1997), who have examined 10 years of Zugspitze data for beryllium 7 activity, relative humidity, and ozone, unambiguously identified 195 events, whereas at a neighboring mountain Wank (1776 m a. s. l.), only 85 events were detected. The highest frequency of intrusions was found to occur in winter and spring.

Figure 5.11 shows the seasonal variation of ozone represented by the daily maxima, observed in measurement series extending for at least 3 years. The first was obtained in Arosa, a high alpine Swiss resort town. The data were discussed by Junge (1962), who first noted the cyclic annual behavior. The maximum occurs in spring, 2–3 month after the late winter maximum of

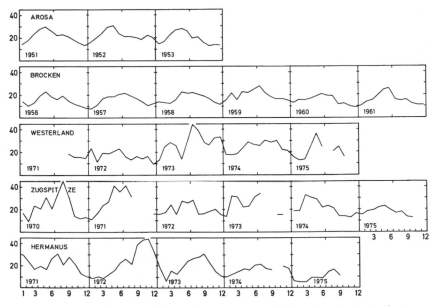

FIGURE 5.11 Monthly means of daily 1-h ozone maxima (daily average for Brocken); observations over at least 3 years at several stations in Europe and South Africa. Data for Arosa, Switzerland (1800 m a. s. l.), are from Junge (1962); those for Brocken, Germany (1140 m a. s. l.), are from Warmbt (1964); the others are from Fabian and Pruchniewicz (1977).

total ozone in the stratosphere. Junge interpreted the seasonal variation as arising from the injection of stratospheric ozone into the troposphere, the flux being proportional to the amount of total ozone, and a phase shift resulting from the transport of ozone in the troposphere until it reaches the surface layer. The balloon sounding data in Figure 5.5 and radioactive tracer observations (for example, of strontium 90; Staley, 1982) show that in the Northern Hemisphere the maximum of such a transport does indeed occur in the spring. The measurements at the Brocken (a middle-range mountain in Germany, 1140 m above sea level), which represent daily average ozone mixing ratios rather than diurnal maxima, display a similar seasonal variation. The same behavior is observed at other mountain stations in the Northern Hemisphere (see Fig. 5.12, to be discussed further below).

After it was recognized that the stratosphere is not the only source of ozone in the troposphere, but that photochemical sources cannot be neglected, other ideas have been put forth as an explanation of the spring maximum. Penkett and Brice (1986) noted that PAN (peroxyacetyl nitrate) also features a spring maximum in the background air over Europe, suggesting that thermal decomposition of PAN and intensifying photochemistry may enhance ozone formation in spring. Liu et al. (1987) proposed that ozone accumulates during winter because of a longer photochemical lifetime in that season. Honrath and Jaffe (1992) discussed the wintertime accumulation of ozone precursors in the Arctic as a potential source of enhanced ozone production in spring. None of these suggestions appears to hold. Recently, Wang et al. (1998a, b) have used a three-dimensional global photochemistry–transport model to simulate the tropospheric O_3–NO_x–hydrocarbon chemistry. The results have greatly clarified the situation. Transport of ozone from the stratosphere contributes 20–60% of tropospheric ozone in mid-latitudes in winter and 5–40% in summer. The spring maximum reflects a superimposition of the seasonal maximum of ozone from the stratosphere, which peaks in later winter, and ozone produced photochemically in the troposphere, which peaks in late spring. Although the ozone production is strongest in summer, long-range transport of ozone to the mid-latitudes of the Northern Hemisphere is more effective in spring, when the lifetime of ozone is longer than in summer.

The remaining data in Figure 5.11 are results from the meridional network of Fabian and Pruchniewicz (1977). The stations selected here are the ones that were characterized earlier as typical coastline, mountain, and flatland sites. As Figure 5.11 indicates, the results are more ragged than those from Arosa and the Brocken. Particularly surprising is the great variability at Zugspitze Mountain, because the distance to Arosa is only 130 km, and at an elevation of 2960 m the Zugspitze should be even better suited for observations of tropospheric ozone than other sites in Europe. Unfortunately, neither Zugspitze nor Westerland can be classified as remote sites, as both are tourist attractions and may suffer from local contamination. Westerland is additionally subject to long-distance pollution transport across the North Sea from industrial centers in the British Isles. It is possible, therefore, that anthropogenic effects spoil the data at both locations. The results from Hermanus, a fairly clean air site, are more regular in exhibiting the seasonal cycle. Note that Hermanus lies near 34° S, so that the annual maximum of surface ozone due to stratospheric injection is expected to occur in the second half of the year rather than the first. It is questionable, however, whether the process would be observable at a flatland station such as Hermanus.

Since these data were taken, photochemical air pollution in the Northern Hemisphere has worsened, and even a high altitude site such as Arosa is affected by it. Data from the period 1954–1958 discussed by Staehelin *et al.* (1994) still resemble those shown in Figure 5.11, whereas data from a more recent period, 1989–1991, indicated an increase in O_3 mixing ratio by about 25 nmol mol^{-1} throughout the year, while preserving the seasonal variation featuring a spring maximum. Routine monitoring of surface ozone at many stations within national air quality networks has shown that the seasonal cycle in rural areas often shows a maximum in late summer rather than in spring, thus indicating the influence of regional photochemical air pollution (Singh *et al.*, 1978; Janach, 1989; Logan, 1989). The data sometimes exhibit maxima both in spring and in late summer, and the record at each site must be carefully analyzed to determine the origin of the seasonal variation.

Oltmans and Levy (1994) discussed the diurnal and seasonal behaviors of surface ozone at several background stations, and Figure 5.12 illustrates some of their results. Mauna Loa (Hawaii) and Izaña (Canary Islands) are high-elevation sites, at 3397 and 2360 m, respectively. The data at these stations exhibit the mountain patterns discussed above, with a typical spring maximum and a late summer minimum. At Izaña, the summer minimum is not nearly as deep as that at Mauna Loa, and the seasonal maximum extends into the summer, indicating some anthropogenic influence. Schmidt *et al.* (1988b) noted an enhancement in the transport of ozone from Europe during this period.

Bermuda, Barbados, and Samoa are island stations that show patterns typical of subtropical and low-latitude sites. Ozone reaches a maximum during the early morning hours and a minimum in the afternoon, contrary to rural continental sites. This type of diurnal variation appears to be the result of photochemical ozone destruction during the day in a region with small losses by dry deposition to the sea's surface and relatively low concentrations of nitrogen oxide. The second precondition will be discussed in more detail in Section 5.5. At Bermuda the diurnal variation is seasonally dependent,

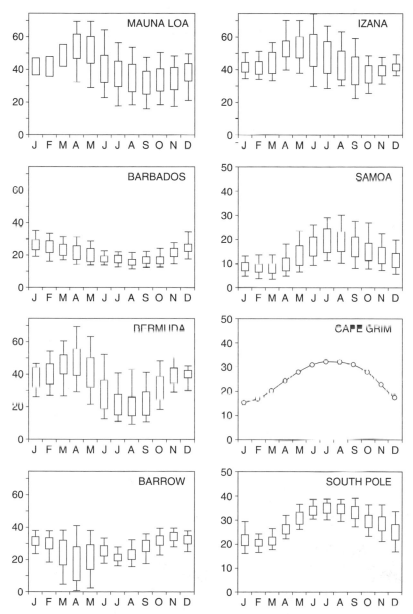

FIGURE 5.12   Monthly mean ozone mixing ratios (nmol mol^{-1}) averaged over a number of years for several baseline stations. Mauna Loa, Hawaii (20° N, 156° W); Izaña, Tenerife (28° N, 16° W); Bermuda (32° N, 64° W); Barbados (13° N, 59° W); Samoa (14° S, 171° W); Cape Grim, Tasmania (41° S, 145° E); South Pole, Antarctica (90° S). Reprinted from *Atmospheric Environment* 28 S. J. Oltmans and H. Levy, Surface ozone measurements from a global network, 9–24 (1994) with permission from Elsevier Science.

with a higher amplitude in summer. The seasonal cycle also has a large amplitude, with a maximum in spring, when daily average values sometimes exceed 70 nmol mol^{-1}. The highest values always occur with the advection of air from the North American continent (Oltmans and Levy, 1992), whereas low concentrations are observed with advection of air from tropical latitudes. During the summer, the influence of air from the continent is restricted by a quasi-stationary anticyclone (Bermuda High), and the advection of pure marine air is dominant. Barbados, being located closer to the equator than Bermuda, shows lower ozone levels throughout the year with a weak winter maximum, which appears to be caused by transport from higher latitudes. At Samoa (14° S), the annual cycle is similar in amplitude to that at Barbados (13° N), except for the 6-month phase shift; that is, the maximum occurs in July during the Southern Hemisphere winter.

At the South Pole, which is a high-elevation site (2835 m a. s. l.), the seasonal variation is similar to that at other sea-level locations in the Southern Hemisphere, such as Cape Grim, Tasmania. In all cases the data exhibit a winter maximum and a summer minimum. Over Antarctica this pattern has prevailed over the last 30 years (Oltmans and Komhyr, 1976), despite a 0.7% annual decrease in the mixing ratio. The Southern Hemisphere, therefore, behaves slightly differently from the Northern Hemisphere, where the annual maximum occurs in the spring. One of the reasons for the difference may be that the Southern Hemisphere does not suffer from anthropogenic pollution to the same degree as the Northern Hemisphere. But it would also be commensurable with observations, recently reconfirmed by Appenzeller *et al.* (1996), that the mass transport across the tropopause, which is responsible for the influx of ozone from the stratosphere, varies annually in such a manner that in the Southern Hemisphere the maximum occurs in winter, whereas in the Northern Hemisphere it occurs in spring.

The behavior of surface ozone in the Arctic differs appreciably from that in other regions. At Barrow, Alaska, there is a distinct spring minimum, which has also been noted elsewhere in the Arctic (Barrie *et al.*, 1988). Ozone levels fluctuate appreciably, and episodes of widespread boundary layer ozone depletion following polar sunrise have been observed at several Arctic measurement sites, including Alert, Canada (Bottenheim *et al.*, 1990; Barrie *et al.*, 1994); Barrow, Alaska (Sturges *et al.*, 1993); Ny-Ålesund, Spitsbergen (Solberg *et al.*, 1996); and Thule, Greenland (Rasmussen *et al.* 1997). The ozone depletion events occur during periods of stable stratification at altitudes up to 2 km, coinciding with the temperature inversion layer. Halogens were suspected to be involved in causing the phenomenon, because the early observations showed a striking anticorrelation between ozone and bromine compounds collected on a cellulose filter. Recent long-path optical absorption measurements have demonstrated that BrO does indeed show up when ozone levels are low (Hausmann and Platt, 1994; Tuckerman *et al.*, 1997). The reactions responsible for ozone destruction would be similar to those occurring in the stratosphere. Bromine has a higher potential for ozone destruction than chlorine under such conditions, because Cl atoms react preferentially with hydrocarbons, whereas the major fate of bromine atoms is reaction with ozone. This forms BrO, which may react with another BrO molecule or with ClO to regenerate Br atoms and thereby continue the chain, or it may react with HO_2 to form HOBr in a chain termination process. The origin of bromine is not yet fully understood. Pure gas-phase reactions are not capable of explaining the explosive increase in the concentration of active halogen and the accompanying sudden ozone loss, and heterogeneous processes either on sea-salt particles or on the Arctic ice surface covered with sea salt have been postulated. Vogt *et al.* (1997) suggested that HOBr reacts with chloride in the aqueous phase of the marine aerosol to form the mixed halogen BrCl, which may react further with bromide to form Br_2; both halogens are released to the gas phase, where they undergo photolysis:

$$HOBr + Cl^- + H + \rightleftharpoons BrCl + H_2O$$

$$BrCl + Br^- \rightarrow Br_2 + Cl^-$$

$$BrCl\,(aqu) \rightarrow BrCl\,(gas)$$

$$Br_2(aqu) \rightarrow Br_2\,(gas)$$

$$BrCl + h\nu \rightarrow Br + Cl$$

$$Br_2 + h\nu \rightarrow Br + Br$$

$$Br + O_3 \rightarrow BrO + O_2$$

$$BrO + BrO \rightarrow Br + Br + O_2$$

$$BrO + HO_2 \rightarrow HOBr$$

Ultimately, the chain is interrupted by the reaction of BrO with HO_2, which reestablishes HOBr, whereby the cycle is closed. The sum of the reactions can be represented by

$$HOBr + 2HO_2 + H^+ + 2O_3 + Br^- + h\nu \rightarrow 2HOBr + 4O_2 + H_2O$$

Two molecules of HOBr are formed for each HOBr spent, so that the process is autocatalytic at the expense of Br^-, which is converted to HOBr. In the gas phase, however, HOBr also undergoes photolysis, whereby bromine atoms are liberated to reenter the ozone destruction cycle. While the reaction scheme contains some speculative elements, it can indeed explain the sudden Arctic ozone depletion events.

Figure 5.13 compares annual averages of surface ozone mixing ratios and their distribution with geographic latitude determined at the baseline stations discussed by Oltmans and Levy (1994) with the earlier data of Fabian and Pruchiewicz (1977). While good agreement exists between the two data sets in the equatorial region and in the Southern Hemisphere, the recent data from the baseline stations in the Northern Hemisphere are higher. The three mountain stations, which are indicated by flags, stick out and indicate the high ozone mixing ratios in the middle troposphere compared to ground-level values. Winkler (1988) has conducted a series of shipboard measurements over the Atlantic Ocean, from which he derived the annual average as a function of latitude. Figure 5.13 includes his results to show that they fit reasonably well into the overall picture. All of the data indicate low ozone mixing ratios in the tropics and a rise toward higher latitudes, in agreement with the distribution shown in Figure 5.6. Johnson *et al.* (1990) have confirmed this behavior by measurements of surface ozone over the Pacific and Indian Oceans during four research cruises. In some cases, mixing ratios of less than 3 nmol mol^{-1} were observed in the central equatorial Pacific. Characteristically, the diurnal variation showed a peak before sunrise and a minimum shortly before sunset, indicative of photochemical destruction

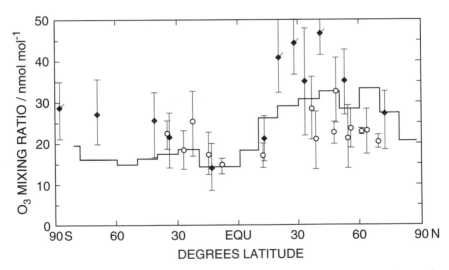

FIGURE 5.13 Latitudinal distribution of the annual average ozone mixing ratio. Diamonds: Measurements at baseline stations (Oltmans and Levy, 1994). Bars give annual range; mountain stations are indicated by flags. Open circles: Network of stations in Europe and Africa (Fabian and Pruchniewicz, 1977). Bars indicate 2σ standard deviation. Solid line: Measurements onboard ships over the Atlantic Ocean (Winkler, 1988).

during daylight. This kind of diurnal variation is also observed at equatorial island stations such as at Barbados and Samoa (Oltmans and Levy, 1994). Singh *et al.* (1996a) considered the possibility that chlorine atoms from photoactive chlorine compounds released from sea salt may contribute to the ozone loss, but they found the abundance of Cl^- atoms to be too small. The concentration of chlorine atoms was estimated from day−night measurements of selected alkanes.

5.4. A MINIMUM TROPOSPHERIC OZONE BUDGET

In this section it will be assumed that the photochemical production of ozone is negligible on a global scale. For such conditions the budget will be dominated by the injection of ozone from the stratosphere and losses at the ground. Injection and loss rates are derived below. For steady-state conditions, the two rates should balance, and if they do not, the existence of additional sources or sinks will be indicated. The discussion will show, however, that within a rather wide margin of error, the two rates are indeed compatible.

5.4.1. INFLUX OF OZONE FROM THE STRATOSPHERE

As discussed in Section 1.5, the principal air exchange mechanisms between stratosphere and troposphere are the seasonal adjustment of the tropopause level, large-scale organized mean air motions in the lower stratosphere, turbulent transport associated with jet streams, and small-scale eddy diffusion. The first three of these processes transfer air in bulk at a rate of about 3×10^{17} kg year^{-1} in the Northern Hemisphere. The contribution of the fourth process is determined by the vertical gradient of O_3 mixing ratios immediately above the tropopause. From Figure 5.5 an annual average gradient of 90 nmol mol^{-1} km^{-1} is estimated for the region poleward of 30° northern latitude. In the Tropics, downward eddy diffusion is counteracted by the upward branch of the Hadley motion.

The ozone flux due to bulk air exchange is obtained by multiplying the exchange rate with an appropriate ozone mass mixing ratio in the lower stratosphere. Figures 3.4 and 5.5 suggest an average value of 1 μg g^{-1}. Both quantities are subject to seasonal variations, however. From observations of radioactive tracers such as strontium 90 it has been shown that the maximum injection rate in spring is about three times the minimum rate in the fall (Danielsen, 1968; Reiter, 1975; Staley, 1982). Similarly, ozone levels in the lower stratosphere reach a maximum in spring, with concentrations about 30% higher than the annual average, and a minimum in the fall, with concentrations 30% lower. Accordingly, the annual flux of ozone into the troposphere of the Northern Hemisphere is estimated as

$$F(O_3) = x(O_3) F_{air}$$

$$= [(1.3 \times 10^{-6})(2 \times 10^{17})] + [(0.7 \times 10^{-6})(1 \times 10^{17})]$$

$$= 3.3 \times 10^{11} \text{ kg year}^{-1}$$

The contribution of small-scale eddy diffusion that must be added is

$$F(O_3) = K_z \rho_{tr}(M_{O_3}/M_{air})[dx(O_3)/dz](A/4)\gamma$$

$$= 0.64 \times 10^{11} \text{ kg year}^{-1}$$

where $K_z = 0.3 \text{ m}^2 \text{ s}^{-1}$ is the eddy diffusion coefficient, M_{O_3} and M_{air} are the molar masses of ozone and air, respectively, and $A = 5 \times 10^{14} \text{ m}^2$ is the surface of the earth. A factor $\gamma = 3.15 \times 10^7$ must be applied to convert seconds to years. The addition of the two fluxes gives 394 Tg year^{-1}. Mohnen (1977) derived a similar estimate from ozone intrusions that occur in association with tropopause folding events in the vicinity of jet streams. He obtained 470–560 Tg year^{-1} for an ozone mixing ratio of 1.3 μg g^{-1} in the lower stratosphere.

Another method of estimating the influx of ozone into the troposphere is based on the observed gradient of ozone mixing ratios in the middle and upper troposphere in conjunction with the eddy diffusion equation. The data of Hering and Borden (1967) as analyzed by Chatfield and Harrison (1977b) may be exploited for this purpose. As Figure 5.4 shows, the average gradients increase with latitude from a typical value of $1 \times 10^{-12} \text{ m}^{-1}$ in the tropics to $7 \times 10^{-12} \text{ m}^{-1}$ at high latitudes. It is assumed that the gradients are zonally representative. Zonal fluxes are calculated by multiplying the gradients with the vertical eddy diffusion coefficient ($K_z = 17 \text{ m}^2 \text{ s}^{-1}$ in the troposphere), the appropriate air density at the altitude considered, and the area of the latitude belt. The results are shown in Table 5.4. The flux above each station should be independent of altitude, but this condition is met only within a fairly wide range of scatter. But the values for the total hemispheric flux, derived by summation over all latitude zones, agree quite well for both altitude layers adopted for the calculation. The total flux also agrees well with the estimates given above.

TABLE 5.4   Downward Flux of Ozone in the Middle Troposphere of the
Northern Hemisphere[a]

Latitude zone	Area $(10^{13}$ m$^2)$	Station	Gradient 5.5–6.5 km $(10^{-12}$ m$^{-1})$	Flux (Tg year$^{-1})$	Gradient 6.5–7.5 km $(10^{-12}$ m$^{-1})$	Flux (Tg year$^{-1})$
50–90°	5.85	Goosebay (53.3°)	7.0	235	10.0	299
40–50°	3.08	Bedford (42.5°)	2.9	51	7.2	113
30–40°	3.57	Wallops (37.5°)	5.4	110	2.7	49
25–30°	1.93	Grand Turk (28.5°)	1.9	21	2.3	22
0–25°	10.56	Canal Zone (9°)	1.3	79	0.9	48
Total flux				497		532

[a] Estimated from average gradients established by the North American Ozonesonde Network (Hering and Broden, 1967; Chatfield and Harrison, 1977b), assuming $K_z = 17$ m^2 s^{-1}.

Table 5.5 summarizes several estimates for the influx of ozone from the stratosphere into the troposphere. Aside from the two methods indicated above, other procedures have been used. One is based on the flux of potential vorticity. In Section 1.5 it was pointed out that a good correlation exists between the ozone mixing ratio and potential vorticity. Mahlman and

TABLE 5.5   Estimates for the Flux of Ozone from the Stratosphere into the Troposphere (Annual Average in Tg year^{-1})

Investigators	Northern Hemisphere	Southern Hemisphere	Remarks
Mohnen (1977), Danielsen and Mohnen (1977)	320–470	—	Stratospheric tropospheric air mass exchange, tropopause folding events plus large-scale circulation
Present (see text)	390	—	Based on the data of Table 1.10
Nastrom (1977)	250	—	Aircraft measurements of O_3 near the tropopause, only for 30° N
Gidel and Shapiro (1980)	310	157	From the flux of potential vorticity in the stratosphere
Present (see text)	490–530	—	From the gradient in the upper troposphere, Table 5.4
Fabian and Pruchniewicz (1977)	425	275	From the phase shift of maxima between total and surface ozone observed at 27 stations
Murphy et al. (1993)	160–545	80–270	From the correlation between NO_y and O_3 in the lower stratosphere (partitioned NH:SH = 2:1)
Beekmann et al. (1997)	230–530	—	From a statistical analysis of the frequency of tropopause folds

Moxim (1978) and Browell *et al.* (1987) showed that any quasi-conservative tracer in the lower stratosphere eventually assumes a positive correlation with potential vorticity, even if the tracer originated from an instantaneous point source. Gidel and Shapiro (1980) and Mahlman *et al.* (1980) made use of this relationship to estimate the flux of ozone on the basis of a general circulation model. The average rates computed by Gidel and Shapiro (1980) for the injection of ozone into the Northern and Southern Hemispheres were 310 and 157 Tg year^{-1}, respectively. The latter value is only one-half that of the former. This result agrees with the aircraft observations (see Fig. 5.7) of ozone in the upper troposphere, which had indicated a reduced injection rate in the region of the southern subtropical jet stream compared to the northern. Murphy *et al.* (1993, 1994) have used the correlation between O_3 and NO_y in the lower stratosphere and the production rate of NO_y from N_2O to calculate the rate of O_3 transfer. NO_y is the sum of all reactive nitrogen oxides ($NO + NO_2 + N_2O_5 + HNO_3 + PAN$).

The estimate of Fabian and Pruchniewicz (1976, 1977) is based on a model for the seasonal behavior of ozone first proposed by Junge (1962). The model assumes that the flux of ozone across the tropopause is proportional to total ozone and that the spring maximum of the flux is responsible for the annual maximum of tropospheric ozone near the Earth's surface. The second maximum occurs somewhat later in the year than the first because of the finite time required for the downward transport of ozone. Ozone injection rates above the station where the observations are made can be calculated from the amplitudes and the phase shift between both (sinusoidal) functions. Although the method would fail if photochemical sources and sinks of tropospheric ozone were important, the results of Fabian and Pruchniewicz are in accord with the other data in Table 5.5. They also indicate a smaller O_3 injection rate in the Southern Hemisphere. The average rates of ozone transfer from the stratosphere to the troposphere, as suggested by the data in Table 5.5, are 400 Tg year^{-1} in the Northern Hemisphere and 210 Tg year^{-1} in the Southern.

5.4.2. DESTRUCTION OF OZONE AT THE GROUND SURFACE

The removal of a trace gas from the atmosphere by absorption and/or destruction at the ground surface was treated in Section 1.6. The flux toward a surface may be expressed in terms of a deposition velocity, which is determined by a series of transfer resistances associated with aerial transport and uptake of material at the surface. Data specific to ozone have been reviewed by Galbally and Roy (1980); these are summarized in Table 1.12.

Resistances to the uptake of ozone by soil, grass, water, and snow were first measured by Aldaz (1969). He observed the rate of disappearance of ozone inside a large box placed open face down on the test surface. The deposition velocities derived (in cm s^{-1}), 0.6 for land, 0.04 for water, and 0.02 for snow and ice, are upper limits because the measurement method does not include the resistance associated with aerial transport. Fabian and Junge (1970) developed a procedure based on the neutral stability model discussed in Section 1.6 by which they corrected the deposition velocities of Aldaz to include the resistances due to turbulent transport in the atmospheric boundary layer and surface roughness. The correction is most significant for continental conditions because uptake rates for ozone by bare and plant-covered soils are fairly high. The solubility of ozone in water is low, so the uptake resistance for seawater is higher than the aerial resistance.

Galbally and Roy (1980) first pointed out that the uptake of ozone by vegetation occurs primarily via the leaf stomata. The surface resistance thus undergoes diurnal variations. Closure of the stomata at night increases the surface resistance from a daytime average of 100 s to 300 s m^{-1} (see Fig. 1.16). Armed with this new information, Galbally and Roy recalculated the global rate of ozone destruction at the earth surface. The results are shown in Table 5.6. Unlike Fabian and Junge (1970), who allowed for different boundary conditions but took the mixing ratio to be that of the free troposphere at 2 km altitude, Galbally and Roy (1980) incorporated the resistance associated with turbulent transport in the atmosphere into the ozone mixing ratio 1 m above the ground. Accordingly, their mixing ratio is lower over the continents than over the ocean. Galbally and Roy also adopted a larger uptake rate of ozone by seawater, because agitation of the water surface was found to increase the uptake rate by a factor of 2 over that reported by Aldaz (1969). This agreed with values derived by Tiefenau and Fabian (Tiefenau and Fabian, 1972, as analyzed by Regener, 1974) from profiles of O_3 mixing ratios over the North Sea.

Table 5.7 presents global ozone destruction rates derived by several investigators. The results are quite consistent, except for the early results of Aldaz (1969), which are too high for the reasons discussed above. The results of Fishman and Crutzen (1978) were based on a less detailed evaluation compared with the estimates of either Galbally and Roy (1980) or Fabian and Pruchniewicz (1977). The last authors applied the procedure of Fabian and Junge (1970) but used updated deposition rates similar to those employed by Galbally and Roy. The accuracy claimed in these studies is about 50%. All authors obtained higher deposition rates in the Northern Hemisphere compared with the Southern. The ratio is approximately 3:2. Fishman and

TABLE 5.6   Ozone Destruction Rates at the Earth's Surface in Different Latitude Zones, Subdivided According to the Type of Surfaces and the Corresponding Average Deposition Velocity v_d and Ozone Mixing ratio $x(O_3)$[a]

Latitude belt	Surface type	Surface area (%)	Surface area (10^{12} m^2)	$x(O_3)$ (μg m^{-3})	v_d (cm s^{-1})	Flux (Tg year^{-1})
60–90° S	Snow/ocean	7	35.7	50	0.10	56.2
30–60° S	Ocean	17	86.7	50	0.10	136.6
	Grassland	0.9	4.6	35.6	0.48	24.7
	Forests	0.1	0.6	35.6	0.63	3.5
15–30° S	Ocean	9	45.9	37.6	0.10	54.4
	Grassland	2.5	12.8	26.4	0.43	45.8
	Forests	0.5	2.6	26.4	0.63	13.6
0.15° S	Ocean	10	51.0	29.6	0.10	47.5
	Grassland	1.4	7.1	20.8	0.43	20.0
	Forests	1.6	8.2	20.8	0.63	34.0
Total flux in the Southern Hemisphere						436.3
0–15° N	Ocean	10	51.0	28.0	0.10	44.9
	Grassland	1.8	9.2	19.6	0.43	25.0
	Forests	1.2	6.1	19.6	0.63	22.7
15–30° N	Ocean	8	40.8	38.4	0.10	49.3
	Grassland	3.8	19.4	26.4	0.43	69.4
	Forests	0.2	1.0	26.4	0.63	5.2
30–60° N	Ocean	9	45.9	44.8	0.10	64.8
	Grassland	5.8	29.6	31.6	0.47	138.5
	Forests	3.2	16.3	31.6	0.63	102.2
60–90° N	Ocean	3.8	19.4	47.2	0.10	28.8
	Grassland	2.2	11.2	33.2	0.47	55.0
	Forests	1.0	5.1	33.2	0.63	33.6
Total flux in the Northern Hemisphere						639.4

[a] Data from Galbally and Roy (1980).

TABLE 5.7   Estimates for the Global Rate of Ozone Deposition in Each Hemisphere (Tg year^{-1})

Authors	Northern Hemisphere	Southern Hemisphere
Aldaz (1969)	905–1308	255–658
Fabian and Junge (1970)	285–491	159–242
Fabian and Pruchniewicz (1977)	302–554	176–378
	425 average	275 average
Fishman and Crutzen (1978)	784	270
Galbally and Roy (1980)	640	436

Crutzen found a higher ratio of about 3:1. The difference derives partly from their use of a lower destruction rate over the oceans, but also from an overestimate of the loss rate in the Northern Hemisphere.

The ozone deposition rates in each hemisphere compare well with the corresponding ozone injection rates shown in Table 5.5. The latter are somewhat smaller than the former, but the rates for the transport of ozone across the tropopause are more difficult to estimate and hence carry larger uncertainties. Within the ranges of uncertainty the two rates are equivalent, making it appear that the budget of ozone in the troposphere is fairly well balanced by these two fluxes alone. Note that the asymmetry between the two hemispheres occurs in both data sets, that is, deposition as well as influx rates are reduced in the Southern Hemisphere. The data of Fabian and Pruchniewicz (1977) are biased in this regard because in their model the two rates are artificially balanced.

5.5. PHOTOCHEMICAL PRODUCTION AND LOSS OF OZONE IN THE UNPERTURBED TROPOSPHERE

5.5.1. REACTION MECHANISM

The assumption made in the preceding section that ozone is photochemically stable in the troposphere is basically incorrect, and this aspect is finally considered. It is true that the photodissociation of ozone leading to the formation of ^3P oxygen atoms does not cause losses, because their subsequent addition to molecular oxygen regenerates ozone. In Section 4.2 it was shown, however, that a part of the $O(^1D)$ atoms produced in the long-wavelength tail of the Hartley absorption band ($\lambda < 320$ nm) reacts with water vapor to form OH radicals, and this process represents a loss of tropospheric ozone. Such losses are most severe in the surface boundary layer, where water vapor concentration are high, whereas in the upper troposphere the losses would be much smaller. The magnitude of the loss may be estimated from the OH production rates reported in the literature. For example, Fishman et al. (1979) calculated column OH production rates of 10.5×10^{14} and 5.0×10^{14} molecule m^{-2} s^{-1} for the Northern and Southern Hemispheres, respectively. The associated ozone loss rates are 668 and 321 Tg year^{-1}, respectively, indicating a very severe perturbation of the minimum ozone budget discussed previously. Because the production of OH radicals maximizes in the Tropics, the loss of ozone would also maximize there.

Additional losses occur because of the reactions

$$OH + O_3 \rightarrow O_2 + HO_2$$

$$HO_2 + O_3 \rightarrow 2O_2 + OH$$

The second of these reactions is more important than the first, since the OH radicals are converted to HO_2, mainly by reacting with methane and carbon monoxide rather than with ozone. All of these losses are at least partly retrieved by the formation of ozone by the photodissociation of NO_2, when NO is converted to NO_2 in smog-like reactions involving the oxidation of methane and carbon monoxide, as was first pointed out by Crutzen (1973). In fact, under favorable conditions, more ozone may be formed in this manner than is lost by the conversion of $O(^1D)$ to OH radicals. The oxidation of CH_4 and CO was discussed in Section 4.2. Here, we will reconsider it from the viewpoint of its ozone-generating potential. If it is assumed that peroxy radicals react only with nitric oxide, the reaction of OH with methane induces the following reaction sequence:

$$CH_4 + OH \rightarrow CH_3 + H_2O$$
$$CH_3 + O_2 + M \rightarrow CH_3O_2 + M$$
$$CH_3O_2 + NO \rightarrow CH_3O + NO_2$$
$$CH_3O + O_2 \rightarrow HCHO + HO_2$$
$$HO_2 + NO \rightarrow NO_2 + OH$$
$$\left. \begin{array}{c} NO_2 + h\nu \rightarrow NO + O \\ O + O_2 + M \rightarrow O_3 + M \end{array} \right\} \times 2$$

net: $$CH_4 + 4O_2 \rightarrow HCHO + H_2O + 2O_3$$

In this scheme two molecules of ozone are formed for each molecule of methane consumed. The subsequent photooxidation of formaldehyde produces equivalent amounts of carbon monoxide, which also undergoes oxidation by OH radicals:

$$CO + OH \rightarrow CO_2 + H$$
$$H + O_2 + M \rightarrow HO_2 + M$$
$$HO_2 + NO \rightarrow NO_2 + OH$$
$$NO_2 + h\nu \rightarrow NO + O$$
$$O + O_2 + M \rightarrow O_3 + M$$

net: $$CO + 2O_2 \rightarrow CO_2 + O_3$$

Here, one ozone molecule is formed for each CO molecule that is oxidized. Altogether, the methane oxidation mechanism is capable of generating three ozone molecules for each methane molecule undergoing oxidation. Still not considered in this account is the formation of HO_2 radicals by the photooxidation of formaldehyde, which convert further NO molecules to NO_2, thereby increasing the yield of ozone.

There should be no doubt that it is quite unrealistic to assume that CH_3O_2 and HO_2 radicals react solely with NO. Radical termination reactions must compete with chain propagation reactions, lest the chain continue indefinitely. Termination reactions divert CH_3O_2 and HO_2 into other reaction channels, so that the yield of ozone does not attain maximum values. The principal termination reactions are

$$HO_2 + HO_2 \rightarrow H_2O_2 + O_2$$

$$OH + HO_2 \rightarrow H_2O + O_2$$

$$HO_2 + CH_3O_2 \rightarrow CH_3OOH + O_2$$

They are in competition with the ozone production and loss reactions:

$$HO_2 + NO \rightarrow NO_2 + OH$$

$$HO_2 + O_3 \rightarrow 2O_2 + OH$$

A production of ozone occurs only by that fraction of HO_2 radicals entering into reaction with NO, whereas the fraction entering into reaction with O_3 causes a loss of ozone. The relative probability of each of the above reaction steps is determined by the steady state concentrations of the corresponding HO_2 reaction partners. Apart from the photochemical production rate, the concentrations of the radicals depend critically on the extent of the recycling reactions of HO_2 with NO and O_3, that is, on the concentrations of nitric oxide and ozone. Since the concentration of ozone may be considered fixed, it is the concentration of NO that is the crucial parameter. To provide a more quantitative assessment of the situation, the concentration of NO will be used as a running parameter, and the radical concentrations as far as needed will be taken from the numerical data of Fishman et al. (1979), who calculated concentrations and reaction probabilities for representative boundary-layer conditions.

Figure 5.14 shows the fraction of HO_2 radicals undergoing reaction with NO and O_3 as a function of NO number density. The first reaction predominates when NO concentrations are high, and the second when they are low. The cross-over point is reached when the two reactions proceed at equal rates according to the condition

$$n(NO) = (k_{59}/k_{60})n(O_3) = (1.8 \times 10^{-15}/8.0 \times 10^{-12})8 \times 10^{11}$$

$$= 1.8 \times 10^8 \text{ molecule cm}^{-3}$$

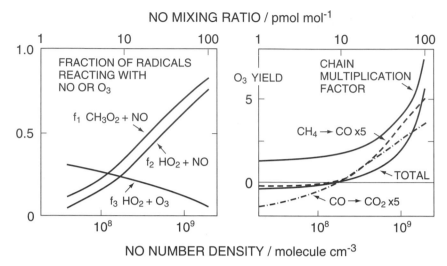

FIGURE 5.14   Left: Fraction of CH_3O_2 and HO_2 radicals reacting with NO, and of HO_2 reacting with O_3 as a function of NO number density. The corresponding NO mixing ratio at ground level is shown on the upper scale. Right: Net production of ozone in numbers of O_3 molecules formed per molecule of CH_4 oxidized to CO and per CO molecule oxidized to CO_2, respectively, both expanded by a factor of 5, and the total number of O_3 molecules formed per OH radical reacted, taking chain multiplication into account. Negative values indicate ozone loss. The chain multiplication factor is also shown.

where the NO concentration corresponds to a mixing ratio of about 10 pmol mol^{-1} (for $x(O_3) = 32$ nmol mol^{-1}). The probability of CH_3O_2 radicals entering into reaction with nitric oxide can be estimated in a similar way from the two competing reactions of CH_3O_2 with NO and HO_2. The reaction of CH_3O_2 with itself is slower and can be ignored. Figure 5.14 shows, as a function of $n(NO)$, the relative probability f_1 for the reaction of CH_3O_2 with NO, and the fractions f_2 and f_3 for the reactions of HO_2 with NO and O_3, respectively.

In determining the yield of ozone from reactions of OH with methane and carbon monoxide, it is instructive to trace the pertinent reaction pathways by means of the flow diagram shown in Figure 5.15. A yield of 0.8 is assumed for HO_2 radicals resulting from the photodissociation of formaldehyde. The mixing ratios of CH_4 and CO are taken to be 1.7 and 0.12 μmol mol^{-1}, respectively, corresponding to conditions in the Northern Hemisphere. About 23% of the OH radicals react with methane, 72% with carbon monoxide, and the rest with HO_2 radicals. The appropriate branching factors are indicated in the figure. The number of ozone molecules resulting from the

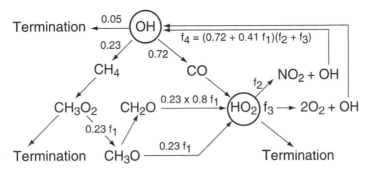

FIGURE 5.15 Flow chart for the formation of ozone from reactions of OH with CH_4 and CO present with mixing ratios of 1.7 and 0.12 μmol mol^{-1}, respectively. The term f_1 is the relative probability for the reaction of CH_3O_2 with NO, f_2 that of HO_2 reacting with NO, and f_3 that of HO_2 reacting with ozone; f_4 is the fraction of OH radicals regenerated in each cycle.

oxidation of one CO molecule is $f_2 - f_3$, and the number of ozone molecules resulting from the oxidation of methane to CO is $1.8f_1(f_2 - f_3)$. The right-hand side of Figure 5.14 shows these individual ozone yields as a function of NO number density. The average yield of ozone—that is, the yield obtained for one OH radical starting the reaction chain—is given by $(0.72 + 0.23 \times 1.8f_1)(f_2 - f_3)p$, where p is a multiplier determined by the effective chain length. The fraction of OH radicals regenerated in each cycle is $f_4 = (0.72 + 0.23 \times 1.8f_1) \times (f_2 + f_3)$. The total number of OH radicals that have reacted after the cycle is repeated an infinite number of times is given by the geometric progression

$$p = \sum_{s}^{\infty} f_4^s = 1/(1 - f_4),$$

and the total yield of ozone from one primary OH radical is

$$Y(O_3) = (0.72 + 0.23 \times 1.8f_1)(f_2 - f_3)/(1 - f_4)$$

The dependence on the multiplier p and the total yield on the NO number density are shown on the right-hand side of Figure 5.14. Both increase strongly with the NO concentration.

As two OH radicals are originally formed for each $O(^1D)$ atom having reacted with H_2O, the loss of ozone due to this reaction is compensated for when the ozone yield from the reaction of OH with methane is 0.5. A yield of 0.5 is obtained for an NO concentration of 4×10^8 molecule cm^{-3}, corresponding to an NO mixing ratio of about 20 pmol mol^{-1}. At lower NO concentrations the formation of OH from $O(^1D)$ represents an ozone loss, whereas higher concentrations cause a net production of ozone. For steady-state conditions, as explained previously, the NO/NO_2 ratio should be given by

$$n(NO)/n(NO_2) = j(NO_2)/[k_{66}n(O_3) + k_{80}n(CH_3O_2) + k_{60}n(HO_2)]$$

With the numerical values $j(NO_2) = 0.005$, $n(O_3) = 8 \times 10^{11}$, $n(CH_3O_2)$ $= 3 \times 10^8$, and $n(HO_2) = 6 \times 10^8$ molecule cm^{-3} gleaned from Fishman et al. (1979), one obtains $n(NO)/n(NO_2) = 0.29$ or $n(NO)/[n(NO) + n(NO_2)] = 0.225$. Accordingly, approximately one-quarter of the total amount of nitrogen oxides is present as NO under daylight conditions. The total concentration of nitrogen oxides needed to reach the break-even point between photochemical production and destruction of ozone then is $4 \times 10^8/0.225 = 1.8 \times 10^9$ molecule cm^{-3} or 90 pmol mol^{-1}. The value corresponds roughly to the mixing ratio of NO_x occurring in tropospheric background air of the Northern Hemisphere (see Section 9.4.2). It thus appears that photochemical production and loss rates for ozone are approximately balanced.

Figure 5.14 should not be taken to indicate that the ozone production rate increases indefinitely with rising NO concentration. The reaction mechanism considered above does not yet include reactions of radicals with NO_2. While these reactions can be neglected for NO_2 mixing ratios less than 200 μmol mol^{-1}, they become increasingly important as the NO_2 mixing ratio rises further. The reaction of foremost importance is the termination process

$$OH + NO_2 \rightarrow HNO_3$$

but other reactions, such as the addition of HO_2 to NO_2 to form peroxynitric acid, also impede chain propagation in that they temporarily trap radicals. Figure 5.16 shows the general behavior of the steady-state ozone mixing ratio resulting from the oxidation of methane and carbon monoxide in the boundary layer. With increasing NO_x mixing ratio, ozone first rises as discussed above, then reaches a maximum of about 26 nmol mol^{-1} when $x(NO_x) \approx 500$ μmol mol^{-1}, and finally declines again for still higher NO_x concentrations. For low NO_x mixing ratios, steady-state ozone is controlled by loss reactions of radicals among themselves, whereas for high NO_x mixing ratios the control is exerted by radical termination with NO_2. The results of Stewart et al. (1977), which are shown in Figure 5.16, and those of Fishman et al. (1979) were based on slightly different mechanisms and rate coefficients, and the two groups derived numerically different radical concentrations. Yet they found a qualitatively similar behavior. The maximum ozone mixing ratio of 26 nmol mol^{-1} evident in Figure 5.16 lies in the range of mixing ratios frequently observed in the continental boundary layer in

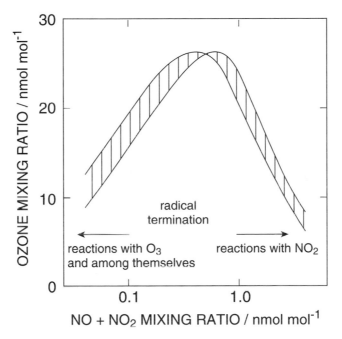

FIGURE 5.16 Photochemical steady state ozone mixing ratios resulting from the oxidation of CH_4 and CO in the boundary layer, calculated as a function of the $NO_x \equiv NO + NO_2$ mixing ratio by Stewart *et al.* (1977). The range of O_3 mixing ratios indicated by the hatched area corresponds to relative humidities between 20% and 80%.

middle latitudes of the Northern Hemisphere. Over the continents, where the NO_x mixing ratio is sufficiently high, it obviously is not possible to distinguish from mixing ratios alone whether the ozone that one observes has its origin in the stratosphere or has resulted from *in situ* photochemical production.

5.5.2. COMBINED TROPOSPHERIC OZONE BUDGET

Fishman *et al.* (1979) used a two-dimensional model to calculate ozone production and equivalent loss rates. They found 1260 Tg year^{-1} for the Northern and 690 Tg year^{-1} for the Southern Hemisphere, about three times higher than the influx of ozone from the stratosphere or destruction at the ground. The budget depends critically on the average NO_x mixing ratio in the troposphere. Fishman *et al.* used a prescribed NO_x altitude profile with a mixing ratio at the ground of approximately 100 pmol mol^{-1} and a total tropospheric column density of about 3×10^{18} molecule m^{-2}. Crutzen and Zimmermann (1991) and Wang *et al.* (1998a, b) have developed three-dimensional global photochemistry–transport models, in which the source distribution of important trace gases such as NO$_x$ is prescribed. The ozone budget derived from these models is dominated by photochemical ozone production and destruction, especially in the Southern Hemisphere. However, the net chemical source, that is, the excess of production over destruction, has a magnitude similar to that of the influx of ozone from the stratosphere and its destruction at the Earth's surface. Table 5.8 presents results obtained by Crutzen and Zimmermann (1991) based on CH$_4$-CO-NO$_x$ chemistry. They have also considered the preindustrial troposphere, which was characterized by smaller sources of CH$_4$, CO, and NO$_x$. In this case, as Table 5.8 indicates, photochemical production and loss rates were significantly smaller, although still dominant in the Southern Hemisphere. The net chemical production rate was smaller than the influx from the stratosphere or destruction at the ground by dry deposition. Wang *et al.* (1998a) included nonmethane hydrocarbons in their model, with about 50% of the emissions being due to isoprene. Wang *et al.* (1998a) found that the addition of nonmethane hydrocarbons raised ozone production by about 10%. The overall photochemical production and loss rates were nevertheless quite similar to those shown in Table 5.8.

Finally we consider the residence time of ozone in the atmosphere. Owing to the uneven distribution of ozone, the tropospheric mass content can be derived only approximately. For the present purpose use may be made of the

TABLE 5.8   Global Ozone Budgets (Tg year^{-1}) for the Present and the Preindustrial Troposphere[a]

	Present troposphere			Preindustrial troposphere		
	NH	SH	Global	NH	SH	Global
Sources						
Stratosphere	335	145	480	335	145	480
HO$_2$ + NO	1970	1150	3120	770	620	1390
CH$_3$O$_2$ + NO	480	335	815	195	195	390
Sinks						
O(^{1}D) + H$_2$O	1055	770	1825	575	480	1055
HO$_2$/OH + O$_3$	865	480	1345	335	240	575
Deposition	865	380	1245	390	240	630
Net chemical source	530	235	765	55	95	150

[a] From Crutzen (1995). NH, Northern Hemisphere; SH, Southern Hemisphere.

data in Figure 5.6 to estimate average mixing ratios of 30 and 20 nmol mol^{-1} for the tropical regions north and south of the equator, and 30 and 45 nmol mol^{-1} poleward of 30° S and 30° N, respectively. Tropospheric air masses in the 0–30° and 30–90° latitude regions are 1.13×10^{18} kg and 9.9×10^{17} kg. Thus, the mass of ozone in the Northern Hemisphere is about 56 + 74 = 130 Tg, while that in the Southern Hemisphere is about 38 + 49 = 87 Tg. The corresponding deposition rates suggested in Table 5.8 are 865 and 380 Tg $year^{-1}$, respectively, which leads to a residence time of 55 days in the Northern Hemisphere and 85 days in the Southern. Within the uncertainty of the estimate, the residence time is about 2 months in either hemisphere. As the residence time involves only the final loss of ozone at the ground surface and not the more frequent photochemical production and loss processes, the turnover of ozone is faster and its lifetime correspondingly shorter, 17 days in the Northern and 19 days in the Southern Hemisphere. In the model simulation performed by Wang *et al.* (1989a), the mass of ozone in the troposphere (actually below the 150-hPa level, which includes the high-latitude stratosphere) was calculated to be 180 Tg in the Northern Hemisphere and 130 Tg in the Southern. Their data lead to annual residence times of 115 and 164 days, and lifetimes of 23 and 28 days, respectively. The photochemical lifetimes depend on the season (that is, they are longer in winter), and the residence time is also expected to show some seasonal dependence, at least at middle latitudes, because of changes in the rate of deposition on vegetation.

CHAPTER 6

Hydrocarbons, Halocarbons, and Other Volatile Organic Compounds

Hydrocarbons, as the name suggests, are compounds consisting of the elements hydrogen and carbon. Methane, the simplest and most abundant hydrocarbon in the atmosphere, was discussed in Section 4.3. Here, we consider other members of the family as far as they are volatile enough to reside primarily in the gas phase. This emphasizes compounds with low molar mass. The substances are often referred to as *nonmethane hydrocarbons* (NMHC) or, when other organic compounds are included, as *nonmethane volatile organic compounds* (NMVOC). The latter are mainly oxygenated organic compounds, such as aldehydes, ketones, alcohols, esters, and organic acids. Many of them are products resulting from the oxidation of hydrocarbons, but direct natural emissions occur as well. Hydrocarbons are emphasized here, but biogenic emissions of some other organic volatiles are also considered. Formic and acetic acids are discussed in more detail. Halocarbons are derivatives of hydrocarbons in which the hydrogen atoms are replaced by fluorine, chlorine, bromine, or iodine. Compounds in which hydrogen is only partly substituted by halogens are also included under this heading.

6.1. HYDROCARBONS AND SOME OTHER ORGANIC VOLATILES

The number of hydrocarbons in the atmosphere is potentially very large. Vapor pressures are favorable, and the heavier species admit many isomers. In urban areas several hundred different hydrocarbons have been identified by gas chromatography (Appel *et al.*, 1979; Louw *et al.*, 1977), including saturated compounds (alkanes), unsaturated species with one carbon-carbon double bond (alkenes), unsaturated species with two such double bonds (alkadienes), acetylene type compounds (alkynes), and benzene derivatives or aromatic compounds (arenes). To separate that many compounds requires sophisticated gas chromatographic techniques. Even more difficult are the sampling and analysis of hydrocarbons in the remote atmosphere, because it is necessary to first accumulate the material and then transfer it to the laboratory for analysis. This requires elaborate precautions against contamination or degradation of the sample during transport and in the laboratory as well.

6.1.1. HYDROCARBON LIFETIMES

All hydrocarbons except methane react rapidly with OH radicals, and these reactions are the principal removal processes for hydrocarbons in the atmosphere. Unsaturated hydrocarbons also react with ozone at rates that make such reactions competitive with OH reactions in the atmosphere. Some unsaturated hydrocarbons, especially the terpenes, react additionally with NO_3, but these reactions are important only at night. Table 6.1 lists for a number of hydrocarbons the rate coefficients of their reactions with OH and O_3. The corresponding atmospheric reaction rates may be estimated on the assumption that typical number densities for OH and O_3 are 8×10^5 and 6.5×10^{11} molecule cm^{-3}, respectively. Atmospheric lifetimes are then obtained by summing the individual rates and taking the inverse value. Lifetimes thus calculated are included in Table 6.1. The exercise shows that hydrocarbon lifetimes in the troposphere generally are on the order of days. Owing to their reactivity, hydrocarbons are expected to develop larger gradients of mixing ratios between the source regions and the remoter troposphere. As the sources occur mainly at the surface of the continents, the gradients will be directed in the vertical direction and from the continents toward the ocean. Figure 6.1 shows vertical profiles for ethane, ethene, and propane over Europe to demonstrate that the expected behavior is actually observed.

TABLE 6.1   Hydrocarbon Reactivities: Rate Coefficients (at 298 K) and Rates for Reactions with OH Radicals and Ozone, and the Corresponding Lifetimes in the Troposphere[a, b]

Compound	$10^{12} \, k_{OH}$ $(cm^3 \, molecule^{-1} \, s^{-1})$	$10^{17} \, k_{O_3}$	$10^6 \, R_{OH}$ (s^{-1})	$10^6 \, R_{O_3}$ (s^{-1})	τ (days)
Alkanes					
Ethane, C_2H_6	0.26	c	0.2	—	56
Propane, C_3H_8	1.15	c	0.9	—	12
n-Butane, C_4H_{10}	2.54	c	2.0	—	5.7
Isobutane, C_4H_{10}	2.33	c	1.9	—	6.2
n-Pentane, C_5H_{12}	3.94	c	3.1	—	3.7
Isopentane, C_5H_{12}	3.90	c	3.1	—	3.7
n-Octane, C_8H_{18}	8.68	c	6.9	—	1.7
Alkenes, alkadienes and alkynes					
Ethene, C_2H_4	8.52	0.16	6.8	1.2	1.4
Propene, C_3H_6	26.3	1.0	21.0	6.5	0.4
1-Butene, C_4H_8	31.4	0.96	25.1	6.2	0.4
cis/trans-2-Butene, C_4H_8	60.2	15.7	48.1	102	0.08
1-Pentene, C_5H_{10}	31.4	1.0	25.1	6.5	0.4
2-Methyl-2-Butene, C_5H_{10}	86.9	40.0	69.5	260	0.04
1,3-Butadiene, C_4H_6	66.6	0.63	53.3	4.1	0.2
Isoprene, C_5H_8	101	1.28	80.8	8.3	0.1
Acetylene, C_2H_2	0.8	0.0008	0.65	0.005	18
Aromatic compounds					
Benzene, C_6H_6	1.23	c	1.0	—	12
Toluene, $C_6H_5CH_3$	6.0	c	4.8	—	2.4
o-Xylene, $C_6H_4(CH_3)_2$	13.7	c	11.0	—	1.0
m-Xylene, $C_6H_4(CH_3)_2$	23.6	c	18.9	—	0.6
p-Xylene, $C_6H_4(CH_3)_2$	14.3	c	11.4	—	1.0
Ethylbenzene, $C_6H_5C_2H_5$	7.1	c	5.7	—	2.0
Monoterpenes,[e] $C_{10}H_{16}$					
α-Pinene	53.7	8.7	42.9	56.5	0.1
β-Pinene	78.9	1.5	63.1	9.7	0.16
Limonene	171	20	137	130	0.04
Myrcene	215	47	172	305	0.02
3-Carene	80	3.7	64	24	0.13

(Continues)

Another interesting aspect is the filter effect associated with long-range transport. Because reactive hydrocarbons are removed at a faster rate than less reactive ones, the abundance spectrum changes in favor of the less reactive species. The longest-lived compounds are ethane and acetylene. They have lifetimes exceeding 1 month, so they are expected to spread around the globe, irrespective of the uneven distribution of their sources. At the other end of the reactivity scale are several alkenes, and especially the

TABLE 6.1 (*Continued*)

Compound	$10^{12} k_{OH}$ $10^{17} k_{O_3}$ (cm³ molecule⁻¹ s⁻¹)		$10^6 R_{OH}$ (s⁻¹)	$10^6 R_{O_3}$ (s⁻¹)	τ (days)
Aldehydes and ketones					
Formaldehyde, HCHO	9.4	*c*	7.5	—	1.5[d]
Acetaldehyde, CH₃CHO	15.8	*c*	12.6	—	0.9[d]
Benzaldehyde, C₆H₅CHO	13.0	*c*	10.4	—	1.1[d]
Acetone, CH₃COCH₃	0.22	*c*	0.17	—	66[d]
Ethyl methyl ketone, C₂H₅COCH₃	11.5	*c*	9.2	—	1.3[d]
Methacrolein	33.5	0.11	26.8	0.74	0.4
Methyl vinyl ketone	18.8	0.46	15.0	3.0	0.6
Alcohols					
Methanol, CH₃OH	0.94	*c*	0.75	—	15
Ethanol, C₂H₅OH	3.3	*c*	2.6	—	4.4
1-Butanol, C₄H₉OH	8.6	*c*	6.8	—	1.7
Carboxylic acids					
Formic acid	0.48	*c*	0.38	—	30
Acetic acid	0.74	*c*	0.59	—	20

[a] Assuming $n(OH) = 8 \times 10^5$, $n(O_3) = 6.5 \times 10^{11}$ molecule cm⁻³.
[b] Rate coefficients from Atkinson (1994).
[c] Rate coefficients for these reactions are less than $\sim 1 \times 10^{-21}$ cm³ molecule⁻¹ s⁻¹.
[d] Upper limit lifetimes, as aldehydes and ketones are subject to photodecomposition; this applies especially to formaldehyde and acetone.
[e] The structures of individual terpenes are shown in Figure 6.2.

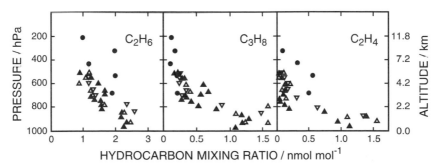

FIGURE 6.1 Vertical distribution of ethane, propane, and ethene over Europe according to Ehhalt *et al.* (1986).

terpenes, which have lifetimes of hours and cannot spread very far from the source regions.

Table 6.1 includes rate coefficients and lifetimes for a few oxygenated compounds of interest. The aldehydes and ketones are products resulting from the oxidation of hydrocarbons, whereas the alcohols are not. Our knowledge of alcohols in the atmosphere is too fragmentary to discuss them in detail here. Methanol is one of the substances known to be emitted from plants (Puxbaum, 1997). The lifetimes of alcohols are slightly shorter than those of the corresponding alkanes. The lifetimes of aldehyde and ketones are relatively short, except that of acetone. These compounds are additionally subject to photodecomposition, and the lifetimes in Table 6.1 represent upper limits. The case of formaldehyde was treated in Section 4.4. The rate of photodissociation of acetone is similar to that of reaction with OH radicals (Meyrahn *et al.*, 1986), so that that the lifetime is about 30 days.

While rate coefficients for reactions with OH radicals with the lower members of the alkane and alkene series are well known from laboratory measurements, only a few data exist for the higher homologues. It is possible, however, to derive rate coefficients by an extrapolation of existing data where needed, and this makes it possible to estimate lifetimes for compounds for which OH rate coefficients have not been specifically deter-mined. Greiner (1970) first showed that the overall rate coefficient for hydrogen abstraction from an alkane can be expressed by the sum of rate constants for the abstraction of primary, secondary and tertiary hydrogen atoms. Primary H-atoms are associated with CH_3 groups, secondary H-atoms are attached to $-CH_2-$ links in the hydrocarbon chain, and tertiary H-atoms occur at chain branching points. The C-H bond dissociation energy decreases when going from primary to secondary, and then on to tertiary H-atoms, and the activation energy for the corresponding abstraction reaction decreases as well. Therefore, one can write

$$k_{OH} = N_P k_P + N_S k_S + N_T k_T$$

$$= N_P A_P \exp(-E_P/R_g T) + N_S A_S \exp(-E_S/R_g T)$$

$$+ N_T A_T \exp(-E_T/R_g T) \tag{6.1}$$

where N_P, N_S, and N_T, respectively, are the number of primary, secondary, and tertiary hydrogen atoms associated with the alkane, A_n are the preexpo-nential factors, and E_n are the activation energies of the individual rate coefficients. Greiner (1970) deduced the following values

$$k_P = 1.02 \times 10^{-12} \exp(-823/T) \tag{6.1a}$$

$$k_S = 2.34 \times 10^{-12} \exp(-428/T) \tag{6.1b}$$

$$k_T = 2.09 \times 10^{-12} \exp(95.6/T) \tag{6.1c}$$

The accuracy of the formula for C_2–C_6 hydrocarbons is about 20%, and this is also expected to hold for the higher homologues. Over a wider temperature range, the A-factors are not fully independent of temperature. This can be accommodated by an additional T^m dependence. Although the expression derived by Greiner (1970) is still useful, the larger data base that subsequently became available led Atkinson (1987) to a modified approach, in which he made allowance for the influence of neighboring hydrocarbon groups on the individual rate coefficients. He derived the following formulae for the rate coefficients (units: cm^3 $molecule^{-1}$ s^{-1})

$$k(CH_3\text{-}X) = k_{prim} F(X) \tag{6.2a}$$

$$k(X\text{-}CH_2\text{-}Y) = k_{sec} F(X) F(Y) \tag{6.2b}$$

$$k\left(X\text{-}CH{<}_Z^Y\right) = k_{prim} F(X) F(Y) F(Z) \tag{6.2c}$$

$$k_{prim} = 3.97 \times 10^{-13} (T/298)^2 \exp(-303/T)$$

$$k_{sec} = 3.84 \times 10^{-13} (T/298)^2 \exp(233/T)$$

$$k_{tert} = 1.68 \times 10^{-13} (T/298)^2 \exp(711/T)$$

where k_{prim}, k_{sec}, and k_{tert} are the OH radical rate constants per -CH_3, -CH_2-, and >CH-group, respectively, for X = Y = Z = -CH_3 as standard substituent group, and $F(X)$, $F(Y)$, and $F(Z)$ are correction factors for actually occurring substituent groups. As Atkinson (1987) showed, $F(\text{-}CH_3)$ = 1, and $F(\text{-}CH_2\text{-}) = F(>CH) = F(>C<) = \exp(76/T)$. For a temperature of 298 K, $\exp(76/T) \approx 1.3$, and $k_{prim} = 1.44 \times 10^{-13}$, $k_{sec} = 8.39 \times 10^{-13}$, and $k_{tert} = 1.83 \times 10^{-12}$. For cycloalkanes, an additional factor must be applied to account for the ring strain involved. The above formulae also allow the calculation of the distribution of isomeric alkyl radicals that result from the OH reaction with a given alkane.

The reaction of OH radicals with alkenes (and aromatic compounds) occurs predominantly by addition to the double bond, so that the above formulae are not applicable. This does not preclude additional hydrogen abstraction from long-chained alkenes and aromatic compounds with long side chains.

6.1.2. ANTHROPOGENIC SOURCES OF ATMOSPHERIC HYDROCARBONS

Hydrocarbons are known to be emitted from anthropogenic sources, from vegetation and soils, and from the oceans. Anthropogenic emissions of hydrocarbons were first reviewed by Tuesday (1976) with specific reference to the situation in the United States. Subsequently, Duce *et al.* (1983) reviewed global-scale emission rates. They considered mostly bulk emissions and not individual compounds. Information on the distribution of individual hydrocarbons being emitted is still fragmentary. Ehhalt *et al.* (1986) prepared a more detailed global source estimate for C_2–C_5 hydrocarbons. Atlas *et al.* (1993a) discussed many other aspects, including anthropogenic polyaromatic compounds and polychlorinated hydrocarbons.

Anthropogenic sources dominate in the cities. Table 6.2 shows mixing ratios of major hydrocarbons that were observed in the 1970s in several North American cities and in Sidney, Australia. The data for Los Angeles in Table 5.1 may be added to the list. A notable feature is the similarity of hydrocarbon compositions in different urban regions. This is not really surprising, however, because source distributions in large cities are much the same. Major sources are vehicle exhaust emissions, gasoline evaporation and spillage, leaks of commercial natural gas, emissions from petroleum manufacturing plants and refineries, and chemical solvents. To this may be added municipal waste, either from incineration or from rotting deposits. Mayrson and Crabtree (1976), who carried out a source apportionment for Los Angeles, estimated the following relative contributions (in mass percentage): car exhaust, 48.8; gasoline, 16.3; gasoline evaporation, 13.3; commercial natural gas, 5.3; and natural gas from oil fields and other production operations, 15.3. Industrial and solvent-related emissions were not considered. For Sidney, Australia, Nelson *et al.* (1983) derived the following source apportionment (in mass percentage): car exhaust, 35.8; gasoline, 16.3; gasoline evaporation, 16.0; solvents, 22.9; commercial natural gas, 3.7; and industrial process emissions, 5.4. Thus, approximately 73% of all emissions arose from the operation of automobiles and their supply with fuel. In the meantime, emission control measures have reduced such emissions, at least in the United States. Data assembled by the Environmental Protection Agency (EPA, 1987; Müller, 1992) show that in the United States volatile hydrocarbon emissions due to road traffic have diminished from 12.3 Tg

TABLE 6.2   Mixing Ratios (nmol mol^{-1}) of Major Hydrocarbons in Several U.S. Cities (Arnts and Meeks, 1981; Sexton and Westberg, 1984) and in Sidney, Australia (Nelson *et al.*, 1983)

Compound	Houston	Philadelphia	Boston	Tulsa	Milwaukee	Sidney
Ethane	12.5	6.5	4.0	4.5	4.5	7.5
Propane	17.0	9.7	3.0	3.1	3.3	5.9
n-Butane	16.0	11.5	7.2	12.5	7.0	7.4
Isobutane	8.2	5.2	3.0	3.1	1.7	4.7
n-Pentane	7.6	5.4	3.2	8.2	2.4	5.0
Isopentane	13.4	8.4	7.0	13.2	4.6	9.0
2-Methylpentane	3.3	2.7	2.0	3.6	1.2	2.6
3-Methylpentane	2.5	1.8	1.3	2.2	0.8	1.6
Ethene	—	—	—	3.4	—	12.5
Propene	5.7	3.3	1.3	1.0	1.0	7.4
1-Butene	1.0	0.25	0.2	—	0.25	1.0
Isobutene	1.5	0.25	0.2	1.1	0.5	1.4
trans-2-Butene	1.2	0.25	0.25	1.1	0.25	1.1
cis-2-Butene	—	—	—	0	—	1.0
Acetylene	7.5	3.0	4.5	4.5	2.5	10.1
Benzene	3.0	2.2	1.3	—	0.7	2.6
Toluene	6.8	4.1	4.0	2.1	2.3	8.9
Ethylbenzene	1.9	0.75	0.5	0.35	0.4	1.3

year^{-1} in 1970 to 7.2 Tg year^{-1} in 1984, while other emissions have remained essentially unchanged: combustion 2.6, industry 8.4, waste disposal 0.6, and miscellaneous uncontrollable emissions 2.7 Tg year^{-1}, totaling 21.5 Tg year^{-1} in 1984. Here, road traffic contributed about 34% and industry 39%.

Table 6.3 gives a breakdown of sources for individual hydrocarbons according to estimates of Mayrson and Crabtree (1976) and Nelson *et al.* (1983). The major source of ethane is natural gas, either from the commercial network or from the direct geological exploitation of resources. Propane appears to derive from natural gas as well as from petrochemical industries. The principal sources of butane and pentane are automotive exhaust and gasoline, although contributions from natural gas and industrial processes are not entirely negligible. The higher alkanes are mostly associated with solvent emissions. The alkenes, especially ethene, derive primarily from automotive exhaust. Nelson *et al.* estimated, however, that 50% of propene is released from industrial processes. The sole source of acetylene is combustion. For this reason, it has been suggested that acetylene might be useful as

TABLE 6.3   Percentage Contribution of Various Sources to Hydrocarbons in (a) Los
Angeles, California (Mayrson and Crabtree, 1976) and (b) Sidney Australia
(Nelson et al., 1983)

Compound	Location	Car exhaust	Gasoline spillage	Gasoline evaporation	Natural gas	Industrial processes	Solvents
Ethane	a	7.9	—	—	90.9[a]	—	—
	b	18.2	—	—	82.2	—	—
Propane	a	—	—	3.6	96.4[a]	—	—
	b	1.2	—	7.9	26.6	64.4	—
n-Butane	a	24.0	7.0	40.5	28.2[a]	—	—
	b	14.6	9.3	59.3	4.0	12.8	—
Isobutane	a	16.3	3.5	33.3	46.9[a]	—	—
	b	11.4	6.1	56.2	4.3	22.0	—
n-Pentane	a	47.5	13.3	23.3	15.9[a]	—	—
	b	26.6	24.2	43.7	1.7	—	2.9
Isopentane	a	37.5	14.0	37.3	10.7[a]	—	—
	b	22.6	22.3	53.6	0.9	—	0.9
2-Methylpentane	b	31.3	29.8	21.8	0.8	—	15.7
3-Methylpentane	b	33.1	30.4	20.4	0.8	—	17.3
n-Hexane	b	32.9	28.3	15.0	1.3	—	22.6
n-Nonane	b	7.6	6.5	—	—	—	85.1
n-Decane	b	18.1	8.7	—	—	—	73.1
Ethene	b	98.9	—	—	1.4	—	—
Propene	b	49.9	—	—	0.3	49.8	—
1-Butene	b	67.3	3.3	29.4	—	—	—
Isobutene	b	77.4	2.6	17.7	—	—	—
trans-2-Butene	b	23.3	10.6	65.8	—	—	—
cis-2-Butene	b	22.7	11.3	63.9	—	—	—
Acetylene	a	100	—	—	—	—	—
	b	100	—	—	—	—	—
Benzene	b	77.0	17.8	6.0	—	—	—
Toluene	b	38.7	16.3	1.7	—	—	43.0
Ethylbenzene	b	45.4	17.5	1.1	—	—	33.7

[a]Includes both natural gas emanating from the ground before processing and commercial natural
gas.

a tracer for automotive emissions. Small amounts of acetylene, however, are
produced in burning of agricultural wastes and other plant materials. For the
aromatic compounds, Table 6.2 suggests that three-quarters of benzene arises
from automobile exhaust and the rest from the evaporation and spillage of
fuels. Toluene and ethylbenzene originate partly from car exhaust and
gasoline and partly from solvent emissions.

Emission inventories of hydrocarbons and some other organic compounds are now more regularly prepared for the developed countries, although the data are largely buried in unpublished reports. Table 6.4 shows a global summary presented by Pacyna and Graedel (1995) in slightly greater detail but without discussion. The data were taken from an agency report (Watson et al., 1991) published in parts by Piccot et al. (1992). According to these data road traffic contributes about 40% in the developed countries, and 25% worldwide. Solvent use is the next most significant activity in the developed countries, where it contributes 25% versus 14% on a global scale. Combustion of wood for heating and cooking in developing countries represents a substantial source of hydrocarbons and other organic volatiles that contributes about 32% to the emissions in these regions, and 18% worldwide. The combustion of crop residues is also a substantial source of hydrocarbons. The total of all emissions is about 142 Tg year^{-1}. The individual estimates as well as the total are rather uncertain. Müller (1992) estimated a total of 98 Tg year^{-1}, but he did not specify the individual contributions, except for natural gas, for which he reported 14 Tg year^{-1}, much higher than the value in Table 6.4. Piccot et al. (1992) presented a detailed account of anthropogenic hydrocarbon emissions summing to a total of 110 Tg year^{-1} worldwide. Their data indicate that about 45% of all emissions are alkanes, 35% alkenes, and 17% aromatic compounds.

TABLE 6.4 Emissions of Nonmethane Volatile Organic Compounds from Anthropogenic Sources (in Tg year^{-1})[a]

Process	Developed[b] countries	Other countries	World	%[c]
Liquid fuel production/distribution	8.6	6.8	15.4	11
Natural gas	1.6	0.4	2.0	1
Road traffic[d]	23.7	12.3	36.0	25
Combustion of wood	4.0	20.8	25.0	18
Combustion of crop residue	4.6	9.8	14.5	10
Combustion of coal, charcoal, etc.[e]	0.9	7.9	8.8	6
Chemical industry	1.5	0.5	2.0	1
Solvent use	15.0	5.0	20.0	14
Miscellaneous[f]	—	—	18.0	13
Total	60.0	63.5	~ 142	

[a] For the year 1985, mainly hydrocarbons, from Pacyna and Graedel (1995).
[b] Mainly OECD and East European countries.
[c] Percentage of global emission rate.
[d] Apportioned like NO_x as given by Müller (1992).
[e] Small sources.
[f] Burning of wastes, and other nonspecified processes.

As Table 6.4 suggests, biomass combustion contributes more than 30% to hydrocarbon emissions on a global scale. Stephens and Burleson (1969) were among the first to investigate the composition of hydrocarbons generated by brush fires. Their data showed the production of alkanes (mainly ethane and propane), acetylene, and alkenes, with ethene and propene as the dominant species. Greenberg *et al.* (1984) examined hydrocarbon emissions from fires in Brazilian grassland and tropical forest regions. The compositions of the hydrocarbons emitted from the two regions were similar. Ethane contributed about 15 mass percent of total nonmethane hydrocarbons, propane 4–15%, ethene 25%, propene 12%, alkynes 1–10%, benzene 8%, toluene 3.5%, furan 4.4%, and 2-methylfuran 1.4%. Altogether these compounds comprised 81–86% of total nonmethane hydrocarbons released from biomass burning. An average of 1.8% was estimated to be emitted relative to CO_2. This compares well with earlier data of Darley *et al.* (1966), Bouble *et al.* (1969), and Sandberg *et al.* (1975) for the burning of grass stubble, straw, agricultural wastes, and brushwood. Lobert *et al.* (1991) conducted laboratory studies that distinguished between flaming and smoldering conditions of open fires. Most of the hydrocarbons detected had maximum emissions early during the smoldering stage, but acetylene was emitted primarily during the flaming stage together with CO_2, NO_x, and N_2O (see Fig. 4.7). About two-thirds of total hydrocarbon emissions occurred during the smoldering stage. Lobert *et al.* (1991) found emission factors for C_2–C_8 hydrocarbons in the range 0.14–3.19%, with an average of 1.18%, expressed as carbon relative to carbon in the biomass fuel, in good agreement with the results of Greenberg *et al.* (1984). Ethene, propene, and 1,3-butadiene were prominent unsaturated hydrocarbons that contributed about 28% to the total nonmethane hydrocarbon mass balance. Benzene and toluene contributed about 3%.

Table 4.6 lists global scale amounts of biomass burned annually and the associated emissions of CO_2. The data discussed above suggest a mean mass ratio of 0.014 for the release of hydrocarbons relative to the carbon released as CO_2. Assuming this emission ratio to be independent of biomass burning activity, one calculates emission rates of 6.6–19.8 Tg year^{-1} for shifting agriculture and deforestation, 10.9–33.3 for savanna burning, 5.5–21.9 for the use of biomass as fuel, and 2.6–9.4 Tg year^{-1} for the burning of agricultural waste in fields. Comparison shows that the entries in Table 6.4 for the combustion of wood and the burning of agricultural waste are at the upper end of the estimates. The total emission of hydrocarbons from biomass burning is 25.6–84.2 Tg year^{-1}, to which the agricultural practices in the Tropics such as savanna burning and deforestation contribute the largest share. Greenberg *et al.* (1984) derived a global emission rate of 34 Tg year^{-1} as carbon, which corresponds to about 40 Tg year^{-1} in terms of hydrocarbon mass. Lobert *et al.* (1991) derived a mean value of 42.6 Tg year^{-1} (as carbon), corresponding to a hydrocarbon mass of nearly 50 Tg year^{-1} with a range similar to that indicated above. On a global scale, the emission of hydrocarbons from biomass burning is evidently comparable to that of other anthropogenic sources.

6.1.3. NATURAL SOURCES OF HYDROCARBONS AND OTHER ORGANIC COMPOUNDS

Went (1960a) pioneered the idea that plants may release significant amounts of hydrocarbons and other organic compounds to the atmosphere, partly by the volatilization of essential oils. From the amount of leaf oils contained in sagebrush, Went (1960b) estimated a global emission rate from this source of 175 Tg year^{-1}. Rasmussen and Went (1965) revised the rate to 200–400 Tg year^{-1} after they had demonstrated the ubiquitous presence of terpenes in the air surrounding natural vegetation and senescent leaves. Later, Sanadze and Kalandadze (1966) and Rasmussen (1970) directed their attention to isoprene (2-methyl-1,3-butadiene), which was found to be the dominant hydrocarbon emitted from deciduous trees such as oak, poplar, sycamore, cottonwood, willow, and eucalyptus (Rasmussen, 1972; Westberg, 1981; Isidorov *et al.*, 1985). Other plants, especially coniferous trees, emit primarily monoterpenes.

Monoterpenes generally are cyclic $C_{10}H_{16}$ hydrocarbons containing at least one double bond. Figure 6.2 shows the chemical structures for nine monoterpenes frequently encountered: α-pinene, β-pinene, myrcene, Δ_3-

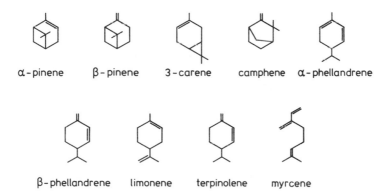

FIGURE 6.2   Skeletal structures of several monoterpenes that have been identified in the atmosphere.

carene, *d*-limonene, terpinolene, α-phellandrene, β-phellandrene, and camphene. Westberg (1981) found the first six plus β-phellandrene to account for most of the mass of volatile organic compounds emitted from Ponderosa pine. Table 6.5 compares the relative contributions to total emissions of isoprene and monoterpenes from several forest species. Pine trees emit mainly α- and β-pinene (in addition to *d*-limonene and myrcene), and deciduous species emit both isoprene and various monoterpenes, with sabinene often making a substantial contribution, in addition to α-pinene and *d*-limonene. Red oak, however, primarily emits isoprene.

Isoprene and monoterpenes derive from the same intermediates in the biosynthetic pathway (Croteau, 1987). The central intermediate is mevalonic acid, which is converted to isopentenyl-diphosphate (IPP). This compound isomerizes enzymatically to dimethylallyl-diphosphate (DMAPP), which is the unique precursor of isoprene. The condensation of IPP and DMAPP, in turn, leads to geranyl-diphosphate, a C_{10} compound, which is the precursor of all monoterpenes. The emission of isoprene from higher plants is light-dependent, so that it must be related to the photosynthetic process. However, a direct correlation between CO_2 assimilation and isoprene emission is frequently not evident, so that CO_2 fixation in itself is not the controlling step in isoprene production. Recent reviews by Fehsenfeld *et al.* (1992), Steinbrecher (1997), and Riba and Torres (1997) have discussed the physiological aspects of isoprene and monoterpene emissions in somewhat greater detail. Despite much progress in this field, it is not known why some plants emit substantial amounts of isoprene and waste up to 20% of fixed carbon, while others, even within the same plant family, do not emit isoprene at all.

Relative and absolute emissions of both isoprene and monoterpenes vary not only among plant species, but also with environmental conditions. Following up on the work of Rasmussen and Jones (1973), Tingey *et al.*

TABLE 6.5   Percentage Distribution of Major Volatile Hydrocarbons Emitted from Trees[a]

Compound	Loblolly pine	Short-leaf pine	Scots pine	Elm	Maple	Sweet gum	Red oak
Isoprene	—	—	—	0.4	1.5	24.1	97.1
α-Pinene	25.5	65.9	42.0	18.0	14.8	16.7	0.3
β-Pinene	28.8	50.8	21.0	7.0	12.6	11.7	0.1
d-Limonene	34.3	27.3	20.0	22.2	2.5	20.9	0.1
Myrcene	7.5	11.3	4.8	3.0	3.0	11.7	< 0.1
Sabinene	0.3	0.7	—	41.7	61.5	6.7	0.1
Camphene	2.0	2.0	6.7	0.3	—	4.6	—
β-Phellandrene	—	—	5.0	—	1.4	—	—

[a] From Khalil and Rasmussen (1992c), except Scots pine, which is from Isidorov (1985).

(1979, 1980, 1981) first explored emission rates as a function of light intensity and temperature. The emission rate of isoprene from oak saplings increased with increasing light intensity and temperature. In the dark, the emission rate was low regardless of temperature. Emissions of monoterpenes from slash pine, in contrast, were almost independent of light intensity, but increased log-linearly with rising temperature. This result is consistent with the behavior of monoterpene vapor pressures. More recently, Guenther *et al.* (1991, 1993) have explored the functional relationship between isoprene emission fluxes and light intensity on the one hand and temperature on the other, for a variety of species and under controlled conditions. Figure 6.3 shows the general behavior. The functional dependence of the emission rate on photosynthetically active radiation intensity is that of a saturation curve, whereas the dependence on temperature goes through a maximum at about 40°C. This effect shows that the functional dependence on temperature is not caused by an increase in volatilization, but rather by that of enzyme activity. It follows that the temperature dependence can be described by an enzyme activation equation that includes denaturation at high temperatures.

In recent years, a great number of studies have been conducted to explore emissions of isoprene, monoterpenes, and other volatile organic compounds from forest species, with two aims: one was to develop suitable algorithms for modeling emission rates; the other was to obtain a sufficiently large number of samples for regional emission estimates. The number of measurements is too large to allow a detailed discussion here. Fehsenfeld *et al.* (1992), Lamb *et al.* (1987, 1993), and Guenther *et al.* (1993, 1995) have reviewed laboratory and field data as well as model development. Isoprene emissions typically are in the range of 1–50 mg h^{-1}/kg leaf biomass, and monoterpenes in the range of 1–10 mg h^{-1}/kg leaf biomass. Many experiments have utilized the dynamic enclosure technique. A branch or twig is enclosed in a plastic bag and flushed with background air, and the effluent is analyzed. The emission rate is given by the difference in concentrations between in-going and out-going air and the volume of the enclosure. The results are related either to leaf area or dry mass. Care must be exercised not to disturb the normal physiological conditions of the plant. In addition, as the technique allows only spot measurements, it must be supplemented by a detailed biomass survey of the forest region. The flux over an area can also be determined from the vertical concentration gradient above the canopy in combination with meteorological measurements to obtain the eddy diffusivity. Such measurements cover a greater area, but they are difficult to conduct properly. A third method involves the controlled release of SF$_6$ to provide a tracer of forest emissions.

The first natural hydrocarbon emissions inventory was constructed by Zimmerman *et al.* (1978) for the contiguous United States. The region was

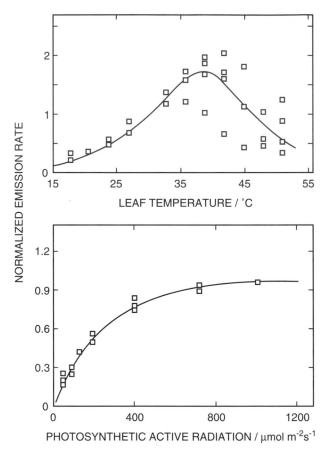

FIGURE 6.3   Isoprene emission from Sweetgum as a function of leaf temperature and photosynthetically active radiation (PAR). The emission rate was normalized to a light flux of 1000 μmol m^{-2} s^{-1} PAR. (Adapted from Guenther *et al.* (1993).)

subdivided into four geographic latitude zones, for which area, total leaf biomass, the average mixture of vegetation, and representative monthly temperature ranges were estimated. Differences in emissions during daytime and nighttime, summer and winter seasons, were taken into account. For example, isoprene emissions were set to zero during winter, except in the southernmost latitude zone, because deciduous trees shed their leaves during the cold season. This procedure gave annual rates of 20 and 50 Tg year^{-1} for isoprene and monoterpene emissions, respectively. To derive an estimate for the global emission rate, Zimmerman *et al.* (1978) first extrapolated their results for the United States to other continental regions in temperature latitude zones. This resulted in 96 Tg year^{-1} for isoprene and 230 Tg year^{-1} for the terpenes. Regarding the Tropics, Zimmerman *et al.* (1978) argued that the emissions should resemble most closely those in the southern United States, where isoprene contributed about 50% to the total emission rate. Then they compared their results for the United States to annual net primary productivity, that is, the fraction of carbon resulting from the uptake of atmospheric CO_2 that is stored by plants and not quickly respired. The comparison suggested a ratio of hydrocarbon emission to net primary productivity close to 0.7%, which was used to extrapolate hydrocarbon emissions to global conditions. Globally, the net primary productivity is about 55 Pg year^{-1} as carbon, or about 120 Pg year^{-1} of plant matter, which leads to a global hydrocarbon emission rate of 830 Tg year^{-1}, with 500 Tg year^{-1} of it occurring in the Tropics. Isoprene was estimated to contribute 350 Tg year^{-1} and monoterpenes 480 Tg year^{-1}, quite similar in magnitude to the estimate derived earlier by Rasmussen and Went (1965).

More recently, Müller (1992) used the temperature parameterization of Lamb (1987) in combination with the net primary productivity assessment of Lieth (1975) and a 5° × 5° land-surface ecosystem classification to determine leaf biomass densities. Observed averaged precipitation and temperature fields were applied to calculate monthly mean emission rates. The overall global rate of nonmethane hydrocarbon emissions was found to be 491 Tg year^{-1}, with the Tropics contributing about 75% to the total. Estimates for the individual contributors were: isoprene 250, monoterpenes 147, alkanes 52, and aromatic compounds 42 Tg year^{-1}. Guenther *et al.* (1995) worked with a 0.5° × 0.5° spatial grid and parameters similar to those of Müller (1992), but included ecosystem type, satellite data base vegetation index, precipitation, temperature, and cloudiness. In addition, they used a canopy radiative transfer model to account for the shading of leaves, which would affect mainly isoprene emissions. Guenther *et al.* (1995) classified emissions of organic volatiles into four groups: isoprene, monoterpenes, other reactive organic compounds such as 2-methyl-3-butene-2-ol, and other less reactive organic species such as methanol. Global emission rates deduced from their model were 503 Tg year^{-1} isoprene, 127 Tg year^{-1} monoterpenes, and 257 Tg year^{-1} each of other reactive and less reactive organic compounds. Compared to the results of Müller (1992), the estimates for isoprene and other volatile organic compounds emissions were higher. The difference appears to arise from the broader empirical emission data base used by Guenther *et al.* (1995). Both estimate emissions of monoterpenes that are smaller than the early estimate of Zimmerman *et al.* (1978). The assessment of Guenther *et al.* (1995) suggests that woods and shrubs contribute about 88% and crops about 11% to the total emissions of 1144 Tg year^{-1}. The fraction of isoprene is about 44%, that of monoterpenes 11%, and that of other organic volatiles 45%.

It has already been noted that a great number of organic compounds other than isoprene and monoterpenes are released from vegetation. Altshuller (1983a) compiled emission data that included C_2-C_6 alkanes, various alkenes, and other volatile C_6-C_{12} compounds. Practically every deciduous plant and all of the grasses studied emitted alkanes, with ethane and propane dominating the mixture. Alkanes often contributed 20% or more to the hydrocarbon emissions from deciduous foliage, and nonisoprene alkenes another 20%, regardless of whether isoprene was emitted.

Zimmerman et al. (1978) included in their estimates emissions from leaf litter and pastures, for which they gave a rate of 162 μg m^{-2} h^{-1} (as carbon). According to Table 11.5, roughly 60×10^{12} m^2 of the continental surface may be characterized as grassland or land covered by grasses and other vegetation. If one assumes that the emissions are negligible during the winter or the dry season, the global rate of hydrocarbons from this source would be 47 Tg year^{-1}. As 40–50% of hydrocarbons from grasses are alkanes, about 20 Tg year^{-1} of alkanes might be produced from this source. Steinbrecher et al. (1997) studied the behavior of ethane and some other hydrocarbons over wheat plants. Emissions occurred at rates similar to those reported by Zimmerman et al. (1978) when temperatures exceeded 20°C, whereas at lower temperatures deposition was preponderant. Accordingly, a global alkane emission rate of 20 Tg year^{-1} from grasses should be an upper limit estimate.

The soil underneath plants is not a source of hydrocarbons. Van Cleemput et al. (1981) found the production of hydrocarbons in various samples of incubated soil small compared with that of methane, 0.07% by volume for ethane, 0.04% for propane, and 0.5% for ethene. Adopting from Table 4.7 a rate of 380 Tg year^{-1} for the global production of methane from microbial sources, one obtains the following production rates: ethane 0.5, propane 0.4, and ethene 3.3 Tg year^{-1}. Except for ethene, these rates are negligible compared to other sources.

Recently, more attention has been directed to emissions from plants of oxygenated organic compounds such as aldehydes, ketones, alcohols, and esters. Graedel et al. (1986a) have prepared an extensive literature survey for the occurrence of such compounds in the atmosphere. The major problem is to isolate compounds that contribute significantly to the great variety of all emissions. Isidorov et al. (1985) identified many nonhydrocarbon organic substances in forest air, Arey et al. (1991) and Winer et al. (1992) described similar emissions from agricultural and natural vegetation in California's central valley, and König et al. (1995) determined emission rates from grape, rye, rape, beech, birch, oak, and grassland. These studies were reviewed by

Puxbaum (1997). MacDonald and Fall (1993) observed significant emissions of methanol from a variety of trees. Table 6.6 shows compounds that have been observed in at least two independent studies. An interesting observation is the ubiquitous occurrence of cis-3-hexen-1-ol and cis-3-hexenylacetate, commonly referred to as leaf alcohol and leaf ester, respectively. These compounds, in addition to cis-3-hexenal and nonanal, were identified previously by Kibe and Kagura (1976) as volatiles released from grass silage. Arey et al. (1991) showed that for some crops the emission of leaf alcohol and leaf ester was higher than that of isoprene and monoterpenes. Linalool (3,7-dimethyl-1,6-octadien-3-ol) is another unsaturated alcohol that appears to be widely emitted from vegetation. It is isomeric with geraniol (3,7-dimethyl-2,6-octadien-1-ol); both are well-known fragrances. Thus, it is perhaps not surprising that Arey et al. (1991) found a 10-fold increase in the emission of linalool from orange trees during the blossoming period. Because their double bonds are easily attacked by hydroxyl radicals, these volatiles must be short-lived in the atmosphere. The great variety of volatile organic compounds emitted from vegetation in addition to isoprene and monoterpenes and the associated variability of fluxes make it difficult to estimate global source strengths. Reported emission rates are in the range of 0.5–5 mg h^{-1}/kg dry leaf biomass. Guenther et al. (1994, 1995) took a geometric mean value of 1.5 mg h^{-1} kg^{-1} to derive the global estimate of 520 Tg $year^{-1}$ discussed above. This figure does not include methanol. Organic acids are not included either; they are not easily analyzed by means of gas

TABLE 6.6   Compounds Widely Observed to be Emitted from Agricultural and Natural Plants in Addition to Isoprene and Monoterpenes[a]

Alcohols		Ketones	
2-Methyl-2-butanol	B85, K95	Butanone	I85, K95
3-Methyl-1-butanol	B85, K95	2-Pentanone	I85, B85, K95
1-Penten-3-ol	B85, K95	3-Pentanone	I85, K95
cis-3-Hexen-1-ol	I85, B85, A91, W92, K95	Camphor	I85, K95
1-Octanol	B85, K95		
Linalool	B85, A91, W92, K95	Esters	
		Hexylacetate	B82, K95
		cis-3-Hexylacetate	I85, B85, A91, K95
Aldehydes		Methylsalicylate	B82, A91
n-Butanal	I85, K95		
n-Hexanal	A91, W92, K95	Alkenes	
trans-2-Hexenal	B85, A91, W92, K95	1-Dodecene	A91, W92, K95

[a] Only those compounds are included that have been observed in least two independent studies. Key: I85 Isidorov et al. (1985), B82 Buttery et al. (1982), B85 Buttery et al. (1985), A91 Arey et al. (1991), W92 Winer et al. (1992), K95 König et al. (1995).

chromatography, and their determination requires a different analytical technique. The major organic acids in the atmosphere, formic and acetic, will be discussed further below.

Finally, the ocean represents a source of nonmethane hydrocarbons. Lamontagne et al. (1974) and Swinnerton and Lamontagne (1974b) found ethane, propane, ethene, and propene to be dissolved in marine surface waters at concentrations high enough to generate a flux to the atmosphere. For certain ocean areas the fluxes can be estimated. Rudolph and Ehhalt (1981) combined their own measurements of atmospheric mixing ratios with the aqueous concentrations reported by Swinnerton and Lamontagne (1974b) and the solubility data of McAuliffe (1966) to derive the following emission rates: ethane 0.3, propane 0.4, ethene 0.6, and propene 0.5 μg m^{-2} h^{-1}. These values referred to a region of the north Atlantic Ocean near 70° northern latitude. Linear extrapolation to the entire ocean surface area was used to obtain global emission rates of 1.0, 1.3, 1.9, and 1.6 Tg $year^{-1}$ for ethane, propane, ethene, and propene, respectively (Ehhalt et al., 1986). Subsequently, Bonsang et al. (1988) measured light hydrocarbons in the air and in surface waters of the Indian Ocean, Donahue and Prinn (1993) made similar measurements in the central Pacific, and Plass et al. (1992) obtained additional data for the Atlantic Ocean. Finally, Plass-Dülmer et al. (1995) evaluated the whole data set to derive the following global emission rates (in Tg $year^{-1}$): ethane 0.32, propane 0.2, butanes 0.11, ethene 0.9, propene 0.5, and butenes 0.5, summing to a total of 2.5 Tg $year^{-1}$. Plass-Dülmer et al. (1995) estimate an upper limit of 1.1 Tg $year^{-1}$ for C_2-C_4 alkanes and 5 Tg $year^{-1}$ for C_2-C_4 alkenes. The values are low in comparison to other sources, so that the ocean plays a minor role in the global budget of these hydrocarbons.

The atmosphere above the ocean contains C_9-C_{28} n-alkanes in addition to low-weight hydrocarbons. Eichmann et al. (1980) estimated that the ocean might emit 26 Tg of higher alkanes each year. The value is based on the measured total atmospheric concentration for C_9-C_{28} n-alkanes, which is 300–400 ng m^{-3} over the Indian Ocean and the north Atlantic Ocean, combined with an assumed average n-alkane residence time of 5 days resulting from reaction with OH radicals. From measurements at Enewetak Atoll (Marshall Islands), Duce et al. (1983) reported atmospheric concentrations of $C_{12}-C_{28}$ n-alkanes that were an order of magnitude lower than those found by Eichmann et al. (1980). It is questionable, therefore, whether their data can be extrapolated to the whole ocean area, so that the emission rate may be much lower than 26 Tg $year^{-1}$.

Table 6.7 summarizes the global emission rates of hydrocarbons and other volatile organic compounds that were discussed above. Emissions from the foliage of vegetation clearly is the largest source. Anthropogenic sources are

TABLE 6.7 Summary of Global Emissions of Hydrocarbons and Other Organic Volatiles from Various Sources

Type of source	Emission rate ($Tg\ year^{-1}$)	Remarks
Anthropogenic sources		
Petroleum-related sources and chemical industry	36–62	Mainly automobiles, alkanes, alkenes, and aromatic compounds
Natural gas	2–14	Mainly light alkanes
Organic solvent use	8–20	Higher alkanes and aromatic compounds
Biomass burning	25–80	Mainly light alkanes and alkenes
Total anthropogenic sources	71–175	
Biogenic sources		
Emissions from foliage		
Isoprene	175–503	
Monoterpenes	127–480	
Other organic compounds	510	Higher alkanes, alkenes, alcohols, aldehydes, ketones, esters
Grasslands	< 26	
Soils	< 3	
Ocean waters	2.5–6	Light alkanes and alkenes
	< 26	C_9–C_{28} alkanes
Total biogenic sources	015 1540	

significant, but on a global scale they contribute only about one tenth of the total.

6.1.4. HYDROCARBON MIXING RATIOS OVER THE CONTINENTS

Table 6.8 presents mixing ratios of C_2–C_5 alkanes at rural measurement sites in North America, Europe, Brazil, and Africa. Table 6.9 adds similar data for several alkenes, acetylene, and a few aromatic compounds. Mixing ratios range from about 2 nmol mol^{-1} for ethane and acetylene to less than 0.1 nmol mol^{-1} for the higher hydrocarbons. The alkanes, and to some extent the alkenes as well, show a distribution in which mixing ratios decrease with increasing carbon number. This differs from the distribution observed in the cities (Table 6.2). Two factors may be responsible for this pattern: a different source distribution, and the greater reactivity of compounds with higher carbon numbers, which would lead to a decrease in the concentration of such hydrocarbons from distant sources such as urban centers. Recent measurements made over periods of more than a year (Lindskog and

TABLE 6.8 Average Mixing Ratios of Alkanes (nmol mol^{-1}) in the Continental Boundary Layer at Rural Measurement Sites

Authors	Location	C_2H_6	C_3H_8	n-C_4H_{10}	i-C_4H_{10}	n-C_5H_{12}	i-C_5H_{12}	Others[a]	Total alkanes[a]
Lonneman et al. (1978)	Florida, May 1976, citrus groves	1.50	1.03	1.87	0.65	0.73	1.80	16.0	48.6
	Everglades	1.15	0.13	0.15	0.07	0.08	0.18	11.1	16.0
Arnts and Meeks (1981)	Great Smoky Mountain, Tennessee, Sept. 1978	5.70	2.72	1.20	0.64	0.59	0.97	8.9	48.2
	Rio Blanco, Colorado, July 1978	2.60	1.63	0.30	0.25	0.40	0.38	30.4	48.1
Sexton and Westberg (1984)	Belfast, Maine, summer 1975	1.75	0.67	0.50	0.12	0.20	0.20	n.g.	10.0
	Miami, Florida, summer 1976	2.00	0.67	0.12	0.10	0.20	0.30	n.g.	9.5
Greenberg and Zimmerman (1984)	Central Brazil, 1979/1980	2.09[c]	0.45	0.24	n.g.	n.d.	n.g.	1.6	8.0
		2.97[d]	0.42	0.34	n.g.	n.d.	n.g.	1.6	12.3
Colbeck and Harrison (1985)	Lancaster, England, early summer 1983	3.20[e]	2.20	0.33	0.20	0.74	n.g.	2.6	21.4
		14.70[f]	4.97	0.85	3.52	0.98	n.g.	17.0	83.8
Greenberg et al. (1985)	Kenya, Lake Baringo, Marigat, July/August 1983	0.68	0.12	0.09	0.05	0.04	0.03	0.7	2.7
		1.85	0.32	0.07	0.03	0.03	n.d.	1.6	6.9
Rudolph and Khedim (1985)	Germany, Schauinsland Mt.	2.33	0.54	0.4	0.20	0.12	0.27	0.3	10.9
Lindskog and Moldanova (1994)	Rörvik, Sweden, Jan. 1990	2.64	1.29	0.76	0.42	0.43	0.42	n.g.	18.2
	July 1990	0.75	0.17	0.18	0.10	0.07	0.12	n.g.	4.1
Bottenheim and Shepherd (1995)	Egbert, Ontario, Feb. 1991	3.84	2.29	1.17	0.50	0.30	0.39	1.92	26.6
	August 1991	1.36	0.49	0.26	0.12	0.11	0.20	1.55	8.81

[a] In nmol mol^{-1} of carbon.
[b] n.d., Not detected; n.g., not given.
[c] Below the canopy.
[d] Above the canopy, determined by aircraft.
[e] On days when ozone levels were near background values.
[f] On days when ozone levels exceeded background values.

TABLE 6.9 Average Mixing Ratios ($nmol\ mol^{-1}$) of Several Alkenes, Acetylene, and Some Aromatic Compounds in the Continental Boundary Layer at Rural Measurement Sites[a]

Authors	Location	Ethene	Propene	Butenes	Alkadienes	Acetylene	Benzene	Toluene	Ethyl benzene
Lonneman et al. (1978)	Florida, May 1976, citrus groves	1.85	0.32	0.25	0.10[g]	2.22	n.g	1.11	0.16
	Everglades	0.20	0.20	n.d.	n.d.	0.60	n.g.	0.36	0.25
Arnts and Meeks (1981)	Great Smoky Mountain, Tennessee, Sept. 1978	2.14	0.74	0.23	0.60[f]	2.39	0.89	0.96	0.12
	Rio Blanco, Colorado, July 1978	0.63	0.30	n.d.	0.75[f]	1.00	n.g.	0.84	0.12
Sexton and Westberg (1984)	Belfast, Maine, summer 1975	1.0	0.17	0.12	n.g.	0.25	n.g.	n.g.	n.g.
	Miami, Florida, summer 1976	0.25	0.17	0.12	n.g.	0.25	n.g.	n.g.	n.g.
Greenberg and Zimmerman (1984)	Central Brazil, 1979/1980	1.89[b]	0.31	0.47	2.44[f]	0.88	0.50	0.12	0.08
		2.70[c]	0.68	1.08	2.50[f]	1.08	1.15	0.14	0.15
Colbeck and Harrison (1985)	Lancaster, England, early summer 1983	0.73[d]	0.93	0.40	n.g.	0.25	n.g.	n.g.	n.g.
		6.03[e]	0.70	0.57	n.g.	2.70	n.g.	n.g.	n.g.
Greenberg et al. (1985)	Kenya, Lake Baringo, Marigat, July/August 1983	3.13	0.84	0.40	0.11[f]	1.14	0.27	0.15	0.04
		0.87	0.18	0.32	0.03[f]	0.53	0.17	0.14	0.06
Rudolph and Khedim (1985)	Germany, Schauinsland Mt.	0.45	0.08	n.g.	n.g.	0.55	0.10	0.11	n.g.
Bottenheim and Shepherd (1995)	Egbert, Ontario, Feb. 1991	0.65	0.10	0.09	0.00[f]	1.14	n.d.	n.d.	n.d.
	August 1991	0.50	0.07	0.08	0.25[f]	0.36	n.d.	n.d.	n.d.
Hagerman et al. (1997)	Yorkville, Georgia, winter 1992/1993, summer	2.85	1.09	2.06	0.04[f]	3.16	1.79	3.01	0.41
		0.90	1.61	1.26	9.80[f]	0.72	1.22	4.75	0.95
	Oak Grove, Mississippi, winter 1992/1993, summer	1.58	0.56	1.02	0.06[f]	1.98	1.36	1.25	0.21
		0.57	0.61	1.72	11.19[f]	0.45	0.69	3.63	0.45

[a] n.d., not detected; n.g., not given.
[b] Below the canopy.
[c] Above the canopy, determined by aircraft.
[d] On days when ozone levels were near background values.
[e] On days when ozone levels exceeded background values.
[f] Mainly isoprene.
[g] Mainly butadiene.

Moldanova, 1994; Bottenheim and Shepherd, 1995) have revealed a significant influence of anthropogenic sources. One piece of evidence is the
weekday/weekend effect, which can be distinguished for acetylene and the
C_3–C_5 alkanes, but not for ethane and the alkenes. Acetylene was found to
correlate well with carbon monoxide, which is not surprising, as both
originate predominantly from vehicular traffic. Back-trajectory analysis
showed that elevated concentrations always occur in air masses originating in
regions of high population and/or traffic density. Another interesting result,
which is apparent from the data in Tables 6.8 and 6.9 and has also been
reported by Boudries *et al.* (1994) and Hagerman *et al.* (1997), is the
observation of a seasonal cycle featuring a winter maximum and summer
minimum. This effect is best defined for the C_2–C_4 alkanes and acetylene.
Figure 6.4 illustrates the seasonal variation of propane as observed at four
sites across Canada. If automobile traffic is the main source of these
hydrocarbons, the source strength will not vary much with the seasons.
Hence, it appears that most of the seasonal variation arises from the annual
variation in the average concentration of OH radicals, which by reacting with
hydrocarbons provide the major removal mechanism.

Rudolph and Johnen (1990) and Parrish *et al.* (1992) have suggested a
procedure to determine whether the variation in the hydrocarbon mixing
ratio has a photochemical origin. In principle, the decrease in concentration
of a hydrocarbon in an air parcel by reaction with OH radicals may be

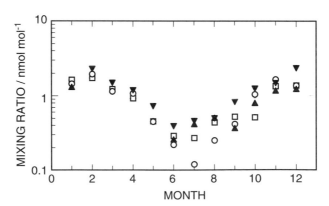

FIGURE 6.4   The seasonal variation of propane at four sites located along the southern border
of Canada: ▲, Saturna Island, British Columbia; ▼, Egbert, Ontario; ○, Lac la Flamme,
Quebec; □, Kemjimkujik National Park, Nova Scotia. Reprinted from Atmospheric Environment
29 J. W. Bottenheim and M. F. Shepherd, C_2–C_6 hydrocarbon measurements at four rural
locations in Canada, 647–664 (1995) with permission from Elsevier Science.

represented by

$$[HC] = [HC]_0 \exp(-k_{HC}[OH]_{av} t) \qquad (6.3)$$

where [HC] is the concentration at the sampling site, $[HC]_0$ is the concentration at the source, t is the time of transport, and $[OH]_{av}$ is the temporal and spatially averaged concentration of OH radicals. By using the ratios $[HC_1]/[HC_3]$ and $[HC_2]/[HC_3]$ of three suitably selected hydrocarbons, it is possible to eliminate the unknown $[OH]_{av}$ as well as dilution effects to obtain the equation

$$\ln([HC_1]/[HC_3]) = A + B \ln([HC_2]/[HC_3]) \qquad (6.4)$$

where the intercept is given by

$$A = \ln([HC_1]_0/[HC_3]_0) - B \ln([HC_2]_0/[HC_3]_0) \qquad (6.4a)$$

and the slope is determined by

$$B = (k_1 - k_3)/(k_2 - k_3) \qquad (6.4b)$$

where the subscript numbers refer to the corresponding hydrocarbons. Using a suitable choice of hydrocarbon concentration ratios, a plot of $\ln([HC_1]/[HC_3])$ versus $\ln([HC_2]/[HC_3])$ should yield a straight line with a slope determined solely by the reaction rate constants. Figure 6.5 shows such plots from measurements at several sites in Canada, one for the hydrocarbon pair n-butane/propane, the other for n-butane/isobutane, relative to ethane in both cases. Linear relationships are observed in both cases. The slopes of the linear regression lines are $B = 1.42$ and $B = 0.75$, respectively. Rudolph and Johnen (1990) and Parrish et al. (1992) likewise worked with n-butane/propane relative to ethane and found good linear relationships with slopes of 1.66 and 1.47 for marine air masses, in good agreement with the data in Figure 6.5. Jobson et al. (1994), who measured C_2–C_6 hydrocarbons at Alert, Canada, considered both reactant pairs and found slopes of 1.44 and 0.97, respectively. The slopes calculated from the rate constants in Table 6.1 are 2.6 for n-butane/propane and 0.95 for n-butane/isobutane. The true values are temperature-dependent and may be different by 20%, but not more. It has been suggested that the deviation from the kinetic data is caused by the effect of mixing air parcels of different ages, an effect that was substantiated by model calculations (McKeen and Liu, 1993). Jobson et al. (1994) therefore argued that it might be better to work with hydrocarbon pairs of similar reactivity. The two butane isomers react with OH radicals at almost the same rate, and the results for this reactant pair are indeed much closer to the theoretical value.

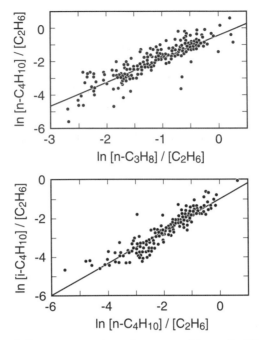

FIGURE 6.5   Plots of alkane concentration ratios as a test of the applicability of Equation (6.4). Upper part: *n*-Butane against propane; lower part: isobutane against *n*-butane. (Adapted from Bottenheim and Shepherd (1995).)

The concentration ratio between winter maxima and summer minima has been found to increase with increasing reactivity of the compound, as expected, but on a relative scale this increase appears to be smaller than one would calculate if a variation of the average OH concentration were the sole cause of the seasonal oscillation. In mid-latitudes the boundary layer contracts in winter and expands in the summer, and the exchange of air with the free troposphere also is more favorable in summer. This meteorological cycle affects the extent of dilution. The enhancement of dilution alone would be able to describe the seasonal variation of the mixing ratio of those trace gases that react slowly with OH radicals. A smaller reaction time is then needed to explain the additional reduction of trace gases that react with OH more rapidly. The treatment outlined above corrects for the dilution effect.

Acetylene has its origin primarily in combustion processes. In the Northern Hemisphere, acetylene provides a good tracer for automotive emissions; in the Tropics it also arises from biomass burning. Vegetation does not emit acetylene. Jobson *et al.* (1994) and Bottenheim and Shepherd (1995) have applied the above-described technique, found the relationship between

$[C_2H_2]/[C_2H_6]$ and $[C_3H_8]/[C_2H_6]$ to be linear, and determined identical values for the slopes of the regression line, $B = 0.66$, quite close to the theoretical value of 0.72. Benzene, toluene, and ethylbenzene are well established to derive primarily from anthropogenic emissions. These compounds react with OH radicals at rates comparable to those of the light alkanes, so that they should exhibit a similar seasonal variation. At locations within the Arctic circle, Rasmussen and Khalil (1983b) and Hov et al. (1984) have observed lower mixing ratios of benzene and toluene during summer. On the other hand, the data of Hagerman et al. (1997) for the southeastern United States indicated an increase in mixing ratios in summer. This behavior has remained unexplained. Ciccioli et al. (1997) discussed the possible emission of aromatic compounds from vegetation. Although some of the data are controversial, the majority of observations argue against such emissions. The only well-identified aromatic compound emitted from vegetation is 1-methyl-4-isopropyl-benzene, also known as p-cymene. This compound is synthesized by plants along the same route as the monoterpenes, so that it is usually considered a member of this group.

Alkenes have atmospheric lifetimes shorter than the light alkanes. After 2 days of transport, propene and the butenes are reduced to a few percent of their original abundances relative to ethane or acetylene. The mixing ratios observed at rural sites for these compounds usually are low and seasonal variations are difficult to discern. Ethene, which is less reactive, shows a slight seasonal variation with higher levels in winter, although not of the sinusoidal type characteristic of the alkanes. Mixing ratios are fairly constant at a low level throughout most of the year and rise during the period of November–February. Isoprene, which has a natural origin, as it is emitted from plants, shows a typical summer maximum, but this behavior is not usually observed for the alkenes. It appears that alkenes derive primarily from automobile traffic, with a dominance of local sources.

Table 6.10 shows atmospheric mixing ratios of isoprene and some monoterpenes observed at rural measurement sites in or near forests. Concentrations of these compounds generally are low compared with the large emission rates that have been projected to arise from vegetation, especially in view of the fact that most measurements were made in summer at times when the emissions maximize. However, the reactions with OH radicals and ozone are very rapid, and the lifetimes of the compounds are on the order of hours. Accordingly, one expects a fairly rapid decline of mixing ratios with distance from the source. Several observers have indeed reported measurable concentrations of isoprene and monoterpenes beneath the canopy, but much lower values, frequently below the detection limit, outside and downwind of forests (Whitby and Coffey, 1977; Yokouchi et al., 1983).

TABLE 6.10 Average Mixing Ratios (nmol mol^{-1}) of Isoprene and Monoterpenes in the Continental Boundary Layer at Rural Measurement Sites, Mainly Underneath Forest Canopies

Authors	Location	Isoprene	α-Pinene	β-Pinene	Δ$_3$-Carene	Camphene	d-Limonene	Others
Holdren et al. (1979)	Moscow Mt. Idaho 1976/1977, pine and fir	Trace	0.113	0.086	0.064	n.g.[a]	0.01	n.g.
Arnts and Meeks (1981)	Great Smoky Mt. National Park, Tennessee, Sept. 1978 deciduous	0.60	0.10	n.g.	n.g.	n.g.	n.g.	n.g.
Roberts et al. (1983)	Niwot Ridge, Colorado 1981/1982, fir and spruce	n.g.	0.054	0.097	0.051	0.038	0.03	n.g.
Shaw et al. (1983)	Mt. Kanobili, Abastumani forest, Georgia (former Soviet Union), July 1979, pine and spruce	1.4	0.8	0.43	0.90	0.09	0.08	0.68[b]
Ciccioli et al. (1984)	Outskirts of Rome, Italy pine forest	n.g.	1.50	0.18	0.06	n.g.	0.04	0.046[c]
Greenberg and Zimmerman (1984)	Central Brazil, below canopy 1979/1980, above canopy	2.40 / 2.27	n.g. / n.g.	0.27 / 0.15	0.24 / 0.82	n.g. / n.g.	n.g. / n.g.	[d] / [d]
Fehsenfeld et al. (1992)	Kinterbish, Alabama, morning summer 1990, afternoon	0.90 / 6.30	0.75 / 0.30	0.40 / 0.17	n.g. / n.g.	0.062 / 0.012	n.g. / n.g.	0.015[c] / 0.008[c]

[a] n.g., Not given.
[b] Myrcene.
[c] p-Cymene.
[d] Below/above canopy: delta-4-carene 0.39/2.08; myrcene 0.19/1.09; a-phellandrene 0.18/1.17; a-terpinene 0.49/0.12; plus others.

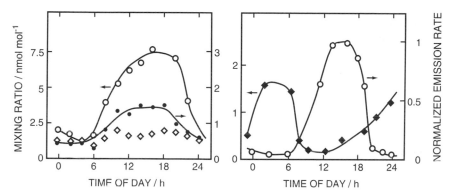

FIGURE 6.6 Left: Diurnal variation of the mixing ratios of isoprene and its oxidation products at Kinterbish, Alabama. $\bigcirc$, isoprene; $\bullet$, methyl vinyl ketone; $\diamondsuit$, methacrolein (adapted from Montzka et al. 1993). Right: Diurnal variation of α-pinene at Jädraås, Sweden. Mixing ratio ($\blacklozenge$), left scale, and normalized emission flux ($\bigcirc$), right scale. (Adapted from Janson (1992, 1993) and Johansson and Janson (1993).)

Diurnal cycles have been observed for the mixing ratios of both isoprene and monoterpenes, and Figure 6.6 shows a typical behavior. The mixing ratio of isoprene often shows a maximum in the early afternoon. Because isoprene emission requires photoactive solar radiation, it occurs only during the day. The diurnal variation, however, is caused primarily by an increase in the emission rate with rising temperature, and less by the dependence on solar intensity. When the air concentrations are normalized to remove the dependence on temperature, most of the variation in isoprene concentration observed during the daytime hours is removed as well. The behavior of monoterpenes is different. Their mixing ratios generally exhibit maxima at night and minima during the day. The emission of monoterpenes occurs during the entire diurnal cycle, but more strongly during the day because of the increasing temperatures. The build-up of nocturnal temperature inversion layers tends to trap the emitted substances, leading to an increase in their concentrations. The inversion layer breaks up in the morning, resulting in rapid vertical mixing and dispersion of the accumulated substances. Monoterpenes react rapidly with OH radicals, with ozone, and with NO_3 radicals, with the latter being effective only at night, because NO_3 is easily photolyzed by sunlight. However, even NO_3 concentrations of 2 pmol mol^{-1} are sufficient to reduce the lifetime of monoterpenes to a few hours. Nevertheless, the increased photochemical activity in daylight must contribute significantly to the removal of monoterpenes to reduce their concentrations despite the increase in emission rates.

Fehsenfeld *et al.* (1992) presented mixing ratios for organic compounds observed at a rural site in the southeastern United States for two times of the day, in the early morning (4 a.m.) and in the afternoon (4 p.m.). In the morning isoprene and the monoterpenes represented about 30% and in the afternoon about 50% of total organic carbon present in the atmosphere. In the afternoon, the relative contributions were alkanes, 10%, aromatic compounds 2%, natural nonmethane hydrocarbons 48%, alcohols 18%, and carbonyl compounds 22% of organic carbon. The carbonyl compounds included acetaldehyde, acetone, methyl vinyl ketone, and methacrolein, which are products resulting from the oxidation of biogenic hydrocarbons in the atmosphere. The alcohols consisted mainly of methanol and ethanol. Their concentrations increased during the day, indicating that they were locally produced. Propane was the dominant alkane (ethane was not reported), and benzene and toluene were the dominant aromatic compounds. Their concentrations decreased slightly during the day.

6.1.5. HYDROCARBONS IN THE MARINE ATMOSPHERE

Measurements of hydrocarbons over the oceans include light alkanes, alkenes, and acetylene. Data for the heavier hydrocarbons refer mainly to *n*-alkanes. Compounds with carbon number up to 32 have been observed. Table 6.11 gives an overview of total concentrations. Figure 6.7 shows the distribution of alkane concentration with carbon number. Individual concentrations can vary over several orders of magnitude, from several μg m^{-3} or about 5 nmol mol^{-1} for ethane to several ng m^{-3} or less than 0.1 pmol mol^{-1} for C_{30} alkanes. The steepest decrease in relative abundance occurs in the C_2–C_{12} range. A slight maximum appears to exist for C_{15}–C_{21} hydrocarbons. Much lower concentrations of C_{12}–C_{30} alkanes were observed at Enewetak Atoll in the tropical northern Pacific Ocean or at Cape Grim, Tasmania. A similar difference was observed for *n*-alkanes associated with aerosol particles (see Section 7.5.2).

The light hydrocarbons have the highest abundances, so they contribute most to the total hydrocarbon mass budget, even though they span only a small range of carbon numbers. Figures 6.8 to 6.10 show meridional distributions of ethane, propane, and acetylene. The data were acquired on ships, on islands, and at coastal stations. The three compounds show a similar behavior in that low mixing ratios are observed in the Southern Hemisphere and higher ones in the Northern, maximizing near 40° N latitude. The behavior is qualitatively similar to that of carbon monoxide, indicating that all of these substances have their major sources in the Northern Hemisphere.

TABLE 6.11 Concentrations of Hydrocarbons (μg m^{-3} except where noted) in the Marine Atmosphere Near the Ocean Surface

Authors	Location	Carbon numbers	Average	Range	Remarks
Robinson et al. (1973)	Point Reyes, California	C_4–C_7	8.8	—	Also found: 2.2 μg m^{-3} acetone
	Brethway, Washington	C_4–C_7	~3	—	
Rasmussen (1974, as reported by Duce et al., 1983)	Hawaii and North Atlantic	C_2–C_{12}	8[a]	4–16[a]	
Wade and Quinn (1974)	Bermuda	C_{14}–C_{37}	0.18[a]	0.05–0.5[a]	
Eichmann et al. (1979)	Loop Head, Ireland	C_6–C_{17}	0.06	0.04–0.16	Only n-alkanes
		C_{13}–C_{29}	0.2	—	
Eichmann et al. (1980)	Cape Grim, Tasmania	C_9–C_{28}	0.34	—	Only n-alkanes in pure marine air, except C_{-7}, where branched species are included
Rudolph and Ehhalt (1981)	North Atlantic Ocean	C_2–C_5	7.1	2.3–28.4	
	Equatorial Atlantic	C_2–C_5	2.5	1.8–33.3	
Singh and Salas (1982)	East Pacific Ocean, N. H.	C_2–C_5	7.58	3.9–9.6	
	East Pacific Ocean, S. H.	C_2–C_5	3.39	2.1–4.2	
Duce and Gagosian (1982, from E. Atlas and C. Giam, unpubl.)	Enewetak Atoll, Pacific	C_{13}–C_{27}	0.0025	—	n-Alkanes
Duce et al. (1983, from O. C. Zafiriou et al., unpubl.)	Enewetak Atoll, Pacific	C_{21}–C_{32}	0.0003	—	n-Alkanes
Bonsang et al. (1988)	Pacific Ocean	C_2–C_6	14.7	2.3–27.8	C_4–C_6 alkanes, C_2–C_6 alkenes

[a] In μg m^{-3} of carbon.

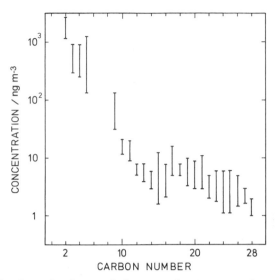

FIGURE 6.7   Abundance distribution of alkanes in the air next to the ocean surface as a function of carbon number. Data for C_2-C_5 alkanes from Rudolph and Ehhalt (1981), Atlantic Ocean, data for C_9-C_{29} alkanes from Eichmann *et al.* (1979, 1980), North Atlantic and South Indian Oceans.

Reaction with OH radicals is the dominant sink. Moreover, their tropospheric lifetimes are rather similar, which largely explains the similarity in the distributions. Ethane has been studied most thoroughly. In the Northern Hemisphere, its mixing ratio shows a seasonal variation with higher values in winter and lower ones in summer, similar to that observed on the continents. In the Southern Hemisphere the seasonal variation and the latitudinal gradient are small. Average mixing ratios are 380 and 1330 pmol mol^{-1} in the Southern and Northern Hemispheres, respectively, with a tropospheric average of 860 pmol mol^{-1} (Rudolph, 1995). Propane and acetylene have been less extensively studied, but the data from the continents indicate that they also should undergo a seasonal variation. The total loss rates can be estimated from the calculated concentration field for OH radicals in the atmosphere and the known rate coefficients for the reactions with OH radicals. In this manner, Kanakidou *et al.* (1991) derived a total rate of 16 Tg year^{-1} for ethane and 23 Tg year^{-1} for propane. Rudolph (1995), who considered only ethane, found a total rate of 14.7 Tg year^{-1}, with an additional loss of 0.8 Tg year^{-1} by transport to the stratosphere. By estimating the loss rate of ethane relative to that of methane, Blake and Rowland (1986) had earlier found 13 ± 3 Tg year^{-1}, in good agreement. Rudolph *et al.* (1996) considered the two hemispheres separately and reported loss rates of 11.8 and 3.7 Tg year^{-1} in the Northern and Southern Hemispheres,

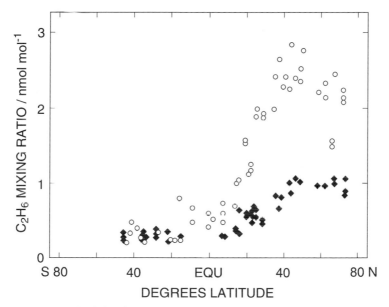

FIGURE 6.8 Latitudinal distribution of ethane in marine surface air of the Pacific Ocean. Open points: December 1984; closed points: June 1985. (Adapted from Blake and Rowland (1986).)

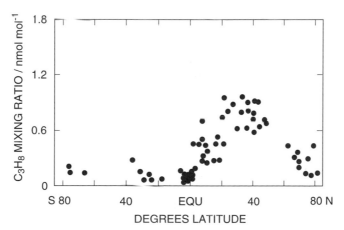

FIGURE 6.9 Latitudinal distribution of propane in the surface air over the Atlantic Ocean (Adapted from Ehhalt *et al.* (1986).)

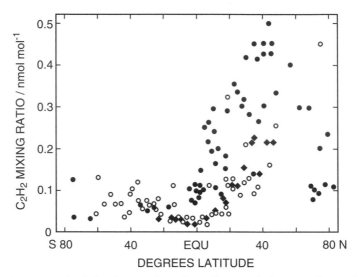

FIGURE 6.10   Latitudinal distribution of acetylene. ○, Aircraft data over the Pacific Ocean from Rasmussen (Robinson, 1978); ●, shipboard data over the Atlantic Ocean (Ehhalt *et al.*, 1986); ◆, shipboard data over the Pacific Ocean (Singh *et al.*, 1988).

respectively, with 1.4 Tg year^{-1} entering the Southern Hemisphere from the Northern Hemisphere because of interhemispheric exchange processes. This means that $3.7 - 1.4 = 2.3$ Tg year^{-1} of ethane must be produced in the Southern Hemisphere, about 10 times more than is emitted from the oceans. As mentioned previously, losses in the exploitation of natural gas represent a significant source of ethane; biomass burning makes another significant contribution. Rudolph (1995) estimated natural gas losses to contribute about 6 Tg year^{-1}, with a range of 2.6–11 Tg year^{-1}, and biomass burning 6.4 Tg year^{-1}, with a range of 2.4–12 Tg year^{-1}. Emissions from grasslands may also contribute. Biomass burning, however, is concentrated in the Tropics, whereas the sources of ethane are mainly located in the Northern Hemisphere. Obviously, the budget derived from the loss rate is much better defined than that from source estimates. The same conclusion applies to propane, which also arises mainly from anthropogenic sources.

Acetylene may be assumed to originate entirely from combustion processes, mainly automobile emissions and biomass burning. The mixing ratio over the Atlantic Ocean appears to be higher than that over the Pacific. In the Southern Hemisphere the mixing ratios of acetylene average about 65 pmol mol^{-1}. The rate coefficient for the reaction of acetylene with OH radicals is fairly insensitive to temperature changes. The lifetime of acetylene shown in Table 6.1 is 18 days, only one-third compared to that for ethane, 56 days. The mixing ratio of acetylene is about 17% of that observed for ethane.

Comparison therefore suggests that the loss rate of acetylene in the Southern Hemisphere is $(65/380) \times (56/18) \times 3.5 = 1.8$ Tg year^{-1}. In the Northern Hemisphere the acetylene mixing ratio is about 4 times higher than in the Southern, so that the global loss rate should be about 9 Tg year^{-1}. According to Table 6.4, anthropogenic emissions of hydrocarbons from road traffic, industries, and biomass burning sum to 84 Tg annually. Hydrocarbons produced by combustion processes contain about 10% acetylene, so that acetylene emissions should amount to 8.4 Tg year^{-1} globally. This agrees with the loss rate. As in the case of ethane, however, biomass burning is estimated to contribute a large share, more than 50% of all acetylene emissions. This raises doubts about the source apportioning, because the latitudinal distribution indicates that the sources should be concentrated in the Northern Hemisphere.

Fewer data exist for butanes and pentanes in the marine atmosphere. Table 6.12 shows a number of measurements. Only minute amounts of the two butane isomers are released from the ocean. These fluxes cannot sustain the observed concentration levels, so that continental sources must contribute. Rudolph and Johnen (1990) applied the plotting method of hydrocarbon pairs discussed in the previous section and found good correlation between n-butane/ethane and propane/ethane, with a linear regression slope in agreement with the known ratios of rate coefficients for reactions with OH radicals. This provides good evidence for a continental origin of n butane and propane. Rudolph and Johnen (1990) also found that n-butane and isobutane are well correlated with each other.

Table 6.12 includes mixing ratios observed for ethene and propene. In contrast to the alkanes, the alkenes are fairly evenly distributed between the two hemispheres, but their mixing ratios can fluctuate considerably. Ethene and propene are released from the oceans at global rates of 0.9 and 0.5 Tg

TABLE 6.12   Average Mixing Ratios (pmol mol^{-1}) of Ethene, Propene, and C_4 and C_5 Alkanes in the Marine Atmosphere near the Ocean Surface

Location	n-C_4H_{10}	i-C_4H_{10}	n-C_5H_{12}	i-C_5H_{12}	C_2H_4	C_3H_8
Equatorial Atlantic[a]	75	23	29	16	250	120
North Atlantic and Barents Sea[a,b]	98	54	57	40	180	110
Eastern Pacific, NH[c]	197	530	232	370	95	545
Eastern Pacific, SH[c]	100	153	143	273	73	207
Indian Ocean[d]	276	43	460	70	1440	1810

[a] Rudolph and Ehhalt (1981).
[b] Two data sets with higher mixing ratios are omitted.
[c] Singh and Salas (1982).
[d] Bonsang et al.(1991).

year^{-1}, respectively. Their atmospheric lifetimes are short, making it unlikely that they would survive long-range transport at travel times exceeding 2 days, so that the observed mixing ratios should be locally derived. Alkenes in ocean waters are formed by the photochemical degradation of dissolved organic matter exuded by algae (Ratte *et al.*, 1993). The chemical or biological removal of alkenes in seawater appears to be slow, so that most of the alkenes produced in ocean surface waters are released to the atmosphere. Production and release rates are largely determined by the rate of net primary production. Because of their short lifetime, alkenes are expected to exhibit a rapid decrease of mixing ratio with altitude. From aircraft measurements over the northwestern Pacific Ocean, Blake *et al.* (1996) found that the decrease was much less than expected. This indicated a rapid vertical transport, so that the hydrocarbon is withdrawn from destruction in the marine boundary layer, where the rate of reaction with OH radicals would be higher than in the free troposphere.

Finally, the behavior of higher hydrocarbons in the marine atmosphere will be considered. Figure 6.11 shows data for the $C_{10}-C_{28}$ *n*-alkanes obtained at Loop Head, Ireland, and at Cape Grim, Tasmania. At both locations, trajectory analysis was used to select pure marine air masses that had no immediate contact with land before reaching the measurement sites. In addition, ethyl benzene and xylene were used as convenient indicators of contaminated air. As Figure 6.11 shows, the concentrations are similar in the mid-latitudes of the Northern and Southern Hemispheres. Somewhat lower concentrations have been observed in the central Atlantic Ocean. At Enewetak Atoll in the tropical north Pacific Ocean, concentrations were lower by up to two orders of magnitude (Duce *et al.*, 1983). Figure 6.11 shows further that the hydrocarbons occurring in the air are also present in seawater at appreciable concentrations. A screen was used to collect material present as a film at the sea surface. A comparison with water samples taken 15 cm below the sea surface indicated an enrichment of hydrocarbons at the air–sea interface. Wade and Quinn (1975), who determined total hydrocarbon concentrations of 13–239 μg liter^{-1} (subsurface) in the Sargasso Sea, had found a similar enrichment. Roughly 11% of the material could be extracted and resolved into the *n*-$C_{13}-C_{32}$ fraction. Barbier *et al.* (1973) found that $C_{14}-C_{32}$-soluble hydrocarbons occurred at depths down to 4500 m off the coast of Africa, with concentrations of 10–140 μg liter^{-1}. Thus, the higher hydrocarbons appear to be ubiquitous in the oceans. Because of the enrichment in the surface microlayer and the incomplete understanding of its physicochemical significance to the transfer of hydrocarbons from the ocean to the atmosphere, it is currently not possible to calculate transfer rates with any confidence.

Hydrocarbons originating from terrestrial vegetation, such as the higher *n*-alkane fraction of plant waxes, show a pronounced predominance of odd

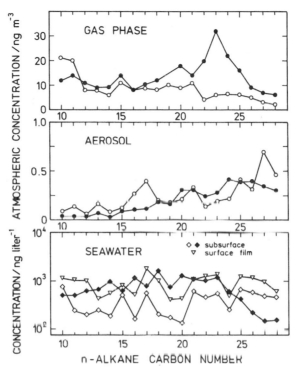

FIGURE 6.11   Concentration of n-alkanes in pure marine air masses and in seawater observed at Loop Head, Ireland (●, ◆), and at Cape Grim, Tasmania (○, ◇, ▽), according to Eichmann et al. (1979, 1980).

over even carbon numbers. In certain cases a preference of odd to even carbon numbers by a factor of 10 has been observed. The gas phase n-alkanes over the oceans do not show this behavior (even though Fig. 6.11 indicates a possible enhancement of the C_{15} component). The observation combined with the ubiquitous presence of n-alkanes in seawater point to a marine rather than terrestrial origin of the compounds. Additional support for this viewpoint is the relatively short lifetime of the higher n-alkanes with regard to their reaction with OH radicals. Eichmann et al. (1980) estimated lifetimes on the order of 1–2 days. The seawater concentrations show little variation with the size of the n-alkanes and no pronounced odd-to-even carbon number preference, indicating the ocean to be the main source of these hydrocarbons in the marine atmosphere.

Finally, it is of interest to compare the observed gas phase concentrations of n-alkanes with their saturation vapor pressures. If saturation were approached, an appreciable fraction of the n-alkanes would become associated with aerosol particles. Vapor pressures of n-alkanes are available only for the temperature range $200-400°C$, so that the data have to be extrapolated to ambient temperatures. Eichmann et al. (1979) estimated vapor pressures at $20°C$ for C_{10}, C_{20}, and C_{28} n-alkanes of 3.6, 2.7×10^{-4}, and 6.7×10^{-8} hPa, respectively. The corresponding saturation concentrations are 2.1×10^{10}, 3.3×10^6, and 1.2×10^3 ng m^{-3}. The observed concentrations, even for n-C_{28}, are much lower, so that a substantial partitioning of n-alkanes in favor of aerosol particles is not expected. Figure 6.11 shows that a small fraction of n-alkanes is indeed associated with the aerosol. The fraction increases with carbon number and approaches about 10% at the high end of the scale.

6.1.6. FORMIC AND ACETIC ACID

The anions of formic acid and acetic acid are the dominant organic components in cloud water and precipitation (Keene et al., 1983). Table 6.13 presents concentrations in rainwater assembled by Keene and Galloway (1986) and cloud-level gas phase mixing ratios calculated from these data to demonstrate that both acids are ubiquitous in the atmosphere, occurring at globally similar abundances. The gas phase reaction rates of formic and acetic acids with OH radicals are moderate, leading to lifetimes of $20-30$ days, as Table 6.1 shows. Pyruvic acid has also been observed, although at far lower levels (Andreae et al., 1987; Helas et al., 1992). Pyruvic acid is subject to photodecomposition (Grosjean, 1983a; Berges and Warneck, 1992), and its atmospheric lifetime is only a few hours. The main decomposition products are acetaldehyde and CO_2. Dicarboxylic acids, which feature lower vapor pressures, occur mainly in association with aerosol particles and not much in the gas phase. Conversely, the concentration of formic and acetic acid associated with aerosol particles is low compared to that in the gas phase. Chebbi and Carlier (1996) have provided a detailed review of carboxylic acids in the troposphere.

Table 6.14 shows gas phase mixing ratios of formic and acetic acids determined at several continental sites and over the ocean, primarily at the surface or in the planetary boundary layer. These data confirm the conclusions drawn from the rainwater analyses. Concentrations in the marine atmosphere are lower than over the continents. The results of Arlander et al. (1990) for the Pacific Ocean indicate a significant influence of air masses of continental origin, and the mixing ratios decrease with increasing distance from the shores. The lowest values were found in regions least subject to such perturbations, that is, in the southern and central Pacific Ocean, where

TABLE 6.13 Average Formic and Acetic Acid Concentrations in Precipitation at Various Locations of the World and the Corresponding Gas-Phase Mixing Ratios x_0 at Cloud Level[a]

	HCOOH			CH$_3$COOH		
	Aqueous[b] (μmol dm^{-3})	Gaseous[c] (nmol mol^{-1})	x_0 (nmol mol^{-1})	Aqueous[b] (μmol dm^{-3})	Gaseous[c] (nmol mol^{-1})	x_0 (nmol mol^{-1})
Continental						
Poker Flat, Alaska	4.3	0.07	0.29	1.2	0.07	0.13
Charlottesville, Virginia						
April–September	11.6	0.55	1.15	4.4	0.39	0.62
October–March	2.7	0.08	0.22	1.6	0.12	0.20
Lago Colado, Brazil	19.0	0.30	1.37	8.8	0.58	1.04
Torres del Paine, Chile	0.5	<0.02	0.05	0.4	0.01	0.03
Katherine, Australia	10.5	0.16	0.70	4.2	0.24	0.46
Marine						
Mauna Loa, Hawaii	2.0	0.03	0.13	0.6	0.03	0.06
North Pacific, *Discoverer*[d]	3.4	0.05	0.23	1.9	0.10	0.20
90 Mile Beach, New Zealand	1.0	<0.02	0.08	0.2	0.01	0.02
Amsterdam Island	2.2	0.01	0.12	0.6	0.02	0.05
High Point, Bermuda	2.2	0.03	0.14	1.3	0.06	0.13
North Atlantic, *Knorr*[e]	4.2	0.05	0.27	1.2	0.14	0.28
Adrigole, Ireland	1.5	0.02	0.10	—	—	—
Cape Point, South Africa	1.8	0.02	0.10	0.6	0.01	0.04

[a] Calculated to exist in the absence of clouds from the data given by Keene and Galloway (1986), assuming a liquid water volume fraction of 2×10^{-6} for raining clouds as discussed in Section 8.4.

[b] Volume-weighted aqueous concentration of anions and undissociated acid combined.

[c] Gas-phase mixing ratio in equilibrium with the aqueous phase as calculated by Keene and Galloway (1986) from the observed anion and hydrogen ion concentrations and the Henry's law coefficients.

[d] From measurements during a cruise of the research ship *Discoverer* in 1982.

[e] From measurements during a cruise of the research ship *Knorr* in 1984.

TABLE 6.14   Gas Phase Mixing Ratios of Formic Acid and Acetic Acid (nmol mol^{-1}) in the Boundary Layer

Author	Location	HCOOH	CH$_3$COOH
Continental			
Dawson et al. (1980)	Rural Arizona	1.2–3.3	1.1–6.2
Dawson and Farmer (1988)	Southwest U. S.	0.7–3.0	0.6–4.0
Talbot et al. (1988)	Rural Virginia, growing season	1.9 ± 1.2	1.3 ± 0.9
	Nongrowing season	0.7 ± 0.4	0.7 ± 0.4
Puxbaum et al. (1988)	Rural Austria	1.6 ± 0.6	1.0 ± 0.4
Hartmann et al. (1989)	Germany, continental	1.04 ± 1.08	1.24 ± 0.72
	Marine influence	0.17 ± 0.06	0.72 ± 0.08
Talbot et al. (1992)	North America, 50–60° N		
	Boundary layer	0.28 ± 0.13	0.21 ± 0.09
	Free troposphere	0.19 ± 0.07	0.22 ± 0.07
	Arctic > 60° N, boundary layer	0.27 ± 0.1	0.28 ± 0.05
	Free troposphere	0.16 ± 0.11	0.21 ± 0.05
Helas et al. (1992)	Congo basin, dry season, surface	0.5 ± 2.1	0.6 ± 5.3
	Boundary layer	3.7 ± 1.0	2.7 ± 0.9
	Free troposphere	0.9 ± 0.3	0.7 ± 0.1
Andreae et al. (1988)	Amazon basin, dry season	1.6 ± 0.6	2.2 ± 1.0
	Wet season, boundary layer	0.37 ± 0.24	0.33 ± 0.15
Talbot et al. (1990)	Amazon basin, wet season	0.15 ± 0.09	0.18 ± 0.07
	Free troposphere		
Sanhueza et al. (1992)	Venezuela, cloud forest, dry	1.70 ± 0.30	1.40 ± 0.60
	wet season	0.79 ± 0.24	0.54 ± 0.20
Hartmann et al. (1991)	Venezuela, savanna, dry season	1.44 ± 0.50	0.72 ± 0.55
	Wet season	1.28 ± 0.34	0.64 ± 0.19
Sanhueza et al. (1996)	Venezuela, savanna, dry season	3.28 ± 0.93	1.64 ± 0.39
	Wet season	0.82 ± 0.28	0.67 ± 0.39
	Coastal site, dry season	0.66 ± 0.31	0.78 ± 0.51
	Wet season	0.62 ± 0.34	0.53 ± 0.34
Marine			
Arlander et al. (1990)	Pacific Ocean, 0–40° N	0.80 ± 0.30	0.78 ± 0.32
	Pacific Ocean, 0–40° S	0.22 ± 0.13	0.28 ± 0.18
	Equatorial region, 110–180° E	0.40 ± 0.50	0.45 ± 0.43
	Indian Ocean, 0–30° S	0.19 ± 0.17	0.29 ± 0.16
Norton et al. (1992)	Mauna Loa, Hawaii, average	0.45	0.37
	Nighttime average	0.063	0.094

mixing ratios sometimes were below 0.1 nmol mol^{-1} and averaged about 0.25 nmol mol^{-1} for both acids. Similar values were observed by aircraft measurements in the free troposphere over the continents.

Mixing ratios in continental surface air range from about 0.1 to 6 nmol mol^{-1}. In contrast to marine air, where the two acids occur in comparable concentrations, the concentration ratio of acetic/formic acid over the continents is quite variable, although generally smaller than unity, approximately 0.7. The data given in Table 6.14 for the Tropics distinguish between wet and dry seasons. Mixing ratios during the wet season usually are lower than during the dry season, presumably because of an increased uptake of the acids by clouds and precipitation. Biomass burning activities toward the end of the dry season provide a source of both acids that may contribute to the seasonal cycle. A seasonal cycle also occurs in the continental air of the Northern Hemisphere. The effect is mainly evident in the annual variation of formate and acetate concentrations in precipitation, as observed in Charlottesville, Virginia (Keene and Galloway, 1986), where the highest levels occurred in August–September and the lowest levels in January–February.

Both acids exhibit a diurnal variation, with maxima occurring generally in midday or in the early afternoon. Figure 6.12 shows some results of Sanhueza *et al.* (1996). It is currently not clear whether the diurnal cycle is due to a photochemical production from a suitable precursor, emission from vegetation, enhancement of deposition processes at night, or all such processes acting together. Sanhueza *et al.* (1996) showed from measurements in a Venezuelan cloud forest that the highest concentrations of formic acid occurred in the daytime, whereas the highest concentrations of acetic acid

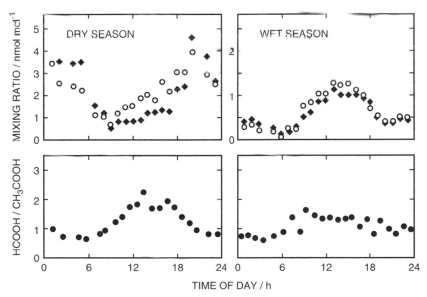

FIGURE 6.12   Top: Diurnal variation of formic acid (O) and acetic acid (●, ♦) mixing ratios in the savanna region of Calabozo, Venezuela, during the dry and wet season. Bottom: the corresponding [HCOOH]/[CH₃COOH] ratios. (Adapted from Sanhueza *et al.* (1996).)

occurred at night. In this case, dry deposition of HCOOH to wet forest surface elements exceeded that of CH_3COOH. Sanhueza et al. (1992) and Helas et al. (1992) have discussed the uptake of acids by dew. This process would contribute to the diurnal variation in ground-level air, if evaporation of moisture in the morning releases the acids again. Any depletion in the concentration occurring by dry deposition at night would be balanced during the next day by the downward mixing of air that has not experienced such losses.

In savanna regions, during the dry season, the diurnal variation was different in that nighttime mixing ratios often were higher than daytime averages, which suggests a nocturnal production process. In addition, the $[HCOOH]/[CH_3COOH]$ ratio varied considerably, showing a maximum in the afternoon, whereas during the wet season it scattered around unity. Sanhueza et al. (1996) suggested that the reaction of ozone with unsaturated hydrocarbons may have produced the acids. Fairly high ozone and hydrocarbon concentrations were observed during the dry season because of biomass burning in the savanna region.

Aircraft measurements of formic and acetic acids indicated a decrease in concentration above the continental boundary layer, showing again that the continents are a source of the acids, whereas in marine air with good vertical mixing little variation in mixing ratio was observed up to 3 km altitude (Hartmann et al., 1989; Talbot et al., 1990, 1992; Helas et al., 1992). In the free troposphere, because of the photochemically long lifetime, both acids can be transported over large distances. In the boundary layer they will undergo wet and dry deposition, which are the main removal processes. Estimates for the corresponding global loss rates provide an insight into the global budgets; Table 6.15 summarizes our current understanding of such budgets.

The data of Keene and Galloway (1986, 1988) may be used to estimate average concentrations of formic and acetic acid in precipitation at various places in the world (see Table 6.13). In the marine atmosphere the data center at 2.3 and 1.1 μmol dm^{-3}. Over the continents the volume-weighted concentrations center at 8.1 and 3.4 μmol dm^{-3}, respectively. The total rate of precipitation worldwide is about 4.4×10^{17} dm^3 (Chahine, 1992), with about 80% occurring over the oceans. The combination of these data leads to the following wet deposition rates for formic acid: 0.71 Tmol year^{-1} over the continents and 0.80 Tmol year^{-1} over the oceans; the corresponding values for acetic acid are 0.40 and 0.30 Tmol year^{-1}. The total wet deposition rates for formic and acetic acids amount to 1.5 and 0.7 Tmol year^{-1}, respectively.

The loss rates due to dry deposition may be estimated in a similar manner. Because of the alkalinity of seawater, the ocean will act as a sink of formic and acetic acids, with a dry deposition velocity similar to that of SO_2,

TABLE 6.15 Estimates of the Global Strengths of Sources and Sinks of Formic and Acetic Acids (Gmol year^{-1})

Type of source	HCOOH	CH$_3$COOH
Sources		
Direct emissions		
Road traffic	0.03	0.1
Biomass burning	63 (8–170)	210 (34–550)
Savanna soils	36	24
Formicine ants	< 13	—
Vegetation	90 (14–256)	70 (9–94)
Sum of direct emissions	189 (71–475)	304 (67–668)
Atmospheric reactions		
Oxidation of isoprene	735	~ 735
Oxidation of selected terpenes	80	—
Total production rate	1004 (944–1752)	1039 (802–1403)
Sinks		
Dry deposition, continents	2800	1700
Ocean surface	630	630
Wet deposition, continents	710	400
To the oceans	800	300
Total removal rate	4640	3030

$v_d \approx 5 \times 10^{-3}$ m s^{-1} (Table 1.13). For mixing ratios above the ocean surface of about 0.25 nmol mol^{-1}, the deposition rate for standard conditions would be $v_d c_{acid} = v_d x_{acid} \rho_{air}/M_{air} \approx 5 \times 10^{-3}$ (0.25 $\times$ 10^{-9} $\times$ 1.293/0.029) = 5.5 $\times$ 10^{-11} mol m^{-2} s^{-1}. The ocean surface is 3.6 $\times$ 10^{14} m^2, so the global dry deposition rate is 0.63 Tmol year^{-1} for each acid. Dry deposition to the continental surface overgrown with vegetation is more difficult to estimate because of the great variability in deposition velocities. Hartmann (1990), Helas *et al.* (1992), and Sanhueza *et al.* (1996) have derived dry deposition velocities in the range of (0.7–1.3) $\times$ 10^{-2} m s^{-1}, or approximately 0.01 m s^{-1}, from measurements of the nocturnal decrease in acid concentrations in savanna and tropical rain forest ecosystems. An order of magnitude estimate for the dry deposition rates may be obtained by taking mixing ratios of 1.2 and 0.8 nmol mol^{-1} to be representative for formic and acetic acid, respectively. This choice leads to deposition rates of 5.4 $\times$ 10^{-10} and 3.6 $\times$ 10^{-10} mol m^{-2} s^{-1}. The corresponding loss rates on the continents are 2.5 and 1.7 Tmol year^{-1}, which is higher than the total wet deposition rates derived above. Finally, by combining wet and dry deposition rates, one obtains order of magnitude fluxes for the removal of formic and acetic acids from the atmosphere. The values derived are 4.6 and 3.0 Tmol year^{-1}, or 213 and 182 Tg year^{-1}, respectively.

A variety of sources of acids in the atmosphere have been discussed in the literature. Among them are direct emissions from automobiles, biomass burning, and vegetation, as well as secondary production in the atmosphere by chemical reactions. Alkanoic acids in automobile exhaust were explored by Kawamura et $al.$ (1985). They found fractions of approximately 9.3 nmol mol^{-1} for formic acid and 31.8 nmol mol^{-1} for acetic acid, corresponding to about 1×10^{-3} and 4×10^{-3} %, respectively, of total hydrocarbons emitted from automobiles (about 0.1% by volume). These values may be combined with the global emission rate of hydrocarbons from road traffic, 36 Tg annually (Table 6.4), to estimate global emission rates of about 0.001 Tg $year^{-1}$ for formic acid and 0.006 Tg $year^{-1}$ for acetic acid, corresponding to 0.03 and 0.1 Gmol $year^{-1}$. The rates are almost negligible compared to those from other sources.

Talbot et $al.$ (1988) first noted that biomass burning in the Tropics is a source of acids. From measurements in the plumes of biomass burning activities in the Congo, Helas et $al.$ (1992) found a good correlation between formic and acetic acids and carbon monoxide. The ratios were 0.018 and 0.012, respectively. Hartmann (1990) conducted laboratory studies by burning straw stubble to determine emission factors relative to CO: $(2.6 \pm 2.0) \times 10^{-3}$ for formic acid and $(8.7 \pm 6.0) \times 10^{-3}$ for acetic acid (amount fraction). The ratio of acetic to formic acid was 3.9 ± 2.5. A considerable fraction of the acids, 35–55%, was associated with smoke particles. Rogers et $al.$ (1991), who conducted both field and laboratory experiments, also found appreciable amounts of formate and acetate on particles resulting from biomass burning. Further acidification of these particles by sulfuric and nitric acid is bound to release the weaker organic acids in the atmosphere. Table 4.6 shows that the global emission rate of carbon monoxide from biomass burning is 680 Tg $year^{-1}$ (range 350–1040), or 24.3 Tmol $year^{-1}$ (range 12.5–37.1). The global emission rate of formic acid from biomass burning thus is 0.063 (range 0.008–0.17) Tmol $year^{-1}$ and that of acetic acid 0.21 (0.034–0.55) Tmol $year^{-1}$.

Sanhueza and Andreae (1991) found that savanna soils emit organic acids during the dry season. The estimated release rates were 0.036 Tmol $year^{-1}$ of formic acid and 0.024 Tmol $year^{-1}$ of acetic acid, on a global scale. Kesselmeier et $al.$ (1997a) also showed that dry leaf litter represents a source of formic and acetic acids, confirming the observations of Sanhueza and Andrea (1991). Moist leaf litter, however, represented a sink. Graedel and Eisner (1988) drew attention to the possibility that formicine ants may be a source of HCOOH in the atmosphere, and they estimated a source strength of 0.013 Tmol $year^{-1}$. This must be considered an upper limit.

The seasonal variation of organic acids in precipitation led Keene and Galloway (1986, 1988) to suggest that terrestrial vegetation provides a source of formic and acetic acids, either directly or by the emission of suitable precursors that are further oxidized to form acids in the atmosphere. Talbot et al. (1990) first showed that formic and acetic acids are emitted from the leaves of tropical trees. Emission rates were 1.9 and 5.0 μg h^{-1}/m^2 leaf area for formic acid and acetic acid, respectively. More recently, Kesselmeier et al. (1997b) explored the emission of organic acids from Mediterranean oak and pine trees at a site near Rome, Italy, and found the emissions to correlate well with the rate of transpiration, that is with stomata conductance. Relative to assimilated carbon dioxide, the emission rates from Quercus ilex were 15.6 μmol mol^{-1} of formic acid and 11.8 μmol mol^{-1} of acetic acid; the rates from Pinus pinea were 20.6 and 17.2 μmol mol^{-1}, respectively. As discussed by Bode et al. (1997), formic and acetic acids are involved in various metabolic pathways, so that their emission from plants in small amounts should be common. The rates found by Kesselmeier et al. (1997b) can be combined with the annual global net primary productivity of 55 Pg (as carbon) to derive global emission rates of about 0.09 Tmol of formic acid and 0.07 Tmol of acetic acid per year. Kesselmeier et al. (1997a) also studied young spruce, beech, and ash trees under conditions excluding the formation of the acids by photochemical processes in the leaf enclosures and found emissions per leaf area in the range of 12–60 nmol m^{-2} h^{-1} for formic acid and 6–24 nmol m^{-2} h^{-1} for acetic acid. From these data they estimated global emission rates from vegetation of 0.014–0.256 Tmol year^{-1} for formic acid and 0.009–0.094 Tmol year^{-1} for acetic acid. The values agree with those obtained above in a different way.

Global emission rates based on the processes discussed so far sum to about 200 Gmol year^{-1} for formic acid and 300 Gmol year^{-1} for acetic acid. This is only a fraction of the amounts that are annually removed from the atmosphere by dry and wet deposition. Additional sources must exit, and these are thought to be hydrocarbon oxidation reactions. In Section 6.2.4 it will be shown that ozone, in reacting with the terminal double bond in alkenes and alkadienes, produces CH_2OO radicals, which when stabilized by collisions react with water vapor to produce formic acid. Neeb et al. (1997, 1998) have measured the yields and found for the reaction with ethene about 50%, and for isoprene at least 30%. Table 6.7 shows isoprene and monoterpenes to be significant sources of unsaturated hydrocarbons. Table 6.1 shows that for isoprene the rate of reaction with ozone contributes only about 10% to the total consumption rate, which is dominated by reaction with OH radicals, while for several monoterpenes featuring terminal double bonds, the two reactions display comparable rates. The global emissions of monoterpenes, however, are smaller than that of isoprene. Methacrolein (MAC) and methyl vinyl ketone (MVK) are major products resulting from the oxidation of isoprene by OH as well as by ozone. Again, as Table 6.1 indicates, these compounds react more rapidly with OH radicals than with ozone. The results of Paulson *et al.* (1992), which are in agreement with earlier studies, suggest that the reaction of isoprene with OH radicals produces about 29% MAC and 42% MVK, if organic nitrates are deducted, and the reaction with ozone produces 67% MAC, 26% MVK, and 7% propene. With these data and the rates given in Table 6.1, and assuming that only CH_2OO radicals but not other Criegee intermediates are formed, the yield of these radicals is estimated to be about 10% from isoprene and another 10% from methacrolein and methyl vinyl ketone relative to the total consumption of isoprene reacting with OH and O_3. Of the CH_2OO radicals thus formed, 50% are stabilized by collisions and react further with water vapor to form formic acid; the rest disintegrates to form other products. As Table 6.7 shows, at least 175 Tg (but more likely 500 Tg or 7.35 Tmol) of isoprene is emitted from vegetation each year. On the basis of the higher emission rate, the production rate of formic acid is 735 Gmol year^{-1}. In a similar way one may estimate the contribution of monoterpenes to formic acid formation. Table 6.5 shows that β-pinene, *d*-limonene, and myrcene, all of which contain terminal double bonds, are widely emitted and contribute 13–21%, 20–25%, and 5–7%, respectively, to total monoterpene emissions. According to Table 6.7 the global emission is at least 130 Tg annually, or about 1 Tmol year^{-1}. Making use of the data in Table 6.1 and similar assumptions for the oxidation of isoprene, the reactions of ozone with β-pinene, *d*-limonene, and myrcene are estimated to produce 70–90 Gmol year^{-1} of formic acid. The production rates derived are rather uncertain because of the many simplifying assumptions made. Reaction rates certainly are quite variable because the concentrations of OH radicals and ozone fluctuate considerably. Accordingly, the production rates are order of magnitude estimates only.

The reactions discussed above primarily generate formic acid. Pyruvic acid may be formed in the reaction of ozone with methyl vinyl ketone, but there is no obvious route to acetic acid, except in the reaction of ozone with propene, which is a minor product. Criegee radical precursors that would lead to the formation of acetic acid are expected to arise from reactions of ozone with 2-alkenes (and propene). Few such compounds are known to be emitted in significant amounts from vegetation. Madronich *et al.* (1990) pointed out that acetic acid is produced together with acetyl hydroperoxide when acetylperoxy and hydroperoxy radicals enter into the reaction

$$CH_3C(O)OO + HO_2 \rightarrow CH_3(CO)OOH + O_2 \qquad 67\%$$

$$\rightarrow CH_3COOH + O_3 \qquad 33\%$$

under NO_x-lean conditions. Hydroperoxy radicals are generated in the oxidation of formaldeyde, a principal product resulting from isoprene oxidation, whereas acetylperoxy radicals are produced in the oxidation of methyl vinyl ketone and methacrolein, either directly or by the photolysis of methylglyoxal, which is one of the oxidation products. In addition, acetaldehyde emitted from vegetation (Kesselmeier et al., 1997b) generates acetyl and acetylperoxy radicals when it reacts with OH radicals. Jacob and Wofsy (1988) studied the photochemistry of isoprene emissions from a tropical forest by means of computer simulations and showed that 0.5 nmol mol^{-1} of formic acid is produced from isoprene oxidation. These calculations were extended by Madronich et al. (1990), who showed that similar mixing ratios of acetic acid can be obtained by including the above reactions. Apparently, acetic acid can be formed in amounts equivalent to formic acid, and the entry in Table 6.15 makes use of this assumption. Acetylperoxy radicals are also intermediates in the oxidation of acetaldehyde and acetone, which are products resulting from the oxidation of ethane and propane, respectively. In the last section it was shown that the global sources of ethane and propane are about 500 Gmol year^{-1} each. If all of the acetylperoxy radicals arising from the oxidation of ethane and propane were to undergo the above reaction with HO$_2$, 330 Gmol of acetic acid would be produced. This rate clearly is an upper limit, as a greater fraction of the radicals will enter into other reactions, but the process may nevertheless add to the rate of acetic acid formation from the oxidation of isoprene.

As Table 6.15 shows, direct emissions and atmospheric hydrocarbon oxidation reactions indicate total source strengths of approximately 1 Tmol year^{-1} for each of the acids, approaching the magnitude of the removal rate due to wet and dry deposition. The budget evidently is not well balanced, and other production mechanisms should be considered. One process that may be important is the oxidation of formaldehyde by OH radicals in the aqueous phase of clouds, originally suggested by Chameides and Davis (1982). As discussed in Section 8.6.1, the reaction is rapidly followed by the destruction of HCOOH. Which of the two processes is more significant depends on the initial concentration of formic acid entering into the aqueous phase.

6.2. HYDROCARBON OXIDATION MECHANISMS

A general scheme for the oxidation of hydrocarbons in the atmosphere was discussed in Section 5.2 in conjunction with the formation of photochemical smog. Here, we shall treat the oxidation of several individual hydrocarbons in more detail. The oxidation may be initiated by reactions with OH radicals, with NO_3 radicals, or with ozone. The oxidation then proceeds via alkylperoxy radicals (ROO·) and alkoxy radicals (RHO·) toward aldehydes and ketones as products. The presence of nitric oxide is required to convert alkylperoxy to alkoxy radicals. In urban and regionally polluted air the abundance of NO_x is usually sufficient to sustain propagation of the oxidation chain. In the unpolluted, remote troposphere the concentration of nitrogen oxides is low and the effectiveness of NO_x as a catalyst may be greatly reduced by competing reactions. Accordingly, before discussing individual reaction mechanisms, it will be useful to consider first the chemical behavior of alkylperoxy and alkoxy radicals.

6.2.1. ALKYLPEROXY RADICALS

The first product resulting from the abstraction of a hydrogen atom from an alkane is an alkyl radical. In the laboratory, alkyl radicals may undergo a variety of reactions, depending on experimental conditions; in the atmosphere, however, because of the overwhelming concentration of oxygen, alkyl radicals combine rapidly with O_2 to produce the corresponding alkylperoxy radicals. As mentioned previously, tertiary hydrogen atoms are abstracted more easily than secondary H atoms, and their abstraction, in turn, is more facile that that of primary H atoms. In the higher hydrocarbons the number of secondary H atoms usually exceeds that of primary or tertiary ones, so that secondary alkyl and alkylperoxy radicals are most frequently formed.

$$\begin{array}{c} R_1 \\ \diagdown \\ \diagup \\ R_2 \end{array} CH_2 \xrightarrow{OH} H_2O + \begin{array}{c} R_1 \\ \diagdown \\ \diagup \\ R_2 \end{array} CH \xrightarrow{O_2} \begin{array}{c} R_1 \\ \diagdown \\ \diagup \\ R_2 \end{array} CHOO$$

From rate coefficients for the addition of molecular oxygen to CH_3 and C_2H_5 radicals (see Table A.4), it can be estimated that the lifetime for alkyl radicals in air at ground-level pressure is about 100 ns. For methyl and ethyl radicals, these addition reactions are in the fall-off regime between second- and third-order kinetics at atmospheric pressure and at room temperature; for higher alkyl radicals they are at the high pressure limit. Occasionally, an addition-abstraction reaction of the type

$$H_3C\dot{C}H_2 + O_2 \rightarrow H_2C{=}CH_2 + HO_2$$

has been invoked for ethyl and other small alkyl radicals, because such reactions appear to be prominent in low-temperature flames (Pollard, 1977).

However, the abstraction reaction requires an activation energy, whereas the addition reaction does not. At low temperatures and high pressures excess energy made available by the addition process is rapidly quenched. For example, at atmospheric pressure and 298 K, the yield of $C_2H_2 + HO_2$ from the reaction $C_2H_5 + O_2$ is about 0.04% (compare Table A.4). Thus, abstraction of an H atom is essentially negligible under atmospheric conditions.

Alkylperoxy radicals in the atmosphere undergo either reactions with other atmospheric constituents or internal rearrangement. The latter possibility exists only for larger alkylperoxy radicals, because a cyclic intermediate is required. In discussing reactions with other species, the methyl peroxy radical, CH_3O_2, may serve as a guide to selecting the probable reaction pathways that alkylperoxy radicals can undergo in the atmosphere. Table 6.16 shows the pertinent reactions, the associated rate coefficients, and relative rates calculated for low and medium levels of NO. Boundary-layer and daylight conditions are adopted.

As Table 6.16 shows, the reactions of CH_3O_2 with NO and NO_2 are predominant. When NO_x concentrations are low, the reaction with HO_2 also becomes important, whereas the reaction of CH_2O_2 with itself is less significant. Alkylperoxy radicals may also be scavenged by aerosol particles. The scavenging rate is determined to some extent by the coefficient of accommodation, which in most cases is not known. The collision frequencies shown in Table 6.16, which are upper-limit values, indicate that this process is potentially important.

TABLE 6.16 Comparison of Methyl Peroxy Radical Reactions in the Atmosphere[a]

Reaction with reactant X		$10^{13}k$	(A)		(B)	
			$n(X)$	$kn(X)$	$n(X)$	$kn(X)$
$CH_3O_2 + HO_2$	→ $CH_3OOH + O_2$	49	5 (8)	2.5 (−3)	2 (7)	1.0 (−4)
$CH_3O_2 + CH_3O_2$	→ $2CH_3O$ (30%)	4.5	1 (8)	9.0 (−5)[b]	3 (6)	2.7 (−6)[b]
	→ $HCHO + CH_3OH$					
$CH_3O_2 + NO$	→ $CH_3O + NO_2$	76	2 (8)	1.5 (−3)	1 (10)	7.6 (−2)
$CH_3O_2 + NO_2$	→ $CH_3O_2NO_2$	41	1 (9)	4.1 (−3)	4 (10)	1.6 (−1)
$CH_3O_2 + SO_2$	→ $CH_3O_2SO_2$	< 5 (−4)	5 (9)	2.5 (−7)	5 (10)	2.5 (−6)
$CH_3O_2 + $ aerosol particles	→ Scavenging products	—	Marine	9.0 (−4)	Rural	9.0 (−3)

[a] Rate coefficients k at 298 K and atmospheric pressure (units: cm^3 molecule^{-1} s^{-1}). Relative rates $kn(X)$ (s^{-1}) are given for two atmospheric conditions: (A) low and (B) medium number densities of NO, NO_2 and SO_2. The last line shows collision frequencies with marine and rural continental aerosol particles according to Table 7.5. Powers of ten are shown in parentheses.
[b] $= 2kn(X)$.

Higher alkylperoxy radicals undergo reactions analogous to those of CH_3O_2, but a few additional comments are in order regarding the products. Reference is made to reviews of Lightfoot *et al.* (1992) and Atkinson (1990, 1994).

(1) The reactions with nitric oxide produce alkyl nitrates in addition to RO· radicals

$$ROO· + NO \xrightarrow{a} NO_2 + RO·$$

$$\xrightarrow{b} RONO_2$$

Although the yield of alkoxy radicals is always preponderant, that of alkyl nitrate formation increases with the carbon number of the alkylperoxy radical. The yield depends on temperature and pressure, increasing with decreasing temperature and increasing pressure. Table 6.17 presents results from laboratory studies to indicate the magnitude of the $k_b/(k_a + k_b)$ ratio observed in several alkylperoxy radical reactions. Secondary alkylperoxy radicals lead to a higher yield than primary alkylperoxy radicals, up to 30% in several cases.

(2) The addition of NO_2 to alkylperoxy radicals leads to alkylperoxy nitrates, $ROONO_2$, which are unstable and revert to the original reactants by thermal decomposition. Zabel *et al.* (1989) determined decomposition rates for several higher alkylperoxy nitrates that indicate lifetimes of 1 min or less at 273 K and higher. Accordingly, alkylperoxy nitrates provide only a

TABLE 6.17   Fractional Yield of Alkyl Nitrates in Selected RO_2 + NO Reactions[a]

Alkyl R	$k_b/(k_a + k_b)$	Alkyl R	$k_b/(k_a + k_b)$
Primary radicals		Secondary radicals	
Ethyl	≤ 0.014	2-Propyl	0.049
1-Propyl	0.020	2-Butyl	0.083
1-Butyl	0.041	2-Pentyl	0.129
2-Methyl-1-propyl	0.075	2-,3-Pentyl	0.125
1-Pentyl	0.061	2-Methyl-2-butyl	0.044
2-Methyl-1-butyl	0.040	2-Methyl-3-butyl	0.109
3-Methyl-1-butyl	0.043	2-Hexyl	0.220
Neo-pentyl	0.051	3-Hexyl	0.220
1-Hexyl	0.121	2-Methyl-2-pentyl	0.035
1-Heptyl	0.195	3-Methyl-2-pentyl	0.140
1-Octyl	0.36	2-Heptyl	0.324
		3-Heptyl	0.312
		4-Heptyl	0.290

[a] Excerpt from the compilation of Lightfoot *et al.* (1992).

temporary storage reservoir of alkylperoxy radicals, except perhaps at the low temperatures of the upper troposphere and lower stratosphere.

(3) Relatively few rate coefficients have been measured for the reactions of alkylperoxy radicals with HO_2, but the existing data show that such reactions are quite rapid. As the reaction leads to stable alkylhydroperoxides as the main product, it is a radical chain termination process. Table 6.16 indicates that in the remote troposphere it competes well with chain propagation by reaction with NO.

(4) A number of self-reactions of alkylperoxy radicals have been studied in addition to the self-reaction of methyl peroxy radicals. There is a tendency for rate constants of secondary alkylperoxy radical self-reactions to be smaller by two orders of magnitude relative to those of primary alkylperoxy radicals (Lightfoot et al., 1992) Self-reactions of tertiary alkylperoxy radicals are even slower. Madronich and Calvert (1990) suggested that values of rate coefficients of cross-combination reactions involving two different alkylperoxy radicals occur midway between the rate constants for the two self-recombination reactions. The data in Table 6.16 indicate that the self-reaction of methyl peroxy radicals is fairly unimportant relative to other reactions of this radical. In the unpolluted remote atmosphere, the CH_3O_2 radical is the most abundant of all alkylperoxy radicals, so that CH_3O_2 will be the main participant in cross-combination reactions with other alkylperoxy radicals As rate constants for such reactions generally are smaller than that for the CH_3O_2 self-reaction, such reactions are expected to be fairly unimportant.

In considering a possible isomerization of alkylperoxy radicals, it should be recognized that internal hydrogen abstraction, exemplified by

$$H_3C-\underset{\underset{O\cdot}{\overset{|}{O}}}{\overset{\overset{H}{|}}{C}}-CH_2-\underset{\overset{|}{H}}{\overset{\overset{H}{|}}{C}}-CH_3 \rightarrow H_3C-\underset{\underset{OH}{\overset{|}{O}}}{\overset{\overset{H}{|}}{C}}-CH_2-\underset{}{\overset{\overset{H}{|}}{C}}-CH_3$$

is the most frequent mode of isomerization, leading to the formation of hydroperoxy-alkyl radicals. These would add another oxygen to form hydroperoxy-alkylperoxy radicals, which by reacting with HO_2 would produce dihydroperoxides. Internal hydrogen abstraction was first postulated by Rust (1957) to explain the production of dihydroperoxides in the low-temperature oxidation of certain hydrocarbons. Internal hydrogen abstraction proceeds via a cyclic transition state, with a six-membered transitory ring structure being favored over other configurations. Abstraction of a hydrogen atom by the peroxy group is slightly endothermic, and the associated activation energies are considerably higher than those for abstraction by OH radicals. Such reactions are usually neglected in atmospheric chemistry, and rate coefficients for the gas phase are not available. The experimental data of Allara *et al.* (1968) and Mill *et al.* (1972) on the oxidation of *n*-butane and isobutane may be utilized to estimate activation energies for the abstraction of primary, secondary, and tertiary H atoms by peroxy radicals. The corresponding values are 77.3, 66.0, and 57.6 kJ mol^{-1}. Ring strain energies must be added to obtain activation energies associated with intramolecular hydrogen abstraction. Laboratory studies of the oxidation of 2,4-dimethylpentane by Mill and Montorsi (1973) and of *n*-pentane and *n*-nonane by VanSickle *et al.* (1973) provide rate coefficients for the internal abstraction of hydrogen atoms that suggest a ring strain energy of about 12 kJ mol^{-1}, if one adopts a value of 10^{11} s^{-1} for the preexponential factor. The combination of these data leads to the following values for the rate coefficients (unit: s^{-1})

$$k_{\text{prim}} = 1 \times 10^{11} \exp(-10{,}770/T)$$

$$k_{\text{sec}} = 1 \times 10^{11} \exp(-9{,}418/T)$$

$$k_{\text{tert}} = 1 \times 10^{11} \exp(-8{,}408/T)$$

At ambient temperatures in the ground-level atmosphere, the rate coefficients assume the following values: $k_{\text{prim}} = 1.4 \times 10^{-5}$, $k_{\text{sec}} = 1.4 \times 10^{-3}$, and $k_{\text{tert}} = 4.2 \times 10^{-2}$ s^{-1}. The values would have been raised by a factor of 2 had we assumed a value of 2×10^{10} s^{-1} for the preexponential factor. A comparison of these data with the rates shown in Table 6.16 for the other processes shows that the internal abstraction of primary hydrogen atoms is unimportant in the atmosphere, but that internal abstraction of secondary and tertiary H atoms may be competitive with other reactions of alkylperoxy radicals.

6.2.2. ALKOXY RADICALS

Alkoxy radials that arise from the reaction of NO with alkylperoxy radicals can also enter into several competing processes. Four types of reactions are generally considered: thermal decomposition, isomerization, reaction with oxygen, and addition to either NO or NO$_2$. Table 6.18 summarizes information on rate coefficients and projected rates in the atmosphere for several small alkoxy radicals.

(1) Decomposition of alkoxy radicals usually leads to the formation of a stable aldehyde. Few direct measurements of such processes exist, and most of the current knowledge was either obtained by extrapolation of high-temperature data to atmospheric conditions or calculated from Arrhenius activa-

TABLE 6.18 Rate Coefficients k and Rates R for Different Types of Reactions of Alkoxy Radicals in the Atmosphere[a,b]

Radical	k_{dec}	k_{isom}	k_{O_2}	k_{NO_2}	R_{dec}	R_{isom}	R_{O_2}	R_{NO_2}
$CH_3O \cdot$	—	—	1.9(−15)	1.3(−11)	—	—	9.0(3)	3.5(−1)
$CH_3CH_2O \cdot$	2(−1)	—	9.5(−15)	2.3(−11)	2(−1)	—	4.8(4)	7(−1)
$C_2H_5CH_2O \cdot$	3(−1)	—	9.5(−15)	3.5(−11)	3(−1)	—	4.6(4)	8.7(−1)
$CH_3CH(O \cdot)CH_3$	2(2)	—	8(−15)	5.5(−11)	2(2)	—	4(4)	8.7(−1)
$HOCH_2CH_2O \cdot$	8.6(4)	—	6(−15)	—	8.6(4)	—	3.1(4)	—
$CH_3(CH_2)_2CH_2O \cdot$	1.6(2)	6.7(4)	7(−15)	3.8(−11)	1.6(2)	6.7(4)	3.5(4)	9.5(−1)
$C_2H_5CH(O \cdot)CH_3$	4.9(3)	—	1(−14)	3.8(−11)	4.9(3)	—	5(4)	9.5(−1)
$C_2H_5CH(O \cdot)CH_2OH$	1.2(5)	—	7(−15)	—	1.2(5)	—	3.5(4)	—
$CH_3CH(O \cdot)CH_2C_2H_5$	9(3)	6.7(4)	9(−15)	3.8(−11)	9(3)	6.7(4)	4.5(4)	9.5(−1)
$C_2H_5CH(O \cdot)C_2H_5$	2.7(4)	—	8(−15)	3.8(−11)	2.7(4)	—	4(4)	9.5(−1)

[a] Alkoxy radical decomposition and isomerization (unit of rate constants: s^{-1}) and reactions with oxygen and nitrogen dioxide (units of rate constants: cm^3 molecule^{-1} s^{-1}). Rates (unit: s^{-1}) were calculated with $n(O_2) = 5(18)$, $n(NO_2) = 2.5(10)$. Rate data were taken from Atkinson (1994) and le Bras (1997). In some cases they were estimated as suggested by Atkinson (1994)

[b] Numbers in parentheses indicate powers of ten.

tion energies that are assumed to rise linearly with the endothermicity of the decomposition process (Baldwin *et al.*, 1977; Carter and Atkinson, 1985; Atkinson and Carter, 1991). Radicals with a hydroxyl and the alkoxy group positioned on neighboring carbon atoms decompose rapidly by carbon-carbon bond cleavage, for example,

$$HOCH_2\text{-}CH_2O\cdot \rightarrow \overset{\centerdot}{C}H_2OH + HCHO$$

1,2 Hydroxy-alkoxy radicals are generally formed in the oxidation of alkenes by the sequential addition of an OH radical to one side and oxygen to the other side of the double bond.

(2) Isomerization of alkoxy radicals may occur by internal hydrogen abstraction in the same manner as discussed above for alkylperoxy radicals. Isomerization of alkoxy radicals is energetically more favorable, however, because the reaction is exothermic. Activation energies tabulated by Gray *et al.* (1967) for H-atom abstraction reactions involving the CH_3O radical suggest values of 29.7, 19.7, and 15.5 kJ mol^{-1} for primary, secondary, and tertiary H atoms, respectively. The ring strain energy must be added. Baldwin *et al.* (1977) assumed ring strain energies of about 2.5 kJ mol^{-1} for a six-membered ring (1,5 H-shift) and 27.2 kJ mol^{-1} for a five-membered ring (1,4 H-shift). The first value may be too low in view of the results for isomerization of alkylperoxy radicals discussed earlier. More recently, Atkinson (1994) suggested a revised activation energy for primary H-atom abstraction of 32.5 kJ mol^{-1}, which leads to a rate coefficient of 6.7×10^4 s^{-1} for 1,5 H-shift for the 1-butoxy radical, in good agreement with experimental data. For the analogous isomerization of the 2-pentoxy radical, Benkelberg *et al.* (1997) found a rate coefficient close to 1×10^5 s^{-1}, which is similar in magnitude. The internal abstraction of a secondary H-atom in 1-butoxy, that is, a 1,4 hydrogen shift, requires a five-membered ring structure with a correspondingly higher ring strain energy. The activation energy of this process is at least $19.7 + 27.2 = 46.9$ kJ mol^{-1}, which makes the process less likely in comparison. Higher alkoxy radicals, starting with 1-pentoxy, admit the possibility of abstraction of a secondary H atom via formation of a six-membered ring. The rate coefficient for this process is calculated to be $k_{isom} > 10^6$ s^{-1}, which would cause the process to overwhelm all others. Eberhard *et al.* (1995) confirmed this prediction from the product yield of the isomerization of the 2-hexoxy radical.

Isomerization by internal hydrogen abstraction creates a new moiety to which oxygen can add, so that it leads to the formation of δ-hydroxy-alkylperoxy radicals, and eventually to the formation of δ-hydroxy-carbonyl compounds. Eberhard *et al.* (1995) detected 5-hydroxy-hexane-2-one to arise from the 2-hexoxy radical, and Kwok *et al.* (1996) identified similar compounds in reactions of OH radicals with C_4–C_8 n-alkanes. Baldwin et al. (1977) envisioned that internal hydrogen abstraction can repeat itself because of the successive formation of alkoxy groups, and they postulated the formation of compounds with the general structure $(OH)_x CH_{3-x}(CH_2)_2$-$CH_{3-y}(OH)_y$. However, compounds with more than one hydroxyl group attached to the same carbon atom are inherently unstable, as they tend to split off water to establish a carbonyl group. In fact, the abstraction of a hydrogen atom from a -CH_2OH group to form -$CHOH$ already strengthens the carbon-oxygen bond, making a further addition of oxygen unlikely. Instead, the simultaneous weakening of the O-H bond enables the hydrogen atom to be abstracted by oxygen. For the simplest radical of this type, CH_2OH, Radford (1980) found the abstraction process

$$\dot{C}H_2OH + O_2 \rightarrow HCHO + HO_2$$

to be very rapid. In the same vein, Carter et al. (1979), who had studied the photooxidation of ethanol and 2-butanol, concluded from the product distribution that H-atom abstraction from the OH group also supersedes oxygen addition for the higher homologues of CH_2OH. The successive addition of oxygen to the 1-butoxy radical thus terminates with the step

$$HO\dot{C}H(CH_2)_2CH_2OH + O_2 \rightarrow OCH(CH_2)_2CH_2OH + HO_2$$

(3) Although alkoxy radicals react with oxygen fairly slowly, the process is important in the atmosphere because of the high concentration of oxygen. The reaction proceeds by abstraction of a hydrogen atom from the -$CH_2O\cdot$ or >$CHO\cdot$ group. Tertiary alkoxy radicals do not contain an abstractable hydrogen atom, so that they do not react with oxygen. Absolute rate constants have been determined for the reactions of methoxy, ethoxy, and isopropoxy radicals. Except for the reaction of methoxy radicals, the rate coefficients of all alkoxy radicals with oxygen are fairly similar: $kO_2 = (8 \pm 2) \times 10^{-15}$ cm^3 molecule^{-1} s^{-1}. Baldwin et al. (1977) and Balla et al. (1985), Atkinson and Carter (1991), and Atkinson (1994) have discussed relationships between rate constants and the exothermicities of the reactions to develop an algorithm for calculating rate constants for the general case.

(4) Association reactions of alkoxy radicals with NO and NO_2 are rapid. Absolute rate constants have been measured for methoxy, ethoxy, and isopropoxy radicals, and relative rate data for several other cases, mostly at higher temperatures, as reviewed by Batt (1987). Extrapolation has led to recommended values for 298 K limiting high-pressure rate constants of $(2-4) \times 10^{-11}$ cm^3 molecule^{-1} s^{-1} (Atkinson, 1994). In the unpolluted atmosphere the concentration of NO_2 exceeds that of NO. Since the rate

coefficients have the same magnitude, only the reaction with NO_2 needs to be considered. Table 6.18 shows, however, that these reactions are essentially unimportant in the atmosphere, even for concentrations of NO_x much higher than assumed, because they cannot compete with the parallel reaction with oxygen. It should be noted that the reactions NO and NO_2 admit the possibility of hydrogen atom abstraction, for example

$$RCH_2O\cdot + NO_2 \rightarrow RCHONO_2$$

$$\rightarrow RCHO + HNO_2$$

which leads to the formation of nitrous acid in addition to an alkyl nitrate. However, the data available for CH_3O, C_2H_5O, and $(CH_3)_2CHO$ radicals indicate only a minor probability for the abstraction pathway, on the order of a few percent.

6.2.3. Oxidation of Alkanes

The preceding discussion of alkylperoxy and alkoxy radical reactions will be used as a guide in presenting in the following some examples of specific oxidation mechanisms, starting with the alkanes.

Ethane

The oxidation of ethane follows a mechanism similar to that of methane and involves the following steps (Aikin *et al.*, 1982):

$$C_2H_6 + OH \rightarrow H_2O + C_2H_5$$

$$C_2H_5 + O_2 \rightarrow C_2H_5OO$$

$$C_2H_5OO + NO \rightarrow C_2H_5O + NO_2$$

$$C_2H_5OO + HO_2 \rightarrow C_2H_5OOH + O_2$$

$$C_2H_5O + O_2 \rightarrow CH_3CHO + HO_2$$

The further oxidation of acetaldehyde then follows the reaction scheme outlined in Section 5.2.

Propane

This alkane contains six primary and two secondary hydrogen atoms. On the basis of Equation (6.1) one estimates that the probability of H-atom abstraction by OH radicals is about 75% for secondary H-atoms and 25% for primary H-atoms. Equation (6.2) leads to similar values of 79% and 21%, respectively. The oxidation mechanism following primary hydrogen atom abstraction is similar to that of ethane and leads to the formation of propanal. The reaction sequence induced by the abstraction of a secondary H-atom is

$$CH_3CH_2CH_3 + OH + (O_2) \rightarrow (CH_3)_2CHOO$$

$$(CH_3)_2CHOO + NO \rightarrow (CH_3)_2CHO + NO_2$$

$$(CH_3)_2CHO + O_2 \rightarrow CH_3COCH_3 + HO_2$$

which gives rise to the formation of acetone. As the abstraction of secondary hydrogen atoms by OH radicals is preponderant, the oxidation of propane is an important source of acetone in the atmosphere.

n-Butane

The oxidation mechanism for n-butane is shown in Figure 6.13. n-Butane contains six primary and four secondary hydrogen atoms. From either Equation (6.1) or Equation (6.2) one estimates that about 15% of all reactive encounters with OH radicals lead to the abstraction of primary and 85% to the abstraction of secondary H-atoms. This opens up two pathways of oxidation that lead to the products butanal and methyl ethyl ketone, respec-

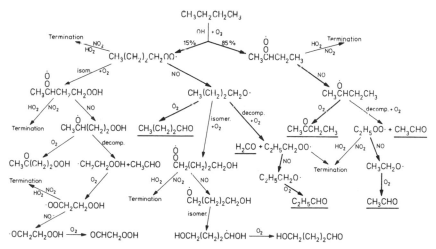

FIGURE 6.13 Oxidation mechanism for n-butane. The dominant reaction channels are indicated by bold arrows.

tively. The data in Table 6.18 show that decomposition of the intermediate alkoxy radicals is less important than reaction with oxygen. Nevertheless, decomposition of *sec*-butoxy leads to the formation of two molecules of acetaldehyde, which is a major product in addition to methyl ethyl ketone observed in studies of *n*-butane oxidation in environmental chambers, showing that decomposition of *sec*-butoxy must be significant. The decomposition of *prim*-butoxy is less important, yet it provides the only route to propanal, which is an observed product in the laboratory oxidation of *n*-butane (Carter *et al.*, 1979). The primary butoxy radical may isomerize and Table 6.18 shows that this process is competitive with the reaction with oxygen. Figure 6.13 indicates that isomerization leads to the formation of 1,4-hydroxy-butanal. The mass spectrometric study of Kwok *et al.* (1996) suggests that this product comprises 7–14% of the entire yield of carbonyl compounds, indicating that a larger fraction of *prim*-butoxy radicals enters this reaction pathway. Heiss and Sahetchian (1996) reported that one of the isomerization products is 4-hydroperoxy butanal. Figure 6.13 includes on the left-hand side an isomerization pathway resulting from internal abstraction of *prim*-butylperoxy radicals to indicate the potential for the formation of products containing both carbonyl and hydroperoxy groups. This reaction may become significant only in the remote troposphere under conditions where butylperoxy radicals are sufficiently long-lived.

Isobutane (2-Methyl Propane)

The oxidation scheme for isobutane is sketched in Figure 6.14. The initial attack of OH radicals and the subsequent addition of oxygen lead to primary isobutylperoxy radicals in 26% of all reactive encounters, whereas 76% lead to tertiary butylperoxy radicals. Both radicals are then converted to the corresponding butoxy radicals. *Prim*-Isobutoxy has a choice of either reacting with oxygen to form isobutanal or decomposing to form formaldehyde and propyl radicals. The addition of oxygen leads to propylperoxy radicals. These are converted to *sec*-propoxy radicals, which react mainly with oxygen to produce acetone. Decomposition of propoxy radicals is a minor pathway, although the only one yielding acetaldehyde. The principal oxidation pathway for isobutane proceeds via tertiary butoxy radicals. They are unstable and undergo decomposition, forming acetone and methyl radicals. The second species enters into the methyl oxidation chain, which leads to formaldehyde and methyl hydroperoxide as the final products. The major products resulting from the oxidation of isobutane thus are acetone and formaldehyde.

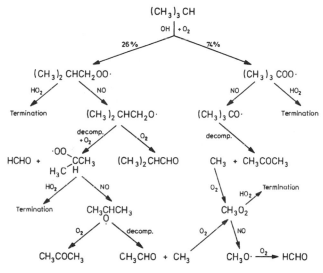

FIGURE 6.14   Oxidation mechanism for isobutane.

6.2.4. OXIDATION OF ALKENES

In reacting with alkenes OH radicals have a choice of either attaching to the C—C double bond or abstracting a hydrogen atom. The generally negative temperature dependence of the rate coefficients for these reactions in addition to the observed product distributions provides ample evidence for the preponderance of the addition reaction, at least for the smaller alkenes and alkadienes. This behavior is explained by the requirement of an activation energy for abstraction in contrast to the addition process. Accordingly, longer side chains are needed before H-atom abstraction can compete with addition. In the following, the oxidation of ethene, propene, and isoprene will be considered as examples.

Ethene

The addition of OH to the double bond of ethene creates a free moiety to which an oxygen molecule attaches under atmospheric conditions. This results in the formation of an α-hydroxyperoxy radical, which undergoes the usual reactions. The following mechanism is based on the suggestions of

Niki *et al.* (1978, 1981). The initial reactions

$$H_2C{=}CH_2 \xrightarrow{OH} \underset{\overset{|}{H}}{\overset{\overset{H}{|}}{HOC}}{-}\dot{C}H_2 \xrightarrow{O_2} \underset{\overset{|}{H}}{\overset{\overset{H}{|}}{OHC}}{-}\underset{\overset{|}{H}}{\overset{\overset{H}{|}}{COO}}{\cdot}$$

are followed by termination as well as O-atom abstraction,

$$HOCH_2CH_2OO + HO_2 \rightarrow HOCH_2CH_2OOH + O_2$$

$$HOCH_2CH_2OO + NO \rightarrow HOCH_2CH_2O + NO_2$$

The hydroxy alkoxy radical can then undergo either reaction with oxygen or cleavage:

$$HOCH_2CH_2O + O_2 \rightarrow HOCH_2CHO + HO_2$$

$$HOCH_2CH_2O \rightarrow H_2COH + HCHO$$

$$H_2COH + O_2 \rightarrow HCHO + HO_2$$

In the first case glycolaldehyde is formed as product. In the second case the products are formaldehyde and a CH_2OH radical, which reacts rapidly with oxygen to form a second formaldehyde molecule. The laboratory experiments indicate that at atmospheric pressure about 75% of the hydroxy ethoxy radicals undergo decomposition, and the rest enter into a reaction with oxygen. In the atmosphere, glycolaldehyde reacts further with OH, whereby additional formaldehyde is formed. The β-hydroxy alkoxy radicals formed from the higher alkenes, starting with propene, mainly undergo decomposition, and the reaction with oxygen plays a subordinate role.

Ethene, like the other alkenes, also reacts with ozone in the atmosphere. Criegee (1957, 1962, 1975), who explored the ozonolysis of alkenes in solution, suggested that ozone adds to the double bond, forming a primary ozonide as an unstable intermediate, which then decomposes to a carbonyl compound and a zwitterion fragment, for example:

$$H_2C{=}CH_2 + O_3 \rightarrow \underset{\overset{|}{O}\ \ \overset{|}{O}}{H_2C{-}CH_2} \rightarrow$$

with the bridging structure $\overset{O}{\underset{O\ \ O}{\triangle}}$ above, giving

HCHO + $H_2C^+OO^-$ liquid phase

HCHO + $H_2\dot{C}OO\cdot$ gas phase

In the gas phase, the zwitterion would be replaced by the corresponding diradical, $H_2\dot{C}OO\cdot$, that is, a radical endowed with two free valences. Such radicals often contain an excess of internal energy, and they undergo internal rearrangement, usually resulting in a further decomposition. In the case of ethene, evidence for the gas phase formation of such a Criegee intermediate has been presented by Lovas and Suenram (1977), who used microwave spectroscopy to identify dioxirane as an intermediate product, and by Martinez *et al.* (1977), who detected the same product by mass spectrometry. Heron and Huie (1977) found in addition that one molecule of formaldehyde was formed for each molecule of ethene consumed in the reaction. Because of the high internal energy content (which may be denoted by an asterisk), the Criegee radical decomposes rapidly unless it is stabilized by collisions:

$$H_2COO^* \rightarrow \text{decomposition products}$$

$$H_2COO^* + M \rightarrow CH_2OO + M$$

At pressures near atmospheric, 37–47% of the Criegee radicals are stabilized (Su *et al.* 1980; Horie and Moortgat, 1991; Atkinson, 1997; Neeb *et al.*, 1997). Recent results of Neeb *et al.* (1998) indicate a value close to 50%. The observed products resulting from decomposition of the energy-rich Criegee intermediate, CO_2, CO, HCOOH, and H_2, established the existence of more than one decomposition channel:

$$H_2COO^* \rightarrow H_2C\begin{matrix} O \\ \diagdown \\ O \end{matrix} \rightarrow \left(HC\begin{matrix} OH \\ \diagdown \\ O \end{matrix} \right)^* \quad \begin{array}{ll} \rightarrow CO_2 + H_2 & \text{(a) } 27\% \\ \rightarrow CO_2 + 2H & \text{(b) } 19\% \\ \rightarrow CO + H_2O & \text{(c)} \\ \rightarrow HCO + OH & \text{(d)} \end{array} \Big\} 46\% \\ \rightarrow HCOOH \quad \text{(e) } 8\%$$

The percentage probabilities for the individual channels were taken from Horie and Moortgat (1991) and Neeb *et al.* (1998). Reaction channel (d), first proposed by Martinez *et al.* (1981), is a source of OH radicals. For the reaction of ozone with ethene, Atkinson *et al.* (1992) have determined an overall OH yield of 12%. Other alkenes reacting with ozone also produce OH radicals, some of them with a much higher yield (Atkinson, 1997). As these reactions also occur in the atmosphere during the night, they provide a nighttime source of OH radicals. In laboratory studies of alkene-ozone reactions, the OH radicals produced react further with the alkene, which complicates the determination of product yields from the ozone reaction. The addition of a scavenger for OH radicals, such as cyclohexane, to the reaction mixture has been used to suppress this effect (Atkinson and Aschmann, 1993).

The stabilized Criegee radical is known to react with aldehydes, SO_2, CO, and H_2O (Hatakeyama and Akimoto, 1994) and is believed to react with NO.

Reactions with formaldehyde and formic acid resulting as products from the ozone-ethene reaction have complicated understanding of this system, because longer-lived and difficult to analyze intermediates are formed. In the atmosphere the reaction with water vapor would be dominant. The reaction leads to the formation of hydroxymethane hydroperoxide, which subsequently decomposes to form formic acid (Horie *et al.*, 1994a; Neeb *et al.*, 1997):

$$CH_2OO + H_2O \rightarrow HOCH_2OOH$$

$$HOCH_2OOH \rightarrow HCOOH + H_2O$$

As nearly 50% of the Criegee radicals from the gas phase reaction of ozone with ethene are stabilized by collisions at atmospheric pressure, the yield of formic acid is also nearly 50%. To this must be added the 8% yield of formic acid resulting from the decomposition of the Criegee radical that remains nonstabilized, so that formic acid is the major product of the reaction.

Propene

The oxidation mechanism for propene initiated by reaction with OH radicals is shown in Figure 6.15. The addition of OH occurs preferentially at the terminal carbon atom of the double bond. Sixty-five percent of the reaction goes this way, whereas addition to the central carbon atom occurs 35% of the time. Hydrogen abstraction is negligible. As in the case of ethene,

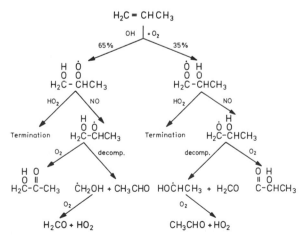

FIGURE 6.15   Oxidation mechanism for propene according to Carter *et al.* (1979).

the peroxy radicals generated by the sequential addition of OH and O_2 to propene may either react with HO_2 to form termination compounds or undergo reaction with NO to yield hydroxy-alkoxy intermediates. Their reaction with oxygen produces 2-hydroxypropanal and hydroxyacetone, whereas decomposition leads to acetaldehyde and formaldehyde, respectively. The relative importance of decomposition versus reaction with oxygen is still not well known. For the 1-hydroxy-2-propoxy radical, a large value of the decomposition rate coefficient has been estimated, so that in this case decomposition should predominate. Decomposition of the 2-hydroxy-1-propoxy radical should be slower, so that its reaction with oxygen becomes more important. Carter et al. (1979) made use of the decomposition rates estimated by Baldwin et al. (1977). By adjusting rates to fit the results of smog chamber experiments, they concluded that $HOCH_2CH(O \cdot)CH_3$ largely decomposes (to form acetaldehyde and CH_2OH), whereas $CH_3CH_2OHCH_2O \cdot$ reacts mostly with oxygen.

The reaction of propene with ozone leads to the formation of the Criegee radicals $H_2\overset{\bullet}{C}OO \cdot$ and $CH_3\overset{\bullet}{C}HOO \cdot$ together with acetaldehyde and formaldehyde, respectively, as primary products. In the presence of a scavenger for OH radicals, Grosjean et al. (1996) found yields of 0.52 for acetaldehyde and 0.78 for formaldehyde, whereas Tuazon et al. (1997) reported yields of 0.65 and 0.45, respectively. The sum of the yields is slightly greater than unity, so that some formaldehyde appears to occur as a product from reactions of the Criegee radicals. The average yield of acetaldehyde is 0.48, suggesting that the two Criegee radicals are produced with approximately equal probability. In contrast, Horie and Moortgat (1991) concluded from a total product study that the yield of $CH_3\overset{\bullet}{C}HOO \cdot$ is about 50% greater than that of $H_2\overset{\bullet}{C}OO \cdot$. The latter radical will enter into the reaction pathways discussed for ethene. The fate of the other Criegee radical is difficult to assess from the propene–ozone reaction as the system becomes too complex, but the $CH_3H\overset{\bullet}{C}OO \cdot$ radical may also be obtained from the reaction of ozone with 2-butene, and in this case it is the only Criegee radical produced. Both cis-2-butene and trans-2-butene have been studied (Niki et al., 1977; Horie and Moortgat, 1991; Horie et al., 1994b, 1997; Tuazon et al., 1997). The observed product distributions are not fully identical, however, indicating that the internal energy content of the Criegee radicals generated from the two 2-butenes is not the same. The following reaction channels have been postulated:

$$CH_3\overset{\bullet}{C}HOO \cdot^* \rightarrow CH_4 + CO_2 \qquad \text{(a)} \quad 15 \ (12-18)\%$$
$$\rightarrow CH_3 + CO + OH \qquad \text{(b)} \quad 19 \ (14-24)\%$$
$$\rightarrow CH_3 + CO_2 + H \qquad \text{(c)} \quad 19 \ (17-21)\%$$

$$\rightarrow HCO + CH_3O \qquad (d)\ 0$$

$$\rightarrow CH_3OH + CO \qquad (e)\ 7\ (6-8)\%$$

$$\rightarrow CH_2CO + H_2O \qquad (f)\ traces$$

$$+ M \rightarrow CH_3CHOO + M \qquad (g)\ 40\ (40-48)\%$$

Probabilities for the individual channels (ranges in parentheses) were taken from the results of Horie and Moortgat (1991) and Horie *et al.* (1994b). Atkinson and Aschmann (1993) derived higher yields for OH radical formation, 41–64% (uncertain by a factor of 1.5), from the yield of cyclohexanone plus cyclohexanol obtained when cyclohexane was added to the reaction mixture. The fate of the stabilized $CH_3\dot{C}HOO\cdot$ radical is not precisely known. Reaction with water vapor is expected to form acetic acid, in analogy with the Criegee radical derived from ethene. Grosjean *et al.* (1994, 1996) found only 10% acetic acid to result from the ozonolysis of *trans*-2-butene and 10% from propene, but these experiments were not designed to study reactions of the stabilized $CH_3\dot{C}HOO\cdot$ radical. If this radical were fully converted to acetic acid, one would expect the yield to be about 40% from 2-butene and about 25% from propene.

The formation of hydrogen atoms in the reactions of ozone with ethene, propene, and butene explains in a natural way the emission of the OH Meinel bands, which were first reported by Finlayson *et al.* (1972, 1974) to occur in these reaction systems. The Meinel bands arise from the reaction $H + O_3 \rightarrow O_2 + OH^*$, which produces excited OH radicals. The excess energy is released as near-infrared radiation. Finally it should be noted that decomposition of the $CH_3CH(O\cdot)CH_2OO\cdot$ and the $CH_3\dot{C}HOO\cdot$ Criegee intermediates is expected to be faster than their reactions with NO, which would abstract the peroxidic O-atom. Accordingly, the oxidation of alkenes by ozone does not require the presence of nitric oxide, in contrast to oxidation by OH radicals.

Isoprene (2-Methyl-1,3-butadiene)

The mechanistic aspects of the oxidation of isoprene under atmospheric conditions have been discussed by Paulson and Seinfeld (1992) and by Carter and Atkinson (1996). Major products of the OH radical initiated oxidation of isoprene are formaldehyde, methyl vinyl ketone, methacrolein, and a number of incompletely identified carbonyl compounds (Paulson *et al.*, 1992; Tuazon and Atkinson, 1990a; Miyoshi *et al.*, 1994). In the presence of NO, methyl vinyl ketone and methacrolein account for about 33% and 25% of total products, and the yield of formaldehyde is an equivalent 60%. In addition, about 5% of 3-methylfuran is formed, as is about 12% of organic nitrates. The formation of 3-methylfuran is unique to this system and does not occur in the reaction of isoprene with ozone. The OH radical-initiated oxidation of isoprene is complicated by the fact that the OH radical may attach to isoprene at four different positions. Addition at the two terminal positions is favored, however. The initial radicals derived isomerize to some extent by shifting the remaining double bond to the center, so that the radical moiety is shifted to the opposite end. Accordingly, six types of radicals are formed that will attach to oxygen to form the corresponding six peroxy radical isomers. Figure 6.16 shows their skeletal structures. Reaction with NO followed by decomposition of the hydroxy-alkoxy radicals formed will cause radicals A and B to produce methyl vinyl ketone, radicals E and F to produce methacrolein, whereas C and D undergo ring formation and hydrogen abstraction by oxygen, which leads to 3-methylfuran.

Methyl vinyl ketone, methacrolein, and formaldehyde are also major products resulting from the reaction of isoprene with ozone, with yields of about 17%, 39–44%, and 90%, respectively (Paulson et al., 1992; Grosjean et al., 1993; Aschmann and Atkinson, 1994). In contrast to the reaction of isoprene with OH, the reaction with ozone produces more methacrolein than methyl vinyl ketone. The reaction of isoprene with ozone can form two types of primary ozonides, which partly split off oxygen to form epoxides in small yield, but mainly decompose to form three types of Criegee radicals. Neeb et al. (1997) have determined the yield of the stabilized $H_2\overset{\bullet}{C}OO\cdot$ Criegee radical to be about 30%, and if the results for ethene are applicable, where 50% of the $H_2\overset{\bullet}{C}OO\cdot$ radicals initially formed are stabilized, their total yield will be about 60%. Paulson et al. (1992) have found that the ozone isoprene reaction also yields 7% propene. This product presumably results from the

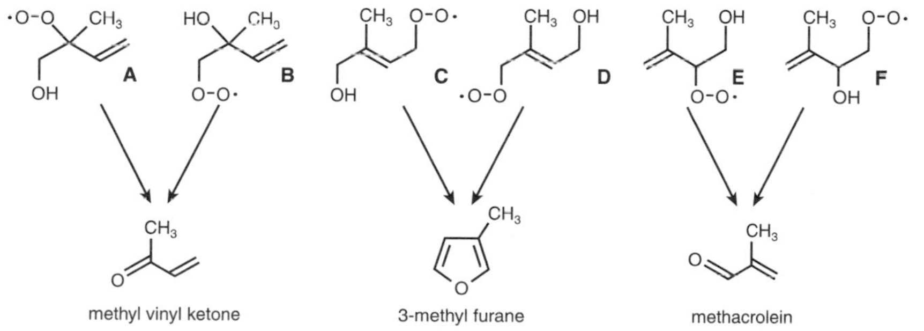

FIGURE 6.16 Skeletal structures of the six hydroxy peroxy radical isomers formed in the reaction of OH radicals with isoprene in air and the major products resulting from their subsequent reactions.

decomposition process $CH_2=C(CH_3)CH\overset{\bullet}{C}OO \rightarrow CH_3CH=CH_2 + CO_2$. The yield of OH radicals as determined by Atkinson *et al.* (1992) is about 27%. Following an earlier suggestion of Niki *et al.* (1987), Aschmann and Atkinson (1994) proposed the $CH_2=CH\overset{\bullet}{C}(CH_3)OO\cdot$ radical to rearrange to form an intermediate hydroperoxide species $CH_2=CHC(OOH)=CH_2$, which then decomposes to form OH and $CH_2=CH\text{-}CO\text{-}\overset{\bullet}{C}H_2$. The reaction of ozone with isoprene is not overly rapid, however, so that it is of secondary importance compared to reaction with OH radicals.

Isoprene reacts also with NO_3 radicals. Although this reaction would be important only at night because of the rapid photodissociation of NO_3 during the day, it may be more efficient than that between isoprene and ozone, provided concentrations of NO_2 (from which NO_3 derives) are favorable. The products of the NO_3 reaction with isoprene are the least understood. The nitrate radical is known to add to unsaturated organic compounds in a manner similar to that of the OH radical, forming a carbon-oxygen bond. The infrared absorption spectroscopy study of Skov *et al.* (1992) indicated that NO_3 adds preferentially at the terminal position, and that the nitrate-carbonyl compound $O_2NOCH_2C(CH_3)=CHCHO$ is the major product.

Methyl vinyl ketone and methacrolein, which are the major products resulting from the oxidation of isoprene, are unsaturated carbonyl compounds that react rapidly with OH radicals and at a lesser rate with ozone. For the reaction of OH with methyl vinyl ketone, Tuazon and Atkinson (1989) have found the products glycolaldehyde and methylglyoxal with yields of about 65% and 25%, respectively. About 70% of the reaction is expected to proceed by addition of OH to the outer position of the double bond, and 30% at the inner position. This leads to $HOCH_2\overset{\bullet}{C}HCOCH_3$ and $\overset{\bullet}{C}H_2CH(OH)COCH_3$ radicals. The subsequent addition of O_2 and abstraction of an oxygen atom by reaction with NO produces the corresponding hydroxy-alkoxy radicals, which undergo decomposition:

$$HOCH_2\overset{\overset{\displaystyle \overset{\bullet}{O}}{|}}{C}HCOCH_3 \rightarrow HOCH_2CHO + CH_3\overset{\bullet}{C}O \overset{O_2}{\rightarrow} CH_3(CO)OO\cdot$$

glycolaldehyde   peroxyacetyl

$$\rightarrow CH_3COCHO + \cdot CH_2OH \overset{O_2}{\rightarrow} HCHO + HO_2$$

methylglyoxal

$$\overset{\displaystyle OH}{\underset{\displaystyle |}{CH_3COCHCH_2O\cdot}} \rightarrow HCHO + CH_3CO\overset{\cdot}{C}HOH \overset{O_2}{\rightarrow} CH_3COCHO + HO_2$$

$$\text{methylglyoxal}$$

According to this mechanism, appreciable amounts of peroxyacetyl radicals are produced, which either form peroxyacetyl nitrate by reacting further with NO_2 or continue the reaction by reacting with NO to form CO_2 and CH_3 radicals. In the atmosphere, glycolaldehyde and methylglyoxal are rapidly degraded further by photodecomposition and reaction with OH radicals, with methylglyoxal leading to an additional production of peroxyacetyl radicals.

The reaction of OH radicals with methacrolein can proceed either by addition of OH to the C=C double bond, which occurs again preferentially at the terminal position, or by abstraction of the aldehydic hydrogen atom. In the first case the study of Tuazon and Atkinson (1990b) showed that hydroxyacetone was the major product, so that the hydroxy-alkoxy radical resulting from the successive addition of OH and O_2, followed by O-abstraction due to reaction with NO, decomposes predominantly by the first of the following two possible pathways:

$$\overset{\displaystyle \overset{\cdot}{O}}{\underset{\displaystyle |}{HOCH_2C(CH_3)CHO}} \rightarrow HOCH_2COCH_3 + HCO \overset{O_2}{\rightarrow} CO + HO_2$$

$$\rightarrow CH_3COCHO + CH_2OH \overset{O_2}{\rightarrow} HCHO + HO_2$$

Hydrogen abstraction, however, leaves the double bond initially intact and leads to the formation of a methacrylperoxy radical:

$$OH + CH_2{=}C\overset{\textstyle CH_3}{\underset{\textstyle CHO}{\diagup\!\!\diagdown}} \overset{O_2}{\rightarrow} H_2O + CH_2{=}C\overset{\textstyle CH_3}{\underset{\textstyle OCOO\cdot}{\diagup\!\!\diagdown}}$$

which may temporarily associate with NO_2 to form peroxymethacrylic nitrate, or react with NO to form NO_2, CO_2 and the vinyl radical $CH_2{=}CH$. Peroxymethacrylic nitrate has been prepared in the laboratory (Roberts and Bertman, 1992) and was found to decompose fairly rapidly at atmospheric pressure and 298 K. The vinyl radical also reacts rapidly with oxygen, leading to HCHO and HCO as the products. Hydrogen abstraction by OH radicals reacting with methacrolein accounts for about 50% of the overall reaction, whereas OH addition according to the data of Tuazon and Atkinson (1990b) yields 41% hydroxyacetone and about 8% methylglyoxal.

6.2.5. AROMATIC COMPOUNDS

The oxidation of aromatic compounds in the atmosphere proceeds primarily by reaction with OH radicals, which may occur either by addition to the benzene ring or by abstraction of a hydrogen atom from the side chain. The first type of reaction is unique, leading either to a permanent addition of the OH group to the benzene ring or to ring opening. The second type of reaction may be modeled in analogy to that of OH radicals reacting with alkanes, or alkenes if double bonds are present. The reaction with toluene may serve as an example. Atkinson (1990, 1994) and Zetzsch (1994) have reviewed the laboratory data. The main conclusions are outlined below.

The principal oxidation pathways for toluene are shown in Figure 6.17. The relative importance of H-atom abstraction versus OH addition may be determined from the yield of benzaldehyde, which is the main terminal product from the abstraction pathway in the presence of NO. Several studies reviewed by Atkinson (1994) suggest that about 7.5% of the reaction enters this route. The product resulting from the addition of OH to the aromatic ring, which is the dominant pathway, is unstable and tends to revert to OH and toluene within about 5 s. In the laboratory, the adduct has been observed to react with oxygen, nitric oxide, and nitrogen dioxide at increasing rates (Knispel *et al.*, 1990). In the atmosphere, only the reaction with oxygen is competitive with decomposition. Oxygen is frequently assumed to add to the OH-toluene complex to form various isomers of the hydroxy methyl cyclohexadienyl peroxy radical. It is doubtful, however, whether the character of a peroxy radical is maintained, because neither hydroperoxides nor peroxy-nitrates, expected to be formed in the interactions with HO_2 and NO_2, have been observed. Zetzsch *et al.* (1997) added small amounts of NO

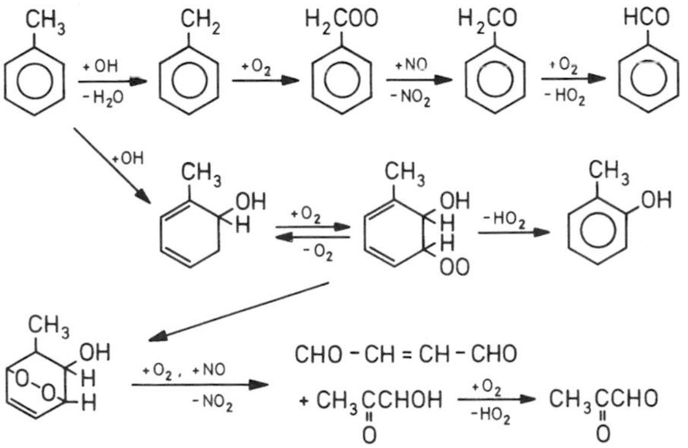

FIGURE 6.17   Oxidation mechanism for toluene. The pathway leading to ring opening is not fully understood; one of several possibilities is indicated.

to the system and found that OH is regenerated, indicating that HO_2 is formed on a rapid time scale. Initial studies of Kenley et al. (1978) showed that ortho-, para-, and meta-cresols are major products, with ortho-cresol being favored. This has been confirmed in subsequent studies. The removal of the excess hydrogen atoms by oxygen from the hydroxy methyl cyclohexa-dienyl radical represents a logical pathway for the formation of the cresols. Their yield is only about 35%, however. Other products are formaldehyde, acetaldehyde, methylglyoxal, glyoxal, unsaturated 1,4-dicarbonyl compounds, and carbon monoxide (Besemer, 1982; Gery et al., 1985; Tuazon et al., 1986; Dumdei et al., 1988; Seuwen and Warneck, 1995). These products must be interpreted to result from ring-opening processes. At least 40% of the reaction enters this pathway. Ring opening has been found to occur in the absence as well as in the presence of NO_x, so that it cannot be triggered by oxygen abstraction from the radical intermediate by reaction with NO. Atkinson et al. (1980) and Atkinson and Lloyd (1984) proposed that the peroxy radical undergoes cyclization to create a new moiety at the carbon atom next to the CH_3 group, which then allows the addition of another O_2 molecule. If this peroxy group loses an oxygen atom, the radical disintegrates to form 1,4-butenal and a hydroxy-acetonyl radical, which by reacting further with oxygen produces methylglyoxal. This pathway is indicated in Figure 6.17. Although the course of the reaction is appealing, other possibilities exist, and additional work is required for confirmation.

6.3. HALOCARBONS

The detection and routine observation of halocarbons at unprecedented low levels of abundances was made possible by the invention of the electron-capture detector by Lovelock (1961) and its uniquely high sensitivity for halogen compounds when used in combination with gas chromatography. Lovelock (1971) himself discovered the presence of CCl_3F and SF_6 in the atmosphere with volume mixing ratios on the order of 10^{-11} and 10^{-14}, respectively. Subsequently, Lovelock et al. (1973) and Wilkness et al. (1973) explored the interhemispheric distribution of CCl_3F and reported the first measurements of CCl_4, while Su and Goldberg (1973) demonstrated the presence of CCl_2F_2 in the air. The fully halogenated compounds CCl_3F, CCl_2F_2, and CCl_4 are essentially inert in the troposphere, but they undergo photodecomposition in the stratosphere. This prompted Molina and Rowland (Molina and Rowland, 1974; Rowland and Molina, 1975) to look into the effects of anthropogenic chlorine compounds on the chemistry of the stratospheric ozone layer (see Section 3.4.3). Once their importance to stratospheric chemistry had been demonstrated, a search for other chlorine compounds in the atmosphere was undertaken to establish the total burden of chlorine.

TABLE 6.19 Halocarbons in the Troposphere: Approximate Abundances, Tropospheric Mass Content G_T, Residence Time τ, Major Sources and Sinks[a]

	Mixing ratio[b] (pmol mol^{-1})	NH-SH[b] ratio	G_T (Tg)	τ (yr)	Sources	Sinks	Remarks
CH_3Cl	610	n.g.[c]	4.5	1.3	Ocean, biomass burning	Reaction with OH	Uniform distribution
CH_2Cl_2	30	~2	0.37	0.46	Anthropogenic	Reaction with OH	
$CHCl_3$	15	~2	0.26	0.6	Anthropogenic Ocean emission	Reaction with OH	
CCl_4	103	1.06	2.9	42	Anthropogenic	Photolysis in the stratosphere	No longer rising
$CHClF_2$ (CFC-22)	117	1.15	1.5	13	Anthropogenic	Reaction with OH	Growth rate ~5% year^{-1}
CCl_3F (CFC-11)	272	1.05	5.4	50	Anthropogenic	Photolysis in the stratosphere	No longer rising
CCl_2F_2 (CFC-12)	532	1.03	8.8	105	Anthropogenic	Photolysis in the stratosphere	Still rising slightly
$CClF_3$ (CFC-13)	3	n.g.	0.05	640	Anthropogenic	Photolysis in the stratosphere	
CF_4	75	n.g.	0.96	~5×10^4	Anthropogenic	Photolysis in the upper atmosphere	Very long-lived Rise rate ~2% year^{-1}
CF_3CH_2F (HCFC-134a)	1.6	~1.8	0.024	14	Anthropogenic	Reaction with OH	CFC replacement, rising
CCl_2FCClF_2 (CFC-113)	84	1.14	2.2	85	Anthropogenic	Photolysis in the stratosphere	No longer rising
$CClF_2CClF_2$ (CFC-114)	20	n.g.	0.5	300	Anthropogenic	O(^{1}D) and photolysis in the stratosphere	
CF_3CClF_2 (CFC-115)	5	n.g.	0.1	1700	Anthropogenic	O(^{1}D) and photolysis in the upper atmosphere	
C_2F_6 (CFC-116)	4	n.g.	0.08	~1×10^4	Anthropogenic	Photolysis in the upper atmosphere	Very long-lived

Species	Lifetime	b			Source	Sink	Comments
CH_3CH_2Cl	12	~4	0.1	0.2	Anthropogenic	Reaction with OH	
CH_3CCl_3	109	1.42	2.5	5.4	Anthropogenic	Reaction with OH	No longer rising
CH_2ClCH_2Cl	25	2.6	0.36	0.4	Anthropogenic	Reaction with OH	
$CHClCCl_2$	8	~4	0.15	0.03	Anthropogenic	Reaction with OH	
CCl_2CCl_2	12	~4	0.3	0.04	Anthropogenic	Reaction with OH	
CH_3CFCl_2 (HCFC-141b)	3.5	~2	0.06	9	Anthropogenic	Reaction with OH	CFC replacement, rising
CH_3CF_2Cl (HCFC-142b)	7.2	~1.3	0.1	20	Anthropogenic	Reaction with OH	CFC replacement, rising
CH_3Br	12	1.2	0.17	1.7	Ocean, anthropogenic	Reaction with OH	
CH_2Br_2	2	n.d.	0.05	0.2	Ocean	Reaction with OH	
CF_3Br	2	n.g.	0.07	65	Anthropogenic	Photolysis in the stratosphere	Used as fire extinguisher Rise rate ~7% year^{-1}.
$CClF_2Br$	3	1.1	0.17	21	Anthropogenic	Photolysis in the stratosphere	Used as fire extinguisher Rise rate ~2% year^{-1}
CH_3I	1-2	n.g.	0.04	0.02	Ocean	Photolysis	

[a] Compiled from Singh (1995), Graedel and Keene (1995), and Montzka et al. (1996).

[b] Average global mixing ratio and approximate north–south abundance ratio.

[c] n.g., No gradient; n.d., not determined.

Table 6.19 gives an overview of a number of halocarbons that have been observed in the troposphere, their approximate abundances, and the dominant sources and sinks as far as known. The data have been reviewed by Fabian (1986), Singh (1995), and Graedel and Keene (1995). Many halocarbons have an anthropogenic origin, and their emissions can, in principle, be brought under control. Only the longer-lived species have been monitored regularly for the past two decades, especially the group of chlorofluorocarbons that harbor the largest potential for stratospheric ozone depletion. It was also recognized that many halocarbons are efficient absorbers of infrared radiation in the infrared atmospheric window (7–9 μm wavelength) and thereby may contribute to global warming (Ramanathan *et al.*, 1985).

6.3.1. METHYL CHLORIDE

Methyl chloride (CH_3Cl), the most abundant chlorocarbon in the atmosphere, was first detected by Grimsrud and Rasmussen (1975). Despite its high abundance, the substance is difficult to analyze because the electron capture detector is less sensitive to CH_3Cl than to other halocarbons. Singh *et al.* (1979b, 1983b) found mixing ratios of 615 $\pm$ 100 pmol mol^{-1} over the Pacific Ocean, which were essentially independent of geographic latitude in both hemispheres. Aircraft measurements by Rasmussen *et al.* (1980), also over the Pacific Ocean, confirmed the uniform distribution with latitude, but they suggested slightly higher mixing ratios in the boundary layer compared to the free troposphere, namely 780 versus 620 pmol mol^{-1}. Mixing ratios over the North American continent, however, were independent of altitude. Lovelock (1975) and Singh *et al.* (1979b, 1983b) showed that the surface waters of the seas are supersaturated with methyl chloride, thereby demonstrating that the ocean provides a source of CH_3Cl. More recent measurements by Atlas *et al.* (1993b) and Koppmann *et al.* (1993) gave mixing ratios similar to those observed previously, indicating that there is no obvious trend with time.

Methyl chloride reacts with OH radicals by hydrogen abstraction in the same way as methane, but the rate coefficient is about six times greater. From the rate coefficient at 277 K, the weighted average tropospheric temperature for OH reactions, and $\bar{n}(OH) = 8 \times 10^5$ molecule cm^{-3} the lifetime of CH_3Cl is inferred to be about 1.3 years. The uniform distribution of methyl chloride would be incompatible with such a relatively short lifetime, if the substance had exclusively anthropogenic origins and the sources were located in the Northern Hemisphere. Thus, it can be concluded that CH_3Cl arises primarily from natural sources that are equally distributed in both hemispheres.

The source strength for CH_3Cl can be calculated by dividing the total mass content in the troposphere (Table 6.19) by the tropospheric lifetime, which gives 3.7 Tg year^{-1}. The contribution of the oceans to atmospheric CH_3Cl can be estimated from the concentrations in seawater. Lovelock (1975) found values of 23 ± 17 ng dm^{-3} near the English coast, whereas Singh et al. (1979b) found 27 ± 31 ng dm^{-3} at various locations in the Pacific Ocean. The variability is considerable, even though the average values are nearly identical. Singh et al. (1979b) used the stagnant film model of Liss and Slater (1974) discussed in Section 1.6 to derive by extrapolation to the total ocean surface area a global CH_3Cl flux of 3 Tg year^{-1}. The value depends on the usual assumptions for the film thickness and the diffusion coefficient. In a more recent assessment, Tait et al. (1994) derived 2.4–3.4 Tg year^{-1}. The main source of methyl chloride in seawater is the production by algae. Wuosmaa and Hager (1990) discussed a possible route for the biosynthesis of halometabolites in marine algae. Zafiriou (1975) pointed out that the reaction $CH_3I + Cl^- \rightarrow CH_3Cl + I^-$ represents a source of CH_3Cl in seawater, but from the rate data reported by Zafiriou (1975), it appears that the reaction is too slow to exert much influence on the flux of methyl chloride to the atmosphere. Singh et al. (1983b) looked for a correlation between seawater concentrations of CH_3I and CH_3Cl but did not find any, indicating that the process must be insignificant.

On the continents, CH_3Cl may be produced by microbes and by anthropogenic activities. Cowan et al. (1973) reported that six species of wood mold of the genus Fomes produce substantial amounts of methyl chloride. The fungi are quite common and may generate appreciable quantities of CH_3Cl, but a reliable source estimate is not available. The existence of anthropogenic sources may be inferred from the observation of higher than average CH_3Cl mixing ratios in urban areas (Singh et al., 1977a, b). It is likely that the contribution is mostly due to the incineration of municipal wastes, because biomass burning releases chlorine present in plants almost exclusively as CH_3Cl. The chlorine contained in polyvinyl chloride is largely converted to HCl. Andreae (1991) has estimated the global rate of CH_3Cl released from biomass burning activities, occurring primarily in the Tropics, to be 0.7 Tg year^{-1} or about 20% of the total source strength. This shows that emissions from the oceans and biomass burning are the dominant contributors to the budget of CH_3Cl in the troposphere, and other sources must be minor.

6.3.2. CHLOROFLUOROCARBONS

Chlorofluorocarbons (CFC) are methane and ethane derivatives that contain only chlorine and fluorine and no hydrogen atoms. They do not exist in nature as such and are synthesized by the chemical industry for commercial applications, namely as refrigerants (e.g., CCl_2F_2), foam-blowing agents (CCl_3F), solvents ($C_2Cl_3F_3$), and aerosol spray propellants. Their main advantages are chemical stability, nonflammability, and absence of toxicity. These compounds have no known sinks in the troposphere and must rise to the stratosphere, where they are consumed mainly by photolysis.

The compounds CFC-11 (CCl_3F) and CFC-12 (CCl_2F_2) have the greatest impact on the stratosphere, and they have received most of the attention. During the period 1950–1970, their production rates rose almost exponentially, but in the mid-1970s they leveled off. During the 1980s the annual production rates were approximately 0.27 and 0.36 Tg, but more recently they have declined in response to regulatory requirements designed to protect the ozone layer. Use patterns also have changed greatly. Applications as propellants were the easiest to control, and these fell from about 70% in 1974 to 18% in 1991 (Singh, 1995). In the same period, their use in foam blowing appears to have increased from 0.05 to 0.2 Tg year^{-1}, whereas their use in refrigeration has remained largely constant at about 0.2 Tg year^{-1}. The estimate once made by McCarthy *et al.* (1977) that about 85% of CFC-11 and CFC-12 is released in the year of production is no longer applicable. Not only has the direct release from spray cans been banned, but greater efforts are being made in the refrigeration business to contain or retrieve chlorofluorocarbons. However, CFCs contained in foams, which are widely used as thermal insulators, cannot be so controlled, and their gradual release into the atmosphere will continue. Figure 6.18 shows the rise of CFC-11 and CFC-12 mixing ratios observed in the background atmosphere. Over the last decades the atmospheric abundances of CFC-11, CFC-12, and CFC-113 have been rising at average rates of 9, 18, and 6 pmol mol^{-1} year^{-1} respectively, or 3.9%, 3.2%, and 7.1% annually. As Figure 6.19 shows, these trends have now been broken as a consequence of the emission restrictions imposed by the Montreal Protocol and its amendments (Montzka *et al.*, 1996). The mixing ratios of CFC-11 and CFC-113 have stabilized, whereas that of CFC-12 is still growing slightly. But the levels are expected to decline in the future.

Because of their long atmospheric residence times, one expects the chlorofluorocarbons to be uniformly distributed in the troposphere. However, as about 95% of the emissions occur in the Northern Hemisphere, and transport to the Southern Hemisphere takes time, an imbalance exists

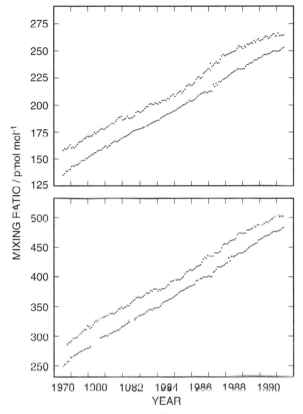

FIGURE 6.18 Rise of CFCl$_3$ (top) and CF$_2$Cl$_2$ (bottom) in the troposphere during the period 1978–1990. The upper and lower curves refer to the Northern and Southern Hemispheres, respectively. (Adapted from Cunnold *et al.*, 1992).)

between the two hemispheres. During the initial period of rapid growth the background mixing ratios in the Southern Hemisphere were considerably smaller than those in the Northern Hemisphere. In 1980 the north–south excess for CFC-11 had decreased to about 10%, and subsequently it fell to about 3%.

The loss of chlorofluorocarbons in the stratosphere requires transport across the tropical tropopause and eddy diffusion toward the 30–35 km altitude region. The atmospheric lifetimes may be assessed from destruction rates in the stratosphere based on absorption cross sections measured in the laboratory. Thus, lifetimes of 50 and 110 years have been estimated for CCl$_3$F and CCl$_2$F$_2$, respectively (WMO, 1995). Cunnold *et al.* (1978, 1983a) have shown that it is possible to use the measured rise in the atmospheric

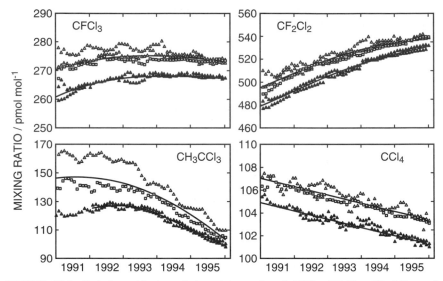

FIGURE 6.19   Variation with time of the mixing ratios of $CFCl_3$, CF_2Cl_2, CH_3CCl_3, and CCl_4 in the troposphere. □, △, Northern Hemisphere; ■, ▲, Southern Hemisphere; □, ■, at Mauna Loa Observatory. Solid lines represent fits to monthly means. (Adapted from Montzka *et al.* (1996).)

mixing ratio in conjunction with known release rates to derive chlorofluoro-
carbon residence times in the troposphere. Even if the release rates are not
accurately known, lifetimes can still be estimated from the observed increase
of the mixing ratios. Toward the end of the 1970s, four monitoring stations,
two in each hemisphere, were established at island and coastal sites to
determine with sufficient precision the abundances of chlorofluorocarbons in
tropospheric background air for the purpose of establishing their atmospheric lifetimes (Cunnold *et al.*, 1983a, b). A more recent analysis (Cunnold
et al., 1992), based on the 1978–1991 time series, has led to equilibrium
lifetimes of 44 ($+17/-10$) years for CCl_3F and of 180 ($+820/-81$) years
for CCl_2F_2, and good agreement was noted between the estimates obtained
by the trend and inventory techniques. The results compare reasonably well
with the lifetimes based on the measured ultraviolet absorption cross sections.

The phase-out of chlorofluorocarbons mandated under the Montreal Protocol has led to the development of alternative compounds that contain
hydrogen atoms. This substitution causes them to react with OH radicals in
the troposphere and shortens their lifetimes, to provide a better protection of
the stratosphere. Temporary substitutes are $CHClF_2$ (HCFC-22), which has

been in use for some time as a blowing agent and has a lifetime of about 13 years, CH_2FCF_3 (HCFC-134a), which is a possible replacement for CFC-12 in refrigeration applications and has a lifetime of 14 years, and CH_3CCl_2F (HCFC-141b) and CH_3CClF_2 (HCFC-142b), which are replacement blowing agents and have lifetimes of 9 and 20 years, respectively. HCFC-22 was first observed in the atmosphere in the late 1970s. In 1995, it had risen to 125 pmol mol^{-1} in the Northern Hemisphere and 110 pmol mol^{-1} in the Southern, and is rising further at a rate of about 4.8% per year. The other compounds have entered the atmospheric cycle only recently and their abundance is still at or below the 10 pmol mol^{-1} level (Montzka et al., 1996), although the mixing ratios are rising rapidly.

6.3.3. Fully Fluorinated Compounds

The fully fluorinated compounds carbon tetrafluoride (CF_4) and carbon hexafluoride (C_2F_6) are of interest because their photolysis requires extreme ultraviolet radiation of less than 100 nm wavelength, so that they must rise to the mesosphere and beyond to undergo photodissociation. Reaction with $O(^1D)$ atoms is slow and results mainly in deactivation (of the excited atom) rather than consumption of CF_4 or C_2F_6. Thus, these substances have extremely long lifetimes in the atmosphere (Cicerone, 1979; Ravishankara et al., 1993). Fabian et al. (1981a, 1987) measured stratospheric altitude profiles and found their mixing ratios to change little up to 33 km altitude, indicating that CF_4 and C_2F_6 are indeed very long-lived. Ravishankara et al. (1993) noted that gaseous fluorides present in air are broken down at the high temperatures associated with the combustion of fossil fuels in electric power stations and internal combustion engines, and they showed that the lifetimes due to these processes are shorter than any other known or presumed processes in the atmosphere but still larger than 10,000 years. The fully fluorinated compounds feature not only long lifetimes, but also strong infrared absorption bands, making them efficient greenhouse gases.

From elevated levels of CF_4 and C_2F_6 in the plumes of aluminum plants, Penkett et al. (1981) concluded that the aluminum industry must be a major source. The electrolytic reduction of aluminum oxide melts at high temperatures is facilitated by the admixture of Na_3AlF_6, which produces a eutectic melt at lower temperatures than for pure aluminum. CF_4 is produced by reactions at the carbon electrodes. The global source of CF_4 amounts to about 30 Gg year^{-1}, as estimated from the current rate of increase in the atmosphere. Aluminum production is the principal source, accounting for about 20 Gg year^{-1}. Emissions of C_2F_6 are lower, but this compound is also synthesized, serving as a dielectric in the electric power industry. The measurements conducted by Penkett *et al.* (1981), Fabian *et al.* (1981, 1987), and Khalil and Rasmussen (1985b) in the late 1970s and mid-1980s established mixing ratios for CF_4 and C_2F_6 of about 70 and 2 pmol mol^{-1}, respectively. Khalil and Rasmussen (1985b) found CF_4 to increase at a rate of about 2%/year; Harnisch *et al.* (1996) reported CF_4 and C_2F_6 to have risen at rates of 1.0 and 0.084 pmol mol^{-1} $year^{-1}$, respectively, during 1982–1995.

6.3.4. METHYL CHLOROFORM AND CARBON TETRACHLORIDE

The presence of 1,1,1-trichloroethane (methyl chloroform, CH_3CCl_3) in the atmosphere was first reported by Lovelock (1974). This compound has been widely used as an industrial solvent, as it is less toxic than other solvents such as C_2HCl_3 and C_2Cl_4. Methyl chloroform production and release increased substantially during the 1970s, until an output of about 5.9×10^8 kg $year^{-1}$ was reached during the 1980s (Midgley, 1989; Prinn *et al.*, 1992). Atmospheric mixing ratios have risen in parallel with the emissions. In 1980, mixing ratios in the 40–50° latitude belts of the Northern and Southern Hemisphere had reached approximately 100 and 65 pmol mol^{-1}. During the 1980s, the mixing ratios increased further at a rate averaging 4.4% annually. In 1991, they attained maximum values of about 155 and 115 pmol mol^{-1}, respectively. These values are lower than those reported from earlier measurements because improved calibration techniques were applied, which changed the absolute calibration factor (Prinn *et al.*, 1995). Figure 6.19 shows that starting in 1991, the abundance of CH_3CCl_3 in the troposphere has been decreasing. In 1995 the global average was 109 pmol mol^{-1} (Montzka *et al.*, 1996). The decrease undoubtedly followed the restrictions imposed by the Montreal Protocol and its amendments on all nonessential uses of the compound. The north–south gradient of mixing ratios for CH_3CCl_3 has been one of the largest among many halocarbons. In addition, seasonal cycles are apparent, similar to those observed for some of the hydrocarbons. This behavior arises from the relatively short lifetime of CH_3CCl_3 in the atmosphere, resulting from its reaction with OH radicals. The recent data of Montzka *et al.* (1996) show that the mixing ratio declines faster in the Northern Hemisphere compared to the Southern Hemisphere and that the gradient between the two is gradually disappearing. In 1991 the mixing ratio in the north was 40% higher than in the south, and by 1995 the excess was reduced to 10%. This behavior is consistent with expectation for a compound, which has its sources predominantly in the Northern Hemisphere, when the sources are suddenly shut off.

Methyl chloroform reacts readily with OH radicals, so that it is destroyed primarily in the troposphere. From the known reaction rate constant and a concentration $\bar{n}(OH) = 8 \times 10^5$ molecule cm^{-3}, one estimates a tropospheric residence time of about 6 years. Small amounts of CH_3CCl_3 are lost to the stratosphere ($\tau_{strat} \approx 45$years), and hydrolysis in the oceans represents another small sink ($\tau_{ocean} \approx 75$years; Butler et al., 1991), so that the true residence time is reduced to about 5 years. Prinn et al. (1983, 1992, 1995) have used the measurements performed at remote sites in the troposphere to derive the residence time of methyl chloroform in various ways. The trend analysis of the data set for the period 1978–1990 gave a lifetime of 4.8 years, while the rise in global tropospheric content combined with emission data gave 6.1 years (Prinn et al., 1992). The trend analysis is independent of absolute calibration errors but sensitive to emission trends. The content method is additionally sensitive to errors in the absolute calibration of the instruments. The new calibration technique applied by Prinn et al. (1995) corrected the data base and reduced the lifetime value derived by the content method to 4.5 years. As discussed in Section 4.2, methyl chloroform has been a key substance for estimating the global average of the OH concentration in the troposphere.

The presence of carbon tetrachloride, CCl_4, in the atmosphere originates almost entirely from anthropogenic emissions. Lovelock et al. (1973) considered it necessary to postulate a natural source of CCl_4 because of the nearly uniform distribution between the hemispheres, and because anthropogenic emission rates could not account for the high mixing ratio of about 125 pmol mol^{-1} observed during the 1970s. Galbally (1976), Singh et al. (1976), and Parry (1977) examined the history of industrial CCl_4 production and its applications more closely and showed that the high mixing ratio was a cumulative effect having resulted from prior emissions. Before 1950 the main market for CCl_4 was in the United States, where it was used as an industrial solvent, in dry cleaning, as a fire extinguisher, and to a limited extent as a grain fumigant. These applications must have led to appreciable emissions to the atmosphere. From 1950 on, carbon tetrachloride was increasingly used as a feed-stock in the production of CCl_2F_2 and CCl_3F, whereas dispersive applications declined, largely for reasons of its toxicity. In recent years the release of CCl_4 must have occurred inadvertently during handling in the production of chlorofluorocarbons.

Simmonds et al. (1983, 1988) have discussed the increase in atmospheric mixing ratios since 1978, when the network of background monitoring stations was established. In 1980 the average mixing ratio of CCl_4 was about 118 nmol mol^{-1}, with a north–south ratio of 1.06. During the period 1978–1985 the annual rate of increase was 1.3%. Both the absolute value and the trend were consistent with release rates supplied by the industry and the approximately known lifetime of 50 years. Simmonds *et al.* (1988), however, noted difficulties with the absolute calibration standards and had to adjust their data slightly to obtain overall consistency. In recent years, the abundance of CCl_4 in the atmosphere has been decreasing as a consequence of restrictions on the use of the compound (see Figure 6.19). In 1995, according to Montzka *et al.* (1996), the average mixing ratio was 103 pmol mol^{-1}, and it has decreased at a rate of 0.8 pmol mol^{-1} $year^{-1}$. Unlike methyl chloroform, however, a decrease in the difference in mixing ratios between the hemispheres has not yet been observed. This behavior is only partly associated with the longer lifetime of CCl_4 compared to that of CH_3CCl_3. It indicates mainly that the release of carbon tetrachloride to the atmosphere is still continuing, although on a lower level.

The early studies of Lovelock *et al.* (1973) of CCl_4 in seawater showed a decline in concentration with depth that was steeper than that for CCl_3F, suggesting a consumption of atmospheric CCl_4 in seawater. The observations were subsequently confirmed by Singh *et al.* (1979b), who calculated a global flux due to absorption in seawater of 32 Gg $year^{-1}$. At the time of the measurements the mass content of CCl_4 in the troposphere was 2.7 Tg, so that the residence time due to the marine sink alone was projected to be 85 years, with a considerable margin of uncertainty. Hydrolysis of CCl_4 in seawater is slow (Jeffers *et al.*, 1989). Singh *et al.* (1979b) suggested that CCl_4 is mainly lost by absorption into the fatty tissue of marine biota. The reaction of carbon tetrachloride with OH radicals in the troposphere is too slow to be important as a removal process in the atmosphere. It is generally agreed that the principal mechanism for the removal of CCl_4 from the atmosphere is transport to the stratosphere and photolysis. The role of ocean as a sink remains uncertain, especially with regard to the mechanism.

6.3.5. METHYL BROMIDE AND METHYL IODIDE

Methyl bromide (CH_3Br) was first observed by Lovelock (1975) in seawater and by Singh *et al.* (1977b) in the air. The compound occurs at mixing ratios of approximately 10 pmol mol^{-1} in the atmosphere and appears to be the main reservoir of bromine in the troposphere (Singh *et al.*, 1983b; Penkett *et al.* 1985; Cicerone *et al.*, 1988; Khalil *et al.*, 1993b; Lobert *et al.*, 1995). A CH_3Br excess of about 25% occurs in the Northern Hemisphere, suggesting a contribution by anthropogenic sources. Mixing ratios observed in various cities are also markedly higher than background values. Khalil *et al.* (1993b) noted a possible annual increase in the mixing ratio on the order of 3% $year^{-1}$. The recent measurements indicated somewhat lower mixing ratios than those reported in the 1980s, which may have resulted from calibration problems. Penkett et al. (1981) had found 10 pmol mol^{-1} in the upper troposphere. Above the tropopause, the mixing ratio of CH_3Br declines rapidly to very low values (Fabian et al., 1981, Kourtidis et al., 1998). In the troposphere, methyl bromide is removed by reaction with OH radicals. This loss reaction leads to a residence time in the troposphere of 1.8 years. For an average mixing ratio of 10 pmol mol^{-1}, the global content of CH_3Br in the troposphere is 0.14 Tg. A residence time of 1.8 years corresponds to a global CH_3Br flux of 76 Gg year^{-1}. To this must be added the observed growth rate, corresponding to about 4 Gg year^{-1}, so that the total source strength is at least 80 Gg year^{-1}. A still greater source strength would be needed, if sinks other than reaction with OH radicals existed. An important possibility for the removal of CH_3Br from the atmosphere is dry deposition to the ocean and to soils.

The surface waters of the ocean are supersaturated with CH_3Br. Singh et al. (1983b) found that CH_3Br and CH_3Cl in seawater are significantly correlated, which points to a common source. Their estimate for the global CH_3Br emission rate from the ocean to the atmosphere was 0.3 Tg year^{-1}. Khalil et al. (1993b), from new measurements, derived an emission rate of about 35 Gg year^{-1}; and Singh and Kanakidou (1993) revised the earlier measurements by comparison with CH_3Cl to obtain 40–80 Gg year^{-1}. These results were based on the Liss and Slater (1974) liquid film model and on the assumption that the ocean is supersaturated everywhere. More recent measurements in the eastern Pacific Ocean (Lobert et al., 1995) revealed saturation anomalies that indicated the existence of regions of CH_3Br consumption. Because methyl bromide can hydrolyze in seawater and react with chloride and organic constituents, loss by dry deposition would be followed by aqueous reactions. Butler (Butler, 1994; Yvon and Butler, 1996) explored the partial tropospheric residence time of methyl bromide following from its degradation in the ocean and derived a value of 2.7 years, with a range of 2.4–6.5 years. The corresponding deposition flux is about 5 Gg year^{-1}, which is only a fraction of the oceanic emission estimates.

The loss rate of methyl bromide in soils is highly uncertain, but the compound is known to undergo hydrolysis and decomposition, so that soils provide a potential sink for CH_3Br in the atmosphere. Shorter et al. (1995) estimated that CH_3Br may be lost to soils at a rate of 42 ± 32 Gg year^{-1}, corresponding to a tropospheric residence time of 3.3 years. By summing the inverse of the partial residence times for the three processes discussed and taking the reciprocal of the sum, one obtains the overall residence time of methyl bromide in the troposphere. The value is 0.8 years, with a considerable margin of uncertainty. The flux of methyl bromide through the troposphere that corresponds to this short residence time and the observed abundance would be 170 Gg year^{-1}. The residence time and budget of methyl bromide in the troposphere thus have ranges of 0.8–1.8 years and 80–170 Gg year^{-1}, respectively.

As summarized by Singh and Kanakidou (1993), the industrial production of CH_3Br has been rising during recent decades to a current level of about 65 Gg year^{-1}. Methyl bromide is primarily used in various applications of soil, grain, and space fumigation, with less than 5% of total production finding other applications. As methyl bromide decomposes in soils, only 30–60% of the CH_3Br applied will be released to the atmosphere. The corresponding release rate would be 20–40 Gg year^{-1}. Manö and Andreae (1994) discussed the production of methyl bromide by biomass burning. The emission factor is about 1% of that for methyl chloride. The release rate was estimated to be 10–50 Gg year^{-1}, but the smaller value appears to be more realistic. These figures show that human-made sources are large enough to rival the emission of CH_3Br from the ocean. But if the total budget of methyl bromide were as large as 170 Gg year^{-1}, the known sources would be insufficient to compensate for the losses.

The presence of methyl iodide in the atmosphere was discovered by Lovelock *et al.* (1973) in marine air samples. These authors also measured CH_3I concentrations in seawater, found a large difference, and concluded that the ocean represents a major source. The origin of CH_3I is, at least partly, the production by marine algae and phytoplankton. It is known, for example, that kelp contains and releases methyl iodide (Geschwend *et al.*, 1985; Manley and Dastoor, 1988), and marine phytoplankton cultures have also been found to produce methyl iodide (Manley and de la Cuesta, 1996). However, Moore and Zafiriou (1994) presented laboratory evidence that CH_3I can be produced photochemically from organic compounds in filtered seawater, and the field measurements of Happell and Wallace (1996) gave support for this hypothesis. In the atmosphere, methyl iodide undergoes photodecomposition. Lovelock *et al.* (1973) assumed an atmospheric lifetime of 2 days and estimated a global source strength of 40 Tg year^{-1}. This value is too high, however, for several reasons. One is that the lifetime of CH_3I due to photodecomposition is longer than 2 days—namely closer to 5 days when nighttime dark periods are included (Zafiriou, 1974). Chameides and Davis (1980) estimated a lifetime of 10 days. The photolysis rate further depends on solar elevation and total ozone overhead, so that the lifetime increases toward higher latitudes. Another reason for doubting the high emission rate deduced by Lovelock *et al.* (1973) is that it would be about 100 times greater than the iodine flux in wet precipitation, determined already by Miyake and Tsunogai (1963). Subsequent measurements of CH_3I in various regions of the world ocean have confirmed the initial observation of a significant supersaturation of CH_3I in ocean waters. Rasmussen *et al.* (1982b), Singh *et al.* (1983b), and Reifenhäuser and Heumann (1992) have used the measurements to derive emission rates of 1.3, 0.5, and 0.8 Tg year^{-1}, respectively.

The global measurements of methyl iodide by Rasmussen *et al.* (1982b) and Singh *et al.* (1983b) showed mixing ratios near 2 pmol mol^{-1} in the marine boundary layer. The mixing ratios exhibited a considerable scatter, but the bulk of the data fell in the range of 1–3 pmol mol^{-1}, decreasing with increasing altitude. The average mixing ratio in the troposphere has been estimated to be approximately 0.8 pmol mol^{-1}, which corresponds to 0.017 Tg CH_3I. This would be in agreement with a source strength of 0.6–1.2 Tg year^{-1} and a residence time of 5–10 days.

CHAPTER 7

The Atmospheric Aerosol

7.1. INTRODUCTION

The concept of air as a colloid and the term *aerosol* for air containing an assembly of suspended particles were originally introduced by Schmauss and Wigand (1929). Colloids are inherently stable because fine particles are subject to Brownian motion and resist settling by sedimentation. The individual aerosol particles may be solid, liquid, or of a mixed variety, and all types are found in the atmosphere. Solid particles are called *dust*. They are formed by the erosion of minerals on the Earth's surface and are kicked up in the air by wind force. Sea spray over the ocean provides a prolific source of liquid drops, which upon evaporation produce sea salt crystals or a concentrated solution thereof. Solid and liquid particles also arise from the condensation of vapors when the vapor pressure exceeds the saturation point. For example, smoke from the open and often incomplete combustion of wood or agricultural refuse is at least partly a condensation aerosol.

For studies of the atmospheric aerosol, the particles are collected on filters or on plates of inertial impactors. Although the original definition refers to particles and carrier medium taken together, it is customary to apply the term *aerosol* more broadly and to include also deposits of particulate matter. We thus speak of "aerosol samples," "aerosol mass," etc., a usage justified by the fact that the deposited material was formerly airborne.

Aerosol particles in the atmosphere usually carry with them some moisture. The amount of water associated with the aerosol depends on the prevailing relative humidity. Increasing the relative humidity condenses more and more water onto the particles, until finally, when the relative humidity exceeds the 100% mark, a certain number of particles grow to produce fog or cloud drops. Meteorologists call these particles *condensation nuclei*, or simply *nuclei*. Fogs and clouds are treated separately and are not included in the normal definition of an aerosol, even though as an assembly of particles suspended in air they represent an atmospheric colloid. The smoothness of transition from an assembly of aerosol particles to one of cloud elements makes it difficult to define a boundary line between the two colloids. Because of the overlap of size ranges of particles in the two systems, any division will be rather arbitrary.

The atmospheric aerosol is not confined to the region adjacent to the Earth's surface, where most of the sources are located. Over the continents, convection currents carry particles to the upper troposphere, where they spread horizontally to fill the entire tropospheric air space. On a global scale, Junge (1963) distinguished three types of aerosol: the continental, the marine, and the tropospheric background aerosol. By chemical composition, the first two contain mainly materials from the nearby surface sources, somewhat modified by the coagulation of particles of different origin and by condensation products resulting from gas phase reactions. The third type represents an aged and much diluted continental aerosol. This fraction would be present also at the ocean's surface. In coastal regions, continental and marine aerosols are blended in that, depending on wind direction, a continental dust plume may travel far over the ocean, or sea salt may be carried inland for hundreds of kilometers.

As a polydisperse system, the aerosol cannot be fully described without taking into account the particle size spectrum. This aspect will be discussed in the next section. Two important bulk parameters associated with the size distribution are the total number density of particles and the total mass concentration. Table 7.1 presents typical values for both quantities of the ground-level aerosol in different tropospheric regions. Concentrations increase as one goes from remote toward urban regions, but mass and particle concentrations are not linearly correlated. The reason is a shift of the median particle size toward smaller values, as the last column of Table 7.1 shows. It is instructive to compare observed aerosol mass concentrations with those of

TABLE 7.1   Typical Mass and Particle Concentrations for Different Tropospheric Aerosols near the Earth's Surface and the Corresponding Mean Particle Radius, Assuming a Mean Density of 1.8 kg dm^{-3} and Spherical Particles

Location	Mass concentration (μg m^{-3})	Particle concentration (particle cm^{-3})	Mean radius (μm)
Urban	~100	10^5–10^6	0.03
Rural continental	30–50	~15,000	0.07
Marine background	~10^a	200–600	0.16
Arctic (summer)	~1	25	0.17

^{a} Includes 8 μg m^{-3} sea salt.

atmospheric trace gases, and Table 7.2 contains a few examples. They indicate that the mass concentration of particulate matter is equivalent to that of trace gases in mixing ratios in the nmol mol^{-1} range.

7.2. PARTICLE SIZE DISTRIBUTION

The size classification of aerosol particles is greatly facilitated if the particles are assumed to have a spherical shape. The size is then defined by an aerodynamically equivalent radius or diameter. The assumption is clearly an idealization, as crystalline particles come in various geometrical but definitely nonspherical shapes, and amorphous particles are rarely perfectly spherical. At sufficiently high relative humidity, however, water-soluble particles turn into concentrated-solution droplets, which are essentially spherical, and therein lies an additional justification for the assumption of spherical particles.

TABLE 7.2   Comparison of Mass Concentration of Several Trace Gases with That of the Aerosol

Trace constituent	Concentration (μg m^{-3})
Hydrogen (0.5 μmol mol^{-1})	40
Ozone (30 nmol mol^{-1})	64
NO_2 (0.03–10 nmol mol^{-1})	0.06–20
CH_3Cl (0.5 nmol mol^{-1})	1
Ethane (0.5–2 nmol mol^{-1})	0.8–3.2
Aerosol	1–100

A size distribution may refer to number concentration, volume, mass, or any other property of the aerosol that varies with particle size. The left-hand part of Figure 7.1 shows somewhat idealized size spectra for the number concentration associated with the marine and rural continental aerosols. The distribution of sea salt particles that contribute to the marine aerosol is added for comparison. Particle radii and concentrations range over many orders of magnitude. In view of the need for presenting the data on a logarithmic scale, it has been found convenient to use the decadic logarithm of the radius as a variable and to define the distribution function by $f(\log r) = dN/d(\log r)$. Conversion to the more conventional form $f(r)$, if needed, can be done by means of the rule $d(\log r) = dr/(r \ln 10)$ with $\ln 10 = 2.302$, which yields

$$dN/dr = f(r) = f(\log r)/2.302\,r$$

Traditionally, particles falling within the size range below 0.1 μm are called *Aitken* nuclei. The adjacent region between 0.1 and 1.0 μm contains *large* particles, whereas components with radii greater than 1 μm are called *giant* particles. Alternatively, one often finds in the literature a distinction between *coarse* particles with radii greater than 1 μm and *fine* particles smaller than 1 μm. Sedimentation is important for particles greater than 20 μm. Such giant particles are still observed in the atmosphere (Delany et al., 1967; Jaenicke and Junge, 1967), although they are rare, occurring with concentrations of less than one particle per cubic meter. Number densities are found to maximize at the other end of the size scale, that is, in the Aitken region. Measurement techniques developed to determine the distribution of Aitken particles show that the size spectrum extends all the way down to the size of large molecules. The observation of a secondary maximum in the Aitken region is common.

No single instrument is capable of spanning the entire size range from Aitken to giant particles, so that a determination of the full size spectrum is a laborious task. The right-hand side of Figure 7.1 shows how a full size spectrum may be obtained by combining several measurement techniques. The book edited by Willeke and Baron (1993) contains detailed descriptions of the various techniques in use, including instruments currently on the market. Particles larger than 0.1 μm can be separated by means of inertial impactors (Marple and Willeke, 1976). When air is forced through a nozzle, particles larger than a predetermined cutoff radius are deposited by inertia on a plate positioned behind a nozzle, whereas smaller particles are carried along with the air stream. Two such units in series, each with its own characteristic cutoff radius, will collect on the second stage only particles within a well-defined size range. These particles can be counted with the help of a microscope. Direct examination and sizing by microscopy is

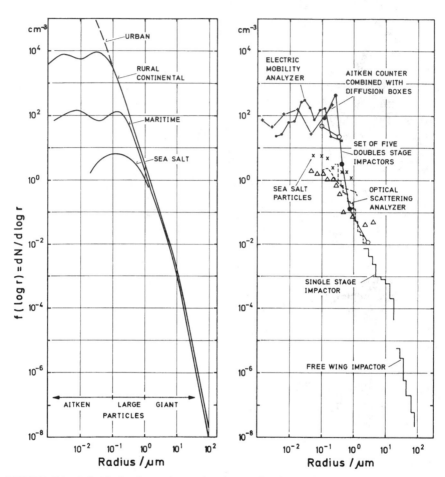

FIGURE 7.1 Left: Idealized particle size distributions for the rural continental and the marine aerosols. The distribution of sea salt particles that contribute to the marine aerosol is shown separately. The dashed line indicates the transition from the rural to the urban aerosol. Right: Determination of the remote tropospheric particle size distribution by a combination of instrumental techniques. Data obtained with single-stage and free wing impactors, set of five double-stage impactors, and single-particle optical scattering analyzer are from Abel *et al.* (1969), observatory Izaña, Tenerife, 2370 m elevation. Data obtained with Aitken counter plus diffusion battery, electric mobility analyzer, and set of double-stage impactors are from ship-board measurements of Jaenicke (1978a) and Haaf and Jaenicke (1980). Sea salt particles were observed by sodium resonance fluorescence (Radke, 1981) and by electron microscopy (Mészarós and Vissy, 1974).

possible for particles larger than 1 μm, and for all particles if electron microscopy is employed. Optical (Mie) scattering techniques are limited to the size range 0.3–3 μm approximately, because for smaller particles the cross section is inadequate and larger particles are not sufficiently numerous.

The condensation nuclei counter (Aitken, 1923), an instrument resembling a Wilson cloud chamber, is an important direct detector suitable for the size range below 0.1 μm. The fact that larger particles are counted as well is not critical, since they add little to the total number concentration. Humidity and volume expansion ratio are adjusted so as to achieve a saturation factor 1.25, which activates essentially all aerosol particles but excludes small ions and molecules with radii below 1.2×10^{-3} μm (Haaf and Jaenicke, 1977; Liu and Kim, 1977). The activated particles grow to a size observable by means of optical techniques. The size classification of Aitken particles requires the application of mobility analyzers. One method of separation is by diffusion losses of particles in narrow channels of various lengths. The size resolution unfortunately is poor, and iterative numerical procedures are required to evaluate the observational data (Maigné et al., 1974; Jaenicke, 1978a). Another method separates charged particles in an electric field (Whitby, 1976; Haaf, 1980). Electric mobility analyzers are endowed with an adequate resolution but the fraction of particles that can be charged decreases rapidly with decreasing size below 0.01-μm radius. The limiting radius is 0.003 μm. Currently available instruments cover the size range 0.005–0.8 μm, approximately.

Size distributions for particle surface and volume can be derived from that of number concentration by applying the appropriate weights:

surface: $dA/d(\log r) = 4\pi r^2 \, dN/d(\log r)$
volume: $dV/d(\log r) = (4\pi/3)r^3 \, dN/d(\log r)$
mass: $dm/d(\log r) = (4\pi/3)r^3 \rho(r) \, dN/d(\log r)$

Integration of these equations yields the total particle surface, volume, or mass per unit volume of air. The last quantity is independently accessible by weighing particulate matter deposited on filter mats or impactor plates. The density of particles as a function of size is difficult to measure and is not generally known. For most purposes the mass distribution is taken to be similar to that of volume, and an average density is applied. Table 7.3 gives aerosol densities measured by Hänel and Thudium (1977) at several locations. Bulk densities for important salts and quartz are shown for comparison. The average density of the dry continental aerosol is about 2×10^3 kg m^{-3}. At relative humidity $< 75\%$ the water content of the continental aerosol is essentially negligible. At higher relative humidity, salts undergo deliquescence, the water content increases, and the density is gradually

TABLE 7.3   Bulk Densities of Continental Aerosol Samples[a]

Sampling location	$\bar{\rho}$ (g cm^{-3})	Size range (μm)	Remarks
Mizpeh Ramon, Israel	2.59–2.72	0.15–5.0	Desert aerosol
Jungfraujoch, Switzerland	2.87	0.15–5.0	3600 m elevation
Mace Head, Ireland	1.93	0.15–5.0	Marine air masses
Mainz, Germany	1.99	0.15–5.0	Marine air masses advected by
(population 200,000)	2.25	> 0.1	strong winds from the Atlantic ocean
	1.87	0.15–5.0	Urban Pollution, inversion layer
Deuselbach, Germany	2.57	0.07–1.0	Winds from rural regions, pre
(rural community)	3.32	1.0–10	dominance of soil erosion particles
	1.82	0.07–1.0	High-pressure situation, weak winds,
	1.92	1.0–10	some showers, some fog
Sodium chloride	2.165		
Sea salt	2.25		
Ammonium sulfate	1.77		
Quartz (SiO$_2$)	2.66		

[a] Selected from measurements of Hänel and Thudium (1977). Size ranges were identified by means of inertial impactors. The bulk densities of some inorganic compounds believed to occur in the aerosols are included for comparison.

lowered to 1×10^3 kg m^{-3}. A similar situation applies to the marine atmosphere.

Figure 7.2 shows surface and volume distributions for the maritime background (plus sea salt) aerosol, derived from the corresponding distributions of number concentration in Figure 7.1. In contrast to number density, which is dominated by Aitken particles, aerosol volume and mass are determined mainly by large and giant particles. Less than 5% of total volume lies in the Aitken range. The volume distribution is inherently bimodal. The shaded area in Figure 7.2, which indicates the sea salt fraction, shows that coarse particle mode represents mainly sea salt. The second mode has its maximum near 0.3 μm, fairly close to that of surface area. We shall see later that whenever the aerosol accumulates material by condensation from the gas phase or coagulation of Aitken particles, the material is deposited in the size range of greatest surface area. Willeke et al. (1974) thus designated the peak near 0.3 μm the accumulation mode. In addition to the coarse mode and the accumulation mode, Figure 7.2 shows a third peak in the distribution of number concentration, centered near 10^{-2} μm in the Aitken range. Later discussion will show that this mode originates as a transient from nucleation processes, that is, the formation of new particles by gas phase chemical reactions resulting in condensable products. As Figure 7.2 illus-

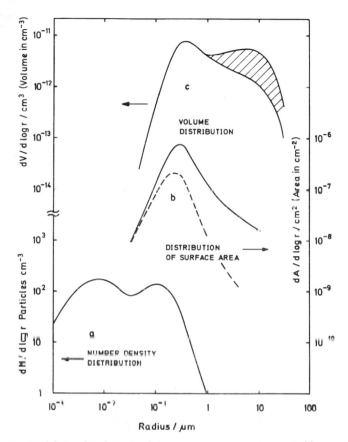

FIGURE 7.2 Model size distributions of the marine background aerosol: (a) particle number density, (b) surface area, (c) volume. The contribution of sea salt to the volume distribution is indicated by the shaded area; arrows indicate the appropriate scale. By integration one obtains a total number density $N = 290$ particle cm^{-3}, a total surface area $A = 5.8 \times 10^{-7}$ cm^2 cm^{-3}, and a total volume $V = 1.1 \times 10^{-11}$ cm^3 cm^{-3}. For an average density 1 g cm^{-3}, the mass concentration is 11 μg m^{-3} (5 μg m^{-3} sea salt). The dashed curve gives the distribution of surface area that is effective in collisions with gas molecules. Diffusion limitation lowers the rate for the larger particles.

trates, the nucleation mode has practically no influence on the size distributions of surface area and volume.

The continental aerosol exhibits similar size distributions. Thus, the aerosol as a whole features a size distribution that is basically trimodal, although it is rare for all three modes to show up simultaneously in any particular size representation (Whitby, 1978; Whitby and Sverdrup, 1980). Figure 7.3 shows a number of examples for averaged size distributions of

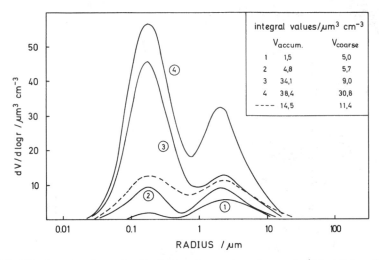

FIGURE 7.3   Average volume size distributions for continental aerosols. (Adapted from Whitby and Sverdrup (1980).) The measurement data were smoothed and idealized by fitting with additive log-normal distributions. (1) Background aerosol, very clean. (2) Normal background aerosol. (3) Background aerosol disturbed by an urban plume (these data from measurements at Goldstone, California). (4) Average urban aerosol (data taken at Minneapolis, Minnesota, Denver, Colorado, and various locations in California). The dashed curve gives the volume distribution resulting from the number density distribution for the rural continental aerosol shown in Figure 7.1. Integrated volumes, given by the area underneath the curves, are shown in the inset.

continental aerosols. The linear volume scale suppresses the nucleation mode but accentuates the two peaks associated with fine and coarse particles. On the continents, the coarse particle mode is due to dust from the wind-driven erosion of soils and released plant fragments. The origin of this mode thus is similar to that of sea salt over the ocean in that it derives from materials present at the Earth's surface. Fine particles arise partly from condensation processes, but a mineral component is still present in the submicrometer size range. Curves 1–3 in Figure 7.3 were obtained at Goldstone, California, in the Mojave desert. Whitby and Sverdrup (1980) refer to curve 1 as clean continental background aerosol. It was obtained under somewhat unusual conditions, namely after a period of rainfalls, and was accompanied by low Aitken nuclei counts (< 1000 cm^{-3}). Curve 2 shows the average distribution at Goldstone. In this case, the Aitken nuclei count was about 6000 cm^{-3}. An average density of 2×10^3 kg m^{-3} may be assumed to estimate the total mass concentration from the integrated volume size distribution. The value obtained is 20 μg m^{-3}. The dashed curve in Figure 7.3 gives the volume distribution of the rural continental aerosol, whose distribution of number density is shown in Figure 7.1. The population of Aitken particles in this case is 1.5×10^4 cm^{-3}, while the total mass concentration is 50 μg m^{-3}. Curve 3 describes the advection of an aged and somewhat diluted urban plume. The term *aging* implies that subsequent to their formation the Aitken particles had sufficient time to coagulate among themselves and with larger particles, so that their number concentration declined to normal background values. In this case, the accumulation mode is considerably enhanced. Curve 4 shows for comparison the volume size distribution of a typical urban aerosol, which is characterized by high Aitken counts, exceeding 10^5 cm^{-3}. Again, the nucleation mode does not really show up in the volume size distribution, except in the vicinity of sources, for example, adjacent to roads with dense automobile traffic. Cars also stir up dust, whereby the concentration of coarse particles also is elevated. The total mass concentration of the aerosol represented by curve 4 is estimated to be 140 μg m^{-3}.

Finally, it should be noted that the initial concept regarding the origin of the accumulation mode has been extended to include physicochemical processes other than coagulation and condensation. Foremost among the additional processes that add new material to an existing aerosol particle are chemical reactions, especially reactions in the aqueous phase. These may occur either in the aqueous shell surrounding the particle or in cloud drops that revert to aerosol particles when the cloud dissipates. John *et al* (1990) and Hering *et al.* (1997) have found evidence for the existence of two submodes in the size range of the accumulation mode. Kerminen and Wexler (1995) applied a growth model to show that the production of sulfate in the aqueous phase of clouds and fogs in addition to condensation can indeed explain the appearance of two submodes within the total accumulation mode.

7.3. THE PHYSICAL BEHAVIOR OF PARTICLES

7.3.1. COAGULATION AND CONDENSATION

Aerosol particles tend to coalesce when colliding with each other. Since in the atmosphere the particles generally are sheathed with moisture, they are sticky and collisions lead to the formation of a new particle with larger size. This process, called *coagulation*, causes a shift in the size distribution in favor of larger particles. Coagulation must be distinguished from *condensation*, which describes the deposition of vapor phase material on particulate matter. In the absence of preexisting particles, condensation leads to the formation of new (Aitken) particles, provided the saturation vapor pressure of the substance is low and the gas phase concentration high. The last process is called *homogeneous nucleation* or *gas-to-particle conversion*. A more detailed discussion of it is given in Section 7.4.3.

The theory of coagulation has been extensively treated (Fuchs, 1964; Hidy and Brock, 1970; Twomey, 1977); only the most salient features will be described here. In the absence of external forces, aerosol particles undergo collisions with each other because of thermal (Brownian) motion. The mathematical description of thermal coagulation goes back to the classical work of Smoluchowski (1918) on hydrosols. Application to aerosols seems to have first been made by Whitlaw-Gray and Patterson (1932). Let $dN_1 = f(r_1)\, dr_1$ and $dN_2 = f(r_2)\, dr_2$ describe the number concentrations of particles in the size intervals $r_1 + dr_1$ and $r_2 + dr_2$, respectively. The collision frequency of these particles is then given by the product

$$K(r_1, r_2) f(r_1) f(r_2)\, dr_1\, dr_2 \tag{7.1}$$

where $K(r_1, r_2)$ is called the *coagulation function*. For particles much smaller than the mean free path of air molecules, $\lambda = 6.5 \times 10^{-2}$ μm under standard atmospheric conditions, the coagulation function is given by the laws of gas kinetics,

$$K(r_1, r_2) = \pi(r_1 + r_2)^2 (\bar{v}_1^2 + \bar{v}_2^2)^{1/2} \tag{7.2}$$

where $\bar{v}_1$ and $\bar{v}_2$ are the mean thermal velocities of the particles, $\bar{v}_i = [8k_B T / \pi m(r_i)]^{1/2}$ v_i^3, $k_B = 1.38 \times 10^{-23}$ kg m^2 s^{-2} K^{-1} is the Boltzmann constant, T is the absolute temperature, and $m_i = 4\pi\rho_i/3$ is the mass of a particle. For particles with radii $r \gg \lambda$, whose motion is hampered by friction, the probability of binary encounters is controlled by diffusion. The collision frequency is then calculated from the theory of Smoluchowski by considering the flow of particles of radius r_2 to a fixed particle of radius r_1, with the result that

$$K(r_1, r_2) = 4\pi(r_1 + r_2)(D_1 + D_2) \tag{7.3}$$

where D_1 and D_2 are the diffusion coefficients associated with the two types of particles. For the transition regime toward small particles where $r \approx \lambda$, Fuchs (1964) has argued that the fixed particle is centered in a concentric sphere $r_a \geq r_1 + r_2$ such that outside this sphere the flow of particles is diffusion-controlled, whereas inside the sphere the particles travel freely according to their mean thermal velocities without undergoing collisions with gas molecules. On this basis, Fuchs derived the correction factor

$$\xi = \left[\frac{r_1 + r_2}{r_a} + \frac{4r_a(D_1 + D_2)}{(r_1 + r_2)^2(\bar{v}_1^2 + \bar{v}_2^2)^{1/2}} \right]^{-1} \tag{7.4}$$

to the Smoluchowski expression for the coagulation function. The problem of selecting the most appropriate value of r_a has produced an animated discussion in the literature. Walter (1973) has summarized the various arguments. He could show that for radii larger than 10^{-3} μm it is allowed to set $r_a = r_1 + r_2$ with reasonable accuracy. This approach has the virtue that in the limit of molecular dimension the second term in the denominator of ξ becomes large compared to the first term, whereby the gas kinetic collision rate is recovered. The coagulation function then has the form

$$K(r_1, r_2) = \frac{4\pi(r_1 + r_2)(D_1 + D_2)}{1 + 4(D_1 + D_2)/(r_1 + r_2)(\bar{v}_1^2 + \bar{v}_2^2)^{1/2}} \tag{7.5}$$

The diffusion coefficient of a particle is connected with its mobility b_i via the Einstein relation $D_i = k_B T b_i$. For spherical particles the dependence of the mobility on r and λ has been derived empirically from careful measurements by Knudsen and Weber (1911) and by Millikan (1923) in the form

$$b(r) = \frac{1}{6\pi\eta r}\left\{ 1 + \frac{\lambda}{r}[A + B\exp(-Cr/\lambda)] \right\} \tag{7.6}$$

where $A = 1.246$, $B = 0.42$, and $C = 0.87$ are constants and $\eta = 1.83 \times 10^{-5}$ kg m^{-1} s^{-1} is the viscosity of air (20°C). For particles with $r \gg \lambda$, the mobility reduces to the well-known law of Stokes. Table 7.4 contains values of mobilities, diffusion coefficients, thermal velocities, and mean free paths for particles with radii in the range of 10^{-3} to 10 μm. The last column gives Fuchs' correction factor for collisions of like particles.

The collision of particles with masses m_1 and m_2 leads to the formation of a new particle with mass $m_3 = m_1 + m_2$. If the new particle is assumed to be spherical, its radius is $r_3 = (r_1^3 + r_2^3)^{1/3}$. To ensure mass conservation, the mathematical description of the coagulation process requires an appropriate weighing of the coagulation function, and the delta function is suitable for this purpose. We define

$$W(r_1, r_2/r_3) = K(r_1, r_2)\delta\left[(r_1^3 + r_2^3)^{1/3} - r_3 \right] \tag{7.7}$$

TABLE 7.4   Mobilities $b(r)$, Diffusion Coefficients $D(r)$, Average Thermal Velocities $\bar{v}(r)$, Mean free Paths $\lambda(r)$, and Fuchs' Correction factor $\xi(r)$ for Like particles, as a Function of Radius for Spherical Particles in Air[a]

r (μm)	b (s kg^{-1})	D (m^2 s^{-1})	$\bar{v}$ (m s^{-1})	λ (m)	ξ
1(−3)	3.172(14)	1.291(−6)	4.975(1)	6.608(−8)	0.0134
1(−2)	3.344(12)	1.361(−8)	1.573	2.203(−8)	0.290
1(−1)	5.467(10)	2.225(−10)	4.975(−2)	1.139(−8)	0.887
1.0	3.135(9)	1.276(−11)	1.573(−3)	2.065(−8)	0.977
10	2.922(8)	1.190(−12)	4.975(−5)	6.091(−8)	0.993

[a]The mean free path of air molecules is $\lambda_{air} = 6.53 \times 10^{-8}$ m at standard pressure and temperature. Powers of 10 are given in parentheses.

as the probability that two particles with radii r_1 and r_2 will form a third particle with radius r_3. The time variation of the distribution function then can be expressed by a combination of integrals over all possible collisions.

$$\frac{\partial f(r,t)}{\partial t} = \frac{1}{2}\int_0^\infty \int [W(r_1, r_2/r)f(r_1,t)f(r_2,t)]\, dr_1\, dr_2$$

$$-\int_0^\infty \int [W(r, r_2/r_3)f(r,t)f(r_2,t)]\, dr_2\, dr_3 \qquad (7.8)$$

The first term on the right describes the gain of particles in the size range $r + dr$ due to coagulation of smaller particles. The factor $\frac{1}{2}$ is required because the integration counts each collision twice. The second term describes the loss of particles in the size range $r + dr$ due to collisions with particles in all other size ranges. After the coagulation probability W is inserted, the double integrals can be integrated once, the first over r_2 (by setting $r_1^3 + r_2^3 = x$) and the second over r_3. Thereby one obtains

$$\frac{\partial f(r,t)}{\partial t} = \frac{1}{2}\int_0^r K(r_1, y)f(r_1,t)f(y,t)\frac{r\, dr_1}{y^2}$$

$$-\int_0^\infty K(r, r_1)f(r,t)f(r_1,t)\, dr_1 \qquad (7.9)$$

where $y = (r^3 - r_1^3)^{1/3}$, and the radii in the second integral are renumbered. The range of the first integral can be reduced to $0-r$, because for $r_1 > r$ the integral is zero. Moreover, because of the symmetry of $K(r_1, r_2)$, the factor $\frac{1}{2}$ can be dropped if the integration limit is replaced by $r/\sqrt[3]{2}$. Twomey (1977) has given a lucid discussion of the properties of these integrals. There are no simple solutions to this integrodifferential equation, and numerical procedures must be applied in solving it. Note that Equation (7.9) describes the temporal change of the distribution function $f(r, t)$ when sources of particles are absent. A source term must be added to the right-hand side, if there is a production of new particles.

Model calculations based on the coagulation equation have yielded several important results. One is the change in the distribution function for the number concentration. Figure 7.4 shows the variation of an assumed size distribution with a maximum particle concentration initially at 0.03-μm radius. In the course of time, the maximum shifts toward larger sizes, while the concentration of Aitken particles decreases. The simultaneous growth of large particles is hardly noticeable, however, as comparatively little volume is added to this size range. The lifetime of particles with radii below 0.01 μm is fairly short. If such particles are found in the atmosphere, they must be

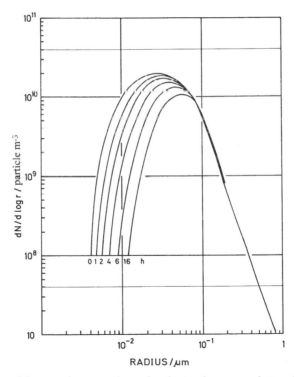

FIGURE 7.4 Modification of an aerosol size distribution due to coagulation of small particles as a function of time. The distribution of large particles is not significantly altered by the process. (Adapted from Junge and Abel (1965).)

young! This conclusion was reached quite early by Junge (1955) and formed an essential piece of support for his viewpoint that the tropospheric background aerosol as far as it derives from the continental aerosol must have undergone aging by coagulation. The data in Figure 7.4 suggest lifetimes on the order of 1 h for particles about 0.01 μm in size. The calculation assumed a total number density of 2×10^4 cm^{-3}, which is representative of the rural continental aerosol. Lifetimes at least an order of magnitude greater are obtained, when the initial number concentration is reduced to that characteristic of the marine aerosol.

Walter (1973) has studied the evolution of the particle size spectrum by coagulation for the case where new particles are continuously generated by nucleation processes. Some of his results are shown in Figure 7.5. The left-hand side shows the evolution of the size spectrum when other particles are initially absent. Starting with 1.2×10^{-3} μm particles, a secondary maximum develops, which moves toward the range of larger particles, until after about a day it has reached the 0.1-μm size range. At this time the further changes are minor, indicating that a quasi-steady state is reached. For such conditions, the particles in the size range > 0.1 μm are recipients of newly created particle mass, so that the peak near 0.2 μm can be identified with the accumulation mode.

The right-hand side of Figure 7.5 shows the effect upon such a steady-state distribution when the population of large particles is varied. As their number density is increased from 270 cm^{-3} to five times that number, a deep gap develops in the Aitken range, whereby the spectrum is split into two parts. The new particles then coagulate preferentially with particles in the accumulation mode rather than undergoing coagulation among themselves. The gap can be partly filled by an appropriate increase in the production rate of new particles. The results are not greatly altered if one assumes a smaller size of new particles formed or an additional condensation of water vapor on the coagulating particles up to equilibrium with the environment (relative humidity $\leq 90\%$) as Hamill (1975) has done. The existence of a gap such as that appearing in Figure 7.5 contrasts with our knowledge of the atmospheric aerosol, which displays a relative minimum in number density near 0.03-μm radius but no cleft (cf. Figure 7.1).

Walter (1973) has not taken into account that in the atmosphere new condensable material such as H_2SO_4 arising from the oxidation of SO_2 is produced by photochemical reactions, so that the rate of new particle formation undergoes a diurnal cycle. A variation in the Aitken size spectrum is inevitable in view of the time constants associated with coagulation (cf. Figure 7.16). The approximate change in the size distribution that will have taken place after a 12-h intermission of new particle production is shown in Figure 7.5 by the dashed curve. The result is a bimodal size distribution,

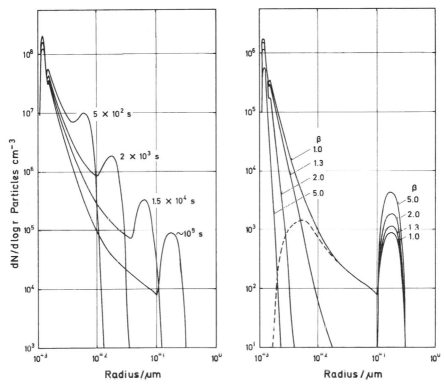

FIGURE 7.5 Coagulation behavior of particles produced by nucleation. (Adapted from Walter (1973).) A continuous generation of embryos with 1.2×10^{-3} μm radius is assumed. Left: Variation of the size distribution with time in the absence of preexisting particles; production rate $q = 10^6$ cm^{-3} s^{-1}. Right: Steady-state distributions for different concentrations of preexisting particles ($r > 0.1$ μm); production rate $q = 10^2$ cm^{-3} s^{-1}; number density of preexisting particles $N_0 = 270\beta$. The dashed curve for ($\beta = 1$) is obtained from the steady-state distribution 12 h after termination of the source. The apparent lifetime of Aitken particles due to coagulation is greater than that indicated in Figure 7.4 because of the smaller number density of particles.

which in addition to the accumulation mode now includes a transient peak caused by the incomplete coagulation of Aitken particles. As discussed previously, this peak is called the *nucleation mode*. The ensuing size spectrum gives a better representation of the natural aerosol, even though the resemblance to the size spectra in Figure 7.1 is still marginal. The change in size distribution resulting from coagulation can be observed in the environment under suitable conditions. Hoppel *et al.* (1985), for example, have followed this change and the associated decay of Aitken particles for 12 h in an air mass moving from the east coast of the United States out over the Atlantic Ocean.

Hoppel *et al.* (1986) suggested that when aerosol particles are cycled through nonprecipitating clouds, in-cloud oxidation of SO_2 to H_2SO_4 would add mass and shift the distribution of those particles that had acted as cloud condensation nuclei to larger sizes (see Section 7.3.2). This would result in a separation in the size distribution between larger particles that were active as cloud nuclei and smaller particles that were not. Hoppel *et al.* (1994) have assembled observational evidence for such a process.

The rate of condensation, that is, the process by which vapor phase material is deposited directly on preexisting particles, is determined by the number of condensable molecules striking the aerosol particles per unit time. A mass accommodation coefficient $\alpha \leq 1$ must be applied to account for the possibility that the number of molecules retained on the surface is smaller than that striking it. If the collision is followed by a chemical reaction with surface materials, α should be replaced by an uptake coefficient, which also may be smaller than unity. The expression for the collision frequency is similar to Equation (7.1). For the purpose of later comparison we use here the aerosol size distribution in the form $dn_2 = f(r_2)d(\log r_2)$. Let n_1 denote the number concentration of condensable gas molecules and n_s that for equilibrium when condensation and reevaporation occur at equal rates (saturation equilibrium). The condensation rate is proportional to $n_1 - n_s$:

$$K(r_1, r_2)(n_1 - n_s)f(r_2)\,d(\log r_2) \qquad (7.10)$$

Note further that for molecules compared with particles, $\bar{v}_1 \gg \bar{v}_2$ and $D_1 \gg D_2$, because $r_1 \ll r_2$. This allows us to set $\bar{v}_1 = \bar{v}$, $D_1 = D$, and $r_2 = r$. These simplifications reduce $K(r_1, r_2)$ to the form

$$K^*(r) = \alpha\pi r^2\bar{v}/(1 + \alpha r\bar{v}/4D) \qquad (7.11)$$

The asterisk is used to distinguish this collision function from the original coagulation function, Equation (7.1). The diffusion coefficient can be expressed by the gas kinetic equivalent $D_1 = (\pi/8)\bar{v}\lambda$, where λ is the mean free path of the condensing molecules in air. The collision cross section πr^2 of a particle may be replaced by the surface area $A(r) = 4\pi r^2$. With these substitutions one obtains

$$K^*(r) = \frac{1}{4}\frac{\alpha A(r)\bar{v}}{(1 + 2\alpha r/\pi\lambda)} = \frac{1}{4}\alpha A_{\text{eff}}\bar{v} \qquad (7.12)$$

The numerator on the right-hand side is the gas kinetic collision term. The denominator reduces the surface area available for collisions to an effective surface area $A_{\text{eff}}(r)$. As long as the particle radius is small so that the term $2\alpha r/\pi\lambda$ is much smaller than unity, the molecules collide with particles at gas kinetic collision rates. For larger particles, when $2\alpha r/\pi\lambda > 1$, the collision rate becomes diffusion-controlled and increases with r rather than with r^2. Setting $\lambda \approx \lambda_{air}$, one finds for $\alpha \approx 1$ that the boundary between the two regimes lies near 0.1-μm radius.

It is instructive to compare the distribution of the effective surface area,

$$dA_{eff}(r)/d(\log r) = \frac{4\pi r^2}{(1 + 2\alpha r/\pi\lambda)} f(\log r) \qquad (7.13)$$

with the real surface area distribution of the natural aerosol. For this purpose, the function $dA_{eff}(r)/d(\log r)$ is included in Figure 7.2 for the maritime background aerosol ($\alpha \approx 1$). The dashed curve makes apparent that the effective surface area is greatly reduced compared with the real one when the radius exceeds 0.3 μm. The maximum of the distribution curve is shifted only slightly toward smaller radii. When condensation takes place, collision of condensable molecules with aerosol particles transfers material predominantly to the 0.1-0.5-μm size range, that is, in the accumulation range. This is so because the number distribution of the particles favors that range. However, similar results are also obtained for the rural continental and the urban aerosol.

The total rate of condensation is obtained by integrating over the entire size range

$$dn_1/dt = -(\alpha\bar{v}/4)(n_1 - n_s)\int A_{eff}(r)f(r)d(\log r)$$

$$= -(\alpha\bar{v}/4)(n_1 - n_s) A_{eff,tot} \qquad (7.14)$$

As in the case of coagulation, a source term must be added to the right-hand side if condensable molecules are continuously produced. In the absence of sources, Equation (7.14) can be integrated to give the variation of n_1 with time:

$$n_1 - n_s = (n_{10} - n_s)\exp(-t/\tau_{cond})$$

with

$$\tau_{cond} = 4/\alpha\bar{v}A_{eff,tot} \qquad (7.15)$$

The associated mass increase of particulate matter is

$$\Delta m(t) = (M_1/N_A)(n_{10} - n_s)[1 - \exp(-t/\tau_{cond})] \qquad (7.16)$$

where M_1 is the molar mass of the condensate and N_A is Avogadro's number. The distribution of the mass increment with particle size,

$$d[\Delta m(t)]/d(\log r) = [\Delta m(t)/A_{eff,tot}] A_{eff}(r)f(r)/d(\log r) \qquad (7.17)$$

is proportional to the rate of condensation in each size range and corresponds to the distribution of $A_{\text{eff}}(r)$. Here, we are mainly interested in the time constant for condensation. Table 7.5 shows values for the integrated surface area, the total effective surface area, and the associated time constants for the three important types of atmospheric aerosol, calculated with the assumption that $\alpha = 1$. The time constants must be considered lower limit values in view of the possibility that the mass accommodation coefficient may be smaller than unity.

If $n_s \ll n_1$, the difference $n_1 - n_s$ is large, and most of the material in the gas phase condenses on the aerosol to stay there. Sulfuric acid is a pertinent example, because it is mostly neutralized by ammonia. The vapor density of the ensuing salt is essentially negligible, so that it cannot reevaporate. If, on the other hand, $n_1 \approx n_s$, only a fraction of the available material is transferred to the aerosol. The amount is controlled by the saturation vapor concentration n_s, which is a function of temperature. In this case, however, the situation may be complicated by the interaction of several condensable substances, an effect that the above treatment did not include. Water vapor is the most abundant condensable constituent in the atmosphere. The fraction of water present on the aerosol is determined to a large degree by its interaction with water-soluble particulate matter. This effect will be discussed next.

7.3.2. INTERACTION OF AEROSOL PARTICLES WITH WATER VAPOR

The great importance of aerosol particles as condensation nuclei in the formation of clouds and fogs requires a more detailed examination of the interaction of the aerosol with water vapor. For individual particles the

TABLE 7.5   Total Surface Area and Total Effective Surface Area Available for Collisions with Gas Molecules for Three Types of Aerosols, and the Associated Time Constants for Condensation of Molecules with Molar Mass $M = 0.1 \text{ kg mol}^{-1}$ ($\bar{v}_1 = 2.5 \times 10^2 \text{ m s}^{-1}$)

Aerosol type	A_{tot} (m^2)	A_{efftot} (m^2 m^{-3})	τ_{cond}^a (s)
Marine background aerosol	5.8×10^{-5}	1.4×10^{-5}	1.15×10^3
Rural continental aerosol	3.1×10^{-4}	1.4×10^{-4}	1.15×10^2
Urban aerosol	1.6×10^{-3}	1.0×10^{-3}	1.6×10^1

^{a} For $\alpha = 1$.

interaction may range from that of partial wetting of an insoluble dust particle to the complete dissolution of a salt crystal, such as sea salt. In the course of time, coagulation, condensation, and in-cloud modification processes cause even an insoluble siliceous particle to acquire a certain share of water-soluble material. Although there will always be particles that have retained their source characteristics, one may assume that many individual aerosol particles contain a mixture of both water-soluble and insoluble matter. Junge (1963), who has promoted the concept of internally mixed particles, assembled evidence from a variety of studies, including examination of particles by electron microscopy, to show that mixed particles dominate the continental aerosol. Winkler (1973) also has discussed internal and external mixtures. We shall come back to this aspect at the end of this section.

The key to understanding the behavior of aerosol particles in a humid environment was first provided by Köhler (1936) on the basis of thermodynamic principles. Dufour and Defay (1963) and Pruppacher and Klett (1997) have given detailed treatments of all of the aspects involved. The main features arise from two opposing effects: the increase in water vapor pressure due to the curvature of the droplet's surface, and the decrease in the equilibrium vapor pressure with increasing concentration of a solute. Let p_s denote the saturation vapor pressure above a plane surface of pure water at temperature T (in Kelvin). The actual vapor pressure above the surface of a drop with radius r is derived from

$$\ln p/p_s = \frac{2 \sigma V_m}{R_g T r} + \ln a_w \qquad (7.18)$$

where σ is the surface tension, V_m is the partial molar volume of water, R_g is the gas constant, and a_w is the somewhat temperature-dependent activity of water in the solution. The first term on the right describes the influence of surface curvature according to Kelvin, and the second term accounts for the influence of solutes in the form of a generalized Raoult's law. For a single solute, the activity of water,

$$a_w = \gamma_w x_w = \gamma_w \frac{\nu_w}{\nu_w + \nu_s} = \gamma_w \left(1 + \frac{m_s M_w}{m_w M_s} \right)^{-1} \qquad (7.19)$$

is proportional to the mole fraction x_w, which is determined by the number of moles of water and solute in the mixture. These can be replaced by the corresponding masses divided by the molar masses. The activity coefficient γ_w is an empirical parameter that is introduced to account for the nonideal behavior of the solution, either because it is not sufficiently dilute to follow

Raoult's law, or because it contains electrolytes that dissociate to form ions. For nondissociating solutes, one often finds $\gamma_w < 1$, although γ_w tends toward unity for dilute solutions.

Equation (7.19) can also be applied to mixtures of solutes, if one replaces r, m, and M by

$$\nu_s = \sum_i \nu_{si} \quad m_s = \sum_i m_{si} \quad \overline{M}_s = m_s / \nu_s$$

No general theory exist that relates activity coefficients for mixtures to those of the individual components, so that γ_w must be determined for each specific mixture that one might wish to consider. If the aerosol particle contains only soluble material, its radius is related to its mass by

$$m = 4\pi\bar{\rho}r^3/3 = \nu_s\overline{M}_s + \nu_w M_w = m_s + m_w \qquad (7.20)$$

where, $\bar{\rho}$, the average density, is determined with the usual and quite realistic assumption (Hänel, 1976) of molar volume additivity,

$$\bar{\rho} = (m_s + m_w)\left/\left[\frac{m_s}{\rho_s} + \frac{m_w}{\rho_w}\right]\right. \qquad (7.21)$$

By combining the above equations, one obtains for a particle in equilibrium with the environment the ratio of the actual to the saturation vapor pressure of water:

$$\frac{p}{p_s} = \gamma_w\left(1 + \frac{m_s M_w}{m_w \overline{M}_s}\right)^{-1} \exp\left(\frac{2\sigma V_m}{R_g Tr}\right)$$

$$\frac{p}{p_s} = \gamma_w\left(1 + \frac{M_w}{\overline{M}_s}\frac{\rho_0 r_0^3}{\bar{\rho}r^3 - \rho_0 r_0^3}\right)^{-1} \exp\left(\frac{2\sigma V_m}{R_g Tr}\right) \qquad (7.22)$$

Here r_0 and ρ_0 denote the particle's dry radius and density, respectively. Note that p/p_s represents the relative humidity. Equation (7.22) thus provides an implicit equation for the droplet's radius r as a function of relative humidity. For pure water, $\sigma_w = 7.42 \times 10^{-2}$ N m^{-1} and $V_m = 1.8 \times 10^{-5}$ m^3 mol^{-1}. If r is given in μm and $T = 283$ K, the exponential assumes the value $1.13 \times 10^{-3}/r$, which shows that for $r > 0.1$ μm the Kelvin effect is small. Surface-active compounds can reduce the surface tension considerably, making the effect even smaller. Strong electrolytes as solutes may either increase or decrease the surface tension, but their influence amounts to less than 20% within the concentrations permitted before the solutions saturate.

The validity of Equation (7.22) has been tested and confirmed experimentally by Orr et al. (1958a, b) and by Tang et al. (1977) for a number of different salt nuclei with sizes in the large particle range. Figure 7.6 demonstrates the behavior for sodium chloride particles. The curves were calculated and the points were taken from the work of Tang et al. (1977). At very low relative humidity the particle is dry and crystalline. Increasing the relative humidity initially causes some water to be adsorbed to the particle's surface, but the amount is insufficient to dissolve the material, the particle remains largely crystalline, and the growth in particle size is small. This portion of the growth curve does not obey Equation (7.22), of course. At a certain critical relative humidity the crystal deliquesces—that is, it takes up enough water to form a saturated solution. The size of the particle then changes abruptly. The deliquescence of NaCl occurs at 75% relative humidity. The increase in size with rising relative humidity follows the growth curve described by Equation (7.22). Decreasing the relative humidity again to a value below the deliquescence point causes the solution to be supersaturated. Recrystallization does not occur as spontaneously as deliquescence, so

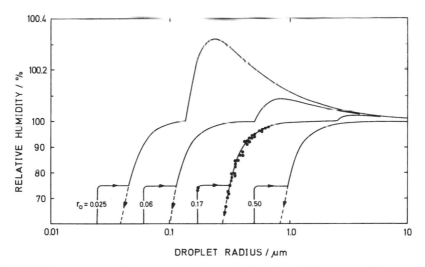

FIGURE 7.6 Variation of particle size with relative humidity (Köhler diagram) for sodium chloride particles with different dry radii. Deliquescence occurs at 75% r. h. Note the hysteresis effect when the humidity is raised or lowered beyond the critical value. Curves for humidities greater than 75% were calculated. The scale above 100% is expanded. The experimental points refer to observations of Tang et al. (1977) on submicrometer-sized monodisperse sodium chloride particles.

that the particle size moves along the extended Köhler curve for a while before it suddenly shrinks to a size near that of the dry particle. This hysteresis branch of the curve has been observed for airborne particles as well as for bulk material (Winkler and Junge, 1972), but it generally is not very reproducible.

An important feature displayed by each particle growth curve is the maximum in the domain of relative humidity > 100%. Low degrees of supersaturation suffice to reach the peak, so that in Figure 7.6 an expanded scale is used to bring out this feature more clearly. The rising branch on the curve to the left of the peak results from the Raoult term in Equation (7.22), whereas the descending branch to the right is due to the Kelvin term. Once a particle has grown to a size just beyond the maximum, it enters a region of instability and must grow further, provided the water supply in ambient air is sufficient. A particle that has passed over the peak is said to have become "activated." This is the mechanism of cloud formation. Note that the height of the maximum decreases with increasing droplet size, so that large particles are preferentially activated. In a cloud produced by adiabatic cooling of moist air, the supersaturation of water vapor first rises because of cooling, but as droplets grow by condensation they eventually offset the rise, and the supersaturation subsequently declines. For giant particles the maximum in the Köhler diagram, although surmountable, occurs at large drop sizes, so that it usually cannot be reached within the time available for condensation. While such particles usually are not activated, they cannot be distinguished from small cloud drops. For particles in the Aitken size range the barrier imposed by the maximum is too high, and these particles also cannot be activated. The number of aerosol particles that become activated depends on the level of water vapor saturation reached. The number density of cloud drops frequently is $100-300$ cm^{-3}, stemming from aerosol particles in the size range $0.1-5$ μm. As the transport of water vapor to the growing droplets occurs by molecular diffusion, the growth rate is determined by the effective surface area, which favors the larger particles. The rate of growth in radius, however, is larger for smaller drops because the relative volume increase is greater. After a short time, all of the droplets approach a rather uniform radius of about 5 μm.

Aitken particles that are not activated remain present as an interstitial aerosol. The smallest and therefore most mobile of the Aitken particles attach to cloud drops by way of collisional scavenging. Although after cloud evaporation these particles have added material to the accumulation and coarse particle range, the resulting change in the original particle size distribution is small. On the other hand, as the cloud ages there are processes that can indeed change the original size distribution, namely, the coalescence of cloud drops among themselves, the chemical conversion of dissolved gases toward nonvolatile products (for example SO_2 to sulfate), and partial rain-out. These processes can be simulated in models. For example, Flossmann et al. (1987) studied the redistribution of particles in a convective cloud. After 30 min of cloud development, coalescence among cloud drops had shifted the size distribution markedly toward giant particles, while at the same time the oxidation of sulfur dioxide had added sulfate to the accumulation mode. The relative importance of processes contributing to in-cloud modification of the aerosol size spectrum clearly depends on the vertical extent and the lifetime of the cloud.

Growth curves of natural aerosol particles were first measured by Junge (1952a, b) at relative humidities below 100%. He found the growth much less compared to that of pure salts and attributed the reduction to the presence of insoluble matter. Winkler and Junge (1972) and Hänel et al. (1976) studied the mass increase with rising humidity of bulk samples of aerosol by means of a gravimetric technique. The procedure is justified, because when the relative humidity is less than 95%, the Kelvin term can be neglected and the relative mass increase m_w/m_0 of the single particles will be identical to that of bulk material of the same composition. Figure 7.7 shows results for the urban, the rural continental, and the maritime aerosol.

Growth curves for continental aerosols are fairly similar. They are much smoother than those for pure salts, and the abrupt mass increase associated with deliquescence is absent. Hysteresis loops were still observed for individual samples, although not to the same extent as for pure salts. A large fraction of the water-soluble material present in the aerosol consists of sulfates, chlorides, and nitrates, the major cations being NH_4^+, Na^+, K^+, Ca^{2+}, and Mg^{2+}. Table 7.6 lists deliquescence humidities for a number of salts containing these ions. The wide range of deliquescence points, the formation of hydrates, and the interaction of the various constituents in a concentrated solution contribute to explain the smoothness of the growth curves. Details of composition are important, of course. To show that the observed growth behavior is due to a mixture of substances, Winkler and Junge (1972) prepared an artificial salt mixture with a composition similar to that often found in continental aerosols. The growth behavior resembled that of the natural aerosol quite well, provided allowance was made for the influence of water-insoluble material.

Figure 7.7 shows further that the marine aerosol features a more pronounced growth with relative humidity than the continental aerosol. The reason is the greater content of water-soluble material. For giant particles, the growth curve bears a close resemblance to that of sea salt. This observation is easily understood, as the mass of sea salt is concentrated in the size range $r_0 > 1$ μm. The growth behavior of sea salt, in turn, is determined primarily by sodium chloride, which is the dominant component. In the size

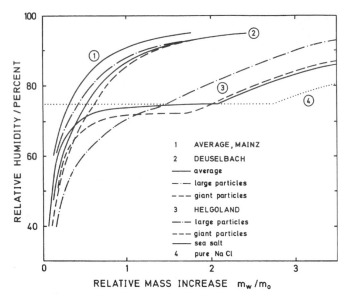

FIGURE 7.7   Growth with relative humidity of aerosol bulk samples as observed by Winkler and Junge (1972). (1) Urban aerosol, average of 28 samples collected at Mainz, Germany. (2) Rural continental aerosol, average of 14 samples collected at Deuselbach, Germany. (3) Marine aerosol, average of 10 samples collected at Helgoland Island (about 50 km off the German coast in the North Sea). In the last two cases, the size ranges $0.1 < r < 1.0$ and $r > 1.0$ were separated with a double-stage impactor, and the average results for each range are shown. Growth curves for sea salt and sodium chloride are shown for comparison.

range $0.1 < r_0 < 1$ μm, the growth curve of the maritime aerosol is more similar to that of the continental aerosol, indicating the increasing influence of the tropospheric background aerosol.

Finally, we may consider the effect of water-insoluble material on the particle growth curve. Let m_u be the mass of the insoluble fraction of the

TABLE 7.6   Deliquescence Humidities (%) for Selected Inorganic Salts at 25°C

$MgSO_4$	88	NH_4NO_3	62
$Na_2SO_4 \cdot 10H_2O$	87	$Mg(NO_3)_2 \cdot 6H_2O$	51
KCl	84	$Mg(NO_3)_2 \cdot 4H_2O$	36
$(NH_4)_2SO_4$	80	NH_4HSO_4	39
NH_4Cl	77	$Ca(NO_3)_2 \cdot 4H_2O$	50
$NaCl$	75	$Ca(NO_3)_2 \cdot 2H_2O$	28
$NaNO_3$	75	$MgCl_2$	33
$(NH_4)_3H(SO_4)_2$	69		

particle's dry mass m_0, and m_s as before the mass of the water-soluble fraction. The total mass of the particle thus is

$$m = m_s + m_u + m_w = m_0 + m_w \tag{7.23}$$

The ratio m_s/m_w appearing in Equation (7.19) can be written

$$m_s/m_w = (m_s/m_0)/(m_0/m_w) = \varepsilon(m_0/m_w) \tag{7.24}$$

where $0 < \varepsilon < 1$. With this relation, Equation (7.22) changes to

$$p/p_s = \gamma_w \left[1 + \varepsilon \frac{M_w}{M_s} \frac{\rho_0 r_0^3}{(\bar{\rho} r^3 - \rho_0 r_0^3)} \right]^{-1} \exp\left(\frac{2\sigma V_m}{R_g T r} \right) \tag{7.25}$$

The form of the equation is the same as that of Equation (7.22), except that the growth of the particle is reduced, because for any fixed relative humidity the mass of water present in the solution is determined by m_s, and hence by ε, but not by $m_0 = m_s + m_u$.

Junge and McLaren (1971) have studied the effect that the presence of water-insoluble material has on the capacity of aerosol particles to serve as cloud condensation nuclei. Using Equation (7.25) they calculated the supersaturation needed for an aerosol particle to grow to the critical radius at the peak of the Köhler curve, where spontaneous formation of cloud drops becomes feasible. The results, which are shown in Figure 7.8a, indicate that the difference is less than a factor of 2 in radius for particles with a water-soluble fraction $\varepsilon \geq 0.1$. The majority of particles appear to meet this condition (see Fig. 7.21). By assuming particle size distributions similar to those of Figure 7.1 for the continental and maritime aerosols, Junge and McLaren (1971) also calculated cloud nuclei spectra as a function of critical supersaturation and compared them with observational data. Their results, which are shown in Figure 7.8b, clearly demonstrate that the presence of water-insoluble material in aerosol particles occurring together with soluble substances does not seriously reduce their capacity to act as cloud condensation nuclei.

In the preceding discussion it was assumed that individual particles consist of a mixture of soluble and insoluble components. This situation describes an *internal* mixture, whereas an assembly of particles that contain exclusively either soluble or insoluble materials represents an *external* mixture. Recent observations have shown that the atmospheric aerosol almost always contains a fraction of essentially nonhygroscopic particles that grow little with increasing relative humidity (Zhang et al., 1993c; Svenningsson et al., 1994). Such particles occur even in the Aitken range, where one would

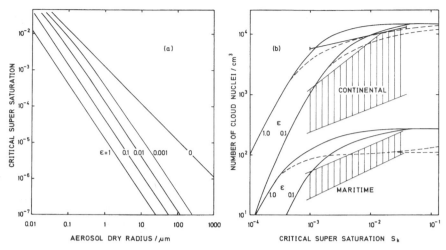

FIGURE 7.8   Influence of the mass fraction ε (water-soluble material/particle dry mass) on the activation of cloud condensation nuclei. Left: Critical supersaturation of aerosol particles as a function of particle dry radius. Right: Cloud nuclei spectra calculated for $\varepsilon = 0.1$ and 1 on the basis of two size distributions each for the continental and maritime aerosols (solid and dashed curves, respectively). (Adapted from Junge and McLaren (1971).) The curves for the marine cloud nuclei spectra are displaced downward from the original data to normalize the total number density to 300 cm^{-3} instead of 600 cm^{-3} used originally. The curves for $\varepsilon = 1$ give qualitatively the cumulative aerosol size distributions, from larger to smaller particles ($s_k = 1 \times 10^{-4}$ corresponds to $r_0 \geq 0.26$ μm, $s_k = 3 \times 10^{-3}$ to $r \geq 0.025$ μm). Similar results were subsequently obtained by Fitzgerald (1973, 1974). The hatched areas indicate the ranges of cloud nuclei concentrations observed in diffusion chambers with material sampled mainly by aircraft, as summarized by Junge and McLaren (1971). The bar represents the maximum number density of cloud nuclei observed by Twomey (1963) in Australia.

expect condensation and coagulation processes to be most effective in producing mixed particles. By means of a tandem mobility analyzer, Svenningsson *et al.* (1994) found at the Kleiner Feldberg mountain station near Frankfurt/Main that in clear air the number fraction of nonhygroscopic particles decreased with increasing size from 64% to 37% in the size range 0.025–0.15 μm. In the presence of clouds, these particles formed part of the interstitial aerosol. The fraction then increased with size from 65% to 87% because an increasing number of other, more hygroscopic particles were lost in the formation of cloud drops. Zhang *et al.* (1993) have shown by chemical analysis that the less hygroscopic fraction sampled in the Grand Canyon consisted of carbon and mineral particles. Eichel *et al.* (1996) sampled aerosol particles at the Kleiner Feldberg with a double stage impactor, dissolved the material in water, and determined the volume distribution of insoluble matter, thereby extending the data set of Svenningsson *et al.* (1994)

toward the size range 0.4–2.3-μm radius. They found three groups of particles with water-soluble fractions of about 9%, 50%, and 88%, respectively. The relative number distributions were similar to those derived from the tandem mobility analyzer experiments. Internal mixing and external mixing evidently exist side by side in the atmospheric aerosol.

7.4. AEROSOL SOURCES AND GLOBAL PRODUCTION RATES

This section deals with source mechanisms. The production of mineral dust, sea salt, and gas-to-particle conversion is described, then estimates of global production rates are discussed.

7.4.1. MINERAL SOURCES

Dust emissions from soils caused by wind erosion have been extensively investigated by Gillette (1974, 1978) and collaborators (Gillette and Goodwin, 1974; Gillette and Walker, 1977; Gillette et al., 1980). The following discussion is mainly based on two review articles (Gillette, 1979, 1980).

Soils are formed by the weathering of crustal material of the earth. Rocks, stones, and pebbles slowly disintegrate through the action of water, chemically by the leaching of water-soluble components, and mechanically by the freeze–thaw cycle of water entering pores and cracks. In this manner, igneous rocks are transformed into clay minerals, carbonates, and quartz grains (sand). Once the material has been broken down to a grain size less than 1 mm, it can be moved by wind force. With increasing wind speed, mobile particles will first creep and roll before they are temporarily lifted off the ground. The ensuing leaping motion of coarse-grained particles has the effect of sandblasting. Fine particles encrusted on coarser grains are loosened, and break off, and a fraction of them are thrown into the air. Most of these particles return quickly to the ground by gravitational settling. Only particles with radii smaller than about 100 μm can remain airborne for a longer period of time, provided they escape the surface friction layer because of turbulent air motions.

Initiation of the motion of soil particles requires wind velocities in excess of a threshold value. Bagnold (1941), Chepil (1951), and Greely et al. (1974), among others, have described experimental studies of threshold velocities for soils consisting of beds of loose monodisperse particles. The results have indicated that threshold velocities are at a minimum for particles in the size

range 25–100 μm. To remove larger particles requires greater wind forces
because the particles are heavier, whereas smaller particles adhere better to
other soil constituents, so that larger pressure fluctuations are needed to
break them loose.

Natural soils are polydisperse systems of particles, which are rarely
present in the form of loose beds. Nevertheless, if loose particles are
available, they will be preferentially mobilized. Other factors that are important are the coverage of the soil surface with roughness elements such as
pebbles, stubbles, bushes, etc., which partly absorb momentum; coherence
forces between soil particles due to clay aggregation, organic material, or
moisture content; and soil texture, that is, the composition of the soil in
terms of particle size classes (see Table 7.7).

Gillette (1980) and Gillette et al. (1980) have presented measured threshold friction velocities for a number of dry soils of different types. The lowest
velocities, 0.2–0.4 m s^{-1}, were found on disturbed soils with less than 50%
clay content and less than 20% pebble cover, or for tilled bare soils. At the
upper end of the range (≥ 1.5 m s^{-1}) were soils with more than 50% clay
content and surface crusts or a cover of coarse (> 5 cm) pebbles. The
corresponding wind speeds when measured 2 m above the ground were 4–8
and 33 m s^{-1}, respectively.

Once the threshold velocity is surpassed, the flux of saltating particles
increases rapidly with wind speed. Measurements (Gillette, 1974, 1978) show
that the horizontal flux of particles through a plane perpendicular to both
ground surface and wind direction increases with $u_*^2(u_* - u_{*0})$, where u_*
is the friction velocity and u_{*0} is the threshold value. For large velocities the
expression tends toward u_*^3, a result that was first derived by Bagnold (1941)
by dimensional arguments, along with the assumption that all of the momen-

TABLE 7.7   International Standard Classification
System for Size Ranges of Soil Particles[a]

Diameter (μm)	Nomenclature
< 2	Clay[b]
2–20	Silt
20–200	Fine sand
200–2000	Coarse sand
> 2000	Gravel

[a] From the "Handbook of Chemistry and Physics"
(Lide, 1995/96). Other classifications use slightly
different ranges.
[b] The term is here used to indicate a size class, not a
type of mineral.

tum is transferred to the ground surface by saltating sand grains. The corresponding flux of kinetic energy delivered to the soil surface is ρu_*^3. The population of particles at some distance from the ground is fed from the friction layer by eddy diffusion with a flux that under neutral stability conditions, when a logarithmic wind profile exists, is approximately proportional to u_*^2. Accordingly, the vertical flux of aerosol particles is expected to grow with the fifth power of the friction velocity, at least for sufficiently loose sandy soils. Figure 7.9 shows vertical fluxes of particles in the $1-10$-μm size range measured over a variety of soils. The data indicate that a fifth power law is indeed observed in a number of cases. For at least one soil, however, the dependence on friction velocity was more pronounced.

Idealized size distributions developing from this mechanism of aerosol generation are summarized in Figure 7.10. A linear $dM/d \log r$ scale of normalized mass distributions is used. The distribution obtained within the dust layer closest to the surface—that is, within the first 2 cm (frame b)—is still fairly similar to that of the parent soil, although it is much narrower. This result is mainly due to the fact that threshold velocities are at a minimum for $50-100$-μm particles. One meter above the ground (frame c), the mass distribution is shifted toward finer particles. Here, the distribution depends much on soil texture and the prevailing wind force. Two examples are shown. One refers to loamy fine sand with a distribution centered at

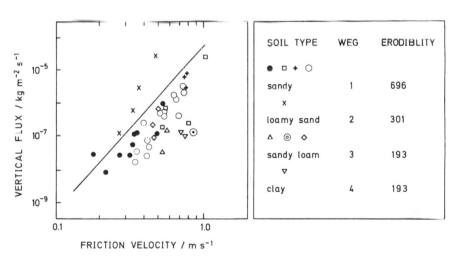

FIGURE 7.9 Vertical mass flux of soil-derived aerosol particles versus the wind friction velocity for different soil types. (Adapted from Gillette (1980).) The solid line indicates a fifth power dependence. The key on the right correlates soil type with wind erodibility groups (WEG) (after Hayes, 1972) and erodibility in Mg hm^{-2} yr^{-1} (after Lyles, 1977).

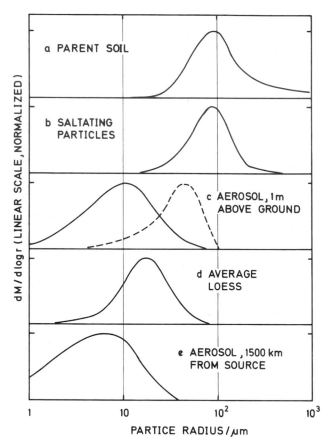

FIGURE 7.10   Idealized representation of aerosol formation from mineral sources, based on measurements of Gillette and Walker (1977), Schütz and Jaenicke (1974), Jaenicke and Schütz (1978), and Junge (1979).

10-μm radius. The other was obtained during a sandstorm in the Saharan Desert and obviously is weighted toward 50-μm particles. Mixed, bimodal distributions are not uncommon, indicating that the assumption of a smooth size distribution of the parent soil is not always warranted. Frame d of Figure 7.10 shows the average distribution of loess taken from data assembled by Junge (1979). Loess is an aeolian deposit mainly composed of particles in the 5–50-μm range of radii. The particles are just small enough to be carried to distances of several hundred kilometers before they are redeposited. Many loess deposits were formed in the final stage of the last glaciation period up to 300 km from the fringes of the great ice sheets (Pye, 1987). Sedimentation of giant particles causes the size distribution of the remaining aerosol to shift further toward smaller particles. The last frame of Figure 7.10 shows, as an example, the mass distribution of the Saharan aerosol observed at Cape Verde Island after 4 days of travel time with the trade winds of the Northern Hemisphere.

Figure 7.11 compares several number density size distributions of soil and aerosol particles obtained at locations in North Africa to show the dominance of submicrometer particles in all populations studied. Maximum number concentrations occur in the size range near 0.1-μm radius in all samples, and particles as small as 0.02 μm were detected. In the aerosol, dust and wind conditions caused an enhancement in the concentration of particles greater

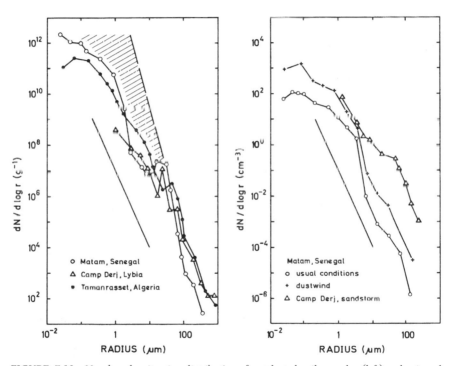

FIGURE 7.11 Number density size distributions for selected soil samples (left) and mineral aerosols 2–3 m above the soils (right) (data from d'Almeida and Jaenicke (1981) and Schütz and Jaenicke (1974)). Left: O, Matam, Senegal; △, Camp Derj, Lybia; ●, Tamarasset, Algeria. Right: Matam, Senegal: O, usual conditions; +, dust-wind; △, Camp Derj, Lybia: sandstorm. The solid lines show a Junge (r^{-3}) power-law distribution. Characterization of soil samples: Matam, Senegal, alluvial flood plains; Tamarasset, Algeria, gravelly soil; Camp Derj, Lybia, desert soil, rock-covered. The hatched area indicates the loss of fine particles to the aerosol according to the interpretation of d'Almeida and Jaenicke (1981).

than 3 μm, in accordance with expectation from the preceding discussion. There is, however, an enhancement of concentration also in the 0.01–1-μm size range relative to that of particles near 1-μm radius. This feature is unexpected and points to an increase, with increasing wind velocity, of the rate at which small particles are liberated by the sandblasting effect.

Another important point is the similarity of size distributions among different soil samples. It made little difference whether these were taken from sand dunes in Dar Albeida, Mali; alluvial soils in Matan, Senegal; or gravel-covered soils from the Nubian Desert in the Sudan or Tamarasset, Algeria. A secondary maximum near 20-μm radius also was frequently observed. D'Almeida and Jaenicke (1981) hypothesized that the gap between this coarse size fraction and submicrometer particles, indicated in Figure 7.11 by the shaded area, is due to the winnowing process of a continuous loss of fine particles to the atmosphere. It is an unresolved question whether this loss is permanent, leading to a gradual depletion of fine particles in the soil, or whether these are continuously replenished by the mechanical disintegration of larger particles.

The mineral dust generated by wind erosion in the great desert regions undergoes long-range transport above the tropospheric boundary layer. Dust plumes from the Sahara–Sahel zone in North Africa and the Gobi–Taklamakan region in central Asia are carried across the Atlantic and Pacific, respectively, as far as 5000 km from the source region. These plumes are conspicuous in satellite images (Durkee et al., 1991; Moulin et al., 1997; Husar et al., 1997), allowing global study of seasonal and annual variations. Mineral dust now has been recognized to affect the radiative budget of Earth by absorbing and scattering solar and terrestrial radiation (Andreae, 1996), which has led to a revival of interest in the problem. Global transport models of mineral dust require a suitable parameterization scheme for dust emission. Marticorena and Bergametti (1995) and Marticorena et al. (1997) have developed emission algorithms that incorporate not only a dependence on wind power but also a variable threshold function accounting for differences in soil texture in the source regions. This model has been successfully applied to desert regions with known surface characteristics. Agreement with satellite data was obtained.

7.4.2. SEA SALT

The production of sea salt aerosol is due to the agitation of the sea surface by wind force, and in this regard its formation is similar to that of dust aerosol. The mechanism is unique, however, in that sea salt particles arise from the bursting of bubbles when they reach the ocean surface. The mechanism is depicted in Figure 7.12. The surface free energy of the collapsing bubble is converted to kinetic energy in the form of a jet of water, which, depending on the size of the bubble, ejects between 1 and 10 drops up to 15 cm above the sea surface. Additional drops are produced from the bursting film of water covering the bubble. A portion of the film drops moves in the direction perpendicular to the jet axis and will be scavenged by the sea. Others are torn from the toroidal rim of the bubble by the escaping air and move upward. The number of film drops increases rapidly with bubble size (Day, 1964), so that bubbles a few millimeters in diameter can produce several hundred film drops. The distinction between film and jet drops is of some importance because surface-active materials such as certain organic compounds, bacteria, etc. concentrate at the air-sea interface of the bubble to disperse with the film drops. A certain enrichment of such material also takes place in jet drops. MacIntyre (1968, 1972) has shown that the jet receives water from the surface layers surrounding the bubble. The uppermost jet drop derives from the innermost shell, the next drop from the adjacent outer shell, and so on. Accordingly, one expects the first jet drop to become endowed with most of the surface-active material. Figure 7.12 illustrates this mechanism.

Seawater contains sea salt to about 3.5% by weight, of which 85% is sodium chloride. It can be safely assumed that the sea salt contents of jet and film drops are similar. As the drops are carried upward in the atmosphere, they experience increasingly lower relative humidity and dry up until their water content is in equilibrium with the environment. The particle radius then is about one-quarter of the parent drop's radius. Although parts of sodium chloride may crystallize, one should not expect the particles to dry up completely, because the deliquescence point of magnesium chloride, which is present in sea salt, lies at 33% relative humidity (see Table 7.6). Such a low relative humidity is not reached in the marine atmosphere.

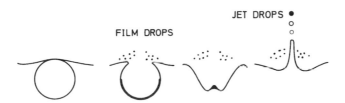

FIGURE 7.12   Sea spray formation from a bursting bubble; schematic representation according to observations of Day (1964) and MacIntyre (1968, 1972). The shading of the uppermost jet drop is to indicate enrichment with material from the interior surface of the bubble. The size of jet drops is about 15% of bubble diameter (Wu, 1979); film drops are smaller.

Bubbles are most numerous in the whitecaps associated with breaking waves, where they are formed by the entrainment of air into the water during the breaking wave motion. Whitecaps begin to appear at wind speeds of about 3 m s^{-1}. At wind speeds near 8 m s^{-1}, approximately 1% of the ocean surface is covered with whitecaps (Monahan, 1971; Toba and Chaen, 1973). Bubble sizes in breaking waves range from perhaps a few micrometers to more than several millimeters in diameter. The exact limits are not known. The concentration of giant bubbles is too low to be observable, and the smallest bubbles not only are difficult to detect, but also go rapidly into solution because of surface-curvature effects. Field measurements by Blanchard and Woodcock (1957), Johnson and Cooke (1979), and others showed bubble size spectra with concentrations maximizing near 100-μm diameter, which was fairly independent of depth, and increasing in concentration about inversely to the fifth power of the bubble diameter. Total concentrations decreased quasi-exponentially with depth below the sea surface (scale height ~ 1 m).

The flux of particles can be estimated from the rate at which bubbles rise to the sea surface. The rise velocity is a function of the bubble size, since it is determined by the buoyancy of the bubble and the drag forces acting upon it. The rise velocity increases roughly with the square of the bubble diameter. Blanchard and Woodcock (1980) assumed that each bubble produces five jet drops and estimated a production rate of about 2×10^6 m^{-2} s^{-1} in the whitecap regions of the ocean. The production rate is heavily weighted toward small drop sizes. Several studies reviewed by Blanchard (1963) and Wu (1979) showed that the mean diameter of jet drops is about 15% of that of the parent bubble. Using this information together with a bubble size distribution $(r/r_0)^{-5}$, with $r_0 = 50$ μm, and the rise velocities given by Blanchard and Woodcock (1957), one can calculate the production rate of jet drops as a function of drop size. The results are shown in Figure 7.13 by the solid line. Monahan (1968) has estimated production rates by combining measured concentration-size distributions of jet drops above the water surface with calculated ejection velocities. Figure 7.13 includes results from his own measurements and from data originally presented by Woodcock et al. (1963) from the coastal surf region of the Hawaiian beach. The flux spectra of the two estimates agree reasonably well. An agreement of the absolute production rates is not to be expected because they depend too much on the prevailing wind speed.

Figure 7.13 also shows size distributions for the concentration of sea salt particles about 6 m above the ocean surface, as derived by Blanchard and Woodcock (1980) from the mass distributions measured by Chaen (1973) onboard ships. More recently, De Leeuw (1986) has obtained quite similar data. The comparison in Figure 7.13 is intended to show that the size

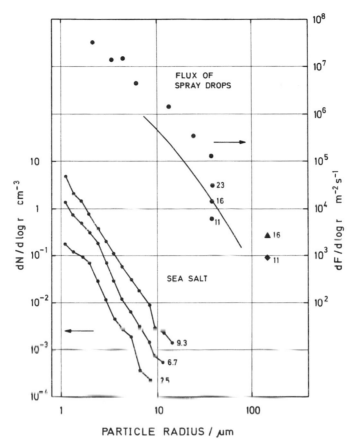

FIGURE 7.13   Vertical flux of sea spray droplets according to Monahan (1968), compared to the number density distribution of sea salt particles 6 m above the ocean surface (Chaen, 1973, as reported by Blanchard and Woodcock, 1980). Both follow a Junge (r^{-3}) power law. Numbers next to points indicate wind speed (m s^{-1}). The solid line for the droplet flux was calculated from ejection velocities given by Wu (1979).

spectrum of sea salt particles corresponds to that of sea spray production, indicating that in the size range below 10 μm the particle size distribution is determined primarily by the production mechanism. The removal rate of sea salt particles thus must be size-independent. This assumption breaks down for particles larger than 10 μm, because these are increasingly subject to gravitational settling. According to Wu (1979), drops greater than 100 μm in size spend less than 0.5 s in the air before they return to the sea under their own weight.

While the data in Figure 7.13 cover only particles with radii larger than 1 μm, it will be clear that the size spectrum extends toward smaller particles. Sea salt crystals as small as 0.05 μm have been detected by electron microscopy (Mészarós and Vissy, 1974) and flame scintillation photometry (Radke, 1981; Radke et al., 1976). Although these small particles contribute little to the total aerosol number density (see Fig. 7.1), their origin remains in doubt. The problem is that bubbles with radii much smaller than 50 μm dissolve too rapidly in seawater. Blanchard and Woodcock (1980) suggest that the absorption of organic matter on the bubble surface stabilizes some of the bubbles in the 10-μm size range to prevent their dissolution. Thus it is possible that such bubbles are responsible for the production of sea salt particles in the submicrometer size range.

Figure 7.14 summarizes current knowledge about the vertical distribution of sea salt particles in the marine atmosphere. Between 10 and 500 m elevation, the concentration changes comparatively little, but once the cloud level is reached, there is a continuous decline toward negligible concentrations in the 2–3-km altitude region. Toba (1965) has pointed out that the concentration decreases nearly exponentially with a scale height of 0.5 km

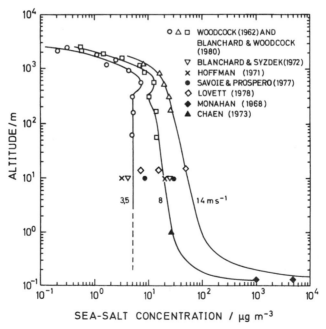

FIGURE 7.14   Distribution of sea salt over the ocean as a function of altitude for different wind velocities. (Adapted from Blanchard and Woodcock (1980).)

for particles lighter than 1 ng ($<$ 7 μm radius). For larger particles the scale height is smaller. The decline is attributed to the incorporation of sea salt particles in cloud drops, which in precipitating clouds is followed by rain-out. Figure 7.14 indicates the occurrence of a sea salt inversion layer below the cloud base. This phenomenon appears to arise from evaporating cloud drops that by coalescence had grown sufficiently to undergo fall-out. As Figure 7.14 further shows, the mass concentration of sea salt in the subcloud boundary layer increases with wind speed. This is expected because coverage of the sea surface with whitecaps increases with wind speed. The data of Woodcock (1953), Blanchard and Syzdet (1972), and Lovett (1978) indicate an exponential increase in the sea salt concentration with wind speeds in the range of 5–35 m s^{-1}, whereas others prefer a linear least square fit to the data (Fitzgerald, 1991). The vertical distribution of sea salt appears to adjust quickly to any given wind speed and to the then existing flux of sea salt from the ocean surface into the atmosphere. One should not assume, however, that the distribution represents a steady state. According to Junge (1957), who numerically solved the eddy diffusion equation, the time constant for the adjustment to steady state is on the order of 12 h, that is, the same magnitude as that of meteorological changes.

7.4.3. GAS-TO-PARTICLE CONVERSION

Atmospheric gas phase reactions may lead to the formation of condensable products that subsequently associate with the atmospheric aerosol. The best-known reaction of this kind is the oxidation of SO_2 to H_2SO_4 and its neutralization by ammonia to form sulfate salts.

Condensation may either cause the formation of new particles in the Aitken range (homogeneous nucleation) or deposit material on preexisting particles (heterogeneous condensation). In laboratory studies of gas-to-particle conversion, one usually starts with air freed from particles by filtration. The development of the particle size spectrum then goes through three successive stages, dominated by nucleation, coagulation, and heterogeneous condensation, in that order (Husar and Whitby, 1973; Friedlander, 1977). In the atmosphere, all three processes take place concurrently. The generation of new particles then requires conditions that allow the growth of molecular clusters by homogeneous nucleation in the face of competition from heterogeneous condensation. These processes are treated below.

Note that gas-to-particle conversion also occurs in nonprecipitating clouds, because the oxidation of gases dissolved in cloud water may form nonvolatile products that remain attached to the particles generated when the cloud dissipates. The oxidation of SO_2 to H_2SO_4 in the aqueous phase of clouds, which is discussed in Section 8.6, is again a good example.

Molecular clusters are formed because of weakly attractive forces between molecules, the so-called van der Waals forces. Except under conditions of low temperature, it is difficult to observe and study in the laboratory clusters containing more than a few molecules, so that details about their properties are sparse. Our understanding of nucleation relies on concepts based largely on the principles of statistical mechanics. The theory of homogeneous nucleation developed by Volmer and Weber (1926), Flood (1934), Becker and Doering (1935), and Reiss (1950) assumes that certain thermodynamic properties, such as molar volume or surface tension, that can be determined for bulk materials remain valid in the molecular regime. The assumption holds for clusters containing at least 20 molecules (Sinanoglu, 1981). Water clusters of single ions attain bulk properties with 15–20 water molecules (Holland and Castleman, 1982; Shin et al., 1993). The error at smaller sizes must be tolerated. The theory shows that a steady-state distribution of clusters of various sizes is set up, conforming to a quasi-equilibrium between clusters having less than a certain critical size. In this size range, the abundance of clusters is proportional to

$$\exp\left[-\Delta G(i)/k_B T\right]$$

where $\Delta G(i)$, the free energy of formation of the cluster, is a function of the number of molecules i harbored in the cluster. At constant temperature, $\Delta G(i)$ is composed of two terms. The first is the difference in the thermodynamic potentials μ_e and μ_g, for the liquid and gas phases, respectively, which is related to the ratio p/p_s of actual to saturation vapor pressure of the compound under consideration. The second term is the surface free energy (Kelvin) term:

$$\Delta G(i) = i(\mu_e - \mu_g) + 4\pi\sigma r^2$$
$$= -k_B Ti \ln(p/p_s) + \sigma(36\pi)^{1/3}(V_m i/N_A)^{2/3} \quad (7.26)$$

Here, k_B is the Boltzmann constant, σ is the surface tension, and $V_m = 4\pi r^3 N_A/3i$ is the molar volume of the condensed phase. When $p/p_s < 1$, the first term is positive and $\Delta G(i)$ grows monotonously with i. When supersaturation occurs, $p/p_s > 1$, the first term becomes negative, and the free energy has a maximum for clusters with radius $r^* = 2\sigma V_m/k_B TN_A \ln(p/p_s)$. The maximum forms a barrier to nucleation, but once a cluster has acquired a sufficient number of molecules to pass over the hill, it enters a region of instability and grows further. The situation is similar to the formation of cloud drops depicted in Figure 7.6. Clusters that have reached the critical size are called *embryos*. It is assumed that the concentration of embryos is still determined by the equilibrium distribution of clusters on the upward slope. The nucleation rate can then be calculated from the equilibrium number density of embryos combined with the net rate at which vapor molecules impinge on them. For the mathematical details of the calculation the reader is referred to the books of Frenkel (1955), Zettlemoyer (1969), Friedlander (1977), and Pruppacher and Klett (1997).

Generally, the homogeneous nucleation of a single compound is effective only at fairly high degrees of supersaturation. Water vapor, for example, requires values in excess of $p/p_s = 5$ (Pruppacher and Klett, 1997). Such high values are not reached in the atmosphere, because the presence of aerosol particles gives rise to heterogeneous nucleation at much lower saturation levels ($p/p_s = 1.05$; see Fig. 7.6). In 1961, in a study of the system H_2SO_4-H_2O, Doyle called attention to the conucleation of two vapors, which he found to be more effective than that of a single component, leading to nucleation even when both vapors are undersaturated. In this case, the equation for the free energy of a cluster contains two pressure terms, one for each component, and $\Delta G(i, j)$ is a function of the numbers of both types of molecules, i and j, present in the cluster:

$$\Delta G(i, j) = -k_B T\left[i \ln(p_1/p_{1s}) + j \ln(p_2/p_{2s})\right] + 4\pi\sigma r^2 \quad (7.27)$$

The free energy can be represented by a surface in three-dimensional space, and the barrier to nucleation now has the shape of a saddle point that the embryos must pass to reach the region of instability for further growth. Doyle's (1961) work was continued by Kiang *et al.* (1973), who introduced a simplified prefactor; by Mirabel and Katz (1974), who corrected sign errors and misprints in Doyle's paper and provided a more complete set of nucleation rates under various conditions; and by Shugard *et al.* (1974), who refined the model by taking into account the effect of H_2SO_4 hydrates. In the troposphere, more than 90% of sulfuric acid molecules are hydrated with one to five water molecules (Jaecker-Voirol *et al.*, 1987; Jaecker-Voirol and Mirabel, 1988, 1989), leading to a reduction of the nucleation rate compared with hydrates being absent. The nucleation rate depends primarily on the concentration of sulfuric acid. The high concentration of water molecules in the atmosphere allows the assumption that the growing embryos are in equilibrium with water vapor. The results depend critically on the choice of the saturation vapor pressure of sulfuric acid. The first reliable measurements were made by Roedel (1979), who found 2.5×10^{-5} torr (3.33×10^{-3} Pa) at 296 K over 99 wt% acid; Ayers *et al.* (1980) determined the temperature dependence in the range 338–445 K and reported $\ln p_s = -10156/T + 16.259$ for pure sulfuric acid, which leads to 1.4×10^{-5} torr (1.9×10^{-3}

Pa) at 298 K. The saturation mixing ratio in air is about 20 nmol mol^{-1} at atmospheric pressure. These measurements supersede the earlier estimates by Gmitro and Vermeulen (1964).

Figure 7.15 illustrates the dependence of nucleation rates on sulfuric acid and water vapor activities and compares calculations with measurements. The nucleation rate increases strongly with the partial pressure of H_2SO_4. At 60% relative humidity, a ratio $p/p_s = 10^{-3}$ corresponds to an H_2SO_4 number density of approximately 5×10^8 at 298 K. The corresponding nucleation rate according to Figure 7.15 is about 10^3 cm^{-3} s^{-1}. Raising the number density 10-fold increases the nucleation rate to about 10^9 cm^{-1} s^{-1}, that is, by six orders of magnitude. To maintain such rates in the atmosphere requires a steady production of H_2SO_4 in the face of competition by heterogeneous condensation of H_2SO_4 onto the surface of the background aerosol. The rate of the second process can be estimated from the rate at which H_2SO_4 molecules strike the effective aerosol surface area, as discussed in Section 7.3.1. Table 7.8 compiles production rates required to maintain the indicated steady-state H_2SO_4 number densities and the loss rates due to nucleation and condensation, respectively. With a mass accommodation coefficient $\alpha = 1$, the first process would become dominant when $n(H_2SO_4)$ changes from 4.6×10^8 to 4.6×10^9 molecule cm^{-3}. However, nucleation would be more competitive with condensation at lower H_2SO_4 concentrations, if α were significantly smaller than unity. Van Dingenen and Raes (1991) have determined $0.02 \leq \alpha \leq 0.3$. Setting $\alpha \approx 0.02$ under marine background conditions would make $R_{cond} \approx 8 \times 10^3$ versus $R_{nucl} = 1 \times 10^3$ for $n(H_2SO_4) = 4.6 \times 10^8$ molecule cm^{-3}, which is still marginal (because accounting for the presence of H_2SO_4 hydrates would lower the nucleation rate even further). Such rates can be sustained in the marine boundary layer in daylight because of the gas phase oxidation of SO_2 by OH radicals, whereas much higher rates would require a more efficient mechanism. Covert et al. (1992) have obtained observational evidence for the competition between homogeneous nucleation and heterogeneous condensation in the marine boundary layer. They found a sharp rise of ultrafine particles in the marine boundary layer when the SO_2 concentration doubled and the surface area concentration of the aerosol decreased. Figure 7.15 shows that the nucleation rate rises significantly with increasing relative humidity. At 90% relative humidity, $R_{nucl} \approx 1 \times 10^6$, so that $R_{nucl} > R_{cond}$. Hegg et al. (1990) have observed an enhanced production of new particles in the vicinity of marine clouds, which they interpret as being derived from the local increase in relative humidity and enhancement of the actinic flux of solar radiation. Table 7.8 includes results for urban conditions obtained by Middleton and Kiang (1978) on the basis of a model that incorporated nucleation, coagulation, and heterogeneous condensation on both new and preex-

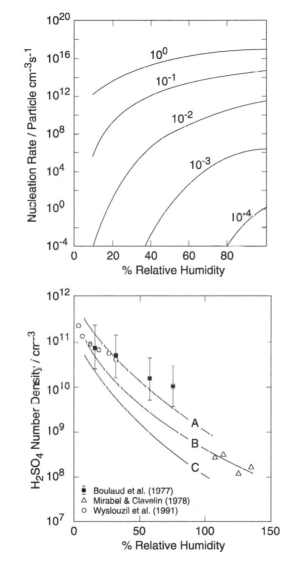

FIGURE 7.15 Top: Nucleation rate in the binary system H_2SO_4-H_2O at 298 K as a function of relative humidity and H_2SO_4 activity p/p_s as a parameter (from calculations of Mirabel and Katz (1974)). Bottom: Steady-state H_2SO_4 number density required to maintain a nucleation rate of 1 particle cm^{-3} s^{-1} at 298 K as a function of relative humidity. Curves represent calculations: (A) Mirabel and Katz (1974); (B) (includes hydrates) Jaecker-Voirol and Mirabel (1988, 1989); (C) Yue and Hamill (1979). Data points are experimental results: squares with error bars, Boulaud *et al.* (1977); triangles, the first four points of Mirabel and Clavelin (1978); open points, Wyslouzil *et al.* (1991).

TABLE 7.8 Steady-State H_2SO_4 Number Density, Associated Production, and
Simultaneous Loss Rates for Nucleation and Heterogeneous Condensation onto
Various Preexisting Aerosols[a]

Aerosol type	r.h (%)	A_{efftot} (m² m⁻³)	$n(H_2SO_4)$ (molecule cm⁻³)	$q(H_2SO_4)$	R(nucl) (molecule cm⁻³ s⁻¹)	R(cond)
Marine background	60	1.4(−5)	4.6(8)	4(5)	1(3)[b]	4(5)[c]
			4.6(8)	9(3)	1(3)[b]	8(3)[d]
			4.6(9)	1(9)	1(9)[b]	4(6)[c]
Rural continental	60	1.4(−4)	4.6(8)	4(6)	1(3)[b]	4(6)[c]
			4.6(9)	1(9)	1(9)[b]	4(7)[c]
Urban[e]	50	1.0(−3)	[f]	1.2(7)	7.5(5)	1.2(7)[g]
			[f]	1.2(8)	1.1(8)	2.4(8)[g]

[a] Values in parentheses indicate powers of 10.
[b] From Figure 7.15; assumes embryos to contain 10 H_2SO_4 molecules.
[c] From collision rate, mass accommodation coefficient $\alpha = 1$.
[d] Mass accommodation coefficient $\alpha = 0.02$.
[e] According to Middleton and Kiang (1978).
[f] Not given.
[g] Reevaporation included.

isting particles, including reevaporation of H_2SO_4, which reduced the rate of condensation onto the preexisting aerosol. Kreidenweis and Seinfeld (1988) and Wyslouzil *et al.* (1991) have compared homogeneous nucleation with methanesulfonic acid as a nucleating agent. This system is less efficient than sulfuric acid.

Despite the effort that has gone into the calculations, it cannot be claimed that the binary system H_2SO_4-H_2O gives a realistic description of the situation existing in the atmosphere. Weber *et al.* (1997) have observed new particle formation in the continental atmosphere at H_2SO_4 concentrations several orders of magnitude lower than those indicated in Table 7.8, which they tentatively ascribe to the interaction with ammonia. Aerosol sulfate indeed occurs mainly as NH_4HSO_4 and $(NH_4)_2SO_4$ and not much in the form of sulfuric acid. Ammonium sulfates arise from the neutralization of H_2SO_4 by ammonia, which may occur either before or after sulfuric acid has entered the particulate phase. However, the interaction of ammonia with sulfuric acid–water aggregates would greatly facilitate homogeneous nucleation because the lower vapor pressure of ammonium bisulfate compared to that of sulfuric acid would substantially reduce the barrier to embryo formation. In fact, one would expect NH_3 to react with sulfuric acid hydrates directly at the molecular level, for example, via

$$H_2SO_4(H_2O)_n + NH_3 \rightarrow NH_4HSO_4(H_2O)_{n-1} + H_2O$$

In other words, the nucleation of sulfuric acid in the presence of water and ammonia represents a ternary rather than binary system, and it should be treated as such. The vapor pressures of NH_3 and H_2SO_4 in equilibrium with ammonium sulfates are much lower than that of sulfuric acid in the presence of water (Scott and Cattell, 1979), so that the critical cluster size is lowered and the rate of nuclei formation is enhanced.

Laboratory and field experiments indicate that the formation of new Aitken particles in outdoor air is very slow in the dark, whereas irradiation by natural or artificial sunlight increases the rate to observable levels (Bricard et al., 1968; Husar and Whitby, 1973; Haaf and Jaenicke, 1980). Ionizing radiation has a similar effect (Vohra et al., 1970). Figure 7.16 shows data illustrating this behavior. While the observations demonstrate a photochemical origin of homogeneous nucleation, they do not give any information on the chemistry involved. A number of observations have shown that the

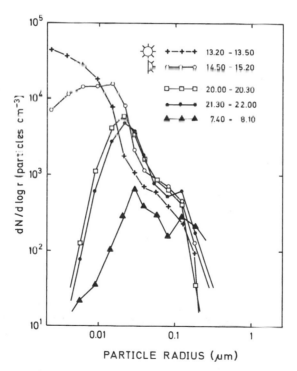

FIGURE 7.16 Particulate size distributions at the summit of Mt. Schauinsland (1250 m a.s.l.), May 16–17, 1978. The variation in the Aitken size range is indicative of gas-to-particle conversion during periods of sunlight, and absence of the process at night. (Adapted from Haaf and Jaenicke, (1980).)

oxidation of SO_2 to sulfate is mediated by sunlight. An example for field observations is the diurnal variation of the SO_2 oxidation rate in power plant plumes discussed in Section 10.3.3. However, the direct photooxidation of SO_2 in air freed from other trace gases is a slow process at wavelengths above 300 nm, because the excited SO_2 molecules initially formed by absorption of solar radiation lose their energy by collisional deactivation instead of undergoing reaction with O_2. This has been shown by laboratory studies (Cox, 1972, 1973; Friend et al., 1973; Sidebottom et al., 1972; Smith and Urone, 1974), as well as by mechanistic considerations (Calvert et al., 1978). The photodissociation rate increases significantly at wavelengths below the SO_2 dissociation limit near 240 nm (Driscoll and Warneck, 1968; Friend et al., 1973), but such short wavelengths do not reach the troposphere. As a consequence, one must turn to other reactions, namely those of SO_2 with transient species that are photochemically generated. Calvert and Stockwell (1984) have reviewed the various possibilities.

Potential oxidants of SO_2 are radicals such as OH, HO_2 and RO_2, NO_3, the Criegee intermediates, and ozone. The reactions of SO_2 with HO_2, CH_3O_2, and NO_3, however, are very slow, and thus they are ineffective (Calvert and Stockwell, 1984). The oxidation of SO_2 by OH radicals is now well understood to proceed via an addition product

$$OH + SO_2 \rightarrow HOSO_2$$

$$HOSO_2 + O_2 \rightarrow SO_3 + HO_2$$

$$SO_3 + H_2O \rightarrow H_2SO_4$$

The rate coefficient for hydrogen abstraction from $HOSO_2$ by oxygen is 4.4×10^{-13} cm^3 $molecule^{-1}$ s^{-1} at 298 K (Martin et al., 1986; Gleason et al., 1987). At the high concentration of oxygen in the atmosphere this step is rapid, so that the reaction of OH with SO_2 (rate coefficient 1.5×10^{-12} cm^3 $molecule^{-1}$ s^{-1} in the high pressure limit) is rate determining in transforming SO_2 to SO_3. The subsequent reaction of SO_3 with water vapor is slower. Reiner and Arnold (1994) found a rate coefficient of 1.2×10^{-15} cm^3 $molecule^{-1}$ s^{-1} independent of pressure. Kolb et al. (1994) and Lovejoy et al. (1996) also have found SO_3 loss to be independent of pressure, but they observed a second order dependence on the H_2O concentration, indicating a more complex mechanism. Even in this case the rate coefficient was sufficient to allow SO_3 to be fairly rapidly converted to H_2SO_4 in the troposphere.

Recently, it has become possible to measure simultaneously the concentrations of gas phase sulfuric acid and OH radicals in the atmosphere. Thus, Weber et al. (1997) demonstrated that H_2SO_4 follows the diurnal variation in OH concentration, and that the concentration of Aitken particles 3–4 nm in size tracks that of H_2SO_4. Figure 7.17 shows the observed diurnal variations of H_2SO_4 and Aitken particles. Although the rise in H_2SO_4 concentration was observed at sunrise together with that of OH radicals, the production of particles followed after an hour delay. This was interpreted to indicate the time required for nuclei to grow to the observed particle size.

The direct oxidation of SO_2 by ozone is slow. Cox and Penkett (1972) demonstrated, however, that SO_2 is nevertheless converted to H_2SO_4 aerosol at an appreciable rate, when SO_2 is added to an ozone–alkene mixture. This suggests that an intermediate of the ozone–alkene reaction is responsible for the oxidation of SO_2 under these conditions. The intermediate might be either an adduct between ozone and the alkene or the intermediate formed from it after aldehyde split-off, so that the reactions oxidizing SO_2 are

$$\underset{\substack{| \qquad\quad | \\ RCH-HCR}}{\overset{\displaystyle \overset{O}{\diagup \ \diagdown}}{O \qquad\quad O}} + SO_2 \rightarrow 2RCHO + SO_3$$

$$R\overset{\bullet}{C}HOO\cdot + SO_2 \rightarrow RCHO + SO_3$$

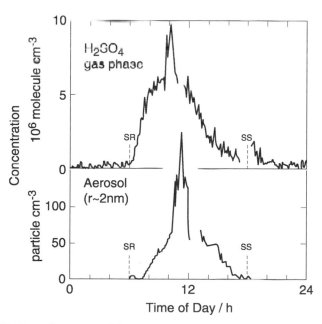

FIGURE 7.17 Diurnal variation of the concentrations of gaseous sulfuric acid and ultrafine particles at Idaho Hill, Colorado. Adapted from measurements of Weber *et al.* (1997).

Niki *et al.* (1977) showed that propylene ozonide formed by the reaction of ozone with *cis*-butene in the presence of formaldehyde was quenched by the addition of SO_2. In addition, the rearrangement of the Criegee intermediate $CH_3\dot{C}HOO \cdot$ to acetic acid was prevented, indicating that one or both of the above reactions did indeed occur. Cox and Penkett (1972), however, also observed an inhibition of H_2SO_4 formation with increasing concentration of water vapor, which is known to react with the Criegee intermediate. Hatakeyama *et al.* (1981) had found earlier that $H\dot{C}HOO\cdot$ produced in the reaction of ozone with ethene interacted with $^{18}H_2O$ to form labeled formic acid. Suto *et al.* (1985) have determined for the reaction of this Criegee intermediate with water a rate coefficient of 1.6×10^{-17}, whereas Becker *et al.* (1990) reported 5.8×10^{-17} cm^3 $molecule^{-1}$ s^{-1}. This is to be compared with a rate coefficient of 7×10^{-14} cm^3 $molecule^{-1}$ s^{-1} given by Atkinson and Lloyd (1984) for the reaction of $H\dot{C}HOO\cdot$ with SO_2. In the troposphere the concentration of H_2O is at least four orders of magnitude greater than that of SO_2, so that for the $H\dot{C}HOO \cdot$ radical the reaction with water vapor takes preference. If other Criegee radicals behaved similarly, the chances would be slim that SO_2 would become oxidized in this manner. This leaves the reaction with OH radicals as the major gas phase process causing SO_2 oxidation.

The oxidation of certain hydrocarbons can also produce aerosols in the absence of SO_2. Grosjean (1977) has reviewed organic particulate formation. Smog chamber experiments have shown that straight-chain alkanes, alkynes, and carbonyl compounds do not generate aerosols. The trend for alkenes, alkadienes, and aromatic compounds is indicated in Table 7.9. Despite their reactivity, most alkenes are not efficient aerosol producers when SO_2 is absent. The ability of alkenes to produce aerosols increases when the carbon number exceeds 6. More effective are alkadienes, cyclic alkenes, and terpenes, the latter being represented in Table 7.9 by α-pinene. For these compounds it makes little difference whether SO_2 is present. For aromatic compounds there are conflicting reports.

The exceptional ability of cyclic alkenes to form aerosols led Grosjean and Friedlander (1980) to study the chemical composition of the particulate products formed. The results for cyclopentene and cyclohexene are listed in Table 7.10. The products are linear, difunctional compounds bearing carboxylic acid, carbonyl, hydroxyl, and nitrate ester functional groups. Equilibrium vapor pressures for these compounds are still essentially unknown. One can nevertheless estimate from the boiling-point temperatures, as far as available, that difunctional compounds have vapor pressures far lower than the corresponding monofunctional species. All of the substances listed in Table 7.10 are well soluble in water, so that binary homogeneous nucleation is expected to occur. The formation of these compounds can be understood

TABLE 7.9 Gas Phase Reactivity and Aerosol-Formation Efficiency for Selected Hydrocarbons in Smog Chambers, in the Presence and Absence of SO$_2$, Relative to Cyclohexane (= 100)

Hydrocarbon	Gas-phase reactivity[a]		Aerosol-forming efficiency		Initial conditions (μmol mol^{-1})				Reference
	(a)	(b)	Without SO$_2$	with SO$_2$	HC	NO	NO$_2$	SO$_2$	
Alkenes									
Ethene	12	49	2.8	65	10	0	5	2	Prager et al. (1960)
1-Butene	46	83	1.4	81	10	0	5	2	
2-Butene	83	202 (cis)	1.4	86	10	0	5	2	
1-Pentene	46	60	2.8	87	40	0	5	2	
2-Pentene	96	187	0	85	100	5	2	2	
1-Hexene	53	49	1.4	96	10	0	5	2	
3-Heptene	—	134 (trans)	12.6	96	10	0	5	2	
Dienes									
1,3-Butadiene	98	123	33	91	10	0	5	2	
1,5-Hexadiene	—	—	104	116	10	0	5	2	
Cyclic alkenes									
Cyclopentene	99	657	75	—	10	0	5	2	
Cyclohexene	100	100	100	100	10	0	5	2	
α-Pinene	49	—	140	137	4	2	0	4	Groblicki and Nebel (1971)
Aromatics									
Toluene	13	37	8.5	8.6	3	0.37	0.37	0.1	Wilson et al. (1973)
Mesitylene	85	146	9	9	3	0.37	0.37	0.1	

[a] (a) k_{OH} from Atkinson (1994); (b) NO-NO$_2$ conversion efficiency form Glasson and Tuesday (1970).

TABLE 7.10   Aerosol Products Obtained from the Photooxidation of Cyclopentene and Cyclohexene[a]

Particulate product	Formula	b.p. (K)[b]	p_s(hPa)
From cyclopentene			
Glutaraldehyde (major)	$OHC(CH_2)_3CHO$	461 (1000)/345 (13.3)	8.5×10^{-1} [b]
5-Oxopentanoic acid	$OHC(CH_2)_3COOH$		
Glutaric acid	$HOOC(CH_2)_3COOH$	576 (1000)/473 (26.7)	1.5×10^{-5} [b]
5-Nitratopentanoic acid	$O_2NOCH_2(CH_2)_3COOH$		
4-Hydroxybutanoic acid	$HOCH_2(CH_2)_2COOH$		
4-Oxobutanoic acid	$OHC(CH_2)_2COOH$		
1,4-Butanedial	$OHC(CH_2)_2CHO$	442 (1000)	1.5^b
4-Nitratobutanal	$O_2NOCH_2(CH_2)_2CHO$		
From cyclohexene			
Adipic acid (major)	$HOOC(CH_2)_4COOH$	538 (133)/478 (13.3)	3×10^{-5} [b,c]
6-Nitratohexanoic acid (major)	$O_2NOCH_2(CH_2)_4COOH$		
6-Oxohexanoic acid	$OHC(CH_2)_4COOH$		
6-Hydroxyhexanoic acid	$HOCH_2(CH_2)_4COOH$		
Glutaric acid	$HOOC(CH_2)_3COOH$	576 (1000)/473 (26.7)	1.5×10^{-5} [b]
5-Nitratopentanoic acid	$O_2NOCH_2(CH_2)_3COOH$		
5-Oxopentanoic acid	$HOC(CH_2)_3COOH$		
5-Hydroxypentanoic acid	$HOCH_2(CH_2)_3COOH$		
Glutaraldehyde	$OHC(CH_2)_3CHO$	461 (1000)/345 (13.3)	8.5×10^{-1} [b]

[a] From Grosjean and Friedlander (1980).
[b] Boiling points, where available, are in Kelvin (at pressures in hPa). They were used to obtain by extrapolation order-of-magnitude estimates for vapor pressures at ambient temperature.
[c] Davies and Thomas (1960) give 1×10^{-7} by extrapolation of the sublimation pressure; Heisler and Friedlander (1977) estimate 2×10^{-5} hPa.

to result from ring opening. The oxidation of cyclohexene by ozone is expected to produce 6-oxohexanoic acid. Figure 7.18 shows a likely mechanism of formation.

Schuetzle et al., (1975) first identified a number of compounds similar to those shown in Table 7.10 in Californian smog aerosols. The occurrence of dicarboxylic acids in association with the atmospheric aerosol is now well documented. Grosjean et al. (1978), Yokouchi and Ambe (1986), and Kawamura and Kaplan (1987) have detected C_2–C_9 dicarboxylic acids, with oxalic acid being the most prominent. Sempéré and Kawamura (1994) and Kawamura et al. (1996), who have used a derivatization technique, have additionally found C_3–C_9 ketoacids in urban and remote aerosols.

Went (1960a, 1964) hypothesized that the blue haze often observed over forested regions is due to aerosols resulting from the photooxidation of

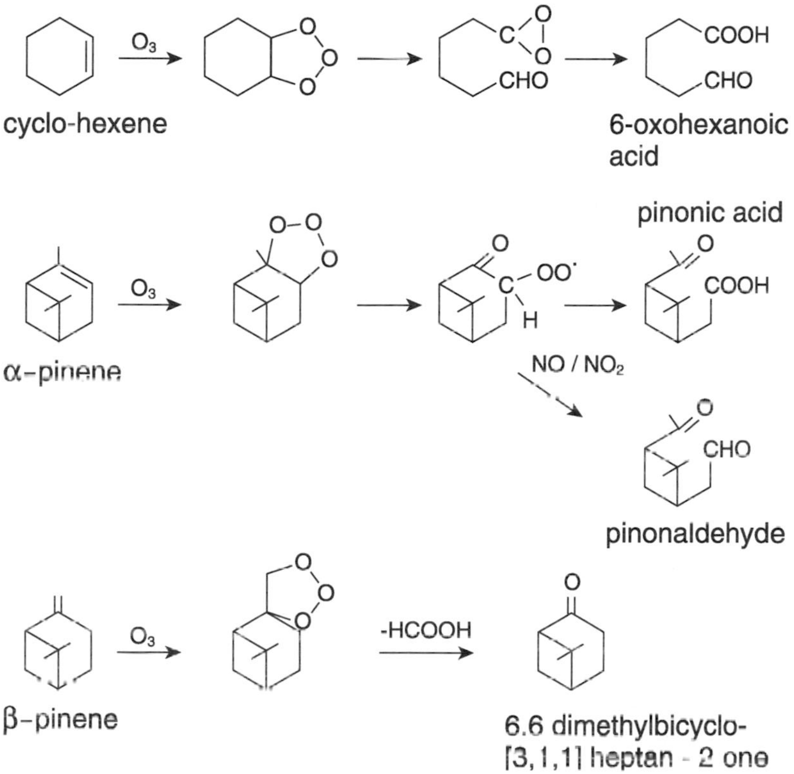

FIGURE 7.18 Pathways for the oxidation of cyclohexene, α-pinene, and β-pinene (after Grosjean and Friedlander, 1980; and Hatakeyama *et al.*, 1989).

terpenes and other organic compounds emitted from vegetation. Schwartz (1974) and Hull (1981) have studied particulate products from the oxidation of α-pinene. Schuetzle and Rasmussen (1978) have examined limonene and terpinolene. Ring opening was observed in all cases. Figure 7.19 shows oxidation products from α-pinene. Prominent products are *cis*-pinonic acid, which is a solid at 298 K, and pinonaldehyde. Figure 7.18 includes pathways for their formation from the reaction with ozone, which is similar to that for the oxidation of cyclohexene. Yokouchi and Ambe (1985) and Hatakeyama *et al.* (1989) have shown that 6,6-dimethyl-bicyclo-[3,1,1]heptan-2-one is the major product from the oxidation of β-pinene. Figure 7.18 shows a probable mechanism of formation. The vapor pressure of this compound is expected to be higher, that is, less favorable for particle formation, than that of pinonic acid. However, particle formation has also been observed with β-pinene.

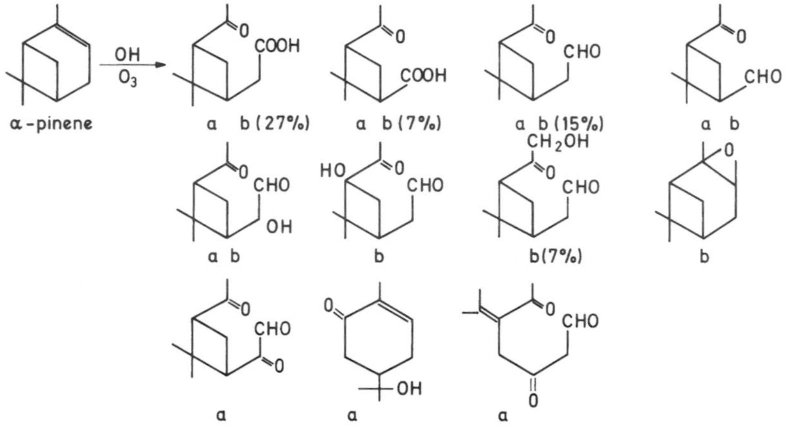

FIGURE 7.19 Compounds identified in the aerosol derived from the oxidation of α-pinene. (a) Sunlight in the presence of nitrogen oxides, (b) by ozonolysis; yields in percent are shown in parentheses. (From Schwartz (1974), as reported by Grosjean (1977) and Hull (1981), respectively.)

Table 7.11 presents yields of particulate material derived from the oxidation of several terpenes in environmental chambers. The dark reaction with ozone appears to be more effective in particle production than reaction with OH radicals produced in sunlight with NO_x present, yet appreciable yields have been observed in both cases, indicating the great potential of such reactions for atmospheric aerosol formation. This may also include reaction of

TABLE 7.11 Yield of Particulate Material from the Oxidation of Several Monoterpenes in Environmental Chambers

Hydrocarbon	x(HC) (nmol mol^{-1})	x(NO$_x$) (nmol mol^{-1})	Aerosol yield %	Reactant	Reference
α-Pinene	10–120	—	18.3	O_3	a
	20–136	113–240	1.2–12.0	OH, O_3	b
	41–168	—	14–23	O_3	b
β-Pinene	10–80	—	13.8	O_3	a
	95	204	30.2	OH, O_3	b
	85	—	32.1	O_3	b
d-3-Carene	112	210	27	OH, O_3	b
	96	—	75.9	O_3	b
Limonene	89–159	134–205	35.7–40.5	OH, O_3	b

[a] Hatakeyama et al. (1989).
[b] Hoffmann et al. (1997).

monoterpenes with the NO_3 radical at night. Its reaction with α-pinene, for example, is rapid and leads to pinonaldehyde as the major product (Wängberg et al., 1997).

7.4.4. MISCELLANEOUS SOURCES

Biogenic particles are released from plants in the form of seeds, pollen, spores, leaf waxes, resins, etc., ranging in size from perhaps 1 to 250 μm. Delany et al. (1967) have used nylon meshes to collect airborne particles greater than 1 μm at Barbados and concluded that most of the material originated on the European and African continents. Besides mineral components, they identified various biological objects, such as microbes and fragments of vascular plants, among them fungus hyphae, marine organisms, and freshwater diatoms, which are also found in Atlantic sediments. Parkin (1972) similarly found fragments of humus, dark plant, and fungus debris. The size distribution of biogenic particles in the range 0.3–50-μm radius is similar to that of the total atmospheric aerosol. The volume fraction is about 15% (Matthias-Maser and Jaenicke, 1994).

The aerial transport of pollen and microorganisms has received some attention (Gregory, 1973, 1978; R. Campbell, 1977). Bacteria (size < 1 μm) are difficult to discern directly, and their study requires cultural growth techniques. In contrast to fungi spores, they usually attach to other aerosol particles because they are mobilized together with dust. Concentrations number several hundred per cubic meter in rural areas and several thousand in the cities. Air is not their natural habitat, so that multiplication does not take place. On filters they remain viable only if provided with nutrients and moisture (Rüden and Thofern, 1976; Rüden et al., 1978; Maus et al., 1997).

The concentration of pollen and spores can change rapidly with locality, season, time of day, and meteorological conditions. On the European continent, concentrations average 12,500 m^{-3} during summer (Gregory and Hirst, 1957; Stix, 1969). The population is dominated by the spores of common moulds, whereas pollen account for less than 1% of the total. Like other aerosol particles, pollen and spores are carried to heights of at least 5 km (Gregory, 1978), and horizontally to distances of several hundred kilometers. The spreading of virulent spores such as those of wheat rusts is important to agriculture.

Volcanism constitutes another source of particles in the atmosphere. The 1963 eruption of Mt. Agung on Bali injected so much particulate material into the stratosphere that the temperature in the 16–20-km altitude region suddenly rose by 6 K in the 30° latitude belt around the equator (Newell, 1970), decaying slowly to pre-Agung values during the subsequent years.

Similar effects were seen following the eruptions of El Chichon in 1982 and Pinatubo in 1991 (Angell, 1993; Fiocco *et al.*, 1996).

Cadle and Mroz (1978), Hobbs *et al.* (1977, 1978), and Stith *et al.* (1978) have used aircraft to study the effluents from the Alaskan volcano St. Augustine during a period of intensified activity in 1976. The results indicated that in periods between eruptions the particles consist of finely divided lava coated with sulfuric acid containing dissolved salts. Sulfur is emitted from volcanoes largely as SO_2, which is subsequently oxidized to H_2SO_4. Paroxysmal events resulting from magmatic movements increase the mass of volcanic ash ejected so much that the acid coating is greatly reduced. The mass of the particles is concentrated in the submicrometer size range. Number densities maximize at radii below 0.1 μm, with the usual fall-off toward larger particle size. Only paroxysmal events enhance the concentration of particles in the size range of 10–100 μm.

Volcanic emissions cause an enrichment of so-called volatile elements in the atmospheric aerosol compared to the relative abundance of such elements found in bulk material of the Earth's crust (Cadle *et al.*, 1973; Mroz and Zoller, 1975; Lepel *et al.*, 1978; Buat-Menard and Arnold, 1978). Figure 7.20 summarizes results obtained by Buat-Menard and Arnold from samples

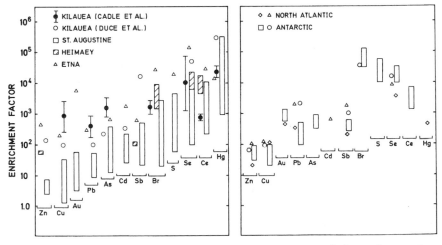

FIGURE 7.20 Left: Enrichment factors for volatile elements in aerosols from volcanoes. Data were compiled from Cadle *et al.* (1973) and Duce *et al.* (1976) for the volcano Kilauea; Mroz and Zoller (1975) for Heimaey; Lepel *et al.* (1978) for St. Augustine; Buat-Menard and Arnold (1978) for Etna. Right: Enrichment factors for volatile elements in background aerosols. Data are from Duce *et al.* (1976) (triangles); Buat-Menard and Chesselet (1979), (diamonds); Zoller et al. (1974) (circles), and Maenhaut *et al.* (1979) (vertical bars).

of the plume of Mt. Etna, Sicily, and compares them with similar data for aerosols collected at remote locations. Enrichment factors shown in Figure 7.20 are defined by the concentration ratios EF = $[X/Al]_{aerosol}/[X/Al]_{crust}$, with aluminum serving as the reference element. Nonvolatile elements have enrichment factors in the vicinity of unity. The values for the elements considered in Figure 7.20 are all much greater than unity. Sulfur, chlorine, and bromine are emitted mainly in gaseous form as SO_2, HCl, and HBr and subsequently become attached to aerosol particles, so that their enrichment is qualitatively understandable. Mercury is a special case in that its vapor pressure is already appreciable. The enrichment of the other elements cannot be so explained, however. It appears that the high-temperature environment of volcanic emissions is responsible for the enhanced volatilization of these elements. The right-hand side of Figure 7.20 shows that remote aerosols feature a similar enrichment. This similarity led Lepel et al. (1978), Mroz and Zoller (1975), and Maenhaut et al. (1979) to suggest that emissions from volcanoes might be responsible for the enrichment of trace metals in the natural background aerosol. Other processes that must be considered as a potential cause of the enrichment are biological volatilization mechanisms, for example, the methylation of trace metals by microorganisms (Wood, 1974; Ridley et al., 1977), and anthropogenic high-temperature combustion processes (Linton et al., 1976; Bolton et al., 1975). Boultron (1980), however, has been able to show from samples taken from the Antarctic ice shield that in the past 100 years the concentrations of and the enrichment factors for Pb, Cd, Cu, Zn, and Ag have remained essentially constant. This confirms that global pollution in the case of these elements is actually negligible in the remote Southern Hemisphere. In the Northern Hemisphere, in contrast, data from the Greenland ice shield clearly indicate an increase in the concentration of lead during the period 1750–1950 (Murozumi et al., 1969). The dominant sources were shown to be lead smelters. Automotive fuel lead additives contributed comparatively little. Today, lead concentrations in Greenland snow are well over 500 times higher than the natural level.

Forest fires, like volcanoes, tend to emit particles into a buoyant plume reaching far into the troposphere (not, however, into the stratosphere). In the United States, according to Brown and Davis (1973), some 10^4 km^2, or about 0.4% of the total, is destroyed annually by wildfires. Seiler and Crutzen (1980) estimated that 3.8×10^4 km^2 of temperate and boreal forests are subject to wildfires each year. These still are fairly isolated events compared with the more regular or recurrent agricultural burning practices in the Tropics, which are mainly savanna fires (Hao and Liu, 1994). As discussed in Section 4.3, the combustion of wood and other biomass proceeds in three stages: the first is pyrolysis of the fuel, leading to the volatilization of organic compounds; the second is the combustion of organic vapors in a turbulent, high-temperature flame; and the third is smoldering combustion, which sets in when the rate of volatile formation decreases so much that it can no longer sustain a flame. The combustion zone then retreats to the surface of the residual (highly charred) fuel, and oxygen must be transported to the combustion site by molecular diffusion. Particle formation occurs during all three stages by the mechanical release of charcoal particles and by the condensation of organic vapors. The char residue left after the fire dies out is subsequently exposed to wind erosion, which brings additional carbonaceous material into the atmosphere. The production rate of smoke particles depends on the type of fire, with flaming fires being characterized by fairly low emissions of 5–10 g kg^{-1} of carbon dioxide carbon, and smoldering fires by higher emissions of 15–50 g kg^{-1}. Savanna fires, which burn hot and efficiently, thus tend to have lower emissions relative to the amount of fuel consumed than deforestation fires that burn less efficiently (Ward and Hardy, 1991).

The analysis of particulate matter from open fires by Gerstle and Kemnitz (1967) and McMahon and Ryan (1976) indicated that about 50% by weight is benzene-soluble organic compounds, 40% is carbonaceous material, and 10% is inorganic components. More recent observations in the Amazon region (Andreae et al., 1988) show that the carbon content of smoke aerosol is 60–70%, the remainder being mostly hydrogen and oxygen, indicating that the material consists of partly oxygenated organic matter. Black carbon from biomass burning, a substance consisting of polymeric carbonaceous materials with a molar H/C content ≤ 0.2, has come under closer scrutiny recently (Cachier et al., 1995; Kuhlbusch and Crutzen, 1995; Kuhlbusch et al., 1996). Black carbon is formed by the pyrolysis of biomass and occurs in particulate matter as well as in the ashes. The content of black carbon in aerosol derived from biomass burning is quite variable. In smoldering fires it is about 4%, whereas in intensely flaming fires it can reach 40%. This makes savanna fires an important source of black carbon in the atmosphere. Prominent among the inorganic components are the cations potassium and ammonium, and the anions chloride and sulfate (Andreae et al., 1996). The presence of potassium in submicron-sized particles can be used as a diagnostic tracer for particles of pyrogenic origin, as mineral-derived potassium is concentrated in larger particles.

Smoke particles from open fires are concentrated in the submicrometer size range. Radke et al. (1978, 1991), Stith et al. (1981), and Andreae et al. (1996) have described size distributions of aerosol particles arising from forest fires. Number density and volume were found to peak near 0.1 and 0.3 μm, respectively, indicating the dominance of particles in the accumulation mode. Eagan et al. (1974) have noted that forest fires are prolific sources of cloud condensation nuclei. Combustion aerosols are only partly water-soluble, but as Radke *et al.* (1978) have pointed out, they enhance the number of large particles normally present in the atmosphere. As these are efficient cloud condensation nuclei, it appears that it is the favorable size distribution of combustion aerosols that makes them effective as cloud condensation nuclei.

7.4.5. GLOBAL AEROSOL PRODUCTION RATES

Table 7.12 summarizes global emission and production rates of particulate matter in the troposphere. Most estimates were derived in the early 1970s, as summarized by Bach (1976). Andreae (1996) recently provided an update, which is included. The data generally refer to all particles that are not immediately returned to the Earth's surface by gravitational settling. We shall briefly indicate the methods that were used in deriving the estimates.

The production rate for sea salt goes back to a detailed study of Ericksson (1959). He calculated the rate of dry fall-out of sea salt particles from a vertical eddy diffusion model and sea salt concentrations existing over the ocean. This led to a global rate for dry deposition of 540 Tg annually. Ericksson then argued that a similar amount would be removed by wet precipitation. It is now known, however, that wet precipitation is more effective than dry deposition in removing aerosol particles from the atmosphere, so that Ericksson's (1959) value must be an underestimate. From the rate of wet deposition of sulfate in the marine atmosphere, discussed in Section 10.3.5, one obtains an upper limit of about 5200 Tg year^{-1} for the wet deposition of sea salt. Estimating the production rate directly from the mechanism of sea salt formation presents formidable difficulties, and no attempt appears to have been made to solve this problem.

The global importance of mineral dust is well documented by dust plumes from desert regions showing up in satellite data (Husar and Stowe, 1994). Soil-derived particle fluxes, however, are critically dependent on local conditions and the prevailing wind force. Direct measurements thus are difficult to extrapolate to global conditions. Peterson and Junge (1971) adopted an estimate of Wadleigh (1968) for the emission of wind-blown dust in the United States that included both natural and agricultural sources. Other estimates made use of the measured dust accumulation in glaciers (Windom, 1969) and deep-sea sediments (Goldberg, 1971). The former gives an underestimate because mountain glaciers are mostly located 2000–3000 m above sea level, where the concentration of aerosol particles is reduced compared to that in the lower atmosphere. The latter may be an overestimate because sediment accumulation is assumed to result exclusively from dust fall. Schütz (1980) has made a detailed study of the dust plume carried from the Sahara

TABLE 7.12 Estimates for Global Production Rates of Particulate Matter from Natural and Anthropogenic Sources (Tg year^{-1})

Source type	Peterson and Junge (1971) All sizes	Peterson and Junge (1971) r < 2.5 μm	Hidy and Brock (1971)	SMIC (1971)	Andreae (1995)	Others
Natural sources						
Direct emissions						
Sea salt	1000	500	1095	300	1300	1300,[a] 5200[b]
Mineral dust	500	250	7–365	100–500	1500	500,[c] 260,[d] 2000,[j] 8000[a]
Volcanoes	25	25	4	25–150	33	—
Forest fires	35	5	146	3–150	—	72–117[e]
Meteorite debris	10	—	0.02–0.2	—	—	—
Biogenic material	—	—	—	—	50	80,[f] 26[g]
Subtotal	1540	780	1610	428–1100	2883	—
Secondary production						
Sulfate	244	226	37–365	130–200	102	3000[a]
Nitrate	75	60	600–620	140–200	22	—
Hydrocarbons	75	75	182–1095	75–200	55	—
Subtotal	394	355	2080	345–1100	179	—
Total natural	1964	1135	3690	773–2200	3062	—

Anthropogenic sources						
Direct emissions						
Transportation	2.2	1.8				
Stationary sources	43.4	9.6				
Industrial processes	56.4	12.4			130	
Solid waste disposal	2.4	0.4				
Miscellaneous	28.8	5.4				
Biomass burning	—	—		10–90	80	62,[h] 130–289,[e] 79,[i]
Subtotal	133.2	29.6	37–110		210	98[g]
Secondary production						
Sulfate	220	200	139.5	130–200	140	
Nitrate	40	35	23	30–35	36	
Hydrocarbons	15	15	27	15–90	10	
Subtotal	275	250	159.5	175–325	186	
Total anthropogenic	408	280	269	185–415	396	
Sum total	2372	1415	3959	958–2615	3450	

[a] Petrenchuk (1980).
[b] See Section 10.3.5.
[c] Goldberg (1971).
[d] Schütz (1980) (Sahara only).
[e] Seiler and Crutzen (1980).
[f] Jaenicke (1978b).
[g] Duce (1978).
[h] Bach (1976).
[i] Penner et al. (1993).
[j] d'Almeida (1986).

Desert westward across the Atlantic Ocean. His estimate for the annual mass transport of dust particles in the plume is 260 Tg year^{-1}. About 80% of it is deposited in the Atlantic Ocean, the rest reaches the Caribbean Sea. There is an additional transport to the Gulf of Guinea, so that the above estimate is again a lower limit. There is a similar transport of dust from the Gobi Desert via the Pacific Ocean into the Arctic Basin and further on to Spitsbergen (Rahn et al., 1977; Griffin et al., 1968). A more recent estimate for the mineral dust emission is 1800–2000 Tg year^{-1} globally (d'Almeida, 1986).

The value derived by Peterson and Junge (1971) for the rate of particulate emissions from volcanoes is based on the long-term burden of particles in the stratosphere combined with an assumed stratospheric residence time of 14 months. This gives a lower limit of 3.3 Tg year^{-1}. If 10% of volcanic particles, on average, reach the stratosphere, the total emission rate will be 33 Tg year^{-1}. Goldberg (1971) took instead the rate of accumulation of montmorillonite in deep-sea sediments as an indicator of average volcanic activity. His estimate of 150 Tg year^{-1} must be an upper limit. The estimate of 10 Tg year^{-1} adopted by Peterson and Junge (1971) for meteorite debris deposited in the stratosphere is due to Rosen (1969).

All estimates for particle production from biomass burning are based on experimental emission factors combined with statistical data for the global consumption of biomass by fires. The study of Seiler and Crutzen (1980) suggested about 100 Tg year^{-1} each for emissions caused by deforestation and agricultural burning practices. Recent estimates are slightly lower. Andreae (1993) reported 104 Tg year^{-1} globally, to which biomass burning in the Tropics contributes 90 Tg year^{-1} and black carbon 17 Tg year^{-1}. Penner et al. (1993) derived a similar estimate of 79 Tg year^{-1}.

The direct release of organic particles from the biosphere in the form of pollen, leaf cuticles, etc., which was not considered by Peterson and Junge (1971), has been estimated by Jaenicke (1978b) to be 80 Tg year^{-1} worldwide, whereas Duce (1978) suggested 26 Tg year^{-1}, and Andreae (1995) chose an intermediate value.

In estimating the rate of gas-to-particle conversion involving SO_2 from anthropogenic sources, Peterson and Junge (1971) assumed that 66% is converted to sulfate and the rest is removed by dry deposition. In addition, is was assumed that sulfate is completely neutralized to ammonium sulfate. An annual emission rate of 160 Tg SO_2 from the combustion of fossil fuels then gives a production of particulate sulfate on the order of 220 Tg year^{-1}. In 1971 the rate of natural emissions of sulfur compounds was less well known than it is today. Peterson and Junge (1971) adopted a tentative estimate of 98 Tg year^{-1} of sulfur from Robinson and Robbins (1970) to derive a production rate of 244 Tg year^{-1}. Both estimates must be revised. The discussion in Section 10.4 will show that dry deposition over the continents removes about

45% of SO_2, and about 25% over the oceans. The global production rate of SO_2 from fossil fuel combustion has not changed much but is uncertain by 20%, whereas the largest source of natural sulfur, namely dimethylsulfide from the oceans, is 16–50 Tg year^{-1}. This would lead to 140–227 Tg year^{-1} from anthropogenic sources and 50–154 Tg year^{-1} from natural sources.

The fraction of NO_2 that is converted to aerosol nitrate is extremely uncertain, because a large fraction of HNO_3 remains in the gas phase and undergoes dry or wet deposition. Peterson and Junge (1971) made use of observational data collected by Ludwig *et al.* (1970) at 217 urban and nonurban stations in the United States. The average ratio of sulfate to nitrate was 5.5, and this value was combined with the rate of sulfate production to obtain a crude estimate for the formation of nitrate. Equally uncertain is the rate of gas-to-particle conversion involving organic compounds. Went (1960b, 1966) had estimated that the rate of hydrocarbon emissions from vegetation, mainly terpenes and hemiterpenes, is 154 Tg year^{-1}. Peterson and Junge (1971) assumed 50% gas-to-particle conversion efficiency, which suggested an aerosol production rate from this source of about 75 Tg year^{-1}. The yield of particulate matter depends on the hydrocarbon, however, and Table 7.12 indicates that it generally is smaller than 50%. On the other hand, the data in Section 6.1.2 show that the global rate of hydrocarbon emissions from vegetation is several times larger than that initially estimated by Went (1966). The oxidation of isoprene, which represents about 50% of the total emission, does not lead to the formation of particulate matter. Andreae (1995) assumed 700 Tg year^{-1} for the global rate of biogenic nonmethane hydrocarbon emissions. With an aerosol yield of 10%, and a carbon content of 66% in the partly oxidized products, he found 55 Tg year^{-1} for the global production rate of particulate matter from this source.

The total rate of aerosol production is on the order of 3000 Tg year^{-1}. Some comparisons are of interest. Anthropogenic sources, direct as well as indirect, contribute about 15% to the global production rate. The percentage of direct emissions from anthropogenic sources ranges from 22% to 53% of the total contribution, so that gas-to-particle conversion is dominant. Among the natural sources, however, direct emissions are more important than secondary particle production.

7.5. THE CHEMICAL CONSTITUTION OF THE AEROSOL

The chemical composition of particulate matter in the atmosphere is complex and not fully known. For many purposes it is useful to distinguish three broad categories of components: (1) water-soluble inorganic salts (electrolytes); (2) water-insoluble minerals of crustal origin; and (3) organic compounds, both water-soluble and insoluble. To determine fractions of materials present in each group, Winkler (1974) performed extraction experiments with bulk samples of European aerosols, using several organic solvents in addition to water. His results are summarized in Table 7.13, which lists averages for urban and rural background aerosols. The large water-soluble fraction comprises about equal parts of organic and inorganic materials. The contribution from water-insoluble organic compounds is small, and the fraction of insoluble minerals is about 30%. The composition of individual aerosol samples varies widely. For example, the frequency distribution of the water-soluble fraction ranges from 30% to 80% of total mass. Winkler (1974) further showed that heating the aerosol samples to 150°C caused a mass decrease, which he attributed to the loss of volatile organic compounds in addition to some inorganic salts such as NH_4HSO_4. Similar effects had been observed earlier by Twomey (1971) for continental cloud nuclei and by Dinger *et al.* (1970) for the marine aerosol.

Other investigators have combined extraction and combustion techniques to determine the water-soluble fraction of particulate (noncarbonate) carbonaceous matter. For example, Cadle and Groblicki (1982) examined the urban aerosol in Denver, Colorado, and found $27 \pm 2\%$ of total organic carbon to be water-soluble; Mueller *et al.* (1982) reported 30–67% in Los Angeles, California; and Sempéré and Kawamura (1994) found 28–55% (average 37%) in Tokyo. In the latter case, the total carbonaceous fraction of

TABLE 7.13   Group Classification of European Continental Aerosols: (A) Mainz and (B) Deuselbach, Germany[a]

Component	Mass fraction (%)	
	(A)	(B)
Insoluble mineral component	35	25
Water-soluble component	58	68
Organic solvent-soluble[b]	40	42
Water-soluble organic fraction[c]	28	25
Water-insoluble organic fraction	5	6
Water-soluble inorganic salts	30	43

[a] From Winkler (1974).
[b] Total organic material soluble in acetone, cyclohexane, ether, and/or methanol.
[c] Taking into account that methanol also dissolves a number of inorganic salts.

the aerosol was only about 12% (range 6.8–19.4). Appel *et al.* (1983) had earlier found about 30% in southern California, and Hashimoto *et al.* (1991) had reported about 20% for Seoul, Korea. In the urban environment combustion products (both primary and secondary) are believed to contribute heavily to organic particulate matter. As the extraction efficiencies of organic solvents are limited, a considerable fraction of organic particulate material may not be recoverable by this technique. In Tokyo, Sempéré and Kawamura (1994) observed that the total carbonaceous fraction as well as the water-soluble organic fraction were four to seven times more abundant in summer than in winter, which suggests that much of the material has a photochemical origin and constitutes secondary aerosol.

Figure 7.21 shows for the rural aerosol studied by Winkler (1974) how the principal chemical component groups are distributed with particle size. Although only a coarse size classification was achieved, water-soluble matter shows up in all size ranges. Giant particles contain roughly 50% each of water-soluble compounds and insoluble minerals. The fraction of water-soluble material increases with decreasing particle size at the expense of the mineral component. If it is allowed to extrapolate the trend into the Aitken size range, more than 90% of all particles with radii less than 0.1 μm will be water-soluble.

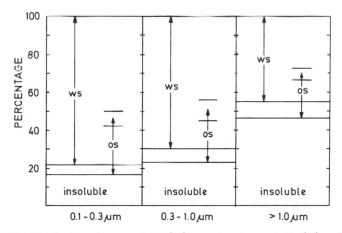

FIGURE 7.21   Distribution of water-soluble (ws), organic solvent-soluble (os), and insoluble mass fractions associated with the rural continental aerosol at Deuselbach, Germany, according to measurements of Winkler (1974). The organic fraction comprises materials soluble in cyclohexane, ether, acetone, and a part of the methanol-soluble fraction. The uncertainty range assumes that 0–40% of the methane-soluble fraction contains organic compounds, and the remainder inorganic salts. The water-soluble fraction averages about 0.66 for all size ranges combined.

7.5.1. THE INORGANIC FRACTION, SOLUBLE AND INSOLUBLE

Wet chemical analysis of aqueous extracts of aerosol samples have established the presence of anions such as sulfate, nitrate, and the halides, and of cations such as ammonium and ions of the alkali and alkaline earth elements. Table 7.14 shows selected data to illustrate the abundances of important inorganic components in the urban, continental, arctic, and marine aerosols. Included for comparison are the concentrations of silicon, aluminum, and iron, which are the major elements of crustal origin. They occur in oxidized form, such as in aluminosilicates, which are practically insoluble. Taken together, the elements listed in Table 7.14 account for 90% of all inorganic constituents of the atmospheric aerosol.

A reasonably complete analysis of the inorganic chemical composition of the aerosol requires much effort and involves, in addition to wet chemical analytical methods, instrumental techniques such as neutron activation analysis, atomic absorption spectroscopy, or proton-induced X-ray emission (PIXE). These latter techniques yield the elemental composition. They furnish no direct information on the chemical compounds involved, although auxiliary data from mineralogy, chemical equilibria, etc. usually leave little doubt about the chemical form in which the elements occur. Thus, sulfur is present predominantly as sulfate, and chlorine and bromine as Cl^- and Br^-, respectively, whereas sodium, potassium, magnesium, and calcium show up as Na^+, K^+, Mg^{2+}, and Ca^{2+}. Reiter et al. (1976) reported a fraction of calcium as insoluble, and Table 7.14 makes allowance for it. Stelson and Seinfeld (1981) concluded from consideration of equilibria and leaching experiments that CaO is more easily converted to its hydroxide than is MgO, so that it should occur predominantly as Ca^{2+}. Heintzenberg et al. (1981) preferred the PIXE technique in determining calcium. In both cases it was assumed that calcium is completely soluble.

The study of Reiter et al. (1976) at the Wank Mountain observatory distinguished days with strong and weak vertical exchange by means of Aitken nuclei counts and concentrations of radon decay products. The main difference between the two meteorological situations was a shift in total aerosol concentration. Variations in chemical composition were minor. This indicates that at the elevation of the Wank observatory (1780 m a.s.l.), the continental aerosol is well homogenized. Gillette and Blifford (1971) and Delany et al. (1973) have reached the same conclusions from aircraft samples over the North American continent.

In the following discussion we shall use Table 7.14 as a reference in delineating the origins of the various inorganic components of continental and marine aerosols. In this discussion one should keep in mind that the aerosol represents a mixture of substances from several sources, and that any specific component may have more than one origin. Sea salt, continental soils, and gas-to-particle conversion represent the major sources. To these must be added various anthropogenic contributions. Comparison of aerosol and source compositions provides important clues to the individual contributions. Below, we discuss first the anions and ammonium, then soluble and insoluble elements.

Sulfate is the most conspicuous constituent of all aerosols considered in Table 7.14 except the marine aerosol, whose mass is dominated by sodium chloride. Mass fractions of SO_4^{2-} range from 22–45% for continental aerosols to 75% in the Arctic region. Similarly high fractions of sulfate occur in the Antarctic (Maenhaut and Zoller, 1977). The origin of sulfate over the continents is primarily gas-to-particle conversion of SO_2. The sulfur content of the Earth's crust is too low for soils to provide a significant source of sulfate, except perhaps locally in the vicinity of gypsum beds. Figure 7.22 shows mass size distributions for sulfate and several other water-soluble constituents of the rural continental aerosol. Sulfate is concentrated in the

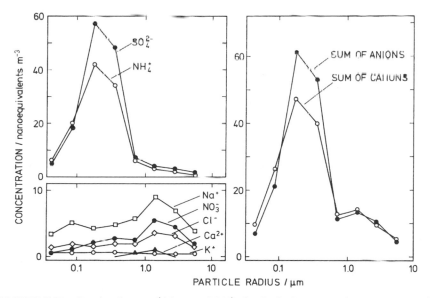

FIGURE 7.22   Size distributions ($\Delta \log r = 0.303$) of individual anions and cations associated with rural continental aerosol at Deuselbach, West Germany (nanoequiv m^{-3} = nmol $m^{-3} \times$ ionic charge number). Left: Individual contributions. Right: Balance between anions and cations. An apparent cation deficit is balanced by hydrogen ion (data from Mehlmann (1986)).

TABLE 7.14 Mass Concentrations ($\mu g\ m^{-3}$) of Inorganic Components in Aerosols at Several Locations[a]

Element or group	Urban aerosol		Continental aerosol Wank, central Europe			Arctic aerosol, Ny-Ålesund		Marine aerosol central Atlantic
	Tees-side	West Covina	Average	High	Low	With sea influence	Without sea influence	
SO_4^{2-}	13.80	16.47	3.150	4.90	0.546	1.95	2.320	2.577
NO_3^-	3.00	9.7	0.920	1.335	0.412	0.022	0.055	0.050
Cl^-	3.18	0.73	0.112	0.137	0.076	0.174	0.013	4.625
Br^-	0.07	0.53	—	—	—	—	—	0.015
NH_4^+	4.84	6.93	1.295	1.960	0.351	0.152	0.226	0.162
Na^+	1.18	3.10	0.053	0.084	0.024	0.209	0.042	2.910
K^+	0.44	0.90	0.062	0.121	0.025	0.050	0.023	0.108
Ca^{2+}	1.56	1.93	0.155	0.303	0.042	0.073 (av.)		0.168
Mg^{2+}	0.60	1.37	—	—	—	0.071	0.032	0.402
Al_2O_3	3.63	6.43	0.223	0.389	0.077	—		—
SiO_2	5.91	21.10	0.663	1.250	0.167	0.235 (av.)		—
Fe_2O_3	5.32	3.83	0.145	0.365	0.035	0.091 (av.)[b]		0.065
CaO	[b]	[b]	0.104	0.182	0.057	[b]		—
Sum	43.53	75.70	6.90	11.03	1.810	3.07 (av.)		11.08

Selected mass fractions and ratios

SO_4^{2-} (%)	29.5	21.8	45.6	44.4	30.2	63.5	75.5	22.6
NO_3^- (%)	6.3	12.8	13.3	12.1	22.7	0.72	1.80	0.44
NH_4^+/SO_4^{2-} [c]	1.9	2.2	2.2	2.1	3.4	0.43[d]	0.53	0.47[d]
$NH_4^+/(NO_3^- + 2SO_4^{2-})$ [c]	0.8	0.8	0.9	0.9	1.0	0.20[d]	0.30	0.30[d]
Si/Al	1.44	2.90	2.63	2.84	1.94	—	—	—
Fe/Al	1.94	0.79	0.86	1.14	0.60	—	—	—
Na/Al	0.61	0.91	0.45	0.41	0.59	—	—	—
K/Al	0.23	0.26	0.52	0.58	0.61	—	—	—
Ca/Al	0.81	0.57	1.94	2.10	2.03	—	—	—

[a] (1) Tees-side, a conglomerate of towns in the industrial area of Middlesbrough in northeast England. Data from June–October 1967, average of five sampling stations (Eggleton, 1969). (2) West Covina, greater Los Angeles, California. Data from three individual days, July, August 1973, averages from three sites (Stetson and Seinfeld, 1981). (3) Wank, a mountain station in the northern Alps (1780 m a.s.l.), near Garmisch, Germany observation period Oct. 1969 to Nov. 1973. Total averages and values for days of strong and minimum vertical exchange (high and low) are shown (Reiter et al., 1975, 1976, 1978). (4) Ny Ålesund, Spitsbergen, data from April and May 1979 (Heinzenberg et al. 1981). Averages for periods with and without air mass advection from the sea are shown. (5) Composite of data taken onboard ships in the central Atlantic ocean. Na, K, Ca, Mg, SO_4^{2-}, Cl from Buat-Menard et al. (1974), Na, K, Ca, Mg, Fe_2O_3 from Hoffman et al. (1974), Na, SO_4^{2-}, NH_4^+, NO^{3-} from Gravenhorst (1975), wind speed range 5–10 m s⁻¹. B⁻ is taken from aircraft data from Duce et al. (1965) in the area of Hawaii, extrapolated to the ocean surface.

[b] All calcium assumed to be soluble.

[c] Molar ratio.

[d] Corrected for sea salt.

submicrometer size range. This important fact was first recognized by Junge (1953, 1954), who had separated the aerosol into a coarse and a fine fraction. The distribution peaks at radii near 0.3 μm, that is, in the size range where the accumulation of material by condensation and coagulation processes is most efficient (cf. Section 7.3.1). There is also a contribution from in-cloud oxidation of SO_2 (see Section 8.6). The observed size distribution thus provides further evidence for the idea that gas-to-particle conversion of SO_2 is the principal source of sulfate over the continents. There are numerous reports in the literature on size distributions of aerosol sulfate similar to that in Figure 7.22; examples can be found in Lundgren (1970), Kadowaki (1976), Hoff et al., (1983), Wall et al. (1988), and Barrie et al. (1994).

The data in Table 7.14 and Figure 7.22 show further that ammonium is the principal cation associated with sulfate in the continental aerosol. The observed size distribution, maximizing in the accumulation mode, leaves no doubt that NH_4^+ arises from a gas-to-particle conversion process, namely, the neutralization of sulfuric acid by gaseous ammonia. Molar NH_4^+/SO_4^{2-} ratios are expected to range from 1 to 2, corresponding to a composition intermediate between NH_4HSO_4 and $(NH_4)_2SO_4$. Table 7.15 lists a number of observations that confirm this expectation. These data lead to values that exceed a value of 2 in some cases. This excess must be assigned to the presence of nitric acid, which also binds some ammonia. If nitrate is included in the ion balance, NH_4^+ is found to compensate about 80% of the acids in the continental aerosol. The right-hand side of Figure 7.22 shows the charge

TABLE 7.15 Some observed Molar NH_4^+/SO_4^{2-} Ratios for the Continental Aerosol

Authors	Location	NH_4^+/SO_4^{2-}	Comment
Kadowaki (1976)	Nagoya, Japan	1.3–1.5	Winter
		1	Summer
Moyers et al. (1977)	Tucson, Arizona	1.5	Average
Penkett et al. (1977c)	Gezira, Sudan	1.2	Average
Brosset (1978)	Gothenburg, Sweden	0.9–2.3	1.5 average
Tanner et al. (1979)	New York City	1.9	February
		1.2–1.5	August
Pierson et al. (1980)	Allegheny Mountains	0.5–2.3	Summer
		0.87	Average
Keller et al. (1991)	Newton, Connecticut	1.72	Summer average
Sirois & Fricke (1992)	Longwood, Ontario	2.15	Four-year average
	Algoma, Ontario	1.41	Four-year average
	Cree Lake, Saskatchewan	1.03	Four-year average
	Bay d'Espoir, Newfoundland	0.67	Four-year average
Mehlmann and Warneck (1995)	Deuselbach, Germany	1–2	Decreasing with rising $[SO_4^{2-}]$; average 1.18

balance between anions and cations. A cation deficit is evident in the size region where SO_4^{2-} and NH_4^+ have their maxima. The major cation missing in the analysis is H^+. Brosset (1978), Tanner et $al.$ (1979), and Pierson et $al.$ (1980) have used the Gran (1952) titration procedure on aqueous extracts of continental aerosols to determine concentrations of protons in addition to those of other cations. They found that including H^+ greatly improves the cation balance. This demonstrates that it is mainly uncompensated acid that accounts for the ammonium deficit, and not so much the presence of other cations. The degree of neutralization reached depends on the supply of ammonia relative to the rate of formation of H_2SO_4. Table 7.15 shows this most clearly in the data from Canada. At sites in agricultural and livestock areas (Longwood), the NH_4^+/SO_4^{2-} ratio was about 2; at rural sites further away from farming activities (Algoma) the ratio fell to between 1 and 2. The lowest ratios occurred at sites remote from human activities (Cree Lake, Bay d'Espoir). As Sirois and Fricke (1992) noted the NH_4^+/SO_4^{2-} ratio underwent an annual cycle, with larger monthly median values in summer. The cycle was mainly due to the variation in NH_4^+ concentration.

Sulfate in the Arctic aerosol is also assumed to derive from the oxidation of SO_2. Rahn and McCaffrey (1980) have studied the annual variation of sulfate at Barrow, Alaska, and found a pronounced winter maximum. This contrasts with the behavior of the continental aerosol, which shows a maximum in summer (cf. Section 10.3.4). From the covariance of sulfate with vanadium, which derives primarily from oil combustion, from the concentration of radon and its decay product ^{210}Pb, and from considerations of air mass trajectories, Rahn and McCaffrey (1980) concluded that the winter Arctic aerosol is associated with aged air masses originating in the middle latitudes, mainly in Europe and Russia. Their analysis suggests that sulfate is formed from SO_2 during transport to the Arctic. Note that the NH_4^+/SO_4^{2-} ratio according Table 7.14 is much lower in the Arctic than over the continental mainland. Evidently, there is even less ammonia available in these remoter regions to balance sulfuric acid. At Spitsbergen, however, Heintzenberg et $al.$ (1981) found $(H^+ + NH_4^+)/SO_4^{2-}$ ratios close to 2 when marine influence was absent, which indicates that also in remote regions ammonia is an important substance neutralizing sulfuric acid. The Antarctic, in contrast to the Arctic, is not polluted by anthropogenic sources. Indeed, Maenhaut and Zoller (1977) have shown that vanadium is not enriched compared with its crustal abundance. And yet, the concentration of sulfate in the Antarctic is as high as in the Arctic summer aerosol. In the Antarctic, sulfate has a natural source. It originates from the oxidation of gaseous sulfur compounds emitted in the surrounding marine environment.

Figure 7.23 shows mass size distributions of various components of the marine aerosol. The ionic composition is quantitatively different from that of

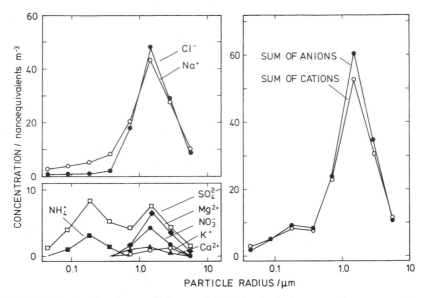

FIGURE 7.23   Size distributions ($\Delta \log r = 0.303$) of individual anions and cations associated with the marine aerosol collected over the Atlantic Ocean (nanoequiv m^{-3} = nmol m^{-3} × ionic charge number). Left: Individual contributions. Right: Balance between anions and cations (data from Mehlmann (1986)).

the continental aerosol. Sulfate in the marine environment exhibits a bimodal size distribution. The coarse particle fraction is associated with sea spray, whereas the submicrometer size fraction may again be assigned to gas-to-particle conversion. Table 7.16 summarizes the major and some minor constituents of seawater. The main components are sodium, chloride, magnesium, sulfate, potassium, and calcium. With the exception of sulfate, these components are all concentrated in the coarse particle mode of the marine aerosol, as expected. The fraction of sulfate originating from sea salt may be estimated by reference to sodium. In seawater the mass ratio of sulfate to sodium is 0.252. The molar equivalence ratio is $[\frac{1}{2}SO_4^{2-}]/[Na^+] = 0.12$. The concentration of sulfate in the coarse-particle mode of the data in Figure 7.23 is somewhat higher, but it varies with magnesium and potassium, as required for sea salt being the major source. Sulfate in the submicrometer size range is associated with ammonium. The charge balance between cations and anions is good, even though NH_4^+ contributes less than one-half to the cation balance in the submicrometer size range. The remainder is made up of sodium. The marine aerosol evidently is less acidic than the continental aerosol.

TABLE 7.16 Major and Some Minor Constituents of Seawater and Marine Aerosols[a]

Element or ion	Seawater Abundance ($mg\ kg^{-1}$)	$(X)/(Na)$	Marine aerosol $(X)/(Na)$	EF
Cl^-	19,344	1.795	1.590	0.88
SO_4^{2-}	2,712	0.252	0.885	3.51
Br^-	67	0.0062	0.0051	0.82
HCO_3^-	142	0.0132	—	—
Na^+	10,773	1	1	1
Mg^{2+}	1,294	0.119	$0.138\ (0.127)^b$	1.16
Ca^{2+}	412	0.0384	$0.058\ (0.046)^b$	1.50
K^+	399	0.0370	$0.037\ (0.036)^b$	1.00
Sr^{2+}	7.9	0.00073	$0.0014^b\ (0.0006)^b$	1.93^b
Si	3	2.8×10^{-4}	—	—
Al	$\sim 10^{-2}$	9.3×10^{-7}	—	—
Fe	$\sim 10^{-2}$	9.3×10^{-7}	1.5×10^{-2}	1.7×10^4
F^-	1.3	1.2×10^{-4}	—	—
I^-	5.4×10^{-2}	5.0×10^{-6}	—	—

[a] Seawater abundances from Wilson (1975), aerosol data from Table 7.14. The $(X)/Na$) mass ratio serves to determine the enrichment factor $EF = (X)/Na)_{aerosol}/(X)/Na)_{seawater}$.
[b] From Hoffman et al. (1974).
Values in parentheses are linear regression data.

The major gaseous precursor of sulfate in the unperturbed marine environment is dimethyl sulfide, a biogenic compound emanating from the sea's surface. Oxidation pathways for dimethyl sulfide will be discussed in Section 10.2.2. A large fraction of it is converted to SO_2, which is subsequently further oxidized to sulfuric acid. Figure 7.23 indicates a deficit of chloride in the submicrometer size range relative to sodium. The charge neutrality observed despite the shortage of ammonium suggests that sulfuric acid liberates some chloride and protons in the form of gaseous HCl. The submicrometer chloride loss was first reported by Martens et al. (1973), but the total deficit had been observed earlier in filter samples (Wilkness and Bressan, 1972). The effect is now well documented. Graedel and Keene (1995) have summarized the data to show that the total chloride deficit may reach 40%. Kolaitis et al. (1989) have analyzed single particles for several elements. The Cl/Na mass ratio in submicrometer-sized particles was about 0.5, whereas coarse particles had a Cl/Na mass ratio near 1.7, close to that of sea salt (1.79). On the other hand, Barrie et al. (1994) noted that part of the submicrometer loss may be an artifact occurring in low-pressure cascade impactors. By means of filters impregnated with LiOH, which absorbs HCl but allows organic chlorine compounds to pass, Rahn et al. (1976) demonstrated the presence of inorganic chlorine in marine and continental air.

Kritz and Rancher (1980) used the same technique to confirm the presence of inorganic gaseous chlorine in the Gulf of Guinea. The observed average concentration was 1 μg m^{-3}, or about one-fifth of that present in particulate form. The corresponding mixing ratio of HCl would have been 0.6 nmol mol^{-1}. Kritz and Rancher (1980) assumed all of it to derive from sea salt and, adopting the steady-state hypothesis, estimated flux rates and residence times (3 days for sea salt, $2\frac{1}{2}$ days for HCl). More recently, Keene et al. (1990) observed proportionality between gaseous HCl concentration and the chloride deficit on particles, especially under anthropogenic influence at a coastal station. Vierkorn-Rudolph et al. (1984) have used an organic derivative technique to determine the HCl mixing ratio in the troposphere west of the European continent. They found about 0.5 nmol mol^{-1} in the boundary layer and 0.05 nmol mol^{-1} in the upper troposphere. Although hydrochloric acid probably is the most abundant inorganic gaseous chlorine species in the marine environment, other compounds have also been considered, including Cl_2 (Cauer, 1951; Finlayson-Pitts et al., 1989; Keene et al., 1990; Pszenny et al., 1993), NOCl (Junge, 1963; Finlayson-Pitts, 1983), and $ClNO_2$ (Finlayson-Pitts et al., 1989; Behnke et al., 1992, 1994). These gases undergo photolysis at wavelengths above 300 nm. Photodecomposition liberates Cl atoms, which sooner or later react with methane or other hydrocarbons to form HCl as a stable end product.

The possibility that the depletion of chloride in the marine aerosol is due to fractionation during the formation of sea salt particles by bursting bubbles can be discounted. Laboratory studies of Chesselet et al. (1972) and Wilkness and Bressan (1972) have shown no deviations of the Cl^-/Na^+ mass ratio in the bubble-produced sea salt particles compared to that of seawater. It may be mentioned that bromide in the marine aerosol exhibits a deficit similar to that of chloride, whereas iodide is present in excess. The latter observation is attributed to chemical enrichment at the sea's surface as well as scavenging of iodine from the gas phase. A portion of iodine is released from the ocean as methyl iodide. In the atmosphere the substance is attacked by OH radicals, and it undergoes photodecomposition, thereby liberating iodine that ultimately is scavenged by aerosol particles. Duce and Hoffman (1976) have reviewed this process. In continental aerosols, chloride and bromide are partly remnants of sea salt, but there is also a contribution from the gas phase. Hydrochloric acid, for example, is released from coal-fired power plants, and it is the end product of oxidation of methyl chloride and other organic chlorine compounds.

The last ion to be discussed is nitrate. Savoie and Prospero (1982) have shown that nitrate in the marine aerosol is associated mainly with coarse particles, and Figure 7.23 confirms this observation. Seawater contains only small amounts of nitrate, so that the nitrate content of the marine aerosol must derive from the gas phase. Gaseous nitric acid is one precursor whose interaction with sea salt is known to form sodium nitrate, liberating HCl in the process (Brimblecombe and Clegg, 1988). Another precursor is N_2O_5, which occurs as a product of the reaction of ozone with NO_2 at night (cf. Section 9.4.1). N_2O_5 is the anhydride of HNO_3, and it reacts with aqueous aerosol particles to form either HNO_3 directly or with aqueous chloride to form $ClNO_2$ and NO_3^- (Finlayson-Pitts et al., 1989; Behnke et al., 1994). Both processes thus contribute to the chloride deficit in the marine aerosol. If the interaction with chloride were dominant, nitrate would become associated primarily with the sea salt fraction, which represents the most effective surface. This is indeed observed (Savoie and Prospero, 1982). Nitric acid does not appear to condense in significant amounts in the size region of the accumulation mode as one would expect, presumably because the volatility of nitric acid is greater than that of H_2SO_4, which dominates this size range.

Nitrate in the continental aerosol is distributed over the entire $0.1-10$-μm size range. A bimodal size distribution is often observed. The first maximum occurs in the accumulation mode, where NO_3^- is associated with ammonium; the second appears in the coarse-particle mode, where sodium, potassium, and earth alkaline elements are the principal cations. The phenomenon was first described by Junge (1963) but has repeatedly been observed since (see, for example, Kadowaki, 1977; Wolff, 1984; Wall et al., 1988, Mehlmann and Warneck, 1995). Kadowaki (1977) observed a seasonal variation in the urban area of Nagoya, Japan, with a clear preference for the coarse-particle mode in summer and for the accumulation mode in winter. In coastal areas, the preference for coarse particles appears to derive from the association of nitrate with sea salt. However, nitrate in coarse particles has also been observed in continental aerosols in the absence of any marine influence (Wolff, 1984). The origin of nitrate in the coarse-particle size range must be the same in the two cases, that is, the interaction with nitric acid or N_2O_5 displaces HCl or $ClNO_2$ in the case of sea salt, and CO_2 from carbonates in mineral dust particles. This leads to stable nitrate salts that remain permanently attached to the aerosol particles. Ammonium nitrate, in contrast, particularly when dry, is more volatile and tends to establish an equilibrium with the vapor phase precursors nitric acid and ammonia (Stelson and Seinfeld, 1982; Mozurkewich, 1993). For particulate NH_4NO_3 to coexist with HNO_3 and NH_3, the product of their partial pressures must be greater than the equilibrium constant $P_{HNO_3} P_{NH_3} \geq K_{equ}$. The equilibrium constant is strongly temperature dependent. Thus, particulate ammonium nitrate will occur either when the equilibrium vapor pressures are high or when temperatures are low, or when both conditions exist. The low abundance of ammonia often is a limiting factor causing NH_4NO_3 to disappear from the accumulation mode. Figure 7.24 shows data from a rural measurement site

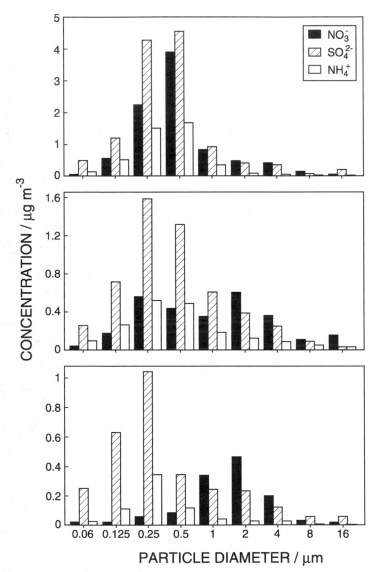

FIGURE 7.24 Size distributions of sulfate, nitrate, and ammonium for three aerosol mass concentration ranges observed in Deuselbach, West Germany. (Adapted from Warneck (1994).) Note the shift in the bimodal distribution of nitrate with increasing concentration.

to demonstrate that nitrate in the accumulation mode grows in as the aerosol loading increases. It appears, therefore, that nitrate in the coarse-particle mode is always present, albeit in fairly low concentration, whereas nitrate in the accumulation mode is an indicator of more polluted atmospheric conditions.

We turn next to consider the nonvolatile alkali and alkaline earth elements and the insoluble components of mineral origin. Their major natural sources are the Earth's crust and the ocean, respectively. The chemical composition of the aerosol is expected to reflect the relative contributions of elements from both reservoirs, provided other contributions from volcanic or anthropogenic sources are negligible. In Section 7.4.4 it has been noted, however, that this premise does not hold for all constituents of the aerosol. Some trace elements are considerably enriched compared to their crustal abundances. It is appropriate, therefore, to inquire whether the observations confirm our expectations at least for the major elements listed in Table 7.14, or whether deviations also occur in these cases. As Rahn (1975a, b) has shown, the problem may be approached in two ways, either by calculating enrichment factors defined by

$$EF(X) - (X)/(Ref)_{aerosol}/(X)/(Ref)_{source} \qquad (7.28)$$

where X is the element under consideration and Ref an appropriate reference element, or by constructing so called scatter diagrams in which the observed data for the element X are plotted versus those for the reference element. Examples of both procedures will be given.

An appropriate choice of reference elements is required, as well as a table for the elemental composition of the source materials. Elements that are useful as a reference for crustal material include silicon, aluminum, iron, and titanium. All of them are minor constituents of seawater, whereas they are abundant in rocks. Silicon presents some analytical difficulties, so that it is not regularly determined. For this reason, aluminum is most frequently used as the reference element for the mineral component. The best reference element for seawater is sodium. Chlorine has been used occasionally, but in the aerosol it is subject to modification, as we have seen, so it should be avoided. We consider the marine aerosol first and the continental aerosol subsequently.

Table 7.16, which was introduced earlier, compares relative element abundances in the marine aerosol with that of seawater and shows the resulting enrichment factors. The deficit for chlorine and bromine and the enrichment of sulfate have been discussed previously. The relative abundances of magnesium, calcium, potassium, and strontium show some enrichment. Early observations indicated even higher enrichments than those shown in Table 7.16, and reports of EF(X) ≈ 10 were not uncommon. This led to the unfortunate and very confusing suggestion that alkali and alkaline earth elements undergo fractionation during the process of sea salt formation. The early studies, however, were conducted at coastal sites without proper allowance for contamination by nonmarine sources. Hoffman and Duce (1972) first showed from measurements on Oahu, Hawaii, that the enrichment is due to the admixture of materials from the Earth's crust. By classifying the data according to wind direction, they found much higher EF(X) values when winds were variable and had partly passed the island compared to strictly on-shore wind conditions. In the latter case the mean element abundance ratios matched those of seawater, and calcium was the only element that showed a statistically significant positive deviation. Figure 7.25 shows scatter diagrams for magnesium and calcium according to Hoffman and Duce (1972). A linear regression analysis leads to the equations

$$c(\text{Mg}) = 0.002 + 0.117 \, c(\text{Na})$$
$$c(\text{Ca}) = 0.021 + 0.037 \, c(\text{Na})$$

The slope of each line corresponds closely to the respective seawater ratio. For calcium, a positive intercept of the regression line with the ordinate indicates the presence of a nonmarine component. It should not be assumed to be constant with time, but it appears that the variation is buried in the statistical scatter of the data. It must be further emphasized that the points in the scatter diagram do not necessarily fall on a straight line. A linear

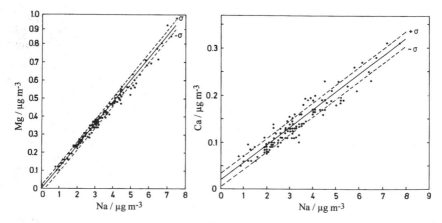

FIGURE 7.25   Scatter diagrams for magnesium and calcium versus sodium in marine aerosols (Adapted from Hoffman and Duce (1972).) Note that the slopes of the regression lines are similar to the mass ratios in seawater (Mg/Na = 0.119, Ca/Na = 0.038), but that the regression line for calcium does not meet the origin of the plot.

correlation will be obtained only if the contribution from one source exceeds that of all others.

Hoffman *et al.* (1974) found the same procedure applicable to ship-board data obtained over the central Atlantic Ocean. Table 7.16 includes average (X)/(Na) ratios from their study. Shown in parentheses are the values derived from the slopes of regression lines. They are distinctly lower than the averaged data. Hoffman *et al.* (1974) also measured the abundance of iron in the marine aerosol. Since the samples were taken in a region partly affected by fall-out of Saharan dust, iron serves as a convenient indicator for the contribution from continental sources. Not surprisingly, the enrichment of the elements Mg, Ca, K, and Sr was well correlated with the iron content. The (X)/(Na) ratios approached those of seawater only when the iron concentrations were very low. These results demonstrate that materials from both marine and continental sources are present over the open ocean. In addition, they confirm the absence of a significant fractionation of alkali and alkaline earth elements in the production of sea salt.

The contribution of crustal elements to the continental aerosol is not easily determined because the composition of material resulting from the wind erosion of soils is not precisely known. Usually a surrogate composition is used. Possibilities include the average composition of crustal rock, bulk soil, or the aerosol-size fraction of the soil. Although the last choice would have advantages, there are not enough data available to construct a useful average. Most investigations have used either globally averaged rock or soil, but neither is entirely satisfactory as a reference source. Rock has not yet undergone the weathering process, and the composition of bulk soil differs from that of particles in the size range of the aerosol. Rahn (1975b) also considered shale as a reference material, with the argument that clays and silts might better approximate soil substances susceptible to aerosol formation. Table 7.17 shows relative abundances of the most frequent elements in crustal rock, bulk soil, and average shale, in addition to fly ash from the combustion of fossil fuels, which is believed to represent an important source of particulate material in densely populated regions. Compared to crustal rock, soils and shales are somewhat depleted with regard to water-soluble elements; otherwise they are fairly similar. Fly ash is depleted in silicon, but not in the other elements.

The third section of Table 7.17 shows the relative composition of two continental aerosols to provide a direct comparison with the data in the second part of the table. The first aerosol represents an average over samples taken at remote sites in Europe, North America, and Africa (Rahn, 1975b). The second aerosol is a 1974 average obtained by Moyers *et al.* (1977) at two sites in rural Arizona. The (X)/(Al) ratios for the elements Si, Fe, Na, K, and Ca in the aerosols presented in Table 7.14 are shown near the bottom of that

TABLE 7.17 Average Absolute and Relative Abundances of Major Elements in Crustal Rock, Soil, and Shale; Relative Abundances of Elements in Fly Ash from Coal and Fuel Oil Combustion; and Relative Abundances of Major Elements in the Remote Continental Aerosol, with Enrichment Factors $EF = (X)/(Al)_{aerosol} / (X)/(Al)_{crustal\ rock}$ [a]

| Element | Elemental abundance (mg kg^{-1}) | | | | Relative composition (X)/(Al) | | | Fly ash (e) | | Remote continental aerosol | | | |
	Soil (a)	Crustal rock (b)	(c)	Shale (d)	Soil (a)	Rock (c)	Shale (d)	Coal	Oil	(X)/(Al) (f)	(g)	EF (f)	(g)
Si	330,000	277,200	311,000	281,900	4.628	4.018	2.591	1.43	0.48	2.7	3.25	0.67	0.81
Al	71,300	81,300	77,400	108,800	1	1	1	1	1	1	1	—	—
Fe	38,000	50,000	34,300	48,400	0.533	0.443	0.445	0.5	0.5	1.3	0.55	2.9	1.24
Ca	13,700	36,300	25,700	22,500	0.192	0.332	0.207	0.29	0.08	0.80	0.66	2.4	1.99
Mg	6,300	20,900	33,000	11,700	0.088	0.426	0.107	0.057	0.06	0.25	0.15	0.58	0.35
Na	6,300	28,300	31,900	9,720	0.088	0.426	0.089	0.029	0.30	0.20	0.23	0.48	0.56
K	13,600	25,900	29,500	32,000	0.191	0.381	0.294	—	0.02	0.65	0.44	1.70	1.15
Ti	4,600	4,400	4,400	4,440	0.064	0.057	0.041	0.064	6×10^{-3}	0.064	0.083	1.12	1.45
Mn	850	950	670	420	0.012	8.6×10^{-3}	3.8×10^{-3}	1.7×10^{-3}	6×10^{-4}	0.025	0.017	2.9	1.98
Cr	200	100	48	80	2.8×10^{-3}	6.2×10^{-4}	7.3×10^{-4}	2.1×10^{-3}	0.024	8×10^{-3}	2.6×10^{-3}	12.9	9.2
V	100	135	98	106	1.4×10^{-3}	1.3×10^{-3}	9.7×10^{-4}	5.7×10^{-3}	0.5	3.5×10^{-3}	—	2.7	—
Co	8	25	12	13	1.1×10^{-4}	1.5×10^{-4}	1.2×10^{-4}	6.4×10^{-4}	0.03	6.5×10^{-4}	5.8×10^{-4}	4.3	3.9

[a] Sources: (a) Vinogradov (1959); (b) Mason (1966); (c) Turekian (1971); (d) Flanagan (1973), model A; (e) Winchester and Nifong (1971); (f) Rahn (1975b); (g) Moyers et al. (1977).

table. Figure 7.26 shows a number of scatter diagrams to further illustrate the behavior of important elements in the atmospheric aerosol. The combination of these data is discussed below.

Table 7.17 demonstrates that the relative abundances of the major elements in the aerosol do not differ greatly from those in bulk soil, crustal rock, or average shale—that is, the elements are neither greatly enriched nor seriously depleted, although a good match with any of the three reference materials is not obtained. The differences are greater than conceivable analytical errors. Consider silicon, for example. The average Si/Al ratio indicated in Tables 7.14 and 7.17 is 2.7, which is lower than that for either bulk soil or crustal rock and is more like that in shales. Fly ash exhibits a particularly low Si/Al ratio. It is possible that the low Si/Al ratio in the aerosol of heavily industrialized Tees-side (Table 7.14) is caused by a mixture of natural and combustion aerosols, but this explanation cannot be extended to the remote continental aerosol. A more likely explanation for the deficiency of silicon is the size distribution of the Si/Al ratio in soil particles. The very coarse quartz particles, which are rich in silicon, are not readily mobilized. Since only the fine fraction of soil particles contributes to aerosol formation, the Si/Al ratio in the aerosol will be determined by that of silts and clays (see Table 7.7 for definitions). Common clay minerals include kaolinite, $Al_4Si_4O_{10}(OH)_8$; montmorillonite, $Al_4(Si_4O_{10})_2(OH)_4 \cdot H_2O$; muscovite, $K_2Al_4(Si_6Al_2)O_{10}(OH)_4$, and chlorite, $Mg_{10}Al_2(Si_6Al_2)O_{20}(OH)_6$. The corresponding Si/Al mass ratios are 1.04, 2.08, 1.04, and 1.56, respectively. Frequently, however, aluminum is partly replaced by other elements, whereby the Si/Al ratio is raised again. Schütz and Rahn (1982) have essentially confirmed this concept. They investigated the size distribution of the Si/Al ratio in various soils and found particles with radii larger than 100 μm enriched by a factor of 10, whereas toward smaller sizes the enrichment decreased steadily until a constant value of about 0.3 was reached in the submicrometer size range.

Iron, in contrast to silicon, is slightly enriched in the atmospheric aerosol, EF(Fe) $\approx$ 1.8. The Fe/Al ratio at Tee-side shown in Table 7.14 is higher and must be attributed to the vicinity of steelworks in that region. The scatter diagram for iron shown in Figure 7.26 includes data from continental, coastal, and strictly marine locations. With few exceptions, the points are seen to parallel the lines representing Fe/Al ratios in soil and crustal rock. Occasionally a site is found where Fe and Al occur in crustal proportions, especially in dust samples from the Sahara Desert (Guerzoni and Chester, 1996). At other locations the scatter of the data must arise from pollution or the variance of iron in different soils. The average Fe/Al ratios in soil, crustal rock, shale and fly ash are nearly equal, however. The same situation applies to titanium. In this case, however, the scatter diagram in Figure 7.26

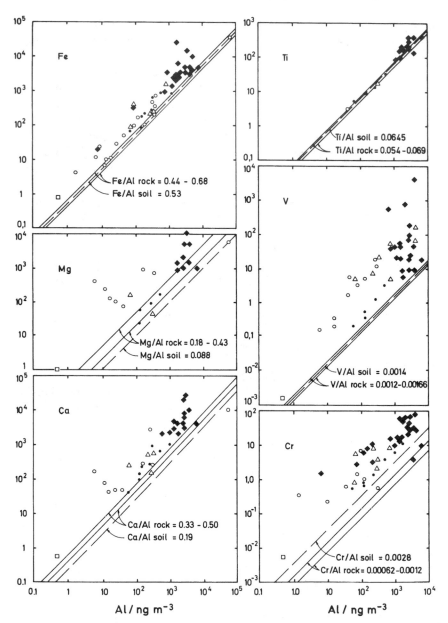

FIGURE 7.26 Scatter diagrams for iron, titanium, magnesium, calcium, vanadium, and chromium in the atmospheric aerosol versus aluminum as reference element (concentrations in ng m^{-3}). Symbols indicate the type of sampling site: $\bigcirc$, marine; $\triangle$, marine-influenced; $\blacklozenge$, urban-continental; $\bullet$, remote continental; $\square$, Antarctic. The range of X/Al ratios in crustal rock is shown by the solid lines, and X/Al ratios in soil by the dashed lines. (Adapted from Rahn (1975b).)

shows a much narrower distribution, and the points follow closely the line representing the crustal Ti/Al ratio. Titanium obviously would serve as a good crustal reference element. Unfortunately, it presents analytical difficulties, and the data carry larger uncertainties than those for many other elements.

Table 7.17 shows that the elements Na, K, Ca, and Mg in soils and shales are depleted in comparison with crustal rock. The depletion is higher for sodium and magnesium than for potassium and calcium. The abundance of these elements in the aerosol matches none of the reference materials, however. The relative abundance of K and Ca is higher than and those for Na and Mg are between those in soil and crustal rock. It is interesting to note that the ratios Ca/Mg and K/Na in soils are approximately preserved. This result suggests a common soil aerosol enrichment process. Schütz and Rahn (1982), who investigated elemental abundances as a function of soil particle size for several desert soils, found no evidence for an increase in the ratios Na/Al, K/Al, Mg/Al, or Ca/Al in particles smaller than 20 μm when compared to larger particles. Clay minerals cannot be responsible for the enrichment. The composition of shales also argues against this possibility. Rahn (1975b) has pointed out that soil humus, which usually is neglected in such considerations, might influence the elemental composition of the aerosol. Humus is a ubiquitous, albeit variable, constituent of soils. The active surface of humus exceeds that of clay minerals; both have a great capacity for binding cations. Being generated from decaying plant debris, humus receives the elements in proportions similar to those in which plants incorporate them from the soil. Since the roots of plants reach down into subsurface layers of the soil, whereas aerosol particles are formed from the top layer, a certain fractionation and enrichment are expected. The desert soils explored by Schütz and Rahn (1982) presumably contained fairly little humus, so that their results should not be generalized.

Figure 7.26 includes scatter diagrams for magnesium and calcium as examples for two elements that can be leached from soils. These data are selected to provide comparison with Figure 7.25 and Table 7.14. Airborne magnesium enables a clean distinction between marine and continental sites. The former fall to the left of the line representing Mg/Al ratios for continental soils and crustal rock; the latter scatter around these lines. Sodium, which is not included in Figure 7.26, shows an almost identical behavior. The scatter diagram for calcium differs from that for magnesium in that marine and continental data points intermingle, making it difficult to identify sites where the marine source dominates. As discussed earlier, calcium is enriched compared with its abundance in sea salt, even in mid-ocean regions. The strong correlation of calcium with aluminum demonstrates the significance of continental sources in the atmospheric background aerosol. The average

Ca/Al ratio in Figure 7.26 is 0.8. Similar values occur at Tees-side and West Covina (see Table 7.14), whereas the Ca/Al ratio at the Wank Mountain site is much higher. This station must be influenced by the preponderance of limestone in the Alpine region.

The remaining elements listed in Table 7.17 have crustal abundances, relative to aluminum, of 1% or less, so that they must be considered trace elements. All of the other elements not listed in the table, with the exception of oxygen, fall into the same category. Vanadium and chromium may be briefly discussed. Vanadium is of interest because fly ash from fuel oil combustion contains disproportionately high amounts. The V/Al scatter diagram indicates two bands of points, each paralleling the lines representing V/Al ratios of crustal rock and soil. In the upper band, vanadium is by an order of magnitude more enriched than in the lower band. The data classified as remote continental lie close to the crustal ratio, even though they are slightly higher. The enrichment factor in this case is 2.7, which is similar to that for other nonvolatile elements such as iron. The high enrichment of vanadium in the upper band must be ascribed to pollution. The greatest effects clearly occur in the cities, but the observation of high enrichment at marine sites of the Northern Hemisphere demonstrates that vanadium from anthropogenic sources also affects the tropospheric background aerosol. The Arctic aerosol is influenced by pollution, as Rahn and McCaffrey (1980) have shown. The Antarctic, however, is not thus affected (Zoller et al., 1974), indicating that neither anthropogenic nor volcanic sources cause an enrichment of vanadium there.

The case of chromium is of interest because the Cr content of soils is higher by a factor of at least 2 than that of crustal rock. The Cr/Al ratio observed in aerosols is also consistently higher than that of either soil or crustal rock. The data scatter broadly without distinction between marine or continental sites. Some of the high values are found in regions with iron and steel industries as well as in the cities. Moyers et al. (1977) have found markedly higher values of chromium, cobalt, and nickel in urban Tucson, Arizona, compared to the surroundings. Data from remote continental sites indicate an average enrichment factor of 12.9, which is the highest for any of the elements listed in Table 7.17, although it is not as high as in urban regions. If, however, soil were used as the reference material, the enrichment factor would reduce to 2.8, comparable to that for the other elements discussed here. In this respect, chromium evidently does not behave exceptionally. The behavior of manganese and cobalt is very similar to that of chromium and needs no further discussion.

Various other elements contained in the atmospheric aerosol are considerably more enriched than the elements treated here. Figure 7.20 shows large enrichment factors for the so-called volatile elements in the remote troposphere not subject to anthropogenic pollution. The arid soils studied by Schütz and Rahn (1982) revealed enrichment factors up to 10 for elements such as Cu, Zn, As, and Sb, that are fairly independent of soil particle size. Ag and Au were the only elements for which they found enrichment factors increasing with decreasing particle size, with values approaching 500 and 50, respectively, in the submicrometer size range. The enrichment in the aerosol is greater. Silver in the aerosol is the only element whose enrichment factor comes closest to being explicable in terms of submicrometer soil particle enrichment. Soil differentiation clearly does not provide a mechanism for the abnormally high enrichment of these elements in the aerosol.

Rahn (1975a, b) has summarized data for many highly enriched elements. From a limited set of mass–size distributions, Rahn (1975a) and subsequently Milford and Davidson (1985) have calculated mass median diameters for each element. This quantity defines the size for which one-half of the mass of an element is associated with smaller and the other half with larger particles. Elements occurring in nearly crustal proportions were found to have mass median diameters in the size range beyond 2 μm, as expected for soil-derived elements. Most of the elements occurring with abnormally high enrichments were found in the submicrometer size range, among them Cu, Zn, As, Sb, Sn, In, Pb, Cd, Se, S, and Hg. This distribution suggests that highly enriched elements have passed through the gas phase at some point in their history and entered the aerosol phase by condensation. As discussed previously, volcanoes and combustion processes are prominent sources of volatile elements. Another process discussed in Section 7.4.4 is biological enrichment. As Wood (1974) and Ridley et al. (1977) have pointed out, trace metals such as As, Se, S, Pd, Tl, Pb, Pt, Au, Sn, and Hg serve as acceptors of methyl groups in microbiological pathways leading to the emission of metal alkyls into the atmosphere. The gas phase oxidation of these compounds then forms metal oxides, which attach to the atmospheric aerosol.

This section may be summarized as follows. The inorganic compositions of the marine and the continental aerosols differ considerably from the underlying source materials. Over the ocean, the sea salt aerosol is modified by condensation processes and by the admixture of the tropospheric background aerosol. One finds a deficit of chloride and bromide, and a surplus of sulfate, ammonium, and nitrate in addition to certain elements of crustal origin. The continental aerosol contains large amounts of sulfate ammonium, nitrate, and possibly chloride, all resulting from gas-to-particle conversion. In contrast to sea salt, which essentially represents the composition of seawater, the elemental composition of the soil-derived fraction of the continental aerosol differs appreciably from that of crustal rock, average soil, or shale. The relative abundances of the major elements differ individually by factors of about 3. A number of minor elements are enriched by several orders of

magnitude. The behavior is globally widespread, but an unambiguous interpretation is not at hand.

7.5.2. The Organic Fraction

Solvent extraction combined with chromatographic techniques is used to
study organic compounds associated with the atmospheric aerosol. Combustion to CO_2 is additionally used to determine total organic carbon. This
includes carbon present as black carbon or soot. Because of the limited
efficiency of solvent extraction and because many compounds showing up in
the chromatograms either are unresolved or cannot been identified, our
knowledge of the composition of the organic fraction is still unsatisfactory,
even though hundreds of different compounds have been detected. The data
of Rogge et al. (1993a), who examined the organic fraction of aerosols
collected at four urban and suburban sites near Los Angeles, may illustrate
the difficulties. At West Los Angeles, for example, total fine particle mass
averaged 24.5 μg m^{-1}. About 40% of it was ammonium, sulfate, and nitrate;
17% represented mineral and other inorganic components; 15% was elemental carbon; and 29% was organic matter. About 53% of the last fraction
could be extracted and eluted, but only 25% of it, that is, 13% of the total
organic material, was resolved in the chromatograms. Although more than 80
compounds were identified, 25% of the components in the resolved fraction
remained unidentified.

Table 7.18 compares concentrations of ether-extractable organic matter in
the aerosol with that of total particulate matter at several locations. The
efficiency of ether extraction is about 50%. The organic compounds are
broadly characterized into five groups: aliphatic, aromatic, and polar neutral
compounds; organic acids; and organic bases. The locations may be classified
as urban, rural, remote continental, and marine. While the absolute concentrations decrease as one goes from urban to remote continental and/or
marine regions, the relative composition of the individual fractions remains
remarkably uniform. Neutral components contribute about 56%, acids 36%,
and bases 8% to the total material. Our knowledge is still inadequate to
determine whether this uniformity is due to a common origin of particulate
organic matter or processes in the atmosphere that produce or modify the
organic content of the aerosol.

Table 7.19 shows concentrations of organic material associated with the
marine aerosol. The total concentration is rather uniform with an average of
0.42 μg m^{-3}. As the residence time of the aerosol in the troposphere is
about 1 week, Eichmann et al. (1980) concluded that continental sources
alone are inadequate to maintain such a uniform concentration, and that

TABLE 7.18   Average Concentrations ($\mu g\ m^{-3}$) of Total Particulate Matter (TPM), Ether-Extractable Organic Matter (EEOM), and Groups of Neutral Compounds, Organic Acids, and Organic Bases, from Measurements at Various Locations (Hahn, 1980; Eichmann et al., 1980)

| Location | TPM | EEOM | Neutral compounds | | | Organic acids | Organic bases |
			Aliphatic	Aromatic	Polar		
Mainz, Germany	150	27.1	6.3	3.1	4.9	10	2.0
Deuselbach, Germany	12	2.0	0.53	0.26	0.35	0.7	0.2
Jungfraujoch, Switzerland	17.6	0.9	0.26	0.12	0.2	0.2	0.03
North-Atlantic Ocean (shipboard)	16.7	0.8	0.2	0.16	0.1	0.23	0.03
Loop Head, Ireland	20.5	0.73		0.38[a]		0.25	0.10
Cape Grim, Tasmania	31	0.5		0.23[a]		0.18	0.08

[a] Sum of neutral compounds.

TABLE 7.19   Concentrations of Organic Carbon Associated with the Marine Aerosol[a]

Location	Sample number	Concentration ($\mu g\ m^{-3}$)	Authors
Bermuda	8	0.29 (0.15–0.47)	Hoffman and Duce (1974)
Bermuda	8	0.37 (0.15–0.78)	Hoffman and Duce (1977)
Tropical North Atlantic	5	0.59 (0.33–0.93)	Ketseridis et al. (1976)
Hawaii	7	0.39 (0.36–0.43)	Hoffman and Duce (1977)
Eastern Tropical North Pacific	3	0.49 (0.22–0.74)	Barger and Garrett (1976)
Samoa	9	0.22 (0.13–0.41)	Hoffman and Duce (1977)
Eastern Tropical South Pacific	9	0.32 (0.07–0.53)	Barger and Garrett (1976)
Loop Head, Ireland	6[b]	0.57 (0.28–0.86)	Eichmann et al. (1979)
Cape Grim, Tasmania	6[b]	0.53 (av.)	Eichmann et al. (1980)
North Pacific	21	0.05–2.5[c]	Rau and Khalil (1993)
South Pacific	5	~0.50 (av.)	Rau and Khalil (1993)

[a] Taken partly from the compilation of Duce (1978).

[b] Samples taken in pure marine air.

[c] Some anthropogenic influence.

these should be supplemented by marine sources or gas-to-particle conversion processes. A uniform distribution, however, would be in harmony with the concept of a uniform background aerosol, whatever its source. Hoffman and Duce (1977) and Chesselet et al. (1981) have shown that the size distribution of organic carbon in the marine aerosol favors the submicrometer size range, in agreement with the idea that the material stems from gas-to-particle conversion. Chesselet et al. (1981) have studied not only the size distribution, but the $^{13}C/^{12}C$ isotope ratio as well. For the smallest particles the ratio was close to that for carbon of continental origin ($\delta^{13}C = 26 \pm 2‰$); that for the largest particles was lower ($\delta^{13}C = 21 \pm 2‰$), indicating that this component derived from the ocean and presumably was associated with sea salt. Chesselet et al. (1981) concluded that about 80% of organic matter in the marine aerosol was due to gas-to-particle conversion involving precursors originating on the continent. The data of Chesselet et al. (1981) were obtained from samples taken in the Sargasso Sea and at Enewetak Atoll. Both sites, especially Enewetak, are influenced by long-range transport of continental aerosols.

Straight-chain organic compounds from terrestrial and marine sources may be distinguished by the relative abundance of members containing odd and even numbers of carbon atoms. The n-alkane fraction released from land-based vegetation features a distinct preference for odd carbon numbers in the C_{19}–C_{35} size range (Simoneit, 1989), whereas n-alkanes from marine biota do not exhibit any carbon number preference. Petroleum-derived hydrocarbons from oil spills and other anthropogenic sources also do not show any carbon number preference. The detection of this feature thus is an indicator for material originating from terrestrial vegetation. Eichmann *et al.* (1979, 1980) have studied *n*-alkanes in pure marine air masses at Loop Head, Ireland, and Cape Grim, Tasmania. The results show some odd-over-even carbon number preference in the aerosol phase but not in the gas phase (see Fig. 6.7). The terrestrial contribution was estimated as 10–25%. Simoneit (1977) found a much higher contribution near the West African coast in particles larger than 1 μm, which evidently derived from the continent. Marty and Saliot (1982), who studied the aerosol size distribution of *n*-alkanes, found an increase in the odd-to-even carbon number index in particles larger than 1 μm. This suggests that larger particles over the Atlantic Ocean are of continental origin, but such a conclusion would contrast with that of Chesselet *et al.* (1981) for the Pacific Ocean, derived from the $^{13}C/^{12}C$ isotope ratio. Gagosian *et al.* (1982) have studied $C_{21}-C_{36}$ *n*-alkanes, in addition to aliphatic alcohols and acids, associated with the aerosol collected at Enewetak Atoll. Here again the observed carbon number preference indicated a considerable fraction of continental origin. In this case it was shown that the terrestrial fraction correlated well with other indicators for continental air mass, such as the elements Al and ^{210}Pb. From air-mass trajectory analyses it appears that Enewetak was in the range of the Asian dust plume at the time the aerosol samples were taken.

It should be noted that *n*-alkanes in the marine atmosphere occur mainly in the gas phase. The data of Eichmann *et al.* (1979, 1980) suggest that no more than 5% of the *n*-alkanes with carbon numbers below C_{28} are associated with the marine aerosol. The abundance is so low that actual vapor pressures do not reach the saturation point.

Table 7.20 gives a condensed summary of organic compounds that have been identified in rural continental and marine aerosols. Table 7.21 shows the most abundant compounds in urban aerosols. The urban situation will not be discussed in detail. Reference is made to the reviews of Daisey (1980), Simoneit and Mazurek (1981), and the recent study of Rogge *et al.* (1993a), which describes the situation in Los Angeles. Urban aerosols contain organic material emitted locally in addition to aged material advected into the urban area. Dicarboxylic acids and nitrate esters are considered to be typical aerosol components in urban air, although they occur also in nonurban areas. Polycyclic aromatic hydrocarbons such as naphthalene and the higher homologues have received much attention because some of them, especially the five-ring benzo[a]pyrene, are carcinogens. These substances are combustion products. In the atmosphere they are photochemically degraded, but they also undergo long-range transport and are present on a global scale.

TABLE 7.20 Identification of Organic Compounds Associated with Continental and Marine Aerosols

Type of compound		Location	Concentration (ng m^{-3})	Authors
Continental				
Alkanes	C_{21}–C_{25}	Ohio River Valley, Indiana	0.2–2	Barkenbus et al. (1983)
	C_{22}–C_{35}	Various locations, Arizona	2.0–8.8	Mazurek et al. (1991)
		Central Africa, Nigeria	110–1700	Simoneit (1989)
		Southeastern Australia	16–80	Simoneit (1989)
		Manaus, Amazon	260–810	Simoneit et al. (1990)
Hydrocarbons[a]	C_{16}–C_{35}	Lake Tahoe, California	53–158	Simoneit and Mazurek (1982)
Polycyclic aromatics	C_{15}–C_{35}	National Park locations, Oregon and California	0.01–0.16	Simoneit (1984)
		Ringwood, New Jersey	0.55 (av. summer) 5.77 (av. winter)	Greenberg et al. (1985)
Nitropolycyclic aromatics		Near Copenhagen, Denmark	9.3 (av.)	Nielsen et al. (1984)
		Near Copenhagen, Denmark	0.18 (av.)	Nielsen et al. (1984)
Phtalate esters		Point Barrow, Alaska	0.1–20	Weschler (1981)
Formaldehyde		Deuselbach, Germany	39 (av.)	Klippel and Warneck (1980)
Aliphatic alcohols	C_{10}–C_{34}	Lake Tahoe, California Simoneit and Mazurek (1982)	198–524	
		Western United States	200–1390	Simoneit (1989)
		Central Africa, Nigeria	230–2200	Simoneit (1989)
		Southeastern Australia	70–380	Simoneit (1989)
	C_{12}–C_{30}	Manaus, Amazon region	10–110	Simoneit et al. (1990)

	Carbon range	Location	Concentration range	Reference
Continental				
Alkanoic acids	$C_{12}-C_{32}$	Lake Tahoe, California	140–332	Simoneit and Mazurek (1982)
	$C_{14}-C_{26}$	Ohio River Valley, Indiana	67	Barkenbus et al. (1983)
		Western United States	90–300	Simoneit (1989)
		Central Africa, Nigeria	80–960	Simoneit (1989)
		Southeastern Australia	30–110	Simoneit (1989)
	$C_{12}-C_{34}$	Manaus, Amazon region	200–620	Simoneit et al. (1990)
Dicarboxylic acids	C_3-C_5	Ohio River Valley, Indiana	30–50	Barkenbus et al. (1983)
	C_2-C_{11}	Alert, Canada	4.3–97	Kawamura et al. (1996)
Marine				
Alkanes	$C_{15}-C_{35}$	Atlantic Ocean	0.1–12	Simoneit (1977)
	$C_{10}-C_{30}$	Atlantic Ocean	14 (av.)	Hahn (1980)
	$C_{10}-C_{30}$	Cape Grim, Tasmania	4.7 (av.)	Eichmann et al. (1980)
	$C_{15}-C_{32}$	Tropical Atlantic ocean	6–13	Marty and Saliot (1982)
	$C_{13}-C_{32}$	Enewetak Atoll	0.07–0.24	Gagosian et al. (1982)
Aromatics	Xylenes	Loop Head Ireland	10.6 (av.)	Eichmann et al. (1979)
	Xylenes	Cape Grim, Tasman a	0.9 (av.)	Eichmann et al. (1980)
Polycyclic aromatics		Tropical Atlantic	0.1–0.2	Marty et al. (1984)
Formaldehyde		Loop Head Ireland	4.9 (av.)	Klippel and Warneck (1980)
Aliphatic alcohols	$C_{10}-C_{34}$	Atlantic Ocean	0.001–30	Simoneit (1977)
		Atlantic Ocean	0.01–10	Simoneit and Mazurek (1982)
	$C_{13}-C_{32}$	Enewetak Atoll	0.07–0.24	Gagosian et al. (1982)
Sterols	Cholesterol	Off the coast, Peru	0.032	Schneider and Gagosian (1985)
Carboxylic acids	$C_{13}-C_{32}$	Enewetak Atoll	0.04–0.38	Gagosian et al. (1982)
	$C_{12}-C_{32}$	Atlantic Ocean	0.001–30	Simoneit and Mazurek (1982)

[a] Includes branched and cyclic compounds

TABLE 7.21 Organic Compounds Associated with Urban Aerosols[a]

Type of compound		Location	Concentration (ng m^{-3})	Authors
Alkanes	$C_{18}-C_{50}$	217 U.S. urban stations	1000–4000	McMullen et al. (1970)
Alkenes		217 U. S. urban stations	2000	McMullen et al. (1970)
Alkylbenzenes		West Covina, California	80–680	Cronn et al. (1977)
Naphthalenes		Pasadena, California	40–500	Schuetzle et al. (1975)
Polycyclic aromatics		Los Angeles, California	0.09–0.18	Simoneit (1984)
Nitropolycyclic aromatics		Southern California	0.09–0.25	Pitts et al. (1985)
Aromatic esters		Antwerp, Belgium	29–132	Cautreels et al. (1977)
Aromatic acids		Pasadena, California	90–380	Schuetzle et al. (1975)
Alphatic alcohols	$C_{10}-C_{34}$	Los Angeles, California	1360–2016	Simoneit and Mazurek (1982)
Monocarboxylic acids	$C_{10}-C_{32}$	Los Angeles, California	217–308	Simoneit and Mazurek (1982)
		Antwerp, Belgium	37	Cautreels et al. (1977)
Ketoacids	C_3-C_9	Tokyo, Japan	100–180	Sempéré and Kawamura (1994)
Dicarboxylic acids	C_3-C_7	Pasadena, California	40–1350	Schuetzle et al. (1975)
	C_2-C_9	Tokyo, Japan	422–1001	Sempéré and Kawamura (1994)
Organic nitrates		Pasadena, California	40–4010	Schuetzle et al. (1975)
Alkylhalides		Pasadena, California	20–320	Schuetzle et al. (1975)
Arylhalides		Pasadena, California	0.5–3	Schuetzle et al. (1975)
Chlorophenols		Antwerp, Belgium	5.7–7.8	Cautreels et al. (1977)

[a] Adapted from Daisey (1980).

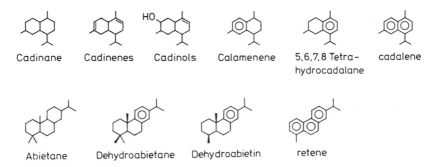

Cadinane   Cadinenes   Cadinols   Calamenene   5,6,7,8 Tetra-hydrocadalane   cadalene

Abietane   Dehydroabietane   Dehydroabietin   retene

FIGURE 7.27 Molecular markers for hydrocarbons originating from plant waxes (Simoneit and Mazurek, 1982).

Organic compounds associated with the rural continental aerosol have been extensively studied by Simoneit and Mazurek (1982), Simoneit (1984, 1989), and Mazurek et al. (1991). The initial study already showed that the material may be regarded as a composite derived from two major sources: waxes from vascular plants and petroleum residues originating from road traffic. Epicuticular plant waxes are abundant on the leaves of many land plants (Kolattukudy, 1976). They frequently occur as microcrystals or hollow tubes with micrometer to submicrometer dimensions (Hall and Donaldson, 1963; Baker and Parsons, 1971) and play a physiological role in water retention (Hall and Jones, 1961). Plant waxes also enter soils via incorporation of senescent vegetation as well as by the erosion of leaf waxes on healthy plants due to natural weathering. Thus, as soils are subject to wind erosion, the wax particles are resuspended in the air together with soil particles, and this explains their presence in the great dust plumes carried from the continents over the oceans.

Apart from the n-alkanes discussed above, epicuticular wax hydrocarbons contain sesqui- and diterpenoids. These compounds are based on the structural skeletons of cadinane and abietane, respectively, which are shown in Figure 7.27. The sesquiterpenoids recovered by Simoneit and Mazurek (1982) in the rural aerosol were calamenene, tetrahydrocadalene, and cadalene. These compounds presumably are degradation products of cadinane derivatives (various isomers of cadinenes and cadinols), which are ubiquitous in essential oils of many higher plants (Simonsen and Barton, 1961). The major diterpenoid hydrocarbons observed in aerosol samples were dehydroabietane, dehydroabietin, and retene. The main sources of abietane derivatives are coniferous resins. The parent compounds dehydrate fairly rapidly to yield the more stable hydrocarbons found in aerosols. These may then serve as markers for hydrocarbons arising from vegetation, in addition to the odd/even carbon number preference in the *n*-alkanes.

The petroleum-derived component associated with the rural aerosol consists of *n*-alkanes without any carbon number predominance, as well as many complex branched and cyclic hydrocarbons that cannot be resolved by gas chromatography, so that they appear as a hump in the C_{15}–C_{30} size range. Although the hump itself is indicative of the presence of petroleum residues, one may additionally use pristane and phytane as molecular markers. These quadruply branched C_{19} and C_{20} hydrocarbons are diagenetic products of phytol and are not primary constituents of terrestrial biota (Didyk *et al.*, 1978). They occur in crude petroleum as well as in lubricating oils and automobile engine exhaust, but not in gasoline.

Rogge *et al.* (1993b) examined automobile exhaust aerosols from noncatalyst, catalyst-equipped, and diesel engines. About 10–15% of the material was resolved. In the first case, about 70% of the compounds in this mass fraction were identified; the majority were *n*-alkanes and polycyclic aromatic hydrocarbons. In the second case, about 60% of the compounds in the resolved mass fraction were identified, with *n*-alkanes, *n*-alkanoic acids, and benzoic acids comprising more than 70%. In the third case, about 40% of the compounds in the resolved mass fraction were identified, with *n*-alkanes and *n*-alkanoic acids being dominant. Essentially all of the particulate *n*-alkanes emitted come from unburned engine oil. The higher alkanoic acids are formed either during the combustion process or by catalytic oxidation; polycyclic aromatic hydrocarbons are formed during combustion by pyrolysis mechanisms (Bartok *et al.*, 1990).

In addition to *n*-alkanes, Simoneit and Mazurek (1982) found the rural continental aerosol to contain alkanoic acids in the range of C_{10}–C_{30}. A strong even-to-odd carbon number preference indicated a biological origin. The alkanoic acid distribution in waxes from composite plant samples did not compare well with that in the aerosol, however. Straight-chain fatty acids in plant waxes, which occur mainly as esters, fall in the C_{22}–C_{32} range of carbon numbers, with maxima between C_{26} and C_{30}, whereas the aerosol constituents exhibit the principal maximum at C_{16} (palmitic acid), with a secondary maximum at C_{22} or C_{24}. Similar distributions have been observed by Ketseridis and Jaenicke (1978) and Gagosian *et al.* (1982) in marine aerosols, by van Vaeck *et al.* (1979) and Matsumoto and Hanya (1980) in continental aerosols, and by Meyers and Hites (1982) in Indiana rainwaters. The origin of the distribution is obscure. Simoneit and Mazurek (1982) suggest that the homologues below C_{20} arise from the degradation of plant waxes inhabited by microorganisms. Soil microbes may additionally contribute to the acids associated with the aerosol.

Apart from alkanoic acids, Simoneit and Mazurek (1982) observed low concentrations of alkenoic acids (range $C_{14}-C_{17}$); α-hydroxy alkanoic acids (range $C_{10}-C_{24}$) that are known components of grass wax; dicarboxylic acids (range $C_{10}-C_{24}$) that possibly arise from the direct biodegradation of hydroxy alkanoic acids; and diterpenoidal acids occurring as diagenetic products of diterpenoids from coniferous resins.

Simoneit and Mazurek (1982), Simoneit (1989), and Mazurek et al. (1991) also described n-alkanols as constituents of rural continental aerosols, ranging from C_{10} to C_{34}, with a strong even-to-odd carbon number preference. The distribution favored the higher homologues, with a maximum in the range of $C_{26}-C_{30}$. This is characteristic of vascular plant waxes, so that, again, leaf waxes are a major source of the material. The fraction below C_{20}, which is not present in plant waxes, may arise from the degradation by microbial activity. Molecular markers for alcohols are the phytosterols. Their distribution in the aerosol is similar to that of the phytosterol content of many plants.

Dicarboxylic acids are less volatile than the corresponding monocarboxylic acids, so that even the lowest homologues are attached to the aerosol. In fact, oxalic acid, $(COOH)_2$, is present in the highest concentration within the C_2-C_9 group, contributing about 30% to the total mass of dicarboxylic acids. As Table 7.21 shows, urban concentrations are 1–3 $\mu g\ m^{-3}$. In the Arctic the concentrations were found to range from about 10 to 100 ng m^{-3}, depending on the season (Kawamura et al., 1996). Kawamura and Usukura (1993) also detected these compounds in aerosol samples of the north Pacific, and Sempéré and Kawamura (1996) showed that they are present in rainwater of the western Pacific ocean, so that they appear to be ubiquitous in the atmosphere. Among the compounds discussed above, dicarboxylic acids are the only ones that are fully water-soluble. In Tokyo, according to Sempéré and Kawamura (1994), they make up less than 5% of the total organic fraction, whereas other data discussed earlier had indicated a higher percentage of water-soluble organic compounds. The major sources of dicarboxylic acids in the urban environment are thought to be the gas phase oxidation of aromatic hydrocarbons and cyclic olefins. In the remote environment, other organic compounds presumably act as precursors. By analyzing aerosol samples from West Los Angeles and from Rubidoux (Riverside) to the east of Los Angeles, Rogge et al. (1993a) demonstrated that the percentage of dicarboxylic acids in the resolved organic fraction increased by about 50% after the air had passed over the city. Here again, dicarboxylic acids contributed less than 5% to the total organic content of the aerosol.

7.6. GLOBAL DISTRIBUTION, PHYSICAL REMOVAL, AND RESIDENCE TIME OF THE TROPOSPHERIC AEROSOL

As for other constituents of the atmosphere, it is possible to set up a mass budget of the aerosol and to calculate its residence time. The main problem is to characterize the global distribution of particulate matter to determine the total mass in the troposphere. One can then apply the emission rates of Table 7.11 to calculate the residence time τ_A with the help of Equation (4.11). This approach will be discussed in the first part of this section. Subsequently, an independent method will be considered that is based on the use of radioactive tracers. Finally, the removal of aerosol particles by sedimentation and impact at the Earth's surface will be discussed.

7.6.1. VERTICAL AEROSOL DISTRIBUTION AND GLOBAL BUDGET

A summary of our knowledge about the vertical distribution of aerosol particles over the continents of the Northern Hemisphere is shown in Figure 7.28. The left-hand side displays the behavior of Aitken particles as determined with condensation nuclei counters; the right-hand side shows the distribution of large particles as determined with optical techniques. In each case the range of observations is shown, and a single altitude profile is included for comparison. The concentration of large particles may be assumed to represent the mass concentrations, whereas that of Aitken particles is more indicative of the production of new particles. The number densities of both types of particles decline rapidly with altitude in the lower troposphere, with a scale height of roughly 1 km. In the upper troposphere the decline corresponds to that of air density—that is, the mass mixing ratio is approximately constant. In the stratosphere, the concentration of Aitken particles declines further, while that of large particles increases again, reaching a peak at altitudes near 20 km, because of the production of sulfuric acid, as discussed in Section 3.4.4 (the so-called Junge sulfate layer).

The observation that in the troposphere the number densities of large and Aitken particles decline with altitude in a similar manner suggests that the shape of the size distribution is approximately preserved. Size distributions of aerosol particles as a function of altitude have been studied by Blifford and Ringer (1969) with single stage impactors and by Patterson et al. (1980) and

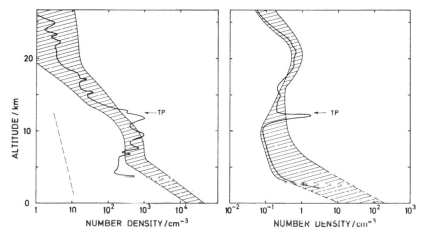

FIGURE 7.28 Vertical distributions of the aerosol over the continents. Left: Profile of condensation nuclei, range of results obtained by Junge (1961, 1963), Weickmann (1957), Hoppel *et al.* (1973), Cadle and Langer (1975), Käselau *et al.* (1974), and Rosen and Hofmann (1977, 1978). The single profile is from Rosen and Hofmann (1977). The location of the tropopause is indicated. The dashed line is the profile for a constant mixing ratio in the troposphere. Right: Vertical distribution of large particles, range of results obtained by Hofmann *et al.* (1975), Patterson *et al.* (1980), and Blifford and Ringer (1969); for earlier results see Junge (1963). The single profile is from Hofmann *et al.* (1975).

Kim *et al.* (1993) with optical techniques. The results indicate that in the free troposphere the volume size distribution maintains its bimodal shape. The slope in the (double logarithmic) number density distribution decreases somewhat, so that the distribution becomes flatter compared to its appearance in the boundary layer. If the size distribution of particles greater than 0.1 μm in radius is assumed to be approximately independent of altitude, one may combine ground-level concentrations with the altitude profile in Figure 7.28 to derive the column density of the aerosol mass in the continental troposphere. The vertical distribution may be visualized to contain two contributions: a uniform background evident in the upper troposphere, and a continental aerosol decreasing from the ground on up with a scale height of 1 km. The background aerosol would also be present in the marine troposphere. Here, the lower air space receives a second contribution from sea salt, whose vertical distribution may be gleaned from Figure 7.14, and from gas-to-particle conversion processes.

The admittedly rather crude model used here is based on the ground-level concentrations given in Table 7.1. The vertical distributions of continental and marine aerosols are illustrated in Figure 7.29. The decline of concentra-

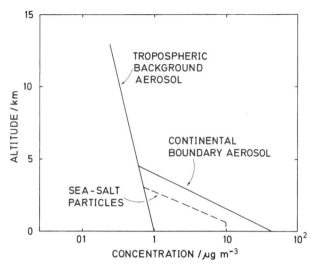

FIGURE 7.29 Model for the vertical distribution of particulate matter in the troposphere. The model assumes a superposition of tropospheric background aerosol with boundary layer aerosol over the continents and sea salt aerosol over the oceans.

tion with increasing altitude follows from the adopted scale heights and is expressed as

continental $(0 < z < z_T)$:

$$c_A = c_1^0 \exp(-z/h_1) + c_2^0 \exp(-z/h_2)$$

marine $(0 < z < 600 \text{ m})$:

$$c_A = c_3^0 + c_2^0 \exp(-z/h_2)$$

$(600 \text{ m} < z < z_T)$:

$$c_A = c_3^0 \exp(-(z - 0.6)/h_3) + c_2^0 \exp(-(z - 0.6)/h_2)   \quad (7.29)$$

with $c_1^0 = 45$ μg m^{-3}, $c_2^0 = 1$ μg m^{-3}, $c_3^0 = 10$ μg m^{-3}, $h_1 = 1$ km, $h_2 = 9.1$ km, and $h_3 = 0.9$ km.

By integration over the vertical extent of the troposphere, assuming an average tropopause height of 12 km, one obtains

continental:

$$W_{AC} = c_1^0 h_1 + c_2^0 h_2 [1 - \exp(-z_T/h_2)] = 5.17 \times 10^{-5} \text{ kg m}^{-2}$$

marine:

$$W_{AS} = 600 c_1^0 + c_3^0 h_3 + c_2^0 h_2 [1 - \exp(-z_T/h_2)] = 2.1 \times 10^{-5} \text{ kg m}^{-2}$$

The total burden of particulate matter in the troposphere is then obtained by applying the surface areas $A_{cont} = 1.49 \times 10^{14}$ m^2 and $A_{sea} = 3.61 \times 10^{14}$ m^2, with the result

$$G_A = W_{AC} A_{cont} + W_{AS} A_{sea} = 1.52 \times 10^{10} \text{ kg}$$

In a similar manner, one calculates the mass content in the troposphere of sea salt alone as

$$G_{seasalt} = 5.4 \times 10^9 \text{ kg}$$

In this calculation, the fraction of sea salt transported from the ocean to the continents is neglected.

The estimates in Table 7.11 indicated a global source strength of particles of $(2.3–3.9) \times 10^{12}$ kg year^{-1} or $(0.63–1.0) \times 10^{10}$ kg day^{-1}. The values include a production rate of sea salt of about 1×10^{12} kg year^{-1} or 2.74×10^9 kg day^{-1}. The corresponding residence times are

$$\tau_A = G_A/Q_A = 1.5–2.5 \text{ days}$$

$$\tau_{seasalt} = G_{seasalt}/Q_{seasalt} = 2 \text{ days}.$$

Table 7.11 admits the possibility of a five times larger rate of sea-salt production. If this rate were adopted, the residence times for the total aerosol and for sea salt alone would be reduced to 0.7 and 0.4 days, respectively.

One should not expect this procedure to provide more than order of-magnitude estimates. On the one hand, the emission rates in Table 7.11 are rather crude estimates, and on the other hand, the tropospheric mass content of particles predicted by the model must also be considered to be approximate. The further discussion below and that in Section 8.3 will show, however, that other methods of determining τ_A lead to similar results, that is, aerosol residence times of a few days. Note that the residence times of the total aerosol and sea salt are almost identical, despite differences in the mass–size distributions, with sea salt favoring the giant particle size range. This suggests that the processes responsible for the removal of particulate matter from the troposphere do not discriminate much between large and giant particles.

7.6.2. RESIDENCE TIMES OF AEROSOL PARTICLES

Independent of budget considerations, one can estimate the residence time of particles by means of radioisotopes as tracers, because these attach to aerosol particles and are removed with them when they undergo precipitation scavenging or dry fallout. One of the earliest estimates of this kind was made by Stewart *et al.* (1956) from the decay of fission products following their dispersal in the Northern Hemisphere from the nuclear weapons tests in Nevada in 1951. The results indicated a residence time of about 1 month. While Stewart *et al.* (1956) had assumed the debris to be confined to the troposphere, it was later shown by Martell and Moore (1974) that the radioactive material from three of the four events had, in fact, penetrated the tropopause. Accordingly, the value of 4 weeks was more likely applicable to the lower stratosphere, so that it represents an upper limit to the aerosol residence time in the troposphere.

Schemes that one may apply to deduce aerosol residence times from various radioactive elements have been reviewed by Junge (1963), Martell and Moore (1974), and Turekian *et al.* (1977). The data fall into two groups of values averaging 6 and 35 days, respectively. Martell and Moore (1974), after a critical review of the older data, concluded that the high values were due to the contribution of stratospheric aerosols, apart from misinterpretations of some data, while the lower values represent the true tropospheric residence time. More recently, Giorgi and Chameides (1986) have summarized a variety of estimates derived from measurements of nuclides of stratospheric debris origin that lie in the range of 25–50 days.

Especially suitable for the determination of the residence time of the tropospheric aerosol are the longer-lived radionuclides within the radon decay sequence, specifically bismuth 210 and lead 210. The major routes for nuclei conversion within the radium decay scheme are shown in Figure 7.30. The direct decay product of radium 226, an alpha emitter, is radon 222, which escapes from continental soils at an average rate of 1×10^4 atom m^{-2} s^{-1} (Turekian *et al.*, 1977) under nonfreezing conditions. Soil emissions may vary spatially and temporally by several orders of magnitude, however. The contribution from the sea surface is essentially negligible. Since the half-life-time of radon 222 is only 3.8 days, its distribution in the troposphere is rather uneven. Over the continents the mixing ratio declines with increasing

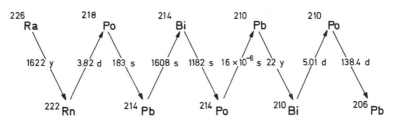

FIGURE 7.30   Major routes within the radium decay scheme. Half-lifetimes are shown in years (y), days (d), or seconds (s).

altitude (see Fig. 1.9). Over the oceans, the vertical gradient is reversed. The immediate decay products of radon 222 are short-lived. The first longer-lived species is lead 210 with a half-lifetime of 22 years. The subsequent products are bismuth 210 and polonium 210, with half-lifetimes of 5 and 138 days, respectively. If one assumes that all of the radium-daughter activities in the atmosphere arise from the decay of radon 222, and that their production and removal are in a steady state, one has

$$dn_2/dt = n_1\lambda_1 - n_2\left(\lambda_2 + \tau_R^{-1}\right) = 0 \qquad (7.30)$$

where n_1 and n_2 are the number concentrations and λ_1 and λ_2 the radioactive decay constants of the parent and daughter isotopes, respectively. Since lead, bismuth, and polonium are metals that become attached to atmospheric particles, the removal term τ_R^{-1} is essentially identical to that for the accumulation mode of the tropospheric aerosol, and $\tau_R \approx \tau_A$. There are three ratios of parent to daughter isotopes that may be utilized for the determination of τ_A: ^{222}Rn/^{210}Pb, ^{210}Pb/^{210}Bi, and ^{210}Po/^{210}Pb. By successive application of Equation (7.30), one obtains for the corresponding residence times

$$^{222}\text{Rn}/^{210}\text{Pb}: \qquad \tau_R = \left(\lambda_{Rn} n_{Rn}/n_{Pb} - \lambda_{Pb}\right)^{-1}$$

$$^{210}\text{Pb}/^{210}\text{Bi}: \qquad \tau_R = \left(\lambda_{Pb} n_{Pb}/n_{Bi} - \lambda_{Bi}\right)^{-1}$$

$$^{210}\text{Pb}/^{210}\text{Po}: \qquad \tau_R^{-1} = \left[-b + \left(b^2 - 4a\right)^{\frac{1}{2}}\right]/2 \qquad (7.31)$$

where

$$a = \lambda_{Bi}\left(\lambda_{Pb} n_{Pb}/n_{Po} + \lambda_{Po}\right) \quad \text{and} \quad b = \left(\lambda_{Po} + \lambda_{Bi}\right)$$

From what has been said above, the model applies only to purely continental conditions, and the interference by marine air should be avoided. The measurements of Poet *et al.* (1972) and Moore *et al.* (1973) were conducted over the central United States. Results for the first two ratios of radionuclides indicated residence times in the ranges of 1.8–9.6 and 3.7–9.2 days, respectively, without evidence of a systematic variation with altitude (in the troposphere). Averaged values were 3.5 and 7.5 days. The ^{210}Pb/^{210}Po ratios, in contrast, gave tropospheric residence times falling in the range of 12–49 days, and the vertical profile of ^{210}Po also differed appreciably from that expected from the decay of radon 222. Lehmann and Sittkus (1959), Burton and Stewart (1960), Pierson *et al.* (1966), and Marenco and Fontan (1972) had earlier measured ^{210}Pb/^{210}Po ratios and obtained similar values. The discrepancy appears to arise from surface sources. Marenco and Fontan

(1972) have considered coal burning and forest fires, and Lambert *et al.* (1979) have shown that volcanoes are a source of radon daughters, especially polonium 210. Because of the relatively long lifetime of polonium 210, surface sources affect the ^{210}Pb/^{210}Po ratio most strongly, making this nuclide pair unsuitable for estimating the aerosol residence time. Poet *et al.* (1972), for example, showed that 85% of ^{210}Po in the atmosphere is of terrestrial origin. In contrast, the ratios ^{222}Rn/^{210}Pb and ^{210}Pb/^{210}Bi are only moderately affected, so that correction factors may be applied. Taking the terrestrial component into account would lower the effective residence time derived from ^{210}Pb/^{210}Bi, whereas that derived from ^{222}Rn/^{210}Pb would be raised. The apparent residence times obtained from these nuclide ratios, 7.5 and 3.4 days, respectively, should thus bracket the true residence time of particulate matter in the troposphere. Values on the order of 5 days were deduced early by Blifford *et al.* (1952) and by Haxel and Schumann (1955) from the ratio of short-lived to long-lived radon daughters in surface air. Lambert *et al.* (1982) considered the inventory of ^{222}Rn and long-lived daughters in the atmosphere and derived 6.5 days as an average for the tropospheric aerosol residence time. Graustein and Turekian (1986) used ^{210}Pb and ^{137}Cs deposition rates and inventories in core samples of soils in the eastern and midwestern United States and reported residence times of 4.8 and 5.5 days, respectively.

A residence time of about 5 days confirms the order of magnitude value derived in the previous section from budget considerations. The aerosol residence time thus is appreciably shorter than the time of vertical transport by turbulent mixing, which takes about 4 weeks to bridge the distance between surface and tropopause. The disparity has led many authors to suggest that the residence time of the aerosol in successive layers of the troposphere increases with altitude (see, for example, Marenco and Fontan, 1972). The data of Moore *et al.* (1973) suggest, however, that values deduced from both ratios, ^{222}Rn/^{210}Pb and ^{210}Pb/^{210}Bi, are essentially independent of altitude, except in the boundary layer. As the decay of radon is essentially a gas-to-particle conversion process, one expects the radon decay products to attach primarily to particles in the size range of accumulation mode. Accordingly, it is primarily the large particles that have a tropospheric residence time of about 5 days.

The finding that the aerosol residence time is shorter than the time constant for vertical transport requires an efficient process for the removal of particles from the troposphere. Precipitation scavenging is thought to be the most efficient process for large particles (see Section 8.3). In this regard it is interesting to note that the ratio of ^{210}Pb/^{210}Bi in rainwater, first studied by Fry and Menon (1962) and subsequently by Poet *et al.* (1972), led to an aerosol residence time of 6–7 days, which compares well with that obtained for the dry aerosol. The good agreement demonstrates that precipitation is indeed important in removing particles from the tropospheric air space. If precipitation is the principal removal process, it follows that the average precipitation rate will be important in determining the local aerosol residence time. Balkanski *et al.* (1993) have applied a three-dimensional global circulation model to study the transport and residence time of ^{210}Pb produced from the radioactive decay of radon 222. They found tropical residence times of 10–15 days, in contrast to 5 days in the middle latitudes, although one would expect a shorter residence time in the Tropics because of higher precipitation rates there. It appears that the longer tropical residence time is due to frequent convective updrafts that bring ^{222}Rn to high altitudes faster.

We come back to the problem of time constants for the removal of individual particles. Jaenicke (1978c, 1980) has studied this aspect in some detail, and Figure 7.31 summarizes his findings. Three processes come into play.

1. Aitken particles are removed mainly by coagulation with other particles. The corresponding time constant is obtained from the decline in concentration to $1/e$ of the initial value, provided sources are absent. Since coagulation is a volume process, the time constant is equivalent to the particle lifetime. As discussed in Section 7.3.1, the rate of coagulation depends on the mobility of Aitken particles, as well as on the number concentration of the entire aerosol population. Figure 7.31 shows that the lifetime of Aitken particles increases essentially with the particle radius squared. The lifetime of background aerosol particles in the Aitken range is almost two orders of magnitude greater than that of the continental aerosol particles, because the number density of the background aerosol is much lower.

2. The main removal process for particles in the 0.1–10 μm size range is wet precipitation. Recall from Section 7.3.2 that most particles in this size range serve as cloud condensation nuclei. Toward small radii, the cut-off limit is determined by the nucleation barrier, which depends on the supersaturation reached in the cloud, on the salt content, and to some extent on the aerosol number density. The cut-off limit lies near 0.1 μm for the marine background aerosol, whereas for the continental aerosol, which features higher number densities, the cut-off limit occurs near 0.2-μm radius. This covers a major portion of all large and giant particles and comprises most of the aerosol mass.

3. For particles exceeding 10 μm in size, sedimentation becomes increasingly important as a removal process. The solid curve shown on the right-hand side of Figure 7.31 was calculated by Jaenicke (1978c) from the time required for the sedimentation of particles starting at 1.5 km altitude, which represents an average height if one assumes an exponential decrease in

446   Chapter 7. The Atmospheric Aerosol

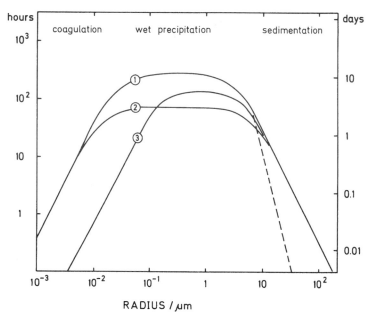

FIGURE 7.31   Combined residence lifetimes of aerosol particles as a function of size. (Adapted from Jaenicke (1978c, 1980).) Important removal processes, active in various size ranges, are indicated. Coagulation and sedimentation time constants were calculated; the time constant for wet removal is the residence time derived from the ^{210}Bi/^{210}Pb and ^{222}Rn/^{210}Pb ratios (Martell and Moore, 1974). Curves 1 and 2 represent the background for τ_{wet} equal to 12 and 3 days, respectively. Curve 3 represents the continental aerosol with $\tau_{wet} = 6$ days. The dashed line was calculated for sedimentation equilibrium, as described in Section 7.6.3.

number density with 1.5-km scale height. While this is a valid approach in the absence of surface sources, the assumption of a common scale height applies only to the bulk of the aerosol. In the presence of sources, particles large enough to undergo sedimentation will establish a balance between fluxes directed upward and downward, so that each size group adjusts to its own scale height. The dashed line in Figure 7.31 shows the corresponding residence times in the case of sedimentation equilibrium.

The residence time of the aerosol has important implications with regard to the transport and distribution of substances associated with particulate matter. For the bulk of the aerosol, the residence time is sufficient to permit particles to spread around the globe within the mean zonal flow, but in latitudinal direction the average distance to transport must be smaller, and any interhemispheric exchange of aerosol particles can have only local significance in the vicinity of the tropical convergence zone. Particles deposited on continental soils will be subject to resuspension into the air, whereby the material is spread over a larger distance after all. Mineral dust deposited on the ocean surface is irretrievably lost from the atmosphere and contributes to the formation of deep-sea sediments. According to Griffin *et al.* (1968), more than 50% of the carbon-free mass of ocean sediments consists of clay minerals, which except in regions directly influenced by river effluents derive from the long-range atmospheric transport of terrestrial matter. Duce (1986) and Duce *et al.* (1991) have pointed out that atmospheric deposition of mineral aerosol to the ocean is also an important source of nutrient elements, such as iron and phosphorus, for the marine biosphere.

7.6.3. SEDIMENTATION AND DRY DEPOSITION

The sedimentation of particles—that is, their downward motion due to gravitational settling—follows Stokes' law. Upward motion by turbulent mixing represents an opposing force, and sedimentation equilibrium is established when the two forces cancel. Assuming a constant production rate of particles at the Earth's surface for each size group and a balance of upward and downward fluxes in the atmosphere leads to the following equation:

$$-K_z n_M \, d[n(r)/n_M]/dz - v_s(r)n(r) = 0 \qquad (7.32)$$

Here, n_M is the number concentration of all molecules, $n(r)$ that of particles with radius r, and $v_s(r)$ is the sedimentation velocity. For an isothermal atmosphere, $n_M = n_M^0 \exp(-z/H)$, where H is the scale height. For simplicity we neglect the variation of K_z with height in the boundary layer and assume a constant $K_z = 20 \text{ m}^2 \text{ s}^{-1}$. The above equation then has the solution

$$n_M = n_0(r) \exp\left[-(v_s(r)/K_z + 1/H)z\right]$$
$$= n_0(r) \exp(-z/h) \qquad (7.33)$$

The scale height of the exponential distribution of $n(r)$,

$$h = H/(1 + v_s(r)H/K_z)$$

decreases with increasing sedimentation velocity. This quantity is given by

$$v_s(r) = b(r) \, gV_p(r)(\rho_p - \rho_{\text{air}}) \approx b(r) \, gV_p(r)\rho_p \qquad (7.34)$$

where $b(r)$ is the mobility of a particle, g is the acceleration due to gravity, $V_p(r)$ is the volume of the particle, and ρ_p is its density, against which the density of air ρ_{air} can be neglected. Inserting $V_p(r) = 4\pi r^3/3$ and $b(r)$ from Equation (7.6) in the approximate form for larger particles $b(r) = (6\pi\eta r)^{-1}$, one obtains

$$v_s(r) = (2g\rho_p/9\eta)r^2 = (2.38 \times 10^{-4})r^2 \text{ m s}^{-1}.$$

Here, $\eta = 1.83 \times 10^{-5}$ kg m^{-1} s^{-1} denotes the viscosity of air, the unit of r is μm, and a particle density of 2×10^3 kg m^{-3} is assumed. The column mass content in the troposphere for particles with radius r is

$$W_p(r) = V_p(r)\rho_p\int_0^{z_T}n(r)\,dz = V_p(r)\rho_p hn_0(r)(1 - \exp - z/h)$$

$$\approx V_p(r)\rho_p hn_0(r) \tag{7.35}$$

The approximation is valid for particles with radii greater than 10 μm. The corresponding mass flux returning to the Earth's surface due to sedimentation is

$$F_p(r) = V_p(r)\rho_p hn_0(r)v_s(r) \tag{7.36}$$

The ratio of the last two equations is equivalent to the local residence time of particles undergoing sedimentation:

$$\tau_p(r) = W_p(r)/F_p(r) = h/v_s(r) = \frac{H/v_s(r)}{[1 + v_s(r)H/K_z]} \tag{7.37}$$

Again, the unit of r is μm. For particles with radii greater than 10 μm, one has $v_s(r)H/K_z \gg 1$, so that Equation (7.37) simplifies to

$$\tau_p(r) = K_z/v_s^2(r) \tag{7.38}$$

The resulting residence times, which are shown in Figure 7.31 by the dashed line, should still be considered upper limits in view of the fact that values for K_z in the planetary boundary layer generally are smaller than 20 m^2 s^{-1}.

In Section 1.6 the concept of dry deposition of gases to the ground surface was discussed. The loss from the atmosphere sets up a gradient of the mixing ratio or concentration, so that a downward flux of material is maintained because of turbulent transport. This flux was expressed as the product of a deposition velocity and the concentration of the substance at a reference height. The similarity of Equations (7.36) and (1.32) shows that sedimentation is a dry deposition process, with the dry deposition velocity being equal to the sedimentation velocity. For particles in sedimentation equilibrium, the concentration gradient is directed upward because a uniform source was assumed to be active at the ground surface. The gradient reverses, however, as soon as the source is shut off. This raises the question of the extent to which particles smaller than 10 μm that are not efficiently removed by sedimentation can undergo dry deposition. Inasmuch as the particles are sticky, they are removed from the atmosphere directly by contact with vegetation and other surface elements. Turbulent mixing then establishes a downward flux, which may be expressed by a deposition velocity similar to that of particles undergoing sedimentation.

Figure 7.32 shows deposition velocities as a function of particle size. The data on the left, which refer to grass and water surfaces, were determined in laboratory and field experiments. McMahon and Denison (1979) and Sehmel (1980) have reviewed these and other data. For particles with radii greater than 1 μm, the deposition rate is similar to the sedimentation velocity. At higher wind speeds the deposition rate is enhanced by the increase in atmospheric turbulence. For submicrometer-sized particles, the deposition velocity reaches a relative minimum of some 10^{-4} m s^{-1} in the 0.1–1.0-μm size range, whereas in the Aitken size range the deposition velocity rises again because of the increased mobility of particles, which causes an increase in the collision frequency with surface elements.

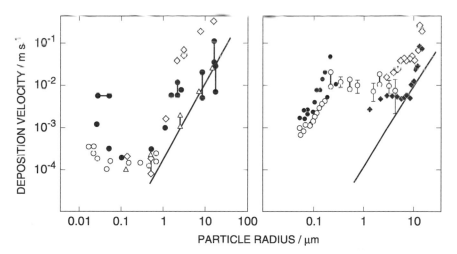

FIGURE 7.32 Dry deposition velocities for particles as a function of size. Left: Deposition to grass (solid symbols) and water surface (open symbols); data are from Chamberlain (1953, 1966b), Möller and Schumann (1970), Sehmel and Sutter (1974), Clough (1975), Little and Wiffen (1977), and Wesley et al. (1977). Wind speed: triangles $\approx$ 2 m s^{-1}, circles $\approx$ 8 m s^{-1}, diamonds $\approx$ 14 m s^{-1}. Right: Deposition to forests (mainly spruce), adapted from Gallagher et al. (1997) and references therein. Diamonds refer to cloud and fog drops. The solid lines in both graphs indicate the sedimentation velocity.

High growing vegetation is more effective in collecting aerosol particles in that stems and leaves of bushes and trees act as a filter by intercepting particles that impact on their surfaces. For many years, the subject has been one of controversy because of difficulties in measuring the dry deposition of particles to forest stands. Many of the difficulties appear to have recently been overcome, and the right-hand side of Figure 7.32 shows some results compiled by Gallagher *et al.* (1997). The data for submicron particles were obtained by eddy correlation techniques and through-fall measurements; the data for supermicron particles were obtained in part from measurements on cloud or fog drops. Dry deposition to high-growing vegetation evidently is important as a removal mechanism, at least in the boundary layer, but it has also been shown that forests can shed particles under turbulent conditions, and these particles become resuspended in the atmosphere. For submicrometer particles there is a strong increase in the dry deposition rate with the wind friction velocity. The situation is different from that of grass in that the structure of high-growing vegetation allows significant air flow within and underneath the canopy.

It is instructive to calculate residence times of aerosol particles in the troposphere resulting from dry deposition. We confine the calculation to particles in the size range 0.1–1.0 μm, where the deposition velocity is roughly size-independent. Over the continents this size range contains approximately one-half of the total mass of all aerosol particles. The residence time is given by the tropospheric column content of such particles divided by the flux to the ground,

$$\tau_p \approx \frac{1}{2}\left(\frac{W_{AC}}{F_p}\right) = \frac{1}{2}\left[\frac{W_{AC}}{\left(c_1^0/2\right)v_d}\right]$$

By inserting $W_{AC} = 5.17 \times 10^{-5}$ kg m^{-2} and $c_1^0 = 45$ μg m^{-3} from Section 7.6.1, and $v_d \approx 1.5 \times 10^{-4}$ m s^{-1} taken from Figure 7.32 (left), the residence time is calculated to be 86 days, which is about 20 times the value derived from radon products. If, on the other hand, the deposition velocity to forest stands is used, $v_d \approx 1 \times 10^{-2}$ m s^{-1} taken from Figure 7.32 (right), the residence time is about 1 day, which is much shorter than that derived from radon products. However, the continents are not fully overgrown with forest, and turbulent conditions do not always exist in the boundary layer, so that the average dry deposition flux generally will be smaller and the residence time greater.

CHAPTER 8

Chemistry of Clouds
and Precipitation

The condensation of water vapor and its precipitation from the atmosphere in the form of rain, snow, sleet, or hail is important to atmospheric chemistry because it removes atmospheric constituents that have a pronounced affinity for water in the condensed state. Cloud and precipitation elements may incorporate both aerosol particles and gases. The uptake mechanisms are discussed in this chapter, together with the inorganic composition of cloud water and rainwater that they determine. These processes are, in principle, well understood. Another subject requiring discussion is the occurrence of chemical reactions in the liquid phase of clouds, especially those involved in the oxidation of dissolved SO_2. In this regard, considerable progress has been made in recent years.

8.1. THE WATER CYCLE

The basic features of the hydrological cycle, described in this section, were established some time ago. Chahine (1992) has discussed details needed to assess perturbations of the global climate system. The column density of water in the troposphere is determined largely by the vertical distribution of the vapor. Liquid water and ice crystals represent only a minor fraction of the total abundance. Even in dense clouds the mass of water in the vapor phase predominates over that in the condensed state. Clouds are instrumental, of course, in the removal of water from the atmosphere by precipitation.

The maximum burden of water vapor in the air is determined by the local saturation vapor pressure p_s. This quantity is a strong function of temperature (Table A.2). An air parcel rising from the ground surface experiences a decrease in temperature and a corresponding lowering of p_s. Condensation sets in when the saturation point is reached. Above this level the air becomes increasingly drier, as more and more water condenses and is left behind. Conversely, sinking dry air eventually picks up water vapor as it mixes with moister air at lower altitudes. At the same time, a subsiding air parcel encounters rising temperatures, causing condensed water to evaporate. This mechanism of convective transport, coupled with the requirement of material balance, demands that at altitudes above the saturation temperature level the water vapor pressure adjusts to approximately one-half the saturation vapor pressure, or a relative humidity of 50%.

Radiosonde data may be used to verify this prediction. The usual instrumentation is not suitable for measuring the low abundance of water in the upper troposphere and stratosphere, so that the data are confined to altitudes below 7 km. Oort and Rasmussen (1971) have compiled zonally averaged, mean monthly specific humidities (H_2O mass mixing ratios) as a function of altitude for the Northern Hemisphere. Figure 8.1 shows average water vapor pressures as a function of temperature calculated from their data. If as a precaution one uses only data for altitudes above the 850-hPa level, the points are found to scatter reasonably well around the 50% relative humidity curve. Average surface H_2O pressures, which in Figure 8.1 are included for comparison, correspond to relative humidities of 75–80%. Such values are typical for surface air over the ocean. Generally, the relative humidity is quite variable, even at higher altitudes. In addition to the expected statistical scatter, there are regional influences such as the meridional Hadley circulation, which causes subtropical latitudes to be drier than the more humid tropics at practically all altitudes.

The radiosonde data also provide the total amount of precipitable water in the atmosphere. Zonal mean values are shown in Figure 8.2. The data are based on the maps of Bannon and Steele (1960) and Newell et al. (1972). The amount of precipitable water decreases from the equator to the Poles because of the poleward decrease in temperature in the troposphere.

Water enters the atmosphere by evaporation from the ocean surface, lakes and rivers, vegetation, and the uppermost layer of soils. On the continents,

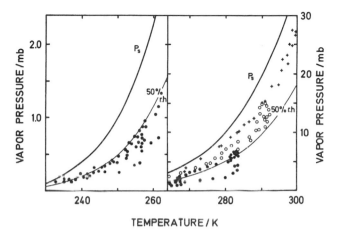

FIGURE 8.1   Zonally averaged partial pressures of water vapor for January and July, calculated from specific humidity data (Oort and Rasmussen, 1971; Newell et al., 1972) for latitudes between 10° S and 75° N. Data for pressure levels 400, 500, and 700 hPa are shown by dots, for 850 hPa by open circles, for 1000 hPa by crosses.

the local balance of surface water is determined by evaporation ($\sim 7.1 \times 10^4$ Pg year^{-1}), precipitation ($\sim 11.1 \times 10^4$ Pg year^{-1}), and losses due to the runoff of water into rivers and underground reservoirs ($\sim 4 \times 10^4$ Pg year^{-1}) and hence to the oceans. A similar balance applies to the oceans, except that here the runoff from the continents represents a gain rather than a loss. To the benefit of the continents, the oceans evaporate more water than they receive by precipitation, although on a global scale the two must balance.

Figure 8.2 depicts zonally averaged, mean annual rates of evaporation and precipitation as a function of latitude. Precipitation rates peak at the equator and in the middle latitudes. The first peak results from the convergence and updraft of moist air advected with the Hadley circulation. The secondary maxima are caused by the high cyclone activity in the westerlies. The subtropical high-pressure regions are much drier because of subsiding dry air in the downward branch of the Hadley cell. Here, and in contrast to the situation in other latitude zones, the rate of evaporation exceeds that of precipitation. Balancing the zonal water budget requires the transport of moisture from the subtropical source regions toward higher and lower latitudes. The poleward flux is maintained by eddy transport and the gradient of precipitable water with latitude. In the Tropics the flow of water follows

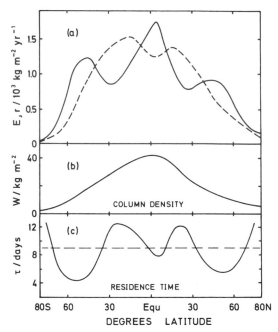

FIGURE 8.2 (a) Zonally averaged, mean annual rates of evaporation E and precipitation R (solid line) versus latitude as given by Sellers (1965). (b) Mean annual column densities W for precipitable water vapor, data from Sellers (1965) and Newell *et al.* (1972). (c) Mean residence time of water vapor, $\tau = W/R$, calculated from (a) and (b). The broken line gives the global mean value.

the direction of the Hadley circulation. The transport is most effective in the lowest strata of the troposphere, where the highest H_2O mixing ratios are found.

A closer inspection of the global water budget shows that the evaporation rates are higher in the Southern Hemisphere compared with the Northern, whereas the integrated precipitation rates are nearly equal in the two hemispheres. The budget is balanced by a flux of water vapor across the equator ($\sim 1.4 \times 10^4$ Pg year^{-1} according to Newell *et al.*, 1972), which is about 6% of the annual precipitation in each hemisphere (2.2×10^5 Pg year^{-1}). The difference in evaporation rates is due, on the one hand, to the greater ratio of ocean to land area in the Southern Hemisphere and, on the other hand, to the greater extent of subtropical desert areas in the Northern

Hemisphere, where the evaporation rates are correspondingly diminished. Transport of water vapor across the equator takes place in both directions, depending on the seasonal position of the Hadley cells and the interhemispheric tropical convergence zone. The strongest transequatorial H_2O flux occurs in the northward direction and is accomplished by the southern Hadley circulation during the period of April–November. In the remaining months of the year the H_2O flux is directed southward, but with lower intensity.

The column densities and precipitation rates in Figure 8.2 also allow an estimation of the residence time of water in the troposphere. The lower section of Figure 8.2 shows the variation of the residence time with latitude. The average residence time is 9 days; the range is 4.5–12.5 days. These data are approximate because they neglect the meridional transport of water vapor from net source to net sink regions. While it seems reasonable that the residence time has a maximum in the dry subtropics, it is from these regions that up to 50% of water actually being precipitated is transferred toward adjacent latitudes. The true residence is lowered accordingly, and the maxima near 25° latitude would probably disappear if appropriate corrections were made. The increase in the residence time toward the Poles must be considered real, since the high latitudes represent sink rather than source regions. Therefore, one estimates H_2O residence times in the troposphere of about 8 days in the ±30° latitude region, 5 days in the 15–65° latitude belts, and more than 12 days in the polar regions.

8.2. CLOUD AND RAIN FORMATION

Processes leading to the formation of clouds and precipitation have been treated in detail in the books of Fletcher (1962), Mason (1971), and Pruppacher and Klett (1997). The following brief summary is intended to provide some background information for the subsequent discussion of chemical constituents of cloud water and rainwater.

Clouds are formed when adiabatic or radiation cooling raises the relative humidity above the moisture saturation point. Adiabatic cooling is associated with rising air in convection cells, large cyclones, and frontal systems and along the windward slope of mountains. Small cumulus clouds formed by local convection frequently have their base about 800 m above the ground and their top a few hundred meters above the base. Entrainment of fresh air at the top combined with the heat release from condensation may revive convection, causing the cloud to rise further. Stratus clouds are formed by radiative cooling below temperature inversion layers, which are caused by subsiding air masses in high-pressure domains at altitudes of about 1 km, or in the vicinity of the ground level by radiative cooling of the Earth's surface, mainly at night. Recall from Section 7.3.2 that condensation nuclei are needed for the formation of cloud drops. The onset of supersaturation in a rising air parcel first activates the largest cloud condensation nuclei in the aerosol size spectrum. Some of the smaller and more numerous particles are activated later as the air rises further and the supersaturation increases. The supersaturation reaches a maximum when the rate of condensation equals the rate at which condensable moisture is made available by adiabatic cooling. The number of particles that become activated and the concentration of cloud drops is determined at this point. Subsequently, supersaturation decreases while condensation continues. The growth rate of cloud drops and the liquid water content of cloud-filled air are highly correlated with the vertical velocity of rising air. The maximum achievable liquid water content is rarely reached, however, because of the entrainment of drier air at the cloud's periphery. This effect is most pronounced in clouds of small width. Generally, the liquid water content increases with height above the cloud base and attains a maximum somewhere in the upper half of the cloud.

Table 8.1 summarizes for several kinds of clouds typically observed quantities such as the number density of cloud elements, the range of radii, and the liquid water content. Cloud drop formation initially leads to a fairly narrow drop size distribution centered near a radius of 5 μm. Droplet growth is mediated by transport of water vapor by molecular diffusion, and the growth rate of the drop's radius dr/dt is proportional to $1/r$. Accordingly, the rate slows down as the drop size increases. This makes condensation inefficient as a mechanism for the formation of drops much larger than 5–10 μm within the lifetime of a cloud. Large drops are thought to form from smaller ones by collision and coalescence in a manner resembling the coagulation of aerosol particles (see Section 7.3.1), except that the collision probabilities are modified by hydrodynamic and gravitational effects. In the initial assembly of cloud drops there are always a few that exceed the average size. They have a somewhat larger gravitational velocity than the others, so that they sweep up smaller drops in their path (sedimentation coagulation). The settling of large cloud elements is partly counterbalanced by the rising air motion within the cloud. Since the settling velocity and consequently the accretion rate increase with the size of the collector drop, its growth rate accelerates with time. This makes deep clouds more effective in the production of large cloud drops compared to the rather shallow fair-weather cumuli. Observed drop size spectra support this view.

Another important aspect of cloud formation is the formation of ice particles. Spontaneous freezing of liquid water drops in clouds is rare unless the temperature approaches $-20°C$, which corresponds to an altitude of about 6 km. In the middle troposphere, therefore, supercooled liquid water

TABLE 8.1 Typical Drop Sizes and Liquid Water Contents of Fogs and Several Types of Clouds[a]

Cloud type		Drop number concentration (cm⁻³)	Drop radius			Liquid water content (g m⁻³)	Vertical extent (m)
			Range (μm)	Most frequent (μm)	Average (μm)		
Fog		2000[b]	0.5–30	0.8	—	0.1–0.5	—
Fair-weather cumulus							
Continental	Nonprecipitating	300	2–30	5	5	0.1–0.4	750
Marine		60	2–25	12	15	0.4–0.5	
Cumulus congestus (continental)		100	2–70	5	20	0.5–1.0	1200–2000
Stratus							
Continental		350	2–25	12	12	0.1–0.9	400
Marine		150	10–50	15	20	0.3	—
Cumulonimbus (continental)	Precipitating	75	2–100	5	25	2.0	5000

[a] Data from Mason (1971), Pruppacher and Klett (1997), and Heymsfield (1993).
[b] Mostly haze particles, that is, nonactivated moist aerosol particles.

clouds are a common phenomenon. The probability of ice formation increases with decreasing temperature, so that at 250 K all clouds contain a certain population of ice particles. Mixed clouds consisting of supercooled water drops and ice particles are unstable because the H_2O equilibrium pressure over ice is lower than over liquid water at the same temperature. Thus the ice crystals grow by sublimation at the expense of the liquid phase. The generation of ice particles takes place in much the same way as the formation of a liquid water drop, in that a solid particle serves as a nucleus for crystallization regardless of whether ice formation proceeds by the freezing of supercooled liquid water or by condensation of water on the nucleus directly from the vapor phase. Compared to cloud condensation nuclei, a much smaller fraction of particles within the total aerosol population is capable of serving as ice nuclei. It is this feature that makes the occurrence of supercooling so common in clouds. Ice nuclei generally are water-insoluble. Clay mineral particles, volcanic ash, and fragments of decaying plant leaves have been found to be particularly effective as ice nuclei. In addition to growing by vapor deposition, ice particles may also collide with and accrete supercooled water drops, which then freeze to the ice particles in a process called *riming*. In the course of time, riming produces graupel or, in the extreme case, hailstones. Similar to the mechanism of liquid drop growth by coalescence, ice particles must reach a critical size before riming becomes effective as a growth process, but thereafter the growth rate increases rapidly.

Most clouds do not lead to precipitation; they evaporate. In longer-lasting deep clouds, a fraction of cloud elements will grow by coalescence to a size that subjects them to gravitational sedimentation. When they descend to levels below the cloud base they enter a region of relative humidity of less than 100%, which causes them to evaporate. A water drop must have a minimum size if it is to survive the sojourn to earth's surface without evaporating completely. Figure 8.3 indicates how much the size of a raindrop is reduced by evaporation as it traverses a stagnant air layer of up to 2 km depth. Only drops with an initial radius greater than about 500 μm will reach the ground surface. Observed size spectra for raindrops extend from perhaps 100 μm in radius (drizzle) up to 4–5 mm. The number frequency decreases approximately exponentially with increasing size because the larger drops suffer collisional breakup.

Two mechanisms are thought to be operative in the formation of rain: the Wegener–Bergeron–Findeisen process and the warm rain process. The first of these is germane to high-reaching continental clouds and starts with the growth of ice particles in the upper part of the cloud. Under favorable conditions, it will take on the order of 30 min for such ice particles to grow to a size sufficient for removal by sedimentation. On their way down they enter warmer layers of the atmosphere and melt. Wegener (1911), Bergeron

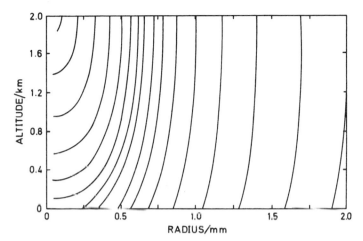

FIGURE 8.3 Decrease in the size of a raindrop by evaporation during the time of descent below the cloud base. (From Kortzeborn and Abraham (1970) with permission.)

(1935), and Findeisen (1939) have been the principal proponents of this mechanism of cold rain formation, so that it now bears their names. It cannot be the only mechanism, however, as in the Tropics and over the oceans one can observe warm precipitating clouds with a vertical extent such that their top remains definitely below the altitude level where the tempera ture is less than 273 K. In such clouds the coalescence of liquid cloud elements is the only mechanism for the formation of millimeter-sized rain-drops. Marine cumulus clouds generally consist of larger drops present at lower concentrations than do continental clouds (cf. Table 8.1). This feature results from the comparatively lower concentration of condensation nuclei in marine air. Thus, larger drops with radii of about 25 μm are formed earlier in the development of the cloud than in continental clouds. This would facilitate the growth of such drops to the millimeter size range during the lifetime of the cloud.

For an assessment of the global effects of clouds on the chemistry of gases, it would be desirable to have data on the average spatial distribution of clouds and the average time an air parcel spends in clouds of various types. Such data are still not widely available. Present satellite observations cannot provide information on the height of the cloud base and the vertical structure of clouds. Junge (1963) discussed the vertical distribution of clouds over Europe obtained from aircraft observations that were evaluated by DeBary and Möller (1960). More recently, Lelieveld *et al.* (1989) have combined visual data for cloud cover frequencies over the continents (Warren *et al.*,

1986) and over the oceans (Hahn *et al.*, 1982) with the vertical extent of different cloud types given by Telegadas and London (1954) and Paltridge and Platt (1976) to derive the zonally averaged volume fraction of cloud-filled air at altitudes up to about 6 km. Figure 8.4 shows results extracted from their data as well as the average column density of liquid water obtained. Clouds fill about 10–16% of the total air space with little indication for variation with latitude, whereas the liquid water column density shows a definite maximum in the Tropics. At latitudes south of 60° S the data base was too scarce to allow evaluation. Integrated amounts of liquid water from the satellite data of Njoku and Swanson (1983) and Prabhakara *et al.* (1983) indicated slightly smaller values. Lelieveld *et al.* (1989) also calculated average residence times of air in clouds, which fell into the approximate range of 2–4 h. The fraction of time that an air parcel spent in a cloud compared to the time inside and outside a cloud was about 15%, on average, with an approximate range of 8–24%.

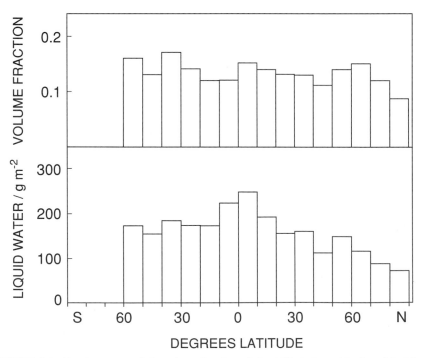

FIGURE 8.4   Zonal averages of the volume fraction of cloud-filled air at altitudes of 0–6 km (upper panel) and the associated column densities of liquid water (lower panel) for June–August (Extracted from Lelieveld *et al.*, 1989.)

A different quantity, which is important in tracing the fate of aerosol particles undergoing in-cloud processing, is the number of condensation–evaporation cycles such particles undergo before they are finally precipitated to the ground. Junge (1964) estimated small convective clouds to have life times of 0.5–1 h before they evaporate, whereas higher-reaching deep clouds may exist for 2–3 h. This led him to estimate that a cloud condensation nucleus undergoes about 10 cycles of condensation–evaporation during its time of residence in the troposphere. In-cloud chemical reactions such as the oxidation of SO_2 lead to nonvolatile products that remain associated with aerosol particles when they are set free after the cloud dissipates.

8.3. THE INCORPORATION OF PARTICULATE MATTER INTO CLOUD AND RAINDROPS

Hydrometeors acquire trace components by scavenging processes occurring within clouds as well as below clouds. In the older literature these are distinguished as rainout and washout, respectively. The term "washout," however, has also been used to describe precipitation scavenging generally. Slinn (1977a) consequently advocated the less ambiguous expressions "in-cloud scavenging" and "below-cloud scavenging."

8.3.1. NUCLEATION SCAVENGING

Owing to their function as cloud condensation nuclei, aerosol particles are naturally imbedded in cloud drops during cloud formation. Maximum values of supersaturation are about 0.1% for stratiform clouds and 0.3–1% for cumuliform clouds. In continental clouds with drop number densities of 100–300 cm^{-3}, essentially all particles with radii greater than 0.2 μm are expected to undergo nucleation scavenging. This expectation has now been confirmed by measurement during a mountain cloud experiment, and Figure 8.5 shows the scavenged fraction as a function of particle size. Even though a substantial fraction of less hygroscopic particles may not be activated, it appears that 60–80% of total aerosol mass undergoes nucleation scavenging. In marine clouds the situation is even more favorable, as the number density of the particles over the ocean is smaller and all of them are fully water-soluble. Once the cloud has formed, the nonactivated fraction of the aerosol population undergoes thermal coagulation with other aerosol particles and with cloud drops. This process removes mainly the smallest and most mobile

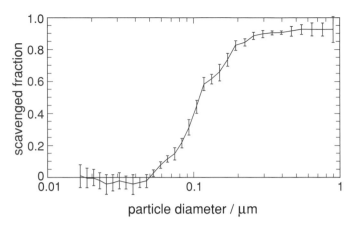

FIGURE 8.5   Fraction of aerosol particles scavenged in cloud drops, derived from observations of total and interstitial aerosol before and after the onset of cloud formation, during the Kleiner Feldberg Cloud Experiment. (Adapted from Svenningsson *et al.*, 1994.)

particles in the Aitken range. Their incorporation in cloud drops adds little extra mass, however. Gravitational scavenging of particles by descending cloud drops also discriminates against particles in the 0.1-μm size range, so that these particles remain present as an interstitial aerosol.

Let c_A (μg m^{-3}) $= c_A^0 \exp(-z_{cb}/h)$ denote the aerosol concentration at the cloud condensation level z_{cb} (cloud base). Here c_A may refer either to an individual constituent of the aerosol or to its entire mass concentration. The aqueous concentration c_w resulting from in-cloud scavenging can then be estimated from a relation first formulated by Junge (1963),

$$c_w = \varepsilon_A c_A/L = (\varepsilon_n + \varepsilon_B + \varepsilon_c)/L^* \qquad \text{mg kg}^{-1} \qquad (8.1)$$

where ε_A (with $0 < \varepsilon_A < 1$) is the mass fraction of material incorporated into cloud water, and L^* (g m^{-3}) is the average liquid water content of the cloud. Although the overall efficiency combines the effects of nucleation scavenging ε_n, attachment to cloud drops by Brownian motion ε_B, and collisional capture ε_c, the first term is the dominant one. We assume values of $\varepsilon_A \approx 0.75$ for continental clouds and $\varepsilon_A \approx 0.95$ for marine clouds. For precipitating clouds in the middle latitudes, the normal range of liquid water content is 1–3 g m^{-3}. For nonprecipitating clouds the values shown in Table 8.1 may be used.

Precipitation removes from the atmosphere particulate matter incorporated into cloud water. The associated average mass flux is

$$F = 10^{-3} c_w \bar{R} = 10^{-3} \varepsilon_A c_A \bar{R}/L^* \qquad (8.2)$$

where $\bar{R}$ (kg m^{-2} year^{-1}) is the average precipitation rate. Locally, the occurrence, rate, and duration of rainfall are distributed with time in a stochastic manner. It is necessary to average over a larger region to have precipitation occur somewhere in that region, and over a sufficiently long period of time. In the middle latitude belt, $\bar{R} \approx 800$ kg m^{-2} year^{-1}. The regional residence time of particulate matter being removed from the troposphere by rain-out is defined by the ratio of the aerosol column density W to the mass flux F. For the vertical distribution of the aerosol we may use the model presented in Section 7.6. In the lower troposphere over the continents, the particle concentration declines quasi-exponentially with altitude with a scale height $h \approx 1$ km. The cloud condensation level also is located at about 1 km altitude. Finally, setting $L^* = 2 \times 10^{-3}$ kg m^{-3} for a raining cloud, one obtains for the residence time

$$
\begin{aligned}
\tau_{\text{rainout}} &= \frac{W}{F} = \frac{c_1^0 h + c_2^0 H[1 - \exp(-z_T/H)]}{\bar{R} \varepsilon_A c_1^0 \exp(-z_{cb}/h)/L^*} \\[2mm]
&- \frac{5.2 \times 10^{-5} \times 2 \times 10^{-3}}{800 \times 0.75 \times 1.7 \times 10^{-5}} = 1 \times 10^{-2} \text{ or } 3.7 \text{ days} \qquad (8.3)
\end{aligned}
$$

This value has the same magnitude as the residence time for the bulk aerosol derived in Section 7.6 in a different manner. The congruence of results shows that in-cloud scavenging and precipitation are the dominant processes for the removal of particle matter from the troposphere.

8.3.2. BELOW - CLOUD SCAVENGING OF PARTICLES

Descending precipitation elements incorporate particles by collisional capture. For falling water drops the usual assumption that particles and drops merge upon contact has been experimentally confirmed by Weber (1969). Consider then a single raindrop of radius r_1 moving with a velocity v. It would sweep out a cylinder with cross section πr_1^2 and collect particles contained therein, were it not for the air flow around the nearly spherical drop, which causes the particles to follow the hydrodynamic streamlines. The situation is depicted schematically in Figure 8.6 to indicate how the

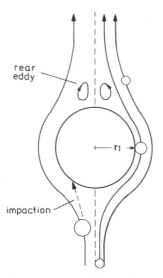

FIGURE 8.6   Schematic representation of the air flow around a falling sphere.

hydrodynamic flow pattern modifies the capture cross section. Formally, the effect is taken into account by an efficiency factor $E_c(r_1, r_2)$, which is a function of the radii of both collision partners. From Figure 8.6 it would seem that E_c generally is smaller than unity, but some experimental results have indicated values $E_c > 1$ under certain conditions. Effects that increase the normal collision efficiency include drop oscillations and electric charges. An important phenomenon is the standing eddy developing in the lee of falling water drops with radii in the range of 200–600 μm. Particles escaping frontal collisions may then become trapped in the eddy and undergo rear capture. Figure 8.7 shows experimental and theoretical evidence for this effect. For drops sizes larger than 600 μm, the collision efficiency declines as turbulence in the drop's wake causes the eddies to break loose at an increasing rate. Rear capture then becomes less significant.

Figure 8.8 shows how the collision efficiency varies with the size of aerosol particles. Three size regions must be distinguished. Small particles with radii less than 0.05 μm attach to water drops by Brownian motion, whereas giant particles with radii above 1 μm undergo inertial impaction. Greenfield (1957) first noted the gap between the two regimes. A later investigation by Slinn and Hales (1971) revealed the influence of phoretic forces, which partly closes the Greenfield gap. Most important is thermophoresis, which arises from the heat flux toward the water drop because of cooling by evaporation when the ambient relative humidity is less than 100%. Figure 8.7 shows this effect in that the collection efficiency rises

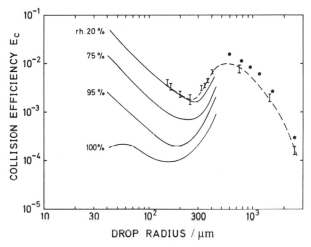

FIGURE 8.7 Collision efficiency E_c for the capture of particles ($r_2 = 0.25$ μm) by raindrops as a function of drop radius r_1 and ambient relative humidity. Solid curves were calculated by Grover *et al.* (1977) from hydrodynamic theory; bars show experimental results of Wang and Pruppacher (1977); points show results of Kerker and Hampl (1974) and of Lai *et al.* (1978). The latter data were partly corrected for terminal drop velocity and particle size. The maximum near $r_1 = 500$ μm is due to rear capture of particles in the standing wake eddy for drops sizes 200–600 μm. The fall-off region for larger drops indicates increasing wake eddy shedding.

sharply as the relative humidity drops from 100% to 20%. Coulomb forces will also be important if the water drop and aerosol particles both carry electric charges. Curves a and b in Figure 8.8 show results from numerical computations by Wang *et al.* (1978) for charged and uncharged water drops, respectively. Calculations for large water drops are only approximate because it is not known how to include the phenomenon of rear eddy shedding. Slinn (1976, 1977b) has derived a semiempirical expression for the collision efficiency, based in part on the theoretical work of Zimin (1964) and Beard and Grover (1974). Slinn's results are included in Figure 8.8. The figure also includes several laboratory results to enable comparison with theoretical predictions. These data are in reasonable agreement with the calculations.

The collection efficiency of falling ice particles and snowflakes for various aerosols has been repeatedly studied, for example, by Sood and Jackson (1970), Knutson *et al.* (1976), Murakami (1985), and Mitra *et al.* (1990). The general behavior of the capture efficiency with the size of aerosol particles was shown to be quite similar to that of water drops. Quantitative differences arise from the different shape of the ice crystals and the corresponding changes in hydrodynamic flow patterns. The collection efficiency of an ice

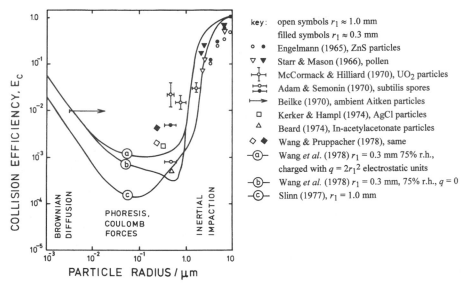

FIGURE 8.8   Collision efficiency for the capture of particles by raindrops as a function of particle radius. Solid curves indicate calculations, points are laboratory results. Three size regimes may be distinguished, depending on the dominant type of capture process.

crystal rises toward a maximum near 0.5 mm nominal radius and then decreases by almost two orders of magnitude toward a radius of 4 mm. However, snowflakes (that is, ice crystal aggregates) are endowed with a scavenging efficiency larger than that of single crystals, which does not increase with an increase in size. Mitra *et al.* (1990) suggest that this effect is due to the air flow through the aggregates rather than around them.

To evaluate the scavenging rate, one has to integrate over all possible collisions. Let $dN_1/dr_1 = f_1(r_1)$ represent the space density size distribution of raindrops (the drop size spectrum), and consider aerosol particles with radii between r_2 and $r_2 + dr_2$. The fraction of such particles that are removed from the air space in unit time is given by

$$\Lambda(r_2) = \int \pi r_1^2 E_c(r_1, r_2) v(r_1) f_1(r_1)\, dr_1 \qquad (8.4)$$

(Chamberlain, 1960; Engelmann, 1968). The quantity $\Lambda(r_2)$ is called the *washout coefficient*. It depends on r_2 in much the same fashion as E_c. Figure 8.9 shows sample calculations for two drop size spectra with maxima at radii of 0.2 and 1 mm, respectively, according to Zimin (1964) and Slinn and Hales (1971). For a given particle size, the washout coefficient is a constant only if the raindrop spectrum does not change with time. This ideal situation

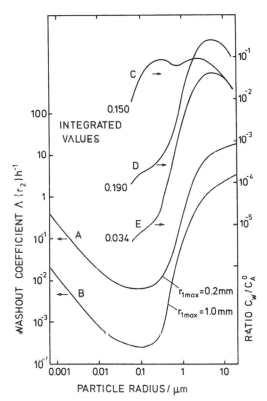

FIGURE 8.9   Curves A and B (left-hand scale) show washout coefficients according to Slinn and Hales (1971), based on raindrop size spectra of Zimin (1964), with $r_{1\,max} = 0.2$ and 1 mm, respectively, and a precipitation rate of 10 mm h^{-1}(10 kg m^{-2} h^{-1}). Curve C represents the first term and curves D and E the second term in Equation (8.6) in nonintegrated form (right-hand scale). These curves are based on the mass size distribution of the rural continental aerosol (Fig. 7.3). Curve C was calculated with $\varepsilon_A(r_2) = 1$ for $r_2 > 0.5\ \mu$m and $\varepsilon_A(r_2) < 1$ for $r_2 < 0.5\ \mu$m, decreasing linearly toward zero at $r_2 = 0.06\ \mu$m. This leads to $\varepsilon_A = 0.8$. Curves D and E were obtained by using the washout coefficients of curves A and B, respectively. Note that below-cloud scavenging (curves D and E) affects only giant particles, whereas nucleation scavenging (curve C) also incorporates submicrometer particles.

is rarely met in nature because of the usual variation in the rainfall rate. If this assumption is nevertheless made, the contribution of below-cloud scavenging to the total concentration of particulate matter in rainwater will be

$$c_w = 10^{-3} c_A^0 \overline{\Lambda} h \left[1 - \exp(-z_{cb}/h) \right] / R^* \qquad \text{mg kg}^{-1} \qquad (8.5)$$

with

$$\overline{\Lambda} = \int \Lambda(r_2) f_m(r_2) \, d \log(r_2)$$

where h is the scale height for the vertical distribution of the aerosol up to the cloud base at the altitude level z_{cb}, R^* is the rainfall rate (kg m^{-2} s^{-1}), and $f_m(r_2)$ represents the normalized aerosol mass size distribution, which is assumed to be height-independent. The total concentration of particulate matter in rainwater is given by the sum of Equations (8.1) and (8.5):

$$c_w = c_A^0 \{ \varepsilon_A \exp(-z_{cb}/h)/L + 10^{-3} \overline{\Lambda} h [1 - \exp(-z_{cb}/h)]/R^* \} \qquad (8.6)$$

This concentration is proportional to c_A^0. The relative importance of the two terms in the brackets is indicated in Figure 8.9 by curves C–E, which are based on the mass–size distribution for the rural continental aerosol shown in Figure 7.3. The comparison shows that below-cloud scavenging primarily affects particles in the 2–15-μm size range, whereas nucleation scavenging includes all particles with radii above 0.2 μm. Below-cloud scavenging thus enhances such elements in precipitation that are associated with the mineral fraction of the aerosol. Elements residing in the accumulation size range are removed from the atmosphere almost exclusively by in-cloud scavenging. Integration leads to the values 0.15 and 0.19 for the two terms in the brackets of Equation (8.6), respectively, assuming a raindrop size spectrum maximizing at $r_1 = 1$ mm. The two scavenging processes evidently contribute equivalent amounts of particulate matter to the total concentration in rainwater. In the alternative case with $r_{1\,max} = 0.2$ mm, the second term in Equation (8.6) has the value 0.034, so that in this case below-cloud scavenging is less significant compared to in-cloud scavenging.

8.4. THE SCAVENGING OF GASES BY CLOUD AND RAINDROPS

Cloud drops and raindrops absorb gases to the extent that these are soluble in water. The resulting aqueous solution is said to be saturated when the gas phase and the liquid phase are in thermodynamic equilibrium. Uptake and release rates are governed by molecular diffusion. As diffusion coefficients for common gases and water vapor are similar in magnitude, they are incorporated into cloud drops at nearly the same rate. Accordingly, one expects most gases in the atmosphere to equilibrate quickly with the aqueous phase while the cloud forms. Raindrops in the region below clouds, however, may experience gas concentrations different from those in clouds, so that the aqueous solution must adjust to the different conditions. The following discussion deals first with equilibria in clouds, then with the time required for the adjustment to equilibrium when conditions change.

8.4.1. Gas - Liquid Equilibria in Clouds

For ideal solutions in which solute interactions can be neglected, the mole fraction x_s of a substance dissolved in water and its equilibrium vapor pressure p_s in the gas phase above the solution are related by Henry's law,

$$p_s = H x_s = H \nu_s / (\nu_w + \Sigma \nu_i) \approx H \nu_s / \nu_w \tag{8.7}$$

where H is a coefficient that is independent of concentration but varies with temperature approximately as $\exp(-A/T)$, ν_s denotes the amount of the solute of interest, ν_w is the total amount of liquid water, and $\Sigma \nu_i$ that of all solutes present. Textbooks usually treat only binary mixtures, but Henry's law applies to multicomponent systems as well. Cloud water represents an almost ideal, dilute solution, so that $\nu_w \gg \Sigma \nu_i$. The mole fraction approximates to ν_s/ν_w and is proportional to concentration. This leads to alternative expressions for Henry's law:

$$p_s = H_c c_s \tag{8.8a}$$

$$c_s = K_H p_s \tag{8.8b}$$

where H_c and K_H are appropriate coefficients. The form of Equation (8.8b) in particular has been a favorite among atmospheric chemists, with concentration and pressure units of mol liter^{-1} and atmospheres frequently employed (1 liter = 1 dm^3; 1 atm = 1.01325×10^5 Pa). However, the atmosphere as a pressure unit does not conform to international standards and should be replaced. An alternative unit for solute concentration is mol kg^{-1} (molality scale) in the place of mol dm^{-3} (molarity scale). If the former is denoted by c_s and the latter by $[s]$, they are related by $[s] = \rho_w c_s$, where $\rho_w \approx 1$ kg dm^{-3} is the density of water. Its variation with temperature causes the molarity scale to depend on temperature, whereas the molality scale does not, so that the latter is preferable. In the temperature range 0–25°C, however, the density of water differs from unity by less than 0.3%, so that the two units are essentially equivalent. From the definitions in Equations (8.7) and (8.8), the coefficients H, H_c, and K_H are related by

$$H = H_c / M_w = (M_w K_H)^{-1} \tag{8.9}$$

where M_w (kg mol^{-1}) is the molar mass of water. Table 8.2 shows Henry's law coefficients for a number of gases of interest and other related data.

TABLE 8.2 Henry's Law Coefficients $K_H^\ominus$ for Several Atmospheric Gases,[a] Enthalpies of Solution ΔH Divided by the Gas Constant, in-Cloud Scavenging Efficiencies ϵ, Concentrations c_s in the Aqueous Phase for Initial Gas Phase Mixing Ratios x_0, and Residence Time τ for Rainout

Constituent	$K_H^\ominus$ (mol/kg Pa)	$\Delta H/R_g$ (K)	ϵ^b	x_0	c_s^b (mol/kg)	τ^b	Reference
N_2	6.4(−9)	1300	2.0(−8)	0.78	5.9(−4)	4.0(5) years	Wilhelm et al. (1977)
O_2	1.3(−8)	1500	3.9(−8)	0.21	3.2(−4)	3.8(5) years	Wilhelm et al. (1977)
O_3	1.2(−7)	2300	4.0(−7)	4(−8)	6.3(−10)	3.7(4) years	Kosak-Channing and Helz (1983)
CH_4	1.4(−8)	1700	4.4(−8)	1.7(−6)	2.9(−9)	3.4(5) years	Wilhelm et al. (1977)
CH_3Cl	9.9(−7)	2800	3.8(−6)	5(−10)	7.4(−11)	3.9(3) years	Wilhelm et al. (1977)
N_2O	2.4(−7)	2600	8.2(−7)	3(−7)	1.0(−8)	1.8(4) years	Wilhelm et al. (1977)
NO	1.8(−8)	1700	4.4(−8)	5(−10)	1.1(−12)	3.4(5) years	Wilhelm et al. (1977)
NO_2	6.9(−8)	—	1.6(−7)	1(−9)	6.3(−12)	9.3(4) years	Lee and Schwartz (1981)
H_2S	8.6(−7)	2100	2.9(−6)	7(−11)	8.0(−12)	5.1(3) year	Wilhelm et al. (1977)
OCS	2.1(−7)	3000	8.3(−7)	5(−10)	1.6(−11)	1.8(4) years	Wilhelm et al. (1977)
CH_3SCH_3	4.7(−6)	3100	1.9(−5)	6(−11)	4.5(−11)	7.9(2) years	De Bruyn et al. (1995)
$HCHO$	3.0(−2)	7200	2.0(−1)	5(−10)	3.8(−6)	16 days	Betterton and Hoffmann (1988a)
H_2O_2	9.9(−1)	6300	8.8(−1)	5(−10)	1.7(−5)	5.9 days	Lind and Kok (1994)
CH_3OOH	3.0(−3)	5300	1.7(−2)	5(−10)	3.3(−7)	158 days	Lind and Kok (1994)

[a] Powers of 10 are indicated in parentheses. $K_H^\ominus$ refers to 298.15 K, except for NO_2, 295 K.

[b] Scavenging efficiency and aqueous concentration for $L = 1 \times 10^{-6}$; residence time for $L = 2 \times 10^{-6}$; $T = 283.15$ K, $p_{air} = 910$ hPa in all cases.

To determine the gas–liquid partitioning of a water-soluble atmospheric component in a cloud, consider a fixed volume V_c of cloud-filled air containing the amount ν_0 of the substance of interest, of which ν_s resides in solution and ν_g in the gas phase. Let $L = V_L/V_c$ be the liquid volume fraction. We make use of the ideal gas law, Equation (1.1), and Henry's law, Equation (8.8b), with c_s in mol kg^{-1} (and ρ_w in kg m^{-3}, $R_g = 8.31$ N m) to derive

$$\nu_0/V_c = \nu_s/V_c + \nu_g/V_c = L\rho_w c_s + (1 - L)c_g$$

$$= L\rho_w c_s + (1 - L)p_s/R_g T = L\rho_w c_s + (1 - L)c_s/K_H R_g T$$

$$= (\nu_s/V_c)\left[1 + (1 - L)/(L\rho_w K_H R_g T)\right] \qquad (8.10)$$

The fraction of the substance residing in solution is

$$\varepsilon = \nu_s/\nu_0 = 1/\left(1 + (1 - L)\left(\rho_w LR_g TK_H\right)^{-1}\right) \qquad (8.11)$$

Note that in the atmosphere $L \ll 1$, typically $L \approx 1 \times 10^{-6}$, so that $(1 - L) \approx 1$. The corresponding concentration in the aqueous phase is

$$c_s = \frac{\nu_s}{\rho_w LV_c} = \frac{\varepsilon \nu_0}{\rho_w LV_c} = \frac{\varepsilon x_0 p_{air}}{\rho_w LR_g T} \qquad (8.12)$$

Here, p_{air} is the air pressure at the condensation level, and x_0 is the mixing ratio of the substance in the gas phase before condensation takes place. For $\varepsilon \ll 1$, one has $(\rho_w LR_g TK_H)^{-1} \gg 1$, and the concentration simplifies to

$$c_s = K_H x_0 p_{air} \qquad (8.12a)$$

which is equal to Equation (8.8b). Table 8.2 includes scavenging efficiencies ε. The values are small in many cases, but are significant for atmospheric constituents such as H_2O_2 or formaldehyde. Concentrations are entered in Table 8.2 for $L = 1 \times 10^{-6}$, which is typical for rain clouds. Concentrations depend on the product εx_0, so that nitrogen and oxygen occur in high concentrations, even though the scavenging efficiencies are small. Oxygen is somewhat more soluble than nitrogen, so that the O_2/N_2 ratio in solution becomes 0.54 instead of 0.27 in the gas phase.

The above equations may be used to derive the residence time of a water-soluble trace gas in the troposphere due to rain-out. For simplicity we assume that the concentration c_s in cloud water is approximately preserved during precipitation—that is, we ignore below-cloud processes such as the adjustment of temperature in and evaporation of water from falling raindrops. The average mass flux for the substance being removed by precipitation is

$$F_s = c_s M_w \bar{R} \tag{8.13}$$

where M_w is the molar mass of the solute and $\bar{R}$ is the average annual precipitation rate within the region considered. The argument is the same as that used in Section 8.3.1 for the removal of particulate matter from the atmosphere. The residence time (years) is obtained from the ratio of the column density of the constituent in the troposphere and the mass flux as

$$\tau_{\text{rainout}} = \frac{W_s}{F_s} = \frac{(M_s/M_{\text{air}})\bar{x}W_T}{c_s M_s \bar{R}} = \frac{\bar{x}W_T \rho_w LR_g T}{x_0 \varepsilon M_{\text{air}} p_{\text{air}} \bar{R}}$$

$$= (1.5 \times 10^{-2})\frac{\bar{x}}{x_0}\left(\frac{1}{\varepsilon}\right) \tag{8.14}$$

where W_T is the column density of air and $\bar{x}$ is the height-averaged mixing ratio of the substance in the troposphere. Approximate averages for one hemisphere are $W_T = 8.3 \times 10^3$ kg m^{-2}, $\bar{R} = 10^3$ kg m^{-2}, $M_{\text{air}} = 29 \times 10^{-3}$ kg mol^{-1}, and $p_{\text{air}} \approx 9.1 \times 10^4$ Pa. The liquid water fraction has been set at $L = 2 \times 10^{-6}$ to derive the numerical factor shown. For constituents that are fairly evenly distributed with altitude, one has $\bar{x} = x_0$, so that the residence time is independent of the ratio $\bar{x}/x_0$. In this case, the residence time depends only on the fraction of the substance that is dissolved in rainwater.

Residence times based on these assumptions are included in Table 8.2 to indicate the orders of magnitude obtained. For sparingly soluble gases, the time scales are too long to have any significance. Permanent gases, moreover, are not retained by the ground surface and are released soon after their deposition. A residence time of 1 year or less requires an in-cloud scavenging efficiency of at least 2%. This condition is met for H_2O_2 and HCHO. The residence time for a scavenging efficiency of unity is 5.5 days. This should not be taken as an absolute lower limit, because clouds then provide an efficient sink and $\bar{x}/x_0$ will be smaller than unity. In the considerations leading to Equations (8.3) and (8.14) it was convenient to use a precipitation rate averaged over a larger region. Locally, there may be dry periods longer than a week, that is, longer than the average residence time for those trace gases that are incorporated into clouds with a high efficiency. In such cases, considerable fluctuations of residence times and mixing ratios must occur. Rodhe and Grandell (1972) have dealt with this aspect.

A number of atmospheric gases and vapors dissolve in water to form ions. Most important among these are CO_2, SO_2, NH_3, HCl, and HNO_3. The interaction with water leads to chemical equilibria, which are summarized in Table 8.3. Rate coefficients for forward and reverse reactions are included as far as they are known, to show that the reactions are very fast so that the equilibria are rapidly established. The dissolution of these gases may be pictured to occur in two steps. The first is the physical dissolution process including hydration; the second is the formation of ions from the hydrates. A third step exists for acids that may donate another proton to the solution. Consider CO_2 as an example. In this case the following chemical equilibria are involved:

$$K_H p(CO_2) = [CO_2 \cdot H_2O]$$
$$K_1[CO_2 \cdot H_2O] = [H^+][HCO_3^-] \qquad (8.15)$$
$$K_2[HCO_3^-] = [H^+][CO_3^{2-}]$$

Here, it is again assumed that cloud water represents an ideal solution. For nonideal systems the concentrations (mol dm^{-3}, shown by brackets) would have to be replaced by the corresponding activities $a_s = \gamma_s[s]$, where the activity coefficient γ_s is a correction factor accounting for interactions among solute molecules (Robinson and Stokes, 1970). Departures from ideal conditions become significant for concentrations greater than 10^{-2} mol dm^{-3}, that is, for evaporating cloud drops. In Table 8.3 and above the hydrate of CO_2 is written in the form of an adduct, because it is difficult to distinguish analytically between a CO_2 molecule trapped in a solvent cage and carbonic acid, H_2CO_3. Both are present in aqueous solution and may undergo ion pair formation. Accordingly, K_1 is a composite equilibrium constant, which refers to the sum of CO_{2aq} and H_2CO_3 (Stumm and Morgan, 1981). A similar situation exists for SO_2. Note further that aqueous hydrogen ions always occur in the form of H_3O^+ and that the notation H^+ is merely a convention.

The total concentration of carbonate species following from the dissolution of CO_2 in water is

$$[CO_2]_{tot} = [CO_2 \cdot H_2O] + [HCO_3^-] + [CO_3^{2-}]$$
$$= [CO_2 \cdot H_2O]\{1 + K_1/[H^+] + K_1 K_2/[H^+]^2\}$$
$$= f_{CO_2} K_{H1} p_{CO_2} \qquad (8.16)$$

where f_{CO_2}, which represents the expression in the curved brackets, is a function solely of hydrogen ion concentration. Similar results are obtained for other weak acids and for ammonia (Warneck, 1986). The result allows us to define a modified Henry's law coefficient,

$$K_H^* = f_a K_H \qquad (8.17)$$

TABLE 8.3 Equilibrium Constants for Reactive Gases and the Ions Involved[a]

Number	Reaction	Coefficient[b]	K_{298}[b]	K_{283}[b]	k_f[b]	k_r[b]
w	$H_2O \rightleftarrows H^+ + OH^-$	K_w	1.0(−14)	3.0(−15)	1.3(−3)	1.3(11)
H1	$CO_{2g} + H_2O \rightleftarrows CO_2 \cdot H_2O$	K_{H1}	3.4(−7)	5.2(−7)		
A1	$CO_2 \cdot H_2O \rightleftarrows H^+ + HCO_3^-$	K_1	4.6(−7)	3.6(−7)	0.04	5.6(4)
A2	$HCO_3^- \rightleftarrows H^+ + CO_3^{2-}$	K_2	4.5(−11)	3.3(−11)	~2.5	~5(10)
H2	$SO_{2g} + H_2O \rightleftarrows SO_2 \cdot H_2O$	K_{H2}	1.2(−5)	2.2(−5)		
A3	$SO_2 \cdot H_2O \rightleftarrows H^+ + HSO_3^-$	K_3	1.3(−2)	1.9(−2)	3.4(6)	2.0(8)
A4	$HSO_3^- \rightleftarrows H^+ + SO_3^{2-}$	K_4	6.4(−8)	8.3(−8)	3.0(3)	5.0(10)
H3	$NH_{3g} \rightleftarrows NH_3 \cdot H_2O$	K_{H3}	5.8(−4)	1.2(−3)		
A5	$NH_3 \cdot H_2O \rightleftarrows NH_4^+ + OH^-$	K_5	1.8(−5)	1.7(−5)	6.0(5)	3.4(10)
H4	$HNO_{2g} \rightleftarrows HNO_{2aq}$	K_{H4}	4.9(−4)	1.1(−3)		
A6	$HNO_{2aq} \rightleftarrows H^+ + NO_2^-$	K_6	6.0(−4)	4.4(−4)	3(5)	5(10)
H5	$HCOOH_g \rightleftarrows HCOOH_{aq}$	K_{H5}	3.8(−2)	1.0(−1)		
A7	$HCOOH_{aq} \rightleftarrows H^+ + HCO_2^-$	K_7	1.8(−4)	1.8(−4)	8.6(6)	~5(10)
H6	$HCl_g \rightleftarrows H^+ + Cl^-$	K_{H6}	2.9(1)	1.5(2)		
H7	$HNO_{3g} \rightleftarrows H^+ + NO_3^-$	K_{H7}	3.2(1)	1.5(2)		
A8	$HSO_4^- \rightleftarrows H^+ + SO_4^{2-}$	K_8	~1.0(−2)	1.6(−2)	~1(9)	~1(11)
A9	$HCHO_{aq} + HSO_3^- \rightleftarrows CH_2(OH)SO_3^-$	K_9	3.6(6)	1.2(7)	1.4(3)[c]	3.9(−4)[c]

Powers of 10 are in parentheses.

[a] Calculated mostly from thermochemical data (Wagman et al., 1982). Data for SO_2 are from Maahs (1982), data for K_{H6} are from Fritz and Fuget (1956), data for K_{H4} are from Park and Lee (1988), data for K_{H7} are from Schwartz and White (1981), data for reaction A9 are from Deister et al. (1986). Rate coefficients are as summarized by Graedel and Weschler (1981), except for reaction A8 from Eigen et al. (1964) and reaction A9 from Bell (1966), Boyce and Hoffmann (1984), and Kok et al. (1986).

[b] Units: dm^3, mol, Pa, s^{-1}.

[c] For pH 4–7.

which depends on hydrogen ion concentration as well as on temperature. The modified coefficient must then be inserted in Equation (8.11) to determine the degree of partitioning of a substance in clouds. Figure 8.10 shows for several gases of interest ε as a function of pH (pH = $\log_{10} a_{H^+}$). The acid-forming gases are poorly absorbed by cloud water at low pH. Ammonia shows the opposite behavior.

The procedure just outlined fails for the strong acids HCl and HNO$_3$, because it is difficult to measure K_H and the dissociation constant separately. Only their product is known, so that

$$K_{HX} = K_{diss} K_H = [H^+][X^-]/p_{HX}$$

where X is either Cl or NO$_3$. The concentration of X$^-$ in solution is

$$[X^-] = (K_{HX}/[H^+]) p_{HX}$$

so that the modified Henry's law coefficient is

$$K_{HX}^* = K_{HX}/[H^+] \tag{8.18}$$

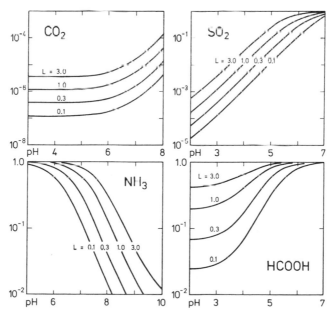

FIGURE 8.10   Gas–liquid scavenging efficiencies ε, for CO$_2$, SO$_2$, NH$_3$, and HCOOH in clouds with a liquid water content $\rho_w L$ (g m^{-3}) as a function of the pH. (From Warneck (1986), with permission.)

Table 8.4 lists, in a fashion similar to that of Table 8.2, modified Henry's law coefficients, in-cloud scavenging efficiencies, concentrations in the aqueous phase, and the associated residence times for the substances due to rain-out. For pH 4.5, a typical value in cloud water, NH_3, HCl, and HNO_3 are essentially completely scavenged from the gas phase. HCl and HNO_3 have such a high affinity for the aqueous phase that the liquid water content of the system would have to be appreciably reduced to release the vapors from solution. In fact, it is necessary to go all the way back to the aerosol stage ($L \approx 10^{-8}$) to return the acids to the vapor phase.

The above treatment of reactive gases assumes that cloud water represents a uniform solution. This assumption is unrealistic. Each cloud drop has its own chemical history. The pH in a cloud drop is determined by the composition of the aerosol particle that had served as the condensation nucleus, by the dissolution of reactive gases, and by the size of the water drop. This may result in a rather nonuniform dissolution of acids and ammonia in cloud drops. Field studies, in which the gas–liquid distribution was inferred from measurements of concentrations in the gas phase and in bulk samples of cloud water, have in some cases revealed serious discrepancies with Henry's law coefficients determined in the laboratory, especially with regard to the predicted pH dependence. Major disagreements have been observed, for example, for formic acid and ammonia (Winiwarter *et al.*, 1994). Both are conservative species. Bulk cloud water represents a volume-weighted average over individual drops. Pandis and Seinfeld (1991) have discussed the resulting effects on the apparent Henry's law coefficient.

While it is possible to calculate the distribution of solute concentration among cloud drops, including pH, starting from an assumed chemical size

TABLE 8.4   Modified Henry's Law Coefficients for Reactive Gases, in-Cloud Scavenging Efficiencies ε, Concentrations c_s in the Aqueous Phase for Initial Gas Phase Mixing Ratios x_0, and Residence Times τ for Rainout[a]

Constituent	pH	K_H^*	ε	x_0	c_s (mol kg^{-1})	τ
CO_2	< 5	5.2 (−7)	1.2 (−6)	3.3 (−4)	1.6 (−5)	6.2 (3) years
SO_2	4.5	1.3 (−2)	3.0 (−2)	1 (−9)	1.2 (−6)	94 days
HNO_2	4.5	1.6 (−2)	3.7 (−2)	5 (−10)	7.5 (−7)	76 days
HCOOH	4.5	6.7 (−1)	6.1 (−1)	5 (−10)	3.0 (−5)	7.2 days
NH_3	4.5	2.1 (2)	9.9 (−1)	1 (−9)	1.9 (−5)	5.5 days
HCl	> 1	> 1.4 (2)	~ 1	5 (−10)	9.6 (−6)	5.5 days
HNO_3	> 1	> 1.5 (2)	~ 1	5 (−10)	9.6 (−6)	5.5 days

[a] Scavenging efficiency for $L = 1 \times 10^{-6}$; aqueous concentration and residence time for $L = 2 \times 10^{-6}$; $T = 283.15$ K, $p_{air} = 910$ hPa in all cases.

distribution of aerosol particles, the results of such calculations cannot yet be compared with observations because of the lack of measurement techniques that would make it possible to determine the pH of individual droplets. For nonvolatile components, Ogren et al. (1989, 1992) have used a counterflow virtual impactor to show that in stratocumulus clouds the concentration increases with increasing drop size in the range of 10–18 μm diameter, while ground fogs feature a continuous decrease in concentration with increasing drop size. Other studies of mountain clouds have applied a segregation of cloud drops into two size fractions to confirm that both types of behavior can be observed (Munger et al., 1989; Collet et al., 1995; Schell et al., 1997). In the initial stages of cloud formation, there are always drops derived from giant aerosol particles that have not had the time to equilibrate with the rising supersaturation of water vapor, and this fraction shows an increase in concentration with increasing drop size in a certain size range. When the cloud ages, the difference disappears because of the increasing dilution, so that all drops show a decline in concentration with increasing size. This behavior, however, is not expected to extend to hydrogen ion concentration, because of the influence of ion dissociation equilibria and the dissolution of reactive gases. Because of the uncertainty regarding the distribution of pH among individual cloud drops, it is currently impossible to derive appropriate averages over cloud water as a whole, so that the preceding description of the scavenging of reactive gases remains an approximation.

Simple Henry's law considerations also ignore interactions between solutes. Laj et al. (1997) have shown, for example, that departures from Henry's law equilibrium in the case of sulfur dioxide and hydrogen peroxide, observed in a cap cloud, can partly be explained by reactive losses. On the other hand, reversible reactions may be incorporated into the equilibrium concept of in-cloud scavenging. An interesting case is the formation of hydroxymethane sulfonate (HMS), an adduct arising from the interaction between formaldehyde and sulfite. (Warneck et al., 1978; Jacob and Hoffmann, 1983; Richards et al., 1983; Munger et al., 1984; Boyce and Hoffmann, 1984; Deister et al., 1986; Kok et al., 1986). In aqueous solution formaldehyde is largely present as a hydrate, and HMS formation is governed by the rate at which formaldehyde is dehydrated. The rate of HMS formation is faster in the alkaline pH range because SO_3^{2-} reacts more rapidly than HSO_3^-. The reverse reaction, that is, the dissociation of HMS, is catalyzed by hydroxyl ion:

$$H_2C(OH)_2 = HCHO + H_2O$$

$$HCHO + HSO_3^- (SO_3^{2-}) = CH_2(OH)SO_3^- (+OH^-)$$

For atmospheric concentrations the forward and reverse reactions are fairly slow, so that in-cloud equilibria between gas and liquid phases are not readily established, but the stability of the adduct is greatest in the pH range of 3–6, which is typical of clouds. The formation of HMS not only protects aqueous SO_2 from oxidation by ozone or H_2O_2, it also brings more formaldehyde and SO_2 into aqueous solution than Henry's law would permit. Figure 8.11 shows calculated in-cloud scavenging efficiencies for formaldehyde and SO_2 due to the formation of hydroxymethane sulfonate, assuming equilibria for gas–liquid exchange and reactions to be maintained. Depending on which of the two components is present in excess, the scavenging efficiency for either SO_2 or HCHO is considerably enhanced. In the marine atmosphere, for example, the residence time of SO_2 due to rain-out may be four times lower than in the absence of formaldehyde.

8.4.2. TIME CONSTANTS FOR THE ADJUSTMENT TO EQUILIBRIUM

Any perturbation of the gas–liquid equilibrium described by Henry's law sets up a flux of material tending to reequilibrate the system to the new conditions. Perturbations are due to uplift in the cloud by the change in temperature or, for a raindrop, by the change in gas phase concentrations in

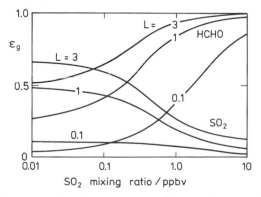

FIGURE 8.11   Enhancement of in-cloud scavenging efficiencies for formaldehyde and SO_2 due to their interaction in aqueous solution; the initial HCHO mixing ratio assumed in the calculation is 0.2 nmol mol^{-1}. (Adapted from Warneck, 1989.)

the air below a cloud. The mass flux is determined by molecular diffusion, inside and outside the water drop, by the transport resistance at the gas–liquid interface, and by the rate of hydrolysis and other chemical reactions in the aqueous phase. Schwartz and Freiberg (1981) and Schwartz (1986) have thoroughly analyzed the problem. Table 8.5 presents expressions for the time constants associated with individual processes. Some numerical values are entered in the fourth column of the table, and Figure 8.12 illustrates the dependence on drop size for diffusion-limited flow.

Hydrolysis and ion-dissociation equilibria are attained very swiftly in most cases, so that these processes are not rate determining. Molecular diffusion in the aqueous phase is rate-determining for slightly soluble gases and small drop sizes ($r < 30$ μm). The characteristic time for diffusion given in Table 8.5 follows from the mathematical solution of Equation (1.20). Diffusion coefficients in the aqueous phase are on the order of 1.8×10^{-9} m^2 s^{-1}. Figure 8.12 shows the resulting time constants as a function of drop size. Larger drops experience frictional drag and develop an internal circulation (Pruppacher and Klett, 1997), so that transport by mixing becomes faster than molecular diffusion. In this case other time constants are more important.

Diffusion coefficients for the gas phase are about four orders of magnitude greater than for the liquid phase. An additional enhancement occurs because the air current surrounding the drop is turbulent. This effect may be taken into account by a semiempirical factor first employed by Frössling (1938) in treating the evaporation of water from raindrops. The effective diffusion coefficient thus is

$$D_g^* = D_g(1 + 0.4\,\mathrm{Re}^{1/2}\,\mathrm{Sc}^{1/3}) \tag{8.19}$$

where D_g is the gas phase diffusion coefficient in the absence of turbulence, $\mathrm{Re} = r_1 v/\eta^*$ is the Reynolds number, $\mathrm{Sc} = \eta^*/D_g$ is the Schmidt number, v is the terminal fall velocity of the water drop, and $\eta^* = \eta/\rho_{\mathrm{air}}$ is the kinematic viscosity of air. The Frössling factor becomes noticeably different from unity for water drops with radii beyond 30 μm. Raindrops 1 mm in size create enough turbulence to enhance molecular diffusion by a factor of 10.

For slightly soluble gases, the gas phase concentration far from a single water drop remains essentially constant, and the concentration gradient in the vicinity of the drop follows the concentration change in the aqueous phase. The drop is assumed to be internally well mixed, and Equation (1.21)

TABLE 8.5 Absorption of Atmospheric Gases by Cloud and Raindrops: Time Constants for the Adjustment to Gas–Liquid Equilibrium

Process	Parameters	Expression for time constant	Examples	References
Diffusion inside spherical drop	D_L	$r_2/\pi^2 D_L$	See Fig. 8.12 curve 1 for $D_L = 1.8 \times 10^{-9}$ m^2 s^{-1}	Postma (1970), Crank (1975)
Approach to ion-dissociation equilibrium and hydrolysis	$k_f, k_r, [H^+]$	$1/(k_f + k_r[H^+] + [X^-])$	SO$_2$, 3×10^{-7}; CO$_2$, 10^{-7}; NH$_3$, 10^{-6} (pH 5)	Eigen et al. (1964)
Adjustment of phase equilibrium at the gas–liquid interface for diffusion limited transport inside the drop	$D_L, \bar{v}, \alpha, K_H^*$	$D_L(4\rho_w K_H^* R_g T/\alpha \bar{v})^2$	SO$_2$, 5×10^{-3}; CO$_2$, 5×10^{-13}; H$_2$O$_2$, 5.8; HNO$_3$, 5×10^7;	Danckwerts (1970)
Gas phase diffusion, drop internally well mixed, constant concentration at large distance from the drop	$D_g^*, \bar{v}, \alpha, K_H^*, r$	$(\rho_w K_H^* R_g T)/k_t$	See Fig. 8.12, curves 2, 3, 4 k_t as defined in Eq. (8.23)	Hales (1972)
Gas phase diffusion, highly soluble gases, substantial reduction of gas phase concentration	$D_g^*, \bar{v}, \alpha, K_H^*, L, r$	$(1-L)\varepsilon/Lk_t$	See Fig. 8.12, curve 5 k_t as defined in Eq. (8.23)	—

r, drop radius; D_L and D_g^* diffusion coefficients for the aqueous and gas phase, respectively; k_f and k_r, forward and reverse rate coefficients for the equilibrium reactions in Table 8.3; $\bar{v}$, gas phase molecular velocity; α, mass accommodation coefficient; K_H^*, modified Henry's law coefficient; R_g, gas constant; T, temperature; ρ_w, density of water.

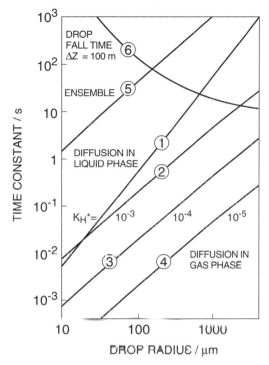

FIGURE 8.12 Time constants for the approach to the generalized Henry's law solution equilibrium for cloud drops and raindrops. Single drops: (1) mixing by diffusion inside the drop; (2)–(4) calculated from Equation (8.24) with $T = 283$ K and the Henry's law coefficients shown. Ensemble of drops: (5) calculated from Equation (8.26) with $\varepsilon = 0.5$ and $L = 1 \times 10^{-6}$. Raindrops: (6) Free-fall time over a distance of 100 m.

is applied. The amount of substance transferred in unit time is

$$dv/dt = A(D_g^*/r)(c_g - c_{g\,\text{surf}}) \qquad \text{due to gas phase diffusion}$$

(8.20a)

$$dv/dt = A(\alpha \bar{v}/4)(c_{g\,\text{surf}} - c_s/K_H^* R_g T) \qquad \text{due to mass accommodation}$$

(8.20b)

where $A = 4\pi r_2$ is the surface area of the drop, $\bar{v}$ is the average gas phase velocity of the molecules, $\alpha \le 1$ is the mass accommodation coefficient, $c_{g\,\text{surf}}$ and c_g are gas phase concentrations at the surface and far from the drop, respectively, and $c_s/K_H^* R_g T$ is the gas phase concentration at the surface corresponding to the aqueous concentration c_s in accordance with Henry's law. Equations (8.20a) and (8.20b) can be combined:

$$\frac{1}{k_1}\frac{dv}{dt} = \left(\frac{r}{D_g^*} + \frac{4}{\alpha\bar{v}}\right)\frac{dv}{dt} = A\left(c_g - c_s/K_H^* R_g T\right) \qquad (8.21)$$

so that one can derive the corresponding change in aqueous concentration:

$$dv/dt = (\rho_w V)\,dc_s/dt = Ak_1\left(c_g - c_s/K_H^* R_g T\right)$$

$$\rho_w\,dc_s/dt = k_t\left(c_g - c_s/K_H^* R_g T\right) \qquad (8.22)$$

where $V = (4/3)\pi r^3$ is the volume of the water drop, $A/V = 3/r$, and $k_t = (A/V)k_1 = 3k_1/r$ is the transfer coefficient. Comparison with Equation (8.21) shows

$$k_t = \left(\frac{r^2}{3D_g^*} + \frac{4r}{3\alpha\bar{v}}\right)^{-1} \qquad (8.23)$$

As the gaseous concentration c_g is assumed to remain constant, Equation (8.22) represents a first-order differential equation. The characteristic time for the quasi-exponential adjustment to equilibrium,

$$\tau = \left(\rho_w K_H^* R_g T\right)/k_t \qquad (8.24)$$

combines individual time constants for gas phase diffusion, for mass accommodation, and for filling the aqueous reservoir with the amount of substance required by Henry's law. Table 8.6 summarizes mass accommodation coef-

TABLE 8.6   Experimentally Determined Mass Accommodation Coefficients for the Gas–Liquid Transport of Several Atmospheric Gases[a]

Constituent	α	Constituent	α
O_3	0.002–0.05	HNO_2	0.15–0.05
H_2O_2	0.08–0.11	N_2O_5	0.02–0.05
HO_2	> 0.2	NO_2	> 0.0015
CO_2	~ 2×10^{-4}	PAN[b]	> 0.001
SO_2	0.09–0.13	HNO_3	0.11
HCOOH	0.015	$ClNO_2$	> 0.01
NH_3	0.04–0.09	HCl	0.08

[a] From the summary of Warneck et al. (1996); $\alpha(HO_2)$ is from Mozurkewich et al. (1987).
[b] Peroxyacetyl nitrate.

ficients for several atmospheric gases. In many cases, $\alpha \approx 0.01$ or more. The resistance due to mass accommodation then is smaller than that due to gas phase diffusion. Figure 8.12 shows time constants for several Henry's law coefficients calculated with $\alpha = 0.02$.

For well-soluble compounds, the gas phase concentration far from a drop will not remain constant, and the presence of many such drops enhances the uptake rate. Consider an ensemble of cloud drops or raindrops of uniform size with liquid water fraction L. Equation (8.10) is used to replace c_g in Equation (8.22) to obtain

$$\rho_w \, dc_s/dt = k_t \left(c_0/(1 - L) - \rho_w L c_s/(1 - L) - c_s/K_H^* R_g T \right)$$

$$dc_s/dt = k_t \left[c_0/\rho_w L - \left(1 + (1 - L)/\rho_w L K_H^* R_g T \right) c_s \right] L/(1 - L)$$

$$(8.25)$$

with $c_0 = \nu_0/V_c$. Note that the factor associated with c_s on the right-hand side is equal to $1/\varepsilon$, that is, the inverse of the fractional amount of substance residing in the aqueous phase at equilibrium, as defined in Equation (8.10). The solution to this first-order differential equation indicates an exponential approach to equilibrium. The corresponding time constant is

$$\tau = (1 - L)\varepsilon/Lk_t$$ $$(8.26)$$

Note that this time constant becomes equal to that shown in Equation (8.24) in the limit $\varepsilon \ll 1$. Figure 8.12 includes a curve calculated for $\varepsilon = 0.5$ and $L = 1 \times 10^{-6}$ to show that the time constants can be appreciable for highly soluble gases. Figure 8.12 includes the time required for a raindrop to traverse a 100-m layer. Comparison makes evident that the adjustment to Henry's law equilibrium can be appreciably longer.

The data in Table 8.5 and Figure 8.12 show that gas–liquid partitioning of small water drops occurs within a fraction of a second, so that for liquid water clouds Henry's law equilibrium is essentially always established. Large raindrops also adjust fairly rapidly to changing external conditions in the case of slightly soluble species. The concentration in the aqueous phase then is given by Equation (8.12a). Comparatively long time constants are observed for the equilibration of well-soluble species such as SO_2, NH_3, or HNO_3. Waltrop et al. (1991), Mitra et al. (1992), Mitra and Hannemann (1993), and Hannemann et al. (1995, 1996) have conducted laboratory studies of the uptake and desorption of SO_2, NH_3, and CO_2 by raindrop-sized water drops and found good agreement with rates predicted from the convective diffusion model. In the region below a cloud, falling raindrops experience rising concentrations of these species, so that equilibrium generally is not reached.

Raindrops thus scavenge only a fraction of the material that they are potentially capable of absorbing. Prolonged rainfalls will nevertheless be effective in eventually depleting gas phase concentrations.

8.5. INORGANIC CHEMICAL COMPOSITION AND pH OF CLOUD AND RAIN WATER

The major ions commonly determined in precipitation are the anions sulfate (SO_4^{2-}), chloride (Cl^-), and nitrate (NO_3^-) and the cations of the alkali and alkaline earth metals $(Na^+, K^+, Mg^{2+}, Ca^{2+})$, in addition to ammonium (NH_4^+). The hydrogen ion concentration $[H^+]$ is inferred from measurement of the pH by virtue of the relation $pH = -\log_{10}(a_{H^+})$ where $a_{H^+} = \gamma[H^+]$ is the proton activity. For sufficiently dilute solutions such as cloud water and rainwater, the activity coefficient γ is close to unity.

Table 8.7 provides an overview on concentrations of major ions in rainwater observed at various locations. Table 8.8 adds similar information on cloud and fog waters. In coastal regions and over the ocean, sodium chloride contributes the largest fraction of all ions. Some of the other ions usually are somewhat enriched in comparison with their abundance in sea salt. The enrichment of potassium and calcium occurs because of the admixture of condensation nuclei from continental sources, and that of sulfate arises from the oxidation of gaseous precursors such as dimethyl sulfide and SO_2. On the continents, the presence of sea salt in precipitation diminishes with distance from the coast, by about 80% within the first 10 km and then more slowly, as indicated by the chloride content of rainwater. Remnants of sea salt still occur hundreds of kilometers inland, as the data from the former Soviet Union in Table 8.6 attest. Junge (1963) discussed the geographical distribution of chloride in precipitation over Europe and the United States, and Hingston and Gailitis (1976) and Hutton (1976) presented similar data for Australia. Gaseous HCl, scavenged by cloud drops and raindrops, also contributes to the chloride content of precipitation, but this fraction is difficult to estimate.

The major fraction of potassium, magnesium, and calcium in continental rainwater is due to the mineral component of aerosol particles originating from soils. Sea salt contributes only small amounts of these components, in contrast to sodium. Nitrate and ammonium enter cloud water as constituents of condensation nuclei as well as by gas-liquid scavenging of HNO_3 and NH_3. The relative input is quite variable and depends strongly on the supply. Precipitation on the continents contains higher amounts of NO_3^- and NH_4^+ than that in marine regions. Sulfate and ammonia are linked in that NH_3

TABLE 8.7 Inorganic Ion Composition (μmol dm^{-3}) of Rainwater at Various Locations

	Continental						Marine			
	North Sweden[a] 1969	Belgium[a] 1969	East Europe[b] 1961–1964	New Hampshire[c] 1963–1974	California[d] 1978–1979	San Carlos, Venezuela[e] 1979–1980	Katherine, Australia[f] 1980–1984	Cape Grim, Tasmania[g] 1977–1981	Amsterdam Island[e] 1980–1987	Bermuda[e] 1979–1980
SO_4^{2-}	21	63	82	29.8 ± 1.2	19.5	1.6 ± 1.2	2	79.1	37.2	24.4 ± 23.8
Cl^-	11	55	60	14.4 ± 2.5	28	4.3 ± 3.4	8	1349	317.6	264 ± 337
NO_3^-	5	36	17.7	23.1 ± 1.7	31	3.5 ± 3.6	4.1	5	1.6	7.9 ± 9.1
HCO_3^-	21	—	71.7	—	0.5	—	—	—	—	—
HCO_3^-[h]	—	0.15	1.73	0.077		0.33	0.32	5.6	0.67	0.27
NH_4^+	6	25	52.8	12.1 ± 0.5	21	17.0 ± 10	2.9	2	2.4	4.8 ± 7.1
Na^+	13	42	65.2	5.4 ± 0.6	24	2.7 ± 2.0	4.5	1297	268.8	221 ± 282
K^+	5	6	17.3	1.9 ± 0.5	2	1.1 ± 1.7	1.1	32.3	5.9	6.5 ± 8.0
Ca^{2+}	16	33	38.1	4.3 ± 0.6	3.5	0.25 ± 0.35	0.95	42.9	12.1	7.2 ± 7.1
Mg^{2+}	5	15	52.5	1.8 ± 0.4	3.5	0.4 ± 0.45	0.7	122	60.3	24.6 ± 30.4
H^+	—	38	3.3	73.9 ± 3.2	39	17 ± 10	16.9	1	8.4	21.1 ± 26.9
pH (av.)	—	4.42	5.48	4.13	4.41	4.77	4.77	5.99	5.08	4.68
Sum of anions[i]	79	217	313	97.1	98	11	16.1	1512	397	321
Sum of cations[i]	66	207	320	105	93	39	28.7	1662	430	273
Number of samples	180	180	n.g.[k]	n.g.	n.g.	14	147	56	179	67
Annual precip.[j]	360	648	—	1310	697	4000	104	1120	1120	1131

[a] Granat (1972).
[b] Former U.S.S.R., Petrenchuk and Selezneva (1970).
[c] Hubbard Brook, Likens et al. (1977).
[d] Liljestrand and Morgan (1980), Morgan (1982).
[e] Galloway and Gaudry (1984), Moody et al. (1991).
[f] Likens et al. (1987).
[g] Ayers (1932).
[h] Calculated from pH.
[i] Sum of charges.
[j] Average annual rate, mm.
[k] n.g., Not given.

TABLE 8.8 Concentrations (μmol dm^{-3}) of Major Inorganic Ions in Cloud and Fog Waters

	East Europe[a]		Southern England[b] nonprecipitating	Whiteface Mt., New York[c] intercepted clouds	Kleiner Feldberg, Germany[d] intercepted clouds	Brocken, Germany[e] mountain fog	Pasadena, California[c] ground fog	Laegeren Mt., Switzerland[f] ground fog
	Frontal precipitating	Non-precipitating						
SO_4^{2-}	29	117	40	26–70	61.7	387	240–472	40–1719
Cl^-	22.2	76.8	94	1.7–3.1	84.7	205	480–730	68–4316
NO_3^-	3.2	16.1	18.6	140–215	183.7	225	1220–3250	44–4420
HCO_3^-	11.5	15.4	—	—	—	—	—	—
HCO_3^- [g]	1.1	0.06	0.14–91	0.023–0.045	0.03	0.72	0.005–0.4	0.004–57
NH_4^+	28.8	11.1	22.1	32–89	185.6	710	1290–2380	102–9215
Na^+	17	29.8	95.2	2.3–11	46.3	295	320–500	14–789
K^+	5.1	20.4	12.5	13–20	8.7	85	33–53	9–424
Ca^{2+}	10	29.3	33.2	5–10	6.0	110	70–265	7–11230
Mg^{2+}	12.3	40.0	12.3	1.1–3.1	8.7	—	45–160	0–135
H^+	5.0	94.4	0.06–40	126–251	168	7.9	14–1200	0.1–1380
pH	5.3	4.02	4.4–7.2	3.6–3.9	3.77	5.1	2.92–4.85	2.9–7.1
Sum of anions[h]	95	327	193	194–358	407	1429	2180–4924	192–12074
Sum of cations[h]	100	294	220	185–397	438	1323	1887–4983	139–34538
Number of samples	125	194	23	Not given	8	19	4	83

[a] Former U.S.S.R., Petrenchuk and Selezneva (1970).
[b] Oddie (1962).
[c] Munger et al. (1983).
[d] Mrose (1966).
[e] Average for eight clouds, Wobrock et al. (1994).
[f] Joos and Baltensperger (1991).
[g] Estimated from pH.
[h] Sum of charges.

partly neutralizes sulfuric acid formed in the oxidation of SO_2, so that condensation nuclei supply SO_4^{2-} and NH_4^+ in molar ratios in the range of 0.5–1. Some nitrate also may be present as ammonium nitrate. The incorporation into cloud water and rainwater of HNO_3 and NH_3 adds further ammonium and nitrate. The data in Table 8.7 indicate that in the Northern Hemisphere the SO_4^{2-}/NH_4^+ ratio usually exceeds unity by some margin, showing that the supply of ammonia is insufficient to balance the flux of excess sulfur from anthropogenic emissions during the observation period. The situation is different for the Southern Hemisphere. As Table 8.7 shows, sulfate is a relatively minor component at remote sites such as San Carlos, Venezuela, and Katherine, Australia.

The total ion concentrations at San Carlos and Katherine are lower than elsewhere. At the tropical rain forest site, the low ion concentration probably is associated with the very high annual precipitation rate. Although there is considerable scatter in the data from all stations, the concentrations usually are inversely correlated with precipitation rate, even at Katherine, where the annual precipitation rate is low. Likens *et al.* (1984) have presented detailed evidence for this relation at Hubbard Brook, New Hampshire. At many locations there is a seasonal variation in ion concentration because of the variation in precipitation rate. In convective showers the intensity of rainfall frequently goes through a maximum, while at the same time the ion concentration goes through a minimum (Gatz and Dingle, 1971; Kins, 1982). This behavior contrasts with frontal precipitation events, where the initial high concentration diminishes with time toward a steady level that is fairly independent of precipitation rate. The latter effect is explained by the evaporation of raindrops during the initial stage of precipitation, and by the gradual cleansing of the atmosphere from material by washout.

Table 8.8 shows that ion concentrations in cloud water are fairly similar to those observed in rainwater. In fog water, however, they are much higher. The phenomenon must be caused by the higher concentration of particulate matter in the ground-level atmosphere compared to cloud levels, according to Equation (8.1), as the liquid water content L does not differ much between fog and cloud systems. The great variability in ion concentrations observed in ground fogs as well as in intercepted mountain clouds is caused not only by the variability of the liquid water content, but also by that in aerosol concentration. This feature is highlighted by considering the product of liquid aqueous concentration and liquid water content. Joos and Baltensperger (1991) found for fogs in Switzerland that the concentrations of NH_4^+, SO_4^{2-}, NO_3^-, and Cl^-, which represented 90% of the ion sum, were well correlated with their concentrations in the precursor aerosol.

The major cations and anions in precipitation are not always well balanced. In Table 8.7, the imbalance is particularly evident in the data from

San Carlos and Katherine, which show a deficit of anions. A simple calculation based on Equation (8.12a) reveals that the concentration of bicarbonate resulting from atmospheric CO_2 is below 1 μmol dm^{-3} as long as the solution has a pH $\leq$ 5.3. The missing concentration of HCO_3^- thus cannot account for the deficit. Galloway *et al.* (1982), Likens *et al.* (1983), and Keene *et al.* (1983) have called attention to the importance of weak organic acids such as formic acid and acetic acid. Both acids may contribute appreciably to hydrogen ion concentration. At Katherine, Likens *et al.* (1987) found that the anions of these acids were dominant. At Hubbard Brook, carboxylic acids contributed about 16% by mass to total dissolved organic matter, but their concentration of 4.3 μmol dm^{-3} represented only a small fraction of total anion activity (Likens *et al.*, 1983). At Amsterdam Island the average concentrations of formic and acetic acid were 3.7 μmol dm^{-3} (Galloway and Gaudry, 1984). At San Carlos, the concentrations were higher, making the contribution to total anion activity more important. Sources of formic and acetic acid in the atmosphere are discussed in Section 6.1.6.

Other minor anions not listed in the tables are the halides F$^-$, Br$^-$, and I$^-$. Nitrite has also been observed. The review of Lammel and Cape (1996) suggests that NO_2^- concentrations range from 0.5 to 50 μmol dm^{-3}, whereas in fogs they are 10 times higher. Hydrogen sulfite, (HSO_3^-) and sulfite (SO_3^{2-}) are also present because of the absorption of gaseous SO_2 by cloud water and rainwater. Richards *et al.* (1983) have explored sulfite in Californian cloud waters and found aqueous concentrations much higher than expected from Henry's law and the acid dissociation constants. High concentrations of formaldehyde were simultaneously observed, suggesting that both compounds occurred as hydroxymethane sulfonate, as discussed in Section 8.4.1. The gas phase mixing ratio of HCHO in unpolluted regions of the lower atmosphere is about 0.2 nmol mol^{-1}. The corresponding concentration of $CH_2(OH)SO_3^-$ is 1–3 μmol dm^{-3}, which is similar to the concentration of formaldehyde in rainwater (Klippel and Warneck, 1978; Thompson, 1980; Zafiriou *et al.*, 1980). The concentrations observed by Richards *et al.* (1983) were an order of magnitude higher.

Hydrogen peroxide in cloud water and rainwater has been a focus of interest, because in addition to being scavenged from the gas phase, it may be produced in the aqueous phase, and it undergoes reaction with sulfite. Gunz and Hoffmann (1990) have reviewed analysis techniques, gas phase and aqueous phase concentrations, and H_2O_2 source mechanisms. The highest concentrations, in both gas and aqueous phases, were found in heavily polluted areas, such as Los Angeles, and during the summer months. This agrees with the notion that H_2O_2 arises from the self-reaction of HO_2 radicals of photochemical origin. Typical concentrations at Whitetop Mountain, Virginia, were gas phase 0.02–2.6 nmol mol^{-1}, cloud water 0.04–247

μmol dm^{-3}, rainwater 0.04–40 μmol dm^{-3} (Olszyna *et al.*, 1988). Several investigators have looked for and found an inverse relationship between H_2O_2 and SO_2—for example, Daum *et al.*, (1984) in an aircraft study, and Munger *et al.* (1989) and Clark *et al.* (1990) in mountain clouds. However, Mohnen and Kadlecek (1989) found at Whiteface Mountain, New York, that either H_2O_2 or S(IV), that is, the sum of all SO_2-related aqueous species, were present in cloud water, but not both at the same time. Richards *et al.* (1983) observed the coexistence of H_2O_2 and S(IV) in the presence of S(IV) aldehyde adducts.

Tables 8.7 and 8.8 indicate that cloud water and rainwater are moderately acidic, on average. The pH generally lies in the range of 3–7. Hydrogen ions derive from the dissociation of acids, which serve as proton donors, whereas bases act as proton acceptors. The associated ion equilibria are summarized in Table 8.3. Strong acids such as HNO_3 are fully dissociated, even at H^+ concentrations approaching 1 mol dm^{-3}. For weak acids, the degree of dissociation usually is smaller than unity. Cloud water and rainwater are aqueous mixtures of strong and weak acids, which are partly neutralized by alkalis. The proton activity and the pH in individual cloud drops and raindrops, just as in bulk solution, adjust to a value governed by the condition of charge balance between positive and negative ions, in combination with the pertinent ion dissociation and Henry's law equilibria. The resulting set of equations enables one to calculate the H^+ concentration of the solution, provided the concentrations of different acids and bases in the system are known. For atmospheric conditions the input parameters usually are incompletely defined, making it necessary to use simplifying assumptions. One assumption appears to hold for polluted regions, where the oxidation of SO_2 and NO_2 is a major source of the strong acids H_2SO_4 and HNO_3, respectively. If these acids dominate over the weak ones, the proton concentration will be determined primarily by the excess of total anion over total cation equivalents. Cogbill and Likens (1974), for example, have applied this principle in comparing predicted and observed H^+ concentrations in rainwaters of the northeastern United States. The found a significant linear correlation, indicating that strong acids were indeed dominant. Another assumption is often made for marine conditions. Here, sea salt is the major contributor to the ion content of atmospheric waters. In aqueous solutions of sea salt, the anions and cations of strong acids and bases are almost perfectly in balance. In this case, the hydrogen ion concentration is determined by the weak acids, first of all by dissolved carbon dioxide. For a CO_2 mixing ratio near 360 μmol mol^{-1}, the pH of a solution in equilibrium with the ground-level atmosphere is about 5.6, depending somewhat on pressure and temperature. In the past, this value has been considered a reference level for the remote atmosphere. However, the uptake of even small concentrations of gaseous ammonia, sulfur dioxide, organic acids, etc., existing in the remote troposphere, may cause the pH level to shift considerably away from the value represented by CO_2 alone. And indeed, the data in Table 8.7 for marine rainwaters clearly show pH values distinctly different from pH 5.6. The data from Cape Grim in Tasmania are slightly higher, whereas the pH values at Amsterdam Island and at Bermuda are lower by almost one unit. Bermuda is episodically subject to pollution from the North American continent (Jickells et al., 1982), but Amsterdam Island is not thus affected.

Similar to the concentrations of other ions, proton concentration and pH fluctuate widely among individual rainwater samples. Figure 8.13 shows frequency distributions of pH values observed at several locations. The data from Bad Reinerz are mainly of historical interest, as they are among the first ever obtained (Ernst, 1938). The measurements at Dresden, 200 km further to the west, were made 20–25 years later, but they show a similar, perhaps

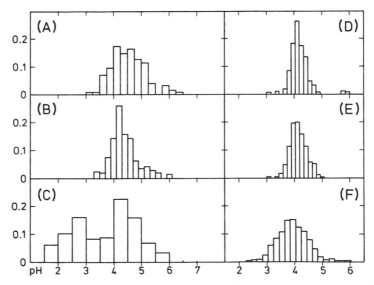

FIGURE 8.13 Frequency distribution of pH values observed in hydrometeors. (A) Ernst (1938), 47 bulk rain samples collected in Bad Reinerz, combined with 80 samples, showing a similar distribution collected in Oberschreiberhau (both locations in upper Silesia, central Europe), 1937–1938. (B) Mrose (1966), 206 bulk rain samples collected at Dresden-Wahnsdorf, Germany, 1957–1964. (C) Esmen and Fergus (1976), about 200 individual raindrops collected during a single rainstorm in Delaware, 1974. (D, E) Likens et al. (1984), weekly samples of bulk rainwater collected at Hubbard Brook Experimental Forest Station, New Hampshire, 1964–1968 (D), 1975–1979 (E). (F) Falconer and Falconer (1980), 824 measurements of cloud water collected continuously at Whiteface Mountain, New York, 1977.

slightly narrower and thus more peaked distribution (Mrose, 1966). The data from Hubbard Brook feature an even narrower distribution compared with that at Dresden. In particular, the tail at values beyond pH 6 is nearly absent. Likens *et al.* (1984) consider it significant that the isolated precipitation events with pH ≈ 6 during 1964–1968 disappeared a decade later. The average proton concentration and pH have remained essentially the same, however. In the same period the annual average of the concentration of sulfate declined, whereas that of nitrate increased.

Figure 8.13 includes the distribution of cloud water collected at the summit of Whiteface Mountain in upper New York state. The distribution is broader than that of rainwater at Hubbard Brook, and it extends toward values of pH < 3. The average pH in nonprecipitating clouds in 1977 was 3.55. The onset of precipitation increased the pH abruptly because the mixing of cloud drops and raindrops at the collector surface, so that the average pH in precipitating clouds was higher by about half a unit compared to nonprecipitating clouds. Esmen and Fergus (1976) have explored the pH of individual raindrops. Figure 8.13 shows one of the distributions observed during four separate rainstorms. All of the data display two maxima, one in the pH range of 2–3, the other one in the pH range of 4–5. The total range covered extends from pH 2 to pH 8, but the highest values are rather infrequent. Bulk collection of rainwater represents a volume-averaging process that discriminates against small drops. If smaller drops were more acidic than larger ones, bulk collection would not only narrow the pH distribution, but would also remove the left-hand tail of pH < 3, because small drops merging with larger ones undergo dilution, causing the lowest and highest pH values to disappear.

Georgii and Wötzel (1970) have used a set of rotating slotted discs to discriminate between raindrops of different sizes and found that the concentrations of several ions other than H^+ decreased with increasing drop size. This behavior may be explained by the increase in concentration when raindrops shrink in size by below-cloud evaporation. Adams *et al.* (1986) have used a similar device to determine the pH as a function of drop size. They observed a decrease with size for drops larger than about 0.7 mm. Below that radius the pH increased again, so that a minimum in pH, that is a maximum of H^+ concentration, occurred near 0.7-mm radius. Bächmann *et al.* (1993, 1996) have collected individual raindrops in liquid nitrogen, separated seven size fractions with radii in the range 0.1–0.8 mm, and subjected them to chemical analysis. Their data confirmed the inverse relationship between concentration and drop size, but only for the initial 10 min of rainfall. Longer precipitation events eventually developed a concentration maximum in raindrops about 0.2 mm in size for various species analyzed, including H^+. The behavior was not the same for all species, however, which suggested that processes other than drop evaporation contributed to the concentration–size dependence. From the nature of rain formation one would expect at the cloud base a fairly even distribution of concentration among raindrops of all sizes. This has also been demonstrated by modeling studies (Flossmann *et al.*, 1987, 1991). Below-cloud concentration changes are due primarily to the scavenging of aerosol particles. Figure 8.7 shows that raindrops in the 0.5-mm size range are more efficient than others in scavenging aerosol particles because of capture in the wake eddy, and Bächmann *et al.* (1993) suggested this to be the most likely process responsible for the concentration maximum. By collecting raindrops at two levels of elevation along the slope of a mountain, Bächmann *et al.* (1996) confirmed that the concentration was size-independent at 550-m elevation, whereas a distinct concentration maximum appeared at the 150-m level for the usual anions and cations.

Trends in precipitation acidity have been under study ever since the first long-term sampling networks of stations were established in the 1950s. Odén (1968, 1976) first called attention to a decrease in the annual average pH of precipitation at a number of Scandinavian stations coupled with the simultaneous rise in sulfate concentration, which was suggested to stem from the increase in anthropogenic SO_2 converted to H_2SO_4. Likens (Likens, 1976; Likens *et al.*, 1979) discussed a similar development for the northeastern United States. Kallend *et al.* (1983) examined European network data and found that among 120 sampling sites, 29 showed a statistically significant trend of increasing hydrogen ion concentration, typically by a factor of 3–4 within 20 years. At the same time the concentrations of nitrate had increased substantially, whereas the rise of sulfate was less conspicuous. In the early 1980s, electric power companies in western Europe and elsewhere undertook efforts to reduce SO_2 emissions from power plants, partly in response to new legal requirements. In 10 western European countries, for example, a 50% reduction was achieved. This has led to a reversal in the trend of precipitation acidity. Figure 8.14 shows annual averages for sulfate, nitrate, and pH at Deuselbach, West Germany, for the period 1982–1995. During this period, the concentration of sulfate decreased by a factor of 2, that of nitrate by about 40%, and the pH rose from 4.2 to about 4.8, which corresponds to reduction in the deposition of hydrogen ion by more than a factor of 2. In the eastern and midwestern United States, where the reduction in SO_2 emissions was about 20% (Butler and Likens, 1991), the concentration of sulfate in precipitation was found to have decreased at 42 of 58 measurement sites during the period 1980–1992 (Lynch *et al.*, 1995). Concurrent with the trend in sulfate, base cations (NH_4^+, Ca^{2+}, Mg^{2+}, Na^+, K^+) have decreased, most notably Ca^{2+} and Mg^{2+}, whereas the decrease in hydrogen ion concentration was less conspicuous. Hedin *et al.* (1994) have suggested that

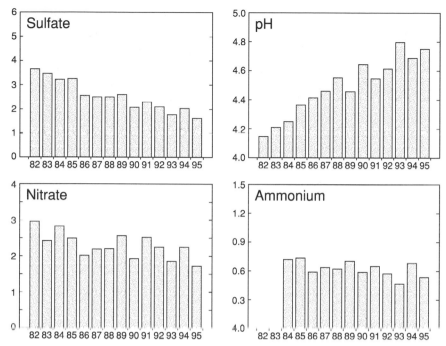

FIGURE 8.14   Trends in the annual average concentrations (in mg dm^{-3}) of sulfate, nitrate, ammonia, and pH in precipitation at Deuselbach, a station of the German Environmental Administration (Umweltbundesamt), during the period 1982–1995 (Fricke *et al.*, 1997).

the combined effects of changes in anion and cation concentrations have offset one another to varying degrees. In fact, as Lynch *et al.* (1995) noted, a significant decrease in hydrogen ion concentration concurrent with that of sulfate has occurred at only 17 of 42 sites.

8.6. CHEMICAL REACTIONS IN CLOUD AND FOG WATERS

Chemical reactions in cloud and fog waters have been considered important ever since the subject of tropospheric chemistry was first developed in the 1950s. Junge and Ryan (1958) called attention to the great potential of cloud water for the oxidation of dissolved SO_2 by transition metals as catalysts, and for many years the process appeared to be the only process capable of oxidizing SO_2 in the atmosphere. When the concept of OH radical reactions gained ground, SO_2 was recognized to also be oxidized by gas phase reactions. Penkett *et al.* (1979b) called attention to ozone and hydrogen peroxide as strong oxidants of SO_2 in the aqueous phase. Chameides and Davies (1982) postulated OH radicals to play a role in the aqueous phase of clouds similar to that in the gas phase. Jacob (1986) and Lelieveld and Crutzen (1991) have pursued this concept further. Graedel *et al.* (1986b) pointed out that HO_2 radicals would react with transition metals in such systems, and recent laboratory studies summarized by Warneck *et al.* (1996) have led to further progress in this direction. Thus, atmospheric aqueous phase chemistry has developed into a subject of considerable complexity. The following presentation will first consider marine clouds, where low concentrations of many atmospheric constituents simplify the system, before other species are discussed.

8.6.1. Basic Photochemistry in Clouds

Sunlit clouds are characterized by reactions in the gas phase as well as in the aqueous phase. However, gas–liquid transfer of strongly water-soluble gases such as formaldehyde and hydrogen peroxide significantly reduces their concentrations in the gas phase. Figure 8.15 shows principal gas phase reactions on the left and the resulting aqueous phase chemistry on the right (for rate coefficients see Table A.6). The main source of OH radicals in the gas phase is the photodissociation of ozone into $O(^1D)$ and its reaction with water vapor. Although this process would also occur in the aqueous phase, it is rather ineffective there because of the much lower concentration of ozone. Photolysis of hydrogen peroxide is not a good source of OH radicals in the gas phase, because of low H_2O_2 absorption cross section at wavelengths greater than 300 nm. The process is more important in the aqueous phase, where H_2O_2 becomes concentrated. The principal source of aqueous OH and HO_2 is intrusion from the gas phase. Most significant in this regard is the fact that HO_2 acts as an acid, with O_2^- as the conjugate anion ($pK_a = 4.88$), which reacts rapidly with dissolved ozone to form the ozonide anion O_3^-. This species is stable only in very alkaline solution. In neutral and acidic solutions it reacts readily with hydrogen ion to generate OH radicals. Hydrated aqueous formaldehyde, $CH_2(OH)_2$, reacts with OH to produce formic acid. In contrast to HCHO, $CH_2(OH)_2$ does not absorb ultraviolet light, so that it is not photolyzed; and carbon monoxide, which in the gas phase is the principal product from formaldehyde, is not produced in the liquid phase. Formic acid reacts further with OH radicals to ultimately produce CO_2. Except in very acidic solution, it is the formate anion ($pK_a = 3.74$) that reacts most rapidly with OH, whereas undissociated acid reacts more slowly. In each of these reactions an HO_2/O_2^- radical is formed

GAS PHASE	GAS-LIQUID EXCHANGE	AQUEOUS PHASE
$NO_2 + h\nu \rightarrow NO + O$		
$O + O_2 + M \rightarrow O_3 + M$	$\longleftarrow O_3 \longrightarrow$	$O_3 + h\nu \rightarrow O(^1D) + O_2$
$NO + O_3 \rightarrow NO_2 + O_2$		$O(^1D) + H_2O \rightarrow 2\,OH$
$O_3 + h\nu \rightarrow O(^1D) + O_2$		
$O(^1D) + M \rightarrow O + M$		
$O(^1D) + H_2O \rightarrow 2\,OH$	$\longleftarrow OH \longrightarrow$	$HO_2 \rightleftharpoons H^+ + O_2^-$
$OH + O_3 \rightarrow HO_2 + O_2$	$\longleftarrow HO_2 \longrightarrow$	$O_2^- + O_3 \rightarrow O_3^- + O_2$
$HO_2 + O_3 \rightarrow HO + 2\,O_2$		$H^+ + O_3^- \rightarrow OH + O_2$
$OH + CH_4 \rightarrow CH_3 + H_2O$		
$CH_3 + O_2 \rightarrow CH_3O_2$	$\leftarrow CH_3O_2 \rightarrow$	$CH_3O_2 + HO_2 \rightarrow CH_3O_2H + O_2$
$HO_2 + CH_3O_2 \rightarrow CH_3O_2H + O_2$	$\leftarrow CH_3O_2H \rightarrow$	$CH_3O_2H + OH \rightarrow CH_3O_2 + H_2O$
$CH_3O_2 + NO \rightarrow CH_3O + NO_2$		$HCHO + H_2O \rightleftharpoons CH_2(OH)_2$
$CH_3O + O_2 \rightarrow HCHO + HO_2$		$OH + CH_2(OH)_2 \rightarrow H_2O + CH(OH)_2$
$HCHO + h\nu \rightarrow CO + H_2$	$\leftarrow HCHO \rightarrow$	$CH(OH)_2 + O_2 \rightarrow HCOOH + HO_2$
$HCHO + h\nu \rightarrow HCO + H$		$HCOOH \rightleftharpoons HCOO^- + H^+$
$HCO + O_2 \rightarrow CO + HO_2$		$OH + HCOO^- \rightarrow H_2O + CO_2^-$
		$CO_2^- + O_2 \rightarrow CO_2 + O_2^-$
		$H_2O_2 + OH \rightarrow H_2O + HO_2$
$OH + CO \rightarrow CO_2 + H$		$HO_2 + HO_2 \rightarrow H_2O_2 + O_2$
$H + O_2 + M \rightarrow HO_2 + M$		$HO_2 + O_2^- + H^+ \rightarrow H_2O_2 + O_2$
$HO_2 + HO_2 \rightarrow H_2O_2 + O_2$	$\longleftarrow H_2O_2 \longrightarrow$	$H_2O_2 + h\nu \rightarrow 2\,OH$

FIGURE 8.15 Photochemically induced reaction pathways in marine clouds: gas phase reactions on the left, aqueous phase reactions on the right.

for an OH radical spent, so that a chain reaction is set up similar to that in the gas phase, except that nitrogen oxides are not involved. Chain termination occurs mainly by the recombination of HO_2/O_2^- radicals, leading to the formation of H_2O_2. Hydrogen peroxide reacts with OH radicals in solution as it does in the gas phase, with HO_2 and H_2O as products. In sunlit marine clouds, aqueous phase steady-state OH and HO_2/O_2^- concentrations adjust to about 0.6 pmol dm^{-3} and 10 nmol dm^{-3}, respectively. Methyl hydroperoxide, CH_3O_2H, undergoes similar reactions in cloud water as H_2O_2. The solubility of CH_3O_2H is lower, however, so that the fraction produced by aqueous recombination of CH_3O_2 with HO_2 largely escapes to the gas phase. Methane, carbon monoxide, and the nitrogen oxides are only slightly soluble in water. Aqueous phase concentrations are low, and their reactions can be neglected. The concentration of bicarbonate, HCO_3^-, is appreciable, but the rate coefficient for reaction with OH is two orders of magnitude smaller than that for formate anion, which is present at a comparable concentration. Moreover, the carbonate radical, CO_3^-, produced from the reaction of OH with HCO_3^-, reacts itself rather slowly except by recombination with HO_2/O_2^- to form HCO_3^- back. Reactions of HCO_3^- therefore do not exert much influence.

Lelieveld and Crutzen (1990) suggested that the above reactions may lead to a significant loss of ozone from the troposphere, not only because of its reaction with O_2^- in cloud water, but also because the gas phase rate of formation is reduced, which is due to the reaction of HO_2 with NO followed by photodissociation of the NO_2 product. The generation of formic acid is only a temporary source of cloud acidity, because the oxidation of formate anion to CO_2 removes a hydrogen ion as well. In addition, more formic acid is present in marine air than is generated in clouds. In the air over the continents, the much higher abundance of formic acid (~ 1 nmol mol^{-1}), released from vegetation, causes this compound to be the dominant reactant with OH radicals. In fact, the rapid rate of this reaction leads to transport limitation of OH radicals that are imported from the gas phase. Continental clouds also differ from marine clouds in that reactions involving dissolved SO_2 can no longer be ignored, and that cloud condensation nuclei supply the aqueous phase with transition metals, which scavenge HO_2/O_2^- radicals, so that OH production from the reaction of O_2^- with ozone is greatly reduced.

8.6.2. REACTIONS OF NITROGEN COMPOUNDS

The ammonium cation is essentially immune to chemical reactions. Ammonia reacts with OH in the aqueous phase as it does in the gas phase. The reaction is slow, however. In addition, as cloud drops usually are acidic, ammonia is largely converted to NH_4^+, so that the concentration of NH_3 in clouds is small in both phases. Gas-phase reactions of nitrogen oxides are discussed in Section 9.4.1. Important species are selected here to show in Table 8.9 the major reactions pertinent to clouds. The oxidation of NO_2 occurs along two pathways. One reaction involves OH to form HNO_3 directly. This gas-phase reaction occurs only during the day. The other pathway involves ozone to generate the NO_3 radical, which then associates with another NO_2 to form nitrogen pentoxide, N_2O_5. This compound is the anhydride of nitric acid. It does not react with water vapor at any measurable rate, but it is rapidly hydrolyzed by liquid water to form HNO_3 essentially upon contact. The gas–liquid uptake coefficient is $\gamma \approx 0.02$ (Table 8.6), so that the reaction is not transport-limited by gas phase diffusion. In daylight,

TABLE 8.9 Reactions and Percentage Contributions to the Oxidation of Nitrogen Oxides in a Sunlit Cloud[a]

Gas phase reactions	Gas–liquid exchange	Aqueous phase reactions	Percentage contribution
$O_3 + NO \rightarrow NO_2 + O_2$	$\leftarrow NO_2 \rightarrow$	$NO_2 + NO_2 \rightleftarrows N_2O_4$ $N_2O_4 + H_2O \rightarrow HNO_2 + NO_3^- + H^+$	$\ll 0.1$
$OH + NO_2 \rightarrow HNO_3$	$\leftarrow HNO_3 \rightarrow$	$HNO_3 \rightarrow H^+ + NO_3^-$ $NO_3^- + h\nu + H^+ \rightarrow NO_2 + OH$	59
$O_3 + NO_2 \rightarrow NO_3 + O_2$ $NO_3 + h\nu \rightarrow NO_2 - O$	$\leftarrow NO_3 \rightarrow$	$NO_3 + Cl^- \rightarrow NO_3^- + Cl$ $NO_3 + HSO_3^- \rightarrow NO_3^- + SO_3^- + H^+$	1.0
$NO_3 + NO_2 \rightleftarrows N_2O_5$	$\leftarrow N_2O_5 \rightarrow$	$N_2O_5 + H_2O \rightarrow 2H^+ + 2NO_3^-$	11
$OH + NO \rightarrow HNO_2$ $HNO_2 + h\nu \rightarrow NO + OH$	$\leftarrow HNO_2 \rightarrow$	$HNO_2 \rightleftarrows NO_2^- + H^+$ $NO_2^- + O_3 \rightarrow NO_3^- + O_2$ $HNO_2 + h\nu \rightarrow NO + OH$ $NO_2^- + h\nu + H^+ \rightarrow NO + OH$ $NO_2^- + OH \rightarrow OH^- + NO_2$	< 0.1
$HO_2 + NO_2 \rightleftarrows HOONO_2$	$\leftarrow HOONO_2 \rightarrow$	$HOONO_2 \rightleftarrows H^+ + NO_4^-$ $NO_4^- \rightarrow NO_2^- + O_2$ $HOONO_2 + HSO_3^- \rightarrow NO_3^- + SO_4^{2-} + 2H^+$	30

[a] Calculated assuming initial gas phase mixing ratios, in nmol mol^{-1}: NO, NO$_2$, 2.0 each; O$_3$ 35; CH$_4$ 700; CO 250; pH ~ 4.5, $L = 0.17 \times 10^{-6}$.

NO_3 is rapidly photolyzed, so that this pathway of nitric acid formation occurs preferentially at night. In the aqueous phase, the NO_3 radical reacts primarily with chloride and with hydrogen sulfite. The reaction with chloride is fairly slow, but at a high concentration of chloride, as in the marine atmosphere, the reaction provides the dominant loss process for NO_3. The reaction of NO_3 with HSO_3^- is rapid and becomes more important in the continental atmosphere. The direct interaction of NO_2 with cloud water is slow under atmospheric conditions and unimportant, because NO_2 is only slightly soluble in water.

Gaseous nitric acid is largely scavenged by cloud drops, where it dissociates to form the nitrate anion. Although photolysis of NO_3^- is a source of OH radicals, the photodissociation coefficient is fairly small ($j \approx 3 \times 10^{-7}$ s^{-1}), and the process contributes significantly only at a high concentration of nitrate. Nitrous acid is more easily photolyzed, both in the gas phase ($j \approx 2 \times 10^{-3}$ s^{-1}) and in the aqueous phase ($j \approx 6 \times 10^{-4}$ s^{-1}), but it is only moderately soluble in water, so that less HNO_2 enters into solution compared to HNO_3. Dissociation causes the presence of the nitrite anion, NO_2^-, except at low pH ($pK_a \approx 3.4$). Ozone rapidly oxidizes nitrite to nitrate, NO_3^-, and this is the major loss reaction. Photolysis of nitrite ($j \approx 4 \times 10^{-5}$ s^{-1}) provides a minor source of OH radicals. Peroxynitric acid, which is formed by the association of HO_2 radicals with NO_2, is thermally unstable. At the altitude level of clouds, however, the temperature is already low enough to favor the formation of $HOONO_2$. The Henry's law coefficient is similar to that of H_2O_2, so that peroxynitric acid is readily absorbed by cloud water. In the aqueous phase, $HOONO_2$ dissociates to form the NO_4^- anion ($pK_a \approx 5$), which is unstable, however, decomposing to form nitrite and oxygen with a time constant of about 1 s (Lammel *et al.*, 1990a; Logager and Sehested, 1993). At pH values greater than 5, the reaction thus provides a major source of nitrite in the aqueous phase. In more acidic solutions peroxynitric acid is stable and undergoes reaction with hydrogen sulfite. This reaction oxidizes NO_2 and SO_2 simultaneously to create nitrate and sulfate.

Table 8.9 includes percentage contributions of processes that lead to the formation of nitrate. The individual contributions were determined by model calculations for conditions typical of sunlit continental fair-weather clouds. The results indicate that gas phase oxidation by OH radicals and transfer of peroxynitric acid to the aqueous phase, followed by reaction with hydrogen sulfite, are the main contributors to the formation of nitric acid. The total rate is about 7×10^{-9} mol dm^{-3} s^{-1} in the aqueous phase or 1.2×10^{-12} mol m^{-3} s^{-1} in both phases together. At night, the production of N_2O_5 by the reaction between ozone and NO_2 leads to a similar rate through this reaction alone.

8.6.3. The Role of Transition metals

The mineral fraction of the aerosol inherits all of the elements present in the Earth's crust, many of which are metals, and these are incorporated into and partly dissolve in cloud water when aerosol particles act as condensation nuclei. The metals most frequently considered in cloud chemistry are iron, manganese, and copper, because these are able to catalyze the oxidation of sulfur dioxide in aerated aqueous solution. These elements are transition metals, that is, they occur in two oxidation states and represent one-electron redox species. Iron is one of the most abundant elements in the environment. It occurs mostly in the higher oxidation state, Fe(III), and forms oxides or hydroxides that are practically insoluble in water, except at low pH. Fe(III) can be brought into solution by the formation of strong complexes, such as hydroxo- or oxalato complexes, or by reduction to Fe(II), which is well soluble in water. The formation of Fe(III)-oxalato complexes in particular appears to be very effective in this regard, because they undergo photolysis in sunlight (Zuo and Hoigné, 1992, 1994), whereby oxalate is converted to CO_2 and Fe(III) is reduced to Fe(II). The somewhat surprising discovery of Fe(II) in fog water by Behra and Sigg (1990) and in aerosol particles by Zhuang et al. (1992) demonstrates the occurrence of such processes in the atmosphere. Manganese occurs in nature mainly in the form of water-soluble Mn(II). The higher oxidation state Mn(III) is a strong oxidant, which in the laboratory is not easily handled, because it disproportionates to form Mn(II) and Mn(IV). Copper (II) also is well soluble in water. It can be reduced to Cu(I), which as a strong reducing agent is readily oxidized back to Cu(II). Table 8.10 shows concentration ranges for the three metals observed in rain, cloud, and fog waters at several locations. The concentrations of manganese and copper are smaller by an order of magnitude than that of iron, but the dissolved fractions are frequently equivalent. The data in Table 8.10 also indicate that dissolved and reduced iron are often present in rather similar concentrations. Little information exists on the speciation of the other two transition metals. Xue et al. (1991) reported on the presence of Cu(I) in addition to Cu(II) in the aqueous phase of fog.

In the region of pH > 2, complexes of Fe(III) with the hydroxyl ion are dominant, so that Fe(III) occurs as $FeOH^{2+}$ (or higher complexes such as $Fe(OH)^{2+}$ that are not considered here). The reaction of iron with HO_2 and O_2^- radicals leads to the following oxidation–reduction sequence:

$$HO_2 + Fe^{2+} (+H_2O) \rightarrow FeOH^{2+} + H_2O_2 \qquad k = 1.2 \times 10^6$$

$$O_2^- + Fe^{2+} (+H_2O, H^+) \rightarrow FeOH^{2+} + H_2O_2 \qquad k = 1.0 \times 10^7$$

TABLE 8.10 Recent Observations of Iron (Total, Dissolved, and Fe(II)), Manganese, and Copper in Rain, Fog, and Cloud Waters[a]

Fe_{tot}	Fe_{diss}	Fe(II)	Mn	Cu	Location	Reference
Rainwater						
9–670	—	—	1.1–21	0.3–13	England	Balls (1989)
0.20–3.0	—	—	0.04–0.28	0.01–0.27	Sweden[c]	Ross (1990)
0.003–1.81	—	—	0.02–0.11	0.006–0.09	Sweden[c]	Ross (1990)
0.63 ± 0.02	0.41 ± 0.04	~ 0.08	0.054 ± 0.01	0.033 ± 0.005	Germany[b, c]	Hofmann et al. (1991)
0.44 ± 0.07	0.18 ± 0.11	~ 0.04	0.044 ± 0.02	0.025 ± 0.003	Germany[b, c]	Hofmann et al. (1991)
Cloud water						
0.45–3.41	—	—	0.04–1.15	—	Germany	Wobrock et al. (1994)
1.7–9.9	0.05–1.20	0.03–0.6	—	0.004–0.011	England	Sedlak et al. (1997)
Fog water						
1.61–37.6	—	—	0.33–2.91	0.02–0.24	California	Munger et al. (1983)
0.3–91	—	—	0–7.6	0–7.0	Switzerland	Joos and Baltensperger (1991)
0.5–42	—	—	0.1–11	—	Po Valley	Fuzzi et al. (1992)
1.0–11.3	0.5–11.3	0.4–8.0	—	—	California[c]	Erel et al. (1993)
2.2–27	2.3–11.6	0.9–5.0	—	—	California[c]	Erel et al. (1993)

[a] Concentrations in μmol dm^{-3}.
[b] Standard deviation of analysis of the rain sample.
[c] Two different sites.

$$HO_2 + FeOH_2 + \rightarrow Fe^{2+} + O_2 + H_2O \quad k = 1.0 \times 10^4$$

$$O_2^- + FeOH_2 + \rightarrow Fe^{2+} + O_2 + OH^- \quad k = 1.5 \times 10^8$$

where the rate coefficients refer to 298 K (Rush and Bielski, 1985). Similar reactions occur for the other transition metal ions, but only those of copper have been well studied (Bielski et al., 1985; von Piechowski et al., 1993). At pH $\approx$ 5, that is in the vicinity of the pK_a value of HO_2/O_2^-, the two radicals are present in nearly equal concentrations, but as the rate coefficients for the reactions of O_2^- are greater than those for HO_2, one may neglect the latter to a first approximation. The two iron species are kept in a steady state. The above scheme suggests that the Fe(II)/Fe(III) concentration ratio is approximately equal to the inverse ratio of the corresponding rate coefficients $[Fe(II)]/[Fe(III)] = k_{Fe(III)}/k_{Fe(II)} = 15$. This keeps the concentration of dissolved Fe(III) low compared to that of Fe(II). For copper ions, which react more rapidly with HO_2/O_2^-, the steady-state concentration ratio is $[Cu(I)]/[Cu(II)] \approx 0.5$. Sedlak and Hoigné (1993) noted that the reaction of Cu(I) with Fe(III) increases the ratio of Fe(II) to Fe(III) while reducing that of Cu(I) to Cu(II).

The reactions of HO_2/O_2^- with transition metals compete with that of O_2^- with ozone. As the concentrations of metal ions are much higher than that of ozone, reactions with metal ions take preference. The consequence is that HO_2/O_2^- concentrations are reduced by almost two orders of magnitude compared with the situation in which transition metal ions are absent. The production of OH radicals from the reaction with ozone is likewise reduced to a rate that is small compared with other production mechanisms. Transition metal ions in concentrations such as those indicated in Table 8.10 do not, however, affect the principal OH loss reactions with formaldehyde and formic acid, dissociated as well as undissociated, which occur at higher concentrations and react more rapidly.

A number of other reactions must be considered in addition to those discussed above. H_2O_2 reacts with the reduced metal species to produce OH radicals. Ozone also reacts with Fe(II) to produce the ferryl ion, $FeO_2 +$, which largely hydrolyses to produce OH and $FeOH_2 +$ but may also undergo other reactions (Logager et al., 1992). Similar reactions occur with Mn(II), which are slow and may be neglected, and with Cu(I), which must be taken into account. $FeOH^{2+}$ also represents a source of OH radicals due to photolysis. Table 8.11 lists reactions involved in the production of OH under conditions typical of continental fair-weather clouds and compares calculated production rates for the case where iron and copper are assumed to be present with those in marine clouds, that is, in the absence of transition metals. Under steady-state conditions, the concentration of Fe(III) is kept

502   Chapter 8.   Chemistry of Clouds and Precipitation

low because of reaction with Cu(I). The reaction of Fe(II) with H_2O_2 contributes appreciably to OH formation despite the small rate constant, and the reaction of Fe(II) with ozone is also significant. Photolysis of $FeOH^{2+}$ is rapid ($j \approx 5 \times 10^{-3}$ s^{-1}), and for this reason it was once postulated to be a major source of OH in continental clouds (Warneck, 1992). Under steady-state conditions, however, the rate of Fe(II) back-oxidation is not fast enough to maintain sufficiently high Fe(III) concentrations, although it may provide a strong OH source initially when the cloud develops. In comparison to marine clouds, where the main sources of aqueous OH are the reaction of O_2^- with ozone and intrusion of OH from the gas phase, continental clouds feature multiple sources of OH, with contributions strongly depending on actual conditions. Table 8.11 indicates that intrusion of OH from the gas phase still makes an important contribution. The combined rate of OH formation in the presence of transition metals is only slightly lower than when they are absent. The steady-state aqueous OH concentration under these conditions, which is calculated to be approximately 0.02 pmol dm^{-3}, is mainly determined by the concentration of formic acid.

TABLE 8.11   Aqueous Phase OH Radical Production Rates from Various Processes in a Sunlit Continental Fair-Weather Cloud in the Absence (a) and in the Presence (b) of Dissolved Transition Metals[a]

Reaction	k_{298}	Production rate (mol dm^{-3} s^{-1}) (a)	(b)	%[b]
$H_2O_2 + h\nu \rightarrow OH + OH$	1.5 (-5)	2.0 (-10)	2.7 (-10)	10.7
$NO_3^- + H^+ + h\nu \rightarrow NO_2 + OH$	5.6 (-7)	1.7 (-10)	1.7 (-10)	6.6
$NO_2^- + H^+ + h\nu \rightarrow NO + OH$	3.7 (-5)	4.4 (-13)	2.8 (-13)	< 0.1
$HNO_2 + h\nu \rightarrow NO + OH$	6.8 (-4)	8.4 (-13)	4.4 (-13)	< 0.1
$FeOH^{2+} + h\nu \rightarrow Fe^{2+} + OH$	5.2 (-3)	—	1.9 (-11)	0.8
$O_3 + O_2^- \rightarrow OH + OH^- + 2O_2$	1.5 (9)	3.38 (-9)	3.4 (-11)	4.1
$Fe^{2+} + H_2O_2 \rightarrow FeOH^{2+} + OH$	7.5 (1)	—	6.7 (-10)	27.4
$Fe^{2+} + O_3(+H_2O) \rightarrow FeOH^{2+} + OH + O_2$	8.2 (5)	—	4.6 (-10)	18.5
$Cu^+ + H_2O_2 \rightarrow Cu^{2+} + OH + OH^-$	1.0 (2)	—	1.6 (-11)	0.6
$Cu^+ + O_3(+H_2O) \rightarrow Cu^{2+} + OH + OH^-$	1.0 (7)	—	1.0 (-10)	4.1
$HO_g \rightarrow OH_a$[c]		1.64 (-9)	7.7 (-10)	30.6
Total rate		5.39 (-9)	2.50 (-9)	

[a] Steady-state concentrations (μmol dm^{-3}), case (b): Fe(III) 4 (-3), Fe(II) 0.99, Cu(II) 0.032, Cu(I) 0.018, H_2O_2 8.7, O_3 5.7 (-4), NO_3^- 295, NO_2^- 7.5 (-3), HNO_2 6.5 (-4), O_2^- 4 (-5), HCO_2^-/HCOOH 50, pH 4.4; numbers in parentheses indicate powers of ten; For rate coefficients see Table A.6.
[b] Percentage contribution for case (b).
[c] Input due to direct gas–liquid phase transfer.

8.6.4. THE OXIDATION OF SULFUR DIOXIDE IN AQUEOUS SOLUTION

As discussed in Section 8.4.1, the dissolution of SO_2 in water yields SO_{2aq}, HSO_3^-, and SO_3^{2-}, in addition to undissociated H_2SO_3, which is indistinguishable from SO_{2aq}. The sum of these species is designated S(IV). The oxidation of S(IV) occurs by ozone, by H_2O_2 and other peroxides, by OH radicals, and by oxygen catalyzed by transition metals.

Penkett (1972) first recognized the importance of ozone as an oxidant of S(IV) in clouds. The rate of the reaction rises with increasing pH, but all studies were confined to the pH range 1–5, approximately, because in alkaline solution ozone undergoes a chain decomposition initiated by the hydroxyl ion, with HO_2/O_2^- and OH radicals acting as chain carriers (Staehelin et al., 1984; Sehested et al., 1991). In acidic solutions the chain is not very effective, and its influence can be further reduced by the addition of radical scavengers (Hoigné et al., 1985). Penkett et al. (1979b) and Lagrange et al. (1994) found rate laws depending approximately inversely on the root of $[H^+]$ concentration, and they interpreted this to arise from a chain reaction propagated by SO_3^- and SO_5^- radicals. On the other hand, the pH dependence can be explained equally well by the shift in dissociation equilibrium between HSO_3^- and SO_3^{2-} with rising pH, combined with a faster rate of the O_3 reaction with SO_3^{2-} compared with HSO_3^-. With this interpretation, Erickson et al. (1977) and Hoigné et al. (1985) have independently derived concordant rate coefficients, which have been widely accepted by the atmospheric science community. The reaction does not appear to represent a simple O-atom transfer, however, because Esphenson and Taube (1965) found by isotope labeling that two oxygen atoms rather than one are incorporated into the product, sulfate.

Penkett et al. (1979b) first suggested hydrogen peroxide to be important as an oxidant of S(IV) in the aqueous phase of clouds. Other hydroperoxides also occur in the atmosphere as terminal products resulting from the recombination of alkylperoxy radicals with HO_2, including methyl hydroperoxide, CH_3OOH, and hydroxymethyl hydroperoxide, $HOCH_2OOH$ (Hellpointer and Gäb, 1989). The significance of peroxynitric acid, $HOONO_2$, has been discussed above. Jacob (1986) has postulated the occurrence of peroxomonosulfate, $HOOSO_3^-$, in clouds because this compound is an intermediate in the OH radical-induced S(IV) chain oxidation. Hydroperoxides react rapidly with HSO_3^- at rates that decrease markedly with increasing pH in the range of 2–7. The reactions with SO_3^{2-} generally are slower, except that of $HOOSO_3^-$, and they are almost independent of pH. In reacting with HSO_3^-, hydroperoxides have been found to follow a common mechanism, first formulated by Hoffmann and Edwards (1975) for H_2O_2. Drexler *et al.* (1991) have demonstrated that the reactions are general acid-catalyzed, that is, all acids including water are catalysts in addition to hydrogen ion. Laboratory studies mandate the elimination of buffer effects; in dilute cloud water such effects are small, but catalysis by H_2O still needs to be considered. In the reaction of ROOH with HSO_3^-, the hydroperoxide acts as a nucleophile. The initial step is the formation of a peroxysulfurous acid anion in a fast preequilibrium (Hoffmann and Edwards, 1975; Martin and Damschen, 1981), which subsequently rearranges into $ROSO_3^-$. This is the general acid-catalyzed step. Finally, the product hydrolyzes to yield ROH and sulfate:

$$ROOH + HSO_3^- \rightleftarrows ROOSO_2^- + H_2O$$

$$ROOSO_2^- (+H^+) \rightarrow ROSO_3^- (+H^+)$$

$$ROSO_3^- + H_2O \rightarrow ROH + SO_4^{2-} + H^+$$

In some cases the hydrolysis step is slow enough to be experimentally observable, for example, in the reaction of HSO_3^- with HOONO (Drexler *et al.*, 1991). Taking into account that the hydroperoxide may dissociate, the general rate expression is

$$-d[S(IV)]/dt = [ROOH]\{k_a[HSO_3^-] + k_b[SO_3^{2-}]\}$$

$$+ k_c[ROO^-][SO_3^{2-}] \qquad (8.27)$$

where

$$k_a = k_H[H^+] + k_{HOH}[H_2O]$$

For hydrogen peroxide $k_H = 3.9 \times 10^7$ at 298 K, $k_{HOH} = 0.16$ or $k_{HOH}[H_2O] \approx 8.9$, $k_b = 0.19$, and $k_c = 0.001$ (units: mol, dm^3, s). Proton catalysis clearly dominates in the pH region of 2–6. At pH values in the vicinity of $pK_a(HSO_3^-) \approx 7.2$, catalysis by water takes over before the term $k_b[SO_3^{2-}]$ gains strength. The rate coefficient then reaches a plateau value with rising pH until the last term in Equation (8.27) takes hold at pH > $pK_a(H_2O_2) \approx 11.3$. At pH < 2, the concentration of HSO_3^- decreases because it is replaced by aqueous SO_2, which does not react, so that the reaction slows down. The pH range of 2–6 is the main region of interest to atmospheric aqueous chemistry. In this region the reactions of S(IV) with ozone and H_2O_2 show an opposite pH dependence. As the oxidation of S(IV) releases hydrogen ions and the solution acidifies, the oxidation rate due to ozone is gradually reduced, while that due to H_2O_2 is maintained. The concentration of S(IV) actually decreases because acidification forces some

SO_2 into the gas phase, but this effect is nearly counterbalanced by the increase in the rate coefficient for the reaction with hydrogen peroxide.

Alkylhydroperoxides oxidize S(IV) in a manner similar to that of H_2O_2. The rate of S(IV) oxidation by peroxomonosulfate is similar to that of H_2O_2 in acidic solutions, but it deviates above pH 5. In this case, values for the individual rate coefficients are $k_H = 6.2 \times 10^6$ at 285 K, $k_{HOH}[H_2O] \approx 189$, $k_b = 2.3 \times 10^3$, and $k_c = 16$ (units: mol, dm^3, s). The pK_a for the second dissociation of $HOOSO_3^-$ is $pK_a \approx 9.1$, so that the overall rate coefficient exhibits a secondary maximum near pH 8 (Elias et al., 1994). The oxidation of S(IV) by peroxynitric acid contrasts with other reactions of hydroperoxides in that it shows no pH dependence at all in the pH region 1–5. In this case it appears that catalysis by water is the dominant term: $k_{HOH}[H_2O] \approx 2 \times 10^5$ dm^3 mol^{-1} s^{-1} at 285 K (Amels et al., 1996). At pH > 5, peroxynitric acid dissociates to form the NO_4^- anion, which is unstable and decomposes.

Chameides and Davis (1982) and McElroy (1986) have discussed the possibility that the reaction of S(IV) with OH radicals may contribute to the oxidation of SO_2 in clouds. Laboratory studies (Buxton et al., 1996a) show that reactions of OH with HSO_3^- and SO_3^{2-} are rapid and that they initiate a chain reaction propagated by SO_3^-, SO_5^-, and SO_4^- radicals

$$OH + HSO_3^-/SO_3^{2-} \rightarrow SO_3^- + H_2O/OH^- \qquad k = 2.7 \times 10^9 / 1.6 \times 10^9$$

$$SO_3^- + O_2 \rightarrow SO_5^- \qquad k = 2.5 \times 10^9$$

$$SO_5^- + HSO_3^-/SO_3^{2-} \rightarrow SO_3^- + HSO_5^-/SO_5^{2-} \qquad k = 8.6 \times 10^3 / 2.1 \times 10^5$$

$$SO_5^- + HSO_3^-/SO_3^{2-} \rightarrow SO_4^- + HSO_4^-/SO_4^{2-} \qquad k = < 4.0 \times 10^2 / 5.5 \times 10^5$$

$$SO_4^- + HSO_3^-/SO_3^{2-} \rightarrow SO_3^- + HSO_4^-/SO_4^{2-} \qquad k = 6.8 \times 10^8 / 3.1 \times 10^8$$

$$SO_5^- + SO_5^- \rightarrow SO_4^- + SO_4^- + O_2 \qquad 2k \approx 5.0 \times 10^8$$

$$SO_5^- + SO_5^- \rightarrow S_2O_8^{2-} + O_2 \qquad 2k = 9.6 \times 10^7$$

(rate coefficients in units of dm^3 mol^{-1} s^{-1}). The peroxomonosulfate anion HSO_5^- reacts further with S(IV) in the manner discussed above. The propagation steps are fairly slow, yet Buxton et al. (1996a), using a steady-state radiolysis technique, observed chain lengths up to 15,000 in alkaline and up to 200 in acidic solutions. The chain length is considerably reduced by radical scavengers such as alcohols, which interfere with the chain in several ways. Sulfate radicals, SO_4^-, react with organic compounds quite rapidly and in a matter similar to that of OH radicals, whereas SO_5^- radicals show little reactivity. In the presence of oxygen, reactions of SO_4^- with organic compounds generate HO_2/O_2^- radicals, for example

$$SO_4^- + CH_3CH_2OH \rightarrow CH_3CHOH + SO_4^{2-} + H^+$$

$$CH_3CHOH + O_2 \rightarrow CH_3CHO + HO_2$$

which provide a more efficient termination pathway for SO_5^- radicals, because the reactions

$$HO_2/O_2^- + SO_5^- (+H^+) \rightarrow HSO_5^- + O_2 \qquad k = 1.8 \times 10^9/2.3 \times 10^8$$

are faster than termination by recombination of SO_5^- radicals (Buxton *et al.*, 1996b; Fischer and Warneck, 1996). This enhances chain termination in comparison to propagation, so that the chain length is markedly reduced. Thus, in the presence of HO_2/O_2^-, Deister and Warneck (1990) observed chain lengths near 300 in alkaline solution, and Fischer and Warneck (1996) found the chain length to be reduced to unity in acidic solution. Weinstein-Lloyd and Schwartz (1992) and Sedlak and Hoigné (1994) also confirmed earlier notions that HO_2 and O_2^- do not undergo any direct reactions with S(IV) species.

The dominant scavenger for SO_4^- radicals in clouds is the Cl^- anion (McElroy, 1990). The Cl atoms formed in this reaction associate with another Cl^- to enter into an equilibrium with the Cl_2^- radical ion (Jayson *et al.*, 1973; Adams *et al.*, 1995):

$$SO_4^- + Cl^- \rightarrow SO_4^{2-} + Cl \qquad k = 3.3 \times 10^8$$

$$Cl + Cl^- \rightleftarrows Cl_2^- \qquad k = 2.1 \times 10^{10}/1.1 \times 10^5$$

$$HO_2/O_2^- + Cl_2^- \rightarrow 2Cl^- + O_2 (+H^+) \qquad k = 1.3 \times 10^{10}/6 \times 10^9$$

$$Cl + H_2O \rightarrow ClOH^- + H^+ \qquad k = 1.3 \times 10^3$$

$$ClOH^- \rightarrow Cl^- + OH \qquad k = 6.1 \times 10^9$$

$$ClOH^- + H^+ \rightarrow Cl + H_2O \qquad k = 2.1 \times 10^{10}$$

(rate coefficients in units of $dm^3 \ mol^{-1} \ s^{-1}$). Although the Cl_2^- radical reacts with many species, most of the rate coefficients have low values (Neta *et al.*, 1988), and the reactions are insignificant. The termination reactions of Cl_2^- with HO_2 and O_2^- are rapid and would be important, but the Cl atom that is in equilibrium with Cl_2^- also reacts rapidly with water, and this reaction has the highest probability. As a consequence, the SO_4^- radicals generated in the S(IV) chain oxidation reaction are largely converted to OH radicals when chloride is present. Although the reaction of OH with S(IV) would continue the chain, S(IV) must compete for OH with formaldehyde and formic acid. The latter reactions lead to the formation of HO_2/O_2^- as radical products, which do not propagate the S(IV) oxidation chain. Only in regions severely polluted with SO_2 are the concentrations of S(IV) high enough to favor chain oxidation in competition with other OH reactions. This circumstance as well as the enhancement of chain termination by HO_2/O_2^- radicals make S(IV) chain oxidation reaction rather inefficient in continental clouds under normal conditions.

Sulfur dioxide dissolved in aerated water is slowly oxidized to sulfuric acid. Titoff (1903) first showed that the reaction is catalyzed by ions of transition metals such as copper, iron, or manganese, and that it is difficult to eliminate traces of such metals from aqueous solutions. In fact, the reaction can be quenched by the addition of metal-complexing agents (Tsunogai, 1971a; Huss et al., 1978), so that the claim of some investigators to have observed the noncatalyzed reaction must be viewed with caution. Connick et al. (1995) recently reassessed the problem and presented new evidence for a noncatalyzed pathway. The observed rate was too slow, however, to be of any significance to clouds. The majority of studies exploring the metal-catalyzed oxidation of S(IV) have focused on iron or manganese as catalysts. Hegg and Hobbs (1978) and Martin (1984) have reviewed empirical rate laws derived from laboratory studies and discussed their relevance to clouds. Brandt and van Eldik (1995) provided an updated, extensive literature review covering many other aspects as well. Penkett et al. (1979a) and Clarke and Radojevic (1987) have explored the oxidation of S(IV) in rainwater and found the oxidation rates to correlate significantly with iron, less with manganese.

While recent laboratory studies have greatly improved our understanding of the mechanistic aspects of metal ion catalysis, only the role of iron is now reasonably well understood. Other catalysts are frequently assumed to behave similarly. As noted earlier, Fe(III) is fully soluble in water only at pH < 4, so that most laboratory studies with iron as catalyst were carried out in this pH region. Martin et al. (1991) and Novic et al. (1996) were able to extend their studies toward the range pH > 5 to show that the reaction depends on the amount of iron in solution rather than the total amount of iron present. Martin et al. (1991) also discussed ionic strength effects, such as inhibition by sulfate and various organic molecules. Iron provides an advantage in that both redox species are accessible to analysis by modern ion chromatography. Grgic et al. (1993), Ziajka et al. (1994), and Novic et al. (1996) have exploited this feature to show that a steady state is established between Fe(III) and Fe(II) during the reaction. Table 8.12 presents a minimum mechanism for the iron-catalyzed oxidation of S(IV) as a basis for the following discussion. Bäckström (1934) first suggested that the metal-catalyzed oxidation of S(IV) occurs by a chain reaction involving sulfite radicals as the principal chain carriers. His idea has been substantiated in studies where the chain was initiated by other means, for example, by OH radicals as discussed above; and Table 8.12 contains this part of the mechanism. The presence of metal ions, however, greatly increases the complexity of the system, making it necessary to include additional reactions that involve both redox species of iron.

All of the metals that act as catalysts are one-electron redox species. The higher oxidation states usually possess a capacity for the formation of complexes by a variety of ligands. In the case of iron, the rapid formation of S(IV) complexes has been demonstrated by the observation of the associated optical spectra (Conklin and Hoffmann, 1988; Kraft and van Eldik, 1989). At least two Fe(III)-sulfito complexes showing different reactivities can be distinguished (Kraft and van Eldik, 1989a; Betterton, 1993). On an extended time scale, the complexes were found to decompose, and sulfate was formed (Kraft and van Eldik, 1989b). Brandt et al. (1994) observed oxygen to be consumed at a rate suggesting the occurrence of a chain reaction, which they assumed was triggered by sulfite radicals generated in the decomposition process, with Fe(II) as the second product. Brandt et al. (1994) also showed that at a critical point in time, when the consumption of oxygen was completed, the decay rate increased abruptly to that observed in the absence of oxygen, which indicated on the one hand that decomposition of the complexes is independent of oxygen, and on the other hand that a back-reaction exists, such as that between Fe(II) and SO_5^-, which reduces the apparent

TABLE 8.12   Minimum Reaction Mechanism for the Iron-Catalyzed Oxidation of Sulfur(IV)[a]

Reaction	k_{298}
$FeOH^{2+} + HSO_3^- \rightleftarrows [FeOHSO_3H]^+$	$600 \text{ dm}^3 \text{ mol}^{-1}$ [b]
$[FeOHSO_3H]^+ \rightarrow Fe^{2+} + SO_3^- + H_2O$	0.065 s^{-1}
$SO_3^- + O_2 \rightarrow SO_5^-$	2.5×10^9
$Fe^{2+} + SO_5^- (+H_2O) \rightarrow FeOH^{2+} + HSO_5^-$	$\sim 7.0 \times 10^5$
$Fe^{2+} + HSO_5^- \rightarrow FeOH^{2+} + SO_4^-$	3.5×10^4
$Fe^{2+} + SO_4^- (+H_2O) \rightarrow FeOH^{2+} + H^+ + SO_4^-$	3.0×10^8
$SO_5^- + HSO_3^- \rightarrow SO_3^- + HSO_5^-$	8.6×10^3
$SO_5^- + HSO_3^- \rightarrow SO_4^- + HSO_4^-$	$\leq 4.0 \times 10^2$
$SO_4^- + HSO_3^- \rightarrow SO_3^- + HSO_4^-$	6.8×10^8
$SO_5^- + SO_5^- \rightarrow SO_4^- + SO_4^- + O_2$	2.5×10^8
$SO_5^- + SO_5^- \rightarrow S_2O_8^{2-} + O_2$	4.8×10^7

[a] Adapted from Warneck and Ziajka (1995); units of rate coefficients $dm^3 mol^{-1} s^{-1}$ unless otherwise noted.
[b] Value of equilibrium constant.

rate of decomposition when oxygen is present. Ziajka *et al.* (1994) have applied suitable radical scavengers to demonstrate that the SO_4^- radical (but not the OH radical) participates in the iron-catalyzed oxidation of S(IV). Generally, organic compounds may inhibit the reaction in two ways, either by acting as a scavenger for SO_4^- radicals, or by the formation of a complex with Fe(III), which suppresses the further addition of S(IV) to Fe(III). Oxalate, which forms strong complexes with Fe(III), is such an inhibitor.

If the reaction is started with Fe(II), the ion must first be converted to Fe(III) for S(IV) oxidation to become effective. The oxidation of Fe(II) to Fe(III) by oxygen alone is slow. Bal Reddy and van Eldik (1992) observed that the addition of S(IV) to an aerated solution of Fe(II) greatly accelerated the process. The reaction was further shown to be autocatalytic, that is, the oxidation rate accelerated with time (Bal Reddy and van Eldik, 1992; Ziajka *et al.*, 1994). A similar behavior has been observed with Co(II) (van Eldik *et al.*, 1992) and with Mn(II) (Berglund *et al.*, 1993) as catalysts. In each case, the phenomenon of autocatalysis was suppressed when a small amount of the catalyst in its higher oxidation state was added to the solution. Autocatalysis usually is associated with branching chain reactions, that is, the mechanism involves a reaction that generates two chain carriers for the one being consumed. The chain-branching reaction in the mechanism shown in Table 8.12 is the interaction of Fe(II) with peroxomonosulfate, HSO_5^-, which produces Fe(III) and a sulfate radical, SO_4^-. Both species react further with S(IV) to produce SO_5^-, which regenerates SO_5^- by rapid reaction with oxygen.

When the initial [S(IV)]/[Fe] concentration ratio is large, Fe(III) is quickly converted to Fe(II). The [Fe(II)]/[Fe(III)] ratio remains high as long as S(IV) is in excess over iron. In this case, the reaction is dominated by the main S(IV) oxidation chain. After the consumption of most of the S(IV), when the [S(IV)]/[Fe] ratio approaches unity, sulfuroxy radicals react mainly with Fe(II). This is the situation usually encountered in clouds. Here, as outlined earlier, the Fe(II)/Fe(III) ratio is determined by other reactions, and the S(IV) oxidation chain is largely quenched as the SO_4^- radicals are scavenged by chloride, and the OH radicals ultimately produced from the reaction of chlorine atoms are scavenged by formaldehyde and formic acid. All of these processes make the iron-catalyzed oxidation of SO_2 in the aqueous phase of clouds rather inefficient.

Table 8.13 summarizes relative rates of the most important S(IV) oxidation reactions under conditions typical of a sunlit fair-weather cloud. The three leading processes involve ozone, hydrogen peroxide, and peroxynitric acid. The reaction of ozone with SO_3^{2-} is more effective than that with HSO_3^-, even though the concentration ratio $[SO_3^{2-}]/[HSO_3^-] \approx 2 \times 10^{-3}$ at pH $\sim$ 4.5. Among the hydroperoxides, the reaction with methyl hydroperox-

TABLE 8.13   Comparison of Reactions and Their Relative Contributions to the Oxidation of Sulfur Dioxide in the Aqueous Phase of a Sunlit Fair-Weather Cloud[a]

Reaction	k_{298}	Reaction rate (mol dm^{-3} s^{-1})	%
$O_3 + HSO_3^- \rightarrow HSO_4^- + O_2$	3.2 (5)	3.5 (−10)	3.6
$O_3 + SO_3^{2-} \rightarrow SO_4^{2-} + O_2$	1.5 (9)	2.12 (−9)	21.8
$H_2O_2 + HSO_3^- \rightarrow HSO_4^- + H_2O$	3.9 (7)[H$^+$]	3.57 (−9)	36.7
$CH_3OOH + HSO_3^- \rightarrow HSO_4^- + CH_3OH$	1.7 (7)[H$^+$]	3.4 (−15)	$\ll$ 0.1
$HOONO_2 + HSO_3^- \rightarrow HSO_4^- + NO_3^- + H^+$	3.1 (5)^c	1.93 (−9)	19.9
$HOONO + HSO_3^- \rightarrow HSO_4^- + HNO_2$	4.0 (4) + 2.0 (7)[H$^+$]	3.4 (−15)	< 0.1
$HSO_5^- + HSO_3^- \rightarrow 2HSO_4^-$ (2 ×)	189 + 6.2 (6)[H$^+$]^c	2.7 (−11)	0.3
$SO_5^- + SO_3^{2-} \rightarrow SO_4^{2-} + SO_4^-$ (2 ×)	5.5 (5)	1.5 (−12)	< 0.1
$SO_5^- + SO_5^- \rightarrow 2SO_4^- + O_2$ (2 ×)	3.3 (8)	8.9 (−11)	0.9
$Fe^{2+} + HSO_5^- \rightarrow FeOH^{2+} + SO_4^-$	3.0 (4)	4.9 (−10)	5.0
$OH^+SO_{2g} + O_2 \rightarrow SO_3 + HO_2$^{b}	2.0 (−12)^b	1.15 (−9)	11.8
Total rate of sulfur dioxide oxidation		9.73 (−9)	

[a] Calculated assuming initial gas phase mixing ratios, in nmol mol^{-1}: NO$_2$, SO$_2$ 2.0 each; O$_3$ 35; CH$_4$ 1700; CO 250; pH ~ 4.5, $L = 0.17 \times 10^{-6}$. Steady-state concentrations in the aqueous phase are as in Table 8.11, case (b). Numbers in parentheses indicate powers of 10.
[b] Gas phase reaction scaled to liquid phase to allow comparison with the other processes. In this case the rate constant is in units cm^3 s^{-1}.
[c] Rate constant for 285 K.

ide turns out to be completely negligible, and that of HSO$_5^-$ with S(IV) makes only a small contribution. The dominant fate of HSO$_5^-$ is reaction with Fe(II), which contributes about 5% to the total rate. The immediate product from this reaction as well as others involved in the chain oxidation of S(IV) is the sulfate radical, SO$_4^-$. This species, however, reacts further in various ways to produce sulfate. The contribution of S(IV) chain oxidation to the total S(IV) oxidation rate is less than 7% under these conditions. Note that the gas phase oxidation of SO$_2$ still contributes about 12%. Sulfur(IV) oxidation catalyzed by iron is unimportant. The concentration of Fe(III) is kept too small by reactions with HO$_2$/O$_2^-$ radicals and with Cu(I) as well as photolysis. It is possible that the iron-catalyzed oxidation of S(IV) is more important at night, when H$_2$O$_2$ oxidizes Fe(II) back to Fe(III). The role of manganese cannot yet be assessed for lack of information on several important reactions. The reader is referred to Berglund and Elding (1995) for a review of laboratory data and mechanisms for the manganese-catalyzed oxidation of S(IV) in aqueous solution.

CHAPTER 9

Nitrogen Compounds in the Troposphere

The better-known among a great number of nitrogen compounds in the atmosphere are ammonia, several nitrogen oxides, and nitric acid. These gases are discussed in the present chapter. NH_3, N_2O, and, to some extent, NO and NO_2 as well are natural constituents of air as they emanate from the biosphere. Although atmospheric chemistry views the biosphere as a source (or sink) of trace gases, it will be helpful for the subsequent discussion to present initially a summary of biochemical processes that are responsible for the release of nitrogenous volatiles from soils and aquatic environs. The individual pathways of nitrogen in the biosphere and in the atmosphere represent portions of a larger network of fluxes involving all geochemical reservoirs. This aspect will be considered in Section 12.3 from a different viewpoint.

9.1. BIOCHEMICAL PROCESSES

The following description of the biological nitrogen cycle emphasizes the production of volatile compounds that can escape from the biosphere to the atmosphere. Delwiche (1970) has given a popular account of the nitrogen cycle. A compilation of review articles edited by Clark and Rosswall (1981) contains more details and numerous literature citations. Conrad (1996a) recently reviewed the role of microorganisms in the release of trace gases from soils. Textbooks on soil microbiology provide additional information (e.g., Paul and Clark, 1996).

Figure 9.1 shows pathways of nitrogen utilization in the soil–plant ecosystem. With appropriate modifications the scheme may also be applied to aquatic ecosystems. Nitrogen enters the biosphere largely by way of bacterial nitrogen fixation, a process by which N_2 is reduced and incorporated directly into the living cell. A limited number of bacteria are fitted with the special enzyme system necessary for this task. Examples include phototrophic cyanobacteria in natural waters, heterotrophs such as *Clostridia* and *Azotobacter* in soils, and symbionts associated with plants, such as *Rhizobia* living in the root nodules of legumes. Burns and Hardy (1975) and, more recently, Sprent and Sprent (1990) have reviewed details of nitrogen fixation. It is an energetically costly process, and the biosphere as a whole tends to preserve its store of fixed nitrogen by extensively recirculating it within the ecosystem. The term *fixed nitrogen* is used to describe nitrogen contained in chemical compounds that all plants and microorganisms can use. Figure 9.1 may serve as a guide for biochemical transformations of fixed nitrogen among various compounds.

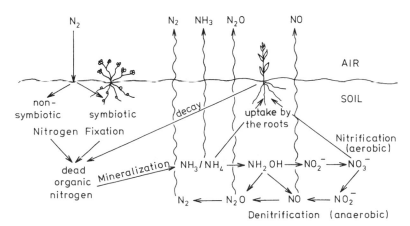

FIGURE 9.1 The biological nitrogen cycle in soils and its connection with the atmosphere. Contributions of nitrogen from rainwater, dry deposition, and fauna are not shown.

Dead organic matter undergoes decomposition by a host of different bacteria. In this manner, organic nitrogen is mineralized to ammonium unless it is assimilated. NH_4^+ and NH_3 are in aqueous equilibrium. If environmental factors such as pH, temperature, and buffer capacity of the soil are favorable, ammonia will be released to the atmosphere. Another important source of ammonia is urea arising from animal excreta (not shown in Fig. 9.1).

In well-aerated soils a number of specialized bacteria derive their energy needs from the oxidation of ammonia to nitrous acid and then further to nitrate. This process is called *nitrification*. Focht and Verstraete (1977), Belser (1979), and Schmidt (1982), among others, have reviewed various aspects of nitrification. The most common nitrifying bacteria in soils are *Nitrosomonas*, which converts ammonium to nitrite, and *Nitrobacter*, which oxidizes nitrite to nitrate. As chemoautotrophs, both species feature nitrification rates several thousand times faster than those of the less populous heterotrophic nitrifiers. As Figure 9.1 shows, hydroxylamine is the only identifiable intermediate product in the oxidation chain. Other intermediates have been postulated, especially $(HNO)_2$, to explain the formation of N_2O as a side product, but a real need for such an intermediate is not obvious. NH_2OH can be oxidized to NO_2^- directly, and N_2O can be formed by the interaction of NH_2OH with NO_2^- (Minami and Fukushi, 1986). Since the process is enzyme-controlled, the true intermediate would be enzyme-bound anyway. More enigmatic is the production of NO observed during nitrification. Several processes appear to contribute. In acidic soils, nitrite is partly present as nitrous acid, which is chemically unstable and decomposes: $2HNO_2 \rightarrow NO + NO_2 + H_2O$ (Chalk and Smith, 1983; McKenney *et al.*, 1982). Calculations of thermodynamic equilibria by Van Cleemput and Baert (1976) show that aerobic conditions favor NO_2 as a decomposition product, whereas anaerobic conditions lead to NO, N_2O, and N_2. Lipschultz *et al.* (1981), experimenting with *Nitrosomonas* in pure culture (pH 7.5), demonstrated that NO and N_2O evolve concurrently with NO_2^-, the latter accumulating at a constant or slightly accelerating rate. Poth and Focht (1985) and Remde and Conrad (1990) have shown that nitrifying bacteria can use NO_2^- as an electron acceptor to produce N_2O and NO when oxygen becomes limiting. However, N_2O and NO are also produced by bacterial reduction of nitrate (see below), and it is difficult to separate the different pathways in microbial communities (Conrad, 1996b).

Nitrate is the major end product of nitrification. Most plants (except certain bog species) can utilize nitrate as well as ammonium to satisfy their nitrogen needs. The process involves uptake through the roots, assimilatory reduction of NO_3^- to NH_3, and metabolic conversion to organic nitrogen compounds. Nitrogen reenters the soil when materials shed by plants (senescent leaves, dead wood, etc.) undergo bacterial degradation. In this manner, fixed nitrogen is circulated through the ecosystem. A certain fraction of nitrogen is leached from the soil together with soluble cations and enters the groundwaters. Yet another fraction undergoes bacterial reduction toward N_2. This process, which is termed *denitrification*, closes the cycle.

Focht and Verstraete (1977), Knowles (1981), and Firestone (1982) have reviewed the microbiology of denitrification. As Figure 9.1 shows, the reduction of NO_3^- toward N_2 proceeds via NO_2^-, NO, and N_2O as intermediates. While the role of N_2O in addition to that of NO_2^- as an obligatory intermediate was well established, it has been shown only recently that NO is also an obligatory intermediate (Ye *et al.*, 1994; Zumft, 1993). Denitrification requires a habitat that is best characterized as being almost but not completely anaerobic. Almost all denitrifying bacteria are aerobic species that turn to oxides of nitrogen as a source of oxygen when the level of O_2 falls to low values. In soils, pore moisture is a major factor regulating the diffusive supply of oxygen, and anaerobic conditions frequently arise from water-logging. In aerated soils, anaerobic microniches are often established in the interior of soil crumbs, where the supply of oxygen is reduced because of consumption in the outer layers. Thus, aerobic and anaerobic domains may exist side by side. The probability of finding anaerobic conditions generally increases as one goes deeper into the soil.

A great many genera of bacteria have the capacity to denitrify, and it is difficult to identify those organisms that are functionally most important. According to Gamble *et al.* (1977) the most commonly isolated bacteria are *Pseudomonas* and *Alicaligenes*, but the authors acknowledge that this observation may simply reflect an inherent selectivity of the cultural media used for growth, which cannot simulate the complexity of the soil environment. Since most denitrifiers are chemoheterotrophs, the availability of organic carbon compounds in the soil as an energy source is an important factor controlling the rate at which denitrification takes place. Another critical parameter is soil pH. Bacterial reduction of nitrate is most facile in the pH range 6–8, whereas in naturally acidic soils denitrification is inhibited.

The ultimate product resulting from denitrification is N_2. The process represents a reversal of nitrogen fixation, as it returns fixed nitrogen from the biosphere to the pool of atmospheric N_2. The release of N_2O also constitutes a loss of fixed nitrogen, as the main fate of N_2O in the atmosphere is photodecomposition by ultraviolet sunlight, with N_2 and O atoms as products. The release of NH_3 or NO, in contrast, does not lead to a loss of fixed nitrogen. Ammonia is not very reactive in the atmosphere and returns to the biosphere by wet and dry deposition; NO is oxidized to NO_2, and ultimately to HNO_3, which also return to Earth's surface by wet and dry deposition.

Owing to the complexity of soil biochemistry, the emissions of nitrogenous gases to the atmosphere are influenced by many parameters, and the emissions are highly variable. The variability makes it difficult to derive representative averages for global source estimates. An important parameter is temperature, because rates of biochemical reactions rise exponentially with temperature, at least in the range 288–308 K. Soil temperature undergoes seasonal and diurnal variations, causing the emissions to follow similar cycles. Diurnal and/or seasonal oscillations in the concentration of a trace gas in continental surface air, with maxima occurring in the late afternoon and/or in the summer months, are indicative of the importance of soil biology as a source of the trace gas under study.

9.2. AMMONIA

The French scientist DeSaussure is credited with the discovery of ammonia in rainwater in 1804. For many years, interest in atmospheric NH_3 has focused on its role as a source of fixed nitrogen for soil and plants. In the 1950s it was recognized, primarily through the work of Junge (1954, 1956), that ammonia is *the* single trace gas capable of neutralizing the acids produced by the oxidation of SO_2 and NO_x. The ammonium salts of sulfuric and nitric acid thus formed are incorporated into the atmospheric aerosol. In Chapter 7 it was shown that acids and ammonia become associated with particulate matter by gas-to-particle conversion, which causes ammonia to be concentrated in the submicrometer size range of particles. Giant particles generated by wind force from materials at the Earth's surface contain practically no ammonium.

For a complete understanding of atmospheric ammonia, it is necessary to differentiate between gaseous NH_3 and particulate NH_4^+. In this regard, many of the existing data are unsatisfactory because of difficulties in separating the two fractions. The older measurements, summarized by Eriksson (1952), did not distinguish gaseous from particle-bound NH_3, so that they provide at best the sum of both. One method of differentiation is to place upstream of the NH_3 sampling device a filter that removes particulate ammonium. During long sampling times, ammonium-rich particles may interact with calcareous soil-derived particles on the filter, causing the release of NH_3 and an overestimate of the concentration of gaseous NH_3 (Wiebe *et al.*, 1990). Another device is the so-called denuder tube, in which NH_3 undergoes diffusion toward acid-coated walls, whereas particles pass through the tube to be collected on a back-up filter. Wet-chemical, colorimetric techniques such as Berthelot's indophenol blue method were often used to determine the amounts of NH_3 and NH_4^+ collected. Today, ion chromatography is the standard technique. Annular denuders consisting of two concentric tubes increase the sampling rate. Continuous-flow annular denuders steadily replenish the acidic solution coating the walls. This system, when combined with modern analytical methods (Wyers *et al.*, 1993), currently meets most of the requirements for automatic monitoring of gaseous NH_3 in the atmosphere. Few data, however, have been obtained so far with this new technique. The detection of atmospheric NH_3 by long-path infrared absorption with the sun as the background source has been accomplished fairly late (Murcray *et al.*, 1978). A spectral resolution better than 0.1 cm^{-1} is required for the identification of NH_3. This explains the failure to detect it in earlier spectra of lesser resolution (Migeotte and Chapman, 1949; Kaplan, 1973).

9.2.1. DISTRIBUTION IN THE TROPOSPHERE

Table 9.1 compiles results from ground-level measurements of gaseous and particulate ammonia in Europe, North America, and over the oceans. Typical total concentrations of $NH_3 + NH_4^+$ in continental air are about 290 nmol m^{-3} or 5 $\mu g\ m^{-3}$. Average background mixing ratios of gaseous NH_3 range from about 0.3 nmol mol^{-1} (0.2 $\mu g\ m^{-3}$) in forest stands to more than 10 nmol mol^{-1} (8 $\mu g\ m^{-3}$) over grasslands. NH_3 exhibits both seasonal and diurnal variations. Summer values are greater than winter values. The diurnal trend, with a maximum in midday and a minimum during the early morning hours, is most pronounced during the summer season. A strong correlation with ambient temperature generally exists. Figure 9.2 illustrates this for two sites where the mean mixing ratios differ in magnitude, yet similar temperature dependencies are observed, suggesting that soils and/or vegetation contribute much to atmospheric ammonia. Comparison with particulate NH_4^+ shows that the NH_3/NH_4^+ ratio generally is smaller than unity. The two phases are poorly correlated, however. Langford *et al.* (1992) have shown that the partitioning depends strongly on the concentration of aerosol sulfate. For sulfate concentrations below about 20 nmol m^{-3} ($\sim 2\ \mu g\ m^{-3}$), most of the ammonia is present in the gas phase during the summer. When the sulfate level exceeds 200 nmol m^{-3} ($\sim 20\ \mu g\ m^{-3}$), less than 5% of NH_3 remains in the gas phase. This demonstrates the strong binding force of sulfuric acid.

By means of small aircraft, Georgii and Müller (1974) and Georgii and Lenhard (1978) have explored the vertical distribution of $NH_3 + NH_4^+$ in the tropospheric boundary layer over Europe up to 3 km altitude. Figure 9.3 shows that the concentrations of both species decrease with altitude. The decrease is equivalent to a scale height of 1–2 km, depending on the season. At the Jungfraujoch, a mountain station in the Swiss Alps 3600 m above sea level, Georgii and Lenhard found $x(NH_3) = 0.26$ nmol mol^{-1}, in agreement

TABLE 9.1 Abundance of Gaseous NH_3 and Particulate NH_4^+ in the Surface Air of Europe and North America and over the Oceans[a]

Authors	Location	Season	NH_3 (nmol m^{-3})	NH_4^+ (nmol m^{-3})	NH_3/NH_4^+	Remarks
Continental atmosphere, rural						
Georgii and Lenhard (1978)	Western Germany	Summer	323	333	0.9	Extrapolated averaged profiles from aircraft measurements
		Winter	100	144	0.6	
	Swiss Alps	Summer	118	—	—	Jungfraujoch, 3600 m a.s.l.
Ferm (1979)	Gothenburg, Sweden	Autumn	12	62	0.33	Denuder tubes
Lenhard and Gravenhorst (1980)	Western Germany	Summer	282 ± 47	272 ± 44	1.0	100 m above ground
		Winter	112 ± 35	144 ± 50	0.73	
Goethel (1980)	Schlüchtern, western Germany	Summer	206 ± 105	206 ± 106	1.0	Seasonal variation studied
		Winter	118 ± 65	194 ± 100	0.6	
Lewin et al. (1986)	State College, PA	Jan.–Mar.	0.9–23.6	7.6–305	0.02–0.15	
Allen et al. (1988)	Colchester, U.K.	annual	139 ± 130	219 ± 188	0.71	19 rural sites, 1-year study
		annual	1430 ± 1420	264 ± 214	5.0	Next to livestock farms
Harrison and Allen (1991)	Colchester, U.K.		171 ± 102	264 ± 200	0.65	
Forested regions						
Tjepkema et al. (1981)	Petersham, Massachusetts	Summer	7.6 ± 2.9	117 ± 55	0.07	Denuder tubes, 1-year study
		Winter	1.0 ± 0.7	54 ± 20	0.02	
Pierson et al. (1989)	Allegheny Mountain	August	< 5	173	< 0.09	
Keeler et al. (1991)	Newton, Connecticut	Summer	34 ± 28	112 ± 116	0.3	Denuder tube
Mountain areas						
Langford et al. (1992)	Boulder, Colorado	Summer	230	44	5.2	Denuder tubes
		Winter	60	29	2.1	
	Niwot Ridge, Colorado	Summer	15	14	1.0	
		Winter	6	9	0.7	
Marine atmosphere						
Tsunogai (1971)	North Pacific		68	29	2.2	
	Tropical Pacific		33	13	2.4	
	South Pacific		7.6	28	0.26	
Georgii and Gravenhorst (1977)	Central Atlantic		2	—	—	
	Sargasso Sea		353	83	4	
Ayers and Gras (1980)	Cape Grim, Tasmania		3.5 ± 1.8	—	—	
Gras (1983)	Antarctic		0.94	2.8	0.3	
Quinn (1990)	Pacific Ocean, 30°–50° N		0.37 ± 0.45	16 ± 7	0.02	
	15°–29° N		0.32 ± 0.38	5.3 ± 2.2	0.06	
	11° S–14° N		0.78 ± 1.3	5.6 ± 1.7	0.14	

[a] 1 nmol mol^{-1} ≈ 44.6 nmol m^{-3} ≈ 0.76 μg m^{-3} [NH_3]; ≈ 0.80 μg m^{-3} [NH_4^+].

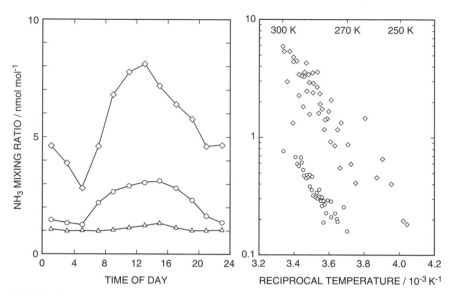

FIGURE 9.2   Left: Diurnal variation of the NH_3 mixing ratio at two locations in Colorado, Boulder ($\diamond$, summer), and Niwot Ridge ($\bigcirc$, summer; $\triangle$ winter; $\times 5$). Right: NH_3 mixing ratios at these locations plotted against reciprocal temperature to show the Arrhenius-type behavior. (Adapted from Langford *et al.* (1992).)

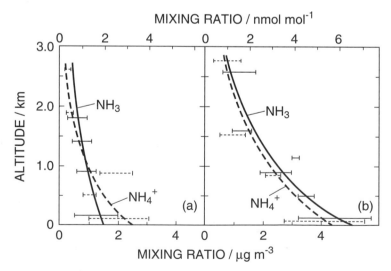

FIGURE 9.3   Vertical distribution of gaseous ammonia and particulate ammonium over Europe in winter (left) and summer (right) according to Lenhard (1977) and Georgii and Lenhard (1978). The lower scale indicates the mass mixing ratio (unit: $\mu g\ m^{-3}$ STP, that is, normalized to 1013.25 hPa and 273.15 K); the upper scale indicates the molar mixing ratio (unit: nmol mol^{-1}) for NH_3 only.

with the aircraft data. The abundance of NH_3 in the upper troposphere is expected to be still smaller. Column abundances estimated from infrared solar spectra by Kaplan (1973) for the air over France and by Murcray et al. (1978) for the air over Colorado suggest average tropospheric mixing ratios of < 0.08 and 0.5 nmol mol^{-1}, respectively. From aircraft-assisted mass spectrometer measurements of gas phase ammonium water cluster ions, Ziereis and Arnold (1986) inferred NH_3 mixing ratios of less than about 30 pmol mol^{-1} above 4 km altitude over Germany.

Table 9.1 includes a few data from marine regions. The exploratory study of Tsunogai (1971) may be open to criticism with regard to the sampling procedure, but it established the general trend. The mixing ratio of NH_3 declines with increasing distance from the shoreline until it reaches a background value of about 0.1 nmol mol^{-1}. The background level determined by Ayers and Gras (1980) for the Indian Ocean was 0.08 nmol mol^{-1}. Even lower concentrations were observed by Quinn et al. (1988, 1990) over the eastern Pacific Ocean. These values indicate that the continents are major source regions, and that the sea would act as a receptor of NH_3 and NH_4^+ of continental origin. Seawater also contains ammonium, which is biologically controlled in concentrations ranging from 0.05 to 0.2 mmol m^{-3}, with values from the open ocean tending toward the lower end of the range. As NH_4^+ is in equilibrium with NH_3, and seawater is slightly alkaline, ammonia can escape to the atmosphere. Several authors have assumed that the low background of NH_3 over the remote ocean reflects a quasi-equilibrium with seawater. The early calculations of Buch (1920) suggested an equilibrium mixing ratio of 0.5 nmol mol^{-1} or less. Georgii and Gravenhorst (1977) assumed average $NH_3 + NH_4^+$ concentrations of 0.3 mmol m^{-3} and calculated aerial equilibrium mixing ratios in the range 0.03–0.4 nmol mol^{-1} for surface water temperatures varying from 5°C to 25°C. More recently, Quinn et al. (1988, 1990) have simultaneously measured NH_4^+, the saturation value of NH_3 in seawater (0.5 ± 0.3 mmol m^{-3} and 10 ± 7 nmol m^{-3}, respectively), and the concentration of NH_4^+ and NH_3 in the air above the ocean (about 5.3 ± 2.2 and 0.67 ± 0.65 nmol m^{-3}, respectively), from which they inferred a flux of ammonia averaging 7 ± 5.6 (range 1.8–15) μmol m^{-2} day^{-1}. The data indicated a rapid conversion of NH_3 to ammonium sulfate in the atmosphere and the subsequent return to the sea surface in a closed cycle. In fact, the consistent observation of ammonium sulfate in the non-sea salt fraction of the marine aerosol (Covert, 1988) suggests that sufficient ammonia emanates from the oceans to compensate for the formation of sulfuric acid. If the fluxes observed by Quinn et al. (1988, 1990) in the north Pacific were representative of marine conditions elsewhere, a simple extrapolation would indicate a global NH_3 source strength of 15 Tg year^{-1}, with a range of 4–34 Tg year^{-1}.

The observational data provide a coarse estimate of the tropospheric mass content of nitrogen present as NH_3 and NH_4^+, if we adopt a procedure similar to that used in Section 7.6.1 for aerosols. We must assume that the concentrations found in Europe and North America can be extrapolated to other continents as well. To describe the concentrations of NH_4^+ in the upper troposphere we adopt concentrations at the top of the boundary layer that decline in accordance with the atmospheric scale height. The concentration of gaseous NH_3 in the upper troposphere is assumed to be negligible. Both species are removed mainly by precipitation. By using the residence times for rain-out of aerosol particles and NH_3, derived in Chapters 7 and 8, respectively, one obtains an estimate for the flux of nitrogen associated with NH_3 through the atmosphere. The data in Table 9.2 suggest a turnover rate of 49 Tg year^{-1}. The following evaluation of the budget of ammonia will show that the flux has the right order of magnitude.

9.2.2. SOURCES AND SINKS OF ATMOSPHERIC NH_3

Table 9.3 lists processes contributing to the tropospheric budget of NH_3 and estimates of the global flux according to several authors. The sources, which are discussed first, may be classified as combustion, bacterial decomposition of animal excreta, and emanation from soils and vegetation.

The nitrogen content of coal is 1–2% by mass, present mainly in organic compounds. Heating releases a large fraction of it as NH_3. The coking of coal once provided the principal source of industrial ammonia before it was replaced by Haber–Bosch synthesis. Wood contains 0.2–0.5% nitrogen; leaves and newly grown twigs contain more, typically 1–2% by mass. Miner (1969) compiled NH_3 emission factors for the combustion of various materials: 1 g kg^{-1} for coal, 0.12 kg m^{-3} for fuel oil, 10 mg m^{-3} for natural gas, 1.2 g kg^{-1} for wood, and 0.15 g kg^{-1} for NH_3 released from forest fires (see also National Research Subcommittee on Ammonia, 1979). The value for coal refers to domestic use and must be taken as an upper limit, because high-temperature combustion units oxidize ammonia further toward NO. According to United Nations (1978) statistics, the world consumption of hard coal is about 2000 Tg year^{-1}, of which 400 Tg year^{-1} is coked and must be deducted, and that of lignite is 750 Tg year^{-1}. Combining these figures gives an NH_3 release rate from coal combustion of less than 2 Tg year^{-1}. The combustion of fuel oil and natural gas produces much less and can be neglected. NH_3 emissions from automobiles likewise are small. The estimate in Table 9.3 is based on an NH_3 emission factor of 30 mg km^{-1} and a world car population of about 300 million. Biomass burning also produces ammonia, primarily by smoldering combustion (Lobert et al., 1991). Accord-

TABLE 9.2. Mass Content of NH_3 and NH_4^+ Nitrogen in the Troposphere and Turnover Rate Resulting from Known Residence Times[a]

Region	A (10^{12} m^2)	c_0 (μg m^{-3})	h (km)	z (km)	$G(N)$ (Tg)	τ (days)	Nitrogen flux (Tg year^{-1})
NH_3							
Continents	149	1.8	1.2	0–5	0.26		
Oceans	361	0.02	1.5	0–3	0.01		
Sum					0.27	5.7	17.3
NH_4^+							
Continents	149	2.5	1.2	0–5	0.36		
Background	149	0.07	9.1	5–12	0.02		
Oceans	361	0.11	1.5	0–3	0.04		
Background	361	0.02	9.1	3–12	0.02		
Sum					0.44	5.0	32.1
$G(NH_3) + G(NH_4^+)$					0.67	Total flux	49.4

[a] The mass content is the product of integrated column density and surface area A, assuming an exponential decrease with scale height h within the boundary layer at altitudes up to z, and a background concentration of NH_4^+ in the upper troposphere decreasing with a scale height of 9.1 km up to the tropopause level at 12 km. The background concentrations of NH_3 is assumed to be negligible.

TABLE 9.3   Estimates of Sources and Sinks of Ammonia (Tg year $^{-1}$ as nitrogen)

Process	Söderlund and Swensson (1976)	Böttger et al. (1978)	Stedman and Shetter (1983)	Warneck (1988)	Schlesinger and Hartley (1992)	Dentener and Crutzen (1994)
Sources						
Coal combustion	4–12	0.03	< 2	≤ 2	2	—
Automobiles	—	0.2–0.3	—	0.2[e]	0.2	—
Biomass burning	—	—	—	2–8	5	2
Domestic animals	20–35	20–30	23	22	32	22
Wild animals	2–6	—	3	4[e]	—	2.5
Human excrements	[a]	—	1.5	3	4	—
Soil/plant emissions	—	1	(51)[b]	15	10	5.1
Fertilizer losses	—	1.2–2.4	3.5	3	9	6.4
Oceans	—	—	—	—	13	7
Sum of sources	26–53	22–34	83	54	75	45
Sinks						
Wet precipitation						
On the continents	20–80	15 ± 7	50[c]	30[e]	30	13.6[f]
Over the oceans	8–26	6 ± 6	10	8[e]	16	16[f]
Dry deposition (land)	69–151	[d]	14	10	10	13.6[f]
Reaction with OH	3–8	3	9	1	1	1.8[f]
Sum of sinks	100–265	24 ± 13	83	49	57	45

[a] Included in the figure for domestic animals.
[b] Adopted to balance the budget.
[c] Obtained by extrapolation, not by integration over latitude belts.
[d] Included in the figure for wet precipitation.
[e] Adopted as representative from previous work.
[f] Derived from three-dimensional model calculations.

ing to Table 4.7 the amounts of biomass burned annually range from 9 to 30 Pg (4.1–13.5 Pg of carbon). With the emission factor of Miner (1969) for wood as fuel (0.15 g kg^{-1}), one calculates ammonia emissions of 1.4–4.5 Tg year^{-1}. Values reported by Andreae (1991) and Laursen et al. (1992) are similar, 5.2 and 3.7 Tg year^{-1} of nitrogen, respectively.

As discussed by Eriksson (1952) and Georgii (1963), coal combustion was thought for many years to provide a prominent source of ammonia, at least in the cities. In 1970, Healy et al. showed that this source is dwarfed by NH_3 released from animal urine due to the bacterial decomposition of urea to NH_3 and CO_2. Healy et al. (1970) reported the following urea excretion rates: cattle 140, pigs 60, sheep 45, and poultry 3, in g urea per animal per day, corresponding approximately to 23, 10, 7.4, and 0.5 kg N per animal per year. Healy et al. (1970) assumed that 10% of urea nitrogen is volatilized as NH_3, but most field experiments, especially those of Watson and Lapins (1969) and Denmead *et al.* (1974), indicate higher percentages of up to 50%. The contribution of feces must be added. Their nitrogen content is similar to that of urine, according to the data assembled by Böttger *et al.* (1978), although it is not as readily volatilized by bacteria. Emission factors are expected to depend very much on the nitrogen content of the feed, the housing system, climatic factors, and the usage of manure. European emission factors reported by Asman (1992) for dairy cows, cattle on range, pigs, sheep, horses, and poultry are, respectively, 33, 10, 12, 4.4, 10, and 0.2 kg N per animal and year. These values correspond to about 50% of those given by Healy *et al.* (1970) for urea production, but they include manure. Animals in developing countries generally are less well fed, so that the excretion rate per head is lower, perhaps half as much. Dentener and Crutzen (1994) have employed these data to improve earlier global NH_3 emission estimates. Using FAO (1991) animal world population statistics, they derived annual emissions of 5.5, 8.7, 2.8, 1.2, 2.5, and 1.3 Tg N for dairy cows, beef cattle/buffaloes, pigs, horses/mules/asses, sheep/goats, and poultry, respectively. The sum is 22 Tg N.

The human body discharges about 30 g of urea daily. Healy *et al.* (1970) assumed that most of it ends up in sewers before it is mineralized. While this assumption may be true for the cities and many smaller communities in developed countries, it should not be generalized. Some NH_3 may even escape from sewer systems. If one-half of the ammonia from human urea were released to the atmosphere, an emission factor of 30% and a world population of 3.6 billion would represent a NH_3 nitrogen source of 2.7 Tg year^{-1}.

Söderlund and Svensson (1976) estimated the contribution to atmospheric ammonia from wild animals as 2–6 Tg N annually. The value is based on the ratio of urea excretion to plant matter ingested by herbivores, 6 mg g^{-1} as given by Loehr and Hart (1970), on an estimated 3–10% consumption by wild animals of annual new growth of green-plant material (primary productivity), which totals about 60 Pg year^{-1} (see Table 11.4), and on the 10% figure adopted by Healy *et al.* (1970) for the conversion of urea to ammonia.

Whereas the various estimates for NH_3 production from animals are reasonably consistent, it is evident from Table 9.3 that the global rate of emissions from soils and vegetation is rather uncertain. Schlesinger and Hartley (1992) provided a comprehensive tabulation of NH_3 flux measurements from different terrestrial ecosystems. Emission rates range over three orders of magnitude (2–4000 μg N m^{-2} h), with little differentiation between grasslands, forests, desert regions, and flooded areas. The diversity of results precludes any reasonable extrapolation to a global scale, so that a different approach is required.

Emissions of NH_3 and other nitrogen compounds from N-fertilized soils have received much study by agronomists interested in keeping such losses small. Studies have been carried out on fields dressed with ammonium salts, urea, manure, and sewage sludge, and on dairy fields and cattle feedlots. The subject has been reviewed by Allison (1955, 1966), by Freney *et al.* (1981), and recently by Asman (1992). Field losses were observed to range from 5% to 50%. High temperature, high wind speeds, and low relative humidities favor such losses, whereas rainfall minimizes them. Asman (1992) reported volatilization fractions of 15%, 2%, 8%, and 3% for urea, ammonium nitrate, ammonium sulfate, and other nitrogen fertilizers, respectively. The annual world production of fertilizers currently is about 80 Tg N, which leads to NH_3 emissions of about 6.5 Tg N year^{-1}, slightly lower than the estimate of Schlesinger and Hartley (1992), 8.5 Tg year^{-1}.

The above-mentioned studies are relevant to natural soils in that they have identified parameters influencing the NH_3 exhalation rate, the most important of which are temperature, pH, and moisture content of the soil. Ammonia is pictured to interact with soil water according to the equilibrium $NH_3 + H_2O \rightleftharpoons NH_4^+ + OH^-$. Increasing the water content shifts the equilibrium toward the right, causing NH_3 to be more tightly bound. Increasing the pH, that is the OH^- concentration, shifts the equilibrium toward the left, causing NH_3 to be released. The magnitude of the equilibrium constant favors the escape of NH_3 from soils with pH ≥ 6. More acidic soils are not expected to release NH_3 except in the autumn, when topped with decaying plant litter. Raising the temperature of the soil enhances the rate at which ammonia is made available by microbiological activity, and it increases the partial pressure of NH_3 in the soil by the evaporation of water.

Dawson (1977) attempted to determine the global rate of NH_3 emissions from unfertilized soils by means of a model incorporating physicochemical and microbiological processes involved in the production and release of ammonia. The value obtained by him, 47 Tg year^{-1} (38 Tg year^{-1} of nitrogen), must be considered an overestimate for several reasons discussed in detail by Langford *et al.* (1992). One problem, which their study highlights, is that measured pH values are consistently lower (by almost one pH unit) than those predicted by Dawson's empirical relationship. Dawson's model underestimates soil hydrogen ion concentration (thus overpredicting soil NH_3 vapor pressure) by a factor of 7 for grasslands and by a factor of 40–400 for forest soils. Another serious limitation of the model is the omission of the interaction of NH_3 with the plant canopy above the soil. Porter *et al.* (1972) and Hutchinson *et al.* (1972) first demonstrated that growing plants assimilate ammonia. Denmead *et al.* (1976) observed significant plant canopy absorption of NH_3 emitted from the soil or decaying organic matter underneath a rye grass/clover pasture. Goethel (1980) studied the NH_3 flux from twin plots, one of which had the plant cover removed, and found emission twice as high from the bare plot compared to the covered one.

Farquhar et al. (1980) first showed that plants, similar to soils (see Section 1.6), establish a compensation point for NH_3, that is, a concentration above which the plant acts as a sink of NH_3 and below which it acts as a source. At the compensation point, uptake and emission fluxes just balance. Micrometeorological measurements indicate compensation points in the range of 2–9 nmol mol^{-1} for grass-clover (Denmead et al., 1976), agricultural grassland (Sutton et al., 1993), barley (Schjørring et al., 1993), and soybeans (Harper et al., 1989). Even higher compensation points were observed for spring wheat (Morgan and Parton, 1989) and winter wheat (Harper et al., 1987). On the other hand, canopies of pine, spruce, fir, and aspen feature compensation points less than 1 nmol mol^{-1}, frequently around 0.4 nmol mol^{-1} (Langford and Fehsenfeld, 1992; Kesselmeier et al., 1993a). This suggests that forests commonly are sinks rather than sources of ammonia. Agricultural crops show higher compensation points, because the leaves of fertilized plants have a higher nitrogen content. The production of NH_3 is due to the enzymatic breakdown of proteins as well as the reduction of cell nitrate. The emission occurs exclusively via the stomata, whereas a considerable fraction of NH_3 uptake occurs via the cuticle. Distinct diurnal cycles can be observed over agricultural crops with NH_3 emissions occurring during the day, when leaf stomata are open, and uptake occurring during the night, when they are closed.

To estimate global scale emissions of ammonia from vegetation, Dentener and Crutzen (1994) adopted the concept of the compensation point and calculated emission and deposition fluxes from the expression $F = (C - C_0)/R_c$, where R_c is the canopy resistance, C is the ambient concentration, and C_0 is the compensation point. The daytime exchange velocity was scaled with the amount of vegetation, which was taken from a global vegetation data base, and ranged from 3 cm s^{-1} in the Tropics to vanishing values in desert areas. To account for the reduced stomata activity at night, the exchange velocities were lowered by a factor of 5 compared to daytime values. The model results indicate that about 5 Tg year^{-1} of ammonia is emitted from vegetation. This is smaller than most previous estimates, but the procedure employed appears to be the best currently available.

The predominant fate of ammonia in the atmosphere is either dry deposition or conversion to NH_4^+ and its return to the ground surface by wet precipitation. In the latter case, the annual deposition rate is determined by the concentration of NH_4^+ in rainwater and the cumulative amount of rainfall. The two parameters are not entirely independent of each other, and they fluctuate considerably at any given location, so that extended measurement series are needed to derive average deposition rates. Eriksson (1952a), Steinhardt (1973), and Böttger et al. (1978) have compiled the earlier data. The first authors included in their compilation observations from the second half of the past century. Böttger et al. (1978) considered only the period 1950–1977. More recent data are available from the stations currently cooperating in the Background Air Pollution Monitoring Network (BAPMoN) established by the World Meteorological Organization (Whelpdale and Kaiser, 1996). Dentener and Crutzen (1994) have summarized wet deposition rates observed at stations outside Europe and North America for comparison with model results. The agreement generally was within a factor of 2. Regional maps for the deposition of NH_4^+ were first prepared by Angström and Högberg (1952) and by Emanuelson et al. (1954) for Sweden. Junge (1963) subsequently discussed similar maps for several rainwater constituents in western Europe and in the United States. Langford et al. (1992) presented a newer map for the United States, and van Leeuwen et al. (1996) provided a recent map for Europe. Figure 9.4 illustrates the global distributions of NH_4^+ concentrations in rainwater and the corresponding deposition rates according to Böttger et al. (1978). The two quantities are not distributed in the same way, but the general behavior is similar in that the deposition is concentrated on the continents and decreases toward the oceans. Thus, NH_3 is removed from the atmosphere essentially in the regions of origin, and this is in accord with the short residence time of NH_3 of only few days.

Global deposition rates on the continents can be obtained by integration over land areas, extrapolating the data wherever necessary. The integrated rates obtained by Söderlund and Svensson (1976), Dawson (1977), and Böttger et al. (1978) amount to 20–80, 35, and 8–22 Tg year^{-1} of nitrogen, respectively. The first estimates are based mainly on the compilation of Eriksson (1952), which included many data from the past century, whereas that of Böttger et al. (1978) covered only the more recent decades. Brimblecombe and Stedman (1982) have shown that the annual deposition of NH_4^+ at several places in Europe and North America has remained fairly constant from 1850 to 1980. The wet deposition of NO_3^-, in contrast, has risen markedly during the same period. Accordingly, the earlier data should be given equal weight, so that the result of Böttger et al. (1978) may be too low. On the other hand, the older measurements do not differentiate between wet and dry deposition. Modern rain samplers cover the container with a lid during nonprecipitation periods, so that the contribution from dry deposition is excluded. Dawson (1977) and Böttger et al. (1978) have used the data to also infer the latitudinal distribution of the NH_4^+ wet deposition rate and

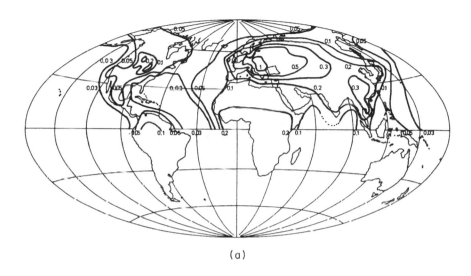

(a)

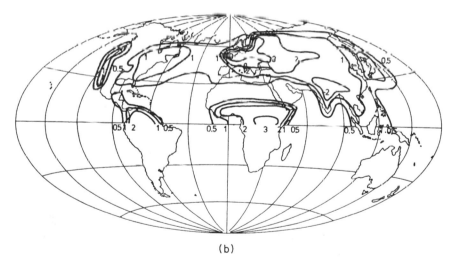

(b)

FIGURE 9.4   Global distribution of wet NH_4^+ deposition according to Böttger *et al.* (1978), derived from measurements during the period 1950–1977. (a) The NH_4^+ nitrogen content of rainwater (unit: mg kg^{-1}). (b) The nitrogen deposition rate (unit: 100 mg m^{-2} year^{-1}). (With permission of Kernforschungsanlage Jülich.)

noted that the distribution is similar to that of the sources. Both distributions reflect primarily the distribution of the land masses.

Duce *et al.* (1991) have used measured deposition fluxes of coastal and island stations to derive the total $NH_3 + NH_4^+$ deposition rate over the oceans. The annual deposition is 16.4 Tg N, of which 91% is due to wet deposition of NH_3 and NH_4^+ combined; the rest is dry deposition of NH_4^+. The three-dimensional modeling study of Dentener and Crutzen (1994) indicated, however, that about 50% of NH_4^+ that is wet deposited to the oceans has its origin on the continents.

Dry deposition of NH_3, and to a lesser extent of NH_4^+, is an important removal process, which is difficult to quantify, however. Deposition of excess NH_3 is efficient, with deposition velocities ranging up to 2 cm s^{-1} (Hanson and Lindberg, 1991). The quantity is highly variable, depending on the resistance to stomatal uptake and cuticular adsorption, the former undergoing diurnal variation, the latter being enhanced when the foliage is wet or when acids are codeposited on the surface. Dry deposition velocities for particulate NH_4^+ are smaller, about 0.2 cm s^{-1}.

A greater number of studies summarized by Höfken *et al.* (1981) have shown that the concentration of many trace substances in rainwater are higher underneath the forest canopy than in clearings at the same location. The effect is caused partly by dry-deposited material being washed off the foliage and partly by leaching. From their own measurements in 1980, Höfken *et al.* (1981) found the enrichment of NH_4^+ to average 1.7 for beech and 2.7 for spruce trees. The relative contribution of NH_4^+ from leaching increased with the amount of precipitation and may reach 40% of the total. The results showed that particulate NH_4^+ can contribute significantly to total intercepted ammonia, contrary to an earlier conclusion of Söderlund and Svensson (1976). If these results were representative of the whole world, the total deposition onto woodlands should be 70% greater than wet deposition alone. Since forests and brushland cover approximately 50% of the continental surface, losses due to interception would augment wet deposition by one-third, or about 10 Tg year^{-1}, if wet deposition were 30 Tg year^{-1} of nitrogen. In the model of Dentener and Crutzen (1994), who worked with deposition velocities scaled to the height of vegetation, the wet deposition flux was 27.5 Tg year^{-1} and dry deposition contributed about 36% of the total global deposition rate of 43.2 Tg year^{-1}; dry deposition of particulate NH_4^+ amounted to 20% of the total dry deposition rate of nitrogen associated with NH_3 and NH_4^+ combined. The data indicated a higher ratio of dry to wet deposition over land than earlier estimates. In agricultural areas with high ammonia emission densities, dry deposition of ammonia increases in importance. Asman and van Jaarsveld (1992) calculated 58% dry versus 42% wet deposition in Europe, and Harrison and Allen (1991) estimated dry deposition of NH_3 and NH_4^+ in England to contribute 64% and 7%, respectively, with the remaining 29% being removed by wet deposition.

Finally we consider the reaction of ammonia with OH radicals as a loss process:

$$NH_3 + OH \rightarrow NH_2 + H_2O$$

The rate coefficient for this reaction is endowed with a moderate temperature dependence (see Table A.4), so that the value for $T = 283$ K, $k = 1.3 \times 10^{-13}$ cm^3 molecule^{-1} s^{-1}, may be taken to be representative for the whole troposphere. The lifetime of NH_3 in the troposphere due to this reaction alone can be estimated with an average tropospheric OH concentration of 8×10^5 molecule cm^{-3}:

$$\tau = \left[k_{127} n(\text{OH}) \right]^{-1} = 9.4 \times 10^6 \text{ s or } 109 \text{ days}$$

This time constant is about 20 times larger than that associated with precipitation of NH_3. Accordingly, the reaction with OH is relatively unimportant. The absolute sink strength can be determined from the mass of NH_3 in the troposphere estimated in Table 9.2. A rate of 1 Tg year^{-1} of nitrogen is indicated. Dentener and Crutzen (1994) obtained 1.8 Tg year^{-1} from three-dimensional model calculations.

The reaction of the NH_2 radical with oxygen is immeasurably slow (Lesclaux and DeMissy, 1977; Cheskis and Sarkisov, 1979). The laboratory results of Hack et al. (1982) suggested that an O_2-NH_2 addition complex is formed that reverts to the original reactants without yielding new products. On the other hand, the reactions of NH_2 with O_3, NO, and NO_2 proceed rapidly with rate coefficients (at 300 K) of 1.7×10^{-13}, 1.6×10^{-11}, and 1.9×10^{-11} cm^3 molecule^{-1} s^{-1}, respectively (Atkinson et al., 1989). NO and NO_2 are favored as reactants unless the NO_2 mixing ratio falls below 220 pmol mol^{-1}, where ozone becomes important. This situation exists over the oceans and in the upper troposphere. The reaction with nitric oxide

$$NH_2 + NO \rightarrow N_2 + H_2O$$

is well known to reduce NO to molecular nitrogen, thus providing a sink for NO as well as for NH_3. The reaction with NO_2 has been shown to proceed similarly (Atkinson et al., 1989):

$$NH_2 + NO_2 \rightarrow N_2O + H_2O$$

The model results of Dentener and Crutzen (1994) suggest that this process may generate up to about 1 Tg year^{-1} of N_2O (as nitrogen). It should be noted, however, that in the laboratory experiments N_2O may have resulted from the reaction sequence

$$NH_2 + NO_2 \rightarrow 2HNO$$

$$2HNO \rightarrow N_2O + H_2O$$

and if the latter sequence were applicable, HNO in the atmosphere would transfer the odd hydrogen atom to other receptors such as oxygen, rather than react with itself. If this were the case, the conversion of NH_2 to N_2O would be much smaller. The fate of the NH_2O radical produced in the reaction of NH_2 with ozone,

$$NH_2 + O_3 \rightarrow NH_2O + O_2$$

is not precisely known. The NH_2O radical presumably reacts further with oxygen in a sequence leading ultimately to the formation of NO.

9.3. NITROUS OXIDE

Atmospheric nitrous oxide (N_2O) was discovered by Adel in 1938 as an infrared absorption feature in the solar spectrum. For the next 30 years, N_2O aroused little interest because it is not a hazardous pollutant and it does not display any particular chemical activity. In fact, no gas phase reaction is known that would remove it from the troposphere. In the stratosphere, N_2O undergoes photodecomposition, and it reacts with $O(^1D)$. The second reaction is the major source of higher nitrogen oxides in that region, and because they catalyze the destruction of ozone, N_2O is an important agent in controlling the stratospheric ozone balance (see Chapter 3). The recognition of this relationship by Crutzen (1970, 1971) and McElroy and McConnell (1971) stimulated many researchers to take a closer look at the behavior and budget of N_2O in the troposphere. In recent years, interest in N_2O has widened to include its role as an atmospheric greenhouse gas (Houghton *et al.*, 1995) and the anthropogenic contributions to the budget. The loss rate of N_2O in the stratosphere has been estimated to range from about 11 to 21 Tg year^{-1}, with an average of 16.5 Tg year^{-1}.

The older concepts regarding tropospheric N_2O have been reviewed by Hahn and Junge (1977) and Pierotti and Rasmussen (1977). The early gas chromatographic measurements, particularly those of Goody (1969) and Schütz *et al.* (1970), had indicated a large variability in the mixing ratio (± 40 nmol mol^{-1}), which in light of more recent results must have resulted from errors associated with the preconcentration procedures then used to overcome the insufficient sensitivity of the thermal conductivity detector.

The difficulties of instrument calibration, influence of temperature on sample processing, etc., also were underrated at that time. The preferred analytical technique today is gas chromatography based on the electron capture detector. Automated sampling procedures allow a precision of 0.02% to be achieved (Prinn et al., 1990). Calibration errors of instruments deployed at different locations can be minimized by the use of a common calibrated standard, and the differences arising from the change of calibrating gas can be further reduced by scaling data from different studies during periods of overlap (Khalil and Rasmussen, 1992a). Absolute calibration is cumbersome and requires a series of comparison measurements.

Table 9.4 summarizes some measurements made during the 1970s at various locations, which demonstrate the rather uniform global distribution in the troposphere with a mixing ratio somewhat exceeding 300 nmol mol^{-1}. In 1978 the problems of instrument calibration were largely solved, and the data converged to a value close to 300 nmol mol^{-1}. At the same time, several monitoring stations were set up in the Northern and Southern Hemispheres. The last entries in Table 9.4 show some results from the 10-year monitoring period of 1978–1988 to indicate the occurrence of a persistent slight excess in the Northern Hemisphere, about 0.7 nmol mol^{-1}. During this period the N$_2$O mixing ratio in the atmosphere was increasing.

TABLE 9.4 Measurements of N$_2$O Mixing Ratios in the Troposphere[a]

Investigators	Location	Years		Mixing ratio (nmol mol^{-1})
Pierotti and Rasmussen (1977)	NE Pacific (31° N–14° S)	1976		332 ± 9
	Pullman, Washington	1976		328.5 ± 0.7
	Flights between New Zealand and Alaska (80° N–59° S)	1976		330 ± 5
Tyson et al. (1978b)	Flight between New Zealand and Alaska	1976	NH	306 ± 21
			SH	314 ± 39
Goldan et al. (1978)	Various locations, mainly Boulder, Colorado	1976–1977 Corrected[b]		325.8 ± 1.2 / 304.5 ± 1.2
Singh et al. (1979a)	Various locations in the world	1975–1978		311 ± 2.8
Weiss (1981)	Various locations in the world	1976–1980		300.2 ± 0.1
	South pole	1978		299.2 ± 0.2
Prinn et al. (1990)	First 6 months of data from	1981	NH	300.3 ± 0.8
	2 years selected from a	1981	SH	299.7 ± 0.7
	10-year monitoring series at five	1988	NH	307.3 ± 1.2
	island and coastal stations	1988	SH	306.6 ± 1.0

[a] All measurements used gas chromatography and electron capture detector.
[b] After correction of calibration by Golden et al. (1981).

On the basis of the older data, and using the inverse relationship between residence time and variance of mixing ratio discussed in Section 4.1, Hahn and Junge (1977) had estimated the residence time of N_2O in the troposphere to be 4–12 years, whereas Singh et al. (1979a), Pierotti and Rasmussen (1977), and Weiss (1981) concluded from the much smaller variability observed by them that the residence time exceeds 30 years. Current estimates fall in the range of 110–170 years for the whole atmosphere (Cicerone, 1989; Prinn et al., 1990; Ko et al. 1991; Minschwaner et al., 1993). While its great chemical stability causes N_2O to be long-lived in the troposphere, it is doubtful whether the inverse relationship between residence time and variability really applies in this case. As Junge (1974) pointed out, the concept is based on an uneven distribution of the sources and sinks of the trace gas under consideration. In the case of N_2O the surface sources are spread fairly evenly over the entire globe, and as measurements by necessity are made in source regions, there is a tendency to observe fluctuations of the source strength rather than of the background mixing ratio.

Long-term measurements made since 1976 (Weiss, 1981; Prinn et al., 1990) have shown that the atmospheric N_2O concentration increases, about 0.27% annually, although with considerable year to year variation in the trend. In 1994 the atmospheric N_2O level was about 312 nmol mol^{-1}, corresponding to a tropospheric N_2O reservoir of about 2000 Tg (Houghton et al., 1995). The atmospheric abundance of N_2O in preindustrial times has been inferred from air trapped in the great ice sheets. The analysis of ice core data by Khalil and Rasmussen (1992a) showed that the N_2O level before 1800 A.D. was approximately constant at 286 nmol mol^{-1}, except during the little ice age in the seventeenth century, when concentrations were slightly lower. The secular increase began around 1890. The difference suggests that anthropogenic activities are responsible for the recent increase. Khalil and Rasmussen (1992a) have performed a box model–time series analysis based on the growth of N_2O from the preindustrial to the present level and an assumed 150-year lifetime of N_2O in the atmosphere to determine the anthropogenic contribution. They found a source of 7 ± 1 Tg $year^{-1}$ over the decade 1977–1987, compared to 15 Tg $year^{-1}$ for natural emissions in preindustrial times. For a brief period, the anthropogenic source responsible for the increase in atmospheric N_2O was thought to be fossil fuel combustion (Weiss, 1981; Hao et al., 1987), but the discovery of an artifact in sampling flue gases from electric power plants (Muzio and Kramlich, 1988) and its subsequent elimination led to a reassessment, which made emissions from power plants insignificant compared to natural sources (Linak et al., 1990; Yokoyama et al., 1991; Khalil and Rasmussen, 1992b).

The principal natural sources of N_2O are microbial processes in soils and in the ocean. For some time it was thought that N_2O production in soils is primarily due to denitrification, despite long-standing laboratory evidence, reviewed by Bremner and Blackmer (1981), showing that nitrifying bacteria also produce N_2O. In temperate agricultural soils, the importance of nitrification has been convincingly demonstrated by the observation of much higher N_2O emission rates in the case of fertilization with urea or ammonium as opposed to nitrate (Breitenbeck et al., 1980). In tropical soils, the opposite behavior has also been observed (Keller et al., 1988; Livingston et al., 1988; Sanhueza et al., 1990). Table 9.5 presents N_2O exhalation rates for a variety of ecosystem soils, both fertilized and unfertilized. Most studies employed chambers covering the soil surface, and the flux was determined from the rise of the N_2O level in the chamber. Formerly, it was believed that N_2O originates in the more anaerobic deeper strata of the soil, that it diffuses toward the surface, that diffusion sets up a vertical gradient of N_2O concentrations in the soil, and that one can calculate the flux from the gradient and a suitably chosen diffusion coefficient (Albrecht et al., 1970; Burford and Stefanson, 1973). This concept had to be revised. The field observations of Seiler and Conrad (1981) clearly showed that surface emission rates and vertical gradients are entirely unrelated. Surface emissions arise from bacterial processes in the uppermost layer of the soil, whereas the gradient is set up by the local balance of production and losses involving both nitrification and denitrification, the latter occurring in anaerobic niches of the soil (Rosswall, 1981). Figure 1.15 illustrates that a compensation level exists for soil derived N_2O. Soils provide a source of atmospheric N_2O when the compensation level is higher than the concentration in the atmosphere, and a sink when it is lower. As Table 9.5 demonstrates, most soils were found to be a net source of N_2O. However, uptake of N_2O from the atmosphere has occasionally been observed (Ryden, 1981; Slemr et al., 1984; Donoso et al., 1993).

The N_2O flux increases with rising temperature and moisture content of the topsoil. This dependence leads to diurnal and seasonal patterns of the flux, with maxima in midday and in summer, and to a fairly drastic enhancement of the release rate when the topsoil is wetted by rainfall or irrigation. Conrad et al. (1983) tried to quantify the temperature dependence in the form of an Arrhenius expression. The apparent activation energy was quite variable, however. The dependence on soil moisture in aerated soils was also studied and could be expressed by a third-order power law. Other factors controlling the N_2O flux (Williams et al., 1992a) are soil porosity, the rate at which NH_4^+ is made available for nitrification, the content of organic carbon serving as substrate for denitrifiers, and the pool of nitrate. Emissions from wetland soils frequently are lower compared to aerated soils, because the migration of gases is hampered under waterlogged conditions, so that a larger percentage of N_2O is further reduced to N_2 by denitrification. Tropical

TABLE 9.5 Some Measurements of N_2O Emission Rates from Soils (as Nitrogen) with the Chamber Technique

Authors	Soil type and cover	Observation period	Nitrogen emission ($\mu g\ m^{-2}\ h^{-1}$)	Remarks
Agricultural fields and grass land				
Bremner et al. (1980)	Various Iowa soils used to grow corn and soybeans	12 months April–April	3.6–22.3 av. 13.7	No fertilization, seasonal variation with maximum in June and July
Mosier and Hutchinson (1981)	Nunn clay loam planted to corn	4 months, May–Sept.	85.6	Fertilized with 200 kg N hm^{-2} NH$_3$
Bremner et al. (1981)	Webster, Canisteo, and Harp soil planted to corn in 1978, to soybeans in 1979	139 days following June 22, 1979	50.2–74.2 365–582	Unfertilized 250 kg N hm^{-2} NH$_3$ applied
Ryden (1981)	Wickham clay loam sown to perennial rye grass	Aug–Dec. (1980) Mar. 1980–Mar. 1981	−8.0[a] 37.0	Unfertilized 250 kg N hm^{-2} NH$_4$NO$_3$
Seiler and Conrad (1981)	Loess loam, grass Aeolian sand, grass	Intermittently over a 3-year period	0.5–3.0 1.9–12.6	Unfertilized Unfertilized
Duxbury et al. (1982)	Mineral soil, weed grasses Covered with alfalfa Organic soils, bare and fallow St. Augustine grass or sugarcane	May 1979–May 1981	10.3–19.4 26.3–47.9 674–3832 80–1107	Unfertilized Unfertilized Unfertilized Unfertilized
Sanhueza et al. (1990)	Scrub grass savanna, sandy loam	October 1988, wet season	−3.0–11.7[a,b]	Application of NO$_3^-$ increased the flux, whereas NH$_4^+$ did not
Forests				
Duxbury et al. (1982)	Mineral soil, New York state	Calendar year 1980	av. 10.3	Unfertilized
Schmidt et al. (1988a)	Six deciduous temperate forest sites, loess loam and sand	July 1981–Aug. 1982	1.0–91.3 av. 2.3–8.0	High spatial variability noted
Keller et al. (1986)	Undisturbed Amazon forest	Dec. 1983–Mar. 1984	av. 14.6	
Livingston et al. (1988)	Amazon forest, infertile oxisol or ultisol	August 1985, dry season	4.3–18.3 av. 13.0	Application of NO$_3^-$ increased the flux, whereas NH$_4^+$ did not
Sanhueza et al. (1990)	Tropical deciduous forest	October 1988, wet season	4.8–20.9	

[a] Negative sign means uptake of N_2O.
[b] Fluxes were 0.2 kg hm^{-2} year^{-1} during the dry season but increased upon irrigation.

forest soils feature comparatively high emission rates, apparently because of a high organic matter turnover and mineralization potential (Keller *et al.*, 1986, 1988; Matson and Vitousek, 1987; Sanhueza *et al.*, 1990)

The great spatial and temporal variability of N_2O fluxes observed in the field is a major obstacle to global-scale extrapolation of soil emission data. The uncertainty may be illustrated by the following global source flux estimates: 7–16 Tg year^{-1} (Bowden, 1986), 3–25 Tg year^{-1} (Banin, 1986), 39 Tg year^{-1} (Seiler and Conrad, 1987), 2.8–7.7 Tg year^{-1} (Watson *et al.*, 1992). Recent extrapolation strategies therefore combine soil type and ecosystem representation with process-oriented models to estimate the global production rate. For example, Bouwman *et al.* (1993) have constructed a model based on nitrogen cycling through the soil–plant–microbial biomass system with five major controls: input of organic matter, soil fertility, soil moisture status, temperature, and soil oxygen status. The satellite-derived normal difference vegetation index was used as a substitute for net primary productivity. Control factors were combined with a $1° \times 1°$ resolution global grid defining soil type, soil texture, vegetation index, and climate. Model results, when compared with measurements at 30 sites, were found to explain 60% of the observed variability in the N_2O fluxes. The results further showed that spatial and seasonal distributions of N_2O emission are very similar to climate patterns. Low temperatures during the cold season limit N_2O production in temperate regions, whereas temperature is not a limiting factor in the Tropics. The highest rates occur in the wet Tropics, where organic matter input is high and both soil moisture and temperature are favorable. The global N_2O emission rate thus calculated was 10.7 Tg year^{-1}, with 79% of it occurring in the Tropics ($\pm 30°$ latitude).

Nevison *et al.* (1996) have used a similar model, which made use of the correlation between N_2O flux and gross mineralization rate, first described by Matson and Vitousek (1987). The study of Nevison *et al.* (1996) focused on changes in N_2O emissions during the period 1860–1990. Anthropogenic perturbations included changes in land use, inputs of nitrogen from fertilizer, livestock manure, and fossil fuel-derived nitrogen oxides. The results indicated that global N_2O emissions associated with soil nitrogen mineralization decreased slightly from 9.3 to 8.9 Tg year^{-1} because of land clearing, whereas emissions due to volatilization and leaching of excess mineral nitrogen increased sharply from 0.7 to 5.2 Tg year^{-1}, with the difference arising from all anthropogenic perturbations. In 1990, emissions attributed to fertilizer, livestock manure, land clearing, and atmospheric deposition were estimated to be 2.2, 1.1, 0.63, and 0.13 Tg year^{-1}, respectively, summing to a total of about 4 Tg year^{-1}.

Fertilization enhances the N_2O output. The field measurements indicate, however, that fertilization increases the flux only for a limited period of time;

thereafter the flux falls back to background values. By integrating the excess flux over the period of enhancement, one can determine the percentage of fertilizer nitrogen that is lost as N_2O. Table 9.6 shows results for a number of such measurements. The lowest value is found for nitrate as fertilizer, and the highest when anhydrous (liquid) ammonia is applied. Eichner (1990) and Harrison *et al.* (1995) have summarized the available data. The majority of applications are in the form of NH_4NO_3. In this case, an average of 0.5% of nitrogen by mass is released as N_2O. The current annual production of 80 Tg year^{-1} (as nitrogen; FAO, 1991) would correspond to a global emission of 0.63 Tg year^{-1} from this source. Bouwman *et al.* (1995) discussed the associated uncertainties. By considering different countries, they estimated global fertilizer-induced N_2O emissions in the range of 0.6–1.6 Tg year^{-1} compared to 0–1.4 Tg year^{-1} from unfertilized arable land. This result is smaller than that calculated by Nevison *et al.* (1996). As the use of nitrogen fertilizers is expected to increase in the future to secure the food supply for a growing world population, N_2O emission rates from agricultural soils also will increase. Concern has been voiced (Crutzen, 1976b; McElroy *et al.*, 1977a) about the prospect that the flux of N_2O to the stratosphere might rise to the point where the UV shielding capacity of the ozone layer is impaired. It appears, however, that other anthropogenic sources are currently more important in this regard.

The world ocean is a significant source of atmospheric nitrous oxide. The surface waters of the sea are usually slightly supersaturated with N_2O relative to atmospheric concentrations. The early data, which were reviewed by Hahn (1981), had indicated an average saturation value of 125%, with the consequence that the global N_2O flux from the ocean was considerably overestimated (~ 40 Tg year^{-1}). High saturation values are found especially in

TABLE 9.6 Fraction (Average Percent) of Fertilizer Nitrogen Released from Soils as N_2O

Authors	Amount of N applied (kg hm^{-2})	Time interval of observation (days)	Type of fertilizer			
			NH_4^+	NO_3^-	Urea	$NH_3{}^a$
Breitenbeck *et al.* (1980)	125	Up to 96	0.11	0.02	0.08	—
Bremner *et al.* (1981)	250	139	—	—	—	5.4
Conrad and Seiler (1980d)	100	72	0.09	0.01	—	—
Conrad *et al.* (1983)	100	20	0.14	0.03	—	—
Mosier and Hutchinson (1981)	200	86	—	—	—	1.3
Slemr *et al.* (1984)	100	30	—	—	0.18	—

a "Anhydrous" ammonia.

estuaries and coastal up-welling regions, whereas in other coastal regions and in the open ocean the supersaturation generally is slight. Numerous measurements from the past two decades now provide an extensive data set. Its recent evaluation by Nevison *et al.* (1995) and by Bange *et al.* (1996) has led to a more realistic assessment of the marine N_2O source. Nevison *et al.* (1995) have globally extrapolated the expedition data and coupled them with air-to-sea gas transfer coefficients as a function of water temperature and wind speed. The parameters were obtained with $2.8° \times 2.8°$ spatial resolution from a 2-year run of the National Center for Atmospheric Research (Boulder, Colorado) climate model. Results for the global N_2O source strength depended on the formula used for the transfer coefficient. The algorithm of Liss and Merlivat (1986) gave 4.6 Tg year^{-1}, that of Wanninkhof (1992) 6.1 Tg year^{-1}, and that of Erickson (1993) 8.2 Tg year^{-1}. The average was 6.3 ± 4.4 Tg year^{-1}. Maxima of zonally averaged N_2O fluxes were predicted to occur at mid-latitudes, 40°–60°, and in the equatorial up-welling zones. Bange *et al.* (1996) have mainly studied emissions of N_2O from estuaries and coastal waters. Their compilation of data indicates that estuaries, while covering only 0.4% of the total ocean area, may contribute as much as 33% to the marine N_2O source, as compared to 25% from coastal waters and 40% from the open ocean. The global source estimate was based on a classification of oceanic provinces, somewhat modified from Menard and Smith (1966), and gave 10.9 or 17.3 Tg year^{-1}, depending on whether the transfer coefficients were based on the formulas of Liss and Merlivat (1986) or Erickson (1993). The contributions from the open ocean were 4.2 and 6.8 Tg year^{-1}, respectively, in good agreement with the results of Nevison *et al.* (1995).

It is still not precisely known how much of the N_2O escaping from the open ocean to the atmosphere is produced in near-surface waters and how much is brought up from deeper levels by diffusive transport. In surface waters, denitrification is inhibited by high levels of oxygen, and nitrification has to compete with the high demand of phytoplankton for NH_4^+ as nutrient (Schlesinger, 1997). The distribution of N_2O with depth is shown in Figure 9.5. The N_2O concentration generally maximizes in a zone where O_2 concentrations are at a minimum, 500–1000 m below the surface. In this region O_2 is consumed by bacterial oxidation of organic matter, and N_2O and nitrate are positively correlated (Cohen and Gordon, 1979; Yoshida *et al.*, 1989; Kim and Craig, 1990). This correlation suggests that N_2O arises primarily from nitrification. The turnover rate of fixed nitrogen between the pools of organic nitrogen in the biota and inorganic nitrogen (mainly nitrate) dissolved in the sea can be derived from the net rate of carbon fixation by photosynthesis (35 Pg year^{-1}, see Chapter 11) and the average C:N ratio in plankton (5.7 by weight; Redfield *et al.*, 1963) as about 6000 Tg year^{-1} N_2.

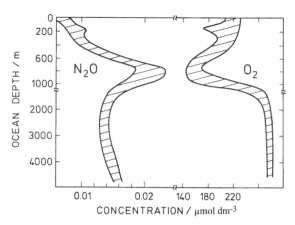

FIGURE 9.5   Vertical concentration profiles of N_2O and O_2 in the Atlantic Ocean (Gulf Stream area and Sargasso Sea) in 1971–1972, according to Yoshinari (1976).

If the fraction of NH_4^+ converted to N_2O in seawater were similar to that in soil systems, which is about 0.1% according to Table 9.6, one would obtain 9.4 Tg year^{-1} for the N_2O production rate from nitrification alone (molar mass ratio $N_2O:N_2 \approx 44:28 = 1.57$). Similar considerations led Najjar (1992) to estimate the production rate from nitrification to be 6.3 Tg N_2O year^{-1}.

Denitrification in zones of low oxygen concentrations may also be a source of marine N_2O. Incubation experiments with seawater samples taken from various depths have shown that nitrification and denitrification occur simultaneously in the same strata (Gundersen et al., 1972). In the open ocean the O_2 concentration never falls to values so low as to establish truly anaerobic conditions. However, anaerobic niches may occur in particles of organic matter. In deep waters, N_2O is enriched with ^{15}N and ^{18}O compared to atmospheric N_2O, and NO_3^- is similarly enriched with ^{15}N. These observations have been interpreted in terms of both nitrification and denitrification (Yoshida et al., 1989; Kim and Craig, 1990) without resolving the controversy. Anaerobic conditions are definitely established in the sediments. Christensen et al. (1987) have suggested that denitrification occurs primarily in continental shelf sediments. Under such conditions, nitrogen is expected to be the main product of denitrification. A limited number of observations indicate a N_2O/N_2 ratio on the order of 0.1% (Seitzinger et al., 1988).

Table 9.7 shows the global budget of N_2O in the atmosphere according to the source estimates of Khalil and Rasmussen (1992a) and Bouwman et al. (1995). In addition to the natural sources discussed above, many anthropogenic sources of small magnitude have been identified. Emissions from the combustion of fossil fuels occur primarily from automobiles equipped with

TABLE 9.7 Tropospheric Budget of Nitrous Oxide (Tg year^{-1})

	1992[a]	1995[b]
Natural sources		
Tropical forests	11.6 (5.3–17.9)	5.2
Temperate forests	1.2 (1.6–0.8)	0.8
Grasslands	0.16	2.3
Arable land	—	1.5
Total soils	12.3 (5.7–18.9)	10.7 (10.4–11.0)
Oceans	3.1 (1.6–4.7)	5.7 (4.4–8.9)
Total natural	15.4 (7.3–23.6)	16.4 (14.8–19.9)
Anthropogenic sources		
Fossil fuel combustion	0.8 (0.1–2.2)	0.47 (0.16–0.94)
Adipic acid Production	0.7	0.47 (0.30–0.61)
Nitric acid production	—	0.3 (0.12–0.47)
Nitrogen fertilizer	1.0 (0.4–3.0)	1.6 (0.64–2.6)
Domestic animal excreta	0.5 (0.3–1.0)	1.6
Biomass burning	1.6 (0.2–3.0)	0.36
Land use change	0.7	0.6
NH_3 conversion[c]	—	0.63 (0.47–1.9)
Total anthropogenic[d]	7.9 (5–10)	9.4 (4.3–8.9)
Total identified sources	23.4 (12.2–33.6)	25.8 (19.1–28.8)
Stratospheric sink	16.5 (11.8–21.2)	19.3 (14.1–25.1)
Atmospheric increase	5.5 (4.7–6.3)	6.1 (4.9–7.4)

[a] Khalil and Rasmussen (1992a); natural sources, stratospheric sink, and atmospheric increase from McElroy and Wofsy (1986).
[b] Bouwman et al., (1995); stratospheric sink and atmospheric increase from Houghton et al. (1995).
[c] Oxidation of ammonia by OH radicals in the atmosphere.
[d] Includes additional small sources.

catalytic converters, whereas emissions from electric power plants may be ignored. The 0.47 Tg year^{-1} emission rate reported by Bouwman et al. (1995) is based on 10 g of N_2O per GJ being emitted in industrialized countries. Adipic acid is used primarily in the production of nylon. One of the steps in the manufacture of adipic acid is the oxidation of a mixture of cyclohexanone and cyclohexanol by nitric acid, leading to N_2O as a by-product. The global production capacity in 1990, according to Bouwman et al., (1995), was 2.18×10^6 kg year^{-1}, and the estimated emission factor is 300 g N_2O per kg adipic acid in production plants with unabated emissions, 6 g kg^{-1} for one plant with 98% abatement (15% of total production), and 231 g kg^{-1} for other plants with partial emission controls (32% of total production). The resulting N_2O emission rate is 0.45 Tg year^{-1}. N_2O also occurs in the manufacture of nitric acid, where it is a by-product of the oxidation of NH_3 over a hot catalyst. Nitric acid is the main feedstock in nitrogen fertilizer production. The global production rate of nitric acid is about 50 Tg year^{-1}, estimated N_2O emission factors are 2.4–9.4 g per kg of nitric acid, which yields 0.12–0.47 Tg year^{-1} N_2O.

Industrial processes and fossil fuel combustion contribute fairly little to all anthropogenic N_2O sources, whereas agricultural activities appear to be more important. N_2O emissions associated with artificial fertilizers were discussed above, together with emissions from soils. This does not yet include emissions from animal excreta. According to the estimate in Section 9.2.2, the amount of nitrogen excreted annually by domestic animals, much of it in the form of urea, is 22 Tg year^{-1}. The fraction of nitrogen converted to N_2O may differ considerably, depending on whether animals are kept in stables or in open fields. The range is about 0.5–1.5%. If the fraction were 1%, the emission rate would be $0.01 \times 22 \times (44/28) = 0.34$ Tg year^{-1}. Bouwman et al. (1995) adopted for the global nitrogen excretion from animals a rate of 100 Tg year^{-1}, which led to nitrous oxide emissions of 1.6 Tg year^{-1} from this source. Biomass burning is another human activity related to agriculture, primarily in the Tropics. The combustion of biomass generates nitrous oxide mainly during the flaming stage, and production rates correlate well with CO_2. Lobert et al. (1991) have measured the average molar emission factor relative to fuel nitrogen content and found 0.77% (range 0.34–1.2%). The global rate of nitrogen emitted from the burning of biomass was estimated from the molar N/C ratio in different materials (forest 1%, savanna grass 0.6%, fuel wood 0.5%, agricultural waste 1%) and the amount of biomass burned (Crutzen and Andreae, 1990). This procedure gave 28–72 Tg N released each year. Global N_2O emissions thus amount to 0.34–0.87 Tg year^{-1} (average 0.67 Tg year^{-1}). Hao et al. (1990) and Cofer et al. (1991) used the ratio of N_2O released to carbon burned, 0.01–0.03%, and obtained 0.8–3.4 Tg year^{-1}. Bouwman et al. (1995) used emission factors presented by Crutzen and Andreae (1990) and obtained a lower value, 0.31 Tg year^{-1}.

The conversion of tropical forests to pastures and fields also leads to an increase in the global emissions of N_2O. Matson and Vitousek (1990) estimated a global emission of 1.1 Tg year^{-1} for an area of 2.2×10^6 km^2 converted to pasture since 1950. This estimate does not consider the decline in emission with pasture age (Keller et al., 1993). Bouwman et al. (1995) used 1990 statistics, which had indicated that 2.4×10^4 km^2 was converted to pasture and 1.26×10^5 km^2 was converted to arable land. They assumed the N_2O emission to be five times that of tropical forest for the first year after land clearance, with a linear decline for the following 10 years to values typical for arable land and grassland. This resulted in an emission of 0.6 Tg year^{-1}. The last source listed in Table 9.7 is the partial conversion of ammonia to N$_2$O in the atmosphere, which was discussed in Section 9.2.2. The source estimate of 0.63 Tg year^{-1} is the anthropogenic contribution (Dentener and Crutzen, 1994). The total source strength including natural emissions may be as high as 2.8 Tg year^{-1}.

Table 9.7 shows that the N$_2$O budget is reasonably well balanced. Stratospheric destruction currently balances mainly the natural sources, 15–16 Tg year^{-1}. Because of the long atmospheric lifetime of N$_2$O, the anthropogenic input has not yet been accommodated by the stratospheric sink, and it shows up as an increase of N$_2$O in the troposphere, 5–6 Tg year^{-1}. Anthropogenic sources sum to about 8 Tg year^{-1}, but the various contributions, resulting primarily from changing agricultural practices, are difficult to estimate.

Removal processes other than photochemical destruction of N$_2$O in the stratosphere have not been found. Bates and Hays (1967) had thought that photolysis of N$_2$O would occur even in the troposphere, but a careful reinvestigation of the long-wavelength tail of the absorption spectrum of N$_2$O by Johnston and Selwyn (1975) as well as measurements of the ground-level photodissociation rate by Stedman et al. (1976) established beyond doubt that this process is insignificant at wavelengths above 300 nm. Rebbert and Ausloos (1978) have shown that photodecomposition occurs at wavelengths greater than 300 nm when N$_2$O is adsorbed on silica or dry desert-type sands. Heat treatment enhances destruction, whereas the presence of moisture reduces it. Schütz et al. (1970), Junge et al. (1971), and Pierotti et al. (1978) all found a depression in the N$_2$O mixing ratio in air masses originating from the Sahara Desert. It is possible that N$_2$O is destroyed on the surface of hot, dusty desert areas, but the magnitude of the effect has not been quantified. Other processes that have been investigated and were found unimportant include the reaction with OH radicals (Biermann et al., 1976) and reactions with negative ions (Fehsenfeld and Ferguson, 1976). Seiler et al. (1978a) showed that growing plants are neither sources nor sinks of atmospheric N$_2$O. Photochemical destruction in the stratosphere thus remains the only well-identified loss process.

9.4. NITROGEN DIOXIDE, NO$_2$, AND RELATED NITROGEN COMPOUNDS

The higher nitrogen oxides are very reactive, and for this reason they hold a prominent position in tropospheric chemistry. Six oxides are potentially important. They fall into two groups: NO, NO$_2$, NO$_3$, and N$_2$O$_3$, N$_2$O$_4$, N$_2$O$_5$. The last three compounds arise from the corresponding species in the first group by association with NO_2. They are thermally unstable, however. Table 9.8 lists association–dissociation equilibrium constants for these compounds at high and low temperatures and product/reactant ratios for a range of NO_2 partial pressures to show that the abundances of N_2O_3 and N_2O_4 are negligible in comparison with NO and NO_2, respectively, whereas N_2O_5 predominates over NO_3 when temperatures are low and/or NO_2 concentrations are high. Accordingly, we may ignore N_2O_3 and N_2O_4, but not N_2O_5.

Table 9.8 includes similar data for the association products of NO_2 with peroxy radicals. Peroxyacetyl nitrate (PAN) is thermally the most stable of these adducts. PAN is important in all regions of the troposphere. Peroxynitric acid and methyl peroxy nitrate are less stable but may be important in the upper troposphere, where temperatures are low. Model calculations suggest that as much as 50% of NO_2 may be present as $HOONO_2$ at higher altitudes (e.g., Logan et al., 1981; Singh et al., 1996b). Observational data are lacking, so that the contribution of these addition products to the total reservoir of NO_2 in the troposphere remains to be established.

The oxidation of NO_2 leads to the formation of nitric acid and aerosol nitrate, which are deposited at the Earth's surface. The relevant oxidation pathways are indicated in Figure 9.6. The following discussion deals first with observations of reaction intermediates, then with tropospheric abundances of NO_2, PAN and HNO_3/aerosol nitrate, and finally with the budget of nitrogen oxides and their oxidation products in the troposphere.

9.4.1. NITROGEN OXIDE CHEMISTRY

Figure 9.6 summarizes chemical pathways involving nitrogen oxides in the troposphere. Photochemically induced reaction pathways are indicated by bold arrows. These processes are active only during the day, whereas the others occur at all times. Table 9.9 is added to show time constants associated with the individual reaction steps.

The photochemical interconversion of NO and NO_2 was discussed in Section 5.2. Photodissociation of NO_2 generates NO and O atoms that quickly attach to molecular oxygen to form ozone. Back-reactions of NO with ozone and with peroxy radicals produced by the oxidation of hydrocarbons establish a steady state between NO and NO_2. In the absence of local sources, their molar ratio should be given by the steady-state equation

$$x(NO_2)/x(NO) = [k_{66}n(O_3) + \Sigma k_x n_x(RO_2)]/j(NO_2) \quad (9.1)$$

where the summation sign denotes the combined effect of all peroxy radicals, including HO_2. In the early studies the reactions of peroxy radicals were

TABLE 9.8 Equilibrium Constants and Relative Abundances for Nitrogen Oxides and Peroxides Originating as Addition Products from NO_2[a]

Reaction	T (K)	K_{eq} (Pa^{-1})	Ratio $p(x \cdot NO_2)/p(x) = K_{eq} \times p(NO_2)$[b]				
			(−7)	(−8)	(−9)	(−10)	(−11)
$NO + NO_2 \rightleftharpoons N_2O_3$	300	4.55(−6)	4.6(−8)	4.6(−9)	4.6(−10)	4.6(−11)	4.6(−12)
$NO_2 + NO_2 \rightleftharpoons N_2O_4$	200	1.63(−2)	1.6(−4)	1.6(−5)	1.6(−6)	1.6(−7)	1.6(−8)
	300	5.21(−5)	5.2(−7)	5.3(−8)	5.3(−9)	5.3(−10)	5.3(−11)
$NO_3 + NO_2 \rightleftharpoons N_2O_5$	200	5.02	5.1(−2)	5.1(−3)	5.1(−4)	5.1(−5)	5.1(−6)
	300	5.46(3)	5.5(1)	5.5	5.5(−1)	5.5(−2)	5.5(−3)
	250	1.00(7)	1.0(5)	1.0(4)	1.0(3)	1.0(2)	1.0(1)
$HO_2 + NO_2 \rightleftharpoons HO_2NO_2$	300	3.14(3)	3.2(1)	3.2	3.2(−1)	3.2(−2)	3.2(−3)
	250	5.25(5)	5.3(4)	5.3(3)	5.3(2)	5.3(1)	5.3
$CH_3O_2 + NO_2 \rightleftharpoons CH_3O_2NO_2$	300	5.14(2)	5.2	5.2(−1)	5.2(−2)	5.2(−3)	5.2(−4)
	250	1.07(6)	1.1(4)	1.1(3)	1.1(2)	1.1(1)	1.1
$CH_2(CO)O_2 + NO_2 \rightleftharpoons PAN$	300	4.02(6)	4.1(4)	4.1(3)	4.1(2)	4.1(1)	4.1
	250	6.12(10)	6.2(8)	6.2(7)	6.2(6)	6.2(5)	6.2(4)

[a] Powers of 10 in parentheses. Equilibrium constants as given by DeMore et al. (1997) and converted to pressure units.

[b] Nitrogen dioxide pressure in atm = 1.01325×10^5 Pa.

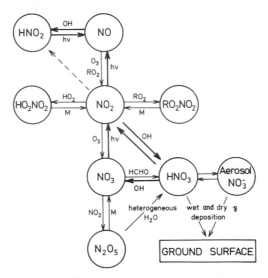

FIGURE 9.6 Oxidation scheme for nitrogen oxides and related compounds. Photochemical processes are indicated by bold arrows.

neglected, and it was assumed that the photostationary state involved only NO_2, NO, and O_3 $[k_{66} n(O_3) \gg \Sigma k_x n_x(RO_2)]$. On clear-sky summer days one may take $j(NO_2) \approx 5 \times 10^{-3}$ s^{-1}, $k_{66} = 1.8 \times 10^{-14}$ molecule cm^{-3} s^{-1}, $n(O_3) = 7.5 \times 10^{11}$ molecule cm^{-3}, corresponding to about 30 nmol mol^{-1}, to estimate $x(NO_2)/x(NO) \approx 2.7$. The first measurements of NO_2/NO ratios designed to check on the above relation were made by O'Brien (1974) and Stedman and Jackson (1975) in polluted city air containing high ozone concentrations. The observed ratios were in reasonable agreement with those predicted from simultaneous measurements of $j(NO_2)$ and $n(O_3)$, and there was no need to include RO_2 reactions. Subsequent observations under clean air conditions revealed NO_2/NO ratios exceeding the predicted values by a considerable margin. For example, ratios in the range 4–10 were found by Ritter et al. (1979) at a rural site in Michigan, and by Kelly et al. (1980) at Niwot Ridge, a mountain site west of Boulder, Colorado. Helas and Warneck (1981), who did not investigate the photostationary state, found NO_2/NO ratios of 10 or greater in marine air masses of the northern Atlantic, and Stedman and McEwan (1983) made similar observations at a mountain site in New Zealand. Parrish et al. (1986a) conducted an extensive study of the deviation from the NO_x-O_3 photostationary state, again at Niwot Ridge, Colorado, by simultaneous measurement of the concentrations of NO, NO_2, O_3, and $j(NO_2)$. They defined the excess

TABLE 9.9 Reactions involving Nitrogen Oxides and Their Oxidation Products and the Associated Time Constants [a]

Reaction	Reactant	x^b	n^b	k, j^c	τ (s)
$O_3 + NO \rightarrow NO_2 + O_2$	O_3	30	7.5(11)	1.8(-14)	7.5(1)
$RO_2 + NO \rightarrow NO_2 + RO$	RO_2	0.1	2.5(9)	8.0(-12)	5.0(1)
$OH + NO \rightarrow HNO_2$	OH	—	2(6)	5.0(-12)	1.0(5)
$HNO_2 + h\nu \rightarrow OH + NO$	$h\nu$	—	—	8.4(-4)	1.2(3)
$NO_2 + h\nu \rightarrow NO + O$	$h\nu$	—	—	3.5(-3)	2.8(2)
$O_3 + NO_2 \rightarrow NO_3 + O$	O_3	30	7.5(11)	3.2(-17)	4.2(4)
$NO_3 + h\nu \rightarrow NO_2 + O$	$h\nu$	—	—	1.6(-1)	6.2
$NO_3 + CH_3CHO \rightarrow HNO_3 + CH_3CO$	CH_3CHO	1.0	2.5(10)	2.4(-15)	1.7(4)
$NO_3 + NO_2 \rightarrow N_2O_5$	NO_2	6.0	1.5(11)	1.2(-12)	5.6
$N_2O_5 + h\nu \rightarrow$	$h\nu$	—	—	6.0(-6)	1.7(-5)
$N_2O_5 + H_2O \rightarrow 2HNO_3$	H_2O	2.5%	6.2(17)	<2(-21)	>8(2)
$N_2O_5 + H_2O \xrightarrow{\text{hetero}} 2HNO_3$	Rural aerosol	—	—	—	1.2(2)
$OH + NO_2 \rightarrow HNO_3$	OH	—	2(6)	1.3(-11)	3.8(4)
$OH + HNO_3 \rightarrow H_2O + NO_3$	OH	—	2(6)	1.5(-13)	3.3(6)
$HNO_3 + h\nu \rightarrow OH + NO_2$	$h\nu$	—	—	1.0(-7)	1.0(7)
$HNO_3 \rightarrow$ wet deposition	—	—	—	—	4.7(5)[d]
$HNO_3 \rightarrow$ dry deposition	—	—	—	—	3.3(5)[d]
$2NO_2 + H_2O \rightarrow HNO_2 + HNO_3$	H_2O	2.5%	6.2(17)	8.0(-38)	7(7)[e]
$NO + NO_2 + H_2O \rightarrow 2$	HNO_2	2.5%	6.2(17)	—	
	NO_2	6.0	1.5(11)	4.4(-40)	2(10)[e]

[a] Orders of magnitude are shown in parentheses. The data are based partly on a compilation of Ehhalt and Drummond (1982).
[b] Mixing ratios in nmol mol^{-1} (except for water vapor in percent), and the corresponding concentrations in molecule cm^{-3}.
[c] Rate coefficients or photodissociation coefficients, in units of cm^3, molecule, s.
[d] Assumes uniform vertical distribution.
[e] Half-lifetime.

oxidant concentration by $n(Ox) = \Sigma k_x n_x (RO_2)/k_{66}$ and showed that it was correlated with solar intensity, undergoing a diurnal cycle with a midday maximum and declining to zero at night. This behavior demonstrated the photochemical origin of the oxidant. Cantrell *et al.* (1984, 1993) developed a measurement technique for HO_2 and RO_2 radicals based on their rapid reaction with NO and chemical amplification by the chain reaction occurring in the presence of excess carbon monoxide:

$$RO_2 + NO \rightarrow NO_2 + RO$$
$$RO + NO \rightarrow HO_2 + R'CO$$
$$\left.\begin{array}{l} HO_2 + NO \rightarrow OH + NO_2 \\ OH + CO + O_2 \rightarrow HO_2 + CO_2 \end{array}\right\} \text{ chain amplification of } NO_2 \text{ formation}$$
$$OH + NO \rightarrow HNO_2$$

followed by the detection of NO_2. Under suitable conditions, chain lengths of several hundred can be achieved. Figure 9.7 shows the diurnal variation of peroxy radicals determined by Cantrell *et al.* (1992) with this technique. The intake of the instrument was placed 10 m above ground in a forest clearing primarily of loblolly pine in west-central Alabama. The simultaneous measurement of important hydrocarbons, mainly isoprene and its oxidation products and α-pinene, provided a basis for a chemical reaction model used to calculate RO_2 radical concentrations. The results are included in Figure 9.7. The good agreement achieved between theory and measurement demon-

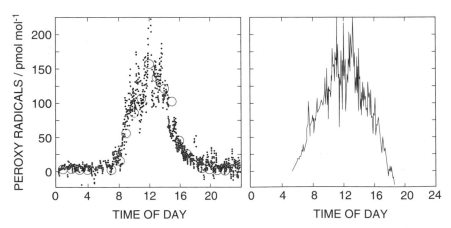

FIGURE 9.7   Diurnal variation of the mixing ratio of RO_2 radicals determined on July 11, 1990, 10 m above ground in a clearing of a forest preserve in west-central Alabama. Left: Direct measurement by means of a chemical amplifier technique. The large open circles indicate values calculated from measured trace gas concentrations. Right: Mixing ratios inferred from the measured deviation from the photostationary state between NO, NO_2 and O_3. (Adapted from Cantrell *et al.* (1992).)

strates that the key processes responsible for the production and losses of RO$_2$ radicals were accounted for. Finally, the measured peroxy radical concentration was compared with the observed deviation from the photostationary state $[x(NO_2)/x(NO)]j(NO_2)/k_{66}n(O_3)$, which should be given by $\Sigma k_x n_x(RO_2)/k_{66}n(O_3)$. As Figure 9.7 shows, good agreement was found between the two quantities, in support of the concepts embodied in Equation (9.1). Mihelcic *et al.* (1978, 1985, 1993) have combined cryogenic matrix isolation with electron paramagnetic resonance detection for the determination of peroxy radical concentrations. This is a discontinuous technique, but it can differentiate between HO$_2$ and alkylperoxy radicals. Its use has indicated peroxy radical concentrations on the same order of magnitude as those observed by Cantrell *et al.* (1992).

As Figure 9.6 indicates, NO$_2$ and NO are also involved in the formation of nitrous acid, HNO$_2$. One pathway to its formation is the association of OH with NO occurring during the day. HNO$_2$ is rapidly photolyzed again to OH and NO, and a photostationary state ensues with low HNO$_2$ concentrations. Tyndall *et al.* (1995) explored the conceivable reaction NO$_2$ + HO$_2$ → HNO$_2$ + O$_2$ and showed that it is negligible. More important is a nighttime process of HNO$_2$ formation, which was first observed by Perner and Platt (1979) by means of long-path optical absorption (see Fig. 5.3). Although the process is still incompletely understood, it has been shown to involve NO$_2$ as a precursor (Kessler and Platt, 1984; Lammel *et al.*, 1990b; Kitto and Harrison, 1992; Harrison *et al.*, 1996). Figure 9.6 indicates this by the broken arrow. Because of the nighttime production and daylight destruction, atmospheric concentrations of HNO$_2$ display a typical diurnal variation, with a nocturnal maximum and low concentrations during the day. In addition to optical techniques, denuder and scrubber systems have been developed to quantify nitrous acid in the air (Sjödin and Ferm, 1985; Kanda and Taira, 1990). A review of measurements by Lammel and Cape (1996) indicates that concentrations can range up to 10 nmol mol^{-1} in urban and rural areas, whereas in remote continental regions they generally fall below 100 pmol mol^{-1}. All data refer to air in the vicinity of the ground surface.

Kessler and Platt (1984) have explored combustion processes as sources of HNO$_2$. They concurred with Pitts *et al.* (1984a) that the most effective source of this type is the automobile engine. Exhaust gases contain up to 0.15% HNO$_2$ relative to NO, which is the major nitrogen oxide emitted from this source. In the atmosphere NO is largely converted to NO$_2$, but the ratio of HNO$_2$ to NO$_2$ observed in the air ranges up to 5%, so that automobiles can account only for a small fraction of nocturnal HNO$_2$. Rondon and Sanhueza (1989), however, have pointed out that HNO$_2$ also occurs at high levels during biomass burning in the tropical savanna.

In the laboratory, nitrous acid has long been known to occur in equilibrium with NO_2, NO, and H_2O according to the reactions

$$NO + NO_2 + H_2O \rightleftharpoons 2HNO_2$$

$$2NO_2 + H_2O \rightleftharpoons HNO_2 + HNO_3$$

although in the gas phase the reactions are slow, as Table 9.9 shows (Kaiser and Wu, 1977). Moreover, the rate of HNO_2 formation exhibits a second-order dependence on the concentration of NO_x, unlike that observed in the atmosphere. Kessler *et al.* (1982) therefore suggested that HNO_2 arises from reactions of NO_2 with liquid water present on aerosol particles, in fog droplets, or on materials at the Earth's surface. The idea that HNO_2 might arise from a surface reaction was reinforced by the observation that NO_2 interacts with water adsorbed on the walls of reaction chambers, leading to gaseous nitrous acid as a product (Sakamaki *et al.*, 1983). Under laboratory conditions, the NO_2 concentrations usually are higher than those in the atmosphere, but the experiments showed, in agreement with the field observations, that the rate of HNO_2 formation was always first order in NO_2 (Pitts *et al.*, 1984b, Svensson *et al.*, 1987; Jenkin *et al.*, 1988) and proportional to the wall surface-to-volume ratio of the chamber. Generally, the rate of HNO_2 formation was insufficient to explain the field observations. Mertes and Wahner (1995) found that the uptake of NO_2 by an aqueous surface was about seven times greater than calculated from the known Henry's law partition coefficient, which also indicates the occurrence of a surface reaction. A number of field studies have found the $[HNO_2]/[NO_2]$ ratio to be correlated with the surface area of aerosol particles (Lammel *et al.*, 1990b; Notholt *et al.*, 1992; Andrès-Hernández *et al.*, 1996). Nitrous acid formation appeared to be particularly effective under hazy conditions. Lammel and Perner (1988) and Li (1994) have also measured concentrations of nitrite associated with the atmospheric aerosol and found them partly higher than calculated from simultaneously measured HNO_2 concentrations in the gas phase and the effective aqueous solution Henry's law constant. The results provide some support for a heterogeneous formation of HNO_2 on the particles.

If nitrous acid can be generated on the water-covered surface of aerosol particles, one would also expect it to be produced at the ground surface. Here, however, the process would have to compete with losses due to dry deposition. Harrison and Kitto (1994) and Harrison *et al.* (1996) have studied the exchange of nitrous acid with the land surface and found both upward and downward fluxes, depending on the abundance of NO_2. Upward fluxes dominated when NO_2 mixing ratios exceeded 10 nmol mol^{-1}; downward fluxes were observed below this value. The average deposition velocity was 0.025 m s^{-1}. HNO_2 production rates ranged up to 25 ng m^{-2} s^{-1}.

Harrison *et al.* (1996) also applied box model calculations based on known gas phase reaction rates and surface production rates to simulate HNO_2 mixing ratios. Good agreement was found between measured and simulated abundances and their diurnal behavior. These results confirm the idea of substantial nitrous acid production on land surfaces from the reaction of nitrogen dioxide with water. The importance of HNO_2 as a source of OH radicals in photochemical smog formation was discussed in Section 5.2. The continuing production of HNO_2 at the ground surface during the day may greatly augment OH production from other sources in nonurban regions.

The oxidation of NO_2 to nitric acid can occur in two ways, either directly by reaction with OH radicals, or indirectly by reaction with ozone, which first forms the NO_3 radical and subsequently N_2O_5 by the association of NO_3 with NO_2. Nitrogen pentoxide ultimately reacts with liquid water to form HNO_3:

$$OH + NO_2(+M) \rightarrow HNO_3$$

$$NO_2 + O_3 \rightarrow NO_3 + O_2$$

$$NO_3 + NO_2 \rightleftharpoons N_2O_5$$

$$N_2O_5 + H_2O \text{ (liquid)} \rightarrow HNO_3$$

The nitrate radical NO_3 is photolyzed more rapidly than NO_2, so that during the day the reaction sequence leading from NO_2 to N_2O_5 is interrupted. The situation changes at dusk when photolysis of NO_2 and NO_3 ceases to be important. The concentration of NO_3 can then build up. Noxon *et al.* (1980) and Platt *et al.* (1980a, 1981) have detected the NO_3 radical by optical absorption and have followed the increase in NO_3 during the night. Figure 9.8 gives two examples for such observations. In the absence of loss reactions, the concentration of NO_3 would steadily increase until dawn, when the rising sun leads to its photodissociation and destruction. Such a behavior is approximated by the data in Figure 9.8 (left). Usually, however, the mixing ratio of NO_3 goes through a maximum and then drops off, as in Figure 9.8 (right). This behavior indicates that loss reactions cannot be neglected. As the association equilibrium between NO_3 and N_2O_5 is reached quickly, the concentrations of the two species are coupled, and the total loss may be due to reactions of either NO_3 or N_2O_5.

The nitrate radical reacts very rapidly with NO, so that NO_3 concentrations remain low as long as NO is present. In the rural atmosphere, after sunset, NO generally is rapidly converted to NO_2 by reacting with ozone. Reactions of NO_3 with inorganic constituents of air other than NO are negligible. NO_3 reacts slowly with alkanes, at moderate rates with aldehydes, alkenes, and aromatic compounds, and quite rapidly with a number of

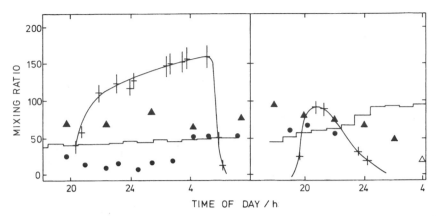

FIGURE 9.8   Two examples for the night-time behavior of NO_3 in ground-level air observed near Deuselbach, western Germany. Mixing ratios are shown for O_3 ($\blacktriangle$, nmol mol^{-1}), NO_2 ($\bullet$, 100 pmol mol^{-1}), NO_3 ($+$, pmol mol^{-1}), and relative humidity (%, stepped curve). (Redrawn from data of Platt et al. (1981).)

terpenes (Wayne et al., 1991). In such reactions NO_3 shows a behavior similar to that of the OH radical. Reactions with alkanes and aldehydes occur by hydrogen abstraction to produce HNO_3 and the corresponding organic radical. As formaldehyde is one of the more abundant aldehydes present, its reaction with NO_3 is included in Figure 9.7 for illustration. The reactions with alkenes (and terpenes) proceed by addition and lead to the formation of organic nitrates. The mechanism is thought to follow a route similar to

$$NO_3 + R_1CH{=}CHR_2 \rightarrow R_1CH(ONO_2)\dot{C}HR_2$$

$$R_1CH(ONO_2)\dot{C}HR_2 + O_2 \rightarrow R_1CH(ONO_2)CHR_2OO\cdot$$

$$R_1CH(ONO_2)CHR_2OO\cdot + NO_2 \rightleftharpoons R_1CH(ONO_2)CHR_2OONO_2$$

$$R_1CH(ONO_2)CHR_2OO\cdot + NO_3 \rightarrow R_1CH(ONO_2)CHR_2O\cdot + NO_2 + O_2$$

$$R_1CH(ONO_2)CHR_2O\cdot + NO_2 \rightarrow R_1CH(ONO_2)CHR_2ONO_2$$

$$R_1CH(ONO_2)CHR_2O\cdot \rightarrow R_1CHO + R_2CHO + NO_2$$

$$R_1CH(ONO_2)CHR_2O\cdot + O_2 \rightarrow R_1CH(ONO_2)COR_2 + HO_2$$

The formation of unstable nitroxy alkyl peroxynitrates has been observed for a number of alkenes, isoprene, and several terpenes (Barnes et al., 1990). The reaction between organic peroxy radicals and NO_3 is fairly rapid, and it may replace the daytime reaction with NO to convert ROO· to RO· at night when the concentration of NO is low. The main products expected are nitroxy aldehydes or nitroxy ketones. For example, Skov *et al.* (1992) showed that the reaction of NO$_3$ with isoprene gave 3-methyl-4-nitroxy-2-butenal as the main product. Mellouki *et al.* (1988) found that the reaction of NO$_3$ with HO$_2$ occurs by two channels:

$$NO_3 + HO_2 \rightarrow HNO_3 + O_2 \qquad 20\%$$

$$\rightarrow OH + NO_2 + O_2 \qquad 80\%$$

The OH radical thus generated is expected to react with organic compounds present to start another reaction chain. The discovery that a large fraction of the NO$_3$ + HO$_2$ reaction causes chain propagation rather than chain termination led Platt *et al.* (1990) to postulate a nighttime chemistry in which NO$_3$ takes on the role of a catalyst similar to NO during the day. Nighttime radical concentrations estimated from such reactions are significant when NO$_2$ concentrations are high, and the radicals can approach concentrations typical of daylight conditions.

Nitrogen pentoxide, N$_2$O$_5$, exists in equilibrium with NO$_3$. The reaction of N$_2$O$_5$ with liquid water results in rapid hydrolysis, in contrast to the gas phase reaction with water vapor, which is immeasurably slow (Sverdrup *et al.*, 1987). In the troposphere, N$_2$O$_5$ is expected to react primarily heterogeneously with liquid water associated with aerosol particles. The mass accommodation coefficient is about 0.02 (see Table 8.6), so that the uptake of N$_2$O$_5$ by liquid water in the atmosphere is not limited by gas phase diffusion. The NO$_3$ radical may also enter the liquid phase of aerosol particles, where it would react mainly with chloride to produce chlorine atoms and nitrate anion (see Table 8.9). This is different from the situation in continental clouds, where the major fate of NO$_3$ in the aqueous phase is reaction with HSO$_3^-$. The direct interaction of NO$_3$ with water is negligible. As discussed in Section 8.6.4, chlorine atoms attach to Cl$^-$ to form the Cl$_2^-$ radical, which is in equilibrium with the precursors. Chlorine atoms react with water, and Cl$_2^-$ reacts with OH$^-$, to produce OH radicals in both cases (Jayson *et al.*, 1973; Jacobi *et al.*, 1997). These would be scavenged by organic components of the continental aerosol. At high chloride concentrations, as in sea salt particles, N$_2$O$_5$ also would react with chloride. This reaction produces partly ClNO$_2$ (Finlayson-Pitts *et al.*, 1989; Behnke *et al.*, 1992), which is sufficiently volatile to escape to the gas phase. In sunlight, ClNO$_2$ is rapidly photolyzed and NO$_2$ is reconstituted, so that this pathway does not lead to the formation of nitric acid.

Heintz *et al.* (1996) have used long-path optical absorption spectroscopy to study the behavior of nocturnal NO$_3$ concentrations for a period longer than a year at a site near the Baltic Sea. They found the atmospheric lifetime of NO_3 to decrease with increasing NO_2 concentration, indicating that the main loss process proceeds by reactions involving N_2O_5. The loss rate was also proportional to the relative humidity, when r.h. > 60%, supporting the notion advanced by Platt et al. (1984) that the reaction of N_2O_5 occurs with water condensing on deliquescent aerosol particles. The average NO_3 mixing ratio was 7.8 pmol mol^{-1}; that of N_2O_5 was calculated to be about 1 nmol mol^{-1}. The lifetime of NO_3 ranged from about 10 to 5600 s, with an average of about 250 s. Lifetimes inferred for N_2O_5 ranged from 1000 to 10,000 s, with a frequency distribution that peaked at 1550 s. Heintz et al. (1996) calculated that about 75% of the N_2O_5 produced is converted to HNO_3 during the night; the remainder undergoes thermal decay and photolysis of NO_3 following sunrise.

The two processes converting NO_2 to nitric acid, namely reaction with OH radicals during the day and reaction with ozone during the night, determine the average lifetime of NO_2 in the troposphere. Accepting the estimate of Heintz et al. (1996) that 75% of N_2O_5 is converted to HNO_3 by reacting with liquid water, the lifetime of NO_2 at night is determined largely by the rate of NO_3 formation. The average lifetime of NO_2 then is

$$\overline{\tau}(NO_2) = 1/[2 \times 0.75 f_{dark} k_{68} \overline{n}(O_3) + k_{122} \overline{n}(OH)] \qquad (9.2)$$

where f_{dark} denotes the fraction of darkness during a full day, and $\overline{n}(O_3)$ and $\overline{n}(OH)$ are the diurnal averages for the number densities of ozone and OH radicals, respectively. The factor 2 in the first term in the denominator on the right of Equation (9.2) is to take into account that two molecules of NO_2 are consumed in forming N_2O_5. For near-surface conditions in mid-latitudes, one may take $\overline{n}(O_3) \approx 6 \times 10^{11}$ and $\overline{n}(OH) \approx 8 \times 10^5$ molecule cm^{-3}. With $f_{dark} = 1/3$ for summer conditions, one finds $\overline{\tau}(NO_2) \approx 1$ day. In summer, the two reactions contribute about equally to the total loss of NO_2. In winter, $\overline{n}(OH)$ decreases, whereas the dark period increases, so that then the reaction with ozone is the dominant loss process. The average lifetime remains almost unchanged. Diurnal variations are considerable, however. The midday lifetime in summer may be as short as 6 h, when OH number densities reach peak values of 4×10^6 molecule cm^{-3}.

In a number of studies, field measurements have been utilized to derive NO_2 lifetimes. For this purpose it is necessary to compare the behavior of NO_2 with that of a tracer of lower reactivity. Chang et al. (1979) examined data from Los Angeles and from St. Louis in this manner and found NO_2 lifetimes of 1–2 days by two independent methods. Spicer (1980, 1982) made aircraft measurements in the urban plumes of Phoenix, Arizona; Boston, Massachusetts; and Philadelphia, Pennsylvania. He found daylight lifetimes of more than 1 day in the first study, and 4–8 h in the other two studies. His measurements confirmed that NO_2 is converted to HNO_3, particulate nitrate, and peroxyacetyl nitrate. The field studies thus support the estimate given above. In the upper troposphere, the lifetime of NO_2 is considerably longer, because the concentration of OH radicals and aerosol particles is lower than in the planetary boundary layer and the air is drier, so that the chemical reactions responsible for the oxidation of NO_2 are slower.

Nitric acid is the stable end product of NO_2 oxidation in the troposphere. Table 9.9 shows that rates for the removal of HNO_3 by photodissociation and reaction with OH are smaller than loss rates due to wet and dry deposition. A portion of HNO_3 becomes associated with aerosol particles. Neutralization may occur by ammonia or by alkaline components present in soil and sea salt particles. The reaction of nitric acid with elements such as calcium or sodium forms nonvolatile nitrates that remain permanently attached to the aerosol. Ammonium nitrate, however, is thermally unstable and occurs in significant amounts only when HNO_3 and NH_3 concentrations are high, or when temperatures are low (see Section 7.5.1). A quasi-equilibrium between gaseous and particulate nitrate is established, which causes a significant fraction of nitric acid to remain in the gas phase. In Figure 9.6 this relation is indicated by the double arrow. The partitioning also depends on the water content of the aerosol. With increasing relative humidity more liquid water condenses on the aerosol particles, so that more nitric acid can dissolve in liquid water (Tang, 1980). The relative humidity frequently increases during the early morning hours because of a decline in temperature. The partitioning of HNO_3 between gas and particulate phases thus undergoes a diurnal variation with higher particulate concentrations in the early morning, which decline again in the late afternoon, as observed, for example, by Mehlmann and Warneck (1995). Table 1.12 shows that the dry deposition velocity of nitric acid is appreciable. As a consequence, a considerable fraction of gaseous nitric acid in the planetary boundary layer is removed by dry deposition.

9.4.2. THE TROPOSPHERIC DISTRIBUTION OF NO_x

Because of the photochemical coupling between nitric oxide (NO) and nitrogen dioxide (NO_2), the sum of their mixing ratios is usually treated as a single variable, which is designated NO_x. The concentrations of NO_3 and N_2O_5 do not generally reach those of NO_x, so that they can be ignored in this context. Peroxyacetyl nitrate (PAN) is an important association product that serves as a temporary reservoir of NO_2. However, PAN can be determined separately, so that it is not included in the definition of NO_x. Although a large fraction of NO_x is originally injected into the atmosphere in the form of NO, high NO/NO_2 ratios are found only in the source regions, that is, mainly in the cities. Here the abundance of NO_x may reach 100 nmol mol^{-1}. Outside the source regions, NO eventually reacts with ozone, and NO_2 becomes the dominant component of NO_x. Nitric oxide can be measured at a level of a few pmol mol^{-1} by analyzers based on the chemiluminescence accompanying the $NO + O_3$ reaction (Fontijn et al., 1970). The determination of NO_2 by the same technique requires a reduction of NO_2 to NO prior to analysis. Hot catalytic converters (for example, heated molybdenum), $FeSO_4$ crystals at ambient temperatures, or photodissociation of NO_2 have been used. Unfortunately, each conversion technique may add NO from the breakup of other nitrogen compounds such as PAN. Even HNO_3 can cause an interference if hot catalytic converters are used. In addition, chemiluminescence analyzers are plagued with a variable background chemiluminescence arising from reactions of ozone with impurities in the gas supplies or at the walls of the reactor. The photolytic converter (Kley and McFarland, 1980; Ridley et al., 1988) is least affected by interference problems, but conversion efficiencies are only about 50%. A gold tube operated at 300°C with CO as the reducing agent converts all oxidized nitrogen species to NO (Fahey et al., 1985). This signal is designated NO_y to indicate that it includes not only NO and NO_2, but also other compounds such as NO_3, N_2O_5, HO_2NO_2, HNO_2, HNO_3, particulate nitrate (unless it is filtered out), PAN, organic nitrates, and perhaps other convertible oxidized nitrogen compounds—but not, however, NH_3, HCN, and CH_3CN. Other methods for measuring NO_2 include long-path differential optical absorption (Platt et al., 1979), tunable diode laser absorption spectroscopy (Walega et al., 1984), and chemiluminescence arising from the interaction of NO_2 with luminol (Wendel et al., 1983; Schiff et al., 1986). Intercomparison of instruments (Fehsenfeld et al., 1987, 1990) has exposed the weak points of each technique, but the results generally were quite favorable. Wet-chemical colorimetric analysis of NO_2, such as the diazotization procedures of Saltzman (1954) and Jacobs and Hocheiser (1958), have also been used, but they are not sufficiently sensitive for measurements in remote tropospheric regions.

Table 9.10 summarizes ground-based measurements of NO_x mixing ratios in rural areas. The background level in well-populated regions is on the order of 3 nmol mol^{-1} in summer and 10 nmol mol^{-1} in winter. Toward remote continental regions, the level is gradually reduced to values in the low pmol mol^{-1} range, as the data from Sweden and Canada show. All sites show a great variability caused by the advection of NO_x richer air from urban centers alternating with cleaner air from other regions, as well as by the varying degree of vertical exchange between the planetary boundary layer and the free troposphere. The seasonal variation observed in Europe and

TABLE 9.10 Some Ground-Based Measurements of NO$_x$ at Continental Rural and Mountaintop Sites

Authors	Location and measurement period	Mixing ratio[a] (nmol mol^{-1})	Remarks[b]
Ritter et al. (1979)	Rural Michigan, 10 days in June 1977	0.3–0.5	CL, FeSO$_4$
Kelly et al. (1980)	Niwot Ridge, Colorado, 2910 m elevation 7 days in April 1979	0.25 av.	CL, FeSO$_4$
	10 days in January 1979	0.21 av.	
Harrison and McCartney (1980)	Rural northwest England May–Sept. 1977	10 ± 8	CL, hot molybdenum
Martin and Barber (1981)	Bottesford, England, 20 km east of Nottingham, 1978–1979	10 av. summer 21 av. winter	CL, seasonal variation, large contribution of NO
Bollinger et al. (1984)	Niwot Ridge, Colorado, 2910 m elevation	0.30 av. summer 0.24 av. winter	CL, photolysis cell; high variability due to advection of polluted air from Denver
Stedman and McEwan (1983)	Mt. John, New Zealand, 1020 m elevation March–April 1981	0.127 av.	CL, FeSO$_4$
Shaw and Paur (1983)	Ohio River Valley, May 1980–Aug. 1981	NO$_2$ 8.2 av. (4.5–6) NO 1.9 av. (0.4–12.5)	Seasonal variation with winter maximum Average NO/NO$_x$ = 0.22
Broll et al. (1984)	Deuselbach, Germany, 1980–81	4.5 av. summer 16.3 av. winter	CL, FeSO$_4$
Johnston and McKenzie (1984)	Rural New Zealand, 29 nighttime measurements	NO$_2$ 0.02–1.36	DOAS (9.2 km), 420–450 nm, low values in windy nights, high values underneath inversion layers
Sjödin and Grennfelt (1984)	Sweden, 1981–1983, Ekered, rural area	2.05 av. summer 4.63 av. winter	WC. NO$_2$ only, daily averages measured
	Velen, forested region	0.68 av. summer 1.78 av. winter	
Poulida et al. (1994)	Candor, North Carolina, Aug. 1991	1.36 av. (0.4–7.2)	CL, photolytic converter
Bakwin et al. (1994)	Shefferville, Quebec, Canada June–Aug. 1990	0.049 ± 0.030	Remote taiga woodland
Munger et al. (1996)	Petersham, Massachusetts, 1990–1994 Harvard Forest	1.25 ± 0.99 summer 9.48 ± 9.25 winter	CL, photolytic converter, data for southwesterly flow are shown here
Williams et al. (1997)	Idaho Hill, Colorado, 3070 m elevation Sept.–Oct. 1993	1.16 av. (0.038–21.3)	CL, photolysis cell; higher values during upslope flow, lower during downslope flow

[a] Sum of NO + NO$_2$, or the measured species is indicated.
[b] Code for analytical techniques: WC, wet chemical; CL (NO + O$_3$) chemiluminescence, method for NO$_2$ to NO conversion indicated; DOAS, differential optical absorption spectroscopy.

North America, leading to higher values in winter and lower ones in summer, probably has several causes, among which are increased emissions due to residential heating during the cold season and a more frequent occurrence of low-lying temperature inversion layers in winter compared to summer conditions. Figure 9.9 shows for a rural community in western Germany that mixing ratios of NO_x and SO_2 are well correlated. The discussion in Section 10.3.4 will show that a large part of the seasonal cycle of SO_2 arises from the greater photochemical oxidation rate in summer. For NO_2, the known oxidation mechanisms do not predict a similar variation in the rate of its conversion to HNO_3, so that the origin of the seasonal cycle of NO_2 is less well understood.

Table 9.10 includes data from several mountain sites. Because of local circulation patterns, these sites receive either uprising air from the valleys or clean air from aloft. All of the mountain data indicate that NO_x mixing ratios decrease with increasing altitude. The difference between summer and winter values is also greatly reduced. For example, at Niwot Ridge, Colorado, average NO_x mixing ratios were about 0.25 nmol mol^{-1} regardless of season, and the values observed at other continental mountain stations were in a similar range. The abundance in the free troposphere is still lower. Figure 9.10 shows results from early aircraft explorations over the continents, which indicated a decline of NO_x with an apparent scale height of about 1 km toward background mixing ratios in the range of 10–100 pmol mol^{-1} in the middle troposphere. More recent aircraft-assisted measurements, reported,

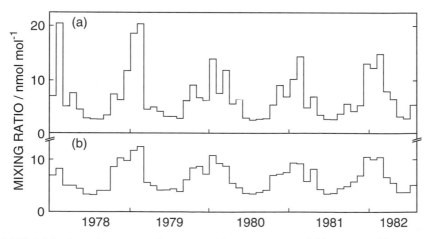

FIGURE 9.9   Annual variation of mean monthly mixing ratios of (a) SO_2 and (b) NO_2 in Deuselbach, a rural community of western Germany. Data were taken from regular reports of the German Environmental Administration (Umweltbundesamt).

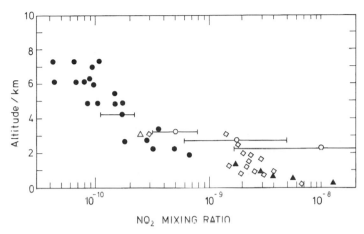

FIGURE 9.10 Vertical distribution of NO$_x$ in the continental troposphere derived from aircraft observation. Filled circles, Wyoming; open circles, over Denver, Colorado (Kley et al., 1981); triangles, over southern Germany in May 1964; diamonds, several flights over Germany in the winter of 1962–63 (Georgii and Jost, 1964).

for example, by Torres and Buchan (1988) over the Amazon region or by Carroll et al. (1990) and Ridley et al. (1994) over the United States, have confirmed the decline of NO$_x$ with altitude in the transition regime from the planetary boundary layer to the free troposphere over the continents. In the upper troposphere, the mixing ratio rises again, as will be discussed further below.

Table 9.11 summarizes NO$_x$ mixing ratios observed in marine air and in the free troposphere. Measurements onboard ships and at coastal stations require careful screening to eliminate contamination by other ships and/or the influence of sources on land. With these precautions it is then possible to show that the abundance of NO$_x$ in pure marine air masses is on the order of 50 pmol mol^{-1}. Broll et al. (1984) have explored the latitudinal distribution of (nighttime) NO$_2$ in surface air over the Atlantic Ocean and found a broad maximum in the Northern Hemisphere near 40° latitude, which suggested that anthropogenic NO$_x$ from the North American continent spreads with the westerlies into the source-free region of the marine atmosphere, and that remnants are still noticeable west of Europe. The lowest mixing ratios, with values in the range of 20–30 pmol mol^{-1}, were encountered near the equator and in polar regions. Noxon (1983) employed optical spectroscopy to determine the NO$_2$ mixing ratio in the 3-km altitude region near Mauna Loa, Hawaii, and found values of about 30 pmol mol^{-1}. These observations suggested a fairly uniform background of NO$_x$ in the remote troposphere,

TABLE 9.11 Some Observations of NO_x Mixing Ratios in the Marine and Remote Troposphere

Authors	Location and measurement period	Mixing ratio[a] (pmol mol^{-1})	Remarks[b]
Surface-bound observations, marine atmosphere			
McFarland et al. (1979)	Tropical Pacific Ocean	~ 4 NO	CL, photolysis cell; diurnal cycle, maximum at noon
Helas and Warneck (1981)	Loop Head Ireland, June 1979	87 ± 47 NO_2 37 ± 6 NO_2 nighttime	CL, $FeSO_4$; diurnal variation noted; all data trajectory selected pure air
Platt and Perner (1980)	Loop Head, Ireland	90–300 NO_2	DOAS, pure marine air
Liu et al. (1983)	Tropical Pacific Ocean	2.8–5.7 NO	CL, photolysis cell; noon ± 3 h
Noxon (1981, 1983)	Mauna Loa, Hawaii, 2.5–3.5 km	30 ± 10 NO_2	DOAS, against the setting sun
Broll et al. (1984)	Atlantic Ocean 60° N–60° S	20–70 NO_2	CL, $FeSO_4$; nighttime data only, maximum at 40° N
Ridley and Robinson (1992)	Mauna Loa, Hawaii, 3600 m elevation May–June 1988	32 ± 13 43 ± 21	Downslope conditions Upslope conditions
Torres and Thompson (1993)	Equatorial Pacific Ocean, Feb.–Mar. 1990	0.3–6.4 NO	CL
Atlas and Ridley (1996)	Mauna Loa, Hawaii, 3600 m elevation, 1991–1992, four seasons	23.8 ± 9.3 NO_2 37.7 ± 15.1 NO_2	Sept.–Oct. 1991, free troposphere April–May 1992, free troposphere
Wang et al. (1996)	Sable Island, Nova Scotia, July–Sept. 1993	153 av.	CL, photolytic converter

Aircraft observations in the free troposphere

Roy et al. (1980)	Australia,	100–300 NO	8–10 km	CL
Davis et al. (1987)	Pacific Ocean, 15°–42° N, Nov. 1983	av. 21 NO	6.1 km	LIF,
		av. 4.4 NO	0.3 km	
Ridley et al. (1987)	Pacific Ocean, autumn 1984	av. 1.7 NO	~1 km	CL
		av. 8.3 NO	1–3 km	
		av. 11.5 NO	5–6.4 km	
		av. 19.5 NO	7.6–10 km	
Carroll et al. (1990)	Continental U.S., Aug–Sept 1986	118.5 ± 68.2	1–3 km	CL, photolytic converter; TDLA
		65.0 ± 46.9	3–6.5 km	
	Eastern Pacific Ocean	18.5 ± 12.5	0–1 km	
		24.1 ± 16.1	1–3 km	
		35.9 ± 25.1	3–6.5 km	
Sandholm et al. (1992)	Alaska, July 1988	30 ± 15	0–6 km	LIF, photolytic converter, little variation with altitude
Singh et al. (1994)	Goose Bay, Canada, July 1990	36 ± 19	0–2 km	LIF, photolytic converter
		36 ± 13	2–4 km	
		44 ± 14	4–6 km	
Singh et al. (1996b)	Tropical Pacific Ocean and Hawaii Sept.–Oct. 1991, 25° N–42° N	44.9 ± 42.4 NO_2	0–3 km	LIF, NO was about 10% of NO_2 in the boundary layer, but increased significantly with increasing altitude
		46.6 ± 15.2 NO_2	3–7 km	
		65.3 ± 24.5 NO_2	7–13 km	
	0°–25° N	21.2 ± 25.1 NO_2	0.3 km	
		19.4 ± 9.8 NO_2	3–7 km	
		33.7 ± 14.9 NO_2	7–13 km	

[a] Sum of NO + NO_2, or the measured species as indicated altitude range for aircraft observations.

[b] Code for analytical techniques: CL, chemiluminescence, method for NO_2 to NO conversion indicated; DOAS, differential optical absorption spectroscopy; LIF, photo-fragment laser-induced fluorescence; TDLA, tunable diode laser absorption.

and this conclusion has essentially been confirmed by more recent aircraft measurements, listed in Table 9.11.

A number of aircraft explorations were confined to daytime measurements of NO. Values near the ocean surface reach only a few pmol mol^{-1}, as Table 9.11 shows. Drummond *et al.* (1988; see also Ehhalt *et al.*, 1992) have mapped the distribution of NO at altitudes of up to 12 km by flights carried out in June 1984 along the coastlines of North America, South America, eastern North Africa, and Europe, covering the latitude range 60° S–70° N. The results indicated significantly elevated NO mixing ratios in the upper troposphere of the Northern Hemisphere ($z > 9$ km), with values of up to 200 pmol mol^{-1}, well above those in the middle troposphere. Because NO arises as a photodissociation product from NO_2, higher concentrations of NO_2 would have to exist in the same region. In the upper troposphere of the Southern Hemisphere, in contrast, the concentration of NO was not elevated.

More recent aircraft missions devoted to measurements of both NO and NO_x have confirmed the increase in NO_x in the upper troposphere. Figure 9.11 shows results of measurements made by Ridley *et al.* (1994) over the continental southern United States. Starting from the boundary layer, both NO and NO_x decline with increasing altitude, attain low values in the 3–6-km altitude region, and then increase again in the upper troposphere. In the middle troposphere, NO contributes about 25% to NO_x; in the upper troposphere the contribution is significantly greater, rising to more than 60%. Aircraft measurements over the Pacific Ocean, described by Singh *et al.* (1996b), have also provided data on the vertical distribution of NO_x and related species in the 25°–42° N and 0°–25° N latitude bands. In the

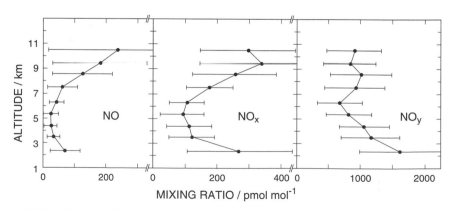

FIGURE 9.11   Vertical distribution of NO, NO_x, and NO_y during July and August 1989 over New Mexico. (Adapted from Ridley *et al.* (1994).)

boundary layer, the mixing ratio of NO was low, in agreement with shipboard measurements close to the sea surface. The level of NO$_x$ remained fairly constant up to about 6 km. With increasing altitude, the mixing ratios of NO$_x$ and NO$_y$ rose about threefold up to 12 km, whereas that of NO rose more steeply and attained a level similar to that of NO$_2$. This behavior and the magnitude of the mixing ratios were quite similar to those observed by Ridley *et al.* (1994) over New Mexico. The abundances of NO$_x$, NO$_y$, and PAN showed a distinct dependence on latitude, increasing from the equator to 40° N at all altitudes, in agreement with the earlier conclusions of Drummond *et al.* (1988).

The relative increase of NO in the upper troposphere can be qualitatively understood to arise from differences in setting up the photostationary state. In the boundary layer, the NO/NO$_x$ ratio is low, because reactions of OH radicals generate high concentrations of peroxy radicals that reduce the NO concentration compared to that of NO$_2$. In the upper troposphere, the formation rate of OH followed by that of peroxy radicals is much lower, so that the NO/NO$_x$ ratio rises toward the photostationary state established in the absence of peroxy radicals. The origin of the enhancement of NO$_x$ in the upper troposphere is not well known. The direct influx of odd nitrogen from the stratosphere is generally thought to be insufficient, whereas production of NO$_x$ by lightning is considered more important. PAN represents a significant reservoir of NO$_2$ in the upper troposphere, which may be tapped as a source of NO$_x$. Singh *et al.* (1996b) concluded that surface sources contribute only about 20% of NO$_x$ in the upper troposphere by convective updraft and that sources in the free troposphere are dominant.

In summary, the mixing ratio of NO$_x$ shows high levels in the boundary layer over the continents, where most of the sources are located. A background of 30–40 pmol mol^{-1} exists over the oceans and in the middle troposphere. In the upper troposphere of the Northern Hemisphere the NO$_x$ mixing ratio increases to values of 100–150 pmol mol^{-1}.

9.4.3. PEROXYACETYL NITRATE

As discussed previously, PAN arises as a secondary product in the oxidation of hydrocarbons in the presence of nitrogen oxides. This compound and its higher homologues were discovered by Stephens *et al.* (1956) in laboratory studies designed to investigate the formation of photochemical smog. Scott *et al.* (1957) then demonstrated the occurrence of PAN in Los Angeles air. The chemical structure of the substance was initially unknown until Stephens *et al.* (1961) established it to be that of the mixed anhydride of peroxoacetic acid and nitric acid. Martinez (1980), who discussed the terminology of organic nitrates, pointed out that peroxyacetyl nitrate is a misnomer, and that it should properly be called *ethaneperoxoic nitric anhydride*. Roberts (1990) suggested *peroxyacetic nitric anhydride* as a compromise to preserve the familiar acronym PAN. In the present text, the term *peroxyacetyl nitrate* will be retained because of its widespread usage. In photochemical smog, PAN is one of the substances responsible for eye irritation and plant damage (Taylor, 1969). The customary technique of measuring PAN in air is gas chromatography with electron capture detection. Although other organic peroxy nitric anhydrides exist, PAN is the most abundant species of the family. The next higher homologue, peroxypropionyl nitrate (PPN), shows an identical behavior but usually occurs in quantities only a few percent of that of PAN (Singh and Salas, 1989).

Temple and Taylor (1983) and Roberts (1990) have reviewed ground-based observations of PAN in Europe, Japan, and the United States. Altshuller (1993) discussed urban PAN measurements, and Grosjean (1984) presented supplemental data for the Los Angeles basin. This smog-affected region is burdened with high PAN mixing ratios. In downtown Los Angeles, for example, values as high as 65 nmol mol^{-1} have been observed (Lonneman *et al.*, 1976), although daily averages rarely exceed 10 nmol mol^{-1}. In rural areas the mixing ratio commonly is in the range of 0.2–0.6 nmol mol^{-1}, but it may rise significantly above this level during smog episodes. Distinct diurnal variations are observed in urban as well as rural environments, with maxima appearing in midafternoon, frequently concurrently with the daily maximum of ozone. The correlation is similar to that observed in smog chambers (cf. Fig. 5.1). It results from the fact that in photochemically active air masses both ozone and PAN are products of the NO_x-catalyzed oxidation of hydrocarbons. A linear relationship between PAN and ozone has been observed at many locations (e.g., Altshuller, 1983b; Singh *et al.*, 1985; Ridley *et al.*, 1990; Shepson *et al.*, 1992; Roberts *et al.*, 1995), but the slope of the PAN/O_3 correlation was found to depend on the extent of air pollution. At high concentrations of NO_2 the formation of ozone is less efficient than that of PAN, so that the slope increases over that observed in less burdened areas. Similar to ozone, PAN has been suggested to represent a useful indicator of photochemical air pollution (Nieboer and van Ham, 1976; Penkett *et al.*, 1977). Unlike ozone, however, PAN is thermally unstable and can decay during transport unless it is continuously produced or enters a colder region of the atmosphere. PAN levels in the remote troposphere exhibit little diurnal variation but are more strongly influenced by the long-range history of the air mass.

In the absence of sources of acetyl peroxy radicals, the lifetime of PAN is determined by the reaction sequence

$$CH_3(CO)OONO_2 \rightleftharpoons CH_3(CO)OO + NO_2 \qquad k_a, k_{-a}$$

$$CH_3(CO)OO + NO \rightarrow CH_3 + CO_2 + NO_2 \qquad k_b$$

The thermal decomposition reaction is highly temperature dependent, leading to lower limit lifetimes of $1/k_a \approx 1$ h at 298 K, 2 days at 273 K, and 148 days at 250 K (Singh, 1987). The extent of the reverse reaction depends on the competition between NO and NO$_2$ for acetyl peroxy radicals and, thus, on the prevalent concentration ratio of nitrogen oxides [NO$_2$]/[NO]. The actual lifetime assumes approximately twice the above values when [NO$_2$]/[NO] $\approx$ 2.3.

Several studies have addressed the annual variation of monthly mean PAN mixing ratios. Figure 9.12 shows data from two locations: Riverside, California, and Harwell, England. At Riverside two time series exist, one based on 24-h measurements, the other limited to daylight hours. The annual averages were 3.4 and 4.4 nmol mol^{-1}, respectively. The 30% difference presumably reflects the diurnal variation, but the seasonal distribution of the former data set appears to be more uniform than the latter, which peaks during the

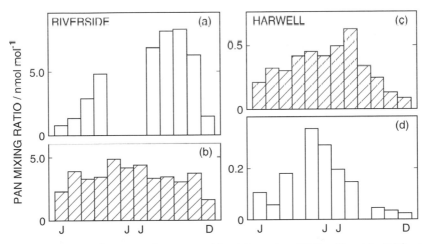

FIGURE 9.12 Annual variation of mean monthly mixing ratios of PAN at Riverside, California (left), and at Harwell, England (right). (a) Values shown combine data from 1967–1968 and 1980; measurements were confined to daytime (Temple and Taylor, 1983). (b) Twenty-four-hour measurements in 1975–76 (Grosjean, 1984). (c) Values shown combine data from 1974–1975 and 1980 (Brice et al., 1984). (d) Data selected for remote air masses (Penkett and Brice, 1986).

second half of the year. The two data sets agree in that the lowest monthly mean values occur at the turn of the year. At Harwell, the PAN mixing ratios were smaller by an order of magnitude than in California, averaging 0.36 nmol mol^{-1} annually. When all of the data are considered, a summer maximum is observed, similar to that in California. However, when the data are selected by the cleanliness of air, a different pattern emerges, showing higher values in spring. Guicherit (1988) observed a similar pattern in Delft, the Netherlands, during a 12-year period of measurements. Brice *et al.* (1988) and Bottenheim *et al.* (1994) also found a spring maximum in Kejimkujik, Nova Scotia, from a 5-year measurement period. At Harwell and Kejimkujik the springtime maxima of PAN were accompanied by higher concentrations of aerosol sulfate and nitrate, and by nitrate and nitric acid, respectively, indicating that well-aged photochemically polluted air masses were sampled. Under these conditions, PAN accounts for a significant fraction of all of the observed oxidized nitrogen species. As ozone also shows a characteristic spring maximum, Penkett and Brice (1986) suggested that tropospheric photochemistry might be responsible for both elevated PAN and ozone levels, following from the accumulation of hydrocarbons in the Northern Hemisphere due to anthropogenic emissions during the winter months. Wang *et al.* (1998), who have studied hydrocarbon chemistry and ozone formation by three-dimensional model calculations, confirmed that photochemical production of ozone contributes to the spring maximum, but found no support for enhanced photochemistry due to the accumulation of ozone precursors such as PAN. The phenomenon appears to be largely explicable by long-range transport.

PAN is not only a product of air pollution, but exists as a natural background in the troposphere. Singh and Hanst (1981) originally pointed out that such a background should arise from the oxidation of ethane and propane, which are the most abundant nonmethane hydrocarbons in the troposphere. Their reactions with OH radicals generate acetaldehyde and acetone, respectively, which then undergo either photodecomposition or further reactions with OH radicals to produce acetyl peroxy radicals and finally PAN. The reaction pathways are shown in Figure 9.13. The photolysis of acetaldehyde is the only process that does not lead to acetyl peroxy radicals, but this pathway is relatively unimportant.

Warneck (1988) discussed early aircraft observations exploring the abundance and vertical distribution of PAN in the continental and marine atmosphere, which indicated mixing ratios of about 50 pmol mol^{-1} in the air over the oceans, and a decline in the mixing ratio from the surface toward higher altitudes over the continents. The early studies have been followed by a great number of measurements revealing the ubiquity of PAN in the atmosphere. Singh *et al.* (1986) showed that values of about 50 pmol mol^{-1}

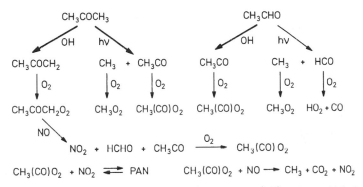

FIGURE 9.13 Routes to the formation of PAN from acetone (left) and acetaldehyde (right). Important pathways are accentuated. The formation of PAN by the association of acetyl peroxy radicals with NO$_2$ is partly compensated for by losses due to the reaction of acetyl peroxy radicals with NO, as shown by the reactions in the last line

are typical of the northern hemispheric marine atmosphere, whereas in the southern troposphere the values are much lower, and that this behavior correlates with that of ethane and propane, from which acetyl peroxy radicals originate as the precursors of PAN. Rudolph *et al.* (1987), who studied the distribution of PAN at altitudes of up to 12 km by flights covering the latitude range 60° S 70° N, found mixing ratios of 7–10 pmol mol^{-1} in the Southern Hemisphere with little variability up to 10 km altitude, where-

as in the Northern Hemisphere the PAN mixing ratios ranged up to 100 pmol mol^{-1}, with a much higher variability.

Table 9.12 shows data from aircraft measurements that indicate the altitude distribution of PAN in the Northern Hemisphere. Over the American continent, at mid-latitudes, the mixing ratios are approximately 150 pmol mol^{-1}, which is fairly independent of altitude above the boundary layer. In remote continental regions the mixing ratios are lower, but they increase with altitude toward values of a few 100 pmol mol^{-1} in the upper troposphere. The measurements over the Pacific Ocean were categorized according to region and air mass origin. In the maritime boundary layer, the PAN mixing ratios are low, especially in the tropical latitude region. With increasing altitude the mixing ratio increases to values of about 20 pmol mol^{-1} in the upper troposphere, in clean air masses. The advection of air from land increases the mixing ratio significantly to essentially continental levels. Under these conditions, the mixing ratio shows relatively little dependence on altitude. All of the measurements showed a considerable variability resulting from the mixing of air masses of different origin. The observations confirm the early conclusions that PAN concentrations are similar to those of

TABLE 9.12   Mean PAN Mixing Ratios in the Northern Hemispheric Troposphere

Air mass category		Altitude range (km)			
		0–1	1–4	4–7	7–12
Tropical marine, 0°–15°	a	2.0	2.0	16.2	12.7
Marine, 15°–40°					
West Pacific	a	2.0	2.0	32.5	22.1
East Pacific	a	2.0	9.7	88.5	20.1
Continental, mid-latitude	a	178.7	117.2	138.4	106.3
Continental U.S.	b	—	174.0	158.5	175.0
Continental, high latitude	c	63	226	310	—

[a] From measurements over the Pacific Ocean, analyzed by Kondo et al. (1996).
[b] From measurements over Colorado (Singh et al., 1986).
[c] From measurements over Goose Bay, Canada (Singh et al., 1994).

NO_x, sometimes even larger, and that they can tie up significant quantities of NO_x, especially in the colder regions of the upper troposphere.

9.4.4. NITRIC ACID AND PARTICULATE NITRATE

The recognition that gaseous nitric acid (HNO_3) and nitrate bound to aerosol particles (NO_3^-) can coexist in the atmosphere has led to a reappraisal of filter techniques for the sampling of particulate nitrate. A widely employed procedure uses Teflon membrane filters, which are immune to HNO_3 and other nitrogen compounds, to collect aerosol nitrate without much interference, followed by a nylon back-up filter for the absorption of nitric acid. Another technique is the use of a wall-coated diffusion-denuder tube for the absorption of gaseous nitric acid and the collection of nitrate-containing particles on a back-up filter. Neither technique prevents losses of particulate nitrate as HNO_3 under conditions of low relative humidity or by the interaction with acidic sulfate particles if the filters are overloaded. Joseph and Spicer (1978) and Kelly et al. (1979) have described modified chemiluminescence analyzers for the determination of ambient HNO_3 mixing ratios from the difference of signals derived from $NO_x + HNO_3$, on the one hand, and NO_x after the removal of HNO_3, on the other. Optical techniques, such as infrared tunable diode laser absorption spectroscopy (Schiff et al. 1983) or laser photolysis fragment fluorescence (Papenbrock and Stuhl, 1989), have also been used as for the determination of HNO_3. Spicer et al. (1982), Anlauf et al. (1985), Hering et al. (1988), Tanner et al. (1989), and Gregory et al. (1990) have conducted field intercomparison tests for a number of these methods, especially filter and denuder techniques, with generally reasonable agreements.

Order-of-magnitude levels for aerosol nitrate in various parts of the world are shown in Table 7.14. Here, the discussion is confined to simultaneous measurements of gaseous HNO$_3$ and particulate nitrate. Although numerous such data exist, many were taken only during short periods. Table 9.13 summarizes some of the data, emphasizing measurements made during different seasons. By and large, the atmospheric concentrations of nitric acid and nitrate are comparable. In the boundary layer, the ratio is on the order of unity, although it fluctuates considerably. Nitric acid is subject to a very high deposition velocity of $2-3 \times 10^{-2}$ m s^{-1} (Huebert and Roberts, 1985; Hanson and Lindberg, 1991), which is at least an order of magnitude greater than that for aerosol particles, so that the HNO$_3$ level and the HNO$_3$/NO$_3$ ratio will be strongly influenced by losses of HNO$_3$ due to dry deposition to vegetation. In the free troposphere, the HNO$_3$/NO$_3$ ratio is greater than unity, indicating that free nitric acid rather than HNO$_3$ bound to particles is the preferred form of nitrate.

Many investigators have noted a pronounced diurnal variation of nitric acid featuring a maximum in the middle of the day and a minimum at night. Several factors may contribute to this behavior. The diurnal cycle of HNO$_3$ is anticorrelated with that of relative humidity, which normally rises during the night because of the decline in temperature. This relation is illustrated in Figure 9.14 (left). The nocturnal rise in relative humidity causes the liquid water content of the aerosol particles to grow (see Fig. 7.7), which, in turn, increases the uptake capacity of the particles with regard to HNO$_3$. The decline in temperature and the increasing liquid water content of the particles also favor the formation of ammonium nitrate from the gas phase precursors NH$_3$ and HNO$_3$. The acidification of the aqueous phase due to the uptake of nitric acid can be partly compensated for by the formation of bisulfate from sulfate, which acts as a buffer. These processes, however, cannot fully explain the observed diurnal behavior of HNO$_3$, and boundary layer meteorology must also be considered. Erisman et al. (1988) and Piringer et al. (1997) have shown that the build-up of a nocturnal inversion layer, which interrupts vertical transport, establishes a vertical concentration gradient over flat terrain such that the HNO$_3$ concentration 100 m above ground undergoes little change, whereas that close to the ground decays to low values because of continuing dry deposition. The next day, when atmospheric turbulence picks up and the vertical air exchange resumes, the ground-level reservoir is filled up again and the HNO$_3$ mixing ratio rises. The diurnal variation in the degree of turbulence that is responsible for the vertical air exchange may be the most important cause of the diurnal variation of the HNO$_3$ mixing ratio. Remember that HNO$_3$ is produced

TABLE 9.13 Simultaneous Observations of Gaseous Nitric Acid and Particulate Nitrate in the Atmosphere[a]

Authors	Location and observation period	Season	HNO$_3$ (nmol m^{-3})	NO$_3^-$ (nmol m^{-3})	Ratio[b]	Remarks
Continental ground-based measurements						
Cadle et al. (1982)	Abbeville, Louisiana, rural 1979	Summer	31.8	14.5	2.2	Diurnal variation noted
Cadle (1985)	Warren, Michigan, June 1981–June 1982	Summer	42.7	49.4	0.87	Suburban location
		Winter	28.2	44.8	0.63	
Meixner et al. (1985)	Jülich, Germany, 1982–1983	Summer	45.7	74.2	0.62	Regionally polluted
		Winter	17.1	124	0.14	
Pierson et al. (1989)	Allegheny Mountain, Pennsylvania Laurel Hill, Pennsylvania, Aug. 1983		71	8	< 8.9	Particle diameter < 1.5, upper limit for HNO$_3$/NO$_3^-$
			85	10	< 8.5	
Benner et al. (1991)	Page, Arizona, Jan.–Feb. 1986		16.8 ± 10.0	2.6 ± 3.6	6.5	12-h samples
Grosjean and Bytnerowicz (1993)	Tanbark Flat, California, summer 1991	Day	179 ± 65	82 ± 56	2.2	12-h samples
		Night	33 ± 22	11 ± 7.5	3.0	
Williams et al. (1997)	Idaho Hill, Colorado, 3070 m elevation, Sept. 1993		5.6	3.1	1.8	
Matsumoto and Okita (1998)	Nara, Japan, 1994–1995		25.6	33.7	0.8	Annual averages
Kasper and Puxbaum (1998)	Mt. Sonnblick, Austria, 3106 m elevation, 1991–1993	Summer	5.0 ± 5.0	6.4 ± 8.1	0.78	High Alpine region
		Winter	1.9 ± 1.8	0.52 ± 0.88	3.65	
Marine atmosphere and aircraft measurements						
Huebert (1980)	Pacific Ocean, 9° S–7° N		1.7	3.5	0.48	Average
Norton et al. (1992)	Mauna Loa, Hawaii, 3400 m elevation, May, 1988	Upslope	5.8 ± 2.7	2.7 ± 2.8	2.2	
		Downslope	4.6 ± 2.5	1.4 ± 2.0	3.3	
Atlas and Ridley (1996)	Mauna Loa, Hawaii, 1991	Fall	2.9 ± 1.3	0.7 ± 1.3	4.0	Data selected for samples
	Jan.–Feb. 1992	Winter	2.8 ± 2.1	0.6 ± 1.1	4.4	from the free troposphere
	April–May 1992	Spring	6.0 ± 3.0	1.6 ± 1.5	3.7	
	July–Aug. 1992	Summer	3.5 ± 2.3	0.7 ± 0.8	4.7	
Huebert and Lazrus (1980a)	Boundary layer, marine, 0–30° N		3.2	6.5	0.49	
	Boundary layer, marine, 0–55° S		3.2	2.7	1.16	0–3 km altitude
	Boundary layer, continental, 30–50° N		13.8	13.5	1.02	
	Free troposphere marine, 55° S–30° N		4.9	2.1	2.3	3–8 km altitude
	Free troposphere, continental, 30–50° N		7.1	2.9	2.5	
Talbot et al. (1992)	Barrow, Alaska, 1988 high Arctic (> 60° N)	Summer	2 ± 0.9	0.5 ± 0.4	4.1	Boundary layer, < 2 km
			2 ± 1.1	1.0 ± 0.5	2.0	Free troposphere, 2–6 km
	Bethel, Alaska 1988 sub-Arctic (50–60° N)	Summer	2.6 ± 1.1	1.3 ± 0.8	2.0	Boundary layer, < 2 km
			2.6 ± 1.4	0.7 ± 0.3	3.9	Free troposphere, 2–6 km

[a] 1 nmol mol^{-1} ≈ 44.6 nmol m^{-3} ≈ 2.81 μg m^{-3} [HNO$_3$] ≈ 2.77 μg m^{-3} [NO$_3^-$].
[b] Ratio = HNO$_3$/NO$_3^-$.

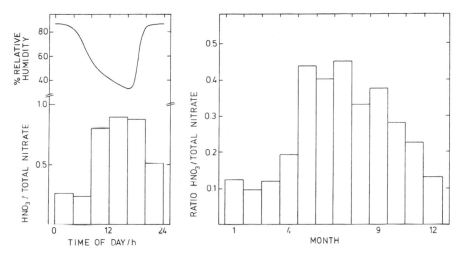

FIGURE 9.14 Diurnal and annual behavior of gaseous nitrate in ground-level air relative to total nitrate (gaseous plus particulate nitrate). Left: Averaged diurnal variation of HNO$_3$ and relative humidity at Claremont, California. (Adapted from Grosjean (1983b).) Right: Monthly mean values of HNO$_3$ in Jülich, Germany, for the period 1982–1983. (Adaped from Meixner et al. (1985) and Meixner (1994).)

during the day by reaction of NO$_2$ with OH radicals, and during the night by the reaction of N$_2$O$_5$ with liquid water on particles. As the two occur at comparable rates, HNO$_3$ production processes can account for only a part of the diurnal variation of HNO$_3$, which should be most effective in summer. The situation can be different at mountain stations when air masses having undergone long-range transport are sampled. At Niwot Ridge, a Colorado mountain site, Parrish et al. (1986b) were able to explain the observed diurnal variation of the HNO$_3$/NO$_x$ ratio by assuming that conversion by reaction with OH radicals is the dominant process generating HNO$_3$ and that dry deposition to the ground surface is the major loss process.

A number of ground-based measurement series have now confirmed that in addition to diurnal variations, the abundance of nitric acid undergoes an annual cycle with a maximum in summer and a minimum in winter. Table 9.14 contains several examples. Figure 9.14 (right) shows the seasonal variation observed by Meixner (1994) in a regionally polluted rural area of Germany. To separate the seasonal variation of HNO$_3$ from that of nitrate, it is convenient to consider the ratio [HNO$_3$]/([HNO$_3$] + [NO$_3^-$]), and Figure 9.14 (right) presents the data in this manner. The seasonal variation appears to be confined to the continental boundary layer. A series of intensive measurements in 1991–1992 at Mauna Loa, Hawaii, summarized by Atlas and Ridley (1996), showed little if any seasonal variation of the $[HNO_3]/([HNO_3] + [NO_3^-])$ ratio when the data were filtered to represent the free troposphere. The seasonal cycle in the boundary layer presumably arises from the same processes that are responsible for the diurnal cycle, that is, the change in temperature and relative humidity, in addition to changes in boundary layer conditions, such as a comparatively greater frequency of temperature inversion layers in winter and the accompanying decrease in atmospheric turbulence. The data from the Mauna Loa observatory for the free troposphere indicate that in contrast to the ratio of nitric acid to total nitrate, the sum of both components undergoes a seasonal variation with a maximum in spring. This feature may be caused by the increased level of ozone during that season and an enhancement of nitric acid formation via the formation of N_2O_5. Alternatively, the data may just indicate a seasonally stronger influence of long-range transport of air from the Asian continent. The ground-based measurements on the continents do not show much variation of total nitrate during the year.

The measurements in the free troposphere over the North American continent and over large parts of the Pacific Ocean indicate a fairly uniform distribution of total nitrate, in the range of 3–10 nmol m^{-3}, with little difference between the Northern and Southern Hemispheres and between continental and marine regions. The partitioning favors nitric acid. Its contribution to the total concentration is 70–85%, considerably larger than that in the boundary layer. The data from Mauna Loa, Hawaii, may be used to compare the abundance of total nitrate with that of NO_x and related compounds. The annual average of total nitrate corresponds to about 105 pmol mol^{-1}; that of NO_x is about 30 pmol mol^{-1}. These values may be assumed to represent steady-state mixing ratios in the free troposphere. According to the discussion in Chapter 8, the residence times for the wet removal of gaseous and particulate nitrate are nearly the same, namely about 6 days. As HNO_3 and its product NO_3^- derive from the oxidation of NO_2, the residence time of NO_x under steady-state conditions will be $\tau_{NO_x} = \tau_{nitrate} \, x(NO_x)/x(nitrate) = 6 \times (30/100) \approx 2$ days, which would be an annual average. A residence time of 2 days is not significantly greater than the value discussed earlier for the continental boundary layer in summer. However, the inclusion of other NO_x reservoir species would increase the residence time considerably. The addition of PAN, which at Mauna Loa was present with a mixing ratio similar to that of NO_x, increases the residence time of NO_x by a factor of 2 to almost 4 days. Total oxidized nitrogen, NO_y, presumably contains additional NO_x reservoir species. The average abundance of NO_y during the 1991–1992 measurement period summarized by Atlas and Ridley (1996) was $x(NO_y) = 233$ pmol mol^{-1}, from which total nitrate must be subtracted. This gives about 130 pmol mol^{-1}, including NO_x

and PAN. The corresponding residence time of NO$_x$ present in free form and contained bound in all reservoir species would be 7 days, and this value sets an upper limit. Accordingly, if HNO$_3$ arises only from the oxidation of NO$_2$ and the other NO$_2$ reservoir species such as PAN are nonreactive, the residence time of NO$_x$ in the free troposphere will be in the range of 2–7 days.

The column density of nitrogen corresponding to a uniform mixing ratio of 100–110 pmol mol^{-1} total nitrate is about 400 μg m^{-2}, and the associated wet deposition flux of nitrogen in the remote troposphere is $(365/6) \times 400 \times 10^{-6} = 24.3$ mg m^{-2} year^{-1}. This flux agrees approximately with that observed for nitrate deposited by precipitation at remote marine locations. For example, the following wet deposition rates were reported: 40 mg m^{-2} year^{-1} on a weather ship in the north Atlantic Ocean (Buijsman et al., 1991), 70 mg m^{-2} year^{-1} at Bermuda (Galloway et al., 1989); 21 mg m^{-2} year^{-1} at Amsterdam Island (Galloway and Gaudry, 1984); 16 mg m^{-2} year^{-1} at Samoa (Dayan and Nelson, 1988); and 30 mg m^{-2} year^{-1} at Cape Grim, Tasmania (Ayers and Ivey, 1988). Larger deposition rates occur in coastal regions of the continents.

9.4.5. TOTAL OXIDIZED NITROGEN

In the literature, the sum of all NO$_x$-related compounds, NO$_y$, is sometimes called *total reactive nitrogen*, in analogy to total odd nitrogen in the stratosphere, where HNO$_3$ must be included in the count because NO$_x$ can be reactivated from HNO$_3$ by photodecomposition. In the troposphere, the intensity of ultraviolet radiation causing the photodissociation of HNO$_3$ is so weak that it is not a significant source of NO$_x$. Other reactions that might convert HNO$_3$ back to NO$_x$ are not known. According to present knowledge, therefore, the term *reactive nitrogen* should be reserved for NO$_x$ and its association products, whereas HNO$_3$ and NO$_3^-$ are essentially stable and represent the final oxidation products. The virtue of NO$_y$ lies in the fact that as a sum of all NO$_x$-related species, it can be used to check on the budget of oxidized nitrogen compounds. Most important in this regard are separate measurements of NO$_y$ and the various species contributing to it, because they provide a check on whether the sum of oxidized nitrogen has been accounted for. Table 9.14 shows a number of data obtained for this purpose. The early measurements at Niwot Ridge, Colorado, a mountain site receiving clean continental air with air flow from the west, but upslope pollution from Denver with winds from the east, showed that in summer the sum of NO$_x$, PAN, HNO$_3$, and NO$_3^-$ accounted for only 60% of the total NO$_y$, whereas in the late fall the balance was closer to 90%. The imbalance suggested either

TABLE 9.14 Average Ratios of NO_x, PAN, HNO_3 and Particulate Nitrate to Total Oxidized Nitrogen NO_y

Authors	Location	Period	$R(NO_x)$	$R(PAN)$	$R(HNO_3)$	$R(NO_3^-)$	$R(\Sigma NO_y)$	Remarks
Fahey et al. (1986)	Niwot Ridge, Colorado	July–Aug. 1984	0.29	0.16	0.11	0.02	0.58	
		Oct.–Nov. 1984	0.40	0.30	0.18	—	0.88	
Parrish et al. (1993)	Scotia, Pennsylvania	July–Aug. 1988	0.38	0.21	0.29	—	0.89	
	Whitetop, Virginia	Aug.–Sept. 1988	0.56	0.17	0.34	—	1.09	
	Bondville, Illinois	Aug. 1988	0.38	0.12	0.13	—	0.80	
	Egbert, Ontario	Aug.–Sept. 1988	0.48	0.19	0.28	—	0.94	
Atlas et al. (1992)	Mauna Loa, Hawaii	May–June 1988	0.14	0.07	0.43	0.15	0.79	Downslope winds
			0.17	0.11	0.51	0.24	1.03	Upslope winds
Bottenheim and Gallant (1989)	Western Atlantic	April 1986	0.32	0.08	—	0.30[a]	0.70	Boundary layer
			0.27	0.11	—	0.18[a]	0.56	Free troposphere
Williams et al. (1997)	Idaho Hill, Colorado 3070 m elevation	Sept.–Oct. 1993	0.36	0.26	0.10	0.06	0.77	Down-slope winds
			0.75	0.14	0.06	0.05	1.00	Upslope winds
Singh et al. (1994)	Goose Bay, Labrador	July–Aug. 1990	0.21	0.39	0.52	0.06	1.13	0–2 km
			0.15	0.77	0.29	0.04	1.19	2–4 km
			0.06	0.38	0.09	0.04	0.54	4–6 km
Atlas and Ridley (1996)	Mauna Loa, Hawaii	Sept.–Oct. 1991	0.13	0.09	0.30	0.12	0.65	Free troposphere
		Jan.–Feb. 1992	0.15	0.28	0.23	0.04	0.70	
		April–May 1992	0.10	0.12	0.36	0.13	0.71	
		July–Aug. 1992	0.14	0.07	0.35	0.13	0.69	
Singh et al. (1996b)	Western Pacific Ocean 25°–42°N	Sept.–Oct. 1991	0.19	0.05	0.38	0.11	0.62	0–3 km
			0.18	0.20	0.25	0.13	0.63	3–7 km
			0.35	0.26	0.12	0.04	0.71	7–12 km

[a] Total inorganic nitrate, $HNO_3 + NO_3^-$.

that species other than the one measured contributed significantly to total oxidized nitrogen, or that systematic errors occurred in the concentration measurements. Appreciable experimental uncertainty is, in fact, expected in view of the difficulty of simultaneously measuring five individual species in addition to NO$_y$ with good accuracy. Table 9.14 shows, however, that a balance of close to 100% was achieved in several more recent studies, in which an intercomparison of standards had improved the reliability of the measurements. On the other hand, the possibility that the measurements had missed an important component contributing to NO$_y$ led to a search for suitable candidate species that might fill the gap. A study of Buhr et al. (1990), which indicated that the missing NO$_y$ was correlated with the degree of photochemical aging, focused attention on alkyl nitrates and multifunctional organic nitrates.

As discussed in Section 6.2.1, alkyl nitrates are formed in small yields during the oxidation of hydrocarbons in reactions of alkylperoxy radicals with NO. The formation of multifunctional organic nitrates results from the reaction of NO$_3$ radicals with organic compounds (Section 9.4.1). Although alkyl nitrates are photolyzed at a moderate rate, they are observable in the atmosphere. Indeed, almost every possible alkyl nitrate in the carbon range C_2–C_8 has been found, in addition to methyl nitrate (Atlas, 1988; Buhr et al., 1990; Ridley et al., 1990; Flocke et al., 1991; Atlas et al., 1993; Shepson et al., 1993). All studies showed that alkyl nitrates contributed only a few percent to total NO$_y$, so that alkyl nitrates alone cannot be responsible for the imbalance between total NO$_y$ and the sum of measured species contributing to it. For example, O'Brien et al. (1995) determined 12 C_3–C_6 alkyl nitrates, four C_2–C_4 hydroxy nitrates, and 1,2-dinitrooxybutane during August 1992 at a rural site in southern Ontario. The sum of their mixing ratios ranged from 12 to 140 pmol mol^{-1}; the average contribution of alkyl nitrates was 82%, that of hydroxy nitrates 16%, and the dinitrate contributed 2%. Yet, the total concentration of organic nitrates contributed only 0.5–3.0% to total oxidized nitrogen. The diurnal variation in relatively clean air showed a maximum in the late afternoon, whereas a more complex behavior occurred during an oxidant pollution episode. All of the data for total RONO$_2$ exhibited a good correlation with ozone, indicating that the organic nitrates are photochemically produced. The distribution of alkyl and hydroxy nitrates was reasonably consistent with computed relative production rates.

The origin of methyl nitrate in the atmosphere is uncertain. The branching ratio for the reaction $CH_3OO + NO \rightarrow CH_3ONO_2$ versus $CH_3O + NO_2$ is immeasurably small, so that methyl nitrate must arise from a different source. Senum et al. (1986) obtained evidence for the formation of methyl nitrate from the thermal decomposition of PAN, but the route is kinetically much less favorable than the decomposition to peroxyacetyl radicals and NO_2. Ridley et al. (1990) found from simultaneous measurements of ozone, PAN, PPN, and methyl nitrate at Boulder, Colorado, that whereas the first three components were well correlated, methyl nitrate did not correlate with either of them, and its mixing ratio stayed essentially constant at about 2 pmol mol^{-1}. The data obtained at Mauna Loa, Hawaii, selected for air from the free troposphere, indicated a similar abundance without any obvious seasonal cycle. At the same site, Walega et al. (1992) had found that C_3–C_5 alkyl nitrates correlated well with each other and with C_2Cl_4, but not with methyl nitrate, indicating that higher alkyl nitrates have a northern hemispheric continental source, whereas methyl nitrate has a different origin.

The data in Table 9.14 show that the imbalance between total NO_y and the sum of the observed individual species contributing to it is particularly poor in the upper troposphere compared to ground level. The difference is shown particularly well by comparing upslope and downslope conditions at Mauna Loa, Hawaii. It appears that in the upper troposphere one or more important species contributing to NO_y have not yet been identified.

9.4.6. SOURCES AND SINKS OF OXIDIZED NITROGEN IN THE TROPOSPHERE

The principal routes to the production of NO_x are combustion processes, nitrification and denitrification in soils, and lightning discharges. The major removal mechanism is oxidation to HNO_3, followed by dry and wet deposition. Ehhalt and Drummond (1982) and Logan (1983) first worked out a global budget in which the fluxes of NO_x into the troposphere and the rate of deposition of nitrate were approximately balanced. A more recent assessment by Lee et al. (1997) focused on refining the emission estimates. Table 9.15 presents a summary of the individual processes and the associated fluxes.

The annual production of NO_x from the combustion of fossil fuels may be estimated from empirical emission factors for various combustion processes combined with the worldwide consumption of coal, oil, and natural gas. Logan (1983) provided a convenient tabular summary of emission factors. Because of unavoidable fluctuations, the values must be considered uncertain by about 30%. Table 9.16 shows a breakdown of global emission estimates for NO_x according to fuel consumption rates. The estimates of Logan (1983) actually were somewhat more detailed than indicated. She found a larger production rate from fossil fuel combustion than that reported by Ehhalt and Drummond (1982), with the largest difference occurring in NO_x from

TABLE 9.15 Global Budget of NO_x in the Troposphere (Tg nitrogen year^{-1})[a]

Type of source or sink	Ehhalt and Drummond (1982)	Logan (1983)	Lee et al. (1997)
Production			
Fossil-fuel combustion	13.5 (8.2–18.5)	19.9 (14–28)	22 (13–31)
Biomass burning	11.5 (5.6–16.4)	12.0 (4–24)	7.9 (3–15)
Lightning discharges	5.0 (2–8)	8.0 (2–20)	5.0 (2–20)
Release from soils	5.5 (1–10)	8.0 (4–16)	7.0 (4–12)
NH_3 oxidation	3.1 (1.2–4.9)	Uncertain (0.10)	0.9 (0.3–1.2)
Stratosphere[b]	0.6 (0.3–0.9)	0.5	0.64 (0.4–1.0)
High-flying aircraft	0.3 (0.2–0.4)	—	0.85
Total production rate	39 (19–59)	48.4 (25–99)	44 (23–81)
Removal			
Wet deposition, land	17 (10–24)	19 (8–30)	—
Wet deposition, oceans	8 (2–14)	8 (2–12)	—
Wet deposition, combined	24 (15–33)	27 (12–42)	—
Dry deposition[c]	—	16 (12–22)	—
Total removal rate	24 (15–40)	43 (23–64)	—

[a] From estimates of Ehhalt and Drummond (1982), Logan (1983), and Lee et al. (1997).
[b] The inflow from the stratosphere consist predominantly of HNO_3.
[c] Dry deposition of NO_2 and HNO_3 over the continents, and of HNO_3 over the ocean.

transportation. The former estimate was based on the amount of fuel used, whereas the latter relied on the world population of automobiles. In the last two decades, attempts have been made to control NO_x emissions from motor vehicles. Logan (1983) pointed out that in the United States, where the control is mandatory, these measures have had little effect on the total emission. The global NO_x emission rate of 19.9 Tg year^{-1} of nitrogen derived by Logan (1983) has been largely confirmed in more recent studies. Dignon (1992) obtained 22 Tg year^{-1} for 1980, and Benkovitz et al. (1996) reported 21 Tg year^{-1} for 1985. Dignon and Hameed (1989) discussed the increase in NO_x emissions that has occurred since 1930. Greenland ice core samples analyzed by Neftel et al. (1985) show some correspondence with estimated increases in NO_x emissions since preindustrial times. In Europe and the United States, the increase in emission rates has leveled off. Currently, the rates are about 6.1 and 6.2 Tg year^{-1} (nitrogen), respectively. The emissions in Asia are continuing to rise at a rate of 4% annually. In 1987 they reached 4.7 Tg year^{-1} and were estimated to have reached 5.7 Tg year^{-1} in 1991 (Kato and Akimoto, 1992). Global emissions are expected to rise further in the future.

TABLE 9.16 Global NO$_x$ Emissions from the Burning of Fossil Fuels and Biomass[a]

Source type	Annual consumption			Emission factors		Global NO$_x$ emission (Tg year^{-1} (as nitrogen))		
	E&D	L	present[b]	E&D	L	E&D	L	present[b]
Fossil fuels								
Hard coal	2150 Tg	2696 Tg	—	1.0–2.8 g kg^{-1}	2.7 g kg^{-1}	3.9 (1.9–5.8)	6.4	—
Lignite	810 Tg		—	0.9–2.7 g kg^{-1}		1.6 (0.8–2.3)		—
Light fuel oil	300 Tg	1.39 Tm^3	—	1.5–3.0 g kg^{-1}	2.2 g m^{-3}	0.7 (0–5–0.9)	3.1	—
Heavy fuel oil	470 Tg		—	1.5–3.1 g kg^{-1}		1.1 (0–7–1.5)		—
Natural gas	1.04 Tg	1.2 Tm^3[c]	—	0.6–3.0 g kg^{-1}	1.9 g m^{-3}	1.9 (0.6–3.1)	2.3	—
Industrial sources	—		—	[c]	—		1.2	—
Automobiles	(4.1–5.4) × 10^{12} km	1.0 Tm^3	—	0.9–1.2 g km^{-1}	8.0 g m^{-3}	4.3 (3.7–6.4)	8.0	—
Total						13.5 (8.2–18.5)	19.9	—
Biomass burning								
Savanna	0.5–1.9 Pg^d	2.0 Pg	3.52 Pg	1.6–2.2 g kg^{-1}	1.7 g kg^{-1}	3.1 (1.8–4.3)	3.4	4.6
Forest clearings	1.6–3.7 Pg^d	4.1 Pg	2.12 Pg	0.5–1.0 g kg^{-1}	2.0 g kg^{-1}	2.1 (0.8–3.4)	8.2	1.3
Biomass fuel	0.5–1.5 Pg^d	850 Tg	2.20 Pg	2.0 g kg^{-1}	0.5 g kg^{-1}	2.0 (1–3)	0.4	0.9
Agricultural waste	1.7–2.1 Pg^d	15 Tg	0.95 Pg	1.0–2.8 g kg^{-1}	1.6 g kg^{-1}	4.0 (2–6)	0.02	1.2
Total						11.2 (5.6–16.4)	12.0	8.0

[a] Estimates according to Ehhalt and Drummond (1982) and Logan (1983) (L). Emission factors refer to gram nitrogen per unit of fuel consumed.
[b] Averages from Table 4.6. Emission factors adopted were (in g kg^{-1}) savanna 1.3, forest clearings 0.6, biomass fuel 0.4, agricultural waste 1.3.
[c] Petroleum refining and manufacture of nitric acid and cement; global emissions were obtained by scaling U.S. emissions for each industrial process.
[d] Ehhalt and Drummond (1982) adopted the data assembled by Crutzen et al. (1979).

Biomass burning as a source of atmospheric trace gases was discussed in Chapter 4. In natural fires and the open burning of wood, savanna grass, agricultural wastes, etc., the temperatures reached are rarely high enough to promote the oxidation of nitrogen present in the air, so that the NO_x emission originates essentially entirely from the nitrogen content of the fuel. Logan (1983) reviewed experimental determinations of emission factors showing that the yield of nitrogen released as NO_x is highest for grass and agricultural refuse fires (1.3 g kg^{-1}), less for prescribed forest fires (0.6 g kg^{-1}), and still lower for the burning of fuel wood in stoves and fireplaces (0.4 g kg^{-1}). The values reflect roughly the differences in the nitrogen content of the materials burned. Grass, leaves, and twigs contain approximately 1% by mass, whereas fuel wood contains only 0.2%. Only a fraction of fuel nitrogen is oxidized to NO_x, however. Figure 4.7 shows that NO_x is released primarily during the flaming stage of the fire. Thus, it can be scaled to CO_2, which is also produced mainly during the flaming stage. Other nitrogen compounds, such as ammonia, hydrogen cyanide, or acetonitrile and other nitriles are released mainly during the smoldering stage. The controlled burning experiments discussed by Lobert et al. (1991) showed that an average of 13.6% of fuel nitrogen occurred as NO_x. Ammonia, cyanogen (C_2N_2), HCN, CH_3CN, and other nitriles contributed 8%, N_2O 0.77%, various amines 0.28%, and gaseous nitrates (predominantly HNO_3) 1.1%, and roughly 10% remained in the ash. Up to 60% of fuel nitrogen could not be accounted for and was assumed to be released as N_2. Kuhlbusch et al. (1991) subsequently confirmed that $46 \pm 15\%$ of fuel nitrogen was volatilized as N_2 during flaming combustion.

The average molar NO_x/CO_2 ratio suggested by Andreae (1991) is 2.1×10^{-3}, with a range of $(0.7-1.6) \times 10^{-3}$ determined in several laboratory studies and $(2-8) \times 10^{-3}$ in field studies. Table 4.6 shows that the amount of CO_2 carbon released annually from biomass burning is 1.5–5.1 Pg $year^{-1}$. The associated amount of nitrogen emitted as NO_x would fall in the approximate range of 3–25 Tg $year^{-1}$, with an average near 7 Tg $year^{-1}$. Alternatively, one may combine emission factors for each category shown in Table 4.6 with the amounts of biomass burned annually. In this case, the global rate of NO_x nitrogen emission obtained is 8 Tg $year^{-1}$ (range 3.8–12.3 Tg $year^{-1}$), in good agreement with the above value (see Table 9.15). Andreae (1991) derived 8.5 Tg $year^{-1}$ without indicating the exact method of accounting, whereas Ehhalt and Drummond (1982) and Logan (1983) had earlier obtained about 12 Tg $year^{-1}$. From the data shown in Table 9.15, it appears that about 57% of the global NO_x from biomass burning is released from savanna fires, about 16% from permanent deforestation and shifting agriculture, and another 16% from the burning of agricultural wastes in the field. Only 11% is due to the use of biomass as fuel.

Thunderstorm electricity has been considered a major source of NO_x in the atmosphere ever since Liebig (1827) proposed that it provides a natural mechanism of nitrogen fixation. Several authors have looked for an increase of nitrate in thunderstorm precipitation, compared to normal rainwater, as evidence for the production of NO_x in lightning discharges. The results reported by Hutchinson (1954), Viemeister (1960), Visser (1961, 1964), Reiter (1970), and Gambell and Fisher (1964) were mainly negative or, at best, inconclusive. The largest effect is expected in the equatorial region, where thunderstorm activity is higher than elsewhere. Yet Visser (1961), who worked in Kampala, Uganda, and made daily observations of NO_3^- in rainwater over a period of a year, failed to detect any correlation between the NO_3^- concentration and the number of lightning flashes. In retrospect it appears that a positive correlation is unlikely because of the time delay between NO_x production and HNO_3 formation, which should precede the incorporation of nitrate into cloud water and rainwater. Inherent in all studies was the assumption that NO_x reacts speedily with cloud water to form HNO_3. However, as discussed in Section 8.6.2 (see Table 8.9), the reaction $2NO_2 + H_2O \rightarrow HNO_2 + HNO_3$ is slow and essentially negligible under atmospheric conditions, because the solubility of NO_2 in water is low and the reaction rate depends on the square of the NO_2 concentration.

Electric discharges in air generate NO_x by the thermal dissociation of N_2 following ohmic heating within the discharge channel and shock-wave heating of the surroundings. Noxon (1976, 1978) first studied the increase of NO_x in the air during a thunderstorm. By relating it to the frequency of cloud-to-ground flashes, he computed a production rate of about 1×10^{26} NO_x molecules per lightning flash. Chameides et al. (Chameides et al., 1977; Chameides, 1979) and Levine et al. (1981) have conducted laboratory studies of small-energy sparks in chambers. The results indicated a yield of NO_x of about 5×10^{16} molecules per joule of spent energy. Borucki and Chameides (1984) applied a simple theory to derive a yield of $9 \pm 2 \times 10^{16}$ molecule J^{-1}. However, Goldenbaum and Dickerson (1993) pointed out that the yield would be sensitive not only to energy density but also to ambient pressure, leading to a narrow region of maximum yield of 25×10^{16} and a broad range of $10–15 \times 10^{16}$ molecule J^{-1}. These results have provided the basis for all estimates of global NO_x production from thunderstorm electricity. Table 9.17 lists a number of such estimates.

Price et al. (1997a) have reviewed the various aspects that must be considered in estimating the energy dissipated in different types of lightning flashes and the global frequency of such events. Ground-to-cloud and within-cloud lightning discharges must be considered. A typical cloud-to-ground lightning discharge begins with a preliminary breakdown channel, extending from the cloud toward the ground. Close to the ground, it is met

TABLE 9.17 Estimates of the Production of NO_x by Lightning[a]

| Authors | NO_x production rate | | Energy deposition | | Global NO_x nitrogen production rate |
	10^{16} Molecule J^{-1}	10^{25} molecules per flash or stroke	10^{-4} J m^{-2} s^{-1}	Strokes s^{-1}	(Tg year^{-1})
Laboratory experiments					
Chameides et al. (1977)	3–7	—	16	—	20–40
Levine et al. (1981)	5 ± 2	—	1–10	—	2–18
Peyrous and Lapeyre (1982)	1.6–2.6	—	15	—	9–15
Atmospheric observations					
Noxon (1976), Dawson (1980)	—	10	—	100	7
Drapcho et al. (1983)	—	40	—	100	30
Ridley et al. (1996)	—	5	—	b	4–5
Model calculations					
Tuck (1976)	—	1.1	—	500	4
Griffing (1977)	—	12–20	—	100	9–15
Chameides (1979)	8–17	—	16	—	47–90
Hill et al. (1980)	—	6	—	100	4
Price et al. (1997)	10	—	—	70–100	12

[a] Based partly on the compilation of Levine (1984).
[b] The observations were made in the anvil and upper core of thunderstorms and refer to the outflow at the top.

by an upward propagating streamer discharge. When the two meet, a large current pulse—the return stroke—occurs, propagating from the ground back up to the cloud along the ionized leader channel. This event may be followed by one or more subsequent strokes. Flashes within clouds occur between two centers of opposite charge. In this case the leader stage of the discharge is the principal process by which charge is transferred. As the leader progresses, it may encounter higher localized concentrations of opposite charge, causing the current to increase to values in the range of 1–4 kA. This is less than that encountered in return strokes of cloud-to-ground flashes, which range from 2 to 35 kA. However, lightning strokes within clouds are more frequent than cloud-to-ground flashes. Price *et al.* (1997a) made use of observations from a lightning detection network in the United States (Orville, 1991) and an empirical model of the lightning discharge to calculate an average energy of cloud-to-ground flashes of 6.7×10^9 J, with a range of 10^9 to 10^{10} J. The energy dissipated by within-cloud flashes is an order of magnitude lower.

The global frequency of lightning flashes may be obtained from satellite observations (Orville and Spencer, 1979; Orville and Henderson, 1986; Turman and Edgar, 1982). The data suggested a frequency of $40–120 \text{ s}^{-1}$, with a distribution maximizing in the tropics. Many estimates have used 100 flashes s^{-1} to calculate the global rate of NO_x production from lightning, but Price *et al.* (1997a) pointed out that 70% of the flashes observed by satellites are discharges within clouds that produce only 10% of the NO_x generated by cloud-to-ground discharges. Some authors have further taken into account that each flash consists of several forward and return strokes, but again, as Hill (1979), Dawson (1980), and Price *et al.* (1997a) have noted, most of the flash energy is dissipated in the first return stroke. The last authors therefore used a different approach, namely that of Price and Rind (1992), which linked lightning activity to the updraft intensity of individual thunderstorms as a parameter, as a basis for calculating lightning frequencies. The results were combined with convective cloud climatology provided by satellite data (Rossow and Schiffer, 1991). The parameterization differentiates between continental and marine clouds. The updraft intensity of the latter is much weaker, leading to low lightning frequencies compared to continental clouds. Integration over the globe resulted in a lightning frequency of 71 s^{-1} in January and 101 s^{-1} in July 1988. The fraction of cloud-to-ground flashes was about 30%. The global rate of NO_x production for the period 1983–1991 averaged 12 Tg nitrogen year^{-1} with a range of $5–20$ Tg year^{-1}. The meridional distribution favors low latitudes, but shows also that 64% of lightning activity occurs in the Northern Hemisphere because of the greater land masses there. Price *et al.* (1997b) pointed out that the global atmospheric electric circuit constrains the total amount of energy available from lightning and thus NO$_x$ production, for which they determined a range of 5–25 Tg year^{-1} (as nitrogen).

Convection within a thunderstorm cell redistributes a large portion of the NO$_x$ from the boundary layer to the upper troposphere. Ridley et al. (1996) have measured the output of NO$_x$ generated in two thunderstorms by probing the anvil and upper core regions of the cloud system by aircraft. The yield was 250–305 kg N. Global-scale extrapolation indicated that 4.1–4.9 Tg of nitrogen produced by lightning discharges is released annually as NO$_x$ to the upper troposphere. This would be about 40% of the total NO$_x$ production.

The emission of NO$_x$ from soils, occurring primarily as NO, is associated with the biological nitrogen cycle discussed in Section 9.1 and follows from the production of nitrogen oxides as intermediates during processes of nitrification and denitrification. Galbally and Roy (1978), who first used the chamber technique to measure emission rates of NO from soils, observed nitrogen fluxes from grazed pastures averaging 12.6 μg m^{-2} h^{-1}. Emission rates from ungrazed plots were about 50% lower. Slemr and Seiler (1984) then showed that the release of NO$_x$ from soils depends critically on the temperature and moisture content of the soil. They found an average release rate of 20 μg m^{-2} h^{-1} for bare natural soils. A grass cover reduced the escape flux, whereas fertilization enhanced it. Ammonium fertilizers were about five times more effective than nitrate fertilizers, indicating that nitrification as a source of NO$_x$ from soils is more important than denitrification. Davidson et al. (1993) showed that in dry tropical forest soils NO emissions and nitrification rates are indeed well correlated. In addition, Williams and Fehsenfeld (1991) found that the concentration of nitrate in soil may be used as a parameter to explain differences in NO emission rates from several ecosystem types in the United States. Generally, as discussed by Williams et al. (1992a) and Conrad (1996), the individual rates of NO production and consumption associated with bacterial processes of nitrification and denitrification, respectively, are regulated differently and depend on substrate availability and local soil microbial communities.

Table 9.18 summarizes results from a variety of field measurements of NO emissions to show that the fluxes can vary considerably. However, grassland and savanna soils tend to be stronger sources than forest soils at the same latitude. Emissions from agricultural fields are proportional to fertilizer application. For example, Slemr and Seiler (1991) found 0.05–0.2% of nitrogen applied as ammonium to be lost as NO$_x$; Harrison et al. (1995), who worked with NH$_4$NO$_3$, found losses of 0.4%; whereas Shepherd et al. (1991) inferred much larger losses of 11%. Slemr and Seiler (1991) and others also noted a compensation point for NO, so that NO undergoes dry deposition when the NO mixing ratio in the air exceeds about 1 nmol mol^{-1}.

TABLE 9.18   Some Measurements of Nitrogen Oxide Emission Rates from Soils (as Nitrogen)[a]

Authors	Site description	Observation period	Nitrogen emission (μg m^{-2} h^{-1})	Remarks
Agricultural fields and grassland				
Galbally and Roy (1978)	Pasture, Australia	July–Aug. 1982	1.6 (0.6–2.6)	Pasture grazed
		Sept.–Oct. 1982	3.5 (1.5–7.3)	Grass cover, unfertilized
Slemr and Seiler (1984)	Loess pararendzina, Mainz, Germany		– 21–51	Bare soil, unfertilized
	Loamy sand, Utrera, Spain		32.8 (–8–385)	40 g m^{-2} NH$_4$NO$_3$ applied
			128 (40–468)	
Williams et al. (1987)	Grassland, Boulder, Colorado	Aug.–Nov. 1985	10.8 (0.1–234)	
Anderson and Levine (1987)	Sandy loam, Jamestown, Virginia planted to corn and barley	June 1984–July 1985	12.8 (0.01–45)	53 mg m^{-2} year^{-1} unfertilized
			26.0 (0.04–242)	208 mg m^{-2} year^{-1}, fertilized
Johansson et al. (1988)	Sandy clay loam, savanna, Venezuela	Feb. 1987	28.4 (11.2–54)	Dry season
Sanhueza et al. (1990)	Sandy clay loam, savanna, Venezuela	Oct.–Nov. 1988	2.4 (0–25)	Rainy season
Shepherd et al. (1991)	Sandy loam, Ontario, Canada field planted to beans	April–Sept. 1988	0.7–19.4	Bare soil, unfertilized
			1.4–272	10 g m^{-2} NH$_4$NO$_3$ applied
Slemr and Seiler (1991)	Sandy loam, grass cover, near Mainz, Germany	Aug.–Sept. 1983	–0.4	Unfertilized
			10.5	~ 40 g m^{-2} NH$_4$Cl applied
Kim et al. (1994)	Fallow unmanaged pasture Candor, North Carolina	June–July 1992	6.4 (0.47–24)	NO compensation point noted
Forests				
Johansson (1984)	Podzolic soil, Sörentorp, Sweden	June 1983	1.26 (0.36–2.74)	Mixed forest
	Podzolic soil, Jädraås, Sweden	Aug.–Sept. 1983	0.83 (0.36–2.0)	Scots pine plus underbrush
Johansson et al. (1988)	Cloud forest, Altos de Pipe, Venezuela	Feb. 1987	3.7 (0.7–7.2)	
Kaplan et al. (1988)	Tropical forest, Manaus, Brazil	July–Aug. 1985	43.6 (29–58)	
Sanhueza et al. (1990)	Semideciduous forest, sandy clay loam, Chaguarama Venezuela	Oct. 1988	1.8 (0.57–3.0)	Small forest near savanna
Williams and Fehsenfeld (1991)	Deciduous forest, Oak Ridge, Tennessee	Summer and fall, 1988	1.0 av. (0.2–4.0)	Oak, hickory
Davidson et al. (1991)	Deciduous forest, Chamela, Mexico	Apr.–May 1990	2.3 (0.58–4.0)	Dry season
		Sept.–Oct. 1990	5.5 (2.27–10.5)	Rainy season

[a] Average and/or range.

Williams *et al.* (1992b) proposed a parameterization scheme that relates emissions to soil temperature, the nitrate content of the soil, and measured fluxes from four different ecosystem types: grasslands, forests, wetlands, and agricultural areas. The latter were subdivided according to principal crops, and the applied amounts of fertilizer were taken into account. For the United States, the method showed agricultural fields to account for 66.4% of all emissions, grassland 28.4%, and forests 5.2%. The total annual emission was calculated to be 314 Gg N, of which 202 Gg, that is 64%, occurred in summer. Williams *et al.* (1992b) combined their algorithm with a global data set of ecosystem types and mean surface temperatures to estimate monthly emissions on a global scale with $1° \times 1°$ grid resolution, from which they derived a total annual emission of 12 Tg N. More recently, Potter (1996) applied a model of nitrogen mineralization rates and soil inundation to estimate the global nitrogen emission of NO from soils. He found 10 Tg year^{-1}. Yienger and Levy (1995) used the Williams *et al.* (1992b) algorithm, modified to simulate pulsing emissions after rainfall, and calculated the global nitrogen emission of NO from soils to be 10.2 Tg year^{-1}.

Although the results are consistent, they define upper limit values because ozone in the air oxidizes NO to NO_2, which subsequently is partly taken up by the plant canopy. The uptake of NO_2 by vegetation has been thoroughly studied by enclosure and eddy correlation techniques; Voldner *et al.* (1986) and Hanson and Lindberg (1991) have reviewed the data. For example, Skärby *et al.* (1981), Johansson (1987), and Hanson *et al.* (1989) showed for a range of deciduous and coniferous trees that the uptake of NO_2 occurs primarily via the stomata and that deposition on cuticular surfaces is practically negligible. Figure 1.16 showed an example for the diurnal variation of the NO_2 flux caused by soil emissions of NO, its conversion to NO_2, and deposition on the canopy of a coniferous forest stand. The uptake of NO_x by the foliage depends on the extent of ventilation underneath the canopy and appears to be greater in dense grass swards than in open forests. Lee *et al.* (1997) accordingly applied approximate canopy reduction factors ranging from 0.2 for grassland to 0.8 for tropical rainforest. When the reduction factors were applied to global areas of ecosystem types, the nitrogen emission as NO_x was 7 Tg year^{-1} versus 11 Tg year^{-1}, calculated without canopy depletion. Yienger and Levy (1995) also modeled the conversion of NO to NO_2 followed by canopy depletion and found the global emission to be reduced to 5.5 Tg year^{-1}.

The oceans are a negligible source of NO_x. Zafiriou and McFarland (1981) observed seawater to be supersaturated with NO in upwelling regions with relatively high concentrations of nitrite. Here, the excess of NO must be due to photodecomposition of nitrite by sunlight. Logan (1983) estimated a local source strength of 1.3×10^{12} molecule m^{-2} s^{-1} under these conditions.

Linear extrapolation to the whole ocean surface, which maximizes this source, leads to an NO flux of 0.35 Tg year^{-1} (as nitrogen).

The oxidation of ammonia initiated by reaction with OH radicals was discussed in Section 9.2. The reaction consumes ammonia at a rate of 1–2 Tg year^{-1} (nitrogen) on a global scale. The yield of NO resulting from the oxidation sequence is not precisely known, because the NH_2 radical reacts partly with NO_2 to produce N_2O, which causes a loss of NO_x. The results of the three-dimensional model calculations of Dentener and Crutzen (1994), who were mainly interested in ammonia oxidation as a source of N_2O, suggest a global NO_x production rate of 0.9 Tg year^{-1} of nitrogen, with an uncertainty range of 0.4–1.2 Tg year^{-1}. This source is small compared with the direct emissions.

Table 9.15 includes the stratosphere as a source of tropospheric NO_x. In the stratosphere, NO_x arises from the oxidation of N_2O (see Section 3.4.1), but the final product returning to the troposphere is predominantly nitric acid. According to Table 3.2 the annual production rate is approximately 0.7 Tg N. In the troposphere, excess NO_x is converted to nitric acid and returns to the Earth's surface in this form. Although the stratosphere adds material to the total budget of oxidized nitrogen in the troposphere, the input is small compared to that of other natural sources.

Comparison of the sources listed in Table 9.15 shows that 50% of the NO_x budget is associated with emissions from fossil fuel combustion, and another 18% results from biomass burning. The natural budget of NO_x contributes approximately 14 Tg nitrogen year^{-1}, with contributions from lightning discharges and soil emissions being most significant.

Wet precipitation provides an efficient mechanism for the removal of gaseous and particulate nitrate from the atmosphere. Like that of ammonia, the precipitation of nitrate provides an important contribution of fixed nitrogen to the terrestrial biosphere, and until 1930 essentially all studies of nitrate in rainwater have dealt with the contribution of fixed nitrogen to agricultural soils. Eriksson (1952a), Steinhardt (1973), and Böttger *et al.* (1978) have compiled most of these data. More recent data are available from national networks of monitoring stations, mainly in Europe (van Leeuwen *et al.*, 1996) and in the United States (Sisterton, 1990), and from the global precipitation network (Whelpdale and Kaiser, 1996). Dentener and Crutzen (1994) have listed wet deposition rates observed at 52 stations in various parts of the world. Despite the wealth of information, it remains difficult to derive a global average for the deposition of nitrate, because of an uneven global coverage of the data and unfavorably short measurement periods at many locations (especially in Africa).

Ehhalt and Drummond (1982) relied on the detailed evaluation of Böttger *et al.* (1978), whose analysis used measurements from the period 1950–1977. The geographic distribution of NO_3^- deposition rates is shown in Figure

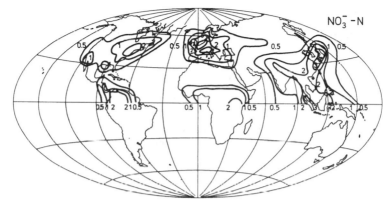

FIGURE 9.15 Global distribution of the wet deposition rate of NO_3^- nitrogen for the Northern Hemisphere (units: 100 mg m^{-2} year^{-1}), derived by Böttger *et al.* (1978) from measurements during the period 1955–1977. (Reproduced with permission of Kernforschungsanlage Jülich.)

9.15. The global deposition rate was derived by integration over 5° latitude belts. Logan (1983) considered more recent network data from North America and Europe, in addition to newer measurements of nitrate in precipitation at remote locations (Galloway *et al.*, 1982). Both estimates gave wet NO_3 deposition rates in the range of 2–14 Tg year^{-1} of nitrogen on the ocean surface, and 8–30 Tg year^{-1} on the continents. An earlier appraisal by Söderlund and Svensson (1976) had led to quite similar values, namely 5–16 and 13–30 Tg year^{-1}, respectively, although it was based primarily on Eriksson's (1952) compilation of data from the period 1880–1930. This remarkable consistency contrasts with the much larger scatter of individual estimates for the wet deposition of NH_4^+ (see Table 9.3) from a similar data base.

Dry deposition is an important sink for those gases that are readily absorbed by materials covering the Earth's surface. In the budget of oxidized nitrogen, NO_2 and HNO_3 are most readily affected by dry deposition. The deposition velocity of NO is too small, and the concentration of PAN generally is not high enough for a significant contribution. According to Table 1.12, dry deposition velocities for NO_2 over bare soil, grass, and agricultural crops fall in the range $(3–8) \times 10^{-3}$ m s^{-1}, but over water the velocities are significantly smaller, so that losses of NO_2 to the ocean can be ignored in comparison to those of HNO_3. The absorption of nitric acid by soil, grass, canopies, and water is significant (Hanson and Lindberg, 1991), so that dry deposition velocities are near the theoretical limit of that allowed by the aerodynamic resistance. While dry deposition of HNO_3 is important, the global flux is difficult to estimate because HNO_3 surface mixing ratios are not globally available. Logan (1983) adopted HNO_3 mixing ratios of 50 pmol mol^{-1} over the oceans and 100 pmol mol^{-1} over the continents. The mixing ratios assumed to exist for NO_2 were 100 and 400 pmol mol^{-1}, respectively. Allowance was made for higher mixing ratios in industrialized areas affected by air pollution. Logan further included the deposition of particulate nitrate over the oceans with a settling velocity of 3×10^{-3} m s^{-1}. This process contributed 2 Tg $year^{-1}$ and dry deposition of HNO_3 about 2.8 Tg $year^{-1}$ to the total annual deposition rate of 12–22 Tg N.

More recently, Duce et al. (1991) reviewed the data base available for an assessment of the deposition of oxidized nitrogen species to the various ocean provinces, and they calculated the corresponding fluxes. For the wet deposition of nitrate originating from the scavenging of particulate nitrate and gaseous nitric acid, they found a global nitrogen flux of 9.6 Tg $year^{-1}$, with a range of 7.6–18.1 Tg $year^{-1}$, which is slightly larger than that of Logan (1983). The rate of dry deposition was estimated to be 3.8 Tg nitrogen $year^{-1}$, with a range of 2.6–12.7 Tg $year^{-1}$, to which particulate nitrate contributed about 25%. This is, again, not much different from Logan's 1983 estimate.

On the continents one should also consider the interception of aerosol particles by high-growing vegetation. The process was discussed previously in Section 7.6.3 and in conjunction with the dry deposition of ammonium. The interception of particulate nitrate may be more effective than that of ammonium. Owing to the bimodal size distribution of NO_3^-, a larger fraction of it resides on giant particles, which are more efficiently removed by impaction. Lovett and Lindberg (1993) found that coarse particles contribute significantly to the dry deposition of nitrate to forest stands, whereas Butler and Likens (1995) found that dry deposition of nitric acid is the dominant process. It appears that dry deposition of particulate nitrate cannot compete with the very high deposition velocity of nitric acid.

If losses by dry deposition of NO_2 and HNO_3 are included in the total budget of oxidized nitrogen, the result is a reasonable balance between sources and sinks, as Table 9.15 shows. Ehhalt and Drummond (1982) felt that an appreciable part of dry deposition is already included in the wet deposition rate, because in the past rain gauges usually were left open continuously, so that the collection of nitrate would occur all the time during wet and dry periods. For NO_2 they estimated a nitrogen dry deposition rate of 7 Tg $year^{-1}$, which they considered rather uncertain. For this reason they included it in the error bounds and not in the mean value of total deposition of oxidized nitrogen. The good agreement between source and sink strengths should not be taken to indicate that the global budget of oxidized nitrogen is well defined. While the behavior of NO_x is made evident, a considerable effort will be required to improve the global description in view of the short residence times of the species involved.

CHAPTER 10

Sulfur Compounds in the Atmosphere

10.1. INTRODUCTORY REMARKS

Large quantities of sulfur dioxide enter the atmosphere each year from anthropogenic sources, mainly the combustion of fossil fuels and the smelting of metals. Sulfur dioxide is indisputably a prominent pollutant, so that a large research effort during the last 40 years has focused on problems such as the dispersal of SO_2 from power stations and urban centers, its conversion to sulfuric acid, the formation of sulfate aerosol, and the deposition of particulate sulfate and SO_2 at the ground surface.

The existence of natural sources of atmospheric sulfur has always been acknowledged. Known processes include volcanic emissions of SO_2 (and some H_2S), the aeolian generation of sulfate particles (e.g., in the form of sea salt), and the emanation of reduced sulfur compounds from the biosphere. The classical representative of the last group is H_2S, which is copiously produced in anaerobic marshlands and tidal flats. In the 1970s a number of additional sulfides were discovered in the atmosphere, namely carbonyl sulfide (OCS) (Hanst *et al.*, 1975), carbonyl disulfide (CS_2) (Sandalls and

Penkett, 1977), dimethyl sulfide (CH_3SCH_3) (Maroulis and Bandy, 1977), and methyl mercaptan (CH_3SH) and dimethyl disulfide ($CH_3S_2CH_3$) (Hill et al., 1978). Various other mercaptans have been identified by Burnett (1969), Stephens (1971), and others over cattle feedlots, where they arise from rotting manure. These discoveries have shifted the focus away from H_2S toward other reduced sulfur compounds and have led to a much better understanding of biogenic sulfur emissions.

The geochemical cycle of sulfur involves mobilization of the element by the weathering of rocks, followed by river transport to the oceans (mainly in the form of dissolved sulfate) and deposition in marine sediments. Tectonic uplift of sedimentary deposits eventually replaces the weathered material on the continents, thus completing the cycle (cf. Fig. 11.3). It is a curious observation that even after correcting for anthropogenic contributions, the rivers carry more sulfate to the oceans than the weathering of rocks can supply. An estimate by Ivanov et al. (1983) for the natural flux of sulfur (as sulfate) to the oceans is 100 Tg year^{-1}, whereas weathering is estimated to provide 33–42 Tg year^{-1} (Berner, 1971; Granat et al., 1976; Lein, 1983). Volcanic emissions cannot close the gap. While the problem has remained unresolved, it has suggested to many authors the existence of a marine source of volatile sulfur whose aerial transport toward the continents might balance the account. Traditionally, the role of the volatile has been assigned to H_2S, with the implicit assumption that H_2S is formed by bacterial reduction of sulfate in marine muds. Yet Östlund and Alexander (1963) had shown that H_2S is quickly oxidized in aerated waters. A recent result for the half-lifetime in seawater is 26 h at 25°C (Millero et al., 1987). Little H_2S is expected to escape from the deep ocean to the atmosphere, but shallow coastal waters would provide a source. In 1972 Lovelock et al. reported the presence of dimethyl sulfide (DMS) in seawater, and they suggested that this compound might be a more important volatile than H_2S. The concept gained ground when Maroulis and Bandy (1977) detected DMS in marine air, Nguyen et al. (1978) demonstrated its widespread occurrence in the world oceans, and Barnard et al. (1982) showed DMS to be ubiquitous in the air over large parts of the Atlantic Ocean. In the atmosphere, DMS is rapidly oxidized to sulfur dioxide and methane sulfonate, and SO_2 is further converted to sulfuric acid and particulate sulfate. These compounds then return to the ocean surface by dry and wet deposition.

Charlson et al. (1987) have called attention to the possibility that DMS affects the global climate through its role as a precursor of cloud condensation nuclei in the marine atmosphere. In Chapter 7 it was shown that particles in the submicrometer size range are more numerous than coarser particles, so that the former generate more cloud condensation nuclei than the latter. The composition of marine submicrometer particles is dominated by sulfate arising from the oxidation of DMS. At the cloud top, visible radiation is reflected because of multiple scattering of light by cloud drops, and this process is important in establishing the radiation balance. The fraction of light that is reflected increases with the number density of cloud drops. The number density, in turn, is determined by the number of condensation nuclei involved in the formation of a cloud. If it were to change, because the DMS flux and the associated rate of particle formation increased or decreased, on average, it would affect both the average cloud albedo and the global radiation budget. Charlson *et al.* (1987) hypothesized that the emission of DMS from the ocean might establish a negative feedback such that it regulates the radiation balance at the ocean surface. Large-scale climate change would control phytoplankton growth and thereby close the feedback loop. Despite a large research effort stimulated by the hypothesis, it is still not possible to represent it in a process-based predictive model. As discussed by Andreae and Crutzen (1997), even the overall sign of the feedback cannot be deduced with certainty because it is not known whether a warming climate would increase or decrease DMS emissions.

10.2. REDUCED SULFUR COMPOUNDS

This section deals with hydrogen sulfide, dimethyl sulfide, carbonyl sulfide, and carbonyl disulfide, which represent the most important reduced sulfur compounds in the atmosphere. We consider, in turn, atmospheric abundances and sources, atmospheric reactions, and exchange processes with the terrestrial biosphere.

10.2.1. ATMOSPHERIC ABUNDANCES, BEHAVIOR, SOURCES AND SINKS

Table 10.1 summarizes a number of measurements made during the last two decades. The currently preferred analytical technique is gas chromatography with flame photometric detection, combined with cryogenic preconcentration of samples. Low levels of H_2S have been determined by fluorescence quenching of fluorescein mercuric acetate after trapping of H_2S on $AgNO_3$-impregnated filters (Natusch *et al.*, 1972; Jaeschke and Herrmann, 1981). Data obtained for H_2S prior to 1970, such as those discussed by Junge (1963), are not included because the analytical methods were of doubtful

TABLE 10.1 Observed Mixing Ratios of Gaseous Sulfides in the Atmosphere (pmol mol^{-1})

Authors	Location	Average	Range	Remarks
Hydrogen sulfide, continental atmosphere				
Breeding et al. (1973)	Jacksonville, Illinois	130	120–140	Rural areas
	Bowling Green, Missouri	320	290–340	
	Athensville. Illinois	100	80–150	
Jaeschke et al. (1978)	Germany, Kleiner Feldberg,	230	—	Daytime average
	820 m elevation	53	—	Nighttime average
	Wetlands, Bavaria	718	—	
	Meadows, Bavaria	33	—	
Jaeschke et al. (1980)	Germany, Kleiner Feldberg	145	—	Annual average 1977
Delmas et al. (1980)	Toulouse, France	65	0–122	Urban, winter months
	Toulouse, residential	29	15–50	May
		76	36–151	June
	Landes, Loire Valley	56	28–121	December
	Pine Forest	44	39–51	April
	Pic du Midi (2980 m a.s.l.)	5	1.3–13	Free atmosphere
	Ivory Coast, tropical forest	4253	—	Anoxic soil region
		459	—	aerobic soils
Jaeschke and Hermann (1981)	Frankfurt/Main, city	164	—	October, daytime
		810	—	nighttime mean
Andreae et al. (1990)	Florida, various locations	187	66–336	Swampy regions
	Amazon Basin, boundary layer	47 ± 21	—	0–1 km Altitude
	Free troposphere	7 ± 7	—	3–5 km altitude
Hydrogen sulfide, marine atmosphere				
Slatt et al. (1978)	Miami, Florida	58 ± 23	41–91	Marine air
	Barbados	7 ± 8	0–45	Marine air
	Sal Island, Cape Verdes	4	0–12	Marine air
Delmas and Servant (1982)	Gulf of Guinea	14	5.5–45	
Herrmann & Jaeschke (1984)	North and Central Atlantic	26	10–110	
Saltzman and Cooper (1988)	Caribbean and Gulf of Mexico	8.5 ± 5.3	2–15	Background data
Andreae et al. (1993)	Tropical Atlantic Ocean,	7.8 ± 1.4	8–12	Night
	boundary layer	4.5 ± 1.8	8–12	Daytime
Cooper and Saltzman (1993)	North Atlantic offshore U.S.	8.4 ± 1.2	—	Marine air
	boundary layer	117 ± 139	—	Continental influence

Dimethyl sulfide, continental atmosphere				
Maroulis and Bandy (1977)	Wallops Island, Virginia	58 ± 2	40–110	Nighttime maximum
	Cape Henry, Virginia	30	—	Marine air influence
Bürgermeister (1984)	Frankfurt/Main, city	32 ± 55	2–349	Seasonal variation, winter maximum
	Kleiner Feldberg, (820 m a.s.l.) Germany	6.5	2.5–20	Highest values under inversion layer
	Wiesbaden, city	106	9–440	Pollution by paper industry
Andreae et al. (1990)	Amazon Basin, ground level	16 ± 10	—	
	Mixed layer	9.1 ± 6.7	—	Below 1 km altitude
	Free troposphere	1.1 ± 0.8	—	3–5 km Altitude
Dimethyl sulfide, marine atmosphere				
Barnard et al. (1982)	Atlantic Ocean	4.3	0.7–31	No diurnal variation
Andreae and Raemdonck (1983)	Remote Pacific Ocean	60 ± 24	42–70	Diurnal cycle, nighttime maximum
	Peru Shelf	14.6 ± 6.8	—	
Ferek et al. (1986)	Atlantic Ocean	73	40–102	
Berresheim (1987)	Drake Passage, subantarctic	108 ± 43	38–186	
	Antarctic coastal shelf	52 ± 11	34–63	
Saltzman and Cooper (1988)	Caribbean and Gulf of Mexico	57 ± 29	25–88	Background data
Bürgermeister et al. (1990)	Atlantic Ocean, 30° S–48° N	23 ± 15	1–68	
Nguyen et al. (1990)	Amsterdam Island	120 ± 74	22–775	Annual average
Staubes and Georgii (1993)	Atlantic Ocean, 50° S–50° N	55	4–704	Diurnal variation noted
	North Atlantic, 70–82° N	560	26–3440	
Cooper and Saltzman (1993)	North Atlantic offshore U.S.	118 ± 13	—	Boundary layer, marine air
	Tropical Atlantic Ocean	34 ± 17	—	Boundary layer
Yvon et al. (1996)	Equatorial Pacific Ocean	453 ± 93	280–650	Diurnal variation studied

(Continues)

TABLE 10.1 (*Continued*)

Authors	Location	Average	Range	Remarks
Carbon disulfide				
Sandalls and Penkett (1977)	Harwell, England	190 ± 110	—	Relatively clean air
Maroulis and Bandy (1980)	Philadelphia, Pennsylvania	37 ± 10	—	Considered polluted
		190 ± 114	65–340	Marine air influence
Bandy et al. (1981)	Wallops Island, Virginia	39 ± 8	—	5 km altitude
Jones et al. (1983)	North American continent	≤ 3	—	Average for clean air
Carroll (1985)	Harwell, England	15		6 km altitude
	North American continent	115 ± 36	65–166	7.3 km altitude
		23 ± 7	11–33	5–6 km altitude
Tucker et al. (1985)	Mainly southeast Pacific	5.7 ± 1.9	1–21	Shipboard measurements
Kim and Andreae (1987)	Northwest Atlantic, 25–44° N	1.3	1.1–1.6	
	Eastern U.S. shelf areas	13 ± 11	5–31	
	Estuaries	69 ± 63	7–184	
Cooper and Saltzman (1993)	Tropical Atlantic Ocean	0.9 ± 0.7	—	Boundary layer
	North Atlantic offshore U.S.	6.4 ± 1.3	—	Boundary layer
Bandy et al. (1993)	West Atlantic, 6° S–35° N	1.1 ± 0.2	0.4–1.8	6 km Altitude
Carbonyl sulfide				
Sandalls and Penkett (1977)	Harwell, England	510 ± 50	—	
Maroulis et al. (1977)	Philadelphia, Pennsylvania	434 ± 56	—	No significant pollution
	Wyoming	454 ± 28	—	
	Oklahoma	510 ± 35	—	
Torres et al. (1980)	Continental U.S.	509 ± 63	—	5–6 km altitude
	Pacific Ocean, 57° S–70° N	518 ± 74	—	Boundary layer, 1–2 km altitude
	Northern Hemisphere	523 ± 63	—	
	Southern hemisphere	498 ± 71	—	
Carroll (1985)	Eastern United States	517 ± 65	—	6–8 km altitude
Bingemer et al. (1990)	Atlantic Ocean, 37° S–51° N	537 ± 104	—	Excess in Northern Hemisphere
Johnson et al. (1993)	North Atlantic east of U.S.	489 ± 6	—	About 5 km altitude
	Tropical Atlantic	419 ± 1	—	About 5 km altitude
Weiss et al. (1995)	Pacific Ocean, 70° S–55° N	470 ± 55	—	Even latitudinal distribution
Thornton et al. (1996a)	North Pacific Ocean	495 ± 9	—	2–10 km altitude

validity, and the values reported often are incompatibly high compared to recent data. All sulfides except OCS react rapidly with OH radicals, and their atmospheric lifetimes are short, on the order of days (see Table 10.3). As a consequence, the mixing ratios fluctuate considerably and decrease strongly with distance from the sources and with elevation.

Mixing ratios of H_2S on the continents, in regions not directly influenced by pollution, range from 30 to about 100 pmol mol^{-1}. Higher values are encountered in the vicinity of cities or larger settlements, thus indicating the importance of anthropogenic sources. Higher values are also observed over anaerobic swamps, salt marshes, tidal flats, and similar regions of enhanced biogenic H_2S production. Jaeschke *et al.* (1978, 1980) have explored the vertical distribution of H_2S over Germany by means of aircraft. The mixing ratio decreased with altitude, in accordance with a scale height of 700–2000 m, depending on meteorological conditions and the strength of the underlying sources. The strongest gradients were found over industrially polluted regions. As Table 10.1 shows, H_2S is present in marine air masses, not only at coastal stations but also over the open ocean. Here, the biochemical sources must be located in near-surface waters, as H_2S from deeper strata would be oxidized before it could escape to the atmosphere. Hydrolysis of OCS, which leads to H_2S, may be a source. Based on simultaneous measurements of H_2S in seawater and air, T. W. Andreae *et al.* (1991) estimated that the oceanic emission rate is less than 0.3 Tg $year^{-1}$ globally.

The importance of dimethyl sulfide (DMS) in the marine environment has led to numerous measurements of its concentrations in ocean surface waters and in the overlying atmosphere. DMS is the predominant volatile sulfur compound in seawater, where it is associated with phytoplankton production. The main precursor of DMS in algae, according to Challenger (1951), is dimethylsulfonium propionate, $(CH_3)_2S^+CH_2CH_2COO^-$ (DMSP), a ternary sulfonium compound thought to be formed from methionine. Although DMSP is known to undergo enzymatic cleavage in the living algal cell, little DMS appears to be released directly. Many observations suggest that DMS in aquatic environments derives largely from the bacterial decomposition of DMSP leaked from aged cells or from zooplankton grazing on these cells. For example, as discussed by Bremner and Steele (1978), very high concentrations of DMS have been observed during the bacterial putrefaction of algae following algal blooms.

Mixing ratios in the air over the open ocean range from 20 to 400 pmol mol^{-1}. The concentrations of DMS in surface air are not necessarily correlated with those in seawater. In coastal regions, comparatively low values are observed despite a high biological activity, presumably because the influence of continental air leads to a more rapid oxidation of DMS. In all regions, regardless of concentration, seawater is found to be appreciably supersaturated with DMS relative to its aerial abundance. The supersaturation gives rise to a continuous flux of DMS from the ocean into the atmosphere. The magnitude of the flux can be estimated from the stagnant-film model discussed in Section 1.6 or an equivalent transport representation. The global mean DMS concentration in seawater is about 100 ng dm^{-3}. Estimates for the global sulfur flux associated with DMS range from 16 to 56 Tg $year^{-1}$ (Nguyen et al., 1978; Barnard et al., 1982; Andreae and Raemdonck, 1983; Andreae and Jaeschke, 1992; Bates et al., 1987, 1992a; Staubes and Georgii, 1993a). The early values, which tended toward the high side, may be criticized in that coastal data were given too much weight and that the seasonal variation at extratropical latitudes was not taken into account. The more recent data of Bates et al. (1987, 1992a) from the Pacific Ocean and Staubes and Georgii (1993a) from the Atlantic Ocean did discriminate between summer and winter half-years. The annually averaged global sulfur fluxes derived were 16 and 27 Tg $year^{-1}$, respectively. The difference in appraisal is mainly due to the use of different transport velocities in the liquid-film model rather than differences in the concentrations. Andreae and Jaeschke (1992) noted that the dependence on season, with low values in winter and high values in summer, is partly compensated for by the opposite behavior of the transfer coefficients, which maximize in winter when wind speeds are higher. Andreae et al. (1994) have compared different algorithms (Liss and Merlivat, 1986; Erickson, 1993) for calculating transfer coefficients as a function of wind speed and found differences on the order of 10–20%. Their preferred global flux rate of sulfur emitted as DMS is 35 Tg $year^{-1}$ (1.1 Tmol $year^{-1}$).

Several investigators (e.g., Maroulis and Bandy, 1977; Andreae and Raemdonck, 1983; Saltzman and Cooper, 1988; Suhre et al., 1995; Yvon et al., 1996) have reported diurnal variations of atmospheric DMS, featuring a build-up during night and a decline during the day. This phenomenon, which appears to be characteristic of clean, remote marine air, was predicted by Graedel (1979) from the reaction of DMS with OH radicals in the sunlit atmosphere. In coastal regions, where DMS mixing ratios are lower, diurnal variations are rarely observed, even though the concentration of DMS in seawater is not significantly different from that in high-productivity zones, and calculated air–sea fluxes are similar in magnitude. The lower mixing ratios and the absence of a diurnal cycle suggest an additional removal process, such as the reaction of DMS with NO_3 radicals that become active at night.

Simultaneous measurements of the DMS flux and its atmospheric concentration allow one to calculate the lifetime of DMS, provided some assumptions are made about the height of the atmospheric mixing layer. The results indicate lifetimes on the order of a day, but generally shorter than that calculated from the reaction with OH radicals alone. This provides additional evidence for the participation of a reaction involving yet another oxidant. Under stable meteorological conditions, the mixing depth of DMS over the open ocean is about 1 km, as found by aircraft observations (Ferek et al., 1986; Andreae et al., 1988c; Luria et al., 1989); in the free troposphere the concentration is < 3 pmol mol^{-1}, except when convection currents carry the substance to the middle troposphere.

The early measurements of Maroulis and Bandy (1977) of atmospheric DMS at a coastal site of the eastern United States indicated little dependence on wind direction, implying the existence of both marine and continental sources. Bürgermeister (1984), who made measurements in the densely populated Rhine-Main area of Germany, observed the highest values (up to 900 pmol mol^{-1}) on winter days, when a well-developed inversion layer restricted the vertical exchange of air, and in an industrial district. Anthropogenic sources of DMS obviously were dominant. In the hills north of Frankfurt/Main, the mixing ratio was reduced to 6 pmol mol^{-1}. Kesselmeier et al. (1993b), while exploring the exchange of sulfides between the atmosphere and tropical vegetation in southern Cameroon, found ground-level DMS mixing ratios in the range of 20–160 pmol mol^{-1}, whereas at the canopy level the mixing ratio rose to 3.5 nmol mol^{-1} at midday. Vegetation evidently provides a source of DMS in the continental atmosphere. Its contribution to the global emission rate, however, is considered minor (< 4 Tg sulfur year^{-1}) compared to that from the marine source.

The abundance of carbon disulfide in the atmosphere is low and variable. The few measurements that exist indicate an appreciable anthropogenic contribution. Jones et al. (1983) analyzed the frequency distribution of CS_2 mixing ratios at Harwell, England. By comparing this with similar data for $CFCl_3$, they concluded that the average CS_2 level in clean air is about 15 pmol mol^{-1}. Maroulis and Bandy (1980) correlated their data with wind direction and found mixing ratios of about 37 pmol mol^{-1} in relatively clean continental air masses. Kim and Andreae (1987) found mixing ratios in the range of 1–20 pmol mol^{-1} over the open Atlantic Ocean, but the values increased significantly in coastal regions. This behavior, which was also observed by Cooper and Saltzman (1991) and Staubes and Georgii (1993b), suggests that continental sources dominate. Bandy et al. (1993) showed from measurements over the eastern United States that local anthropogenic sources of CS_2 are significant. Tucker et al. (1985) made aircraft measurements to show that CS_2 is ubiquitous in the 45° S–45° N latitude belt, with an average of 5.7 ± 1.9 pmol mol^{-1} at 5–6 km altitude.

The ocean provides a source of CS_2, because seawater is slightly supersaturated with CS_2 compared with its concentration in the overlying air. Kim and Andreae (1987) estimated from their data a global sulfur flux in the range of 0.11–0.44 (average 0.23) Tg year^{-1}. Emissions from soils and plants may represent another source of CS_2, but they are quite variable and thus are difficult to evaluate. Chin and Davis (1993) have collected data to estimate the global annual emission rate of CS_2 from soils and marshes to be about 0.02 Tg S, that is, an order of magnitude lower than that from the ocean. Chin and Davis (1993) also discussed anthropogenic sources. The principal industrial use of CS_2 (75%) is in the manufacture of regenerated cellulose rayon and cellophane. Lay *et al.* (1986) estimated that in 1984 the worldwide production of CS_2 was 1.1 Tg. If 25–40% of it were released into the atmosphere, the source strength would be about 0.35 Tg CS_2 year^{-1} or 0.3 Tg sulfur year^{-1}. Other anthropogenic sources, such as sulfur recovery from oil refinery processes or the paper industry, appear to contribute less.

Carbonyl sulfide is fairly evenly distributed throughout the troposphere, with a mixing ratio of about 500 pmol mol^{-1}. Thus, OCS is the most abundant sulfide in the atmosphere. Bandy *et al.* (1992a) have examined a time series of measurements since 1977 and found neither temporal trends nor seasonal variations. A slight excess of about 5% appears to exist in the Northern Hemisphere compared to the Southern (Torres *et al.*. 1980; Bingemer *et al.*, 1990; Johnson *et al.*, 1993; Thornton *et al.*, 1996a). The uniform distribution points toward a residence time larger than 1 year. Thornton *et al.* (1996a) have used Junge's (1974) relation (see Eq. 4.13) to estimate from the relative standard deviation of the OCS mixing ratio a residence time of 7 years. The mass content of OCS in the troposphere is close to 4.6 Tg, so that global sources and sinks, if balanced, would amount to about 0.65 Tg year^{-1} each. In the stratosphere OCS undergoes photodecomposition by ultraviolet radiation, and the mixing ratio declines with increasing altitude. Inn *et al.* (1981) obtained the first evidence for the decline of OCS above the tropopause. By fitting their data to a one-dimensional diffusion profile, they determined the upward flux of OCS, about 1×10^{11} molecule m^{-2} s^{-1} or 0.16 Tg year^{-1} globally. Chin and Davis (1995) have evaluated several other OCS altitude profiles to estimate the global flux to the stratosphere as 0.048 Tg year^{-1}, about one-third of the former value. The measurements of Engel and Schmidt (1994) that were not taken into account by Chin and Davis indicated a range of 0.047–0.11 Tg year^{-1}. Thus, the stratosphere takes up 7–25% of the total source strength. The next section will show that reactions of OCS with OH radicals and oxygen atoms are slow, leading to an OCS lifetime in the troposphere of about 25 years as the upper limit. This sink contributes another 28% to the total. Brown and Bell (1986) and, subsequently, Goldan *et al.* (1988) have pointed out that the uptake of OCS by plants represents another sink. Goldan *et al.* (1988) and Chin and Davis (1993) estimated the global annual uptake of OCS by vegetation, using CO_2

net primary productivity as a scaling factor, and found 0.24–0.59 and 0.16–0.91 Tg, respectively. This essentially closes the gap. The exchange of OCS with vegetation will be discussed further in Section 10.2.3.

Concentrations of OCS in surface seawater are in the range of 0.03–1.0 nmol dm^{-3}, with lower values occurring in open ocean areas and higher ones in coastal and shelf areas (Rasmussen et al., 1982a; Ferek and Andreae, 1984; Turner and Liss, 1985; Andreae and Ferek, 1992). The observed concentrations were generally in the regime of supersaturation, indicating a net sea-to-air flux. However, Weiss et al. (1995) have found regions in the open ocean that were undersaturated with OCS, and they concluded that the open ocean represents no significant source. Diurnal variations of the OCS concentration in seawater suggest a photochemical origin. Laboratory experiments with seawater and distilled water showed that OCS is produced by photosensitized oxidation of dissolved organic sulfur compounds commonly found in biological materials (Ferek and Andreae, 1984; Zepp and Andreae, 1994). The presence of sulfate was not necessary. Removal of OCS from the water column is believed to occur both by sea–air exchange and by hydrolysis (Johnson, 1981; Elliot et al. 1989). Estimates for the global flux of OCS from the ocean range from 0.2 to 0.9 Tg year^{-1} (Rasmussen et al., 1982a; Ferek and Andreae, 1984; Andreae and Jaeschke, 1992). In contrast to DMS, the flux of OCS is dominated by the high-productivity areas, especially coastal and shelf regions.

Continental soils and marshes may provide another source of atmospheric OCS. Chin and Davis (1993) have examined the available data and derived a global flux in the range of 0.14–0.52 (average 0.27) Tg year^{-1}. Some observations, however, suggest that soils may absorb OCS rather than release it (Castro and Galloway, 1991; de Mello and Hines, 1994), so that the importance of soils may be overrated. Chin and Davis (1993) also discussed volcanoes, biomass burning, and various anthropogenic activities as sources of OCS in the atmosphere. The most significant of these appears to be biomass burning, which they estimated to contribute about 0.14 Tg year^{-1}. Andreae (1991) derived a similar estimate, 0.17 Tg year^{-1}. This is only a minor fraction of the total sulfur flux from fires, which occurs mainly in the form of SO_2. As discussed further below, the photochemical oxidation of CS_2 in the atmosphere produces equal amounts of SO_2 and OCS. Thus, carbon disulfide makes a significant contribution to the tropospheric budget of OCS, about 0.34 Tg year^{-1}. The OH-induced oxidation of dimethyl sulfide also produces OCS in a side reaction (0.17 Tg year^1), as Barnes et al. (1994b) discovered.

Table 10.2 summarizes estimates for global sources and sinks of carbonyl sulfide. The sources exceed the sinks by more than a factor of 2, indicating

TABLE 10.2   Global Sources and Sinks of Carbonyl Sulfide (Tg year $^{-1}$ of OCS)[a]

Sources		Sinks	
Ocean (coastal)	0.32 (0.17−0.47)	Stratosphere	0.04 (0.029−0.066)
Soil and marsh	0.27 (0.14−0.52)	Open ocean	0.03 (0.010−0.054)
Biomass burning	0.14 (0.04−0.26)	Reaction with OH	0.13 (0.02−0.80)
Anthropogenic	0.04 (0.027−0.059)	Vegetation	0.43 (0.16−0.91)
Volcanoes	0.02 (0.006−0.09)		
CS_2 oxidation	0.34 ((0.17−0.61)		
DMS oxidation	0.17 (0.10−0.28)		
Total	1.30 (0.67−2.5)	Total	0.63 (0.22−1.9)

[a] From Weiss *et al.* (1995)

significant uncertainties. A much better agreement would exist if the role of soils were reversed, so that they represented a sink rather than a source.

10.2.2. ATMOSPHERIC REACTIONS AND LIFE TIMES

The oxidation of sulfides in the atmosphere is initiated mainly by reaction with OH radicals. Table 10.3 lists the pertinent rate coefficients for two conditions: in air at atmospheric pressure, and in the absence of oxygen. The data show that O_2 accelerates the reactions in several cases, most notably in the reaction with CS_2. This effect is explained by the incipient formation of an OH addition complex, which then has three options: redissociation toward the original reactants, decomposition to new products, or interaction

TABLE 10.3   Rate Coefficients k_{298} for Reactions of OH Radicals at Atmospheric Pressure, O-Atoms, and Ozone with Various Sulfides[a]

Compound	OH O_2 absent	OH O_2 present	O_3	O	Lifetime in the troposphere
H_2S	4.8	4.8	2×10^{-20}	0.022	3.0 days
CH_3SH	33	33	—	1.9^b	0.4 days
CH_3SCH_3	4.4	6.3	1×10^{-18}	50	2.2 days
CH_3SSCH_3	200	200	—	130	0.1 days
CS_2	< 0.007	2	—	3.6	7.2 days
OCS	0.002	—	—	0.014	25 years[c]

[a] Unit of rate coefficients: 10^{-12} cm^3 molecule^{-1} s^{-1}; data from Atkinson *et al.* (1989, 1997).
[b] Slagle *et al.* (1976).
[c] For $T = 277$ K taken to represent a tropospheric average.

with O_2 to form a different set of products. The individual reaction pathways are discussed below.

Sulfides also react with oxygen atoms and with ozone. The corresponding rate coefficients are included in Table 10.3 as far as available. Laboratory studies of ozone reactions are complicated by the observation of fractional orders, autocatalytic behavior, and chemiluminescence, all features indicating the occurrence of radical chain processes triggered by a slow initiation step. The rate coefficient for the direct reaction of O_3 with H_2S is less than 2×10^{-20} cm^3 molecule^{-1} s^{-1} (Becker et al., 1975); that for the bimolecular reaction of O_3 with CH_3SCH_3 is less than 1×10^{-16} cm^3 molecule^{-1} s^{-1} (Martinez and Herron, 1978). In the atmosphere, ozone would begin to compete with OH if the reactions of ozone with H_2S and CH_3SCH_3 had rate coefficients greater than about 3×10^{-18} and 3×10^{-17} cm^3 molecule^{-1} s^{-1}, respectively. The laboratory evidence therefore suggests that the reactions with ozone are relatively unimportant, and we shall not consider them in detail.

The reactions of sulfides with oxygen atoms are quite rapid, as Table 10.3 shows, with rate constants comparable to those of the OH reactions in many cases. However, concentrations of O-atoms in the troposphere are in the range of 10^4 to 10^5 atom cm^{-3}—that is, at least an order of magnitude less than those for OH radicals. Accordingly, the latter is the preferred reagent.

Oxidation mechanisms for the reactions of OH with H_2S, CS_2, and OCS are reasonably well established (see Table 10.4). With regard to H_2S, the absence of a rate-accelerating effect of oxygen and the identification of HS as an intermediate product by Leu and Smith (1982) demonstrated that the reaction

$$OH + H_2S \rightarrow H_2O + HS$$

proceeds mainly by abstraction. The reaction of HS with oxygen is very slow, but HS reacts quite rapidly with ozone as well as with NO_2, and these reactions will be important in the atmosphere. The laboratory observations of Glavas and Toby (1975) and Becker et al. (1975) on the H_2S-O_3 reaction system indicated that it involves HS as a chain carrier, and that both channels of the HS-O_3 reaction shown in Table 10.4 may be involved. The first channel leading to HSO is exothermic by 287 kJ mol^{-1}. The excess energy occurs partly as electronic excitation of the HSO product and is released as chemiluminescence. The second reaction channel is slightly endothermic. Either this reaction or the equivalent reaction of HSO with ozone is required as a chain-branching step to explain the autocatalytic behavior of the H_2S-O_3 reaction. A third conceivable reaction channel, namely HS + O_3 → H + SO + O_2, can be ignored because the subsequent

TABLE 10.4 Overview of Reactions Involved in the Oxidation of H_2S, CS_2, and OCS in the Atmosphere[a]

Reaction		Heat of reaction (kJ mol^{-1})	k_{298} (cm^3 molecule^{-1} s^{-1})
Hydrogen sulfide			
OH + H_2S	$\rightarrow$ HS + H_2O	-120	4.8×10^{-12}
HS + O_2	$\rightarrow$ OH + SO	-96	$\leq 4 \times 10^{-19}$
	$\rightarrow$ SO_2 + H	-219	
HS + O_3	$\rightarrow$ HSO + O_2	-287	3.6×10^{-12}
	$\rightarrow$ OH + SO + O	$+10$	
HS + NO_2	$\rightarrow$ HSO + NO	-88	5.8×10^{-11}
HSO + O_2	$\rightarrow$ HO_2 + SO	$+24$	$\leq 2 \times 10^{-17}$
HSO + O_3	$\rightarrow$ OH + SO + O_2	-94	1.1×10^{-13}
	$\rightarrow$ HS + 2O_2	$+2$	
HSO + NO_2	$\rightarrow$ HSO_2 + NO	-161	9.6×10^{-12}
HSO_2 + O_2	$\rightarrow$ SO_2 + HO_2	-60	3.0×10^{-13}
SO + O_2	$\rightarrow$ SO_2 + O	-53	6.7×10^{-17}
SO + O_3	$\rightarrow$ SO_2 + O_2	-444	8.9×10^{-14}
Carbon disulfide			
OH + CS_2	$\rightleftharpoons$ SCSOH		$(< 0.007$–$2.0) \times 10^{-12}$
SCSOH + O_2	$\rightarrow$ COS + SO_2 + H	-377^b	
Carbonyl sulfide			
OH + OCS	$\rightarrow$ CO_2 + HS	-150	2.0×10^{-15}
O + OCS	$\rightarrow$ CO + SO	-213	1.4×10^{-14}
OCS + $h\nu$	$\rightarrow$ CO + S(^{3}P, ^{1}D)		^{c}
S(^{1}D) + N_2	$\rightarrow$ S(^{3}P) + N_2	-110	$> 1.2 \times 10^{-11\,d}$
S(^{3}P) + O_2	$\rightarrow$ SO + O	-23	2.3×10^{-12}
SO + O_2	$\rightarrow$ SO_2 + O	-53	6.7×10^{-17}

[a] Enthalpy data and rate coefficients from Atkinson et al. (1989).
[b] Overall reaction enthalpy.
[c] See Table 2.6.
[d] Little et al. (1972).

reaction of H-atoms with O_3 would produce excited OH radicals, but the associated emission of the Meinel bands has not been observed. The bands do occur, however, in the O_3 + CH_3SCH_3 reaction. The reaction of HSO with oxygen is endothermic, so that HSO must enter another reaction path in the atmosphere. Reactions with ozone and NO_2 would convert HSO to SO and HSO_2, respectively, which then react with oxygen to form SO_2 as the final product.

The OH-induced oxidation of CS_2 in air at atmospheric pressure has been studied by Jones et al. (1982), Barnes et al. (1983), and Hynes et al. (1988). Equal amounts of SO_2 and OCS were observed as products. The rate-accelerating effect of oxygen then suggests the incipient formation of an $HOCS_2$

addition complex, which reacts with oxygen to generate the final products SO_2 and COS, as shown in Table 10.4. The precise mechanism is not known. The overall process is sufficiently exothermic to generate H-atoms directly. Alternatively, one may envision the abstraction of H by oxygen to yield the products HO_2, OCS, and S, or formation of an HSO_2 intermediate, which would react further with O_2 to form SO_2 and HO_2.

The reaction of OH with OCS is again thought to proceed via an OCS-OH addition complex, but in the absence of any rate-accelerating effect of O_2, the main fate of the complex must be reversal to the original reactants. The actual forward reaction yielding new products is a minor pathway showing bimolecular behavior. By means of mass spectrometry, Leu and Smith (1982) identified HS as one of the products, so that the reaction takes the course OH + OCS → CO_2 + HS; this is indicated in Table 10.4. All other conceivable product channels are endothermic and may be ignored. The further oxidation reactions of HS follow the scheme outlined in Table 10.4. The reaction of OCS with oxygen atoms, which supplements the OH + OCS reaction in the upper troposphere and stratosphere, leads to the formation of SO and CO.

In the stratosphere, at altitudes above about 20 km, OCS is exposed to ultraviolet radiation of wavelengths shorter than 250 nm. Carbonyl sulfide has its first absorption band in this spectral region, giving rise to photodissociation. The photofragments are CO and a sulfur atom. As discussed by Okabe (1978), a major portion of the sulfur atoms thus produced are excited to the metastable ^1D state, but the excess energy is rapidly removed by collisions with N_2 (Little et al., 1972). Sulfur atoms in the ^3P ground state, and presumably in the excited ^1D state as well, react readily with oxygen and are then oxidized to SO_2. Table 10.4 includes this reaction sequence.

The reaction of OH with dimethyl sulfide proceeds by abstraction of a hydrogen atom as well as by the addition of OH to the sulfur atom. The rate-accelerating effect in the presence of oxygen is moderate, suggesting that abstraction is the dominant pathway:

$$OH + CH_3SCH_3 \rightarrow H_2O + CH_3SCH_2$$

$$\rightleftharpoons CH_3S(OH)CH_3$$

The $CH_3S(OH)CH_3$ radical undergoes rapid reverse decomposition, so that in the absence of O_2, only the rate coefficient for abstraction is observed. When oxygen is present, the addition product reacts with O_2, whereby new product channels are opened. The difference in the rate coefficients suggests that in air at atmospheric pressure about 75% of the reaction follows the abstraction pathway. The mechanism of DMS oxidation induced by OH radicals is still poorly understood, and the complete reaction sequence is uncertain. The early studies of Niki *et al.* (1983) and Hatakeyama *et al.* (1982, 1985) were performed in the presence of NO_x. Under these conditions the major sulfur-containing products were methanesulfonic acid (MSA, $\sim$ 50%) and SO_2 ($\sim$ 21%), and a significant fraction of methyl was converted to formaldehyde ($\sim$ 100%). The same products were observed in the oxidation of CH_3SH and CH_3SSCH_3 with similar yields. More recent results on the OH-induced oxidation of DMS under NO_x-free conditions (Barnes *et al.*, 1996) indicated that dimethyl sulfoxide (DMSO, CH_3SOCH_3) and SO_2 are major products (about 15% and 80%, respectively), whereas the yield of MSA is fairly small ($\sim$ 4%). The occurrence of DMSO is most readily explained to follow from reaction of the OH-DMS addition product with oxygen:

$$CH_3S(OH)CH_3 + O_2 \rightarrow CH_3SOCH_3 + HO_2$$

Sulfur dioxide as a product must then originate primarily as a consequence of the abstraction pathway. Under laboratory conditions, DMSO undergoes a rapid secondary reaction with OH radicals, which presumably leads to the formation of dimethylsulfone, $(CH_3)_2SO_2$, although the oxidation pathways are still uncertain. Both species, DMSO and $DMSO_2$, have been detected together with DMS in marine air (Pszenny *et al.*, 1990; Berresheim *et al.*, 1993).

The abstraction pathway is thought to generate the CH_3S radical as an intermediate (Yin *et al.*, 1990) as follows:

$$OH + CH_3SCH_3 \rightarrow H_2O + CH_3SCH_2$$

$$CH_3SCH_2 + O_2 \rightarrow CH_3SCH_2OO$$

$$CH_3SCH_2OO + NO \rightarrow CH_3SCH_2O + NO_2$$

$$CH_3SCH_2O \rightarrow CH_3S + HCHO$$

Support for this scheme is seen in the observation of HCHO and CH_3SNO in OH-initiated reactions (Hatakeyama and Akimoto, 1983; Niki *et al.*, 1983), and of $CH_3SCH_2OONO_2$ and CH_3SNO_2 in NO_3-initiated reactions (Jensen *et al.*, 1992) when NO_x is present. The CH_3S radical has been shown to react with ozone and with NO_2; the rate coefficients are 5.2×10^{-12} and 6.1×10^{-11} cm^3 molecule^{-1} s^{-1}, respectively (Tyndall and Ravishankara, 1989a, b). The product is CH_3SO in both cases. Oxygen interacts with CH_3S by forming a reversible addition product, $CH_3S + O_2 \rightleftharpoons CH_3SOO$ (Turnipseed *et al.*, 1993); 20–80% of CH_3S was estimated to occur as CH_3SOO under typical atmospheric conditions. The photolysis of CH_3SSCH_3 has been exploited as a source of CH_3S (Barnes *et al.*, 1994a) to study the products resulting from its oxidation in the absence of NO_x. Under near-atmospheric conditions the yield of SO_2 was 90% and that of MSA about 10%. The major carbon-containing product was HCHO, in addition to small amounts of CH_3OOH and CH_3OH. This suggests that CH_3SOO reacts further with O_2 to form CH_3OO and possibly SO_2 directly. In the presence of NO, CH_3SOO would produce CH_3SO and NO_2. Thus, in the presence of NO_x, CH_3S would be converted to CH_3SO. Since the presence of NO_x was observed to increase the yield of methanesulfonic acid, CH_3SO appears to be an intermediate to its formation. The pathway to MSA formation has remained speculative, however.

Under NO_x-lean conditions the CH_3SCH_2OO radical may also react with HO_2 to form a hydroperoxide or methyl thiolformate:

$$CH_3SCH_2OO + HO_2 \rightarrow CH_3SCH_2OOH + O_2$$

$$\rightarrow CH_3SCHO + H_2O + O_2$$

For the reaction of the analogous CH_3OCH_2OO radical with HO_2, Wallington *et al.* (1993) have observed the two channels to occur with approximately equal yields. Barnes *et al.* (1994b) discovered OCS as a product of the OH-induced oxidation of DMS, which they found to occur with a yield of $0.7 \pm 0.2\%$ The original suggestion was that methyl thiolformate would serve as a source of OCS via

$$CH_3SCH_2O + O_2 \rightarrow CH_3SCHO + HO_2$$

$$CH_3SCHO + OH \rightarrow CH_3 + H_2O + OCS$$

or

$$CH_3SCHO + h\nu \rightarrow H + CH_3 + OCS$$

The presence of NO_x, however, reduces the yield of OCS from DMS to zero, so that this pathway of OCS formation does not seem to hold. Barnes *et al.* (1996) therefore considered the CH_3S radical as a possible intermediate. Their experiments, based on CH_3SSCH_3 photolysis as a source of CH_3S in the presence of oxygen, indicated a OCS yield of 0.4% per radical.

10.2.3. TERRESTRIAL BIOSPHERE–ATMOSPHERE EXCHANGE

Not too long ago it was thought that microbial activity in soils was mainly responsible for the occurrence of sulfides in the continental atmosphere. Many studies in recent years have shown, however, that vegetation contributes significantly to the exchange of sulfides between the biosphere and the atmosphere. The corresponding change in outlook will permeate much of the discussion in this section.

Sulfur is a component of the essential amino acids methionine, cysteine, and cystine, which the living cell must either synthesize (plants, algae, microorganisms) or procure through the food supply (animals). Biological utilization causes sulfur to enter a complex chain of oxidation–reduction processes whereby the element is circulated through various reservoirs of organic and inorganic sulfur compounds. The biological sulfur cycle combines three principal pathways: (1) Organisms capable of synthesizing sulfur-containing amino acids do so by intracellular reduction of sulfate absorbed from the environment. Most of the reduced sulfur is fixed in amino acids and other organic compounds; a minor fraction may be released in the form of volatile sulfur compounds. In vascular plants in particular such emissions are thought to provide a regulatory step to balance the sulfur pool in the plant. (2) Dead organic matter is subject to microbial degradation, which liberates sulfur to the environment, mainly as H_2S but also in the form of other sulfides. It is during this stage that volatile sulfides may escape from the soil to the atmosphere. Sulfides are unstable in aerobic media, and they are reoxidized to sulfate by a variety of microorganisms, which exploit the oxidation process as a source of energy. Thiobacilli are the best known and presumably most important among the sulfur oxidizers. (3) In anaerobic media, sulfate and other sulfur oxides provide a source of oxygen for respiration, and they undergo dissimilatory reduction to H_2S. A small family of strict anaerobes (the desulfuricants) is able to reduce sulfate, whereas many other microorganisms use oxygen contained in sulfites or thiosulfate. Dissimilatory sulfate reduction is most important in marine ecosystems, including salt marshes, because of the high concentration of sulfate in seawater. In normal soils the fraction of sulfur present as sulfate is less than 10% of the total, as sulfur occurs mostly in organic compounds, either carbon-bound or in the form of sulfate esters. In anaerobic habitats, ferric iron (Fe^{3+}) is reduced to ferrous (Fe^{2+}) before H_2S appears, so that the latter is precipitated as FeS. This reaction is considered the major route to inorganic sulfur fixation in anaerobic domains. In aerobic strata, iron sulfide is unstable and undergoes reoxidation. Although the rate of H_2S production by dissimilatory sulfate reduction in shallow near-shore marine sediments and inundated marshlands will be appreciable, the escape of H_2S to the atmosphere is limited by bacterial reoxidation in the overlying aerated strata. Further details on the sulfur cycle, especially on the microorganisms involved, can be found in Alexander (1977). Anderson (1990) has reviewed aspects of the assimilation of sulfur in plants.

Information on the biochemical precursors of volatile sulfur compounds in soils has been obtained from artificial culture studies, reviewed by Kadota and Ishida (1972), and from incubation studies of natural and amended soils, reviewed by Bremner and Steele (1978). Table 10.5 summarizes our knowledge about the origin of sulfides, which have been observed to emanate as degradation products under aerobic as well as anaerobic conditions. Rennenberg (1991) has reviewed sources of sulfides emitted by plants. In plants, cysteine and methionine contain the major fraction of stored sulfur. These amino acids are involved in the metabolism leading to the emission of H_2S and CH_3SH, respectively, whereas S-methyl-methionine is thought to be an important precursor of DMS. Hydrogen sulfide is also produced directly by the assimilatory reduction of sulfate and, in addition by the reduction of SO_2 absorbed by plant tissue, in a light-dependent process.

Table 10.6 presents results from field studies attempting to quantify the flux of H_2S and other sulfides from coastal wetlands. The dynamic enclosure method was used in most cases. Emission rates vary considerably because of spatial, diurnal, and tidal factors. An unexpected discovery of the early studies was the unusual diurnal behavior of H_2S in inundated source regions. The highest emission rates occurred at night and in the early morning hours, whereas during the day the rate decreased by several orders of magnitude (Hansen et al., 1978; Jaeschke et al., 1980). It appears that under nocturnal stagnant conditions the oxygen concentration in the water falls to low values, so that H_2S can escape to the atmosphere, whereas during the day, when oxygen is photosynthesized by algae, H_2S undergoes bacterial oxidation. The H_2S mixing ratio in the air above high-production regions shows a similar diurnal variation (Goldberg et al., 1981), because meteorological conditions favor the accumulation of H_2S underneath nocturnal

TABLE 10.5 Biochemical Origin of Volatile Sulfides Produced in Soils by the Microbial Degradation of Organic Matter under Aerobic and Anaerobic Conditions[a]

Volatile	Biochemical precursors
H_2S[b]	Proteins, polypeptides, cystine, cysteine, gluthadione
CH_3SH	Methionine, methionine sulfoxide, methionine sulfone, S-methyl cysteine
CH_3SCH_3	Methionine, methionine sulfoxide, methionine sulfone, S-methyl cysteine homocysteine
CH_3SSCH_3	Methionine, methionine sulfoxide, methionine sulfone, S-methyl cysteine
CS_2	Cysteine, cystine, homocysteine, lanthionine, djencolic acid
COS	Lanthionine, djencolic acid

[a] From Bremner and Steele (1978).
[b] H_2S can also be formed by anaerobic dissimilatory sulfate reduction, as indicated in the text.

TABLE 10.6 Average Biogenic Emissions of Sulfides from Salt Marshes and Tidal Flats (in mg m^{-2} year^{-1} as Sulfur)

Authors	Location	Sulfur emission rate						Remarks
		H_2S	DMS	DMDS	CH_3SH	CS_2	OCS	
Hansen et al. (1978)	Coastal lagoons	17,600	—	—	—	—	—	Release occurs mainly at night
	Denmark, two sites	442,000	—	—	—	—	—	
Jaeschke et al. (1978)	Tidal flats, Denmark	42	—	—	—	—	—	
Hill et al. (1978)	Salt marsh, New York	550	150	18	64	—	—	
Aneja et al. (1979)	Salt marshes, N. Carolina	40/190	180/1,310	—	—	—	—	Spartina alterniflora Mud flat zone
	Two sites	120/410	<40	—	—	—	—	
Goldberg et al. (1981)	Salt marsh, Virginia	3,000–16,000	—	—	—	—	—	Strong build-up at night
Aneja et al. (1981)	Salt marsh, N. Carolina	<10	400	<50	<50	150	30	S. alterniflora zone,
		500	<10	<50	<50	<50	30	Mud flat zone
Adams et al. (1981)	Salt marshes, N. Carolina	67	538	0.5	0.3	35	12	(5/7/10)
	Delaware	96	480	0.5	—	70	13	
	Virginia	—	1,870	40	220	1,380	30	
Steudler and Peterson (1985)	Saltmarsh, Massachusetts	2,000	2,870	420	—	160	300	Annual cycle observed
Cooper et al. (1987a)	Saltmarsh, Florida	25	448	13	—	31	—	Over s. alterniflora
		756	55	10	—	19	—	Adjacent sand plot
		21	—	—	—	—	—	Tidal mud flat
Cooper et al. (1987b)	Coastal wetlands, Florida	73–1,332	5–202	0.9–13	—	1.7–15	—	Distichlis spicata
		0.9–25	0.9–56	<0.3	—	<0.5	—	Juncus roemerianus
		5–8.7	8.7–61	0.9–4.4	—	0.9–2.6	—	Batis maritima
Lamb et al. (1987)	Salt marsh, N. Carolina	87	82	0.45	—	2.1	16	Spartina alterniflora

atmospheric inversion layers, followed by dilution during the day due to enhanced vertical mixing.

Sulfur emissions from tidal mud flats and salt marshes include DMS, OCS, CS_2, and some CH_3SH and CH_3SSCH_3 (DMDS), in addition to H_2S. The flux of DMS is comparable to that of H_2S; the other sulfides are less significant. The release of DMS appears to be related to marsh grasses and algae, in contrast to H_2S, which derives mainly from anaerobic strata in the soil. The two components show different diurnal and seasonal patterns. Steudler and Peterson (1985), who have studied a complete annual emission cycle at a New England site covered with *Spartina alterniflora*, found a total average sulfur emission rate of 0.18 mol m^{-2} $year^{-1}$. This rate is significantly higher than that observed by several other investigators at similar sites. Cooper *et al.* (1987a, 1989), for example, made measurements in April, May, October, and January at a *Spartina alterniflora* vegetated site in Florida. They found 0.026 mol m^{-2} $year^{-1}$, on average, and even lower fluxes at other Florida sites with different vegetation. Extrapolation to the total area of salt marshes in the world (3.8×10^{11} m^2) suggests a global sulfur emission rate from such areas of about 10 Gmol $year^{-1}$, or 0.3 Tg $year^{-1}$. Because of the great variability of emission rates, the range of uncertainty is very large.

Table 10.7 contains selected results for rates of sulfur emissions from inland soils. The most comprehensive study was that of Adams *et al.* (1981a, b), who determined fluxes of H_2S, DMS, OCS, CS_2, CH_3SH, and DMDS from soils at 35 sites in the eastern and southeastern United States. Their data exhibited a large variability in the type of sulfur species, as well as in the individual fluxes. Hydrogen sulfide, which was the dominant component at most sites, contributed about 60%, CS_2 and OCS about 15% each, and the rest was DMS and traces of DMDS; only salt marshes emitted substantial amounts of CH_3SH. Temperature is an important parameter controlling sulfur emissions, as Delmas *et al.* (1980) had shown earlier for H_2S. Adams *et al.* (1981a) have studied the temperature dependence in some detail and found the data to correlate with ambient temperature as well as with soil temperature. The former was then used as a parameter, and linear regression analysis was employed to group the data according to soil order. The results were used to estimate the total annual sulfur emission rate in the study area and on a global scale. The global estimate obtained, 64 Tg $year^{-1}$, was based on unrealistic assumptions and, as discussed previously (Warneck, 1988), is much too high. Goldan *et al.* (1987) and Lamb *et al.* (1987) have subsequently revisited several of the sites explored by Adams *et al.* (1981a, b) and observed lower emission rates, especially of H_2S. Their studies also compared bare soils with those covered by vegetation. Plants were found to modify the mixture of sulfides emitted as well as the total emission rate. To make the comparison possible, the temperature dependence of the emissions

TABLE 10.7 Average Biogenic Emissions of Sulfides from Inland Soils (in mg m^{-2} year^{-1} as sulfur)

Authors	Location and soil type	Sulfur emission rate					Remarks
		H$_2$S	DMS	CS$_2$	OCS	Total	
Delmas et al. (1980)	France, Toulouse	44	—	—	—	—	Grass cover, 3–13°C
		58	—	—	—	—	15–20°C
		230	—	—	—	—	20–25°C
	Loire valley	18	—	—	—	—	3–9°C
	Landes	38	—	—	—	—	9–12°C
	Pine forest	22	—	—	—	—	9–12°C
	Ivory Coast, littoral forest	300	—	—	—	—	
Adams et al. (1981)	Ames, Iowa, mollisol	147	3	16	17	180	Sampling between crop rows
	Celleryville, Ohio, histosol	47	3	6	12	68	Bare soil
	Belle Valley, Ohio, inceptisol	72	4	10	4	94	Pasture grass
Staubes et al. (1986)	Frankfurt/Main, Germany, acrisol	—	1.86	2.6	2.4	6.9	May–October, 9–22.5°C
	Histosol	—	16.9	12.3	1.14	30.4	May, 7–26°C
Goldan et al. (1987)	Ames, Iowa, mollisol	0.32	0.47	0.32	1.47	2.6 ± 0.8	Bare soil, corrected to 25°C
		1.58	3.2	0.32	1.0	6.2 ± 2.3	Oat grass, at 25°C
		1.2	1.84	0.21	0.89	4.2 ± 1.4	Gramma grass, at 25°C
		0.53	2.79	0.26	0.37	3.9 ± 2.0	Soy plants, 25°C
	Celleryville, Ohio, histosol	2.4	0.32	0.74	3.6	7.1 ± 1.7	Bare soil, corrected to 22°C
		3.5	3.9	0.63	0.63	8.7 ± 1.8	Purple clover, 22°C
		3.0	1.4	0.74	1.2	6.4 ± 1.0	Orchard grass, 22°C
		2.3	2.9	0.42	0.58	6.2 ± 1.8	Quack grass, 22°C
Lamb et al. (1987)	Ames, Iowa, mollisol	0.2	0.5	0.2	2.9	1.6	Bare soil, 25°C
	Celleryville, Ohio, histosol	9.1	0.4	0.04	14.7	26.4	Bare soil, 22°C
	Ames, Iowa, corn	10.9	62.9	—	—	30.9	Soil emissions subtracted, 28°C
Andreae and Andreae (1988)	Manaus, Brazil, tropical soil above the canopy	1.2	6.7	—	—	8.4 ± 1.4	Dry season only, CH$_3$SH included in total flux
		< 26	13.7	—	—	21	
Andreae et al. (1990)	Manaus, Brazil, tropical forest	18.4	3.4	—	—	22	Wet season, above canopy

from the bare soils was determined separately for each sulfide, and the fluxes were corrected to the temperature observed in the experiments with vegetated soils. As shown by the data in Table 10.7, the presence of vegetation greatly reduced the flux of OCS, whereas it enhanced that of DMS. This demonstrates that plants can absorb OCS and emit DMS. Andreae and Andreae (1988) and Andreae et al. (1990) studied soil/plant sulfur emissions in the Amazon forest by a gradient flux technique. They found that the fluxes of H_2S and DMS above the canopy were larger than those emitted from the soil underneath. In the case of DMS the vertical profiles suggested that emission from the forest took place only in daytime, and that the canopy acted as a sink during the night.

The recognition that soils and vegetation must be considered together and that plants may be more important emitters of sulfides than soils has changed the direction of research. A laboratory study of Fall et al. (1988) on corn, alfalfa, and wheat showed DMS to be the main sulfur species emitted, followed by variable amounts of H_2S, CH_3SH, and CS_2. The emission rates rose logarithmically with increasing temperature, and with increasing illumination. This behavior has been confirmed in more recent studies on wheat and various mid-latitude trees (spruce, beech, ash, oak, Kesselmeier et al., 1997c) and on tropical trees (Kesselmeier et al., 1993). Specifically, the emission rate for DMS by an oak species (Quercus petrea) was shown to correlate closely with that of CO_2 assimilation and thus with stomatal conductance. The uptake of OCS is also coupled to CO_2 assimilation. OCS is metabolized to H_2S by carbonic anhydrase, whose affinity for OCS is a 1000 times higher than that for CO_2 (Protoschill-Krebs et al., 1992, 1996) Although many field measurements indicated an uptake of OCS (Goldan et al., 1987; Fall et al., 1988; Bartell et al., 1993), plants have also been observed to release it (Lamb et al., 1987) or showed no evidence of uptake (Berresheim and Vulcan, 1992). A study of the exchange of OCS with a wheat field and with corn and rape seed leaves has revealed the existence of a compensation point in each case (Schröder, 1993; Kesselmeier and Merk, 1993). Uptake occurs when the aerial mixing ratio is higher than the compensation point, whereas emission occurs when it is lower. The value of the compensation point depends on numerous factors such as light intensity, temperature, and plant development stage. For young plants of corn and rape, Kesselmeier and Merk (1993) found compensation points of about 144 pmol mol^{-1} and 90 pmol mol^{-1}, respectively; for wheat, Schröder (1993) found 160 pmol mol^{-1}; for soybeans, the data of Goldan et al. (1988) suggest a compensation point of about 200 pmol mol^{-1}. These values are significantly lower than the mixing ratio of OCS in the atmosphere, indicating that the plants should normally assimilate OCS.

The problem of how these data can be globally extrapolated has not yet been satisfactorily solved. Bates *et al.* (1992a) have used the emission data of Goldan *et al.* (1987) and Lamb *et al.* (1987) obtained with the enclosure technique, applied a suitable temperature algorithm, and combined them with land cover data compiled by Henderson-Sellers (1986) to derive a global inventory that resulted in 11 Gmol year^{-1} of total sulfur (0.35 Tg year^{-1}). The contribution of DMS was about 50%, that of H_2S 37%, and that of CS_2 7%, with more than 60% of the emissions occurring in the Tropics. This estimate, which includes a fraction of about 15% of sulfur released from soils, appears to be conservative. The field data of Andreae *et al.* (1990) and Bingemer *et al.* (1992), taken in the tropical forests of the Amazon and the Congo, indicated emission rates for DMS and H_2S in the ranges of 0.3–1.0 and 0.9–1.5 μmol m^{-2} days^{-1}, respectively, which if extrapolated to the full area of tropical rainforests ($\sim 15 \times 10^{12}$ m^2), would indicate 6.3–13.7 Gmol year^{-1} of total sulfur to be emitted from this source alone. Because sulfur emissions from plants correlate well with CO_2 assimilation, net primary productivity may provide a suitable scaling parameter for extrapolation. Tropical rainforests contribute 17–30% to the total net primary productivity of the world (see Table 11.4), so that the global sulfur emission rate from vegetation would be 20–81 Gmol year^{-1} (0.6–2.5 Tg year^{-1}). DMS is estimated to contribute about 30%, and H_2S about 70%, to the emission. This assessment may indicate the range of uncertainty in the global emission rate of sulfur from vegetation.

The estimates for the global uptake rate of OCS by plants discussed in Section 10.2.1 (Goldan *et al.*, 1988; Chin and Davis, 1993) were also based on net primary productivity as a scaling parameter. The uptake of OCS, however, is followed by metabolization to H_2S. Accordingly, the full rate of CO_2 assimilation should be applied rather than net primary productivity (assimilation minus respiration). The potential for destruction of OCS by vegetation therefore appears to have been underestimated.

10.2.4. SUMMARY OF NATURAL SOURCES OF SULFUR

Table 10.8 summarizes our current knowledge on the strength of natural sources of sulfides and sulfur dioxide. The comparison shows that DMS from the oceans is the dominant source, and that terrestrial sources, that is, emissions from soils and vegetation, cannot compete with it, even in the face of uncertainties in the global emission rates. The global budget of H_2S clearly is less satisfactory than that of the other sulfides. Emissions of SO_2 from

TABLE 10.8 Summary of Natural Sources of Sulfur in the Atmosphere
(Gmol year $^{-1}$ of Sulfur)

Source	H_2S^a	DMS^a	$CS_2{}^b$	OCS^c	$SO_2{}^a$
Oceans	< 9	500–1300	2.4–9.5	2.8–7.8	—
Coastal wetlands	0.2–30	0.2–18	0.2–1.2	2.3–8.7	—
Soils and plants	2–56	3–24	0.4	—	—
Volcanoes	16–47	—	0.2–2.4	0.1–1.5	230–300
Biomass burning	—	—	—	0.7–4.3	81
Other	—	—	—	4.5–14.8^d	—
Sums	18–133	503–1342	3.3–14.1	10.4–37.1	311–381

^{a} Data from Andreae and Jaeschke (1992), Bates *et al.* (1992a), and text.
^{b} Chin and Davis (1993).
^{c} From Table 10.2.
^{d} Reaction of OH with CS_2 and DMS.

natural sources show that volcanoes are dominant and that they emit sulfur in amounts comparable to DMS from the oceans. Emissions of sulfur dioxide from volcanoes will be discussed further below.

10.3. SULFUR DIOXIDE AND PARTICULATE SULFATE

Sulfur dioxide (SO_2) enters the atmosphere by direct emissions from volcanic and anthropogenic sources, as well as by the oxidation of organic sulfides, as described in Section 10.2.2. Owing to a rapid further oxidation to sulfuric acid, SO_2 also accounts for a good deal of sulfate associated with the atmospheric aerosol. The two components, therefore, are appropriately discussed together. We consider, in turn, the magnitude of human-made and natural emissions, the rate at which SO_2 is converted to sulfate, the distribution of SO_2 and SO_4^{2-} in the troposphere, and the removal of both components by wet and dry deposition. The observational data are then used to construct, in Section 10.4, regional and global sulfur budgets.

10.3.1. ANTHROPOGENIC EMISSIONS

Table 10.9 shows a breakdown of processes causing the release of SO_2 from anthropogenic sources. According to these data, the combustion of coal contributes about 60% to all such emissions, that of petroleum and its products 28%, and the smelting of nonferrous ores as well as miscellaneous

TABLE 10.9   The Most Important Sources of Human-made Sulfur Emissions
in 1976–1980

Source type	Production and use $(Tg\ year^{-1})^a$	Sulfur emission factor $(g\ kg^{-1})^a$	Emission rate $(Tg\ year^{-1})$ [a]	[b]	Percentage of total[a]
Hard coal combustion	1830	24.1	44.1	44.6	
Lignite combustion	930	17.8	16.6	10.2	60.0
Coal coked	473	2.7	1.3	0.9	
Petroleum refining	3672.8	1.0	3.7		
Motor spirit	854	0.36	0.3		
Light fuel oils	854.3	2.23	1.9	25.9	28.1
Residual fuel oils	1273	18	22.9		
Petroleum coked	45.2	6.75	0.3		
Copper smelted	7.9	1000	7.9	5.4	
Copper refined	8.7	175	1.5		10.4
Lead smelted	3.4	235	0.8	0.3	
Zinc smelted	5.1	100	0.5	1.1	
Sulfuric acid	105.7	12	1.3	—	
Pulp and paper	240.1	1	0.3	—	1.5
Total emissions			103.4	88.4	

[a] From Cullis and Hirschler (1980), for 1976.
[b] According to Spiro et al. (1992), for 1980.

industrial processes take up the remainder. The individual emission estimates are derived as usual by combining statistical production data with
emission factors. Fossil fuels contain sulfur mainly in the form of organic
sulfur compounds. Combustion converts them to SO_2, which is vented with
the flue gases. The ashes retain very little sulfur, so that emission factors for
coals correspond closely to their sulfur contents. Crude oil is processed
before use, and a certain fraction of sulfur escapes from the refineries.
Distillation leaves most of the sulfur in the residual oil. Gasoline contains
very little sulfur, and diesel and light fuel oils contain about 0.2% by mass.
The major source of SO_2 from petroleum is the combustion of heavy residual
fuel oils.

   Nonferrous ores occur mainly in the form of pyrites. The large emission
factors associated with nonferrous metal production derives from the fact
that sulfur contained in ores escapes mostly as SO_2, in spite of control
measures. The most significant contribution to SO_2 emissions from industrial
processes lies in the manufacture of sulfuric acid. The conversion of pulp to
paper leads to emissions of H_2S and other organic sulfides, but the magni-

tude is comparatively small. The combustion of natural gas, which is another important source of energy, causes negligible sulfur emissions, so that it is not even listed in Table 10.9. This fuel has a low sulfur content to begin with, and almost all of it is removed before use.

Anthropogenic sulfur emissions have increased steadily in the past because of the rise in fossil fuel consumption. Figure 10.1 indicates the historic development. Assessments of global-scale emissions of SO_2 were begun by Robinson and Robbins (1970b), who derived 73 Tg year^{-1} (as sulfur) for 1970. The value of about 103 Tg year^{-1} given by Cullis and Hirschler (1980) for 1976, which is shown in Table 10.9, appears to have been an overestimate. Möller (1984) reported a global sulfur emission rate of 69 Tg year^{-1}, and Varhelyi (1985) estimated 79 Tg year^{-1}, partly because they used lower emission factors. Dignon and Hameed (1989) obtained 63 Tg year^{-1} from the combustion of fossil fuels alone. Spiro et al. (1992), who used a lower emission factor for lignite and accounted for sulfur recovery in the smelting of iron ores, found 88 Tg year^{-1}. Table 10.9 includes their data for comparison. These values are reasonably consistent. By using national inventories rather than United Nations statistics, Spiro et al. (1992) refined the

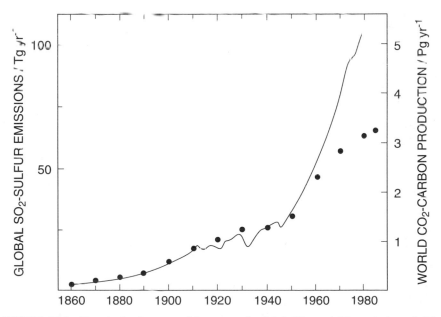

FIGURE 10.1   Historic development of human-made global CO_2 and SO_2 emissions. Solid curve, data for CO_2 (Bolin et al., 1981); circles, data for SO_2 (Dignon and Hameed, 1989); last point, estimate for 1985 (Benkovitz et al., 1996).

global estimate and derived 77.6 Tg year^{-1}, with about 26% from Europe, 16% from the former Soviet Union, 19% from North America, and 38% from the rest of the world. The need for global gridded inventories has led to new estimates, but the range of uncertainty is still appreciable. For the year 1985, Dignon (1992) found 64 Tg year^{-1}, Müller (1992) reported 92 Tg year^{-1}, and Benkovitz et al. (1996) derived 65 Tg year^{-1}. Further uncertainties arise from efforts to curb SO_2 emissions by energy conservation measures and by the installation of flue gas cleaning equipment. In West Germany, the latter type of control measures has decreased emissions by 70% since 1980 (Fricke and Beilke, 1993). This kind of emission reduction is not taken into account in the above estimates, which are based on fuel consumption statistics. Reductions have also occurred in other regions of Europe and in the United States and Japan. The collapse of the political and economic system in eastern Europe has led to a decrease of about 25% in this region during the period 1990–1993 (Tuovinen et al., 1994). On the other hand, the increased energy demand in China has led to an increase of SO_2 emissions in that region by about 10% since 1985, according to Kato and Akimoto (1992). In view of the emission reductions, the low values given above probably are more realistic than the higher ones. Accordingly, the global SO_2 sulfur emission rate probably ranges between 65 and 70 Tg year^{-1}. More than 90% of the emissions occur in the Northern Hemisphere. The great improvement of air quality in several regions of western Europe that followed the installation of flue gas cleaning devices gives rise to the hope that in the future all new power plants will be fitted with the necessary equipment, so that air pollution by SO_2 will be largely overcome.

In combustion chambers, a small fraction of SO_2 is further oxidized to SO_3, which reacts with water vapor and is emitted as sulfate. Ryaboshapko (1983) estimated the direct emission of particulate sulfate to reach 12 Tg sulfur year^{-1}. Robinson and Homolya (1983) showed, however, that the average ratio of SO_4^{2-} to SO_2 in power-plant flue gases is close to 0.03. The corresponding global release rate of sulfur as sulfate then is less than 3 Tg year^{-1}, a comparatively minor fraction of total sulfur emissions.

10.3.2. SULFUR EMISSIONS FROM VOLCANOES

Volcanic gases contain sulfur in the form of SO_2 and H_2S. Thermodynamic equilibrium calculations of Heald et al. (1963) indicate that in the anoxic environment of the magma SO_2 is dominant at high temperatures, whereas H_2S is preferred at low temperatures. Accordingly, most active volcanoes are expected to release sulfur mainly as SO_2. Hot volcanic gases react with oxygen. Thus, upon entering the atmosphere, both compounds are oxidized; H_2S is partly converted to SO_2, and SO_2 is converted to sulfuric acid.

A first estimate for the global rate of sulfur emissions from volcanoes was made by Kellog et al. (1972). Their assessment was based on the total volume of lava and ashes ejected during the time period from 1500 to 1914 according to Sapper (1927), the estimate by MacDonald (1955) of the weight of the gases that evolved from fresh lava on Hawaii, and the analysis of Shepherd (1938), who found 10% by mass of the gases in the form of SO_2. The global flux of volcanic SO_2 derived in this manner was 0.75 Tg sulfur year^{-1}. Actually, this is a lower limit, because it accounts only for the SO_2 released during eruptions, whereas the much longer quiescent periods of volcanic activity are not considered. The same deficiency was inherent in later attempts by Friend (1973) and Cadle (1975) to improve on the estimate of Kellog et al. (1972). Subsequently, more direct information has been obtained by means of aircraft samples of volcanic plumes, and by remote SO_2 sensing techniques. Table 10.10 presents a compilation of observational data prepared by Berresheim and Jaeschke (1983). To determine average emission rates from these data requires a correlation with the intensity of volcanic activity at the time of observation. Berresheim and Jaeschke (1983) distinguished five categories of volcanic activity: eruptive, preeruptive, intraeruptive, posteruptive, and extraeruptive phases. The first category was further subdivided into nine groups of differing intensities. The corresponding definitions are given in the footnote to Table 10.10. The authors then considered the worldwide frequency of eruptions in the period between 1961 and 1979, appropriately classified by the strength of individual eruptions. An average of 55 volcanoes undergo eruptions each year. This is about one-tenth of all active volcanoes. By summing over the entire amount of SO_2 released within each intensity class, Berresheim and Jaeschke derived an average emission rate of 0.5 Tg year^{-1} of sulfur within the time period considered. The value differs little from that of Kellog et al. (1972). A much greater total emission rate was obtained for the steadier, noneruptive activities. Among 520 live volcanoes are about 365 that may be classified as posteruptive, and an additional 100 in the extraeruptive phase. Volcanoes in the latter category may be neglected to a first approximation because they emit only a small fraction of sulfur released by others. The preeruptive and intraeruptive phases are relatively short and may also be neglected.

According to Table 10.10, the average SO_2 flux for a single volcano in the posteruptive activity class is roughly 3×10^5 kg SO_2 day^{-1}. The application of this value leads to a global sulfur emission rate of 20 Tg year^{-1}. Berresheim and Jaeschke (1983) adopted a more cautious attitude. On the basis of MacDonald 's (1972) description of 516 active volcanoes, they assigned a source strength of the magnitude suggested by the average of the

TABLE 10.10   Measured or Estimated Rates of SO$_2$ Emissions from Volcanoes[a]

Authors	Volcano	SO$_2$ Emission rate (10^3 kg day^{-1})	Category
Italy			
Haulet et al. (1977)	Etna	3,740	Eruptive (I)
Zettwoog and Haulet (1978)	Etna	3,325	Preeruptive
Zettwoog and Haulet (1978)	Etna	1,130	Posteruptive
Malinconico (1979)	Etna	1,000	Intraeruptive
Jaeschke et al. (1982)	Etna	142	Posteruptive
Jaeschke et al. (1982)	Etna	955/1,600	Preeruptive
Nicaragua			
Stoiber and Jepsen (1973)	Telica	20	Posteruptive
Stoiber and Jepsen (1973)	Momotombo	50	Posteruptive
Stoiber and Jepsen (1973)	San Cristobal	360	Posteruptive
Stoiber and Jepsen (1973)	Masaya	180	Preeruptive
Taylor and Stoiber (1973)	Cerro Negro	120[b]	Eruptive (I)
Taylor and Stoiber (1973)	Cerro Negro	2,000[b]	Eruptive (II–IV)
Guatemala			
Stoiber and Jepsen (1973)	Pacaya	260	Eruptive (I)
Stoiber and Bratton (1978)	Pacaya	300	Posteruptive
Stoiber and Jepsen (1973)	Santiaguito	420	Eruptive
Stoiber and Jepsen (1973)	Santiaguito	300/1,500	Eruptive (IV)
Rose et al. (1973)	Fuego	10,000[b]	Eruptive (II–IV)
Stoiber and Jepsen (1973)	Fuego	40	Posteruptive
Crafford (1975)	Fuego	55,000[c]	Eruptive (V)
Crafford (1975)	Fuego	423	Intraeruptive
Stoiber and Bratton (1978)	Fuego	300/1,500	Eruptive (I)
Hawaii			
Naughton et al. (1975)	Kilauea	1,280[d]	Eruptive (I–V)
Naughton et al. (1975)	Kilauea	280	Undetermined
Stoiber and Malone (1975)	Mauna Ulu	30	Extraeruptive
Stoiber and Malone (1975)	Sulfur Banks	7	Extraeruptive

(*Continues*)

data in Table 10.10 to no more than 90 volcanoes. The remaining 275 volcanoes with posteruptive activity were taken to produce 5×10^4 kg SO$_2$ day^{-1} each. The global sulfur flux is then reduced to 7 Tg year^{-1}. This rate is still an order of magnitude greater than that associated with erupting volcanoes. Berresheim and Jaeschke also used their own measurements of H$_2$S in the plume of Mt. Etna to estimate the rate of H$_2$S emissions from all volcanoes with post- and extraeruptive activities as about 1 Tg sulfur year^{-1}. Cadle (1980) considered OCS and CS$_2$ as effluents from volcanoes and found their contributions to be negligible by comparison.

TABLE 10.10   (*Continued*)

Authors	Volcano	SO$_2$ Emission rate (10^3 kg day^{-1})	Category
Alaska			
Stith *et al.* (1978)	St. Augustine	86,400[c]	Eruptive (V)
Stith *et al.* (1978)	St. Augustine	432[c]	Posteruptive
Stith *et al.* (1978)	St. Augustine	8,640	Eruptive (II–IV)
Stith *et al.* (1978)	St. Augustine	86	Posteruptive
Stith *et al.* (1978)	Mt. Martin	3	Extraeruptive
Washington state			
Hobbs *et al.* (1981)	St. Helens	10/50	Preeruptive
Casadevall *et al.* (1981)	St. Helens	1,000/1,900	Preeruptive
Hobbs *et al.* (1981)	St. Helens	1,300/2,400	Eruptive (V)
Casadevall *et al.* (1981)	St. Helens	900	Intraeruptive
Casadevall *et al.* (1981)	St. Helens	600/3,600	Posteruptive
Japan and former Soviet Union			
Okita and Shimozuru (1975)	Mihara	345	Posteruptive
Okita and Shimozuru (1975)	Asama	142	Posteruptive
Okita and Shimozuru (1975)	Asama	780	Posteruptive
Crafford (1975)	Karimski	173	Undetermined

[a] Classification of activity: eruptive (intensity), intensity scale after Tsuya (1955) in terms of total volume of material ejected, I ($= 10^{-5}$ km^3) to IX ($= 10^2$ km^3); preeruptive, intensification of activity before an eruption; intraeruptive, phase of repose between two paroxysmal eruptions; posteruptive, permanent fuming or fumarolic activity; extraeruptive, exclusively fumarolic and solfataric activity (from Berresheim and Jaeschke, 1983).
[b] Calculated from the SO$_4^{2-}$ content of ash particles.
[c] Estimated value.
[d] Calculated from the SO$_2$ content of lava.

While total volcanic sulfur emissions during any specific year may reach or even exceed 20 Tg, the geochemical data that will be discussed in Chapter 12 also support a lower average release rate. The continuous degassing of the Earth's interior since the formation of the planet 4.5 billion years ago has virtually depleted the upper mantle of juvenile volatiles. As acidic gases, CO$_2$ and SO$_2$ are responsible for the chemical weathering of virgin rocks. Over geological time, most of the CO$_2$ released was transformed to carbonate and SO$_2$ was converted to sulfate. In this form, carbon and sulfur accumulated in the oceans and eventually came to rest in marine sediments. Cycles of sedimentary rocks will be discussed in Section 11.3.1 in conjunction with the geochemistry of carbon dioxide (see Fig. 11.3.) Current losses of carbon and sulfur from the upper mantle, caused by volcanic emissions to the atmosphere, will have to be replaced if these emissions are to continue. The replacement is accomplished by the subduction of sediments in those tectonic zones, where marine and continental crustal plates meet and, by their counteracting motion, force crustal material downward into regions of the upper mantle. Assuming that steady-state conditions are approximately established, modern volcanoes should emit volatile elements in roughly the same proportions as they occur in average sedimentary rock. Table 12.5 shows that the mass ratio of sulfur and carbon residing in the sediments is $1.2/6.2 = 0.19$, whereas Table 11.8 suggests that the mass flux of carbon associated with volcanic emissions or the subduction of sedimentary rocks into the upper mantle is 23 Tg $year^{-1}$. Taken together, both values suggest that the mass flux of sulfur subducted into the upper mantle and reemitted by volcanoes is 4.5 Tg $year^{-1}$. The rate agrees surprisingly well with the estimate derived from direct observations of volcanic sulfur emissions.

10.3.3. CHEMICAL CONVERSION OF SO_2 TO PARTICULATE SULFATE

The oxidation of SO_2 in the atmosphere may occur in the gas phase, in the aqueous phase of clouds and fogs, and on the surface of aerosol particles. Reaction mechanisms for the gas phase and the aqueous phase were discussed in Sections 7.4.3 and 8.6.4, respectively. Here, the discussion will focus on the overall oxidation rate and the associated lifetime of SO_2 with regard to its conversion to H_2SO_4. Data are available from both laboratory and field observations.

Table 10.11 summarizes processes that have been studied in the laboratory, including the inferred atmospheric conversion rates and time constants. The direct gas phase photooxidation of SO_2 due to its absorption of solar radiation is slow because the excitation energy is mainly lost in collisions with air molecules rather than utilized to convert SO_2 to SO_3. More important are oxidation reactions initiated by radicals. The best known and most effective is the reaction of SO_2 with OH. It leads first to the formation of an adduct, then to SO_3, which ultimately reacts with water vapor to produce H_2SO_4. Peroxy radicals do not react with SO_2 at significant rates. Even the Criegee intermediates produced in the reaction of ozone with alkenes are fairly ineffective as oxidants, because their reaction with water vapor quenches the reaction with SO_2.

In the presence of clouds and fogs, SO_2 dissolves to some extent in the liquid phase, where it forms HSO_3^- and SO_3^{2-}. The anions are oxidized by ozone and by all hydroperoxides at significant rates. Hydrogen peroxide is usually the most effective hydroperoxide present. The direct reaction of sulfite anions with oxygen is slow and can be neglected, but transition metals

TABLE 10.11 Summary of SO_2 Oxidation Processes Studied in the Laboratory, Application to Atmospheric Conditions, and Estimates of Rates and SO_2 Lifetimes

Process	Rate (% h⁻¹)	Time constant (days)	Remarks
Direct photooxidation[a]			
$SO_2 + h\nu \rightarrow SO_2^* \xrightarrow{O_2} SO_2 + O$	< 0.04	> 100	SO_2^* involves both singlet and triplet states. Quenching by N_2 makes this process unimportant in the atmosphere (Calvert et al., 1978)
Reactions involving radicals			
$SO_2 + OH \xrightarrow{O_2} HOSO_2 \rightarrow SO_3 + HO_2$	0.35	12	Based on $n(OH) = 1 \times 10^6$ molecule cm⁻³. Rate undergoes diurnal, seasonal, and latitudinal variation.
$SO_2 + HO_2 \rightarrow$ products	< 10⁻⁴	> 4 × 10⁴	Based on $n(HO_2) = 2.5 \times 10^8$ molecule cm⁻³, and $k \leq 1 \times 10^{-18}$ cm³ molecule⁻¹ s⁻¹ (Atkinson et al., 1989).
$SO_2 + RO_2 \rightarrow$ products	< 0.3	> 14	Based on $n(RO_2) = 7 \times 10^7$ molecule cm⁻³ assuming CH_3O_2 to be the principal species; $k \leq 1 \times 10^{-14}$ cm³ molecule⁻¹ s⁻¹ (Calvert and Stockwell, 1984).
$SO_2 + R\dot{C}HOO \rightarrow RCHO + SO_3$	1 × 10⁻⁵	4 × 10⁵	Estimated for the Criegee radical resulting from ethene; the discussion in Section 7.4.3 showed that the radical reacts mainly with water vapor. Calvert and Stockwell (1984) suggested a greater efficiency.
Aqueous reactions in clouds[b,c]			
$S(IV) + \frac{1}{2}O_2 \xrightarrow{Fe^{3+}} SO_4^{2-}$	(6)[c] 1.2	(0.7)[c] 1.5	$[Fe^{3+}] = 2\ \mu M$, $k = 1 \times 10^9 [Fe^{3+}]^2$ s⁻¹ (Martin and Hill, 1987; Martin et al., 1991).
$S(IV) + \frac{1}{2}O_2 \xrightarrow{Mn^{2+},Fe^{3+}} SO_4^{2-}$	(9)[c] 1.8	(0.7)[c] 3	$[Fe^{3+}] = 2\ \mu M$ (serves as initiator) $[Mn^{2+}] = 0.5\ \mu M$, $k = 1.6 \times 10^3 [Mn^{2+}]$ s⁻¹ (Berglund et al., 1993).
$S(IV) + O_3 \rightarrow SO_4^{2-} + O_2$	(0.3)[c] 0.06	(5)[c] 75	$p(O_3) = 4 \times 10^{-3}$ Pa (~ 40 nmol mol⁻¹), rate coefficients taken from Table A.6.
$S(IV) + H_2O_2 \rightarrow SO_4^{2-} + H_2O$	(21)[c] 4	(0.2)[c] 1	$[H_2O_2] = 15\ \mu M$ corresponding to $x(H_2O_2) \approx 0.1$ nmol mol⁻¹ before cloud condensation. For the rate coefficient, see Table A.6.
Oxidation of SO_2 on particles			
Fly-ash particles in power plant plumes	22	0.2	Initial oxidation rate 0.37 m³ g⁻¹ min⁻¹ (Dlugi and Jordan, 1982) The assumed particle concentration is 10 mg m⁻³.
Natural aerosol of the same composition	0.1	40	Prorated to a particle concentration of 50 μg m⁻³.

[a] See Tables 2.4 and 2.6.

[b] Rate estimates are based on clouds with a liquid water content of 0.1 g m⁻³, pH 4, and a fraction of SO_2 in the aqueous phase, as in Table 8.4. Mass transport limitations and rate changes due to acidification by product sulfuric acid are neglected.

[c] Values in parentheses refer to cloud-filled air; the other data are based on the assumption that clouds occur in only 20% of the entire tropospheric air space.

act as catalysts and may promote the oxidation under favorable conditions. Traditionally, iron and manganese are assumed to be most effective as catalysts because they are the most abundant transition metals. The role of iron is now quite well understood, and that of manganese is similar, in principle. The discussion in Section 8.6.4 showed, however, that at least during the day, because of the presence of HO_2 radicals in cloud water, Cu(II) is reduced to Cu(I), and the concentrations of the catalytic active species, Fe(III) and Mn(III), are lowered so much that this pathway of SO_2 oxidation becomes rather inefficient. The possibility remains that the process is more favorable at night. In daylight, OH radicals also are present, and they may trigger the chain oxidation of sulfite in aqueous solution, but even this pathway is inhibited by other reactions competing for OH radicals. This situation leaves the reactions of dissolved SO_2 with ozone and hydrogen peroxide as the dominant oxidation reactions in liquid water.

In calculating SO_2 residence times in the atmosphere due to losses in clouds, one must take into account the fact that clouds fill only a portion of the tropospheric air space. The values in Table 10.11 are based on a fraction of 20% in the lower troposphere. Reaction rates also depend on the pH of the solution. As SO_2 is converted to H_2SO_4, the pH is lowered and the gas–liquid partitioning of SO_2 changes in favor of the gas phase. This effect may be partly compensated for by neutralization with ammonia. As the pH decreases, the oxidation rate due to ozone declines, whereas that due to hydrogen peroxide is maintained. In daylight, H_2O_2 is produced in the gas phase and incorporated into cloud water, so that a continuous supply exists. Nevertheless, it will be clear that the conversion rates in Table 10.11 are coarse estimates.

The last entry in Table 10.11 gives the rate of SO_2 oxidation on the surface of fly-ash particles emitted from coal-fired electric power plants. The material consists largely of the oxides of silicon, aluminum, iron, and calcium. It may thus serve as a surrogate for the mineral fraction of the natural aerosol. Laboratory studies have shown that the rate of SO_2 oxidation on such particles increases with increasing relative humidity. As this parameter determines the amount of liquid water associated with the particles, it appears that the oxidation mechanism is mainly that of heavy metal ion catalysis in the liquid phase. Initially only a few percent of solid fly ash is water-soluble, but after some H_2SO_4 is produced it dissolves additional material. The capacity of the particles for H_2SO_4 formation is limited by the increasing acidity of the developing aqueous solution. The initial conversion rates listed in Table 10.11 indicate that the process is important in the plumes of power plants in the vicinity of the stacks, where particle concentrations are high. Concentrations of natural aerosol particles generally are too low to make the process significant. Dentener *et al.* (1996) pointed out that fresh mineral aerosol particles generated from deserts, owing to their alkaline nature, may attract SO_2 and moisture, which would allow ozone to oxidize SO_2 to sulfate until the rising acidity stops the process.

According to the data in Table 10.11, SO_2 is converted to H_2SO_4 at similar rates in the two phases, although all processes are not necessarily active at the same time. In the gas phase, the total rate is about 0.4% h^{-1}, whereas in the liquid phase it should be less than 7% h^{-1}. The laboratory data, therefore, suggest a lifetime of SO_2 with respect to conversion to H_2SO_4 on the order of a few days.

Field studies of SO_2 oxidation have been performed mostly in the downwind regions of sufficiently isolated sources. The favored objects were stacks of electric power plants, whose plumes are often identifiable at distances up to 300 km from the source, and larger cities. Sulfur dioxide and particulate sulfate were measured, and the data were analyzed in terms of the concentration ratio $[SO_4^{2-}]/([SO_2] + [SO_4^{2-}])$ as a function of time or a combination of wind speed and distance. The sampling was accomplished with the help of aircraft, although urban plumes have also been studied with ground-based measurements. Table 10.12 presents some results from such studies.

With regard to plumes from electric power generating stations, the evidence indicates that at least in summer the SO_2-to-SO_4^{2-} conversion rate undergoes a diurnal cycle, with maximum values at midday and minimum rates at night. Wilson (1981) has analyzed data from many observations in the plumes of nine different power plants, in addition to some metal smelters. Only data from plumes with ages greater than 1 h were included, to limit the influence of fly-ash catalyzed processes. Wilson found the daytime conversion rate to correlate well with the solar radiation dose. In a similar analysis of eight field studies at four coal-fired power stations, Meagher et al. (1983) have determined the seasonal variation of the conversion rate, which showed a maximum in summer and a minimum in winter. These results clearly established the importance of photochemical conversion mechanisms.

Most of the observations referred to dry atmospheric conditions. A few data obtained in the presence of moisture were treated by Gillani and Wilson (1983), who attempted to separate the effects of gas and liquid phase processes by applying an empirical parameterization of the dry conversion rate. The results confirm the idea that the two types of processes are about equally important. The majority of studies were conducted in the plumes of coal-fired power plants. Wilson (1981) included in his analysis one oil-fired unit and found similar results, whereas Garber et al. (1981) found lower conversion rates in the plumes of an oil-fired power plant in Northport, New York. The reasons for the difference are unknown.

TABLE 10.12 SO₂ to H₂SO₄ Conversion Rates Derived from Field Observations

Authors	Rate ($\%\ h^{-1}$) Range	Average	Time constant (days)	Remarks
Observations in power-plant plumes				
Cantrell and Whitby (1978)	0.41–4.9	1.5	2.9	Coal-fired power station, Labadie, Missouri, July and August, 1974
Husar et al. (1978)	0.1–4.8	1.5	2.7	1976; diurnal variation of conversion rate noted.
Hegg and Hobbs (1980)	0–5.7	1.1	3.6	Data from three coal-fired power stations, midwestern U.S., June 1978; review of previous data.
Gillani et al. (1981)	0.3–3.1	1.4	3.0	Coal-fired power station, Labadie, Missouri, July 1976.
	0.1–5.8	2.0	2.0	Coal-fired power station, Cumberland, Tenessee, August 1978.
Forrest et al. (1981)	0.2–7.0	2.2	1.9	Coal-fired power station, Cumberland, Tenessee, August 1978.
Wilson (1981), night	0.1–1.3	0.5	8.3	Analysis of 230 measurements in plumes of nine power plants
Wilson (1981), whole day	0–8.4	1.5	2.8	and two metal smelters; plume age > 1 h; strong dependence on solar radiation dose.
Meagher et al. (1983), winter	—	0.15	28	Averages from measurements in the plumes of five coal-fired
Meagher et al. (1983), summer	—	1.3	3.2	power plants; seasonal variation described.
Gillani and Wilson (1983), dry	0.2–2.1	0.97	4.3	Analysis of plume data from three coal-fired power plants under
Gillani and Wilson (1983), wet	0.3–8.0	1.77	2.3	dry and wet conditions.
Urban plumes and regional budget studies				
Eliassen and Saltbones (1975)	0.3–1.7	0.8	5	Comparison of trajectory transport calculations with observations at six stations in western Europe, based on an area emission inventory.
Breeding et al. (1976)	2.6–14[a]	8.1[a]	0.5[a]	Ground-based measurements in the plume of St. Louis, Missouri.
Mészarós et al. (1977)	5.3–32	16.4	0.3	Ground-based measurements in a suburban area of Budapest, Hungary;
	7.2–35	—	—	correlation analysis with wind speed, temperature, data grouping.
Alkezweeny (1978)	8–11.5	9.8	0.4	Urban plume of St. Louis, Missouri; aircraft observations during 3 days in August 1975.
Elshout et al. (1978)	0.6–4.4	1.7	2.4	Ground-based measurements in Arnshem, The Netherlands, of polluted air originating in the Ruhr valley; four selected days in winter 1972.
Forrest et al. (1979)	0–4	2	2	Urban plume of St. Louis, Missouri, followed by a manned balloon on a day in June 1976.
Miller and Akezweeny (1980)	1–9	4	1	Urban plume of Milwaukee, Wisconsin, over Lake Michigan, on 2 days in August 1976, 1 day in July 1977.
McMurry and Wilson (1983), dry	0–5	—	—	Urban plume of Columbus, Ohio; ground-based
McMurry and Wilson (1983), r. h. > 75%	≤ 12	—	—	measurements in July and August 1980.
Husain and Dutkiewicz (1992)	—	3	1.4	Selenium was used as a tracer of sulfur oxidation.

[a] Total removal rate includes dry deposition of SO₂ to the ground surface.

Husain and Dutkiewicz (1992) have used selenium as a tracer to determine the lifetime of SO_2 against gas phase oxidation by measuring sulfate particle concentrations. Selenium-containing particles have the same sinks as sulfate aerosols, but no atmospheric sources. Therefore, as sulfur dioxide is oxidized, the ratio of sulfate to selenium increases. Husain and Dutkiewicz conducted aircraft measurements over upstate New York and found SO_2 oxidation rates of 3% hour^{-1}, in agreement with other studies.

Measurements in urban plumes have in many cases given conversion rates considerably in excess of those deduced from studies of power-plant plumes. Because of the greater spatial extent, urban plumes come into contact with the ground surface more readily than power-plant plumes, so that losses of SO_2 due to dry deposition to the ground are unavoidable. This effect shows up most markedly in the data from ground-based observations, while airborne studies should be less influenced. One regional budget study is included in Table 10.12 to indicate that in polluted regions dry deposition can be more significant for the removal of SO_2 from the atmosphere than conversion to sulfate. Regional and global budgets of SO_2 will be discussed in Section 10.4.1.

Dry deposition cannot account for the great variation in the conversion rate revealed by some of the aircraft studies. Specifically, the measurements of Miller and Alkezweeny (1980) in the urban plume of Milwaukee, Wisconsin, indicated average rates of 9% and 0.7%, respectively, on two consecutive days characterized by nearly identical meteorological conditions. Hydrocarbon mixing ratios were measured along with several other pollutants. While the total nonmethane hydrocarbon reactivity toward OH radicals differed little between the two days, alkenes were considerably more abundant on the first day, when the SO_2-to-SO_4^{2-} conversion rate was high, compared to the second day, when it was low. Thus, the data may possibly confirm the importance of SO_2 oxidation by Criegee intermediates produced in the oxidation of alkenes. Miller and Alkezweeny (1980) suggested that the dramatic difference in the SO_2-to-SO_4^{2-} conversion rate was caused by the different quality of the upwind air masses. On the first day the air moved into the city from the southwest. High levels of ozone and sulfate were typical of well-aged polluted air (with Chicago as a possible point of origin). On the second day, in contrast, the wind direction was from the northwest, and ozone and sulfate concentrations were low. This leads to the conclusion that the SO_2 oxidation rate is enhanced by the presence of oxidants or oxidant precursors carried along with polluted air.

In general, and despite the uncertainties noted, the observational data of Table 10.12 confirm the predictions of the laboratory studies. The combination of all of the data suggests an SO_2 oxidation rate of 1–2% h^{-1} and a corresponding lifetime of SO_2 of 2–4 days in the lower troposphere.

10.3.4. DISTRIBUTION OF SO_2 AND SO_4^{2-}
IN THE TROPOSPHERE

The combustion of fossil fuels is still the dominant source of sulfur dioxide in the industrialized regions of the world. Before 1950, coal provided the major fuel for energy generation, steel production, and domestic applications. The comparatively high sulfur content of many coals gave rise to urban SO_2 levels often exceeding 100 nmol mol^{-1} (Meetham *et al.*, 1964). In England the former practice of heating homes by burning coal in open fireplaces further burdened the atmosphere with high concentrations of carbonaceous particles. The development of radiation fogs under stagnant and humid weather conditions (usually during the winter months), if persisting for several days, led to incidents of the infamous London-type smog, in which smoke, fog, SO_2, and H_2SO_4 combined to cause bronchial irritation, vomiting, and in some cases death due to heart failure. The worst episode on record occurred in London during 4 days in December 1952, when SO_2 and smoke concentrations rose to levels well above 1 mg m^{-3}. During this episode the mortality reached twice the normal rate (Wilkins, 1954). After 1950, coal for domestic heating was gradually replaced by light fuel oil and natural gas. Both are low-sulfur fuels. New electric power plants were constructed outside the cities. They were fitted with taller stacks, which discharged the effluents higher into the atmosphere, thus spreading SO_2 more over the countryside. Altshuller (1980) and Husar and Patterson (1980) have discussed this development for the United States. Between 1963 and 1973, urban SO_2 mixing ratios in the northeastern United States declined by a factor of 5 to about 10 nmol mol^{-1}, while nonurban values remained fairly constant around 4 nmol mol^{-1}. A similar though less extensive reduction, starting from lower levels, occurred in midwestern cities in spite of rising SO_2 emissions from new power stations in that region. In the British Isles, the average urban SO_2 concentration decreased by a factor of 2 from 1960 to 1975 (Department of the Environment, 1983). The average SO_2 level in London at that time was 23 nmol mol^{-1}. In the city of Frankfurt (Main), Germany, the annual average declined by a similar factor to about 35 nmol mol^{-1} during the same period (Georgii, 1982).

In the last decade, following the implementation of further abatement strategies, the situation has substantially improved both in North America and in central Europe. Figure 10.2 shows, as an example, the development in West Germany since 1980. Selected for comparison are annual emission rates in the whole region, annual averages of SO_2 concentrations in a major city (Frankfurt, Main), and annual averages of rural SO_2 and particulate sulfate concentrations at a measurement station (Deuselbach) representative of the

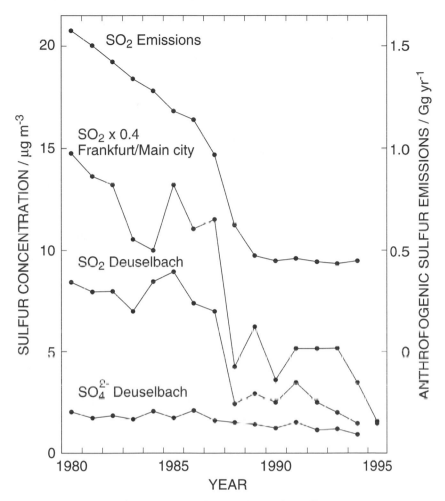

FIGURE 10.2   Decline of concentrations of SO_2 and particulate sulfate in Germany since 1980; total anthropogenic emissions and annual averages of concentrations at two locations: in the city of Frankfurt (Main), and at the station of the German Environmental Service in Deuselbach, a rural community about 150 km west of Frankfurt.

whole region. The major emission sources are electric power stations and industrial heating plants. During the first 5 years, the emission rate declined by about 20%, but this had relatively little influence on the concentrations of SO_2 and sulfate, in Frankfurt or in Deuselbach. In 1987–1988, following stricter regulations, most power plants were equipped with flue gas cleaning devices, which reduced total SO_2 emissions by almost 60%. At this time, the concentrations of SO_2 responded quickly and dropped by a similar percentage, in the cities as well as in rural areas. The concentration of particulate sulfate followed more slowly, because the region was still influenced by emissions that continued in the neighboring former Eastern Bloc countries. Other countries in western Europe have been similarly successful in the reduction of SO_2 emissions, and the SO_2 concentrations dropped accordingly. As discussed in Section 8.5, the reduction in SO_2 emissions has led to a significant decrease in precipitation acidity. In the United States, tighter controls of SO_2 emissions mandated by the 1990 amendments to the Clean Air Act have also led to significant reductions in the concentrations of SO_2 and sulfate. Holland *et al.* (1999), who analyzed trend data from 34 rural stations in the eastern United States, showed that from 1989 to 1995 the level of SO_2 declined by about 35% across all sites, and that of sulfate 26%. The most significant emission reductions occurred along the Ohio River, where numerous electric power plants are located (Benkovitz, 1982), and this led to a corresponding reduction in the adjacent midwestern region.

The distribution of SO_2 over the European continent was first estimated by de Bary and Junge (1963), who used data from a network of monitoring stations set up in the 1950s at nonurban sites. A more recent study of Ottar (1978) combined measurements at 70 ground stations, aircraft observations, an emission inventory, and advection models to derive the distribution of SO_2 and particulate sulfate occurring in the early 1970s. Figure 10.3 shows the results. The distribution is similar to that derived earlier by de Bary and Junge (1963). The highest SO_2 levels appeared in and around industrial centers. In Scandinavia and toward the Atlantic Ocean, the mixing ratios decrease to low values.

In North America, the rural background of SO_2 initially received less attention than urban SO_2. Altshuller (1976) listed only six nonurban measurement sites, all located in the northeast, as opposed to 48 urban stations. The annual average SO_2 level at the six rural sites during the period 1968–1972 was about 3.5 nmol mol^{-1}. In the late 1970s a concerted effort was made in the Sulfate Regional Experiment to determine SO_2 and particulate sulfate in a larger region east of the Mississippi River. The program involved 54 ground stations and supplemental aircraft observations (Mueller *et al.*, 1980). Additional results were incorporated from six Canadian stations operated during 1978–1979 (Whelpdale and Barrie, 1982). Figure 10.4 shows the distribution of SO_2 and particulate sulfate in northeast America as derived from these measurements. The situation resembled that in Europe, with SO_2 levels peaking in heavily industrialized regions. High concentrations also occurred along the Ohio River because of the concentration of electric power plants in that region.

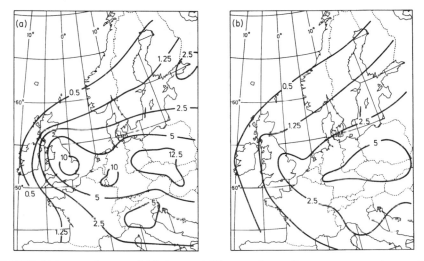

FIGURE 10.3 Distribution of (a) SO_2 and (b) particulate sulfate over Europe during the 1970s; sulfur concentrations units are $\mu g \ m^{-3}$ (STP). (Adapted from Ottar (1978).)

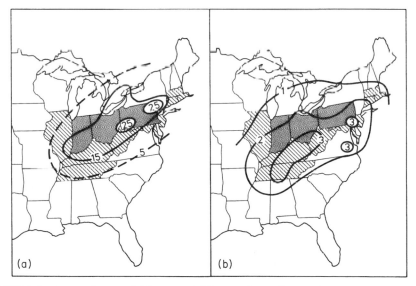

FIGURE 10.4 Distribution of (a) SO_2 and (b) particulate sulfate in the northeastern United States during the 1970s; sulfur concentration units are $\mu g \ m^{-3}$ (STP). (Adapted from Mueller et al. (1980).) Sulfur emission densities in various states are indicated by heavy stipples, 20–30 g $m^{-2} \ year^{-1}$; light stipples, 10–20 g $m^{-2} \ year^{-1}$; no stipples, less than 10 g $m^{-2} \ year^{-1}$ (from Husar and Patterson, 1980).

There can be no doubt that the change in absolute concentrations of SO_2 and particulate sulfate during the last decade have also changed the distribution patterns. At least in Europe the situation is currently in a state of flux. Remaining centers of high emissions are expected to eventually be removed when the old equipment is replaced by new technology. Thus, although the distributions shown in Figures. 10.3 and 10.4 are no longer valid, it will take some time before the situation stabilizes and a new steady state can be presented. The data in these figures can nevertheless be used to derive an important conclusion. The distribution of particulate sulfate on both continents is similar to, though somewhat broader than, that of sulfur dioxide. This may indicate either that emissions of SO_4^{2-} and SO_2 occur in the same regions, that is, from the same sources, or that the oxidation of SO_2 to SO_4^{2-} is rapid compared to long-distance transport. The discussion of anthropogenic emissions has shown, however, that direct emissions of particulate sulfate are minor compared to those of gaseous sulfur dioxide. This leads to the conclusion that SO_2 is the major precursor of SO_4^{2-} and that conversion is fairly rapid. Further support is obtained from the mass–size distribution of sulfate associated with continental aerosol particles (see Fig. 7.22 or 7.24). As outlined in Section 7.4.3, gas-to-particle conversion channels material primarily into the submicrometer size range, whereas wind-generated particles such as sea salt or mineral dust show up mainly in the size range above 1 μm radius. The continental aerosol carries its burden of sulfate mostly in the submicrometer size fraction. Exceptions are found in coastal areas, where the influence of sea salt is dominant, and in regions where mineral deposits of sulfate are exposed to aeolian weathering.

At the rural monitoring stations in Europe and North America, the average SO_2 mixing ratio generally follows an annual cycle, with high values in winter and lower ones in summer. The trend showed up already in the early data treated by de Bary and Junge (1963). Figure 9.9 presented a record from a rural station in Germany. A similar behavior was observed at other rural stations in Europe and in America. Figure 10.5 shows the seasonal variation of both SO_2 and SO_4^{2-} at three stations in the Ohio River valley as reported by Shaw and Paur (1983). The sum of both components exhibits a winter maximum such as one would expect from the increased consumption of fossil fuels during the cold season of the year combined with the depth of the continental boundary layer, which tends to increase concentrations. The concentration of sulfate, however, varies in a manner opposite to that of SO_2, going through a minimum in winter and a maximum in summer. The existence of such a seasonal cycle had been noted earlier by Hidy et al. (1978) and by Husar and Patterson (1980) from observations at several rural as well as urban sites. The summer maximum of SO_4^{2-} (and the simultaneous minimum of SO_2) clearly indicates an increase in the rate of SO_2 oxidation

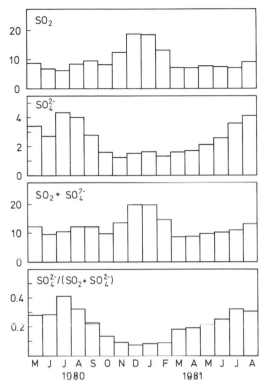

FIGURE 10.5 Annual variation of monthly means for gaseous SO_2, particulate SO_4^{2-}, their sums, and their ratios, in the Ohio River valley; units: $\mu g \ m^{-3}$ (STP) of sulfur. The plots show averages from three stations located in the western, central, and eastern valley regions. (Data from Shaw and Paur (1983).

during the warmer and brighter period of the year and may be ascribed to enhanced photochemical activity, especially the rise in the concentration of OH radicals during the summer months. The data in Figure 10.5 thus provide further support for the results of Wilson (1981) and Meagher et al. (1983), which had demonstrated diurnal and seasonal cycles for the rate of SO_2-to-SO_4^{2-} conversion in power plant stack plumes.

The continental background of SO_2 in regions not directly influenced by anthropogenic emissions has remained largely unexplored. Further below it will be shown that the mixing ratio of SO_2 in the middle and upper troposphere ranges from about 20 to 100 pmol mol^{-1}, and this range may provide a guide to the SO_2 mixing ratio expected in clean continental air. In the absence of direct sources of SO_2 in the continental boundary layer, sulfur dioxide may arise from the oxidation of reduced sulfur compounds emitted locally from soils and plants, as well as by downward mixing of SO_2 from the upper troposphere.

The vertical distribution of SO_2 in the troposphere has been explored by means of aircraft ascents. Figure 10.6 shows results from several individual flights. The SO_2 mixing ratio decreases with altitude in the first few thousand meters above the ground and then reaches an almost constant level in the upper troposphere. Individual altitude profiles are strongly influenced by the concentrations at ground level and their seasonal variation, the vertical stability of the atmosphere, the presence of inversion layers, and other meteorological factors. The selection of data in Figure 10.6 suggests an average scale height of 1250 ± 500 m in the continental boundary layer, where SO_2 is lost by reaction with OH radicals and by oxidation processes in clouds. A simple one-dimensional eddy diffusion model incorporating a constant sink term, but no volume sources, yields the observed scale height, provided the life of SO_2 due to oxidation is about 4 days. In the upper troposphere the concentration of OH is smaller than that in the boundary layer, and the scavenging effect of clouds is also less important. Accordingly, one expects the lifetime of SO_2 to increase considerably, and this may be the reason for the almost constant SO_2 background. Over the Atlantic Ocean west of Europe, the SO_2 mixing ratio is essentially independent of altitude, as Figure 10.6 (middle) shows. The values are similar to those in the continental upper troposphere, suggesting the presence of a fairly uniform background of SO_2 in the entire troposphere. While this conclusion has

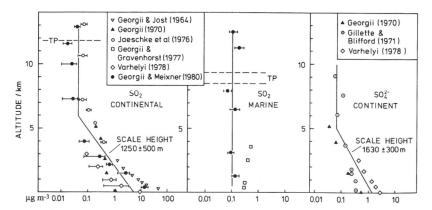

FIGURE 10.6   Vertical distribution of SO_2 in the continental atmosphere (left), and over the Atlantic Ocean (center), and the vertical distribution of sulfate over the continents (right). Sulfur concentrations shown on the abscissa are reduced to standard temperature and pressure (STP). Dashed horizontal lines indicate the tropopause level.

received support from numerous aircraft-assisted measurements (Table 10.13), the discussion below will show that the mixing ratio of SO_2 in the marine boundary layer is somewhat smaller than that in the middle and upper troposphere.

The right-hand side of Figure 10.6 includes the vertical distribution of particulate sulfate observed over the continents. Similar to the behavior of SO_2, the data indicate a decrease in concentration with increasing altitude toward a nearly constant level of about 70 ng m^{-3} (STP) of sulfur, or about 50 pmol mol^{-1}. The SO_4^{2-} concentration declines with altitude more slowly than that of SO_2, so that the average scale height of 1630 ± 300 m is slightly larger. The difference must be ascribed to the conversion of SO_2 to SO_4^{2-} during vertical transport. In polluted regions the ground-level concentration of SO_4^{2-} is usually smaller than that of SO_2, whereas in the upper troposphere the two concentrations are approximately equal. A crossover of concentrations at an intermediate altitude has been observed (Georgii, 1970; Varhelyi, 1978). The scale height for SO_4^{2-} in the boundary layer is also larger than that for total aerosol mass, which according to Figure 7.28 is on the order of 1 km. With increasing altitude, therefore, the aerosol must become enriched with sulfate.

Table 10.13 presents data from aircraft observations in the free troposphere to show that the distributions of SO_2 and particulate sulfate are globally quite uniform. For SO_2, this conclusion was initially reached by Maroulis et al. (1980), who measured SO$_2$ in the 5-6 km altitude band over the Pacific Ocean and found that the mixing ratio showed little dependence on latitude. Maroulis et al. (1980) explored also the free troposphere over North America, where SO_2 mixing ratios were somewhat higher, but the flight level of 5-6 km probably was not high enough to preclude perturbations by the meteorology of the boundary layer. Evidence for such perturbations may be seen in the greater variability of their data over the continent compared to those over the ocean. Maroulis et al. (1980) argued that the uniform background of SO_2 in the free troposphere might arise from the oxidation of reduced sulfur compounds. Carbonyl sulfide would be a suitable candidate because of its uniform distribution in the troposphere. Unfortunately, its reactivity against OH radicals or other oxidants is too low to provide an SO_2 source of the required magnitude. On the other hand, the oxidation of other reduced sulfur compounds, such as H_2S, CS_2, or CH_3SCH_3, is so rapid that they are lost predominantly in the boundary layer. The fraction that escapes destruction and is transported to the upper troposphere by turbulent mixing is small. Chatfield and Crutzen (1984) proposed a rapid transfer of reduced sulfur compounds in the strong updraft regions of the Tropics. Sulfur dioxide would be subject to the same transport mechanism. Over the ocean, DMS would provide the most significant source of

TABLE 10.13 Sulfur Dioxide and Particulate Sulfate in the Remote Troposphere (above the Boundary Layer)

Authors	Location	Mixing ratio (pmol mol^{-1})	(nmol m^{-3})[a]	Remarks
Sulfur dioxide				
Maroulis et al. (1980)	Continental U.S.	160 ± 100	7.1 ± 4.5	5–6 km altitude
	Pacific ocean, 57° S–37° N	85 ± 28	3.8 ± 1.2	No significant interhemispheric gradient
Georgii and Meixner (1980)	Bay of Biscayne	95 ± 28	4.2 ± 1.2	One-flight average, 1–12 km altitude
Meixner (1984)	Continental Europe	44 ± 15	2.0 ± 0.7	Three-flight average above 4 km altitude
Andreae et al. (1990)	Amazon Basin, dry season	18 ± 16	0.81 ± 0.71	Free troposphere, 3–5 km altitude
	Amazon Basin, wet season	15 ± 8	0.67 ± 0.36	Free troposphere, 3–5 km altitude
Luria et al. (1990)	North Atlantic, offshore U.S.	110 ± 50	4.9 ± 2.2	About 2.5 km altitude
	Bermuda	60 ± 35	2.9 ± 1.8	About 2.5 km altitude
Berresheim et al. (1990)	West of Tasmania	16 ± 3	0.71 ± 0.13	Five-flight average, 2–3 km altitude
Thornton et al. (1993)	Atlantic Ocean, offshore U.S.	89 ± 11	4.0 ± 0.5	Free troposphere, 5 km altitude
	Atlantic Ocean, offshore Brazil	100 ± 77	4.4 ± 3.4	Free troposphere, 5 km altitude
Wolz and Georgii (1996)	South America	10 ± 8	0.45 ± 0.36	South of 40° S, 2–8 km altitude
	55° S–65° N, 0°–120° W	20 ± 10	0.89 ± 0.45	Average of 47 points, 6–9 km altitude
Particulate sulfate				
Gillette and Blifford (1971)	Scottsbluff, Nebraska		2.44 ± 0.34	4–9 km altitude
	Death Valley, California		2.94 ± 0.31	
	Pacific Ocean, west of U.S.		1.38 ± 0.81	
Adams et al. (1977)	Chacaltaya Mountain, Bolivia		3.69 ± 1.28	5,200 m elevation
Maenhaut and Zoller (1977)	South Pole		1.53 ± 0.31	2,800 m elevation
Lezberg et al. (1979)	Continental U.S.		2.03 ± 1.59	Near tropopause
			4.84 ± 1.59	Above tropopause
Huebert and Lazrus (1980b)	Continental U.S.		4.06 av.	5–6 km altitude
	Pacific Ocean		2.44 av.	5–6 km altitude
Cunningham and Zoller (1981)	South Pole, winter		0.91 ± 0.31	2,800 m elevation
	South Pole, summer		2.38 ± 0.75	
Berresheim et al. (1990)	West of Tasmania		1.4 ± 0.3	Five-flight average, 2–3 km altitude

[a] Concentration reduced to standard temperature and pressure (STP).

SO_2. Chatfield and Crutzen used their model to explain the higher SO_2 mixing ratios reported by Maroulis et al. (1980) in the 5–6 km altitude regime, as opposed to that in the marine boundary layer (see Table 10.14). However, in view of the overlapping variances of the two data sets the difference does not appear to be statistically significant. Andreae et al. (1988c) have explored the vertical distribution of DMS and its oxidation products over the northeast Pacific Ocean and found that the concentration of sulfur dioxide increases with altitude in the boundary layer, whereas that of DMS decreases to low values in the first 2 km, in accordance with expectations. While biogenic sulfur emissions accounted for most of the sulfur budget in the boundary layer, trajectory analyses and radon measurements led to the conclusion that long-range transport was responsible for the elevated levels of SO_2 and excess sulfate in the free troposphere. More recently, Thornton et al. (1996b) concluded from aircraft measurements over the western and central Pacific Ocean that the SO_2 in the upper troposphere arises mainly from anthropogenic sources and long-range transport, in addition to some return of volcanic SO_2 from the stratosphere. The most plausible explanation for the uniform background of SO_2 in the upper troposphere is its longer lifetime in that region combined with fast horizontal transport. In the marine atmosphere the scheme works only because the ocean is an indirect source of SO_2, which compensates for losses due to dry deposition.

Table 10.13 includes concentrations of particulate sulfate in the free troposphere above 4 km. Although the data come from various parts of the world, the mixing ratios are fairly uniform, in the range of 1–4 nmol m^{-3}, which is similar to that of sulfur dioxide. The average is near 2.4 nmol m^{-3}, or 230 ng m^{-3} of sulfate at STP. Thus, sulfate is an important constituent of the tropospheric aerosol. In Section 7.6, the mixing ratio for the background aerosol was taken to be typically 1 μg m^{-3} STP. Comparison shows that sulfate contributes about 25% by mass to the total aerosol background of the troposphere.

Finally, we consider sulfur dioxide and particulate sulfate in the air above the ocean surface. Tables 10.14 and 10.15 list data from a number of measurements, most of which were made onboard ships. In near-coastal areas the atmosphere is often burdened with SO_2 and/or SO_4^{2-} of continental origin. The standard procedure in eliminating such perturbations from the data set is the calculation of air mass trajectories and a selection of air masses that have had no land contact for at least 3–4 days. Bonsang et al. (1980) and others have made use of radon 222 as an indicator for continental air. This element emanates from the earth's crust and has a half-lifetime of 3.8 days, which is similar to that of SO_2 in the lower atmosphere. Bonsang et al. (1987) reviewed results from a number of such studies and concluded

TABLE 10.14 Sulfur Dioxide Mixing Ratios in the Remote Marine Troposphere (Surface Air and Boundary Layer)

Authors	Location	Mixing ratio (pmol mol^{-1})	Mixing ratio (nmol m^{-3})[a]	Remarks
Nguyen et al. (1974)	Subantarctic Pacific Ocean	89.6 ± 83.1	4.0 ± 3.7	Some continental influence
	South Indian Ocean	71.4 ± 75.6	3.2 ± 3.4	(Australia)
	Antarctic	65.4 ± 17.3	2.9 ± 0.8	
Prahm et al. (1976)	Faroe Islands	65 ± 18	2.9 ± 0.8	Trajectory-selected clean air masses
Bonsang et al. (1980)	Central Pacific Ocean	36 ± 26	1.6 ± 1.1	
	Central Atlantic Ocean	22.6 ± 10.9	1.0 ± 0.5	
Maroulis et al. (1980)	Pacific Ocean, 57° S–43° N	54 ± 19	2.4 ± 0.8	0–2 km altitude
Delmas and Servant (1982)	Gulf of Guinea	28.7 ± 10.4	1.3 ± 0.5	
Ryaboshapko (1983)	Central Pacific Ocean	24.5 ± 1.5	1.1 ± 0.1	Trajectory-selected, no land contact
Nguyen et al. (1983)	North Atlantic Ocean	87 ± 47	3.9 ± 2.1	
	Central Atlantic Ocean	51.5 ± 68.4	2.3 ± 3.0	
	South Indian Ocean	33.6 ± 38.1	1.5 ± 1.7	
Herrmann and Jaeschke (1984)	Atlantic Ocean, 30–45° N	27 ± 8	1.2 ± 0.4	Trajectory-selected, no land contact
Berresheim (1987)	Drake Passage, subantarctic	8.5 ± 4.3	0.38 ± 0.19	
	Antarctic coastal shelf	10.5	0.47	
Putaud et al. (1992)	Amsterdam Island	16.8 ± 7.3	0.76 ± 0.33	Seasonal cycle with summer maximum
Berresheim et al. (1990)	West of Tasmania	13 ± 7	0.59 ± 0.31	Boundary layer, 0–1 km altitude
Quinn et al. (1990)	Equatorial Pacific Ocean	27 ± 16	1.2 ± 0.7	
Bandy et al. (1992b)	Northeast Pacific Ocean	22.4 ± 11.2	1.0 ± 0.5	
Huebert et al. (1993)	Equatorial Pacific Ocean	24.7 ± 20.2	1.1 ± 0.9	Low SO$_2$/DMS ratio observed
Yvon and Saltzman (1996)	Equatorial Pacific Ocean	71 ± 56	3.2 ± 2.5	DMS is dominant source of SO$_2$

[a] Concentration reduced to standard temperature and pressure (STP).

TABLE 10.15 Particulate Sulfate in the Remote Marine Troposphere (Surface Air and Boundary Layer)

Authors	Location	Mixing ratio (nmol m^{-3})[a]			Remarks
		Range	Average		
Mézáros and Vissy (1974)	South Atlantic Ocean	—	2.3	Excess	
Prahm et al. (1976)	North Atlantic Ocean	2.5–3.3	2.7	Total	Trajectory selected for pure marine air
		0.3–2.4	1.46	Excess	
Gravenhorst (1975)	North Atlantic Ocean,	1.0–3.0	2.4	Total	Data selected to exclude continental
	Meteor cruise 23 and 32	1.0–2.4	1.77	Excess	influence
		2.1–7.3	4.6	Total	
		1.2–3.9	2.81	Excess	
Bonsang et al. (1980)	Central Atlantic Ocean	4.4–6.7	5.63	Total	
		0.8–3.0	2.5	Excess	
	South Indian Ocean	2.9–87	19.7	Total	
		1.1–16	10.7	Excess	
Huebert and Lazrus (1980b)	Pacific Ocean	8.3–1.4	3.8	Total	0–2 km altitude
Heintzenberg et al. (1981)	Spitzbergen	5.2–8.7	6.8	Total	Marine air masses
Horvath et al. (1981)	Indian Ocean	0.7–8.7	5.94	Total	
		0–6.5	3.0	Excess	
Maenhaut et al. (1983)	Central Pacific Ocean	—	0.48	Excess	Two-Month average (particles < 1 μm)
Saltzman et al. (1985)	Fanning Island, north Pacific	—	7	Excess	Annual average
	American Samoa	—	4.3	Excess	
Ayers et al. (1986)	Cape Grim, Tasmania	—	2.9	Excess	
Huebert et al. (1993)	Equatorial Pacific Ocean	—	7	Excess	
Yvon et al. (1996)	Equatorial Pacific Ocean	10–18	14	Excess	

[a] Concentration reduced to standard temperature and pressure (STP).

that in the marine boundary layer the residence time of SO_2 derived from continental sources is 15 ± 3 h. The elimination of sulfate in the air advected from the continents takes longer. Bonsang et al. (1980) suggested 10 days. The north Atlantic Ocean is particularly affected by continental sulfate carried along with the Saharan dust plume. Ito et al. (1986) made a careful study at Chichi-Jima Island, more than 1000 km from the Japanese mainland, using trajectory analyses combined with radon and chlorofluoro-carbon measurements. They found residence times of 3.7–7.4 days for particulate sulfate, depending on the size of the particles.

The data in Table 10.14 indicate that the distribution of SO_2 over the ocean is globally quite uniform, with mixing ratios in the range of 20–80 pmol mol^{-1}. Especially noteworthy is the absence of a significant latitudinal gradient, which is similar to the even distribution of SO_2 in the free troposphere. Seawater, which is slightly alkaline, provides an almost perfect sink for SO_2, and dry deposition to the ocean surface is an efficient process. The even distribution of SO_2 and the difference in mixing ratios between the marine boundary layer and the free troposphere suggest that a fraction of anthropogenic sulfur dioxide, upon entering the upper troposphere over the continents, becomes globally well distributed and is subsequently lost by dry deposition to the ocean. The process would be augmented by the oxidation of SO_2 to sulfate and its removal by precipitation. Because of the uneven distribution of land masses between the hemispheres, it is necessary to transport SO_2 to the Southern Hemisphere, where the ocean area is larger, and this feature may provide an explanation for the even distribution of SO_2 despite the dominance of anthropogenic sources in the north. Over the oceans, however, the release of dimethyl sulfide from the surface waters provides another source of SO_2, which adds to the downward flow from the free troposphere. The impact of DMS oxidation on local SO_2 mixing ratios may be assessed by means of the steady-state hypothesis, with the assumption that reaction of DMS with OH radicals is the dominant source of SO_2. For such conditions, the SO_2 mixing ratio above the ocean surface should be

$$x(SO_2) = \varepsilon \left(\tau_{SO_2} / \tau_{DMS} \right) x(DMS)$$

where the factor $\varepsilon \approx 0.8$ is the yield of SO_2 from the oxidation of DMS indicated by the laboratory results for NO_x-lean conditions, and τ_{SO_2} and τ_{DMS} are the residence times of SO_2 and DMS in the mixed layer of the atmosphere. For DMS, the reaction with OH radicals establishes a lifetime of approximately 2 days (Table 10.3). The residence time of SO_2 is largely determined by dry deposition, whereas losses due to oxidation by OH radicals are less important. However, the uptake of SO_2 by deliquescent sea salt particles and subsequent oxidation by ozone may contribute appreciably to the loss rate, as Chameides and Stelson (1992) have shown. The height of the mixed layer is on the order of 1 km, and the deposition velocity is about 5 mm s^{-1} (Table 1.13). These data suggest an upper limit for the residence time of SO_2 of about 2 days. The lower limit is given by the field measurements reviewed by Bonsang et al. (1987), about 0.7 days. Accordingly, the residence times of SO_2 and DMS are comparable, and one expects similar mixing ratios. Table 10.1 showed that the majority of the observed DMS mixing ratios fall in the range of 30–80 pmol mol^{-1}. Higher values occur in polar regions. Table 10.14 indicates that the SO_2 mixing ratios are in the same range, 20–80 pmol mol^{-1}. Thus, within a fairly wide margin of variability, the data do support the notion that SO_2 in the atmospheric layer adjacent to the ocean surface derives largely from DMS. Additional evidence for DMS being an important precursor of SO_2 in the marine atmosphere has been obtained from studies of the seasonal variation of both components.

The seasonal variation of DMS in the atmosphere was first described by Nguyen et al. (1990) at Amsterdam Island in the Indian Ocean and by Ayers et al. (1991) at Cape Grim, Tasmania. The seasonal cycle follows that of the DMS concentration in seawater (Turner et al., 1988) and features a summer maximum and winter minimum. Putaud et al. (1992) conducted a 19-month measurement series of atmospheric DMS and SO_2 mixing ratios at Amsterdam Island and demonstrated that the two species were highly correlated, provided the SO_2 data were selected by means of the radon technique to exclude a long-range continental component. Gillett et al. (1993) reported comparable results for a 2-year period at Cape Grim.

Ayers et al. (1991) and Gillett et al. (1993) studied the seasonal cycles of methanesulfonic acid (MSA), a product of DMS oxidation, and non-sea-salt sulfate associated with aerosol particles, as well as the concentration of cloud condensation nuclei (CCN), and compared them to that of DMS. All species showed a distinct seasonal cycle, with similar phase and amplitude. Concentrations increase in austral spring, peak during austral summer, and decrease to a minimum in austral winter. The similarities suggest a linkage between DMS, MSA, SO_2, excess sulfate, and CCN, which is largely in agreement with the known parts of the DMS oxidation mechanism.

As noted earlier, because of reaction with OH radicals, DMS undergoes a diurnal cycle with a maximum in the morning hours and a minimum in the afternoon. Yvon and Saltzman (1996) have studied the relation between DMS and SO_2 in the equatorial region of the Pacific Ocean by simulating the observed diurnal cycle with a time-dependent photochemical box model of the marine boundary layer. The yield of SO_2 from DMS oxidation that was consistent with the measurements was found to range from 27% to 54%, which is less than that suggested by the laboratory data. Huebert et al. (1993) also found low SO_2/DMS ratios in the same ocean region. Nevertheless, even with low yields, Yvon and Saltzman (1996) were able to show that DMS was the dominant source of SO_2. The major sinks of SO_2 were dry deposition to the ocean surface and uptake by alkaline sea salt aerosol followed by oxidation with ozone, which contributed 58% and 28%, respectively, to the total loss rate. Oxidation by OH radicals contributed only 5%, and oxidation in clouds 9%. Taken together, these results show that SO_2 in the remote marine atmosphere arises primarily from natural sources.

Table 10.15 shows concentrations of particulate sulfate observed over the open ocean under conditions eliminating continental effects as far as possible. It was noted earlier that the marine aerosol contains sulfate from sea salt as well as sulfate arising from the oxidation of sulfur compounds (mainly sulfur dioxide). The latter fraction is called non-sea-salt or excess sulfate. As described in Section 7.5, the excess can be determined either by size fractionation of the aerosol, as it is concentrated in the submicrometer size range, or from the sulfate-to-sodium mass ratio in excess of that found in seawater. However, the submicrometer size fraction frequently provides an incomplete sample because some of the excess sulfate derives from the oxidation of SO_2 absorbed in deliquescent sea salt particles. The data in Table 10.15 suggest that excess sulfate contributes 40–75% to total sulfate. The concentrations of sulfate in the two categories are quite variable, however. The concentration of sea salt aerosol depends on the wind force, whereas the flux of gaseous sulfur compounds that act as precursors to excess sulfate varies with biological productivity. Bonsang et al. (1980) have studied the latter aspect and found that the concentration of excess sulfate was linearly correlated with that of SO_2. Its concentration, in turn, followed the strength of biological activity. As discussed above, Ayers et al. (1991) and Gillett et al. (1993) have confirmed the close correlation between DMS and excess sulfate, both showing a summer maximum and winter minimum. Sulfate, in addition to methanesulfonic acid (MSA), would be the final product resulting from the oxidation of DMS. In mid-latitudes, the rates of DMS production and oxidation maximize in summer, but the amplitude of the seasonal cycle for excess sulfate was found to be somewhat smaller than that of the DMS and MSA curves, which may imply the existence of another source of excess sulfate in addition to DMS that dampens the amplitude. The admixture of particles advected from the continents by long-range transport would provide such a source, especially during the winter season. As noted above, the residence time of particles in the marine atmosphere is longer than that of SO_2. On the other hand, much of the detailed structure in the variation of MSA with time, such as temporal spikes, is reproduced in the variation of excess sulfate. As MSA is a unique product of DMS oxidation, the observations clearly demonstrate that large parts of excess sulfate originate from DMS. Yvon et al. (1996) have estimated the steady-state concentration of excess sulfate resulting from the oxidation of SO_2 that is produced from DMS and found somewhat less than they measured. Because SO_2 is not the only product resulting from the oxidation of DMS, it is possible that there are other routes to the formation of sulfate with DMS as precursor. Suhre et al. (1995) have also modeled the generation of excess sulfate by the oxidation of DMS and found good agreement of the results with measurements performed during a cruise in the tropical south Atlantic Ocean.

Since Saltzman et al. (1983) first demonstrated the presence of MSA in the marine aerosol, a greater number of observations have confirmed its ubiquity in the marine atmosphere. MSA is a unique product of DMS oxidation, so that the ratio of MSA to non-sea-salt sulfate is of interest. Table 10.16 lists some of the measurements in this form. The majority of data from remote tropical and subtropical ocean areas indicate a ratio of MSA to excess sulfate in the range of 0.02–0.06. Considerably larger ratios have been found at high latitudes The difference has been interpreted to arise from a temperature dependence in the DMS oxidation mechanism (Berresheim, 1987). The laboratory data have not yet clarified this aspect, because the precise route to MSA production remains to be established. One possibility is the production of MSA via the OH addition pathway of DMS oxidation, which Hynes et al. (1986) have shown to increase in importance compared to the abstraction pathway as the temperature is lowered. However, Ayers et al. (1991) found that the ratio of MSA to excess sulfate increases from 0.06 in winter to 0.18 in summer, in contrast to the predicted behavior. If MSA were produced in a direct oxidation pathway and sulfate mainly via SO_2, the ratio would depend greatly on the relative rates of SO_2 losses by dry deposition and the efficiency of the various oxidation mechanisms involved. MSA appears to be fairly stable in the atmosphere, as it occurs in cloud water and rainwater (Ayers and Ivey, 1990; Berresheim et al., 1990; Galloway et al., 1990) and in polar ice (Ivey et al., 1986; Legrand et al., 1991; Whung et al., 1994).

10.3.5. WET AND DRY DEPOSITION OF SULFATE

Sulfate in rainwater has been studied for over a century, and Eriksson (1952b, 1960, 1966), Granat et al. (1976), and Ryaboshapko (1983) have summarized these observations. Many of the studies have addressed ecological problems such as the input of sulfur and other elements to agricultural soils. Smith (1872) was probably the first to note the relation between rainwater chemistry and air pollution, but his work was not immediately followed up. In the 1950s, when interest in the fate of pollutants revived, networks of monitoring stations in Europe and in the United States established for the first time regional distribution patterns for the concentrations

TABLE 10.16 Methane Sulfonate Relative to Non-Sea-Salt Sulfate in Marine Surface Air

Authors	Location		$[\text{MSA}]/[\text{nss-SO}_4^{2-}]$ (mol mol^{-1})	
Saltzman et al. (1985)	Fanning Island		0.068 ± 0.016	
	American Samoa		0.066 ± 0.019	
Pszenny et al. (1989)	Drake Passage/Gerland Strait		0.53 ± 0.05	
Bates et al. (1990)	Northeast Pacific Ocean		0.07 ± 0.03	
Quinn et al. (1990)	Pacific Ocean, 14° S–14° N		0.03 ± 0.01	
	15° N–30° N		0.012 ± 0.011	
Ayers et al. (1991)	Cape Grim, Tasmania	winter	0.06	(range 0.02–0.10)
		summer	0.18	(range 0.12–0.24)
Bürgermeister and Georgii (1991)	Atlantic Ocean, 17–29° S		0.047	
Savoie et al. (1992)	Mawson, Antarctica		0.21 ± 0.08	(seasonal cycle)
Huebert et al. (1993)	Pacific Ocean, 10° S–15° N		0.67 ± 0.18	
Suhre et al. (1995)	Atlantic Ocean, 19° S		0.072 ± 0.032	

of sulfate and other ions in precipitation. The results were reviewed by Junge (1963). Fifteen years later, Granat (1978) discussed wet precipitation of sulfate again. He had available a record of European network data spanning two decades, a period long enough to expose trends. The rate of deposition of sulfate was found to have increased with time, although not to the extent expected from the simultaneous rise in the anthropogenic SO_2 emissions. More recently, as discussed in Section 8.5, the rate of wet deposition of sulfate (and of acidity) has declined substantially as a consequence of reductions in SO_2 emissions, both in Europe and in America. Thus, the regional distribution patterns of the wet deposition of sulfate are changing. In southeast Asia, a strong economic development has led to a considerable burden of regional uncontrolled air pollution, which has been followed by a similar increase in the rate of wet deposition of sulfate in that region. According to the data assembled by Whelpdale and Kaiser (1996), Southeast Asia—except Japan—now experiences two to three times the rate of wet deposition of sulfate compared to that in Europe or North America.

The mid-1970 distribution in Europe discussed by Ottar (1978) and Eliassen (1978) resembled that of particulate sulfate in surface air, which was shown in Figure 10.3. Lindberg and Lovett (1992) presented the distribution of the wet deposition of sulfate in the United States for the years 1986–1989 derived from data collected by the National Atmospheric Deposition Program (NADP). The distribution also was rather similar to that of particulate sulfate in Figure 10.4, showing a high deposition rate in the northeast. The wet deposition rate is the product of the concentration of sulfate in precipitation and the local annual precipitation rate. While both factors can vary considerably with time and location, the product often has a narrower range of variance. The spatial distribution of wet deposition results from the distribution of the anthropogenic sources, the lifetime of SO_2 for conversion to sulfate, the time scale of horizontal transport, and the residence time of sulfate particles in the atmosphere due to wet precipitation. Similarities in the distribution patterns of particulate sulfate and sulfate in wet precipitation indicate that the precipitation rates are high enough to remove much of the sulfate within a short distance from the region of origin. The recent emission reductions have led to considerable changes in the deposition patterns. Not only have concentrations of sulfate declined, but their distribution has also changed. But the distribution of precipitation rates has remained unchanged. Van Leeuwen et al. (1996) have prepared wet deposition maps for Europe based on 1989 data. They noted that because of the reduction of sulfur emissions, the climatology of precipitation as well as orographic effects become increasingly important in determining the deposition rate. High deposition rates for sulfate still occurred in regions where anthropogenic sources were dominant, such as the Czech Republic, Poland, and Ukraine, whereas in the mountainous regions in the countries formerly known as Yugoslavia, high SO_4^{2-} deposition rates were due to large long-term mean precipitation fluxes. In most areas of western Europe, the wet deposition rate of sulfate is now about 2 g m^{-2} year^{-1}, in Scandinavia it generally is less than 1 g m^{-2} year^{-1}, but in some regions of eastern Europe it still exceeds 10 g m^{-2} year^{-1}. In Germany the range of wet sulfate deposition in 1995 was 1.0–2.6 g m^{-2} year^{-1}, with an average of 1.78 $\pm$ 0.65 g m^{-2} year^{-1} (Fricke *et al.*, 1997), determined at eight background stations of the German Federal Environmental Administration.

Marquardt *et al.* (1996) and Acker *et al.* (1998) have studied the decline of sulfate in precipitation at Seehausen, a station near the former border between East and West Germany. During the period following German unification, 1989–1996, the aqueous concentration of sulfate dropped from 7.2 to 2.4 mg dm^{-3}. Back-trajectory analysis was used to separate contributions from the eastern and western sectors. Sulfate in precipitation associated with air advected from the west largely followed the trend in emission reductions in West Germany. In contrast, sulfate in precipitation from the eastern sector showed a dramatic decline during the first 4 years after 1989, corresponding to a similar decline in SO_2, which followed the collapse of the economic structure, the decrease in energy consumption in the region, and the decommissioning of several old power stations. About one-half of the precipitation events from the eastern sector were associated with air that had traveled a distance of less than 500 km (that is, only within Germany) during the 24 h preceding the event. The remaining precipitation events from the eastern sector were connected with air masses that had a more distant origin. Sulfur dioxide emissions in the neighboring countries, Poland and the Czech Republic, underwent little change during the period covered, but these emissions evidently have had much less influence on wet sulfate deposition at the observation site. The results are remarkable, as they clearly demonstrate the geographic range contributing to sulfate deposition at any sampling station.

One of the problems arising in the collection of precipitation for chemical analysis is that open samplers, which are often used for simplicity, overestimate wet deposition because dry deposition of SO_2 and of sulfate particles adds material. Modern automated equipment can differentiate between the two forms of deposition. The ratio of wet-only to bulk precipitation can be determined by using either dry-only/wet-only samplers or wet-only samplers and bulk samplers in parallel. Van Leeuwen *et al.* (1996) have summarized a number of such studies, indicating that for sulfate the ratio is 0.85 $\pm$ 0.09, which can be used as a correction factor. At coastal sites, as Losno *et al.* (1998) have shown, even wet-only samples may be contaminated with sea salt entering the bucket with the wind.

More significant than dry deposition onto flat terrain probably is the interception of aerosol particles by high-growing vegetation. The arguments are the same as those presented in Section 9.2 for particulate ammonium. Höfken et al. (1981) have studied the filtering effect of forests in some detail. They found, for example, that the concentration of particulate sulfate is up to 35% lower underneath the canopy than above it. Rain rinses the dry-deposited material off the foliage, so that rainwater collected below the canopy is enriched with sulfate and other trace substances compared to rainwater collected in forest clearings. From the observed enrichments, Höfken et al. (1981) calculated ratios of dry to wet deposition rates for sulfate of 0.9 for beech and 4.1 for spruce trees. The corresponding ratios for ammonium were 0.4 and 0.8, respectively. As in the polluted regions of western Europe, sulfate and ammonium are closely associated, the much higher dry to wet deposition rates for SO_4^{2-}, especially in spruce stands, suggests that a large part of the enrichment was caused not by sulfate particles, but by dry deposition of gaseous SO_2 that was subsequently oxidized to sulfate.

Lindberg and Lovett (1992) have conducted a detailed study of atmospheric fluxes of airborne sulfur compounds above and below the canopies at 13 forest sites, both coniferous and deciduous, in different parts of the United States. The fraction of wet to total deposition averaged 67%, with a range of 47–86%. At high elevation sites, fog and cloud interception contributed significantly to wet total deposition. Dry deposition of SO_2 contributed 19%, on average, with a range of 3–37%, depending on ambient SO_2 concentrations. Dry deposition of particulate sulfate contributed 16%, on average, with coarse particles being most important except at high elevation sites, where fine particles assumed a larger role. The deposited sulfur behaved conservatively in the forest systems studied, and little net uptake by the canopy was observed.

Three-dimensional global transport models are now capable of simulating the essential features of the distribution of sulfur compounds and their wet deposition. Such computer models are based on generalized global circulation schemes and parameterized convective cloud formation and precipitation scavenging routines. The strengths and spatial distributions of anthropogenic sources (SO_2) and natural sources (DMS) are prescribed, and the oxidation chemistry is highly simplified. Figure 10.7 shows typical results for the wet deposition of sulfate obtained by means of a three-dimensional computer simulation. In this case, the source strengths adopted for sulfur emitted as SO_2 and DMS were 70 and 16 Tg year^{-1}, respectively. The three regions most affected by pollution are made evident by the high deposition rates in regions close to the main emissions, in Europe, in North America, and in southeast Asia. However, the effects of anthropogenic emissions are also felt in marine regions of the Northern Hemisphere, which receive a

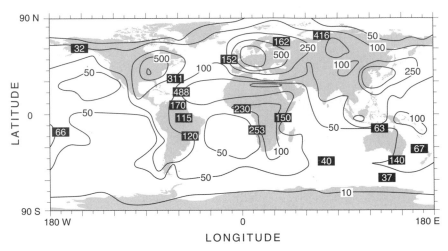

FIGURE 10.7   Global distribution of the wet deposition of non-sea-salt sulfate derived from a three-dimensional computer model simulation. Isolines show wet deposition rates of sulfur in units of mg m^{-2} year^{-1}. Numbers in boxes show observed deposition rates at the respective locations. (Reprinted with permission from *J. Atmos. Chem.*, 13 (1991), 225–263, J. Langner and H. Rodhe, with kind permission from Kluwer Academic Publishers.)

higher wet sulfate deposition than comparable regions in the Southern Hemisphere. Figure 10.7 includes annual wet sulfate deposition rates from observations at a number of sites for comparison with the calculated deposition field. Although there is good agreement at many stations, large deviations are evident in several cases, especially at the stations in Nigeria and east Africa. The agreement is much better in marine regions.

The wet deposition of sulfate onto the oceans has been studied mainly at island stations. In a few cases rainwater samples have been collected onboard ships. Table 10.17 presents average concentrations and annual deposition rates from a number of such studies. The fraction of sulfate originating from sea salt is usually determined by reference to sodium and the SO_4^{2-}/Na mass ratio in seawater. The concentration of sea-salt-related sulfate in marine rainwater is fairly constant (except at Cape Grim) and averages about 10 μmol dm^{-3}, or 0.32 mg dm^{-3} as sulfur. According to Baumgartner and Reichel (1975) the annual precipitation rate over the ocean is 3.84×10^{17} kg. Thus, the annual wet deposition rate of sulfate originating from sea salt amounts to about 3.8 Tmol year^{-1}, or 125 Tg sulfur year^{-1}. A global deposition of 125 Tg year^{-1} of sea-salt sulfur would correspond to a total mass flux of sea salt of 4.8 Pg year^{-1}, almost five times the rate that had been widely accepted (see Table 7.12) for some time after Eriksson (1959, 1960)

TABLE 10.17 Concentrations of Sulfate in Marine Precipitation and Annual Wet Sulfate Deposition Rates at Various Locations

Authors	Location	Sulfur concentration (mg dm^{-3})			Excess sulfur deposition rate (g m^{-2} year^{-1})	
		Total	Sea salt	Excess		
Eriksson (1957)	Hawaii	0.64	0.44	0.10	0.11	Near sea level, R = 550 mm year^{-1}
Bruyevich and Kulik (1967)	Pacific Ocean	1.18	0.47	0.71	—	142–154° W, 1° S–31° N
Gravenhorst (1975b)	Atlantic Ocean	0.69	0.39	0.30	—	Equatorial region
Nyberg (1977)	North Atlantic	0.84	0.43	0.41	—	All data
		0.55	0.44	0.11	—	Trajectory-selected, less polluted
Mészáros (1982)	Samoa	0.58	0.25	0.32	0.34	R = 1070 mm year^{-1}
Pszenny et al. (1982)	Samoa	0.19	0.15	0.03	0.07	
Jickells et al. (1982)	Bermuda	0.69	0.34	0.34	0.50	R = 1502 mm year^{-1}
Galloway et al. (1982)	Bermuda	0.58	0.20	0.39	0.43	R = 1525 mm year^{-1}
Galloway and Gaudry (1984)	Amsterdam Island	0.47	0.39	0.38	0.10	R = 1120 mm year^{-1}
Ayers and Ivey (1988)	Cape Grim, Tasmania	2.43	2.25	0.18	0.14	High sea salt loading
Berresheim et al. (1990)	South Pacific	0.10	0.04	0.06	—	Rainwater
		0.36	0.19	0.17	—	Cloud water, same day
Buijsman et al. (1991)	North Atlantic	0.86	0.73	0.13	0.10	Shipboard, 57° N, 20° W, 4-year average

provided an initial estimate. Mészarós (1982) and Várhelyi and Gravenhorst (1983) first noted the discrepancy. The latter authors used actual precipitation measurements to derive a range of 94–177 Tg year^{-1} for the wet deposition of sea-salt sulfur. More recently, Erickson and Duce (1988) calculated global wet and dry deposition fields, using a sea-salt mass size distribution appropriate for an elevation of 15 m above the mean sea level. Combining rainfall distribution data with a range of empirical scavenging ratios, they obtained a global wet deposition flux for sea salt of 1.5–4.5 Pg year^{-1}. This is smaller than the above value and leads to 45–120 Tg year^{-1} for the wet deposition of sea-salt sulfur. According to their calculations, however, wet deposition contributes only about 25% to the total deposition of sea salt and dry deposition the larger share. Várhelyi and Gravenhorst (1983), in contrast, found wet deposition to be three times more effective than dry deposition.

Excess sulfate in marine precipitation arises from the oxidation of sulfides emanating from the sea's surface as well as from the overflow of anthropogenic sulfur dioxide emitted on the continents. The North Atlantic Ocean in particular is affected by pollution carried out to sea from North America. In such regions, the great variability of excess sulfate derives primarily from the fluctuations in the air mass flow patterns. The observations of Jickells *et al.* (1982) at Bermuda provide a pertinent example of the influence of the synoptic weather situation. High SO_4^{2-} concentrations (and low pH values) in rainwater are correlated with air masses originating in the United States, whereas low SO_4^{2-} concentrations (and higher pH values) occur when the subtropical high pressure system blocks the advection of continental air to the Bermuda site. Nyberg (1977) made measurements onboard weather ships west of Europe and found the lowest values of excess sulfate in air masses originating in the region of the Azores. At Amsterdam Island in the south Indian Ocean, the influence of pollution is almost imperceptible. The lowest concentrations, 0.07–0.14 mg dm^{-3} of sulfur, may be taken to indicate the natural background level of excess SO_4^{2-} in marine precipitation. Galloway (1985) has reviewed data on excess sulfate wet deposition rates at remote locations and derived an average of 0.11 ± 0.06 g m^{-2} year^{-1} from 20 values ranging from 0.03 to 0.3 g m^{-2} year^{-1} (as sulfur). Extrapolation to global conditions leads to a total wet deposition rate for excess sulfur of 35 Tg year^{-1}. Although this is a coarse estimate, it agrees approximately with that for the emission of sulfides from the ocean (16–42 Tg year^{-1}; see Table 10.8). Actually, it should be smaller, because not all of the sulfur flux emanating from the ocean is converted to particulate sulfate, and some of it is lost by dry deposition.

10.4. TROPOSPHERIC SULFUR BUDGETS

Although much progress has been made in delineating the source strengths and the fate of natural sulfur emissions, it is still not possible to describe the behavior of sulfur compounds in the atmosphere in sufficient detail. Difficulties arise from the inhomogeneous concentration patterns of many compounds, the problems of quantifying dry and wet deposition rates, and uncertainties connected with biogenic source strengths, especially on the continents, where the interaction with vegetation must be taken into account. In view of the short residence times of all sulfur compounds except OCS, the local behavior of atmospheric sulfur is best described by regional budgets. For many years, the focus of attention has been on regions heavily polluted by SO_2 and its oxidation products. The aim was to establish the magnitude of the fluxes and to achieve a reasonable balance between sources and sinks. Such regional budgets have been discussed by Rodhe (1972, 1976) and Mészáros *et al.* (1978) for western Europe, and by Galloway and Whelpdale (1980) for the northeastern United States. Ryaboshapko (1983) presented estimates for the atmospheric sulfur balance in polluted, clean continental, dusty continental, and marine environments. Kritz (1982) first worked out a local budget for the marine atmosphere. The results may be presented as simple box models in which the fluxes are derived by arithmetic sums and differences. Examples of this approach will be given further below. Computer model studies based on Lagrangian or trajectory schemes were also used to overcome the limitations inherent in the flux estimate approach (e.g., Eliassen, 1978; Shannon, 1981; Renner, 1985). The rapid advances in computer technology have now made it possible to study the distribution and the local and global budgets of sulfur species in three-dimensional space by means of Eulerian transport models that solve the continuity equation at each grid point of the domain (Langner and Rodhe, 1991; Feichter *et al.* 1996; Chin *et al.*, 1996; Wojcik and Chang, 1997, Restad *et al.*, 1998). The technique allows to follow variations of concentrations in time anywhere in the study area for comparison of the results with observations. The results obtained have finally brought within reach a truly global description of the atmospheric sulfur budget.

10.4.1. THE REGIONALLY POLLUTED CONTINENTAL ATMOSPHERE

The main purpose of this section is to show how one may construct a simple sulfur budget for the regionally polluted continental troposphere based on data given in the preceding text. Figure 10.8 presents a flux diagram for the disposal of sulfur from anthropogenic sources in the continental atmosphere of the Northern Hemisphere. Biogenic and volcanic emissions are relatively minor and are ignored. The continent is subdivided into urban, regionally polluted, and remote regions. The input consists of 100 Tg sulfur annually, mainly as SO_2. This high rate is unrealistic, as the global anthropogenic source of sulfur dioxide is not all injected into one continental region. In fact, it is already difficult to accommodate the input with the removal rates, which are determined by the concentrations and residence times chosen. The ground-level concentrations adopted are indicated at the bottom of Figure 10.8. Also shown are the scale heights in the lower atmosphere, which were estimated from the vertical profiles shown in Figure 10.6. The flux scheme starts with the injection of SO_2 and some SO_4^{2-} into the urban atmosphere. This is an oversimplification, of course. Transport then carries SO_2 and SO_4^{2-} into the regionally polluted troposphere, and from there to the remoter regions of the continents. In each region, SO_2 is partly converted to

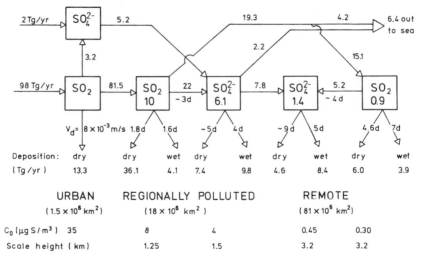

FIGURE 10.8   Flux diagram for the disposal of anthropogenic sulfur emissions in the continental troposphere of the Northern Hemisphere. The unit for fluxes is Tg year^{-1} (sulfur). Numbers in boxes indicate the column densities (unit: mg m^{-2} as sulfur). They are derived from the adopted ground-level concentrations and scale heights below 5.5 km altitude. Above this level, mixing ratios are constant, with $x(SO_2) = 45$ pmol mol^{-1} and $x(SO_4^{2-}) = 49$ pmol mol^{-1}. Chemical conversion of SO_2 to SO_4^{2-} is assumed to occur only in the lower troposphere. Deposition velocities adopted are 8 mm s^{-1} for SO_2 and 0.15 mm s^{-1} for SO_4^{2-} in flat terrain, and 5 mm s^{-1} for SO_4^{2-} interception by forests, which are assumed to occupy 50% of the continental area.

SO_4^{2-}, and both components are removed by wet and dry deposition. Wet deposition rates for SO_2 make allowance for the enhancement resulting from the interaction with hydrogen sulfite with formaldehyde in clouds, as discussed in Section 8.4.1. The dry deposition velocity is about 8 mm s^{-1}, according to Table 1.12. The rate of wet deposition of sulfate is determined primarily by the residence time of aerosol particles, which is approximately 5 days. In the polluted atmosphere the rate should be slightly higher because of below-cloud scavenging of particles by falling raindrops. The rate of dry deposition of particulate sulfate is uncertain. Estimates used in Figure 10.8 are based on the process of particle interception by forest canopies.

An important result immediately apparent from the box model of Figure 10.8 is the great effect of dry deposition of SO_2, which takes up about 50% of the entire flux in urban and polluted regions. Rodhe (1976), Garland (1977, 1978), Mészáros et al. (1978), and Eliassen (1978) reached similar conclusions when they considered the sulfur budget of western Europe. Here, the source strength adopted by the various authors was 15–20 Tg year^{-1}, the total area considered was $(4-9) \times 10^6$ km^2. The rate of SO_2 removal by dry deposition was estimated to range from 20% to 55% of the total budget. These authors also estimated losses due to advection from the European territory to other domains and found this fraction to comprise 9–43% of the total sulfur budget. In Figure 10.8, the flux of sulfur from the polluted to the remote continental atmosphere is 22.9% of the input. Another 6.4% is carried out to sea. A flux of 6.4 Tg year^{-1} is a relatively small fraction of anthropogenic sulfur, but it is substantial compared with marine or coastal emissions from the biosphere, which are on the order of 10 Tg year^{-1} in the Northern Hemisphere.

The box model further predicts a wet deposition rate of sulfur contained in SO_2 and sulfate in the polluted regions reaching 770 mg m^{-2} year^{-1}, or 14% of total anthropogenic sulfur. The first value compares well with observed deposition rates. The authors cited above reported average wet deposition rates in the range of 620–1100 mg m^{-2} year^{-1}. The corresponding percentages in their total budgets are 19–48. The fluxes were derived in most cases from actual measurements of sulfate in rainwater collected at various stations of European sampling networks. The fluxes are higher than that shown in Figure 10.8. The reason for the disparate relative contributions of wet sulfur deposition to the total budgets is not obvious, but it should be noted that dry deposition of particulate sulfate in the model presented in Figure 10.8 is almost as significant as the wet deposition rate.

Wojcik and Chang (1997) discussed the sulfur budget in the northeastern United States. Here, the annual input rates were estimated to range from 10 to 16 Tg. Dry deposition of SO_2 was found to contribute 17–31%, wet deposition 30–48%, and export to other domains 8–43% to the total budget.

The Regional Acid Deposition Model applied by Wojcik and Chang, with input data representative of the early 1980s, made it possible to differentiate between the formation of sulfate by gaseous and aqueous phase reactions. The results of this model indicated that about 50% of the total input was mainly exported out to sea. The remaining fraction split into losses of 27% due to dry deposition of SO_2, 7.5% due to dry deposition of particulate sulfate, and 65% due to wet deposition, primarily as sulfate. The latter fraction contained 53% from the scavenging of aerosols by clouds and 12% from the immediate wet removal of SO_2 conversion in clouds. Particulate sulfate arose to 70% from aqueous phase oxidation (mainly by H_2O_2) and to 30% from gas phase oxidation (mainly by OH radicals), similar to predictions from results obtained with global models discussed further below.

10.4.2. The Circulation of Sulfur in the Unperturbed Marine Atmosphere

The transformation of reduced sulfur compounds to SO_2 and excess sulfate in the marine atmosphere can be quantified by means of a steady-state local circulation model. An appropriate flow diagram is shown in Figure 10.9. Bonsang et al. (1980) and Kritz (1982) initially assembled the elements necessary for constructing the marine sulfur cycle. Bates et al. (1990), Suhre et al. (1995), and Yvon and Saltzman (1996) refined the quantitative aspects of the model. The column densities for SO_2 and excess particulate sulfate shown in Figure 10.9 are based on the observed concentrations discussed in Section 10.3.4. The flux and the concentration of dimethyl sulfide (DMS) were taken from the range of data discussed in Section 10.2.2 to represent approximately tropical conditions. The contribution of H_2S to the marine sulfur cycle can be neglected in comparison to that of DMS. The oxidation of DMS, primarily by reaction with OH radicals, was taken to yield 7.5% methane sulfonic acid and 30% SO_2. This fraction was chosen to accommodate the results of Yvon and Saltzman (1996), although the laboratory results indicate that the yield of SO_2 is greater. The remaining products, such as dimethyl sulfoxide and dimethyl sulfone, are not specified but are assumed to be oxidized further to form particulate sulfate. A significant fraction of SO_2 undergoes dry deposition (wet deposition is ignored). The other fraction is oxidized to sulfuric acid, which is neutralized by ammonia as well as by sodium and ends up as excess sulfate. As noted earlier, the main pathway of oxidation is absorption of SO_2 by deliquescent sea-salt particles and aqueous phase reaction with ozone. Oxidation by OH radicals in the gas phase and by H_2O_2 in clouds is less important. The residence time of sulfate in particles is

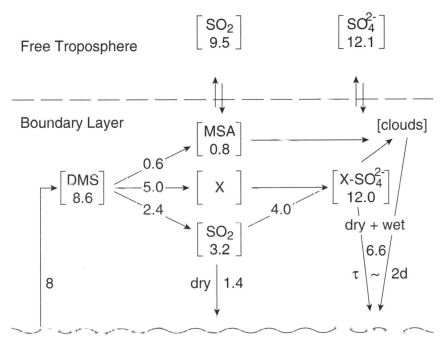

FIGURE 10.9   Flux diagram for sulfur in the unperturbed marine atmosphere. The unit for fluxes is $\mu g\ m^{-2}\ days^{-1}$ (sulfur). Numbers in boxes indicate column densities (unit: $\mu g\ m^{-2}$ as sulfur). DMS, dimethyl sulfide; MSA, methane sulfonate (associated primarily with the marine aerosol); X, unspecified intermediate compound. The mixing ratio of SO_2 is 42 pmol mol^{-1}, independent of altitude; the mixing ratio of excess SO_4^{2-} is 196 pmol mol^{-1} in the boundary layer and 55 pmol mol^{-1} in the free troposphere. Contrary to the model of Kritz (1982), the fluxes are confined to the boundary layer, and there is no significant net flux into or out of the free troposphere.

determined primarily by wet deposition. The residence time of 2 days indicated in Figure 10.9 for particulate matter is less than that derived in Section 8.3 for the tropospheric aerosol because the precipitation rate in the tropics is higher than elsewhere.

The oxidation of DMS and the conversion of SO_2 to SO_4^{2-} occur largely in the boundary layer. Sulfur dioxide in the free troposphere is comparatively stable. The cycle of sulfur in the marine boundary layer is also much faster than the exchange of air with the upper troposphere. However, vertical mixing of air is still faster than interhemispheric air exchange, so that the sulfur cycles in the two hemispheres are decoupled. In the model depicted in Figure 10.9, SO_2 and SO_4^{2-} are allowed to communicate freely with the upper troposphere, but the exchange is not considered to produce significant net fluxes in either the upward or the downward direction. Kritz (1982) included nonnegligible downward fluxes of both SO_2 and SO_4^{2-} in his model. These would have to be maintained by fast horizontal transport in the upper troposphere. For excess SO_4^{2-}, however, a comparison of the data in Tables 10.13 and 10.15 shows that the mixing ratios near the ocean surface are higher than in the upper troposphere, so that an upward rather than a downward flux is indicated. An upward flux is needed to bring SO_4^{2-} to the cloud level to allow its removal by rainout. In the free troposphere, the fractions of SO_2 and SO_4^{2-} that penetrate the cloud layer intermingle with SO_2 and SO_4^{2-} of continental origin. Thus, both continental and marine sources will contribute to the general background of sulfur in the troposphere.

10.4.3. THE GLOBAL TROPOSPHERIC SULFUR BUDGET

In the past, the tropospheric sulfur budget has been mostly a subject of speculation. The overview given by Warneck (1988) on budget estimates of several authors between 1960 and 1985 showed the great uncertainty in the derived individual fluxes. In the meantime, source strength estimates of the most important sulfur compounds, namely anthropogenic sulfur dioxide and biogenic dimethyl sulfide, have greatly improved, and global distributions have been determined. The data provided an input for three-dimensional global chemical transport models, which have finally led to reasonably reliable results regarding the fluxes of individual sulfur compounds.

Table 10.18 compares the results obtained by Langner and Rodhe (1991), Pham et al. (1995, 1996), Feichter et al. (1996), and Chin et al. (1996). In these studies, the source strength of DMS was chosen to be in the range of 500–680 Gmol year^{-1}, corresponding to a sulfur flux of 16–21.8 Tg year^{-1}, which is toward the lower end of the possible range indicated in Table 10.8. Terrestrial emissions of biogenic reduced sulfur compounds were ignored. The range of anthropogenic SO_2 emissions was 2030–2875 Gmol year^{-1}, corresponding to 65–92 Tg sulfur year^{-1}; volcanoes were taken to emit sulfur as SO_2 in the range of 210–290 Gmol year^{-1}, or 3.5–9.2 Tg year^{-1}. The emission rates agree with the range of estimates discussed in Sections 10.3.1 and 10.3.2. All studies include SO_2 emissions from biomass burning, with a contribution of about 2.5 Tg sulfur annually. The major new development in these models is the incorporation of oxidant concentrations that allow the computation of oxidation rates. Langner and Rodhe (1991) considered DMS to be oxidized exclusively by OH radicals, whereas Pham et al.

TABLE 10.18 The Global Atmospheric Sulfur Budget as Determined from Three-Dimensional Models (Fluxes in Gmol year $^{-1}$)

			LR91[a]	P95[b]	F96[c]	C96[d]
DMS[e]	Source		500	625	528	681
SO$_2$	Sources	Anthropogenic	2078	2875	2425	2034
		Volcanoes	266	288	109	209
		Biomass burning	78	91	78	72
		Oxidation of DMS	531	584	528	672
		Total	2953	3838	3140	2987
	Sinks	Gas-phase oxidation	244	203	525	234
		Oxidation in clouds	1312	1734	1078	1300
		Dry deposition	953	1719	1256	831
		Wet deposition	444	182	281	622
		Residence time[f]	1.2	0.6	1.5	1.3
SO$_4^{2-}$	Sources	Anthropogenic	109	—	—	—
		Gas-phase production	244	203	525	234
		In-cloud production	1312	1734	1078	1300
		Total	1665	1937	1603	1534
	Sinks	Dry deposition	268	531	209	175
		Wet deposition	1397	1406	1394	1359
		Residence time[f]	5.3	4.7	4.3	3.9
MSA[e]	Source	Oxidation of DMS	—	38	—	34
	Sinks	Dry deposition	—	16	—	2
		Wet deposition	—	22	—	32
		Residence time[f]	—	6.1	—	6.2

[a] Languer and Rodhe (1991).
[b] Pham et al. (1995, 1996).
[c] Feichter et al. (1996).
[d] Chin et al. (1996).
[e] DMS, dimethyl sulfide; MSA, methane sulfonate.
[f] Residence time in days.

(1995) and Feichter et al. (1996) included the reaction with NO$_3$ radicals and found that it contributed another 16%. Chin et al. (1996) found it necessary to include an additional oxidation pathway (oxidant unknown) to bring calculated and observed DMS concentrations into agreement. This problem was also made evident in the field studies of Suhre et al. (1995) and Yvon et al. (1996). All studies assumed that SO$_2$ was the major oxidation product. Pham et al. (1995) and Chin et al. (1996) also accounted for the additional formation of methane sulfonate. This product comprises only a small fraction, however.

The oxidation of SO$_2$ was assumed to occur both in the gas phase by reaction with OH radicals and in the aqueous phase of clouds by reaction with dissolved H$_2$O$_2$. The latter reaction was assumed to be rapid, so that it titrates the limiting reagent within the course of the time step, usually a few hours. In polluted continental regions, the concentration of H_2O_2 is limiting, whereas in other regions H_2O_2 is present in excess. The fraction of air filled with clouds is prescribed by the models. As Table 10.18 shows, 67–89% of SO_2 oxidation (average 81%) was found to occur in the aqueous phase. This agrees with the conclusions of Wojcik and Chang (1997) regarding the regional sulfur budget in northeast America. All of the global studies agree that the oxidation of SO_2 to SO_4^{2-} contributes close to 50% to the total losses of sulfur dioxide. The remaining fraction undergoes dry and wet deposition. Dry deposition is preponderant, showing that 28–45% of SO_2 is removed in this manner. The fraction of wet deposition varies with assumptions on the scavenging mechanism. Langner and Rodhe (1991) took into account that formaldehyde assists in dissolving SO_2 in cloud water, in accordance with the discussion in Section 8.4.1. Chin et al. (1996) and Feichter et al. (1996) accounted for much scavenging of SO_2 in wet convective updrafts. Pham et al. (1995) used a higher dry deposition velocity for SO_2 than the other investigators; wet deposition was correspondingly reduced.

Direct anthropogenic emissions of sulfate are small; only one model took them explicitly into account. Sea salt as a source of sulfate is also excluded. The predominant source of non-sea-salt sulfate is the oxidation of SO_2. The main removal process is wet deposition; 73–89% of sulfate is removed in this way. Chin and Jacob (1996) used their model to study the origin of sulfate in various regions of the world. Anthropogenic sources represent the principal contributor to SO_4^{2-} in the surface air over the continents. In marine areas of the Northern Hemisphere, anthropogenic SO_2 and DMS contribute about equally, whereas in the Southern Hemisphere DMS is the dominant source of excess sulfate. In the middle and upper troposphere the anthropogenic influence on sulfate is considerably reduced compared to that at the surface because of efficient scavenging of SO_2 and SO_4^{2-} in wet convective updrafts. DMS can be injected into high altitudes without being scavenged, so that much of the sulfate present in the upper troposphere may originate from the oxidation of DMS. Volcanoes also inject SO_2 and sulfate into the middle troposphere and they contribute a large fraction of SO_4^{2-}, especially at high latitudes. Volcanic emissions also account for 20–40% of sulfate in the surface air over the north Pacific Ocean.

Table 10.18 includes residence times of SO_2, SO_4^{2-}, and MSA that were calculated from the concentration fields and removal rates derived in the models. For SO_2, the main result is a residence time of 1.2–1.5 days, which is largely in agreement with the field studies summarized in Table 10.12. Dry deposition and in-cloud oxidation are the processes contributing most to the short residence time. The residence times calculated for sulfate are 4–5 days, which agrees quite well with the time constant derived in Section 8.3.1 for the removal of particles by nucleation scavenging. The residence time for MSA is somewhat longer, about 6 days, which is unexpected, as MSA should be removed by the same mechanisms as sulfate. Chin *et al.* (1996) suggested that the difference occurs because a larger fraction of MSA resides in the free troposphere, where precipitation is less frequent. However, the models ignore the presence of sea salt in the marine boundary layer. A large fraction of MSA is associated with sea salt particles that do not reach the free troposphere. Accordingly, the removal rate of MSA appears to have been underestimated.

Table 10.18 does not include the cycle of sulfur associated with sea-salt particles, which is essentially independent of that of excess sulfate. The generation of sea spray is difficult to estimate on a global scale, and knowledge about the annual flux of sea salt is obtained from estimates of the dry and wet deposition rates. Várhelyi and Gravenhorst (1983) proceeded from observations of sulfur in marine aerosols and precipitation. After separating sea-salt and excess sulfate they estimated the global annual production rate of sulfur associated with sea salt to be 130–275 Tg, of which about 75% is removed by wet deposition. The corresponding molar flux is in the range 4–9 Tmol year^{-1}. Erickson and Duce (1988) considered global wet and dry deposition fields, using a sea-salt mass size distribution appropriate for an elevation of 13 m above the mean sea level. The equations describing dry deposition as a function of wind speed were coupled with a global wind speed climatology to determine the dry deposition field. Wet deposition was estimated from the annual amount of rainfall and an empirical average scavenging ratio. Integration led to a global flux of 10–30 Pg year^{-1} of sea salt, to which dry deposition contributed about 75%. According to Table 7.16 the sulfur content in fresh sea salt should be about 2.6% by mass, so that the associated flux of sulfur would be 260–770 Tg year^{-1}, or 8–24 Tmol year^{-1}, much larger than that of excess sulfate discussed above. Most of the sea salt generated from the ocean also returns to the ocean. The fraction deposited on the continents is less than 10%.

CHAPTER 11

Geochemistry of Carbon Dioxide

11.1. INTRODUCTION

Atmospheric carbon dioxide is chemically quite inert except at high altitudes, where it undergoes photodissociation. Carbon dioxide also occurs dissolved in the oceans and in the surface waters of the continents. Here it participates in several geochemically important reactions, such as the weathering of rocks and the formation of limestone deposits. Another important process involving CO_2 is its assimilation by plants. Carbon is a key element of life, and atmospheric CO_2 provides the principal source of it. These interactions lead to a complex system of carbon fluxes connecting a number of well-differentiated geochemical reservoirs: the atmosphere, the biosphere, the oceans, and the sediments of the Earth's crust. Reservoirs and fluxes combined describe the geochemical carbon cycle. The atmosphere cannot be isolated from the rest of the system, because the abundance of carbon dioxide in the air is determined by the behavior of the other reservoirs to which the atmosphere is coupled and by the associated exchange processes. To elucidate what controls atmospheric CO_2 thus requires a detailed discussion of the major reservoirs and their interactions.

The current great interest in carbon dioxide derives from the observed rise of concentration in the atmosphere and a growing concern about the prospect of climatic changes if the trend continues. Callendar (1938) appears to have first noted the increase since the beginning of the century, but systematic measurements were not begun until 1957. Since then, a unique record of the atmospheric CO_2 mixing ratio has been obtained at Mauna Loa, Hawaii, and somewhat less extensively at the South Pole (see Fig. 1.2). These data document the annual increase over a period of the first 25 years. Following Callendar, the effect has generally been attributed to the combustion of fossil fuels by humans. Bolin (1977) and others have subsequently shown that the destruction of forests accompanying the expansion of arable land areas during the past 150 years must have liberated additional nonnegligible quantities of CO_2 to the atmosphere.

As an infrared-active molecule, CO_2 assumes a significant role in determining the heat balance of the atmosphere. By intercepting thermal radiation emitted from the Earth's surface, CO_2 raises the temperature in the troposphere (the so-called greenhouse effect), while at the same time it serves as a cooling agent in the upper atmosphere by radiating heat away toward space. Around 1900, Arrhenius (1896, 1903) estimated that the surface temperature would rise by 9°C if the abundance of atmospheric CO_2 were tripled. Recent studies carried out with the aid of high-speed electronic computers have revealed the complexity of quantifying this effect. Plass (1956) provided the first realistic model for the 15 μm absorption band of CO_2; Gebhart (1967) included the absorption of solar radiation in the near infrared; Möller (1963) took into account the overlap of H_2O and CO_2 absorption and the feedback by the increase of water evaporation due to the rise in surface temperature (a very significant effect, as water vapor is the main infrared absorber); and Manabe and Weatherald (1967) considered the convective readjustment of the troposphere accompanying the change in heat fluxes. Furthermore, Manabe and Weatherald (1975, 1980) developed a three-dimensional circulation model that incorporates radiative, convective, and advective heat transport as well as the water vapor balance resulting from evaporation and precipitation. They found a temperature rise of 2.9 K, on average, when the CO_2 level is doubled from 300 to 600 μmol mol^{-1}, with a stronger warming in the polar regions. Similar one-dimensional radiative-convective models reviewed by Schneider (1975) and Ramanathan and Coakley (1978) gave a temperature increase of 2–3 K, depending on assumptions on the cloud-top behavior. Dickinson (1986) has discussed such models in some detail. In the meantime, the global warming potential of other greenhouse gases, such as methane, nitrous oxide, and the halocarbons, as well as the cooling effects of sulfate aerosols, has been explored, but CO_2 and the water vapor feedback remains the major contributor. Current global climate studies are based on three-dimensional general circulation models that couple heat and moisture fluxes between the atmosphere and the ocean and between the atmosphere and continental soils and vegetation. Such models and their capabilities are discussed in depth in a book edited by Trenberth (1992) and in an assessment report of the Intergovernmental Panel on Climate Change (Houghton *et al.*, 1996).

11.2. THE MAJOR CARBON RESERVOIRS

In describing the natural carbon cycle, it will be useful to commence with a characterization of the major reservoirs. Table 11.1 presents an overview. We shall have to differentiate between organic (or reduced) carbon and inorganic (or oxidized) carbon. The former includes living biomass as well as decay products of the biosphere, such as plant debris, soil humus, and metamorphic organic compounds. Inorganic carbon refers to CO_2, bicarbonate, and carbonate.

TABLE 11.1   Geochemical Carbon Reservoirs

Reservoir	Mass content of carbon (Pg)	Remarks and references
Atmosphere		
Present level	7.6×10^2	$x(CO_2) = 358 \ \mu mol \ mol^{-1}$
Preindustrial	6.0×10^2	Corresponds to $x(CO_2) = 282 \ \mu mol \ mol^{-1}$
Oceans		
Total dissolved CO_2	3.74×10^4	Bolin *et al.* (1981); see Table 11.2
CO_2 in the mixed layer	6.70×10^2	Depth of mixed layer, 75 m.
Living biomass carbon	3	Mainly plankton; Mopper and Degens (1979)
Dissolved organic carbon	1.0×10^3	Average conc. 0.7 g m^{-3}; Williams (1975)
Sediments		
Carbonates, continental, and shelf	2.7×10^7	Carbonates: total of 5×10^7 Pg (Garrels and Perry, 1974), subdivided according to Hunt
carbonates, oceanic	2.3×10^7	(1972), see Table 11.3
Organic carbon, continental, and shelf	1.0×10^7	Total 1.2×10^7 Pg (Schidlowski, 1982); subdivided according to Hunt (1972)
organic carbon, oceanic	0.2×10^7	
Biosphere		
Terrestrial biomass carbon	6.5×10^2	Includes deadwood and plant litter; Ajtay *et al.* (1979)
Soil organic carbon	2×10^3	Estimate of Ajtay *et al.* (1979)
Oceanic organic carbon	1×10^3	From Williams (1975), see above

11.2.1. CARBON DIOXIDE IN THE ATMOSPHERE

The average mixing ratio of carbon dioxide in the atmosphere in 1994 was 358 μmol mol^{-1}. The distribution is fairly uniform, so that one can immediately calculate the total atmospheric content as 2780 Pg. The corresponding mass of inorganic carbon is 760 Pg. The contribution due to other carbon-containing compounds can be neglected. The preindustrial level of atmospheric CO_2—that is, the level that existed in the years around 1850—was about 280 μmol mol^{-1}. Information on the CO_2 mixing ratio 150 years ago can be obtained in several ways. One is to assess the reliability of measurements made in the second half of the nineteenth century; another involves back-extrapolation of CO_2 values based on an estimate of total fossil fuel consumption coupled with auxiliary assumptions about the fraction that has remained in the atmosphere; and a third method relies on CO_2 preserved in ancient ice samples. The first method was applied by Callendar (1958) and by Bray (1959). Their analyses gave 290–295 μmol mol^{-1}. The second approach made use of the data assembled by Keeling (1973a) and Rotty (1981, 1983) for the integrated amount of CO_2 released from combustion and cement manufacture since 1860. The cumulative input by the end of 1977 was 150 Pg. Bacastow and Keeling (1981) have also shown that the increase in atmospheric CO_2 for the period 1958–1978 was equivalent to 54% of the cumulative emissions from fossil fuel combustion during the same period. If one assumes that this fraction has stayed constant since 1850, one obtains the preindustrial carbon content of the atmosphere by difference: 700 − (0.54 × 150) = 619 Pg, which corresponds to 292 μmol mol^{-1}. This calculation does not take into account the CO_2 released from forest clearings, so that the true value must be lower. Rather detailed information on atmospheric CO_2 during the last millennium has been obtained from polar ice cores. Neftel et al. (1983) have studied gas inclusions in Greenland ice cores and reported an atmospheric CO_2 level of 271 ± 9 μmol mol^{-1} 600 years ago. The analysis of Antarctic ice cores by Neftel et al. (1985), Raynaud and Barnola (1985), and Pearman et al. (1986) also showed that during the period 1000–1800 the atmospheric CO_2 mixing ratio ranged from 270 to 290 μmol mol^{-1}. Figure 11.1 shows a reconstruction of the CO_2 increase during the past two centuries from an ice core taken at Siple Station, Antarctica. The record shows that CO_2 started to increase during the first half of the nineteenth century, and that the ice core data coincide with the direct measurements that started in the late 1950s.

Most ground-based measurements of CO_2 in air have revealed diurnal and seasonal variations due to the assimilation and respiration of CO_2 by land

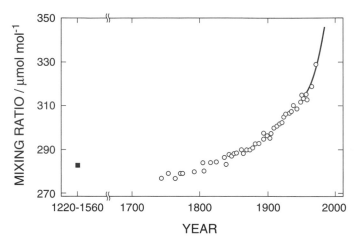

FIGURE 11.1   Reconstructed atmospheric CO_2 mixing ratio from an ice core at Siple Station, Antarctica (open points, except square). The solid line shows the increase of atmospheric CO_2 observed at Mauna Loa. (Reprinted with permission from Neftel *et al.* (1985) and Friedli *et al.* (1986), MacMillan Magazine, Ltd.)

plants (see, for example, Figs. 1.1, 1.2, and Fig. 5 in Junge (1963)). The diurnal variations are rapidly dampened with height above the ground, whereas the seasonal variations permeate the entire troposphere and do not disappear before one transcends the tropopause. Bischof (1977, 1981) has collected an extensive set of data on tropospheric CO_2 variations from measurements by aircraft in the Northern Hemisphere. Bolin and Bischof (1970) have analyzed such data and found them to be consistent with our notions about the large-scale mixing processes in the atmosphere, particularly with a vertical exchange time of about 1 month. In the Southern Hemisphere, the amplitude of the seasonal oscillations exhibits a phase shift of 6 months. In qualitative terms, both observations are readily understood to derive from the decoupled hemispheres, the smaller land area available for support of land biota in the Southern Hemisphere, and the displacement of the seasonal cycle by half a year. The quantitative aspects will be discussed in Section 11.3.3.

11.2.2. THE OCEANS

The total volume of the world ocean is 1.35×10^{18} m^3. The value is obtained by combining the areal extent of the oceans with a mean depth of

3730 m, derived by Menard and Smith (1966) from hypsometric charts of all ocean basins.

The vertical structure of the ocean features a decrease in temperature with increasing depth and a correspondingly stable stratification. A shallow surface layer of 50–100 m thickness is vertically well mixed by agitation by wind force. This portion represents only a small subvolume, but it is of crucial importance to the exchange of CO_2 with the atmosphere. Compared with the bulk of the ocean, the mixed layer responds quickly to changes in the atmosphere and must be treated as a separate reservoir. The depth of the mixed layer is variable. Bolin et al. (1981) recommended a seasonal average depth of 75 m derived by Bathen (1972) from measurements in the Pacific Ocean.

Intermediate between the mixed layer and the deep sea, at depths between 100 and 1000 m, lies the main thermocline, a region where mixing is imperfect and the exchange with the upper layer is slow but nevertheless faster than in the denser and cooler waters of the deep ocean. At high latitudes a mixed layer often cannot be discerned, and the surface waters mix directly with the deeper strata. Circulation in the main body of the ocean is accomplished by the downward motion of cold surface waters near the Poles and a slow up-drift at low latitudes. Estimated rates of the downward flow are 30×10^6 m^3 s^{-1}, as given by Munk (1966) and Gordon (1975) for the Antarctic, and $(10–30) \times 10^6$ m^3 s^{-1}, as derived by Broecker (1979) for the north Atlantic. The resulting turnover time of deep ocean waters is 700–1000 years.

The total amount of carbon dioxide dissolved in the ocean is known from measurements at various locations in the three main ocean basins and the Antarctic. Takahashi et al. (1980, 1981) have provided summaries of the data. Measurement techniques involved infrared gas analysis and potentiometric titration by the Gran (1952) procedure, the latter in conjunction with the determination of total alkalinity. The two methods gave concordant results. Inconsistencies noted by Broecker and Takahashi (1978) in the data from the Pacific and Indian Oceans were later traced to the nonideal behavior of the glass electrodes used in the titration, and appropriate corrections were applied. Table 11.2 lists mean values of total CO_2 obtained for the three depth ranges in seven areas of the world ocean. Figure 11.2 is added to show the average distribution with depth together with other data of interest. The average concentration of CO_2 in the surface layer is 2.0 mmol kg^{-1}, or 2.05 mol m^{-3} if one adopts a value of 1025 kg m^{-3} for the density of seawater. The total mass of inorganic carbon in the mixed layer then amounts to 670 Pg, which is approximately the same as that residing in the atmosphere. The average concentration of dissolved CO_2 for the entire ocean of 2.25 mmol kg^{-1} leads to a total mass content of 37,400 Pg inorganic carbon.

TABLE 11.2 Arithmetic Means of Total CO_2 Concentrations (μmol kg^{-1}) in Seawater in Three Depth Ranges of Seven Ocean Areas, and Averages for the World Ocean[a]

Depth range	North Atlantic	South Atlantic	North Pacific	South Pacific	North Indian	South Indian	Antarctic	Average
0.50 m	1944 ± 25	1961 ± 34	1985 ± 77	1971 ± 12	1933 ± 13	1936 ± 25	2172 ± 49	2002
50–1200 m	2124 ± 96	2142 ± 94	2227 ± 140	2146 ± 112	2211 ± 83	2160 ± 109	2230 ± 47	2182
> 1200 m	2181 ± 21	2212 ± 38	2356 ± 32	2309 ± 31	2330 ± 16	2298 ± 17	2263 ± 12	2288
Percentage area	11.9	9.7	21.9	21.4	3.4	13.6	18.2	—
Whole ocean (Volume-weighted arithmetic mean)								2254

[a] From Takahashi et al. (1981). The data are normalized to 34.78% mean salinity of the world ocean.

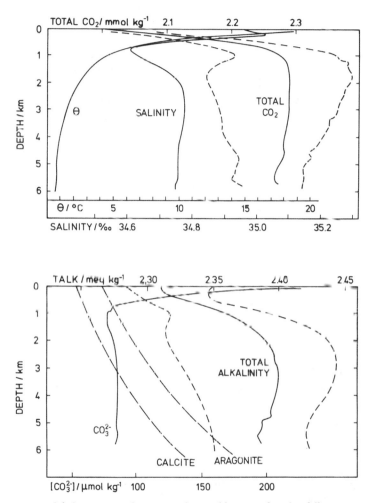

FIGURE 11.2 Global mean conditions in the world ocean for the following quantities: potential temperature Θ (i.e., temperature corrected for adiabatic heating), salinity, concentrations of total CO_2 and CO_3^{2-}, and total alkalinity. (Adapted from Takahashi *et al.* (1981).) Dashed curves indicate the spread of total CO_2 and alkalinity; long dashed curves show the critical dissolution regimes for calcite and aragonite according to Broecker and Takahashi (1978).

The dissolution of CO_2 in seawater involves a set of ion equilibria, which can be expressed by the sequence

$$CO_{2gas} \rightleftharpoons CO_{2aq} \rightleftharpoons HCO_3{-} \rightleftharpoons CO_3^{2-} \rightleftharpoons CaCO_{3solid}$$

The first three relations were discussed in Section 8.4 in conjunction with the dissolution of CO_2 in cloud water. The last reaction is specific to conditions existing in the ocean and leads to solid calcium carbonate as a precipitate. For a concentrated salt solution such as seawater, the effects of complex formation and other ionic interactions cannot be neglected, and ion activities must be inserted in the equilibrium relations (Eq. (8.15)) rather than concentrations. Thus, HCO_3^- forms a complex with Na^+, and a major portion of CO_3^{2-} occurs as a complex with Mg^{2+}. The situation was discussed by Stumm and Brauner (1975). A rigorous description of all of the ion equilibria is not necessary, however, to quantify the carbon dioxide system. For practical purposes it suffices to express the effects by apparent equilibrium constants. The relative concentrations of CO_{2aq}, HCO_3^-, and CO_3^{2-} resulting from Equation (8.15) then refer to the stoichiometric concentrations of free ions plus those bound to other species. The effective equilibrium constants depend on the salinity (salt content) of seawater in addition to temperature, and to some extent on pressure. Edmond and Gieskes (1970), Disteche (1974), Skirrow (1975), and Kennish (1989) have reviewed the data and tabulated them. The salinity of seawater is close to 35%, and the pH lies in the vicinity of 8. One then finds that about 0.6% of dissolved carbon dioxide is present as CO_{2aq}, 90% as bicarbonate, and 9% as carbonate ion. The amount of suspended $CaCO_3$ is small in comparison.

From equilibrium constants and observations of total CO_2, temperature, pressure, salinity, and alkalinity, Takahashi et al. (1981) calculated the concentration of CO_3^{2-} as a function of depth at various locations. Averaged values are shown in Figure 11.2. It is apparent that the concentration of CO_3^{2-} decreases with depth, although that of total CO_2 increases. The behavior is mainly due to the decrease in temperature. Figure 11.2 also shows the critical dissolution curves proposed by Broecker and Takahashi (1978) for the two principal crystalline forms of $CaCO_3$ in the ocean: calcite and aragonite. The solubility of $CaCO_3$ increases with depth because of the decrease in temperature and the increase in pressure. Near the surface, CO_3^{2-} is found to be supersaturated relative to calcium carbonate, whereas at greater depths it is undersaturated.

Our knowledge of carbonate sedimentation was reviewed by Cloud (1965), Bathurst (1975), and Holland (1978). Despite the apparent supersaturation of CO_3^{2-} in the upper strata of the sea, there is little evidence for the direct, inorganic chemical precipitation of $CaCO_3$, except in the warm and shallow waters of the Tropics. Today as well as during the recent history of the Earth, the predominant mode of $CaCO_3$ production takes place by shell-forming organisms such as protozoa (foraminifers), algae (coccolithophores), mollusks, corals, etc., all of them residing in the sunlit surface waters. The majority of carbonate sediments are formed in the nutrient-rich shelf regions and mingle with detritus of continental origin advected by the rivers. In the open sea, the calcareous remains of organisms, while settling toward the ocean floor, eventually enter the region of undersaturation and slowly dissolve. The rate of dissolution increases with depth until at a critical level, called the carbonate compensation depth, the rate of dissolution equals the flux of solid carbonate from above. Below this depth calcium carbonate dissolves rapidly. According to Gieskes (1974), as much as 80% of precipitating $CaCO_3$ redissolves in the deep ocean. Only those skeletal remains that have undergone transformation in the intestines of predators and that as fecal pellets are shielded by an organic coating can escape dissolution and reach deeper strata. From sedimentary evidence assembled by Pytkowicz (1970) and Bathurst (1975), the carbonate compensation level has been shown to occur at depths near 5000 m, although as a kinetic boundary its position must be quite variable.

Organic carbon in the ocean has been the subject of detailed reviews by Menzel (1974), Williams (1975), Parsons (1975), and Mopper and Degen (1979). It is customary to distinguish between particulate and dissolved organic carbon by filtration with a 0.45 μm filter. For both fractions the concentrations fluctuate widely in various parts of the ocean. In addition, there are discrepancies between different analytical techniques, so that estimates of the average content of organic carbon in seawater involve large uncertainties. Concentrations reported for waters below the mixed layer range from zero to 120 mg m^{-3} for particulate and from 0.3 to 1.7 g m^{-3} for dissolved organic carbon. Following Williams (1975), average concentrations of 20 mg m^{-3} and 0.7 g m^{-3}, respectively, may be adopted, which give 30 and 10^3 Pg for the whole ocean. The mass of living biomass, which is concentrated in plankton residing in near-surface waters, has been estimated to amount to 3 Pg; that of bacteria is an order of magnitude smaller.

The dominant source of organic carbon in seawater is the photosynthetic fixation of CO_2 by unicellular algae (phytoplankton) in the photic zone. Their growth by cell division is rapid, but the population is kept in balance by grazing species (zooplankton). Techniques for determining the rate of primary production of marine biota are mainly based on the uptake of radiocarbon (de Vooys, 1979). Estimates for the global rate fall in the range of 23–80 Pg year^{-1}, with a probable value of about 35 Pg year^{-1}. The contribution of kelps, rockweeds, and other macrophytes that grow in coastal regions and feature high local production rates is not globally significant.

Dissolved organic carbon appears to arise mainly from phytoplankton, both directly, through exudation, and indirectly, by the decay of dead cells. The chemical composition of the exudate is complex. Amino acids, carbohydrates, hydrocarbons, and aromatic compounds have been identified. The compilation of Kennish (1989) contains an extensive list of observed substances. A good deal of the material is scavenged by bacteria and other heterotrophs. The remainder appears to undergo polymerization. The molar mass has been found to increase with depth from values near 1500 in the mixed layer to more than 10,000 at depth of 5000 m (Mopper and Degens, 1979). By means of radiocarbon dating, Williams *et al.* (1969) estimated the average age of dissolved organic carbon in deep waters to be 3400 years. The result should be considered a minimum age, owing to the possibility of contamination by ^{14}C generated by atomic bomb tests. But the radiocarbon age provides an upper limit of 0.3 Pg year^{-1} for the flux of dissolved organic carbon required to maintain a steady-state reservoir of 10^3 Pg in the deep ocean. Removal processes probably include metabolic oxidation and conversion to particulate organic carbon, but the details of organic carbon recycling are uncertain. An even smaller fraction of primary organic carbon enters the sediments to be trapped there. Estimates for this fraction range from 10^{-4} to 4×10^{-3} (Garrels and Perry, 1974; Walker, 1974; Mopper and Degens, 1979), depending on location. Low values apply to deep sea sediments; higher values are associated with continental shelf regions. The rate of carbon deposition is difficult to measure, because again, much of the organic material entering the sediment–water interface is subject to microbial utilization and decays.

11.2.3. CARBON IN SEDIMENTARY ROCKS

The planet Earth is chemically differentiated into a core, a mantle, and a crust, which is rich in calcium, aluminum, sodium, potassium, and several other elements that are easily melted out of rock and rise to the surface. Following Anderson (1989), if the earth is likened to a blast furnace, the crust is the lighter slag and the core is the iron at the bottom of the kiln. The crust averages about 20 km in thickness. It is thin underneath the ocean and reaches 60 km under some of the higher mountain ranges. The bulk of the earth's crust consists of igenous rocks that represent primary material of magmatic origin. Exposure to the atmosphere and the action of surface waters cause a small fraction of crustal rocks to undergo weathering and erosion. The debris is washed into the continental and oceanic basins, where it is deposited and compacted. In the course of time, this secondary material is converted to sedimentary rocks.

The total mass of the sediments that has been generated from igneous rocks since the formation of our planet amounts to approximately $G_{sed} = 2.4 \times 10^{21}$ kg, excluding volcanic intercalations, or about 8% of the total mass of the crust. The estimate can be derived in two ways. The direct method makes use of the observed spatial extent of continental sedimentary shields combined with seismic evidence about the depth of the deposits and deep drillings on the continents and in the oceans. Thus, Ronov and Yaroshevskiy (1969) obtained a mass of 2.24×10^{21} kg, excluding volcanogenics. An indirect estimate is obtained from geochemical mass balancing based on the redistribution of elements during the transformation of igneous rocks by weathering. In this manner Garrels and MacKenzie (1971) calculated a mass of 2.34×10^{21} kg, whereas Li (1972) estimated 2.4×10^{21} kg.

The carbon content of igneous rocks (~ 100 mg kg^{-1}) is minor compared to that of the sediments. Table 11.3 shows representative data for the distribution of the three major sedimentary rock types: limestones, shales, and sandstones, and the weight percentages of carbon in each. Garrels and MacKenzie (1971) emphasized the difficulties encountered in estimating the relative proportions of limestones, shales, and sandstones for the totality of all sediments because of uncertainties about the individual contributions from Phanerozoic (younger than $\sim 600 \times 10^6$ years) and Precambrian (older than $\sim 600 \times 10^6$ years) sediments. If one accepts the mass ratios of 15:74.11 that these authors derived from geochemical mass balances, the mass percentage of carbonate carbon is 2.1%, on average, for all three rock types. The resulting mass of inorganic carbon in the sediments is 0.021 $G_{sed} = 5 \times 10^{19}$ kg. Based on the data presented by Ronov and Yaroshevskiy (1989), Hunt (1972) derived a detailed breakdown of inorganic and organic carbon by sedimentary rock type. His summary is included in Table 11.3. The subdivision into oceanic and continental plus shelf sediments gives masses of 0.8×10^{21} kg and 1.4×10^{21} kg, respectively, with inorganic carbon contents of 3×10^{19} kg and 3.4×10^{19} kg. The combined total is 6.4×10^{19} kg. It is interesting to note that Hunt's distribution of inorganic carbon between limestones, shales, and sandstones, which is 80:14:6, compares well with that computed from the data of Garrels and MacKenzie (1971), despite differences in the ratios of sedimentary rock types adopted for the calculation.

The value 5×10^{19} kg for the total mass of inorganic carbon in the sediments was chosen for presentation in Table 11.1. Here one finds the largest pool of carbon, harboring an amount 1000 times greater than that in the atmosphere–ocean system. If all of the carbonate now buried in the sediments were volatilized and the CO_2 were released to the atmosphere, the pressure would rise to about 40 times the current total pressure, and the composition of the atmosphere would resemble that existing on the

TABLE 11.3 Distribution of Carbon in Sedimentary Rocks, Excluding Volcanogenics

Sedimentary rock type	Mass (10⁷ Pg)	Distribution (%)	Carbonate carbon Weight percent	Carbonate carbon Mass (10⁷ Pg)	Carbonate carbon Distribution (%)	Organic carbon Weight percent	Organic carbon Mass (10⁷ Pg)	Organic carbon Distribution (%)
Continental shelf and slope[a]								
Clays and shales	83	59	0.84	0.70	21	0.99	0.82	83
Carbonates	25	18	9.40	2.35	69	0.33	0.08	8
Sandstones	32	23	1.05	0.35	10	0.28	0.09	9
Sum	140			3.40			0.99	
Oceanic[a]								
Clays and shales	34	40	0.68	0.23	8	0.22	0.07	33
Carbonates	35	42	7.90	2.76	91	0.28	0.10	48
Sandstones	15	18	0.28	0.04	1	0.26	0.04	19
Sum	84			3.03			0.21	
Total sediments[a]								
Clays and shales	117	52	0.79	0.93	14	0.76	0.89	74
Carbonates	60	27	8.50	5.11	80	0.30	0.18	15
Sandstones	47	21	0.83	0.39	6	0.27	0.13	11
Sum	224			6.43			1.20	
Total sediments[b]								
Clays and shales	178	75	0.42	0.75	15	0.76	1.35	88
Carbonates	36	15	11.4	4.10	81	0.30	0.11	7
Sandstones	26	11	0.71	0.18	4	0.27	0.07	5
Sum	240			5.03			1.53	

[a] From Hunt (972); inorganic carbon contents originally from Ronov and Yaroshevskiy (1969, 1971).
[b] From data assembled by Garrels and McKenzie (1971) and Hunt (1972).

planet Venus. As will be considered in more detail in Chapter 12, the atmospheres of the terrestrial planets in the solar system originated from thermal outgassing of volatiles from virgin planetary matter, with water vapor and CO_2 providing the principal degassing products. The difference in evolutionary trends between Venus and Earth rose from the more favorable conditions for the condensation of water on Earth, a prerequisite for the formation of an ocean and the formation of carbonate sediments.

As was pointed out earlier, the sediments also inherit a small fraction of reduced carbon as a residue of biological activity. Once trapped, the organic compounds eventually undergo stabilization by polymerization reactions and are diagenetically converted to a product called *kerogen* (Durand, 1980), a geopolymer defined by its insolubility in the usual organic solvents as opposed to soluble bitumen. The formation of kerogen must be viewed as a continuous disproportionation process that goes hand in hand with an increase in the degree of aromaticity and a progressive elimination of lighter organic compounds. Kerogen is essentially immobile, whereas the lighter fraction tends to migrate and may accumulate to form economically exploitable deposits.

Evidence for the biological origin of reduced carbon in sedimentary rocks comes from two sources. One is the existence of chemofossils, that is, characteristic remnants of biologically important compounds more resistant to chemical degradation than others. Notable examples are the isoprenoids pristane (2,4,6,10-tetramethyl pentadecane) and phytane (2,4,6,10-tetramethyl hexadecane), which arise from the decay of chlorophyll (Eglinton and Calvin, 1967; Didyk *et al.*, 1978; McKirdy and Hahn, 1982, Hahn, 1982). The second kind of evidence is obtained from the $^{13}C/^{12}C$ isotope ratio of sedimentary reduced carbon. Schidlowski (1983a) and Schidlowski *et al.* (1983) have reviewed this aspect in some detail. The various assimilatory pathways by which carbon enters the biosphere always favors the light ^{12}C isotope over the heavier ^{13}C carbon, so that the latter is somewhat depleted in organic compounds compared with the inorganic carbon reservoir. The $^{13}C/^{12}C$ ratio in the contemporary biosphere is 2–3% smaller than that of oceanic bicarbonate, which represents the largest pool of inorganic carbon accessible to the biosphere. In the sediments, the isotopic compositions of both organic and inorganic carbon are preserved with only minor alterations. Figure 11.3 shows a synopsis of $^{13}C/^{12}C$ data in the conventional notation, expressing the isotope ratio of a sample by its permille (‰) deviation from that of a standard. The left-hand side shows the spread of ^{13}C values in sedimentary carbonates (C_{carb}) and reduced carbon (C_{org}) over geological times, the record going back almost 3.8×10^9 years into the past. The right-hand side gives recent ^{13}C values observed in various autotrophic organisms and the marine, atmospheric, and sedimentary environments.

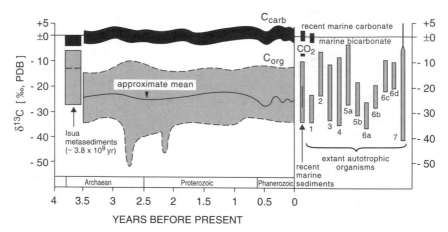

FIGURE 11.3   Left: Geological record for the ^{13}C content (in permille PDB standard) of carbonates (C_{carb}) and organic carbon (C_{org}) in the sediments. The spread of values is indicated. Right: Carbon 13 content in atmospheric CO_2, marine carbonates, organic material of modern sediments, and various organisms: (1) C_3 plants, (2) C_4 plants, (3) CAM (crassulacean acid metabolism) plants, (4) eucaryotic algae (black bar, range for marine plankton, (5) cyanobacteria from (a) natural and (b) culture communities, (6) several nonoxygenic photosynthetic bacteria, and (7) methanogenic bacteria. (Reprinted from Schidlowski (1987), with permission, from the Annual Review of Earth and Planetary Science, Volume 15, © 1987, by Annual Reviews.)

Figure 11.3 demonstrates that the ^{13}C/^{12}C ratio of carbonates is tied to the zero permille line, whereas that of organic carbon is clearly displaced toward negative values scattering around 25‰. The large spread must be taken to reflect variations in the isotope effect by individual organisms and environmental factors.

The constancy of the isotope shift is remarkable. It implies that biogenic organic carbon has been incorporated into the sediments since the earliest Precambrian period, when marine bacteria and blue-green algae were the sole agents of biological carbon fixation. This interpretation is supported by the morphological fossil record, which, if one includes the microfossils and stromatolites (see Section 12.4.3), extends back to 3.5×10^9 years ago. The possibility of a nonbiological formation of reduced carbon has been considered. Schidlowski *et al.* (1983) pointed out, however, that the magnitude of the observed isotope shift is hard to match by nonbiological reactions, and a convincing reaction scheme yielding an effect of the required size has yet to be found.

Since organic and inorganic carbon in the sediments derives from a common source, namely primordial carbon dioxide, the negative ^{13}C isotope shift of C_{org} must be compensated by a corresponding positive isotope shift of inorganic carbon relative to the ^{13}C/^{12}C ratio associated with geochemically undifferentiated primary carbon. This condition simply follows from the requirement of mass balance for ^{12}C and ^{13}C. Expressed in quantitative terms, one obtains an equation relating the ratio of organic to inorganic carbon in the sediments to the isotope shift in the two fractions. Using the conventional notation,

$$\delta^{13}C = 10^3\left[(^{13}C/^{12}C)_{sample}/(^{13}C/^{12}C)_{standard} - 1\right]$$

in permille and setting $f = C_{org}/(C_{org} + C_{carb})$, the mass balance equation can be cast in the form

$$\delta^{13}C_{prim} = (1 - f)\delta^{13}C_{carb} + f\delta^{13}C_{org} \qquad (11.1)$$

Here, $\delta^{13}C_{prim}$ denotes the isotopic composition of the primordial, that is initial, carbon input to the atmosphere–ocean system, commonly adopted as being close to $-5\%o$. With $\delta^{13}C_{carb} = \pm0\%o$ and $\delta^{13}C_{org} = -25\%o$, one finds by solving the above equation that $f \approx 0.2$. The value implies a partitioning of total sedimentary carbon between C_{org} and C_{carb} in the ratio $C_{org}/C_{carb} = f/(1 - f) = 1:4$. The mass of reduced carbon in sedimentary rocks is one-quarter of that of carbonate carbon, or 1.2×10^{19} kg. By adding the two fractions one obtains for the total mass of sedimentary carbon the value 6.2×10^{19} kg.

Direct measurements also provide fairly reliable information on the content of reduced carbon in sedimentary rocks. The data have been reviewed and evaluated by Ronov and Yaroshevskiy (1969), Hunt (1972), and Schidlowski (1982). Table 11.3 presents distributions among sedimentary rock types as given by Hunt (1972). Continental clays and shales are found to contain the highest weight percentages of organic carbon, with an average of about 1%, whereas the averages for carbonates and sandstones are about 0.2%. Marine sediments contain between 0.1 and 0.3%, except those in proximity to the continents. A reliable estimate for the average content of reduced carbon in all of the sediments depends again on the frequency distribution of limestones, shales, and sandstones. Fortunately, the errors are less severe here than for the carbonates, because shales not only represent the largest mass fraction of sedimentary rocks; they also contain the highest fraction of reduced carbon. The detailed breakdown of the data in Table 11.3 leads to an average content of organic carbon in the sediments of 0.54–0.62%. The corresponding mass of reduced carbon is $(1.3–1.5) \times 10^{19}$ kg for all sediments. Only 1% of it occurs in economically exploitable amounts as coal or petroleum.

The last result is remarkably close to that obtained from the $^{13}C/^{12}C$ ratio. It may be combined with the range of estimates for the total mass of inorganic carbon, $(5.0–6.4) \times 10^{19}$ kg, to derive for the ratio C_{carb}/C_{org} a value of 4 ± 0.8, in good agreement with that deduced from the ^{13}C and ^{12}C mass balance. In Table 11.1 a ratio of 4 was adopted and the mass of organic carbon was adjusted accordingly.

11.2.4. THE TERRESTRIAL BIOSPHERE

The last reservoir to be discussed is organic matter accrued by the terrestrial biosphere. Here again, several subreservoirs must be distinguished. One comprises the living biomass of plants (and animals); another leaf litter, deadwood, and other debris; yet another soil humus. The following account is largely based on the review of Ajtay et al. (1979). As in the marine environment, it is photosynthesis that provides the terrestrial biosphere with organic carbon. Net primary productivity is defined as the rate of storage of organic carbon in plant tissue due to the uptake of CO_2 in excess of that released again by respiration. Net primary productivity frequently is given in terms of dry weight of organic matter accumulated per year, and a suitable conversion factor must be used to determine the mass of carbon that is fixed. The carbon content of the biomass is variable but usually higher than that of pure hexose. For living biomass, a conversion factor of 0.45 is commonly employed. For plant litter and soil humus slightly higher values of 0.5 and 0.6, respectively, are required.

Methods for the assessment of terrestrial primary productivity and biomass were reviewed in Lieth and Whittaker (1975). The techniques include harvests from small sample plots, dimensional analysis and census of forest stands, growth relations between different plant tissues (for example, leaf or twig dry weight versus stem dry weight), and gas-exchange measurements. Primary productivity and biomass depend heavily on the type of vegetation considered, and on environmental factors such as water and nutrient supply, temperature, duration of the photoproductive period, etc. Estimation of the world's total productivity and biomass thus involves a classification of the biosphere according to ecosystem types and a detailed accounting of land areas occupied, combined with data characterizing the desired quantities for each ecosystem. Of the numerous estimates reported in the literature as summarized by Whittaker and Likens (1973), only the more recent ones are based on a sufficiently detailed differentiation of vegetation types to be considered reliable.

Table 11.4 compares two detailed estimates of global net primary productivity and biomass. The compilation of Whittaker and Likens (1973, 1975) makes use of the classification and area assignments developed by Lieth (1975) and, as the authors emphasize, reflects the situation existing in 1950. The assessment by Ajtay et al. (1979) followed the same principles but was

TABLE 11.4 Comparison of Two Estimates for Terrestrial Biomass and Net Primary Production Rates

Ecosystem type	Land area[a] (10¹² m²)	Mean biomass[a] (kg m⁻²)	Mass of carbon[a] (Pg)	Land area[b] (10¹² m²)	Mean biomass[b] (kg m⁻²)	Mass of carbon[b] (Pg)	Net primary productivity[a] (kg m⁻² year⁻¹)	Net primary productivity[a] (Pg year⁻¹)	Net primary productivity[b] (kg m⁻² year⁻¹)	Net primary productivity[b] (Pg year⁻¹)
Tropical rain forests	17.0	45	344	10	42	189	2.2	16.8	2.3	10.4
Tropical seasonal forests	7.5	35	117	4.8	25	54	1.6	5.4	1.6	3.4
Evergreen forests	5.0	35	79	3	30	41	1.3	2.9	1.5	2.0
Deciduous forests	7.0	30	95	3	28	38	1.2	3.8	1.3	1.8
Boreal forests	12.0	20	108	10.5	22	105	0.8	4.3	0.93	4.4
Woodlands and shrubs	8.5	6	22	4.5	12	24	0.7	2.7	1.1	2.2
Savanna	15.0	4	27	22.5	6.5	66	0.9	6.1	1.75	17.7
Temperate grassland	9.0	1.6	6	12.5	1.6	9	0.6	2.4	0.78	4.4
Tundra and alpine	8.0	0.6	2	9.5	1.4	6	0.14	0.5	0.22	0.9
Semi-desert shrubs	18.0	0.7	6	21	0.8	7	0.09	0.7	0.14	1.3
Cultivated land	14.0	1	6	16	0.5	3	0.65	4.1	3.6	3.3
Swamps and marshes	2.0	15	14	2	15	12	3.0	2.7	0.94	6.8
Miscellaneous	26	0.08	1	30	0.4	6	0.4	0.4	0.10	1.3
Sums	149		327	149		560		52.8		59.9

[a] From Whittaker and Likens (1975).
[b] From Ajtay et al. (1979).

based on newer data and a slightly more detailed subdivision according to ecosystem types. With regard to primary productivity, the two estimates are in reasonable agreement and are consistent with earlier estimates of Olson (1970) and SCEP (1970). The data suggest a global productivity of about 55 Pg year^{-1}. The Russian school (Bazilevich et al., 1971) came up with a higher value of 77 Pg year^{-1}. It was based on a more detailed accounting, but appears to have overestimated the productivities in some ecosystems and thus should be considered an upper limit. All results indicate that the net primary productivity on land exceeds that in the oceans in spite of the smaller area available to terrestrial life.

With regard to living biomass, the two estimates in Table 11.4 agree that forests harbor more than 80% of the world's biomass. In contrast to primary productivity, however, a disparity exists for total biomass. The difference arises mainly from the smaller area of forest stands adopted by Ajtay et al. (1979) compared to that of Whittaker and Likens (1975). Specifically the area of tropical rain forests is smaller, and it is compensated for by a greater area of savanna and grasslands. There is no obvious reason for the difference, so it must mainly reflect the uncertainties inherent in such estimates. The global biomass presumably declined since 1950 because of a great deal of deforestation in recent years, especially in the Tropics. The loss of carbon resulting from changes in tropical land use is currently estimated as 1.6 ± 1.0 Pg year^{-1} (Schimel et al., 1995), which would accumulate to a total of 50 ± 30 Pg over a period of 30 years, which is much less than the difference between the two estimates in Table 11.4, so that it cannot be the source of the discrepancy.

Various other assessments of the global biomass may be cited for comparison. Bowen (1966) reported 518 Pg of carbon, Bolin (1970) 450 Pg, Baes et al. (1976) 680 Pg, and Bazilevich et al. (1971) 1080 Pg. The last estimate refers to a reconstructed plant cover of the Earth without corrections for agricultural areas and forest cuts, assuming an optimum of vegetation for each of the bioclimatic zones considered. Bazilevich et al. (1971) were mainly interested in the potential of resources, and their value should be considered an upper limit. Terrestrial animals represent only a small fraction of total living biomass and, according to the estimates of Bowen (1966) and Whittaker and Likens (1973), may be neglected in this context.

The preceding estimates do not yet include deadwood and plant litter and must be amended. Data on dead phytomass still attached to living plants are scarce and extrapolation is difficult. Ajtay et al. (1979) collected a number of data for above ground dead biomass in various ecosystems, from which they concluded that total standing dead material equals roughly 5% of living biomass.

Dead material shed by plants is called *litter*. Reiners (1973) estimated total litter fall by four different procedures and derived values ranging from 37 to

64 Pg carbon year^{-1}. The results were largely based on the estimates of Whittaker and Likens (1973) for the areal extent, primary productivity, and mean biomass of different ecosystems. Ajtay *et al.* (1979) repeated Reiners' calculations using their own data and obtained for the annual litter fall values between 37 and 49 Pg carbon year^{-1}. One of their detailed estimates is shown in Table 11.5. These results demonstrate that the global primary production of the terrestrial biosphere, about 55 Pg carbon year^{-1}, turns mainly to plant litter. Ajtay *et al.* further estimated a consumption by herbivores of about 6 Pg year^{-1}, including consumption by domestic livestock. This leaves about 10% of net primary production to be incorporated into the living biomass annually.

Observational data for the mass of litter are available mainly for forest ecosystems but are scarce for savanna and grasslands, and estimates are correspondingly tenuous. The data of Ajtay *et al.* (1979) in Table 11.5 suggest a value of 60 Pg for the total mass of carbon contained in plant litter of all types. The amount represents about 10% of carbon in living biomass. If one adds the 5% assumed to exist in standing dead material, the total mass of carbon in the biosphere, both living and dead, increases to 650 Pg, and this value is entered in Table 11.1.

Finally, carbon occurs in organic matter contained in soils. Again, it is necessary to extrapolate from field samples and world maps of soil types to obtain an estimate for the global mass of carbon present in soils as humus. Soil carbon is particularly difficult to assess because it generally decreases exponentially with depth, and it is not surprising that estimates vary widely. The earlier results of Waksman (1938), Bolin (1970), and Baes *et al.* (1976) indicated 400–1080 Pg of carbon. More recent studies have led to higher values. Bohn (1976) and subsequently Ajtay *et al.* (1979) presented detailed data based on UNESCO world soil maps. They determined the global mass of soil carbon as 2950 and 2070 Pg, respectively. Schlesinger (1977) noted the close relationship between humus accumulation and temperature and wetness as environmental factors. From his data, and using the humus content of soils according to ecosystem types, Ajtay *et al.* (1979) prepared another detailed account from which they derived an independent estimate of 1640 Pg, as Table 11.5 shows. The value chosen here, which is entered in Table 11.1, lies midway between the extremes of the two estimates.

11.3. THE GLOBAL CARBON CYCLES

The carbon reservoirs described in the preceding sections are coupled to each other by a variety of exchange processes. The resulting total network of fluxes is complex, but by considering the nature of the interactions, the fluxes can be arranged into a system of closed cycles that can be dealt with

TABLE 11.5 Rates of Plant Litter Fall, Mass of Plant Litter, and Organic Soil Carbon in Different Ecosystem Types[a]

Ecosystem type	Surface area[b] (10^6 km^2)	Litter fall[c] (kg m^{-2} year^{-1})	Total rate as carbon (Pg year^{-1})	Litter mass[c] (kg m^{-2})	Total mass as carbon (Pg)	Organic soil carbon (kg m^{-2})	Total soil carbon (Pg)
Tropical rain forests	10	1.85	8.3	0.65	3.3	8	80
Tropical seasonal forests	4.5	1.3	2.6	0.85	1.9	9	40.5
Mangroves	0.3	0.6	0.1	10.0	1.5	8	2.4
Temperate forests	7.5	0.86	2.8	2.5	9.4	12	90
Boreal forests	9	0.59	2.4	3.5	15.8	15	135
Temperate woodlands	2	1.22	1.1	2.5	2.5	12	24
Chaparral, maquis, brushland	2.5	1.0	1.1	0.5	0.6	12	30
Savanna grassland	19	1.5	12.8	0.35	3.3	12	228
Savanna woodland	3.5	0.8	1.3	0.35	0.6	10	36
Temperate grassland, dry	7.5	0.55	1.9	0.5	1.3	30	225
Temperate grassland, wet	5	0.9	2.0	0.325	1.2	14	70
Tundra, high arctic	5.1	0.105	0.23	0.36	0.9	6.5	33.2
Scrub tundra	4.4	0.2	0.4	5.0	11.0	20	88
Desert, semidesert	21	0.125	1.2	0.1	1.1	8	168
Extreme desert	9	0.015	0.06	0.015	0.1	2.5	22.5
Swamps, marshes, bog, and peatland	3.5	0.6	0.95	2.5	4.4	64	225
Cultivated land	16	0.425	3.1	0.05	0.4	8	128
Human areas	2	0.3	0.2	0.3	0.2	5	10
Sums			42.6		59.5		1636

[a] Slightly condensed from Ajtay et al. (1979). Conversion factors for carbon in dry organic matter: litter fall 0.45, litter mass 0.5.
[b] Note: 17.5×10^6 km^2 are perpetual ice, lake, and streams, etc., which are not included.
[c] Dry mass

individually. Table 11.6 gives an overview. Associated with each system of cycles is a turnover time (or residence time) of carbon in the atmospheric reservoir. The turnover time provides a useful criterion for the classification of cycles. There are two main groups. One system of cycles, which arises from geochemical processes, has time constants of several thousand years. On this time scale, the atmosphere and the ocean tend to be in equilibrium. The other set of cycles describes the exchange of CO_2 with the biosphere and with the ocean. In this case, the characteristic times are on the order of 10 years. The great difference in time scales shows that the groups of cycles are essentially decoupled and can be treated separately.

11.3.1. THE GEOCHEMICAL CYCLES

The slow carbon cycles are intimately connected with the sedimentary rock cycle, which involves all crustal elements. Although the interest here is centered on carbon, we must discuss several related elements but can do so only superficially. For aspects not covered, especially those of mineralogy, reference is made to the geochemistry texts of Krauskopf (1979) and Garrels and McKenzie (1971).

The salient features of the geochemical (rock) cycles are shown in Figure 11.4 in the form of a simplified box model. The driving forces are volcanism, tectonics, and weathering reactions. Volcanic exhalations supply carbon dioxide from the earth's mantle to the atmosphere, and weathering causes erosion and chemical breakdown of rocks by reactions involving CO_2 dissolved in terrestrial surface waters. With regard to carbon, one may distinguish three types of reactions: (1) the weathering of silicates, (2) the

TABLE 11.6 Overview on Carbon Cycles Involving Atmospheric CO_2 and the Associated Turnover Times in the Atmosphere

Type of cycle	Time constant (years)
Geochemical rock cycles	$(2.4-30) \times 10^3$
(see Table 11.8 for details)	
Exchange with the ocean	
Mixed layer (Table 11.9)	4–10
Deep ocean	20
Exchange with the biosphere	
(see Table 11.10 for specification)	
Short-term storage	15
Long-term storage	75
Soil humus	200

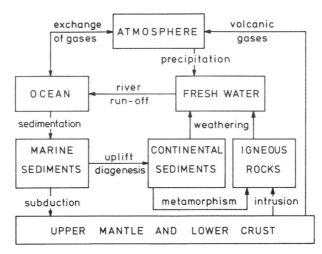

FIGURE 11.4   Geochemical cycles involving igneous and sedimentary rocks. The cycles are driven by the weathering action of atmospheric CO_2 dissolved in terrestrial surface waters, by the supply of CO_2 to the atmosphere from volcanic exhalations, and by the tectonic displacement of crustal material.

weathering of carbonates, and (3) the oxidation of organic carbon in sedimentary rocks by atmospheric oxygen. The inorganic reactions will be treated first.

Feldspars comprise the major fraction of igneous rocks. They consist of aluminosilicates of sodium, potassium, and calcium. Their weathering may be exemplified by the following generic reactions:

$$2KAlSi_3O_8 + 2CO_2 + 3H_2O \rightleftharpoons Al_2Si_2O_5(OH)_4 + 4SiO_2 + 2K^+ + 2HCO_3^-$$

K-feldspar   kaolinite   quartz

$$CaAl_2Si_2O_8 + 2CO_2 + 3H_2O \rightleftharpoons Al_2Si_2O_5(OH)_4 + Ca^{2+} + 2HCO_3^-$$

anorthite   kaolinite

These equations are summations of several individual processes that convert igneous rock to clay minerals, quartz, and dissolved salts. Following a suggestion of Sillén (1963), who noted that the reactions resemble an acid–base titration, most authors (e.g., Siever, 1968; Pytkowitz, 1975) now present them in the form

$$2KAlSi_3O_8 + 2H^+ + 2A^- + H_2O \rightleftharpoons Al_2Si_2O_5(OH)_4 + 4SiO_2 + 2K^+ + 2A^-$$

It is then apparent that the characteristic feature of aluminosilicate weathering is the fixation of hydrogen ions in the place of cations that are liberated. The source of the hydrogen ions is left unspecified, and A^- may be any suitable anion. A we have seen in Chapter 8, the acidity of rainwater is determined by acids stronger than H_2CO_3. Once these are consumed, however, H_2CO_3 takes over. In soils, moreover, the abundance of CO_2 usually is greater than in the atmosphere because of biological decay processes. Analyses of groundwaters beneath granitic and basaltic rocks are shown in Table 11.7. HCO_3^- is the major anion, demonstrating the significance of CO_2 as a weathering agent.

Sedimentary rocks also contain silicates, which are weathered in the same manner as igneous silicates. Of particular importance to the carbon cycle are sedimentary limestones. They occur mainly as calcite ($CaCO_3$), with a smaller fraction of dolomite [$CaMg(CO_3)_2$]. Both carbonates are poorly soluble in water, whereas the corresponding bicarbonates dissolve easily. The weathering of carbonates may be written as

$$CaCO_3 + H^+ + HCO_3^- \rightleftharpoons Ca^{2+} + 2HCO_3^-$$

$$CaMg(CO_3)_2 + 2H^+ + 2HCO_3^- \rightleftharpoons Ca^{2+} + Mg^{2+} + 4HCO_3^-$$

Here, hydrogen ions are used to mobilize carbon as bicarbonate and, as in the case of silicate weathering, an equivalent amount of CO_2 is used up to establish a charge balance.

The weathering products from both igneous and sedimentary rocks eventually enter the rivers and are swept into the ocean. The river load contains suspended particulate material in addition to dissolved salts. The relative proportions are indicated in Table 11.7. Silicon is transported mainly in the quartz and clay mineral particles of the suspended load. Only a very small fraction appears to be dissolved as silicic acid. The alkali and alkaline earth metals are abundant in both fractions. Sodium and chloride are not exclusively weathering products but are in part recycled via the atmosphere as sea salt. Sulfate and magnesium are less affected in this way, but a significant part of sulfate is of anthropogenic origin. Holland (1978) has given a thorough discussion of the provenances of individual river constituents.

In the ocean, the material advected by the rivers participates in the formation of new sediments, whereby the geochemical cycle is closed. The suspended load is redeposited in the continental shelf regions. The dissolved elements are temporarily stored in the ocean reservoir before they, too, are incorporated in sedimentary deposits. In the long run the river influx must be balanced by an appropriate sedimentary output for each element considered. A certain degree of accumulation is nevertheless evident from the higher concentrations of the principal ions in seawater as compared to those in river waters (Table 11.7, columns 6 and 7). Residence times in the ocean

TABLE 11.7 Comparison of Dissolved and Suspended Components in Natural Waters[a]

| Constituent | Molar distribution (%) of dissolved constituents | | | Concentrations (mg kg^{-1}) | | | Residence time in the ocean (10^6 years) |
| | Groundwater | | Rivers | River load | | Ocean, dissolved | |
	Granitic	Basaltic		Dissolved	Suspended		
SiO_2	14.2	12.5	9.1	13.1	244	62	0.014
Al	—	—	—	—	35	0.002	—
Fe	0.2	0.06	0.5	0.7	25	0.002	0.0001
Mg^{2+}	6.3	10.4	7.0	4.1	5.3	1,294	10[c]
Ca^{2+}	16.1	12.3	15.6	15	11	412	0.8
Na^+	10.3	8.9	11.4	6.3	7.6	10,760	95[c]
K^+	1.6	1.6	2.4	2.3	4.2	399	5.0
HCO_3^-	43.1	44.4	39.9	58.4	—	145	0.075
SO_4^{2-}	3.9	1.9	4.9	11.2	—	2,712	8[c]
Cl^-	4.2	7.9	9.2	7.8	—	19,350	225[c]
Average total concentration (mg kg^{-1})	221	334	120	119	332[b]		

[a] Dissolved constituents in groundwaters beneath granitic and basaltic rocks, from White et al. (1963); dissolved constituents in river waters, world average (Livingstone, 1963); suspended load in rivers, composition from Garrels and MacKenzie (1972), combined with estimate of world river efflux, 18 Pg year^{-1} (Holeman, 1968); dissolved constituents in seawater, from Table 7.16; and residence time of elements in the ocean based on river input, 4.6×10^4 Pg year^{-1} (Holland, 1978).

[b] Total load adopted is 400 mg kg^{-1}. The metals usually are reported as oxides, so that the difference is due to bound oxygen.

[c] After correction for the contribution by sea salt, assuming two-thirds of Cl^- in the river load is derived from fallout of sea salt. Note that the aerial transport of aerosol particles from the continents to the oceans (Table 7.11) is negligible compared to that of the suspended load annually discharged by the rivers.

can be computed by means of Equation (4.11) from the ratio of the reservoir's mass content to the observed river influx. Residence times are shown in the last column of Table 11.7. The values must be considered order-of-magnitude estimates because of the implicit assumption that the current continental runoff rates have persisted throughout geological times. The longest residence times are those of sodium and chloride, but even for these elements the values are much shorter than the age of the Earth, indicating that they have been recycled many times.

The fate of alkali and alkaline earth elements in the ocean has intrigued geochemists for some time. Clearly identified removal mechanisms are the deposition of $CaCO_3$, as described earlier, which provides the major loss process for calcium and inorganic carbon, and the formation of salt deposits (evaporites) by the evaporation of seawater in closed-off estuaries. Although evaporites form a relatively minor fraction of all sediments, it appears that chloride is lost mainly by this route, as is sodium to a significant extent.

The removal processes for potassium and magnesium are less obvious. Sillén (1961), who considered seawater to be in equilibrium with the underlying oceanic sediments, was impressed by the potential ion exchange capacity of marine clay minerals. Indeed, in geochemical mass balances most potassium and much magnesium are recovered in shales. Following Sillén's proposal, a number of workers (e.g., MacKenzie and Garrels, 1966; Siever, 1968; Pytkowicz, 1975) have advocated the reconstitution of silicates in the ocean's sediments by reactions such as

$$5Al_2Si_2O_5(OH)_4 + 4SiO_2 + 2K^+ \rightleftharpoons 2KAl_5Si_7O_2(OH)_4 + 2H^+ + 5H_2O$$

kaolinite   quartz   illite

which result in the release of hydrogen ions and thus represent a reversal of weathering. The hydrogen ions would eventually combine with bicarbonate and allow CO_2 to return to the atmosphere. MacKenzie and Garrels (1966) constructed a set of reactions for the purpose of mass balance considerations, which predict that reverse weathering removes magnesium largely as chlorite, potassium as illite, and sodium partly as montmorillonite (all are clay minerals). The transformation must be viewed strictly as model reactions, however. Drever (1974) and Holland (1978) have reviewed the observational evidence for this scheme and found little support for it. Magnesium in particular presents a problem. As discussed by Bathurst (1975) and Holland (1978), the direct deposition of dolomite in the ocean is difficult, in spite of an apparent supersaturation of magnesium, and most dolomite formations have resulted from alterations of calcium carbonates subsequent to their deposition. Holland pointed out that the reaction of seawater with basalts of the midoceanic ridges should not be neglected; here magnesium is preferentially fixed in the place of potassium and calcium, which are liberated. Although the process helps to improve the magnesium balance, it raises problems with the potassium and calcium inputs.

When the impact of these processes on the inorganic carbon budget is considered, the following three cycles can be discerned in Figure 11.4:

1. One cycle of carbon involves the mobilization of carbonate during the weathering of limestones, transfer of Ca^{2+} and HCO_3^- to the ocean, and the redeposition as calcium carbonate. The cycle is closed by the diagenesis (postdeposition change) of the precipitate to carbonate rocks, followed by tectonic uplift to the surface of the continents. The fate of magnesium from dolomites is currently not obvious. On a molar basis, the Mg/Ca ratio in all limestones is 0.25. If magnesium were used to replace calcium in suboceanic sediments, the cycle would be closed because the calcium released would also eventually form carbonate deposits. Otherwise the cycle would not be completely closed.

2. In spite of many uncertainties still inherent in the concept of reverse weathering, it appears reasonable to assume to a first approximation that the amount of CO_2 tied up as bicarbonate during the weathering of silicates is released again to the atmosphere–ocean system when the cations involved are removed by incorporation into the sediments. The corresponding flow of inorganic carbon represents an approximately closed cycle from the atmosphere via rock weathering and river transport into the ocean, and back again into the atmosphere. The carbon flux within this cycle is augmented by the amount of atmospheric CO_2 consumed in the weathering of limestones, as an equivalent portion is set free upon the formation and deposition of $CaCO_3$ in the ocean.

3. An exception to the above scheme must be seen in the weathering of calcium-bearing silicates in igneous rocks. In this case only one-half of the CO_2 consumed in the mobilization of calcium is returned to the atmosphere when calcium is removed from the ocean; the other half enters the $CaCO_3$ reservoir of the sediments. The resulting loss of CO_2 from the atmosphere must be balanced by volcanic exhalations if stationary conditions are to prevail. It is probable that in the long run, the cycle is closed by the subduction of carbonate sediments into regions of the upper mantle due to plate tectonic activity (cf. Figure 11.4). The subduction zones are located in coastal regions, where marine and continental plates meet.

The magnitudes of the fluxes associated with the three cycles can be estimated from the calcium and magnesium data in Table 11.7 and auxiliary information. According to Blatt and Jones (1975), roughly 25% of the continental surface is covered with igneous rocks; in the remainder sedimentary rocks are exposed to weathering. Garrels and MacKenzie (1971) reported the CaO content of igneous rocks to be 4.9% by mass, that of limestones 43%, shales 1.5%, and sandstones 3.2%. With the distribution of these rocks in continental sediments given in Table 11.2, one finds that the ratio of calcium from silicates in igneous rocks to that from sedimentary carbonates is about 12:88. The contribution to dissolved calcium in river waters stemming from evaporite gypsum has been estimated by Garrels and MacKenzie (1972) and Holland (1978) to be 6% and 9%, respectively. The correspondingly adjusted concentration of Ca^{2+} in river water, 14 mg kg^{-1}, leads to a global calcium flux of

$$\left(14 \times 10^{-3}/40\right)\left(4.6 \times 10^{16}\right) = 16.1 \text{ Tmol year}^{-1}$$

where 4.6×10^{16} kg year^{-1} is the world river input into the ocean according to Holland (1978). About 12% of it is associated with the weathering of silicates. The remaining fraction, which is assigned to carbonate rocks, must be increased by about 25% to account for the contribution of magnesium carbonates. Roughly one-half of the magnesium found dissolved in river water is released in this manner. The resulting carbon fluxes in cycles (1) and (3) are 17.8 and 1.9 Tmol year^{-1}, respectively, or 213 and 23 Tg year^{-1}. The average bicarbonate content of river effluents of 58 mg kg^{-1} corresponds to a total carbon flux of 44 Tmol year^{-1} or 530 Tg year^{-1} of carbon. The excess of 295 Tg year^{-1} over the sum of the fluxes in cycles (1) and (3) must be attributed to the carbon flux in cycle (2).

Table 11.8 summarizes the results. Again it should be emphasized that the fluxes derived are current ones and that the uncertainties are considerable, even though Garrels and MacKenzie (1972), Garrels and Perry (1974), and Holland (1978), all using similar procedures, have obtained comparable results. To indicate the range of values, the estimates of the first authors are included in Table 11.8. The fluxes may also be used to calculate the residence times of carbon in the participating reservoirs, and Table 11.8 includes these data. The longest residence times are associated with the sediments because of of the high carbon content of this reservoir. But values of a few hundred million years indicate that carbonates have been recycled at least 10 times since the first deposits occurred after the planet Earth formed 4.5×10^9 years ago.

The inorganic carbon cycles must be supplemented by yet another cycle, cycle (4), which results from the weathering of reduced carbon in sedimentary rocks. The geological record indicates that if occasional fluctuations can be disregarded, the content of kerogen in sedimentary rocks has stayed remarkably constant throughout geological time (Schidlowski, 1982). This observation implies the existence of a stationary state, because sedimentary rocks have been recycled several times, and fresh sediments are always endowed with organic detritus from the marine biosphere. One would therefore expect sedimentary carbon to accumulate with time, were it not

TABLE 11.8 Carbon Fluxes Active in the Geochemical Cycles of Inorganic and Organic Carbon and the Corresponding Residence Times in the Three Participating Reservoirs

Type of cycle	Carbon flux (Tg year^{-1})			Residence time[b] (year)		
	Designation	a	b	Atmosphere	Ocean	Sediments
1. Inorganic carbon from the weathering of carbonate rocks redeposited as $CaCO_3$	F_1	132	213	—	1.7×10^5	2.3×10^8
2. CO_2 used up in the weathering of silicates and carbonate rocks and returned to the atmosphere	F_2	210	295	2.4×10^3	1.3×10^5	—
3. CO_2 used up in the weathering of calcium-bearing silicates in igneous rocks, redeposited as $CaCO_3$	F_3	10	23	3.0×10^4	1.6×10^6	2.2×10^9
4. Oxidation of organic carbon in weathered sedimentary rocks and incorporation of recent organic carbon in new sediments	F_4	30	60	1.2×10^4	6.2×10^5	2.1×10^8

[a] Garrels and MacKenzie (1972); Garrels and Perry (1974).
[b] Present estimates.

removed in the exterior branch of the rock cycle. For this reason it is generally assumed that reduced carbon is oxidized to CO_2 upon exposure to air, even though kerogen is recognized to be chemically rather inert. The mechanism of oxidation is unknown, but it may be mediated by bacteria (Shneour, 1966). Unfortunately, the fate of fossil carbon cannot be traced, because in river waters it is masked by the much larger burden of young biogenic matter from the terrestrial biosphere. Attempts of Erlenkeuser et al. (1974) and by Baxter et al. (1980) to determine by the ^{14}C method the age of surficial organic carbon in recent sediments in bay areas have led to ambiguous results due to contamination by anthropogenic material. The ^{14}C ages in deeper strata cannot be extrapolated to the sediment's surface without using auxiliary assumptions. If a linear extrapolation is allowed, the published data would indicate a contribution of fossil carbon amounting to 10–30% of the total deposition of organic carbon. Thus some accumulation may take place. In the following, however, it is assumed that the recycling of ancient reduced carbon can be neglected.

A rough estimate for the flux of carbon within this cycle is obtained from the average content of C_{org} in sedimentary rocks, which is about 0.5% as discussed in Section 11.2.3, combined with the total annual sediment discharge of the world's rivers. The latter is dominated by suspended material, for the flux of which Holeman's (1968) estimate of $18 \times Pg$ years^{-1} may be used. Adjustment is necessary to account for the contribution from igneous rocks. If one assumes as previously that sedimentary rocks contribute 75% to all weathered material, the flux of organic carbon will be $(5 \times 10^{-3})(0.75 \times 1.8 \times 10^4) = 68$ Tg years^{-1}. As Table 11.8 shows, the flux derived by Garrels and MacKenzie (1972) is smaller by a factor of 2, which demonstrates the great uncertainties involved in such estimates.

The flux of reduced carbon should be consistent with the observed ratio of carbonate carbon to organic carbon in sedimentary rocks. Recall that carbonate carbon is concentrated in limestones, and organic carbon in shales, and assume for simplicity that the two types of rocks are weathered at the same rate. For steady-state conditions, the flux balance equations then read

$$F_1 - aG(C_{carb}) = 0$$

$$F_3 - bG(C_{carb}) = 0$$

$$F_4 - (a + b)G(C_{org}) = 0$$

Here, the flux numbering used in Table 11.8 is retained. We denote by a the nominal exchange coefficient for the weathering of sedimentary rocks, and by b the coefficient associated with the subduction of sediments into regions of the upper mantle (compare Figure 11.4). By combining the three equations one obtains

$$F_4 = (F_1 + F_3)G(C_{org})/G(C_{carb})$$

$$= (213 + 23) \times 0.25 = 60 \text{ Tg year}^{-1}.$$

The result agrees well with that derived above from the total sediment discharge of the world's rivers.

The production of organic carbon by photosynthesis represents in effect a separation of carbon from oxygen in CO_2, and the burial of organic carbon in the sediments accordingly leaves an equivalent amount of oxygen behind. The mass of oxygen in the atmosphere is understood to have been generated in this fashion. This aspect will be followed up in Chapter 12. In the closed carbon cycle, the annual production of O_2 resulting from the incorporation of organic carbon into the sediments is balanced by a corresponding loss due to the weathering of reduced carbon in sedimentary rocks.

11.3.2 EXCHANGE OF CO_2 BETWEEN ATMOSPHERE AND OCEAN

The fact that the Pacific, the Atlantic, and the Indian Oceans are separate entities with mixing characteristics of their own makes it difficult to describe the world ocean in a realistic manner by box models, and rather detailed three-dimensional global general circulation models are now used (e.g., Sarmiento et al., 1992). Nevertheless, to illustrate the most salient features of the exchange of CO_2 between atmosphere and ocean, the approach adopted here is based on a simple two-box model, which subdivides the ocean into a wind-mixed layer and the deep sea. Figure 11.5 shows how they are connected to the atmosphere. The exchange of CO_2 takes place mainly via the mixed layer. The polar currents subsiding from the surface directly into the deep ocean are comparatively small and will be ignored, even though Siegenthaler (1983) first showed that they are not entirely negligible. Keeling (1973b) has given a thorough mathematical analysis of the reservoir system shown in Figure 11.5. The main deficiency of the two-box ocean model is its neglect of the transition zone of the main thermocline, which connects the mixed layer with the deep sea. Oeschger et al. (1975) developed a diffusion model that better accounts for the transport of material in the domain of the thermocline, but compared with the three-dimensional models, the variation of the exchange coefficient with latitude cannot be simulated in such models. The shortcomings of the two-box model are felt most severely when one

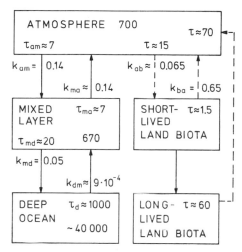

FIGURE 11.5 Exchange of CO_2 between atmosphere and ocean in a two-box model. Reservoir contents are given in Pg of carbon, exchange coefficients in year^{-1}, and residence times in years. For comparison, the exchange between the atmosphere and the biosphere is shown in parallel. The biosphere is represented by only two reservoirs for simplicity. A more detailed network of reservoirs for the biosphere is shown in Figure 11.7.

considers variations on a time scale that is short compared with the turnover of the deep ocean, because the usual assumption of a well-mixed reservoir does not apply. As Keeling (1973b) pointed out, however, the two-box model approach requires only that the outgoing fluxes be proportional to the total content of the reservoir, whereby the usefulness of the model is considerably extended.

Consider first the exchange of CO_2 between the atmosphere and the mixed layer of the ocean. The majority of studies concerned with CO_2 exchange have made use of radiocarbon data. Carbon 14 is formed naturally in the atmosphere by the interaction of cosmic ray-produced neutrons with nitrogen. The global ^{14}C production rate is about 3×10^{26} atom year^{-1} (Lingenfelter, 1963; Lingenfelter and Ramaty, 1970). Temporarily, ^{14}C is stored in the atmosphere, the biosphere, and in the mixed layer of the ocean before it is transferred to the deep sea, where it decays with a half-lifetime of 5730 years. Nuclear weapons tests beginning in 1954 and culminating in the early 1960s caused additional short-term inputs, which resulted in a drastic increase in atmospheric ^{14}C levels. The history of excess ^{14}C in the troposphere is depicted in Figure 11.6. A significant fraction of bomb ^{14}C was injected into the stratosphere. The annual oscillations appearing superimposed on the decline of ^{14}C activity following the 1963 test moratorium must be attributed to the seasonal variation of air exchange across the tropopause. The response of the Southern Hemisphere lags behind by about 1 year because of the equatorial barrier to eddy transport imposed by the interhemispheric tropical convergence zone. Figure 11.6 also shows the rise of excess ^{14}C concentration in the surface waters of the ocean. Here, a new steady state was reached about 4 years after the last major injection of ^{14}C into the atmosphere.

Exchange rates between the troposphere and the mixed layer of the ocean can be derived from the ^{14}C data in several ways: (1) by considering the steady-state flux of natural ^{14}C from the atmosphere through the mixed layer toward the deep ocean; (2) from the inventory of excess bomb ^{14}C in the atmosphere and the ocean; and (3) from the change in the natural $^{14}CO_2/^{12}CO_2$ ratio caused by the addition of inactive CO_2 to the atmospheric reservoir, the dilution resulting from the combustion of fossil fuels (Suess effect). Independently, one may also use the stagnant film model described in Section 1.6 to estimate the CO_2 exchange rate. Table 11.9 lists results obtained by various investigators and the equivalent residence times of CO_2 in the atmosphere. They average $\tau_{am} \approx 7$ years. Below, we shall illustrate the procedure employed to derive k_{am} and k_{md} by the first of the methods indicated above.

The mass contents of ^{12}C in the atmosphere, the mixed layer, and the deep ocean are abbreviated by G_a, G_m, and G_d, respectively. The corresponding quantities for radiocarbon are indicated by an asterisk. Fluxes are

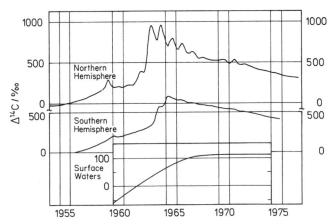

FIGURE 11.6 History of ^{14}C activity in the troposphere and in the surface waters of the ocean (inset). (Adapted from the compilation of Tans (1981).) The data are given as permille deviation from an oxalic acid standard.

TABLE 11.9 Transfer of CO_2 from the Atmosphere to the Mixed Layer of the Ocean[a]

Authors	Method[b]	k_{am} (year^{-1})	τ_{am} (year)
Münnich and Roether (1967)	(a)	0.18	5.4
Bien and Suess (1967)	(a)	0.04	25
Nydal (1968)	(a)	0.1–0.2	5–10
Young and Fairhall (1968)	(a)	0.25	4
Machta (1971)	(a)	0.46	2.15
Revelle and Suess (1957)	(b)	0.1	10
Fergusson (1958)	(b)	0.14–0.5	2–7
Craig (1957)	(c)	0.1–0.25	7 ± 3
Broecker (1963)	(c)	0.1–0.14	7–10
Oeschger et al. (1975)	(c)	0.137	7.3
Peng et al. (1979)	(d)	0.114	8.8
Stuiver (1981)	(a)	0.14	6.8

[a] Transfer coefficients k_{am} and the corresponding residence times of CO_2 in the atmosphere as deduced by various investigators.

[b] Methods: (a) transfer of excess bomb ^{14}C to ocean and biosphere; (b) change of atmospheric ^{14}C activity due to dilution with fossil fuel CO_2; (c) steady state of natural ^{14}C exchange and decay in the ocean; (d) application of stagnant film model independent of ^{14}C.

assumed to be proportional to the mass content of the three reservoirs; the notation for the exchange coefficients is shown in Figure 11.5. The measurements provide activities expressed as $^{14}C/^{12}C$ mole ratios. Their average values in the three reservoirs are denoted by

$$R_a = fG_a^*/G_a \quad R_m = fG_m^*/G_m \quad R_d = fG_d^*/G_d \tag{11.2}$$

where the factor $f = 12/14$ is the mass ratio of the two isotopes. It will later be seen to cancel in the flux balance equations. In thermodynamic equilibrium, the $^{14}C/^{12}C$ ratios of total dissolved CO_2 in seawater and that in air are not identical because the equilibrium constants for the solubility and ion formation differ slightly for the two isotopes. The resulting fractionation has been determined in the laboratory for $^{13}C/^{12}C$ and amounts to 8‰. as discussed by Siegenthaler and Münnich (1981). The fractionation for $^{14}C/^{12}C$ will be twice as large, so that

$$\frac{k_{am}^*}{k_{ma}^*} = \alpha^* \frac{k_{am}}{k_{ma}} \quad \alpha^* = 1.016 \tag{11.3}$$

In the steady state, the radioactive decay of ^{14}C in the ocean must be balanced by the net uptake from the atmosphere:

$$k^*_{am}G^*_a - k^*_{ma}G^*_m = \lambda(G^*_m + G^*_d) \approx \lambda G^*_d \qquad (11.4)$$

Here, $\lambda = 1.21 \times 10^{-4}$ year^{-1} is the ^{14}C decay constant. For ordinary CO_2, equilibrium conditions can be assumed:

$$k_{am}G_a = k_{ma}G_m \qquad (11.5)$$

Dividing Equation (11.4) through by G_a and converting to activities yields

$$k^*_{am}R_a - k^*_{ma}R_m(G_m/G_a) = \lambda R_d(G_d/G_a) \qquad (11.6)$$

which with the help of the preceding equations can be cast in the form

$$k_{am} \approx k^*_{am} = \frac{\lambda(R_d/R_a)}{(1 - R_m/\alpha^* R_a)}\left(\frac{G_d}{G_a}\right) \qquad (11.7)$$

The decay of ^{14}C in the sediments is nearly balanced by the river input of ^{14}C. Both quantities amount to only a few percent of that present in the ocean and can be neglected. From measurements conducted during the period preceding the major weapons tests, Broecker (1963) and Oeschger *et al.* (1975) estimated $R_d / R_a = 0.85 \pm 0.05$ and $R_m/\alpha^* R_a = 0.95 \pm 0.015$. The mass of carbon in the deep sea is $G_d \approx 38,000$ Pg if organic carbon is included, and that of carbon dioxide in the atmosphere was about 650 Pg during the period considered. The insertion of these data into Equation (11.7) yields $k_{am} = 1.333 \pm 0.047$ year^{-1}. The equivalent atmospheric residence time, $\tau_{am} = 1/k_{am} = 7.5$ year, is long compared to the time for mixing in the troposphere. The uncertainty in the value arises primarily from the fact that the denominator of Equation (11.7) represents a difference of two almost equal numbers, and for the same reason one must include the seemingly small correction for isotope fractionation. The coefficient for the reverse flux, k_{ma}, is determined by the equilibrium relation, Equation (11.5). It turns out that $k_{ma} \approx k_{am}$, because the mass of CO_2 dissolved in the mixed layer of the ocean is essentially the same as that in the atmosphere.

Consider now the exchange of inorganic carbon between the mixed layer and the deep ocean. The steady-state conditions for the latter reservoir are

$$k_{ma}G^*_m = (k_{dm} + \lambda)G^*_d$$
$$k_{ma}G_m = k_{dm}G_d \qquad (11.8)$$

In this case the isotope fractionation is expected to be negligible, and the asterisk on the rate coefficients for the exchange of radiocarbon can be dropped. Elimination of k_{dm} and conversion to activities gives

$$k_{md} R_m = k_{dm} R_d + \lambda R_d (G_d/G_m) \qquad (11.9)$$

which after dividing through by R_a can be arranged to yield

$$k_{md} = \frac{\lambda (G_d/G_m)(R_d/R_a)}{R_m/R_a - R_d/R_a} \qquad (11.10)$$

The numerical values given above for the quantities appearing in this expression lead to $k_{md} = 0.05$ year^{-1} or a residence time of inorganic carbon in the mixed layer of 20 years. The range of uncertainty is similar to that for k_{am} and amounts to $\pm 50\%$. In contrast to k_{am}, the values obtained for k_{md} and τ_{md} depend on the assumed thickness of the mixed layer, so that results reported in the literature vary somewhat. Oeschger *et al.* (1975) reported $\tau_{md} = 22.7$ years; Houtermans *et al.* (1973) found $\tau_{md} \approx 30$ years. The validity of the above result can be checked by calculating the equilibrium residence time of CO_2 dissolved in the deep ocean, which is independent of G_m:

$$\tau_{dm} = 1/k_{dm} = \tau_{md} G_d/G_m = 1100 \text{ years}$$

The value is in satisfactory agreement with the turnover time of 700-1000 years resulting from the circulation of deep ocean waters.

Further insight into the behavior of the ocean–atmosphere system can be gained by considering a sudden perturbation of atmospheric ^{14}C (Houtermans *et al.*, 1973) and the subsequent adjustment to new steady-state conditions. We set $G_a^*(t) = \overline{G}_a^* - \Delta G_a^*$ and $G_m^*(t) = \overline{G}_m^* - \Delta G_m^*$, where $\overline{G}_a^*$ and $\overline{G}_m^*$ represent the inventories of ^{14}C in the atmosphere and the mixed layer, respectively, after the new steady state has been reached, and ΔG_a^* and ΔG_m^* are the deviations from these values at time t. The equations governing the temporal change following an input pulse then reduce to

$$-d \Delta G_a^*/dt = k_{am}^* \Delta G_a^* - k_{ma}^* \Delta G_m^* \qquad (11.11a)$$

$$d \Delta G_m^*/dt = (k_{am}^* - k_{dm}) \Delta G_a^* - (k_{ma}^* + k_{md} + k_{dm}) \Delta G_m^* \qquad (11.11b)$$

Here, the interest is confined to a time scale of 50–100 years, so that the radioactive decay of ^{14}C can be ignored. Also neglected for simplicity is the influence of the terrestrial biosphere, although its omission is not really justified (in contrast to treating the steady state), because it acts as a temporary sink for ^{14}C on a time scale of a few decades.

If the initial conditions are chosen to represent a sudden increase ΔG^*_{a0} of ^{14}C in the atmosphere (for example, due to the nuclear weapons tests), the solutions to the above system of differential equations are

$$\Delta G^*_a / \Delta G^*_{a0} = A \exp(-\lambda_1 t) + B \exp(-\lambda_2 t) \qquad (11.12a)$$

$$\Delta G^*_m / \Delta G^*_{a0} = C[\exp(-\lambda_1 t) - \exp(-\lambda_2 t)] \qquad (11.12b)$$

where A, B, and C are appropriate constants with which we shall not be concerned, and λ_1 and λ_2 are the reciprocal time constants for the relaxation of the system to the new steady state. They are give by

$$\tau^{-1}_{1,2} = \lambda_{1,2} = \tfrac{1}{2}(a+d) \pm \tfrac{1}{2}\sqrt{(a-d)^2 + 4bc} \qquad (11.13)$$

with $a = k^*_{am}$, $b = k^*_{ma}$, $c = k^*_{am} - k_{dm}$, and $d = k^*_{ma} + k_{md} + k_{dm}$. By inserting for the exchange coefficients the values discussed previously, one finds

$$\tau_1 = 1/\lambda_1 = 3.3 \text{ years} \qquad \tau_2 = 1/\lambda_2 = 42 \text{ years}$$

The first relaxation time is associated with the equilibrium of excess ^{14}C between the atmosphere and the mixed layer of the ocean; the second relaxation time describes the much slower rate of filling up of the deep ocean reservoir. The relaxation time for the first of these processes agrees with the time constant observed for the rise of ^{14}C in the mixed layer after the last major injection of ^{14}C into the atmosphere by the nuclear weapons tests (see Fig. 11.6 for comparison). Note that the equilibration with the mixed layer is faster than one would expect from the atmospheric residence time under steady-state conditions. After a time span of about 10 years the atmosphere and the mixed layer approach steady state, and they may be combined into a single reservoir. The relaxation then is determined by the second process. The excess flux of ^{14}C into the deep ocean then is

$$k_{md}\,\Delta G^*_m = \frac{1}{\tau_2}(\Delta G^*_m + \Delta G^*_a) = \frac{1}{\tau_2}\left(1 + \frac{k^*_{ma}}{k^*_{am}}\right)\Delta G^*_m \qquad (11.14)$$

or

$$\tau_2 = \frac{k^*_{am} + k^*_{ma}}{k^*_{am}}\,\tau_{md} \approx 2\tau_{md}$$

The fact that the relaxation time τ_2 is about twice the residence time τ_{md} is seen to be a consequence of treating the atmosphere and the mixed layer together as one reservoir. The decline of ^{14}C activity in the atmosphere that has been observed following the nuclear test ban is more rapid than a relaxation time of 42 years would predict. The small additional injections of ^{14}C since 1963 can be ignored in this context, but not the influence of the biosphere, which acts as another absorbing reservoir on the time scale of interest and therefore leads to faster decay of ^{14}C in the atmosphere.

Compared with excess $^{14}CO_2$, the addition of fossil fuel-derived $^{12}CO_2$ to the atmosphere–ocean system forms a much more massive perturbation, which also affects the ion equilibria in the mixed layer. According to Equation (8.15), the absorption capacity of seawater for CO_2 depends on the hydrogen ion concentration. This quantity is largely determined by the charge balance equation,

$$A + [H^+] = [HCO_3^-] + 2[CO_3^{2-}] + [H_2BO_3^-] + [OH^-]$$

where A stands for the net charge of all other ions, both positive and negative, that do not vary significantly with changes of the CO_2 content. Using the substitution procedure in Equation (8.16), the equilibrium relations between CO_{2aq}, HCO_3^- and CO_3^{2-} on the one hand and H_3BO_3 and $H_2BO_3^-$ ion on the other, modify the above equation to

$$A = \frac{K_H K_1}{[H^+]}\left(1 + \frac{2K_2}{[H^+]}\right)P(CO_2)$$

$$+ \frac{[H_2BO_3] + [H_2BO_3^-]}{1 + [H^+]/K_B} + \frac{K_w}{[H^+]} - [H^+] \qquad (11.15)$$

where K_H, K_1, and K_2 are the apparent equilibrium constants for the dissolution of CO_2 in seawater, and K_B and K_w are the dissociation constants of boric acid and water, respectively. The expression shows that the hydrogen ion concentration is a function of the partial pressure of CO_2 in the atmosphere. Solving for $[H^+]$ in terms of $P(CO_2)$ and inserting the result into Equation (8.16) yields a nonlinear relation between $P(CO_2)$ and the total concentration $\Sigma C_{inorg} = [CO_2]_{aq} + [HCO_3^-] + [CO_3^{2-}]$ of inorganic carbon in seawater. A rise in $P(CO_2)$ increases $[H^+]$, thus lowering the capacity of seawater for the additional absorption of CO_2. In other words, the surface layer of the ocean appears to evade the uptake of CO_2. This relation can be expressed in terms of a buffer factor ζ defined by

$$\frac{\delta P(CO_2)}{P^0(CO_2)} = \zeta \frac{\delta \Sigma C_{inorg}}{C^0_{inorg}} \quad \text{or} \quad \frac{G_a}{G_a^0} = \zeta \frac{\delta G_m}{G_m^0} \qquad (11.16)$$

where δ denotes an infinitesimal change in pressure or concentration and the superscript zero indicates preindustrial conditions. The buffer factor can be calculated from the known equilibrium constants, and it has also been measured at various points on the sea surface. The two procedures give concordant results, as Sundquist *et al.* (1979), among others, have shown. The values for ζ range from 8 to 14, depending on the temperature. A common choice for the average ocean is $\zeta \approx 10$ under current conditions. The buffer factor is not independent of ΣC_{inorg}, however, and it is expected to increase with the further rise of atmospheric CO_2 in the future (see Keeling (1973b) for details).

The existence of a buffer factor has the consequence that the exchange coefficient k_{ma} associated with the equilibration of CO_2 between mixed layer and atmosphere must be replaced by ζk_{ma} when the equilibrium is perturbed. The uptake capacity of the mixed layer is reduced to one-tenth of its equilibrium value, and the relaxation time for the transfer of excess CO_2 toward the deep sea, given by Equation (11.14) for a pulse input, is raised to

$$\tau_2 = \frac{k_{am} + \zeta k_{ma}}{k_{am}} \tau_{md} = 11\tau_{md} \approx 220 \text{ years} \qquad (11.14')$$

This important result shows that it will take several centuries to drain from the atmosphere the excess CO_2 injected by the combustion of fossil fuels. It makes little difference that combustion must be represented as a continuous source function, because any continuous function can be expressed by a series of pulses. In the box-diffusion model of the ocean discussed by Siegenthaler and Oeschger (1978), the response to a pulse input leads to a nonexponential decay of atmospheric CO_2, which after equilibration with the mixed layer is somewhat faster than in the two-box model treated here, but the time scales are still very similar.

11.3.3. INTERACTION BETWEEN THE ATMOSPHERE AND THE TERRESTRIAL BIOSPHERE

The biological carbon cycle in the ocean is essentially short-circuited. Despite the high turnover rate, it has little effect on the atmosphere, because marine organisms and detritus contain only a minute fraction of total carbon in the ocean. Accordingly, attention must be directed toward the terrestrial biosphere. Ocean–atmosphere exchange models usually treat the terrestrial biosphere as a delay line in which carbon is temporarily stored before it is returned to the atmosphere from which it was taken (see Fig. 11.5). It was pointed out earlier, however, that the biosphere as a whole does not represent a uniform reservoir with a well-defined residence time, and a subdivision into several compartments, each with its own characteristic behavior, is necessary. How many such compartments are required to provide a realistic description depends on the purpose of the model. Figure 11.7 shows a possible scheme. It follows naturally from our previous discussion and distinguishes four important subreservoirs on which information is available: leaves and other assimilating parts of plants; wood contained in roots, stems, branches, and standing deadwood; leaf and wood litter; and soil humus. In this model, the various ecosystems are lumped together. More precisely, one should treat each ecosystem separately and then sum the equivalent reservoir contents and fluxes. For steady-state conditions or annual averages, the results of the two procedures will be identical. The more general case of time-dependent fluxes for which the results are expected to differ is more complex and cannot be treated here. Bolin (1986) discussed several such models.

The fluxes in Figure 11.7 are annual averages estimated by the methods discussed in Section 11.2.4. If the system is in a stationary state, the fluxes may be taken as constant. The residence times in each reservoir are readily evaluated, and the results are summarized in Table 11.10. Several subcycles of carbon through the system can be identified. One is the direct exchange of CO_2 with the atmosphere, due to assimilation during daylight and respiration at all times. About one-half of the gross primary production appears as plant tissue and constitutes net primary production. Most of this material is returned as CO_2 to the atmosphere when leaves and other assimilating parts wither, turn to litter, and undergo decomposition. This pathway forms the

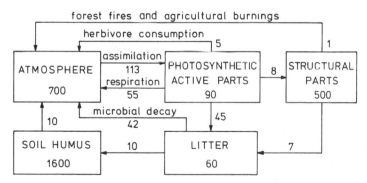

FIGURE 11.7   Four-reservoir model of organic carbon in the terrestrial biosphere. Reservoir contents are given in Pg of carbon, fluxes in Pg year^{-1}. The numbers shown follow from the data presented in Section 11.2.4. The flux due to biomass burning (Seiler and Crutzen, 1980) excludes deforestation.

TABLE 11.10   Carbon Reservoirs of the Terrestrial Biosphere, Fluxes and
Residence Times[a]

Reservoir	Carbon content (Pg)	Net flux of carbon (Pg year^{-1})	Residence time (years)
Atmosphere, short-term storage	~ 700	47	15
Atmosphere, long-term storage		8–10	70–87
Assimilating parts of the biosphere	90	58	1.5
Structural parts including standing deadwood	500	8	63
Stem, branch, twig and leaf litter	60	52	1.2
Soil humus	2000	10	200

[a] For interactions and connections with the atmosphere see Figure 11.7. Individual
data were gleaned from the summary of Ajtay et al. (1979).

second subcycle. The ensuing residence times in both participating compartments are on the order of 1 year. A third subcycle is set up by the small
fraction of organic carbon converted to wood, and a fourth subcycle involves
the fraction of plant litter that enters the soil and turns to humus. This
fraction is eventually metabolized to CO_2 and is expired to the atmosphere.
Wood and soil humus are the major reservoirs of organic carbon. The
residence times associated with these reservoirs are on the order of 63 and
200 years, respectively. Herbivore consumption followed by metabolization
of green plant tissue and biomass burning represent comparatively minor
pathways for the return of CO_2 to the atmosphere.

In considering the effect of the biosphere on atmospheric CO_2, one must
distinguish three time scales. The first is given by the diurnal variation of
primary productivity. It may be locally significant but averages out over the
whole troposphere. The next time scale is determined by the growth and
decay of foliage and other assimilating parts of plants in the second subcycle
described above. The corresponding residence time of CO_2 in the atmosphere is 15 years. This is sufficiently short to produce a seasonal variation of
atmospheric CO_2. The third time scale follows from the slower flux of
carbon through the major reservoirs of the biosphere (wood and soil humus)
and leads to an atmospheric residence time of 70–90 years.

The seasonal oscillation of CO_2 mixing ratios in the troposphere has been
investigated by means of global transport models and is well understood.
Junge and Czeplak (1968) made a first attempt to relate the variations to the
temporal imbalance between biosphere net uptake (by photosynthesis) and

release (by decaying plant litter), using a simple one-dimensional north–south transport model. Machta (1971) and Pearman and Hyson (1981) worked with two-dimensional models. The input data required are the transport coefficients as a function of latitude and altitude and the biospheric source functions. Seasonal variations due to anthropogenic activities such as home heating and temperature variations of ocean surface waters were assumed to be negligible. The early studies employed net primary productivities of Lieth (1965), which have since been revised, together with an estimate of the fraction of organic carbon already suffering decay during the growth period. Net uptake and release were assumed to balance out over a complete annual cycle. In the middle and high latitudes the growth season is the warm period of the year; in subtropical regions it is the rainy season. In the Tropics, where temperature and solar intensities vary little over the year, growth and decay are nearly always balanced and seasonal effects are nil.

Although the calculations of Junge and Czeplak (1968) and of Machta (1971) gave results in reasonable accord with observations, there were still differences with respect to amplitude and phase. Pearman and Hyson (1981) adopted an inverse procedure. By matching calculated and measured variations of atmospheric CO_2, they sought the biosphere source function that best reproduces the observational data. Figure 11.8a shows for the two measurement stations at Mauna Loa and the South Pole the level of agreement that can be reached. Figure 11.8b shows the corresponding seasonal variations of the biosphere source for the three latitude bands of equal area in the Northern Hemisphere. These curves may be interpreted as a superposition of two curves, one describing the release of CO_2 by a continuous decay of plant material and the other representing the net uptake of CO_2 due to

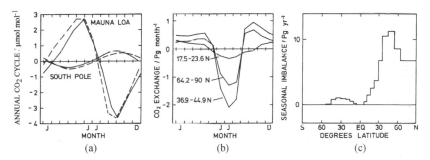

FIGURE 11.8   Exchange of atmospheric CO_2 with the terrestrial biosphere. (a) (solid line) Seasonal variation of atmospheric CO_2 at two stations (Northern and Southern Hemispheres) compared with (dashed line) model calculations by Pearman and Hyson (1981). (b) Rate of CO_2 exchange between biosphere and atmosphere for three latitude bands in the Northern Hemisphere derived from a best fit of the model to observational data. (c) Integrated annual CO_2 net exchange rate versus latitude.

photosynthesis during the growth period. The area underneath each curve gives the amplitude of the imbalance within the chosen latitude band. The resulting distribution with latitude is shown in Figure 11.8c. The largest amplitudes are found in mid-latitudes and, as expected on account of the distribution of land masses, in the Northern Hemisphere. Summation over the entire globe yields a peak-to-peak amplitude for the CO_2 imbalance of 15 Pg year^{-1} (as carbon), which corresponds roughly to 25% of global net primary productivity.

In the meantime, three-dimensional general circulation models have been adapted for study of the seasonal variability of CO_2 on a global scale. These models incorporate the global distribution of all sources and sinks, including chemical sources such as the oxidation of anthropogenic carbon monoxide and biomass burning. The results can be compared with observations at baseline monitoring stations. An important constraint of the simulation is the interhemispheric gradient, which the atmospheric models tend to overpredict. For example, Erickson *et al.* (1996) found that the three-dimensional model used by them reproduced the seasonal variations of CO_2 within 10% of those observed at many stations in the Northern Hemisphere, in both amplitude and phase, but the amplitude of the seasonal cycle in the Southern Hemisphere and the interhemispheric gradient were over-predicted. When the seasonal variation of the air–sea exchange of CO_2 was incorporated into the model, the discrepancy was very much reduced.

11.3.4. Long - Term Atmosphere - Biosphere Interactions and Current CO_2 Budget

Finally, we consider long-range effects of the interaction between the atmosphere and the biosphere. The two main questions are: what is the response of the biosphere to an increase in the atmospheric CO_2 level, and how much has deforestation by humans contributed to the rise of CO_2 in the atmosphere? A third problem that cannot be dealt with here is the response of the biosphere to climate changes caused by the CO_2 increase or other factors.

The dependence of net primary productivity on atmospheric CO_2 concentration has been studied for various plant species, and Figure 11.9 illustrates the observed behavior for two different types of plants. The initial response is linear, but saturation sets in for higher CO_2 concentrations. Similar curves are found for the dependence of primary productivity on light intensity and annual precipitation. The saturation effect is due to the closure of the stomata (leaf breathing pores), which regulate the intercellular CO_2 level and the water transpiration rates. As Figure 11.9 shows, the efficiency of C_4

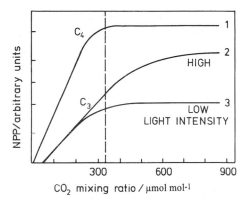

FIGURE 11.9   Net primary productivity (NPP): typified response of plants to changes in atmospheric CO_2 concentrations. (Adapted from Berry (1975).) C_3 and C_4 plants differ in their photosynthetic mechanism. The former generates a three-carbon carboxylic acid (phosphoglyceric acid) as the first identifiable photosynthetic product; the latter forms first a four-carbon dicarboxylic acid (oxaloacetic acid); see Devlin and Barker (1971) or Tolbert and Zelitch (1983) for details. Examples for C_4 plants are tropical grasses such as maize, sorghum, and sugarcane. All forest species are C_3 plants.

plants in utilizing CO_2 is higher than that of C_3 plants because they reach saturation at lower CO_2 partial pressures. At current levels of CO_2 in the atmosphere, C_4 plants already operate near saturation, whereas C_3 plants offer a potential for increasing the net primary productivity with rising CO_2 concentration, provided other factors such as nutrient and water supply are not rate-limiting. This process is called *carbon dioxide fertilization*. Raising the carbon dioxide level in the atmosphere would lead to an increase in primary productivity, so that more carbon is stored in the terrestrial biosphere. The ensuing withdrawal of CO_2 from the atmosphere would dampen the rate of the CO_2 increase.

A great number of experiments in enclosed environments, primarily of crop species, have demonstrated the enhanced growth rate caused by increased carbon dioxide concentration. Kimball (1983), Cure and Acock (1986), and Poorter (1993) have reviewed the extensive literature on the subject. Several open field studies have confirmed the results obtained in controlled environments (Lawlor and Mitchell, 1991; Hendrey, 1993). For C_3 plants, crop yields increased by about 30%, on average, under doubled CO_2 conditions, whereas C_4 plants showed an increase of about 10%. Resource limited or stressed plants showed a higher increase than plants under unstressed conditions. Young seedlings generally show a higher growth rate than plants reaching maturity. However, potted plants suffer root restrictions, which reduces the response to increased CO_2 (Thomas and Strain,

1991). Trees appear to respond to CO_2 fertilization more strongly than most other plants (Idso and Kimball, 1997), but longer-running controlled tree experiments are not yet available, and this problem deserves more study. Körner and Arnone (1992), who studied the effect of CO_2 fertilization on tropical plant communities, found vigorous growth in CO_2-enriched atmospheres, but no significant differences in stand biomass over a period of 100 days.

There is now sufficient evidence to show that uptake of excess CO_2 by the terrestrial biosphere has, in fact, occurred and that it currently helps to dampen the rise of CO_2 in the atmosphere due to fossil fuel combustion. Initial indications came from mass considerations based on atmosphere–ocean box-diffusion model calculations. During the 20-year period 1958–1978, the amount of CO_2 added to the atmosphere was about 54% of that released by fossil fuel combustion during the same period. The box-model calculations showed that another 35% had entered the oceans (Siegenthaler and Oeschger, 1978; Peng et al., 1979). The remaining 11%, which were left unaccounted for, were then assigned to uptake by the biosphere to balance the budget. A larger gap would exist if significant amounts of CO_2 were released from deforestation and land use changes, in addition to fossil fuel combustion.

As discussed above, two-dimensional models can be used to invert observed concentration gradients in the boundary layer to obtain information on the net surface sources of CO_2 as a function of latitude and time. Tans et al. (1990) worked with a three-dimensional transport model in combination with assumptions about the global distribution of CO_2 sources and sinks to simulate the mean annual interhemispheric gradient of CO_2 for comparison with observations. The results showed again that the oceans alone are insufficient to simulate the distribution of atmospheric CO_2 and that a terrestrial sink at temperate latitudes in the Northern Hemisphere must be added to balance the north–south gradient of CO_2.

Recently it became possible to measure the rate of atmospheric oxygen depletion resulting from the combustion of fossil fuels, and these data opened a new pathway for estimating the individual contributions to the current CO_2 budget (Keeling and Shertz, 1992, Bender et al., 1996, Keeling et al., 1996). Following Keeling and Shertz (1992), the annual atmospheric changes and the global transfer rates of carbon dioxide and oxygen are coupled by the equations

$$\Delta G(CO_2)/dt = F_{\text{fuel}} - F_{\text{biota}} - F_{\text{ocean}} \quad (\text{Tmol year}^{-1}) \quad (11.17a)$$

$$\Delta G(O_2)/dt = -R_c F_{\text{fuel}} + R_b F_{\text{biota}} \quad (\text{Tmol year}^{-1}) \quad (11.17b)$$

where F_{fuel} is the production of CO_2 from fossil fuel combustion, F_{biota} is the uptake rate by terrestrial biosphere, and F_{ocean} is the uptake rate by the world ocean; $R_c \approx 1.4$ represents a stoichiometric ratio for the consumption of oxygen by burning coal and fuel oil (the latter contains hydrogen, which consumes oxygen but does not produce CO_2); and $R_b \approx 1$ is the exchange ratio for production and destruction of terrestrial biomass. Rearrangement gives

$$F_{biota} = (\Delta G(O_2)/dt + R_c F_{fuel})/R_b \qquad (11.18a)$$

$$F_{ocean} = F_{fuel} - \Delta G(CO_2)/dt - (\Delta G(O_2)/dt + R_c F_{fuel})/R_b \qquad (11.18b)$$

The terms on the left of Equations 11.17 are measured, and F_{fuel} is assessed from production statistics for the measurement period. As an example, we consider the data reported by Bender et al. (1996) for the time period 1991–1993. The observed average rise of CO_2 was 0.7 μmol mol^{-1} year^{-1}, the loss of O_2 (measured against N_2) was $-(2.5 \pm 0.8)$ μmol mol^{-1} year^{-1}, whereas the fossil fuel CO_2 production rate was 6.1 Pg year^{-1} (as carbon). These data are converted to the global fluxes $\Delta G(CO_2)/dt = 124$, $\Delta G(O_2)/dt = -445$, and $F_{fuel} = 508$, in Tmol year^{-1}. From Equations 11.18 one obtains the transfer rates $F_{biota} = 266$ and $F_{ocean} = 118$ Tmol year^{-1}, which correspond to about 3.2 and 1.4 Pg carbon year^{-1}, respectively, with an uncertainty of 1.7 Pg year^{-1}.

Yet another method for gaining information on the partitioning between ocean and land ecosystem sink strengths is based on the latitudinal dependence of the $^{13}C/^{12}C$ ratio in atmospheric CO_2. The isotope ratio is determined largely by the biosphere. Photosynthesis discriminates strongly against ^{13}C, whereas uptake by the ocean does not to the same degree. Ciais et al. (1995) have used a global network of stations for measurements of the $^{13}CO_2/^{12}CO_2$ ratio. The data were evaluated with a two-dimensional atmospheric transport model dealing with $^{13}CO_2$ and $^{12}CO_2$ separately. The analysis is complicated by the fact that the average $\delta^{13}C$ in the atmosphere has decreased since the industrial revolution because of dilution with fossil carbon. As a consequence, old carbon that is respired today from the biosphere is somewhat enriched in ^{13}C compared to carbon currently incorporated into the biosphere. Respiration associated with green leaf photosynthesis is rapid, but soil humus has a longer residence time, and this causes ^{13}C to be out of balance with the atmosphere. The results indicated the existence of a major CO_2 sink in temperate and boreal forests, a net source in the northern Tropics, and a net sink in the southern Tropics. In the equatorial regions the ocean is a net source due to upwelling waters supersaturated with CO_2, but on the whole the ocean acts as a sink.

Table 11.11 presents a summary of the results. Note that the annual rise of CO_2 in the atmosphere fluctuates. For example, during the period 1991–1993 the growth was slower than either before or afterward. This has been taken as evidence for a higher uptake efficiency by the terrestrial biosphere. However, the evaluation of the observational data is afflicted by large uncertainties, which would obscure any variation of the sink strength associated with the biosphere. Thus, while the results clearly demonstrate that the terrestrial biosphere is capable of absorbing excess CO_2, the effect needs to be better quantified. Currently, the northern land biota and the oceans remove about 30% each of the total fossil fuel-derived CO_2 emissions. Compared with the period 1958–1978 discussed above, a greater share of carbon dioxide now appears to be taken up by the biosphere.

The transfer rates to the terrestrial biosphere shown in Table 11.11 are net fluxes, that is, they are sums of global CO_2 uptake and release. An enhanced uptake may not only be due to increased photosynthesis, but may occur also by the expansion of forest areas in the Northern Hemisphere. The release rate, in turn, is greatly influenced by the clearing of forests. Bolin (1977) first pointed out that the decimation of forests by humans for the purpose of gaining arable land must have released CO_2 in amounts comparable to that from the combustion of fossil fuels. In the past centuries this activity has taken place mainly in the Northern Hemisphere following population growth. The principal sites of deforestation are currently in the tropics.

While Bolin (1977) based his conclusions on an assessment of the biomass change, other studies have used the change in the $^{13}C/^{12}C$ ratio to estimate

TABLE 11.11   Global Budgets for Excess CO_2^a

Author	Fossil source	Land biota[b]	Ocean uptake	$\Delta x(CO_2)/\Delta t$ atmosphere[c]	Remarks[d]
Tans et al. (1990)	5.3	−1.9	−0.4	1.4	Latitudinal CO_2 gradient (1981–1988)
Keeling and Shertz (1992)	5.9	−0.2 ± 1.7	−2.5 ± 1.7	1.3	O_2 depletion (1989)
Ciais et al. (1995)	6.1	−1.5 ± 1.1	−3.1 ± 1.1	0.7	Latitudinal $^{13}C/^{12}C$ gradient (1991–1993)
Bender et al. (1996)	6.1	−3.2 ± 1.6	−1.4 ± 1.6	0.7	O_2 depletion (1991–1993)
Keeling et al. (1996)	6.2	−2.0 ± 0.9	−1.7 ± 0.9	1.1	Latitudinal O_2 and CO_2 gradients (1991–1994)

[a] In Pg year^{-1} as carbon; minus sign indicates uptake.
[b] Net uptake rate.
[c] Atmospheric CO_2 increase in μmol mol^{-1} year^{-1}.
[d] Method used and time period considered.

the transfer of carbon from the terrestrial biosphere to the atmosphere. The history of atmospheric ^{13}C is recorded in tree rings, which can be accurately dated by dendrochronology, and it is recorded in air entrapped in the polar ice sheets, which can be dated by the seasonal cycles of electric conductivity in addition to other techniques. Because of the kinetic isotope effect associated with photosynthesis, the $^{13}C/^{12}C$ ratio in plants is about 18% smaller than that in the atmosphere (see Figure 11.3), and this effect must be taken into account. Stuiver (Stuiver, 1978; Stuiver et al., 1984), Peng et al. (1983), and Freyer (1986) have reviewed measurements of the ^{13}C isotope shift in tree rings. Many of the trees studied showed a decrease of the $^{13}C/^{12}C$ ratio since the middle of the last century, although the data exhibit much scatter due to environmental factors. A dependence on latitude also appears to exist in addition to differences between C_3 and C_4 plants (Körner et al., 1991). To differentiate between the contributions of biomass and fossil fuel burning, Stuiver (1978) corrected the data by means of the Suess effect, that is, the dilution of the natural $^{14}C/^{12}C$ ratio by the addition of fossil fuel-derived CO_2 to the atmosphere. The corrected data showed little net change in the $^{13}C/^{12}C$ ratio between 1945 and 1980. Prior to 1945, however, the change was appreciable, indicating a decrease in biomass since 1850. Other authors have come to similar conclusions. To provide an example, Figure 11.10 shows the history of ^{13}C in tree rings as reported by Peng et al. (1983). These authors tried to minimize environmental factors by selecting nine trees from remote unpolluted sites, trees that were not water-stressed and that had experienced consistent temperatures during the growth season. The individual ^{13}C data were pooled. The release of CO_2 from the use of fossil fuels was taken from the compilation of Rotty (1981), and the uptake of CO_2 by the ocean was modeled in the fashion pioneered by Oeschger et al. (1975). With the help of these auxiliary data, the time history for the CO_2 input from the biosphere was constructed. Figure 11.10 compares the input from the biosphere with that from the combustion of fossil fuels. The biosphere appears to have been a greater source of atmospheric CO_2 during the nineteenth century than fossil fuels. The cumulative release of CO_2 by human manipulation of the biosphere was calculated to have been 240 Pg, which is somewhat higher than other estimates. Peng et al. (1983) did not constrain the preindustrial CO_2 concentration, for which they deduced a value of 243 μmol mol^{-1}. The ice core data show that it was closer to 280 μmol mol^{-1}.

Siegenthaler and Oeschger (1987) have used $^{13}C/^{12}C$ ratios of CO_2 trapped in the polar ice of the Antarctic (Friedli et al., 1986) to determine the transfer of CO_2 from the biosphere to the atmosphere. In this case the preindustrial atmospheric CO_2 mixing ratio was constrained to 278 ± 3 μmol mol^{-1}. The results were similar to those obtained by Peng et al.

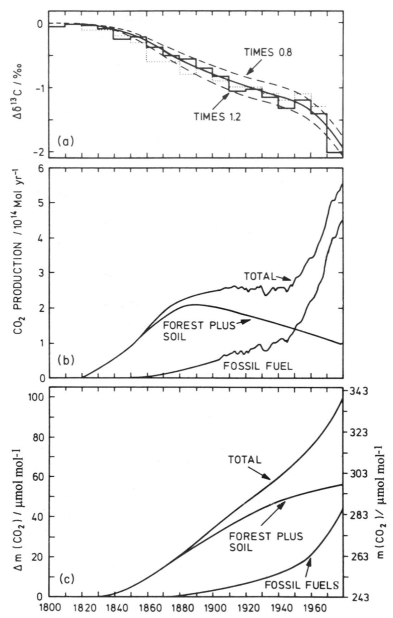

FIGURE 11.10   (a) Atmospheric ^{13}C record in tree rings. (b) Deconvolution of tree ring data to determine the history of CO_2 released to the atmosphere. By deducting from the resulting curve the fossil fuel-derived part one obtains the contribution arising from deforestation and soil manipulations. (c) The corresponding rise of the CO_2 mixing ratio in the atmosphere. (Adapted from Peng *et al.*, 1983), with permission.)

(1983), except that the range of the cumulative CO_2 input was smaller, 90–150 Pg. Table 11.12 shows results obtained by several authors.

In recent times, the destruction of forest has taken place primarily in the Tropics. The burning that follows most forest clearing in the Tropics converts some of the biomass directly to CO_2, whereas the decay of the remaining material and the decline of soil organic matter add additional CO_2 to the atmosphere for a longer time after an area is cleared. In Section 4.3 (see Table 4.6) it was shown that savanna fires and the burning of agricultural waste make a large contribution to the total amount of carbon released as CO_2 from biomass burning. These sources, however, have little influence on atmospheric CO_2, because the CO_2 set free is soon reincorporated into new vegetation during the next growth cycle. Forest clearing, in contrast, liberates carbon that has been stored for a longer time and that is not reincorporated within a short period. The conversion of forests to arable land releases additional CO_2 from the soil subsequently because of cultivation.

The amount of carbon released annually by biomass burning can be assessed from the forest area cleared, the density of biomass in that area, the fraction of biomass exposed to fire, and the burning efficiency. Seiler and Crutzen (1980) first conducted a detailed survey based on United Nations Food and Agriculture (Lanly, 1982; FAO, 1987) statistics and showed that

TABLE 11.12 Estimates of CO_2 Released from the Biosphere to the Atmosphere (as Carbon)[a]

Authors	Present annual input[b] (Pg year^{-1})	Cumulative input (Pg)	Pre-industrial $x(CO_2)$ (μmol mol^{-1})
Bolin (1977)[c]	1.0 ± 0.6	70 ± 30 (1800–1975)	275
Stuiver (1978)[d]	1.2	120 (1850–1950)	268
Freyer (1978)[d]	—	70 (1860–1974)	—
Siegenthaler et al. (1978)[d]	—	133–195 (1860–1974)	—
Wagener (1978)[d]	—	170 (1800–1935)	—
Peng et al. (1983)[d]	1.2	240 (1850–1975)	243
Stuiver et al. (1984)[d]	0.3	150 ± 100 (1600–1975)	270
Peng (1985)[d]	0.5	~ 144 (1860–1980)	266
Bolin (1986), review[c]	1.6 ± 0.8	150 ± 50 (1860–1980)	275
Siegenthaler and Oeschger (1987)[e]	0.9	90–150 (1860–1980)	278

[a] For comparison: the present rate of fossil fuel combustion is about 5.5 Pg year^{-1} and the total release from fossil fuels since the beginning of industrialization is about 200 Pg.
[b] About 1980.
[c] Based on biomass change data.
[d] Based on $^{13}C/^{12}C$ data from tree rings.
[e] Based on $^{13}C/^{12}C$ data from air bubbles in ice cores.

70–80% of the total biomass cleared remains present in the roots, and that about 40% of the remaining material is actually burned and converted to CO_2 (and CO). Their data indicated a total of 2–4 Pg of carbon released annually by biomass burning, but if only burning due to shifting agriculture and deforestation were included, the rate would be reduced to 0.7–1.7 Pg. In the meantime Brown and Lugo (1984) have derived somewhat lower estimates for biomass densities in the Tropics. The data in Table 4.6, which are based on more recent biomass density data, indicated that shifting agriculture and deforestation in the Tropics release carbon at a rate of 0.3–1.1 Pg year^{-1}. According to Hao et al. (1990), 30% of the release occurred in Africa, 23% in Asia, and 47% in Latin America. The above rate does not yet include the use of biomass as fuel, which according to the assessment of Andreae (1991, 1993) is widespread and contributes 0.4–1.6 Pg carbon year^{-1} (see Table 4.6). The total annual emission of CO_2 from the Tropics accordingly is 0.7–2.6 Pg of carbon. The contribution from forest fires in the extratropical regions, at 0.1–0.3 Pg year^{-1}, is comparatively small.

While many other estimates exist for the release of CO_2 by land-use changes in the Tropics, most of them refer to the years around 1980. Detwiler and Hall (1988) have reviewed these data and concluded that the release of carbon from cleared vegetation during that year was 0.3–1.3 Pg, supplemented by another 0.1–0.3 Pg from the conversion of forest soils to agricultural fields and pastures. More recently, Dixon et al. (1994) and Houghton (1996) have reassessed the situation by using newer statistical data (FAO, 1993) for the Tropics, and they derived transfer rates of 1.7 ± 0.4 and 1.6 ± 0.5, respectively.

Whereas data for the land-use change in the Tropics indicate a net transfer of carbon from the biosphere to the atmosphere, similar assessments for temperate zone forests are less certain about how much of the excess CO_2 in the atmosphere these biota have actually absorbed. Two approaches have been used to calculate changes in terrestrial carbon storage. One is based on changes in land use, that is, conversion of forest to agricultural land or the reverse; the other approach is a more direct measurement of forest biomass. Studies based on the first approach (e.g., Melillo et al., 1988; Houghton, 1993) have mostly found the northern temperate and boreal areas approximately balanced with regard to transfer of carbon to or from the atmosphere. Studies of forest inventories (Dixon et al., 1994; Houghton, 1996) have found the northern regions to act as a sink for excess atmospheric CO_2. The difference is that analyses based on land-use change determine the accumulation of carbon only as resulting from deliberate human activity, such as forest regrowth, logging, or abandonment of agriculture, whereas forest inventories measure biomass directly. Accumulations between inventories may occur not only because of additional growth due to forest management, but also because of changing environmental conditions, such as changes in climate, availability of nitrogen, or CO_2 fertilization. This difference should make the second approach more suitable. Table 11.13 compares recent results from two such assessments with the $^{13}C/^{12}C$ data discussed above, confirming the notion that temperate and boreal forests currently provide a significant sink for excess carbon dioxide. Although all of these studies show temperate and boreal forests in the Northern Hemisphere to act as a sink for atmospheric CO_2, only the studies based on $^{13}C/^{12}C$ data indicate the sink to be sufficiently large to balance the excess. However, as the preceding discussion has made clear, all of the assessments of changes in terrestrial biomass carry large errors, which makes it difficult to quantify the effect with sufficient accuracy.

11.4. SUMMARY

The diversity of arguments in the preceding sections calls for a summation. The essential questions in the current debate of rising CO_2 levels in the atmosphere are: what controls the abundance of CO_2 in the atmosphere, and how does the system respond to perturbations? These points can be summarized as follows.

The current atmospheric CO_2 budget is summarized in Table 11.14. The input from human sources amounts to about 7.1 Pg year^{-1}, with 5.5 Pg year^{-1} deriving from the combustion of fossil fuels and 1.6 Pg year^{-1} from the deforestation of the biosphere in the Tropics. About 47% of the CO_2 released by human activities is temporarily stored in the atmosphere, and

TABLE 11.13   Comparison of Recent Estimates of the Net Transfer of Carbon from Terrestrial Ecosystems to the Atmosphere (Pg year^{-1} as Carbon)

Latitude	Ciais et al. (1995)[a]	Dixon et al. (1994)[b]	Houghton (1996)[c]
Temperate and boreal			
(30–60°N)	−3.5 ± 0.9	−0.74 ± 0.1	−0.8 ± 0.4
Tropics			
(0–30°N)	2.0 ± 1.3 ⎱	1.65 ± 0.4	0.8 ± 0.5
(0–30°S)	−0.3 ± 1.0 ⎰		0.8 ± 0.5
Global	−1.8	0.9 ± 0.4	0.8 ± 0.6

[a] From the seasonal and latitudinal variation of the $^{13}C/^{12}C$ ratio in atmospheric CO_2.
[b] From land-use change data.
[c] From land-use change data for the Tropics and forest inventories in temperate and boreal latitudes.

TABLE 11.14   Annual Carbon Budget of CO_2 for the Decade 1980–1990 ($Pg \ year^{-1}$) and Percentage Distribution among Reservoirs

CO_2 sources[a]	
Fossil fuel combustion	5.5 ± 0.5
Tropical land-use changes	1.6 ± 1.0
Total emission	7.1 ± 1.1
Partitioning among reservoirs[a,b]	
Storage in the atmosphere	3.3 ± 0.2 (47%)
Uptake by the ocean	2.0 ± 0.8 (28%)
Northern Hemisphere reforestation	0.8 ± 0.4 (11%)
Other terrestrial sinks (e.g., CO_2 fertilization)	1.0 ± 0.6 (14%)

[a] From Schimel et al. (1995).
[b] Terrestrial sinks according to Table 11.13.

28% is transferred to the ocean. A similar amount, about 25%, is absorbed by the terrestrial biosphere. Storage of excess carbon in the biosphere occurs both by deliberate reforestation and by enhanced CO_2 uptake, that is, CO_2 fertilization.

With regard to CO_2, the atmosphere (and the biosphere as well) represents only a small appendix to the much larger ocean reservoir. In the absence of perturbations and on a time scale sufficiently long to guarantee a well-mixed ocean, the partial pressure of CO_2 in the atmosphere will be governed by Henry's law equilibrium, that is, equilibrium between CO_2 dissolved in seawater and the partial pressure in the gas phase of the atmosphere. On a shorter time scale only the upper layer of the ocean is equilibrated with atmospheric CO_2, so that deviations from equilibrium with the deep ocean are expected. The inorganic carbon content of the ocean is determined by the balance between the weathering of rocks and the formation of calcium carbonate sediments. Table 11.7 shows that the influx of fresh carbon arises mainly from the weathering of limestone. The current rate of biological $CaCO_3$ formation is about five times greater than that of carbonate deposition. The mismatch appears to result from the dissolution of $CaCO_3$ below the carbonate compensation level. Broecker (1973) consequently suggested that it is this level that controls the global $CaCO_3$ deposition rate. Although the role of the carbonate compensation level is still somewhat obscure, it may, in principle, adjust to the amount of CO_2 dissolved in the ocean and thereby provide a feedback mechanism, which by regulating the $CaCO_3$ deposition rate tends to keep the inorganic carbon content of the ocean nearly constant. According to the data in Table 11.7, the time constant for the removal of carbonate is 1.7×10^5 years. The geochemical CO_2 cycle is clearly coupled to that of calcium. Under present conditions, calcium is not the controlling factor, because its concentration in the ocean, in mol kg^{-1}, is about five times greater than that of dissolved CO_2, and the time constant for changes is correspondingly longer. The time constant for the removal of carbonate may be reduced by a factor of 5 under favorable conditions, that is, if the rate of biologically assisted $CaCO_3$ production stayed constant, but all of the calcium carbonate formed were also deposited. Currently, 80% of it is redissolved.

The current perturbation of the system consists of the addition of CO_2 to the atmosphere following the combustion of fossil fuels and changes in tropical land use. Equilibration with the mixed layer of the ocean is achieved quickly within a few years, even though the evasion factor prevents a favorable partitioning of CO_2 between the two reservoirs and more than 90% of the additional CO_2 stays initially in the atmosphere. The excess is eventually transferred to the deep ocean. Complete equilibrium is achieved slowly within a time period on the order of 1000 years. In the interim, the terrestrial biosphere utilizes some of the excess CO_2 to store it as organic carbon in biomass and soil humus. Table 11.10 shows that soil humus provides the longest storage, a few hundred years on average. This is still shorter than the time needed for a complete equilibration between atmosphere and ocean, following a pulse input. The latter process, therefore, determines the long-term behavior of the system.

The exploitable fossil fuel reserves have been estimated as 7000 Pg of carbon (Zimen et al., 1973, 1977). The amount corresponds to nearly 19% of that currently residing in the ocean–atmosphere reservoir. If all of the fossil fuel reserves were burned and the resulting CO_2 transferred to the ocean, the evasion factor would rise from 10 to 15, and the mixing ratio of CO_2 in the atmosphere, at equilibrium, would increase by a factor of 2.7 to about 800 μmol mol^{-1}. Much higher mixing ratios would temporarily occur before the final equilibrium were reached, depending on the rate at which fossil fuel reserves are consumed. In the worst case, if most of the fuel reserves were utilized within a century, the mixing ratio of CO_2 in the atmosphere would rise to 3000 μmol mol^{-1} for some time. The climatic and other effects of CO_2 levels that high are unknown, but they may have deleterious consequences for life on Earth, and the prospect for such a development should give rise to serious concern, more so than any other effect of local air pollution.

The slow geochemical cycles begin to take hold long after the excess CO_2 has been transferred to the deep sea. Because of the inorganic chemistry involved, one would expect a rise in the atmospheric CO_2 level to enhance the weathering rate, but the effect is hard to quantify. The current weathering rate is largely due to the action of soil bacteria and the release of CO_2 (and organic acids) from the roots of vascular plants, which enhances the local CO_2 partial pressure 10–100-fold above the atmospheric level (Schlesinger, 1997). In the oceans, however, the deposition of $CaCO_3$ appears to be controlled by an inorganic process, although the formation of $CaCO_3$ is biologically mediated. Hence one expects an increase of inorganic carbon to enhance the deposition rate. Again, it is difficult to make quantitative predictions, but the time scale for the removal is not expected to change too much, so that the excess of inorganic carbon will remain in the ocean for 100,000 years to come.

CHAPTER 12

The Evolution of the Atmosphere

Two groups of gases remain to be discussed: the noble gases, and the major constituents of air, nitrogen and oxygen. The great stability of their abundances indicates residence times in the atmosphere much longer than those of other constituents, and in discussing these gases we are invariably led to consider the evolution of the atmosphere over a time span reaching back to the formation of the Earth some 4.5 billion years ago. The earth is differentiated into an iron-rich core (about 32% by mass), a silicate mantle (4.1×10^{24} kg) and an outer crust ($\sim 2.5 \times 10^{22}$ kg). The core is chemically isolated from the mantle, and even the lower mantle appears to have remained essentially unchanged since its formation. The upper mantle, in turn, has been depleted of elements that went into the continental crust, and in volatile elements that formed the atmosphere.

Our knowledge of the primitive atmosphere existing on Earth after its creation is quite fragmentary and leaves much room for speculation. Nevertheless, there is sufficient evidence to show that the early atmosphere was a secondary feature, produced by thermal outgassing of virgin planetary matter, in contrast to a primordial atmosphere composed of gases originally

711

present in the solar nebula. The evidence comes from the depletion of rare gases on the Earth relative to their solar abundances, and from the composition of excess volatiles during the formation of the sediments. These topics are discussed in the next two sections before the behavior of nitrogen and oxygen is considered.

12.1. THE NOBLE GASES

Mixing ratios of the noble (or rare) gases are shown in Table 12.1. Included are the corresponding mass contents in the atmosphere and the observed isotopic distributions. The fractions dissolved in the oceans are comparatively small, but those present in portions of the Earth's crust or mantle are not necessarily negligible. Because they are chemically inert, the noble gases are not expected to enter into chemical cycles, so that they must either have accumulated in the atmosphere as a result of terrestrial exhalation, or they represent a remnant of a primordial atmosphere. We shall see that the latter possibility is unlikely. A third possibility, the capture of solar wind particles by the planet, also turns out to be unimportant. By combining the observed extraterrestrial proton flux with known solar rare gas abundances, Axford (1970) and Banks and Kockarts (1973) have shown that the accretion rate is too small by several orders of magnitude to account for the rare-gas content of the atmosphere.

Helium and ^{40}Ar are recognized to have originated as products from radioactive decay processes in the Earth's crust and mantle. Thus, ^{40}Ar

TABLE 12.1   Mixing Ratios, Total Mass Contents, and Isotopic Compositions of Stable Noble Gases in the Atmosphere[a]

Constituent	Mixing ratio (μmol mol^{-1})	G_A (Pg)	Isotopic composition (%)		
He	5.24	3.71	^{4}He ($\sim$ 100)	^{3}He (1.4×10^{-4})	
Ne	18.2	65.0	^{20}Ne (90.5)	^{21}Ne (0.268)	^{22}Ne (9.23)
Ar	9340	6.6×10^4	^{36}Ar (0.337)	^{38}Ar (0.063)	^{40}Ar (99.6)
Kr	1.14	16.9	^{78}Kr (0.347)	^{80}Kr (2.257)	^{82}Kr (11.523)
			^{83}Kr (11.477)	^{84}Kr (57.00)	^{86}Kr (17.398)
Xe	0.087	2.0	^{124}Xe (0.095)	^{126}Xe (0.089)	^{128}Xe (1.919)
			^{129}Xe (26.44)	^{130}Xe (4.07)	^{131}Xe (21.22)
			^{132}Xe (26.89)	^{134}Xe (10.43)	^{136}Xe (8.86)

[a] Volume fractions and isotopic abundances from Ozima and Podosek (1983).

derives from K-shell electron capture in ^{40}K, and ^{4}He is generated by the disintegration of nuclei within the uranium and thorium decay series. These gases, therefore, have definitely accumulated. Argon is entirely retained by gravitation, whereas helium has a tendency to escape slowly to interplanetary space, so that it does not accumulate to the same degree as argon. The rates of production and the total amounts of argon and helium that have been generated since the Earth's formation can, in principle, be calculated from terrestrial abundances of the precursor elements. It is justifiable to consider the Earth's iron-rich core chemically isolated from the silicate-rich mantle and crust. From analyses of midoceanic ridge basalts and crustal rocks, the average abundance of potassium in the mantle and crust was estimated to be 230 mg kg^{-1} (O'Neill and Palme, 1998). The fraction present as ^{40}K is 1.2×10^{-4}, its total radioactive decay constant is 5.54×10^{-10} year^{-1}, and about 10% of the product is ^{40}Ar (Steiger and Jäger, 1977). With the mass of the Earth (5.98×10^{24} kg), and considering that the silicate mantle comprises 68% of the total, one calculates that ^{40}K on Earth 4.5×10^9 years ago was 1.47×10^6 Pg, and that the total production of ^{40}Ar since that time is 1.35×10^5 Pg. The amount is twice that currently residing in the atmosphere. Thus, it appears that one-half the argon produced from ^{40}K has been exhaled, and a comparable fraction still remains in the mantle. This fraction must reside in the lower and less well stirred region of the Earth's mantle, whereas the upper portion is largely outgassed.

The argon isotopes ^{36}Ar and ^{38}Ar are primordial outgassing products. The ^{36}Ar/^{38}Ar ratios in basaltic rocks dredged from the ocean floor, which are regarded as having originating from the Earth's mantle, are essentially indistinguishable from that found in the atmosphere. The ^{40}Ar content shows two signatures, however. Midoceanic ridge basalts, which are related to the upper mantle, contain very high ^{40}Ar/^{36}Ar ratios (($13-25$) $\times 10^4$, with an average of 16.7×10^4, as summarized by Allègre et al. (1987)), whereas ocean-island basalts, which are thought to arise from plumes originating in the lower mantle, have ^{40}Ar/^{36}Ar ratios near 390, which is much closer to the atmospheric value of 296. This finding is interpreted as indicating a fairly rapid outgassing of ^{36}Ar from the upper mantle early in Earth's history because ^{40}Ar is produced by radioactive decay and has accumulated (Hart and Hogan, 1978; Hamano and Ozima, 1978; Allègre et al., 1987).

Similar considerations apply to helium. Zartman et al. (1961) and Wasserburg et al. (1963) have studied helium and argon in natural gases, including geothermal and bedrock gases of varied chemical compositions. In most samples, the radiogenic He/Ar ratio fell into the range 6–25, which compares well with production rates calculated from abundances of uranium, thorium, and potassium in average igneous and sedimentary rocks. The helium content of the gases varied by a factor of 2000, so that the degassing of helium and argon appears to occur at roughly equivalent rates.

The observed $^4\text{He}/^{40}\text{Ar}$ production ratio of about 10, on average, corresponds to a mass ratio of about unity. And yet, as Table 12.1 shows, the ratio in the atmosphere is $1/1800$. Apparently, most of the helium has escaped the gravitational field of the Earth. One can derive an estimate for the residence time of helium in the atmosphere in the following way. The average release rate of argon over geological times to reach the current atmospheric abundance has been $6.6 \times 10^{16}/4.5 \times 10^9 = 1.4 \times 10^7$ kg year^{-1}. The current production rate of ^{40}Ar is $\exp(-\lambda \Delta t) = 0.082$ times the rate 4.5×10^9 years ago, where $\lambda = 5.54 \times 10^{-10}$ year^{-1} is the ^{40}K decay constant, and the average production rate over the period Δt is

$$[1 - \exp(-\lambda \Delta t)]/\lambda \Delta t \exp(-\lambda \Delta t) = 4.45 \qquad (12.1)$$

times the present value. If one accepts the observed average mass ratio $^4\text{He}/^{40}\text{Ar} \approx 1$, one obtains for the current helium release rate and the helium residence time in the atmosphere the values

$$Q(\text{He}) = 1.4 \times 10^7/4.45 = 3.3 \times 10^6 \text{ kg year}^{-1} \approx 8 \times 10^8 \text{ mol year}^{-1}$$

$$\tau(\text{He}) = G(\text{He})/Q(\text{He}) = 3.7 \times 10^{12}/3.3 \times 10^6 \approx 10^6 \text{ years}$$

The time is appreciably shorter than the age of the planet, indicating that helium in the atmosphere is essentially in a steady state. Note that the estimate for the helium production rate does not depend on assumptions about the uranium content of the Earth.

Average abundances of uranium, thorium, and potassium in the Earth's crust and mantle are now quite well known, by virtue of the observation that the K/U abundance ratio is rather constant at 1.27×10^4 in midoceanic ridge basalts (Jochum et al., 1983), which are indicative of upper mantle conditions, whereas in the continental crust it is about 1.0×10^4 (Taylor and McLennan, 1995). Allègre et al. (1987) have used this feature and measurements of rare gas isotopes in rocks of various provenance as well as in deep-sea water to estimate the amounts of uranium, thorium, and potassium present in the continental crust, in the upper depleted portion of the Earth's mantle, and in the lower undisturbed mantle. The data, which are summarized in Table 12.2, may be used to estimate the production rates of helium and argon in and the fluxes from these reservoirs. The two uranium isotopes responsible for the production of helium are ^{238}U and ^{235}U, but the latter contributes only 0.72%, which is almost negligible. Compared to ^{238}U, however, the decay constant for ^{235}U is about six times greater, so that it

TABLE 12.2   Abundances of Radioactive Uranium, Thorium and Potassium in Three Geochemical Reservoirs of the Earth[a]

	Mass fractions (μg kg^{-1})			Total amounts (Pmol)			Production rate[d] (Mmol year^{-1})	
	U[b]	^{232}Th	^{40}K[c]	U[b]	^{232}Th	^{40}K[c]	^{4}He	^{40}Ar
Continents[e]	1700	5850	1985	179	630	1300	415	72
Upper mantle[f]	3.3	8.3	4.9	2	5.1	18	41	10
Lower mantle	22	92	32	25	110	220	650	120

[a] Abundances taken from Allègre et al. (1987).
[b] Sum of ^{238}U (99.28%) and ^{235}U (0.72%).
[c] The K/U mass ratio is 1.27 × 10^4 in the mantle, and 1 × 10^4 in the crust, ^{40}K/(^{39}K + ^{41}K) − 1.167 × 10^{-4}.
[d] The decay constants are λ = 1.551 × 10^{-10} year^{-1} for ^{238}U, λ = 9.849 × 10^{-10} year^{-1} for $_{235}$U, and λ = 5.54 × 10^{-10} year^{-1} for ^{40}K.
[e] The mass of the crust 2.5 × 10^{22} kg.
[f] The mass of the Earth's mantle is 4.1 × 10^{24} kg; the upper depleted portion of the mantle is 35% of the total.

must have played a greater role initially after formation of the planet. In calculating helium production rates, one has to take into account that several helium atoms are produced within each radioactive decay series: the number is 8 for ^{238}U, 7 for ^{235}U, and 6 for ^{232}Th. Table 12.2 includes the corresponding fluxes. The crust acquired practically all of the uranium, thorium, and potassium during the process of separation from the upper mantle, while the lower mantle has retained the original share of material. Accordingly, the production of helium and argon in the continental crust is dominant. The production in the lower mantle is comparable to that in the crust, but little of it can penetrate into the upper mantle region and escape to the atmosphere. Allégre et al. (1987) concluded that roughly 10% of the helium produced but no argon diffuses to the upper mantle. The total ^{4}He production rate derived from Table 12.2 is (415 + 41 + 65) × 10^6 = 5.21 × 10^8 mol year^{-1}, of which 80% is generated within the Earth's crust. This agrees quite well with the early estimate of Wasserburg et al. (1963) discussed above.

The mechanism of the thermal escape of a gaseous constituent from the Earth's gravitational field was first described by Jeans (1916) on the basis of the kinetic theory of gases. Hunten (1973) and Walker (1977) have more recently considered the problem. The loss of a molecule or atom to interplanetary space takes place in the outermost region of the atmosphere, which is called the *exosphere*. Here, the gas densities are so low that collisions between molecules or atoms become rather improbable. For such conditions, an upward-moving molecule or atom will be lost when, during the last collision at altitudes near the exobase, the particle acquires sufficient energy to overcome the back-pull by gravitation. Otherwise it will return to the exobase in a ballistic orbit. The kinetic energy required for escape is $\frac{1}{2}mv^2 > mrg(r)$ where m is the mass of the particle, v is its velocity, $r = r_0 + z$ is the distance from the center and $r_0 =$ the radius of the Earth, and $g(r) = g_0(1 + z/r_0)^{-2}$ is the acceleration due to gravity at the altitude $z = r - r_0$. The minimum velocity for escape then is

$$v_c = \left[2g(r_c)r_c\right]^{\frac{1}{2}} \qquad (12.2)$$

where r_c denotes the critical altitude, essentially the base of the exosphere. At this level, the gas particles are assumed to have a Maxwellian velocity distribution, although it is recognized that the high-energy tail of the distribution is modified somewhat by the loss of fast particles. The error incurred is considered tolerable. After weighting the number density n_c at the critical level with the Maxwellian distribution function and integrating over all directions and velocities greater than v_c, one obtains an expression for the escape flux:

$$F_c = n_c\left(\frac{R_g T}{2\pi M}\right)^{1/2}\left[1 + \frac{Mg(r_c)r_c}{R_g T}\right]\exp\left[-\frac{Mg(r_c)r_c}{R_g T}\right] \qquad (12.3)$$

The overriding factor is the exponential term. In the Earth's atmosphere, only hydrogen and helium have masses small enough and velocities high enough to overcome gravity at temperatures existing in the thermosphere (750–1500 K). For hydrogen the escape flux is so high that the supply by transport cannot keep up with it, so that the escape flux is transport-limited (Hunten, 1973). This is not the case for helium.

The level of the exobase is determined essentially by the average collision cross section σ_c (2×10^{-14} cm^2). The critical altitude is given with sufficient accuracy by the condition

$$\int_{r_c}^{\infty} \sigma_c n_T(r)\, dr = \sigma_c n_T(r_c)H(r_c) = 1 \qquad (12.4)$$

where n_T is the total number density and H is the scale height. Both depend on temperature, which varies with the time of the day (weakly) and with solar activity (strongly). Table 12.3 shows as a function of temperature the altitude of the exobase and other parameters required to evaluate the escape flux. The major constituent at the critical level is atomic oxygen, and helium ranks next in abundance. These conditions follow from the fact that at

TABLE 12.3 Conditions at the Base of the Exosphere, Helium Escape Fluxes, and Annual Helium Loss Rates as a Function of Temperature[a]

	Parameter	Temperature (K)					
		750	1000	1250	1500	1750	2000
z_c	(km)[b]	400	450	550	650	700	800
n(He)	(cm^{-3})	2.0 (6)	2.4 (6)	1.6 (6)	1.3 (6)	1.4 (6)	1.2 (6)
n_T	(cm^{-3})	3.6 (7)	5.2 (7)	3.4 (7)	2.3 (7)	2.9 (7)	2.3 (7)
$g(r)$	(m s^{-2})	8.68	8.55	8.31	8.07	7.96	7.74
F_c(He)	(atom m^{-2} s^{-1})	1.6	2.6 (4)	5.7 (6)	2.2 (8)	3.3 (9)	2.2 (10)
F_c(He)	(kg year^{-1})	1.9 (-4)	3.2	7.2 (2)	2.8 (5)	4.4 (5)	3.0 (6)

[a] Partly based on data compiled by Banks and Kockarts (1973). Powers of ten are shown in parentheses.
[b] Altitude above the Earth's surface.

altitudes above 120 km oxygen is largely dissociated, mixing no longer prevails, and molecular diffusion is the dominant mechanism of transport. Each constituent then establishes its own scale height. Table 12.3 includes the resulting helium escape fluxes. Even with temperatures on the high side, one finds that the loss rate falls far short of that required to accommodate the release of helium to the atmosphere from terrestrial sources. MacDonald (1963, 1964) considered the variations in the escape flux over a complete solar cycle and, taking an average, found the loss rate to be too low by a factor of 30. Although the resulting atmosphere residence time is still short compared with the age of the Earth, it is obvious that Jeans escape does not balance the helium budget.

A variety of suggestions have been made to overcome the difficulty (see Hunten (1973) for details). The most reasonable one appears to be the proposition by Axford (1968) that helium leaves the Earth's atmosphere in the ionized state. Ion flow would be most effective in the polar regions, where magnetic field lines are open, but some doubt exists whether the ionization rate is adequate. The question has not yet been resolved, so that the budget of helium in the atmosphere remains unsatisfactory.

Most rare gases other than ^{40}Ar and ^{4}He are understood to have occurred in the solar nebula prior to the formation of Earth, so that they must represent primordial components. Brown (1949) and Suess (1949) independently noted that the noble gases on Earth are severely underrepresented compared to their solar abundances. Signer and Suess (1963) later showed that similar deficiencies occur for the primordial noble gas components in certain meteorites. Table 12.4 lists relative abundances of neon, ^{36}Ar + ^{38}Ar, krypton, and xenon in the sun, on Earth, and in meteorites. Silicon is used here as a reference element. The left-hand side of Table 12.4 shows that the

TABLE 12.4   Abundances of the Rare Gases in the Sun, in the Earth's Atmosphere, and in Several Meteorites, Relative to Silicon ($X_{Si} = 10^6$)[a]

Constituent	$(X/X_{Si})_{solar}$	$(X/X_{Si})_{Earth}$[b]	DF_{Earth}	$DF_{meteorite}$[c]
He	3.1 (9)	—	—	7.9–10.9
Ne	1.6 (7)	1.0 (−4)	11.2	8.0–11.4
$^{36}Ar + ^{38}Ar$	2.4 (5)	2.1 (−4)	9.05	7.3–8.9
Kr	5.1 (1)	6.7 (−6)	6.90	5.7–7.7
Xe	3.6	4.9 (−7)	6.85	4.2–6.7

[a] Data from Signer and Suess (1963). The deficiency factor is defined as $DF = -\log_{10}(X/X_{Si})_{sample}/(X/X_{Si})_{solar}$.
[b] Assumes a terrestrial mass abundance of silicon of 14.7%.
[c] Assumes an average silicon content of meteorites of 20% by mass.

terrestrial noble gases are indeed significantly depleted by many orders of magnitude. Similar deficiencies occur for several other volatile elements, such as hydrogen and nitrogen (Urey, 1952; Pepin, 1991). In discussing relative abundances of the aerosol, it was found convenient to introduce (on p. 419) a fractionation factor, which in view of the often observed enrichment relative to crustal abundances, was termed the *enrichment factor*. Here, the deficiency of volatile elements can be treated in a similar manner. Because of the large range of values, it is useful to work with the negative logarithm of the fractionation factor and to define the deficiency by

$$DF = -\log_{10}\left[(X/Si)_{sample}/(X/Si)_{solar}\right] \qquad (12.5)$$

Values derived from the noble gases on the Earth are entered in column three of Table 12.4. It is then easy to see that the deficiency increases toward the lighter noble gases. If this feature is interpreted to have resulted from Jeans escape, one finds that the mass of the body from which the escape took place should have been 100 times smaller than that of the present Earth, except for extremely high temperatures. The result may be used to support the argument that any primordial atmosphere that the Earth might have had was removed before the planet acquired its present size.

Noble gases in meteorites have been studied after releasing them from the solid by heating. They have at least four sources: radiogenesis of ^4He and ^{40}Ar; spallation processes induced by the interaction with cosmic rays; implantation of solar wind particles; and the incorporation of primordial gases during the formation of solid matter by condensation from the solar nebula. The last component is the one of interest here. It occurs primarily in dark grains imbedded in lighter texture, so that it can be isolated and studied separately. Deficiencies for this fraction, taken from the compilation of Signer and Suess (1963), are entered in the last column of Table 12.4 to demonstrate that their magnitudes are similar to those of the terrestrial noble gases.

Most meteorites were formed during a narrow time interval (4.54–4.57) $\times 10^9$ years ago, as revealed by standard radiogenic-isotope geochronometers (Dalrymple, 1991). Only for a comparably short fraction of time, namely a few million years, were they exposed to cosmic rays, so that they must be fragments of rocks from the interior of parent bodies that were shattered by collisions. The various classes of meteorites can be understood to have resulted from parent bodies that had experienced differentiation by melting (in part) and by metal segregation, possibly leading to core formation, as well as from undifferentiated parent bodies. Meteorites of the latter class are called *chondrites*. Chemically, the most primitive chondrites are the CI carbonaceous chondrites. They also are rich in volatiles, and they have acquired special significance in that their relative element abundances are similar to those in the solar photosphere, except for the lightest and most volatile elements. Figure 12.1 compares deficiency factors for the rare gases (except helium) observed in CI chondrites with those occurring in the atmospheres of Venus, Earth, and Mars. The similarity of distributions attests to the common origin of the rare gases in all cases. The fact that the

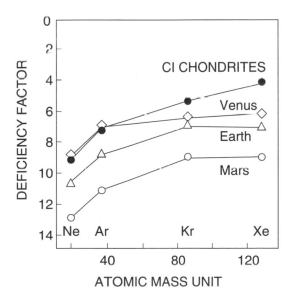

FIGURE 12.1   Abundance deficiencies DF for the rare gases in CI chondrites, on Earth, Venus, and Mars, relative to solar abundances, with silicon as the reference element. (Adapted from Pepin (1991).)

distribution is different from that in the sun shows that the fractionation must have occurred during the formation of the planets or their precursors. This aspect will be briefly considered.

Earth and the other planets in the solar system are believed to have formed by the accretion of dust grains present in the solar nebula or produced by condensation from the plasma phase surrounding the developing sun. The dust grains coagulated to form 1–10-km sized bodies (planetesimals, $\sim 10^{21}$ kg). The subsequent growth was mainly determined by gravitational forces. Computer models (Wetherill, 1990, 1994) show that fast runaway accretion will produce planetary embryos, that is, bodies of some 10^{23} kg (about 2% of the mass of the Earth), within about 10^5 years. At some point in time, during an intense activity of the early sun (T-Tauri phase), the surrounding space was cleared of dust and gases. This event provided a potentially effective mechanism for the fractionation of volatile elements. The formation of the inner planets by the accumulation of embryos lasted about 100 million years. From the distribution of lunar craters it can be inferred that the period of large impacts by accreting bodies ended 3.9×10^9 years ago (Neukum, 1977). Because of radial mixing it appears that embryos formed as far out as in the asteroid belt made a significant contribution to the evolution of Earth.

The fact that the relative distributions of rare gases in chondrites and in the atmospheres of the inner planets are rather similar suggests that the rare gases were acquired by a common mechanism. An important process must have been the implantation of rare gas atoms into the dust grains that ultimately grew to form precursors of the planets. Laboratory studies summarized by Pepin (1991) have shown that noble gases adsorbed on the surfaces of finely divided solid materials exhibit a fractionation such that the heavier species are enriched, and that starting from the solar distribution, a relative composition is obtained that resembles that of noble gases in meteorites and planetary atmospheres. This mechanism provided the planet with a reservoir of fractionated rare gases, which were partly released when the planet's interior was heated. Within the framework of this concept it is not necessary to assume that the Earth ever possessed a primordial atmosphere other than that derived by outgassing. However, to account for differences in specific isotopic distributions of rare gases, especially of neon, it is frequently assumed that the inner planets retained part of the original atmosphere of solar composition, which was then modified by hydrodynamic escape of the lighter volatile elements when (during the T-Tauri phase) a temporarily intense flux of extreme ultraviolet radiation from the young evolving sun heated the atmosphere (Pepin, 1991).

Knowledge about relative isotope abundances of noble gases in the sun, which cannot be measured directly, comes largely from collection foils deployed on the moon for irradiation by the solar wind, or from lunar surface minerals. The data compiled by Pepin (1991) show that argon and krypton isotope ratios in CI meteorites and the atmospheres of the terrestrial planets differ little from each other. For example, the solar $^{36}Ar/^{38}Ar$ ratio is 5.6 ± 0.1, that in CI chondrites 5.30 ± 0.05, Venus 5.56 ± 0.62, Earth 5.320 ± 0.002, and Mars 4.1 ± 0.2. The neon isotope ratio $^{20}Ne/^{22}Ne$, in contrast, shows a larger scatter: solar 13.7, CI chondrites 8.9 ± 1.3, Venus 11.8 ± 0.7, Earth 9.80 ± 0.08, and Mars 10.1 ± 0.7. While the values for CI chondrites, Earth, and Mars are similar within the uncertainties quoted, the solar $^{20}Ne/^{22}Ne$ ratio is distinctly higher, and that for the atmosphere of Venus lies in between. On Earth, $^{20}Ne/^{22}Ne$ ratios higher than atmospheric occur in midoceanic ridge basalts and ocean-island basalts. The former have derived from the upper mantle, whereas the latter are thought to represent material from the deeper mantle brought upward in thermal plumes. The mantle of the Earth evidently contains primordial neon closer to solar composition, which indicates again that fractionation of neon isotopes must have occurred by escape from the atmosphere. The isotope ^{21}Ne, in contrast to ^{20}Ne and ^{22}Ne, is produced partly by the nuclear reactions $^{18}O(\alpha, n)^{21}Ne$ and $^{24}Mg(n, \alpha)^{21}Ne$, which involve α particles and neutrons derived from the decay of uranium and thorium.

12.2. THE PRIMITIVE ATMOSPHERE

It is now widely accepted that the volatiles on Earth, which include water in the ocean, nitrogen in the atmosphere, and carbon dioxide occurring as carbonate deposits in the sediments, resulted from the thermal outgassing of virgin planetary matter during and subsequent to the formation of the planet. From the arguments developed in the preceding section, one expects the degassing products and the atmospheres of the terrestrial planets to be rather similar. Table 12.5 compares abundances of the major constituents in the atmospheres of Venus, Earth, and Mars. CO_2 is the major constituent of Venus and Mars. The much lower abundance of CO_2 in the atmosphere of Earth must be attributed to its removal by limestone formation, which follows from the presence of liquid water on Earth but not on the other planets. If all the CO_2 now buried in the sediments were brought into the atmosphere, CO_2 would be the dominant constituent on Earth. Hence, it appears that CO_2 was a major exhalation product on all three planets, in addition to water vapor. Yet, it is equally apparent that the three atmospheres have undergone evolutionary changes. Venus has lost most of the water vapor initially present, and Earth has developed a considerable amount of oxygen that the other planets are lacking.

TABLE 12.5   Data for the Planets Venus, Earth, and Mars: Physical Parameters and the
Main Constituents of the Atmospheres[a]

Parameter		Venus	Earth	Mars	
Mass of planet (kg)		4.88 (24)	5.98 (24)	6.42 (23)	
Acceleration of gravity (m s^{-1})		8.88	9.81	3.73	
Radius (km)		6053	6371.3	3380	
Surface area (m^2)		4.6 (14)	5.1 (14)	1.44 (14)	
Surface temperature (K)		730	288	218	
Surface pressure (Pa)		9.1 (6)	1.0 (5)	7.0 (2)	
Mass of atmosphere (kg)		4.78 (20)	5.13 (18)	2.5 (16)	
Composition	CO_2	96	0.03	95.3	
of the atmosphere	N_2	3.4	78.08	2.7	
(in percent)	O_2	6.9 (-3)	20.9	0.13	
	H_2O	0.1–0.5	2	0.03	
	^{40}Ar	(2–7) (-3)	0.93	1.6	
Ratios: mass of the	CO_2	9.4 (-5)	3.8 (-5)[b]	4.0 (-8)[c]	> 4 (-8)[d]
volatile to mass of	N_2	2.1 (-6)	8.0 (-7)[b]	6.8 (-10)[c]	4 (-8)[d]
the planet	H_2O	(1–5) (-7)	2.8 (-4)[b]	5 (-12)[c]	> 5 (-6)[d]
	^{40}Ar	(2–7) (-9)	1.1 (-8)[b]	5.6 (-10)[c]	5.6 (-10)[d]

[a] From the compilations of Owen et al. (1977), Oyama et al. (1979), and Pollack and Yung (1980). Orders of magnitude are shown in parentheses.
[b] Includes CO_2 in carbonates, N_2 in shales, and H_2O in the ocean.
[c] Actual values.
[d] Estimates including material in near-surface reservoirs or lost to space, according to McElroy et al. (1977b) and Pollack and Black (1979). H_2O on Venus has probably been lost to space.

The thermal exhalation of volatiles such as water and carbon dioxide is a process occurring today in conjunction with volcanism, with heating provided by radioactive decay. As the level of radioactivity has declined, it is virtually certain that the current degassing rate is only a small fraction of that 4.5×10^9 years ago. During the period of accretion, additional heat was supplied by impacting bodies, and conceivably by the gravitational settling of molten iron to the core. The last point is still controversial (see Ringwood (1979) for details). If accretion had proceeded homogeneously, core formation would have occurred comparatively late and the associated sudden energy release would have been sufficient for a complete melting of the planet (Birch, 1965; Hanks and Anderson, 1969). Alternatively, there are plausible arguments for a stepwise formation of the planet, where the core was formed prior to the addition of siliceous material. As the core is inaccessible to inspection, evidence in favor of one or the other model is by necessity indirect. Ringwood (1966) first noted that the upper mantle is grossly out of chemical equilibrium with the metal that formed the Earth's core. O'Neill and Palme (1998) have reviewed data on metal/silicate partitioning for many siderophile elements (elements having a high affinity for iron). Most of them are overabundant in the bulk silicate Earth. Homogeneous accretion would have stripped the Earth's mantle of all moderately siderophile and highly siderophile elements and transferred them to the core. In the heterogeneous accretion model of O'Neill (1991), 85–90% of the Earth was formed from indigenous, volatile-poor material present in the solar nebula in the vicinity of Earth's present orbit, while 10–15% of exogenous, volatile-rich material having originated at a greater heliocentric distance was added later. If this material had been added by a single giant impact, it would have driven off most of the then existing atmosphere. Such a massive perturbation also would have raised the surface temperature to a point where most of the volatiles contained in the added material would have been released rapidly by catastrophic outgassing, in contrast to the much slower release of volatiles from the interior of the Earth due to heating by radioactivity. The current understanding is that both types of processes have contributed to formation of the atmosphere. Holland (1984) has combined data on the variation of the heat flow with current gas release rates measured at midoceanic ridges to estimate that up to 75% of the atmosphere was released initially by rapid degassing. Allègre et al. (1987) have used a systematic interpretation of rare gas data to estimate the fraction of volatiles that underwent rapid degassing and arrived at a similar conclusion. Thus, it appears that a large portion of the atmosphere was already present 4.4×10^9 years ago.

Evidence from lunar craters and maria tells us that impacting bodies caused a partial melting of materials at the lunar surface, and the same effect must have occurred on Earth. As long as the Earth's atmosphere contained little water vapor and carbon dioxide, excess surface heat was rapidly dissipated into space. Surface heating would not have affected the interior of the planet. On the other hand, the sudden, catastrophic release of large amounts of water and CO_2 combined with elevated temperatures due to a giant impact would have produced a thick steam atmosphere nearly opaque to infrared radiation. As Rasool and DeBergh (1970) demonstrated for the planet Venus, such conditions may have led to an irreversible greenhouse effect, causing temperatures to remain high. Earth, however, appears to have escaped overheating. Figure 12.2 shows for the planets Venus, Earth, and Mars the temperatures expected because of infrared absorption of thermal radiation by water vapor as a function of surface partial pressure. The dashed curve begins on the left at temperatures for radiative equilibrium in the absence of any absorbing atmosphere. Increasing the water vapor pressure causes the temperature to rise until, for Earth and Mars, the temperature curve intersects the saturation pressure curve, whereupon the rise is halted. If the initial temperature is too high, as for Venus, the saturation curve

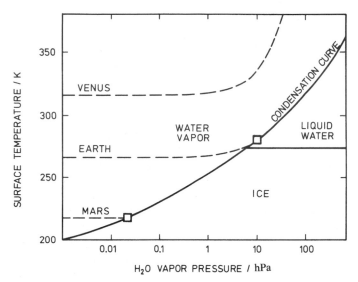

FIGURE 12.2 Surface temperatures expected on the three planets Venus, Earth, and Mars as a function of the water vapor pressure. An increase in the vapor pressure increases the retention of infrared radiation in the atmosphere, raising the temperature via the greenhouse effect. Overlaid is the phase diagram of water. On Earth and Mars the starting (radiation equilibrium) temperatures are low enough for water to condense, when the temperature intersects the condensation curve. On Venus, the temperature rises more rapidly and runs away. (Adapted from Walker (1977); originally modeled by Rasool and DeBergh (1970).)

cannot be reached, and the temperature runs away. These results argue against average temperatures greatly exceeding 300 K on primitive Earth. The overall temperature must have been favorable for water to condense. Figure 12.2 suggests that the principal factor responsible for such conditions is the optimum distance from the sun. Venus, being much closer to the sun than Earth, has temperatures that are too high, whereas the more distant planet Mars has temperatures too low to maintain liquid water for long before it freezes. Earth evidently has escaped the fate of Venus, so that any excess water vapor initially present in the atmosphere must have condensed to form the ocean. Carbon dioxide would have partly dissolved in this water reservoir, thereby also reducing the abundance of CO_2 in the atmosphere.

   Evolutionary models of main-sequence stars indicate that the brightness of the sun has increased by 25% since the formation of the solar system (Newman and Rood, 1977). Accordingly, the energy received from the sun should have been lower 4.5×10^9 years ago. A smaller solar flux would have caused surface temperatures to remain below the freezing point of water for some time, unless sufficient CO_2 and water vapor had existed to raise the temperature by greenhouse warming. The age of the oldest known sediments from the Isua formation is 3.7×10^9 years (Moorbath *et al.*, 1973). The rocks contain well-rounded pebbles, indicating the presence of liquid water at that time, in addition to carbonates and other evidence of weathering processes similar to those occurring today. The presence of liquid water is deemed necessary for the development of life, and for this reason it is unlikely that the Earth has ever been fully glaciated. The ability of CO_2 to compensate for the reduction in solar energy has been explored in models with the result that 100–1000 times the present amount would have been required to prevent global-scale freezing (Kuhn and Kasting, 1983; Kasting *et al*., 1984). In the case of a catastrophic early degassing event, the CO_2 level may have been even higher for some time.

Magmatic gases released from volcanoes today contain water vapor and carbon dioxide as the major components, with smaller contributions of SO_2/H_2S, HCl, CO, H_2, and N_2. A large share of them probably have undergone extensive recycling, so that they should not be considered juvenile degassing products. It is nevertheless possible to derive approximately the average composition of juvenile gases by tracing them to the geochemical reservoirs in which they now reside, and by setting up an inventory of the accumulated masses. Geochemists have long recognized that water in the hydrosphere, CO_2 buried as carbonates in the sediments, chloride and sulfate dissolved in seawater, and nitrogen contained in the atmosphere are far too abundant to have originated from rock weathering alone and, hence, must have been supplied by thermal exhalation. Rubey (1951), who was one of the first to prepare a quantitative account of the volatiles, appropriately termed them *excess* volatiles. Table 12.6 shows an inventory for the most important volatile elements in addition to water, fashioned after that presented by Rubey. More recent data are used in Table 12.6, but the results are very similar to the earlier ones, and the conclusions remain the same. Li (1972) employed a somewhat different method, using mass balance relations for the weathering reactions in combination with ^{18}O, ^{13}C, and ^{34}S isotope data to fix the cumulative masses of individual volatiles. His results are entered in the last row of the table, to show that with regard to the total amounts of volatiles, very little divergence of opinion exists. The entries in Table 12.6 indicate that, indeed, water resides mainly in the oceans, carbon in the sediments, and nitrogen in the atmosphere, whereas chlorine and sulfur are about evenly divided between oceans and sediments. In the latter domain, Cl and S occur mainly in evaporites, sulfur being additionally present in reduced form in shales.

Table 12.7 compares the relative abundances of excess volatiles with the composition of gases escaping from volcanoes and fumaroles, and with gases that can be extracted from solid lava. Only a few selected data are shown.

TABLE 12.6 Estimates for the Masses (kg) of Water and Other Volatiles Residing in the Atmosphere, the Ocean, and the Sediments, with Contributions from the Weathering of Rocks and Excess (Magmatic) Exhalations[a]

Source	H_2O	C	CO_2	N_2	Cl	S
Atmosphere	Negligible	7.0 (14)	2.5 (15)	3.9 (18)	Negligible	Negligible
Hydrosphere	1.4 (21)	3.4 (16)	1.4 (17)	2.2 (16)	2.7 (19)	1.2 (18)
Sedimentary rocks	3.3 (20)[b]	6.2 (19)[c]	2.3 (20)	9.0 (17)[d]	2.2 (19)[e]	1.2 (19)[f]
Total volatiles	1.7 (21)	6.2 (19)	2.3 (20)	4.8 (18)	4.9 (19)	1.3 (19)
Supplied by weathering of igneous rocks[g]	3.0 (19)	6.9 (17)	2.5 (18)	5.9 (16)[d]	6.5 (17)	8.8 (17)
Excess volatiles	1.7 (21)	6.2 (19)	2.2 (20)	4.7 (18)	4.8 (19)	1.2 (19)
Excess volatiles according to Li(1972)[h]	1.7 (21)	5.6 (19)	2.0 (20)	3.9 (18)	4.2 (19)	1.3 (19)

[a] Orders of magnitude are shown in parentheses.
[b] Garrels and McKenzie (1971).
[c] From Table 11.1.
[d] Wlotzka (1972), corrected to a total sediment mass of 2.4 (21) kg.
[e] Sediment distribution of Hunt (1972). Table 11.3, assuming 3% evaporites.
[f] Holser and Kaplan (1966).
[g] 2.4 (21) kg derives from the weathering of 2.15 (21) kg of igneous rock. Element abundances were taken from Horn and Adams (1966).
[h] The estimates of Li (1972) were derived in part from isotope balances combined with mass balances of rock weathering.

TABLE 12.7 Composition of Volcanic Gases (mole %) Compared with the Composition of Excess Volatiles

Location	H_2O	CO_2	N_2	Cl	S	H_2	CO	References
Hawaii, composite[a]	79.3	11.6	1.3	0.05	6.9	0.6	0.4	Eaton and Murata (1960)
Kilauea, Hawaii[b]	97.3	2.3	—	—	0.43	0.23	—	Heald et al. (1963)
Surtsey, Iceland[c]	86	6.0	0.07	0.4	2.7	4.7	0.4	Sigvaldason and Elisson (1968)
Erta'Ale, Ethopia[d]	79.4	10.0	0.18	0.4	1.0	1.5	0.5	Giggenbach and LeGuern (1976)
Lassen Peak, California, gas extracted from lava	93.7	2.1	0.6	0.3	0.9	0.4	0.6	Shepherd (1925)
Mauna Loa, Hawaii, lava	73.2	15.3	5.2	0.2	1.2	4.4	1.4	Shepherd (1938)
Kudryavy, Kuril Islands	93.7	2.3	0.3	0.06	2.3	1.2	0.02	Taran et al. (1995)
Excess volatiles	93.6	4.6	0.2	1.2	0.4	—	—	Last two lines from Table 12.6

[a] Composite from data reported by Shepherd (1938), Jagger (1940), and Naughton and Terada (1954).
[b] Corrected for the presence of air.
[c] Average of three least contaminated samples.
[d] Average of 18 samples.

727

Volcanic gases are notoriously difficult to study, because contamination with air is almost unavoidable, and chemical reactions may change the composition during the time between collection and analysis. Moreover, all of the measurements reveal a remarkable variability in composition, which is partly due to variations in temperature and the accompanying changes in the chemical equilibrium. These difficulties notwithstanding, it can be seen from Table 12.7, and it has been emphasized by Rubey (1951), that the excess volatiles occur in relative proportions comparable to those found for the same elements in volcanic emanations. This implies either that the degassing conditions during the early period of Earth's history were roughly the same as they are today, or that current volcanic exhalations feed largely on volatiles being recycled from the store of sedimentary deposits. Both conclusions probably are applicable. The oxidation state of volcanic gases is mainly determined by the Fe^{3+}/Fe^{2+} ratio in magmas from which the gases evolve. It is interesting to note that Precambrian basalts contain Fe_2O_3 and FeO in essentially the same ratio as recent primary rocks (Eaton and Murata, 1960; Holland, 1978), indicating that the oxidation state has not changed significantly with time.

Free oxygen is not a volcanic exhalation product. The partial pressure of oxygen above basaltic melts is about 1 mPa at 1500 K (Katsura and Nagashima, 1974), and it decreases toward lower temperatures (Heald et al., 1963). The O_2 concentration is low enough for the appearance of H_2 and CO in volcanic gases (see Table 12.7). These gases are present because of the high temperature dissociation of H_2O and CO_2, respectively. The fractions of H_2 and CO found agree approximately with those expected from the thermochemical equilibria with their precursors, provided oxygen pressures 10 mPa or lower are inserted into the calculations (Matsuo, 1978; Holland, 1978).

In the primitive atmosphere, oxygen was essentially absent. The prevalence of anoxic conditions during the first 2.5×10^9 years of the Earth's history is documented by the preservation of ancient detritals of minerals like uranite and pyrite (Schidlowski, 1966, 1970), which are unstable in oxidizing environments and do not occur in later deposits. Authors addressing the problem of the origin of life (e.g., Rutten, 1971; Miller and Orgel, 1974) have emphasized that the initial development of life demanded a reducing habitat. Volcanic gases having a modern composition presumably would have provided a sufficiently reducing environment. The more radical suggestions of Urey (1952) and Miller and Urey (1959) of a primitive atmosphere consisting mainly of methane and ammonia are no longer maintained (Kasting, 1993). Their principal arguments were based on average solar system abundances and were not concerned with outgassing conditions.

The possibility exists, of course, that primitive volcanic gases were more reducing that they are now, but the admissible margin is rather narrow. Suppose for the sake of argument that the 5×10^{19} kg of inorganic carbon residing in the sediments had been issued originally as CO. Its subsequent oxidation to CO_2 would have required 6.7×10^{19} kg of oxygen. As will be shown in the next section, the mass of oxygen derived from photochemical processes is approximately 3.4×10^{19} kg, that is only one-half of that required. Since further amounts of oxygen were needed to oxidize other elements such as sulfur and bivalent iron, no more than a fraction of volatile carbon can have occurred as CO. Even greater difficulties would be encountered, if carbon were assumed to be released in the form of methane. Thus, it appears that not only the relative amounts but also the oxidation state of the ancient degassing products were similar to those of modern volcanic gases.

From the data in Table 12.6 one calculates that the total amount of the major volatiles H_2O, CO_2, and N_2 is about 10^{23} mol, with molar percentage fractions of 9.48, 5.0, and 0.17, respectively. If a major portion of the volatiles had been liberated by an early catastrophic outgassing event, and water had condensed to form an ocean, CO_2 would have been the dominant constituent in the atmosphere. A sudden release of 50% of volatile carbon would raise the CO_2 pressure to 2.1 MPa. About 30% of it would have dissolved in the ocean, reducing the CO_2 partial pressure in the atmosphere to 1.5 MPa. This is still 15 times the present total atmospheric pressure. The partitioning of CO_2 between ocean and atmosphere would have been much less favorable than today, because the dissolution of that much CO_2 in seawater would have decreased the pH to a value near pH 3. Under such conditions, weathering reactions must have been important. The dissolution in seawater of alkaline elements liberated from the evolving crust undoubtedly raised the pH of the ocean, allowing more CO_2 to be dissolved. In addition, CO_2 would have reacted with calcium and magnesium to form carbonate deposits. In this manner, the CO_2 pressure in the atmosphere would have gradually been lowered. Eventually, the CO_2 partial pressure must have decreased to a value below that of nitrogen, and from then on nitrogen became the major constituent of the atmosphere. If, on the other hand, CO_2 were not suddenly released in large amounts, weathering may have kept pace with the rate of outgassing, so that the content of CO_2 in the atmosphere–ocean system would have been lower, although still higher than that today. In that case nitrogen may well have been the major constituent all the time. Kasting (1993) has reviewed ideas about the development of CO_2 in the primitive atmosphere. He suggested a CO_2 pressure range of 0.01–1 MPa soon after Earth's formation, while in the late Precambrian, 700 million years ago, the partial pressure of CO_2 would have decreased to 0.3–30 hPa, corresponding to volume mixing ratios in the range of 0.03–3%.

12.3. NITROGEN

According to the concepts discussed in the preceding section, nitrogen evolved by thermal degassing of virgin planetary matter forming the Earth and accumulated predominantly in the atmosphere. The accumulated mass is 3.9×10^{18} kg. Wlotzka (1972) reviewed the nitrogen contents of other geochemical reservoirs. As Table 12.6 shows, the sediments contain 9×10^{17} kg, mainly as part of organic compounds in shales. The amount represents an outflow of the biosphere and must be counted as part of the total mass of nitrogen liberated. Igneous rocks contain about 25 mg kg^{-1} nitrogen as ammonium and 3 mg kg^{-1} as N_2, summing to 9×10^{17} kg for the total mass of igneous rocks in the Earth's crust. The amount of juvenile nitrogen remaining in the Earth's mantle is unknown, because most of the nitrogen found in volcanic gases presumably is due to the recycling of sediments. If one assumes that only the upper mantle has released its volatiles while the lower mantle has retained them, an amount similar to that in the atmosphere is expected to reside in the lower mantle.

Owing to the utilization by the biosphere, atmospheric nitrogen participates in two slow cycles. One is the exchange of N_2 between the atmosphere and the biosphere via processes of nitrogen fixation and denitrification; the other cycle involves the burial and recovery of organic nitrogen as part of the turnover of the sediments. Table 12.8 lists estimates for the inventories of fixed nitrogen in various compartments of the biosphere and in the oceans. It can be seen that this fraction is small compared with that residing in the atmosphere. Few organisms are capable of utilizing atmospheric nitrogen directly, forcing plants to rely extensively on nitrogen offered by decaying organic matter. As discussed by Rosswall (1976) and Schlesinger (1997), up to 95% of the nitrogen present in terrestrial ecosystems is recycled within the soil and between soil and vegetation. The biochemical pathways operative in converting different nitrogen compounds were treated in Section 9.1. Here, the interest is focused on the rate of nitrogen fixation as the principal supply of nitrogen to the biosphere. There are two routes. In addition to the biochemical reduction of N_2 by bacteria and blue-green algae, one must consider the nonbiotic oxidation to nitrate, which all plants and most microorganisms can use directly. Fixed nitrogen is ultimately lost from the biosphere by denitrification processes, which cause the return of nitrogen to the atmosphere in the form of N_2 or N_2O.

Estimates for the fluxes associated with N fixation processes are listed in Table 12.9. The individual entries carry the usual uncertainties. The global rate of biological nitrogen fixation in terrestrial habitats is the estimate of Burns and Hardy (1975). It represents the sum of rates for agricultural

TABLE 12.8   Global Nitrogen Contents in the Biosphere, in Soils, and in the Ocean[a]

Reservoir	Mass content (Tg)	Remarks
Terrestrial		
Land plants	0.75–1.1 (4)	Based on C/N = 75 and Table 11.4
Litter and dead wood	1.5 (3)	Based on C/N = 60 and 90 Pg of litter and dead wood
Animals	2.0 (2)	Delwiche (1970)
Soil humus, organic fraction	2.0 (5)	Based on C/N = 10 and Table 11.1
Soil, inorganic fraction (mainly fixed, insoluble NH_4^+)	1.6 (4)	Söderlund and Svensson (1976)
Sum	2.3 (5)	
Marine		
Plankton	5.2 (2)	Based on C/N = 5.7 (Redfield et al., 1963) and Table 11.1
Animals	1.7 (2)	Delwiche (1970)
Dead organic matter (particulates)	5.3 (3)	Based on C/N = 5.7 and 20 Pg of particulate carbon (Section 11.2.2)
Dissolved organic matter	3.7 (5)	Based on C/N = 2.7 (Duursma, 1961) and Table 11.1
Inorganic nitrogen	5.7 (5)	Söderlund and Svensson (1976)
Sum	9.4 (5)	
Dissolved N_2	2.2 (7)	From Table 12.6
Total mass of nitrogen	2.3 (7)	

[a] Orders of magnitude shown in parentheses; based partly on data from Chapter 11.

TABLE 12.9   Global Fluxes and Atmospheric Residence Times of Nitrogen Associated with
Biological Nitrogen Fixation, Inorganic Atmospheric Oxidation, Agricultural Fertilizers,
Combustion Processes, and the Sedimentary Cycle[a]

Process	Flux ($Tg\ year^{-1}$)	τ (years)	Remarks
Biological N fixation			
Terrestrial	139		Burns and Hardy (1975)
Marine	70 ± 50		Söderlund and Svensson (1976)
Combined	159–259	1.8 (7)	
Lightning flashes	8	5 (8)	From Table 9.15
Cosmic rays	0.04	9.7 (10)	Warneck (1972)
Industrial fertilizer	80	4.9 (7)	FAO (1992) for the year 1990
Combustion and			
biomass burning	32	1.2 (8)	From Table 9.15
Sedimentary cycle	4.3	9 (8)	Ratio N/C = 0.072; carbon flux From Table 11.8

[a] Powers of 10 are in parentheses.

systems (44 Tg year^{-1}, mainly from legumes), from grasslands (45 Tg year^{-1}), from forests (40 Tg year^{-1}), and from miscellaneous sources (10 Tg year^{-1}). The rate of biological nitrogen fixation in the marine environment is still poorly defined. Söderlund and Svensson (1976) have considered a range of 20–120 Tg year^{-1}. The minimum uptake rate by cyanobacteria in the open ocean is 10 Tg year^{-1}, according to Carpenter and Romans (1991). As in the terrestrial biosphere, fixed nitrogen is rapidly recycled among the biota. As fixed nitrogen is lost by denitrification, one may also estimate the rate of nitrogen fixation via the rate of denitrification. Here again, however, the estimates are rather uncertain. Söderlund and Svenson (1976) estimated the denitrification rate in the deep ocean to be 91 Tg year^{-1}. Christensen *et al.* (1987) suggested a denitrification rate of 110 Tg year^{-1} for the continental shelf areas alone. Estuaries and coastal regions receive fixed nitrogen also with the river input at a rate of about 36 Tg year^{-1} globally (Meybeck, 1982), as well as an undetermined amount by wet and dry deposition of nitrate and ammonium originating on the continents. These fluxes should be deducted to balance the cycle. Accordingly, it is possible that the nitrogen fixation rate in continental shelf areas is rather small. The input of fixed nitrogen from the atmosphere to the open ocean is 30 Tg year^{-1} (Duce *et al.*, 1991), of which about 16.5 Tg year^{-1} is ammonium and 13.5 Tg year^{-1} is nitrate or nitric acid. As discussed in Section 9.2.2, 50% of the input of ammonium derives from the continents; the rest represents the return flow of NH_3 emissions from the ocean. The extra input thus amounts to about 21 Tg year^{-1}. The rate of nitrogen fixation in the open ocean, if balanced by denitrification, therefore amounts to $91 - 21 = 70$ Tg year^{-1}, approximately. Nitrogen fixation in continental shelf areas (< 70 Tg year^{-1}) would have to be added. Together, terrestrial and marine biological nitrogen fixation processes transfer about 210 Tg nitrogen year^{-1} from the atmosphere to the biosphere. Inorganic nitrogen fixation processes are the natural oxidation of N_2 by lightning discharges and by ionizing radiation in the atmosphere, by the inadvertent oxidation of nitrogen in air during high-temperature combustion processes, and by the application of fertilizers in agriculture. The fluxes in Table 12.9 show that biological nitrogen fixation is the dominant process, although the application of fertilizers is expected to grow in the future and may eventually reach a similar magnitude.

Table 12.9 includes the residence times of nitrogen in the atmosphere calculated from the individual fluxes. Biological nitrogen fixation results in a residence time of 18 million years, which is much shorter than the age of the Earth and its atmosphere. Even the oxidation of N_2 in lightning discharges leads to a residence time that is shorter. Lewis and Randall (1923) and Sillén (1966) pointed out that in the presence of free oxygen and liquid water, nitrogen is thermodynamically unstable because the equilibrium

$$N_2 + 2\tfrac{1}{2}O_2 + H_2O \rightleftharpoons 2H^+ + 2NO_3^-$$

favors the right-hand side. If the entire amount of oxygen in the atmosphere were used to oxidize nitrogen to nitrate, the mass of N_2 in the atmosphere would be reduced by 11% to about 3.4×10^{18} kg, and the difference of 4.7×10^{17} kg would have been transferred to occur as nitrate in the ocean (and the sediments). The present amount of NO_3^- nitrogen in the ocean is 5.7×10^{14} kg, as Table 12.8 shows. It is clear that denitrification and the accompanying return of nitrogen to the atmosphere keep the concentration of nitrate in the ocean at a low level. The external biological cycle consisting of nitrogen fixation and denitrification proceeds at a rate at least 10 times faster than NO_3^- formation due to lightning discharges, so that it is the controlling process.

Table 12.9 includes the flux of nitrogen associated with the cycle of the sediments. The rate can be calculated from the flux of organic carbon given in Table 11.6 and the ratio of organic nitrogen to carbon in the sediments. The small rate obtained makes evident that this cycle is trivial compared to that between the atmosphere and the biosphere. The fate of organic nitrogen exposed to the atmosphere upon weathering of the sediments has not been explored. Presumably it becomes part of organic matter in the soil, is utilized by microorganisms, and thus returns to the biosphere from which it originated.

12.4. OXYGEN

12.4.1. SOURCES OF OXYGEN

Because oxygen is not a terrestrial outgassing product, the primitive atmosphere on Earth must have been essentially devoid of oxygen. Under such conditions, the photodissociation of H_2O and CO_2 would have provided the only physicochemical source. In the early treatment of these processes by Berkner and Marshall (1964, 1966) and Brinkmann (1969), it was assumed that the photodecomposition of water vapor was confined to the lowest regions of the atmosphere, that the products were hydrogen and oxygen, that hydrogen underwent Jeans escape so that oxygen accumulated, and that eventually oxygen overtook water vapor as an effective ultraviolet absorber, whereupon H_2O was shielded from further photodecomposition. Because of the emphasis placed on water vapor and an insufficient understanding of the factors controlling the escape of hydrogen, the early studies concentrated on the amount of oxygen required as a radiation shield, but ignored all aspects of reaction kinetics. These are important, however.

Kasting *et al.* (1979) finally clarified the situation by means of a study involving a one-dimensional steady-state diffusion model incorporating the essential photochemical reactions. Figure 12.3 shows some of their results. The overriding source of oxygen at all altitudes is the photodecomposition of

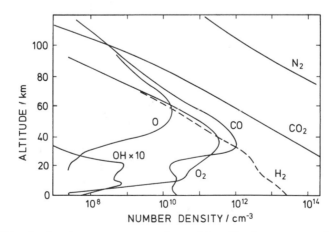

FIGURE 12.3   Steady-state concentrations of N_2, CO_2, CO, O_2, H_2, O, and OH in the prebiotic primitive atmosphere according to model calculations of Kasting *et al.* (1979). These authors assumed a steady flux of $F(H_2) = 1 \times 10^{12}$ m^{-2} s^{-1} through the atmosphere, emanating from volcanoes and leaving by Jeans escape.

CO_2, which produces CO and O atoms. Carbon monoxide is oxidized back to CO_2 by reaction with OH derived from the photolysis of water vapor. Molecular oxygen is formed by the recombination of oxygen atoms ($z \approx 60$ km) and by the reaction $O + OH \rightarrow O_2 + H$ ($z < 30$ km). The steady-state O_2 mixing ratio is a function of altitude. It rises from a value near 10^{-12} at ground level to almost 10^{-5} at 60 km altitude. The small concentration of O_2 in the lower atmosphere is due partly to an efficient catalytic chain in which hydrogen and oxygen combine to produce water. Another reason for the low oxygen level is the presence of H_2 due to volcanic outgassing. Kasting *et al.* (1979) assumed an H_2 flux of 7×10^7 kg year^{-1}, which they considered to be the present rate. Walker (1977) derived a rate about three times higher. Because of more intense volcanic outgassing, the flux may have been even higher 4×10^9 years ago. Such conditions would have reduced the O_2 concentration even further.

In discussing Jeans escape of hydrogen generated by the photodissociation of water vapor, we reiterate that the escape rate is limited by the rate of eddy diffusion of hydrogen-containing compounds in the region, where the atmosphere is still well mixed, that is, in the homosphere. In the lower atmosphere the decomposition of H_2O by ultraviolet radiation is largely suppressed by back-reactions. The study of Kasting *et al.* (1979) showed that these reactions become ineffective at altitudes above 20 km, and at higher levels H_2O is essentially completely dissociated. The supply to these altitudes is determined by the temperature minimum in the middle atmosphere, so that the escape rate of H_2 derived from water vapor should have been similar to that existing today. The present rate of Jeans escape of hydrogen is about 2×10^{12} atom m^{-2} s^{-1}, or 2.6×10^{10} mol year^{-1} (Hunten and Strobel, 1974; Liu and Donahue, 1974). The net loss is somewhat reduced by accretion of hydrogen from the solar wind, but not by much (Walker, 1977). The corresponding rate of oxygen production is 4.2×10^8 kg year^{-1}. If one assumes that the rate has remained unchanged over geological time, the total gain of oxygen in 4.5×10^9 years is 1.9×10^{18} kg. This value compares well with that currently residing in the atmosphere and ocean combined (1.2×10^{18} kg). The comparison is misleading, however, because it neglects losses of oxygen due to the oxidation of reduced compounds at the Earth's surface. In the model of Kasting *et al.* (1979) the gain of oxygen is essentially balanced by losses resulting from the oxidation of hydrogen emanating from volcanoes. Additional losses are important, as will be shown further below, so that Jeans escape cannot account for the mass of oxygen in the atmosphere.

Our current understanding of the evolution of oxygen is founded on the recognition that oxygen is a byproduct of the assimilation of CO_2 by green plants and phytoplankton. Recall from Chapter 11 that the annual exchange of carbon between the atmosphere and the biosphere is almost balanced. The same must hold true for the equivalent exchange of oxygen, as the amount of oxygen generated by photosynthesis is used up again when organic matter reverts to CO_2. However, a tiny fraction of organic carbon is lost each year by incorporation into the sediments, and this process then imposes a similar imbalance on oxygen. Thus, the evolution of oxygen is intimately tied to that of organic carbon imbedded in the sediments. The photosynthetic process may be represented by the reaction

$$nCO_2 + nH_2O \rightarrow (CH_2O)_n + nO_2$$

which shows that 1 mole of oxygen is liberated for 1 mole of carbon that becomes fixed. The relation accordingly determines the mass of oxygen that must have been generated from the mass of carbon buried in the sediments. Table 11.1 shows that the sediments harbor 1.2×10^{19} kg of organic carbon. The corresponding amount of oxygen is 1.2×10^{19} $(32/12) = 3.2 \ 10^{19}$ kg, or 25 times the mass currently residing in the atmosphere. If the amount derived from the photodissociation of H_2O and escape of H_2 is added, a total of 3.4×10^{19} kg of oxygen must be accounted for. Since the atmosphere contains so little of it, it appears that the dominant part has been consumed in oxidation processes, and these must be identified to establish a mass balance.

12.4.2. The Budget and Cycles of Oxygen

Elements or compounds that can be oxidized are made available from two geochemical sources: volcanic emanations yield H_2, CO, and sulfur compounds such as SO_2 and H_2S; and the weathering of igneous rocks supplies bivalent iron in the form of $FeSiO_3$. Other elements occurring in reduced form are of lesser importance. The total quantities of reduced material released during geological times may be estimated from the data assembled by Li (1972). Table 12.10 lists the pertinent reactions, the amounts of material released, and the contributions to the total oxygen budget. Several comments are required. Because the biosphere was initially quite tenuous, practically the entire production of oxygen was used up to oxidize iron and other reduced materials. Thus, oxygen did not immediately accumulate in the atmosphere. Under these initial conditions, the oxidation of hydrogen exhaled from volcanoes cannot have exceeded the rate of oxygen production from H_2O and CO_2 photolysis. Later, when oxygen had accumulated in the atmosphere, all of the hydrogen from geothermal processes was oxidized. When the transition occurred is not well known, but it probably occurred

TABLE 12.10 Geochemical Oxygen Budget, Based Mainly on Data Presented by Li (1972) from Mass Balance Relations Describing Weathering of Igneous Rocks by Primary Magmatic Volatiles.

Material	Reaction	Amount released (10^{21} mol)	Mass of oxygen (10^{18} kg)	Remarks
Production				
C_{org}	$CO_2 + H_2O \xrightarrow{h\nu} CH_2O + O_2$	10	32	Photosynthesis and burial of C_{org} in the sediments
H_2O	$H_2O \xrightarrow{h\nu} \frac{1}{2}O_2 + H_2$	1.2	1.9	Jeans escape of hydrogen, present rate
Sum			33.9	
Consumption				
H_2	$H_2 + \frac{1}{2}O_2 \rightarrow H_2O$	1.5	3.6	Two-thirds of O_2 from H_2O photodissociation plus losses of H_2 released since the advent of O_2 in the atmosphere
CO	$CO + \frac{1}{2}O_2 \rightarrow CO_2$	1.4	2.2	Assuming CO_2/CO mole ratio of 37 in volcanic exhalations
H_2S	$H_2S + 2O_2 \rightarrow H_2SO_4$	2.1	13.4	Assuming sulfur to be entirely released as H_2S; about 50% occurs as SO_4^{2-} in the ocean and in evaporites, the rest as FeS_2 in shales
	$2H_2S + \frac{1}{2}O_2 \rightarrow 2H^+ + S_2^{2-} + H_2O$	2.1	1.7	
$FeSiO_3$	$2FeO + \frac{1}{2}O_2 \rightarrow Fe_2O_3$	12.4	9.9	
Free in the atmosphere		0.4	1.2	Including the amount dissolved in the ocean
Sum			32.0	

not earlier than 2×10^9 years ago. In Table 12.10 the estimate for losses resulting from the oxidation of hydrogen was obtained by extrapolating the present rate of hydrogen release back over 1.5×10^9 years, which gives 2.4×10^{18} kg O_2 year^{-1}. Two-thirds of the amount produced by CO_2 + H_2O photodissociation and presumably consumed by reacting with hydrogen is then added: $(2/3)1.9 \times 10^{18} = 1.2 \times 10^{18}$. The amount of CO that is assumed to have been released with volcanic gases is based on a volume ratio of $CO_2/CO = 37$, according to the thermal equilibrium at 1500 K (Matsuo, 1978), combined with the total carbon content of sedimentary rocks (Table 11.1). Sulfur presents a problem in that modern volcanic emissions contain both SO_2 and H_2S, but the partitioning is uncertain, for the present as well as the past. Here, it is assumed in accordance with Li (1972) that H_2S predominated. This assumption maximizes the consumption of O_2.

Only one-half of the sulfur inventory occurs as sulfate. The other half is present as pyrite in the sediments due to the action of sulfate-reducing bacteria living in marine muds. The reduction is made possible by organic carbon, and the reaction may be written as

$$2Fe_2O_3 + 8SO_4^{2-} + 15CH_2O + 16H^+ \rightarrow 4FeS_2 + 15CO_2 + 23H_2O$$

Here, 3.75 moles of organic matter are consumed for 1 mole of pyrite deposited, and an equivalent quantity of O_2 is left behind in the atmosphere–ocean system. The excess is used up again in the oxidation of pyrite when it is brought back to the surface of the Earth by the sedimentary rock cycle. With regard to the oxygen budget, it is important to note that disulfide represents a higher oxidation stage of sulfur than sulfide. The substitute reaction entered in Table 12.9 demonstrates that the oxidation of 1 mole of sulfide to disulfide consumes 0.25 mole of oxygen. The mass of sulfur occurring as pyrite in the sediments was estimated by Holser and Kaplan (1966) as 6.65×10^{18} kg. The corresponding mass of oxygen that has been consumed is 1.7×10^{18} kg, and this value is entered in the table.

FeO present in igneous rocks is oxidized to Fe_2O_3 by weathering when such rocks are exposed to air. In a way similar as to the reduction of sulfate, however, Fe_2O_3 can be reduced to Fe^{2+} in anaerobic environs such as marine sediments. The reduction is again mediated by organic carbon and leads to the occurrence of bivalent iron in sedimentary rocks in amounts greater than that represented by pyrite. The total mass of FeO that was converted to Fe_2O_3 and remains in that oxidation state can be estimated from the difference in the average contents of FeO and Fe_2O_3 in igneous and sedimentary rocks. Difficulties are encountered because of widely varying iron contents, so that representative averages are not easily established. Holland (1978) cites average abundances of FeO in igneous rocks and in the sediments of 3.5% and 1.6% by mass, respectively. The total mass of the sediments is about 2.4×10^{21} kg. Because of the incorporation of volatiles, the weathering of 1 kg of igneous rock produces 1.12 kg of sedimentary rock. The amount of FeO that has been oxidized to Fe_2O_3 then is

$$\left(\frac{3.5}{1.12} - 1.6 \right) 10^{-2} \frac{2.4 \times 10^{21}}{72 \times 10^{-3}} = 5.1 \times 10^{20} \text{ mol}$$

where $M_{FeO} = 72 \times 10^{-3}$ kg mol^{-1} is the molar mass of FeO. Basaltic rocks often contain a higher percentage of FeO than granitic rocks, on which the 3.5% estimate of Holland is based. Li (1972) assumed instead a mass percentage of 5.8 in average igneous rock, so that he derived a higher value of 12.4×10^{20} mol FeO that was oxidized to Fe_2O_3. This value is entered in Table 12.10 to maximize the consumption of O_2, but it is also rather uncertain.

The total mass of oxygen consumed by the reactions listed, added to that contained in the atmosphere, sums to about 3.1×10^{19} kg. The result represents a reasonable balance between production and loss only because an effort has been made to maximize the consumption of oxygen. On the whole, the budget is still quite uncertain. But the data show that the major reservoirs of oxygen are sulfate in seawater and in evaporites and Fe_2O_3 in sedimentary rocks. Only 4% of the total amount of oxygen produced during Earth's history resides in the atmosphere. One must appreciate the peculiarity of this distribution. As oxidative weathering causes a steady drain on atmospheric O_2, it must be continuously replenished. It appears that production and consumption processes are approximately in balance and keep atmospheric oxygen in a steady state. The subsequent discussion deals with the ensuing cycle of oxygen. We consider first the flux of oxygen through the atmosphere, then the mechanism controlling the abundance, and finally the rise of oxygen during Earth's history.

The weathering of rocks leads to a loss of oxygen, mainly by the oxidation of reduced material that sedimentary rocks inherit as a consequence of biological activities in the marine environment. The resulting rate of oxygen consumption can be determined from the average contents of C_{org}, FeS_2, and FeO in sedimentary rock and the flux of organic carbon associated with the geochemical rock cycle. The latter quantity was discussed in Section 11.3.1 (see Table 11.8). For completeness, one must also consider O_2 losses incurred in the weathering of igneous rocks. The data are summarized in Table 12.11, which shows the abundances of the substances of interest in the two rock types, the associated material fluxes, and the rates of oxygen consumption, calculated for a complete oxidation of C_{org} to CO_2, FeO to Fe_2O_3, and S_2^{2-} to SO_4^{2-}. The oxidation of organic carbon is found to

TABLE 12.11 Oxygen Consumption Rates Associated with the Weathering of Sedimentary and Igneous Rocks[a]

Material	Content (% by mass)	Flux rate (Tg year^{-1})	(Tmol year^{-1}) S	Rate of oxygen consumption (Tg year^{-1})
Sedimentary rocks				
C_{org}	0.5	60	5.0	160
FeO	1.6	192	2.67	21
S_2^{2-}	0.28	34	0.52	62
Igneous rocks				
FeO	5.8	232	3.22	26
Total rate				269

[a] Based on a flux of organic carbon of 60 Tg year^{-1} (Table 11.8), average content of 0.5% C_{org} in sedimentary rocks, and a ratio of 25:75 of igneous to sedimentary rocks being weathered. The amount of oxygen consumed by reduced volatiles exhaled from volcanoes is negligible in comparison.

contribute the major share of the loss, whereas the oxidation of bivalent iron in igneous rocks accounts for only 10%. The total loss rate is about 270 Tg year^{-1}. Other authors have come up with similar estimates. Holland (1973, 1978) derived a loss rate of 400 Tg year^{-1}, and Walker (1974) obtained 300 Tg year^{-1}. Walker also pointed out that reduced volatiles exhaled from volcanoes consume oxygen at a rate that is negligible compared with weathering.

The loss of oxygen due to weathering should be balanced by a similar net production rate associated with the incorporation of organic carbon in marine sediments. Walker (1974) has examined rates for the burial of organic carbon in several aquatic environs. He concluded that the highest rates are found in marine sediments known as "blue muds" occurring on the continental slopes, where carbon contents and accumulation rates are both high. Walker estimated global burial rates ranging from 8 to 22 Tmol year^{-1} for such locations. The equivalent O_2 production rate of 250–700 Tg year^{-1} has the correct magnitude to compensate for the loss due to weathering. Walker further found that the incorporation of carbon into deep sea sediments is an order of magnitude smaller despite the large area covered, and anoxic basins such as the Black Sea are even less effective. Globally negligible, too, is the burial of organic carbon in freshwater lakes.

Some of the organic carbon in continental shelf regions originates from the terrestrial biosphere. Meybeck (1982) estimated that the flux of organic matter transported to the ocean by the rivers totals 400 Tg year^{-1}, with tropical streams contributing the major share. How much of this material survives to be incorporated into marine sediments is quite uncertain, but it is generally thought to be small (Holland, 1978). Griffin and Goldberg (1975) have shown that charcoal derived from forest fires also enters marine sediments. The corresponding carbon fluxes are an order of magnitude smaller than those from the marine biosphere.

A turnover rate of about 300 Tg year^{-1} corresponds to a residence time of oxygen in the atmosphere–ocean reservoir of

$$\tau_{atm}(O_2) = 4 \times 10^6 \text{ years}$$

The weathering of rocks would deplete atmospheric oxygen in a geologically short period of time, were it not replenished by the activity of the biosphere. The latter is the controlling factor, of course, whereas weathering reactions are a secondary phenomenon. But a residence time of 4 million years indicates the time scale on which changes in the atmospheric oxygen level can be expected. Such changes presumably have occurred even after the atmospheric oxygen reservoir was well established.

The above value for $\tau_{atm}(O_2)$ may also be compared with the residence time resulting from the rapid exchange of carbon between the biosphere and CO_2 in the atmosphere–ocean system. In Chapter 11 the net primary productivity for the marine and terrestrial biosphere was given as 35 and 58 Pg year^{-1}, respectively. Correcting for the different molar masses of carbon and oxygen, one has

$$\tau_{atm}(O_2) = \frac{(1.2 \times 10^6)(12)}{(35 + 58)(32)} \approx 5 \times 10^3 \text{ years}$$

The turnover rate of atmospheric oxygen resulting from the interaction with the biosphere is about 1000 times faster that that caused by the geochemical cycle. The same result would have been obtained had we compared the pertinent fluxes directly.

The turnover time of 5000 years corresponds to an annual oscillation of the oxygen mixing ratio in the troposphere of about 50 μmol mol^{-1}. By means of an interferometric technique, Keeling and Shertz (Keeling and Shertz, 1992; Keeling et al., 1996) have recently confirmed the existence of the seasonal variation, relative to nitrogen, and Figure 12.4 shows some of their data. The seasonal cycles in the Northern and Southern Hemispheres are opposite each other, and they are anticorrelated with the cycles of CO_2, thus demonstrating the close relation between the two substances. Figure 12.4 shows that the amplitude of the oscillation is almost twice that expected, but the two hemispheres are decoupled, so that the amplitude in each hemisphere should be enhanced compared with the average. Keeling

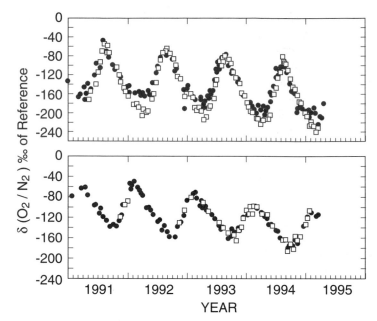

FIGURE 12.4   Seasonal variation of the O_2/N_2 ratio, relative to that of a reference, expressed as $\delta(O_2/N_2) = 10^6\,[(O_2/N_2)/(O_2/N_2)_{ref} - 1]$. Upper panel: Measurements at La Jolla, California (●), and Alert, Canada (□); lower panel: measurements at Cape Grim, Tasmania (●), and South Pole (□). (Adapted from Keeling *et al.* (1996).)

and Shertz (Keeling and Shertz, 1992; Keeling *et al.*, 1993) suggested that an additional effect might result from the annual warming and cooling of the upper layer of the ocean, which modulates the amount of oxygen dissolved in seawater. CO_2 would not be affected because it is tied to bicarbonate and thus is less volatile.

The question of what controls the abundance of oxygen in the atmosphere can now be considered. First, it should be noted that despite the rapid turnover, the fast cycle of carbon cannot be the controlling mechanism, because the size of the biosphere is too small for that purpose. The quantity of oxygen required to convert the entire mass of carbon in the biosphere to CO_2, including dead organic carbon in the soil and in the oceans, amounts to less than 1% of the mass of oxygen in the atmosphere. Accordingly, the control must be exercised by processes associated with the slower geochemical cycle. Broecker (1970, 1971), Holland (1973, 1978), and Walker (1974, 1977) have examined these processes, and the following discussion summarizes their main arguments.

The current high concentration of atmospheric oxygen ensures a fairly rapid oxidation of reduced material in sedimentary rocks upon their exposure to weathering. Certainly this is true for bivalent iron and sulfides, which are both soluble in water but which are not significant in river waters. The possibility that some organic carbon escapes oxidation and undergoes resedimentation was discussed in Section 11.3.1. If this occurs, the rate of oxygen consumption would be somewhat diminished, but the conclusions that follow would not be substantially altered. For simplicity, it may be assumed with Walker (1974) that the oxidation is complete. The rate of oxygen loss is then limited by the rate at which reduced material in rocks is made available by erosion and chemical weathering, and since these processes are independent of the abundance of free oxygen, they cannot stabilize it. The situation was clearly different in the Precambrian, when oxygen started to evolve and its concentration was much lower than today. Under such conditions oxygen consumption would have been independent of the weathering rate. This situation is not considered here.

If, under limiting conditions, weathering reactions are not effective in controlling atmospheric O_2, one must look to the source of it, namely the burial of organic carbon in marine sediments. The efficiency of this process is determined by two opposing factors. One is the supply of organic matter by the biosphere, and the other is the rate at which oxygen dissolved in seawater is made available for the reoxidation of organic material. The details of the oxidation process operating along the food chain and by microbial decay are complicated and are not fully understood. For the purpose of illustration consider the following mechanism. Organic carbon meets aerobic waters while traveling from the photosynthetic production region near the sea's surface toward the sediment layer at the bottom of the ocean. Along the way, organic matter is subject to oxidation, and the fraction reaching the ocean floor will depend on both the depth of the water layer to be traversed and the average concentration of oxygen in it. Somewhere near the ocean water–sediment interface, either above or below it, conditions become anoxic, and after reaching this region organic carbon is shielded from further oxidation. Subsequent reactions such as fermentation, sulfate reduction, etc., do not change the local oxidation state and have no consequences for the oxygen balance. From this description of the situation it will be clear that if the oxygen concentration in the atmosphere–ocean system were lowered, less organic carbon would be oxidized and more would be buried. The result would be an increase in the net production of oxygen. Conversely, if the oxygen level in the atmosphere were raised, more oxygen would dissolve in the ocean, more organic carbon would undergo oxidation, and less of it would be incorporated into the sediments. The net production of oxygen would then decrease. In this manner a feedback mechanism is established that keeps the oxygen concentration at a level sufficiently high for an almost complete oxidation of organic matter. According to this model, the fraction of organic carbon escaping oxidation should be higher in relatively shallow waters as opposed to the deep sea, a feature explaining in part the high rates of carbon burial along the ocean's margins. The model also makes clear that the biosphere regulates primarily oxygen dissolved in the ocean. The role of the atmosphere is that of a buffering reservoir. This is a consequence of the relatively low solubility of oxygen in seawater, which causes the physical partitioning to favor the atmosphere as the main reservoir.

12.4.3. THE RISE OF ATMOSPHERIC OXYGEN

Finally, some notes are in order describing the evolution of free oxygen toward the present abundance, starting from the very low levels in the primitive atmosphere. This process must be viewed as an integral part of the early history of life and the development of photosynthetic carbon fixation. Our knowledge in this regard is quite fragmentary. Compared with the rich macroscopic fossil record covering the last 600 million years (Phanerozoicum), the sediments representing the more distant geological period (Precambrium) contain hardly any evolutionary indicators. The few that exist are summarized in Figure 12.5 and will be discussed below. Supplemental information comes from the realm of microorganisms, especially anaerobic bacteria, some of which represent very ancient forms of life. The technique of nucleotide sequence analysis as applied to certain proteins and nucleic acids has made it possible to establish the genealogy of bacteria with some confidence (Schwartz and Dayhoff, 1978; Fox *et al.*, 1980). The results have greatly benefited our understanding of cellular evolution. A few descent lines are included in Figure 12.5 for illustration, although the time scale for the branching points in the development remain conjectural at present.

Phototrophic bacteria hold a prominent position in the hierarchy of all bacteria. The majority of photosynthetic species are not capable of utilizing water as a hydrogen donor, and they must either make use of hydrogen from organic compounds or from H_2S by means of processes that do not liberate oxygen. Table 12.12 compares a number of simplified reactions for the fixation of carbon as well as the associated energy needs to show that the reaction of CO_2 with water as a hydrogen donor demands the highest energy input. One expects this process to have been adapted subsequent to the exploitation of other reactions with lesser energy requirements. Indeed, the phylogenetic tree in Figure 12.5 bears this out in that cyanobacteria appear later than green sulfur bacteria *Chloroflexus* and *Chlorobium*, although all three species represent very early forms of life. Among the bacteria, the

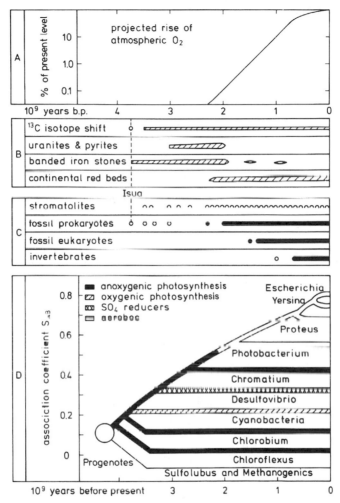

FIGURE 12.5 (A) Projected rise of atmospheric oxygen based on (B) geological evidence (Schidlowski, 1978) and (C) fossil remains (Awramik, 1982; McAlester, 1968; Pflug, 1978; Schidlowski, 1983a,b; Schopf, 1975; Schopf and Oehler, 1976). (D) Evolution of bacteria according to Kandler (1981). A few lines of development are selected for illustration. The phylogenetic tree is based on nucleotide sequences in 16 S ribosomal RNAs (Fox et al., 1980). The association coefficient S_{AB} ($0 < S_{AB} < 1$) is a measure of the overlap of common sequences, with $S_{AB} = 0$ indicating complete difference and $S_{AB} = 1$ full identity. The oldest line is represented by Archebacteria, which include *Sulfolubus*, living in hot, hydrogen sulfide-rich springs, and the methanogenic bacteria. *Chlorobium* and *Chloroflexus* are green sulfur bacteria. Cyanobacteria are the only oxygen producers. Time markers for oxygenic photosynthesis are the $^{13}C/^{12}C$ shift, the occurrence of banded iron stones, and stromatolites; the beginning of the $^{34}S/^{32}S$ shift is an indicator of the evolution of sulfate reducers 3.3×10^9 years ago (Monster et al., 1979); the advent of continental red-beds is taken to indicate that the Pasteur point was reached; aerobic bacteria should have appeared at the same time.

TABLE 12.12 Free-Energy Requirements for Reactions Reducing Carbon Dioxide to the Carbohydrate Level[a]

Number	Process	ΔG (kJ mol^{-1})	Examples of bacteria utilizing the reaction
1	$CO_2 + 2H_2 \rightarrow CH_2O + H_2O$	-4.2	Chemoautotrophs (e.g., methanogenics)
2	$CO_2 + 2H_2S \xrightarrow{h\nu} CH_2O + H_2O + 2S$	50.2	Chlorobacteriaceae (e.g., *Chlorobium*)
3	$CO_2 + H_2O + \frac{1}{2}H_2S \xrightarrow{h\nu} CH_2O + \frac{1}{2}H_2SO_4$	117.2	Thiorodoceae (e.g., *Chromatium*)
4	$CO_2 + 2H_2O^* \xrightarrow{h\nu} CH_2O + H_2O + O_2^*$	470.7	Cyanophytaceae and chloroplasts

[a] Reaction 1 is exoergic; the other reactions require an energy input. Photosynthesis using water as the hydrogen donor, causing the release of free oxygen, has the highest energy demand; the asterisk indicates the oxygen atom transferred.

splitting of water seems to have been mastered only by the cyanobacteria, also known as blue-green algae, and by prokaryotic algal species (*Prochloron*) (Lewin, 1976). The chloroplasts contained in the more advanced eucaryotic cells of green plants are believed to have descended from cyanobacteria or *Prochloron*-like precursors, which entered into a symbiotic relation with the eucaryotic host cell.

The record of reduced carbon in sedimentary rocks and the associated $^{13}C/^{12}C$ isotope shift (see Fig. 11.2) dates back to the oldest deposits of 3.5×10^9 years ago, and even further to 3.7×10^9 years if one makes allowance for the metamorphic alterations of the Isua formation. The biological origin of the isotope shift has been convincingly demonstrated, as reviewed by Schidlowski *et al.* (1983). The implication that life had been in full swing 3.5×10^9 years ago is supported by several lines of fossil evidence: stromatolites, microfossils, and banded iron formations.

1. Stromatolites are the commonest and most conspicuous fossils (Walter, 1976). They consist of bun-shaped or columnar, laminated structures, usually in limestone deposits; since they are essentially indistinguishable from modern analogs forming today by colonies of blue-green algae, stromatolites from the Precambrian generally are taken as evidence for the presence of cyanobacteria during that period, although it should be clear that other species may have produced similar biosedimentary structures. The earliest stromatolites occur in 3.5×10^9-year-old rocks (Lowe, 1980; Walter *et al.*, 1980), and they become quite common in the areally more extensive sediments younger than 2.2×10^9 years.

2. Microfossils, spheroidal or filamentous objects, can be discerned in stromatolites and nonstromatolitic cherts. On the basis of morphology and size, they are ascribed to unicellular organisms, either bacteria or eucaryotic cells, the latter making their appearance in sediments younger than 1.4×10^9 years (Cloud, 1976; Schopf and Oehler, 1976). The biogenicity of microfossils has been firmly established for strata younger than 2.3×10^9 years on the grounds of cellular differentiation or similarity to living organisms combined with carbonaceous composition and careful elimination of contamination by younger intrusions. More debatable are the primitive features occurring in 3.5×10^9-year-old rocks (Dunlop *et al.* 1978) and the still tentative microfossils described by Pflug (Pflug, 1978; Pflug and Jaeschke-Boyer, 1979) in the 3.7×10^9-year-old sediments of the Isua formation.

3. Banded iron stones are sedimentary structures consisting of alternating layers of iron oxides and iron-poor siliceous material. With few exceptions, their occurrence is confined to sediments older than about 2×10^9 years. Since the contemporaneous occurrence of detrital uranites and pyrites indicates the absence of free oxygen at that time, water-soluble ferrous iron was

not immediately immobilized as it is today by being oxidized to insoluble ferric iron, and it was carried by the rivers into the sea, where it accumulated. The currently accepted explanation for the origin of banded iron formations was formulated by Cloud (1972, 1973), who considered it a biosedimentary structure generated by the action of cyanobacteria or their progenitors. The local production of oxygen due to photosynthesis converted bivalent, ferrous iron to ferric hydroxides, which subsequently deposited to form hematite (Fe_2O_3). The presence of ferrous iron in the ancient seas must have kept the oxygen concentration at a low level, permitting oxygen-sensitive organisms to survive. The periodicity of deposition giving rise to alternating iron-rich and iron-poor microbands is still not well understood. It may have resulted from seasonal variations in the supply of vital nutrients. While the last point leaves many questions open, there is support for the biogenicity of banded iron formations from chemical indicators. Fiebiger (1975) studied hydrocarbons in a variety of banded ironstones and found evidence for photosynthetic activity in the presence of phytane and pristane, two stable decomposition products of chlorophyll. This finding suggests that photosynthetic activity was already in existence 3.7×10^9 years ago (banded iron deposits in Isua, West Greenland).

The incipient rise of oxygen in the ancient atmosphere is signaled by the appearance of the first continental red-beds of ferric iron about 2×10^9 years ago. This event essentially coincides with the termination of widespread banded iron formation. The pyrite and uranite detritals disappear at the same time. By then, in a long sequence of evolutionary steps, the early organisms must have learned to cope with the toxicity of free oxygen by having developed a variety of antioxidant enzymes such as cytochromes, carotenoids, and eventually superoxide dismutase. Before this capability existed, the rate of photosynthetic carbon fixation was limited by the rate at which the inorganic oxygen acceptors ferrous iron and hydrogen sulfide were made available by weathering processes and volcanic activity, respectively. The added capacity of tolerating free oxygen gave room for the expansion of primary productivity as far as the nutrient supply permitted it. This development led to an increase in the oxygen production rate and the rise of oxygen in the atmosphere. Now, as oxidative weathering took hold, ferrous iron no longer reached the seas, and bacteria unable to develop defense mechanisms against an aerobic environment had to withdraw to anoxic niches.

At the same time, respiratory metabolism developed. In the utilization of energy stored in organic compounds, respiration is an order of magnitude superior to fermentation, and this pathway opened up a new era in the evolution of life. A number of bacteria are facultative anaerobes, reverting to fermentation when the oxygen level in the environment decreases below the so-called Pasteur point, about 1% of the present abundance of oxygen in the atmosphere. The corresponding O_2 concentration thus indicates the minimum level required for respiration. The advent 1.4×10^9 years ago of the first eucaryotic cells, which is contingent upon the respiratory metabolic pathway, makes it likely that the atmospheric O_2 level had reached the Pasteur point at that time. The possibility exists, of course, that the first eucaryotes lived in locally oxygen-rich niches and that atmospheric oxygen reached the Pasteur point somewhat later. According to McAlester (1968), the earliest metazoa appeared 700 million years ago. This was a soft-bodied fauna dependent on diffusive transport of oxygen to the interior cells, so that by that time the oxygen level in the atmosphere had almost certainly risen beyond the Pasteur point.

Life began to inhabit the land surface in the upper Silurian, about 400 million years ago. During this period and afterward the abundance of atmospheric oxygen must have been high enough to form an ozone layer of sufficient optical depth to shield terrestrial life from lethal ultraviolet radiation. The precise functional relationship between oxygen and ozone is uncertain because it depends on trace gases such as N_2O and CH_3Cl, whose abundances in earlier times are not known. Ratner and Walker (1972) have studied the development of the ozone layer as a function of oxygen concentration for a pure oxygen atmosphere, whereas Hestvedt et al. (1974) allowed for the additional presence of water vapor. Both groups found that an O_2 concentration of 1% of the present level produces enough ozone to shield the Earth's surface from radiation at wavelengths below 290 nm, except for a small contribution in the absorption window near 210 nm. The latter effect is completely eliminated when the O_2 concentration is raised to 10% of the present value. Since according to the preceding arguments the 1% level had been reached 700 million years ago at the latest, it appears likely that 400×10^6 years ago the O_2 concentration was higher and had approached the 10% mark. This value was used to construct the rise curve in Figure 12.5.

The following 100 million years saw the first air-breathing lungfishes and primitive amphibians, indicting that the oxygen level had risen still further. Fossil remains of insects, including giant dragonfly-like species featuring a 70-cm wingspan, first became abundant in the late Coniferous and early Permian strata (about 300×10^6 years ago; McAlester, 1968). As is well known, the respiratory tracts of insects, the tracheae, conduct oxygen passively by diffusion. This kind of respiratory system is considered to require rather high concentrations of oxygen and suggests that modern O_2 levels had been attained at that time. In fact, Schidlowski (1971) argued that giant insects might have needed an O_2 concentration higher than that existing today. On the other hand, the continuous evolution of mammalian species from the Triassic on seems to preclude significant excursions of the atmospheric oxygen level during the past 200×10^6 years.

Although the qualitative nature of the preceding arguments leaves many questions open, there can be little doubt that atmospheric oxygen is a consequence of life. Indeed, Lovelock (Lovelock, 1979; Lovelock and Margulis, 1974) considers oxygen a biological contrivance, specifically developed to further the evolution of life. This viewpoint is not entirely without justification if one recalls that the respiratory metabolism is superior to fermentation in exploiting energy stored in organic compounds. Respiration certainly made possible the tremendous diversification of life forms, starting with the Phanerozoicum.

12.5. ATMOSPHERIC GASES: CUMULATIVE VERSUS CYCLIC BEHAVIOR

The main objective of the present chapter was to show that the Earth's atmosphere developed together with the ocean as a consequence of geochemical processes, primarily by the thermal exhalation of volatiles from the planet's interior. To summarize briefly, the rare gases and nitrogen reside largely in the atmosphere because their solubilities in seawater are small. Carbon dioxide, in contrast, is much more soluble, and it interacted with dissolved calcium to become overwhelmingly trapped in limestone sediments. Oxygen as a biogenic constituent is unique in that it appears to have accumulated in the atmosphere as an excess over the amount required for the oxidation of reduced elements or compounds derived from exhalations or the weathering of igneous rocks. Seemingly, all of these gases must be considered as having accumulated, but they also are subject to geochemical and biological cycles. This raises the question of which of the two processes, accumulation or cycling, actually controls the abundance of these gases in the atmosphere. For CO_2 the discussions in the preceding chapter have shown that cycling exerts the dominant control. For nitrogen and oxygen, the situation is not immediately obvious, so that additional comments concerning this aspect are useful in concluding the present chapter.

Junge (Junge, 1972; see also Junge and Warneck, 1979) has studied the problem and proposed a set of criteria based on the distribution of an atmospheric constituent among the various geochemical reservoirs to facilitate an assignment to one of the two categories. For a better understanding of the criteria, it is instructive to consider initially the exchange of material between two coupled reservoirs A and B. Let G_A and G_B denote the mass contents of the reservoirs, and k_{AB} and k_{BA} the exchange coefficients. In the steady state, the exchange fluxes are equal, so that

$$k_{AB}G_A = k_{BA}G_B \qquad (12.6)$$

The residence times of the material in the reservoirs, τ_A and τ_B, respectively, are reciprocals of the exchange coefficients, so that the above relation can also be written as

$$\tau_A/\tau_B = G_A/G_B$$

Assume now that $G_A \gg G_B$. It is then evident that the residence time τ_A is much greater than τ_B and the substance being considered must accumulate in reservoir A, even if it were initially introduced to reservoir B. Because of the smaller residence time, the content of B is controlled by that of reservoir A because of the exchange of material between the two reservoirs. Two examples discussed previously may illustrate the behavior: (1) the amount of water in the ocean (reservoir A) is much greater than that in the atmosphere (reservoir B), so that the former controls the latter by means of the water cycle in the atmosphere; (2) a similar distribution between ocean and atmosphere exists for CO_2, so that again the ocean determines the atmospheric cycle, provided, of course, that interactions with other geochemical reservoirs are ignored.

For the more general case, one needs to consider the four principal geochemical reservoirs: the sediments, the ocean, the atmosphere, and the biosphere. The total amount of a volatile constituent exhaled during the history of the Earth is

$$G_{tot} = G_{sed} + G_{ocean} + G_{atm} + G_{bio} = G_{sed} + G_{mobile} \qquad (12.7)$$

The distributions of H_2O, N_2, CO_2, and O_2 in the four reservoirs are summarized in Table 12.13. From the geochemical point of view, we may combine the contents of the ocean, the atmosphere, and the biosphere and consider this the mobile fraction, in contrast to the amount of substance fixed in the sediments. The ratio G_{mobile}/G_{sed} then tells us whether the substance has accumulated mainly in the sediments or in the other reservoirs. If the ratio is appreciably greater than unity, the constituent has remained mobile. If the ratio is much smaller than unity, the constituent has entered the sediments and the mobile fraction is governed by material exchange with the lithosphere. We can further form the ratios G_{atm}/G_{ocean}

and G_{atm}/G_{bio} to determine which of the three reservoirs containing the mobile fraction controls the atmospheric reservoir. Table 12.13 includes the three ratios. They provide a desired set of criteria for the categorization of atmospheric constituents. At the same time they furnish a useful summary of the discussion of these constituents and their behavior.

H_2O and N_2 occur mainly in the mobile fraction, as the sediments contain less than one-fifth of G_{tot}. The small value of G_{atm}/G_{ocean} for water indicates accumulation in the ocean reservoir, which controls the amount of water in the atmosphere, as discussed above. For nitrogen, the ratios G_{atm}/G_{ocean} and G_{atm}/G_{bio} are both much greater than unity. This demonstrates that nitrogen has, in fact, accumulated in the atmosphere and that the cycles of nitrogen through the biosphere and the sediments are subordinate features. For CO_2, the total of organic and inorganic carbon is used in Table 12.13. The procedure is chosen to enable a comparison of carbon in the ocean with that in the biosphere. The contents of carbon in the ocean and in the sediments are dominated by inorganic carbon anyway. For a better comparison, however, organic carbon is also entered separately. The three ratios derived from the contents of the individual reservoirs are all smaller than unity. Accordingly, atmospheric CO_2 is governed by the other reservoirs. Which of these exerts the control cannot be decided without considering the individual exchange fluxes. The discussion in Chapter 11 has shown that the ocean and the biosphere control the short-term variations, whereas weathering of sedimentary rocks is responsible for the long-term average level. In the line for organic carbon, only the ratio G_{atm}/G_{bio} is relevant. This value is smaller than unity, indicating that, in accordance with expectation, methane and the other organic compounds in the atmosphere are controlled by the biosphere.

Finally, the behavior of oxygen will be considered. This is a special case, because oxygen bound to sulfate and trivalent iron is permanently fixed and cannot be liberated by weathering. As a consequence, there is no direct exchange between the mobile fraction and the reservoir of oxygen in sedimentary rocks, so that the ratio G_{mobile}/G_{sed} becomes meaningless. Since the source of oxygen is the disproportionation of CO_2 to give C_{org} and O_2, and the sink is the reoxidation of C_{org} to CO_2, it will be more appropriate to relate the mobile fraction of O_2 to the contents of organic carbon (and other reduced material) in the sediments and in the biosphere. When this procedure is employed one obtains the last line of Table 12.13, where the contents of reduced material in the sediments and in the biosphere are expressed as equivalents of oxygen. The ratios now show that $G_{mobile}/G_{sed} \ll 1$, which agrees with the earlier conclusion that the cycle of organic carbon through the sediments controls the behavior of oxygen in the atmosphere. At the

TABLE 12.13 Contents of H_2O, N_2, CO_2 (as Carbon), and O_2 in the Four Principal Geochemical Reservoirs[a]

Constituent	G_{sec}	G_{ocean}	G_{atm}	G_{bio}	G_{mobile}/G_{sed}	G_{atm}/G_{ocean}	G_{atm}/G_{bio}
H_2O	3.3 (8)	1.4 (9)	1.3 (4)	—	4.2	9.0 (−6)	na[h]
N_2	9.0 (5)	2.2 (4)	3.9 (6)	1.3 (3)	4.4	1.7 (2)	3.0 (3)
CO_2 (C_{org} + C_{inorg})	6.2 (7)	3.8 (4)[b]	7.0 (2)	2.5 (3)	6.6 (−4)	1.8 (−2)	2.8 (−1)
C_{org}	1.2 (7)	[b]	3.1[c]	3.5 (3)	2.9 (−4)	na	8.8 (−4)
O_2	6.5 (7)[d]	1.7 (4)[b]	1.2 (6)	—	na	7.0 (1)	na
C_{org} + other reduced material expressed as O_2	4.8 (7)[e]	[b]	(8.3)[c]	9.3 (3)	2.5 (−2)[f]	na	1.3 (2)[g]

[a] Values (in Pg) from Tables 11.1, 12.6, 12.8, and 12.10; orders of magnitude shown in parentheses. Ratios of reservoir contents are used as criteria for establishing control over the atmosphere reservoir—accumulation or cycles. $G_{mobile} = G_{atm} + G_{ocean} + G_{bio}$.

[b] The content of C_{org} in the ocean is included in G_{bio}.

[c] Methane.

[d] Counting only the amount of oxygen fixed as sulfate and trivalent iron, but including the amount of SO_4^{2-} in the ocean.

[e] Includes FeS_2 and FeO.

[f] Mobile O_2 over C_{org} and other reduced material expressed as O_2.

[g] Atmospheric oxygen over C_{org} expressed as O_2.

[h] na, Not applicable.

same time, the ratio G_{atm}/G_{bio} is much larger than unity, and since this is also true for the ratio G_{atm}/G_{ocean}, one finds that interactions with the biosphere or with the ocean are not the controlling factors. In this sense free oxygen must be considered to have accumulated in the atmosphere. The behavior of oxygen thus shows hybrid features. On the one hand, it is controlled by the burial of organic carbon in the sediments as a consequence of biological activity; on the other hand, it behaves like a cumulative constituent of the atmosphere. The reason is the slowness of the sediment cycle, which causes the residence time of oxygen in the atmosphere to be a few million years.

Supplementary Tables

TABLE A.1   Temperature, Pressure, Density and Number Density of Air as a Function of Altitude according to the U.S. Standard Atmosphere (1976)[a]

z (km)	T (K)	p (hPa)	ρ (kg m^{-3})	n (cm^{-3})
0	288.15	1.01 (3)	1.23	2.55 (19)
2	275.15	7.95 (2)	1.00	2.09 (19)
4	262.17	6.17 (2)	8.19 (−1)	1.70 (19)
6	249.19	4.72 (2)	6.60 (−1)	1 37 (19)
8	236.22	3.57 (2)	5.26 (−1)	1.09 (19)
10	223.25	2.65 (2)	4.14 (−1)	8.60 (18)
12	216.65	1.94 (2)	3.12 (−1)	6.49 (18)
14	216.65	1.42 (2)	2.28 (−1)	4.74 (18)
16	216.65	1.04 (2)	1.67 (−1)	3.46 (18)
18	216.65	7.57 (1)	1.22 (−1)	2.53 (18)
20	216.65	5.53 (1)	8.89 (−2)	1.85 (18)
22	218.57	4.05 (1)	6.45 (−2)	1.34 (18)
24	220.56	2.97 (1)	4.69 (−2)	9.76 (17)
26	222.54	2.19 (1)	3.43 (−2)	7.12 (17)
28	224.53	1.62 (1)	2.51 (−2)	5.21 (17)
30	226.51	1.20 (1)	1.84 (−2)	3.83 (17)
32	228.49	8.89	1.36 (−2)	2.82 (17)
34	233.74	6.63	9.89 (−3)	2.06 (17)
36	239.28	4.99	7.26 (−3)	1.51 (17)

(*Continues*)

TABLE A.1 (*Continued*)

z (km)	T (K)	p (hPa)	ρ (kg m^{-3})	n (cm^{-3})
38	244.82	3.77	5.37 (-3)	1.12 (17)
40	250.35	2.87	4.00 (-3)	8.31 (16)
42	255.88	2.20	3.00 (-3)	6.23 (16)
44	261.40	1.70	2.26 (-3)	4.70 (16)
46	266.94	1.31	1.71 (-3)	3.57 (16)
48	270.65	1.03	1.32 (-3)	2.74 (16)
50	270.65	8.00 (-1)	1.03 (-3)	2.14 (16)
52	269.03	6.22 (-1)	8.10 (-4)	1.67 (16)
54	267.56	4.85 (-1)	6.39 (-4)	1.32 (16)
56	258.02	3.77 (-1)	5.04 (-4)	1.04 (16)
58	252.51	2.87 (-1)	3.96 (-4)	8.23 (15)
60	247.02	2.19 (-1)	3.10 (-4)	6.44 (15)
70	219.58	5.22 (-2)	8.28 (-5)	1.72 (15)
80	198.64	1.04 (-2)	1.85 (-5)	3.84 (14)
90	186.87	1.84 (-3)	3.42 (-6)	7.12 (13)

[a] Powers of 10 are in parentheses.

TABLE A.2 Saturation Vapor Pressure (p_s) over Water or Ice as a Function of Temperature[a]

Water		Ice	
T (°C)	p_s (hPa)	T (°C)	p_s (hPa)
-35	3.14 (-1)	-90	9.67 (-5)
-30	5.09 (-1)	-85	2.35 (-4)
-25	8.07 (-1)	-80	5.47 (-4)
-20	1.25	-75	1.22 (-3)
-15	1.91	-70	2.62 (-3)
-10	2.86	-65	5.41 (-3)
-5	4.21	-60	1.08 (-2)
0	6.11	-55	2.09 (-2)
5	8.72	-50	3.93 (-2)
10	1.23 (1)	-45	7.20 (-2)
15	1.70 (1)	-40	1.28 (-1)
20	2.34 (1)	-35	2.23 (-1)
25	3.17 (1)	-30	3.80 (-1)
30	4.24 (1)	-25	6.32 (-1)
35	5.62 (1)	-20	1.03
40	7.38 (1)	-15	1.65
		-10	2.60
		-5	4.02
		0	6.12

[a] From the Smithonian Meteorological Tables (1951); powers of 10 are in parentheses.

TABLE A.3 Average Solar Photon Flux Outside Earth's Atmosphere,
According to Samain and Simon (1976) and Neckel and Labs (1981)

Wavelength interval (nm)	Solar photon flux $(m^{-2} s^{-1} nm^{-1})$	Wavelength interval (nm)	Solar photon flux $(m^{-2} s^{-1} nm^{-1})$
Lyα 121.57	3.0 (15)	156.2–158.7	9.6 (13)
116.3–117.0	9.8 (12)	158.7–161.3	1.1 (14)
117.0–117.6	4.7 (13)	161.3–163.9	1.4 (14)
117.6–118.3	1.0 (13)	163.9–166.7	2.2 (14)
118.3–119.0	9.3 (12)	166.7–169.5	2.8 (14)
119.0–119.8	1.9 (13)	169.5–172.4	4.6 (14)
119.8–120.5	2.0 (13)	172.4–173.9	4.9 (14)
120.5–121.2	9.0 (13)	173.9–175.4	5.9 (14)
121.2–122.0	2.9 (13)	175.4–177.0	7.1 (14)
122.0–122.7	1.8 (13)	177.0–178.0	1.1 (15)
122.7–123.5	1.1 (13)	178.0–179.0	1.3 (15)
123.5–124.2	1.7 (13)	179.0–180.0	1.3 (15)
124.2–125.0	7.4 (12)	180.0–181.0	1.5 (15)
125.0–125.8	8.4 (12)	181.0–182.0	2.0 (15)
125.8–126.6	2.3 (13)	182.0–183.0	1.9 (15)
126.6–127.4	5.7 (13)	183.0–184.0	2.0 (15)
127.4–128.2	1.0 (13)	184.0–185.0	1.7 (15)
128.2–129.0	7.5 (12)	185.0–186.0	1.9 (15)
129.0–129.9	7.9 (12)	186.0–187.0	2.4 (15)
129.9–130.7	8.6 (13)	187.0–188.0	2.7 (15)
130.7–131.6	2.0 (13)	188.0–189.0	2.8 (15)
131.6–132.4	1.0 (13)	189.0–190.0	2.9 (15)
132.4–133.3	3.7 (13)	190.0–191.0	2.9 (15)
133.3–134.2	5.8 (13)	191.0–192.0	3.3 (15)
134.2–135.1	9.5 (12)	192.0–193.0	3.5 (15)
135.1–136.0	2.3 (13)	193.0–194.0	2.5 (15)
136.0–137.0	1.4 (13)	194.0–195.0	4.5 (15)
137.0–137.9	1.5 (13)	195.0–196.0	4.3 (15)
137.9–138.9	1.4 (13)	196.0–197.0	4.9 (15)
138.9–140.8	1.6 (13)	197.0–198.0	4.9 (15)
140.8–142.8	3.8 (13)	198.0–199.0	4.9 (15)
142.8–144.9	3.2 (13)	199.0–200.0	5.5 (15)
144.9–147.0	3.4 (13)	200.0–202.0	7.0 (15)
147.0–149.2	5.0 (13)	202.0–204.1	8.0 (15)
149.2–151.5	5.3 (13)	204.1–206.2	9.9 (15)
151.5–153.8	7.7 (13)	206.2–208.3	1.2 (16)
153.8–156.2	1.2 (14)	208.3–210.5	1.9 (16)

(*Continues*)

TABLE A.3 (*Continued*)

Wavelength interval (nm)	Solar photon flux $(m^{-2} s^{-1} nm^{-1})$	Wavelength interval (nm)	Solar photon flux $(m^{-2} s^{-1} nm^{-1})$
210.5–212.8	3.1 (16)	250.9–253.2	5.7 (16)
212.8–215.0	3.5 (16)	253.2–256.4	7.0 (16)
215.0–217.4	3.5 (16)	256.4–259.7	1.4 (17)
217.4–219.8	4.6 (16)	259.7–263.2	1.2 (17)
219.8–222.2	5.0 (16)	263.2–266.7	3.2 (17)
222.2–224.7	6.6 (16)	266.7–270.3	3.3 (17)
224.7–227.3	4.9 (16)	270.3–274.0	2.9 (17)
227.3–229.9	5.0 (16)	274.0–277.8	2.7 (17)
229.9–232.6	5.6 (16)	277.8–281.7	2.0 (17)
232.6–235.3	4.9 (16)	281.7–285.7	3.7 (17)
235.3–238.1	5.4 (16)	285.7–289.9	5.2 (17)
238.1–241.0	4.6 (16)	289.9–294.1	8.2 (17)
241.0–243.9	7.0 (16)	294.1–298.5	7.7 (17)
243.9–246.9	6.1 (16)	298.5–303.0	7.2 (17)
246.9–250.0	6.1 (16)	303.0–307.7	9.4 (17)

Wavelength (±2.5 nm)		Wavelength (± 2.5 nm)	
310	9.9 (17)	405	3.4 (18)
315	1.2 (18)	410	3.7 (18)
320	1.2 (18)	415	3.9 (18)
325	1.4 (18)	420	3.9 (18)
330	1.7 (18)	425	3.6 (18)
335	1.6 (18)	430	3.3 (18)
340	1.8 (18)	435	4.0 (18)
345	1.7 (18)	440	4.0 (18)
350	1.7 (18)	445	4.4 (18)
355	1.8 (18)	450	4.7 (18)
360	1.6 (18)	455	4.6 (18)
365	2.1 (18)	460	4.8 (18)
370	2.2 (18)	465	4.8 (18)
375	1.9 (18)	470	4.8 (18)
380	2.2 (18)	475	4.9 (18)
385	1.8 (18)	480	5.0 (18)
390	2.4 (18)	485	4.6 (18)
395	1.9 (18)	490	4.8 (18)
400	3.4 (18)	495	5.0 (18)

(*Continues*)

TABLE A.3 (*Continued*)

Wavelength (±2.5 nm)	Solar photon flux (m^{-2} s^{-1} nm^{-1})	Wavelength (±2.5 nm)	Solar photon flux (m^{-2} s^{-1} nm^{-1})
500	4.8 (18)	675	5.2 (18)
505	4.9 (18)	680	5.1 (18)
510	5.0 (18)	685	5.2 (18)
515	4.6 (18)	690	5.0 (18)
520	4.8 (18)	695	5.2 (18)
525	4.8 (18)	700	5.2 (18)
530	5.1 (18)	705	5.0 (18)
535	5.1 (18)	710	5.0 (18)
540	5.0 (18)	715	5.0 (18)
545	5.1 (18)	720	4.9 (18)
550	5.1 (18)	725	5.0 (18)
555	5.1 (18)	730	4.9 (18)
560	5.0 (18)	735	4.9 (18)
565	5.1 (18)	740	4.8 (18)
570	5.2 (18)	745	4.8 (18)
575	5.3 (18)	750	4.8 (18)
580	5.3 (18)	755	4.8 (18)
585	5.4 (18)	760	4.8 (18)
590	5.2 (18)	765	4.7 (18)
595	5.4 (18)	770	4.6 (18)
600	5.3 (18)	775	4.6 (18)
605	5.4 (18)	780	4.7 (18)
610	5.3 (18)	785	4.7 (18)
615	5.2 (18)	790	4.6 (18)
620	5.2 (18)	795	4.6 (18)
625	5.2 (18)	800	4.5 (18)
630	5.2 (18)	805	4.5 (18)
635	5.2 (18)	810	4.4 (18)
640	5.3 (18)	815	4.4 (18)
645	5.2 (18)	820	4.4 (18)
650	5.1 (18)	825	4.4 (18)
655	5.0 (18)	830	4.3 (18)
660	5.1 (18)	835	4.3 (18)
665	5.2 (18)	840	4.2 (18)
670	5.2 (18)		

[a] Powers of 10 are shown in parentheses.

TABLE A.4 Rate Coefficients at 298 K and Arrhenius Parameters for Gas Phase Atmospheric Reactions[a]

Number	Reaction	k_{298} (cm^3 molecule^{-1} s^{-1})[b]	A (cm^3 molecule^{-1} s^{-1})[b]	E_a/R_g (K)	References[c]
Reactions of O(^{3}P) atoms					
1	$O + O_2 + M \rightarrow O_3 + M$	See Table A.5			a
2	$O + O_3 \rightarrow O_2 + O_2$	8.0 (−15)	8.0 (−12)	2,060	b
3	$O + OH \rightarrow O_2 - H$	3.3 (−11)	2.2 (−11)	−120	a
4	$O + HO_2 \rightarrow OH + O_2$	5.8 (−11)	2.7 (−11)	−224	a
5	$O + H_2 \rightarrow OH + H$	9.0 (−18)	2.5 (−16) (T/298)$^{7.8}$	966	c
6	$O + H_2O_2 \rightarrow OH + HO_2$	1.7 (−15)	1.4 (−12)	2,000	b
7	$O + N_2 \rightarrow NO + N$	3.3 (−66)	3.0 (−10)	38,400	d
8	$O + NO + M \rightarrow NO_2 + M$	See Table A.5			
9	$O + NO_2 + M \rightarrow NO_3 + M$	See Table A.5			
10	$O + NO_2 \rightarrow NO + O_2$	9.7 (−12)	6.5 (−12)	−120	a, b
11	$O + NO_3 \rightarrow O_2 + NO_2$	1.0 (−11)	1.0 (−11)	±0	b
12	$O + HNO_3 \rightarrow OH + NO_3$	< 3 (−17)	—	—	b
13	$O + N_2O_5 \rightarrow$ products	< 3 (−16)	—	—	b
14	$O + HO_2NO_2 \rightarrow$ products	8.6 (−16)	7.8 (−11)	3,400	b
15	$O + OCS \rightarrow SO + CO$	1.3 (−14)	2.1 (−11)	2,200	b
16	$O + SO_2 + M \rightarrow SO_3 + M$	See Table A.5			
17	$O + ClO \rightarrow Cl + O_2$	3.8 (−11)	3.0 (−11)	−70	b
18	$O + HCl \rightarrow OH + Cl$	1.5 (−16)	1.0 (−11)	3,300	b
19	$O + HOCl \rightarrow OH + ClO$	1.3 (−13)	1.0 (−11)	1,300	b
20	$O + ClONO_2 \rightarrow$ products	2.0 (−13)	2.9 (−12)	800	b
21	$O + BrO \rightarrow Br + O_2$	4.0 (−11)	1.7 (−11)	−260	b
22	$O + HBr \rightarrow OH + Br$	3.8 (−14)	5.8 (−12)	1,500	b
23	$O + HOBr \rightarrow OH + BrO$	2.5 (−11)	—	—	b
24	$O + CO + M \rightarrow CO_2 + M$	See Table A.5			
25	$O + CH_3 \rightarrow HCHO + H$	1.1 (−10)	1.1 (−10)	±0	b
26	$O + CH_4 \rightarrow OH - CH_3$	5.0 (−18)	8.3 (−12) (T/298)$^{1.56}$	4,270	c
27	$O + C_2H_6 \rightarrow OH - C_2H_5$	4.7 (−16)	8.5 (−12) (T/298)$^{1.5}$	2920	c

No.	Reaction	k(298)	A	E/R	Note
28	$O + C_3H_8 \rightarrow OH + C_3H_7$	7.9(−15)	5.8(−13)$(T/298)^{3.5}$	1,280	e
29	$O + n\text{-}C_4H_{10} \rightarrow OH + n\text{-}C_4H_9$	1.8(−14)	3.5(−13)$(T/298)^{3.9}$	780	e
30	$O + C_2H_4 \rightarrow$ products	8.1(−13)	8.1(−13)$(T/298)^{2.08}$	±0	c, g
31	$O + C_3H_6 \rightarrow$ products	4.8(−12)	1.2(−12)$(T/298)^{2.15}$	−400	e, g
32a	$O + cis\text{-}2\text{-}butene \rightarrow$ products	1.7(−11)	1.5(−12)$(T/298)^{1.94}$	−716	f
32b	$O + trans\text{-}2\text{-}butene \rightarrow$ products	2.0(−11)	1.6(−12)$(T/298)^{1.87}$	−743	f
33	$O + C_2H_2 \rightarrow$ products	1.4(−13)	3.0(−11)	1600	b
34	$O + benzene \rightarrow$ products	2.5(−14)	1.7(−13)$(T/298)^{3.7}$	570	c
35	$O + toluene \rightarrow$ products	7.6(−14)	5.2(−12)$(T/298)^{1.21}$	1260	c
36	$O + HCHO \rightarrow OH + HCO$	1.6(−13)	3.4(−11)	1600	b
Reactions of O(^{1}D) atoms					
37	$O^* + N_2 \rightarrow O + N_2$	2.6(−11)	1.8(−11)	−110	b
38	$O^* + O_2 \rightarrow O + O_2$	4.0(−11)	3.2(−11)	−70	b
39a	$O^* + O_3 \rightarrow O_2 + O_2$	1.2(−10)	1.2(−10)	±0	b
39b	$\rightarrow O_2 + O + O$	1.2(−10)	1.2(−10)	±0	b
40	$O^* + H_2O \rightarrow OH + OH$	2.2(−10)	2.2(−10)	±0	b
41	$O^* + H_2 \rightarrow OH + H$	1.0(−10)	1.0(−10)	±0	b
42a	$O^* + N_2O \rightarrow N_2 + O_2$	4.9(−11)	4.9(−11)	±0	b
42b	$\rightarrow NO + NO$	6.7(−11)	6.7(−11)	±0	b
43	$O^* + CO_2 \rightarrow O + CO_2$	1.1(−10)	7.4(−11)	−120	b
44a	$O^* + CH_4 \rightarrow OH + CH_3$	1.4(−10)	1.4(−10)	±0	a
44b	$\rightarrow HCHO + F_2$	1.5(−11)	1.5(−11)	±0	a
45	$O^* + CCl_4 \rightarrow$ products	3.3(−10)	3.3(−10)	±0	b
46	$O^* + CCl_3F \rightarrow$ products	2.3(−10)	2.3(−10)	±0	b
47	$O^* + CCl_2F_2 \rightarrow$ products	1.4(−10)	1.4(−10)	±0	b
48	$O^* + HCl \rightarrow OH + Cl$	1.5(−10)	1.5(−10)	±0	b
49	$O^* + HF \rightarrow OH + F$	1.4(−10)	1.4(−10)	±0	b
Reactions involving H atoms and HO$_2$ radicals					
50	$H + O_2 + M \rightarrow HO_2 + M$	See Table A.5			
51	$H + O_3 \rightarrow OH + O_2$	2.9(−11)	1.4(−10)	470	b
52a	$H + HO_2 \rightarrow OH + OH$	7.2(−11)	7.2(−11)	±0	a

(Continues)

TABLE A.4 (Continued)

Number	Reaction	k_{298} (cm³ molecule⁻¹ s⁻¹)ᵇ	A (cm³ molecule⁻¹ s⁻¹)ᵇ	E_a/R_g (K)	Referencesᶜ
Reactions involving H atoms and HO₂ radicals					
52b	→ H₂O + O	2.4 (−12)	2.4 (−12)	±0	a
52c	→ H₂ + O₂	5.6 (−12)	5.6 (−12)	±0	a
53	OH + OH → H₂O + O	1.9 (−12)	4.2 (−12)	240	b
54	OH + OH + M → H₂O₂ + M	See Table A.5			
55	OH + HO₂ → H₂O + O₂	1.1 (−10)	4.8 (−11)	−250	a, b
56	HO₂ + HO₂ → H₂O₂ + O₂	1.7 (−12)	2.3 (−13)	−600	a, b
57	HO₂ + HO₂ + M → H₂O₂ + O₂ + M	4.9 (−32) [M]	1.7 (−33) [M]	−1000	b
58	OH + O₃ → HO₂ + O₂	6.8 (−14)	1.6 (−12)	940	b
59	HO₂ + O₃ → OH + 2O₂	2.0 (−15)	1.4 (−14)	600	a, b
60	HO₂ + NO → OH + NO₂	8.3 (−12)	3.4 (−12)	−270	a, b
61	HO₂ + NO₂ → HO₂NO₂	See Table A.5			
Reactions involving N atoms, NOx, and NH₂ radicals					
62	N + O₂ → NO + O	8.5 (−17)	1.5 (−11)	3600	b
63	N + O₃ → NO + O₂	< 1 (−16)	—	—	a, b
64	N + NO → N₂ + O	3.0 (−11)	2.1 (−11)	−100	b
65	N + NO₂ → N₂O + O	1.2 (−11)	5.8 (−12)	−220	b
66	NO + O₃ → NO₂ + O₂	1.8 (−14)	2.0 (−12)	1400	b
67	NO + NO₃ → 2NO₂	2.7 (−11)	1.6 (−11)	−150	a, b
68	O₃ + NO₂ → NO₃ + O₂	3.2 (−17)	1.2 (−13)	2450	a, b
69	NO₃ + NO₂ → N₂O₅	See Table A.5			
70	NO + NO₂ + H₂O → 2HNO₂	See Table A.5			
71	2NO₂ + H₂O → HNO₃ + HNO₂	See Table A.5			
72	N₂O₅ + H₂O → 2HNO₃	< 2 (−21)	—	—	a, b
73	2NO + O₂ → 2NO₂	2.0 (−38)	3.3 (−39)	−530	a
74	NH₂ + O₂ → products	< 6 (−21)	—	—	b

No.	Reaction				
75	NH₂ + O₃ → products	1.9(-13)	4.3(-12)	930	b
76	NH₂ + NO → products	1.7(-11)	3.8(-12)	-450	b
77	NH₂ + NO₂ → products	1.9(-11)	2.1(-12)	-650	b
Methane oxidation reactions					
78	OH + CH₄ → CH₃ + H₂O	6.2(-15)	2.3(-13)(T/298)²·⁵⁸	1082	g
79	CH₃ + O₂ + M → CH₃O₂ + M	See Table A.5			
80	CH₃O₂ + NO → CH₃O + NO₂	7.5(-12)	2.9(-12)	-285	g,h
81a	CH₃O₂ + CH₃O₂ → HCHO + CH₃OH + O₂	2.5(-13)			
81b	→ 2CH₃O + O₂	1.2(-13)	9.1(-14)	-416	a,h
81c	→ CH₃OOCH₃ + O₂	<3(-14)			
82	CH₃O₂ + NO₂ → CH₃O₂NO₂	See Table A.5			
83	CH₃O₂ + HO₂ → CH₃OOH + O₂	5.8(-12)	4.1(-13)	-790	g,h
84	CH₃O + O₂ → HCHO + HO₂	1.9(-15)	7.2(-14)	1080	a
85	CH₃O + NO₂ → CH₃ONO₂	See Table A.5			
86	OH + HCHO → HCO + H₂O	1.1(-11)	1.6(-11)	110	a
87	HCO + O₂ → CO + HO₂	5.5(-12)	3.5(-12)	-140	a,b
Halogen atom and radical reactions					
88	Cl + O₃ → ClO + O₂	1.2(-11)	2.9(-11)	260	b
89	Cl + O₂ + M → ClO₂ + M	See Table A.5			
90	Cl + H₂ → HCl + H	1.6(-14)	3.7(-11)	2300	b
91	Cl + CH₄ → HCl + CH₃	1.0(-13)	1.1(-11)	1400	b
92	Cl + C₂H₆ → HCl + C₂H₅	5.7(-11)	7.7(-11)	90	a,b
93	Cl + C₃H₈ → HCl + C₃H₇	1.5(-10)	1.3(-10)	-40	a,b
94	Cl + CH₃Cl → HCl + CH₂Cl	4.9(-13)	3.3(-11)	1250	b
95	Cl + HCHO → HCl + HCO	7.3(-11)	8.1(-11)	30	a,b
96	Cl + HOCl → products	1.6(-12)	2.5(-12)	130	b
97	Cl + H₂O₂ → HCl + HO₂	4.1(-13)	1.1(-11)	980	b
98	Cl + HNO₃ → products	<2(-16)	—		a,b
99a	Cl + HO₂ → HCl + O₂	3.2(-11)	1.8(-11)	-170	b
99b	→ OH + ClO	9.1(-12)	4.1(-11)	450	b
100a	Cl + ClO₂ → Cl₂ + O₂	2.3(-10)	2.3(-10)	±0	b

(Continues)

TABLE A.4 (*Continued*)

Number	Reaction	k_{298} (cm^3 molecule^{-1} s^{-1})[b]	A (cm^3 molecule^{-1} s^{-1})[b]	E_a/R_g (K)	References[c]
Halogen atom and radical reactions					
100b	→ ClO + ClO	1.2(−11)	1.2(−11)	±0	b
101	Cl + ClONO$_2$ → products	1.2(−11)	6.8(−12)	−160	a, b
102	ClO + NO → NO$_2$ + Cl	1.7(−11)	6.4(−12)	−290	b
103	ClO + NO$_2$ + M → ClONO$_2$ + M	See Table A.5			
104	ClO + HO$_2$ → HOCl + O$_2$	5.0(−12)	4.6(−13)	−710	a
105a	ClO + ClO → Cl$_2$ + O$_2$	4.8(−15)	1.0(−12)	1590	b
105b	→ ClOO + Cl	8.0(−15)	3.0(−11)	2450	b
105c	→ OClO + Cl	3.5(−15)	3.5(−13)	1370	b
106	ClO + ClO + M → Cl$_2$O$_2$ + M	See Table A.5			
107	Br + O$_3$ → BrO + O$_2$	1.2(−12)	1.7(−11)	800	b
108	Br + H$_2$O$_2$ → HBr + HO$_2$	< 5(−16)	—	—	b
109	Br + HCHO → HBr + HCO	1.1(−12)	1.7(−11)	800	b
110	BrO + NO → NO$_2$ + Br	2.1(−11)	8.8(−12)	−260	b
111	BrO + NO$_2$ + M → BrONO$_2$ + M	See Table A.5			
112a	BrO + ClO → Br + OClO	6.8(−12)	1.6(−12)	−430	b
112b	→ Br + ClOO	6.1(−12)	2.9(−12)	−220	b
112c	→ BrCl + O$_2$	1.0(−12)	5.8(−13)	−170	b
113a	BrO + BrO → 2Br + O$_2$	2.1(−12)	4.0(−12)	190	b
113b	→ Br$_2$ + O$_2$	3.8(−13)	4.2(−14)	−660	b
114	F + H$_2$ → HF + H	2.6(−11)	1.4(−10)	500	b
115	F + CH$_4$ → HF + CH$_3$	8.0(−11)	3.0(−10)	400	b
116	F + H$_2$O → HF + OH	1.4(−11)	1.4(−11)	±0	b
117	F + O$_3$ → FO + O$_2$	1.0(−11)	2.2(−11)	230	b
118	FO + NO → F + NO$_2$	2.2(−11)	8.2(−12)	−300	b
Reactions of OH radicals					
119	OH + H$_2$ → H$_2$O + H	6.7(−15)	7.7(−12)	2100	a
120	OH + H$_2$O$_2$ → H$_2$O + HO$_2$	1.7(−12)	2.9(−12)	160	a, b

No.	Reaction	$k(298)$	Arrhenius expression	E/R (K)	Note
58	$OH + O_3 \rightarrow HO_2 + O_2$	$6.8(-14)$	$1.6(-12)$	940	b
121	$OH + NO + M \rightarrow HNO_2 + M$	See Table A.5			
122	$OH + NO_2 + M \rightarrow HNO_3$	See Table A.5			
123	$OH + HNO_2 \rightarrow H_2O + NO_2$	$4.9(-12)$	$1.8(-11)$	390	a,b
124	$OH + HNO_3 \rightarrow H_2O + NO_3$	$1.5(-3)\; p = 10^5$ Pa	$7.2(-15)^d$	-785	a,b
125	$OH + HO_2NO_2 \rightarrow$ products	$4.5(-12)$	$1.3(-12)$	-380	a,b
126	$OH + CH_3CO_3NO_2 \rightarrow$ products	$1.4(-13)$	$1.2(-12)$	650	a
127	$OH + NH_3 \rightarrow H_2O + NH_2$	$1.5(-13)$	$1.7(-12)$	710	b
128	$OH + HCl \rightarrow H_2O + Cl$	$8.0(-13)$	$2.6(-12)$	350	a,b
129	$OH + HOCl \rightarrow H_2O + ClO$	$5.0(-13)$	$3.0(-12)$	500	a,b
130	$OH + ClONO_2 \rightarrow$ products	$3.9(-13)$	$1.2(-12)$	330	a,b
131	$OH + HBr \rightarrow H_2O + Br$	$1.1(-11)$	$1.1(-11)$	± 0	b
132	$OH + CO \rightarrow H + CO_2$	$1.5(-13)[1 + 5.92 \times 10^{-6}\,(p_{air}/Pa)]$		± 0	a
78	$OH + CH_4 \rightarrow CH_3 + H_2O$	$6.2(-15)$	$2.3(-13)(T/298)^{2.58}$	1082	g
133	$OH + C_2H_6 \rightarrow H_2O + C_2H_5$	$2.5(-13)$	$1.3(-12)(T/298)^2$	498	g
134	$OH + C_3H_8 \rightarrow H_2O + C_3H_7$	$1.1(-12)$	$1.4(-12)(T/298)^2$	61	g
135	$OH + n\text{-}C_4H_{10} \rightarrow H_2O + n\text{-}C_4H_9$	$2.4(-12)$	$1.5(-12)(T/298)^2$	-145	g
136	$OH + i\text{-}C_4H_{10} \rightarrow H_2O + i\text{-}C_4H_9$	$2.2(-12)$	$1.0(-12)(T/298)^2$	-225	g
137	$OH + n\text{-}C_5H_{12} \rightarrow H_2O + n\text{-}C_5H_{11}$	$4.0(-12)$	$2.2(-12)(T/298)^2$	-183	g
138	$OH + i\text{-}C_5H_{12} \rightarrow H_2O + i\text{-}C_5H_{11}$	$3.7(-12)$	—	—	g
139	$OH + neo\text{-}C_5H_{12} \rightarrow H_2O + neo\text{-}C_5H_{11}$	$8.5(-13)$	$1.6(-12)$	189	g
140	$OH + C_2H_4 \rightarrow$ products	$8.5(-12)\; p \approx 10^5$ Pa	$2.0(-12)^d$	-438	g
141	$OH + C_3H_6 \rightarrow$ products	$2.6(-11)\; p \approx 10^5$ Pa	$4.9(-12)^d$	-504	g
142	$OH + 1\text{-butene} \rightarrow$ products	$5.6(-11)$	$6.5(-12)$	-467	g
143	$OH + cis\text{-}2\text{-butene} \rightarrow$ products	$5.6(-11)$	$1.1(-11)$	-487	g
144	$OH + trans\text{-}2\text{-butene} \rightarrow$ products	$6.4(-11)$	$1.1(-11)$	-550	g
145	$OH + isobutene \rightarrow$ products	$5.1(-11)$	$9.5(-12)$	-504	g
146	$OH + 1,3\text{-butadiene} \rightarrow$ products	$6.7(-11)$	$1.5(-12)$	-448	g
147	$OH + isoprene \rightarrow$ products	$1.0(-10)$	$2.5(-11)$	-410	g
148	$OH + C_2H_2 \rightarrow$ products	See Table A.5			b
149	$OH - benzene \rightarrow$ products	$1.2(-12)\; p \approx 10^5$ Pa	$2.5(-12)^d$	201	i

(Continues)

TABLE A.4 (*Continued*)

Number	Reaction	k_{298} (cm³ molecule⁻¹ s⁻¹)[b]	A (cm³ molecule⁻¹ s⁻¹)[b]	E_a/R_g (K)	References[c]
Reactions of OH radicals					
150	OH + toluene → products	$6.2\,(-12)\ p \approx 10^5$ Pa	$1.8\,(-12)$[d]	-355	i
151	OH + ethylbenzene → products	$7.1\,(-12)$	—	—	i
152	OH + o-xylene → products	$1.4\,(-11)$	—	—	i
153	OH + m-xylene → products	$2.4\,(-11)$	—	—	i
154	OH + p-xylene → products	$1.4\,(-11)$	—	—	i
155	OH + α-pinene → products	$5.4\,(-11)$	—	—	g
156	OH + β-pinene → products	$7.9\,(-11)$	—	—	g
157	OH + d-limonene → products	$1.7\,(-10)$	—	—	g
158	OH + myrcene → products	$2.2\,(-10)$	—	—	g
159	OH + 3-carene → products	$8.0\,(-11)$	—	—	g
160	OH + sabinene → products	$1.2\,(-10)$	—	—	g
161	OH + camphene → products	$5.3\,(-11)$	—	—	g
162	OH + α-phellandrene → products	$3.1\,(-10)$	—	—	g
163	OH + β-phellandrene → products	$1.7\,(-11)$	—	—	g
164	OH + terpinolene → products	$2.2\,(-10)$	—	—	g
86	OH + HCHO → HCO + H₂O	$9.4\,(-12)$	$3.6\,(-12)\,(T/298)$	-287	i
165	OH + CH₃CHO → CH₃CO + H₂O	$1.6\,(-11)$	$5.6\,(-12)$	-311	i
166	OH + C₆H₅CHO → C₆H₅CO + H₂O	$1.3\,(-11)$	—	—	i
167	OH + CH₃COCH₃ → CH₃COCH₂ + H₂O	$2.2\,(-13)$	$5.3\,(-18)$	230	i
168	OH + CH₃OH → CH₂OH + H₂O	$9.4\,(-13)$	$5.3\,(-13)\,(T/298)^2$	-170	i
169	OH + C₂H₅OH → CH₃CHOH + H₂O	$3.3\,(-12)$	$5.5\,(-13)$	-530	i
170	OH + 1-butanol → products	$8.6\,(-12)$	—	—	i
171	OH + phenol → products	$2.6\,(-11)$	—	—	i
172	OH + o-cresol → products	$4.2\,(-11)$	—	—	i
173	OH + m-cresol → products	$6.4\,(-11)$	—	—	i
174	OH + p-cresol → products	$4.7\,(-11)$	—	—	i
175	OH + HCOOH → CO₂ + H + H₂O	$4.5\,(-13)$	$4.5\,(-13)$	±0	i

No.	Reaction	A	k	E/R	Ref.
176	OH + CH$_3$COOH → products		8.0(−13)	—	—	i
177	OH + CH$_3$Cl → CH$_2$Cl + H$_2$O	4.0(−12)	3.6(−14)	1400	b
178	OH + CH$_2$Cl$_2$ → CHCl$_2$ + H$_2$O	3.8(−12)	1.1(−13)	1050	b
179	OH + CHCl$_3$ → CCl$_3$ + H$_2$O	2.0(−12)	1.0(−13)	900	b
180	OH + CCl$_4$ → products	—	<5.0(−16)	—	b
181	OH + CH$_3$CH$_2$Cl → products	6.8(−13) (T/298)^2	4.1(−13)	150	i
182	OH + CH$_2$ClCH$_2$Cl → products	9.8(−13) (T/298)^2	2.5(−13)	410	i
183	OH + CH$_3$CCl$_3$ → CH$_2$CCl$_3$ + H$_2$C	1.8(−12)	1.0(−14)	1550	b
184	OH + CHCl=CCl$_2$ → products	4.9(−13)	2.2(−12)	−450	b
185	OH + 'Cl$_2$C=CCl$_2$ → products	9.4(−12)	1.7(−13)	1200	b
186	OH + CHClF$_2$ → CClF$_2$ + H$_2$O	1.2(−12)	5.9(−14)	1100	b
187	OH + CH$_3$CFCl$_2$ → CH$_2$CFCl$_2$ + H$_2$O	1.7(−12)	5.7(−15)	1700	b
188	OH + CH$_3$CF$_2$Cl → CH$_2$CF$_2$Cl + H$_2$O	1.3(−12)	3.1(−15)	1800	b
189	OH + CH$_2$FCF$_3$ → CHFCF$_3$ + H$_2$O	1.4(−13) (T/298)^2	4.9(−15)	1000	i
190	OH + CH$_3$Br → CH$_2$Br + H$_2$O	4.0(−12)	2.9(−14)	1470	b
191	OH + CH$_2$Br$_2$ → CHBr$_2$ + H$_2$O	1.9(−12)	1.1(−13)	840	b
192	OH + CH$_3$I → CH$_2$I + H$_2$O	3.1(−12)	7.2(−14)	1120	b
193	OH + CH$_3$ONO$_2$ → CH$_2$ONO$_2$ + H$_2$O		3.5(−13)	—	i
194	OH + C$_2$H$_5$ONO$_2$ → C$_2$H$_4$ONO$_2$ + H$_2$O		4.9(−13)	—	i
195	OH + (CH$_3$)$_2$CHONO$_2$ → products		4.9(−13)	—	i
196	OH + 2-butyl nitrate → products		9.2(−13)	—	i
197	OH + 2-pentyl nitrate → products		1.9(−12)	—	i
198	OH + 3-pentyl nitrate → products		1.1(−12)	—	i
199	OH + 2-hexyl nitrate → products		3.2(−12)	—	i
200	OH + 3-hexyl nitrate → products		2.7(−12)	—	i
201	OH + CH$_3$CO$_3$NO$_2$ → CH$_2$CO$_2$NO$_2$ + H$_2$O		<4(−14)	—	b
202	OH + H$_2$S → H$_2$O + HS	6.3(−12)	4.8(−12)	80	a, b
203	OH + OCS → products	1.1(−13)	2.0(−15)	1200	a, b
204	OH + CS$_2$ (+O$_2$) → products	8.8(−16)	2.0(−12) P ≈ 10^5 Pa	−2300	a, b
205	OH + CH$_3$SH → products	9.9(−12)	3.3(−11)	−360	a, b
206	OH + CH$_3$SCH$_3$ → products	1.2(−11)	5.0(−12)	260	a, b

(Continues)

TABLE A.4 (Continued)

Number	Reaction	k_{298} (cm^3 molecule^{-1} s^{-1})[b]	A (cm^3 molecule^{-1} s^{-1})[b]	E_a/R_g (K)	References[c]
Reactions of OH radicals					
207	OH + CH$_3$SSCH$_3$ → products	2.3 (−10)	2.3 (−10)	−400	a, b
208	OH + SO$_2$ (+M) → HOSO$_2$	See Table A.5			
Ozone reactions (see also reactions 2, 39, 51, 58, 59, 63, 66, 68, 75, 88, 107, 117)					
209	O$_3$ + CH$_4$ → products	< 7 (−24)	—	—	j
210	O$_3$ + CO → CO$_2$ + O$_2$	< 4 (−25)	—	—	k
211	O$_3$ + SO$_2$ → SO$_3$ + O$_2$	< 2 (−22)	—	—	j
212	O$_3$ + C$_2$H$_4$ → products	1.6 (−18)	9.1 (−15)	2580	g
213	O$_3$ + C$_3$H$_6$ → products	1.0 (−17)	5.5 (−15)	1880	g
214	O$_3$ + 1-butene → products	9.6 (−18)	3.4 (−15)	1740	g
215	O$_3$ + cis-2-butene → products	1.2 (−16)	3.2 (−15)	970	g
216	O$_3$ + trans-2-butene → products	1.9 (−16)	6.6 (−15)	1060	g
217	O$_3$ + 1-pentene → products	1.0 (−17)	—	—	g
218	O$_3$ + 2-methyl-2-butene → products	4.0 (−16)	6.5 (−15)	830	g
219	O$_3$ + 1-hexene → products	1.1 (−17)	—	—	g
220	O$_3$ + 1,3-butadiene → products	6.3 (−18)	13.4 (−15)	2280	g
221	O$_3$ + isoprene → products	1.3 (−17)	7.9 (−15)	1910	g
222	O$_3$ + C$_2$H$_2$ → products	7.8 (−21)	—	—	i
223	O$_3$ + benzene → products	1.7 (−22)	1.1 (−11)	7400	i
224	O$_3$ + toluene → products	4.1 (−22)	2.3 (−12)	6695	i
225	O$_3$ + o-xylene → products	1.7 (−21)	2.4 (−13)	5590	i
226	O$_3$ + m-xylene → products	8.5 (−22)	5.4 (−13)	6040	i
227	O$_3$ + p-xylene → products	1.4 (−21)	1.9 (−13)	5590	i
228	O$_3$ + α-pinene → products	8.7 (−17)	1.0 (−15)	730	g, i
229	O$_3$ + β-pinene → products	1.5 (−17)	—	—	g, i
230	O$_3$ + limonene → products	2.0 (−16)	—	—	g, i
231	O$_3$ + myrcene → products	4.7 (−16)	—	—	g, i

232	O_3 + 3-carene → products	3.7 (−17)	—	—	i
233	O_3 + sabinene → products	8.6 (−17)	—	—	g, i
234	O_3 + camphene → products	0.9 (−18)	—	—	g, i
235	O_3 + α-phellandrene → products	3.1 (−10)	—	—	g
236	O_3 + β-phellandrene → products	1.7 (−11)	—	—	g
237	O_3 + terpinolene → products	2.2 (−10)	—	—	g
238	O_3 + acrolein → products	2.8 (−19)	—	—	i
239	O_3 + methacrolein → products	1.1 (−18)	1.4 (−15)	2110	i
240	O_3 + methyl vinyl ketone → products	4.6 (−18)	7.5 (−16)	1520	i
	Reactions of NO_3 radicals (see also reactions 11, 67, 69)				
241	NO_3 + CO → CO_2 + NO_2	< 4 (−19)	—	—	l, b
242	NO_3 + SO_2 → SO_3 + NO_2	< 7 (−21)	—	—	l, b
243	NO_3 + HCHO → HNO_3 + HCO	5.8 (−16)	—	—	a, b
244	NO_3 + CH_3CHO → HNO_3 + CH_3CO	2.8 (−15)	1.4 (−12)	1860	a, b
245	NO_3 + C_2H_4 → products	2.0 (−16)	5.3 (−12)	3100	l, i
246	NO_3 + C_3H_8 → products	9.5 (−15)	4.6 (−13)	1160	l, i
247	NO_3 + 1-butene → products	1.2 (−14)	2.0 (−13)	845	l, i
248	NO_3 + iso-butene → products	3.3 (−13)	—	—	l, i
249	NO_3 + cis-butene → products	3.5 (−13)	—	—	l, i
250	NO_3 + trans-butene → products	3.9 (−13)	—	—	l, i
251	NO_3 + 1,3-butadiene → products	1.0 (−13)	—	—	l, i
252	NO_3 + isoprene → products	6.8 (−13)	3.0 (−12)	450	l, i
253	NO_3 + α-pinene → products	6.2 (−12)	1.2 (−12)	−490	l, i
254	NO_3 + β-pinene → products	2.5 (−12)	—	—	l, i
255	NO_3 + d-limonene → products	1.2 (−11)	—	—	l, i
256	NO_3 + myrcene → products	1.1 (−11)	—	—	l, i
257	NO_3 + 3-carene → products	9.1 (−12)	—	—	l, i
258	NO_3 + sabinene → products	1.0 (−11)	—	—	i

(Continues)

TABLE A.4 (Continued)

Number	Reaction	k_{298} (cm^3 molecule^{-1} s^{-1})[b]	A (cm^3 molecule^{-1} s^{-1})[b]	E_a/R_g (K)	References[c]
Reactions of NO$_3$ radicals (see also reactions 11, 67, 69)					
259	NO$_3$ + camphene → products	6.6 (−13)	—	—	i
260	NO$_3$ + α-phellandrene → products	8.5 (−11)	—	—	l, i
261	NO$_3$ + β-phellandrene → products	8.0 (−12)	—	—	l, i
262	NO$_3$ + terpinolene → products	9.7 (−11)	—	—	l, i
263	NO$_3$ + acrolein → products	5.1 (−15)	—	—	l, i
264	NO$_3$ + toluene → products	6.8 (−17)	—	—	l, i
265	NO$_3$ + o-xylene → products	3.8 (−16)	—	—	l, i
266	NO$_3$ + m-xylene → products	2.3 (−16)	—	—	l, i
267	NO$_3$ + p-xylene → products	4.5 (−16)	—	—	l, f
268	NO$_3$ + phenol → products	3.9 (−12)	—	—	l, f
269	NO$_3$ + o-cresol → products	1.4 (−11)	—	—	l, i
270	NO$_3$ + m-cresol → products	9.7 (−12)	—	—	l, i
271	NO$_3$ + p-cresol → products	1.1 (−11)	—	—	l, i
272	NO$_3$ + CH$_3$SCH$_3$ → products	9.7 (−13)	—	—	i
Additional reactions					
273	CH$_3$ + O$_3$ → CH$_3$O + O$_2$	2.6 (−12)	5.4 (−12)	220	a, b
274	CH$_2$OH + O$_2$ → HCHO + HO$_2$	9.1 (−12)	—	—	a, b
275a	C$_2$H$_5$ + O$_2$ → C$_2$H$_5$O$_2$	See Table A.5			
275b	→ C$_2$H$_4$ + HO$_2$	< 2 (−15)	—	—	a, b

No.	Reaction				Ref.
276	$C_2H_5O + O_2 \rightarrow CH_3CHO + HO$	1.0(-14)	6.3(-14)	550	b
277	$C_2H_5O_2 + NO \rightarrow C_2H_5O + NO_2$	8.7(-12)	8.7(-12)	±0	a, b
278	$C_2H_5O_2 + HO_2 \rightarrow C_2H_5OOH + O_2$	5.8(-12)	6.5(-13)	-650	h
279	$CH_3O_2 + CH_3CO_3 \rightarrow CH_3O + CH_3 + CO_2$	1.5(-11)	5.1(-12)	-270	h
280	$CH_3CO_3 + NO \rightarrow CH_3 + CO_2 + NO_2$	2.-(-11)	2.4(-11)	±0	a, b
281a	$CH_3CO_3 + HO_2 \rightarrow CH_3C(O)OOH + O_2$	4.7(-12)	1.4(-13)	-1040	h
281b	$\rightarrow CH_3C(O)OH + O_3$	9.4(-12)	2.9(-13)	-1040	h
282	$CH_3O_2 + SO_2 \rightarrow$ products	<5(-17)	—	—	a, b
283	$HO_2 + SO_2 \rightarrow$ products	<1(-18)	—	—	b
284	$S + O_2 \rightarrow SO + O$	2.3(-12)	2.3(-12)	±0	a, b
285	$S + O_3 \rightarrow SO + O_2$	1.2(-11)	1.2(-11)	—	a, b
286	$SO + O_2 \rightarrow SO_2 + O$	8.4(-17)	2.6(-13)	2400	b
287	$SO + O_3 \rightarrow SO_2 + O_2$	9.0(-14)	3.6(-12)	1100	a, b
288	$HS + O_2 \rightarrow OH + SO$	<4(-19)	—	—	a, b
289	$HS + O_3 \rightarrow HSO + O_2$	3.5(-12)	9.0(-12)	280	b
290	$HSO + O_2 \rightarrow$ products	<2(-17)	—	—	a, b
291	$HSO + O_3 \rightarrow$ products	1.2(-13)	—	—	a, b
292	$HSO_2 + O_2 \rightarrow SO_2 + HO_2$	3.2(-13)	—	—	a, b
293	$HOSO_2 + O_2 \rightarrow SO_3 + HO_2$	4.4(-13)	1.3(-12)	330	b
294	$ClO + SO \rightarrow Cl + SO_2$	2.8(-11)	2.8(-11)	±0	b
295	$ClO + SO_2 \rightarrow Cl + SO_3$	<4(-18)	—	—	b

[a] The Arrhenius equation is $k_{bim} = A \exp(-E_a/R_g T)$. In several cases the A factor was taken to have a temperature dependence T^n. Powers of 10 are shown in parentheses.

[b] Except for termolecular reactions: cm^6 molecule^{-2} s^{-1}.

[c] References: (a) Atkinson et al. (1989, 1997); (b) DeMore et al. (1997); (c) Baulch et al. (1992) (d) Michael and Lim (1992), Glassman (1996); (e) Cohen and Westberg (1991); (f) Adusei and Fontijn (1994); (g) Atkinson (1997); (h) Lightfoot et al. (1992); (i) Atkinson (1994); (j) Anderson (1976); (k) Arin and Warneck (1972); (l) Wayne et al. (1991).

[d] The reaction is slightly pressure dependent, see DeMore et al. (1997) and Atkinson et al. (1989, 1997).

TABLE A.5 Rate coefficient Parameters for Gas Phase Association Reactions, with Air as the Carrier Gas[a]

Number	Reaction	k_0^{300}	n	k_∞^{300}	m	k_{atm}^{300}
1	$O + O_2 + M$	6.0 (−34)	2.3	—	—	1.5 (−14)
8	$O + NO + M$	9.0 (−32)	1.5	3.0 (−11)	0	1.6 (−12)
9	$O + NO_2 + M$	9.0 (−32)	2.0	2.2 (−11)	0	1.6 (−12)
16	$O + SO_2 + M$	1.3 (−33)	−3.6	—	—	2.2 (−12)
24	$O + CO + M$	3.3 (−36)	−6.3	—	—	8.1 (−17)
50	$H + O_2 + M$	5.7 (−32)	1.6	7.5 (−11)	0	1.2 (−12)
54	$OH + OH + M$	6.2 (−31)	1.0	2.6 (−11)	0	5.9 (−12)
61	$HO_2 + NO_2 + M$	1.8 (−31)	3.2	4.7 (−12)	1.4	1.4 (−12)
69	$NO_3 + NO_2 + M$	2.2 (−30)	3.9	1.5 (−12)	0.7	1.2 (−12)
70	$NO + NO_2 + H_2O$	4.4 (−40)	—	—	—	—
71	$NO_2 + NO_2 + H_2O$	8.0 (−38)	—	—	—	—
79	$CH_3 + O_2 + M$	4.5 (−31)	3.0	1.8 (−12)	1.7	1.1 (−12)
82	$CH_3O_2 + NO_2 + M$	1.5 (−30)	4.0	6.5 (−12)	2.0	4.0 (−12)
85	$CH_3O + NO_2 + M$	1.1 (−28)	4.0	1.6 (−11)	1.0	1.5 (−11)
89	$Cl + O_2 + M$	2.7 (−33)	1.5	—	—	4.9 (−14)
103	$ClO + NO_2 + M$	1.8 (−31)	3.4	1.5 (−11)	1.9	2.3 (−12)
106	$ClO + ClO + M$	2.2 (−32)	3.1	3.5 (−12)	1.0	3.4 (−13)
111	$BrO + NO_2 + M$	5.2 (−31)	3.2	6.9 (−12)	2.9	2.8 (−12)
121	$OH + NO + M$	7.0 (−31)	2.6	3.6 (−11)	0.1	7.3 (−12)
122	$OH + NO_2$	2.5 (−30)	4.4	1.6 (−11)	1.7	8.7 (−12)
148	$OH + C_2H_2$	5.5 (−30)	0	8.3 (−13)	−2.0	7.6 (−13)
208	$OH + SO_2$	3.0 (−31)	3.3	1.5 (−12)	0	8.8 (−13)
277a	$C_2H_5 + O_2$	1.5 (−28)	3.0	8.0 (−12)	0	7.5 (−12)

[a] The second-order compound rate coefficient at ambient temperature $T(K)$ and number density $n(M)$ may be calculated from $k = 0.6^y\, k_0(T)n(M)/[1 + k_0(T)n(M)/k_\infty(T)]$, where $y = \{1 + [\log_{10} k_0(T)n(M)/k_\infty(T)]^2\}^{-1}$. The low pressure rate coefficient is $k_0(T) = k_0^{300}(T/300)^{-n}$, in units of cm^6 $molecule^{-2}$ s^{-1}; the high pressure rate coefficient is $k_\infty(T) = k_\infty^{300}(T/300)^{-m}$, in units of cm^3 $molecule^{-1}$ s^{-1}. The table gives k_0^{300}, n, k_∞^{300}, and m as far as known, and the rate coefficient k_{atm}^{300} at 300 K for atmospheric pressure (1013 hPa). Powers of 10 are indicated in parentheses.

TABLE A.6 Rate Coefficients at 298 K and Activation Energies (as Far as Known) for Reactions in Dilute Aqueous Solutions as in Clouds[a]

		k_{298} (dm^3 mol^{-1}, s^{-1})	E_a / R_g (K)	Ref.[b]
Aqueous photochemical processes				
1	$H_2O_2 + h\nu \rightarrow OH + OH$	1.5(−5)[c]		a
2	$NO_3^-(+H^+) + h\nu \rightarrow NO_2 + OH$	5.6(−7)[c]		a
3	$NO_2^-(+H^+) + h\nu \rightarrow NO + OH$	3.7(−5)[c]		a
4	$HNO_2 + h\nu \rightarrow NO + OH$	6.8(−4)[c]		a
5	$FeOH^{2+} + h\nu \rightarrow Fe^{2+} + OH$	5.2(−3)[c]		a

(*Continues*)

TABLE A.6 (*Continued*)

		k_{298} (dm^3 mol^{-1}, s^{-1})	E_a / R_g (K)	Ref.[b]
Basic chemistry involving HO$_x$ and C$_1$ reactions				
6	HO$_2$ + HO$_2$ → H$_2$O$_2$ + O$_2$	8.3(5)	2720	b
7	HO$_2$ + O$_2^-$(+H$^+$) → H$_2$O$_2$ + O$_2$	9.7(7)	1060	b
8	O$_3$ + O$_2^-$(+H$^+$) → OH + 2O$_2$	1.5(9)		b
9	O$_3$ + OH → HO$_2$ + O$_2$	1.1(8)		b
10	OH + HO$_2$ → O$_2$ + H$_2$O	7.1(9)		b
11	OH + O$_2^-$ → OH$^-$ + O$_2$	1.0(10)		b
12	OH + OH → H$_2$O$_2$	5.5(9)		b
13	OH + H$_2$O$_2$ → HO$_2$ + H$_2$O	2.7(7)		b
14	OH + HCO$_3^-$ → CO$_3^-$ + H$_2$O	1.0(7)		c
15	CO$_3^-$ + O$_2^-$(+H$^+$) → HCO$_3^-$ + O$_2$	6.5(8)		d
16	CO$_3^-$ + HCOO$^-$ → HCO$_3^-$ + CO$_2^-$	1.5(5)		d
17	CO$_3^-$ + H$_2$O$_2$ → HCO$_3^-$ + HO$_2$	4.3(5)		d
18[d]	CH$_3$OO + HO$_2$ → CH$_3$OOH + O$_2$	4.3(5)		b
19[d]	CH$_3$OO + O$_2^-$ → CH$_3$OOH + OH$^-$ + O$_2$	5.0(7)		b
20a[e]	CH$_3$OOH + OH → CH$_3$OO + H$_2$O	2.7(7)		b
20b[e]	CH$_3$OOH + OH → HCHO + OH + H$_2$O	1.1(7)		b
21	OH + CH$_3$OH(+O$_2$) → CH$_2$(OH)$_2$ + HO$_2$	9.7(8)	580	c
22	OH + CH$_2$(OH)$_2$(+O$_2$) → HCOOH + H$_2$O + HO$_2$	1.1(9)	1020	c, e
23	OH + HCOOH + O$_2$ → CO$_2$ + HO$_2$ + H$_2$O	1.2(8)	990	c, e
24	OH + HCOO$^-$ → CO$_2$ + H$_2$O	3.1(9)	1240	c, e
25	CO$_2^-$ + O$_2$ → O$_2^-$ + CO$_2$	2.4(9)		d
Reactions involving sulfur–oxygen species				
26	CH$_2$(OH)$_2$ ⇌ HCHO(+H$_2$O)	5.3(−3)/13	4000	f
27	HCHO + HSO$_3^-$ → HOCH$_2$SO$_3^-$	4.3(2)	2660	f
28	HCHO + SO$_3^{2-}$(+H$^+$) → HOCH$_2$SO$_3^-$	5.4(6)	2530	f
29	HOCH$_2$SO$_3^-$ + OH$^-$ → CH$_2$(OH)$_2$ + SO$_3^{2-}$	4.6(3)	4880	f
30	OH + HOCH$_2$SO$_3^-$ → HOCHSO$_3^-$ + H$_2$O	3.0(8)		a, g
31	HOCHSO$_3^-$ + O$_2$(+H$_2$O) → HCOOH + HSO$_3^-$ + HO$_2$	2.5(9)		g
32	O$_3$ + SO$_2$ + H$_2$O → HSO$_4^-$ + O$_2$ + H$^+$	2.4(4)		h
33	O$_3$ + HSO$_3^-$ → HSO$_4^-$ + O$_2$	3.2(5)	2030	h
34	O$_3$ + SO$_3^{2-}$ → SO$_4^{2-}$ + O$_2$	1.5(9)	3080	h
35	HSO$_3^-$ + H$_2$O$_2$ → HSO$_4^-$ + H$_2$O	4.0(7)[H$^+$]	4100	a, i
36	HSO$_3^-$ + CH$_3$OOH → HSO$_4^-$ + CH$_3$OH	1.7(7)[H$^+$]	3800	a, j
37	HSO$_3^-$ + HOONO$_2$ → HSO$_4^-$ + NO$_3^-$ + H$^+$	3.1(5)	(at 285 K)	a, k
38	HSO$_3^-$ + HOONO → HSO$_4^-$ + HNO$_2$	4.0(4) + 2.0(7)[H$^+$]		a, i
39	HSO$_3^-$ + HSO$_5^-$ → 2HSO$_4^-$	189 + 6.2(6)[H$^+$]	(at 285 K)	a, l
40	OH + HSO$_3^-$ → SO$_3^-$ + H$_2$O	2.7(9)		m
41	OH + SO$_3^{2-}$ → SO$_3^-$ + OH$^-$	4.6(9)		m
42	SO$_3^-$ + O$_2$ → SO$_5^-$	2.5(9)		m
43	SO$_5^-$ + HSO$_3^-$ → HSO$_5^-$ + SO$_3^-$	8.6(3)		m
44	SO$_5^-$ + SO$_3^{2-}$(+H$^+$) → HSO$_5^-$ + SO$_3^-$	2.1(5)		m
45	SO$_5^-$ + SO$_3^{2-}$ → SO$_4^{2-}$ + SO$_4^-$	5.5(5)		m
46	SO$_5^-$ + SO$_5^-$ → S$_2$O$_8^{2-}$ + O$_2$	4.8(7)		m

(*Continues*)

TABLE A.6 (*Continued*)

		k_{298} (dm^3 mol^{-1}, s^{-1})	E_a/R_g (K)	Ref.[b]
47	$SO_5^- + SO_5^- \rightarrow 2SO_4^- + O_2$	2.3(8)		m
48	$SO_5^- + HO_2 \rightarrow HSO_5^- + O_2$	2.0(9)		n
49	$SO_5^- + O_2^-(+H^+) \rightarrow HSO_5^- + O_2$	2.3(8)		m
50	$SO_4^- + HSO_3^- \rightarrow HSO_4^- + SO_3^-$	6.8(8)		m
51	$SO_4^- + SO_3^{2-} \rightarrow SO_4^{2-} + SO_3^-$	3.1(8)		m
52	$SO_4^- + H_2O_2 \rightarrow HSO_4^- + HO_2$	1.2(7)		o
53	$SO_4^- + HCOOH \rightarrow HSO_4^- + H^+ + CO_2^-$	1.4(6)		o
54	$SO_4^- + HCOO^- \rightarrow HSO_4^- + CO_2^-$	1.1(8)		o
55	$SO_4^- + CH_3OH(+O_2) \rightarrow HSO_4^- + H^+$ $+ HCHO + HO_2$	9.0(6)		o
Reactions involving chlorine species				
56	$SO_4^- + Cl^- \rightarrow SO_4^{2-} + Cl$	3.4(8)		o, p
57	$OH + Cl^- \rightarrow Cl + OH^-$	4.3(9)		b
58	$Cl + H_2O \rightarrow ClOH^- + H^+$	1.3(3)		b
59	$ClOH^- \rightarrow Cl^- + OH$	6.1(9)		b
60	$ClOH^- + H^+ \rightarrow Cl + H_2O$	2.1(10)		b
61	$Cl + Cl^- \rightleftharpoons Cl_2^-$	2.7(10)/1.1(5)		b
62	$Cl_2^- + HO_2 \rightarrow 2Cl^- + H^+ + O_2$	1.3(10)		q
63	$Cl_2^- + O_2^- \rightarrow 2Cl^- + O_2$	6.0(9)		q
64	$Cl_2^- + HSO_3^- \rightarrow 2Cl^- + H^+ + SO_3^-$	1.7(8)		q
65	$Cl_2^- + SO_3^{2-} \rightarrow 2Cl^- + SO_3^-$	6.2(7)		q
66	$Cl_2^- + HCOOH \rightarrow 2Cl^- + 2H^+ + CO_2^-$	8.0(4)		q
67	$Cl_2^- + HCOO^- \rightarrow 2Cl^- + H^+ + CO_2^-$	1.3(6)		q
68	$Cl_2^- + H_2O_2 \rightarrow 2Cl^- + H^+ + HO_2$	7.0(5)		q
Reactions involving nitrogen–oxygen species				
69	$NO_3 + OH^- \rightarrow NO_3^- + OH$	8.2(7)	2700	r
70	$NO_3 + Cl^- \rightarrow NO_3^- + Cl$	1.0(7)	4300	r
71	$NO_3 + HSO_3^- \rightarrow NO_3^- + H^+ + SO_3^-$	1.3(9)	2000	r
72	$NO_3 + HCOOH(+O_2) \rightarrow NO_3^- + H^+ + CO_2 + HO_2$	3.8(5)	3400	r
73	$NO_3 + HCOO^-(+O_2) \rightarrow NO_3^- + CO_2 + HO_2$	5.1(7)	2200	r
74	$HOONO_2 \rightleftharpoons H^+ + NO_4^-$	$K_{eq} \approx 3.0(-6)$		s
75	$NO_4^- \rightarrow NO_2^- + O_2$	8.0(-1)		s
76	$O_2^- + NO(+H^+) \rightarrow HOONO$	6.7(9)		t
77	$HOONO \rightarrow NO_3^- + H^+$	7.0(-1)		t
78	$HO_2 + NO_2 \rightarrow HOONO_2$	1.8(9)		t
79	$O_2^- + NO_2 \rightarrow NO_4^-$	4.5(9)		t
80	$NO_2^- + O_3 \rightarrow NO_3^- + O_2$	5.0(5)		u
81	$OH + NO_2^- \rightarrow NO_2 + OH^-$	1.0(10)		c
82	$OH + HNO_2 \rightarrow NO_2 + H_2O$	2.6(9)		t
83	$NO_2 + NO_2(+H_2O) \rightarrow HNO_2 + NO_3^- + H^+$	8.0(7)		t
Reactions involving metal ion species[f]				
84	$OH + Fe^{2+} \rightarrow FeOH^{2+}$	4.3(8)		c
85	$HO_2 + Fe^{2+}(+H_2O) \rightarrow FeOH^{2+} + H_2O_2$	1.2(6)		v

(*Continues*)

TABLE A.6 (*Continued*)

	k_{298} (dm^3 mol^{-1}, s^{-1})	E_a / R_g (K)	Ref.[b]
86 $O_2^- + Fe^{2+}(+H_2O + H^+) \rightarrow FeOH^{2+} + H_2O_2$	1.0(7)		v
87 $HO_2 + FeOH^{2+} \rightarrow Fe^{2+} + O_2 + H_2O$	1.0(4)		v
88 $O_2^- + FeOH^{2+} \rightarrow Fe^{2+} + O_2 + OH^-$	1.5(8)		v
89 $Fe^{2+} + H_2O_2 \rightarrow FeOH^{2+} + OH$	7.5(1)		v
90 $Fe^{2+} + O_3(+H_2O) \rightarrow FeOH^{2+} + OH + O_2$	8.2(5)		w
91 $FeOH^{2+} + HSO_3^- \rightarrow Fe^{2+} + SO_3^- + H_2O$	1.0(1)	8300	x
92 $Fe^{2+} + HSO_5^- \rightarrow FeOH^{2+} + SO_4^-$	3.6(4)		a
93 $Fe^{2+} + SO_4^- (+H_2O) \rightarrow FeOH^{2+} + HSO_4^-$	3.0(8)		a
94 $Fe^{2+} + SO_5^- (+H_2O) \rightarrow FeOH^{2+} + HSO_5$	1.0(6)		x
95 $Cu^+ + HO_2(+H^+) \rightarrow Cu^{2+} + H_2O_2$	1.0(9)		v
96 $Cu^+ + O_2^- (+2H^+) \rightarrow Cu^{2+} + H_2O_2$	9.4(9)		v
97 $Cu^{2+} + HO_2 \rightarrow Cu^+ + H^+ + O_2$	5.0(7)		v
98 $Cu^{2+} + O_2^- \rightarrow Cu^+ + O_2$	8.0(9)		v
99 $Cu^+ + H_2O_2 \rightarrow Cu^{2+} + OH + OH^-$	4.7(2)		y
100 $Cu^+ + O_3(+H_2O) \rightarrow Cu^{2+} + OH + OH$	1.0(7)		z
101 $Cu^+ + FeOH^{2+} \rightarrow Cu^{2+} + Fe^{2+} + OH^-$	3.0(7)		v

[a] Rate coefficients for 3 < pH < 7; numbers in parentheses indicate powers of 10. The reactions were used in model calculations together with the principal gas phase and gas aqueous phase exchange reactions to obtain the results in Tables 8.11 and 8.13. Some of the rate constants are estimates (see remarks).

[b] References: (a) Warneck *et al.* (1996); (b) Sander and Crutzen (1996); (c) Buxton *et al.* (1988); (d) Ross *et al.* (1992); (e) Chin and Wine (1994); (f) Boyce and Hoffmann (1984), Kok *et al.* (1986), Deister *et al.* (1986), Sorensen and Andersen (1970), Schecker and Schulz (1969); (g) Barlow *et al.*, 1997; (h) Hoigné *et al.* (1985), Hoffmann (1986), Botha *et al.* (1993); (i) Drexler *et al.* (1991); (j) Lind *et al.* (1987); (k) Amels *et al.* (1996); (l) Betterton and Hoffmann (1988b), Elias *et al.* (1994); (m) Buxton *et al.* (1996a); (n) Fischer and Warneck (1996a), Buxton *et al.* (1996b); (o) Wine *et al.* (1989); (p) McElroy (1990), Huie and Clifton (1990); (q) Jacobi (1996), Jacobi *et al.* (1996); (r) Exner *et al.* (1992, 1994); (s) Lammel *et al.* (1990), Logager and Sehested (1993); (t) Fischer and Warneck (1996b); (u) Damschen and Martin (1983); (v) Sedlak and Hoigné (1993); (w) Logager *et al.* (1992); (x) Ziajka *et al.* (1994), Warneck and Ziajka (1995), Kraft and van Eldik (1989); (y) Weinstein-Lloyd and Schwartz (1992); (z) von Piechowski (1991). Note: Ref. (a) is a review, references (b)–(d) and (v) contain compilations.

[c] Photodissociation coefficients were calculated by the two-stream method of Brühl and Crutzen (1989), which includes back-scattered light from the ground with albedo = 0.25. The data refer to 0.8 km elevation above sea level and 25° solar zenith angle (summer conditions at 45–50° latitude).

[d] Rate coefficients for these reactions were estimated by Jacob (1986); experimental confirmation is required.

[e] Rate coefficients estimated by Sander and Crutzen (1996); experimental confirmation is required.

[f] The predominant Fe(III) species at pH 3–4 is $FeOH^{2+}$, and above pH 4 it is $Fe(OH)_2^+$. The reactions are written as if $FeOH^+$ were the only Fe(III) species present.

REFERENCES

Abel, N., R. Jaenicke, C. Junge, P. Rodriguez Garcia Prieto, and W. Seiler (1969) Luftchemische Studien am Observatorium Izaña (Teneriffa). *Meteorol. Rundsch.* **22**, 158–167.

Acker, K., D. Möller, W. Marquardt, E. Brüggemann, W. Wieprecht, R. Aurel, and D. Kalass (1998) Atmospheric research program for studying changing emission patterns after German unification. *Atmos. Environ.* **32**, 3435–3443.

Ackerman, M. (1971) Ultraviolet solar radiation related to mesospheric processes, In "Mesospheric Models and Related Experiments" (G. Fiocco, ed.), pp. 149–159. Reidel, Dordrecht, the Netherlands.

Ackerman, M., and F. Biaumé (1970) Structure of the Schumann-Runge bands from the (0-0) to the (13-0) band. *J. Mol. Spectrosc.* **35**, 73–82.

Adam, J. R., and R. G. Semonin (1970) Collection efficiencies of raindrops for submicron particulates. In "Precipitation Scavenging 1970" (R. J. Engelmann and W. G. N. Slinn, coordinators), pp. 151–160. U.S. Atomic Energy Commission, Division of Technical Information, Oak Ridge, TN.

Adams, D. F., S. O. Farwell, E. Robinson, M. R. Pack, and W. L. Bamesberger (1981a) Biogenic sulfur source strengths. *Environ. Sci. Technol.* **15**, 1493–1498.

Adams, D. F., S. O. Farwell, M. R. Pack, and E. Robinson (1981b) Biogenic gas emissions from soils in eastern and southeastern United States. *J. Air. Pollut. Control Assoc.* **31**, 1083–1089.

Adams, D. J., S. Barlow, G. V. Buxton, T. M. Malone, and G. A. Salmon (1995) Evaluation of the stability constant of Cl_2^- in neutral aqueous solution. *J. Chem. Soc. Faraday Trans. I* **91**, 3303–3305.

Adams, F., R. Dams, L. Guzman, and J. W. Winchester (1977) Background aerosol composition on Chacaltaya Mountain, Bolivia. *Atmos. Environ.* **11**, 629–634.

Adams, S. J., S. G. Bradley, C. D. Stow, and S. J. de Mora (1986) Measurement of pH versus drop size in natural rain. *Nature* **322**, 842–844.

Adel, A. (1938) Further detail in the rock-salt prismatic solar spectrum. *Astrophys. J.* **88**, 186–188.

Adusei, G. Y., and A. Fontijn (1994) Comparison of the kinetics of $O(^3P)$ reactions with the four butenes over wide temperature ranges. *J. Phys. Chem.* **98**, 3732–3739.

Aikin, A. C., J. R. Herman, E. J. Maier, and C. J. McQuillan (1982) Atmospheric chemistry of ethane and ethylene. *J. Geophys. Res.* **87**, 3105–3118.

Air Quality Criteria for Photochemical Oxidants (1970) Publication AP-63. U.S. Department of Health, Education and Welfare, Public Health Service, Environmental Health Service, National Air Pollution Control Administration, Washington, DC.

Aitken, J. (1923) "Collected Scientific Papers" (C. Knott, ed.). Cambridge University Press, London and New York.

Ajtay, G. L., P. Ketner, and P. Duvigneaud (1979) Terrestrial primary production and phytomass. In "The Global Carbon Cycle" (B. Bolin, E. T. Degens, S. Kempe, and P. Ketner, eds.). SCOPE 13, 129–181.

Albert-Levy (1878) Analyse de l'air. In "Annulaire de l'Observatoire Montsouris," pp. 495–505. Gauthier-Villars, Paris.

Albrecht, B. C., C. Junge, and H. Zatosek (1970) Der N_2O Gehalt der Bodenluft in drei Bodenprofilen. Z. Pflanzenernähr. Bodenkd. 125, 205–211.

Aldaz, L. (1969) Flux measurements of atmospheric ozone over land and water. J. Geophys. Res. 74, 6934–6946.

Alexander, M. (1977) "Soil Microbiology," 2nd edition, Wiley, New York.

Alkezweeny, A. J. (1978) Measurements of aerosol particles and trace gases in METROMEX. J. Appl. Meteorol. 17, 609–614.

Allara, D. L., T. Mill, D. G. Hendry, and F. R. Mayo (1968) Low temperature gas- and liquid-phase oxidations of isobutane. Adv. Chem. Ser. 76, 40–57.

Allègre, C. J., T. Staudacher, and P. Sarda (1987) Rare gas systematics: formation of the atmosphere, evolution and structure of the earth's mantle. Earth Planet. Sci. Lett. 81, 127–150.

Allen, A. G., R. M. Harrison, and M. T. Wake (1988) A meso-scale study of the behaviour of atmospheric ammonia and ammonium. Atmos. Environ. 22, 1347–1353.

Allen, D. J., P. Kasibhatla, A. M. Thompson, R. B. Rood, B. G. Doddridge, K. E. Pickering, R. D. Hudson, and S.-J. Lin (1996) Transport-induced interannual variability of carbon monoxide determined using a chemistry and transport model. J. Geophys. Res. 101, 28,655–28,669.

Allison, F. E. (1955) The enigma of soil nitrogen balance sheets. Adv. Agron. 7, 213–250.

Allison, F. E. (1966) The fate of nitrogen applied to soils. Adv. Agron. 18, 219–258.

Altshuller, A. P. (1976) Regional transformation of sulfur dioxide to sulfate in the U.S. J. Air Pollut. Control. Assoc. 26, 318–324.

Altshuller, A. P. (1980) Seasonal and episodic trends in sulfate concentrations in the eastern United States. Environ. Sci. Technol. 14, 1337–1349.

Altshuller, A. P. (1983a) Review: natural volatile organic substances and their effects on air quality in the eastern United States. Environ. Sci. Technol. 14, 1337–1349.

Altshuller, A. P. (1983b) Measurements of the products of atmospheric photochemical reactions in laboratory studies and in ambient air-relationships between ozone and other products. Atmos. Environ. 17, 2383–2427.

Altshuller, A. P. (1993) PANs in the atmosphere. J. Air Waste Manag. Assoc. 43, 1221–1230.

Altshuller, A. P., and J. J. Bufalini (1965) Photochemical aspects of air pollution: a review. Photochem. Photobiol. 4, 97–146.

Altshuller, A. P., and J. J. Bufalini (1971) Photochemical aspects of air pollution: a review. Environ. Sci. Technol. 5, 39–64.

Altshuller, A. P., S. L. Kopczynski, W. A. Lonneman, T. L. Becker, and R. Slater (1967) Chemical aspects of photooxidation for the system propene-nitric oxide. Environ. Sci. Technol. 1, 899–914.

Amels, P., H. Elias, U. Götz, U. Steingens, and K. J. Wannowius (1996) Kinetic investigation of the stability of peroxynitric acid and of its reaction with sulfur(IV) in aqueous solution. In "Heterogeneous and Liquid Phase Processes" (P. Warneck, ed.), Transport and Chemical Transformation of Pollutants in the Troposphere, Vol. 2 (P. Borrell, P. M. Borrell, T. Cvitas, K. Kelly, and W. Seiler, series eds.), pp. 77–88. Springer-Verlag, Berlin.

Anastasi, C., and I. W. M. Smith (1976) Rate measurements of OH by resonance absorption. Part 5: Rate constants for $OH + NO_2$ $(+M) \rightarrow HNO_3$ $(+M)$ over a wide range of temperature and pressure. *J. Chem. Soc. Faraday Trans.* 2 **72**, 1459–1468.

Anderson, D. L. (1989) Where on Earth is the crust? *Physics Today* **42** 38–46.

Anderson, I. C., and J. S. Levine (1987) Simultaneous field measurements of biogenic emissions of nitric oxide and nitrous oxide. *J. Geophys. Res.* **92**, 965–976.

Anderson, J. G., W. H. Brune, S. A. Lloyd, D. W. Toohey, S. P. Sander, W. L. Starr, M. Loewenstein, and R. J. Podolske (1989a) Kinetics of ozone destruction by ClO and BrO within the Antarctic vortex: an analysis based on *in situ* ER-2 data. *J. Geophys. Res.* **94**, 11,480–11,520.

Anderson, J. G., W. H. Brune, and M. H. Proffitt (1989b) Ozone destruction by chlorine radicals within the Antarctic vortex: the spatial and temporal evolution of $ClO-O_3$ anticorrelation based on *in situ* ER-2 data. *J. Geophys. Res.* **94**, 11,465–11,479.

Anderson, J. G., D. W. Toohey, and W. H. Brune (1991) Free radicals within the Antarctic vortex: the role of CFCs in Antarctic ozone loss. *Science* **251**, 39–46.

Anderson, J. W. (1990) Sulfur metabolism in plants. *In* "The Biochemistry of Plants" (B. J. Miflin and P. J. Lea, eds.), pp. 327–381. Academic Press, New York.

Anderson, L. G. (1976) Atmospheric chemical kinetics data survey. *Rev. Geophys. Space Phys.* **14**, 151–170.

Anderson, L. G., J. A. Lanning, R. Barrell, J. Miyagishima, R. H. Jonnes, and P. Wolfe (1996) Sources and sinks of formaldehyde and acetaldehyde: an analysis of Denver's ambient concentration data. *Atmos. Environ.* **30**, 2113–2123.

Andreae, M. O. (1991) Biomass burning: its history, use and distribution, and its impact on environmental quality and global climate. *In* "Global Biomass Burning, Atmospheric, Climatic and Biospheric Implications" (J. S. Levine, ed.), pp. 3–21. MIT Press, Cambridge, MA.

Andreae, M. O. (1993) The influence of tropical biomass burning on climate and the atmospheric environment *In* "Biogeochemistry of Global Change. Radiatively Active Gases" (R. S. Oremland, ed.), pp. 113–150. Chapman and Hall, New York.

Andreae, M. O. (1995) Climatic effects of changing atmospheric aerosol levels. *In* World Survey of Climatology, Vol. 16, "Future Climates of the World" (A. Henderson-Sellers, ed.), pp. 341–392. Elsevier, Amsterdam.

Andreae, M. O. (1996) Raising dust in the greenhouse. *Nature* **380**, 389–390.

Andreae, M. O., and T. W. Andreae (1988) The cycle of biogenic sulfur compounds over the Amazon basin. I. Dry season. *J. Geophys. Res.* **93**, 1487–1497.

Andreae, M. O., and P. J. Crutzen (1997) Atmospheric aerosols: biogeochemical sources and role in atmospheric chemistry. *Science* **276**, 1052–1058.

Andreae, M. O., and R. J. Ferek (1992) Photochemical production of carbonyl sulfide in sea water and its emission to the atmosphere. *Global Biogeochem. Cycles* **6**, 175–183.

Andreae, M. O., and W. A. Jaeschke (1992) Exchange of sulphur between biosphere and atmosphere over temperate and tropical regions. *In* "Sulphur Cycling on the Continents" (R. W. Howarth, J. W. B. Stewart and M. V. Ivanov, eds.), pp. 27–61. Wiley, New York.

Andreae, M. O., and H. Raemdonck (1983) Dimethyl sulfide in the surface ocean and the marine atmosphere: a global view. *Science* **221**, 744–747.

Andreae, M. O., and P. Warneck (1994) Global methane emissions from biomass burning and comparison with other sources. *Pure Appl. Chem.* **66**, 162–169.

Andreae, M. O., R. W. Talbot, and S.-M. Li (1987) Atmospheric measurements of pyruvic and formic acid. *J. Geophys. Res.* **92**, 6635–6641.

Andreae, M. O., R. W. Talbot, T. W. Andreae, and R. C. Harris (1988a) Formic and acetic acids over the central Amazon region, Brazil. 1. Dry season. *J. Geophys. Res.* **93**, 1616–1624.

Andreae, M. O., E. V. Browell, M. Garstang, G. L. Gregory, R. C. Harris, G. F. Hill, D. J. Jacob, M. C. Pereira, G. W. Sachse, A. W. Setzer, P. L. Silva Dias, R. W. Talbot, A. L. Torres, and S. C. Wofsy (1988b) Biomass-burning emissions and associated haze layers over Amazonia. *J. Geophys. Res.* **93**, 1509–1527.

Andreae, M. O., H. Berresheim, T. W. Andreae, M. A. Kritz, T. S. Bates, and J. T. Merrill (1988c) Vertical distribution of dimethyl sulfide, sulfur dioxide, aerosol ions, and radon over the northeast Pacific Ocean. *J. Atmos. Chem.* **6**, 149–173.

Andreae, M. O., H. Berresheim, H. Bingemer, D. J. Jacob, B. L. Lewis, S.-M. Li, and R. W. Talbot (1990) The atmospheric sulfur cycle over the Amazon basin. 2. Wet season. *J. Geophys. Res.* **95**, 16,813–16,824.

Andreae, M. O., W. Elbert, and S. J. DeMora (1995) Biogenic sulfur emissions and aerosols over the tropical south Atlantic. 3. Atmospheric dimethyl sulfide, aerosols and cloud condensation nuclei. *J. Geophys. Res.* **100**, 11,335–11,356.

Andreae, M. O., E. Atlas, H. Cachier, W. R. Cofer, G. W. Harris, G. Helas, R. Koppmann, J.-P. Lacaux, and D. F. Ward (1996) Trace gas and aerosol emissions from savanna fires. *In* "Biomass Burning and Global Change" (J. S. Levine, ed.), pp. 278–295. MIT Press, Cambridge, MA.

Andreae, T. W., G. A. Cutter, N. Hussain, J. Radford-Knoery, and M. O. Andreae (1991) Hydrogen sulfide and radon in and over the western north Atlantic ocean. *J. Geophys, Res.* **96**, 18,753–18,760.

Andreae, T. W., M. O. Andreae, H. G. Bingemer, and C. Leck (1993) Measurements of dimethyl sulfide and H$_2$S over the western north Atlantic and tropical Atlantic. *J. Geophys. Res.* **98**, 23,389–23,396.

Andreae, T. W., M. O. Andreae, and G. Schebeske (1994) Biogenic sulfur emissions and aerosols over the tropical south Atlantic. I. Dimethylsulfide in sea water and in the atmospheric boundary layer. *J. Geophys. Res.* **99**, 22,819–22,829.

Andrés-Hernández, M. D., J. Notholt, J. Hjorth, and O. Schrems (1996) A DOAS study on the origins of nitrous acid at urban and non-urban sites. *Atmos. Environ.* **30**, 175–180.

Andrews, D. G., J. R. Holton, and C. B. Leovy (1987) "Middle Atmosphere Dynamics." Academic Press, London.

Aneja, V. P., J. H. Overton, L. T. Cupitt, J. L. Durban, and W. E. Wilson (1979) Direct measurements of emission rates of some atmospheric biogenic sulfur compounds. *Tellus* **31**, 174–178.

Aneja, V. P., J. H. Overton, and A. P. Aneja (1981) Emission survey of biogenic sulfur flux from terrestrial surfaces. *J. Air. Poll. Contr. Assoc.* **31**, 256–258.

Angell, J. K. (1993) Comparison of stratospheric warming following Agung, El Chichon and Pinatubo volcanic eruptions. *Geophys. Res. Lett.* **20**, 715–718.

Angell, J. K., and J. Korshover (1978) Global ozone variations: an update into 1976. *Mon. Weather Rev.* **104**, 63–75.

Angstrom, A., and L. Högberg (1952) On the content of nitrogen in atmospheric precipitation in Sweden 2. *Tellus* **4**, 271–279.

Animoto, S. T., A. P. Force, and J. R. Wiesenfeld (1978/1979) Ozone photochemistry: production and deactivation of O(2^{1}D$_2$) following photolysis at 248 nm. *Chem. Phys. Lett.* **60**, 40–43.

Anlauf, K. G., P. Fellin, H. A. Wiebe, H. I. Schiff, G. I. Mackay, R. S. Braman, and R. Gilbert (1985) A comparison of three methods for measurement of atmospheric nitric acid and particulate nitrate and ammonium. *Atmos. Environ.* **19**, 325–333.

Appel, B. R., E. M. Hoffer, E. L. Kothny, S. M. Wall, M. Haik, and R. L. Knights (1979) Analysis of carbonaceous material in southern Californian aerosols. 2. *Environ. Sci. Technol.* **13**, 98–104.

Appel, B. R., Y. Tokiwa, and E. L. Kothny (1983) Sampling of carbonaceous particles in the atmosphere. *Atmos. Environ.* 17, 1787–1796.

Appenzeller, C., J. R. Holton, and K. H. Rosenlof (1996) Seasonal variation of mass transport across the tropopause. *J. Geophys. Res.* 101, 15,071–15,078.

Arey, J., A. M. Winer, R. Atkinson, S. M. Aschmann, W. D. Long, and C. L. Morrison (1991) The emission of (Z)-3-hexen-1-ol, (Z)-3-hexenylacetate and other oxygenated hydrocarbons from agricultural plant species. *Atmos. Environ.* 25A, 1063–1075.

Arijs, E., D. Nevejans, P. Frederick, and J. Ingels (1982) Stratospheric negative ion composition measurements, ion abundances, and related trace gas detection. *J. Atmos. Terr. Phys.* 44, 681–694.

Arin, L. M., and P. Warneck (1972) Reaction of ozone with carbon monoxide. *J. Phys. Chem.* 76, 1514–1516.

Arlander, D. W., D. R. Cronn, J. C. Farmer, F. A. Menzia, and H. H. Westburg (1990) Gaseous oxygenated hydrocarbons in the remote marine troposphere. *J. Geophys. Res.* 95, 16,391–16,403.

Armerding, W., F. J. Comes, and B. Schülke (1995) $O(^1D)$ quantum yields of ozone photolysis in the UV from 300 nm to its threshold and at 355 nm. *J. Phys. Chem.* 99, 3137–3143.

Arnold, F., R. Fabian, G. Henschen, and W. Joos (1980) Stratospheric trace gas analysis from ions, H_2O and HNO_3. *Planet. Space Sci.* 28, 681–685.

Arnts, R. R., and S. A. Meeks (1981) Biogenic hydrocarbon contribution to the ambient air of selected areas. *Atmos. Environ.* 15, 1643–1651.

Arrhenius, S. (1896) On the influence of carbonic acid in the air upon the temperature of the ground. *Philos. Mag.* 41, 237–276.

Arrhenius, S. (1903) "Lehrbuch der kosmischen Physik." Hirzel, Leipzig.

Aschmann, S. M., and R. Atkinson (1994) Formation yields of methyl vinyl ketone and methacrolein from the gas phase reaction of O_3 with isoprene. *Environ. Sci. Technol.* 28, 1539–1542.

Aselmann, I., and P. J. Crutzen (1989) Global distribution of natural freshwater wetlands and rice paddies, their net primary productivity, seasonality and possible methane emissions. *J. Atmos. Chem.* 8, 307–358.

Asman, W. A. H. (1992) "Ammonia Emission in Europe: Updated Emission and Emission Variations," Report no. 228471008. National Institute of Public Health and Environmental Protection, Bilthoven, the Netherlands.

Asman, W. A. H., and H. A. Jaarsveld (1992) A variable resolution transport model applied to NH_3 in Europe. *Atmos. Environ.* 26A, 445–465.

Atkinson, R. (1986) Kinetics and mechanisms of the gas-phase reactions of the hydroxyl radical with organic compounds under atmospheric conditions. *Chem. Rev.* 86, 69–201.

Atkinson, R. (1987) A structure-activity relationship for the estimation of rate constants for the gas-phase reactions of OH radicals with organic compounds. *Int. J. Chem Kinet.* 19, 799–828.

Atkinson, R. (1990) Gas-phase tropospheric chemistry of organic compounds: a review. *Atmos. Environ.* 24A, 1–41.

Atkinson, R. (1994) Gas-phase tropospheric chemistry of organic compounds. *J. Phys. Chem. Ref. Data Monograph 2*, 1–216.

Atkinson, R. (1997) Gas-phase tropospheric chemistry of volatile organic compounds. 1. Alkanes and alkenes. *J. Phys. Chem. Ref. Data* 26, 215–290.

Atkinson, R., and S. M. Aschmann (1993) OH production from the gas-phase reaction of O_3 with a series of alkenes under atmospheric conditions. *Environ. Sci. Technol.* 27, 1357–1363.

Atkinson, R., and W. P. L. Carter (1991) Reactions of alkoxy radicals under atmospheric conditions: the relative importance of decomposition versus reaction with O_2. *J. Atmos. Chem.* 13, 195–210.

Atkinson, R., and A. C. Lloyd (1984) Evaluation of kinetic and mechanistic data for modeling of photochemical smog. *J. Phys. Chem. Ref. Data* 13, 315–444.

Atkinson, R., K. R. Darnall, A. C. Lloyd, A. M. Winer, and J. N. Pitts, Jr. (1979) Kinetics and mechanisms of the reactions of the hydroxyl radical with organic compounds in the gas phase. *Adv. Photochem.* 11, 375–488.

Atkinson, R., W. P. L. Carter, K. R. Darnall, A. M. Winer, and J. N. Pitts, Jr. (1980) A smog chamber and modeling study of gas-phase NO_x—air photooxidation of toluene and the cresols. *Int. J. Chem. Kinet.* 12, 779–836.

Atkinson, R., D. L. Baulch, R. A. Cox, R. F. Hampson, Jr., J. A. Kerr, and J. Troe (1989) Evaluated kinetic and photochemical data for atmospheric chemistry. Supplement III. *J. Phys. Chem. Ref. Data* 18, 881–1097.

Atkinson, R., D. L. Baulch, R. A. Cox, R. F. Hampson, Jr., J. A. Kerr, and J. Troe (1992a) Evaluated kinetic and photochemical data for atmospheric chemistry. Supplement IV. *J. Phys. Chem. Ref. Data* 21, 1125–1568.

Atkinson, R., S. M. Aschmann, J. Arey, and B. Shorees (1992b) Formation of OH in the gas phase reactions of O_3 with a series of terpenes. *J. Geophys. Res.* 97, 6065–6073.

Atkinson, R., D. L. Baulch, R. A. Cox, R. F. Hampson, Jr., J. A. Kerr, M. J. Rossi, and J. Troe (1997) Evaluated kinetic and photochemical data for atmospheric chemistry: Supplements V and VI. *J. Phys. Chem. Ref. Data* 26, 521–1011; 1329–1499.

Atlas, E. L. (1988) Evidence for $\geq C_3$ alkyl nitrates in rural and remote atmospheres. *Nature* 331, 426–428.

Atlas, E. L., and B. A. Ridley (1996) The Mauna Loa Observatory photochemistry experiment: introduction. *J. Geophys. Res.* 101, 14,531–14,541.

Atlas, E. L., B. A. Ridley, G. Hübler, J. G. Walega, M. A. Carroll, D. D. Montzka, B. J. Huebert, R. B. Norton, F. E. Grahek, and S. Schauffler (1992) Partitioning and budget of NO_y species during Mauna Loa Observatory photochemistry experiment. *J. Geophys. Res.* 97, 10,449–10,462.

Atlas, E. I., S.-M. Li, L. J. Stanley, and R. A. Hytes (1993a) Natural and anthropogenic organic compounds in the global atmosphere. In "Global Atmospheric Chemical Change" (C. N. Hewitt and W. T. Sturges, eds.), pp. 313–381. Elsevier, London and New York.

Atlas, E., W. Pollock, J. Greenberg, L. Heidt, and A. M. Thompson (1993b) Alkyl nitrates, nonmethane hydrocarbons, and halocarbon gases over the equatorial Pacific Ocean during SAGA 3. *J. Geophys. Res.* 98, 16,933–16,947.

Attmannspacher, W., and H. U. Dütsch (1970) International ozone sonde intercomparison at the observatory Hohenpeissenberg. *Ber. Dtsch. Wetterdienst* 16 (120).

Attmannspacher, W., and R. Hartmannsgruber (1973) On extremely high values of ozone near the ground. *Pure Appl. Geophys.* 106–108, 1091–1096.

Awramik, S. M. (1982) The pre-phanerozoic fossil record. In "Mineral Deposits and the Evolution of the Biosphere" (H. D. Holland and M. Schidlowski, eds.), Dahlem-Konferenzen, pp. 67–82. Springer-Verlag, Berlin and New York.

Axford, W. I. (1968) The polar wind and the terrestrial helium budget. *J. Geophys. Res.* 73, 6855–6859.

Axford, W. I. (1970) On the origin of the radiation belt and auroral primary ions. In "Particles and Fields in the Magnetosphere" (B. M. McCormac, ed.), pp. 46–59. Reidel, Dordrecht, the Netherlands.

Ayers, G. P. (1982) The chemical composition of precipitation. A southern hemisphere perspective. In "Atmospheric Chemistry" (E. D. Goldberg, ed.), pp. 41–56, Dahlem Konferenzen. Springer-Verlag, Berlin.

Ayers, G. P., and J. L. Gras (1980) Ammonia gas concentrations over the southern ocean. *Nature* 284, 539–540.

Ayers, G. P., and J. P. Ivey (1988) Precipitation composition at Cape Grim, 1977–1985. *Tellus* 40B, 297–307.

Ayers, G. P., and J. P. Ivey (1990) Methanesulfonate in rainwater at Cape Grim, Tasmania. *Tellus* **42B**, 217–222.

Ayers, G. P., R. W. Gillett, and J. L. Gras (1980) On the vapor pressure of sulfuric acid. *Geophys. Res. Lett.* **17**, 433–436.

Ayers, G. P., J. P. Ivey, and H. S. Goodman (1986) Sulfate and methane sulfonate in the marine aerosol at Cape Grim, Tasmanaia. *J. Atmos. Chem.* **4**, 173–185.

Ayers, G. P., J. P. Ivey, and R. W. Gillett (1991) Coherence between seasonal cycles of dimethyl sulphide, methanesulphonate and sulphate in marine air. *Nature* **349**, 404–406.

Bacastow, R. B., and C. D. Keeling (1981) Atmospheric carbon dioxide concentration and observed airborne fraction. *In* "Carbon Cycle Modelling" (B. Bolin, ed.). SCOPE **16**, 103–112.

Bach, W. (1976) Global air pollution and climate change. *Rev. Geophys. Space Phys.* **14**, 429–474.

Bachelet, D., and H. U. Neue (1993) Methane emissions from wetland rice areas of Asia. *Chemosphere* **26**, 219–238.

Bächmann, K., I. Haag, and A. Röder (1993) A field study to determine the chemical content of individual raindrops as a function of their size. *Atmos. Environ.* **27A**, 1951–1958.

Bächmann, K., P. Ebert, I. Haag, and T. Prokop (1996) The chemical content of raindrops as a function of drop radius. I. Field measurements at the cloud base and below the cloud. *Atmos. Environ.* **30**, 1019–1025.

Bäckström, H. (1934) Der Kettenmechanismus bei der Autoxydation von Natriumsulfit-Lösungen. *Z. Phys. Chem.* **25B**, 122–138.

Baes, C. F., H. E. Goelle, J. S. Olsen, and R. M. Rotty (1976) The global carbon dioxide problem. ORNL-5194, pp. 1–72. Oak Ridge National Laboratory, Oak Ridge, TN.

Bagnold, R. A. (1941) "The Physics of Blown Sand and Desert Dunes." Methuen, London.

Bahe, F. C., W. N. Marx, and U. Schurath (1979) Determination of the absolute photolysis rate of ozone by sunlight, $O_3 + h\nu \rightarrow O(^1D) + O_2(^1\Delta_g)$ at ground level. *Atmos. Environ.* **13**, 1515–1522.

Bahe, F. C., U. Schurath, and K. H. Becker (1980) The frequency of NO_2 photolysis at ground level as recorded by a continuous actinometer. *Atmos. Environ.* **14**, 711–718.

Baker, E. A., and E. Parsons (1971) Scanning electron microscopy of plant cuticles. *J. Microsc.* **94**, 39–49.

Bakwin, P. S., D. J. Jacob, S. C. Wofsy, J. W. Munger, B. C. Daube, J. D. Bradshaw, S. T. Sandholm, R. W. Talbot, H. B. Singh, G. L. Gregory, and D. R. Blake (1994) Reactive nitrogen oxides and ozone above a taiga woodland. *J. Geophys. Res.* **99**, 1927–1936.

Baldwin, A. C., J. R. Barker, D. M. Golden, and D. G. Hendry (1977) Photochemical smog rate parameter estimates and computer simulations. *J. Phys. Chem.* **81**, 2483–2492.

Balkanski, Y. J., D. J. Jacob, G. M. Gardner, W. C. Graustein, and K. K. Turekian (1993) Transport and residence times of tropospheric aerosols inferred from a global three-dimensional simulation of ^{210}Pb. *J. Geophys. Res.* **98**, 20,573–20,586.

Ball, S. M., and G. Hancock (1995) The relative quantum yield of $O_2(a^1\Delta_g)$ from the photolysis of ozone at 227 K. *Geophys. Res. Lett.* **22**, 1213–1216.

Balla, R. J., H. H. Nelson, and J. R. McDonald (1985) Kinetics of the reactions of isopropoxy radicals with NO, NO_2, and O_2. *Chem. Phys.* **99**, 323–335.

Balls, P. W. (1989) Trace metal and major ion composition of precipitation at a North Sea coastal site. *Atmos. Environ.* **23**, 2751–2759.

Bal Reddy, K., and R. van Eldik (1992) Kinetics and mechanism of the sulfite-induced autoxidation of Fe(II) in aqueous solution. *Atmos. Environ.* **26A**, 661–665.

Bandy, A. R., P. J. Maroulis, L. Shalaby, and L. A. Wilner (1981) Evidence for a short tropospheric residence time for carbon disulfide. *Geophys. Res. Lett.* **8**, 1180–1183.

Bandy, A. R., D. C. Thornton, D. L. Scott, M. Lalevic, E. E. Lewin, and A. R. Driedger (1992a) A time series for carbonyl sulfide in the northern hemisphere. *J. Atmos. Chem.* **14**, 527–534.

Bandy, A. R., D. L. Scott, B. W. Blomquist, S. M. Chen, and D. C. Thornton (1992b) Low yields of SO_2 from dimethyl sulfide oxidation in the marine boundary layer. *Geophys. Res. Lett.* **19**, 1125–1127.

Bandy, A. R., D. C. Thornton, and J. E. Johnson (1993) Carbon disulfide measurements in the atmosphere of the western north Atlantic and the northwestern south Atlantic oceans. *J. Geophys. Res.* **98**, 23,449–23,457.

Bange, H. W., S. Rapsomanikis, and M. O. Andreae (1996) Nitrous oxide in coastal waters. *Global Biogeochem. Cycles* **10**, 197–207.

Banin, A. (1986) A global budget of N_2O: the role of soils and their change. *Sci. Total Environ.* **55**, 27–38.

Banks, P. M., and G. Kockarts (1973) "Aeronomy." Academic Press, New York.

Bannon, J. K., and L. P. Steele (1960) Average water-vapour content of the air. Geophysical Memo no. 102. Meteorological Office Air Ministry, London.

Barbier, M., D. Joly, A. Saliot, and D. Tourres (1973) Hydrocarbons from sea water. *Deep Sea Res.* **20**, 305–314.

Barger, W. R., and W. D. Garrett (1976) Surface active organic material in the marine atmosphere. *J. Geophys. Res.* **81**, 3151–3157.

Barkenbus, B. D., C. S. MacDougall, W. H. Griest, and J. E. Caton (1983) Methodology for the extraction and analysis of hydrocarbons and carboxylic acids in atmospheric particulate matter. *Atmos. Environ.* **17**, 1537–1543.

Barker, J. R., L. Brouwer, R. Patrick, M. J. Rossi, P. L. Trevor, and D. M. Golden (1985) N_2O_5 photolysis: products investigated by fluorescence and optoacoustic techniques. *Int. J. Chem. Kinet.* **17**, 991–1006.

Barlow, S., G. V. Buxton, S. A. Murray, and G. A. Salmon (1997) Oxidation of hydroxymethane sulfonate initiated by the hydroxyl radical. *In* "Transport and Transformation of Pollutants in the Troposphere" (P. M. Borrell, P. Borrell, K. Kelly, T. Cvitas, and W. Seiler, eds.), Vol. 1, Clouds, Aerosols, Modelling and Photo-oxidants. Proceedings of EUROTRAC Symposium '96, pp. 361–365. Computational Mechanics Publications, Southampton, UK.

Barnard, W. R., M. O. Andreae, W. E. Watkins, H. Bingemer, and H. W. Georgii (1982) The flux of dimethyl sulfide from the oceans to the atmosphere. *J. Geophys. Res.* **87**, 8787–8793.

Barnes, I., K. H. Becker, E. H. Fink, A. Reimer, F. Zabel, and H. Niki (1983) Rate constant and products of the reaction CS_2 + OH in the presence of O_2. *Int. J. Chem. Kinet.* **15**, 631–645.

Barnes, I., V. Bastian, K. H. Becker, and Z. Tong (1990) Kinetics and products of the reactions of NO_3 with monoalkenes, dialkenes, and monoterpenes. *J. Phys. Chem.* **94**, 2413–2419.

Barnes, I., K. H. Becker, and N. Mihalopoulos (1994a) An FTIR product study of the photooxidation of dimethyl disulfide. *J. Atmos. Chem.* **18**, 267–289.

Barnes, I., K. H. Becker, and I. Patroescu (1994b) The tropospheric oxidation of dimethyl sulfide: a new source of carbonyl sulfide. *Geophys. Res. Lett.* **21**, 2389–2392.

Barnes, I., K. H. Becker, and I. Patroescu (1996) FTIR product study of the OH initiated oxidation of dimethyl sulphide: observation of carbonyl sulphide and dimethyl sulphoxide. *Atmos. Environ.* **30**, 1805–1814.

Barnola, J. M., D. Raynaud, Y. S. Korotkevitch, and C. Lorius (1987) Vostok ice core provides 160,000 year record of atmospheric CO_2. *Nature* **329**, 408–414.

Barrie, L. A., J. W. Bottenheim, R. C. Schnell, P. J. Crutzen, and R. A. Rasmussen (1988) Ozone destruction and photochemical reactions at polar sunrise in the lower Arctic atmosphere. *Nature* **334**, 138–141.

Barrie, L. A., R. Staebler, D. Toom, B. Georgi, G. den Hartog, S. Landsberger, and D. Wu (1994a) Arctic aerosol size-segregated chemical observations in relation to ozone depletion during Polar Sunrise Experiment 1992. *J. Geophys. Res.* **99**, 25,439–25,451.

Barrie, L. A., S. M. Li, D. M. Toom, S. Landsberger, and W. Sturges (1994b) Measurements of aerosol and gaseous halogens, nitrates and sulphur oxides by denuder and filter systems during Polar Sunrise Experiment 1992. *J. Geophys. Res.* **99**, 25,453–25,468.

Bartell, U., U. Hofmann, R. Hofmann, B. Kreuzburg, M. O. Andreae, and J. Kesselmeier (1993) COS and H_2S fluxes over a wet meadow in relation to photosynthetic activity: an analysis of measurements made on 6 September 1990. *Atmos. Environ.* **27A**, 1851–1864.

Barth, C. A., C. B. Farmer, D. E. Siskind, and J. P. Perich (1996) ATMOS observations of nitric oxide in the mesosphere and lower thermosphere. *J. Geophys. Res.* **101**, 12,489–12,494.

Bartholomew, G. W., and M. Alexander (1979) Microbial metabolism of carbon monoxide in culture and in soil. *Appl. Environ. Microbiol.* **37**, 932–937.

Bartholomew, G. W., and M. Alexander (1982) Microorganisms responsible for the oxidation of carbon monoxide in soil. *Environ. Sci. Technol.* **16**, 300–301.

Bartlett, K. B., and R. C. Harriss (1993) Review and assessment of methane emissions from wetlands. *Chemosphere* **26**, 261–320.

Bartok, W., and A. F. Sarofim, eds. (1990) "Fossil Fuel Combustion: A Source Book." Wiley, New York.

Bates, D. R., and P. B. Hays (1967) Atmospheric nitrous oxide. *Planet. Space Sci.* **15**, 189–197.

Bates, T. S., J. D. Cline, R. H. Gammon, and S. R. Kelly-Hansen (1987) Regional and seasonal variations in the flux of oceanic dimethyl sulfide to the atmosphere. *J. Geophys. Res.* **92**, 2930–2938.

Bates, T. S., J. E. Johnson, P. K. Quinn, P. D. Goldan, W. C. Kuster, D. C. Covert, and C. J. Hahn (1990) The biogeochemical sulfur cycle in the marine boundary layer over the northeast Pacific Ocean. *J. Atmos. Chem.* **10**, 59–81.

Bates, T. S., B. K. Lamb, A. Guenther, J. Dignon, and R. E. Stoiber (1992a) Sulfur emissions to the atmosphere from natural sources. *J. Atmos. Chem.* **14**, 315–337.

Bates, T. S., J. A. Calhoun, and P. K. Quinn (1992b) Variations in the methanesulfonate to sulfate molar ratio in sub-micrometer marine aerosol particles over the south Pacific Ocean. *J. Geophys. Res.* **97**, 9859–9865.

Bathen, K. H. (1972) On the seasonal changes in depth of the mixed layer in the north Pacific Ocean. *J. Geophys. Res.* **77**, 7138–7150.

Bathurst, R. G. C. (1975) "Carbonate Sediments and Their Diagenesis," 2nd ed. Elsevier, Amsterdam.

Batt, L. (1987) Reactions of alkoxy and alkylperoxy radicals. *Int. Rev. Phys. Chem.* **6**, 53–90.

Bauer, E. (1974) Dispersion of tracers in the atmosphere and ocean: survey and comparison of experimental data. *J. Geophys. Res.* **79**, 789–795.

Bauer, K., W. Seiler, and H. Giehl (1979) CO Produktion höherer Pflanzen an natürlichen Standorten. *Z. Pflanzenphysiol.* **94**, 219–230.

Bauer, K., R. Conrad, and W. Seiler (1980) Photooxidative production of carbon monoxide by phototrophic microorganisms. *Biochim. Biophys. Acta* **589**, 46–55.

Baulch, D. L., R. A. Cox, R. F. Hampson, Jr., J. A. Kerr, J. Troe, and T. R. Watson (1980) Evaluated kinetic and photochemical data for atmospheric chemistry. *J. Phys. Chem Ref. Data* **9**, 295–471.

Baulch, D. L., R. A. Cox, P. J. Crutzen, R. F. Hampson, Jr., J. A. Kerr, J. Troe, and T. R. Watson (1982) Evaluated kinetic and photochemical data for atmospheric chemistry, Supplement I. *J. Phys. Chem Ref. Data* **11**, 327–496.

Baulch, D. L., R. A. Cox, R. F. Hampson, Jr., J. A. Kerr, J. Troe, and T. R. Watson (1984) Evaluated kinetic and photochemical data for atmospheric chemistry. Supplement II. *J. Phys. Chem Ref. Data* **13**, 1259–1380.

Baulch, D. L., C. J. Cobos, R. A. Cox, C. Esser, P. Frank, T. Just, J. A. Kerr, M. J. Pillling, J. Troe, R. W. Walker, and J. Warnatz (1992) Evaluated kinetic data for combustion modelling. *J. Phys. Chem. Ref. Data* **21**, 411–429.

Baum, F. (1972) CO Emissionen aus Hausbrand Feuerstätten. *Staub* **32**, 54–59.

Baumgartner, A., and E. Reichel (1975) "The World Water Balance." R. Oldenburg Verlag, Munich, Germany.

Baxter, M. S., M. J. Stenhouse, and N. Drudarski (1980) Fossil carbon in coastal sediments. *Nature* **287**, 35–36.

Bazilevich, N. I., L. E. Rodin, and N. N. Rozov (1971) Geographical aspects of biological productivity. *Sov. Geogr. Rev. Transl.* **12**, 293–317.

Beard, K. V. (1974) Experimental and numerical collision efficiencies for submicron particles scavenged by small raindrops. *J. Atmos. Sci.* **31**, 1595–1603.

Beard, K. V., and S. N. Grover (1974) Numerical collection efficiencies for small raindrops colliding with micron size particles. *J. Atmos. Sci.* **31**, 1595–1603.

Beck, L. L., S. D. Piccot, and D. A. Kirchgessner (1993) Industrial sources. *In* "Atmospheric Methane: Sources, Sinks, and Role in Global Change" (M. A. K. Khalil, ed.), NATO ASI Series, Vol. I 13, pp. 399–431. Springer-Verlag, Berlin.

Becker, K. H., A. Inocenncio, and U. Schurath (1975) The reaction of ozone with hydrogen sulfide and its organic derivatives. *Int. J. Chem. Kinet.* **7** (Symp. no. 1), 205–220.

Becker, K. H., K. J. Brockmann, and J. Bechara (1990) Production of hydrogen peroxide in forest air by reaction of ozone with terpenes. *Nature* **346**, 256–258.

Becker, R., and W. Doering (1935) Kinetische Behandlung der Keimbildung in übersättigten Dämpfen. *Ann. Phys. (Leipzig)* **24**, 719–752.

Beekmann, M., G. Ancellet, S. Blonsky, D. De Muer, A. Ebel, H. Elbern, J. Hendricks, J. Kowol, C. Mancier, R. Sladkowic, H. G. J. Smit, P. Speth, T. Trickl, and P. Van Haver (1997) Stratosphere-troposphere exchange: regional and global tropopause folding occurrence. *In* "Tropospheric Ozone Research" (O. Hov, ed.), Transport and Chemical Transformation of Pollutants in the Troposphere, Vol. 6 (P. Borrell, P. M. Borrell, T. Cvitas, K. Kelly, and W. Seiler, Series eds.), pp. 131–151. Springer-Verlag, Berlin.

Begemann, F., and I. Friedman (1968) Isotopic composition of atmospheric hydrogen. *J. Geophys. Res.* **73**, 1139–1147.

Behnke, W., H. U. Krüger, V. Scheer, and C. Zetzsch (1992) Formation of $ClNO_2$ and HONO in the presence of NO_2, O_3, and wet NaCl aerosol. *J. Aerosol Sci.* **23** (Suppl.), 933–936.

Behnke, W., V. Scheer, and C. Zetzsch (1994) Production of $BrNO_2$, Br_2, and $ClNO_2$ from the reaction between sea spray aerosol and N_2O_5. *J. Aerosol Sci.* **25** (Suppl.), 277–278.

Behra, P., and L. Sigg (1990) Evidence for redox cycling of iron in atmospheric droplets. *Nature* **344**, 419–421.

Beilke, S. (1970) Untersuchungen über das Auswaschen atmosphärischer Spurenstoffe durch Niederschläge. Dissertation. Universität Frankfurt (Main), Germany.

Bell, R. P., and P. G. Evans (1966) Kinetics of the dehydration of methylene glycol in aqueous solution. *Proc. Roy. Soc. Lond. Ser. A* **291**, 297–323.

Belser, L. W. (1979) Population ecology of nitrifying bacteria. *Annu. Rev. Microbiol.* **33**, 309–333.

Bender, M., T. Ellis, P. Tans, R. Francey, and D. Lowe (1996) Variability in the O_2/N_2 ratio of southern hemispheric air, 1991–1994; implication for the carbon cycle. *Global Biogeochem. Cycles* **10**, 9–21.

Benkelberg, H.-J., G. Heimann, O. Böge, and P. Warneck (1996) Self-reaction of 2- and 3-pentylperoxy radicals. *In* "Transport and Transformation of Pollutants in the Troposphere" (P. M. Borrell, P. Borrell, T. Cvitas, K. Kelly and W. Seiler, eds.), Vol. 2, Emissions, Deposition, Laboratory Work and Instrumentation, Proceedings of EUROTRAC Symposium '96, pp. 467–470. Computational Mechanics Publications, Southampton, UK.

Benkovitz, C. M. (1982) Compilation of an inventory of anthropogenic emissions in the United States and Canada. *Atmos. Environ.* 16, 1551–1563.

Benkovitz, C. M., M. T. Scholtz, J. Pacyna, L. Tarrasón, J. Dignon, E. C. Voldner, P. A. Spiro, J. A. Logan, and T. E. Graedel (1996) Global gridded inventories of anthropogenic emissions of sulfur and nitrogen. *J. Geophys. Res.* 101, 29,239–29,253.

Benner, C. L., D. J. Eatough, N. L. Eatough, and P. Bhardwaja (1991) Comparison of annular denuder and filter pack collection of HNO_3 (g), HNO_2 (g), SO_2 (g), and particulate-phase nitrate, nitrite, and sulfate in the south-west desert. *Atmos. Environ.* 25A, 1537–1545.

Bergeron, T. (1935) On the physics of clouds and precipitation. Proceedings of the 5th Assem. U.G.G.I, Lisbon, Vol. 2, p. 156.

Berges, M. G. M., and P. Warneck (1992) Product quantum yields for the 350 nm photodecomposition of pyruvic acid in air. *Ber. Bunsenges. Phys. Chem.* 96, 413–416.

Berglund, J., and L. I. Elding (1995) Manganese-catalyzed autoxidation of dissolved sulfur dioxide in the atmospheric aqueous phase. *Atmos. Environ.* 29, 1379–1391.

Berglund, J., S. Fronaeus, and L. I. Elding (1993) Kinetics and mechanism for the manganese-catalyzed oxidation of sulfur(IV) by oxygen in aqueous solution. *Inorg. Chem.* 32, 4527–4538.

Berkner, L. V., and L. C. Marshall (1964) The history of oxygenic concentrations in the Earth's atmosphere. *Discuss. Faraday Soc.* 37, 122–141.

Berkner, L. V., and L. C. Marshall (1966) Limitation on oxygen concentration in a primitive planetary atmosphere. *J. Atmos. Sci.* 23, 133–143.

Berner, R. A. (1971) Worldwide sulfur pollution of rivers. *J. Geophys. Res.* 76, 6597–6600.

Berresheim, H. (1987) Biogenic sulfur emissions from the subantarctic and Antarctic Oceans. *J. Geophys. Res.* 92, 13,245–13,262.

Berresheim, H., and W. Jaeschke (1983) The contribution of volcanoes to the global atmospheric sulfur budget. *J. Geophys. Res.* 88, 3732–3740.

Berresheim, H., and V. D. Vulcan (1992) Vertical distribution of COS, CS_2, DMS and other sulfur compounds in a loblolly pine forest. *Atmos. Environ.* 26A, 2031–2036.

Berresheim, H., M. O. Andreae, G. P. Ayers, R. W. Gillett, J. T. Merril, V. J. Davis, and W. L. Chameides (1990) Airborne measurements of dimethyl sulfide, sulfur dioxide, and aerosol ions over the southern ocean south of Australia. *J. Atmos. Chem.* 10, 341–370.

Berresheim, H., F. L. Eisele, D. J. Tanner, D. S. Covert, L. McInnes, and D. C. Ramsey-Bell (1993) Atmospheric sulfur chemistry and cloud condensation nuclei (CCN) concentrations over the northeastern Pacific coast. *J. Geophys. Res.* 98, 12,701–12,711.

Berry, J. A. (1975) Adaptation of photosynthetic processes to stress. *Science* 188, 644–650.

Besemer, A. C. (1982) Formation of chemical compounds from irradiated mixtures of aromatic hydrocarbons and nitrogen oxides. *Atmos. Environ.* 16, 1599–1602.

Betterton, E. A. (1993) On the pH dependent formation constants of iron(III)–sulfur(IV) transient complexes. *J. Atmos. Chem.* 17, 307–324.

Betterton, E. A., and M. R. Hoffmann (1988a) Henry's law constants for some environmentally important aldehydes. *Environ. Sci. Technol.* 22, 1415–1418.

Betterton, E. A., and M. R. Hoffmann (1988b) Oxidation of aqueous SO_2 by peroxomonosulfate. *J. Phys. Chem.* 92, 5962–5965.

Beyerle, G., R. Neuber, and O. Schrems (1994) Multi-wavelength lidar measurements of stratospheric aerosols above Spitsbergen during winter 1992/93. *Geophys. Res. Lett.* 21, 57–60.

Bidwell, R. G. S., and D. E. Fraser (1972) Carbon monoxide uptake and metabolism by leaves. *Can J. Bot.* 50, 1435–1439.

Bielski, B. H. J., D. E. Cabelli, R. L. Arudi, and A. B. Ross (1985) Reactivity of HO_2/O_2^- in aqueous solution. *J. Phys. Chem. Ref. Data* 14, 1041–1100.

Bien, G., and H. E. Suess (1967) Transfer and exchange of C-14 between the atmsophere and surface waters of the Pacific Ocean. *In* "Radioactive Dating and Methods of Low-Level Counting," Symposium Proceedings, Monaco, STI/PUB/152, pp. 105–115. International Atomic Energy Agency, Vienna.

Biermann, H. W., C. Zetzsch, and F. Stuhl (1976) Rate constant for the reaction of OH with N_2O at 298 K. *Ber. Bunsenges. Phys. Chem.* **80**, 909–911.

Bingemer, H. G., and P. J. Crutzen (1987) The production of CH_4 from solid wastes. *J. Geophys. Res.* **92**, 2181–2187.

Bingemer, H. G., S. Bürgermeister, R. L. Zimmerman, and H. W. Georgii (1990) Atmospheric OCS: evidence for a contribution of anthropogenic sources. *J. Geophys. Res.* **95**, 20,617–20,622.

Bingemer, H. G., M. O. Andreae, T. W. Andreae, P. Artaxo, G. Helas, D. J. Jacob, N. Mihalopoulos, and B. C. Nguyen (1992) Sulfur gases and aerosols in and above the equatorial African rain forest. *J. Geophys. Res.* **97**, 6207–6217.

Birch, F. (1965) Speculations on the earth's thermal history. *Bull. Geol. Soc. Am.* **76**, 133–154.

Bischof, W. (1977) Comparability of CO_2 measurements. *Tellus* **29**, 435–444.

Bischof, W. (1981) The CO_2 content of the upper polar troposphere between 1963–1970. *In* "Carbon Cycle Modelling" (B. Bolin, ed.). *SCOPE* **16**, 113–116.

Blackburn, T. E., S. T. Bairai, and D. H. Stedman (1992) Solar photolysis of ozone to singlet D oxygen atoms. *J. Geophys. Res.* **97**, 10,109–10,117.

Blake, A. J. (1979) An atmospheric absorption model for the Schumann-Runge bands of oxygen. *J. Geophys. Res.* **84**, 3272–3282.

Blake, D. R., and F. S. Rowland (1986) Global atmospheric concentrations and source strength of ethane. *Nature* **321**, 231–233.

Blake, D. R., and F. S. Rowland (1988) Continuing worldwide increase in tropospheric methane, 1978–1987. *Science* **239**, 1129–1131.

Blake, D. R., E. W. Mayer, S. C. Tyler, Y. Makide, D. C. Montagne, and F. S. Rowland (1982) Global increase in atmospheric methane concentrations between 1978–1980. *Geophys. Res. Lett.* **9**, 477–480.

Blake, D. R., T.-Y. Chen, T. W. Smith, C. J.-L. Wang, O. W. Wingenter, N. J. Blake, and F. S. Rowland (1996) Three-dimensional distribution of nonmethane hydrocarbons and halocarbons over the northwestern Pacific during the 1991 Pacific Exploratory Mission (PEM-West A). *J. Geophys. Res.* **101**, 1763–1778.

Blanchard, D. C. (1963) Electrification in the atmosphere. *Prog. Oceanogr.* **1**, 71–202.

Blanchard, D. C., and L. Syzdet (1972) Variations in Aitken and giant nuclei in marine air. *J. Phys. Oceanogr.* **2**, 255–262.

Blanchard, D. C., and A. H. Woodcock (1957) Bubble formation and modification in the sea and its meteorological significance. *Tellus* **9**, 145–158.

Blanchard, D. C., and A. H. Woodcock (1980) The production, concentration and vertical distribution of sea-salt aerosol. *Ann. N.Y. Acad. Sci.* **338**, 330–347.

Blatherwick, R. D., A. Goldman, D. G. Murcray, F. J. Murcray, G. R. Cook, and J. W. van Allen (1980) Simultaneous mixing ratio profiles of stratospheric NO and NO_2 as derived from balloon borne infrared solar spectra. *Geophys. Res. Lett.* **7**, 471–473.

Blatt, H., and R. L. Jones (1975) Proportions of exposed igneous, metamorphic and sedimentary rocks. *Bull. Geol. Soc. Am.* **86**, 1085–1088.

Blaxter, K. L., and J. L. Clapperton (1965) Prediction of the amount of methane produced by ruminants. *Br. J. Nutr.* **19**, 511–522.

Blifford, I. H., and L. D. Ringer (1969) The size and number distribution of aerosols in the continental troposphere. *J. Atmos. Sci.* **26**, 716–726.

Blifford, I. H., L. B. Lockhart, and H. B. Rosenstock (1952) On the natural radioactivity in the air. *J. Geophys. Res.* 57, 499–509.

Bloomfield, P., M. L. Thompson, and S. Zeger (1982) A statistical analysis of Umkehr measurements of 32–46 km ozone. *J. Appl. Meteorol.* 21, 1828–1837.

Blundell, R. V., W. G. A. Cook, D. E. Hoare, and G. S. Milne (1965) Rates of radical reactions in methane oxidation. Tenth International Symposium on Combustion, pp. 445–452. The Combustion Institute, Pittsburgh, Pensylvania.

Bluth, G. J. S., S. D. Doiron, C. C. Schnetzler, A. J. Krueger, and I. S. Walter (1992) Global tracking of the SO_2 cloud from the June 1991 Mount Pinatubo eruptions. *Geophys. Res. Lett.* 19, 151–154.

Bode, K., G. Helas, and J. Kesselmeier (1997) Biogenic contribution to atmospheric organic acids. In "Biogenic Volatile Organic Compounds in the Atmosphere" (G. Helas, J. Slanina, and R. Steinbrecher, eds.), pp. 157–170, SPB Academic Publishing bv, Amsterdam, the Netherlands.

Bodenstein, M. H. (1918) Die Geschwindigkeit der Reaktion zwischen Stickoxyd und Sauerstoff. *Z. Elektrochemie* 24, 183–201.

Bohn, H. L. (1976) Estimate of organic carbon in soils. *J. Am. Soc. Soil Sci.* 40, 468–470.

Bojkov, R. D., L. Bishop, W. J. Hill, G. C. Reinsel, and G. C. Tiao (1990) A statistical trend analysis of revised Dobson total ozone data over the northern hemisphere. *J. Geophys. Res.* 95, 9785–9807.

Bojkov, R. D. V. E. Fioletov, D. S. Balis, C. S. Zerefos, T. V. Kadygrova, and A. M. Shalamjanski (1995) Further ozone decline during the northern hemisphere winter-spring of 1994–1995 and the new record low ozone over Siberia. *Geophys. Res. Lett.* 22, 2729–2732.

Bolin, B. (1970) The carbon cycle. *Sci. Am.* 223, 124–132.

Bolin, B. (1977) Changes of land biota and their importance for the carbon cycle. *Science* 196, 613–615.

Bolin, B. (1986) How much CO_2 will remain in the atmosphere? In "The Greenhouse Effect, Climatic Change, and Ecosystems" (B. Bolin, B. R. Döös, J. Jäger, and R. A. Warrick, eds.) *SCOPE* 29, 93–155.

Bolin, B., and W. Bischof (1970) Variation of carbon dioxide content of the atmosphere in the northern hemisphere. *Tellus* 22, 431–442.

Bolin, B., A. Björkström, C. D. Keeling, R. Bacastow, and U. Siegenthaler (1981) Carbon Cycle Modelling. In "Carbon Cycle Modelling" (B. Bolin, ed.). *SCOPE* 16, 1–28.

Bolle, H.-J., W. Seiler, and B. Bolin (1986) Other greenhouse gases and aerosols. In "Greenhouse Effects, Climatic Change, and Ecosystems" (B. Bolin, B. R. Doos, J. Jager, and R. A. Warrick, eds.). *SCOPE* 29, 157–203.

Bollinger, M. J., C. J. Hahn, D. D. Parrish, P. C. Murphy, D. L. Albritton, and F. C. Fehsenfeld (1984) NO_x measurements in clean continental air and analysis of the contributing meteorology. *J. Geophys. Res.* 89, 9623–9631.

Bolton, N. E., J. A. Carter, J. F. Emery, C. Feldman, W. Fulkerson, L. D. Hulett, and W. S. Lyon (1975) Trace element balance around a coal-fired steam plant. In "Trace Elements in Fuel" (S. P. Babu, ed.) *Adv. Chem. Ser.* 141, 175–187.

Bongartz, A., J. Kames, F. Welter, and U. Schurath (1991) Near UV absorption cross sections and *cis/trans* equilibrium of nitrous acid. *J. Phys. Chem.* 95, 1076–1082.

Bonsang, B., B. C. Nguyen, A. Gaudry, and G. Lambert (1980) Sulfate enrichment in marine aerosols owing to biogenic gaseous sulfur compounds. *J. Geophys. Res.* 85, 7410–7416.

Bonsang, B., B. C. Nguyen, and G. Lambert (1987) Comment on "The residence time of aerosols and SO_2 in the long-range transport over the ocean," by Ito et al. *J. Atmos. Chem.* 5, 367–369.

Bonsang, B., M. Kanakidou, G. Lambert, and P. Monfray (1988) The marine source of C_2–C_6 aliphatic hydrocarbons. *J. Atmos. Chem.* **6**, 3–20.

Bonsang, B., D. Martin, G. Lambert, M. Kanakidou, J. C. Le Roulley, and G. Sennequier (1991) Vertical distribution of non-methane hydrocarbons in the remote marine boundary layer. *J. Geophys. Res.* **91**, 7313–7324.

Borchers, R., P. Fabian, and S. A. Penkett (1983) First measurements of the vertical distribution of CCl_4 and CH_3CCl_3 in the stratosphere. *Naturwissenschaften* **70**, 514–517.

Borucki, W. J., and W. L. Chameides (1984) Lightning: estimates of the rates of energy dissipation and nitrogen fixation. *Rev. Geophys. Space Phys.* **22**, 363–372.

Botha, C. F., J. Hahn, J. J. Pienaar, and R. van Eldik (1995) Kinetics and mechanism of the oxidation of sulfur(IV) by ozone in aqueous solutions. *Atmos. Environ.* **28**, 3207–3212.

Bott, A., and W. Zdunkowski (1987) Electromagnetic energy within dielectric spheres. *J. Opt. Soc. Am. A* **4**, 1361–1365.

Bottenheim, J. W., and A. J. Gallant (1989) PAN over the Arctic: observations during AGASP-2 in April 1986. *J. Atmos. Chem.* **9**, 301–306.

Bottenheim, J. W., and M. F. Shepherd (1995) C_2–C_6 hydrocarbon measurements at four rural locations in Canada. *Atmos. Environ.* **29**, 647–664.

Bottenheim, J. W., L. W. Barrie, E. Atlas, L. Heidt, H. Niki, R. A. Rasmussen, and P. B. Shepson (1990) Depletion of lower tropospheric ozone during Arctic spring: the polar sunrise experiment 1988. *J. Geophys. Res.* **95**, 18,555–18,568.

Bottenheim, J. W., A. Sirois, K. A. Brice, and A. J. Gallant (1994) Five years of continuous observations of PAN and ozone at a rural location in eastern Canada. *J. Geophys. Res.* **99**, 5333–5352.

Böttger, A., D. H. Ehhalt, and G. Gravenhorst (1978) Atmosphärische Kreisläufe von Stickoxiden und Ammoniak, Report no. 1558. Kernforschungsanlage Jülich, Germany.

Bouble, R. W., E. F. Darley, and E. A. Schuck (1969) Emissions from burning grass stubble and straw. *J. Air Pollut. Control Assoc.* **19**, 497–500.

Boudries, H., G. Toupance, and A. L. Dutot (1994) Seasonal variation of atmospheric non-methane hydrocarbons on the western coast of Brittany, France. *Atmos. Environ.* **28**, 1095–1112.

Boulaud, D., G. Madelaine, D. Vigla, and J. Bricard (1977) Experimental study on the nucleation of water vapor sulfuric acid binary system. *J. Chem. Phys.* **66**, 4854–4860.

Boultron, C. (1980) Respective influence of global pollution and volcanic eruptions on the past variations of the trace metal content of Antarctic snow since 1880. *J. Geophys. Res.* **85**, 7426–7432.

Bouwman, A. F., I. Fung, E. Matthews, and J. John (1993) Global analysis of the potential for N_2O production in natural soils. *Global Biogeochem. Cycles* **7**, 557–597.

Bouwman, A. F., K. W. Van der Hoek, and J. G. J. Olivier (1995) Uncertainties in the global source distribution of nitrous oxide. *J. Geophys. Res.* **100**, 2785–2800.

Bowden, W. B. (1986) Gaseous nitrogen emissions from undisturbed terrestrial ecosystems: an assessment of their impacts on local and global nitrogen budgets. *Biogeochemistry* **2**, 249–279.

Bowen, H. J. M. (1966) "Trace Elements in Biochemistry." Academic Press, London.

Boyce, S. D., and M. R. Hoffmann (1984) Kinetics and mechanism of the formation of hydroxymethane sulfonic acid at low pH. *J. Phys. Chem.* **88**, 4740–4746.

Boyer, C. M., J. R. Kelafant, V. A. Kuuskraa, K. C. Manger, and D. Kruger (1990) Methane emissions from coal mining: issues and opportunities for reduction. EPA-400/9-90/008, U.S. Environmental Protection Agency, Office of Air and Radiation, Washington, DC.

Brandt, C., and R. van Eldik (1995) Transition metal-catalyzed oxidation of sulfur(IV) oxides. Atmospheric-relevant processes and mechanisms. *Chem Rev.* **95**, 119–190.

Brandt, C., I. Fábián, and R. van Eldik (1994) Kinetics and mechanism of the iron(III)-catalyzed autoxidation of sulfur(IV) oxides in aqueous solution. Evidence for the redox cycling of iron in the presence of oxygen and modeling of the overall reaction. *Inorg. Chem.* **33**, 687–701.

Brandtjen, R., T. Klüpfel, D. Perner, and B. M. Knudsen (1994) Airborne measurements during the European Arctic Stratospheric Ozone Experiment: observation of OClO. *Geophys. Res. Lett.* **21**, 1363–1366.

Braslau, N., and J. V. Dave (1973), Effect of aerosols on the transfer of solar energy through realistic model atmospheres, Parts I and II. *J. Appl Meteorol.* **12**, 601–619.

Brasseur, G., and S. Solomon (1986) "Aeronomy of the Middle Atmosphere." 2nd Edition. Reidel, Dordrecht, Holland.

Brasseur, G., and C. Granier (1992) Mount Pinatubo aerosols, chlorofluorocarbons, and ozone depletion. *Science* **257**, 1239–1242.

Brasseur, G., and M. Nicolet (1973) Chemospheric processes of nitric oxide in the mesosphere and stratosphere. *Planet. Space Sci.* **21**, 939–961.

Brauers, T., U. Aschmutat, U. Brandenburger, H.-P. Dorn, M. Hausmann, M. Heßling, A. Hofzumahaus, F. Holland, C. Plass-Dülmer, and D. H. Ehhalt (1996) Intercomparison of tropospheric OH radical measurements by multiple folded long-path laser absorption and laser induced fluorescence. *Geophys. Res. Lett.* **18**, 2545–2548.

Bray, J. R. (1959) An analysis of the possible recent change in atmospheric carbon dioxide concentration. *Tellus* **11**, 220–230.

Breeding, R. J., H. B. Klonis, J. P. Lodge, J. B. Pate, D. C. Sheesley, T. R. Englert, and D. R. Sears (1976) Measurements of atmospheric pollutants in the St. Louis area. *Atmos. Environ.* **10**, 181–194.

Breitenbeck, G. A., A. M. Blackmer, and J. M. Bremner (1980) Effect of different nitrogen fertilizers on emissions of nitrous oxide from soil. *Geophys. Res. Lett.* **7**, 85–88.

Bremner, J. M., and A. M. Blackmer (1981) Terrestrial nitrification as a source of atmospheric nitrous oxide. In "Denitrification, Nitrification and Atmospheric Nitrous Oxide" (C. C. Delwiche, ed.), pp. 151–170, Wiley, New York.

Bremner, J. M., and C. G. Steele (1978) Role of microorganisms in the atmospheric sulfur cycle. *Adv. Microb. Ecol.* **2**, 155–201.

Bremner, J. M., S. G. Robbins, and A. M. Blackmer (1980) Seasonal variability in emission of nitrous oxide from soil. *Geophys. Res. Lett.* **7**, 641–644.

Bremner J. M., G. A. Breitenbeck, and A. M. Blackmer (1981) Effect of anhydrous ammonia fertilization on emission of nitrous oxide from soils. *J. Environ. Qual.* **10**, 77–80.

Brewer, A. W. (1949) Evidence for a world circulation provided by the measurements of helium and water vapor distribution in the stratosphere. *Q. J. R. Meteorol. Soc.* **75**, 351–363.

Brewer, A. W., and C. T. McElroy (1973) Nitrogen dioxide concentrations in the atmosphere. *Nature* **246**, 129–133.

Brewer, A. W., and J. R. Milford (1960) The Oxford-Kew Ozonesonde. *Proc. R. Soc. Lond. A* **256**, 470–495.

Bricard, J., F. Billard, and G. J. Madeleine (1968) Formation and evolution of nuclei of condensation that appear in air initially free of aerosols. *J. Geophys. Res.* **73**, 4487–4496.

Brice, K. A., S. A. Penkett, D. H. F. Atkins, F. J. Sandalls, D. J. Bamber, A. F. Tuck, and G. Vaughan (1984) Atmospheric measurements of peroxyacetyl nitrate (PAN) in rural, south-east England: seasonal variations, winter photochemistry, and long-range transport. *Atmos. Environ.* **18**, 2691–2702.

Brice, K. A., J. W. Bottenheim, K. G. Anlauf, and H. A. Wiebe (1988) Long-term measurements of atmospheric peroxyacetyl nitrate (PAN) at rural sites in Ontario and Nova Scotia; seasonal variations and long-range transport. *Tellus* **40B**, 408–425.

Briggs, J., and W. T. Roach (1963) Aircraft observations near jet streams. *Q. J. R. Meteorol. Soc.* **89**, 225–247.

Brimblecombe, P., and S. L. Clegg (1988) The solubility and behaviour of acid gases in the marine aerosol. *J. Atmos. Chem.* **7**, 1–18 (erratum: *J. Atmos. Chem.* **8**, 95).

Brimblecombe, P., and D. H. Stedman (1982) Evidence for a dramatic increase in the contribution of oxides of nitrogen to precipitation acidity. *Nature* **298**, 460–462.

Brinkmann, R. T. (1969) Dissociation of water vapor and evolution of oxygen in the terrestrial atmosphere. *J. Geophys. Res.* **74**, 5355–5368.

Brock, J. C., and R. T. Watson (1980) Laser flash photolysis of ozone: $O(^1D)$ quantum yield in the fall-off region 297–325 nm. *Chem Phys.* **46**, 477–484.

Broecker, W. S. (1963) Radioisotopes and large-scale ocean mixing. *In* "The Sea: Ideas and Observations on Progress in the Study of the Seas" (M. N. Hill, ed.), pp. 88–108. Wiley Interscience, New York.

Broecker, W. S. (1970) A boundary condition on the evolution of atmospheric oxygen. *J. Geophys. Res.* **75**, 3553–3557.

Broecker, W. S. (1971) A kinetic model for the chemical composition of the sea water. *Q. Res. (N.Y.)* **1**, 188–207.

Broecker, W. S. (1973) Factors controlling the CO_2 content in the oceans and atmosphere. *In* "Carbon and the Biosphere" (G. M. Woodwell and E. V. Pecan, eds.). AEC Symposium Series, Vol. 30, pp. 32–50. NTIS U.S. Department of Commerce, Springfield, VA.

Broecker, W. S. (1979) A revised estimate of the radiocarbon age of North Atlantic deep water. *J. Geophys. Res.* **84**, 3218–3226.

Broecker, W. S., and T. H. Peng (1971) The vertical distribution of radon in the Bomex area. *Earth Planet. Sci. Lett.* **11**, 99–108.

Broecker, W. S., and T. H. Peng (1974) Gas exchange rates between air and sea. *Tellus* **20**, 21–35.

Broecker, W. S., and T. Takahashi (1978) The relationship between lysocline depth and in-situ carbonate concentration. *Deep Sea Res.* **25**, 65–95.

Broll, A., G. Helas, K. J. Rumpel, and P. Warneck (1984) NO_x background mixing ratios in surface air over Europe and the Atlantic Ocean. *In* "Physico-chemical Behaviour of Atmospheric Pollutants" (B. Versino and G. Angeletti, eds.). Proceedings of the 3rd European Symposium, Varese, Italy, pp. 390–400. Reidel, Dordrecht, the Netherlands.

Brosset, C. (1978) Water-soluble sulphur compounds in aerosols. *Atmos. Environ.* **12**, 25–38.

Browell, E. V., E. F. Danielsen, S. Ismail, G. L. Gregory, and S. M. Bleck (1987) Tropopause fold structure determined from airborne lidar and in situ measurements. *J. Geophys. Res.* **92**, 2112–2120.

Brown, A. P., and K. P. Davis (1973) "Forest Fires, Control and Use." McGraw-Hill, New York.

Brown, H. (1949) Rare gases and the formation of the earth's atmosphere. *In* "The Atmospheres of the Earth and Planets" (G. P. Kuiper, ed.), pp. 258–266. University of Chicago Press, Chicago.

Brown, K. A., and J. N. B. Bell (1986) Vegetation—the missing sink in the global cycle of COS. *Atmos. Environ.* **20**, 537–540.

Brown, S., and A. E. Lugo (1984) Biomass of tropical forests: a new estimate based on forest volumes. *Science* **223**, 1290–1293.

Brueckner, G. E., J. D. F. Bartoc, O. K. Moe, and M. E. van Hoosier (1976) Absolute solar intensities and their variations with solar activity in the wavelength region 1750–2100 Å. *Astrophys. J.* **209**, 935–944.

Brühl, C., and P. J. Crutzen (1989) On the disproportionate role of tropospheric ozone as a filter against solar U.V. radiation. *Geophys. Res. Lett.* **16**, 703–706.

Brühl, C., S. R. Drayson, J. M. Russell, III, P. J. Crutzen, J. M. McInerney, P. N. Purcell, H. Claude, H. Gernandt, T. J. McGee, I. S. McDermid, and M. R. Gunson (1996) Halogen Occultation Experiment ozone channel validation. *J. Geophys. Res.* **101**, 10,217–10,240.

Brunke, E.-G., H. E. Scheel, and W. Seiler (1990) Trends of tropospheric CO, N_2O and CH_4, as observed at Cape Point, South Africa. *Atmos. Environ.* **24A**, 585–595.

Bruyevich, S. V., and E. Z. Kulik (1967) Chemical interaction between atmosphere and the ocean (salt exchange). *Oceanology (Engl. Transl.)* **7**, 279–293.

Buat-Menard, P., and M. Arnold (1978) The heavy metal chemistry of atmospheric particulate matter emitted by Mount Etna volcano. *Geophys. Res. Lett.* **5**, 245–248.

Buat-Menard, P., J. Morelli, and R. Chesselet (1974) Water-soluble elements in atmospheric particulate matter over the tropical and equatorial Atlantic. *J. Rech. Atmos.* **8**, 661–673.

Buch, K. (1920) Ammoniakstudien an Meer- und Hafenwasserproben. *Kemistsamfundets Meddelanden* **1920**, 59–60.

Bufalini, J. J., B. M. Gay, Jr., and K. L. Brubaker (1972) Hydrogen peroxide formation from formaldehyde photo-oxidation and its presence in urban atmosphere. *Environ. Sci. Technol.* **6**, 816–821.

Buhr, M., F. C. Fehsenfeld, D. D. Parrish, R. E. Sievers, and J. M. Roberts (1990) Contribution of organic nitrates to the total odd-nitrogen budget at a rural eastern U.S. site. *J. Geophys. Res.* **95**, 9809–9816.

Buijs, H. L., G. L. Vail, G. Tremblay, and D. J. W. Kendall (1980) Simultaneous measurements of the volume mixing ratios of HCl and HF in the stratosphere. *Geophys. Res. Lett.* **7**, 205–208.

Buijsman, E., P. J. Jonker, W. A. H. Asman, and T. B. Ridder (1991) Chemical composition of precipitation collected on a weathership on the north Atlantic. *Atmos. Environ.* **25A**, 873–883.

Burford, J. R., and R. C. Stefanson (1973) Measurements of gaseous losses of nitrogen from soils. *Soil. Biol. Biochem.* **5**, 133–141.

Burgermeister, S. (1984) Messung von Bodenemissionen und atmosphärischen Konzentrationen des Dimethylsulfids. Diploma-Thesis Institut für Geologie und Geophysik, Universität Frankfurt (Main).

Bürgermeister, S., and H. W. Georgii (1991) Distribution of methanesulfonate, nss-sulfate and dimethylsulfide over the Atlantic and the North Sea. *Atmos. Environ.* **25A**, 587–595.

Bürgermeister, S., R. L. Zimmermann, H. W. Georgii, H. G. Bingemer, G. O. Kirst, M. Janssen, and W. Ernst (1990) On the biogenic origin of dimethylsulfide: relation between chlorophyll, ATP, organism DMSP, phytoplankton species and DMS distribution in Atlantic surface water and atmosphere. *J. Geophys. Res.* **95**, 20,607–20,615.

Burnett, W. E. (1969) Air pollution from animal wastes. *Environ. Sci. Technol.* **3**, 744–749.

Burns, R. C., and R. W. F. Hardy (1975) "Nitrogen Fixation in Bacteria and Higher Plants." Springer-Verlag, New York.

Burton, W. M., and N. G. Stewart (1960) Use of long-lived natural radioactivity as an atmospheric tracer. *Nature* **186**, 584–589.

Bush, Y. A., A. L. Schmeltekopf, F. C. Fehsenfeld, D. L. Albritton, J. R. McAfee, P. D. Goldan, and E. E. Ferguson (1978) Stratospheric measurements of methane at several latitudes. *Geophys. Res. Lett.* **5**, 1027–1029.

Butler, J. H. (1994) The potential role of the ocean in regulating atmospheric CH_3Br. *Geophys. Res. Lett.* **21**, 185–188.

Butler, J., J. Elkins, T. Thompson, B. Hall, T. Swanson, and V. Koropalav (1991) Oceanic consumption of CH_3CCl_3: implications for tropospheric OH. *J. Geophys. Res.* **96**, 22,347–22,355.

Butler, T. J., and G. E. Likens (1991) The impact of changing regional emissions on precipitation chemistry in the eastern United States. *Atmos. Environ.* **25A**, 305–315.

Butler, T. J., and G. E. Likens (1995) A direct comparison of throughfall plus stemflow to estimates of dry and total deposition for sulfur and nitrogen. *Atmos. Environ.* **29**, 1253–1265.

Buttery, R. G., L. C. Ling, and S. G. Wellso (1982) Oat leaf volatiles: possible insect attractants. *J. Agric. Food Chem.* **30**, 791–792.

Buttery, R. G., C. Xu, and L. C. Ling (1985) Volatile components of wheat leaves (and stems): possible insect attractants. *J. Agric. Food Chem.* **33**, 115–117.

Buxton, G. V., C. L. Greenstock, W. P. Hellman, and A. B. Ross (1988) Critical review of rate constants for reactions of hydrated electrons, hydrogen atoms and hydroxyl radicals in aqueous solution. *J. Phys. Chem. Ref. Data* **17**, 513–886.

Buxton, G. V., S. McGowan, G. A. Salmon, J. E. Williams, and N. D. Wood (1996a) A study of the spectra and reactivity of oxysulphur-radical anions involved in the chain oxidation of S(IV): a pulse and γ-radiolysis study. *Atmos. Environ.* **30**, 2483–2493.

Buxton, G. V., T. N. Malone, and G. A. Salmon (1996b) Pulse radiolysis study of the reaction of SO_5^- with HO_2. *J. Chem. Soc. Faraday Trans.* **92**, 1287–1289.

Cachier, H., C. Liousse, P. Buat-Menard, and A. Gaudichet (1995) Particulate content of savanna fire emissions. *J. Atmos. Chem.* **22**, 123–148.

Cadle, R. D. (1973) Particulate matter in the lower stratosphere. *In* "Chemistry of the Lower Atmosphere" (S. I. Rasool, ed.), pp. 69–120. Plenum Press, New York.

Cadle, R. D. (1975) Volcanic emissions of halides and sulfur compounds to the troposphere and stratosphere. *J. Geophys. Res.* **80**, 1650–1652.

Cadle, R. D. (1980) A comparison of volcanic with other fluxes of atmospheric trace gas constituents. *Rev. Geophys. Space Phys.* **18**, 746–752.

Cadle, R. D., and E. R. Allen (1970) Atmospheric photochemistry. *Science* **167**, 243–249.

Cadle, R. D., and G. Langer (1975) Stratospheric Aitken particles near the tropopause. *Geophys. Res. Lett.* **2**, 329–332.

Cadle, R. D., and E. J. Mroz (1978) Particles in the eruption cloud from St. Augustine volcano. *Science* **199**, 455–456.

Cadle, R. D., A. F. Wartburg, W. H. Pollock, B. W. Gandrud, and J. P. Shedlovsky (1973) Trace constituents emitted to the atmosphere by Hawaiian volcanoes. *Chemosphere* **6**, 231–234.

Cadle, R. D., F. G. Fernald, and C. L. Frush (1977) Combined use of lidar and numerical diffusion models to estimate the quantity and dispersion of volcanic eruption clouds in the stratosphere: Vulcan Fuego 1974 and Augustine 1976. *J. Geophys. Res.* **82**, 1783–1786.

Cadle, S. H. (1985) Seasonal variation in nitric acid, nitrate strong aerosol acidity, and ammonia in urban areas. *Atmos. Environ.* **19**, 181–188.

Cadle, S. H., and P. J. Groblicki (1982) An evaluation of methods for the determination of organic and elemental carbon in particulate samples. *In* "Particulate Carbon—Atmospheric Life Cycle" (G. T. Wolff and R. L. Klimisch, eds.), pp. 89–109. Plenum Press, New York.

Cadle, S. H., R. J. Countess, and N. A. Kelly (1982) Nitric acid and ammonia in urban and rural locations. *Atmos. Environ.* **16**, 2501–2506.

Callendar, G. S. (1938) The artificial production of carbon dioxide and its influence on temperature. *Q. J. R. Meteorol. Soc.* **64**, 223–237.

Callendar, G. S. (1940) Variations of the amount of carbon dioxide in different air currents. *Q. J. R. Meteorol. Soc.* **66**, 395–400.

Callendar, G. S. (1958) On the amount of carbon dioxide in the atmosphere. *Tellus* **10**, 243–248.

Calvert, J. G., and W. R. Stockwell (1984) Mechanisms and rates of the gas-phase oxidation of sulfur dioxide and nitrogen oxides in the atmosphere. *In* "SO_2, NO and NO_2 Oxidation Mechanisms: Atmospheric Considerations" (J. G. Calvert, ed.), Acid Precipitation Series, Vol. 3, pp. 1–62, Butterworth, Boston.

Calvert, J. G., J. A. Kerr, K. L. Demerjian, and R. D. McQuigg (1972) Photolysis of formaldehyde as a hydrogen atom source in the lower atmosphere. *Science* 175, 751–752.

Calvert, J. G., F. Su, J. W. Bottenheim, and O. P. Strausz (1978) Mechanism of the homogeneous oxidation of sulfur dioxide in the troposphere. *Atmos. Environ.* 12, 197–226.

Campbell, R. (1977) "Microbial Ecology." Halsted Press, New York.

Cantrell, B. K., and K. T. Whitby (1978) Aerosol size distribution and aerosol volume formation for a coal-fired power plant plume. *Atmos. Environ.* 12, 323–333.

Cantrell, C. A., D. H. Stedman, and G. J. Wendel (1984) Measurement of atmospheric peroxy radicals by chemical amplification. *Anal. Chem.* 56, 1496–1502.

Cantrell, C. A., J. A. Lind, R. E. Shetter, J. G. Calvert, P. D. Goldan, W. Kuster, F. C. Fehsenfeld, S. A. Montzka, D. D. Parrish, E. J. Williams, M. P. Buhr, H. H. Westberg, G. Allwine, and R. Martin (1992) Peroxy radicals in the ROSE experiment: measurement and theory. *J. Geophys. Res.* 97, 20,671–20,686.

Cantrell, C. A., R. E. Shetter, J. A. Lind, A. H. McDaniel, J. G. Calvert, D. D. Parrish F. C. Fehsenfeld, M. P. Buhr, and M. Trainer (1993) An improved chemical amplifier technique for peroxy radical measurements. *J. Geophys. Res.* 98, 2897–2909.

Carlier, P., H. Hannachi, and G. Mouvier (1986) The chemistry of carbonyl compounds in the atmosphere—a review. *Atmos. Environ.* 20, 2079–2099.

Carpenter, E. J., and K. Romans (1991) Major role of cyanobacterium *Trichodesmium* in nutrient cycling in the north Atlantic Ocean. *Science* 254, 1356–1358.

Carroll, M. A. (1985) Measurements of OCS and CS_2 in the free troposphere. *J. Geophys. Res.* 90, 10,483–10,486.

Carroll, M. A., R. W. Sanders, S. Solomon, and A. L. Schmeltekopf (1989) Visible and near-ultraviolet spectroscopy at McMurdo Station, Antarctica 6. Observations of BrO. *J. Geophys. Res.* 94, 16,633–16,638.

Carroll, M. A., D. R. Hastie, B. A. Ridley, M. O. Rodgers, A. L. Torres, D. D. Davis, J. D. Bradshaw, S. T. Sandholm, H. I. Schiff, D. R. Karecki, G. W. Harris, G. I. Mackay, G. L. Gregory, E. P. Condin, M. Trainer, G. Hübler, D. D. Montzka, S. Madronich, D. L. Albritton, H. B. Singh, S. M. Beck, M. C. Shipham, and A. S. Bachmeier (1990) Aircraft measurements of NO_x over the eastern Pacific and continental United States and implications for ozone production. *J. Geophys. Res.* 95, 10,205–10,233.

Carslaw, K. S., and T. Peter (1997) Uncertainties in reactive uptake coefficients for solid stratospheric particles. 1. Surface chemistry. *Geophys. Res. Lett.* 24, 1743–1746.

Carslaw, K. S., B. P. Luo, S. L. Clegg, T. Peter, P. Brimblecombe, and P. J. Crutzen (1994) Stratospheric aerosol growth and HNO_3 gas phase depletion from coupled HNO_3 and water uptake by liquid particles. *Geophys. Res. Lett.* 21, 2479–2482.

Carslaw, K. S., T. Peter, and S. L. Clegg (1997) Modeling the composition of liquid stratospheric aerosols. *Rev. Geophys.* 35, 125–154.

Carter, W. P. L., and R. Atkinson (1985) Atmospheric chemistry of alkanes. *J. Atmos. Chem.* 3, 377–405.

Carter, W. P. L., and R. Atkinson (1996) Development and evaluation of a detailed mechanism for the atmospheric reactions of isoprene and NO_x. *Int. J. Chem. Kinet.* 28, 497–530.

Carter, W. P. L., A. C. Lloyd, J. L. Sprung, and J. N. Pitts, Jr. (1979) Computer modeling of smog chamber data: progress in validation of a detailed mechanism for the photooxidation of propene and *n*-butane in photochemical smog. *Int. J. Chem. Kinet.* 11, 45–101.

Casadevall, T. J., D. A. Johnston, D. M. Harris, W. I. Rose, L. L. Malinconico, R. E. Stoiber, T. J. Bornhorst, S. N. Williams, L. Woodruff, and J. L. Thompson (1981) SO_2 emission rates at Mount St. Helens from March 29 through December, 1980. *U. S. Geol. Surv. Professional Papers* 1250, 193–200.

Castellano, E., and H. J. Schumacher (1962a) Die Kinetik des photochemischen Zerfalls von Ozon in rotgelbem Licht. *Z. Phys. Chem. N.F.* **34**, 198–212.

Castellano, E., and H. J. Schumacher (1962b) Photochemical decomposition of ozone yellow-red light and the mechanism of its thermal decomposition. *J. Chem. Phys.* **36**, 2238.

Castellano, E., and H. J. Schumacher (1972) The kinetics and the mechanism of photochemical decomposition of ozone with light of 3340 Å wavelength. *Chem. Phys. Lett.* **13**, 625–627.

Castleman, A. W., Jr., H. R. Munkelwitz, and B. Manowitz (1973) Contribution of volcanic sulphur compounds to the stratospheric aerosol layer. *Nature* **244**, 345–346.

Castleman, A. W., Jr., H. R. Munkelwitz, and B. Manowitz (1974) Isotope studies of the sulfur component of the stratospheric aerosol layer. *Tellus* **26**, 222–234.

Castro, M. S., and J. N. Galloway (1991) A conmparison of sulfur-free and ambient enclosure techniques for measuring the exchange of reduced sulfur gases between soils and the atmosphere. *J. Geophys. Res.* **96**, 15,427–15,437.

Cauer, H. (1935) Bestimmung des Gesamtoxidationswertes des Nitrits, des Ozons, und des Gesamtchlorgehaltes in roher und vergifteter Luft. *Z. Anal. Chem.* **103**, 321–324, 385–416.

Cauer, H (1951) Some problems of atmospheric chemistry. *In* "Compendium of Meteorology" (T. F. Malone, ed.), pp. 1126–1136. American Meteorological Society, Boston.

Cautreels, W. K., K. van Cauwenberghe, and L. A. Guzman (1977) Comparison between the organic fraction of suspended matter at background and an urban station. *Sci. Total Environ.* **8**, 79–88.

Chaen, M. (1973) Studies of the production of sea-salt particles on the sea surface. *Mem. Fac. Fish. Kagoshima Univ.* **22**, 49–107.

Chahine, M. T. (1992) The hydrological cycle and its influence on climate. *Nature* **359**, 373–380.

Chalk, P. M., and C. J. Smith (1983) Chemodenitrification. *Dev. Plant Soil Sci.* **9**, 65–89.

Challenger, F. (1951) Biological methylation. *Adv. Enzymol.* **12**, 429–491.

Chamberlain, A. C. (1953) "Aspects of Travel and Deposition of Aerosol and Vapor Clouds," AERE-HP/R 1261. Atomic Energy Research Establishment, Harwell, England.

Chamberlain, A. C. (1960) Aspects of the deposition of radioactive and other gases and particles. *Int. J. Air. Pollut.* **3**, 63–88.

Chamberlain, A. C. (1966a) Transport of gases to and from grass and grass-like surfaces. *Proc. R. Soc. A* **290**, 236–265.

Chamberlain, A. C. (1966b) Transport of *Lycopodium* spores and other small particles to rough surfaces. *Proc. R. Soc. A* **290**, 45–70.

Chameides, W. L. (1979) Effect of variable energy input on nitrogen fixation in instantaneous linear discharges. *Nature* **277**, 123–125.

Chameides, W. L., and D. D. Davis (1980) Iodine: its possible role in tropospheric chemistry. *J. Geophys. Res.* **85**, 7383–7398.

Chameides, W. L., and D. D. Davis (1982) The free radical chemistry of cloud droplets and its impact upon the composition of rain. *J. Geophys. Res.* **87**, 4863–4877.

Chameides, W. L., and A. W. Stelson (1992) Aqueous-phase chemical processes in deliquescent sea salt aerosols: A mechanism that couples the cycles of S and sea salt. *J. Geophys. Res.* **97**, 20,565–20,580.

Chameides, W. L., and A. Tan (1981) The two-dimensional diagnostic model for tropospheric OH: an uncertainty analysis. *J. Geophys. Res.* **86**, 5209–5223.

Chameides, W. L., and J. C. G. Walker (1973) A photochemical theory of tropospheric ozone. *J. Geophys. Res.* **78**, 8751–8760.

Chameides, W. L., and J. C. G. Walker (1976) A time-dependent photochemical model for ozone near the ground. *J. Geophys. Res.* **81**, 413–420.

Chameides, W. L., D. H. Stedman, R. R. Dickerson, D. W. Rush, and R. J. Cicerone (1977) NO_x production in lightning. *J. Atmos. Sci.* 34, 143–149.

Chance, K. V., D. G. Johnson, and W. A. Traub (1989) Measurement of stratospheric HOCl: concentration profiles, including diurnal variation. *J. Geophys. Res.* 94, 11,059–11,069.

Chang, T. Y., J. M. Norbeck, and B. Weinstock (1979) An estimate of the NO_x removal rate in an urban atmosphere. *Environ. Sci. Technol.* 13, 1534–1537.

Chapman, S. (1930) A theory of upper atmospheric ozone. *Q. J. R. Meteorol. Soc.* 3, 103–125.

Chappellaz, J., J. M. Barnola, D. Raynaud, Y. S. Korotkevich, C. Lorius. Ice record of atmospheric methane over the past 160,000 years. *Nature* 345, 127–131.

Charlson, R. J., J. E. Lovelock, M. O. Andreae, and S. G. Warren (1987) Oceanic phytoplankton, atmospheric sulfur, cloud albedo, and climate. *Nature* 326, 655–661.

Chatfield, R. B., and P. J. Crutzen (1984) Sulfur dioxide in remote oceanic air: cloud transport of reactive precursors. *J. Geophys. Res.* 89, 7111–7132.

Chatfield, R., and H. Harrison (1976) Ozone in the remote troposphere, mixing versus photochemistry. *J. Geophys. Res.* 81, 421–423.

Chatfield, R., and H. Harrison (1977a) Tropospheric ozone. 1. Evidence for higher background values. *J. Geophys. Res.* 82, 5965–5968.

Chatfield, R., and H. Harrison (1977b) Tropospheric ozone. 2. Variations along a meridional band. *J. Geophys. Res.* 82, 5969–5976.

Chebbi, A., and P. Carlier (1996) Carboxylic acids in the troposphere, occurrence, sources and sinks: a review. *Atmos. Environ.* 30, 4233–4249.

Chen, Z., L. Debo, K. Shao, and B. Wang (1993) Features of CH_4 emissions from rice paddy fields in Beijing and Nanjing, China. *Chemosphere* 26, 239–246.

Chepil, W. S. (1951) Properties of soils which influence wind erosion. IV. State of dry aggregate soil structure. *Soil Sci.* 72, 387–401.

Cheskis, S. G., and O. M. Sarkisov (1979) Flash photolysis of ammonia in the presence of oxygen. *Chem. Phys. Lett.* 62, 72–76.

Chesselet, R., J. Morelli, and P. Buat-Menard (1972) Variations in ionic ratios between reference sea water and marine aerosols. *J. Geophys. Res.* 77, 5116–5131.

Chesselet, R., M. Fontugné, P. Buat-Menard, V. Ezat, and C. E. Lambert (1981) The origin of particulate organic carbon in the marine atmosphere as indicated by its stable carbon isotope composition. *Geophys. Res. Lett.* 8, 345–348.

Chin, M., and D. D. Davis (1993) Global sources and sinks of OCS and CS_2 and their distribution. *Global Biogeochem. Cycles* 7, 321–337.

Chin, M., and D. D. Davis (1995) A reanalysis of carbonyl sulfide as a source of stratospheric background sulfur aerosol. *J. Geophys. Res.* 100, 8993–9005.

Chin, M., and D. N. Jacob (1996) Anthropogenic and natural contributions to tropospheric sulfate: a global model analysis. *J. Geophys. Res.* 101, 18,691–18,699.

Chin, M., and P. H. Wine (1994) A temperature-dependent competitive kinetics study of the aqueous phase reactions of OH radicals with formate, formic acid, acetate, acetic acid, and hydrated formaldehyde. *In* "Aquatic and Surface Photochemistry" (G. R. Heltz, R. G. Zepp, and D. G. Crosby, eds.), pp. 85–96. Lewis Publications, Boca Raton, FL.

Chin, M., J. D. Jacob, G. M. Gardner, M. S. Foreman-Fowler, and P. A. Spiro (1996) A global three-dimensional model of tropospheric sulfate. *J. Geophys. Res.* 101, 18,667–18,690.

Chou, C. C., J. G. Lo, and F. S. Rowland (1974) Primary processes in the photolysis of water vapor at 174 nm. *J. Chem. Phys.* 60, 1208–1210.

Choularton, T. W., H. Coe, S. Walton, M. W. Gallagher, K. M. Bewick, C. Dore, J. Duyzer, H. Westrate, K. Pilegaard, N. O. Jensen, and P. Hummelshøj (1994) Photochemical modification of ozone and NO_x deposition to forests: results from the Speuderbos and Rivox experiments. *In* "Transport and Transformation of Pollutants in the Troposphere" (P. M. Borrell, P.

Borrell, T. Cvitas, and W. Seiler, eds.). Proceedings of EUROTRAC Symposium '94, pp. 645–650. SPB Academic Publishing bv, Den Haag, the Netherlands.

Christensen, J. P., J. W. Murray, A. H. Devol, and L. A. Codispoti (1987) Denitrification in continental shelf sediments has major impact on oceanic nitrogen budget. *Global Biogeochem. Cycles* 1, 97–116.

Ciais, P., P. P. Tans, J. W. C. White, M. Trollier, R. J. Francey, J. A. Berry, D. R. Randall, P. J. Sellers, J. G. Collatz, and D. S. Schimel (1995) Partitioning of ocean and land uptake of CO_2 as inferred by $\delta^{13}C$ measurements from the NOAA Climate Monitoring and Diagnostics laboratory global air sampling network. *J. Geophys. Res.* 100, 5051–5070.

CIAP (1975) "The Natural Stratosphere of 1974," CIAP Monograph 1 (A. J. Grobecker, chief ed.), Department of Transportation Climatic Impact Assessment Program, Final Report, DOT-TST-75-51. National Technical Information Service, Springfield, VA.

Ciccioli, P., E. Brancaleoni, M. Possanzini, A. Brachetti, and C. Di Palo (1984) Sampling, identification, and quantitative determination of biogenic and anthropogenic hydrocarbons in forested areas. In "Physico-Chemical Behaviour of Atmospheric Pollutants" (B. Versino and G. Angeletti, eds.) Proceedings of the 3rd European Symposium, Varese, Italy, pp. 62–73. Reidel, Dordrecht, the Netherlands.

Ciccioli, P., R. Valentini, S. Cieslik, G. Seufert, and J. Kesselmeier (1997) Volatile arenes from biogenic sources. *In* "Biogenic Volatile Organic Compounds in the Atmosphere" (G. Helas, J. Slanina, and R. Steinbrecher, eds.), pp. 145–155. SPB Academic Publishing bv, Amsterdam, the Netherlands.

Cicerone, R. J. (1979) Atmospheric carbon tetrafluoride: a nearly inert gas. *Science* 206, 59–61.

Cicerone, R. J. (1989) Analysis of sources and sinks of atmospheric nitrous oxide (N_2O). *J. Geophys. Res.* 94, 18,265–18,271.

Cicerone, R. J. and J. L. McCrumb (1980) Photodissociation of isotopically heavy O_2 as a source of atmospheric O_3. *Geophys. Res. Lett.* 7, 251–254.

Cicerone, R. J., and R. Oremland (1988) Biogeochemical aspects of atmospheric methane *Global Biogeochem. Cycles* 2, 299–327.

Cicerone, R.J., J. D. Shetter, and C. C. Delwiche (1983) Seasonal variation of methane flux from a California rice paddy. *J. Geophys. Res.* 88, 11,022–11,024.

Cicerone, R. J., L. E. Heidt, and W. H. Pollock (1988) Measurements of atmospheric methyl bromide and bromoform. *J. Geophys. Res.* 93, 3745–3749.

Clark, F. E., and T. Rosswall, eds. (1981) "Terrestrial nitrogen cycles: processes, ecosystem, strategies and management impacts." *Ecol. Bull. (Stockholm)* 33, Swedish Natural Science Research Council.

Clark, J. H., C. B. Moore, and N. S. Nogar (1978) The photochemistry of formaldehyde: absolute quantum yields, radical reactions, and NO reactions. *J. Chem. Phys.* 68, 1264–1271.

Clark, P. A., G. P. Gervat, T. A. Hill, A. R. W. Marsh, A. S. Chandler, and T. W. Choularton (1990) A field study of the oxidation of SO_2 in cloud. *J. Geophys. Res.* 95, 13,985–13,995.

Clarke, A. G., and M. Radojevic (1987) Oxidation of SO_2 in rainwater and its role in acid rain chemistry. *Atmos. Environ.* 21, 1115–1123.

Cleveland, W. S., T. E. Graedel, and B. Kleiner (1977) Urban formaldehyde: observed correlation with source emissions and photochemistry. *Atmos. Environ.* 11, 357–360.

Cloud, P. E. (1965) Carbonate precipitation and dissolution in the marine environment. *In* "Chemical Oceanography" (J. P. Riley and G. Skirrow, eds.), Vol. 2, pp. 127–158. Academic Press, London.

Cloud, P. E. (1972) A working model of the primitive atmosphere. *Am. J. Sci.* 272, 537–548.

Cloud, P. E. (1973) Paleological significance for the banded iron formation. *Econ. Geol.* 68, 1135–1143.

Cloud, P. E. (1976) Beginnings of biospheric evolution and their biogeochemical consequences. *Paleobiology* 2, 351–387.

Clough, W. S. (1975) The deposition of particles on moss and grass surfaces. *Atmos. Environ.* **9**, 1113–1119.

Cofer, W. R., J. S. Levine, E. L. Winstead, and B. J. Stocks (1991a) Trace gas and particulate emissions from biomass burning in temperature ecosystems. In "Global Biomass Burning: Atmospheric, Climatic, and Biospheric Implications" (J. S. Levine, ed.), pp. 203–208. MIT Press, Cambridge, MA.

Cofer, W. R., J. S. Levine, E. L. Winstead, and B. J. Stocks (1991b) New estimates of nitrous oxide emissions from biomass burning. *Nature* **349**, 689–691.

Coffey, M. T., W. G. Mankin, and A. Goldman (1981) Simultaneous spectroscopic determination of the latitudinal, seasonal and diurnal variability of stratospheric N_2O, NO, NO_2, and HNO_3. *J. Geophys. Res.* **86**, 7331–7341.

Cogbill, C. V., and G. E. Likens (1974) Acid precipitation in the northeastern United States. *Water Resour. Res.* **10**, 1133–1137.

Cohen, N., and K. R. Westberg (1991) Chemical kinetic data sheets for high-temperature reactions. Part II. *J. Phys. Chem. Ref. Data* **20**, 1211–1311.

Cohen, Y., and L. I. Gordon (1979) Nitrous oxide production in the ocean. *J. Geophys. Res.* **84**, 347–353.

Colbeck, I., and R. M. Harrison (1985) The concentration of specific C_2–C_6 hydrocarbons in the air of northwest England. *Atmos. Environ.* **19**, 1899–1904.

Collet, J., R. Iovinelli, and B. Demoz (1995) A three-stage cloud impactor for size-resolved measurement of cloud droplet chemistry. *Atmos. Environ.* **29**, 1145–1154.

Conklin, M. H., and M. R. Hoffmann (1988) Metal ion-sulfur(IV) chemistry. 3. Thermodynamics and kinetics of transient iron(III)-sulfur(IV) complexes. *Environ. Sci. Technol.* **22**, 899–907.

Connell, P., and H. S. Johnston (1979) The thermal dissociation of N_2O_5 in N_2. *Geophys. Res. Lett.* **6**, 553–556.

Connick, R. E., Y.-X. Zhang, S. Lee, R Adamic, and P. Chieng (1995) Kinetics and mechanism of the oxidation of HSO_3^- by O_2. 1. The uncatalyzed reaction. *Inorg. Chem.* **34**, 4543–4553.

Connor, B. J., D. E. Siskind, J. J. Tsou, A. Parrish, and E E. Remsberg (1994) Ground-based microwave observation of ozone in the upper stratosphere and mesosphere. *J. Geophys. Res.* **99**, 16,757–16,770.

Conrad, R. (1988) Biogeochemistry and ecophysiology of atmospheric CO and H_2. *Adv. Microb. Biol.* **10**, 231–283.

Conrad, R. (1996a) Soil microorganisms as controllers of atmospheric trace gases (H_2, CO, CH_4, OCS, N_2O, and NO). *Microbiol. Rev.* **60**, 609–640.

Conrad, R. (1996b) Metabolism of nitric oxide in soil and soil microorganisms and regulation of flux into the atmosphere. In "Microbiology of Atmospheric Trace Gases: Sources, Sinks and Global Change Processes" (J. C. Murrell and D. P Kelly, eds.), pp. 167–203. Springer-Verlag, Berlin.

Conrad, R., and W. Seiler (1980a) Photooxidative production and microbial consumption of carbon monoxide in sea water. *FEMS Microbiol Lett.* **9**, 61–64.

Conrad, R., and W. Seiler (1980b) Role of microorganisms in the consumption and production of atmospheric carbon monoxide by soil. *Appl. Environ. Microbiol.* **40**, 437–445.

Conrad, R., and W. Seiler (1980c) Contribution of hydrogen production by biological nitrogen fixation to the global hydrogen budget. *J. Geophys. Res.* **85**, 5493–5498.

Conrad, R., and W. Seiler (1980d) Field measurements of the loss of fertilizer nitrogen into the atmosphere as nitrous oxide. *Atmos. Environ.* **14**, 555–558.

Conrad, R., and W. Seiler (1981) Decomposition of atmospheric hydrogen by soil microorganisms and soil enzymes. *Soil. Biol. Biochem.* **13**, 43–49.

Conrad, R., and W. Seiler (1982) Utilization of traces of carbon monoxide by aerobic oligotrophic microorganisms in ocean, lake and soil. *Arch. Microbiol.* **132**, 41–46.

Conrad, R., and W. Seiler (1985a) Influence of temperature, moisture and organic carbon on the flux of H_2 and CO between soil and atmosphere: field studies in subtropical regions. *J. Geophys. Res.* 90, 5699–5709.

Conrad, R., and W. Seiler (1985b) Characteristics of abiological carbon monoxide formation from soil organic matter, humic acids, and phenolic compounds. *Environ. Sci. Technol.* 19, 1165–1169.

Conrad, R., W. Seiler, and G. Bunse (1983) Factors influencing the loss of fertilizer nitrogen into the atmosphere as N_2O. *J. Geophys. Res.* 88, 6709–6718.

Conrad, R., W. Seiler, G. Bunse, and H. Giehl (1982) Carbon monoxide in sea water (Atlantic ocean). *J. Geophys. Res.* 87, 8839–8852.

Cooper, D. J., and E. S. Saltzman (1991) Measurements of atmospheric dimethyl sulfide and carbonyl disulfide in the western Atlantic boundary layer. *J. Atmos. Chem.* 12, 153–168.

Cooper, D. J., and E. S. Saltzman (1993) Measurements of atmospheric dimethylsulfide, hydrogen sulfide, and carbon disulfide during GTE/CITE 3. *J. Geophys. Res.* 98, 23,397–23,409.

Cooper, D. J., W. Z. de Mello, W. J. Cooper, R. G. Zika, E. S. Saltzman, J. M. Prospero, and D. L. Savoie (1987a) Short-term variability in biogenic sulfur emissions from a Florida *Spartina alterniflora* marsh. *Atmos. Environ.* 21, 7–12.

Cooper, W. J., D. J. Cooper, E. S. Saltzman, W. Z. de Mello, D. L. Savoie, R. G. Zika, and J. M. Prospero (1987b) Emissions of biogenic sulfur compounds from several wetland soils in Florida. *Atmos. Environ.* 21, 1491–1495.

Cooper, D. J., W. J. Cooper, W. Z. de Mello, E. S. Saltzman, and R. G. Zika (1989) Variability in biogenic sulfur emissions in Florida wetlands. *In* "Biogenic Sulfur in the Environment" (E. S. Saltzman and W. J. Cooper, eds.). ACS Symposium Series vol. 393, pp. 31–43. American Chemical Society, Washington, DC.

Covert, D. S. (1988) North Pacific marine background aerosol: average ammonium to sulfate ratio equals 1. *J. Geophys. Res.* 93, 8455–8458.

Covert, D. S., V. N. Kapustin, P. K. Quinn, and T. S. Bates (1992) New particle formation in the marine boundary layer. *J. Geophys. Res.* 97, 20,581–20,589.

Cowan, M. I., A. T. Glen, S. A. Hutchinson, M. E. MacCartney, J. M. Mackintosh, and A. M. Moss (1973) Production of volatile metabolites by the species of *fomes*. *Trans. Br. Mycol. Soc.* 60, 347–360.

Cox, R. A. (1972) Quantum yields for the photooxidation of sulfur dioxide in the first allowed absorption region. *J. Phys. Chem.* 76, 814–820.

Cox, R. A. (1973) Some experimental observations of aerosol formation in the photooxidation of sulphur dioxide. *J. Aerosol Sci.* 4, 473–483.

Cox, R. A. (1974) Photolysis of gaseous nitrous acid. *J. Photochem.* 3, 175–188.

Cox, R. A., and R. G. Derwent (1976) The ultraviolet absorption spectrum of gaseous nitrous acid. *J. Photochem.* 6, 23–34.

Cox, R. A., and G. D. Hayman (1988) The stability and photochemistry of dimers of the ClO radical and implications for Antarctic ozone depletion. *Nature* 332, 796–800.

Cox, R. A., and S. A. Penkett (1972) Aerosol formation from sulphur dioxide in the presence of ozone and olefinic hydrocarbons. *J. Chem. Soc. Faraday Trans. 1* 68, 1735–1753.

Cox, R. A., and M. J. Roffey (1977) Thermal decomposition of peroxyacetyl nitrate in the presence of nitric oxide. *Environ. Sci. Technol.* 11, 900–906.

Crafford, T. C. (1975) SO_2 emission of the 1974 eruption of volcano Fuego, Guatemala. *Bull. Volcanol.* 39, 536–556.

Craig, H. (1957) The natural distribution of radiocarbon and the exchange time of carbon dioxide between atmosphere and sea. *Tellus* 9, 1–17.

Craig, H., and C. C. Chou (1982) Methane: the record in polar ice cores. *Geophys. Res. Lett.* 9, 1221–1224.

Craig, R. A. (1950) The observations and photochemistry of atmospheric ozone and their meteorological significance. Meteorological Monographs I/2. American Meteorological Society, Boston.

Crank, J. (1975) "The Mathematics of Diffusion," 2nd ed. Oxford University Press, London.

Criegee, R. (1957) The course of ozonization of unsaturated compounds. *Rec. Chem. Prog.* 18, 111–120.

Criegee, R. (1962) Peroxide pathways in ozone reactions. *In* "Peroxide Reaction Mechanisms" (J. O. Edwards, ed.), pp. 29–39, Wiley Interscience, New York.

Criegee, R. (1975) Mechanisms of ozonolysis. *Angew. Chem. Int. Ed. Engl.* 14, 745–751.

Cronn, D. R., R. J. Charlson, R. L. Knights, A. L. Crittenden, and B. R. Appel (1977) A survey of the molecular nature of primary and secondary components of particles in urban air by high resolution mass spectroscopy. *Atmos. Environ.* 11, 929–937.

Croteau, R. (1987) Biosynthesis and catabolism of monoterpenoids. *Chem. Rev.* 87, 929–954.

Crutzen, P. J. (1970) The influence of nitrogen oxides on the atmospheric ozone content. *Q. J. R. Meteorol. Soc.* 96, 320–325.

Crutzen, P. J. (1971) Ozone production rates in an oxygen-hydrogen-nitrogen oxide atmosphere. *J. Geophys. Res.* 76, 7311–7327.

Crutzen, P. J. (1973) A discussion of the chemistry of some minor constituents in the stratosphere and troposphere. *Pure Appl. Geophys.* 106–108, 1385–1399.

Crutzen, P. J. (1974) photochemical reactions initiated by and influencing ozone in unpolluted tropospheric air. *Tellus* 26, 47–57.

Crutzen, P. J. (1976a) The possible importance of CSO for the sulfate layer of the stratosphere. *Geophys. Res. Lett.* 3, 73–76.

Crutzen, P. J. (1976b) Upper limits on atmospheric ozone reductions following increased applications of fixed nitrogen to the soil. *Geophys. Res. Lett.* 3, 169–172.

Crutzen, P. J. (1982) The global distribution of hydroxyl. *In* "Dahlem Konferenzen: Atmospheric Chemistry" (E. D. Goldberg, ed.), pp. 313–328. Springer-Verlag, Berlin.

Crutzen, P. J. (1995) Ozone in the troposphere. *In* "Composition, Chemistry, and Climate of the Atmosphere" (H. B. Singh, ed.), pp. 349–393. Van Nostrand-Reinhold, New York.

Crutzen, P. J., and M. O. Andreae (1990) Biomass burning in the tropics: impact on atmospheric chemistry and biogeochemical cycles. *Science* 250, 1669–1678.

Crutzen, P. J., and F. Arnold (1986) Nitric acid cloud formation in the cold Antarctic statosphere: a major cause for the springtime "ozone hole." *Nature* 324, 651–655.

Crutzen, P. J., and J. Fishman (1977) Average concentrations of OH in the troposphere and the budgets of CH_4, CO, H_2, and CH_3CCl_3. *Geophys. Res. Lett.* 4, 321–324.

Crutzen, P. J., and L. T. Gidel (1983) A two-dimensional photochemical model of the atmosphere. 2. The tropospheric budgets of the anthropogenic chlorocarbons, CO, CH_4, CH_3Cl and the effect of various NO_x sources on tropospheric ozone. *J. Geophys. Res.* 88, 6641–6661.

Crutzen, P. J., and U. Schmailzl (1983) Chemical budgets of the stratosphere. *Planet. Space Sci.* 31, 1009–1032.

Crutzen, P. J., and P. H. Zimmermann (1991) The changing photochemistry of the troposphere. *Tellus* 43AB, 136–151.

Crutzen, P. J., S. A. Isaksen, and G. C. Reid (1975) Solar proton events: stratospheric sources of nitric oxide. *Science* 189, 457–459.

Crutzen, P. J., L. E. Heidt, J. P. Krasnec, W. H. Pollock, and W. Seiler (1979) Biomass burning as a source of atmospheric gases. CO, H_2, N_2O, NO, CH_3Cl and COS. *Nature* 282, 253–256.

Crutzen, P. J., I. Aselmann, and W. Seiler (1986) Methane production by domestic animals, wild ruminants, and other herbivorous fauna, and humans. *Tellus* **38B**, 271–284.

Crutzen, P. J., J.-U. Grooß, C. Brühl, R. Müller, and J. M. Russell, III (1995) A reevaluation of the ozone budget with HALOE UARS data: no evidence for the ozone deficit. *Science* **268**, 705–708.

Cullis, C. F., and M. M. Hirschler (1980) Atmospheric sulfur: natural and man-made sources. *Atmos. Environ.* **14**, 1263–1278.

Cunningham, W. C., and W. H. Zoller (1981) The chemical composition of remote area aerosols. *J. Aerosol Sci.* **12**, 367–384.

Cunnold, D. M., F. N. Alyea, and R. G. Prinn (1978) A methodology for determining the atmospheric life time of fluorocarbons. *J. Geophys. Res.* **83**, 5493–5500.

Cunnold, D. M., R. G. Prinn, R. A. Rasmussen, P. G. Simmonds, F. N. Alyea, C. A. Cardelino, A. J. Crawford, P. J. Fraser, and R. D. Rosen (1983a) The atmospheric life time experiment. 3. Life time methodology and application to three years of CFCl$_3$ data. *J. Geophys. Res.* **88**, 8379–8400.

Cunnold, D. M., R. G. Prinn, R. A. Rasmussen, P. G. Simmonds, F. N. Alyea, C. A. Cardelino, and A. J. Crawford (1983b) The atmospheric life time experiment. 4. Results for CF$_2$Cl$_2$ based on three years of data. *J. Geophys. Res.* **88**, 8401–8414.

Cunnold, D. M., P. J. Fraser, R. Weiss, R. G. Prinn, P. G. Simmonds, F. N. Alyea, and A. J. Crawford (1992) Global trends and annual releases of CFCl$_3$ and CF$_2$Cl$_2$ estimated from ALE/GAGE and other measurements from July 1978 to June 1991. *J. Geophys. Res.* **99**, 1107–1126.

Cure, J. D., and B. Acock (1986) Crop responses to carbon dioxide doubling: a literature survey. *Agric. For. Meteorol.* **8**, 127–145.

Cvetanovic, R. J. (1965) Excited oxygen atoms in the photolysis of N$_2$O and NO$_2$. *J. Chem. Phys.* **43**, 1850–1851.

Czeplak, G., and C. Junge (1974) Studies of interhemispheric exchange in the troposphere by a diffusion model. *Adv. Geophys.* **18B**, 57–72.

Daisey, J. M. (1980) Organic compounds in urban aerosols. *Ann. N.Y. Acad. Sci.* **338**, 50–69.

d'Almeida, G. A. (1986) A model for Saharan dust transport. *J. Clim. Appl. Meteorol.* **25**, 903–916.

d'Almeida, G., and R. Jaenicke (1981) The size distribution of mineral dust. *J. Aerosol Sci.* **12**, 160–162.

d'Almeida, G. A., and L. Schütz (1983) Number, mass and volume distribution of mineral aerosol and soils of the Sahara. *J. Clim. Appl. Meteorol.* **22**, 233–243.

Dalrymple, G. B. (1991) "The Age of the Earth." Stanford University Press, Stanford, CA.

Damschen, D. E., and L. R. Martin (1983) Aqueous phase oxidation of nitrous acid by O$_2$, O$_3$ and H$_2$O$_2$. *Atmos. Environ.* **17**, 2005–2011.

Danckwerts, P. V. (1970) "Gas-Liquid Reactions." McGraw-Hill, New York.

Danielsen, E. F. (1968) Stratospheric-tropospheric exchanged based on radioactivity, ozone and potential vorticity. *J. Atmos. Sci.* **25**, 502–518.

Danielsen, E. F. (1975) The nature of transport processes in the stratosphere. In "The Natural Stratosphere of 1974" (A. J. Grobecker, ed.), CIAP Monograph 1, pp. 6-12–6-22. U.S. Department of Transportation, Washington, DC.

Danielsen, E. F., and V. A. Mohnen (1977) Project Duststorm report: ozone transport, *in situ* measurements and meteorological analysis of tropopause folding. *J. Geophys. Res.* **82**, 5867–5877.

Danielsen, E. F., R. Bleck, J. Shedlovsky, A. Wartburg, P. Haagenson, and W. Pollock (1970) Observed distribution of radioactivity, ozone and potential vorticity associated with tropopause folding. *J. Geophys. Res.* **75**, 2353–2361.

Darley, E. F., F. R. Burleson, E. H. Mateer, J. T. Middleton, and V. P. Osterli (1966) Contribution of burning of agricultural wastes to photochemical air pollution. *J. Air Pollut. Control Assoc.* 11, 685–690.

Darnall, K. R., A. C. Lloyd, A. M. Winer, and J. N. Pitts, Jr. (1976) Reactivity scale for atmospheric hydrocarbons based on reaction with hydropxyl radicals. *Environ. Sci. Technol.* 10, 692–696.

Daum, P. H., T. J. Kelly, S. E. Schwartz, and L. Newman (1984) Measurements of the chemical composition of stratiform clouds. *Atmos. Environ.* 18, 2671–2684.

Davidson, B., J. P. Friend, and H. Seitz (1966) Numerical models of diffusion and rainout of stratospheric radioactive materials. *Tellus* 18, 301–315.

Davidson, E. A., P. M. Vitousek, P. A. Matson, R. Riley, G. Garcia-Mendez, and J. M. Maas (1991) Soil emission of nitric oxide in a seasonally dry tropical forest of Mexico. *J. Geophys. Res.* 96, 15,439–15,445.

Davidson, E. A., P. A. Matson, P. M. Vitousek, R. Riley, K. Dunkin, G. Garcia-Mendez, and J. M. Maass (1993) Processes regulating soil emissions of NO and N_2O in seasonally dry tropical forest. *Ecology* 74, 130–139.

Davies, M., and G. H. Thomas (1960) The lattice energy, infrared spectra and possible crystallization of some dicarboxylic acids. *Trans. Faraday Soc.* 56, 185–192.

Davis, D. D., W. S. Heaps and T. McGee (1976) Direct measurements of natural tropospheric levels of OH via an aircraft-borne tunable dye laser. *Geophys. Res. Lett.* 3, 331–333.

Davis, D. D., M. O. Rogers, S. D. Fisher, and K. Asai (1981) An experimental assessment of the O_3/H_2O interference problem in the detection of OH via LIF. *Geophys. Res. Lett.* 8, 69–72.

Davis, D. D., J. D. Bradshaw, M. O. Rodgers, S. T. Sandholm, and S. KeSheng (1987) Free tropospheric and boundary layer measurements of NO over the central and eastern north Pacific Ocean. *J. Geophys. Res.* 92, 2049–2070.

Dawson, G. A. (1977) Atmospheric ammonia from undisturbed land. *J. Geophys. Res.* 82, 3125–3133.

Dawson, G. A. (1980) Nitrogen fixation by lightning. *J. Atmos. Sci.* 37, 174–178.

Dawson, G. A., and J. C. Farmer (1988) Soluble atmospheric trace gases in the southwest United States. 2. Organic species HCHO, HCOOH, CH_3COOH. *J. Geophys. Res.* 93, 5200–5206.

Dawson, G. A., J. C. Farmer, and J. L. Moyers (1980) Formic and acetic acids in the atmosphere. *Geophys. Res. Lett.* 7, 725–728.

Day, J. A. (1964) Production of droplets and salt nuclei by bursting of air bubble films. *Q. J. R. Meteorol. Soc.* 90, 72–78.

Dayan, U., and D. W. Nelson (1988) Origin and composition of Samoan precipitation. *Tellus* 40B, 148–153.

De Bary, E., and C. E. Junge (1963) Distribution of sulfur and chlorine over Europe. *Tellus* 15, 370–381.

De Bary, E., and F. Möller (1960) Die mittlere vertikale Verteilung von Wolken in Abhängigkeit von der Wetterlage. *Ber. Dtsch. Wetterdienst* 9 (67).

De Bruyn, W. J., E. Swartz, J. H. Hu, J. A. Shorter, P. Davidovits, D. R. Worsnop, M. S. Zahniser, and C. E. Kolb (1995) Henry's law solubilities and Setchenow coefficients for biogenic reduced sulfur species obtained from gas-liquid uptake measurements. *J. Geophys. Res.* 100, 7245–7251.

De Leeuw, G. (1986) Vertical profiles of giant particles close above the sea surface. *Tellus* 38B, 51–61.

De Mello, W. Z., and M. E. Hines (1994) Application of static and dynamic enclosures for determining dimethyl sulfide and carbonyl sulfide exchange in spagnum peatlands; implications for magnitude and direction of flux. *J. Geophys. Res.* 99, 14,601–14,607.

De Vooys, G. G. N. (1979) Primary production in aquatic environments. In "The Global Carbon Cycle" (B. Bolin, E. T. Degens, S. Kempe, and P. Ketner, eds.). *SCOPE* 13, 259–292.

De Zafra, R. L., A. Parrish, P. M. Solomon, and J. W. Barrett (1984) A measurement of stratospheric HO_2 by ground-based millimeter-wave spectroscopy. *J. Geophys. Res.* 89, 1321–1326.

De Zafra, R. L., M. Jaramillo, A. Parrish, P. Solomon, R. Connor, and J. Barrett (1987) High concentrations of chlorine monoxide at low altitudes in the Antarctic spring stratosphere. *Nature* 328, 408–411.

Deister, U., and P. Warneck (1990) Photodissociation of SO_3^{2-} in aqueous solution. *J. Phys. Chem.* 90, 3213–3217.

Deister, U., R. Neeb, G. Helas, and P. Warneck (1986) Temperature dependence of the equilibrium $CH_2(OH)_2 + HSO_3^- = CH_2(OH)SO_3^- + H_2O$ in aqueous solution. *J. Phys. Chem.* 90, 3213–3217.

Delany, A., A. C. Delany, D. W. Parkin, J. J. Griffin, E. D. Goldberg, and B. E. Reimann (1967) Airborne dust collected at Barbados. *Geochim. Cosmochim. Acta* 31, 885–909.

Delany, A. C., W. H. Pollock, and J. P. Shedlowsky (1973) Tropospheric aerosol: the relative contribution of marine and continental components. *J. Geophys. Res.* 78, 6249–6265.

Delmas, R., and J. Servant (1982) The origin of sulfur compounds in the atmosphere of a zone of high productivity (Gulf of Guinea). *J. Geophys. Res.* 85, 7410–7416.

Delmas, R., J. Baudet, J. Servant, and Y. Baziard (1980) Emissions and concentrations of hydrogen sulfide in the air of the tropical forest of the Ivory Coast and of temperate regions in France. *J. Geophys. Res.* 85, 4468–4474.

Delmas, R., A. Marenco, J. P. Tathy, B. Cros, and J. G. R. Baudet (1991) Sources and sinks of methane in the African savanna. CH_4 emissions from biomass burning. *J. Geophys. Res.* 96, 7287–7299.

Delwiche, C. C. (1970) The nitrogen cycle. *Sci. Am.* 223, 137–146.

DeMore, W. B., S. P. Sander, D. M. Golden, R. F. Hampson, M. J. Kurylo, C. J. Howard, A. R. Ravishankara, C. E. Kolb, and M. J. Molina (1997) "Chemical Kinetics and Photochemical data for Use in Stratospheric Modeling, Evaluation No. 12," JPL 97–4. National Aeronautics and Space Administration, Jet Propulsion Laboratory, Pasadena, CA.

Denmead, O. T., J. R. Freney, and J. R. Simpson (1976) A closed ammonia cycle within a plant canopy. *Soil Biol. Biochem.* 8, 161–164.

Denmead, O. T., J. R. Simpson, and J. R. Freney (1974) Ammonia fluxes into the atmosphere from a grazed pasture. *Science* 185, 609–610.

Dentener, F. J., and P. J. Crutzen (1994) A three-dimensional model of the global ammonia cycle. *J. Atmos. Chem.* 19, 331–369.

Dentener, F. J., G. R. Carmichael, Y. Zhang, J. Lelieveld, and P. J. Crutzen (1996) Role of mineral aerosols as a reactive surface in the global troposphere. *J. Geophys. Res.* 101, 22,860–22,889.

Department of the Environment (1983) "Digest of Environmental Protection and Water Statistics," No. 6. HMSO, London.

Derwent, R. G. (1982) On the comparison of global, hemispheric, one-dimensional model formulations of halocarbon oxidation by OH radicals in the troposphere. *Atmos. Environ.* 16, 551–561.

Detwiler, R. P., and C. S. Hall (1988) Tropical forests and the global carbon cycle. *Science* 239, 42–47.

Detwiler, C. R., J. D. Garret, J. D. Purcell, and R. Tousey (1961) The intensity distribution in the solar ultraviolet spectrum. *Ann. Geophys.* (*Paris*) 17, 263–272.

Devlin, R. M., and A. V. Barker (1971) "Photosynthesis." Van Nostrand, New York.

Dickerson, R. R., D. H. Stedman, W. L. Chameides, P. J. Crutzen, and J. Fishman (1979) Actinometric measurements and theoretical calculations of $j(O_3)$, the rate of photolysis of ozone to $O(^1D)$. *Geophys. Res. Lett.* **6**, 833–836.

Dickerson, R. R., D. H. Stedman, and A. C. Delany (1982) Direct measurement of ozone and nitrogen dioxide photolysis rates in the troposphere. *J. Geophys. Res.* **87**, 4933–4946.

Dickinson, R. E. (1986) How will climate change? The climate system and modelling of future climate. *In* "The Greenhouse Effect, Climatic Change, and Ecosystems" (B. Bolin, B. R. Döös, J. Jäger, and R. A. Warrick, eds.). *SCOPE* **29**, 207–270.

Didyk, B. M., B. R. T. Simoneit, S. C. Brassel, and G. Eglinton (1978) Organochemical indicators of paleo-environmental conditions of sedimentation. *Nature* **272**, 216–222.

Dignon, J. (1992) NO_x and SO_2 emissions from fossil fuels: a global distribution. *Atmos. Environ.* **14**, 1263–1278.

Dignon, J., and S. Hameed (1989) Global emissions of nitrogen and sulfur oxides from 1860 to 1980. *J. Air. Pollut. Control Assoc.* **39**, 180–186.

Dinger, J. E., H. B. Howell, and T. A. Wojciechowski (1970) On the source and composition of cloud nuclei in a subsident air mass over the north Atlantic. *J. Atmos. Sci.* **27**, 791–797.

Disteche, A. (1974) The effect of pressure on dissociation constants and its temperature dependence. *In* "The Sea" (E. D. Goldberg, ed.), Vol. 5, pp. 81–120. Wiley Interscience, New York.

Ditchburn, R. W., and P. Young (1962) The absorption of molecular oxygen between 1850 and 2500 Å. *J. Atmos. Terr. Phys.* **24**, 127–139.

Dixon, R. K., S. Brown, R. A. Houghton, A. M. Solomon, M. C. Trexler, and J. Wisniewski (1994) Carbon pools and flux of global forest ecosystems. *Science* **263**, 185–190.

Dlugi, R., and S. Jordan (1982) Heterogeneous SO_2 oxidation: its contribution to cloud condensation nuclei formation. *Időjárás* (*J. Hung. Meteorol. Serv.*) **86**, 82–88.

Dobson, G. M. B. (1973) The laminated structure of the ozone in the atmosphere. *Q. J. R. Meteorol. Soc.* **99**, 599–607.

Dobson, G. M. B., and D. N. Harrison (1926) Measurements of the amount of ozone in the earth's atmosphere and its relation to other geophysical conditions. Part 1. *Proc. R. Soc. Lond.* **110**, 660–693.

Donahue, N. M., and R. G. Prinn (1993) In situ nonmethane hydrocarbon measurements on SAGA 3. *J. Geophys. Res.* **98**, 16,915–16,932.

Donoso, L. R., R. Santana, and E. Sanhueza (1993) Seasonal variation of N_2O fluxes at a tropical savanna site: soil consumption of N_2O during the dry season. *Geophys. Res. Lett.* **20**, 1379–1382.

Dorn, H.-P., U. Brandenburger, T. Brauers, M. Hausmann, and D. H. Ehhalt (1996) In-situ detection of tropospheric OH radicals by folded long-path laser absorption. Results from the POPCORN field campaign in August 1994. *Geophys. Res. Lett.* **18**, 2537–2540.

Doyle, G. J. (1961) Self-nucleation in the sulfuric acid water system. *J. Chem. Phys.* **35**, 795–799.

Drapcho, D. L., D. Sisterson, and R. Kumar (1983) Nitrogen fixation by lightning activity in a thunderstorm. *Atmos. Environ.* **17**, 729–734.

Drever, J. I. (1974) The magnesium problem. *In* "The Sea" (E. D. Goldberg, ed.), Vol. 5, 337–357. Wiley Interscience, New York.

Drexler, C. H. Elias, B. Fecher, and K. J. Wannowius (1991) Kinetic investigation of sulfur(IV) oxidation by peroxo-compounds R-OOH in aqueous solution. *Fresenius J. Anal. Chem.* **340**, 605–615.

Driscoll, J. N., and P. Warneck (1968) Primary processes in the photolysis of SO_2 at 1849 Å. *J. Phys. Chem.* **72**, 3736–3740.

Drummond, J. W., and R. F. Jarnot (1978) Infrared measurements of stratospheric composition II. Simultaneous NO and NO_2 measurements. *Proc. R. Soc. Lond. A* **364**, 237–254.

Drummond, J. W., D. H. Ehhalt, and A. Volz (1988) Measurements of nitric oxide between 0–12 km altitude and 67°N–60°S latitude obtained during STRATOZ III. *J. Geophys. Res.* **93**, 15,831–15,849.

Duce, R. A. (1978) Speculations on the budget of particulate and vapor phase non-methane organic carbon in the global troposphere. *Pure Appl. Geophys.* **116**, 244–273.

Duce, R. A. (1986) The impact of atmospheric nitrogen, phosphorus, and iron species on marine biological productivity. *In* "The Role of Air-Sea Exchange in Geochemical Cycling" (Buat-Menard, ed.), pp. 497–529. Reidel, Norwell, MA.

Duce, R. A., and R. B. Gagosian (1982) The input of atmospheric n-C_{10} to n-C_{30} alkanes to the ocean. *J. Geophys. Res.* **87**, 7192–7200.

Duce, R. A., and E. J. Hoffman (1976) Chemical fractionation at the air/sea interface. *Annu. Rev. Earth. Planet. Sci.* **4**, 187–228.

Duce, R. A., J. W. Winchester, and T. W. van Nahl (1965) Iodine, bromine and chlorine in the Hawaiian marine atmosphere. *J. Geophys. Res.* **70**, 1775–1799.

Duce, R. A., G. L. Hoffman, B. J. Ray, I. S. Fletcher, G. T. Wallace, J. L. Fasching, S. R. Piotrowicz, P. R. Walsh, E. J. Hoffman, J. M. Miller, and J. L. Heffter (1976) Trace metals in the marine atmosphere: sources and fluxes. *In* "Marine Pollutant Transfer" (H. L. Windom and R. A. Duce, eds.), pp. 77–119. Heath, Lexington, MA.

Duce, R. A., V. A. Mohnen, P. R. Zimmerman, D. Grojean, W. Cautreels, R. Chatfield, R. Jaenicke, J. A. Ogren, E. D. Pellizzari, and G. T. Wallace (1983) Organic material in the global atmosphere. *Rev. Geophys. Space Phys.* **21**, 921–952.

Duce, R. A., P. S. Liss, J. T. Merril, E. L. Atlas, P. Buat-Menard, B. B. Hicks, J. M. Miller, J. M. Prospero, R. Arimoto, T. M. Church, W. Ellis, J. N. Galloway, L. Hansen, T. D. Jickills, A. H. Knap, K. H. Reinhardt, B. Schneider, A. Soudine, J. J. Tokos, S. Tsunogai, R. Wollast, and M. Zhou (1991) The atmospheric input of trace species to the world ocean. *Global Biogeochem. Cycles* **5**, 193–259.

Dufour, L., and R. Defay (1963) "Thermodynamics of Clouds." Academic Press, New York.

Dumdei, B. E., D. V. Kenny, P. B. Shepson, T. E. Kleindienst, C. M. Nero, L. T. Cupitt, and L. D. Claxton (1988) MS/MS analysis of the products of toluene photooxidation and measurement of their mutagenic activity. *Environ. Sci. Technol.* **22**, 1493–1498.

Dunlop, J. S. R., M. D. Muir, V. A. Milne, and D. I. Groves (1978) A new microfossil assemblage from the archaean of western Australia. *Nature* **274**, 676–678.

Durand, B., ed. (1980) "Kerogen, Insoluble Matter from Sedimentary Rocks." Editions Technip, Paris.

Durkee, P. A., F. Pfeil, E. Frost, and R. Shema (1991) Global analysis of aerosol particle characteristics. *Atmos. Environ.* **25A**, 2457–2471.

Dutkiewicz, V. A., and L. Husain (1979) Determination of stratospheric ozone at ground level using beryllium-7/ozone ratios. *Geophys. Res. Lett.* **6**, 171–174.

Dütsch, H. U. (1968) The photochemistry of stratospheric ozone. *Q. J. R. Meteorol. Soc.* **94**, 483–497.

Dütsch, H. U. (1971) Photochemistry of atmospheric ozone. *Adv. Geophys.* **15**, 219–322.

Dütsch, H. U. (1980) Ozon in der Atmosphäre: Gefährdet die stratosphärische Verschmutzung die Ozonschicht? Neujahrsblatt der Naturforschenden Gesellschaft in Zürich auf das Jahr 1980. *Vierteljahresschr.* **124**(5), 1–48.

Duursma, E. K. (1961) Dissolved organic carbon, nitrogen, and phosphorus in the sea. *Neth. J. Sea Res.* **1**, 1–147.

Duxbury, J. M., D. R. Bouldin, R. E. Terry, and R. L. Tate (1982) Emission of nitrous oxide from soils. *Nature* **298**, 462–464.

Dye, J. E., D. Baumgardner, B. W. Gandrud, S. R. Kawa, K. K. Kelly, M. Loewenstein, G. V. Ferry, K. R. Chan, and B. L. Gary (1992) Particle size distributions in Arctic polar stratospheric clouds, growth and freezing of sulfuric acid droplets, and implications for cloud formation. *J. Geophys. Res.* 97, 8015–8034.

Fagan, R. C., P. V. Hobbs, and L. F. Radke (1974) Measurements of cloud condensation nuclei and cloud droplet size distributions in the vicinity of forest fires. *J. Appl. Meteorol.* 13, 553–557.

Eaton, J. P., and K. J. Murata (1960) How volcanoes grow. *Science* 132, 925–938.

Ebel, A., H. Hass, H. J. Jakobs, M. Laube, M. Memmesheimer, A. Oberreuter, H. Geiss, and Y. H. Kuo (1991) Simulation of ozone intrusion caused by a tropopause fold and cut-off low. *Atmos. Environ.* 25A, 2131–2144.

Eberhard, J., C. Müller, D. W. Stocker, and J. A. Kerr (1995) Isomerization of alkoxy radicals under atmospheric conditions. *Environ. Sci. Technol.* 29, 232–241.

Edmond, J. M., and J. M. T. M. Gieskes (1970) On the calculation of the degree of saturation of sea water with respect to calcium carbonate under in-situ conditions. *Geochim. Cosmochim. Acta* 34, 1261–1291.

Egerton, A., G. J. Minkoff, and K. C. Salooja (1956) The slow oxidation of methane. *Proc. R. Soc. Lond. A* 235, 158–173.

Eggleton, A. E. J. (1969) The chemical composition of atmospheric aerosols on the Tees-Side and its relation to visibility. *Atmos. Environ.* 3, 355–372.

Eglinton, G., and M. Calvin (1967) Chemical fossils. *Sci. Am.* 216, 32–43.

Ehhalt, D. H. (1974) The atmospheric cycle of methane. *Tellus* 26, 58–70.

Ehhalt, D. H., and J. W. Drummond (1982) The tropospheric cycles of NO_x. In "Chemistry of the Unpolluted and Polluted Troposphere" (H. W. Georgii and W. Jaeschke, eds.), NATO ASI Series, Vol. C 96, pp. 219–251. Reidel, Dordrecht, the Netherlands.

Ehhalt, D. H., and L. E. Heidt (1973a) Vertical profiles of CH_4 in the troposphere and stratosphere. *J. Geophys. Res.* 78, 5265–5271.

Ehhalt, D. H., and L. E. Heidt (1973b) The concentration of molecular H_2 and CH_4 in the stratosphere. *Pure Appl. Geophys.* 106–108, 1352–1360.

Ehhalt, D. H., and U. Schmidt (1978) Sources and sinks of atmospheric methane. *Pure Appl. Geophys.* 116, 452–464.

Ehhalt, D. H., G. Israel, W. Roether, and W. Stich (1963) Tritium and deuterium content of atmospheric hydrogen. *J. Geophys. Res.* 68, 3747–3751.

Ehhalt, D. H., L. E. Heidt, R. H. Lueb, and W. Pollock (1975) The vertical distribution of trace gases in the stratosphere. *Pure Appl. Geophys.* 113, 389–402.

Ehhalt, H. D., U. Schmidt, and L. E. Heidt (1977) Vertical profiles of molecular hydrogen in the troposphere and stratosphere. *J. Geophys. Res.* 82, 5907–5910.

Ehhalt, D. H., J. Rudolph, and U. Schmidt (1986) On the importance of light hydrocarbons in multiphase atmospheric systems. In "Chemistry of Multiphase Atmospheric Systems" (W. Jaeschke, ed.), NATO ASI Series, Vol. G 6, pp. 321–350. Springer-Verlag, Berlin.

Ehhalt, D. H., F. Rohrer, and A. Wahner (1992) Sources and distribution of NO_x in the upper troposphere at northern midlatitudes. *J. Geophys. Res.* 97, 3725–3738.

Ehmert, A. (1949) Ein einfaches Verfahren zur Bestimmung kleinster Jodkonzentrationen, Jod- und Natriumthiosulfatmengen in Lösungen. *Z. Naturforsch.* 4b, 321–327.

Ehmert, A. (1951) Ein einfaches Verfahren zur absoluten Messung des Ozongehaltes der Luft. *Meteorol. Rundschau* 4, 64–68.

Eichel, C., M. Krämer, L. Schütz, and S. Wurzler (1996) The water-soluble fraction of atmospheric aerosol particles and the influence on cloud microphysics. *J. Geophys. Res.* 101, 29499–29510.

Eichmann, R., P. Neuling, G. Ketseridis, J. Hahn, R. Jaenicke, and C. Junge (1979) n-Alkane studies in the troposphere. I. Gas and particulate concentrations in north Atlantic air. *Atmos. Environ.* 13, 587–599.

Eichmann, R., G. Ketseridis, G. Schebeske, J. Hahn, R. Jaenicke, P. Warneck, and C. Junge (1980) n-Alkane studies in the troposphere. II. Gas and particulate concentrations in Indian Ocean air. *Atmos. Environ.* **14**, 695–703.

Eichner, M. J. (1990) Nitrous oxide emissions from fertilized soils: summary of available data. *J. Environ. Qual.* **19**, 272–280.

Eigen, M., W. Kruse, G. Maass, and L. De Maeyer (1964) Rate constants of protolytic reactions in aqueous solution. *Progr. Reaction Kinet.* **2**, 285–318.

Eisele, F. L., and D. J. Tanner (1991) Ion-assisted tropospheric OH measurement. *J. Geophys. Res.* **96**, 9295–9308.

Eisele, F. L., G. H. Mount, F. C. Fehsenfeld, J, Harder, E. Marovich, D. D. Parrish, J. Roberts, and M. Trainer (1994) Intercomparison of tropospheric OH and ancillary trace gas measurements at Fritz Peak Observatory, Colorado. *J. Geophys. Res.* **99**, 18605–18626.

Elbern, H., J. Kowol, R. Sladkovic, and A. Ebel (1997) Deep stratospheric intrusions: a statistical assessment with model guided analyses. *Atmos. Environ.* **31**, 3207–3226.

Elbern, H., J. Hendricks, and A. Ebel (1998) A climatology of tropopause folds by global analyses. *Theor. Appl. Climatol.* **59**, 181–200.

Elias, H., U. Götz, and K. J. Wannowius (1994) Kinetics and mechanism of the oxidation of sulfur(IV) by peroxomonosulfuric acid anion. *Atmos. Environ.* **28**, 439–448.

Eliassen, A. (1978) The OECD study of long range transport of air pollutants: long range transport modeling. *Atmos. Environ.* **12**, 479–487.

Eliassen, A., and J. Saltbones (1975) Decay and transformation rates of SO_2 as estimated from emission data, trajectories and measured air concentrations. *Atmos. Environ.* **9**, 425–429.

Elliot, S., E. Lu, and F. S. Rowland (1989) Rates and mechanisms for the hydrolysis of carbonyl sulfide in natural waters. *Environ. Sci. Technol.* **23**, 458–461.

Elsaesser, H. W., J. E. Harris, D. Kley, and R. Penndorf (1980) Stratospheric H_2O. *Planet. Space Sci.* **28**, 827–835.

Elshout, A. J., J. W. Viljeer, and H. Van Duuren (1978) Sulphates and sulphuric acid in the atmosphere in the years 1971–1976 in the Netherlands. *Atmos. Environ.* **12**, 785–790.

Eluszkiewicz, J., and M. Allen (1993) A global analysis of the ozone deficit in the upper stratosphere and lower mesosphere. *J. Geophys. Res.* **98**, 1069–1082.

Emanuelson, A., E. Eriksson, and H. Egner (1954) Composition of atmospheric precipitation in Sweden. *Tellus* **6**, 261–267.

Engel, A., and U. Schmidt (1994) Vertical profile measurements of carbonyl sulfide in the stratosphere. *Geophys. Res. Lett.* **21**, 2219–2222.

Engelmann, R. J. (1965) Rain scavenging of zinc sulfide particles. *J. Atmos. Sci.* **22**, 719–729.

Engelmann, R. J. (1968) The calculation of precipitation scavenging. In "Meteorology and Atomic Energy 1968" (D. H. Slade, ed.), pp. 208–221. U.S. Atomic Energy Commission, Division of Technical Information, Oak Ridge, TN.

EPA (1976) National air quality and emission trend, EPA-450/1-76-002. U.S. Environmental Protection Agency, Research Triangle Park, NC.

EPA (1987) National air pollution estimates 1940–1985, EPA-450/4-86-018, U.S. Environmental Protection Agency, Research Triangle Park, NC.

Erel, Y., S. O. Pehkonen, and M. R. Hoffmann (1993) Redox chemistry of iron in fog and stratus clouds. *J. Geophys. Res.* **98**, 18423–18434.

Erickson, D. J. (1993) A stability-dependent theory for air-sea gas exchange. *J. Geophys. Res.* **98**, 8471–8488.

Erickson, D. J., and R. A. Duce (1988) On the global flux of atmospheric sea salt. *J. Geophys. Res.* **93**, 14,079–14,088.

Erickson, D. J., P. J. Rasch, P. P. Tans, P. Friedlingstein, P. Ciais, E. Maier-Reimer, K. Six, C. A. Fischer, and S. Walters (1996) The seasonal cycle of atmospheric CO_2: a study based on the NCAR community climate model (CCM2). *J. Geophys. Res.* **101**, 15,079–15,097.

Erickson, R. E., L. M. Yates, R. L. Clark, and D. McEwen (1977) The reaction of sulfur dioxide with ozone in water and its possible atmospheric significance. *Atmos. Environ.* **11**, 813–817.

Eriksson, E. (1952a) Composition of atmospheric precipitation. I. Nitrogen compounds. *Tellus* **4**, 215–232.

Eriksson, E. (1952b) Composition of atmospheric precipitation. II. Sulfur, chloride, iodine compounds; bibliography. *Tellus* **4**, 280–303.

Eriksson, E. (1957) The chemical composition of Hawaiin rainfall. *Tellus* **9**, 509–520.

Eriksson, E. (1959) The yearly circulation of chloride and sulfur in nature: meteorological, geochemical and pedological implications. Part I. *Tellus* **11**, 375–603.

Eriksson, E. (1960) The yearly circulation of chloride and sulfur in nature: meteorological, geochemical and pedological implications. Part II. *Tellus* **12**, 63–109.

Eriksson, E. (1966) Air and precipitation as sources of nutrients. In "Handbuch der Pflanzenernährung und Düngung" (H. Linser, ed.), pp. 774–792. Springer-Verlag, Vienna.

Erisman, J.-W., A. W. M. Vermetten, A. A. H. Asman, A. Waijers-Ijelaan, and J. Slanina (1988) Vertical distribution of gases and aerosols: the behaviour of ammonia and related components in the lower atmosphere. *Atmos. Environ.* **22**, 1153–1160.

Erlenkeuser, H., E. Suess, and H. Wilkomm (1974) Industrialization affects heavy metal and carbon isotope concentrations in recent Baltic Sea sediments. *Geochim. Cosmochim. Acta* **38**, 823–842.

Ernst, W. (1938) Über pH-Wert Messungen von Niederschlägen. *Balneologe* **5**, 545–549.

Ermon, N. A., and R. D. Fergus (1976) Rainwater acidity: pH spectrum of individual drops. *Sci. Total Environ.* **6**, 223–226.

Euphenson, J. C., and H. Taube (1963) Tracer experiments with ozone as oxidizing agent in aqueous solution. *Inorg. Chem.* **4**, 704–709.

Etheridge, D. M., G. I. Pearman, and P. J. Fraser (1992) Changes in tropospheric methane between 1841 and 1978 from high accumulation rate Antarctic ice core. *Tellus* **44B**, 282–294.

Eucken, A., and F. Patat (1936) Die Temperaturabhängigkeit der photochemischen Ozonbildung. *Z. Phys. Chem.* B **33**, 459–474.

Evans, W. F., J. B. Kerr, C. T. McElroy, R. S. O'Brien, B. A. Ridley, and D. I. Wardle (1977) The odd nitrogen mixing ratio in the stratosphere. *Geophys. Res. Lett.* **4**, 235–238.

Exner, M., H. Herrmann, and R. Zellner (1992) Laser-based studies of reactions of the nitrate radical in aqueous solution. *Ber. Bunsenges. Phys. Chem.* **96**, 470–477.

Exner, M., H. Herrmann, and R. Zellner (1994) Rate constants for the reactions of the NO_3 radical with $HCOOH/HCOO^-$ and CH_3COOH/CH_3COO^- in aqueous solution between 278 and 328 K. *J. Atmos. Chem.* **18**, 359–378.

Eyre, J. R., and H. K. Roscoe (1977) Radiometric measurements of stratospheric HCl. *Nature* **226**, 243–244.

Fabian, P. (1974) Comment on a photochemical theory of tropospheric ozone by Chameides. *J. Geophys. Res.* **79**, 4124–4125.

Fabian, P. (1986) Halogenated hydrocarbons in the atmosphere. In "The Handbook of Environmental Chemistry" (O. Hutzinger, ed.), Vol. 4A, pp. 23–51. Springer-Verlag, Berlin.

Fabian, P., and C. E. Junge (1970) Global rate of ozone destruction at the Earth's surface. *Arch. Meteorol. Geophys. Bioklimatol. Ser.* A **19**, 161–172.

Fabian, P., and P. G. Pruchniewicz (1976) Final Report on Project "Troposphärisches Ozon" (Project TROZ), MPAE-W-100-76-21. Max-Planck-Institut für Aeronomie, Katlenberg, Lindau, Germany.

Fabian, P., and P. G. Pruchniewicz (1977) Meridional distribution of ozone in the troposphere and its seasonal variations. *J. Geophys. Res.* **82**, 2063–2073.

Fabian, P., W. F. Libby, and C. E. Palmer (1968) Stratospheric residence time and interhemispheric mixing of strontium 90 from fallout in rain. *J. Geophys. Res.* **73**, 3611–3616.

Fabian, P., R. Borchers, K. H. Weiler, U. Schmidt, A. Volz, D. H. Ehhalt, W. Seiler, and F. Müller (1979) Simultaneously measured vertical profiles of H_2, CH_4, CO, N_2O, $CFCl_3$, and CF_2Cl_2 in the mid-latitude stratosphere and troposphere. *J. Geophys. Res.* **84**, 3149–3154.

Fabian, P., R. Borchers, S. A. Penkett, and N. J. D. Prosser (1981a) Halocarbons in the stratosphere. *Nature* **294**, 733–735.

Fabian, P., R. Borchers, G. Flentje, W. A. Matthews, W. Seiler, H. Giehl, K. Bunse, F. Müller, U. Schmidt, A. Volz, A. Khedim, and F. J. Johnen (1981b) The vertical distribution of stable trace gases at mid-latitudes. *J. Geophys. Res.* **86**, 5179–5184.

Fabian, P., R. Borchers, B. C. Krüger, and S. Lal (1987) CF_4 and C_2F_6 in the atmosphere. *J. Geophys. Res.* **92**, 9831–9835.

Fabry, C., and H. Buisson (1921) Étude de l'extrémité du spectre solaire. *J. Phys.* (*Orsay, France*) **2**, 197–226.

Fahey, D. W., C. S. Eubank, G. Hübler, and F. C. Fehsenfeld (1985) Evaluation of a catalytic reduction technique for the measurement of total reactive odd-nitrogen NO_y in the atmosphere. *J. Atmos. Chem.* **3**, 435–468.

Fahey, D. W., G. Hübler, D. D. Parrish, E. J. Williams, R. B. Norton, B. A. Ridley, H. B. Singh, S. C. Liu, and F. C. Fehsenfeld (1986) Reactive nitrogen species in the troposphere: measurements of NO, NO_2, HNO_3, particulate nitrate, peroxyacetyl nitrate (PAN), O_3, and total reactive odd nitrogen (NO_y) at Niwot Ridge, Colorado. *J. Geophys. Res.* **91**, 9781–9793.

Fahey, D. W., D. M. Murphy, K. K. Kelly, M. K. W. Ko, M. H. Proffitt, C. S. Eubank, G. V. Ferry, M. Loewenstein, and K. R. Chan (1989a) Measurement of nitric oxide and total reactive nitrogen in the Antarctic stratosphere. Observations and chemical implications. *J. Geophys. Res.* **94**, 16,665–16,681.

Fahey, D. W., K. K. Kelly, G. V. Ferry, L. R. Poole, J. C. Wilson, D. M. Murphy, M. Loewenstein, and K. R. Chan (1989b) In situ measurements of total reactive nitrogen, total water, and aerosol in polar stratospheric cloud in the Antarctic. *J. Geophys. Res.* **94**, 11,299–11,315.

Fahey, D. W., K. K. Kelly, S. R. Kawa, A. F. Tuck, M. Loewenstein, K. R. Chan, and L. E. Heidt (1990) Observations of denitrification and dehydration in the winter polar stratospheres. *Nature* **344**, 321–324.

Fahey, D. W., S. R. Kawa, E. L. Woodbridge, P. Tin, J. C. Wilson, H. H. Jonsson, J. E. Dye, D. Baumgardner, S. Borrmann, D. W. Toohey, L. M. Avallone, M. H. Proffitt, J. Margitan, M. Loewenstein, J. R. Podolske, R. J. Salawitch, S. C. Wofsy, M. K. W. Ko, D. E. Anderson, M. R. Schoeberl, and K. R. Chan (1993) In situ measurements constraining the role of sulphate aerosols in mid-latitude ozone depletion. *Nature* **363**, 509–514.

Fairchild, P. W., and E. K. C. Lee (1978) Relative quantum yields of $O(^1D)$ in ozone photolysis in the region 250–300 nm. *Chem. Phys. Lett.* **60**, 36–39.

Fairchild, C. E., E. J. Stone, and G. M. Lawrence (1978) Photofragment spectroscopy of ozone in the UV region 270–310 nm and at 600 nm. *J. Chem. Phys.* **69**, 3632–3638.

Falconer, R. E., and P. D. Falconer (1980) Determination of cloud water acidity at a mountain observatory in the Adirondack Mountains of New York state. *J. Geophys. Res.* **85**, 7465–7470.

Fall, R., D. L. Albritton, F. C. Fehsenfeld, W. C. Kuster, and P. D. Goldan (1988) Laboratory studies of some environmental variables controlling sulfur emissions from plants. *J. Atmos. Chem.* **6**, 341–362.

FAO (1984) "Production Yearbook 1983," Vol. 37. Food and Agricultural Organization of the United Nations, Rome, Italy.

FAO (1987) "Yearbook of Forest Products 1976–1987," Vol. 41. Food and Agricultural Organization of the United Nations, Rome, Italy.

FAO (1991) "Production Yearbook 1990," Vol. 44. Food and Agricultural Organization of the United Nations, Rome, Italy.

FAO (1993) "Forest Resources Assessment 1990, Tropical Countries," Forestry paper 112. Food and Agricultural Organization of the United Nations, Rome, Italy.

FAO (1992) "Fertilizer Use by Crop," Report ESS/MISC 1992-2. Food and Agricultural Organization of the United Nations, Rome, Italy.

Farley, F. F. (1978) Correspondence. Environ. Sci. Technol. 12, 99–100.

Farlow, N. H., D. M. Hayes, and H. Y. Lem (1977) Stratospheric aerosols: undissolved granules and physical state. J. Geophys. Res. 82, 4921–4929.

Farlow, N. H., K. G. Suetsinger, H. Y. Lem, D. M. Hayes, and B. M. Tooper (1978) Nitrogen-sulfur compounds in stratospheric aerosols. J. Geophys. Res. 83, 6207–6212.

Farman, J. C., B. G. Gardiner, and J. D. Shanklin (1985) Large losses of total ozone in Antarctica reveal seasonal ClO_x/NO_x interaction. Nature 315, 207–210.

Farmer, C. B., O. F. Raper, D. Robbins, R. A. Toth, and C. Müller (1980) Simultaneous spectroscopic measurements of stratospheric species: O_3, CH_4, CO, CO_2, N_2O, H_2O, HCl, and HF at northern and southern mid-latitudes. J. Geophys. Res. 85, 1621–1632.

Farquhar, G. D., P. M. Firth, R. Wetselaar, and B. Wier (1980) On the gaseous exchange of ammonia between leaves and the environment: determination of the ammonia compensation point. Plant Physiol. 66, 710–714.

Feely, H. W., H. Seitz, R. J. Lagomarsino, and P. E. Biscaye (1966) Transport and fallout of stratospheric radioactive debris. Tellus 18, 316–329.

Fehsenfeld, F. C., and E. E. Ferguson (1976) Reactions of atmospheric negative ions with N_2O. J. Chem. Phys. 64, 1853–1854.

Fehsenfeld, F. C., R. R. Dickerson, G. Hübler, W. T. Luke, L. J. Nunnermacker, E. J. Williams, J. M. Roberts, J. G. Calvert, C. M. Curran, A. C. Delany, C. S. Eubank, D. W. Fahey, A. Fried, B. W. Gandrud, A. O. Langford, P. C. Murphy, R. B. Norton, K. E. Pickering, and R. A. Ridley (1987) A ground-based intercomparison of NO, NO_x, and NO_y measurement techniques. J. Geophys. Res. 92, 14,710–14,722.

Fehsenfeld, F. C., J. W. Drummond, U. K. Roychowdhury, P. J. Galvin, E. J. Williams, M. P. Buhr, D. D. Parrish, G. Hübler, A. O. Langford, J. G. Calvert, B. A. Ridley, F. Grahek, B. G. Heikes, G. L. Kok, J. D. Shetter, J. G. Walega, C. M. Elsworth, R. B. Norton, D. W. Fahey, P. C. Murphy, C. Hovermale, V. A. Mohnen, K. L. Demerjian, G. I. Mackay, and H. I. Schiff (1990) Intercomparison of NO_2 measurement techniques. J. Geophys. Res. 95, 3579–3597.

Fehsenfeld, F., J. Calvert, R. Fall, P. Goldan, A. B. Guenther, C. N. Hewitt, B. Lamb, S. Liu, M. Trainer, H. Westberg, and P. Zimmerman (1992) Emissions of volatile organic compounds from vegetation and the implications for atmospheric chemistry. Global Biogeochem. Cycles 6, 389–430.

Feichter, J., E. Kjellström, H. Rodhe, F. Dentener, J. Lelieveld, and G. J. Roelofs (1996) Simulation of the tropospheric sulfur cycle in a global climate model. Atmos. Environ. 30, 1693–1707.

Feister, U., and W. Warmbt (1987) Long-term measurements of surface ozone in the German Democratic Republic. J. Atmos. Chem. 5, 1–21.

Fenimore, C. P. (1971) Formation of nitric oxide in premixed hydrocarbon flames. Thirteenth International Symposium on Combustion, pp. 373–379. The Combustion Institute, Pittsburgh, PA.

Ferek, R. J., and M. O. Andreae (1984) The photochemical production of carbonyl sulfide in marine surface waters. Nature 307, 148–150.

Ferek, R. J., R. B. Chatfield, and M. O. Andreae (1986) Vertical distribution of dimethyl sulphide in the marine atmosphere. *Nature* 320, 514–516.

Fergusson, G. J. (1958) Reduction of atmospheric radiocarbon concentrations by fossil fuel carbon dioxide and the mean life of carbon dioxide in the atmosphere. *Proc. R. Soc. Lond. A* 243, 561–574.

Ferm, M. (1979) Method for the determination of atmospheric ammonia. *Atmos. Environ.* 13, 1385–1393.

Fiebiger, W. (1975) Organische Substanzen in präkambrischen Itabiriten und deren Nebengesteinen. *Geol. Rundsch.* 64, 641–652.

Findeisen, W. (1939) Zur Frage der Regentropfenbildung in reinen Wasserwolken. *Meteorol. Z.* 56, 365–368.

Finlayson, B. J., and J. N. Pitts, Jr. (1976) Photochemistry of the polluted troposphere. *Science* 192, 111–119.

Finlayson, B. J., J. N. Pitts, Jr., and H. Akimoto (1972) Production of vibrationally excited OH in chemiluminescent ozone-olefin reactions. *Chem. Phys. Lett.* 12, 495–498.

Finlayson, B. J., J. N. Pitts, Jr., and R. Atkinson (1974) Low pressure gas-phase ozone–olefin reactions. Chemiluminescence, kinetics and mechanisms. *J. Am. Chem. Soc.* 96, 5356–5367.

Finlayson-Pitts, B. J. (1983) Reaction of NO_2 with NaCl and atmospheric implications for NOCl formation. *Nature* 306, 676–677.

Finlayson-Pitts, B. J., and J. N. Pitts, Jr. (1986) "Atmospheric Chemistry." Wiley, Chichester.

Finlayson-Pitts, B. J., M. J. Ezell, and J. N. Pitts, Jr. (1989) Formation of chemically active chlorine compounds by reactions of atmospheric NaCl particles with gaseous N_2O_5 and $ClONO_2$. *Nature* 337, 241–244.

Fiocco, G., D. Fuà, and G. Visconti, eds. (1996) "The Mount Pinatubo Eruption. Effects on the Atmosphere and Climate," NATO ASI Series, Vol. I 42. Springer-Verlag, Berlin.

Firestone, M. K. (1982) Biological denitrification. *Agronomy* 22, 289–326.

Fischer, H., F. Fergg, D. Rabus, and P. Burkert (1985) Stratospheric H_2O and HNO_3 profiles derived from solar occultation measurements. *J. Geophys. Res.* 90, 3831–3843.

Fischer, K., and U. Lüttge (1978) Light-dependent net production of carbon monoxide by plants. *Nature* 275, 740–741.

Fischer, M., and P. Warneck (1996a) Photodecomposition and photooxidation of hydrogen sulfite in aqueous solution. *J. Phys. Chem.* 100, 15,111–15,117.

Fischer, M., and P. Warneck (1996b) Photodecomposition of nitrite and undissociated nitrous acid in aqueous solution. *J. Phys. Chem.* 100, 18,749–18,756.

Fishman, J., and P. J. Crutzen (1978) The origin of ozone in the troposphere. *Nature* 274, 855–858.

Fishman, J., S. Solomon, and P. J. Crutzen (1979) Observational and theoretical evidence in support of a significant in-situ photochemical source of tropospheric ozone. *Tellus* 31, 432–446.

Fishman, J., C. E. Watson, J. C. Larsen, and J. A. Logan (1990) Distribution of tropospheric ozone determined from satellite data. *J. Geophys. Res.* 95, 3599–3619.

Fishman, J., K. Fakhruzzaman, B. Cros, and D. Nganga (1991) Identification of widespread pollution in the southern hemisphere deduced from satellite analyses. *Science* 252, 1693–1696.

Fitzgerald, J. W. (1973) Dependence of the supersaturation spectrum of CCN on aerosol size distribution and composition. *J. Atmos. Sci.* 30, 628–634.

Fitzgerald, J. W. (1974) Effect of aerosol composition on cloud droplet size distribution. A numerical study. *J. Atmos. Sci.* 31, 1358–1367.

Fitzgerald, J. W. (1991) Marine aerosols: a review. *Atmos. Environ.* 25A, 533–545.

Flanagan, F. J. (1973) 1972 values for international geochemical reference samples. *Geochim. Cosmochim. Acta* **37**, 1189–1200.

Fleig, A. J., P. K. Bhartia, C. G. Wellemeyer, and D. S. Silberstein (1986) Seven years of total ozone from the TOMS instrument—a report on data quality. *Geophys. Res. Lett.* **13**, 1355–1358.

Fletcher, N. H. (1962) "The Physics of Rain Clouds." Cambridge University Press, Cambridge, UK.

Flocke, F. A. Volz-Thomas, and D. Kley (1991) Measurements of alkyl nitrates in rural and polluted air masses. *Atmos. Environ.* **25A**, 1951–1960.

Flohn, H. (1961) Meridional transport of particles and standard vector deviation of upper winds. *Pure Appl. Geophys.* **50**, 229–234.

Flood, H. (1934) Tröpfchenbildung in übersättigten Äthylalkohol-Wasserdampfgemischen. *Z. Phys. Chem.* **A 170**, 286–294.

Flossmann, A. I. (1991) The scavenging of two different types of marine aerosol particles calculated using a two-dimensional detailed cloud model. *Tellus* **43B**, 301–321.

Flossmann, A. I., H. R. Pruppacher, and J. H. Topalian (1987) A theoretical study of the wet removal of atmospheric pollutants. Part II: The uptake and redistribution of $(NH_4)_2SO_4$ particles and SO_2 gas simultaneously scavenged by growing cloud drops. *J. Atmos. Sci.* **44**, 2912–2923.

Focht, D. D., and W. Verstraete (1977) Biochemical ecology of nitrification and denitrification. *Adv. Microb. Ecol.* **1**, 135–214.

Fontanella, J. C., A. Girard, L. Gramont, and N. Louisnard (1975) Vertical distribution of NO, NO_2 and HNO_3 as derived from stratospheric absorption infrared spectra, *Appl. Optics* **14**, 025 039.

Fontijn, A., A. J. Sabadell, and R. Ronco (1970) Homogeneous chemiluminescence measurement of nitric oxide with ozone. *Anal. Chem.* **42**, 575–579.

Forrest, J. R., S. E. Schwartz, and L. Newman (1979) Conversion of sulfur dioxide to sulfate during the da Vinci flights. *Atmos. Environ.* **13**, 157–167.

Forrest, J., R. W. Garber, and L. Newman (1981) Conversion rates in power plant plumes based on filter pack data: the coal-fired Cumberland Plume. *Atmos. Environ.* **15**, 2273–2282.

Fowler, D. (1978) Dry deposition of SO_2 on agricultural crop. *Atmos. Environ.* **12**, 369–373.

Fox, G. E., E. Stackebrandt, R. B. Hespel, J. Gibson, J. Maniloff, T. A. Dyer, R. S. Wolfe, W. E. Balch, R. S. Tanner, L. J. Magrum, L. B. Zablen, R. Blakemore, R. Gupta, L. Bonen, B. J. Lewis, D. A. Stahl, K. R. Luehrsen, K. N. Chen, and C. R. Woese (1980) The phylogeny of procaryotes. *Science* **209**, 457–463.

Fraser, P. J., R. A. Rasmussen, J. W. Crefield, J. R. French, and M. A. K. Khalil (1986) Termites and global methane—another assessment. *J. Atmos. Chem.* **4**, 295–310.

Frederick, J. E. (1976) Solar corpuscle emission and neutral chemistry in the Earth's middle atmosphere. *J. Geophys. Res.* **81**, 3179–3185.

Frederick, J. E., and R. D. Hudson (1979a) Predissociation linewidths and oscillator strengths for the (2-0) to (13-0) Schumann-Runge bands of O_2. *J. Mol. Spectrosc.* **74**, 247–258.

Frederick, J. E., and R. D. Hudson (1979b) Predissociation of nitric oxide in the mesosphere and stratosphere. *J. Atmos. Sci.* **36**, 737–745.

Freney, J. R., J. R. Simpson, and O. T. Denmead (1981) Ammonia volatilization. *In* "Terrestrial Nitrogen Cycles" (F. E. Clark and T. Rosswall, eds.). *Ecol. Bull* (*Stockholm*) **33**, 291–302.

Frenkel, J. (1955) "Kinetic Theory of Liquids." Dover, New York.

Freyer, H. D. (1978) Preliminary evaluation of past CO_2 increase derived from ^{13}C measurements in tree rings. *In* "Carbon Dioxide and Society" (J. Williams, ed.), pp. 69–78. Pergamon, New York.

Freyer, H. D. (1986) Interpretation of the northern hemisphere record of $^{13}C/^{12}C$ trends of atmospheric CO_2 in tree rings. In "The Changing Carbon Cycle: A Global Analysis" (J. R. Trabalka and D. E. Reichle, eds.), pp. 125–150. Springer-Verlag, Berlin.

Fricke, W., and S. Beilke (1993) Changing concentrations and deposition of sulfur and nitrogen compounds in central Europe between 1980 and 1992. In "General Assessment of Biogenic Emissions and Deposition of Nitrogen Compounds, Sulphur Compounds and Oxidants in Europe" (J. Slanina, G. Angeletti, and S. Beilke, eds.). Air Pollution Research Report, Vol. 47, pp. 9–30. Commission of the Europena Communities, Guyot, Brussels.

Fricke, W., S. Beilke, E. Bieber, K. Uhse, and M. Wallasch (1997) Results of daily precipitation analyses in Germany from 1982 to 1995 (in German). Texte 10/97. Umweltbundesamt, Berlin.

Friedl, R. R., and S. P. Sander (1989) Kinetics and product studies of the reaction ClO + BrO using discharge flow mass spectrometry. *J. Phys. Chem.* **93**, 4756–4764.

Friedlander, S. K. (1977) "Smoke, Dust and Haze. Fundamentals of Aerosol Behavior." Wiley, New York.

Friedli, H., K. Loetscher, H. Oeschger, U. Siegenthaler, and B. Stauffer (1986) Ice core record of the $^{13}C/^{12}C$ record of atmospheric CO_2 in the past two centuries. *Nature* **324**, 237–238.

Friend, J. P. (1973) The general sulfur cycle. In "Chemistry of the Lower Atmosphere" (S. I. Rasool, ed.), pp. 177–201. Plenum Press, New York.

Friend, J. P., R. Leifer, and M. Trichon (1973) On the formation of stratospheric aerosols. *J. Atmos. Sci.* **30**, 465–479.

Fritz, J. J., and C. R. Fuget (1956) Vapor pressure of aqueous hydrogen chloride solutions. *Ind. Eng. Chem.* **1**, 10–12.

Frössling, N (1938) The evaporation of falling drops. *Gerlands Beitr. Geophys.* **52**, 170–216.

Fry, L. M., and K. K. Menon (1962) Determination of the tropospheric residence time of lead-210. *Science* **137**, 994–995.

Fuchs, N. A. (1964) "The Mechanics of Aerosols." Pergamon, Oxford.

Fushimi, K., and Y. Miyake (1980) Contents of formaldehyde in the air above the surface of the ocean. *J. Geophys. Res.* **85**, 7533–7536.

Fuzzi, S., M. C. Facchini, G. Orsi, J. A. Lind, W. Wobrock, M. Kessel, R. Maser, W. Jaeschke, K. H. Enderle, B. G. Arends, A. Berner, I. Solly, C. Kruisz, G. Reischl, U. Kaminski, P. Winkler, J. A. Ogren, K. J. Noone, A. Hallberg, H. Fierlinger-Oberlinninger, H. Puxbaum, A. Marzorati, H.-C. Hansson, A. Wiedensohler, I. B. Svenningsson, B. G. Martinssson, D. Schell, and H.-W. Georgii (1992) The Po Valley fog experiment 1989, an overview. *Tellus* **44B**, 448–468.

Gaedtke, H., and J. Troe (1975) Primary processes in the photolysis of NO_2. *Ber. Bunsenges. Phys. Chem.* **79**, 184–191.

Gagosian, R. B., O. C. Zafiriou, E. T. Peltzer, and J. B. Alford (1982) Lipids in aerosols from the tropical north pacific: temporal variability. *J. Geophys. Res.* **87**, 11,133–11,144.

Galbally, I. E. (1976) Man-made carbon tetrachloride in the atmosphere. *Science* **193**, 573–576.

Galbally, I. E., and C. R. Roy (1978) Loss of fixed nitrogen from soils by nitric oxide exhalation. *Nature* **275**, 734–735.

Galbally, I. E., and C. R. Roy (1980) Destruction of ozone at the earth's surface. *Q. J. R. Meteorol. Soc.* **106**, 599–620.

Gallagher, M. W., T. W. Choularton, R. M. Dowener, B. J. Tyler, I. M. Stromberg, C. S. Mill, S. A. Penkett, B. Bandy, G. J. Dollard, T. J. Davies, and B. M. R. Jones (1991) Measurements of the entrainment of hydrogen peroxide into cloud systems. *Atmos. Environ.* **25A**, 2029–2038.

Gallagher, M. W., K. M. Beswick, J. Duyzer, H. Westrate, T. W. Choularton, and P. Hummelshøj (1997) Measurements of aerosol fluxes to Speulder forest using a micrometeorological technique. *Atmos. Environ.* **31**, 359–373.

Galloway, J. N. (1985) The deposition of sulfur and nnitrogen from the remote atmosphere. *In* "The Biogeochemical Cycling of Sulfur and Nitrogen in the Remote Atmosphere" (J. N. Galloway, R. J. Charlson, M. O. Andreae, and H. Rodhe, eds.), NATO ASI Series, Vol. C 159, pp. 143–175. Reidel, Dordrecht, the Netherlands.

Galloway, J. N., and A. Gaudry (1984) The composition of precipitation at Amsterdam Island, Indian Ocean. *Atmos. Environ.* **18**, 2649–2656.

Galloway, J. N., and D. M. Whelpdale (1980) An atmospheric sulfur budget for eastern North America. *Atmos. Environ.* **14**, 409–417.

Galloway, J. N., G. E. Likens, W. C. Keene, and J. M. Miller (1982) The composition of precipitation in remote areas of the world. *J. Geophys. Res.* **87**, 8771–8786.

Galloway, J. N., W. C. Keene, R. S. Artz, J. M. Miller, T. M. Church, and A. H. Knap (1989) Processes controlling the concentrations of $SO_4^=$, NO_3^-, NH_4^+, H^+, $HCOO_T$ and CH_3COO_T in precipitation on Bermuda. *Tellus* **41B**, 427–433.

Galloway, J. N., W. C. Keene, A. A. P. Pszenny, D. M. Whelpdale, H. Sievering, J. T. Merrill, and J. F. Boatman (1990) Sulfur in the western north Atlantic Ocean atmosphere: results from a summer 1988 ship/aircraft experiment. *Global Biogeochem. Cycles* **4**, 349–365.

Gambell, A. W., and D. W. Fisher (1964) Occurrence of sulfate and nitrate in rainfall. *J. Geophys. Res.* **69**, 4203–4210.

Gamble, T. N., M. R. Betlach, and J. M. Tiedje (1977) Numerically dominant denitrifying bacteria from world soils. *Appl. Env. Microbiol.* **33**, 926–939.

Ganapathy, R., and D. E. Brownlee (1979) Interplanetary dust: trace element analysis of individual particles by neutron activation. *Science* **206**, 1075–1077.

Garber, R. W., J. Forrest, and L. Newman (1981) Conversion rates in power plant plumes based on filter pack data: the oil fired Northport plume. *Atmos. Environ.* **15**, 2203–2292.

Garcia, R. R., F. Strodal, S. Solomon, and J. T. Kiehl (1992) A new numerical model of the middle atmosphere 1. Dynamics and transport of tropospheric source gases. *J. Geophys. Res.* **97**, 12,967–12,991.

Garcia, R. R., and S. Solomon (1994) A new numerical model of the middle atmosphere 2. Ozone and related species. *J. Geophys. Res.* **99**, 12,937–12,951.

Gardner, E. P., R. D. Wijayaratne, and J. G. Calvert (1984) Primary quantum yields of the photodecomposition of acetone in air under tropospheric conditions. *J. Phys. Chem.* **88**, 5069–5076.

Gardner, E. P., P. D. Sperry, and J. G. Calvert (1987) Primary quantum yields of NO_2 photodissociation. *J. Geophys. Res.* **92**, 6642–6653.

Garland, J. A. (1977) The dry deposition of sulphur dioxide to land and water surfaces. *Proc. R. Soc. Lond. A* **354**, 245–268.

Garland, J. A. (1978) Dry and wet removal of sulphur from the atmosphere. *Atmos. Environ.* **12**, 349–362.

Garland, J. A. (1979) Dry deposition of gaseous pollutants. *In* "WMO Symposium on the Long-Range Transport of Pollutants and Its Relation to General Circulation Including Stratospheric/Tropospheric Exchange Processes," WMO no. 538, pp. 95–103. World Meteorological Organization, Geneva, Switzerland.

Garland, J. A., and J. R. Branson (1977) The deposition of sulphur dioxide to pine forest assessed by a radioactive tracer method. *Tellus* **29**, 445–454.

Garland, J. A., A. W. Elzerman, and S. A. Penkett (1980) The mechanism for dry deposition of ozone to sea water surfaces. *J. Geophys. Res.* **85**, 7488–7492.

Garrels, R. M., and F. T. MacKenzie (1971) "Evolution of Sedimentary Rocks." Norton, New York.

Garrels, R. M., and F. T. MacKenzie (1972) A quantitative model for the sedimentary rock cycle. *Marine Chem.* **1**, 27–61.

Garrels, R. M., and E. A. Perry, Jr. (1974) Cycling of carbon, sulfur, and oxygen through geologic time. In "The Sea" (E. D. Goldberg, ed.), Vol. 5, pp. 303–336. Wiley Interscience, New York.

Gatz, D. F., and A. N. Dingle (1971) Trace substances in rainwater: concentration variation during convective rains and their interpretation. Tellus 23, 14–27.

Gauthier, M., and D. R. Snelling (1971) Mechanism of singlet molecular oxygen formation from photolysis of ozone at 2537 Å. J. Chem. Phys. 54, 4317–4325.

Gebhart, R. (1967) On the significance of short-wave CO_2 absorption in investigations concerning the CO_2 theory of climatic change. Arch. Meteorol. Geophys. Bioklimatol. B 15, 52–61.

Georgii, H.-W. (1963) Oxides of nitrogen and ammonia in the atmosphere. J. Geophys. Res. 68, 3963–3970.

Georgii, H.-W. (1970) Contribution to the atmospheric sulfur budget. J. Geophys. Res. 75, 2365–2371.

Georgii, H.-W. (1982) The atmospheric sulfur budget. In "Chemistry of the Unpolluted and Polluted Troposphere" (H. W. Georgii and W. Jaeschke, eds.), NATO ASI Series, Vol. C 96, pp. 295–324. Reidel, Dordrecht, the Netherlands.

Georgii, H.-W., and G. Gravenhorst (1977) The ocean as a source or sink of reactive trace gases. Pure Appl. Geophys. 115, 503–511.

Georgii, H.-W., and D. Jost (1964) Untersuchung über die Verteilung von Spurengasen in der freien Atmosphäre. Pure Appl. Geophys. 59, 217–224.

Georgii, H.-W., and U. Lenhard (1978) Contribution to the atmospheric NH_3 budget. Pure Appl. Geophys. 116, 385–391.

Georgii, H.-W., and F. X. Meixner (1980) Measurement of tropospheric and stratospheric SO_2 distribution. J. Geophys. Res. 85, 7433–7438.

Georgii, H.-W., and W. J. Müller (1974) On the distribution of ammonia in the middle and lower troposphere. Tellus 26, 180–184.

Georgii, H.-W., and D. Wötzel (1970) On the relation between drop size and concentration of trace elements in rainwater. J. Geophys. Res. 75, 1727–1731.

Gerstle, R. W., and D. A. Kemnitz (1967) Atmospheric emission from open burning. J. Air. Pollut. Control Assoc. 17, 324–327.

Gery, M. W., D. L. Fox, H. E. Jeffries, L. Stockburger, and W. S. Weathers (1985) A continuous stirred tank reactor investigation of the gas-phase reaction of hydroxyl radicals and toluene. Int. J. Chem. Kinet. 17, 931–955.

Gidel, L. T., and M. A. Shapiro (1980) General circulation model estimates of the net vertical flux of ozone in the lower stratosphere and the implications for the tropospheric ozone budget. J. Geophys. Res. 85, 4049–4058.

Gidel, L. T., P. J. Crutzen, and J. Fishman (1983) A two-dimensional photochemical model of the atmosphere. 1. Chlorocarbon emissions and their effect on stratospheric ozone. J. Geophys. Res. 88, 6622–6640.

Gieskes, J. M. (1974) The alkalinity-total carbon dioxide system in sea water. In "The Sea" (E. D. Goldberg, ed.), Vol. 5, pp. 123–151. Wiley Interscience, New York.

Giggenbach, W. F., and F. LeGuern (1976) The chemistry of magmatic gases from Erta'Ale, Ethiopia. Geochim. Cosmochim. Acta 42, 25–30.

Gillani, N. V., and W. E. Wilson (1983) Gas-to-particle conversion of sulfur in power plant plumes. II. Observations of liquid-phase conversion. Atmos. Environ. 17, 1739–1752.

Gillani, N. V., S. Kohli, and W. E. Wilson (1981) Gas-to-particle conversion of sulfur in power plant plumes. I. Parametrization of the conversion rate for dry, moderately polluted ambient conditions. Atmos. Environ. 15, 2293–2313.

Gille, J. C., and J. M. Russell, III (1984) The limb infrared monitor of the stratosphere: experiment description, performance, results. J. Geophys Res. 89, 5125–5140.

Gillett, R. W., G. P. Ayers, J. P. Ivey, and J. L. Gras (1993) Measurement of dimethyl sulfide, sulfur dioxide, methane sulfonic acid and non sea salt sulfate at the Cape Grim Baseline Station. *In* "Dimethylsulphide: Oceans, Atmosphere and Climate" (G. Restelli and G. Angeltti, eds.), pp. 117–128. Kluwer, Dordrecht, the Netherlands.

Gillette, D. A. (1974) On the production of soil wind erosion aerosols having the potential for long-range transport. *J. Rech. Atmos.* **8**, 735–744.

Gillette, D. A. (1978) A wind tunnel simulation of the erosion of soil: effect of soil texture, sandblasting, wind speed and soil condition on dust production. *Atmos Environ.* **12**, 1735–1743.

Gillette, D. A. (1979) Environmental factors affecting dust emission and wind erosion. *In* "Sahara Dust: Mobilization, Transport, Deposition" (C. Morales, ed.). *SCOPE* **14**, 71–91.

Gillette, D. A. (1980) Major contribution of natural primary continental aerosols: source mechanisms. *Ann. N.Y. Acad. Sci.* **338**, 348–358.

Gillette, D. A., and I. H. Blifford (1971) Composition of tropospheric aerosols as a function of altitude. *J. Atmos. Sci* **28**, 1199–1210.

Gillette, D. A, and P. A. Goodwin (1974) Microscale transport of sand-sized soil aggregates eroded by wind. *J. Geophys. Res.* **79**, 4080–4089.

Gillette, D. A , and T. R. Walker (1977) Characteristics of airborne particles produced by wind erosion of sandy soil, high plains of West Texas. *Soil. Sci.* **123**, 97–110.

Gillette, D. A., J. Adams, C. Endo, and D. Smith (1980) Threshold velocities for input of soil particles into the air by desert soils. *J. Geophys. Res.* **85**, 5621–5630.

Giorgi, F., and W. L. Chameides (1986) Rainout life times of highly soluble aerosols and gases as inferred from simulations with a general circulation model, I, *Geophys. Res.* **91**, 14,367–14,376.

Girard, A., J. Besson, R. Giraudet, and L. Gramont (1978/1979) Correlated seasonal and climatic variatons of trace constituents in the stratosphere. *Pure Appl. Geophys.* **117**, 381–394.

Girard, A., G. Fergant, L. Gramont, O. Lado-Bordowsky, J. Laurent, S. LeBoiteux, M. P. Lemaitre, and N. Louisnard (1983) Latitudinal distribution of ten stratospheric species deduced from simultaneous spectroscopic measurements. *J. Geophys. Res.* **88**, 5377–5392.

Glassman, I. (1996) "Combustion," 3rd ed. Academic Press, San Diego, CA.

Glasson, W. A., and C. S. Tuesday (1970a) Hydrocarbon reactivities in the atmospheric photooxidation of nitric oxide. *Environ. Sci. Technol.* **4**, 916–924.

Glasson, W. A., and C. S. Tuesday (1970b) Hydrocarbon reactivity and the kinetics of the atmospheric photooxidation of nitric oxide. *J. Air Pollut. Control Assoc.* **20**, 239–243.

Glavas, S , and S. Toby (1975) Reaction between ozone and hydrogen sulfide. *J. Phys. Chem.* **79**, 779–782.

Gleason, J. F., A. Sinha, and C. J. Howard (1987) Kinetics of the gas-phase reaction $HOSO_2 + O_2 \rightarrow HO_2 + SO_3$. *J. Phys. Chem.* **91**, 719–724.

Glueckauf, E. (1951) The composition of atmospheric air. *In* "Compendium of Meteorology" (T. F. Mahone, ed.), pp. 3–12. American Meteorological Society, Boston.

Gmitro, J. I., and T. Vermeulen (1964) Vapor-liquid equilibria for aqueous sulfuric acid. *Am. Inst. Chem. Eng.* **10**, 741–746.

Goethel, M. (1980) Untersuchung ausgewählter Quellen und Senken des atmosphärischen Ammoniaks mit einem neuen Probenahmeverfahren. Diploma-Thesis. Institut für Meteorologie und Geophysik, Universität Frankfurt (Main), Germany.

Goldan, P. D., Y. A. Bush, F. C. Fehsenfeld, D. L. Albritton, P. J. Crutzen, A. L. Schmeltekopf, and E. E. Ferguson (1978) Tropospheric N_2O mixing-ratio measurements. *J. Geophys. Res.* **83**, 935–939.

Goldan, P. D., W. C. Kuster, D. L. Albritton, and A. L. Schmeltekopf (1980) Stratospheric $CFCl_3$, CF_2Cl_2, and N_2O profile measurements at several latitudes. *J. Geophys. Res.* **85**, 413–423.

Goldan, P. D., W. C. Kuster, A. L. Schmeltekopf, F. C. Fehsenfeld, and D. L. Albritton (1981) Correction of atmospheric N_2O mixing ratio data. *J. Geophys. Res.* **86**, 5385–5386.

Goldan, P. D., W. C. Kuster, D. L. Albritton, and F. C. Fehsenfeld (1987) The measurement of natural sulfur emissions from soils and vegetation: three sites in the eastern United States revisited. *J. Atmos. Chem.* **5**, 439–467.

Goldan, P. D., R. Fall, W. C. Kuster, and F. C. Fehsenfeld (1988) Uptake of OCS by growing vegetation: a major tropospheric sink. *J. Geophys. Res.* **93**, 14,186–14,192.

Goldberg, A. B., P. J. Maroulis, L. A. Wilner, and A. R. Bandy (1981) Study of H_2S emissions from a salt water marsh. *Atmos. Environ.* **15**, 11–18.

Goldberg, E. D. (1971) Atmospheric dust, the sedimentary cycle and man. *Geophysics* **3**, 117–132.

Goldenbaum, G. C., and R. R. Dickerson (1993) Nitric oxide production in lightning discharges. *J. Geophys. Res.* **98**, 18,333–18,338.

Goldman, A., F. G. Fernald, W. J. Williams, and D. G. Murcray (1978) Vertical distribution of NO_2 in the stratosphere as determined from balloon measurements of solar spectra in the 4500 Å region. *Geophys. Res. Lett.* **5**, 257–260.

Goody, R. M. (1969) Time variations in atmospheric N_2O in eastern Massachusetts. *Planet. Space Sci.* **17**, 1319–1320.

Gordley, L. L., J. M. Russell, L. J. Mickley, J. E., Frederick, J. H., Park, K. A. Stone, G. M. Beaver, J. M. McInermey, L. E. Deaver, G. C. Toon, F. G. Murcray, R. D. Blatherwick, M. R. Gunson, J. P. D. Abbatt, R. L. Mauldin, G. H. Mount, B. Sen, and J.-F. Blavier (1996) Validation for nitric oxide and nitrogen dioxide measurements made by the Halogen Occultation Experiment for UARS platform. *J. Geophys. Res.* **101**, 10,241–10,266.

Gordon, A. L. (1975) General ocean circulation. *In* "Numerical Models of Ocean Circulation" (R. O. Reid, ed.), pp. 39–53. National Academy of Science, Washington, DC.

Götz, F. W. P. (1931) Zum Strahlungsklima des Spitzbergensommers: Strahlungs- und Ozonmessungen in der Königsbucht 1929. *Gerlands Beitr. Geophys.* **31**, 119–156.

Götz, F. W. P., A. R. Meetham, and G. M. B. Dobson (1934) The vertical distribution of ozone in the atmosphere. *Proc. R. Soc. Lond. A* **145**, 416–446.

Graedel, T. E. (1979) Reduced sulfur emissions from the open ocean. *Geophys. Res. Lett.* **6**, 329–331.

Graedel, T. E., and T. Eisner (1988) Atmospheric formic acid from formicine ants: a preliminary assessment. *Tellus* **40B**, 335–339.

Graedel, T. E., and W. C. Keene (1995) Tropospheric budget of reactive chlorine. *Global Biogeochem. Cycles* **9**, 47–77.

Graedel, T. E., and C. J. Weschler (1981) Chemistry within aqueous aerosols and raindrops. *Rev. Geophys. Space Phys.* **19**, 505–539.

Graedel, T. E., D. T. Hawkins, and L. D. Claxton (1986a) "Atmospheric Chemical Compounds: Sources, Occurrence, and Bioassay." Academic Press, Orlando, FL.

Graedel, T. E., M. L. Mandich, and C. J. Weschler (1986b) Kinetic model studies of atmospheric droplet chemistry. 2. Homogeneous transition metal chemistry in rain drops. *J. Geophys. Res.* **91**, 5205–5221.

Gran, G. (1952) Determination of the equivalence point in potentiometric titrations, Part II. *Analyst (Lond.)* **77**, 661–671.

Granat, L. (1972) On the relation between pH and the chemical composition in atmospheric precipitation. *Tellus* **24**, 550–560.

Granat, L. (1978) Sulfate in precipitation as observed by the European atmospheric chemistry network. *Atmos. Environ.* **12**, 413–424.

Granat, L. R., O. Hallberg, and H. Rodhe (1976) The sulfur cycle. *In* "Nitrogen, Phosphorus and Sulfur–Global Cycles" (B. H. Svensson and R. Söderlund, eds.). *SCOPE Report 7, Ecol. Bull. (Stockh.)* **22**, 89–134.

Gras, J. L. (1983) Ammonia and ammonium concentrations in the Antarctic atmosphere. *Atmos. Environ.* **17**, 815–818.

Graustein, W. C., and K. K. Turekian (1986) ^{210}Pb and ^{137}Cs in air and soils measure the rate and vertical profile of aerosol scavenging. *J. Geophys. Res.* **91**, 14,355–14,366.

Gravenhorst, G. (1975a) The sulfate component in aerosol samples over the north Atlantic. *Meteor Forschungserg.* B **10**, 22–33.

Gravenhorst, G. (1975b) Der Sulfatanteil im atmosphärischen Aerosol über dem Nordatlantik, Report no. 30. Institut für Meteorologie und Geophysik, Universität Frankfurt (Main), Germany.

Gray, P., R. Shaw, and J. C. J. Thynne (1967) The rate constants of alkoxy radical reactions. *Prog. React. Kinet.* **4**, 65–117.

Greeley, R., J. D. Iverson, J. B. Pollak, N. Udovich, and B. White (1974) Wind tunnel studies of Martian aeolian processes. *Proc. R. Soc. Lond.* A **341**, 331–360.

Greenberg, A., F. Darack, R. Harkov, P. Lioy, and J. Daisey (1985) Polycyclic aromatic hydrocarbons in New Jersey: a comparison of winter and summer concentrations over a two-year period. *Atmos. Environ.* **19**, 1325–1339.

Greenberg, J. P., and P. R. Zimmerman (1984) Nonmethane hydrocarbons in remote tropical, continental and marine atmospheres. *J. Geophys. Res.* **89**, 4767–4778.

Greenberg, J. P., P. R. Zimmerman, L. Heidt, and W. Pollock (1984) Hydrocarbon and carbon monoxide emissions from biomass burning in Brazil, *J. Geophys. Res.* **89**, 1350–1354.

Greenberg, J. P., P. R. Zimmerman, and R. B. Chatfield (1985) Hydrocarbons and carbon monoxide in African savanna air. *Geophys. Res. Lett.* **12**, 113–116.

Greenfield, S. M. (1957) Rain scavenging of radioactive particulate matter from the atmosphere. *J. Meteorol.* **14**, 115–125.

Gregory, G. L., J. M. Hoell, Jr., B. J. Huebert, S. E. Van Bramer, P. J. LeBel, S. A. Vay, R. M. Marinaro, H. I. Schiff, D. R. Hastie, G. I. Mackay, and D. R. Karecki (1990) An intercomparison of airborne nitric acid measurements. *J. Geophys. Res.* **95**, 10,089–10,102

Gregory, P. H. (1973) "The Microbiology of the Atmosphere," 2nd ed. Leonard Hill, Aylesbury.

Gregory, P. H. (1978) Distribution of airborne pollen and spores and their long distance transport. *Pure Appl. Geophys.* **116**, 309–315.

Gregory, P. H., and J. M. Hirst (1957) The summer air spora at Rothamsted in 1952. *J. Gen. Microbiol.* **17**, 135–152.

Greiner, N. R. (1966) Flash photolysis of H_2O vapor in the presence of D_2, Ar, $H_2^{18}O$. *J. Chem. Phys.* **45**, 99–103.

Greiner, N. R. (1967a) Photochemistry of N_2O essential to a simplified vacuum-ultraviolet actinometer. *J. Chem. Phys.* **47**, 4373–4377.

Greiner, N. R. (1967b) Hydroxyl radical kinetics by kinetic spectroscopy. I. Reactions with H_2, CO and CH_4 at 300 K. *J. Chem. Phys.* **46**, 2795–2799.

Greiner, N. R. (1970) Hydroxyl radical kinetics by kinetic spectroscopy. VI. Reactions with alkanes in the range 300–500 K. *J. Chem. Phys.* **53**, 1070–1076.

Grennfelt, P., C. Bengtson, and L. Scärby (1983) Dry deposition of nitrogen dioxide to Scots Pine needles. *In* "Precipitation Scavenging, Dry Deposition and Resuspension" (H. R. Pruppacher, R. G. Semonin, W. G. N. Slinn, eds.), Vol. 2, pp. 753–761. Elsevier, New York.

Grgic, I., V. Hudnik, M. Bizjak, and J. Levec (1993) Aqueous S(IV) oxidation. III. Catalytic effect of soot particles. *Atmos. Environ.* **27A**, 1409–1416.

Griffin, J. J., and E. D. Goldberg (1975) The flux of elemental carbon in coastal marine sediments. *Limnol. Oceanogr.* 20, 456–463.

Griffin, J. J., H. Windom, and E. D. Goldberg (1968) The distribution of clay minerals in the world ocean. *Deep Sea Res.* 15, 433–459.

Griffing, G. W. (1977) Ozone and oxides of nitrogen production during thunderstorms. *J. Geophys. Res.* 82, 943–950.

Griggs, M. (1968) Absorption coefficients of ozone in the ultraviolet and visible region. *J. Chem. Phys.* 49, 857–859.

Grimsrud, E. P., and R. A. Rasmussen (1975) Survey and analysis of halocarbons in the atmosphere by gas chromatography-mass spectrometry. *Atmos. Environ.* 9, 1014–1017.

Groblicki, P., and G. J. Nebel (1971) The formation of aerosols in urban atmospheres. *In* "Chemical Reactions in Urban Atmospheres" (C. S. Tuesday, ed.), pp. 241–263. American Elsevier, New York.

Grosjean, D. (1977) Aerosols. *In* "Ozone and other Photochemical Oxidants," (Committee on Medical and Biological Effects of Environmental Pollution, eds.), pp. 45–125. National Academy of Sciences, Washington, DC.

Grosjean, D. (1983a) Atmospheric reactions of pyruvic acid. *Atmos. Environ.* 17, 2379–2382.

Grosjean, D. (1983b) Distribution of atmospheric nitrogenous pollutants at a Los Angeles receptor site. *Environ. Sci. Technol.* 17, 13–19.

Grosjean, D. (1984) Discussion: world-wide ambient measurements of peroxyacetyl nitrate (PAN) and implications for plant injury. *Atmos. Environ.* 18, 1489–1491.

Grosjean, D. (1991) Ambient levels of formaldehyde, acetaldehyde, and formic acid in southern California: results of a one-year base-line study. *Environ. Sci. Technol.* 25, 710–715.

Grosjean, D., and A. Bytnerovicz (1993) Nitrogenous air pollutants at a southern California mountain forest smog receptor site. *Atmos. Environ.* 27A, 483–492.

Grosjean, D., and S. K. Friedlander (1980) Formation of organic aerosols from cyclic olefins and diolefins. *In* "Character and Origin of Smog Aerosols" (G. M. Hidy, P. K. Mueller, D. Grosjean, B. R. Appel, and J. J. Weslovski, eds.). *Adv. Environ. Sci. Technol.* 9, 435–473.

Grosjean, D., K. Van Cauwenberghe, J. P. Schmid, P. E. Kelly, and J. N. Pitts, Jr. (1978) Identification of C_3–C_{10} aliphatic dicarboxylic acids in airborne particulate matter. *Environ Sci. Technol.* 12, 313–317.

Grosjean, D., R. Swanson, and C. Ellis (1983) Carbonyls in Los Angeles air: contributions of direct emissions. *Sci. Total Environ.* 29, 65–85.

Grosjean, D., E. L. Williams, and E. Grosjean (1993) Atmospheric chemistry of isoprene and its carbonyl products. *Environ Sci. Technol.* 27, 830–840.

Grosjean, D., E. Grosjean, and E. L. Williams (1994) Atmospheric chemistry of olefins: a product study of the ozone–alkene reactions with cyclohexane added to scavenge OH. *Environ. Sci. Technol.* 28, 186–196.

Grosjean, E., J. B. deAndrade, and D. Grosjean (1996) Carbonyl products of the gas phase reaction of ozone with simple alkenes. *Environ. Sci. Technol.* 30, 975–983.

Grover, S. N., H. R. Pruppacher, and A. E. Hamielec (1977) A numerical determination of the efficiency with which spherical aerosol particles collide with spherical water drops due to inertial impaction and phoretic and electrical forces. *J. Atmos. Sci.* 34, 1655–1663.

Gschwend, P. M., J. K. MacFarlane, and K. A. Newman (1985) Volatile halogen and organic compounds released to seawater from temperate marine macroalgae. *Science* 227, 1033–1036.

Gudiksen, P. H., A. W. Fairhall, and R. J. Reed (1968) Roles of mean meridional circulation and eddy diffusion in the transport of trace substances in the lower stratosphere. *J. Geophys. Res.* 73, 4461–4473.

Guenther, A. B., R. S. Monson, and R. Fall (1991) Isoprene and monoterpene emission rate variability: observation with eucalyptus and emission rate algorithm development. *J. Geophys. Res.* **96**, 10,799–10,808.

Guenther, A. B., P. R. Zimmerman, P. C. Harley, R. K. Monson, and R. Fall (1993) Isoprene and monoterpene emission rate variability: model evaluations and sensitivity analysis. *J. Geophys. Res.* **98**, 12,609–12,617.

Guenther, A., P. Zimmerman, and M. Wildermuth (1994) Natural volatile organic compound emission estimates from U.S. woodland landscapes. *Atmos. Environ.* **28**, 1197–1210.

Guenther A., C. N. Hewitt, D. Erickson, R. Fall, C. Geron, T. Graedel, P. Hartley, L. Klinger, M. Lerdau, W. A. McKay, T. Pierce, B. Scholes, R. Steinbrecher, R. Tallamraju, J. Taylor, and P. Zimmerman (1995) A global model of natural volatile organic compound emissions. *J. Geophys. Res.* **100**, 8873–8892.

Guerzoni, S., and R. Chester (eds.) (1996) "The Impact of Desert Dust across the Mediterranean." Kluwer, Dordrecht, the Netherlands.

Guicherit, R. (1988) Ozone on an urban and regional scale: with special reference to the situation in the Netherlands. In "Tropospheric Ozone—Regional and Global Scale Indications" (I. S. A. Isaksen, ed.), NATO ASI Series,Vol. C 227, pp. 49–62. Reidel, Dordrecht, the Netherlands.

Guicherit, R., and H. van Dop (1977) Photochemical production of ozone in western Europe (1971–1978) and its relation to meteorology. *Atmos. Environ.* **11**, 145–155.

Gundersen, K., C. W. Mountain, D. Taylor, R. Ohye, and J. Shen (1972) Some chemical and microbiological observations in the Pacific Ocean off the Hawaiian Islands. *Limnol. Oceanogr.* **17**, 524–531.

Gunson, M. R., C. B. Farmer, R. H. Norton, R. Zander, C. P. Rinsland, J. H. Shaw, and D. G. Gao (1990) Measurement of CH_4, N_2O, CO, H_2O and O_3 in the middle atmosphere by the Atmospheric Trace Molecule Spectroscopy Experiment on Spacelab 3. *J. Geophys. Res.* **95**, 13,867–13,882.

Gunz, D. W., and M. R. Hoffmann (1990) Atmospheric chemistry of peroxides: a review. *Atmos. Environ.* **24A**, 1601–1633.

Haaf, W. (1980) Accurate measurements of aerosol size distributions II. Construction of a new plate condensor electric mobility analyser and first results. *J. Aerosol Sci.* **11**, 201–212.

Haaf, W., and R. Jaenicke (1977) Determination of the smallest particle size detectable in condensation nucleus counters by observation of the coagulation of SO_2 photooxidation products. *J. Aerosol Sci.* **8**, 447–456.

Haaf, W., and R. Jaenicke (1980) Results of improved size distribution measurements in the Aitken range of atmospheric aerosols. *J. Aerosol Sci.* **11**, 321–330.

Hack, W., O. Horie, and H. Gg. Wagner (1982) The rate of the reaction of NH_2 with O_3. *Ber. Bunsen Ges. Phys. Chem.* **85**, 72–78.

Hagen-Smit, A. J. (1952) Chemistry and physiology of Los Angeles smog. *Ind. Eng. Chem.* **44**, 1342–1346.

Hagerman, L. M., V. P. Aneja, and W. A. Lonneman (1997) Characterization of non-methane hydrocarbons in the rural south-east United States. *Atmos. Environ.* **31**, 4017–4038.

Hahn, C. J., S. G. Warren, J. London, R. M. Chervin, and R. Jenne (1982) Atlas of simultaneous occurrence of different cloud types over the oceans. NCAR Technical Note TN-201 + STR. Boulder, CO.

Hahn, J. (1980) Organic constituents of natural aerosols. *Ann. N.Y. Acad. Sci.* **338**, 359–376.

Hahn, J. (1981) Nitrous oxide in the oceans. In "Denitrification, Nitrification and Atmospheric Nitrous Oxide" (C. C. Delwiche, ed.), pp. 191–277. Wiley, New York.

Hahn, J. (1982) Geochemical fossils of a possible archae-bacterial origin in ancient sediments. *Zentralbl. Bacteriol. Mikrobiol. Hyg. Abt. 1 Orig. C* **3**, 40–52.

Hahn, J., and C. Junge (1977) Atmospheric nitrous oxide: a critical review. Z. Naturforsch. **32a**, 190–214.

Hales, J. M. (1972) Fundamentals of the theory of gas scavenging by rain. Atmos. Environ. **6**, 635–659.

Hall, D. M., and L. A. Donaldson (1963) The ultrastructure of wax deposits on plant leaf surfaces. 1. Growth of wax on leaves of tripolium repens. Ultrastruct. Res. **9**, 259–267.

Hall, D. M., and R. L. Jones (1961) Physiological significance of surface wax on leaves. Nature **191**, 95–96.

HALOE (1998) Halogen Occultation Experiment on the Upper Atmosphere Research Satellite (http://haloe data.lavc.nasa.gov/home.html).

Hamano, Y., and M. Ozima (1978) Earth-atmosphere evolution models based on Ar isotopic data. In "Terrestrial Rare Gases" (E. C. Alexander, Jr., and M Ozima, eds.), pp. 155–171. Japan Scientific Society Press, Tokyo.

Hamill, P. (1975) The time-dependent growth of H_2O-H_2SO_4 aerosols by heteromolecular condensation. J. Aerosol Sci. **6**, 475–484.

Hamill, P., R. P. Turco, O. B. Toon, C. S. Kiang, and R. C. Whitten (1982) On the formation of sulfate aerosol particles in the stratosphere. J. Aerosol Sci. **13**, 561–585.

Hamrud, M. (1983) Residence time and spatial variability for gases in the atmosphere. Tellus **35B**, 295–303.

Hänel, G. (1976) The properties of atmospheric aerosol particles as function of the relative humidity at thermodynamic equilibrium with the surrounding moist air. Adv. Geophys. **19**, 73–188.

Hänel, G., and J. Thudium (1977) Mean bulk densities of dry atmospheric aerosol particles: a summary of measured data. Pure Appl. Geophys. **115**, 799–803.

Hanks, T. C., and D. L. Anderson (1969) The early thermal history of the Earth. Phys. Earth Planet. Inter. **2**, 19–29.

Hannemann, A. U., S. K. Mitra, and H. R. Pruppacher (1995) On the scavenging of gaseous nitrogen compunds by large and small rain drops. I. A wind tunnel and theoretical study of the uptake and desorption of NH_3 in the presence of CO_2. J. Atmos. Chem. **21**, 293–307.

Hannemann, A. U., S. K. Mitra, and H. R. Pruppacher (1996) On the scavenging of gaseous nitrogen compunds by large and small rain drops. II. Wind tunnel and theoretical studies of the simultaneous uptake of NH_3, SO_2 and CO_2 by water drops. J. Atmos. Chem. **24**, 271–284.

Hansen, M. H., K. Ingvorsen, and B. B. Jorgensen (1978) Mechanism of hydrogen sulfide release from coastal marine sediments to the atmosphere. Limnol. Oceanogr. **23**, 68–76.

Hanson, D. R., and E. R. Lovejoy (1995) The reaction of $ClONO_2$ with sub-micrometer sulfuric acid aerosol. Science **267**, 1326–1328.

Hanson, D. R., and K. Mauersberger (1988) Vapor pressures of HNO_3/H_2O solutions at low temperatures. J. Phys. Chem. **92**, 6167–6170.

Hanson, D. R., and A. R. Ravishankara (1993) The uptake of HCl and HOCl onto sulfuric acid: solubilities, diffusivities, and reaction. J. Phys. Chem. **97**, 12,309–12,319.

Hanson, D. R., and A. R. Ravishankara (1994) Reactive uptake of $ClONO_2$ onto sulfuric acid due to reaction with HCl and H_2O. J. Phys. Chem. **98**, 5728–5735.

Hanson, D. R., A. R. Ravishankara, and S. Solomon (1994) Heterogeneous reactions in sulfuric acid aerosols: a framework for model calculations. J. Geophys. Res. **99**, 3615–3629.

Hanson, P. J., and S. E. Lindberg (1991) Dry deposition of reactive nitrogen compounds: a review of leaf, canopy and non-foliar measurements. Atmos. Environ. **25A**, 1615–1634.

Hanson, P. J., K. Rott, G. E. Taylor, C. A. Gunderson, S. E. Lindberg, and B. M. Ros-Todd (1989) NO_2 deposition to elements representative of a forest landscape. Atmos. Environ. **23**, 1783–1794.

Hanst, P. L., L. L. Spiller, D. M. Watts, J. W. Spence, and M. F. Miller (1975) Infrared measurements of fluorocarbons, carbon tetrachloride, carbonyl sulfide and other atmospheric trace gases. *J. Air Pollut. Control Assoc.* **25**, 1220–1226

Hanst, P. L., J. W. Spence, and O. Edney (1980) Carbon monoxide production in photooxidation of organic molecules in the air. *Atmos. Environ.* **14**, 1077–1088.

Hao, W.-M., S. C. Wofsy, M. B. McElroy, J. M. Beer, and M. A. Toqan (1987) Sources of atmospheric nitrous oxide from combustion. *J. Geophys. Res.* **92**, 3098–3104.

Hao, W.-M., M.-H. Liu, and P. J. Crutzen (1990) Estimates of annual and regional releases of CO_2 and other trace gases to the atmosphere from fires in the tropics, based on FAO statistics for the period 1975–1980. *In* "Fire in the Tropical Biota: Ecosystem Processes and Global Challenges" (J. G. Goldammer, ed.). *Ecol. Stud.* **84**, 440–462.

Hao, W.-M., and M.-H. Liu (1994) Spatial and temporal distributions of tropical biomass burning. *Global Biogeochem. Cycles* **8**, 495–503

Happel, J. D., and D. W. R. Wallace (1996) Methyl iodide in the Greenland/Norwegian seas and the tropical Atlantic Ocean: evidence for photochemical production. *Geophys. Res. Lett.* **23**, 2105–2108.

Harker, A. B., W. Ho, and J. J. Ratto (1977) Photodissociation quantum yield of NO_2 in the region 375–420 nm. *Chem. Phys. Lett.* **50**, 394–397.

Harnisch, J., R. Borchers, P. Fabian, and M. Maiss (1996) Tropospheric trends for CF_4 and C_2F_6 since 1982 derived from SF_6 dated stratospheric air. *Geophys. Res. Lett.* **23**, 1099–1102.

Harper, L. A., R. R. Sharpe, G. W. Langdale, and J. E. Giddens (1987) Nitrogen cycling in a wheat crop: soil, plant and aerial nitrogen transport. *Agron. J.* **79**, 965–973.

Harper, L. A., J. E. Giddens, and G. W. Langdale (1989) Environmental effects on nitrogen dynamics in soybean under conservation and clean tillage systems. *Agron. J.* **81**, 623–631.

Harries, J. E. (1976) The distribution of water vapor in the stratosphere. *Rev. Geophys. Space Phys.* **14**, 565–575.

Harries, J. E., D. G. Moss, N. R. W. Swann, G. F. Neill, and P. Gildwarg (1976) Simultaneous measurements of H_2O, NO_2, and HNO_3 in the daytime stratosphere from 15 to 35 km. *Nature* **259**, 300–302.

Harris, G. W., G. I. MacKay, T. Iguchi, L. K. Mayne, and H. I. Schiff (1989) Measurements of formaldehyde in the troposphere by tunable diode laser absorption spectroscopy. *J. Atmos. Chem.* **8**, 119–137.

Harris, G. W., D. Klemp, T. Zenker, J. P. Burrows, and B. Mathieu (1992) Tunable diode laser measurements of trace gases during the 1988 Polarstern cruise and inter-comparison with other methods. *J. Atmos. Chem.* **15**, 315–326.

Harrison, R. M., and A. G. Allen (1991) Scavenging ratios and deposition of sulphur, nitrogen and chlorine species in eastern England. *Atmos. Environ.* **25A**, 1719–1723.

Harrison, R. M., and A.-M. N. Kitto (1994) Evidence for a surface source of atmospheric nitrous acid. *Atmos. Environ.* **28A**, 1089–1094.

Harrison, R. M., and H. A. McCartney (1980) Ambient air quality at a coastal site in rural north-west England. *Atmos. Environ.* **14**, 233–244.

Harrison, R. M., J. D. Peak, and G. M. Collins (1996) Tropospheric cycle of nitrous acid. *J. Geophys. Res.* **101**, 14,429–14,439.

Harrison, R. M., S. Yamulki, K. W. T. Goulding, and C. P. Webster (1995) Effect of fertilizer application on NO and N_2O fluxes from agricultural fields. *J. Geophys. Res.* **100**, 25,923–25,931.

Hart, R., and L. Hogan (1978) Earth degassing models, and the heterogeneous vs. homogeneous mantle. *In* "Terrestrial Rare Gases" (E. C. Alexander, Jr., and M Ozima, eds.), pp. 193–206. Japan Scientific Society Press, Tokyo.

Hartley, W. N. (1881) On the absorption of solar rays by atmospheric ozone. *J. Chem. Soc.* **39**, 111–128.

Hartmann, W. R., (1990) Carbonsäuren in der Atmosphäre. Dissertation. Universität Mainz, Germany.

Hartmann, W. R., M. O. Andreae, and G. Helas (1989) Measurements of organic acids over central Germany. *Atmos. Environ.* **23**, 1531–1533.

Hartmann, W. R., M. Santana, M. Hermosos, M. O. Andreae, and E. Sanhueza (1991) Diurnal cycles of formic and acetic acids in the northern part of the Guyana Shield. *J. Atmos. Chem.* **13**, 63–72.

Harvey, R. B., D. H. Stedman, and W. Chameides (1977) Determination of the absolute rate of solar photolysis of NO_2. *J. Air. Poll. Control. Assoc.* **27**, 663–666.

Hashimoto, Y., H. K. Kim, T. Otoshi, and Y. Sekine (1991) Air quality in Seoul, Korea, May–March 1989. *J. Japan Soc. Air. Poll.* **26**, 51–58.

Haszpra, L., I. Szilágyi, A. Demeter, T. Turányi, and T. Bérces (1991) Non-methane hydrocarbon and aldehyde measurmenets in Budapest, Hungary. *Atmos. Environ.* **25**, 2103–2110.

Hatakeyama, S., and H. Akimoto (1983) Reactions of OH radicals with methanethiol, dimethyl-sulfide and dimethyldisulfide in air. *J. Phys. Chem.* **87**, 2387–2395.

Hatakeyama, S., and H. Akimoto (1994) Reactions of Criegee intermediates in the gas phase. *Res. Chem. Intermed.* **20**, 503–524.

Hatakeyama, S., H. Bandow, M. Okuda, and H. Akimoto (1981) Reactions of CH_2OO and $CH_2(^1A_1)$ with H_2O in the gas phase. *J. Phys. Chem.* **85**, 2249–2254.

Hatakeyama, S., M. Okuda, and H. Akimoto (1982) Formation of sulfur dioxide and methane sulfonic acid in the photooxidation of dimethylsulfide. *Geophys. Res. Lett.* **9**, 583–586.

Hatakeyama, S., K. Izumi, and H. Akimoto (1985) Yields of SO_2 and formation of aerosol in the photooxidation of dimethylsulfide under atmospheric conditions. *Atmos. Environ.* **19**, 135–141.

Hatakeyama, S., K. Izumi, T. Fukuyama, and H. Akimoto (1989) Reactions of ozone with α-pinene and β-pinene in air: yields of gaseous and particulate products. *J. Geophys. Res.* **94**, 13,013–13,024.

Haulet, R., P. Zettwoog, and J. C. Sabroux (1977) Sulphur dioxide discharge from Mount Etna. *Nature* **268**, 715–717.

Hausmann, M., and U. Platt (1994) Spectroscopic measurements of bromine oxide and ozone in the high Arctic during Polar Sunrise Experiment 1992. *J. Geophys. Res.* **99**, 25,399–25,414.

Haxel, U., and G. Schumann (1955) Selbstreinigung der Atmosphäre. *Z. Phys.* **142**, 127–132.

Hayes, W. (1972) Designing wind crosion control systems in the midwest region. RT SCS Agronomical Technical Note LI-9. Soil Conservation Service, USDA, Lincoln, NE.

Heald, E. F., J. J. Naughton, and I. L. Barnes (1963) The chemistry of volcanic gases. 2. Use of equilibrium calculations in the interpretation of volcanic gas samples. *J. Geophys. Res.* **68**, 545–557.

Healy, T. V., H. A. C. McKay, A. Pilbeam, and D. Scargill (1970) Ammonia and ammonium sulfate in the troposphere over the United Kingdom. *J. Geophys. Res.* **75**, 2317–2321.

Heath, D. F. (1973) Space observations of variability of solar irradiance in the near and far U.V. *J. Geophys. Res.* **78**, 2779–2792.

Heath, D. F., and M. P. Thekaekara (1977) The solar spectrum between 1200–3000 Å. *In* "The Solar Output and Its Variation" (O. R. White, ed.), pp. 193–212. Colorado Associated University Press, Boulder CO.

Heaton, W. B., and J. T. Wentworth (1959) Exhaust gas analysis by gas chromatography combined with infrared detection. *Anal. Chem.* **31**, 349–357.

Hedin, L. O., L. Granat, G. E. Likens, T. A. Buishand, J. N. Galloway, T. J. Butler, and H. Rodhe (1994) Steep declines in atmospheric base cations in regions of Europe and North America. *Nature* **367**, 351–354.

Hegg, D. A., and P. V. Hobbs (1978) Oxidation of sulfur dioxide in aqueous systems with particular reference to the atmosphere. *Atmos. Environ.* **15**, 1597–1604.

Hegg, D. A., and P. V. Hobbs (1980) Measurements of gas-to-particle conversion in the plumes of five coal-fired electric power plants. *Atmos. Environ.* **14**, 99–116.

Hegg, D. A., L. F. Radke, and P. V. Hobbs (1990) Particle production associated with marine clouds. *J. Geophys. Res.* **95**, 13,917–13,925.

Heicklen, J., K. Westberg, and N. Cohen (1971) Discussion remark. In "Chemical Reactions in Urban Atmospheres" (C. S. Tuesday, ed.), pp. 55–59. American Elsevier, New York.

Heidt, L. E., and D. H. Ehhalt (1980) Corrections of CH_4 concentrations measured prior to 1974. *Geophys. Res. Lett.* **7**, 1023.

Heidt, L. E., J. P. Krasnec, R. A. Lueb, W. H. Pollock, B. E. Henry, and P. J. Crutzen (1980) Latitudinal distribution of CO and CH_4 over the Pacific. *J. Geophys. Res.* **85**, 7329–7336.

Heidt, L. E., R. Lueb, W. Pollock, and D. H. Ehhalt (1975) Stratospheric profiles of CCl_3F and CCl_2F_2. *Geophys. Res. Lett.* **2**, 445–447.

Heintz, F., U. Platt, H. Flentje, and R. Dubois (1996) Long-term observation of nitrate radicals at the Tor Station Kap Arkona (Rügen). *J. Geophys. Res.* **101**, 22,891–22,910.

Heintzenberg, J., H. C. Hanson, and H. Lannefors (1981) The chemical composition of arctic haze at Ny-Alesund, Spitzbergen. *Tellus* **33**, 162–171.

Heisler, S. L., and S. K. Friedlander (1977) Gas to particle conversion in photochemical smog: aerosol growth laws and mechanisms for organics. *Atmos. Environ.* **11**, 157–168.

Heiss, A., and K. Sahetchian (1996) Isomerization reactions of the n-C_4H_9O and n-OOC_4H_8OH radicals in oxygen. *Int. J. Chem. Kinet.* **28**, 531–544.

Helas, G., and P. Warneck (1981) Background mixing ratios in air masses over the north Atlantic Ocean. *J. Geophys. Res.* **86**, 7283–7290.

Helas, G. H. Bingemer, and M. O. Andreae (1992) Organic acids over equatorial Africa: results from DECAFE 88. *J. Geophys. Res.* **97**, 6187–6193.

Hellpointner, E., and S. Gäb (1989) Detection of methyl, hydroxymethyl and hydroxyethyl hydroperoxides in air and precipitation. *Nature* **337**, 631–634.

Henderson-Sellers, A., M. F. Wilson, G. Thomas, and R. E. Dickinson (1986) Current global land-surface data sets for use in climate-related studies, NCAR/TN-272 + STR. Atmospheric Analysis and Prediction Division, National Center for Atmospheric Research, Boulder, CO.

Hendry, G. R. (1993) "Free-Air CO_2 Enrichment for Plant Research in the Field." Smoley, Boca Raton, FL.

Hendry, D. G., and R. A. Kenley (1977) Generation of peroxy radicals from peroxynitrate (RO_2NO_2), decomposition of peroxyacylnitrates. *J. Am. Chem. Soc.* **99**, 3198–3199.

Hering, S. V., D. R. Lawson, I. Allegrini, A. Febo, C. Perrino, M. Possanzini, J. E. Sickles, K. G. Anlauf, A. Wiebe, B. R. Appel, J. John, J. Ondo, S. Wall, R. S. Braman, R. Sutton, G. R. Cass, P. A. Solomon, D. J. Eatough, N. L. Eatough, E. C. Ellis, D. Grojean, B. B. Hicks, J. D. Womack, J. Horrocks, K. T. Knapp, T. G. Ellestad, R. J. Paur, W. J. Mitchell, M. Pleasant, E. Peake, A. MacLean, W. R. Pierson, W. Brachaczek, H. I. Schiff, G. I. Mackay, C. W. Spicer, D. H. Stedman, A. M. Winer, H. W. Biermann, and E. C. Tuazon (1988) The nitric acid shoot-out: field comparison of measurement methods. *Atmos. Environ.* **22**, 1519–1539.

Hering, S., A. Eldering, and J. H. Seinfeld (1997) Bimodal character of accumulation mode aerosol mass distributions in southern California. *Atmos. Environ.* **31**, 1–11.

Hering, W. S., and T. R. Borden (1967) Ozone sonde observations over North America, Vol. 4, AFCRL 64-30 (IV), Environmental Research Paper no. 279. Air Force Cambridge Research Laboratories, Bedford, MA.

Hering, W. S., and H. U. Dütsch (1965) Comparison of chemiluminescent and electrochemical ozone sonde observations. *J. Geophys. Res.* 70, 5483–5490.

Herman, J. R., and J. E. Mentall (1983) O_2 absorption cross sections (187–225 nm) from stratospheric solar flux measurements. *J. Geophys. Res.* 87, 8967–8975.

Herman, J. R., R. Hudson, R. McPeters, R. Stolarski, Z. Ahmad, X.-Y. Gu, S. Taylor, and C. Wellemeyer (1991) A new self-calibration method applied to TOMS/SBUV backscattered ultraviolet data to determine long-term global ozone change. *J. Geophys. Res.* 96, 7531–7545.

Herrmann, J., and W. Jaeschke (1984) Measurements of H_2S and SO_2 over the Atlantic Ocean. *J. Atmos. Chem.* 1, 111–123.

Herron, J. T., and R. E. Huie (1977) Stopped-flow studies of the mechanism of ozone-alkene reactions in the gas phase: ethylene. *J. Am. Chem. Soc.* 99, 5430–5435.

Hestvedt, E., S. Hendriksen, and H. Hjartarson (1974) On the development of an aerobic atmosphere. A model experiment. *Geophys. Norv.* 31, 1–8.

Heymsfield, A. J. (1993) Microphysical structures of stratiform and cirrus clouds. In "Aerosol-Cloud-Climate Interactions" (P. V. Hobbs, ed.), pp. 97–121. Academic Press, San Diego, CA.

Hidalgo, H., and P. J. Crutzen (1977) The tropospheric and stratospheric composition perturbed by NO_x emissions of high-altitude aircraft. *J. Geophys. Res.* 82, 5833–5866.

Hidy, G. M., and J. R. Brock (1970) "The Dynamics of Aerocolloidal Systems." Pergamon, Oxford.

Hidy, G. M., P. K. Mueller, and E. Y. Tong (1978) Spatial and temporal distributions of airborne sulfate in parts of the United States. *Atmos. Environ.* 12, 735–752.

Hill, F. B., V. P. Aneja, and R. M. Felder (1978) A technique for measurements of biogenic sulfur emission fluxes. *J. Environ. Sci. Health A* 13, 199–225.

Hill, R. D. (1979) A survey of lightning energy estimates. *Rev. Geophys. Space Phys.* 17, 155–164.

Hill, R. D., R. G. Rinker, and H. D. Wilson (1980) Atmospheric nitrogen fixation by lightning. *J. Atmos. Sci.* 37, 179–192.

Hilsenrath, E., and B. M. Schlesinger (1981) Total ozone seasonal and inter-annual variations derived from the 7-year Nimbus-4 BUV data set. *J. Geophys. Res.* 86, 12,087–12,096.

Hilsenrath, E., D. F. Heath, and B. M. Schlesinger (1979) Seasonal and interannual variations in total ozone revealed by Nimbus 4 backscattered ultraviolet experiment. *J. Geophys., Res.* 84, 6969–6979.

Hingston, F. J., and V. Gailitis (1976) The geographic variation of salt precipitation over western Australia. *Austr. J. Soil. Res.* 14, 319–335.

Hitchcock, D. R., and A. E. Wechsler (1972) Biological cycling of trace gases, Final Report NASW-2128, pp. 117–154. Littleton, Cambridge, MA.

Hoare, D. E. (1967) The combustion of methane. In "Low Temperature Oxidation" (W. Jost, ed.), pp. 125–167. Gordon and Breach, New York.

Hobbs, P. V., L. F. Radke, and J. L. Stith (1977) Eruption of the St. Augustine volcano: airborne measurements and observations. *Science* 195, 871–873.

Hobbs, P. V., L. F. Radke, and J. L. Stith (1978) Particles in the eruption cloud from St. Augustine volcano. *Science* 199, 456–458.

Hobbs, P. V., L. F. Radke, M. W. Eltgroth, and D. A. Hegg (1981) Airborne studies of the emissions from the volcanic eruptions of Mount St. Helens. *Science* 211, 816–818.

Hoff, R. M., W. R. Leaitch, P. Fellin, and L. A. Barrie (1983) Mass size distributions of chemical constituents of the winter Arctic aerosol. *J. Geophys. Res.* 88, 10,947–10,956.

Hoffman, E. J., and R. A. Duce (1974) The organic carbon content of marine aerosols collected at Bermuda. *J. Geophys. Res.* 79, 4474–4477.

Hoffman, E. J., and R. A. Duce (1977) Organic carbon in marine atmospheric particulate matter, concentration and particle size distribution. *Geophys. Res. Lett.* 4, 449–452.

Hoffman E. J., G. L. Hoffman, and R. A. Duce (1974) Chemical fractionation of alkali and alkaline earth metals in atmospheric particulate matter on the north Atlantic. *J. Rech. Atmos.* 8, 676–688.

Hoffman, G. L., and R. A. Duce (1972) Consideration of the chemical fractionation of alkali and alkaline earth metals in atmospheric particulate matter over the north Atlantic. *J. Geophys. Res.* 77, 5161–5169.

Hoffmann, A., V. Mörs, and R. Zellner (1992) A novel laser-based technique for the time-resolved study of integrated hydrocarbon oxidation mechanisms. *Ber. Bunsenges. Phys. Chem.* 96, 437–440.

Hoffmann, M. R. (1986) On the kinetics and mechanism of oxidation of aquated sulfur dioxide by ozone. *Atmos. Environ.* 20, 1145–1154.

Hoffmann, M. R., and J. O. Edwards (1975) Kinetics of the oxidation of sulfite by hydrogen peroxide in aqueous solution. *J. Phys. Chem.* 79, 2096–2098.

Hoffmann, T., J. R. Odum, F. Bowman, D. Collins, D. Klockow, R. C. Flagan, and J. H. Seinfeld (1997) Formation of organic aerosols from the oxidation of biogenic hydrocarbons. *J. Atmos. Chem.* 26, 189–222.

Höfken, K. D., H. W. Georgii, and G. Gravenhorst (1981) Untersuchungen über die Deposition atmosphärischer Spurenstoffe an Buchen- und Fichtenwald, Report no. 46. Institut für Meteorologie und Geophysik, Universität, Frankfurt (Main), Germany.

Hofmann, D. J., and S. Solomon (1989) Ozone destruction through heterogeneous chemistry following the eruption of El Chichón. *J. Geophys. Res.* 94, 5029–5041.

Hofmann, D. J., J. M. Rosen, T. J. Pepin, and R. G. Pinnick (1975) Stratospheric aerosol measurement I. Time variations at northern midlatitudes, *J. Atmos. Sci.* 32, 1446–1456.

Hofmann, D. J., S. J. Oltmans, J. M. Harris, J. A. Lathrop, G. L. Koenig, W. D. Komhyr, R. D. Evans, D. M. Quincy, T. Deshler, and B. J. Johnson (1994a) Recovery of stratospheric ozone over the United States in the winter 1993–1994. *Geophys. Res. Lett.* 21, 1779–1782.

Hofmann, D. J., S. J. Oltmans, J. M. Harris, J. A. Lathrop, G. L. Koenig, W. D. Komhyr, R. D. Evans, D. M. Quincy, T. Deshler, and B. J. Johnson (1994b) Ozone loss in the lower stratosphere over the United States in 1992–1993: evidence for heterogeneous chemistry on the Pinatubo aerosol. *Geophys. Res. Lett.* 21, 65–68.

Hofmann, H., P. Hoffmann, and K. H. Lieser (1991) Transition metals in atmospheric aqueous samples, analytical determination and speciation. *Fresenius J. Anal. Chem.* 340, 591–597.

Hoigné, J., H. Bader, W. R. Haag, and J. Staehelin (1985) Rate constants for the reactions of ozone with organic and inorganic compounds in water. III. Inorganic compounds and radicals. *Water Res.* 19, 993–1004.

Holdren, M. W., H. H. Westberg, and P. R. Zimmerman (1979) Analysis of monoterpene hydrocarbons in rural atmospheres. *J. Geophys. Res.* 84, 5083–5088.

Holeman, J. N. (1968) The sediment yield of major rivers of the world. *Water Resour. Res.* 4, 737–747.

Holland, D. M., and A. W. Castleman, Jr. (1982) The Thomson equation revisited in the light of ion-clustering experiments. *J. Chem. Phys.* 86, 4181–4188.

Holland, D. M., P. P. Principe, and J. E. Sickles (1999) Trends in atmospheric sulfur and nitrogen species in the eastern United States for 1989–1995. *Atmos. Environ.* 33, 37–49.

Holland, H. D. (1973) Ocean water, nutrients and atmospheric oxygen. In "Hydrogeochemistry" (E. Ingersoll, ed.). Proceedings of the IAGG Symposium on Hydrogeochemistry and Biogeochemistry, Tokyo, 1970, Vol. I, pp. 68–81. The Clark Co., Washington, DC.

Holland, H. D. (1978) "The Chemistry of the Atmosphere and Oceans." Wiley, New York.

Holland, H. D. (1984) "The Chemical Evolution of the Atmosphere and Oceans." Princeton University Press, Princeton, NJ.

Holser, W. T., and I. R. Kaplan (1966) Isotope geochemistry of sedimentary sulfates. *Chem. Geol.* 1, 93–135.

Holton, J. R. (1990) On the global exchange of mass between the stratosphere and troposphere. *J. Atmos. Sci.* 47, 392–395.

Holton, J. R., P. H. Haynes, M. E. McIntyre, A. R. Douglass, R. B. Rood, and L. Pfister (1995) Stratosphere-troposphere exchange. *Rev. Geophys.* 33, 403–439.

Holzapfel-Pschorn, A., and W. Seiler (1986) Methane emission during a cultivation period from an Italian rice paddy. *J. Geophys. Res.* 91, 11,803–11,814.

Honrath, R. E., and D. A. Jaffe (1992) The seasonal cycle of nitrogen oxides in the Arctic troposphere at Barrow, Alaska. *J. Geophys. Res.* 97, 20,615–20,630.

Hoppel, W. A., J. E. Dinger, and R. E. Ruskin (1973) Vertical profile of CCN at various geographic locations. *J. Atmos. Sci.* 30, 1410–1420.

Hoppel, W. A., J. W. Fitzgerald, and R. E. Larson (1985) Aerosol size distribution in air masses advecting off the east coast of the United States. *J. Geophys. Res.* 90, 2365–2379.

Hoppel, W. A., G. M. Frick, and R. E. Larson (1986) Effect of non-precipitating clouds on the aerosol size distribution. *Geophys. Res. Lett.* 13, 125–128.

Hoppel, W. A., G. M. Frick, J. W. Fitzgerald, and R. E. Larson (1994) Marine boundary layer measurements of new particle formation and the effects non-precipitating clouds have on aerosol size distribution. *J. Geophys. Res.* 99, 14,443–14,459.

Horie, O., and G. K. Moortgat (1991) Decomposition pathways of the excited Criegee intermediates in the ozonolysis of simple alkenes. *Atmos. Environ.* 25, 1881–1896.

Horie, O., P. Neeb, S. Limbach, and G. K. Moortgat (1994a) Formation of formic acid and organic peroxides in the ozonolysis of ethene with added water vapor. *Geophys. Res. Lett.* 21, 1523–1526.

Horie, O., P. Neeb, and G. K. Moortgat (1994b) Ozonolysis of *trans*- and *cis*-2-butenes in low part-per-million concentration ranges. *Int. J. Chem. Kinet.* 26, 1075–1094.

Horie, O., P. Neeb, and G. K. Moortgat (1997) The reactions of the Criegee intermediate CH_3CHOO in the gas-phase ozonolysis of 2-butene isomers. *Int. J. Chem. Kinet.* 29, 461–468.

Horn, M. K., and J. A. S. Adams (1966) Computer-derived geochemical balances and element abundances. *Geochim. Cosmochim. Acta* 30, 279–297.

Horowitz, A., and J. G. Calvert (1978) Wavelength dependence of the quantum efficiencies of the primary processes in formaldehyde photolysis at 25°C. *Int. J. Chem. Kinet.* 10, 805–819.

Horowitz, A., and J. G. Calvert (1982) Wavelength dependence of the primary process in acetaldehyde photolysis. *J. Phys. Chem.* 86, 3105–3114.

Horvath, J. J., and C. J. Mason (1978) Nitric oxide mixing ratios near the stratopause measured by a rocket-borne chemiluminescent detector. *Geophys. Res. Lett.* 5, 1023–1026.

Horvath, L., E. Mészáros, E. Antal, and A. Simon (1981) On the sulfate, chloride and sodium concentrations in maritime air around the Asian continent. *Tellus* 33, 382–386.

Houghton, J. T., L. G. Meira Filho, J. Bruce, H. Lee, B. A. Callander, E. F. Haites, N. Harris, and K. Maskell (eds.) (1995) "Climate Change 1994: Radiative Forcing of Climate Change and an Evaluation of the IPCC IS92 Emission Scenarios." Cambridge University Press, Cambridge, UK.

Houghton, J. T., L. G. Meira Filho, B. A. Callander, N. Harris, A. Kattenberg, and K. Maskell (eds.) (1996) "Climate Change 1995: The Science of Climate Change." Cambridge University Press, Cambridge, UK.

Houghton, R. A. (1993) Is carbon accumulating in the northern temperate zone? *Global Biogeochem. Cycles* 7, 611–617.

Houghton, R. A. (1996) Terrestrial sources and sinks of carbon inferred from terrestrial data. *Tellus* 48B, 420–432.

Houtermans, J. C., H. E. Suess, and H. Oeschger (1973) Reservoir models and production rate variations of natural radiocarbon. *J. Geophys. Res.* **78**, 1897–1908.

Hov, O., S. A. Penkett, I. S. A. Isaksen, and A. Semb (1984) Organic gases in the Norwegian Arctic. *Geophys. Res. Lett.* **11**, 425–428.

Hübler, G., D. Perner, U. Platt, A. Tönissen, and D. H. Ehhalt (1984) Ground level OH radical concentration: new measurements by optical absorption. *J. Geophys. Res.* **89**, 1309–1319.

Hubrich, C., C. Zetzsch, and F. Stuhl (1977) Absorptionsspektren von halogenierten Methanen im Bereich von 275 bis 160 nm bei Temperaturen von 298 und 208 K. *Ber. Bunsenges. Phys. Chem.* **81**, 437–442.

Hudson, R. D. (chief ed.) (1982) "The Stratosphere 1981: Theory and Measurements," WMO Global Research and Monitoring Project Report no 11. World Meteorological Organization, Geneva.

Hudson, R. D., and S. H. Mahle (1972) Photodissociation rates of molecular oxygen in the mesosphere and lower thermosphere. *J. Geophys. Res.* **77**, 2902–2914.

Hudson, R. D., V. L. Carter, and J. A. Stein (1966) An investigation of the effect of temperature on the Schumann-Runge absorption continuum of oxygen 1580–1950 Å. *J. Geophys. Res.* **71**, 2295–2298.

Hudson, R. D., V. L. Carter, and E. L. Breig (1969) Predissociation in the Schumann-Runge band system of O_2: laboratory measurements and atmospheric effects. *J. Geophys. Res.* **74**, 4079–4086.

Huebert, B. J. (1980) Nitric acid and aerosol nitrate measurements in the equatorial Pacific region. *Geophys. Res. Lett.* **7**, 325–328.

Huebert, B. J., and A. L. Lazrus (1980a) Tropospheric gas-phase and particulate nitrate measurements. *J. Geophys. Res.* **85**, 7322–7328.

Huebert, B. J., and A. L. Lazrus (1980b) Bulk composition of aerosols in the remote troposphere. *J. Geophys. Res.* **85**, 7337–7344.

Huebert, B. J., and C. H. Robert (1985) The dry deposition of nitric acid to grass. *J. Geophys. Res.* **90**, 2085–2090.

Huebert, B. J., S. Howell, P. Laj, E. Johnson, T. S. Bates, P. K. Quinn, V. Yegorov, A. D. Clarke, and J. N. Porter (1993) Observations of the atmospheric sulfur cycle on SAGA 3. *J. Geophys. Res.* **98**, 16,985–16,995.

Huie, R. E., and C. L. Clifton (1990) Temperature dependence of the rate constants for reactions of the sulfate radical, SO_4^-, with anions. *J. Phys. Chem.* **94**, 8561–8567.

Hull, L. A. (1981) Terpene ozonolysis products. *In* "Atmospheric Biogenic Hydrocarbons" (J. J Bufalini and R. R. Arnts, eds.), pp. 161–186. Ann Arbor Science, Ann Arbor, MI.

Hunt, B. G. (1966a) Photochemistry of ozone in a moist atmosphere. *J. Geophys. Res.* **71**, 1386–1398.

Hunt, B. G. (1966b) The need for a modified photochemical theory of the ozonosphere. *J. Atmos. Sci.* **23**, 2388–2395.

Hunt, J. M. (1972) Distribution of carbon in crust of earth. *Bull. Am. Assoc. Pet. Geol.* **56**, 2273–2277.

Hunten, D. M. (1973) The escape of light gases from planetary atmospheres. *J. Atmos. Sci.* **30**, 1481–1484.

Hunten, D. M. (1975) Estimates of stratospheric pollution by an analytical model. *Proc. Natl. Acad. Sci. USA.* **72**, 4711–4715.

Hunten, D. M., and D. F. Strobel (1974) Production and escape of terrestrial hydrogen. *J. Atmos. Sci.* **31**, 305–317.

Husain, L., and V. A. Dutkiewicz (1992) Elemental tracers for the study of homogeneous gas phase oxidation of SO_2 in the atmosphere. *J. Geophys. Res.* **97**, 14,635–14,643.

Husar, R. B., and D. E. Patterson (1980) Regional scale air pollution sources and effects. *Ann. N.Y. Acad. Sci.* **338**, 399–417.

Husar, R. B., and K. T. Whitby (1973) Growth mechanism and size spectra of photochemical aerosols. *Environ. Sci. Technol.* **7**, 241–247.

Husar, R. B., D. E. Patterson, J. D. Husar, N. V. Gillani, and W. E. Wilson (1978) Sulfur budget of a power plant plume. *Atmos. Environ.* **12**, 549–568.

Husar, R. B., L. L. Stowe, and J. M. Prospero (1997) Satellite sensing or tropospheric aerosols over the oceans with NOAA AVHRR. *J. Geophys. Res.* **102**, 16,889–16,909.

Huss, A., P. K. Kim, and C. A. Eckert (1978) On the uncatalyzed oxidation of sulfur (IV) in aqueous solution. *J. Am. Chem. Soc.* **100**, 6252–6253.

Hutchinson, G. E. (1948) Circular causal systems in ecology. *Ann. N.Y. Acad. Sci.* **50**, 221–246.

Hutchinson, G. E. (1954) The biogeochemistry of the terrestrial atmosphere. In "The Earth as a Planet" (G. P. Kuiper, ed.), pp. 371–433. University of Chicago Press, Chicago.

Hutchinson, G. L., R. L. Millington, and D. B. Peters (1972) Atmospheric ammonia: absorption by plant leaves. *Science* **175**, 771–772.

Hutton, J. T. (1976) Chloride in rainwater in relation to distance from ocean. *Search* **7**, 207–208.

Hynes, A. J., P. H. Wine, and D. H. Semmes (1986) Kinetics and mechanism of OH reactions with organic sulfides. *J. Phys. Chem.* **90**, 4148–4156.

Hynes, A. J., P. H. Wine, and J. M. Nicovich (1988) Kinetics and mechanism of the reaction of OH with CS_2 under atmospheric conditions. *J. Phys. Chem.* **92**, 3846–3852.

Idso, S. B., and B. A. Kimball (1997) Effects of long-term atmospheric CO_2 enrichment on the growth and fruit production of sour orange trees. *Global Change Biol.* **3**, 89–96.

Ingersoll, R. B., R. E. Inman, and W. R. Fisher (1974) Soil's potential as a sink for atmospheric carbon monoxide. *Tellus* **26**, 151–159.

Inman, R. E., R. B. Ingersoll, and E. A. Levy (1971) Soil: a natural sink for carbon monoxide *Science* **172**, 1229–1231.

Inn, E. C. Y., and Y. Tanaka (1953) Absorption coefficient of ozone in the ultraviolet and visible regions. *J. Opt. Soc. Am.* **43**, 870–873.

Inn, E. C. Y., J. F. Vedder, and B. J. Tyson (1979) COS in the stratosphere. *Geophys. Res. Lett.* **6**, 191–193.

Inn, E. C. Y., J. F. Vedder, and D. O'Hara (1981) Measurements of stratospheric sulfur constituents. *Geophys. Res. Lett.* **8**, 5–8.

Inn, E. C. Y., N. H. Farlow, P. B. Russel, M. P. McCormick, and W. P. Chu (1982) Observations. In "The Stratospheric Aerosol Layer" (R. C. Whitten, ed.), pp. 15–68. Springer-Verlag, Berlin.

Isaksen, I. S. A., K. H. Mitboe, J. Sunde, and P. J. Crutzen (1977) A simplified method to include molecular scattering and reflection in calculations of photon fluxes and photodissociation rates. *Geophys. Norv.* **31**, 11–26.

Isodorov, V. A., I. G. Zenkevich, and B. V. Ioffe (1985) Volatile organic compounds in the atmosphere of forests. *Atmos. Environ.* **19**, 1–8.

Ito, T., T. Okita, M. Ikegami, and I. Kanazawa (1986) The characterization and distribution of aerosol and gaseous species in the winter monsoon over the western Pacific Ocean. II. The residence time of aerosols and SO_2 in the long range transport over the ocean. *J. Atmos. Chem.* **4**, 401–411.

Ivanov, M. V., V. A. Grinenko, and A. P. Rabinowich (1983) The sulfur cycle in continental reservoirs. Part II. Sulfur flux from continents to ocean. In "The Global Biogeochemical Sulphur Cycle" (M. V. Ivanov and J. R. Freney, eds.). *SCOPE* **19**, 331–356.

Ivey, J. P., D. M. Davies, V. Morgan, and G. P. Ayers (1986) Methanesulphonate in Antarctic ice. *Tellus* **38B**, 375–379.

Jackman, C. H., J. E. Frederick, and R. S. Stolarski (1980) Production of odd nitrogen in the stratosphere and mesosphere. An intercomparison of sources strengths. *J. Geophys Res.* **85**, 7495–7505.

Jacob, D. J. (1986) Chemistry of OH in remote clouds and its role in the production of formic acid and peroxomonosulfate. *J. Geophys. Res.* **91**, 9807–9826.

Jacob, D. J., and M. R. Hoffmann (1983) A dynamic model for the production of H^+, NO_3^- and SO_4^{2-} in urban fog. *J. Geophys. Res.* **88**, 6611–6621.

Jacob, D. J., and S. C. Wofsy (1988) Photochemistry of biogenic emissions over the Amazon forest. *J. Geophys. Res.* **93**, 1477–1486.

Jacob, D. J., M. J. Prather, S. C. Wofsy, and M. B. McElroy (1987) Atmospheric distribution of ^{85}Kr simulated with a general circulation model. *J. Geophys. Res.* **92**, 6614–6626.

Jacobi, H.-W. (1996) Kinetische Untersuchungen und Modellrechnungen zur troposphärischen Chemie von Radikalanionen und Ozon in wäßriger Phase. Dissertation. Universität Essen, Germany.

Jacobi, H.-W., H Herrmann, and R. Zellner (1996) Kinetic investigation of the Cl_2^- radical in the aqueous phase. *In* "Homogeneous and Heterogeneous Chemical Processes in the Troposphere" (P. Mirabel, ed.). Air Pollution Research Report 57, pp. 172–176. European Commission, Brussels.

Jacobi, H.-W., H. Herrmann, and R. Zellner (1997) A laser flash photolysis study of the decay of Cl-atoms and Cl_2^- radical anions in aqueous solution at 298 K. *Ber. Bunsenges. Phys. Chem.* **101**, 1909–1913.

Jacobs, M B., and S. Hocheiser (1958) Continuous sampling and ultramicrodetermination of nitrogen dioxide in air. *Anal. Chem* **30** 426–428.

Jaecker-Voirol, A., and P. Mirabel (1988) Nucleation rates in a binary mixture of sulfuric acid and water vapor. *J. Phys. Chem.* **92**, 3518–3521.

Jaecker Voirol, A., and P. Mirabel (1989) Heteromolecular nucleation in the sulfuric acid-water system. *Atmos. Environ.* **23**, 2053–2057.

Jaecker-Voirol, A., P. Mirabel, and J. J. Rriss (1987) Hydrates in supersaturated binary sulfuric acid water vapor: a reexamination. *J. Chem. Phys.* **87**, 4849–4053.

Jaenicke, R. (1978a) Aitken particle size distribution in the Atlantic north east trade winds. *Meteor Forschungerg. Reihe B* **13**, 1–9.

Jaenicke, R. (1978b) The role of organic material in atmospheric aerosols. *Pure Appl. Geophys.* **116**, 283–292.

Jaenicke, R. (1978c) Über die Dynamik atmosphärischer Aitkenteilchen. *Ber. Bunsenges. Phys. Chem.* **82**, 1198–1202.

Jaenicke, R. (1980) Natural aerosols. *In* "Aerosols: Anthropogenic and Natural, Sources and Transport" (T. J. Kneip and P. J. Lioy, eds.). *Ann. N Y. Acad. Sci.* **338**, 317–329.

Jaenicke, R., and C. E. Junge (1967) Studien zur oberen Grenzgröße des natürlichen Aerosols. *Beitr. Phys. Atmos.* **40**, 129–143.

Jaenicke, R., and L. Schütz (1978) Comprehensive study of physical and chemical properties of the surface aerosol in the Cape Verde Island region. *J. Geophys. Res.* **83**, 3585–3599.

Jaeschke, W., and J. Herrmann (1981) Measurements of H_2S in the atmosphere. *Int. J. Environ. Anal. Chem.* **10**, 107–120.

Jaeschke, W., H. W. Georgii, and R. Schmitt (1976) Preliminary results of stratospheric SO_2 measurements. *Geophys. Res. Lett.* **3**, 517–519.

Jaeschke, W., H. W. Georgii, H. Claude, and H. Malewski (1978) Contribution of H_2S to the atmospheric sulfur cycle. *Pure Appl. Geophys.* **116**, 465–475.

Jaeschke, W., H. Claude, and J. Herrmann (1980) Sources and sinks of atmospheric H_2S. *J. Geophys. Res.* **85**, 5639–5644.

Jaeschke, W., H. Berresheim, and H. W. Georgii (1982) Sulfur emissions from Mount Etna. *J. Geophys. Res.* **87**, 7253–7261.

Jagger, T. A. (1940) Magmatic gases. *Am. J. Sci.* **238**, 313–353.

Janach, W. E. (1989) Surface ozone: trend details, seasonal variations, and interpretation. *J. Geophys. Res.* **94**, 18,289–18,295.

Janson, R. W. (1992) Monoterpene concentrations in and above a forest of Scots pine. *J. Atmos. Chem.* **14**, 385–394.

Janson, R. W. (1993) Monoterpene emissions from Scots pine and Norwegian spruce. *J. Geophys. Res.* **98**, 2839–2850.

Jayson, G. G., B. J. Parson, and A. J. Swallow (1973) Some simple, highly reactive inorganic chlorine derivatives in aqueous solution. *J. Chem. Soc. Faraday Trans. I* **69**, 1697–1607.

Jeans, J. H. (1916) "The Dynamical Theory of Gases," 4th ed. (republication, 1954). Dover, New York.

Jeffries, H. E., D. L. Fox, and R. Kamens (1976) Photochemical conversion of NO to NO_2 by hydrocarbon in an outdoor chamber. *J. Air. Pollut. Control Assoc.* **26**, 480–484.

Jenkin, M. E., R. A. Cox, and D. J. Williams (1988) Laboratory studies of the kinetics of formation of nitrous acid from the thermal reaction of nitrogen dioxide and water vapor. *Atmos. Environ.* **22**, 487–498.

Jenouvrier, A., B. Coquart, and M. F. Merienne-Lafore (1986) New measurements of the absorption cross sections in the Herzberg continuum of molecular oxygen in the region between 205 and 240 nm. *Planet. Space Sci.* **34**, 253–254.

Jensen, N. R., J. Hjorth, C. Lohse, H. Skov, and G. Restelli (1992) Products and mechanisms of the gas phase reactions of NO_3 with CH_3SCH_3, CD_3SCD_3, CH_3SH and CH_3SSCH_3. *J. Atmos. Chem.* **14**, 95–108.

Jickells, T., A. Knap, T. Church, J. Galloway, and J. Miller (1982) Acid rain on Bermuda. *Nature* **297**, 55–57.

Jobson, B. T., H. Niki, Y. Yokouchi, J. Bottenheim, F. Hopper, and R. Leaitch (1994) Measurements of C_2–H_6 hydrocarbons during the Polar Sunrise 1992 Experiment: evidence for Cl atom and Br atom chemistry. *J. Geophys. Res.* **99**, 25,355–25,368.

Jochum, K. P., A. W. Hofmann, E. Ito, H. M. Seufert, and W. M. White (1983) K, U, and Th in mid-ocean ridge basalt glasses and heat production, K/U and K/Rb in the mantle. *Nature* **306**, 431–436.

Johansson, C. (1984) Field measurements of emission of nitric oxide from fertilized and unfertilized forest soils in Sweden. *J. Atmos. Chem.* **1**, 429–442.

Johansson, C. (1987) Pine forest: a negligible sink for atmospheric NO_2 in Sweden. *Tellus* **39B**, 426–438.

Johansson, C., and R. W. Janson (1993) Diurnal cycle of O_3 and monoterpenes in a coniferous forest: importance of atmospheric stability, surface exchange, and chemistry. *J. Geophys. Res.* **98**, 5121–5133.

Johansson, C., H. Rodhe, and E. Sanhueza (1988) Emission of NO in tropical savanna and a cloud forest during the dry season. *J. Geophys. Res.* **93**, 7180–7192.

John, W., S. M. Wall, J. L. Ondo, and W. Winklmayr (1990) Modes in the size distributions of atmospheric inorganic aerosol. *Atmos. Environ.* **24A**, 2349–2359.

Johnson, B., and R. C. Cooke (1979) Bubble populations and spectra in coastal waters: a photographic approach. *J. Geophys. Res.* **84**, 3763–3766.

Johnson, D. E., T. M. Hill, G. M. Ward, K. A. Johnson, M. E. Branine, B. R. Carmean, and D. W. Lodman (1993) Ruminants and other animals. *In* "Atmospheric Methane: Sources, Sinks, and Role in Global Change" (M. A. K. Khalil, ed.), NATO ASI Series, Vol. I 13, pp. 199–229. Springer-Verlag, Berlin.

Johnson, E. J., R. H. Gammon, J. Larsen, T. S. Bates, S. J. Oltmans, and J. C. Farmer (1990) Ozone in the boundary layer over the Pacific and Indian Oceans: latitude gradients and diurnal cycles. *J. Geophys. Res.* 95, 11,874–11,856.

Johnson, J. E. (1981) The life time of carbonyl sulfide in the troposphere. *Geophys. Res. Lett.* 8, 938–940.

Johnson, J. E., A. R. Bandy, D. C. Thornton, and T. S. Bates (1993) Measurements of atmospheric carbonyl sulfide during the NASA CITE 3 project: implications for the global COS budget. *J. Geophys. Res.* 98, 23,443–23,448.

Johnston, H. S. (1966) "Gas Phase Reaction Rate Theory." Ronald Press, New York.

Johnston, H. S. (1971) Reduction of stratospheric ozone by nitrogen oxide catalysis from SST exhaust. *Science* 173, 517–522.

Johnston, H. S. (1972) Laboratory kinetics as an atmospheric science. In "Climatic Impact Assessment Program, Proceedings of the Survey Conference" (A. E. Barrington, ed.), DOT-TSC-OST-72-13, pp. 90–112. U.S. Department of Transportation, Washington, DC.

Johnston, H. S. (1975) Global balance in the natural stratosphere. *Rev. Geophys. Space Phys.* 13, 637–649.

Johnston, H. S., and J. Podolske (1978) Interpretations of stratospheric photochemistry. *Rev. Geophys, Space Phys.* 16, 491–519.

Johnston, H. S., and G. Selwyn (1975) New cross section for the absorption of near ultraviolet radiation by nitrous oxide (N_2O). *Geophys. Res. Lett.* 2, 549–551.

Johnston, H. S., and G. Whitten (1973) Instantaneous photochemical rates in the global stratosphere. *Pure Appl. Geophys.* 106–108, 1468–1489.

Johnston, H. S., and G. Whitten (1975) Chemical reactions in the atmosphere as studied by the method of instantaneous rates. *Int. J. Chem. Kinet. Symp.* 1, 1 26.

Johnston, H. S., W. A. Bonner, and D. J. Wilson (1957) Carbon isotope effect during oxidation of carbon monoxide with nitrogen dioxide. *J. Chem. Phys.* 26, 1002 1006.

Johnston, H. S., S. G. Chang, and G. Whitten (1974) Photolysis of nitric acid vapor. *J. Phys. Chem.* 78, 1–7.

Johnston, H. S., O. Serang, and J. Podolske (1979) Instantaneous global nitrous oxide photochemical rates. *J. Geophys. Res.* 84, 5077–5082.

Johnston, P. V., and R. L. McKenzie (1984) Long path absorption measurements of tropospheric NO_2 in rural New Zealand. *Geophys. Res. Lett.* 11, 69–72.

Jones, B. M. R., J. P. Burrows, R. A. Cox, and S. A. Penkett (1982) OCS formation in the reaction of OH with CS_2. *Chem. Phys. Lett.* 88, 372–376.

Jones, B. M. R., R. A. Cox, and S. A. Penkett (1983) Atmospheric chemistry of carbon disulphide. *J. Atmos. Chem.* 1, 65–86.

Jones, I. T. N., and K. D. Bayes (1973) Photolysis of nitrogen dioxide. *J. Chem. Phys.* 59, 4836–4844.

Jones, I. T. N., and R. P. Wayne (1970) The photolysis of ozone by ultraviolet radiation IV: effect of photolysis wavelength on primary steps. *Proc. R. Soc. Lond.* A 319, 273–287.

Jones, R. L., J. A. Pyle, J. E. Harries, A. M. Zavody, J. M. Russel, and J. C. Gille (1984) The water vapour budget of the stratosphere studied using LIMS and SAMS satellite data. *Q. J. R. Meteorol. Soc.* 112, 1127–1143.

Joos, F., and U. Baltensperger (1991) A field study on chemistry, S(IV) oxidation rates and vertical transport during fog conditions. *Atmos. Environ.* 25A, 217–230.

Joseph, D. W., and C. W. Spicer (1978) Chemiluminescence method for atmospheric monitoring of nitric acid and nitrogen oxides. *Anal. Chem.* 50, 1400–1403.

Jouzel, J., C. Lorius, J. R. Petit, C. Genthon, N. I. Barkov, V. M. Kotlyakov, and V. M. Petrov (1987) Vostok ice core: a continuous isotope temperature record over the last climatic cycle (160,000 years). *Nature* 329, 403–407.

Judd, A. G., R. H. Charlier, A. Lacroix, G. Lambert, and C. Rouland (1993) Minor sources of methane. *In* "Atmospheric Methane: Sources, Sinks, and Role in Global Change" (M. A. K. Khalil, ed.), NATO ASI Series, Vol. I 13, pp. 432–456. Springer-Verlag, Berlin.

Junge, C. E. (1952a) Gesetzmäßigkeiten in der Größenverteilung des atmosphärischen Aerosol über dem Kontinent. *Ber. Dtsch. Wetterdienst, U.S. Zone* 35, 261–277.

Junge, C. E. (1952b) Das Größenwachstum der Aitkenkerne. *Ber. Dtsch. Wetterdienst, U.S. Zone* 38, 264–267.

Junge, C. E. (1953) Die Rolle der Aerosole und der gasförmigen Beimengungen der Luft im Spurenhaushalt der Troposphäre. *Tellus* 5, 1–26.

Junge, C. E. (1954) The chemical composition of atmospheric aerosols. I. Measurements at Round Hill Field Station June–July 1953. *J. Meteorol.* 11, 223–233.

Junge, C. E. (1955) The size distribution and aging of natural aerosols as determined from electrical and optical data on the atmosphere. *J. Meteorol.* 12, 13–25.

Junge, C. E. (1956) Recent investigations in air chemistry. *Tellus* 8, 127–139.

Junge, C. E. (1957) The vertical distribution of aerosols over the ocean. *In* "Artificial Stimulation of Rain" (J. Weickmann and W. Smith, eds.), Proceedings of the First Conference on the Physics of Cloud Precipitation Particles, pp. 89–96. Pergamon, London.

Junge, C. E. (1961) Vertical profiles of condensation nuclei in the stratosphere. *J. Meteorol.* 18, 501–509.

Junge, C. E. (1962) Global ozone budget and exchange between stratosphere and troposphere. *Tellus* 14, 363–377.

Junge, C. E. (1963) "Air Chemistry and Radioactivity." Academic Press, New York.

Junge, C. E. (1964) The modification of aerosol size distribution in the atmosphere, Final Technical Report, Contract Da 91-591-EVC 2979. Institute of Meteorology, University Mainz, Germany.

Junge C. E. (1972) The cycles of atmospheric gases—natural and man-made. *Q. J. R. Meteorol. Soc.* 98, 711–729.

Junge, C. E. (1974) Residence time and variability of tropospheric trace gases. *Tellus* 26, 477–488.

Junge, C. E. (1979) The importance of mineral dust as an atmospheric constituent. *In* "Sahara Dust: Mobilization, Transport, Deposition" (C. Morales, ed.). *SCOPE* 14, 49–60.

Junge, C. E., and N. Abel (1965) Modification of aerosol size distribution in the atmosphere and development of an ion counter of high sensitivity, Final Technical Report no. Da 91-591-EVC-3404. Institute of Meteorology, University Mainz, Germany.

Junge, C. E., and G. Czeplak (1968) Some aspects of the seasonal variation of carbon dioxide and ozone. *Tellus* 20, 422 434.

Junge, C. E., and E. McLaren (1971) Relationship of cloud nuclei spectra to aerosol size distribution and composition. *J. Atmos. Sci.* 28, 382–390.

Junge, C. E., and T. G. Ryan (1958) Study of the SO_2 oxidation in solution and its role in atmospheric chemistry. *Q. J. R. Meteorol. Soc.* 84, 46–55.

Junge, C. E., and P. Warneck (1979) Composition of the atmosphere. *In* "Review of Research on Modern Problems in Geochemistry" (F. R. Siegel, ed.), pp. 139–165. UNESCO, Paris.

Junge, C. E., C. W. Chagnon, and J. E. Manson (1961) Stratospheric aerosols. *J. Meteorol.* 18, 81–108.

Junge, C. E., B. Bockholt, K. Schütz, and R. Beck (1971) N_2O measurements in air and sea water over the Atlantic. *Meteor. Forschungsergeb. B* 6, 1–11.

Kadota, H., and Y. Ishida (1972) Production of volatile sulfur compounds by microorganisms. *Annu. Rev. Microbiol.* 26, 127–163.

Kadowaki, S. (1976) Size distribution of atmospheric total aerosol, sulfate, ammonium and nitrate particulates in the Nagoya area. *Atmos. Environ.* 11, 671–695.

Kadowaki, S. (1977) Size distribution and chemical composition of atmospheric particulate nitrate in Nagoya area. *Atmos. Environ.* 11, 671–675.

Kaiser, E. W., and C. H. Wu (1977) A kinetic study of the gas phase formation and decomposition of nitrous acid. *J. Phys. Chem.* 81, 1701–1708.

Kajimoto, O., and R. J. Cvetanovich (1979) Absolute quantum yields of $O(^1D_2)$ in the photolysis of ozone in the Hartley band. *Int. J. Chem. Kinet.* 11, 605–612.

Kalabokas, P., P. Carlier, P. Fresnet, G. Mouvier, and G. Toupance (1988) Field study of aldehyde chemistry in the Paris area. *Atmos. Environ.* 22, 149–155.

Kallend, A. S., A. R. W. Marsh, J. H. Pickles, and M. V. Proctor (1983) Acidity of rain in Europe. *Atmos. Environ.* 17, 127–137.

Kanakidou, M., H. B. Singh, K. M. Valentin, and P. J. Crutzen (1991) A two-dimensional study of ethane and propane oxidation in the troposphere. *J. Geophys. Res.* 96, 15,395–15,413.

Kanda, Y., and M. Taira (1990) Chemiluminescent method for continuous monitoring of nitrous acid in ambient air. *Anal. Chem* 62, 1084–2087.

Kandler, O. (1981) Archebakterien und Phylogenie der Organismen. *Naturwissenschaften* 68, 183–192.

Kaplan, L. D. (1973) Background concentration of photochemically active trace constituents in the stratosphere and upper atmosphere. *Pure Appl. Geophys.* 106–108, 1342–1345.

Kaplan, W. A., S. C. Wofsy, M. Keller, and J. M. Costa (1988) Emission of NO and deposition of O_3 in a tropical forest system. *J. Geophys. Res.* 93, 1389–1395.

Käselau, K. H., P. Fabian, and H. Röhrs (1974) Measurements of aerosol concentrations up to a height of 27 km. *Pure Appl. Geophys,* 112, 877–885.

Kasper, A., and H. Puxbaum (1998) Seasonal variation of SO_2, HNO_3, NH_3, and selected aerosol components at Sonnblick (3106 m a s. l.). *Atmos. Environ* 32, 3925–3939.

Kasting, J. F. (1993) Earth's early atmosphere. *Science* 259, 920 –926

Kasting, J. F, S. C. Liu, and T. M. Donahue (1979) Oxygen levels in the primitive atmosphere. *J. Geophys. Res* 84, 3097–3107.

Kasting, J. F., J. B. Pollack, and T. P. Ackerman (1984) Response of the earth's atmosphere to increase in solar flux and implications for loss of water from Venus. *Icarus* 57, 335–355.

Kato, N., and H. Akimoto (1992) Anthropogenic emissions of SO_2 and NO_x in Asia: emission inventories plus errata. *Atmos. Environ.* 26A, 2997–3017

Katsura, T., and S. Nagashima (1974) Solubility of sulfur in some magmas at one atmosphere. *Geochim. Cosmochim. Acta* 38, 517–531.

Kawamura, K., and I. R. Kaplan (1987) Motor exhaust emissions as a primary source for dicarboxylic acids in Los Angeles air. *Environ. Sci. Technol.* 21, 105–110.

Kawamura, K., and K. Usukara (1993) Distributions of low molecular weight dicarboxylic acids in north Pacific aerosol samples. *J. Oceanogr.* 49, 271–283.

Kawamura, K., Lai-Ling Ng, and I. R. Kaplan (1985) Determination of organic acids ($C1–C_{10}$) in the atmosphere, motor exhaust and engine oils. *Environ. Sci. Technol.* 19, 1082–1086.

Kawamura, K., H. Kasukabe, and L. A. Barrie (1996) Source and reaction pathways of dicarboxylic acids, ketoacids and dicarbonyls in Arctic aerosols: one year of observations. *Atmos. Environ.* 30, 1709–1722.

Keeler, G. L., J. D. Spengler, and R. A. Castillo (1991) Acid aerosol measurements at a suburban Connecticut site. *Atmos. Environ.* 25A, 681–690.

Keeling, C. D. (1973a) Industrial production of carbon dioxide from fossil fuels and limestone. *Tellus* 27, 174–198.

Keeling, C. D. (1973b) The carbon dioxide cycle. Reservoir models to depict the exchange of atmospheric carbon dioxide with the ocean and land plants. *In* "Chemistry of the Lower Atmosphere" (S. Rasool, ed.), pp. 251–329. Plenum, New York.

Keeling, C. D., R. B. Bacastow, and T. P. Whorf (1982) Measurements of the concentration of carbon dioxide at Mauna Loa Observatory, Hawaii. In "Carbon Dioxide Review 1982" (S. Rasool, ed.), pp. 251–329. Plenum, New York.

Keeling, R. F., and S. R. Shertz (1992) Seasonal and inter-annual variations in atmospheric oxygen and implications for the global carbon cycle. *Nature* **358**, 723–727.

Keeling, R. F., R. P. Najjar, M. L. Bender, and P. P. Tans (1993) What atmospheric oxygen measurements can tell us about the global carbon cycle. *Global Biogeochem. Cycles* **7**, 37–67.

Keeling, R. F., S. C. Piper, and M. Heimann (1996) Global and hemispheric CO_2 sinks deduced from changes of the atmospheric O_2 concentration. *Nature* **381**, 218–221.

Keene, W. C., and J. N. Galloway (1986) Considerations regarding sources of formic and acetic acids in the troposphere. *J. Geophys. Res.* **91**, 14,466–14,474.

Keene, W. C., and J. N. Galloway (1988) The biogeochemical cycling of formic and acetic acids through the troposphere: an overview of current understanding. *Tellus* **40B**, 322–334.

Keene, W. C., J. N. Galloway, and J. D. Holden, Jr. (1983) Measurements of weak organic acidity in precipitation from remote areas of the world. *J. Geophys. Res.* **88**, 5122–5130.

Keene, W. C., A. A. P. Pszenny, D. J. Jacob, R. A. Duce, J. N. Galloway, J. J. Schultz-Tokos, H. Sievering, and J. Boatman (1990) The geochemical cycling of reactive chlorine through the marine troposphere. *Global Biogeochem. Cycles* **4**, 407–430.

Keller, G. J., J. D. Spengler, and R. A. Castillo (1991) Acid aerosol measurements at a suburban Connecticut site. *Atmos. Environ.* **25A**, 681–690.

Keller, M., T. J. Goreau, S. C. Wofsy, W. A. Kaplan, and M. B. McElroy (1983) Production of nitrous oxide and consumption of methane by forest soils. *Geophys. Res. Lett.* **10**, 1156–1159.

Keller, M., W. A. Kaplan, and S. C. Wofsy (1986) Emissions of N_2O, CH_4 and CO_2 from tropical forest soils. *J. Geophys. Res.* **91**, 11,791–11,802.

Keller, M., W. A. Kaplan, S. C. Wofsy, and J. M. DaCosta (1988) Emission of N_2O from tropical forest soils: response to fertilization with NH_4^+, NO_3^-, and PO_4^{3-}. *J. Geophys. Res.* **93**, 1600–1604.

Keller, M., E. Veldkamp, A. M. Weitz, and W. A. Reiners (1993) Pasture effects on soil-atmosphere trace gas exchange in a deforested area of Costa Rica. *Nature* **365**, 244–246.

Kellog, W. W., R. D. Cadle, E. R. Allen, A. L. Lazrus, and E. A. Martell (1972) The sulfur cycle. *Science* **175**, 587–596.

Kelly, K. K., A. F. Tuck, D. M. Murphy, M. H. Proffitt, D. W. Fahey, R. L. Jones, D. S. McKenna, M. Loewenstein, J. R. Podoslke, S. E. Strahan, G. V. Ferry, K. R. Chan, J. F. Vedder, G. L. Gregory, W. D. Hypes, M. P. McCormick, E. V. Browell, and L. E. Heidt (1989) Dehydration in the lower Antarctic stratosphere during late winter and early spring 1987. *J. Geophys. Res.* **94**, 11,317–11,357.

Kelly, T. J., D. H. Stedman, and G. L. Kok (1979) Measurements of H_2O_2 and HNO_3 in rural air. *Geophys. Res. Lett.* **6**, 375–378.

Kelly, T. J., D. H. Stedman, J. A. Ritter, and R. B. Harvey (1980) Measurements of oxides of nitrogen and nitric acid in clean air. *J. Geophys. Res.* **85**, 7417–7425.

Kenley, R. A., J. E. Davenport, and D. G. Hendry (1978) Hydroxyl radical reactions in the gas phase. Products and pathways for the reaction of OH with toluene. *J. Phys. Chem.* **82**, 1095–1096.

Kennish, M. J. (1989) "Practical Handbook of Marine Science." CRC Press, Boca Raton, FL.

Kerker, M., and V. Hampl (1974) Scavenging of aerosol particles by a falling water drop and calculation of washout coefficients. *J. Atmos. Sci.* **31**, 1368–1376.

Kerminen, V. M., and A. S. Wexler (1995) Growth laws for atmospheric aerosol particles: an examination of the bimodality of the accumulation mode. *Atmos. Environ.* **29**, 3263–3275.

Kerr, J. B., and C. T. McElroy (1976) Measurements of stratospheric nitrogen dioxide from the AES stratospheric balloon program. *Atmosphere* **14**, 166–171.

Kesselmeier, J., and L. Merk (1993) Exchange of carbonyl sulfide (COS) between agricultural plants and the atmosphere: studies on the deposition of COS to peas, corn and rape seed. *Biogeochemistry* 23, 47–59.

Kesselmeier, J., L. Merk, M. Bliefernicht, and G. Helas (1993a) Trace gas exchange between terrestrial plants and atmosphere: carbon dioxide, carbonyl sulfide and ammonia under the rule of compensation points. In "General Assessment of Biogenic Emissions and Deposition of Nitrogen Compounds, Sulfur Compounds and Oxidants in Europe" (J. Slanina, G. Angeletti, and S. Beilke, eds.). Air Pollution Research Report, Vol. 47, pp. 71–80. Commission of the European Communities, Brussels.

Kesselmeier, J., F. X. Meixner, U. Hofmann, A. Ajavon, S. Leimbach, and M. O. Andreae (1993b) Reduced sulfur compound exchange between the atmosphere and tropical tree species in southern Cameroon. *Biogeochemistry* 23, 23–45.

Kesselmeier, J., C. Ammann, J. Beck, K. Bode, R. Gabriel, U. Hofmann, G. Helas, U. Kuhn, F. X. Meixner, T. Rausch, L. Schäfer, D. Weller, and M. O. Andreae (1997a) Exchange of short chained organic acids between the biosphere and the atmosphere. In "Biosphere-Atmosphere Exchange of Pollutants and Trace Substances" (J. Slanina, ed.), Transport and Chemical Transformation of Pollutants in the Troposphere, Vol. 4 (P. Borrell, P. M. Borrell, T. Cvitas, K. Kelly, and W. Seiler, series eds.), pp. 327–334. Springer-Verlag, Berlin.

Kesselmeier, J., K. Bode, U. Hofmann, H. Müller, L. Schäfer, A. Wolf, P. Ciccioli, E. Brancaleoni, A. Cecinato, M. Frattoni, P. Foster, C. Ferrari, V. Jacob, J. L. Fugit, L. Dutaur, V. Simon, and L. Torres (1997b) Emission of short chained organic acids, aldehydes and monoterpenes from *Quercus Ilex* L. and *Pinus pinea* L. in relation to physiological activities, carbon budget and emission algorithms. *Atmos. Environ.* 31, 119–133.

Kesselmeier, J., P. Schröder, and J. W. Erisman (1997c) Exchange of sulfur gases between the biosphere and the atmosphere. In "Biosphere Atmosphere Exchange of Pollutants and Trace Substances" (J. Slanina, ed.), Transport and Chemical Transformation of Pollutants in the Troposphere, Vol. 4 (P. Borrell, P. M. Borrell, T. Cvitas, K. Kelly, and W. Seiler, series eds.), pp. 167–198. Springer-Verlag, Berlin.

Kessler, C., and U. Platt (1984) Nitrous acid in polluted air masses — sources and formation pathways. In "Physico-Chemical Behaviour of Atmospheric Pollutants" (B. Versino and G. Angeletti, eds.). Proceedings of the 3rd European Symposium, Varese, Italy, pp. 412–421. Reidel, Dordrecht, the Netherlands.

Kessler, C., D. Perner, and U. Platt (1982) Spectroscopic measurements of nitrous acid and formaldehyde — implications for urban photochemistry. In "Physico-Chemical Behaviour of Atmospheric Pollutants" (B. Versino and H. Ott, eds.). Proceedings of the 2nd European Symposium, Varese, Italy, pp. 393–399. Reidel, Dordrecht, the Netherlands.

Ketseridis, G., and R. Jaenicke (1978) Organische Beimengungen atmosphärischer Reinluft: ein Beitrag zur Budget-Abschätzung. In "Organische Verbindungen in der Umwelt" (K. Anrand, H. Hässelbarth, E. Lahmann, G. Müller, and W. Niemitz, eds.), pp. 379–390. Erich Schmidt Verlag, Berlin.

Ketseridis, G., J. Hahn, R. Jaenicke, and C. Junge (1976) The organic constituents of atmospheric particulate matter. *Atmos. Environ.* 10, 603–610.

Khalil, M. A. K., and R. A. Rasmussen (1983) Sources, sinks and seasonal cycles of atmospheric methane. *J. Geophys. Res.* 88, 5131–5144.

Khalil, M. A. K., and R. A. Rasmussen (1985a) Causes of increasing atmospheric methane: depletion of hydroxy radicals and the rise of emissions. *Atmos. Environ.* 19, 397–407.

Khalil, M. A. K., and R. A. Rasmussen (1985b) Atmospheric carbon tetrafluoride (CF_4): sources and trends. *Geophys. Res. Lett.* 12, 671–672.

Khalil, M. A. K., and R. A. Rasmussen (1990a) The global cycle of carbon monoxide: trends and mass balance. *Chemosphere* 20, 227–242.

Khalil, M. A. K., and R. A. Rasmussen (1990b) Global increase of atmospheric molecular hydrogen. *Nature* 347, 743–745.

Khalil, M. A. K., and R. A. Rasmussen (1992a) The global sources of nitrous oxide. *J. Geophys. Res.* 97, 14,651–14,660.

Khalil, M. A. K., and R. A. Rasmussen (1992b) Nitrous oxide from coal-fired power plants. *J. Geophys. Res.* 97, 14,645–14,649.

Khalil, M. A. K., and R. A. Rasmussen (1992c) Forest hydrocarbon emissions: relationships between fluxes and ambient concentrations. *J. Air Waste Manage. Assoc.* 42, 810–813.

Khalil, M. A. K., and R. A. Rasmussen (1994) Global decrease of atmospheric carbon monoxide concentration. *Nature* 370, 639–641.

Khalil, M. A. K., and M. J. Shearer (1993) Sources of methane: an overview. *In* "Atmospheric Methane: Sources, Sinks, and Role in Global Change" (M. A. K. Khalil, ed.), NATO ASI Series, Vol. I 13, pp. 180–198. Springer-Verlag, Berlin, Germany.

Khalil, M. A. K., M. J. Shearer, and R. A. Rasmussen (1993a) Methane sinks and distributions. *In* "Atmospheric Methane: Sources, Sinks, and Role in Global Change" (M. A. K. Khalil, ed.), NATO ASI Series, Vol. I 13, pp. 168–179. Springer-Verlag, Berlin.

Khalil, M. A. K., R. A. Rasmussen and R. Gunawardena (1993b) Atmospheric methyl bromide: trends and global mass balance. *J. Geophys. Res.* 98, 2887–2896.

Kiang, C. S., D. Stauffer, V. A. Mohnen, J. Bricard, and D. Viglia (1973) Heteromolecular nucleation theory applied to gas to particle conversion. *Atmos. Environ.* 7, 1279–1283.

Kibe, K., and S. Kagura (1976) Distribution of volatile components of grass silage. *J. Sci. Food Agric.* 27, 726–732.

Kim, D.-S., V. P. Aneja, and W. R. Robarge (1994) Characterization of nitrogen oxide fluxes from soil of a fallow field in the central Piedmont of North Carolina. *Atmos. Envion.* 28, 1129–1137.

Kim, K. H., and M. O. Andreae (1987) Carbon disulfide in seawater and the marine atmosphere over the North Atlantic. *J. Geophys. Res.* 92, 14,733–14,738.

Kim, K.-R., and H. Craig (1990) Two isotope characterization of N_2O in the Pacific Ocean and constraints on its origin in deep water. *Nature* 247, 58–61.

Kim, Y. J., J. F. Boatman, R. L. Gunter, D. L. Wellman, and S. W. Wilkison (1993) Vertical distribution of atmospheric aerosol size distribution over south-central New Mexico. *Atmos. Environ.* 27A, 1351–1362.

Kimball, B. A. (1983) Carbon dioxide and agricultural yield: an assemblage and analysis of 430 prior observations. *Agron. J.* 75, 779–788.

Kins, L. (1982) Temporal variation of chemical composition of rainwater during individual precipitation events. *In* "Deposition of Atmospheric Pollutants" (H.-W. Georgii and J. Pankrath, eds.), pp. 87–96. Reidel, Dordrecht, the Netherlands.

Kirchgessner, D. A., S. D. Piccot, and J. D. Winkler (1993) Estimate of global methane emissions from coal mines. *Chemosphere* 26, 453–472.

Kitto, A.-M. N., and R. M. Harrison (1992) Nitrous and nitric acid measurements at sites in south-east England. *Atmos. Environ.* 26A, 235–241.

Kley, D., and M. McFarland (1980) Chemiluminescence detector for NO and NO_2. *Atmos. Technol.* 12, 63–69.

Kley, D., J. W. Drummond, M. McFarland, and S. C. Liu (1981) Tropospheric profiles of NO_x. *J. Geophys. Res.* 86, 3153–3161.

Kley, D., A. L. Schmeltekopf, K. Kelly, R. H. Winkler, T. L. Thomson, and M. McFarland (1982) Transport of water vapor through the tropical tropopause. *Geophys. Res. Lett.* 9, 617–620.

Kley, D., A. Volz, and F. Mülheims (1988) Ozone measurements in historic perspective. *In* "Tropospheric Ozone" (I. S. A. Isaksen, ed.), pp. 63–72. Reidel, Dordrecht, the Netherlands.

Klippel, W., and P. Warneck (1978) Formaldehyde in rainwater and on the atmospheric aerosol. *Geophys. Res. Lett.* 5, 177–179.

Klippel, W., and P. Warneck (1980) The formaldehyde content of the atmospheric aerosol. *Atmos. Environ.* 14, 809–818.

Knispel, R., R. Koch, M. Siese, and C. Zetzsch (1990) Adduct formation of OH radicals with benzene, toluene and phenol and consecutive reactions of the adducts with NO_x and O_2. *Ber. Bunsenges. Phys. Chem.* 94, 1375–1379.

Knowles, R. (1981) Denitrification. In "Terrestrial Nitrogen Cycles" (F. E. Clark and T. Rosswall, eds.). *Ecol. Bull. (Stockh.)* 33, 315–329.

Knudsen, M., and S. Weber (1911) Luftwiderstand gegen die langsame Bewegung kleiner Kugeln. *Ann. Phys.* 36, 981–994.

Knutson, E. O., S. K. Sood, and J. D. Stockham (1976) Aerosol collection by snow and ice crystals. *Atmos. Environ.* 10, 395–402.

Ko, M. K., N. D. Sze, and D. K. Weisenstein (1991) Use of satellite data to constrain the model-calculated atmospheric life time for N_2O; implications for other trace gases. *J. Geophys. Res.* 96, 7547–7552.

Kockarts, G. (1976) Absorption and photodissociation in the Schumann-Runge bands of molecular oxygen in the terrestrial atmosphere. *Planet. Space Sci.* 24, 589–604.

Köhler, H. (1936) The nucleus in and the growth of hygroscopic droplets. *Trans. Faraday Soc.* 32, 1152–1162.

Kok, G. L., K. R. Darnall, A. M. Winer, J. N. Pitts, Jr., and B. W. Gay (1978) Ambient air measurements of hydrogen peroxide in the Californian southeast air basin. *Environ. Sci. Technol.* 12, 1077–1080.

Kok, G. L., S. N. Gitlin, and A. L. Lazrus (1986) Kinetics of the formation and decomposition of hydroxymethane sulfonate. *J. Geophys. Res.* 91, 2801–2804.

Kolaitis, L. N., F. J. Bruynseels, R. E. Van Grieken, and M. O. Andreae (1989) Determination of methanesulfonic acid and non-sea-salt sulfate in single marine aerosol particles. *Environ. Sci. Technol.* 23, 236–240.

Kolattukudy, P. E. (ed.) (1976) "Chemistry and Biochemistry of Natural Waxes." Elsevier, Amsterdam.

Kolb, C. E., J. T. Jayne, P. R. Worsnop, M. J. Molina, R. F. Meads, and A. A. Viggiano (1994) Gas phase reaction of sulfur trioxide with water vapor. *J. Am. Chem. Soc.* 116, 10,314–10,315.

Komhyr, W. D. (1969) Electrochemical concentration cells for gas analysis. *Ann. Geophys. (Paris)* 25, 203–210.

Komhyr, W. D., R. W. Barrett, G. Slocum, and H. K. Weickmann (1971) Atmospheric total ozone increase during 1960s. *Nature* 232, 390–391.

Kondo, Y., P. Aimedieu, M. Koike, Y. Iwasaka, P. A. Newman, U. Schmidt, W. A. Matthews, M. Hayashi, and W. R. Sheldon (1992) Reactive nitrogen, ozone, and nitrate aerosols observed in the Arctic stratosphere in January 1990. *J. Geophys. Res.* 97, 13,025–13,038.

Kondo, Y., H. Ziereis, M. Koike, S. Kawakami, G. L. Gregory, G. W. Sachse, H. B. Singh, D. D. Davis, and J. T. Merrill (1996) Reactive nitrogen over the Pacific Ocean during PEM-West A. *J. Geophys. Res.* 101, 1809–1828.

König, G., M. Brunda, H. Puxbaum, C. N. Hewitt, S. C. Duckham, and J. Rudolph (1995) Relative contribution of oxygenated hydrocarbons to the total biogenic volatile organic compound emissions of selected mid-European agricultural and natural plant species. *Atmos. Environ.* 29, 861–874.

Koop, T., U. M. Biermann, W. Raber, B. P. Luo, P. J. Crutzen, and T. Peter (1995) Do stratospheric aerosol droplet freeze above the ice frost point? *Geophys. Res. Lett.* 22, 917–920.

Koppmann, R., F. J. Johnen, C. Plass-Dülmer, and R. Rudolph (1993) Distribution of methyl chloride, dichloromethane, trichloroethene, and tetrachloroethene over the north and south Atlantic. *J. Geophys. Res.* **98**, 20,517–20,526.

Kopzcynski, S. T., W. A. Lonneman, F. D. Sutterfield, and P. E. Darley (1972) Photochemistry of atmospheric samples in Los Angeles. *Environ. Sci. Technol.* **6**, 342–347.

Körner, C., and J. A. Arnone, III (1992) Responses to elevated carbon dioxide in artificial tropical ecosystems. *Science* **257**, 1672–1675.

Körner, C., G. D. Farquar, and S. C. Wong (1991) Carbon isotope discrimination by plants follows latitudinal and altitudinal trends. *Oecologia* **88**, 30–36.

Kortzeborn, R. N., and F. F. Abraham (1970) Scavenging of aerosols by rain: a numerical study. In "Precipitation Scavenging 1970" (R. J. Engelmann and W. G. N. Slinn, coordinators), *AEC Symposium Series*, Vol. 30, pp. 433–446. U.S. Atomic Energy Commission, Division of Technical Information, Oak Ridge, TN.

Kosak-Channing, L. F., and G. R. Helz (1983) Solubility of ozone in aqueous solutions of 0–6 M ionic strength at 5–30°C. *Environ. Sci. Technol.* **17**, 145–149.

Kossina, E. (1933) Die Erdoberfläche. In "Handbuch der Geophysik" (B. Gutenberg, ed.), Vol. 2, pp. 869–954. Borntraeger, Berlin.

Kourtidis, K., R. Borchers, and P. Fabian (1998) Vertical distribution of methyl bromide in the stratosphere. *Geophys. Res. Lett.* **25**, 505–508.

Koyama, T. (1963) Gaseous metabolism in lake sediments and paddy soils and the production of atmospheric methane and hydrogen. *J. Geophys. Res.* **68**, 3971–3973.

Kraft, J., and R. van Eldik (1989a) Kinetics and mechanism of the iron(III) catalyzed autoxidation of sulfur(IV) oxides in aqueous solution. 1. Formation of transient iron(III)-sulfur(IV) complexes. *Inorg. Chem.* **28**, 2297–2305.

Kraft, J., and R. van Eldik (1989b) Kinetics and mechanism of the iron(III) catalyzed autoxidation of sulfur(IV) oxides in aqueous solution. 2. Decomposition of transient iron(III)-sulfur(IV) complexes. *Inorg. Chem.* **28**, 2306–2312.

Krall, A. R., and A. Tolbert (1957) A comparison of the light-dependent metabolism of carbon monoxide by barley leaves with that of formaldehyde, formate and carbon dioxide. *Plant Physiol.* **32**, 321–326.

Krauskopf, K. B. (1979) "Introduction to Geochemistry," 2nd ed. McGraw-Hill, New York.

Kreidenweis, S. M., and J. H. Seinfeld (1988) Nucleation of sulfuric acid-water and methanesulfonic acid-water solution particles: implications for the atmospheric chemistry of organic sulfur species. *Atmos. Environ.* **22**, 283–296.

Krey, P. W., and B. Krajewski (1970) Comparison of atmospheric transport model calculations with observations of radioactive debris. *J. Geophys. Res.* **75**, 2901–2908.

Krey, P. W., M. Schomberg, and L. Toonkel (1974) Updating stratospheric inventories to January 1973, Report no. HASL-281. U.S. Atomic Energy Commission, New York.

Krey, P. W., R. J. Lagomarsino, and L. E. Toonkel (1977) Gaseous halogens in the atmosphere in 1975. *J. Geophys. Res.* **82**, 1753–1766.

Kritz, M. A. (1982) Exchange of sulfur between the free troposphere, marine boundary layer and the sea surface. *J. Geophys. Res.* **87**, 8795–8803.

Kritz, M. A., and J. Rancher (1980) Circulation of Na, Cl, and Br in the tropical marine atmosphere. *J. Geophys. Res.* **85**, 1633–1639.

Krueger, A. J. (1969) Rocket measurements of ozone over Hawaii. *Ann. Geophys.* **25**, 307–311.

Krueger, A. J. (1973) Mean ozone distribution from several series of rocket soundings to 52 km at latitudes from 58°S to 64°N. *Pure Appl. Geophys.* **106–108**, 1272–1280.

Kuhlbusch, T. A., and P. J. Crutzen (1995) Toward a global estimate of black carbon in residues of vegetation fires representing a sink of atmospheric CO_2 and a source of O_2. *Global Biogeochem. Cycles* **9**, 491–501.

Kuhlbusch, T. A., J. M. Lobert, P. J. Crutzen, and P. Warneck (1991) Molecular nitrogen emissions from denitrification during biomass burning. *Nature* 351, 135–137.

Kuhlbusch, T. A., M. O. Andreae, H. Cachier, J. G. Goldammer, J.-P. Lacaux, R. Shea, and P. J. Crutzen (1996) Black carbon formation by savanna fires: measurements and implications for the global carbon cycle. *J. Geophys. Res.* 101, 23,651–23,665.

Kuhn, W., and J. F. Kasting (1983) Effects of increased CO_2 concentrations on surface temperatures of the early Earth. *Nature* 301, 53–55.

Kuis, S., R. Simonaitis, and J. Heicklen (1975) Temperature dependence of the photolysis of ozone at 3130 Å. *J. Geophys. Res.* 80, 1328–1331.

Kuwata, K., M. Uebori, Y. Yamasaki, and Y. Kuge (1983) Determination of aliphatic aldehyde in air by liquid chromatography. *Anal. Chem.* 55, 2013–2016.

Kwok, E. S. C., J. Arey, and R. Atkinson (1996) Alkoxy radical isomerization in the OH radical-initiated reactions of C_4–C8 *n*-alkanes. *J. Phys. Chem.* 100, 214–219.

Labitzke, K. (1980) Climatology of the stratosphere and mesosphere. *Philos. Trans. R. Soc. Lond.* A 296, 7–18.

Labitzke, K., and J. J. Barnett (1979) Review of climatological information obtained from remote sensing of the stratosphere and the mesosphere. *COSPAR Space Res.* 19, 97–106.

Lagrange, J., C. Pallares, and P. Lagrange (1994) Electrolyte effects on aqueous atmospheric oxidation of sulfur dioxide by ozone. *J. Geophys. Res.* 99, 14,595–14,600.

Lai, K. Y., N. Dayan, and M. Kerker (1978) Scavenging of aerosol particles by a falling water drop. *J. Atmos. Sci.* 35, 674–682.

Laj, P., S. Fuzzi, M. C. Facchini, J. A. Lind, G. Orsi, M. Preiss, R. Maser, W. Jaeschke, E. Seyffer, G. Helas, K. Acker, W. Wieprecht, D. Möller, D. G. Arends J J, Möls, R. N. Colvile, M. W. Gallagher, K. M. Beswick K, J. Hargreaves, R. L. Storeton-West, and M. A. Sutton (1997) Cloud processing of soluble gases. *Atmos. Environ.* 31, 2589–2598.

Lal, D., and B. Peters (1967) Cosmic ray produced radioactivity on Earth. In "Encyclopedia of Physics" (S. Flügge, ed.), Vol. 46/2, pp. 551–612. Springer Verlag, Berlin and New York.

Lal, D., and Rama (1966) Characteristics of global tropospheric mixing based on ^{14}C, ^4H, and ^{90}Sr. *J. Geophys. Res.* 71, 2865–2874.

Lamarque, J. F., and P. Hess (1994) Cross-tropopause mass exchange and potential vorticity budget in a simulated tropopause folding. *J. Atmos. Sci.* 51, 2246–2269.

Lamb, B., A. Guenther, D. Gay, and H. Westberg (1987a) A national inventory of biogenic hydrocarbon emissions. *Atmos. Environ.* 21, 1695–1705.

Lamb, B., H. Westberg, G. Allwine, L. Bamesberger, and A. Guenther (1987b) Measurement of biogenic sulfur emissions from soils and vegetation: application of dynamic enclosure methods with Natusch filter and GC/FDP analysis. *J. Atmos. Chem.* 5, 469–491.

Lamb, B., D. Gay, H. Westberg, and T. Pierce (1993) A biogenic hydrocarbon emission inventory for the U. S. A. using a simple forest canopy model. *Atmos. Environ.* 27A, 1673–1690.

Lamb, R. G. (1977) A case study of stratospheric ozone affecting ground-level oxidant concentrations. *J. Appl. Meteorol.* 16, 780–794.

Lambert, G., A. Buison, J. Sanak, and B. Ardouin (1979) Modification of the atmospheric polonium 210 to lead 210 ratio by volcanic emissions. *J. Geophys. Res.* 84, 6980–6986.

Lambert, G., G. Polian, J. Sanak, B. Ardouin, A. Buisson, A. Jegou, and C. LeRoulley (1982) Cycle de radon et des ses descendants: application à l'étude des échanges troposphère-stratosphère. *Ann. Geophys.* 38, 497–531.

Lammel, G., and J. N. Cape (1996) Nitrous acid and nitrite in the atmosphere. *Chem. Soc. Rev.* 1996, 361–369.

Lammel, G., and D. Perner (1988) The atmospheric aerosol as a source of nitrous acid in the polluted atmosphere. *J. Aerosol Sci.* 19, 1199–1202.

Lammel, G., D. Perner, and P. Warneck (1990a) Decomposition of pernitric acid in aqueous solution. *J. Phys. Chem.* **94**, 6141–6144.

Lammel, G., D. Perner, and P. Warneck (1990b) Nitrous acid at Mainz: observations and implications for its formation mechanism. *In* "Physico-Chemical Behaviour of Atmospheric Pollutants" (G. Restelli and G. Angeletti, eds.), pp. 469–476. Kluwer, Dordrecht, the Netherlands.

Lamontagne, R. A., J. W. Swinnerton, and V. J. Linnenboom (1974) C_1–C_4 hydrocarbons in the north and south Pacific. *Tellus* **26**, 71–77.

Langford, A. O., and F. C. Fehsenfeld (1992) Natural vegetation as a source or sink for atmospheric ammonia: a case study. *Science* **255**, 581–583.

Langford, A. O., F. C. Fehsenfeld, J. Zachariassen, and D. S. Schimel (1992) Gaseous ammonia fluxes and background concentrations in terrestrial ecosystems of the United States. *Global Biogeochem. Cycles* **4**, 459–483.

Langner, J., and H. Rodhe (1991) A global three-dimensional model of the tropospheric sulfur cycle. *J. Atmos. Chem.* **13**, 225–263.

Lanly, J. P. (1982) Tropical forest resources, FAO Forestry Paper 30. Food and Agricultural Organization, United Nations, Rome, Italy.

Laursen, K. K., P. V. Hobbs, L. F. Radke, and R. A. Rasmussen (1992) Some trace gas emissions from North American biomass fires with assessment of regional and global fluxes from biomass burning. *J. Geophys. Res.* **97**, 20,687–20,701.

Lawlor, D. W., and R. A. C. Mitchell (1991) The effects of increasing CO_2 on crop photosynthesis and productivity: a review of field studies. *Plant Cell Environ.* **14**, 807–818.

Lay, M. D. S., M. W. Sauerhoff, and D. R. Saunders (1986) Carbon disulfide. *In* "Ullmann's Encyclopedia of Industrial Chemistry" (W. Gerhartz, exec. ed.), Vol. A5, pp. 185–195. VCH Verlagsgesellschaft, Weinheim, Germany.

Lazrus, A. L., and B. W. Gandrud (1974) Distribution of stratospheric nitric acid vapor. *J. Atmos. Sci.* **31**, 1102–1108.

Lazrus, A. L., and B. W. Gandrud (1977) Stratospheric sulfate at high altitudes. *Geophys. Res. Lett.* **4**, 521–522.

Leach, P. W., L. J. Leng, T. A. Bellar, J. E. Sigsby, Jr., and A. P. Altshuller (1964) Effects of HC/NO ratios on irradiated auto exhaust. Part II. *J. Air Pollut. Control. Assoc.* **14**, 176–183.

Le Bras, G. (ed.) (1997) "Chemical Processes in Atmospheric Oxidation," Transport and Chemical Transformation of Pollutants in the Troposphere, Vol. 3 (P. Borrell, P. M. Borrell, T. Cvitas, K. Kelly, and W. Seiler, series eds.). Springer-Verlag, Berlin.

Lee, D. S., I. Köhler, E. Grobler, F. Rohrer, R. Sausen, L. Gallardo-Klenner, J. G. J. Oliver, F. J. Dentener, and A. F. Bouwman (1997) Estimates of global NO_x emissions and their uncertainties. *Atmos. Environ.* **31**, 1735–1749.

Lee, L. C., T. G. Slanger, G. Black, and R. L. Sharpless (1977) Quantum yields for the production of $O(^1D)$ from photodissociation of O_2 at 1160–1770 Å. *J. Chem. Phys.* **67**, 5602–5606.

Lee, L. C., G. Black, R. L. Sharpless, and T. G. Slanger (1980) $O(^1S)$ yield from O_3 photodissociation of 1700–2400 Å. *J. Chem. Phys.* **73**, 256–258.

Lee, Y.-N., and S. E. Schwartz (1981) Reaction kinetics of nitrogen dioxide with liquid water at low partial pressure. *J. Phys. Chem.* **85**, 840–848.

Legrand, M., C. Feniet-Saigné, E. S. Saltzman, C. Germain, N. I. Barkov, and V. N. Petrov (1991) Ice-core record of oceanic emissions of dimethylsulphide during the last climate cycle. *Nature* **350**, 144–146.

Lehmann, L., and A. Sittkus (1959) Bestimmung von Aerosol Verweilzeiten aus dem RaD und RaF Gehalt der atmosphärischen Luft und des Niederschlags. *Naturwissenschaften* **46**, 9–10.

Leifer, R., K. Sommers, and S. F. Guggenheim (1981) Atmospheric trace gas measurements with a new clean air sampling system. *Geophys. Res. Lett.* **8**, 1079–1082.

Leighton, P. A. (1961) "Photochemistry of Air Pollution." Academic Press, New York.

Lein, A. Yu. (1983) The sulfur cycle in the lithosphere. Part II. Cycling. In "The Global Biogeochemical Sulphur Cycle" (M. V. Ivanov and J. R. Freney, eds.). SCOPE 19, 95–127.

Lelieveld, J., and P. J. Crutzen (1990) Influences of cloud photochemical processes on tropospheric ozone. Nature 343, 227–233.

Lelieveld, J., and P. J. Crutzen (1991) The role of clouds in tropospheric chemistry. J. Atmos. Chem. 12, 229–267.

Lelieveld, J., P. J. Crutzen, and H. Rodhe (1989) Zonal average cloud characteristics for global atmospheric chemistry modelling, GLOMAC Report UDC 551.510.4, CM-76. International Meteorological Institute, Stockholm.

Lenhard, U. (1977) Messungen von Ammoniak in der unteren Troposphäre und Untersuchung der Quellstärke von Böden. Diploma Thesis. Institut für Meteorologie und Geophysik, Universität Frankfurt (Main), Germany.

Lenhard, U., and G. Gravenhorst (1980) Evaluation of ammonia fluxes into the free atmosphere over western Germany. Tellus 32, 48–55.

Lepel, E. A., K. M. Stefansson, and W. H. Zoller (1978) The enrichment of volatile elements in the atmosphere by volcanic acitivity: Augustine Volcano 1976. J. Geophys. Res. 83, 6213–6220.

Lesclaux, R., and M. DeMissy (1977) On the reaction of the NH_2 radical with oxygen. Nouv. J. Chim. 1, 443–444.

Lettau, H., and W. Schwerdtfeger (1933) Untersuchungen über atmosphärische Turbulenz und Vertikalaustausch vom Freiballon aus. Meteorol. Z. 50, 250–256.

Lettau, H., and W. Schwerdtfeger (1934) Untersuchungen über atmosphärische Turbulenz und Vertikalaustausch vom Freiballon aus. Meteorol. Z. 51, 249–257.

Lettau, H., and W. Schwerdtfeger (1936) Untersuchungen über atmosphärische Turbulenz und Vertikalaustausch vom Freiballon aus. Meteorol. Z. 53, 44.

Leu, M. T., and R. H. Smith (1982) Kinetics of the gas-phase reaction between hydroxyl and carbonyl sulfide over the temperature range 300–517 K. J. Phys. Chem. 86, 73–83.

Levin, I., and V. Hesshaimer (1996) Refining of atmospheric transport model entries by the globally observed passive tracer distributions of ^{85}Kr and sulfur hexafluoride (SF_6). J. Geophys. Res. 101, 16,745–16,755.

Levine, J. S. (1984) Nitrogen fixation by lightning activity in a thunderstorm. Atmos. Environ. 18, 2272–2280.

Levine, J. S., R. S. Rogowski, G. I. Gregory, W. E. Howell, and J. Fishman (1981) Simultaneous measurements of NO_x, NO, and O_3 production in a laboratory discharge: atmospheric implications. Geophys. Res. Lett. 8, 357–360.

Levine, J. S., W. R. Cofer, and J. P. Pinto (1993) Biomass burning. In "Atmospheric Methane: Sources, Sinks, and Role in Global Change" (M. A. K. Khalil, ed.), NATO ASI Series, Vol. I 13, pp. 299–313. Springer-Verlag, Berlin.

Levy, H. (1971) Normal atmosphere: large radical and formaldehyde concentrations predicted. Science 173, 141–143.

Levy, H. (1972) Photochemistry of the lower troposphere. Planet. Space Sci. 20, 919–935.

Levy, H., J. D. Mahlman, and W. J. Moxim (1980) A stratospheric source of reactive nitrogen in the unpolluted troposphere. Geophys. Res. Lett. 7, 441–444.

Lewin, E. E., R. G. de Pena, and J. P. Shimshock (1986) Atmospheric gas and particle measurements at a rural northeastern U.S. site. Atmos. Environ. 20, 59–70.

Lewin, R. A. (1976) Prochlorophyta as a proposed new division of algae. Nature 261, 697–698.

Lewis, G. N., and M. Randall (1923) "Thermodynamics and the Energy of Chemical Substances." McGraw-Hill, New York.

Lezberg, E. A., F. M. Hummenik, and D. A. Otterson (1979) Sulfate and nitrate mixing ratios in the vicinity of the tropopause. *Atmos. Environ.* **13**, 1299–1304.

Li, S.-M. (1994) Equilibrium of particle nitrite with gas phase HONO: tropospheric measurements in the high Arctic during polar sunrise. *J. Geophys. Res.* **99**, 25,479–25,488.

Li, Y. H. (1972) Geochemical mass balance among lithosphere, hydrosphere and atmosphere. *Am. J. Sci.* **272**, 119–137.

Lide, D. R. (1995/1996) "Handbook of Chemistry and Physics," 76th ed. CRC Press, Boca Raton, FL.

Liebig, J. (1827) Une note sur la nitrification. *Ann. Chem Phys.* **35**, 329–333.

Liebl, K. H., and W. Seiler (1976) CO and H_2 destruction at a soil surface. *In* "Microbial Production and Utilization of Gases" (H. Schlegel, G. Gottschalk, and N. Pfennig, eds.), pp. 215–229. Akademie der Wissenschaften, Göttingen, Germany.

Lieth, H. (1965) Versuch einer kartographischen Darstellung der Produktivität der Pflanzendecke auf der Erde. *In* "Geographisches Taschenbuch," pp. 72–79. Franz Steiner Verlag, Wiesbaden.

Lieth, H. (1975) Primary production of the major vegetation units of the world. *In* "Primary Productivity of the Biosphere" (H. Lieth and R. H. Whittaker, eds.). *Ecol. Stud.* **14**, 203–215.

Lieth, H., and R. H. Whittaker, eds. (1975) "Primary Productivity of the Biosphere." *Ecol. Stud.* **14**.

Lightfoot, P. D., R. A. Cox, J. N. Crowley, M. Destriau, G. D. Hayman, M. E. Jenkin, G. K. Moortgat, and F. Zable (1992) Organic peroxy radicals: kinetics, spectroscopy and tropospheric chemistry. *Atmos Environ.* **26A**, 1805–1861.

Likens, G. E. (1976) Acid precipitation. *Chem. Eng. News* **54**, 29–44.

Likens, G. E., F. H. Bormann, R. S. Pierce, J. S. Eaton, and N. M. Johnson (1977) "Biochemistry of a Forested Ecosystem." Springer-Verlag, Berlin.

Likens, G. E., R. F. Wright, J. N. Galloway, and T. J. Butler (1979) Acid rain. *Sci. Am.* **241**, 39–47.

Likens, G. E., E. S. Edgerton, and J. N. Galloway (1983) The composition and deposition of organic carbon in precipitation. *Tellus* **35B**, 16–24.

Likens, G. E., F. H. Bormann, R. S. Pierce, J. S. Eaton, and R. E. Munn (1984) Long-term trends in precipitation chemistry at Hubbard Brook, New Hampshire. *Atmos. Environ.* **18**, 2641–2647.

Likens, G. E., W. C. Keene, J. M. Miller, and J. N. Galloway (1987) Chemistry of precipitation from a remote terrestrial site in Australia. *J. Geophys. Res.* **92**, 13,299–13,314.

Liljestrand, H. M., and J. J. Morgan (1980) Spatial variations of acid precipitation in southern California. *Environ. Sci. Technol.* **15**, 333–339.

Lin, C. L., and W. B. DeMore (1973) $O(^1D)$ production in ozone photolysis near 3100 Å. *J. Photochem.* **2**, 161–164.

Linak, W. P., J. A. McSorley, R. E. Hall, J. V. Ryan, R. S. Srivastava, J. O. L. Wendt, and J. B. Mereb (1990) Nitrous oxide emissions from fossil fuel combustion. *J. Geophys. Res.* **95**, 7533–7541.

Lind, J. A., and G. L. Kok (1994) Correction to Henry's law determinations for aqueous solutions of hydrogen peroxide, methylhydroperoxide and peroxyacetic acid. *J. Geophys. Res.* **99**, 21,119.

Lind, J. A., A. L. Lazrus, and G. L. Kok (1987) Aqueous phase oxidation of sulfur(IV) by hydrogen peroxide, methylhydroperoxide and peroxyacetic acid. *J. Geophys. Res.* **92**, 4171–4177.

Lindberg, S. E., and G. M. Lovett (1992) Deposition and forest canopy interactions of airborne sulfur: results from the integrated forest study. *Atmos. Environ.* **26A**, 1477–1492.

Lindskog, A., and J. Moldanova (1994) The influence of the origin, season and time of the day on the distribution of individual NMHC measured at Rörvik, Sweden. *Atmos. Environ.* 28, 2383–2398.

Lingenfelter, R. E. (1963) Production of carbon-14 by cosmic ray neutrons. *Rev. Geophys.* 1, 35–55.

Lingenfelter, R. E., and R. Ramaty (1970) Astrophysical and geographic variations in C-14 production. In "Radiocarbon Variations and Absolute Chronology" (I. U. Olsson, ed.), pp. 513–537. Almquist and Wiksell, Stockholm; Wiley Interscience, New York.

Linnenboom, V. J., J. W. Swinnerton, and R. A. Lamontagne (1973) The ocean as a source for atmospheric carbon monoxide. *J. Geophys. Res.* 78, 5333–5340.

Linton, R. W., A. Loh, and D. F. S. Natusch (1976) Surface predominance of trace elements in airborne particles. *Science* 191, 852–854.

Lipari, F., J. M. Dash, and W. F. Scruggs (1984) 2,4-Dinitro-phenyl-hydrazine-coated fluorosil sampling cartridges for the determination of formaldehyde in air. *Environ. Sci. Technol.* 19, 70–74.

Lipschultz, F., O. C. Zafiriou, S. C. Wofsy, M. B. McElroy, F. W. Valois, and S. W. Watson (1981) Production of NO and N_2O by soil nitrifying bacteria. *Nature* 294, 641–643.

Liss, P. S. (1973) Process of gas exchange across an air-water interface. *Deep Sea Res.* 20, 221–238.

Liss, P. S., and L. Merlivat (1986) Air-sea gas exchange rates: introduction and synthesis. In "The Role of Air-Sea Exchange in Geochemical Cycling" (P. Buat-Menard, ed.), pp. 113–127, Reidel, Hingham, MA.

Liss, P. S., and P. G. Slater (1974) Flux of gases across the air-sea interface. *Nature* 247, 181–194.

List, R. J., and K. Telegadas (1969) Using radioactive tracers to develop a model of the circulation of the stratosphere. *J. Atmos. Sci.* 26, 1128–1136.

Little, D. J., A. Dalgleish, and R. J. Donovan (1972) Relative rate data for the reaction of $S(3^1D_1)$ using the NS radical as a spectroscopic marker. *Faraday Discuss. Chem. Soc.* 53, 211–216.

Little, P., and R. D. Wiffen (1977) Emission and deposition of petrol engine exhaust Pb-I: deposition of exhaust Pb to plant and soil surfaces. *Atmos. Environ.* 11, 437–467.

Liu, B. Y. H., and C. S. Kim (1977) On the counting efficiency of condensation nuclei counters. *Atmos. Environ.* 11, 1097–1110.

Liu, S. C., and T. M. Donahue (1974) Realistic model of hydrogen constituents in the lower atmosphere and escape flux from the upper atmosphere. *J. Atmos. Sci.* 31, 2238–2242.

Liu, S. C., M. McFarland, D. Kley, O. Zafiriou, and B. J. Huebert (1983) Tropospheric NO and O_3 budgets in the equatorial Pacific. *J. Geophys. Res.* 88, 1360–1368.

Liu, S. C., J. R. McAfee, and R. J. Cicerone (1984) Radon 222 and tropospheric vertical transport. *J. Geophys. Res.* 89, 7291–7297.

Liu, S. C., M. Trainer, F. C. Fehsenfeld, D. D. Parrish, E. J. Williams, D. W. Fahey, G. Hübler, and P. C. Murphy (1987) Ozone production in the rural troposphere and the implications for regional and global ozone distributions. *J. Geophys. Res.* 92, 4191–4207.

Livingston, G. P., P. M. Vitousek, and P. Matson (1988) Nitrous oxide flux and nitrogen transformations across a landscape gradient in Amazonia. *J. Geophys. Res.* 93, 1593–1599.

Livingstone, D. A. (1963) Chemical composition of rivers and lakes. In "Data of Geochemistry," 6th ed. (M. Fleischer, ed.), U.S. Geological Survey Professional Paper, pp. 440–446,

Lloyd, A. C., K. R. Darnall, A. M. Winer, and J. N. Pitts, Jr. (1976) Relative rate constants for the reaction of the hydroxyl radical with a series of alkanes, alkenes and aromatic hydrocarbons. *J. Phys. Chem.* 80, 789–794.

Lobert, J. M., D. H. Scharffe, W.-M. Hao, T. A. Kuhlbusch, R. Seuwen, P. Warneck, and P. J. Crutzen (1991) Experimental determination of biomass burning emissions: nitrogen and carbon containing compounds. *In* "Global Biomass Burning: Atmospheric, Climatic, and Biospheric Implications" (J. S. Levine, ed.), pp. 289–304. MIT Press, Cambridge, MA.

Lobert, J. M., J. H. Butler, S. A. Montzka, L. S. Geller, R. C. Myers, and J. W. Elkins (1995) A net sink for atmospheric CH_3Br in the east Pacific Ocean. *Science* **267**, 1002–1005.

Loehr, R. C., and S. A. Hart (1970) Changing practices in agriculture and their effects on the environment. *Crit. Rev. Environ. Control* **1**, 87–92.

Loewenstein, M., W. J. Borucki, H. F. Savage, J. G. Borucki, and R. C. Whitten (1978a) Geographical variations of NO and O_3 in the lower stratosphere *J. Geophys. Res.* **83**, 1874–1882.

Loewenstein, M., W. J. Starr, and D. G. Murcray (1978b) Stratospheric NO and HNO_3 observations in the northern hemisphere for three seasons. *Geophys. Res. Lett.* **5**, 531–534.

Løgager, T., and K. Sehested (1993) Formation and decay of peroxynitric acid: a pulse radiolysis study. *J. Phys. Chem.* **97**, 10,047–10,052.

Løgager, T., J. Holcman, K. Sehested, and T. Pedersen (1992) Oxidation of ferrous ions by ozone in acidic solutions. *Inorg. Chem.* **31**, 3523–3529.

Logan, J. A. (1983) Nitrogen oxides in the troposphere: global and regional budgets. *J. Geophys. Res.* **88**, 10,785–10,807.

Logan, J. A. (1989) Ozone in rural areas of the United States. *J. Geophys. Res.* **94**, 8511–8532.

Logan, J. A., M. J. Prather, S. C. Wofsy, and M. B. McElroy (1978) Atmospheric chemistry: response to human influence. *Philos. Trans. R. Soc. Lond.* **290**, 187–192.

Logan, J. A., M. J. Prather, S. C. Wofsy, and M. B. McElroy (1981) Tropospheric chemistry: a global perspective. *J. Geophys. Res.* **86**, 7210–7254.

London, J., and S. J. Oltmans (1979) The global distribution of long-term ozone variations during the period 1957–1975. *Pure Appl. Geophys.* **117**, 346–354.

Lonneman, W. A., S. L. Kopczinski, P. E. Darley, and F. D. Sutterfield (1974) Hydrocarbon composition of urban air pollution. *Environ. Sci. Technol.* **10**, 374–380.

Lonneman, W. A., J. J. Bufalini, and R. L. Seila (1976) PAN and oxidant measurements in ambient atmosphere. *Environ. Sci. Technol.* **10**, 374–380.

Lonneman, W. A., R. L. Seila, and J. J. Bufalini (1978) Ambient air hydrocarbon concentrations in Florida. *Environ. Sci. Technol.* **12**, 459–463.

Losno, R., J. L. Collin, L. Spokes, T. Jickells, M. Schulz, A. Rebers, M. Leermakers, C. Meuleman, and W. Baeyens (1998) Non-rain deposition significantly modifies rain samples at a coastal site. *Atmos. Environ.* **32**, 3445–3455.

Louis, J. F. (1975) Mean meridional circulation. *In* "The Natural Stratosphere 1974" (A. J. Grobecker, ed.), CIAP Monograph 1, pp. 6-25–6-31. U.S. Department of Transportation, Washington, DC.

Louw, C. W., J. F. Richards, and P. K. Faure (1977) The determination of volatile organic compounds in city air by gas chromatography combined with standard addition, selective substraction, infrared spectrometry and mass spectrometry. *Atmos. Environ.* **11**, 703–717.

Lovas, F. J., and R. D. Suenram (1977) Identification of dioxirane H_2COO in ozone-olefin reactions via microwave spectroscopy. *Chem. Phys. Lett.* **51**, 453–456.

Lovejoy, E. R., D. R. Hanson, and L. G. Huey (1996) Kinetics and products of the gas-phase reaction of SO_3 with water. *J. Phys. Chem.* **100**, 19,911–19,916.

Lovelock, J. E. (1961) Ionization methods for the analysis of gases and vapors. *Anal. Chem.* **33**, 162–178.

Lovelock, J. E. (1971) Atmospheric fluorine compounds as indicators of air movements. *Nature* **230**, 379.

Lovelock, J. E. (1974) Atmospheric halocarbons and stratospheric ozone. *Nature* **252**, 292–294.

Lovelock, J. E. (1975) Natural halocarbons in the air and in the sea. *Nature* **256**, 193–194.

Lovelock, J. E. (1979) "Gaia. A New Look at Life on Earth." Oxford University Press, London.

Lovelock, J. E., and L. Margulis (1974) Atmospheric homostasis by and for the biosphere: the Gaia hypothesis. *Tellus* **26**, 2–12.

Lovelock, J. E., R. J. Maggs, and R. A. Rasmussen (1972) Atmospheric dimethyl sulfide and the natural sulfur cycle. *Nature* **237**, 452–453.

Lovelock, J. E., R. G. Maggs, and R. J. Wade (1973) Halogenated hydrocarbons in and over the Atlantic. *Nature* **242**, 194–196.

Lovet, G. M., and S. E. Lindberg (1993) Atmospheric deposition and canopy interactions of nitrogen in forests. *Can. J. Forest Res.* **23**, 1603–1616.

Lovett, R. F. (1978) Quantitative measurement of airborne sea-salt in the north Atlantic. *Tellus* **30**, 358–364.

Lowe, D. C., and U. Schmidt (1983) Formaldehyde (HCHO) measurements in the non-urban atmosphere. *J. Geophys. Res.* **88**, 10,844–10,858.

Lowe, D. C., U. Schmidt, and D. H. Ehhalt (1980) A new technique for measuring tropospheric formaldehyde (CH_2O). *Geophys. Res. Lett.* **7**, 825–828.

Lowe, R. D. (1980) Stromatolites: 3400 Myr old from the archean of western Australia. *Nature* **284**, 441–443.

Ludwig, J. H., G. B. Morgan, and T. B. McMullen (1970) Trends in urban air quality. *EOS Trans. Am. Geophys. Union* **51**, 468–475.

Lundgren, D. A. (1970) Atmospheric aerosol composition and concentration as a function of particle size and time. *J. Air Pollut. Control. Assoc.* **20**, 603–608.

Luo, B. P., T. Peter, and P. J. Crutzen (1994) Freezing of stratospheric aerosol droplets. *Geophys. Res. Lett.* **21**, 1447–1450.

Luria, M., C. C. Van Valin, J. N. Galloway, W. C. Keene, D. L. Wellman, H, Sievering, and J. F. Boatman (1989) The relationship between dimethyl sulfide and particulate sulfate in the mid-Atlantic ocean atmosphere. *Atmos. Environ.* **23**, 139–147.

Luria, M., C. C. Van Valin, R. L. Gunter, D. L. Wellman, W. C. Keene, J. N. Galloway, H Sievering, and J. F. Boatman (1990) Sulfur dioxide over the western north Atlantic Ocean during CGE/Case/Watox. *Global Biogeochem. Cycles* **4**, 381–393.

Luther, F. M. (1975) Large-scale eddy diffusion. *In* "The Natural Stratosphere of 1974" (A. J. Grobecker, ed.), CIAP Monograph 1, pp. 6-31–6-38, U.S. Department of Transportation, Washington, DC.

Luther, F. M., and R. J. Gelinas (1976) Effect of molecular multiple scattering and surface albedo on atmospheric photodissociation rates. *J Geophys. Res.* **81**, 1125–1132.

Lyles, L. (1977) Wind erosion: processes and effect on soil productivity. *Trans. ASAE* **20**, 880–884.

Lynch, J. A., J. W. Grimm, and V. C. Bowersox (1995) Trends in precipitation chemistry in the United States: a national perspective, 1980–1992. *Atmos. Environ.* **29**, 1231–1246.

Maahs, H. G. (1982) Sulfur-dioxide/water equilibria between 0° and 50°C, an examination of data at low concentrations. *In* "Heterogeneous Atmospheric Chemistry" (D. R. Schryer, ed.). Geophysical Monograph Series, Vol. 26, pp. 187–195. American Geophysical Union, Washington, DC.

MacDonald, G. A. (1955) Hawaiian volcanoes during 1952. *U.S. Geol. Surv. Bull.* **1021-B**, 15–108.

MacDonald, G. A. (1972) "Volcanoes." Prentice-Hall, Englewood Cliffs, NJ.

MacDonald, G. J. F. (1963) The escape of helium from the Earth's atmosphere. *Rev. Geophys.* **1**, 305–349.

MacDonald, G. J. F. (1964) The escape of helium from the earth's atmosphere. In "The Origin and Evolution of Atmospheres and Oceans" (P. J. Brancazio and A. G. W. Cameron, eds.), pp. 127–182. Wiley, New York.

MacDonald, R., and R. Fall (1993) Detection of substantial emissions of methanol from plants to the atmosphere. Atmos. Environ. 27A, 1709–1713.

Machta, L. (1971) The role of the oceans and the biosphere in the carbon cycle. In "The Changing Chemistry of the Oceans" (F. Dryssen and D. Jagner, eds.), Nobel Symposium 20, pp. 121–145. Almquist and Wikrell, Stockholm.

Machta, L. (1974) Global scale atmospheric mixing. Adv. Geophys. 18 B, 33–56.

MacIntyre, F. (1968) Bubbles: a boundary layer "microtome" for micron thick samples of liquid surface. J. Phys. Chem. 72, 589–592.

MacIntyre, F. (1972) Flow patterns in breaking bubbles. J. Geophys. Res. 77, 5211–5228.

MacKenzie, F. T., and R. M. Garrels (1966) Chemical mass balance between rivers and oceans. Am. J. Sci. 264, 507–525.

Madronich, S. (1987a) Photodissociation in the atmosphere. 1. Actinic flux and effects of ground reflections and clouds. J. Geophys. Res. 92, 9740–9752.

Madronich, S. (1987b) Intercomparison of NO_2 photodissociation and u.v. radiometer measurements. Atmos. Environ. 21, 569–578.

Madronich, S., and J. G. Calvert (1990) Permutation reactions of organic peroxy radicals in the troposphere. J. Geophys. Res. 95, 5697–5715.

Madronich, S., D. R. Hastie, B. A. Ridley, and H. I. Schiff (1983) Measurements of the photodissociation coefficient for NO_2 in the atmosphere. I. Method and surface measurements. J. Atmos. Chem. 1, 3–25.

Madronich, S., R. B. Chatfield, J. G. Calvert, G. K. Moortgat, B. Veyret, and R. Lesclaux (1990) A photochemical origin of acetic acid in the troposphere. Geophys. Res. Lett. 17, 2361–2364.

Maenhaut, W., and W. H. Zoller (1977) Determination of the chemical composition of the South Pole aerosol by instrumental neutron activation analysis. J. Radioanal. Chem. 37, 637–650.

Maenhaut, W., W. H. Zoller, R. A. Duce, and G. L. Hoffman. (1979) Concentration and size distribution of particulate trace elements in the south polar atmosphere. J. Geophys. Res. 84, 2421–2431.

Maenhaut, W., H. Raemdonck, A. Selen, R. Van Grieken, and J. W. Winchester (1983) Characterization of the atmospheric aerosol over the eastern equatorial Pacific. J. Geophys. Res. 88, 5353–5364.

Mahlman, J. D., and W. J. Moxim (1978) Tracer simulation using a global general circulation model: results from a midlatitude instantaneous source experiment. J. Atmos. Sci. 35, 1340–1374.

Mahlman, J. D., H. B. Levy, and W. J. Moxim (1980) Three-dimensional tracer structure and behavior as simulated in two ozone precursor tracer experiments. J. Atmos. Sci. 37, 655–685.

Maigné, J. P., P. Y. Turpin, G. Madelaine, and J. Bricard (1974) Nouvelle méthode de détermination de la granulométrie d'un aerosol moyen d'une battérie de diffusion. J. Aerosol Sci. 5, 339–355.

Maiss, M., L. P. Steele, R. J. Francey, P. J. Fraser, R. L. Langenfelds, N. B. A. Trivett, and I. Levin (1996) Sulfur hexafluoride—a powerful new atmospheric tracer. Atmos. Environ. 30, 1621–1629.

Malinconico, L. L. (1979) Fluctuations in SO_2 emissions during recent eruptions of Etna. Nature 278, 43–45.

Manabe, S., and R. T. Weatherald (1967) Thermal equilibrium of the atmosphere with a given distribution of relative humidity. J. Atmos. Sci. 24, 241–259.

Manabe, S., and R. T. Weatherald (1975) The effects of doubling the CO_2 concentration on climate of a general circulation model. J. Atmos. Sci. 32, 3–17.

Manabe, S., and R. T. Weatherald (1980) On the distribution of climate change resulting from an increase in the CO_2 content of the atmosphere. *J. Atmos, Sci.* **37**, 99–118.

Mankin, W. G., and M. T. Coffey (1983) Latitudinal distributions and temporal changes of stratospheric HCl and HF. *J. Geophys. Res.* **88**, 10,776–10,784.

Mankin, W. G., M. T. Coffey, D. W. T. Griffith, and S. R. Drayson (1979) Spectroscopic measurement of carbonyl sulfide in the stratosphere. *Geophys. Res. Lett.* **6**, 853–856.

Manley, S. L., and M. N. Dastoor (1988) Methyl iodide (CH_3I) production and associated microbes. *Marine Biol.* **98**, 477–482.

Manley, S. L., and J. L. de la Cuesta (1997) Methyl iodide production from marine phytoplankton cultures. *Limnol. Oceanogr.* **42**, 142–147.

Manö, S., and M. O. Andreae (1994) Emission of methyl bromide from biomass burning. *Science* **263**, 1255–1257.

Marenco, A., and J. Fontan (1972) Etude par simulation sur modèles numérique du temps de séjour des aérosols dans la troposphère. *Tellus* **24**, 428–441.

Marenco, A., and F. Said (1989) Meridional and vertical ozone distribution in the background troposphere (70°N–60°S; 0–12 km altitude) from scientific aircraft measurements during the STRATOZ III experiment (June 1984). *Atmos. Environ.* **23**, 201–214.

Marenco, A., M. Macaigne, and S. Prieur (1989) Meridional and vertical CO and CH_4 distributions in the background troposphere (70°N–60°S, 0–12 km altitude) from scientific aircraft measurements during the STRATOZ III experiment (June 1984). *Atmos. Environ.* **23**, 185–200.

Maroulis, P. J., and A. R. Bandy (1977) Estimate of the contribution of biologically produced dimethyl sulfide to the global sulfur cycle. *Science* **196**, 647–648.

Maroulis, P. J., and A. R. Bandy (1980) Measurements of atmospheric concentrations of CS_2 in the eastern United States. *Geophys. Res. Lett.* **7**, 681–684.

Maroulis, P. J., A. L. Torres, and A. R. Bandy (1977) Atmospheric concentrations of carbonyl sulfide in the southwestern and eastern United States. *Geophys. Res. Lett.* **4**, 510–512.

Maroulis, P. J., A. L. Torres, A. B. Goldberg, and A. R. Bandy (1980) Atmospheric SO_2 measurements on project GAMETAG. *J. Geophys. Res.* **85**, 7345–7349.

Marple, V. A., and K. Willeke (1976) Inertial impactors: theory, design and use. In "Fine Particles, Aerosol Generation, Measurement, Sampling and Analysis" (B. Y. H. Liu, ed.), pp. 411–446. Academic Press, New York.

Marquardt, W., E. Brüggemann, and P. Ihle (1996) Trends in the composition of wet precipitation after long-scale transport—effects of atmospheric rehabilitation in East Germany. *Tellus* **48B**, 361–371.

Martell, E. A., and H. E. Moore (1974) Tropospheric aerosol residence times: a critical review. *J. Rech. Atmos.* **8**, 903–910.

Martens, C. S., J. J. Wesolowski, R. C. Harris, and R. Kaifer (1973) Chlorine loss from Puerto Rico and San Francisco Bay area marine aerosols. *J. Geophys. Res.* **78**, 8778–8792.

Marticorena, B., and G. Bergametti (1995) Modelling the atmospheric dust cycle. 1. Design of a soil-derived dust emission scheme. *J. Geophys. Res.* **100**, 16,415–16,430.

Marticorena, B., G. Bergametti, B. Aumont, Y. Callot, C. N. Doumé, and M. Legrand (1997) Modelling the atmospheric dust cycle. 2. Simulation of Saharan dust sources. *J. Geophys. Res.* **102**, 4387–4404.

Martin, A., and F. R. Barber (1981) Sulfur dioxide, oxides of nitrogen, and ozone measured continuously for two years at a rural site. *Atmos. Environ.* **15**, 567–578.

Martin, D., J. L. Jourdain, and G. LeBras (1986) Discharge flow measurement of the rate constant for the reaction $OH + SO_2 + He$ and $HOSO_2 + O_2$ in relation with the atmospheric oxidation of SO_2. *J. Phys. Chem.* **90**, 4143–4147.

Martin, L. R. (1984) Kinetic studies of sulfite oxidation in aqueous solution. In "SO_2, NO and NO_2 Oxidation Mechanisms: Atmospheric Considerations" (J. G. Calvert, ed.), Acid Precipitation Series, Vol. 3, pp. 63–100. Butterworth, Boston.

Martin, L. R., and D. E. Damschen (1981) Aqueous oxidation of sulfur dioxide by hydrogen peroxide at low pH. Atmos. Environ. 15, 1617–1622.

Martin, L. R., and M. W. Hill (1987) The iron-catalyzed oxidation of sulfur: reconciliation of the literature rates. Atmos. Environ. 21, 1487–1490.

Martin, L. R., M. W. Hill, A. E. Tai, and T. W. Good (1991) The iron catalyzed oxidation of sulfur(IV) in aqueous solution. Differing effects of organics at high and low pH. J. Geophys. Res. 96, 3085–3097.

Martinez, R. I. (1980) A systematic nomenclature for peroxyacyl nitrates, the functional and structural misnomers for anhydride derivatives of nitrogen oxo acids. Int. J. Chem. Kinet. 12, 771–775.

Martinez, R. I., and J. T. Herron (1978) Stopped-flow study of the gas-phase reaction of ozone with organic sulfides: dimethyl sulfide. Int. J. Chem. Kinet. 10, 433–452.

Martinez, R. I., R. E. Huie, and J. T. Herron (1977) Mass spectrometric detection of dioxirane, H_2COO and its decomposition products, H_2, CO, from the reaction of ozone with ethylene. Chem. Phys. Lett. 51, 457–459.

Martinez, R. I., J. T. Herron, and R. E. Huie (1981) The mechanism of ozone–alkene reactions in the gas phase. A mass spectrometric study of the reactions of eight linear and branched chain alkenes. J. Am. Chem. Soc. 103, 3807–3820.

Marty, J. C., and A. Saliot (1982) Aerosols in equatorial Atlantic air: n-alkanes as a function of particle size. Nature 298, 144–147.

Marty, J. C., M. J. Tissier, and A. Saliot (1984) Gaseous and particulate polycyclic aromatic hydrocarbons (PAH) from the marine atmosphere. J. Atmos. Environ. 18, 2183–2190.

Marx, W., P. B. Monkhouse, and U. Schurath (1984) "Kinetik und Intensität photochemischer Reaktionsschritte in der Atmosphäre." Gesellschaft für Strahlen- und Umweltforschung, Munich, Germany.

Mason, B. (1966) "Principles of Geochemistry," 3rd ed. Wiley, New York.

Mason, B. J. (1971) "The Physics of Clouds," 2nd ed. Oxford University Press (Clarendon), London

Mateer, C. L., and J. J. DeLuisi (1992) A new Umkehr inversion algorithm. J. Atmos. Terr. Phys. 54, 537–556.

Matson, P. A., and P. M. Vitousek (1987) Cross-system comparisons of soil nitrogen transformations and nitrous oxide flux in tropical forest ecosystems. Global Biogeochem. Cycles 2, 163–170.

Matson, P. A., and P. M. Vitousek (1990) Ecosystem approach to a global nitrous oxide budget. Processes that regulate gas emissions vary in predictable ways. BioScience 40, 667–672.

Matsumoto, G., and T. Hanya (1980) Organic constituents in fallout in the Tokyo area. Atmos. Environ. 14, 1409–1419.

Matsumoto, M., and T. Okita (1998) Long term measurements of atmospheric gaseous and aerosol species using an annular denuder system in Nara, Japan. Atmos. Environ. 32, 1419–1425.

Matsuo, S. (1978) The oxidation state of the primordial atmosphere. In "Origin of Life" (H. Noda, ed.), pp. 21–27. Center for Academic Publishing Japan, Tokyo.

Matthews, E., and I. Fung (1987) Methane emission from natural wetlands: global distribution, area, and environmental characteristics of sources. Global Biogeochem. Cycles 1, 61–86.

Matthews, E., I. Fung, and J. Lerner (1991) Methane emission from rice cultivation: geographic and seasonal distribution of cultivated areas and emissions. Global Biogeochem. Cycles 5, 3–24.

Matthias-Maser, S., and R. Jaenicke (1994) Examination of atmospheric bio-aerosol particles with radii > 0.2 μm. *J. Aerosol Sci.* 25, 1605–1613.

Mauersberger, K., R. Finstad, S. Anderson, and P. Robbins (1981) A comparison of ozone measurements. *Geophys. Res. Lett.* 8, 361–364.

Maus, R., A. Goppelsröder, and H. Umhauer (1997) Viability of bacteria on unused filter media. *Atmos. Environ.* 31, 2305–2310.

Maynard, J. B., and W. N. Sanders (1969) Determination of a detailed hydrocarbon composition and potential atmospheric reactivity of full-range motor gasolines. *J. Air Pollut. Control Assoc.* 19, 505–510.

Mayrson, H., and J. H. Crabtree (1976) Source reconciliation of atmospheric hydrocarbons. *Atmos. Environ.* 10, 137–143.

Mazurek, M. A., G. R. Cass, and B. R. T. Simoneit (1991) Biological input to visibility-reducing aerosol particles in the remote arid southwestern United States. *Environ. Sci. Technol.* 25, 684–694.

McAlester, A. L. (1968) "The History of Life." Prentice Hall, Englewood Cliffs, NJ.

McAuliffe, C. (1966) Solubility in water of paraffin, cyclo-paraffin, olefin, acetylene, cyclo-olefin and aromatic hydrocarbons. *J. Phys. Chem.* 70, 1267–1275.

McCarthy, R. L., F. A. Bower, and J. P. Jesson (1977) The fluorocarbon-ozone theory I: production and release. World production and release of CCl_3F and CCl_2F_2 through 1975. *Atmos. Environ.* 11, 491–497.

McConnell, J. C., M. B. McElroy, and S. C. Wofsy (1971) Natural sources of atmospheric CO. *Nature* 233, 187–188.

McCormack, J. D., and R. K. Hilliard (1970) Scavenging of aerosol particles by sprays. In "Precipitation Scavenging 1970" (R. J. Engelmann and W. G. N. Slinn, coordinators). AEC Symposium Series, 30, 187–204. U.S Atomic Energy Commission, Division of Technical Information, Oak Ridge, TN.

McCormick, M. P., H. M. Steele, P. Hamill, W. P. Chu, and T. J. Swissler (1982) Polar stratospheric cloud sightings by SAM II. *J. Atmos. Sci.* 39, 328–331.

McCormick, M. P., C. R. Trepte, and M. C. Pitts (1989) Persistence of polar stratospheric clouds in the southern polar region. *J. Geophys. Res.* 94, 11,241–11,251.

McElroy, M. B., and J. C. McConnell (1971) Nitrous oxide: a natural source of stratospheric NO. *J. Atmos. Sci.* 28, 1095–1098.

McElroy, M. B., and R. J. Salawitch (1989) Changing composition of the global stratosphere. *Science* 243, 763–770.

McElroy, M. B., and S. C. Wofsy (1986) Tropical forests. Interactions with the atmosphere. In "Tropical Rain Forests and the World Atmosphere" (G. T. Prance, ed.), pp. 33–60. Westview, Boulder, CO.

McElroy, M. B., S. C. Wofsy, and Y. L. Yung (1977a) The nitrogen cycle: perturbations due to man and their impact on atmospheric N_2O and O_2. *Philos. Trans. R. Soc. Lond.* 277, 159–181.

McElroy, M. B., T. Y. Kong, and Y. L. Yung (1977b) Photochemistry and evolution of Mars' atmosphere. A Viking perspective. *J. Geophys. Res.* 82, 4379–4388.

McElroy, M. B., R. J. Salawitch, and S. C. Wofsy (1986a) Antarctic O_3: chemical mechanism for the spring decrease. *Geophys. Res. Lett.* 13, 1296–1299.

McElroy, M. B., R. J. Salawitch, S. C. Wofsy, and J. A. Logan (1986b) Reduction of Antarctic ozone due to synergistic interactions of chlorine and bromine. *Nature* 321, 759–762.

McElroy, M. B., R. J. Salawitch, and K. Minschwaner (1992) The changing stratosphere. *Planet. Space Sci.* 40, 373–401.

McElroy, W. J. (1986) The aqueous oxidation of SO_2 by OH radicals. *Atmos. Environ.* 20, 323–330.

McElroy, W. J. (1990) A laser photolysis study of the reaction of SO_4^- with Cl^- and the subsequent decay of Cl_2^- in aqueous solution. *J. Phys. Chem.* 94, 2435–2441.

McElroy, W. J., and S. J. Waygood (1990) Kinetics of the reactions of the SO_4^- radical with SO_4^-, $S_2O_8^{2-}$, H_2O and Fe^{2+}. *J. Chem Soc. Faraday Trans.* 86, 2557–2564.

McEwen, D. J. (1966) Automobile exhaust hydrocarbon analysis by gas chromatography. *Anal. Chem.* 38, 1047–1053.

McFarland, M., D. Kley, J. W. Drummond, A. L. Schmeltekopf, and R. J. Winkler (1979) Nitric oxide measurements in the equatorial Pacific region. *Geophys. Res. Lett.* 6, 605–608.

McGrath, W. D., and R. G. W. Norrish (1960) Studies of the reactions of excited atoms and molecules produced in the flash photolysis of ozone. *Proc. R. Soc. Lond.* 254, 317–326.

McKay, C., M. Pandow, and R. Wolfgang (1963) On the chemistry of natural radiocarbon. *J. Geophys. Res.* 68, 3929–3931.

McKeen, S. A., and S. C. Liu (1993) Hydrocarbon ratios and photochemical histories of air masses. *Geophys. Res. Lett.* 20, 2363–2366.

McKenney, D. J., K. F. Shuttleworth, J. R. Vriesacker, and W. I. Findlay (1982) Production and loss of nitric oxide from denitrification in anaerobic Brookston clay. *Appl. Environ. Microbiol.* 43, 534–541.

McKirdy, D. M., and J. Hahn (1982) Composition of kerogen and hydrocarbons in Precambrian rocks. *In* "Mineral Deposits and the Evolution of the Biosphere" (H. D. Holland and M. Schidlowski, eds.), pp. 123–154. Springer-Verlag, Berlin.

McMahon, C. K., and P. W. Ryan (1976) Some chemical and physical characteristics of emission from forest fires, Paper no. 76-2.3, presented at the 69th Annual Meeting of the Air Pollution Control Association, Portland, OR, 1976.

McMahon, T. A., and P. J. Denison (1979) Empirical atmospheric deposition parameters—a survey. *Atmos. Environ.* 13, 571–585.

McMullen, T. B., R. B. Faoro, and G. B. Morgan (1970) Profile of pollutant fractions in nonurban suspended particulate matter. *J. Air Pollut. Control Assoc.* 20, 369–376.

McMurry, P. H., and J. C. Wilson (1983) Droplet phase (heterogeneous) and gas phase (homogeneous) contributions to secondary ambient aerosol formation as functions of relative humidity. *J. Geophys. Res.* 88, 5101–5108.

McPeters, R. D. (1992) The atmospheric SO_2 budget for Pinatubo derived from NOAA-11 SBUV/2 spectral data. *Geophys. Res. Lett.* 20, 1971–1974.

Meagher, J. F., E. M. Bailey, and M. Luria (1983) The seasonal variation of the atmospheric SO_2 to SO_4 conversion rate. *J. Geophys. Res.* 88, 1525–1527.

Meetham, A. R., D. W. Bottom, and S. Cayton (1964) "Atmospheric Pollution: Its Origin and Prevention," 3rd ed. Pergamon Press, Oxford.

Mehlmann, A. (1986) Größenverteilung des Nitrats und seine Beziehung zur gasförmigen Salpetersäure. Dissertation. University Mainz, Germany.

Mehlmann, A., and P. Warneck (1995) Atmospheric gaseous HNO_3, particulate nitrate and aerosol size distribution of major ionic species at a rural site in western Germany. *Atmos. Environ.* 29, 2359–2373.

Meinel, A. B. (1950) OH emission bands in the spectrum of the night sky. *Astrophys. J.* 111, 555–564.

Meixner, F. X. (1984) The vertical sulfur dioxide distribution at the tropopause level. *J. Atmos. Chem.* 2, 175–189.

Meixner, F. X. (1994) Surface exchange of odd nitrogen oxides. *Nova Acta Leopoldina* NF 70, 299–348.

Meixner, F. X., P. Müller, G. Aheimer, and K. D. Höfken (1985) Measurements of gaseous nitric acid and particulate nitrate. *In* "Pollutant Cycles and Transport: Modelling and Field

Experiment" (F. A. A. M. De Leeuw and N. D. van Egmond, eds.), pp. 103–119. National Institute of Public Health and Environmental Protection, Bilthoven, the Netherlands.

Melillo, J. M., J. R. Fruci, R. A. Houghton, B. Moore, and D. L. Skole (1988) Land-use change in the Soviet Union between 1850 and 1980: causes of a net release of CO_2 to the atmosphere. *Tellus* **40B**, 116–128.

Mellouki, A., G. Le Bras, and G. Poulet (1988) Kinetics of the reaction of NO_3 with OH and HO_2. *J. Phys. Chem.* **92**, 2229–2234.

Menard, H. W., and S. M. Smith (1966) Hypsometry of ocean basin provinces. *J. Geophys. Res.* **71**, 4305–4325.

Mentall, J. E., B. Guenther, and D. Williams (1985) The solar spectral irradiance between 150–200 nm. *J. Geophys. Res.* **90**, 2265–2271.

Menzel, D. W. (1974) Primary productivity, dissolved and particulate organic matter, and the sites of oxidation of organic matter. In "The Sea" (E. D. Goldberg, ed.), Vol. 5, pp. 659–678. Wiley Interscience, New York.

Menzies, R. T. (1983) A reevaluation of laser heterodyne radiometer ClO measurements. *Geophys. Res. Lett.* **10**, 729–732.

Mertes, S., and A. Wahner (1995) Uptake of nitrogen dioxide and nitrous acid on aqueous surfaces. *J. Phys. Chem.* **99**, 14,000–14,006.

Mészáros, A., and K. Vissy (1974) Concentration, size distribution and chemical nature of atmospheric aerosol particles in remote ocean areas. *J. Aerosol Sci.* **5**, 101–110.

Mészáros, E. (1982) On the atmospheric input of sulfur into the ocean. *Tellus* **34**, 277–282.

Mészáros, E., D. J. Moore, and J. P. Lodge, Jr (1977) Sulfur dioxide-sulfate relationships in Budapest *Atmos. Environ.* **11**, 345–349.

Mészáros, E., G. Varhelyi, and L. Haszpra (1978) On the atmospheric sulfur budget over Europe. *Atmos. Environ.* **12**, 2273–2277.

Meybeck, M. (1982) Carbon, nitrogen, and phosphorus transport by world rivers. *Am. J. Sci.* **282**, 401–450.

Meyers, P. A., and R. A. Hites (1982) Extractable organic compounds in midwest rain and snow. *Atmos. Environ.* **16**, 2169–2175.

Meyers, T. P., B. J. Huebert, and B. B. Hicks (1989) HNO_3 deposition to a deciduous forest. *Boundary Layer Met.* **49**, 395–410.

Meyrahn, H., G. K. Moortgat, and P. Warneck (1982) The photolysis of acetaldehyde under atmospheric conditions. In "Atmospheric Trace Constituents" (F. Herbert, ed.), pp. 65–72. Vieweg and Sohn, Braunschweig, Germany.

Meyrahn, H., J. Pauly, W. Schneider, and P. Warneck (1986) Quantum yields of the photodissociation of acetone in air and an estimate for the life time of acetone in the troposphere. *J. Atmos. Chem.* **4**, 277–291.

Michael, J. V., and K. P. Lim (1992) Rate constants for the N_2O reaction system: thermal decomposition of N_2O; $N + NO \rightarrow N_2 + O$; and implications for $O + N_2 \rightarrow NO + N$. *J. Chem. Phys.* **97**, 3228–3234.

Middleton, P., and C. S. Kiang (1978) A kinetic aerosol model for the formation and growth of secondary sulfuric acid particles. *J. Aerosol Sci.* **9**, 359–385.

Midgley, P. (1989) The production and release to the atmosphere of 1,1,1-trichloroethane (methyl chloroform). *Atmos. Environ.* **23**, 2663–2665.

Migeotte, M. V. (1948) Spectroscopic evidence of methane in the Earth's atmosphere. *Phys. Rev.* **73**, 519–520.

Migeotte, M. V. (1949) The fundamental band of carbon monoxide at 4.7 microns in the solar spectrum. *Phys. Rev.* **75**, 1108–1109.

Migeotte, M. V., and R. M. Chapman (1949) On the question of atmospheric ammonia. *Phys. Rev.* **75**, 1108–1109.

Mihelcic, D., D. H. Ehhalt, G. F. Kules, J. Klomfass, M. Trainer, U. Schmidt, and H. Röhrs (1978) Measurements of free radicals in the atmosphere by matrix isolation and electron paramagnetic resonance. *Pure Appl. Geophys.* 116, 530–536.

Mihelcic, D., P. Müsgen, and D. H. Ehhalt (1985) An improved method of measuring tropospheric NO_2 and RO_2 by matrix isolation and electron spin resonance. *J. Atmos. Chem.* 3, 341–361.

Mihelcic, D., D. Klemp, P. Müsgen, H. W. Pätz, and A. Volz-Thomas (1993) Simultaneous measurement of peroxy and nitrate radicals at Schauinsland. *J. Atmos. Chem.* 16, 313–335.

Milford, J. B., and C. I. Davidson (1985) The size of particulate trace elements in the atmosphere —a review. *J. Air Pollut. Control Assoc.* 35, 1249–1260.

Mill, T., and G. Montorsi (1973) The liquid phase oxidation of 2,4-dimethylpentane. *Int. J. Chem. Kinet.* 5, 119–136.

Mill, T., F. Mayo, H. Richardson, K. Irvin, and D. A. Allara (1972) Gas- and liquid-phase oxidations of *n*-butane. *J. Am. Chem. Soc.* 94, 6802–6811.

Miller, C., D. L. Filkin, A. J. Owens, J. M. Steed, and J. P. Jesson (1981) A two-dimensional model of stratospheric chemistry and transport. *J. Geophys. Res.* 86, 12,039–12,065.

Miller, D. F., and A. J. Alkezweeny (1980) Aerosol formation in urban plumes over Lake Michigan. *Ann. N.Y. Acad. Sci.* 338, 219–232.

Miller, D. F., and C. W. Spicer (1975) Measurement of nitric acid in smog. *J. Air Pollut. Control Assoc.* 25, 940–942.

Miller, R. L., A. G. Suits, P. L. Houston, R. Toumi, J. A. Mack, and A. M. Wodtke (1994) The ozone deficit problem: $O_2(X, \nu \le 26) + O(^3P)$ from 226 nm ozone photodissociation. *Science* 265, 1831–1838.

Miller, S. L., and L. E. Orgel (1974) "The Origin of Life on Earth." Prentice Hall, Englewood Cliffs, NJ.

Miller, S. L., and H. L. Urey (1959) Organic compound synthesis on the primitive Earth. *Science* 130, 245–251.

Millero, J. J., S. Hubinger, M. Fernandez, and S. Garnett (1987) Oxidation of H_2S in sea water as a function of temperature, pH and ionic strength. *Environ. Sci. Technol.* 21, 439–443.

Millikan, R. A. (1923) The general law of fall of a small spherical body through a gas, and its bearing upon the nature of molecular reflections from surfaces. *Phys. Rev.* 22, 1–23.

Minami, K., and S. Fukushi (1986) Emission of nitric oxide from a well-aerated andosol treated with nitrite and hydroxylamine. *Soil. Sci. Plant Nutr.* 32, 233–237.

Miner, S. (1969) Preliminary air pollution survey of ammonia, a literature review, publication APTD 69-25. U.S. Department of Health, Education and Welfare, National Air Pollution Control Administration, Raleigh, NC.

Minschwaner, K., G. P. Anderson, L. A. Hall, and K. Yoshino (1992) Polynomial coefficients for calculating O_2 Schumann-Runge cross sections at 0.5 cm^{-1} resolution. *J. Geophys. Res.* 97, 10,103–10,108.

Minschwaner, K., R. J. Salawitch, and M. B. McElroy (1993) Absorption of solar radiation by O_2: implications for O_3 and lifetimes of N_2O, $CFCl_3$, and CF_2Cl_2. *J. Geophys. Res.* 98, 10,543–10,561.

Mirabel, P., and J. L. Clavelin (1978) On the limiting behavior of binary homogeneous nucleation theory. *J. Aerosol. Sci.* 9, 219–225.

Mirabel, P., and J. L. Katz (1974) Binary homogeneous nucleation as a mechanism for the formation of aerosols. *J. Chem. Phys.* 60, 1138–1144.

Mitra, S. K., and A. U. Hannemann (1993) On the scavenging of SO_2 by large and small rain drops 5. A wind tunnel and theoretical study of the desorption of SO_2 from water drops containing S(IV). *J. Atmos. Chem.* 16, 201–218.

Mitra, S. K., S. Barth, and H. R. Pruppacher (1990) A laboratory study of the efficiency with which aerosol particles are scavenged by snow flakes. *Atmos. Environ.* 24A, 1247–1254.

Mitra, S. K., A. Waltrop, A. Hannemann, A. Flossmann, and H. R. Pruppacher (1992) A wind tunnel and theoretical investigation to test various theories for the absorption of SO_2 by drops of pure water and water drops containing H_2O_2 and $(NH_4)_2SO_4$. *In* "Precipitation Scavenging and Atmosphere—Surface Exchange" (S. E. Schwartz and W. G. N. Slinn, eds.), pp. 123–141. Hemisphere Publishing Corporation, Washington, DC.

Miyake, Y., and S. Tsunogai (1963) Evaporation of iodine from the ocean. *J. Geophys. Res.* 68, 3989–3994.

Miyoshi, A., S. Hatakeyama, and N. Washida (1994) OH radical initiated photooxidation of isoprene: an estimate of global CO production. *J. Geophys. Res.* 99, 18,779–18,787.

Mohnen, V. A. (1977) The issue of stratospheric ozone intrusion. Atmospheric Science Research Center Publication no. 428. New York State University, Albany, NY.

Mohnen, V. A., and J. A. Kadlecek (1989) Cloud chemistry research at Whiteface Mountain. *Tellus* 41B, 79–91.

Molina, M. J., and F. S. Rowland (1974) Stratospheric sink for chlorofluoromethanes: chlorine atom catalyzed destruction of ozone. *Nature* 249, 810–812.

Molina, M. J., A. J. Colussi, L. T. Molina, R. N. Schindler, and T.-L. Tso (1990) Quantum yield of chlorine-atom formation in the photodissociation of chlorine peroxide (ClOOCl) at 308 nm. *Chem. Phys. Lett.* 173, 310–315.

Molina, M. J., R. Zhang, P. J. Wooldridge, J. R. McMahon, J. E. Kim, H. Y. Chang, and K. D. Beyer (1993) Physical chemistry of the $H_2SO_4/HNO_3/H_2O$ system: implications for polar stratospheric clouds. *Science* 261, 1418–1423.

Möller, D. (1984) Estimation of the global man-made sulphur emission. *Atmos. Environ.* 18, 19–27.

Möller, F. (1950) Eine Berechnung des horizontalen Grossaustausches über dem Atlantischen Ozean. *Arch. Meteorol. Geophys. Bioklimatol. Ser. A* 2, 73–81.

Möller, F. (1963) On the influence of changes in the CO_2 concentration in air on the radiation balance at the Earth's surface and on climate *J. Geophys. Res.* 68, 3877–3886.

Möller, U., and G. Schumann (1970) Mechanism of transport from the atmosphere to the Earth's surface. *J. Geophys. Res.* 75, 3013–3019.

Monahan, E. C. (1968) Sea spray as a function of low elevation wind speed. *J. Geophys. Res.* 73, 1127–1137.

Monahan, E. C. (1971) Oceanic whitecaps. *J. Phys. Oceanogr.* 1, 139–144.

Monster, J., P. W. U. Appel, H. G. Thode, M. Schidlowski, C. M. Carmichael, and D. Bridgewater (1979) Sulfur isotope studies in early archaean sediments from Isua, West Greenland: implications for the antiquity of bacterial sulfate reduction. *Geochim. Cosmochim. Acta* 43, 405–413.

Montzka, S. A., M. Trainer, P. D. Goldan, W. C. Kuster, and F. C. Fehsenfeld (1993) Isoprene and its oxidation products, methyl vinyl ketone and methacrolein, in the rural atmosphere. *J. Geophys. Res.* 98, 1101–1111.

Montzka, S. A., J. H. Butler, R. C. Myers, T. M. Thompson, T. H. Swanson, A. D. Clarke, L. T. Lock, and J. W. Elkins (1996) Decline in the tropospheric abundance of halogen from halocarbons: implications for stratospheric ozone depletion. *Science* 272, 1318–1322.

Moody J. L., A. A. P. Pszenny, A. Gaudry, W. C. Keene, J. N. Galloway, and G. Polian (1991) Precipitation composition and its variability in the southern Indian Ocean: Amsterdam island, 1980–1987. *J. Geophys. Res.* 96, 20,769–20,786.

Moorbath, C., R. K. O'Nions, and R. J. Pankhurts (1973) Early archaean age for the Isua iron formation of west Greenland. *Nature* 245, 138–139.

Moore, H. E., S. E. Poet, and E. A. Martell (1973) ^{222}Rn, ^{210}Pb, ^{210}Bi, and ^{210}Po profiles and aerosol residence times versus altitude. *J. Geophys. Res.* **78**, 7065–7075.

Moore, R. M., and O. C. Zafiriou (1994) Photochemical production of methyl iodide in seawater. *J. Geophys. Res.* **99**, 16,415–16,420.

Moortgat, G. K., and C. E. Junge (1977) The role of SO_2 oxidation for the background stratospheric sulfate layer in the light of new reaction rate data. *Pure Appl. Geophys.* **115**, 759–774.

Moortgat, G. K., and E. Kudzus (1978) Mathematical expression for the $O(^1D)$ quantum yields from the O_3 photolysis as a function of temperature. *Geophys. Res. Lett.* **5**, 191–194.

Moortgat, G. K., and P. Warneck (1979) CO and H_2 quantum yields in the photodecomposition of formaldehyde in air. *J. Chem. Phys.* **70**, 3639–3651.

Moortgat, G. K., E. Kudszus, and P. Warneck (1977) Temperature dependence of $O(^1D)$ formation in the near UV photolysis of ozone. *J. Chem. Soc. Faraday Trans. 2* **73**, 1216–1221.

Moortgat, G. K., W. Seiler, and P. Warneck (1983) Photodissociation of HCHO in air: CO and H_2 quantum yields at 220 and 300 K. *J. Chem. Phys.* **78**, 1185–1190.

Mopper, K., and E. Degens (1979) Organic carbon in the ocean, nature and cycling. *In* "The Global Carbon Cycle" (B. Bolin, E. T. Degens, S. Kempe, and P. Ketner, eds.). *SCOPE* **13**, 293–316.

Morgan, J. A., and W. J. Parton (1989) Characteristics of ammonia volatilization from spring wheat. *Crop Sci.* **29**, 726–731.

Morgan, J. J. (1982) Factors governing the pH, availability of H^+, and oxidation capacity of rain. *In* "Atmospheric Chemistry" (E. D. Goldberg, ed.), pp. 17–40. Dahlem Konferenzen. Springer-Verlag, Berlin.

Morris, E. D., and H. Niki (1971) Reactivity of hydroxyl radicals with olefins. *J. Phys. Chem.* **75**, 3640–3641.

Mosier, A. R., and G. L. Hutchinson (1981) Nitrous oxide emissions from cropped fields. *J. Environ. Qual.* **10**, 169–173.

Moulin, C., F. Guillard, F. Dulac, and C. E. Lambert (1997) Long-term daily monitoring of Sahara dust load over marine areas using Meteosat ISCCP-B2 data. *J. Geophys. Res.* **102**, 16,947–16,958.

Mount, G. H., and G. J. Rottman (1981) Solar intensities: the solar spectral irradiance 1200–3184 Å near solar maximum: 15 July 1980. *J. Geophys. Res.* **86**, 9193–9198.

Mount, G. H., and G. J. Rottman (1983) The solar absolute spectral irradiance at 1150–3173 Å May 17, 1982. *J. Geophys. Res.* **88**, 5403–5410.

Mount, G. H., and G. J. Rottman (1985) Solar absolute spectral irradiance 118–300 nm: July 25, 1983. *J. Geophys. Res.* **90**, 13,031–13,036.

Moyers, J. L., L. E. Rauweiler, S. B. Hopf, and N. E. Korte (1977) Evaluation of particulate trace species in southwest desert atmosphere. *Environ. Sci. Technol.* **11**, 789–795.

Mozurkewich, M. (1993) The dissociation constant of ammonium nitrate and its dependence on temperature, relative humidity and particle size. *Atmos. Environ.* **27A**, 261–270.

Mozurkewich, M., P. H. McMurry, A. Gupta, and J. G. Calvert (1987) Mass accommodation coefficient for HO_2 radicals on aqueous particles. *J. Geophys. Res.* **92**, 4163–4170.

Mrose, H. (1966) Measurements of pH and chemical analysis of rain, snow and fog water. *Tellus* **18**, 266–270.

Mroz, E., and W. H. Zoller (1975) Composition of atmospheric particulate matter from the eruption of Heimaey, Iceland. *Science* **199**, 461–464.

Mueller, P. K., K. K. Fung, S. L. Heisler, D. Grosjean, and G. M. Hidy (1982) Atmospheric particulate observations in urban and rural areas of the United States. In "Particulate Carbon —Atmospheric Life Cycle" (G. T. Wolff and R. L. Klimisch, eds.), pp. 343–370. Plenum, New York.

Mueller, T. K., G. M. Hidy, K. Warren, T. F. Lavery, and R. L. Baskett (1980) The occurrence of atmospheric aerosols in the north-eastern United Sates. Ann. N.Y. Acad. Sci. 338, 463–482.

Müller, J. F. (1992) Geographical distribution and seasonal variation of surface emissions and deposition velocities of atmospheric trace gases. J. Geophys. Res. 97, 3787–3804.

Munger, J. W., D. J. Jacob, J. M. Waldman, and M. R. Hoffmann (1983) Fog water chemistry in an urban atmosphere. J. Geophys. Res. 88, 5109–5121.

Munger, J. W., D. J. Jacob, and M. R. Hoffmann (1984) The occurrence of bisulfite aldehyde addition products in fog and cloud water. J. Atmos. Chem. 1, 335–350.

Munger, J. W., J. L. Collet, B. Daube, and M. R. Hoffmann (1989) Chemical composition of coastal stratus clouds: dependence on droplet size and distance from the coast. Atmos. Environ. 23, 2305–2320.

Munger, J. W., S. C. Wofsy, P. S. Bakwin, S.-M. Fan, M. L. Goulden, B. C. Daube, A. H. Goldstein, K. E. Moore, and D. R. Fitzgerald (1996) Atmospheric deposition of reactive nitrogen oxides and ozone in a temperate deciduous forest and subarctic woodland. 1. Measurements and mechanisms. J. Geophys. Res. 101, 12,639–12,657.

Munk, W. H. (1966) Abyssal recipes. Deep Sea Res. 13, 707–730.

Münnich, K. O. (1963) Der Kreislauf des radioaktiven Kohlenstoffs in der Natur. Naturwissenschaften 50, 211–218.

Münnich, K. O., and W. Roether (1967) Transfer of bomb C-14 and tritium from the atmosphere to the ocean. Internal mixing of the ocean on the basis of tritium and C-14 profiles. In "Radioactive Dating and Methods of Low Level Counting," Symposium Proceedings, Monaco, STI/PUB/152, pp. 93–104. International Atomic Energy Agency, Vienna.

Murakami, M., K. Kikichi, and C. Mangono (1985) Experiments on aerosol scavenging by natural snow crystals. Part I. Collection efficiency of uncharged snow crystals for micron and submicron particles. J. Meteorol. Soc. Japan 63, 119–129.

Murcray, D. G., A. Goldman, C. M. Bradford, G. R. Cook, J. W. van Allen, F. S. Bonomo, and F. H. Murcray (1978) Identification of the ν_2 vibration-rotation band of ammonia in ground level solar spectra. Geophys. Res. Lett. 5, 527–530.

Murozumi, M., T. J. Chow, and C. Patterson (1969) Pollutant lead aerosols, terrestrial dusts and sea salt in Greenland snow strata. Geochim. Cosmochim. Acta 33, 1247–1294.

Murphy, D. M., and D. W. Fahey (1994) An estimate of the flux of stratospheric reactive nitrogen and ozone into the troposphere. J. Geophys. Res. 99, 5325–5332.

Murphy, D. M., D. W. Fahey, M. H. Proffitt, S. C. Liu, K. R. Chan, C. S. Eubank, S. R. Kawa, and K. K. Kelly (1993) Reactive nitrogen and its correlation with ozone in the lower stratosphere and upper troposphere. J. Geophys. Res. 98, 8751–8773.

Muzio, L. J., and J. C. Kramlich (1988) An artifact in the measurement of N_2O from combustion sources. Geophys. Res. Lett. 15, 1369–1372.

Najjar, R. G. (1992) Marine biogeochemistry. In "Climate System Modeling" (K. E. Trenberth, ed.), pp. 241–280. Cambridge University Press, New York.

Nastrom, G. D. (1977) Vertical and horizontal fluxes of ozone at the tropopause from the first year of GASP data. J. Appl. Meteorol. 16, 740–744.

National Research Council, Subcommittee on Ammonia (1979) "Ammonia." University Park Press, Baltimore, MD.

Natusch, D., H. Konis, H. Axelrod, R. Teck, and J. Lodge (1972) Sensitive method for the measurement of atmospheric hydrogen sulfide. Anal. Chem. 44, 2067–2070.

Naughton, J. J., and K. Terada (1954) Effect of eruption of Hawaiian volcanoes on the composition and carbon isotope content of associated volcanic and fumarolic gases. *Science* 120, 580–581.

Naughton, J. J., V. A. Lewis, D. Thomas, and J. B. Finlayson (1975) Fume compositions found at various stages of activity at Kilauea volcano, Hawaii. *J. Geophys. Res.* 80, 2963–2966.

Neckel, H., and D. Labs (1981) Improved data of solar spectral irradiance from 0.33 to 1.25 μm. *Solar Phys.* 74, 231–249.

Neeb, P., F. Sauer, O. Horie, and G. K. Moortgat (1997) Formation of hydroxymethyl hydroperoxide and formic acid in alkene ozonolysis in the presence of water vapor. *Atmos. Environ.* 31, 1417–1423.

Neeb, P., O. Horie, and G. K. Moortgat (1998) The ethene-ozone reaction in the gas phase. *J. Phys. Chem.* 102, 6778–6785.

Neely, W. H., and J. H. Ploncka (1978) Estimation of the time-averaged hydroxyl radical concentration in the troposphere. *Environ. Sci. Technol.* 12, 317–321.

Neftel, A., E. Moor, H. Oeschger, and B. Stauffer (1985a) Evidence from polar ice cores for the increase in atmospheric CO_2 in the past two centuries. *Nature* 315, 45–47.

Neftel, A., J. Beer, H. Oeschger, F. Zürcher, and R. C. Finkel (1985b) Sulphate and nitrate concentrations in snow from southern Greenland 1895–1978. *Nature* 314, 611–613.

Neitzert, V., and W. Seiler (1981) Measurement of formaldehyde in clean air. *Geophys. Res. Lett.* 8, 79–82.

Neligan, R. E. (1962) Hydrocarbons in the Los Angeles atmosphere. Comparison between hydrocarbons in automobile exhaust and those found in the Los Angeles atmosphere. *Arch. Environ. Health* 5, 581–591.

Nelson, P. F., S. M. Quigley, and M. Y. Smith (1983) Sources of atmospheric hydrocarbons in Sidney: a quantitative determination using a source reconciliation technique. *Atmos. Environ.* 17, 439–449.

Neta, P., R. E. Huie, and A. B. Ross (1988) Rate constants for reactions of inorganic radicals in aqueous solution. *J. Phys. Chem. Ref. Data* 17, 1027–1284.

Neukum, G. (1977) Lunar cratering. *Philos. Trans. R. Soc. Lond.* A 285, 267–272.

Nevison, C. D., R. F. Weiss, and D. J. Erickson (1995) Global oceanic emissions of nitrous oxide. *J. Geophys. Res.* 100, 15,809–15,820.

Nevison, C. D., G. Esser, and E. A. Holland (1996) A global model of changing N_2O emissions from natural and perturbed soils. *Climatic Change* 32, 327–378.

Newell, R. E. (1970) Stratospheric temperature change from the Mt. Agung volcanic eruption of 1963. *J. Atmos. Sci.* 27, 977–978.

Newell, R. E., J. M. Wallace, and J. R. Mahoney (1966) The general circulation of the atmosphere and its effect on the movement of trace substances, Part 2. *Tellus* 18, 363–380.

Newell, R., D. G. Vincent, and J. W. Kidson (1969) Interhemispheric mass exchange from meteorological and trace substance observations. *Tellus* 21, 641–647.

Newell, R. E., J. W. Kidson, D. G. Vincent, and G. J. Boer (1972) "The General Circulation of the Tropical Atmosphere and Interactions with Extratropical Latitudes," Vol. 1. MIT Press, Cambridge, MA.

Newell, R., G. Boer, and J. Kidson (1974) An estimate of the interhemispheric transfer of carbon monoxide from tropical general circulation data. *Tellus* 26, 103–107.

Newman, M. J., and R. T. Rood (1977) Implications of solar evolution for the Earth's earliest atmosphere. *Science* 198, 1035–1037.

Newman, P. A., L. R. Lait, M. R. Schoeberl, E. R. Nash, K. K. Kelly, D. W. Fahey, R. M. Nagatani, D. W. Toohey, L. M. Avallone, and J. G. Anderson (1993) Stratospheric meteorological conditions on the Arctic polar vortex 1991 to 1992. *Science* 261, 1143–1146.

Nguyen, B. C., B. Bonsang, and G. Lambert (1974) The atmospheric concentration of sulfur dioxide and sulfate aerosols over antarctic, subantarctic areas and oceans. *Tellus* **26**, 241–247.

Nguyen, B. C., A. Gaudry, B. Bonsang, and G. Lambert (1978) Reevaluation of the role of dimethyl sulfide and the natural sulfur cycle. *Nature* **237**, 452–453.

Nguyen, B. C., B. Bonsang, and A. Gaudry (1983) The role of the ocean in the global atmospheric sulfur cycle. *J. Geophys. Res.* **88**, 10,903–10,914.

Nguyen, B. C., N. Mihalopoulos, and S. Belviso (1990) Seasonal variation of atmospheric dimethylsulfide at Amsterdam Island in the southern Indian Ocean. *J. Atmos. Chem.* **11**, 123–141.

Nicolet, M. (1954) Dynamic effects in the high atmosphere. *In* "The Earth as a Planet" (G. P. Kuiper, ed.), pp. 644–712. University of Chicago Press, Chicago, IL.

Nicolet, M. (1975) On the production of nitric oxide by cosmic rays in the mesosphere and stratosphere. *Planet. Space Sci.* **23**, 637–649.

Nicolet, M. (1978) "Etude des Réactions Chimiques de l'Ozone dans la Stratosphère." Institut Royal Météorologique de Belgique, Brussels.

Nicolet, M. (1981) The solar spectral irradiance and its action in the atmospheric photodissociation process. *Planet. Space. Sci.* **29**, 951–974.

Nicolet, M., and W. Peetermans (1980) Atmospheric absorption in the O_2 Schumann-Runge band spectral range and photodissociation rates in the stratosphere and mesosphere. *Planet. Space. Sci.* **28** 85–103.

Niehoer, H., and J. van Ham (1976) Peroxyacetyl nitrate (PAN) in relation to ozone and some meteorological parameters at Delft in the Netherlands. *Atmos. Environ.* **10**, 115–120.

Nielsen, T., B. Seitz, and T. Ramdahl (1984) Occurrence of nitro-PAH in the atmosphere in a rural area. *Atmos. Environ.* **18**, 2159–2165.

Niki, H., E. E. Daby, and B. Weinstock (1972) Mechanisms of smog reactions. *In* "Photochemical Smog and Ozone Reactions" (R. F. Gould, ed.). *Adv. Chem. Ser.* **113**, 116–176.

Niki, H., P. D. Maker, C. M. Savage, and L. P. Breitenbach (1977) Fourier transform IR spectroscopic observation of propylene ozonide in the gas phase reaction of ozone–cis-2-butene–formaldehyde. *Chem. Phys. Lett.* **46**, 327–330.

Niki, H., P. D. Maker, C. M. Savage, and L. P. Breitenbach (1978) Mechanism for the OH radical initiated oxidation of olefin-NO mixtures in ppm concentrations. *J. Phys. Chem.* **82**, 135–137.

Niki, H., P. D. Maker, C. M. Savage, and L. P. Breitenbach (1981) An FTIR study of mechanisms for the OH radical initiated oxidation of C_2H_4 in the presence of NO: detection of glycolaldehyde. *Chem. Phys. Lett.* **80**, 499–503.

Niki, H., P. D. Maker, C. M. Savage, and L. P. Breitenbach (1983) An FTIR study of the mechanism for the reaction $OH + CH_3SCH_3$. *Int. J. Chem. Kinet.* **15**, 647–654.

Niki, H., P. D. Maker, C. M. Savage, L. P. Breitenbach, and M. D. Hurley (1987) FTIR spectroscopic study of the mechanism for the gas-phase reaction between ozone and tetramethylethylene. *J. Phys. Chem.* **91**, 941–946.

Njoku, E. G., and L. Swanson (1983) Global measurements of sea surface temperature, wind speed and atmospheric water content from satellite microwave radiometry. *Mon. Weather Rev.* **111**, 1977–1987.

Norton, R. B. (1992) Measurements of gas-phase formic and acetic acids at the Mauna Loa observatory, Hawaii, during the Mauna Loa observatory photochemical experiment 1988. *J. Geophys. Res.* **97**, 10,389–10,393.

Norton, R. B., M. A. Carroll, D. D. Montzka, G. Hübler, B. J. Huebert, G. Lee, W. Warren, B. A. Ridley, and J. G. Walega (1992) Measurements of nitric acid and aerosol nitrate at the Mauna Loa observatory during the Mauna Loa Observatory Photochemistry Experiment 1988. *J. Geophys. Res.* **97**, 10,415–10,425.

Notholt, J. J. Hjorth, and F. Raes (1992) Formation of HNO_2 on aerosol surfaces during foggy periods in the presence of NO and NO_2. *Atmos. Environ.* **26A**, 211–218.

Novelli, P. C., K. A. Masarie, P. P. Tans, and P. M. Lang (1994) Recent changes in atmospheric carbon monoxide. *Science* **263**, 1587–1590.

Novic, M., I. Grgic, M. Poje, and V. Hudnik (1996) Iron-catalyzed oxidation of S(IV) species by oxygen in aqueous solution: influence of pH on the redox cycling of iron. *Atmos. Environ.* **30**, 4191–4196.

Noxon, J. F. (1975) Nitrogen dioxide in the stratosphere and troposphere measured by ground-based absorption spectroscopy. *Science* **189**, 547–549.

Noxon, J. F. (1976) Atmospheric nitrogen fixation by lightning. *Geophys. Res. Lett.* **3**, 463–465.

Noxon, J. F. (1978) Tropospheric NO_2. *J. Geophys. Res.* **83**, 3051–3057 (correction: *J. Geophys. Res.* **85**, 4560–4561).

Noxon, J. F. (1979a) Nitrogen dioxide in the stratosphere and troposphere measured by ground-based absorption spectroscopy. *Science* **189**, 547–549.

Noxon, J. F. (1979b) Stratospheric NO_2 global behavior. *J. Geophys. Res.* **84**, 5067–5076.

Noxon, J. F. (1981) NO_x in the mid-Pacific troposphere. *Geophys. Res. Lett.* **8**, 1223–1226.

Noxon, J. F. (1983) NO_3 and NO_2 in the mid-Pacific troposphere. *J. Geophys. Res.* **88**, 11,017–11,021.

Noxon, J. F., R. Norton, and G. Marovich (1980) NO_3 in the troposphere. *Geophys. Res. Lett.* **7**, 125–128.

Nyberg, A. (1977) On air-borne transport of sulphur over the North Atlantic. *Q. J. R. Meteorol. Soc.* **103**, 607–615.

Nydal, R. (1968) Further investigation on the transfer of radiocarbon in nature. *J. Geophys. Res.* **73**, 3617–3635.

O'Brien, R. J. (1974) Photostationary state in photochemical smog studies. *Environ. Sci. Technol.* **8**, 579–583.

O'Brien, J. M., P. B. Shepson, K. Muthuramu, C. Hao, H. Niki, D R. Hastie, R. Taylor, and P. B. Roussel (1995) Measurements of alkyl and multifunctional organic nitrates at a rural site in Ontario. *J. Geophys. Res.* **100**, 22,795–22,804.

Oddie, B. C. V. (1962) The chemical composition of precipitation at cloud levels. *Q. J. R. Meteorol. Soc.* **88**, 535–538.

Odén, S. (1968) The acidification of air and precipitation and its consequence on the natural environment. *Swed. Nat. Sci. Res. Counc. Ecol. Comm. Bull.* **1**, 1–86.

Odén, S. (1976) The acidity problem: an outline of concepts. *Water Air Soil Pollut.* **6**, 137–166.

Oeschger, H., U. Siegenthaler, U. Schotterer, and A. Gugelmann (1975) A box diffusion model to study the carbon dioxide exchange in nature. *Tellus* **27**, 168–192.

Ogawa, M. (1971) Absorption cross sections of O_2 and CO_2 continua in the Schumann and far-UV regions. *J. Chem. Phys.* **54**, 2550–2556.

Ogren, J. A., J. Heintzenberg, A. Zuber, K. J. Noone, and R. J. Charlson (1989) Measurements of the size-dependence of solute concentrations in cloud droplets. *Tellus* **41B**, 24–31.

Ogren, J. A., K. J. Noone, A. Hallberg, J. Heintzenberg, D. Schell, A. Berner, I. Solly, C. Kruisz, G. Reisch, B. G. Arends, and W. Wobrock (1992) Measurements of the size dependence of the concentration of non-volatile material in fog droplets. *Tellus* **44B**, 570–580.

Oh, D., W. Sisk, A. Young, and H. S. Johnston (1986) Nitrogen dioxide fluorescence from N_2O_5 photolysis. *J. Chem. Phys.* **85**, 7146–7158.

Okabe, H. (1978) "Photochemistry of Small Molecules." Wiley, New York.

Okita, T., and D. Shimozuro (1975) Remote sensing measurements of mass flow of sulfur dioxide gas from volcanoes. *Bull. Volcanol. Soc. Japan.* 19-3, 153–157.

Olson, J. S. (1970) Geographic index of world ecosystem. *In* "Analysis of Temperate Forest Ecosystems" (D. Reichle, ed.). *Ecol. Stud.* 1, 297–304.

Olszyna, K. J., J. F. Meagher, and E. M. Bailey (1988) Gas-phase, cloud and rain-water measurements of hydrogen peroxide at a high-elevation site. *Atmos. Environ.* 19, 687–690.

Oltmans, S. J., and W. D. Komhyr (1976) Surface ozone in Antarctica. *J. Geophys. Res.* 81, 5359–5364.

Oltmans, S. J., and H. Levy (1992) Seasonal cycle of surface ozone over the western North Atlantic. *Nature* 358, 392–394.

Oltmans, S. J., and H. Levy (1994) Surface ozone measurements from a global network. *Atmos. Environ.* 28, 9–24.

O'Neill, H. St. C. (1991) The origin of the moon and the early history of the earth––a chemical model. *Geochim. Cosmochim. Acta* 55, 1135–1157, 1159–1172.

O'Neill, H. St. C., and H. Palme (1998) Composition of the silicate earth: implications for accretion and core formation. *In* "The Earth's Mantle, Composition, Structure and Evolution" (I. Jackson, ed.), pp. 3–126, Cambridge University Press, Cambridge, UK.

Oort, A. H., and E. M. Rasmussen (1971) Atmospheric circulation statistics, National Oceanic and Atmospheric Administration Professional Paper 5. U.S. Department of Commerce, Rockville, MD.

Orr, C., K. Hurd, and W. J. Corbett (1958a) Aerosol size and relative humidity. *J. Colloid Sci.* 13, 472–482.

Orr, C., K. Hurd, W. Hendrix, and C. E. Junge (1958b) The behavior of condensation nuclei under changing humidities. *J. Metrorol.* 15, 240–242.

Ortgies, G. K., K. H. Gericke, and F. J. Comes (1980) Is UV laser induced fluorescence a method to monitor tropospheric OH? *Geophys. Res. Lett* 7, 905–908.

Orville, R. E. (1991) Lightning ground flash density in the contiguous United States––1989. *Mon. Weather Rev.* 119, 573–577.

Orville, R. E., and R. W. Henderson (1986) Global distribution of midnight lightning: September 1977 to August 1978. *Mon. Weather Rev.* 114, 2640–2653.

Orville, R. E., and D. W. Spencer (1979) Global lightning flash frequency. *Mon. Weather Rev.* 107, 934–943.

Östlund, H. G., and J. Alexander (1963) Oxidation rate of sulfide in sea water, a preliminary study. *J. Geophys. Res.* 68, 3995–3997.

Ottar, B. (1978) An assessment of the OECD study on long range transport of air pollutants (LRTAP). *Atmos. Environ.* 12, 445–454.

Owen, T., K. Biemann, D. R. Rushneck, J. E. Biller, D. W. Horwarth, and A. L. Lafleur (1977) The composition of the atmosphere at the surface of Mars. *J. Geophys. Res.* 82, 4635–4639.

Oyama, V. I., G. C. Carle, F. Woeller, and J. B. Pollack (1979) Venus lower atmospheric composition. Analysis by gas chromatography. *Science* 203, 802–805.

Ozima, M., and F. A. Podosek (1983) "Nobel Gas Geochemistry." Cambridge University Press, Cambridge, UK.

Pacyna, J. M., and T. E. Graedel (1995) Atmospheric emissions inventories: status and prospects. *Annu. Rev. Energy Environ.* 20, 265–300.

Pales, J. C., and C. D. Keeling (1965) The concentration of atmospheric carbon dioxide in Hawaii. *J. Geophys. Res.* 70, 6053–6076.

Paltrige, G. W., and C. M. R. Platt (1976) "Radiative Processes in Meteorology and Climatology." Elsevier, Amsterdam.

Pandis, S. N., and J. H. Seinfeld (1991) Should bulk cloud water or fog water samples obey Henry's law? *J. Geophys. Res.* **96**, 10,791–10,798.

Paneth, J. A. (1937) The chemical composition of the atmosphere. *Q. J. R. Meteorol. Soc.* **63**, 433–438.

Pannetier, R. (1970) Original use of the radioactive tracer gas krypton 85 to study the meridian atmospheric flow. *J. Geophys. Res.* **75**, 2985–2989.

Papenbrock, T., and F. Stuhl (1989) Detection of nitric acid in air by a laser-photolysis fragment-fluorescence (LPFF) method. *In* "Physico-Chemical Behaviour of Atmospheric Pollutants" (G. Restelli and G. Angeletti, eds.), pp. 651–656. Kluwer, Dordrecht, the Netherlands.

Paraskepopoulos, G., and R. J. Cvetanovic (1969) Competitive reactions of the excited oxygen atoms (O^1D). *J. Am. Chem. Soc.* **91**, 7572–7577.

Park, J. H. (1974) The equivalent mean absorption cross sections of the O_2 Schumann-Runge bands: application to the H_2O and NO dissociation rates. *J. Atmos. Sci.* **31**, 1893–1897.

Park, J.-Y., and Y.-N. Lee (1988) Solubility and decomposition kinetics of nitrous acid in aqueous solution. *J. Phys. Chem.* **92**, 6294–6302.

Parkin, D. W., D. R. Phillips, R. A. Sullivan, and L. R. Johnson (1972) Airborne dust collections down the Atlantic. *Q. J. R. Meteorol. Soc.* **98**, 789–808.

Parrish, D. D., M. Trainer, E. J. Williams, D. W. Fahey, G. Hübler, C. S. Eubank, S. C. Liu, P. C. Murphy, D. L. Albritton, and F. C. Fehsenfeld (1986a) Measurements of the NO_x-O_3 photostationary state at Niwot Ridge, Colorado. *J. Geophys. Res.* **91**, 5361–5370.

Parrish, D. D., R. B. Norton, M. J. Bollinger, S. C. Liu, P. C. Murphy, D. L. Albritton, and F. C. Fehsenfeld (1986b) Measurements of HNO_3, and NO_3^- particulates at a rural site in the Colorado mountains. *J. Geophys. Res.* **91**, 5379–5393.

Parrish, D. D., C. J. Hahn, E. J. Williams, R. B. Norton, F. C. Fehsenfeld, H. B. Singh, J. D. Shetter, B. W. Gandrud, and B. A. Ridley (1992) Indications of photochemical histories of Pacific air masses from measurements of atmospheric trace gas species at Pt. Arena, California. *J. Geophys. Res.* **97**, 15,883–15,901.

Parrish, D. D., M. P. Buhr, M. Trainer, R. B. Norton, J. P. Shimshock, F. C. Fehsenfeld, K. G. Anlauf, J. W. Bottenheim, Y. Z. Tang, H. A. Wiebe, J. M. Roberts, R. L. Tanner, L. Newman, V. C. Bowersox, K. J. Olszyna, E. M. Bailey, M. O. Rodgers, T. Wang, H. Berresheim, U. K. Roychowdhury, and K. Demerjian (1993) The total reactive oxidized nitrogen levels and partitioning between the individual species at six rural sites in Eastern North America. *J. Geophys. Res.* **98**, 2927–2939.

Parry, H. D. (1977) Ozone depletion by chlorofluoromethanes? Yet another look. *J. Appl. Meteorol.* **16**, 1137–1148.

Parsons, T. R. (1975) Particulate organic carbon in the sea. *In* "Chemical Oceanography," 2nd ed. (J. P. Riley and G. Skirrow, eds.), Vol. 2, pp. 365–383. Academic Press, London.

Pasquill, F. (1974) "Atmospheric Diffusion," 2nd ed. Ellis Harwood, Chichester.

Patten, K. O., Jr., P. S. Connell, D. E. Kinnison, D. J. Wuebbles, G. Slanger, and L. Froidevaux (1994) Effect of vibrationally excited oxygen on ozone production in the stratosphere. *J. Geophys. Res.* **99**, 1211–1223.

Patterson, E. M., C. S. Kiang, A. C. Delany, A. F. Wartburg, A. C. D. Leslie, and B. J. Huebert (1980) Global measurements of aerosols in remote continental and marine regions: concentrations, size distributions and optical properties. *J. Geophys. Res.* **85**, 7361–7376.

Paul, E. A., and F. E. Clark (1996) "Soil Microbiology and Biochemistry," 2nd ed. Academic Press, San Diego, CA.

Paulson, S. E., and J. H. Seinfeld (1992) Development and evaluation of a photooxidation mechanism for isoprene. *J. Geophys. Res.* **97**, 20,703–20,715.

Paulson, S. E., R. C. Flagan, and J. H. Seinfeld (1992) Atmospheric photooxidation of isoprene. Part I. The hydroxyl radical and ground state atomic oxygen reactions. *Int. J. Chem. Kinet.* 24, 79–101.

Paulson, S. E., R. C. Flagan, and J. H. Seinfeld (1992) Atmospheric photooxidation of isoprene. Part II. The ozone-isoprene reaction. *Int. J. Chem. Kinet.* 24, 103–125.

Pawson, S., B. Naujokat, and K. Labitzke (1995) On the polar stratospheric cloud formation potential of the northern stratosphere. *J. Geophys. Res.* 100, 23,215–23,225.

Pearman, G. I., and P. Hyson (1981) A global atmospheric diffusion simulation model for atmospheric carbon studies. In "Carbon Cycle Modelling" (B. Bolin, ed.). *SCOPE* 16, 227–240.

Pearman, G. I., D. Etheridge, F. de Silva, and P. J. Fraser (1986) Evidence of changing concentrations of atmospheric CO_2, N_2O, and CH_4 from air bubbles in Antarctic ice. *Nature* 320, 248–250.

Peer, R. L., S. A. Thorneloe, and D. L. Epperson (1993) A comparison of methods for estimating global methane emissions from landfills. *Chemosphere* 26, 387–400.

Peirson, R., and I. Cambray (1968) Interhemispheric transfer of debris from nuclear explosions using a simple atmospheric model. *Nature* 216, 755–758.

Peiser, G. D., C. C. Lizada, and S. F. Yang (1982) Dark metabolism of carbon monoxide in lettuce leaf discs. *Plant Physiol.* 70, 397–400.

Peng, T. H. (1985) Atmospheric CO_2 variations based on the tree-ring ^{13}C record. In "The Carbon Cycle and Atmospheric CO_2: Natural Variations Archean to Present" (E. T. Sundquist and W. S. Broecker, eds.), Geophysical Monograph Series Vol. 32, pp. 123–131. American Geophysical Union, Washington, DC.

Peng, T. H., W. S. Broecker, G. G. Mathieu, and Y. H. Li (1979) Radon evasion rates in the Atlantic and Pacific Oceans as determined during the GEOSECS program. *J. Geophys. Res.* 84, 2471–2486.

Peng, T. H., W. S. Broecker, H. D. Freyer, and S. Trumbore (1983) A deconvolution of the tree ring based delta C-13 record. *J. Geophys. Res.* 88, 3609–3620.

Penkett, S. A. (1972) Oxidation of SO_2 and other atmospheric gases by ozone in aqueous solution. *Nature* 240, 105–106.

Penkett, S. A., and K. A. Brice (1986) The spring maximum in photo-oxidants in the Northern Hemisphere troposphere. *Nature* 319, 655–657.

Penkett, S. A., F. J. Sandalls, and B. M. R. Jones (1977) PAN measurements in England, analytical methods and results. *VDI Ber.* 270, 47–54.

Penkett, S. A., B. M. R. Jones, and A. E. J. Eggleton (1979a) A study of SO_2 oxidation in stored rainwater samples. *Atmos. Environ.* 13, 139–147.

Penkett, S. A., B. M. R. Jones, K. A. Brice, and A. E. J. Eggleton (1979b) The importance of atmospheric ozone and hydrogen peroxide in oxidising sulphur dioxide in cloud and rain water. *Atmos. Environ.* 13, 323–337.

Penkett, S. A., F. J. F. Atkins, and M. H. Unswork (1979c) Chemical composition of the ambient aerosol in the Sudan Gezire. *Tellus* 31, 285–327.

Penkett, S. A., R. G. Derwent, P. Fabian, R. Borchers, and U. Schmidt (1980) Methylchloride in the stratosphere. *Nature* 283, 58–60.

Penkett, S. A., N. J. D. Prosser, R. A. Rasmussen, and M. A. K. Khalil (1981) Atmospheric measurements of CF_4 and other fluorocarbons containing the CF_3 grouping. *J. Geophys. Res.* 86, 5172–5178.

Penkett, S. A., B. M. R. Jones, M. J. Rycroft, and D. A. Simmons (1985) An interhemispheric comparison of the concentrations of bromine compounds in the atmosphere. *Nature* 318, 550–553.

Penner, J. E., S. J. Ghan, and J. J. Walton (1993) The role of biomass burning in the budget and cycle of carbonaceous soot aerosols and their climatic impact. *In* "Global Biomass Burning: Atmospheric, Climatic, and Biospheric Implications" (J. S. Levine, ed.), pp. 387–393. MIT Press, Cambridge, MA.

Pepin, R. O. (1991) On the origin and early evolution of terrestrial planet atmospheres and meteoritic volatiles. *Icarus* 92, 2–79.

Perner, D., and U. Platt (1979) Detection of nitrous acid in the atmosphere by differential optical absorption. *Geophys. Res. Lett.* 6, 917–920.

Perner, D., D. H. Ehhalt, H. W. Pätz, U. Platt, E. P. Röth, and A. Volz (1976) OH radicals in the lower troposphere. *Geophys. Res. Lett.* 3, 466–468.

Perner, D., U. Platt, M. Trainer, G. Hübler, J. Drummond, W. Junkermann, J. Rudolph, B. Schubert, A. Volz, and D. D. Ehhalt (1987) Measurements of tropospheric OH concentrations: a comparison of field data with model predictions. *J. Atmos. Chem.* 5, 185–216.

Peter, T. (1997) Microphysics and heterogeneous chemistry of polar stratospheric clouds. *Annu. Rev. Phys. Chem.* 48, 785–822.

Peterson, J. T. (1976) Calculated actinic fluxes (290–700 nm) for air pollution photochemistry application. EPA-600/4-76-025, U.S. Environmental Protection Agency, Research Triangle Park, NC.

Peterson, J. T., and C. E. Junge (1971) Sources of particulate matter in the atmosphere. *In* "Man's Impact on Climate" (W. H. Matthews, W. W. Kellog, and G. D. Robinson, eds.), pp. 310–320. MIT Press, Cambridge, MA.

Petrenchuk, O. P. (1980) On the budgets of sea salt and sulfur in the atmosphere. *J. Geophys. Res.* 85, 7439–7444.

Petrenchuk, O. P., and E. S. Selezneva (1970) Chemical composition of precipitation in regions of the Soviet Union. *J. Geophys. Res.* 75, 3629–3634.

Peyrous, R., and R. M. Lapeyre (1982) Gaseous products created by electrical discharges in the atmosphere and condensation nuclei resulting from gas-phase reactions. *Atmos. Environ.* 16, 959–986.

Pflug, H. D. (1978) Yeast-like microfossils deteected in oldest sediments of the earth. *Naturwissenschaften* 65, 611–615.

Pflug, H. D., and H. Jaeschke-Boyer (1979) Combined structural and chemical analysis of 3800 Myr old microfossils. *Nature* 280, 483–486.

Pham, M., J.-F. Müller, G. Brasseur, C. Granier, and G. Mégie (1995) A three-dimensional study of the tropospheric sulfur cycle. *J. Geophys. Res.* 100, 26,061–26,092.

Pham, M., J.-F. Müller, G. P. Brasseur, C. Granier, and G. Mégie (1996) A 3D model study of the global sulfur cycle: contributions of anthropogenic and biogenic sources. *Atmos. Environ.* 30, 1815–1822.

Piccot, S. D., J. J. Watson, and J. W. Jones (1992) A global inventory of volatile organic compound emissions from anthropogenic sources. *J. Geophys. Res.* 97, 9897–9912.

Pierotti, D., and R. A. Rasmussen (1977) The atmospheric distribution of nitrous oxide. *J. Geophys. Res.* 82, 5823–5832.

Pierotti, D., L. E. Rasmussen, and R. A. Rasmussen (1978) The Sahara as a possible sink for trace gases. *Geophys. Res. Lett.* 5, 1001–1004.

Pierson, D. H., R. S. Cambray, and G. S. Spicer (1966) Lead 210 and polonium 210 in the atmosphere. *Tellus* 18, 428–433.

Pierson, W. R., W. W. Brachaczek, T. J. Truex, J. W. Butler, and T. J. Korniski (1980) Ambient sulfate measurements on Allegheny Mountain and the question of atmospheric sulfate in the northeastern United States. *In* "Aerosols: Anthropogenic and Natural Sources and Transport" (T. J. Kneip and P. J. Lioy, eds.). *Ann. N.Y. Acad. Sci.* 338, 145–173.

Pierson W. R., W. W. Brachaczek, R. A. Gorse, Jr., S. M. Japar, J. M. Norbeck, and G. J. Keeler (1989) Atmospheric acidity measurements on Allegheny mountain and the origins of ambient acidity in the northeastern United States. *Atmos. Environ.* 23, 431–459.

Piringer, M., E. Ober, H. Puxbaum, and H. Kromb-Kolb (1997) Occurrence of nitric acid and related compounds in the northern Vienna basin during summertime anticyclonic conditions. *Atmos. Environ.* 31, 1049–1057.

Pittock, A. B. (1977) Climatology of the vertical distribution of ozone over Aspendale (38°S, 145°E). *Q. J. R. Meteorol. Soc.* 103, 575–585.

Pitts, J. N., Jr., J. H. Sharp, and S. I. Chan (1964) Effects of wavelength and temperature on primary processes in the photolysis of nitrogen dioxide and a spectroscopic-photochemical determination of the dissociation energy. *J. Chem. Phys.* 40, 3655–3662.

Pitts, J. N., Jr., A. C. Lloyd, and J. L. Sprung (1975) Ecology, energy and economics. *Chem. Br.* 11, 247–256.

Pitts, J. N., Jr, A. M. Winer, G. J. Doyle, and K. R. Darnall (1978) Correspondence. *Environ. Sci. Technol.* 12, 100–102.

Pitts, J. N., Jr., H. W. Biermann, A. M. Winer, and E. C. Tuazon (1984a) Spectroscopic identification and measurement of gaseous nitrous acid in dilute auto exhaust. *Atmos Environ.* 18, 847–854.

Pitts, J. N., Jr., E. Sanhueza, R. Atkinson, W. P. L. Carter, A. M. Winer, G. W. Harris, and C. N. Plum (1984b) An investigation of the dark formation of nitrous acid in environmental chambers. *Int. J. Chem Kinet.* 16, 919–939.

Pitts, J. N. Jr., H. W. Biermann, R. Atkinson, and A. M. Winer (1984c) Atmospheric implications of simultaneous nighttime measurements of NO$_3$ radicals and HONO. *Geophys. Res. Lett.* 11, 557–560.

Pitts, J. N., J. N. Sweetman, B. Zielinska, C. Winer, and R. Atkinson (1985) Determination of 2-nitrofluoranthene and 2-nitropyrene in ambient particulate organic matter: evidence for atmospheric reactions. *Atmos. Environ.* 19, 1601–1608.

Plass, C., R. Koppmann and J. Rudolph (1992) Light hydrocarbons in the surface water of the mid-Atlantic. *J. Atmos. Chem.* 15, 235–251.

Plass, G. N. (1956) The carbon dioxide theory of climatic change. *Tellus* 8, 140–154.

Plass-Dülmer, C., R. Koppmann, M. Ratte, and J. Rudolph (1995) Light nonmethane hydrocarbons in sea water. *Global Biogeochem. Cycles* 9, 79–100.

Platt, U., and D. Perner (1980) Direct measurement of atmospheric CH$_2$O, HNO$_2$, O$_3$, NO$_2$, and SO$_2$ by differential optical absorption. *J. Geophys. Res.* 85, 7453–7458.

Platt, U., D. Perner, and H.W. Pätz (1979) Simultaneous measurement of atmospheric CH$_2$O, O$_3$, and NO$_2$ by differential optical absorption. *J. Geophys. Res.* 84, 6329–6335.

Platt, U., D. Perner, G. W. Harris, and J. N. Pitts, Jr. (1980a) Detection of NO$_3$ in the polluted atmosphere by differential optical absorption. *Geophys. Res. Lett.* 7, 89–92.

Platt, U., D. Perner, G. W. Harris, A. M. Winer, and J. N. Pitts, Jr. (1980b) Observations of nitrous acid in an urban atmosphere by differential optical absorption. *Nature* 285, 312–314.

Platt, U., D. Perner, J. Schröder, C. Kessler, and A. Toenissen (1981) The diurnal variation of NO$_3$. *J. Geophys. Res.* 86, 11,965–11,970.

Platt, U., A. M. Winer, H. W. Biermann, R. Atkinson, and J. N. Pitts, Jr. (1984) Measurement of nitrate radical concentrations in continental air. *Environ. Sci. Technol.* 18, 365–369.

Platt, U., M. Rateike, W. Junkermann, J. Rudolph, and D. D. Ehhalt (1988) New tropospheric OH measurements. *J. Geophys. Res.* 93, 5159–5166.

Platt, U., G. Le Bras, G. Poulet, J. P. Burrows, and G. K. Moortgat (1990) Peroxy radicals from night-time reactions of NO$_3$ with organic compounds. *Nature* 348, 147–149.

Poet, S. E., H. E. Moore, and E. A. Martell (1972) Lead 210, bismuth 210 and polonium 210 in the atmosphere: accurate ratio measurements and application to aerosol residence time determinations. *J. Geophys. Res.* 77, 6515–6527.

Pollack, J. B., and D. C. Black (1979) Implications of the gas compositional measurement of Pioneer Venus for the origin of planetary atmospheres. *Science* 205, 56–59.

Pollack, J. B., and Y. L. Yung (1980) Origin and evolution of planetary atmospheres. *Annu. Rev. Earth Planet. Sci.* 8, 425–487.

Pollard, R. T. (1977) Hydrocarbons. In "Chemical Kinetics" (C. H. Bamford and C. F. H. Tipper, eds.), Vol. 17, Gas-Phase Combustion, pp. 249–367. Elsevier, Amsterdam.

Pollock, W. L., L. E. Heidt, R. Lueb, and D. H. Ehhalt (1980) Measurements of stratospheric water vapor by cryogenic collection. *J. Geophys. Res.* 85, 5555–5568.

Poole, L. R., and M. P. McCormick (1988) Airborne lidar observations of arctic polar stratospheric clouds: indications of two distinct growth stages. *Geophys. Res. Lett.* 15, 21–23.

Poorter, H. (1993) Interspecific variation in the growth response of plants to elevated ambient CO_2 concentration. *Vegetatio* 104–105, 77–97.

Porter, L. K., F. G. Viets, Jr., and G. L. Hutchinson (1972) Air containing nitrogen-15 ammonia: foliar absorption by corn seedlings. *Science* 175, 759–761.

Possanzini, M., V. Di Palo, M. Petricca, R. Fratarcangeli, and D. Brocco (1996) Measurements of the lower carbonyls in Rome ambient air. *Atmos. Environ.* 30, 3757–3764.

Postma, A. K. (1970) Effect of solubilities of gases on their scavenging by rain drops. In "Precipitation Scavenging 1970" (R. J. Engelmann and W. G. N. Slinn, coordinators). AEC Symposium Series, Vol. 30, pp. 247–259. U.S. Atomic Energy Commission, Division of Technical Information, Oak Ridge, TN.

Poth, M., and D. D. Focht (1985) ^{15}N kinetic analysis of N_2O production by *Nitrosomonas europaea*: an examination of nitrifier denitrification. *Appl. Environ. Microbiol.* 49, 1134–1141.

Potter, C. S., P. A. Matson, P. M. Vitousek, and E. A. Davidson (1996) Process modeling of controls on nitrogen trace gas emissions from soils worldwide. *J. Geophys. Res.* 101, 1361–1377.

Poulida, O., K. L. Civerolo, and R. R. Dickerson (1994) Observations and tropospheric photochemistry in central North Carolina. *J. Geophys. Res.* 99, 10,553–10,563.

Prabhakara, C., I. Wang, A. T. C. Chang, and P. Gloersen (1983) A statistical examination of Nimbus-7 SMMR data and remote sensing of sea surface temperature, liquid water content in the atmosphere and surface wind speed. *J. Appl. Meteorol.* 22, 2023–2037.

Prager, M. J., E. R. Stephens, and W. E. Scott (1960) Aerosol formation from gaseous air pollutants. *Ind. Eng. Chem.* 52, 521–524.

Prahm, L. P., U. Thorp, and R. M. Stern (1976) Deposition and transformation rates of sulfur oxides during atmospheric transport over the Atlantic. *Tellus* 28, 355–372.

Prather, M., and C. M. Spivakovsky (1990) Tropospheric OH and the lifetimes of hydrochlorofluorocarbons. *J. Geophys. Res.* 95, 18,723–18,729.

Pressman, J., and P. Warneck (1970) The stratosphere as a chemical sink for carbon monoxide. *J. Atmos. Sci.* 27, 155–163.

Preston, K. F., and R. F. Barr (1971) Primary processes in the photolysis of nitrous oxide. *J. Chem. Phys.* 54, 3341–3348.

Price, C., and D. Rind (1992) A simple lightning parameterization for calculating global lightning distributions. *J. Geophys. Res.* 97, 9919–9933.

Price, C., J. Penner, and M. Prather (1997a) NO_x from lightning. 1. Global distribution based on lightning physics. *J. Geophys. Res.* 102, 5929–5941.

Price, C., J. Penner, and M. Prather (1997b) NO_x from lightning 2. Constraints from the global atmospheric electric circuit. *J. Geophys. Res.* 102, 5943–5951.

Prinn, R. G., R. A. Rasmussen, P. G. Simmonds, F. N. Alyea, D. M. Cunnold, B. C. Lane, C. A. Cardelino, and A. J. Crawford (1983) The atmospheric life time experiment 5 Results for CH_3CCl_3 based on three years of data. *J. Geophys. Res.* 88, 8415–8426.

Prinn, R., D. Cunnold, R. Rasmussen, P. Simmonds, F. Alyea, A. Crawford, P. Fraser, and R. Rosen (1987) Atmospheric trends in methylchloroform and the global average for the hydroxyl radical. *Science* 238, 945–950.

Prinn, R., D. Cunnold, R. Rasmussen, P. Simmonds, F. Alyea, A. Crawford, P. Fraser, and R. Rosen (1990) Atmospheric emissions and trends of nitrous oxide from 10 years of ALE-GAGE data. *J. Geophys. Res.* 95, 18,369–18,385.

Prinn, R., D. Cunnold, P. Simmonds, F. Alyea, R. Boldi, A. Crawford, P. Fraser, D. Gutzler, D. Hartley, R. Rosen, and R. Rasmussen (1992) Global average concentration and trend for hydroxyl radicals deduced from ALE/GAGE trichloroethane (methyl chloroform) data for 1978–1990. *J. Geophys. Res.* 97, 2445–2461.

Prinn, R. G., R. F. Weiss, B. R. Miller, J. Huang, F. N. Alyea, D. M. Cunnold, P. J. Fraser, D. E. Hartley and P. G. Simmonds (1995) Atmospheric trends and life time of CH_3CCl_3 and global OH concentrations. *Science* 269, 187–192.

Propero, J. M., D. Savoie, R. T. Ness, R. Duce, and J. Merrill (1985) Particulate sulfate and nitrate in the boundary layer over the north Pacific Ocean. *J. Geophys. Res.* 90, 10,586–10,596.

Protoschill-Krebs, G., and J. Kesselmeier (1992) Enzymatic pathways for the metabolization of carbonyl sulfide. *Botanica Acta* 105, 206–212.

Protoschill-Krebs, G., C. Wilhelm, and J. Kesselmeier (1996) Consumption of carbonyl sulfide by carbonic anhydrase. *Atmos. Environ.* 30, 3151–3156.

Pruchniewicz, P. G., H. Tiefenau, P. Fabian, P. Wilbrandt, and W. Jessen (1974) The distribution of tropospheric ozone from world-wide surface and aircraft observations. In "Structure, Composition, General Circulation of the Upper and Lower Atmospheres and Possible Anthropogenic Perturbations" (P. Goldsmith, A. D. Belmont, N. J. Derco, and E. J. Truhlar, eds.), Proceedings of the International Conference of IAMAP/IAPSO, Melbourne, Australia, Vol. 1, pp. 439–451.

Pruppacher, H. R., and J. D. Klett (1997) "Microphysics of Clouds and Precipitation," 2nd ed. Kluwer, Dordrecht, the Netherlands.

Pszenny, A. A., P. F. McIntyre, and R. A. Duce (1982) Sea salt and the aciditiy of marine rain on the windward coast of Samoa. *Geophys. Res. Lett.* 9, 751–754.

Pszenny, A. A. P., A. J. Castelle, J. N. Galloway, and R. A. Duce (1989) A study of the sulfur cycle in the Antarctic marine boundary layer. *J. Geophys. Res.* 94, 9818–9830.

Pszenny, A. P., G. R. Harvey, C. J. Brown, R. F. Lang, W. C. Keene, J. N. Galloway, and J. T. Merrill (1990) Measurements of dimethylsulfide oxidation products in the summertime north Atlantic marine boundary layer. *Global Biogeochem. Cycles* 4, 367–379.

Pszenny, A. A. P., W. C. Keene, D. J. Jacob, S. Fan, J. R. Maben, M. P. Zetwo, M. Springer-Young, and J. N. Galloway (1993) Evidence of inorganic chlorine gases other than hydrogen chloride in marine surface air. *Geophys. Res. Lett.* 20, 699–702.

Putaud, J. P., N. Mihalopoulos, B. C. Nguyen, J. M. Campin, and S. Belviso (1992) Seasonal variation of atmospheric sulfur dioxide and dimethylsulfide concentrations at Amsterdam Island in the southern Indian Ocean. *J. Atmos. Chem.* 15, 117–131.

Puxbaum, H. (1997) Biogenic emissions of alcohols, esters ethers and higher aldehydes. In "Biogenic Volatile Organic Compounds in the Atmosphere" (G. Helas, J. Slanina, and R. Steinbrecher, eds.), pp. 79–99. SPB Academic Publishing bv, Amsterdam, the Netherlands.

Puxbaum, H., C. Rosenberg, M. Gregori, C. Lanzerstorfer, E. Ober, and W. Winiwarter (1988) Atmospheric concentrations of formic and acetic acid and related compounds in eastern and northern Austria. *Atmos. Environ.* 22, 2841–2850.

Pye, K. (1987) "Aeolian Dust and Dust Deposits." Academic Press, London.

Pytkowicz, R. M. (1970) On the carbonate compensation depth in the Pacific Ocean. *Geochim. Cosmochim. Acta* **34**, 836–839.

Pytkowicz, R. M. (1975) Some trends in marine chemistry and geochemistry. *Earth Sci. Rev.* **11**, 1–46.

Quay, P. D., S. L. King, J. Stutsman, D. O. Wilbur, L. P. Steele, I. Fung, R. H. Gammon, T. A. Brown, G. W. Farwell, P. M. Grootes, and F. H. Schmidt (1991) Carbon isotope composition of atmospheric CH_4: fossil and biomass burning source strengths. *Global Biogeochem. Cycles* **5**, 25–47.

Quinn, P. K., R. J. Charlson, and T. S. Bates (1988) Simultaneous observation of ammonia in the atmosphere and ocean. *Nature* **335**, 336–338.

Quinn, P. K., T. S. Bates, J. E. Johnson, D. S. Covert, and R. J. Charlson (1990) Interactions between the sulfur and reduced nitrogen cycles over the central Pacific Ocean. *J. Geophys. Res.* **95**, 16,405–16,416.

Radford, H. E. (1980) The fast reaction of CH_2OH with O_2. *Chem. Phys. Lett.* **71**, 195–197.

Radke, L. F. (1981) Marine aerosol: simultaneous size distributions of the total aerosol and the sea-salt fraction from 0.1–10 μm diameter. Proceedings of the Ninth International Conference Atmosphere Aerosol, Condensation, Ice Nuclei, Galway, Ireland (A. F. Roddy and T. C. O'Connor, eds.), pp. 487–491.

Radke, L. F., P. V. Hobbs, and J. E. Pinnons (1976) Observations of cloud condensation nuclei, sodium containing particles, ice nuclei and the light scattering coefficient near Barrow, Alaska. *J. Appl. Meteorol.* **15**, 982–995.

Radke, L. F., J. L. Stith, D. A. Hegg, and P. V. Hobbs (1978) Airborne studies of particles and gases from forest fires. *J. Air Pollut. Control Assoc.* **28**, 30–34.

Radke, L. F., D. A. Hegg, P. V. Hobbs, J. D. Nance, J. H. Lyons, K. K. Laursen, R. E. Weiss, P. J. Riggan, and D. E. Ward (1991) Particulate and trace gas emissions from large biomass fires in North America. *In* "Global Biomass Burning, Atmospheric, Climatic and Biospheric Implication" (J. S. Levine, ed.), pp. 209–224. MIT Press, Cambridge, MA.

Rahn, K. A. (1975a) Chemical composition of the atmospheric aerosol. A compilation. I. *Extern.* **4**, 286–313.

Rahn, K. A. (1975b) Chemical composition of the atmospheric aerosol. A compilation. II. *Extern.* **4**, 639–667.

Rahn, K. A., and R. J. McCaffrey (1980) On the origin and transport of winter Arctic aerosol. *In* "Aerosols: Anthropogenic and Natural Sources and Transport" (T. J. Kneip and P. J. Lioy, eds.). *Ann. N.Y. Acad. Sci.* **338**, 486–503.

Rahn, K. A., R. D. Borys, and R. A. Duce (1976) Tropospheric halogen gases: inorganic and organic components. *Science* **192**, 549–550.

Rahn, K. A., R. D. Borys, and G. E. Shaw (1977) The Asian source of Arctic haze bands. *Nature* **268**, 713–715.

Ramanathan, V., and J. A. Coakley, Jr. (1978) Climate modeling through radiative-convective models. *Rev. Geophys. Space Phys.* **16**, 465–689.

Ramanathan, V., R. H. Cicerone, H. B. Singh, and J. Kiehl (1985) Trace gas trends and their potential role in climate change. *J. Geophys. Res.* **90**, 5547–5566.

Randhawa, J. S. (1971) The vertical distribution of ozone near the equator. *J. Geophys. Res.* **76**, 8139–8142.

Raper, O. F., C. B. Farmer, R. A. Toth, and B. D. Robbins (1977) The vertical distribution of HCl in the stratosphere. *Geophys. Res. Lett.* **4**, 531–534.

Raper, O. F., C. B. Farmer, R. Zander, and J. H. Park (1987) Infrared spectroscopic measurements of halogenated sink and reservoir gases in the stratosphere with the ATMOS instrument. *J. Geophys. Res.* **92**, 9851–9858.

Rasmussen, A., S. Kilsholm, J. H. Sørensen, and I. S. Mikkelsen (1997) Analysis of tropospheric ozone measurements in Greenland. *Tellus* **49B**, 510–521.

Rasmussen, R. A. (1970) Isoprene: identified as a forest-type emission to the atmosphere. *Environ. Sci. Technol.* **4**, 669–673.

Rasmussen, R. A. (1972) What do hydrocarbons from trees contribute to air pollution? *J. Air Pollut. Control Assoc.* **22**, 537–543.

Rasmussen, R. A., and C. A. Jones (1973) Emission of isoprene from leaf discs of *Hamamelis. Phytochemistry* **12**, 15–19.

Rasmussen, R. A., and M. A. K. Khalil (1980) Atmospheric halocarbons, measurements and analyses of selected trace gases. In "Atmospheric Ozone: Its Variation and Human Influences" (M. Nicolet and A. C. Aikin, eds.), NATO ASI, FAA-11-80-20, pp. 209–231. U.S. Department of Transportation, Washington, DC.

Rasmussen, R. A., and M. A. K. Khalil (1981) Increase in the concentration of atmospheric methane. *Atmos. Environ.* **15**, 883–886.

Rasmussen, R. A., and M. A. K. Khalil (1983a) Global production of methane by termites. *Nature* **301**, 700–702.

Rasmussen, R. A., and M. A. K. Khalil (1983b) Altitudinal and temporal variations of hydrocarbons and other gaseous tracers of Arctic haze. *Geophys. Res. Lett.* **10**, 144–147.

Rasmussen, R. A., and M. A. K. Khalil (1984) Atmospheric methane in recent and ancient atmospheres: concentrations, trends, and interhemispheric gradient. *J. Geophys. Res.* **89**, 11,599–11,605.

Rasmussen, R. A., and F. W. Went (1965) Volatile organic material of plant origin in the atmosphere. *Proc. Natl. Acad. Sci. USA* **53**, 215–220.

Rasmussen, R. A., L. E. Rasmussen, M. A. K. Khalil, and R. W. Dalluge (1980) Concentration distribution of methyl chloride in the atmosphere. *J. Geophys. Res.* **85**, 7350–7356.

Rasmussen, R. A., M. A. K. Khalil, and S. D. Hoyt (1982a) The oceanic source of carbonyl sulfide. *Atmos. Environ.* **16**, 1591–1594.

Rasmussen, R. A., M. A. K. Khalil, R. Gunawardena, and S. D. Hoyt (1982b) Atmospheric methyl iodide (CH_3I). *J. Geophys. Res.* **87**, 3086–3090.

Rasmussen, R. A., M. A. K. Khalil, A. J. Crawford, and P. J. Fraser (1982c) Natural and anthropogenic trace gases in the southern hemisphere. *Geophys. Res. Lett.* **9**, 704–707.

Rasool, S. I., and C. DeBergh (1970) The runaway greenhouse and accumulation of CO_2 in the Venus atmosphere. *Nature* **226**, 1037–1039.

Ratner, M. J., and J. C. G. Walker (1972) Atmospheric ozone and the history of life. *J. Sci.* **29**, 803–808.

Ratte, M., C. Plass-Dülmer, R. Koppmann, and J. Rudolph (1993) Production mechar C_2-C_4 light hydrocarbons in sea water. Field measurements and experiments. *Biogeochem. Cycles* **7**, 369–378.

Rau, J. A., and M. A. K. Khalil (1993) Anthropogenic contributions to the carbonaceous of aerosols over the Pacific ocean. *Atmos. Environ.* **27A**, 1297–1307.

Ravishankara, A. R., P. H. Wine, C. A. Smith, P. E. Barbone, and A. Torabi (198€ photolysis: quantum yields for NO_3 and $O(^3P)$ *J. Geophys. Res.* **91**, 5355–5360.

Ravishankara, A. R., S. Solomon, A. A. Turnipseed, and R. F. Warren (1993) Atmosph times of long-lived halogenated species. *Science* **259**, 194–199.

Raynaud, D., and J. M. Barnola (1985) An Antarctic ice core reveals atmospheric CO_2 v over the past few centuries. *Nature* **315**, 309–311.

Raynaud, D., J. Chappellaz, J. M. Barnola, Y. S. Korotkevich, and C. Lorius (1988) Cli methane cycle implications of glacial-interglacial CH_4 change in the Vostok ice cor **333**, 655–657.

Rebbert, R. E., and P. J. Ausloos (1975) Photodecomposition of $CFCl_3$ and CF_2Cl_2. *J. Photochem.* **4**, 419–434.

Rebbert, R. E., and P. J. Ausloos (1977) Gas-phase photodecomposition of carbon tetrachloride. *J. Photochem.* **6**, 265–276.

Rebbert, R. E., and P. J. Ausloos (1978) Decomposition of N_2O over particulate matter. *Geophys. Res. Lett.* **5**, 761–764.

Redfield, A. C., B. H. Ketchum, and F. A. Richards (1963) The influence of organisms in the composition of sea water. *In* "The Sea" (M. N. Hill, ed.), Vol. 2, pp. 26–77. Wiley Interscience, New York.

Reed, R. J., and K. E. German (1965) A contribution to the problem of stratospheric diffusion by large-scale mixing. *Mon. Weather Rev.* **93**, 313–321.

Regener, E., and V. H. Regener (1934) Aufnahme des ultravioletten Sonnenspektrums in der Stratosphäre und vertikale Ozonverteilung. *Z. Phys.* **35**, 788–793.

Regener, V. H. (1938) Messung des Ozongehalts der Luft in Bodennähe. *Met. Zeitschr.* **55**, 459–462.

Regener, V. H. (1957) Vertical flux of atmospheric ozone. *J. Geophys. Res.* **62**, 221–228.

Regener, V. H. (1964) Measurement of atmospheric ozone with a chemiluminescent method. *J. Geophys. Res.* **69**, 3795–3800.

Regener, V. H. (1974) Destruction of atmospheric ozone at the ocean surface. *Arch. Meterol. Geophys. Bioklimatol. Ser. A* **23**, 131–135.

Reifenhäuser, W., and L. G. Heumann (1992) Determination of methyl iodide in the Antarctic atmosphere and the south polar area. *Atmos. Environ.* **26A**, 2905–2912.

Reihs, C. M., D. M. Golden, and M. A. Tolbert (1990) Nitric acid uptake by sulfuric acid solutions under stratospheric conditions: determination of Henry's law solubility. *J. Geophys. Res.* **95**, 16,545–16,550.

Reiner, T., and F. Arnold (1994) Laboratory investigation of gaseous sulfuric acid formation via $SO_3 + H_2O + M \rightarrow H_2SO_4 + M$: measurement of the rate constant and product identification. *J. Chem. Phys.* **101**, 7399–7407.

Reiners, W. A. (1973) Terrestrial detritus and the carbon cycle. *In* "Carbon and the Biosphere" (G. M. Woodwell and E. V. Pecan, eds.), AEC Symposium Series, Vol. 30, pp. 303–327, NTIS U.S. Department Commerce, Springfield, VA.

Reinsel, G. C. (1981) Analysis of total ozone data for the detection of recent trends and the effect of nuclear testing during 1960s. *Geophys. Res. Lett.* **8**, 1227–1230.

Reinsel, G. C., G. C. Tiao, R. Lewis, and M. Bobkowski (1983) Analysis of upper stratospheric ozone profile data from the ground-based Umkehr method and the Nimbus-4 BUV satellite experiment. *J. Geophys. Res.* **88**, 5393–5402.

Reinsel, G. C., G. C. Tiao, J. J. DeLuisi, C. L. Mateer, A. J. Miller, and J. E. Frederick (1984) Analysis of upper stratospheric Umkehr ozone profile data and the effects of stratospheric aerosols. *J. Geophys. Res.* **89**, 4833–4840.

Reinsel, G. C., G. C. Tiao, D. J. Wuebbels, J. B. Kerr, A. J. Miller, R. M. Nagatani, L. Bishop, and L. H. Ying (1994) Seasonal trend analysis of published ground-based and TOMS total ozone data through 1991. *J. Geophys. Res.* **99**, 5449–5464.

Reiss, H. (1950) The kinetics of phase transitions in binary systems. *J. Chem. Phys.* **18**, 840–848.

Reiter, E. R. (1975) Stratospheric-tropospheric exchange processes. *Rev. Geophys. Space Phys.* **13**, 459–474.

Reiter, E. R., and J. D. Mahlmann (1965) Heavy radioactive fallout on the southern United States, November 1962. *J. Geophys. Res.* **70**, 4501–4520.

Reiter, E. R., H. J. Kanter, R. Reiter, and R. Sladkovic (1977) Lower-tropospheric ozone of stratospheric origin. *Arch. Met. Geophys. Ser. A* **26**, 179–186.

Reiter, R. (1970) On the causal relation between nitrogen-oxygen compounds in the troposphere and atmospheric electricity. *Tellus* 22, 122–135.

Reiter, R., R. Sladkovic, and K. Pötzl (1975) Die wichtigsten chemischen Bestandteile des Aerosols über Mitteleuropa unter Reinluftbedingungen in 1800 m Seehöhe. *Meteorol. Rundsch.* 28, 37–55.

Reiter, R., R. Sladkovic, and K. Pötzl (1976) Chemical components of aerosol particles in the lower troposphere above central Europe measured under pure-air conditions. *Atmos. Environ.* 10, 841–853.

Reiter, R., R. Sladkovic, and K. Pötzl (1978) Chemische Komponenten des Reinluftaerosols in Abhängigkeit vom Luftmassencharakter und meteorologischen Bedingungen. *Ber. Bunsen-Ges. Phys. Chem.* 82, 1188–1193.

Remde, A., and R. Conrad (1990) Production of nitric oxide in *Nitrosomonas europaea* by reduction of nitrite. *Arch. Microbiol.* 154, 187–191.

Remsberg, E. E., J. M. Russell, III, J. G. Gille, J. L. Gordley, P. L. Bailey, W. G. Planet, and J. E. Harries (1984) The validation of NIMBUS 7 LIMS measurement of ozone. *J. Geophys. Res.* 89, 5161–5178.

Rennenberg, H. (1991) The significance of higher plants in the emission of sulfur compounds from terrestrial ecosystems. *In* "Trace Gas Emissions from Plants" (T. D. Sharkey, E. A. Holland, and H. A. Mooney, eds.), pp. 217–260. Academic Press, San Diego, CA.

Renner, E., U. Ratzlaff, and W. Rolle (1985) A Lagrangian multi-level model of transport, transformation and deposition of atmospheric sulfur dioxide and sulfate. *Atmos. Environ.* 19, 1351–1359.

Restad, K., I. S. A. Isaksen, and T. K. Berntsen (1998) Global distribution of sulphate in the troposphere. A three-dimensional model study. *Atmos. Environ.* 32, 3593–3609.

Revelle, R., and H. E. Suess (1957) Carbon dioxide exchange between atmosphere and ocean and the question of an increase of atmospheric CO_2 during the past decades. *Tellus* 9, 18–27.

Riba, M. L., and L. Torres (1997) Terpenes: biosynthesis and transport in plants; emission and presence in the troposphere. *In* "Biogenic Volatile Organic Compounds in the Atmosphere" (G. Helas, J. Slanina, and R. Steinbrecher, eds.), pp. 115–143. SPB Academic Publishing bv, Amsterdam, the Netherlands.

Richards, L. W., J. A. Anderson, D. L. Blumenthal, and J. A. McDonald (1983) Hydrogen peroxide and sulfur (IV) in Los Angeles cloud water. *Atmos. Environ.* 17, 911–914.

Ridley, B. A., and D. R. Hastie (1981) Stratospheric odd nitrogen: NO measurements at 51°N in summer. *J. Geophys. Res.* 86, 3162–3166.

Ridley, B. A., and E. Robinson (1992) The Mauna Loa Observatory Photochemistry Experiment. *J. Geophys. Res.* 97, 10,285–10,290.

Ridley, B. A., and H. I. Schiff (1981) Stratospheric odd nitrogen nitric oxide measurements at 31°N in autumn. *J. Geophys. Res.* 86, 3167–3172.

Ridley, B. A., M. McFarland, J. T. Bruin, H. I. Schiff, and J. C. McConnell (1977) Sunrise measurements of stratospheric nitric oxide. *Can J. Phys.* 55, 212–221.

Ridley, B. A., M. A. Carroll, and G. L. Gregory (1987) Measurements of nitric oxide in the boundary layer and free troposphere over the Pacific Ocean. *J. Geophys. Res.* 92, 2025–2047.

Ridley, B. A., M. A. Carroll, G. L. Gregory, and W. G. Sachse (1988) NO and NO_2 in the troposphere: technique and measurements in regions of a folded tropopause. *J. Geophys. Res.* 93, 15,813–15,830.

Ridley, B. A., J. D. Shetter, J. G. Walega, S. Madronich, C. M. Elsworth, F. E. Grahek, F. C. Fehsenfeld, R. B. Norton, D. D. Parrish, G. Hübler, M. Buhr, E. J. Williams, E. J. Allwine, and H. H. Westberg (1990) The behavior of some organic nitrates at Boulder and Niwot Ridge, Colorado. *J. Geophys. Res.* 95, 13,949–13,961.

Ridley, B. A., J. G. Walega, J. E. Dye, and F. E. Grahek (1994) Distributions of NO, NO_x, NO_y and O_3 to 12 km altitude during the summer monsoon season over New Mexico. *J. Geophys. Res.* **99**, 25,519–25,534.

Ridley, B. A., J. E. Dye, J. G. Walega, J. Zheng, F. E. Grahek, and W. Rison (1996) On the production of active nitrogen by thunderstorms over New Mexico. *J. Geophys. Res.* **101**, 20,985–21,005.

Ridley, W. P., L. J. Dizikesand, and J. M. Wood (1977) Biomethylation of toxic elements in the environment. *Science* **197**, 329–332.

Ringwood, A. E. (1966) The chemical evolution of terrestrial planets. *Geochim. Cosmochim. Acta* **30**, 41–104.

Ringwood, A. E. (1979) "Origin of the Earth and the Moon." Springer-Verlag, Berlin and New York.

Ritter, J., D. H. Stedman, and T. J. Kelly (1979) Ground level measurements of NO, NO_2, and O_3 in rural air. *In* "Nitrogenous Air Pollutants—Chemical and Biological Implications" (D. Grosjean, ed.), pp. 325–343. Ann Arbor Science Publications, Ann Arbor, MI.

Roberts, J. M. (1990) The atmospheric chemistry of organic nitrates. *Atmos. Environ.* **24A**, 243–287.

Roberts, J. M., and S. B. Bertman (1992) The thermal decomposition of peroxy acetyl nitric anhydride (PAN) and peroxy methacrylic nitric anhydride (MPAN) *Int. J. Chem. Kinet.* **24**, 297–307.

Roberts, J. M., F. C. Fehsenfeld, D. L. Albritton, and R. E. Sievers (1983) Measurement of monoterpene hydrocarbons at Niwot Ridge, Colorado. *J. Geophys. Res.* **88**, 10,667–10,678.

Roberts, J. M., R. L. Tanner, L. Newman, V. C. Bowersox, J. W. Bottenheim, K. G. Anlauf, K. A. Brice, D. D. Parrish, F. C. Fehsenfeld, M. P. Buhr, J. F. Meagher, and E. M. Bailey (1995) Relationships between PAN and ozone at sites in eastern North America. *J. Geophys. Res.* **100**, 22,821–22,830.

Robinson, E. (1978) Hydrocarbons in the atmosphere. *Pure Appl. Geophys.* **116**, 372–384.

Robinson, E., and R. C. Robbins (1968a) Evaluation of CO data obtained on Eltanin cruises 27, 29 and 31. *Antarctic J. U.S.* **3**, 194–196.

Robinson, E., and R. C. Robbins (1968b) Sources, abundances and fate of gaseous atmospheric pollutants. Final Report, Project PR-6755. Stanford Research Institute, Palo Alto, CA.

Robinson, E., and R. C. Robbins (1970a) Atmospheric background concentrations of carbon monoxide. *Ann. N.Y. Acad. Sci.* **174**, 89–95.

Robinson, E., and R. C. Robbins (1970b) Gaseous sulfur pollutants from urban and natural sources. *J. Air Pollut. Control Assoc.* **20**, 233–235.

Robinson, E., and J. B. Homolya (1983) Natural and anthropogenic emission sources. *In* "The Acidic Deposition Phenomenon and Its Effects: Critical Assessment Review Papers" (A. P. Altshuller and R. A. Linthurst, eds.), EPA-600/8-83-016A, U.S. Environmental Protection Agency, Office of Research and Development, Washington, DC.

Robinson, E., R. A. Rasmussen, H. H. Westberg, and M. W. Holdren (1973) Non-urban nonmethane low molecular weight hydrocarbon concentrations related to air mass identification. *J. Geophys. Res.* **78**, 5345–5351.

Robinson, E., R. A. Rasmussen, J. Krasnec, D. Pierotti, and M. Jacubovic (1977) Halocarbon measurements in the Alaskan troposphere and lower stratosphere. *Atmos. Environ.* **11**, 213–223.

Robinson, E., D. Clark, and W. Seiler (1984) The latitudinal distribution of carbon monoxide across the Pacific from California to Antarctica. *J. Geophys. Res.* **86**, 5163–5171.

Robinson, R. J., and R. H. Stokes (1970) "Electrolyte Solutions," 2nd ed. Butterworth, London.

Robson, R. L., and J. R. Postgate (1980) Oxygen and hydrogen in biological nitrogen fixation. *Annu. Rev. Microbiol.* **34**, 183–207.

Rodhe, H. (1972) A study of the sulfur budget for the atmosphere over northern Europe. *Tellus* **24**, 128–138.

Rodhe, H. (1976) An atmospheric sulfur budget for NW Europe. *In* "Nitrogen, Phosphorous, and Sulfur—Global Cycles (B. H. Svensson and R. Söderlund, eds.). *SCOPE 7, Ecol. Bull.* (*Stockh.*) **22**, 123–134.

Rodhe, H., and J. Grandell (1972) On the removal of aerosol particles from the atmosphere by precipitation scavenging. *Tellus* **24**, 442–454.

Roedel, W. (1979) Measurement of sulfuric acid saturation vapor pressure: implications for aerosol formation by heteromolecular nucleation. *J. Aerosol Sci.* **10**, 375–386.

Roehl, C. M., J. J. Orlando, G. S. Tyndall, R. E. Shetter, G. J. Vásquez, C. A. Cantrell, and J. G. Calvert (1994) Temperature dependence of the quantum yield for the photolysis of NO_2 near the dissociation limit. *J. Phys. Chem.* **98**, 7837–7843.

Rogers, C. F., J. G. Hudson, B. Zielinska, R. L. Tanner, J. Hallett, and J. G. Watson (1991) Cloud condensation nuclei from biomass burning. *In* "Global Biomass Burning, Atmospheric, Climatic, and Biospheric Implications" (J. S. Levine, ed.), pp. 431–438. MIT Press, Cambridge, MA.

Rogge, W. F., L. M. Hildemann, M. A. Mazurek, G. R. Cass, and B. R. T. Simoneit (1993a) Sources of fine organic aerosol. 2. Noncatalyst and catalyst-equipped automobiles and heavy-duty diesel trucks. *Environ. Sci. Technol.* **27**, 636–651.

Rogge, W. F., M. A. Mazurek, L. M. Hildemann, G. R. Cass, and B. R. T. Simoneit (1993b) Quantification of urban organic aerosols at a molecular level: identification, abundance and seasonal variation. *Atmos. Environ.* **27A**, 1309–1330.

Rohrer, F., and D. Brüning (1992) Surface NO and NO_2 mixing ratios measured between 30°N and 30°S in the Atlantic region. *J. Atmos. Chem.* **15**, 253–267.

Rondon, A., and E. Sanhueza (1989) High HONO atmospheric concentrations during vegetation burning in tropical savannah. *Tellus* **41B**, 474–477.

Ronov, A. B., and A. A. Yaroshevskiy (1969) Chemical composition of the Earth's crust. *In* "The Earth Crust and Upper Mantle," Geophysical Monograph Series 13, pp. 37–57, American Geophysical Union, Washington, DC.

Roscoe, H. K., J. R. Drummond, and R. F. Jarnot (1981) Infrared measurements of stratospheric composition. III. The daytime changes of NO and NO_2. *Proc. R. Soc. Lond. A* **375**, 507–520.

Rose, W. I., S. Bonis, R. E. Stoiber, M. Keller, and T. Bickford (1973) Studies of volcanic ash from two recent Central American eruptions. *Bull. Volcanol.* **37**, 338–364.

Rosen, J. M. (1969) Stratospheric dust and its relationship to the meteorite influx. *Space Sci. Rev.* **9**, 58–89.

Rosen, J. M. (1971) The boiling point of stratospheric aerosols. *J. Appl. Meteorol.* **10**, 1044–1046.

Rosen, J. M., and D. J. Hofmann (1977) Balloon-borne measurements of condensation nuclei. *J. Appl. Meterol.* **16**, 56–62.

Rosen, J. M., and D. J. Hofmann (1978) Vertical profiles of condensation nuclei. *Amer. Meteorol. Soc.* **17**, 1737–1740.

Rosen, J. M., D. J. Hofmann, J. R. Carpenter, J. W. Harder, and S. J. Oltmans (1988) Balloon-borne Antarctic frost-point measurements and their impact on polar stratospheric cloud theories. *Geophys. Res. Lett.* **15**, 859–862.

Rosenlof, K. H., and J. R. Holton (1993) Estimates of stratospheric residual circulation using the downward control principle. *J. Geophys. Res.* **98**, 10,465–10,479.

Ross, A. B., W. G. Mallard, W. P. Helman, B. H. J. Bielski, G. V. Buxton, D. E. Cabelli, C. L. Greenstock, R. E. Huie, and P. Neta (1992) "NDRL-NIST Solution Kinetics Data Base." National Institute of Standard and Technology, Gaithersburg, MD.

Ross, H. B. (1990) Trace metal wet deposition in Sweden: insight gained from daily wet-only collectors. *Atmos. Environ.* **24A**, 1929–1938.

Rossow, W. B., and R. A. Schiffer (1991) ISCCP cloud data products. *Bull. Am. Meteorol. Soc.* 72, 2–20.

Rosswall, T. (1976) The internal nitrogen cycle between microorganisms, vegetation and soil. *In* "Nitrogen, Phosphorus and Sulphur-Global Cycles" (B. H. Svensson and R. Söderlund, eds.). *SCOPE 7, Ecol. Bull. (Stockh.)* 22, 157–167.

Rosswall, T. (1981) The biogeochemical nitrogen cycle. *In* "Some Perspectives of the Major Biogeochemical Cycles" (G. E. Likens, ed.). *SCOPE* 17, 25–49.

Rotty, R. M. (1981) Data for global CO_2 production from fossil fuels and cement. *In* "Carbon Cycle Modelling" (B. Bolin, ed.). *SCOPE* 16, 121–125.

Rotty, R. M. (1983) Distribution of and changes in industrial carbon dioxide production. *J. Geophys. Res.* 88, 1301–1308.

Rouland, C., A. Brauman, M. Labat, and M. Lepage (1993) Nutritional factors affecting methane emissions from termites. *Chemosphere* 26, 617–622.

Routhier, F., R. Dennet, P. D. Davis, A. Wartburg, P. Haagenson, and A. C. Delany (1980) Free tropospheric and boundary layer airborne measurements of ozone over the latitude range of 58°S to 70°N. *J. Geophys. Res.* 85, 7307–7321.

Rowland, F. S., and M. J. Molina (1975) Chlorofluoromethanes in the environment. *Rev. Geophys. Space Phys.* 13, 1–35.

Roy, C. R., I. E. Galbally, and B. A. Ridley (1980) Measurements of nitric oxide in the stratosphere of the southern hemisphere. *Q. J. R. Meteorol. Soc.* 106, 887–894.

Rubey, W. W. (1951) Geological history of sea water, an attempt to state the problem. *Bull. Geol. Soc. Am.* 62, 1111–1147.

Rüden, H., and E. Thofern (1976) Abscheidung von Schadstoffen und Mikroorganismen in Luftfiltern. *Staub Reinhalt. Luft* 36, 33–36.

Rüden, H., E. Thofern, P. Fischer, and U. Mihm (1978) Airborne microorganisms: their occurrence, distribution and dependence on environmental factors—especially on organic compounds of air pollution. *Pure Appl. Geophys.* 116, 335–350.

Rudolph, J. (1995) The tropospheric distribution and budget of ethane. *J. Geophys. Res.* 100, 11,369–11,381.

Rudolph, J., and D. H. Ehhalt (1981) Measurement of C_2–C_5 hydrocarbons over the north Atlantic. *J. Geophys. Res.* 86, 11,959–11,964.

Rudolph, J., and F. J. Johnen (1990) Measurements of the light hydrocarbons over the Atlantic in regions of low biological activity. *J. Geophys. Res.* 95, 20,583–20,591.

Rudolph, J., and A. Khedim (1985) Hydrocarbons in the non-urban atmosphere: analysis, ambient concentrations and impact on the chemistry of the atmosphere. *Int. J. Environ. Anal. Chem.* 20, 265–282.

Rudolph, J., B. Vierkorn-Rudolph, and F. X. Meixner (1987) Large-scale distribution of peroxy-acetyl nitrate: results from the STRATOZ III flights. *J. Geophys. Res.* 92, 6653–6661.

Rudolph, J., R. Koppmann, and C. Plass-Dülmer (1996) The budgets of ethane and tetra-chloroethene: is there evidence for an impact of reactions with chlorine atoms in the troposphere? *Atmos. Environ.* 30, 1887–1894.

Rush, J. D., and B. H. J. Bielski (1985) Pulse radiolysis studies of the reactions of HO_2/O_2^- with ferric ions and its implication on the occurrence of the Haber-Weiss reaction. *J. Phys. Chem.* 89, 5062–5066.

Russell, J. M., J. C. Gille, E. E. Remsberg, L. L. Gorley, P. L. Bailey, S. R. Drayson, H. Fischer, A. Girard, J. E. Harries, and W. F. Evans (1984) Validation of nitrogen dioxide results mesaured by the Limb Infrared Monitor of the Stratosphere (LIMS) experiment on Nimbus 7. *J. Geophys. Res.* 89, 5099–5108.

Russell, J. M., C. B. Farmer, C. P. Rinsland, R. Zander, L. Froidevaux, G. C. Toon, B. Gao, J. Shaw, and M. Gunson (1988) Measurements of odd nitrogen compounds in the stratosphere by the ATMOS Experiment on Spacelab 3. *J. Geophys. Res.* **93**, 1718–1736.

Russell, J. M., L. L. Gordley, J. H. Park, S. R. Drayson, W. D. Hesketh, R. J. Cicerone, A. F. Tuck, J. E. Frederick, J. E. Harries, and P. J. Crutzen (1993) The Halogen Occultation Experiment. *J. Geophys. Res.* **98**, 10,777–10,797.

Rust, F. (1957) Intra-molecular oxidation: the autoxidation of some dimethyl alkanes. *J. Am. Chem. Soc.* **79**, 4000–4003.

Rutten, M. G. (1971) "The Origin of Life by Natural Causes." Elsevier, Amsterdam, the Netherlands.

Ryaboshapko, A. G. (1983) The atmospheric sulfur cycle. *In* "The Global Biogeochemical Sulfur Cycle" (M. V. Ivanov and J. R. Freney, eds.). *SCOPE* **19**, 203–296.

Ryden, J. C. (1981) N_2O exchange between a grass land soil and the atmosphere. *Nature* **292**, 235–237.

Sagebiel, J. C., B. Zielinska, W. R. Pierson, and A. Gertler (1996) Real world emissions and calculated reactivities of organic species from motor vehicles. *Atmos. Environ.* **30**, 2287–2296.

Sakamaki, F., S. Hatakeyama, and H. Akimoto (1983) Formation of nitrous acid and nitric oxide in the heterogeneous dark reaction of nitrogen dioxide with water vapor in a smog chamber. *Int. J. Chem. Kinet.* **15**, 1013–1029.

Saltzman, B. E. (1954) Colorimetric microdetermination of nitrogen dioxide in the atmosphere. *Anal. Chem.* **26**, 1949–1955.

Saltzman, E. S., and D. J. Cooper (1988) Shipboard measurements of atmospheric dimethylsulfide and hydrogen sulfide in the Caribbean and Gulf of Mexico. *J. Atmos. Chem.* **7**, 191–209.

Saltzman, E. S., D. L. Savoie, R. G. Zika, and J. M. Propero (1983) Methane sulfonic acid in the marine atmosphere. *J. Geophys. Res.* **88**, 10,897–10,902.

Saltzman, E. S., D. L. Savoie, J. M. Prospero, and R. G. Zika (1985) Atmospheric methane sulfonic acid and non sea-salt sulfate at Fanning and American Samoa. *Geophys. Res. Lett.* **13**, 437–440.

Saltzman, E. S., D. L. Savoie, J. M. Prospero, and R. G. Zika (1986) Methanesulfonic acid and non-sea-salt sulfate in Pacific air: regional and seasonal variations. *J. Atmos. Chem.* **4**, 227–240.

Samain, D., and P. C. Simon (1976) Solar flux determination in the spectral range 150–210 nm. *Solar Phys.* **49**, 33–41.

Sanadze, G. A., and A. N. Kalandadze (1966) Light and temperature curves on the evolution of isoprene. *Sov. Plant Physiol.* **13**, 411–413.

Sandalls, F. J., and S. A. Penkett (1977) Measurements of carbonyl sulfide and carbon disulfide in the atmosphere. *Atmos. Environ.* **11**, 197–199.

Sandberg, D. V., S. G. Pickford, and E. F. Darley (1975) Emissions from slash burning and the influence of flame retardant chemicals. *J. Air Pollut. Control Assoc.* **25**, 278–281.

Sander, R., and P. J. Crutzen (1996) Model study indicating halogen activation and ozone destruction in polluted air masses transported to the sea. *J. Geophys. Res.* **101**, 9121–9138.

Sanders, R. W., S. Solomon, J. P. Smith, L. Perliski, H. L. Miller, G. H. Mount, J. G. Keys, and A. L. Schmeltekopf (1993) Visible and near ultraviolet spectroscopy at McMurdo Station, Antarctica 9. Observations of OClO from April to October 1991. *J. Geophys. Res.* **98**, 7219–7228.

Sandholm, S. T., J. D. Bradshaw, G. Chen, H. B. Singh, R. W. Talbot, G.-L. Gregory, D. R. Blake, G. W. Sachse, E. V. Browell, J. D. W. Barrick, M. A. Shipham, A. S. Bachmeier, and D. Owen (1992) Summertime tropospheric observations related to N_xO_y distributions and partitioning over Alaska: Arctic Boundary Layer Expedition 3A. *J. Geophys. Res.* **97**, 16,481–16,509.

Sanhueza, E., and M. O. Andreae (1991) Emission of formic and acetic acids from tropical savanna soils. *Geophys. Res. Lett.* **18**, 1707–1710.

Sanhueza, E., W. M. Hao, D. Scharffe, L. Donoso, and P. J. Crutzen (1990) N_2O and NO emissions from soils of the northern part of the Guayana Shield, Venezuela. *J. Geophys. Res.* **95**, 22,481–22,488.

Sanhueza, E., M. Santana, and M. Hermoso (1992) Gas and aqueous formic and acetic acids at a tropical cloud forest site. *Atmos. Environ.* **26A**, 1421–1426.

Sanhueza, E., L. Figueroa, and M. Santana (1996) Atmospheric formic and acetic acids in Venezuela. *Atmos. Environ.* **30**, 1861–1873.

Santee, M. L., W. G. Read, J. W. Waters, L. Froidevaux, G. L. Manney, D. A. Flower, R. F. Jarnot, R. S. Harwood, and G. E. Peckham (1994) Interhemispheric differences in polar stratospheric HNO_3, H_2O, ClO and O_3. *Science* **267**, 849–852.

Sapper, K. (1927) "Vulkankunde." Engelhorn-Verlag, Stuttgart, Germany.

Sarmiento, J. L., J. C. Orr, and U. Siegenthaler (1992) A perturbation simulation of CO_2 uptake in an ocean general circulation model. *J. Geophys. Res.* **97**, 3621–3645.

Savoie, D. L., and J. M. Prospero (1982) Particle size distribution of nitrate and sulfate in the marine atmosphere. *Geophys. Res. Lett.* **9**, 1207–1210.

Savoie, D. L., J. M. Prospero, R. J. Larsen, and E. S. Saltzman (1992) Nitrogen and sulfur species in aerosols at Mawson, Antarctica, and their relationship to natural radionucleides. *J. Atmos. Chem.* **14**, 181–204.

SCEP (1970) "Man's Impact on the Global Environment," Report of the Study of Critical Environmental Problems (SCEP). MIT Press, Cambridge, MA.

Schecker, H. G., and G. Schulz (1969) Untersuchungen zur Hydrationskinetik von Formaldehyd in wässriger Lösung. *Z. Phys. Chem. N. F.* **65**, 221–224.

Schell, D, W. Wobrock, R. Maser, M. Preiss, W. Jaeschke, H.-W. Georgii, M. W. Gallagher, K. N. Bower, K. M. Beswick, S. Pahl, M. C. Facchini, S. Fuzzi, A. Wiedensohler, H-C. Hansson, and M. Wendisch (1997) The size-dependent chemical composition of cloud droplets. *Atmos. Environ.* **31**, 2561–2576.

Schidlowski, M. (1966) Beiträge zur Kenntnis der radioaktiven Bestandteile der Witwatersrand Konglomerate I–III. *N. Jahrb. Miner. Abh.* **106/107**, 55–71, 183–202, 310–324.

Schidlowski, M. (1970) Untersuchungen zur Metallogenese im südwestlichen Witwatersrand Becken (Oranje-Freistaat Goldfeld, Südafrika). *Beih. Geo. Jahrb.* **85**, 80–126.

Schidlowski, M. (1971) Probleme der atmosphärischen Evoution im Präkambrium. *Geol. Rundsch.* **60**, 1351–1384.

Schidlowski, M. (1982) Content and isotope composition of reduced carbon in sediments. *In* "Mineral Deposits and the Evolution of the Biosphere" (H. D. Holland and M. Schidlowski, eds.), pp. 103–122, Dahlem-Konferenzen, Springer-Verlag, Berlin.

Schidlowski, M. (1983a) Biologically mediated isotope fractionation: biochemistry, geochemical significance and preservation in the earth's oldest sediments. *In* "Cosmochemistry and the Origin of Life" (C. Ponnamperuma, ed.), pp. 277–322. Reidel, Dordrecht, the Netherlands.

Schidlowski, M. (1983b) Evolution of photo-autotrophy and early atmospheric oxygen levels. *Precambriam Res.* **20**, 319–335.

Schidlowski, M. (1987) Application of stable carbon isotopes to early biogeochemical evolution on earth. *Annu. Rev. Earth Planet. Sci.* **15**, 47–72.

Schidlowski, M., J. M. Hayes, and I. R. Kaplan (1983) Isotopic interferences of ancient biochemistry: carbon, sulfur, and nitrogen. *In* "Earth's Earliest Biosphere: Its Origins and Evolution" (J. W. Schopf, ed.), pp 149–186. Princeton University Press, Princeton, NJ.

Schiff, H. I. (1972) Laboratory measurements of reactions related to ozone photochemistry. *Ann. Geophys.* **28**, 67–77.

Schiff, H. I., D. R. Hastie, G. I. Mackay, T. Iguchi, and B. A. Ridley (1983) Tunable diode laser systems for measuring trace gases in tropospheric air. *Env. Sci. Technol.* **17**, 352A–364A.

Schiff, H. I., G. I. Mackay, C. Castledine, G. W. Harris, and Q. Tran (1986) Atmospheric measurements of nitrogen dioxide with a sensitive luminol instrument. *Water Air Soil Pollut.* **30**, 105–114.

Schiller, C., A. Wahner, U. Platt, H.-P. Dorn, J. Callies, and D. H. Ehhalt (1990) Near UV atmospheric absorption measurements of column abundances during Airborne Arctic Stratospheric Expedition, January–February 1989. 2. OClO observations. *Geophys. Res. Lett.* **17**, 501–504.

Schimel, D. S., I. G. Enting, M. Heimann, T. M. I. Wigley, D. Raynaud, D. Alves, and U. Siegenthaler (1995) CO_2 and the carbon cycle. *In* "Climate Change 1994" (J. T. Houghton, L. G. Meira Filho, J. Bruce, H. Lee, B. A. Callander, E. Haites, N. Harris, and K. Maskell, eds.), pp. 35–71. Cambridge University Press, Cambridge, UK.

Schjørring, J. K., A. Kyllingsbaek, J. V. Mortensen, and S. Byskov-Nielsen (1993) Field investigations of ammonia exchange between barley plants and the atmosphere. *Plant Cell Environ.* **16**, 161–178.

Schlager, H., and F. Arnold (1990) Measurements of stratospheric gaseous nitric acid in the winter Arctic vortex using a novel rocket-borne mass spectrometric method. *Geophys. Res. Lett.* **17**, 433–436.

Schlegel, H. G. (1974) Production, modification, and consumption of atmospheric trace gases by microorganisms. *Tellus* **26**, 11–20.

Schlesinger, W. H. (1977) Carbon balance in terrestrial detritus. *Annu. Rev. Ecol. Syst.* **8**, 51–81.

Schlesinger, W. H. (1997) "Biogeochemistry, An Analysis of Global Change," 2nd ed., Academic Press, San Diego, CA.

Schlesinger, W. H., and A. E. Hartley (1992) A global budget for atmospheric NH_3. *Biogeochemistry* **15**, 191–211.

Schmauss, A., and A. Wigand (1929) "Die Atmosphäre als Kolloid." Vieweg und Sohn, Braunschweig

Schmeltekopf, A. L., P. D. Goldan, W. R. Henderson, W. J. Harrop, T. L. Thompson, F. C. Fehsenfeld, H. I. Schiff, P. J. Crutzen, and I. S. A. Isaksen, and E. E. Ferguson (1975) Measurements of stratospheric $CFCl_3$, CF_2Cl_2, and N_2O. *Geophys. Res. Lett.* **2**, 393–396.

Schmeltekopf, A. L., D. L. Albritton, P. J. Crutzen, P. D. Golden, W. J. Harrop, W. R. Henderson, J. R. McAffee, M. McFarland, H. I. Schiff, T. L. Thompson, D. J. Hoffman, and N. T. Kjome (1977) Stratospheric nitrous oxide altitude profiles at various latitudes. *J. Atmos. Sci.* **34**, 729–736.

Schmidt, E. L. (1982) Nitrification in soil. *Agronomy* **22**, 253–288.

Schmidt, J., W. Seiler, and R. Conrad (1988a) Emission of nitrous oxide from temperate forest soils into the atmosphere. *J. Atmos. Chem.* **6**, 95–115.

Schmidt, R., B. Schreiber, and I. Levin (1988b) Effects of long-range transport on atmospheric trace constituents at the Baseline Station Tenerife (Canary Islands). *J. Atmos. Chem.* **7**, 335–351.

Schmidt, U. (1974) Molecular hydrogen in the atmosphere. *Tellus* **26**, 78–90 (erratum *Tellus* **27**, 1).

Schmidt, U. (1978) The latitudinal and vertical distribution of molecular hydrogen in the troposphere. *J. Geophys. Res.* **83**, 941–946.

Schmidt, U., J. Rudolph, F. J. Johnen, D. H. Ehhalt, A. Volz, E. P. Röth, R. Borchers, and P. Fabian (1981) The vertical distribution of CH_3Cl, $CFCl_3$, and CF_2Cl_2 in the mid-latitude stratosphere, Proceedings of the Quadrennial International Ozone Symposium (J. London, ed.), pp. 816–823. National Center of Atmospheric Research, Boulder, CO.

Schneider, J. K., and R. B. Gagosian (1985) Particle size distribution of lipids in the aerosols off the coast of Peru. *J. Geophys. Res.* **90**, 7889–7898.

Schneider, S. H. (1975) On the carbon dioxide-climate confusion. *J. Atmos. Sci.* **32**, 2060–2066.

Schneider, W., G. K. Moortgat, G. S. Tyndall, and J. P. Burrows (1987) Absorption cross sections of NO_2 in the UV and visible region (200–700 nm) at 298 K. *J. Photochem. Photobiol A* **40**, 195–217.

Schönbein, C. F. (1840) Beobachtungen über den bei der Elektrolysation des Wassers und dem Ausströmen der gewöhnlichen Elektrizität aus Spitzen sich entwickelnden Geruch. *Ann. Phys. Chem.* **50**, 616.

Schönbein, C. F. (1854) Über verschiedene Zustände des Sauerstoffs. *Liebigs Ann. Chem.* **89**, 257–300.

Schooley, A. H. (1969) Evaporation in the laboratory and at sea. *J. Marine Res.* **27**, 335–338.

Schopf, J. W. (1975) Precambrian paleobiology: problems and perspectives. *Annu. Rev. Earth Planet. Sci.* **3**, 213–249.

Schopf, J. W., and D. Z. Oehler (1976) How old are the eucaryotes? *Science* **193**, 47–49.

Schröder, P. (1993) Plants as sources of atmospheric sulfur. In: "Sulfur Nutrition and Assimilation in Higher Plants" (L. J. De Kok, I. Stulen, H. Rennenberg, C. Brunold, and W. E. Rauser, eds.), pp. 253–270. SPB Acad. Publishing bv, The Hague, the Netherlands.

Schubert, K. R., and H. J. Evans (1976) Hydrogen evolution: a major factor affecting the efficiency of nitrogen fixation in modulated symbionts. *Proc. Natl. Acad. Sci. USA* **73**, 1207–1211.

Schuetzle, D., and R. A. Rasmussen (1978) The molecular composition of secondary aerosol particles formed from terpenes. *J. Air Pollut. Control Assoc.* **28**, 236–240.

Schuetzle, D., D. Cronn, A. L. Crittenden, and R. J. Charlson (1975) Molecular composition of secondary aerosols and its possible origin. *Environ. Sci. Technol.* **9**, 835–845.

Schumacher, H. J. (1930) The mechanism of the photochemical decomposition of ozone. *J. Am. Chem. Soc.* **52**, 2377–2391.

Schumacher, H. J. (1932) Die Photokinetik des Ozons I. Der Zerfall im roten Licht. *Z. Phys. Chem. B* **17**, 405–416.

Schütz, H., A. Holzapfel-Pschorn, R. Conrad, H. Rennenberg, and W. Seiler (1989) A 3-year continuous record on the influence of daytime, season and fertilizer treatment on methane emission rates from an Italian rice paddy. *J. Geophys. Res.* **94**, 16,405–16,416.

Schütz, K., C. Junge, R. Beck, and B. Albrecht (1970) Studies of atmospheric N_2O. *J. Geophys. Res.* **75**, 2230–2246.

Schütz, L. (1980) Long-range transport of desert dust with special emphasis on the Sahara. In "Aerosols: Anthropogenic and Natural, Sources and Transport" (T. J. Kneip and P. J. Lioy, eds.). *Ann. N.Y. Acad. Sci.* **338**, 515–532.

Schütz, L., and R. Jaenicke (1974) Particle number and mass distribution above 10^{-4} cm radius in sand and aerosol of the Sahara desert. *J. Appl. Meteorol.* **13**, 863–870.

Schütz, L., and K. A. Rahn (1982) Trace element concentrations in erodible soils. *Atmos. Environ.* **16**, 171–176.

Schwartz, R. M., and M. O. Dayhoff (1978) Origins of procaryotes, eucaryotes, mitochondria and chloroplasts. *Science* **199**, 395–403.

Schwartz, S. E. (1986) Mass transport considerations pertinent to aqueous phase reactions of gases in liquid water clouds. In "Chemistry of Multiphase Atmospheric Systems" (W. Jaeschke, ed.), NATO ASI Series, Vol. G 6, pp. 415–471. Springer-Verlag, Berlin.

Schwartz, S. E., and J. E. Freiberg (1981) Mass-transport limitation to the rate of reactions of gases in liquid droplets: application to SO_2 in aqueous solutions. *Atmos. Environ.* **15**, 1129–1144.

Schwartz, S. E., and W. H. White (1981) Solubility equilibria of the nitrogen oxides and oxiacids in dilute aqueous solution. *Adv. Environ. Sci. Technol.* **12**, 1–116.

Schwartz, W. (1974) Chemical characterization of model aerosols. EPA-650/3-76-085. U.S. Environmental Protection Agency, Research Triangle Park, NC.

Scott, W. D., and F. C. R. Cattell (1979) Vapor pressure of ammonium sulfates. *Atmos. Environ.* 13, 307–317.

Scott, W. E., E. R. Stephens, P. L. Hanst, and R. C. Doerr (1957) Further developments in the chemistry of the atmosphere. *Proc. Am. Petrol. Inst.* (*III*) 37, 171–183.

Scranton, M. I., W. R. Barger, and F. L. Herr (1980) Molecular hydrogen in the urban troposphere: measurements of seasonal variability. *J. Geophys. Res.* 85, 5575–5580.

Sedlak, D. L., and J. Hoigné (1993) The role of copper and oxalate in the redox cycling of iron in atmospheric waters. *Atmos. Environ.* 27A, 2173–2185.

Sedlak, D. L., and J. Hoigné (1994) Oxidation of S(IV) in atmospheric water by photooxidants and iron in the presence of copper. *Environ. Sci. Technol.* 28, 1898–1906.

Sedlak, D. L., J. Hoigné, M. M. David, R. N. Colvile, E. Seyffer, K. Acker, W. Wieprecht, J. A. Lind, and S. Fuzzi (1997) The cloud water chemistry of iron and copper at Great Dun Fell, U. K. *Atmos. Environ.* 31, 2515–2526.

Sehested, K., H. Corfitzen, J. Holcman, H. Fischer, and E. Hart (1991) The primary reaction in the decomposition of ozone in acidic aqueous solutions. *Environ. Sci. Technol.* 25, 1589–1596.

Sehmel, G. A. (1980) Particle and gas dry deposition: a review. *Atmos. Environ.* 14, 983–1011.

Sehmel, G. A., and S. L. Sutter (1974) Particle deposition rates on water surface as a function of particle diameter and air velocity. *J. Rech. Atmos.* 8, 911–918.

Seiler, W. (1974) The cycle of atmospheric CO. *Tellus* 26, 116–135.

Seiler, W. (1978) The influence of the biosphere on the atmospheric CO and H_2 cycles. In "Environmental Biochemistry and Geomicrobiology" (W. E. Krumbein, ed.), pp. 773–810. Ann Arbor Science Publications, Ann Arbor, MI.

Seiler, W., and R. Conrad (1981) Field measurements of natural and fertilizer-induced N_2O release rates from soils. *J. Air Pollut. Control Assoc.* 31, 767–772.

Seiler, W., and R. Conrad (1987) Contribution of tropical ecosystems to the global budgets of trace gases, especially CH_4, H_2, CO, and N_2O. In "Geophysiology of Amazonia" (R. Dickinson, ed.), pp. 133–162. Wiley, New York.

Seiler, W., and P. J. Crutzen (1980) Estimates of the gross and natural flux of carbon between the biosphere and the atmosphere from biomass burning. *Clim. Change* 2, 207–247.

Seiler, W., and J. Fishman (1981) The distribution of carbon monoxide and ozone in the free troposphere. *J. Geophys. Res.* 86, 7255–7265.

Seiler, W., and C. E. Junge (1969) Decrease of carbon monoxide mixing ratio above the polar tropopause. *Tellus* 21, 447–449.

Seiler, W., and C. E. Junge (1970) Carbon monoxide in the atmosphere. *J. Geophys. Res.* 75, 2217–2226.

Seiler, W., and U. Schmidt (1974) Dissolved non-conservative gases in sea water. In "The Sea" (E. P. Goldberg, ed.), Vol. 5, pp. 219–243. Wiley Interscience, New York.

Seiler, W., and P. Warneck (1972) Decrease of carbon monoxide mixing ratio at the tropopause. *J. Geophys. Res.* 77, 3204–3214.

Seiler, W., and J. Zankl (1975) Die Spurengase CO und H_2 über München. *Umschau* 75, 735–736.

Seiler, W., H. Giehl, and G. Bunse (1978a) Influence of plants on atmospheric carbon monoxide and dinitrogen oxide. *Pure Appl. Geophys.* 116, 439–451.

Seiler, W., F. Müller, and H. Oeser (1978b) Vertical distribution of chlorofluorocarbons in the upper troposphere and lower stratosphere. *Pure Appl. Geophys.* 116, 554–566.

Seiler, W., A. Holzapfel-Pschorn, R. Conrad, and D. Scharffe (1984a) Methane emission from rice paddies. *J. Atmos. Chem.* 1, 241–268.

Seiler, W., R. Conrad, and D. Scharffe (1984b) Field studies of methane emissions from termite nests into the atmosphere and measurements of methane uptake by tropical soils. *J. Atmos. Chem.* 1, 171–186.

Seiler, W., H. Giehl, E. G. Brunke, and E. A. Halliday (1984c) The seasonality of CO abundance in the southern hemisphere. *Tellus* 36B, 219–231.

Seitzinger, S. P. (1988) Denitrification in freshwater and coastal marine ecosystems: ecological and geochemical significance. *Limnol. Oceanogr.* 33, 702–724.

Sellers, W. D. (1965) "Physical Climatology." University of Chicago Press, Chicago, IL.

Sempéré, R., and K. Kawamura (1994) Comparative distributions of dicarboxylic acids and related polar compounds in snow, rain and aerosols from urban atmosphere. *Atmos. Environ.* 28, 449–459.

Sempéré, R., and K. Kawamura (1996) Low molecular weight dicarboxylic acids and related polar compounds in the remote marine rain samples collected from the western Pacific. *Atmos. Environ.* 30, 1609–1619.

Senum, G. I., R. Fajer, and J. S. Gaffney (1986) Fourier transform infrared spectroscopic study of the thermal stability of peroxyacetyl nitrate. *J. Phys. Chem.* 90, 152–156.

Seuwen, R., and P. Warneck (1995) Oxidation of toluene in NO_x-free air: product distribution and mechanism. *Int. J. Chem. Kinet.* 28, 315–332.

Sexton, K., and H. Westberg (1984) Nonmethane hydrocarbon composition of urban and rural atmospheres. *Atmos. Environ.* 18, 1125–1132.

Shannon, J. D. (1981) A model of regional long-term average sulfur atmospheric pollution, surface removal, and net horizontal flux. *Atmos. Environ.* 15, 689–701.

Shardanand (1969) Absorption cross sections of O_2 and O_4 between 2000 and 2800 Å. *Phys. Rev.* 186, 5–9.

Shardanand and A. D. P. Rao (1977) Collision-induced absorption of O_2 in the Herzberg continuum. *J. Quant. Spectrosc. Radiat. Transfer* 17, 433–439.

Shaw, R. W., and R. J. Paur (1983) Measurements of sulfur in gases and particles during sixteen months in the Ohio River Valley. *Atmos. Environ.* 17, 1431–1438.

Shaw, R. W., A. L. Crittenden, R. K. Stevens, D. R. Cronn, and V. S. Titov (1983) Ambient concentrations of hydrocarbons from conifers in atmospheric gases and aerosol particles measured in Soviet Georgia. *Environ. Sci. Technol.* 17, 389–396.

Shearer, M. J., and M. A. K. Khalil (1993) Rice agriculture: emissions. *In* "Atmospheric Methane: Sources, Sinks, and Role in Global Change" (M. A. K. Khalil, ed.), NATO ASI Series, Vol. I 13, pp. 230–253. Springer-Verlag, Berlin.

Shen, T.-L., P. J. Wooldridge, and M. J. Molina (1995) Stratospheric pollution and ozone depletion. *In* "Composition, Chemistry and Climate of the Atmosphere" (H. B. Singh, ed.), pp. 394–442. Van Nostrand Reinhold, New York.

Shepherd, E. S. (1925) The analysis of gases from volcanoes and from rocks. *J. Geol.* 33, 289–370.

Shepherd, E. S. (1938) The gases in rocks and some related problems. *Am. J. Sci. Ser.* 5 235a, 311–351.

Shepherd, J. G. (1974) Measurements of the direct deposition of sulphur dioxide onto grass and water by the profile method. *Atmos. Environ.* 8, 69–74.

Shepherd, M. F., S. Barzetti, and D. R. Hastie (1991) The production of atmospheric NO_x and N_2O from a fertilized agricultural soil. *Atmos. Environ.* 25A, 1961–1969.

Sheppard, J. C., H. Westberg, J. F. Hopper, K. Ganesan, and P. Zimmerman (1982) Inventory of global methane sources and their production rates. *J. Geophys. Res.* 87, 1305–1312.

Shepson, P. B., D. R. Hastie, H. I. Schiff, M. Polizzi, J. W. Bottenheim, K. Anlauf, G. I. MacKay, and D. R. Karecki (1991) Atmospheric concentrations and temporal variations of C_1-C_3 carbonyl compounds at two rural sites in central Ontario. *Atmos. Environ.* **25A**, 2001–2015.

Shepson, P. B., D. R. Hastie, K. W. So, H. I. Schiff, and P. Wong (1992) Relationships between PAN, PPN, and O_3 at urban and rural sites in Ontario. *Atmos. Environ.* **26A**, 1259–1270.

Shepson, P. B., K. G. Anlauf, J. W. Bottenheim, H. A. Wiebe, N. Gao, K. Muthuramu, and G. I. Mackay (1993) Alkyl nitrates and their contribution to reactive nitrogen at a rural site in Ontario. *Atmos. Environ.* **27A**, 749–757.

Shin, Z., J. V. Ford, S. Wie, and A. W. Castleman, Jr. (1993) Water clusters: contribution of binding energy and entropy to stability. *J. Phys. Chem.* **99**, 8009–8015.

Shneour, E. A. (1966) Oxidation of graphitic carbon in certain soils. *Science* **151**, 991–992.

Shorter, J. H., C. E. Kolb, P. M. Crill, R. A. Kervin, R. W. Talbot, M. E. Hines, and R. C. Harriss (1995) Rapid degradation of atmospheric methyl bromide in soils. *Nature* **377**, 717–719.

Shugard, W. J., R. H. Heist, and J. J. Reiss (1974) Theory of water phase nucleation in binary mixtures of water and sulfuric acid. *J. Chem. Phys.* **61**, 5298–5307.

Sidebottom, H. W., C. C. Badcock, G. E. Jackson, J. G. Calvert, G. W. Reinhardt, and E. K. Damon (1972) Photooxidation of sulfur dioxide. *Environ. Sci. Technol.* **6**, 72–79.

Siegenthaler, U. (1983) Uptake of excess CO_2 by an outcrop diffusion model of the ocean. *J. Geophys. Res.* **88**, 3599–3608.

Siegenthaler, U., and K. O. Münnich (1981) $^{13}C/^{12}C$ fractionation during CO_2 transfer from air to sea. *In* "Carbon Cycle Modelling" (B. Bolin, ed.). *SCOPE* **16**, 249–257.

Siegenthaler, U., and H. Oeschger (1978) Predicting future atmospheric carbon dioxide levels. *Science* **199**, 388–395.

Siegenthaler, U., and H. Oeschger (1987) Biospheric CO_2 emissions during the past 200 years reconstructed by deconvolution of ice core data. *Tellus* **39B**, 140–154.

Siegenthaler, U., M. Heimann, and H. Oeschger (1978) Model responses of the atmospheric CO_2 level and $^{13}C/^{12}C$ ratio to biogenic CO_2 input. *In* "Climate and Society" (J. Williams, ed.), pp. 79–84. Pergamon, New York.

Siever, R. (1968) Sedimentological consequences of a steady state ocean-atmosphere. *Sedimentology* **11**, 5–29.

Signer, P., and H. E. Suess (1963) Rare gases in the sun, in the atmosphere and in meteorites. *In* "Earth Science and Meteorites" (J. Geiss and E. D. Goldberg, eds.), pp. 241–272. North-Holland Publishing, Amsterdam, the Netherlands.

Sigvaldason, G. E., and G. Elisson (1968) Collection and analysis of volcanic gases at Surtsey, Iceland. *Geochim. Cosmochim. Acta* **32**, 797–805.

Sillén, L. G. (1961) The physical chemistry of sea water. *In* "Oceanography" (M. Sears, ed.). *Am. Assoc. Adv. Sci. Publ.* **67**, 549–581.

Sillén, L. G. (1963) How has sea water got its present composition? *Sven. Kem. Tidskr.* **75**, 161–177.

Sillén, L. G. (1966) Regulation of O_2, N_2, and CO_2 in the atmosphere: thoughts of a laboratory chemist. *Tellus* **18**, 198–206.

Simmonds, P. G., F. N. Alyea, C. A. Cardelino, A. J. Crawford, D. M. Cunnold, B. C. Lane, J. E. Lovelock, R. G. Prinn, and R. A. Rasmussen. (1983) The atmospheric life time experiment. 6. Results for carbon tetrachloride based on 3 years data. *J. Geophys. Res.* **88**, 8427–8441.

Simmonds, P. G., D. M. Cunnold, F. N. Alyea, C. A. Cardelino, A. J. Crawford, R. G. Prinn, P. J. Fraser, R. A. Rasmussen, and R. D. Rosen (1988) Carbon tetrachloride lifetimes and emissions determined from daily global measurements during 1978–1985. *J. Atmos. Chem.* **7**, 35–58.

Simon, P. C. (1978) Irradiation solar flux measurements between 120 and 400 nm. Current position and future needs. *Planet. Space Sci.* **26**, 355–365.

Simoneit, B. R. T. (1977) Organic matter in eolian dust over the Atlantic ocean. *Marine Chem.* 5, 443–464.

Simoneit, B. R. T. (1984) Organic matter in the troposphere. III. Characterization and sources of petroleum and pyrogenic residues in aerosols over the western United States. *Atmos. Environ.* 18, 51–57.

Simoneit, B. R. T. (1989) Organic matter of the troposphere. V. Application of molecular marker analysis to biogenic emissions into the troposphere for source reconciliation. *J. Atmos. Chem.* 8, 251–275.

Simoneit, B. R. T., and M. A. Mazurek (1981) Air pollution: the organic components. *CRC Crit. Rev. Environ. Control* 11, 219–276.

Simoneit, B. R. T., and M. A. Mazurek (1982) Organic matter of the troposphere. II. Natural background of biogenic lipid matter in aerosols over the rural western United States. *Atmos. Environ.* 16, 2139–2159.

Simoneit, B. R. T., J. N. Cardoso, and N. Robinson (1990) An assessment of the origin and composition of higher molecular weight organic matter in aerosols over Amazonia. *Chemosphere* 21, 1285–1301.

Simonsen, J., and D. H. R. Barton (1961) "The Terpenes," Vol. III, The Sequiterpenes, Diterpenes and Their Derivatives. Cambridge University Press, London.

Sinanoglu, O. (1981) What size cluster is like a surface? *Chem. Phys. Lett.* 81, 188–190.

Singh, H. B. (1977a) Atmospheric halocarbons: evidence in favor of reduced average hydroxyl radical concentrations in the troposphere. *Geophys. Res. Lett.* 3, 101–104.

Singh, H. B. (1977b) Preliminary estimation of average tropospheric OH concentrations in the northern and southern hemisphere. *Geophys. Res. Lett.* 4, 453–456.

Singh, H. B. (1987) Reactive nitrogen in the atmosphere. *Environ. Sci. Technol.* 21, 320–327.

Singh, H. B. (1995) Halogens in the atmospheric environment. *In* "Composition, Chemistry, and Climate of the Atmosphere" (H. B. Singh, ed.), pp. 216–250. Van Nostrand-Reinhold, New York.

Singh, H. B., and P. L. Hanst (1981) Peroxyacetyl nitrate (PAN) in the unpolluted atmosphere: an important reservoir for nitrogen oxides. *Geophys. Res. Lett.* 8, 941–944.

Singh, H. B., and M. Kanakidou (1993) An investigation of the atmospheric sources and sinks of methyl bromide. *Geophys. Res. Lett.* 20, 133–136.

Singh, H. B., and L. J. Salas (1982) Measurement of selected light hydrocarbons over the Pacific Ocean: latitudinal and seasonal variations. *Geophys. Res. Lett.* 9, 842–845.

Singh, H. B., and L. J. Salas (1989) Measurements of peroxyacetyl nitrate (PAN) and peroxypropionyl nitrate (PPN) at selected urban, rural and remote sites. *Atmos. Environ.* 23, 231–238.

Singh, H. B., P. D. Fowler, and T. O. Peyton (1976) Atmospheric carbon tetrachloride: another man-made pollutant. *Science* 192, 1231–1234.

Singh, H. B., L. J. Salas, and L. A. Cavanagh (1977a) Distribution, sources and sinks of atmospheric halogenated compounds. *J. Air Pollut. Control Assoc.* 27, 332–336.

Singh, H. B., L. J. Salas, H. Shigeishi, and A. Crawford (1977b) Urban-non-urban relationships of halocarbons, SF_6, N_2O and other atmospheric trace constituents. *Atmos. Environ.* 11, 819–828.

Singh, H. B., F. L. Ludwig, and W. B. Johnson (1978) Tropospheric ozone: concentrations and variabilities in clean remote atmospheres. *Atmos. Environ.* 12, 2185–2196.

Singh, H. B., L. J. Salas, and H. Shigeishi (1979a) The distribution of nitrous oxide in the global atmosphere and in the Pacific ocean. *Tellus* 31, 313–320.

Singh, H. B., L. J. Salas, H. Shigeishi, and E. Scribner (1979b) Atmospheric halocarbons, hydrocarbons, sulfur hexafluoride: global distribution, sources and sinks. *Science* 203, 899–903.

Singh, H. B., L. J. Salas, and R. E. Stiles (1983a) Selected man-made halogenated chemicals in the air and oceanic environment. *J. Geophys. Res.* 88, 3675–3683.

Singh, H. B., L. J. Salas, and R. E. Stiles (1983b) Methyl halides in and over the eastern Pacific (40°N–32°S). *J. Geophys. Res.* **88**, 3684–3690.

Singh, H. B., L. J. Salas, B. A. Ridley, J. D. Shetter, N. M. Donahue, F. C. Fehsenfeld, D. W. Fahey, D. D. Parrish, E. J. Williams, S. C. Liu, G. Hübler, and P. C. Murphy (1985) Relationship between peroxyacetyl nitrate and nitrogen oxides in the clean troposphere. *Nature* **318**, 347–349.

Singh, H. B., L. J. Salas, and W. Viezee (1986) Global distribution of peroxyacetyl nitrate. *Nature* **321**, 588–591.

Singh, H. B., W. Viezee, and L. J. Salas (1988) Measurements of selected C_2–C_5 hydrocarbons in the troposphere: latitudinal, vertical and temporal variations. *J. Geophys. Res.* **93**, 15,861–15,878.

Singh, H. B., D. Herlth, D. O'Hara, K. Zahnle, J. D. Bradshaw, S. T. Sandholm, R. Talbot, G. L. Gregory, G. W. Sachse, D. R. Blake, and S. C. Wofsy (1994) Summer time distribution of PAN and other reactive nitrogen species in the northern high-latitude atmosphere of eastern Canada. *J. Geophys. Res.* **99**, 1821–1835

Singh, H. B., G. L. Gregory, B. Anderson, E. Browell, G. W. Sachse, D. D. Davis, J. Crawford, J. D. Bradshaw, R. Talbot, D. R. Blake, D. Thornton, R. Newell, and J. Merrill (1996a) Low ozone in the marine boundary layer of the tropical Pacific Ocean: photochemical loss, chlorine atoms, and entrainment. *J. Geophys. Res.* **101**, 1907–1917.

Singh, H. B., D. Herlth, R. Kolyer, L. Salas, J. D. Bradshaw, S. T. Sandholm, D. D. Davis, J. Crawford, Y. Kondo, M. Koike, R. Talbot, G. L. Gregory, G. W. Sachse, E. Browell, D. R. Blake, F. S. Rowland, R. Newell, J. Merrill, B. Heikes, S. C. Liu, P. J. Crutzen, and M. Kanakidou (1996b) Reactive nitrogen and ozone over the western Pacific: distribution, partitioning, and sources. *J. Geophys. Res.* **101**, 1793–1808.

Singh, O. N., (1994) Indian methane budget and its global perspective. *Pure Appl. Chem.* **66**, 197–200.

Sirois, A., and W. Fricke (1992) Regionally representative daily air concentrations of acid-related substances in Canada: 1983–1987. *Atmos. Environ.* **26A**, 593–607.

Sisterton, D. L. (1990) "Deposition Monitoring, Methods and Results," NASP State of Science and Technology Report no. 6. U.S. National Acid Precipitation Assessment Program, Washington, DC.

Sivaldason, G. E., and G. Ellison (1968) Collection and analysis of volcanic gases at Surtsey, Iceland. *Geochim. Cosmochim. Acta* **32**, 797–805.

Sjödin, A., and M. Ferm (1985) Measurements of nitrous acid in an urban area. *Atmos. Environ.* **19**, 985–992.

Sjödin, A., and P. Grennfelt (1984) Regional background concentration of NO_2 in Sweden. In "Physico-chemical Behaviour of Atmospheric Pollutants" (B. Versino and G. Angeletti, eds.), Proceedings of the Third European Symposium, Varese, Italy, pp. 401–411. Reidel, Dordrecht, the Netherlands.

Skärby, L., C. Bengtson, C.-A. Boström, P. Grennfelt, and E. Troeng (1981) Uptake of NO_x in Scots pine. *Silvae Fennica* **15**, 396–398.

Skirrow, G. (1975) The dissolved gases—carbon dioxide. In "Chemical Oceanography," 2nd ed. (J. P. Riley and G. Skirrow, eds.), Vol. 2, pp. 1–192. Academic Press, London.

Skov, H., J. Hjorth, C. Lohse, N. R. Jensen, and G. Restelli (1992) Products and mechanisms of the reactions of the nitrate radical (NO_3) with isoprene, 1,3-butadiene and 2,3-dimethyl-1,3-butadiene in air. *Atmos. Environ.* **26A**, 2771–2783.

Slagle, I. R., R. E. Graham, and D. Gutman (1976) Direct identification of reactive routes and measurement of rate constants in the reaction of oxygen atoms with methanethiol, ethanethiol and methylsulfide. *Int J. Chem. Kinet.* **8**, 451–548.

Slanger, T. G., L. E. Jusinski, G. Black, and G. E. Gadd (1988) A new laboratory source of ozone and its potential atmospheric implications. *Science* 241, 945–950.

Slatt, B. J., D. F. S. Natush, J. M. Prospero, and D. L. Savoie (1978) Hydrogen sulfide in the atmosphere of the northern equatorial Atlantic Ocean and its relation to the global sulfur cycle. *Atmos. Environ.* 12, 981–991.

Slemr, F., and W. Seiler (1984) Field measurements of NO and NO_2 emissions from fertilized and unfertilized soils. *J. Atmos. Chem.* 2, 1–24.

Slemr, F., and W. Seiler (1991) Field study of environmental variable controlling the NO emission from soil and the NO compensation point. *J. Geophys. Res.* 96, 13,017–13,031.

Slemr, F., R. Conrad, and W. Seiler (1984) Nitrous oxide emissions from fertilized and unfertilized soils in a subtropical region (Andalusia, Spain). *J. Atmos. Chem.* 1, 159–169.

Slemr, J., W. Junkermann, and A. Volz-Thomas (1996) Temporal variations in formaldehyde, acetaldehyde and acetone and a budget of formaldehyde at a rural site in southern Germany. *Atmos. Environ.* 30, 3667–3676.

Slinn, W. G. N. (1976) Some approximations for the wet and dry removal of particles and gases from the atmosphere. *J. Air Water Soil Pollut.* 7, 513–543.

Slinn, W. G. N. (1977a) Proposed terminology for precipitation scavenging. *In* "Precipitation Scavenging (1974)" (R. G. Semonin and R. W. Beadle, coordinators), ERDA Symposium Series 41, CONF-741003, NTIS, pp. 813–818. U.S. Department. of Commerce, Springfield, VA.

Slinn, W. G. N. (1977b) Precipitation scavenging: some problems approximate solultions and suggestions for further research. *In* "Precipitation Scavenging (1974)" (R. G. Semonin and R. W. Beadle, coordinators). ERDA Symposium Series 41, CONF-741003, NTIS, pp. 1–60. U.S. Department of Commerce, Springfield, VA.

Slinn, W. G. N. (1988) A simple model for Junge's relationship between concentration fluctuations and residence times for tropospheric trace gases. *Tellus* 40B, 229–232.

Slinn, W. G. N., and J M. Hales (1971) A reevaluation of the role of thermophoresis as a mechanism of in and below cloud scavenging. *J. Atmos. Sci.* 28, 1465–1471.

SMIC (1971). "Inadvertent Climate Modification," Report of the Study of Man's Impact on Climate, MIT Press, Cambridge, MA.

Smith, J. P., and P. Urone (1974) Static studies of sulfur dioxide reactions, effects of NO_2, C_2H_6 and H_2O. *Environ. Sci. Technol.* 8, 742–746.

Smith, R. A. (1882) "Air and Rain—The Beginnings of a Chemical Climatology." Longmans, Green, London.

Smithsonian Meteorological Tables (1951), 6th ed. Smithsonian Institution, Washington, DC.

Smoluchowski, M. V. (1918) Versuch einer mathematischen Theorie der Koagulationskinetik kolloidaler Lösungen. *Z. Phys. Chem.* 92, 129–168.

Söderlund, R., and B. H. Svensson (1976) The global nitrogen cycle. *In* "Nitrogen, Phosphorus, and Sulphur-Global Cycles" (B. H. Svensson and R. Söderlund, eds.). SCOPE 7, *Ecol. Bull.* (*Stockh.*) 22, 23–73.

Solberg, S., N. Schmidbauer, A Semb, F. Stordal, and O. Hov (1996) Boundary layer ozone depletion as seen in the Norwegian Arctic in spring. *J. Atmos. Chem.* 23, 301–332.

Solomon, S., R. R. Garcia, F. S. Rowland, and D. J. Wuebbles (1986a) On the depletion of Antarctic ozone. *Nature* 321, 755–758.

Solomon, S., J. T. Kiehl, J. B. Kerridge, E. E. Remsberg, and J. M. Russell, III (1986b) Evidence for non-local thermodynamic equilibrium in the mode of mesospheric ozone. *J. Geophys. Res.* 91, 9865–9876.

Solomon, S., R. W. Sanders, M. A. Carroll, and A. L. Schmeltekopf (1989) Visible and near-ultraviolet spectroscopy at McMurdo Station, Antarctica. 5. Observations of the diurnal variations of BrO and OClO. *J. Geophys. Res.* 94, 11,393–11,403.

Solomon, S., R. W. Portmann, R. R. Garcia, L. W. Thomason, L. R. Poole, and M. P. McCormick (1996) The role of aerosol variations in anthropogenic ozone depletion at northern mid-latitudes. *J. Geophys. Res.* **101**, 6713–6727.

Sood, S. K., and M. R. Jackson (1970) Scavenging by snow and ice crystals. In "Precipitation Scavenging 1970" (R. Engelmann and W. G. N. Slinn, coordinators). AEC Symposium Series 30, pp. 121–134. U.S. Atomic Energy Commission, Division of Technical Information, Oak Ridge, TN.

Sørensen, P. E., and V. S. Andersen (1970). The formaldehyde-hydrogen sulphite system in alkaline aqueous solution: kinetics, mechanism and equilibria. *Acta Chem. Scand.* **24**, 1301–1306.

Spicer, C. W. (1980) The rate of NO_x reaction in transported urban air. In "Studies in Environmental Science" (M. M. Benarie, ed.), Vol. 8, pp. 181–186. Elsevier, Amsterdam, the Netherlands.

Spicer, C. W. (1982) Nitrogen oxide reactions in the urban plume of Boston. *Science* **215**, 1095–1097.

Spicer, C. W., J. E. Howes, Jr., T. A. Bishop, and L. H. Arnold (1982) Nitric acid measurement methods: an intercomparison. *Atmos. Environ.* **16**, 1476–1500.

Spiro, P. A., D. J. Jacob, and J. A. Logan (1992) Global inventory of sulfur emissions with a 1° × 1° resolution. *J. Geophys. Res.* **97**, 6023–6036.

Spivakovsky, C. M., R. Yevich, J. A. Logan, S. C. Wofsy, M. B. McElroy, and M. J. Prather (1990) Tropospheric OH in a three-dimensional chemical tracer model: an assessment based on observations of CH_3CCl_3. *J. Geophys. Res.* **95**, 18,441–18,471.

Sprent, J. I., and P. Sprent (1990) "Nitrogen Fixing Organisms—Pure and Applied Aspects." Chapman and Hall, New York.

Staehelin, J., and W. Schmid (1991) Trend analysis of tropospheric ozone concentrations utilizing the 20-year data set of balloon soundings over Payerne (Switzerland). *Atmos. Environ.* **25A**, 1739–1749.

Staehelin, J., R. E. Bühler, and J. Hoigné (1984) Ozone decomposition in water studied by pulse radiolysis. 2. OH and HO_4 as chain intermediates. *J. Phys. Chem.* **88**, 5999–6004.

Staehelin, J., J. Thudium, R. Buehler, A. Volz-Thomas, and W. Graber (1994) Trends in surface ozone concentrations at Arosa (Switzerland). *Atmos. Environ.* **28**, 75–84.

Staley, D. O. (1982) Strontium-90 in surface air and the stratosphere: some interpretations of the 1963–1975 data. *J. Atmos. Sci.* **39**, 1571–1590.

Stallard, R. J., J. M. Edmond, and R. E. Newell (1975) Surface ozone in the south east Atlantic between Dakar and Walvis Bay. *Geophys. Res. Lett.* **2**, 289–292.

Stanford, J. L. (1973) Possible sink for stratospheric water vapor at the winter Antarctic pole. *J. Atmos. Sci.* **30**, 1431–1436.

Stanford, J. L., and J. S. Davis (1974) A century of stratospheric cloud reports: 1870–1972. *Bull. Am. Meteorol. Soc.* **55**, 213–219.

Starr, J. R., and B. J. Mason (1966) The capture of airborne particles by water drops and simulated snow crystals. *Q. J. R. Meteorol. Soc.* **92**, 490–499.

Starr, W. L., and J. F. Vedder (1989) Measurements of ozone in the Antarctic atmosphere during August and September 1987. *J. Geophys. Res.* **94**, 11,449–11,463.

Staubes, R., and H. W. Georgii (1993a) Measurements of atmospheric and sea water DMS concentrations in the Atlantic, the Arctic and Antarctic region. In "Dimethylsulphide, Oceans, Atmosphere and Climate" (G. Restelli and G. Angeletti, eds.), pp. 95–102, Kluwer Academic Publishers, Dordrecht, the Netherlands.

Staubes, R., and H.-W. Georgii (1993b) Biogenic sulfur compounds in sea water and the atmosphere of the Arctic region. *Tellus* **45B**, 127–137.

Staubes, R., G. Ockelmann, and H. W. Georgii (1986) Emissions of COS, DMS and CS_2 from various soils in Germany. *Tellus* **41B**, 305–313.

Stauffer, B., G. Fischer, A. Neftel, and H. Oeschger (1985) Increase of atmospheric methane recorded in Antarctic ice core. *Science* **229**, 1386–1388.

Stauffer, B., E. Lochbronner, H. Oeschger, and J. Schwander (1988) Methane concentration in the glacial atmosphere was only half that of pre-industrial holocene. *Nature* **332**, 812–814.

Stedman, D. H., and J. O. Jackson (1975) The photostationary state in photochemical smog. *Int. J. Chem. Kinet.* **7** (Symp. 1), 493–501.

Stedman, D. H., and M. J. McEwan (1983) Oxides of nitrogen at two sites in New Zealand. *Geophys. Res. Lett.* **10**, 168–171.

Stedman, D. H., and R. E. Shetter (1983) The global budget of atmospheric nitrogen species. *In* "Trace Atmospheric Constituents, Properties, Transformations and Fates" (S. E. Schwartz, ed.) pp. 411–454. Wiley, New York.

Stedman, D. H., W. Chameides, and J. O. Jackson (1975) Comparison of experimental and computed values for $j(NO_2)$. *Geophys. Res. Lett.* **2**, 22–25.

Stedman, D. H., R. J. Cicerone, W. L. Chameides, and R. B. Harvey (1976) Absence of N_2O photolysis in the troposphere. *J. Geophys. Res.* **81**, 2003–2004.

Steele, L. P., P. J. Fraser, R. A. Rasmussen, M. A. K. Khalil, T. J. Conway, A. J. Crawford, R. H. Gammon, K. A. Masarie, and K. W. Thonok (1987) The global distribution of methane in the troposphere. *J. Atmos. Chem.* **5**, 125–171.

Steiger, R. H., and E. Jaeger (1977) Subcommission on geochronology: convention on the use of decay constants in geo- and cosmochronology. *Earth Planet. Sci. Lett.* **36**, 359–362.

Steinbrecher, R. (1997) Isoprene: production by plants and ecosystem-level estimates. *In* "Biogenic Volatile Organic Compounds in the Atmosphere" (G. Helas, J. Slanina, and R. Steinbrecher, eds.), pp. 101–114. SPB Academic Publishing bv, Amsterdam, the Netherlands.

Steinbrecher, R., J. Hahn, K. Stahl, G. Eichstädter, K. Lederle, R. Rabong, A. M. Schreiner, and J. Slemr (1997) Investigations on emissions of low molecular weight compounds (C_2-C_10) from vegetation. *In* "Biosphere-Atmosphere Exchange of Pollutants and Trace Substances" (S. Slanina, ed.), Transport and Chemical Transformation of Pollutants in the Troposphere, Vol. 4 (P. Borrell, P. M. Borrell, T. Cvitas, K. Kelly, and W. Seiler, series eds.), pp. 342–351. Springer-Verlag, Berlin.

Steinhardt, U. (1973) Input of chemical elements from the atmosphere. A tabular review of literature. *Göttinger Bodenkd. Ber.* **29**, 93–132.

Stelson, A. W., and J. H. Seinfeld (1981) Chemical mass accounting of urban aerosol. *Environ. Sci. Technol.* **15**, 671–679.

Stelson, A. W., and J. H. Seinfeld (1982) Relative humidity and temperature dependence of the ammonium nitrate dissociation constant. *Atmos. Environ.* **16**, 983–992.

Stephens, E. R. (1971) Identification of odors from cattle feed lots. *Calif. Agric.* **25**, 10–11.

Stephens, E. R., and F. R. Burleson (1967) Analysis of the atmosphere for light hydrocarbons. *J. Air Pollut. Control Assoc.* **17**, 147–153.

Stephens, E. R., and F. R. Burleson (1969) Distribution of light hydrocarbons in ambient air. *J. Air Pollut. Control Assoc.* **19**, 929–936.

Stephens, E. R., P. L. Hanst, R. C. Doerr, and W. E. Scott (1956) Reactions of nitrogen dioxide and organic compounds in air. *Ind. Eng. Chem.* **48**, 1498–1504.

Stephens, E. R., P. L. Hanst, R. C. Doerr, and W. E. Scott (1958) Auto exhaust: composition and photolysis products. *J. Air Pollut. Control Assoc.* **8**, 333–335.

Stephens, E. R., E. F. Darley, O. C. Taylor, and W. E. Scott (1961) Photochemical reaction products in air pollution. *Int. J. Air Water Pollut.* **4**, 79–100.

Steudler, P., and B. Peterson (1985) Annual cycle of gaseous sulfur emissions from a New England *Spartina alterniflora* marsh. *Atmos. Environ.* **19**, 1411–1416.

Stewart, N. G., R. N. Crooks, and E. M. R. Fisher (1956) The radiological dose to persons in the U.K. due to debris from nuclear test explosions prior to January 1956, AERE HP/R-2017. Atomic Energy Research Establishment, Harwell, England.

Stewart, R. W., S. Hameed, and J. P. Pinto (1977) Photochemistry of the tropospheric ozone. *J. Geophys. Res.* **82**, 3134–3140.

Stimpfle, R. M., and J. G. Anderson (1988) In-situ detection of OH in the lower stratosphere with balloon-borne high-repetition rate laser system. *Geophys. Res. Lett.* **15**, 503–506.

Stimpfle, R. M., L. B. Lapson, P. O. Wennberg, and J. G. Anderson (1989) Balloon-borne detection of OH in the stratosphere from 37 to 23 km. *Geophys. Res. Lett.* **16**, 1433–1436.

Stimpfle, R. M., P. O. Wennberg, L. B. Lapson, and J. G. Anderson (1990) Simultaneous in situ measurement of OH and HO_2 in the stratosphere. *Geophys. Res. Lett.* **17**, 1905–1908.

Stith, J. L., P. V. Hobbs, and L. F. Radke (1978) Airborne particle and gas measurements in the emissions from six volcanoes. *J. Geophys. Res.* **83**, 4009–4017.

Stith, J. L., L. F. Radke, and P. V. Hobbs (1981) Particle emissions and the production of ozone and nitrogen oxides from the burning of forest slash. *Atmos. Environ.* **15**, 73–84.

Stix, E. (1969) Schwankungen des Pollen- und Sporengehaltes der Luft. *Umschau* **19**, 620–621.

Stockwell, W. R., and J. G. Calvert (1978) The near ultraviolet absorption spectrum of gaseous HONO and N_2O_3. *J. Photochem.* **8**, 193–203.

Stoiber, R. E., and G. Bratton (1978) Airborne correlation spectromter measurements of SO_2 in eruption clouds from Guatemalan volcanoes. *EOS Trans. Am. Geophys. Union* **59**, 1222.

Stoiber, R. E., and A. Jepsen (1973) Sulfur dioxide contribution to the atmosphere by volcanoes. *Science* **182**, 577–578.

Stoiber, R. E., and G. B. Malone (1975) SO_2 emission at the crater of Kilauea, at Mauna Ulu and Sulfur Banks, Hawaii. *EOS Trans. Am. Geophys. Union* **56**, 461.

Strobel, D. F. (1971) Odd nitrogen in the mesosphere. *J. Geophys. Res.* **76** 8384–8393.

Strutt, R. J. (1918) Ultraviolet transparency of the lower atmosphere and its relative poverty in ozone. *Proc. R. Soc. Lond. A* **94**, 260–268.

Stuhl, F., and H. Niki (1972) Pulsed vacuum UV photochemical study of reactions of OH with H_2, D_2 and CO using a resonance-fluorescence method. *J. Chem. Phys.* **57**, 3671–3677.

Stuiver, M. (1978) Atmospheric CO_2 increase related to carbon reservoir changes. *Science* **199**, 253–258.

Stuiver, M. (1981) C-14 distribution in the Atlantic Ocean. *J. Geophys. Res.* **85**, 2711–2718.

Stuiver, M., R. L. Burk, and P. D. Quay (1984) $^{13}C/^{12}C$ ratios in tree rings and the transfer of biospheric carbon to the atmosphere. *J. Geophys. Res.* **89**, 11,731–11,748.

Stumm, W., and P. A. Brauner (1975) Chemical speciation. *In* "Chemical Oceanography," 2nd ed. (J. P. Riley and G. Skirrow, eds.), Vol. 1, pp. 173–239. Academic Press, London.

Stumm, W., and J. J. Morgan (1981) "Aquatic Chemistry." Wiley, New York.

Sturges, W. T., R. C. Schnell, G. S. Dutton, S. R. Garcia, and J. A. Lind (1993) Spring measurements of tropospheric bromine at Barrow, Alaska. *Geophys. Res. Lett.* **20**, 201–204.

Su, C. W., and E. D. Goldberg (1973) Chlorofluorocarbons in the atmosphere. *Nature* **245**, 27.

Su, F., J. G. Calvert, and J. H. Shaw (1980) A FTIR study of the ozone-ethene reaction mechanism in O_2-rich mixtures. *J. Phys. Chem.* **84**, 239–246.

Suess, H. E. (1949) Die Häufigkeit der Edelgase auf der Erde und im Kosmos. *J. Geol. Res.* **57**, 600–607.

Suess, H. E. (1966) Some chemical aspects of the evolution of the terrestrial atmosphere. *Tellus* **18**, 207–211.

Suhre, K., M. O. Andreae, and R. Rosset (1995) Biogenic sulfur emissions and aerosols over the tropical south Atlantic. 2. One-dimensional simulation of sulfur chemistry in the marine boundary layer. *J. Geophys. Res.* **100**, 11,232–11,334.

Sullivan, J. O., and P. Warneck (1967) Reaction of ^1D oxygen atoms III. Ozone formation in the 1470 Å photolysis of O_2. *J. Chem. Phys.* **46**, 953−959.

Sundquist, E. T., L. N. Plummer, and T. M. L. Wigley (1979) Carbon dioxide in the ocean surface: a homogeneous buffer factor. *Science* **204**, 1203−1205.

Suto, M., E. R. Manzares, and L. C. Lee (1985) Detection of sulfuric acid aerosols by ultraviolett scattering. *Environ. Sci. Technol.* **19**, 815−820.

Sutton, M. A., D. Fowler, and J. B. Moncrieff (1993a) The exchange of atmospheric ammonia with vegetated surfaces. I. Unfertilized vegetation. *Q. J. R. Meteorol. Soc.* **119**, 1023−1045.

Sutton, M. A., D. Fowler, and J. B. Moncrieff (1993b) The exchange of atmospheric ammonia with vegetated surfaces. II. Fertilized vegetation. *Q. J. R. Meteorol. Soc.* **119**, 1047−1070.

Sutton, O. G. (1953) "Micrometeorology." McGraw-Hill, New York.

Sutton, O. G. (1955) "Atmospheric Turbulence," 2nd ed. Wiley, New York.

Svenningsson, B., H.-C. Hansson, A. Wiedensohler, K. Noone, J. Ogren, A. Hallberg, and R. Colvile (1994) Hygroscopic growth of aerosol particles and its influence on nucleation scavenging in cloud: experimental results from Kleiner Feldberg. *J. Atmos. Chem.* **19**, 129−152.

Svensson, R., E. Ljungström, and O. Lindquist (1987) Kinetics of the reaction between nitrogen dioxide and water vapour. *Atmos. Environ.* **21**, 1529−1539.

Sverdrup, G. M., C. W. Spicer, and G. F. Ward (1987) Investigation of the gas-phase reaction of dinitrogen pentoxide with water vapor. *Int. J. Chem. Kinet.* **19**, 191−205.

Swanson, D., B. Kan, and H. S. Johnston (1984) NO_3 quantum yield from N_2O_5 photolysis. *J. Phys. Chem.* **88**, 3115−3118.

Swinnerton, J. W., and R. A. Lamontagne (1974a) Carbon monoxide in the southern hemisphere. *Tellus* **26**, 136−142.

Swinnerton, J. W., and R. A. Lamontagne (1974b) Oceanic distribution of low-molecular weight hydrocarbons. *Environ. Sci. Technol.* **8**, 657−663.

Swinnerton, J. W., and V. J. Linnenboom (1967) Gaseous hydrocarbons in sea water: determination. *Science* **156**, 119−120.

Swinnerton, J. W., J. Linnenboom, and C. H. Cheek (1969) Distribution of methane and carbon monoxide between the atmosphere and natural waters. *Environ. Sci. Technol.* **3**, 836−838.

Swinnerton, J. W., V. J. Linnenboom, and R. A. Lamontagne (1970) Ocean: a natural source of carbon monoxide. *Science* **167**, 984−986.

Tabazadeh, A., and R. P. Turco (1993) A model for heterogeneous chemical processes on the surface of ice and nitric acid trihydrate particles. *J. Geophys. Res.* **98**, 12,727−12,740.

Tait, V. K., R. M. Moore, and R. Tokarczyk (1994) Measurements of methyl chloride in the northwest Atlantic. *J. Geophys. Res.* **99**, 7821−7833.

Takahashi, T., W. S. Broecker, S. R. Werner, and A. E. Bainbridge (1980) Carbonate chemistry of the surface waters of the world oceans. *In* "Isotope Marine Chemistry" (E. D. Goldberg, Y. Hozibe, and K. Sarnhashi, eds.), pp. 291−326. Uchida Rokakuho, Tokyo.

Takahashi, T., W. S. Broecker, and A. E. Bainbridge (1981) The alkalinity and total carbon dioxide concentration in the world oceans. *In* "Carbon Cycle Modelling" (B. Bolin, ed.). *SCOPE* **16**, 159−199, 271−286.

Talbot, R.W., K. M. Beecher, R. C. Harris, and W. R. Cofer, III (1988) Atmospheric geochemistry of formic and acetic acids at a mid-latitude temperate site. *J. Geophys. Res.* **93**, 1638−1652.

Talbot, R. W., M. O. Andreae, H. Berresheim, D. J. Jacob, and K. M. Beecher (1990) Sources and sinks of formic, acetic, and pyruvic acids over central Amazonia. 2 Wet season. *J. Geophys. Res.* **95**, 17,799−17,811.

Talbot, R. W., A. S. Vijgen, and R. C. Harriss (1992) Soluble species in the Arctic summer troposphere: acidic gases, aerosols, and precipitation. *J. Geophys. Res.* **97**, 16,531−16,543.

Talukdar, R. K., C. A. Longfellow, M. K. Gilles, and A. R. Ravishankara (1998) Quantum yields of $O(^1D)$ in the photolysis of ozone between 289 and 329 nm as a function of temperature. *Geophys. Res. Lett.* 25, 143–146.

Tanaka, Y., E. C. Y. Inn, and K. Watanabe (1953) Absorption coeffcients of gases in the vacuum ultraviolet. Part IV. Ozone. *J. Chem Phys.* 21, 1651–1653.

Tang, I. N. (1980) On the equilibrium partial pressure of nitric acid and ammonia in the atmosphere. *Atmos. Environ.* 14, 819–828.

Tang, I. N., H. R. Munkelwitz, and J. G. Davis (1977) Aerosol growth studies. II. Preparation and growth measurements of monodisperse salt aerosols. *J. Aerosol Sci.* 8, 149–159.

Tang, K. Y., P. W. Fairchild, and E. C. K. Lee (1979) Laser-induced photodecomposition of formaldehyde $(\tilde{A}^1A_2)$ from its single vibronic levels. Determination of quantum yields of H-atom by HNO* chemiluminescence. *J. Phys. Chem.* 83, 569–573.

Tanner, R., W. H. Marlow, and L. Newman (1979) Chemical composition correlations of size-fractionated sulfate in New York City. *Environ. Sci. Technol.* 13, 75–78.

Tanner, R. L., T. J. Kelly, D. A. Dezaro, and J. Forrest (1989) A comparison of filter, denuder, and real-time chemiluminescence techniques for nitric acid determination in ambient air. *Atmos. Environ.* 23, 2213–2222.

Tans, P. (1981) A computation of bomb C-14 data for use in global carbon model calculations. In "Carbon Cycle Modelling" (B. Bolin, ed.). *SCOPE* 16, 131–157.

Tans, P. P., I. Y. Fung, and T. Takahashi (1990) Observational constraints on the global atmospheric CO_2 budget. *Science* 247, 1431–1438.

Taran, Y. A., J. W. Hedenquist, M. A. Korzhinsky, S. I. Tkachenko, and K. I. Shmulowich (1995) Geochemistry of magmatic gases from Kudryavy volcano, Iturup, Kuril Islands. *Geochim. Cosmochim. Acta* 59, 1749–1761.

Taylor, O. C. (1969) Importance of peroxyacetyl nitrate (PAN) as a phytotoxic air pollutant. *J. Air Poll. Control Assoc.* 19, 347–351.

Taylor, P. S., and R. E. Stoiber (1973) Soluble material on ash from active Central American volcanoes. *Geol. Soc. Am. Bull.* 84, 1031–1042.

Taylor, S. R., and S. M. McLennan (1995) The geochemical evolution of the continental crust. *Rev. Geophys.* 33, 241–265.

Telegadas, K. (1972) Atmospheric radioactivity along the HASL ground-levels sampling network, 1968–1970, as an indicator of tropospheric and stratospheric sources. *J. Geophys. Res.* 77, 1004–1011.

Telegadas, K., and J. London (1954) A physical model of the northern hemispheric troposhere for winter and summer, *Air Force Science Report* 19(122)-165. New York University.

Temple, P. J., and O. C. Taylor (1983) World-wide ambient measurements of peroxyacetyl nitrate (PAN) and implications for plant injury. *Atmos. Environ.* 17, 1583–1587.

Thomas, R. B., and B. R. Strain (1991) Root restriction as a factor in photosynthetic acclimation of cotton seedling grown in elevated carbon dioxide. *Plant Physiol.* 96, 627–634.

Thompson, A. M. (1980) Wet and dry removal of atmospheric formaldehyde at a coastal site. *Tellus* 32, 376–383.

Thorne, L., and G. P. Hanson (1972) Species differences in rates of vegetal ozone absorption. *Environ. Pollut.* 3, 303–312.

Thorneloe, S. A., M. A. Barlaz, R. Peer, L. C. Huff, L. Davies, and J. Mangino (1993) In "Atmospheric Methane: Sources, Sinks, and Role in Global Change" (M. A. K. Khalil, ed.), NATO ASI Series, Vol. I 13, pp. 362–398. Springer-Verlag, Berlin.

Thornton, D. C., A. R. Bandy, N. Beltz, A. R. Driedger, and R. Ferek (1993) Advection of sulfur dioxide over the western Atlantic Ocean during CITE 3. *J. Geophys. Res.* 98, 23,459–23,467.

Thornton, D. C., A. R. Bandy, B. W. Blomquist, and B. E. Anderson (1996a) Impact of anthropogenic and biogenic sources and sinks on carbonyl sulfide in the North Pacific troposphere. *J. Geophys. Res.* 101, 1873–1881.

Thornton, D. C., A. R. Bandy, B. W. Blomquist, D. D. Davis, and R. W. Talbot (1996b) Sulfur dioxide as a source of condensation nuclei in the upper troposphere of the Pacific Ocean. *J. Geophys. Res.* 101, 1883–1890.

Tiefenau, H. K., and P. Fabian (1972) The specific ozone destruction at the ocean surface and ist dependence on horizontal wind velocity from profile measurements. *Arch. Meteor. Geophys. Bioklimatol. Ser. A* 21, 399–412.

Tiefenau, H. K., P. G. Pruchniewicz, and P. Fabian (1973) Meridional distribution of tropospheric ozone from measurements aboard commercial airliners. *Pure Appl. Geophys.* 106–108, 1036–1040.

Tingey, D. T. (1981) The effect of environment factors on the emission of biogenic hydrocarbons from live oak and slash pine. *In* "Atmospheric Biogenic Hydrocarbons" (J. J. Bufalini and R. R. Arnts, eds.), Vol. 1, Emissions, pp. 53–79. Ann Arbor Science Publications, Ann Arbor, MI.

Tingey, D. T., M. Manning, L. C. Grothaus, and W. F. Burns (1979) The influence of light and temperature on isoprene emissions from life oak. *Physiol. Plant.* 47, 112–118.

Tingey, D. T., M. Manning, L. C. Grothaus, and W. F. Burns (1980) Influence of light and temperature on monoterpene emissions from slash pine. *Plant. Physiol.* 65, 797–801.

Titoff, A. (1903) Beiträge zur Kenntnis der negativen Katalyse im homogenen System. *Z. Phys. Chem.* 45, 641–683.

Tjepkema, J. D., R. J. Cartica, and H. F. Hemond (1981) Atmospheric concentration of ammonia in Massachusetts and deposition on vegetation. *Nature* 294, 445–446.

Toba, Y. (1965) On the giant sea-salt particles in the atmosphere. *Tellus* 17, 131–145.

Toba, Y., and M. Chaen (1973) Quantitative expression of the breaking of wind waves on the sea surface. *Rec. Oceanogr. Works Japan* 12, 1–11.

Tolbert, N. E., and I. Zelitch (1983) Carbon metabolism. *In* "CO_2 and Plants: The Response of Plants to Rising Levels of Atmospheric Carbons Dioxide" (E. R. Lemon, ed.), pp. 21–64, Westview Press, Boulder, CO.

Toohey, D. W., L. M. Avallone, L. R. Lait, P. A. Newman, M. R. Schoeberl, D. W. Fahey, E. L. Woodbridge, and J. G. Anderson (1993) The seasonal evolution of reactive chlorine in the northern hemisphere stratosphere. *Science* 261, 1134–1136.

Toon, G. C., C. B. Farmer, and R. H. Norton (1986a) Detection of stratospheric N_2O_5 by infrared remote sounding. *Nature* 319, 570–571.

Toon, O. B., P. Hamill, R. P. Turco, and J. Pinto (1986b) Condensation of HNO_3 and HCl in the winter polar stratospheres. *Geophys. Res. Lett.* 13, 1284–1287.

Toon, O. B., E. V. Browell, S. Kinne, and J. Jordan (1990) An analysis of lidar observations of polar stratospheric clouds. *Geophys. Res. Lett.* 17, 393–396.

Torres, A. L., and H. Buchan (1988) Tropospheric nitric oxide over the Amazon Basin. *J. Geophys. Res.* 93, 1396–1406.

Torres, A. L., and A. M. Thompson (1993) Nitric oxide in the equatorial Pacific boundary layer: SAGA 3 measurements. *J. Geophys. Res.* 98, 16,949–16,954.

Torres, A. L., P. J. Maroulis, A. B. Goldberg, and A. R. Bandy (1980) Atmospheric OCS measurements on Project GAMETAG. *J. Geophys. Res.* 85, 7357–7360.

Toumi, R., B. J. Kerridge, and J. A. Pyle (1991) Highly vibrationally excited oxygen as a potential source of ozone in the upper stratosphere and mesosphere. *Nature* 351, 217–219.

Trapp, D., and C. De Serves (1995) Intercomparison of formaldehyde measurements in the tropical atmosphere. *Atmos. Environ.* 29, 3239–3243.

Traub, W. A., D. G. Johnson, and K. V. Chance (1990) Stratospheric hydroperoxyl measurements. *Science* **247**, 446–449.

Trenberth, K. E. (1981) Seasonal variations in global sea level pressure and the total mass of the atmosphere. *J. Geophys. Res.* **86**, 5238–5246.

Trenberth, K. E. (ed.) (1992) "Climate System Modeling." Cambridge University Press, London.

Trenberth, K. E., and C. J. Guillemot (1994) The total mass of the atmosphere. *J. Geophys. Res.* **99**, 23,079–23,088.

Troe, J. (1974) Fall-off curves of unimolecular reactions. *Ber. Bunsen-Ges. Phys. Chem.* **78**, 478–488.

Trolier, M. R., and J. R. Wiesenfeld (1988) Relative quantum yield of $O(^1D_2)$ following ozone photolysis between 275 and 325 nm. *J. Geophys. Res.* **93**, 7119–7124.

Troxler, R. F., and J. M. Dokos (1973) Formation of carbon monoxide and bile pigment in red and green algae. *Plant Physiol.* **51**, 72–75.

Tsunogai, S. (1971a) Oxidation rate of sulfite in water and its bearing on the origin of sulfate in meteoritic precipitation. *Geochem. J.* **5**, 175–185.

Tsunogai, S. (1971b) Ammonia in the oceanic atmosphere and the cycle of nitrogen compounds through the atmosphere and hydrosphere. *Geochem. J.* **5**, 57–67.

Tsuya, H. (1955) Geological and petrological studies of volcano Fuji. *Tokyo Daigaku Jishin Kenkyusho Iho* **33**, 341–382.

Tuazon, E. C., and R. Atkinson (1989) A product study of the gas-phase reaction of methyl vinyl ketone with the OH radical in the presence of NO_x. *Int. J. Chem. Kinet.* **21**, 1141–1152.

Tuazon, E. C., and R. Atkinson (1990a) A product study of the gas-phase reaction of isoprene with the OH radical in the presence of NO_x. *Int. J. Chem. Kinet.* **22**, 1221–1236.

Tuazon, E. C., and R. Atkinson (1990b) A product study of the gas-phase reaction of methacrolein with the OH radical in the presence of NO_x. *Int. J. Chem. Kinet.* **22**, 591–602.

Tuazon, A. C., A. M. Winer, and J. N. Pitts, Jr. (1981) Trace pollutant concentrations in a multiday smog episode in the Californian south coast air basin by long path length Fourier transform infrared spectroscopy. *Environ. Sci. Technol.* **15**, 1232–1237.

Tuazon, E. C., H. MacLeod, R. Atkinson, and W. P. L. Carter (1986) α-Dicarbonyl yields from the NO_x-air photooxidation of a series of aromatic hydrocarbons in air. *Environ. Sci. Technol.* **20**, 383–387.

Tuazon, E. C., S. M. Aschmann, J. Arey, and R. Atkinson (1997) Products of the gas-phase reactions of O_3 with a series of methyl-substituted ethenes. *Environ. Sci. Technol.* **31**, 3004–3009.

Tuck, A. F. (1976) Production of nitrogen oxides by lightning discharges. *Q. J. R. Meteorol. Soc.* **102**, 749–755.

Tuck, A. F., R. T. Watson, E. P. Condon, J. J. Margitan, and O. B. Toon (1989) The planning and execution of ER-2 and DC-8 aircraft flights over Antarctica, August–September 1987. *J. Geophys. Res.* **94**, 11,181–11,222.

Tucker, B. J., P. J. Maroulis, and A. R. Bandy (1985) Free tropospheric measurements of CS_2 over a 45°N to 45°S latitude range. *Geophys. Res. Lett.* **12**, 9–11.

Tuckerman, M., R. Ackermann, C. Gölz, H. Lorenzen-Schmidt, T. Senne, J. Stutz, B. Trost, W. Unold, and U. Platt. (1997) DOAS observation of halogen radical-catalyzed Arctic boundary layer ozone destruction during ARCTOC campaigns 1995 and 1996 in Ny-Alesund, Spitsbergen. *Tellus* **49B**, 533–555.

Tuesday, C. S. (1976) Sources of atmospheric hydrocarbons. In "Vapor-Phase Organic Pollutants," pp. 4–42. National Academy of Sciences, Washington, DC.

Tuovinen, J. P., K. Barrett, and H. Styve (1994) "Transboundary acidifying pollution in Europe: calculated fields and budgets 1985–1993." UN Economic Commission for Europe (ECE) Cooperative Programme for Monitoring and Evaluation of the Long-Range Transmission of Air Pollutants in Europe, Oslo, Norway.

Turco, R. P., R. C. Whitten, O. B. Toon, J. B. Pollack, and P. Hamill (1980) OCS, stratospheric aerosols and climate. *Nature* **283**, 283–286.

Turco, R. P., R. J. Cicerone, E. C. Y. Inn, and L. A. Capone (1981) Long-wavelength carbonyl sulfide photodissociation. *J. Geophys. Res.* **86**, 5373–5377.

Turco, R. P., R. C. Whitten, and O. B. Toon (1982) Stratospheric aerosols: observations and theory. *Rev. Geophys. Space Phys.* **20**, 233–279.

Turco, R. P., O. B. Toon, and P. Hamill (1989) Heterogeneous physico-chemistry of the polar ozone hole. *J. Geophys. Res.* **94**, 16,493–16,510.

Turekian, K. K. (1971) Elements, geochemical distribution of. *In* "Encyclopedia of Science and Technology," 2nd ed., Vol. 4, pp. 627–630. McGraw-Hill, New York.

Turekian, K. K., Y. Nozaki, and L. K. Benninger (1977) Geochemistry of atmospheric radon and radon products. *Annu. Rev. Earth. Planet. Sci.* **5**, 227–255.

Turman, B. N., and B. C. Edgar (1982) Global lightning distribution at dawn and dusk. *J. Geophys. Res.* **87**, 1191–1206.

Turner, N. C., P. E. Waggoner, and S. Rich (1974) Removal of ozone from the atmosphere by soil and vegetation. *Nature* **250**, 486–489.

Turner, S. M., and P. S. Liss (1985) Measurements of various sulphur gases in a coastal marine environment. *J. Atmos. Chem.* **2**, 223–232.

Turner, S. M., G. Malin, P. S. Liss, D. S. Harbour, and P. M. Holligan (1988) The seasonal variation of dimethyl sulfide and dimethylsulfoniopropionate concentrations in nearshore waters. *Limnol. Oceanogr.* **33**, 364–375.

Turnipseed, A. A., and A. R. Ravishankara (1993) The atmospheric oxidation of dimethyl sulfide: elementary steps in a complex mechanism. *In* "Dimethylsulphide: Oceans, Atmosphere and Climate" (G. Restelli and G. Angeletti, eds.), pp. 185–195. Kluwer Academic Publishers, Dordrecht, the Netherlands.

Turnipseed, A. A., J. W. Birks, and J. G. Calvert (1991) Kinetics and temperature dependence of the BrO + ClO reaction. *J. Phys. Chem.* **95**, 4356–4364.

Turnipseed, A. A., G. L. Vaghjiani, J. E. Thompson, and A. R. Ravishankara (1992) Photodecomposition of HNO_3 at 193, 222, and 248 nm: products and quantum yields. *J. Chem. Phys.* **96**, 5887–5895.

Twenhofel, W. H. (1951) "Principles of Sedimentation." McGraw-Hill, New York.

Twomey, S. (1963) Measurements of natural cloud nuclei. *J. Rech. Atmos.* **1**, 101–104.

Twomey, S. (1971) The composition of cloud nuclei. *J. Atmos. Sci.* **28**, 377–381.

Twomey, S. (1977) "Atmospheric Aerosols." Elsevier, Amsterdam, the Netherlands.

Tyndall, G. S., and A. R. Ravishankara (1989a) Kinetics and mechanism of the reactions of CH_3S with O_2 and NO_2 at 298K. *J. Phys. Chem.* **93**, 2426–2435.

Tyndall, G. S., and A. R. Ravishankara (1989b) Kinetics of the reaction of CH_3S with O_3 at 298K. *J. Phys. Chem.* **93**, 4707–4710.

Tyndall, G. S., J. J. Orlando, and J. G. Calvert (1995) Upper limit for the rate coefficient for the reaction $HO_2 + NO_2 \rightarrow HONO + O_2$. *Environ. Sci. Technol.* **29**, 202–206.

Tyson, B. J., J. F. Vedder, J. C. Arvesen, and R. B. Brewer (1978a) Stratospheric measurements of $CFCl_3$ and N_2O. *Geophys. Res. Lett.* **5**, 369–372.

Tyson, B. J., J. C. Arvesen, and D. O'Hara (1978b) Interhemispheric gradients of CF_2Cl_2, $CFCl_3$, CCl_4, and N_2O. *Geophys. Res. Lett.* **5**, 535–538.

U.S. Standard Atmosphere (1976) National Oceanic and Atmospheric Administration, Washington, DC.

UNEP (1987) Montreal Protocol on Substances That May Deplete the Ozone Layer—Final Act. United Nations Environment Programme, Nairobi, Kenya.

United Nations (1978) "Yearbook of Industrial Statistics," 1976 ed., New York.

United Nations (1984) "Demographic Yearbook 1982." New York.

Urey, H. C. (1952) "The Planets." Yale University Press, New Haven, CT.

Vaghjiani, G. L., and A. R. Ravishankara (1990) Photodissociation of H_2O_2 and CH_3OOH at 248 nm and 298 K: quantum yields for OH, $O(^3P)$ and $H(^2S)$. J. Chem. Phys. 92, 996–1003.

Van Cleemput, O., and L. Baert (1976) Theoretical considerations on nitrite self-decomposition reactions in soils. Soil Sci. Soc. Am. J. 40, 322–323.

Van Cleemput, O., A. S. El-Sebaay, and L. Baert (1981) Production of gaseous hydrocarbons in soil. In "Physico-Chemical Behaviour of Atmospheric Pollutants" (B. Versino and H. Ott, eds.). Proceedings of the Third European Symposium, Varese, Italy, pp. 349–355. Reidel, Dordrecht, the Netherlands.

Van Dingenen, R., and F. Raes (1991) Determination of the condensation accommodation coefficient of H_2SO_4 on H_2SO_4-H_2O aerosol. J. Aerosol. Sci. Technol. 15, 93–106.

Van Eldik, R., N. Coichev, K. Bal Reddy, and A. Gerhard (1992) Metal ion catalyzed autoxidation of sulfur(IV) oxides: redox cycling of metal ions induced by sulfite. Ber. Bunsenges. Phys. Chem. 96, 478–481.

Van Leeuwen, E. P., G. P. J. Draaijers, and J. W. Erisman (1996) Mapping wet deposition of acidifying components and base cations over Europe using measurements. Atmos. Environ. 30, 2495–2511.

Van Sickle, D. E., T. Mill, F. R. Mayo, H. Richardson, and C. W. Gould (1973) Intramolecular propagation in the oxidation of n-alkanes. Autoxidation of n-pentane and n-octane. J. Org. Chem. 38, 4435–4440.

Van Vaeck, L., G. Brodddin, and K. van Cauwenberghe (1979) Differences in particle size distributions of major organic pollutants in ambient aerosols in urban, rural and seashore areas. Environ. Sci. Technol. 13, 1494–1502.

Van Velthoven, P. F. J., and H. Kelder (1996) Estimates of stratosphere-troposphere exchange: sensitivity to model formulation and horizontal resolution. J. Geophys. Res. 101, 1429–1434.

Várhelyi, G. (1978) On the vertical distribution of sulfur compounds in the lower troposphere. Tellus 30, 542–545.

Várhelyi, G. (1985) Continental and global sulfur budgets. I. Anthropogenic SO_2 emissions. Atmos. Environ. 19, 1029–1040.

Várhelyi, G., and G. Gravenhorst (1983) Production rate of airborne sea-salt sulfur deduced from chemical analysis of marine aerosols and precipitation. J. Geophys. Res. 88, 6737–6751.

Vasudev, R. (1990) Absorption spectrum and solar photodissociation of gaseous nitrous acid in the actinic wavelength region. Geophys. Res. Lett. 17, 2153–2165.

Vedder, J. F., B. J. Tyson, R. B. Brewer, C. A. Boitnott, and E. Y. C. Inn (1978) Lower stratospheric measurements of variation with latitude of CF_2Cl_2, $CFCl_3$, CCl_4, and N_2O profiles in the northern hemisphere. Geophys. Res. Lett. 5, 33–36.

Vedder, J. F., C. E. Inn, B. J. Tyson, C. A. Boitnott, and D. O'Hara (1981) Measurements of CF_2Cl_2, $CFCl_3$, and N_2O in the lower stratosphere between 2°S and 73°N latitude. J. Geophys. Res. 86, 7363–7368.

Vidal-Madjar, A. (1975) Evolution of solar Lyman-alpha flux during four consecutive years. Solar Phys. 40, 69–86.

Vidal-Madjar, A. (1977), The solar spectrum of Lyman α. In "The Solar Output and Its Variation" (O. R. White, ed.), pp. 213–234. Colorado Associated Press, Boulder, CO.

Viemeister, P. E. (1960) Lightning and the origin of nitrates found in precipitation. J. Meteoro. 17, 681–683.

Vierkorn-Rudolph, B., K. Bächmann, B. Schwarz, and F. X. Meixner (1984) Vertical profiles of hydrogen chloride in the troposphere. *J. Atmos. Chem.* **2**, 47–63.

Viggiano, A. A., and F. Arnold (1983) Stratospheric sulfuric acid vapor. *J. Geophys. Res.* **88**, 1457–1462.

Vinogradov, A. P. (1959) "The Geochemistry of Rare and Dispersed Elements in Soils," 2nd ed. Consultants Bureau, New York.

Visser, S. A. (1961) Chemical composition of rainwater in Kampala, Uganda, and its relation to meteorological and topographic conditions. *J. Geophys. Res.* **66**, 3759–3765.

Visser, S. A. (1964) Origin of nitrate in tropical rainwater. *Nature* **201**, 35–36.

Vogt, R., P. J. Crutzen, and R. Sander (1997) A mechanism for halogen release from sea-salt aerosol in the remote marine boundary layer. *Nature* **383**, 327–330.

Vohra, K. G., K. N. Vasuderan, and P. V. N. Nair (1970) Mechanism of nucleus-forming reactions in the atmosphere. *J. Geophys. Res.* **75**, 2951–2960.

Voldner, E. C., L. A. Barrie, and A. Sirois (1986) A literature review of dry deposition of oxides of sulphur and nitrogen with emphasis on long range transport modelling in North America. *Atmos. Environ.* **20**, 2101–2123.

Volman, D. H. (1963) Photochemical gas phase reactions in the hydrogen-oxygen system. *Adv. Photochem.* **1**, 43–82.

Volmer, M., and A. Weber (1926) Keimbildung in übersättigten Gebilden. *Z. Phys. Chem.* **119**, 277–301.

Volz, A., and D. Kley (1988) Evaluation of the Montsouris series of ozone measurements made in the nineteenth century. *Nature* **332**, 240–242.

Volz, A., D. H. Ehhalt, and R. G. Derwent (1981a) Seasonal and latitudinal variation of ^{14}CO and the tropospheric concentration of OH radicals. *J. Geophys. Res.* **86**, 5163–5171.

Volz, A., U. Schmidt, J. Rudolph, D. H. Ehhalt, F. J. Johnson, and A. Khedim (1981b) Vertical profiles of trace gases at mid-latitudes, Report no. 1742. Kernforschungsanlage, Jülich, Germany.

Von Piechowski, M. (1991) Der Einfluß von Kupferionen auf die Redoxchemie des atmosphärischen Wassers; kinetische Untersuchungen. Dissertation. Eidgenössische Technische Hochschule, Zürich, Switzerland.

Von Piechowski, M., T. Nauser, J. Hoigné, and R. E. Bühler (1993) O_2^- decay catalyzed by Cu^{2+} and Cu^+ ions in aqueous solutions: a pulse radiolysis study for atmospheric chemistry. *Ber. Bunsenges. Phys. Chem.* **97**, 762–771.

Vukovich, F. M., W. D. Bach, B. W. Crissman, and D. J. King (1977) On the relationship between high ozone in the rural surface layer and high pressure systems. *Atmos. Environ.* **11**, 967–983.

Wade, T. C., and J. G. Quinn (1974) Transfer processes to the marine environment. *In* "Pollutant Transfer to the Marine Environment" (R. A. Duce, P. L. Parker, and G. S. Giam, eds.), p. 25. University of Rhode Island, Kingston, RI.

Wade, T. C., and J. G. Quinn (1975) Hydrocarbons in the Sargasso Sea surface microlayer. *Marine Pollut. Bull.* **6**, 54–57.

Wadleigh, C. H. (1968) Wastes in relation to agriculture and forestry, Miscellaneous Publication no. 1065. Department of Agriculture, Washington, DC.

Wagener, K. (1978) Total anthropogenic CO_2 production during the period 1800–1935 from carbon-13 measurements in tree rings. *Radiat. Environ. Biophys.* **15**, 101–111.

Wagman, D. D., W. H. Evans, V. B. Parker, R. H. Schumm, I. Halow, S. M. Bailey, K. L. Churney, and R. L. Nuttal (1982) The NBS tables of chemical thermodynamic properties, selected values for inorganic and C_1 and C_2 organic substances in SI units. *J. Phys. Chem. Ref. Data* **11**, (Suppl. 2), 1–39.

Wahlen, M., N. Tanaka, R. Henry, B. Deck, J. Zeglen, J. S. Vogel, J. Southon, A. Shemesh, R. Fairbanks, and W. Broecker (1989) Carbon-14 in methane sources and in atmospheric methane: the contribution of fossil carbon. *Science* 245, 286–290.

Waksman, S. A. (1938) "Humus, Origin, Chemical Composition and Importance in Nature," 2nd ed. Williams and Wilkins, Baltimore, MD.

Walega, J. G., D. H. Stedman, R. E. Shetter, G. I. Mackay, T. Iguchi, and H. I. Schiff (1984) Comparison of a chemiluminescent and a tunable diode laser absorption technique for the measurement of nitrogen oxide, nitrogen dioxide, and nitric acid. *Environ. Sci. Technol.* 18, 823–826.

Walega, J. G., B. A. Ridley. S. Madronich, F. E. Grahek, J. D. Shetter, T. D. Sauvain, C. J. Hahn, J. T. Merrill, B. A. Bodhaine, and E. Robinson (1992) Observations of peroxyacetyl nitrate, peroxypropionyl nitrate, methyl nitrate and ozone during the Mauna Loa Observatory Photochemistry Experiment. *J. Geophys. Res.* 97, 10,311–10,330.

Walker, J. C. G. (1974) Stability of atmospheric oxygen. *Am. J. Sci.* 274, 193–214.

Walker, J. C. G. (1977) "Evolution of the Atmosphere." Macmillan, New York.

Wall, S. M., W. John, and J. L. Ondo (1988) Measurement of aerosol size distribution for nitrate and major ionic species. *Atmos. Environ.* 22, 1649–1656.

Wallington, J., M. D. Hurley, J. C. Ball, and M. E. Jenkin (1993) FTIR product study of the reaction $CH_3OCH_2O_2 + HO_2$. *Chem. Phys. Lett.* 211, 41–47.

Walter, H. (1973) Coagulation and size distribution of condensation aerosols. *J. Aerosol Sci.* 4, 1–15.

Walter, M. R. (ed.) (1976) "Stromatolites." Elsevier, Amsterdam, the Netherlands.

Walter, M. R., R. Buick, and J. S. R. Dunop (1980) Stromatolites 3400–3500 Myr old from the North Pole area, western Australia. *Nature* 284, 443–445.

Walton, A., M. Ergin, and D. D. Markness (1970) Carbon-14 concentrations in the atmosphere and carbon dioxide exchange rates. *J. Geophys. Res.* 75, 3089–3098.

Waltrop, A. S. K. Mitra, A. Flossmann, and H. R. Pruppacher (1991) On the scavenging of SO_2 by rain and cloud drops. IV. A wind tunnel and theoretical study of the absorption of SO_2 in the ppbv range by water drops containing H_2O_2. *J. Atmos. Chem.* 12, 1–17.

Wang, C. C., and L. I. Davis (1974) Ground- state population distribution of OH determined with a tunable laser. *Appl. Phys. Lett.* 25, 34–35.

Wang, C. C., L. I. Davis, P. M. Selzer, and R. Munoz (1981) Improved airborne measurements of OH in the atmosphere using the technique of LIF. *J. Geophys. Res.* 86, 1181–1186.

Wang, P. K., and H. R. Pruppacher (1977) An experimental determination of the efficiency with which aerosol particles are collected by water drops in sub-saturated air. *J. Atmos. Sci.* 34, 1664–1669.

Wang, P. K., S. N. Grover, and H. R. Pruppacher (1978) On the effect of electric charges on the scavenging of aerosol particles by clouds and small rain drops. *J. Atmos. Sci.* 35, 1735–1743.

Wang, T., M. A. Carroll, G. M. Albercock, K. R. Owens, K. A. Duderstadt, A. N. Markevitch, D. D. Parrish, J. S. Holloway, F. C. Fehsenfeld, G. Forbes, and J. Ogren (1996) Ground-based measurements of NO_x and total reactive oxidized nitrogen (NO_y) at Sable Island, Nova Scotia, during the NARE 1993 summer intensive. *J. Geophys. Res.* 101, 28,991–29,004.

Wang, Y., J. A. Logan, and D. J. Jacob (1998a) Global simulation of tropospheric O_3-NO_x-hydrocarbon chemistry. 2. Model evaluation and global ozone budget. *J. Geophys. Res.* 103, 10,727–10,755.

Wang, Y., J. A. Logan, and D. J. Jacob (1998b) Global simulation of tropospheric O_3-NO_x-hydrocarbon chemistry. 3. Origin of tropospheric ozone and effects of non-methane hydrocarbons. *J. Geophys. Res.* 103, 10,757–10,767.

Wängberg, I., I. Barnes, and K. H. Becker (1997) Product and mechanistic study of the reaction of NO_3 radicals with α-pinene. *Environ. Sci. Technol.* 31, 2130–2135.

Wanninkhof, R. (1992) Relationship between wind speed and gas exchange over the ocean. *J. Geophys. Res.* **97**, 7373–7382.

Ward, D. E., and C. C. Hardy (1991) Smoke emissions from wild-land fires. *Environ. Int.* **17**, 117–134.

Warmbt, W. (1964) "Luftchemische Untersuchungen des bodennahen Ozons 1952–1961, Methoden und Ergebnisse." Abhandlungen Meteorologischen Dienstes der Deutschen Demokratischen Republik, no. 72 (Vol. X). Akademie Verlag, Berlin.

Warneck, P. (1972) Cosmic radiation as a source of odd nitrogen in the stratosphere. *J. Geophys. Res.* **77**, 6589–6591.

Warneck, P. (1974) On the role of OH and HO_2 radicals in the troposphere. *Tellus* **26**, 39–46.

Warneck, P. (1975) OH production rates in the troposphere. *Planet. Space Sci.* **23**, 1507–1518.

Warneck, P. (1986) The equilibrium distribution of atmospheric gases between the two phases of liquid water clouds. *In* "Multiphase Atmospheric Chemistry" (W. Jaeschke, ed.), NATO ASI Series, Vol. G 6, pp. 473–499. Springer-Verlag, Berlin.

Warneck, P. (1988) "Chemistry of the Natural Atmosphere." Academic Press, San Diego, CA.

Warneck, P. (1989) Sulfur dioxide in rain clouds: gas-liquid scavenging efficiencies and wet deposition rates in the presence of formaldehyde. *J. Atmos. Chem.* **8**, 99–117.

Warneck, P. (1992) Chemistry and photochemistry in atmospheric water drops. *Ber. Bunsenges. Phys. Chem.* **96**, 454–460.

Warneck, P. (1994) Clouds, rain and aerosols. *In* "Low-Temperature Chemistry of the Atmosphere" (G. K. Moortgat, A. J. Barnes, G. Le Bras, J. R. Sodeau, eds.), NATO ASI Series, Vol. I 21, pp. 49–68. Springer-Verlag, Berlin.

Warneck, P., and J. Ziajka (1995) Reaction mechanism for the iron(III)-catalyzed autoxidation of bisulfite in aqueous solution: steady state description for benzene as radical scavenger. *Ber. Bunsenges. Phys. Chem.* **99**, 59–65.

Warneck, P., F. F. Marmo, and J. O. Sullivan (1964) Ultraviolet absorption of SO_2: dissociation energies of SO_2 and SO. *J. Chem. Phys.* **40**, 1132–1136.

Warneck, P., C. E. Junge, and W. Seiler (1973) OH radical concentrations in the stratosphere. *Pure Appl. Geophys.* **106–108**, 1417–1430.

Warneck, P., W. Klippel, and G. K. Moortgat (1978) Formaldehyd in troposphärischer Reinluft. *Ber. Bunsen-Ges. Phys. Chem.* **82**, 1136–1142.

Warneck, P., P. Mirabel, G. A. Salmon, R. van Eldik, C. Vinckier, K. J. Wannowius, and C. Zetzsch (1996) Review of activities and achievements of the EUROTRAC subproject HALIPP. *In* "Heterogeneous and Liquid Phase Processes" (P. Warneck, ed.), Transport and Chemical Transformation of Pollutants in the Troposphere, Vol. 2 (P. Borrell, P. M. Borrell, T. Cvitas, K. Kelly, and W. Seiler, series eds.), pp. 7–74. Springer-Verlag, Berlin.

Warren, S. G., C. J. Hahn, J. London, R. M. Chervin, and R. Jenne (1986) Global distribution of total cloud cover and cloud type amounts over land, NCAR Technical Note TN-273 + STR. Boulder, CO.

Washida, N., Y. Mori, and I. Tanaka (1971) Quantum yields of ozone formation from photolysis of oxygen molecule at 1849 and 1931 Å. *J. Chem. Phys.* **54**, 1119–1122.

Wasserburg, G. J., E. Mazor, and R. E. Zartman (1963) Isotopic and chemical composition of some terrestrial natural gases. *In* "Earth Science and Meteorites" (J. Geiss and E. D. Goldberg, eds.), pp. 219–240. North-Holland Publishing, Amsterdam, the Netherlands.

Watanabe, K., E. C. Y. Inn, and M. Zelikoff (1953) Absorption coefficients of oxygen in the vacuum ultraviolet. *J. Chem. Phys.* **21**, 1026–1030.

Waters, J. W., J. C. Hardy, R. F. Jarnot, and H. M. Pickett (1981) Chlorine monoxide radical, ozone and hydrogen peroxide: stratospheric measurements by microwave limb sounding. *Science* **214**, 61–64.

Waters, J. W., L. Froidevaux, W. G. Read, G. L. Manney, L. S. Elson, D. A. Flower, R. F. Jarnot, and R. S. Harwood (1993) Stratospheric ClO and ozone from the microwave limb sounder on the Upper Atmosphere Research Satellite. *Nature* **362**, 597–602.

Watson, E. R., and P. Lapins (1969) Losses of nitrogen from urine on soils from south-west Australia. *Austr. J. Exp. Agric. Anim. Husb.* **9**, 85–91.

Watson, J. J., J. A. Probert, and S. D. Piccot (1991) Global inventory of volatile organic compound emissions from anthropogenic sources, EPA-600/8-91-002 (NTIS PB91-161687). U.S. Environmental Protection Agency, Research Triangle Park, NC.

Watson, R. T., L. G. Meira Filho, E. Sanhueza, and A. Janetos (1992) Sources and sinks. *In* "Climate Change 1992. The Supplementary Report to the IPCC Scientific Assessment" (J. T. Houghton, B. A. Callander, and S. K. Varney, eds.), pp. 25–46. Cambridge University Press, New York.

Wayne, L. G., and J. G. Romanofsky (1961) Rates of reaction of oxides of nitrogen in the photooxidation of diluted automobile exhaust gases. *In* "Chemical Reactions in the Lower and Upper Atmosphere" (R. D. Cadle, ed.), pp. 71–86. Wiley Interscience, New York.

Wayne, R. P. (1987) The photochemistry of ozone. *Atmos. Environ.* **21**, 1683–1694.

Wayne, R. P., I. Barnes, P. Biggs, J. P. Burrows, C. E. Canosa-Mas, J. Hjorth, G. Le Bras, G. K. Moortgat, D. Perner, G. Poulet, G. Restelli, and H. Sidebottom (1991) The nitrate radical: physics, chemistry, and the atmosphere. *Atmos. Environ.* **25A**, 1–203.

Weber, E. (1969) Stand und Ziel der Grundlagenforschung bei der Nassentstaubung. *Staub* **29**, 272–277.

Weber, R. J., J. J. Marti, P. H. McMurry, F. L. Eisele, D. J. Tanner, and A. Jefferson (1997) Measurements of new particle formation and ultrafine particle growth rates at a clean continental site. *J. Geophys. Res.* **102**, 4375–4385.

Webster, C. R., R. D. May, D. W. Toohey, L. M. Avallone, J. G. Anderson, P. A. Newman, L. R. Lait, M. R. Schoeberl, J. W. Elkins, and K. R. Chan (1993) Chlorine chemistry on polar stratospheric cloud particles in the Arctic winter. *Science* **261**, 1130–1134.

Wege, K., H. Claude, and R. Hartmannsgruber (1989) Several results from the 20 years of ozone observations at Hohenpeissenberg. *In* "Ozone in the Atmosphere" (D. Bojkov and P. Fabian, eds.), pp. 109–112. Deepak, Hampton, VA.

Wegener, A. (1911) "Thermodynamik der Atmosphäre." Barth, Leipzig.

Weickmann, H. (1957) Recent measurements of the vertical distribution of Aitken nuclei. *In* "Artifical Stimulation of Rain" (Weickmann and W. Smith, eds.). Proceedings of the First Conference on the Physics of Cloud Precipitation Particles, pp. 81–88. Pergamon, New York.

Weinstein-Lloyd, J. B., and S. E. Schwartz (1992) Free radical reactions in cloud water: the role of transition metals in hydrogen peroxide production and destruction. *In* "Precipitation Scavenging and Atmosphere Surface Exchange" (S. E. Schwartz and W. G. N. Slinn, eds.), pp. 161–175. Hemisphere Publishing, Washington, DC.

Weinstock, B. (1969) Carbon monoxide residence time in the atmosphere. *Science* **166**, 224–225.

Weinstock, B. (1971) Discussion remark. *In* "Chemical Reactions in Urban Atmospheres" (C. S. Tuesday, ed.), pp. 54–55, American Elsevier, New York.

Weinstock, E. M., M. J. Phillips, and J. G. Anderson (1981) *In situ* observations of ClO in the stratosphere: a review of recent results. *J. Geophys. Res.* **86**, 7273–7278.

Weiss, P. S., J. E. Johnson, R. E. Gammon, and T. S. Bates (1995) Reevaluation of the open ocean source of carbonyl sulfide to the atmosphere. *J. Geophys. Res.* **100**, 23,083–23,092.

Weiss, R. F. (1981) The temporal and spatial distribution of tropospheric nitrous oxide. *J. Geophys. Res.* **86**, 7185–7195.

Weiss, W., A. Sittkus, H. Stockburger, and H. Sartorius (1983) Large-scale atmospheric mixing derived from meridional profiles of krypton 85. *J. Geophys. Res.* **88**, 8574–8578.

Welge, K. H. (1974) Photolysis of O_x, HO_x, CO_x, and SO_x compounds. *Can. J. Chem.* **52**, 1424–1435.

Wendel, G. J., D. H. Stedman, C. A. Cantrell, and L. Damrauer (1983) Luminol-based nitrogen dioxide detector. *Anal. Chem.* **55**, 937–940.

Went, F. W. (1960a) Blue hazes in the atmosphere. *Nature* **187**, 641–643.

Went, F. W. (1960b) Organic matter in the atmosphere and its possible relation to petroleum formation. *Proc. Natl. Acad. Sci. USA* **46**, 212–221.

Went, F. W. (1964) The nature of Aitken nuclei in the atmosphere. *Proc. Natl. Acad. Sci. USA* **51**, 1259–1267.

Went, F. W. (1966) On the nature of Aitken condensation nuclei. *Tellus* **18**, 549–556.

Weschler, C. J. (1981) Identification of selected organics in the Arctic aerosol. *Atmos. Environ.* **15**, 1365–1369.

Weseley, M. L., B. B. Hicks, W. P. Dannevick, S. Frissela, and R. B. Husar (1977) An eddy correlation measurement of particle deposition from the atmosphere. *Atmos. Environ.* **16**, 815–820.

Weseley, M. L., J. A. Eastman, D. H. Stedman, and E. D. Yalvac (1982) An eddy-correlation measurement of NO_2 flux to vegetation and comparison to O_3 flux. *Atmos. Environ.* **16**, 815–820.

Westberg, H. H. (1981) Biogenic hydrocarbon measurements. *In* "Atmospheric Biogenic Hydrocarbons" (J. J. Bufalini and R. R. Arnts, eds.), Vol. 2, Ambient Concentrations and Atmospheric Chemistry, pp. 25–49. Ann Arbor Science Publications, Ann Arbor, MI.

Wetherill, G. W. (1990) Formation of the Earth. *Annu. Rev. Earth Planet. Sci.* **18**, 205–256.

Wetherill, G. W. (1994) Provenance of the terrestrial planets. *Geochim. Cosmochim. Acta* **58**, 4513–4520.

Whelpdale, D. M., and L. A. Barrie (1982) Atmospheric monitoring network operations and results in Canada. *Water Air Soil Pollut.* **18**, 7–23.

Whelpdale, D. M., and M. S. Kaiser (eds.) (1996) Global acid deposition assessment. *WMO Global Atmospheric Watch* **106**.

Whitby, K. T. (1976) Electrical measurement of aerosols. *In* "Fine Particles, Aerosol Generation, Measurement, Sampling and Analysis" (B. Y. H. Liu, ed.), pp. 581–624. Academic Press, New York.

Whitby, K. T. (1978) The physical characteristics of sulfur aerosols. *Atmos. Environ.* **12**, 135–159.

Whitby, K. T., and G. M. Sverdrup (1980) California aerosols: their physical and chemical characteristics. *In* "The Character and Origin of Smog Aerosols" (G. M. Hidy, P. K. Mueller, D. Grosjean, B. R. Appel, and J. J. Wesolovski, eds.). *Adv. Environ. Sci. Technol.* **9**, 477–525.

Whitby, R. A., and P. E. Coffey (1977) Measurement of terpenes and other organics in an Adirondack mountain pine forest. *J. Geophys. Res.* **82**, 5928–5934.

White, D. E., J. D. Hem, and G. A. Waring (1963) Chemical composition of subsurface waters. *In* "Data of Geochemistry," 6th ed. (F. Chap, ed.), U.S. Geological Survey Professional Paper 440-F.

Whitlaw-Gray, R., and H. S. Patterson (1932) "Smoke: A Study of Aerial Disperse Systems." E. Arnold, London.

Whittaker, R. H., and G. E. Likens (1973) Carbon in the biota. *In* "Carbon and the Biosphere" (G. M. Woodwell and E. V. Pecan, eds.), AEC Symposium Series 30, pp. 281–302, NTIS. U.S. Deptartment Commerce, Springfield, VA.

Whittaker, R. H., and G. E. Likens (1975) The biosphere and man. *In* "Primary Productivity of the Biosphere" (H. Lieth and R. H. Whittaker, eds.). *Ecol. Stud.* **14**, 305–328.

Whung, P.-Y., E. S. Saltzman, M. J. Spencer, P. A. Mayewski, and N. Gundestrup (1994) Two-hundred-year record of biogenic sulfur in a south Greenland ie core (20D). *J. Geophys. Res.* 99, 1147–1156.

Wiebe, H. A., K. G. Anlauf, E. C. Tuazon, A. M. Winer, H. W. Biermann, B. R. Appel, P. A. Solomon, G. R. Cass, T. G. Ellestad, K. T. Knapp, E. Peake, W. Spicer, and D. R. Lawson (1990) A comparison of measurements of atmospheric ammonia by filter packs, transition-flow reactors, simple annular denuders and Fourier transform infrared spectroscopy. *Atmos. Environ.* 24A, 1019–1028.

Wilcox, R. W., G. D. Nastrom, and A. D. Belmont (1977) Periodic variations of total ozone and its vertical column. *J. Appl. Meterol.* 16, 290–298.

Wilhelm E., R. Battino, and R. J. Wilcock (1977) Low pressure solubilities of gases in liquid water. *Chem. Rev.* 77, 219–262.

Wilkening, M. H. (1970) Radon 222 concentrations in the convective patterns of a mountain environment. *J. Geophys Res.* 75, 1733–1740.

Wilkins, E. T. (1954) Air pollution and the London fog of December 1952. *J. R. Sanit. Inst.* 74, 1–21.

Wilkness, P. E., and D. J. Bressan (1972) Fractionation of the elements F, Cl, Na, and K at the sea-air interface. *J. Geophys. Res.* 77, 5307–5315.

Wilkness, P. E., R. A. Lamontagne, R. E. Larson, J. W. Swinnerton, C. R. Dickson, and T. Thompson (1973) Atmospheric trace gases in the southern hemisphere. *Nature* 245, 45–47.

Wilks, S. S. (1959) Carbon monoxide in green plants. *Science* 129, 964–966.

Willeke, K., and P. A. Baron (eds.) (1993) "Aerosol Measurement: Principles, Techniques, and Applications." Van Nostrand-Reinhold, New York.

Willeke, K., K. T. Whiby, W. E. Clark, and V. A. Marple (1974) Size distribution of Denver aerosols—a comparison of two sites. *Atmos. Environ.* 8, 609–633.

Williams, E. J., and F. C. Fehsenfeld (1991) Measurement of soil nitrogen oxide emissions at three North American ecosystems. *J. Geophys. Res.* 96, 1033–1042.

Williams, E. J., D. D. Parrish, and F. C. Fehsenfeld (1987) Determination of nitrogen dioxide emissions from soils: results from a grassland site in Colorado, United States. *J. Geophys. Res.* 92, 2173–2179.

Williams, E. J., G. L. Hutchinson, and F. C. Fehsenfeld (1992a) NO_x and N_2O emission from soil. *Global Biogeochem. Cycles* 6, 351–388.

Williams, E. J., A. Guenther, and F. C. Fehsenfeld (1992b) An inventory of nitric oxide emissions from soils in the United States. *J. Geophys. Res.* 97, 7511–7519.

Williams, E. J., J. M. Roberts, K. Baumann, S. B. Bertman, S. Buhr, R. B. Norton, and F. C. Fehsenfeld (1997) Variations in NO_y composition at Idaho Hill, Colorado. *J. Geophys. Res.* 102, 6297–6314.

Williams, P. J. le B. (1975) Biological and chemical aspects of dissolved organic material in sea water. In "Chemical Oceanography," 2nd ed. (J. P. Riley and G. Skirrow, eds.), Vol. 2, pp. 301–363. Academic Press, London.

Williams, P. M., H. Oeschger, and P. Kinney (1969) Natural radiocarbon activity of the dissolved organic carbon in the north-east Pacific Ocean. *Nature* 224, 256–258.

Williams, W. J., J. J. Kostus, A. Goldman, and D. G. Murcray (1976) Measurement of the stratospheric mixing ratio of HCl using an infrared absorption technique. *Geophys. Res. Lett.* 3, 383–385.

Wilson, T. R. S. (1975) Salinity and major elements of sea water. In "Chemical Oceanography" 2nd ed. (J. P. Riley and G. Skirrow, eds.), Vol. 1, pp. 365–413. Academic Press, London.

Wilson, W. E., Jr. (1981) Sulfate formation in point source plumes: a review of recent field studies. *Atmos. Environ.* 15, 2573–2581.

Wilson, W. E., Jr., D. F. Miller, A. Levy, and R. K. Stone (1973) The effect of fuel composition on atmospheric aerosol due to auto exhaust. *J. Air Pollut. Control Assoc.* **23**, 949–956.

Winchester, J. W., and G. D. Nifong (1971) Water pollution in Lake Michigan by trace elements from pollution fall-out. *Water Air Soil Pollut.* **1**, 50–64.

Windom, H. L. (1969) Atmospheric dust records in permanent snow fields, implications to marine sedimentation. *Bull. Geol. Soc. Am.* **80**, 761–782.

Wine, P. H., Y. Tang, R. P. Thorn, J. R. Wells, and D. D. Davis (1989) Kinetics of aqueous phase reactions of the SO_4^- radical with potential importance in cloud chemistry. *J. Geophys. Res.* **94**, 1085–1094.

Winer, A. M., K. R. Darnall, R. Atkinson, and J. N. Pitts, Jr. (1979) Smog chamber study of the correlation of hydroxyl radical rate constants with ozone formation. *Environ. Sci. Technol.* **13**, 822–826.

Winer, A. M., J. Arey, R. Atkinson, S. M. Aschmann, W. D. Long, C. L. Morrison, and D. M. Olszyk (1992) Emission rate of organics from vegetation in California's central valley. *Atmos. Environ.* **26A**, 2647–2659.

Winiwarter, W., H. Fierlinger, H. Puxbaum, M. C. Facchini, B. G. Arends, S. Fuzzi, D. Schell, U. Kaminski, S. Pahl, T. Schneider, A. Berner, I. Solly, and C. Kruisz (1994) Henry's law and the behavior of weak acids and bases in fog and cloud. *J. Atmos. Chem.* **19**, 173–188.

Winkler, P. (1973) The growth of atmospheric particles as a function of relative humidity, Part II. Improved concept of mixed nuclei. *J. Aerosol Sci.* **4**, 373–387.

Winkler, P. (1974) Die relative Zusammensetzung des atmosphärischen Aerosols in Stoffgruppen. *Meteorol. Rundsch.* **27**, 129–136.

Winkler, P. (1988) Surface ozone over the Atlantic Ocean. *J. Atmos. Chem.* **7**, 73–91.

Winkler, P., and C. E. Junge (1972) The growth of atmospheric aerosol particles a function of the relative humidity, Part I. Method and measurements at different locations. *J. Rech. Atmos.* **6**, 617–638.

Wlotzka, F. (1972) Nitrogen. *In* "Handbook of Geochemistry" (K. H. Wedepohl, ed.). Springer-Verlag, Berlin and New York.

WMO (1986) "Atmospheric Ozone 1985." Global Ozone Research and Monitoring Project, Report No. 16. World Meteorological Organization, Geneva, Switzerland.

WMO (1990) "Scientific Assessment of Ozone Depletion: 1989," Global Ozone Research and Monitoring Project, Report No. 20. World Meteorological Organization, Geneva, Switzerland.

WMO (1992) "Scientific Assessment of Ozone Depletion: 1991," Global Ozone Research and Monitoring Project, Report No. 25. World Meteorological Organization, Geneva, Switzerland.

WMO (1995) "Scientific Assessment of Ozone Depletion: 1994" (C. A. Ennis, coordinating ed.), Global Ozone Research and Monitoring Project, Report No. 37. World Meteorological Organization, Geneva, Switzerland.

Wobrock, W., D. Schell, R. Maser, W. Jaeschke, H.-W. Georgii, W. Wieprecht, B. G. Arends, J. J. Möls, G. P. A. Kos, S. Fuzzi, M. C. Facchini, G. Orsi, A. Berner, I. Solly, C. Kruisz, I. B. Svenningsson, A. Wiedensohler, H.-C. Hansen, J. A. Ogren, K. J. Noone, A. Hallberg, S. Pahl, T. Schneider, P. Winkler, W. Winiwarter, R. N. Colvile, T. W. Choularton, A. I. Flossmann, and S. Borrmann (1994) The Kleiner Feldberg cloud experiment 1990, an overview. *J. Atmos. Chem.* **19**, 1–35.

Wofsy, S. C. (1978) Temporal and latitudinal variation of stratospheric trace gases: a critical comparison between theory and experiment. *J. Geophys. Res.* **83**, 364–378.

Wofsy, S. C., and M. B. McElroy (1973) On vertical mixing in the upper stratosphere and mesosphere. *J. Geophys. Res.* **78**, 2619–2624.

Wofsy, S. C., and J. A. Logan (1982) Recent developments in stratospheric photochemistry. In "Causes and Effects of Stratospheric Ozone Reduction: An Update" (National Research Council), pp. 167–205. National Academy Press, Washington, DC.

Wofsy, S. C., J. C. McConnell, and M. B. McElroy (1972) Atmospheric CH_4, CO and CO_2. J. Geophys. Res. 77, 4477–4493.

Wofsy, S. C., M. B. McElroy, and Y. L. Yung (1975) The chemistry of atmospheric bromine. Geophys. Res. Lett. 2, 215–218.

Wojcik, G. S., and J. S. Chang (1997) A re-evaluation of sulfur budgets, lifetimes, and scavenging ratios for eastern North America. J. Atmos. Chem. 26, 109–145.

Wolff, G. T. (1984) On the nature of nitrate in coarse continental aerosols. Atmos. Environ. 18, 977–981.

Wolz, G., and H. W. Georgii (1996) Large scale distribution of sulfur dioxide results from the TROPOZ II flights. Időjárás (J. Hung. Meteorol. Serv.) 100, 23–41.

Wood, J. M. (1974) Biological cycles for toxic elements in the environment. Science 183, 1049–1052.

Wood, T. G., and W. A. Sands (1978) The role of termites in ecosystems. In "Production Ecology of Ants and Termites" (M. V. Brian, ed.), pp. 245–292. Cambridge University Press, London.

Woodcock, A. H. (1953) Salt nuclei in marine air as a function of altitude and wind force. J. Meteorol. 10, 362–371.

Woodcock, A. H. (1972) Small salt particles in oceanic air and bubble behavior in the sea. J. Geophys. Res. 77, 5316–5321.

Woodcock, A. H., D. C. Blanchard, and C. G. H. Rooth (1963) Salt-induced convection and clouds. J. Atmos. Sci. 20, 159–169.

Wu, C. H., S. M. Japar, and H. Niki (1976) Relative reactivities of OH-hydrocarbon reactions from smog reactions studies. J. Environ. Health (Part A) Environ. Sci. Eng. A 11, 191–200.

Wu, J. (1979) Sea spray in the atmospheric surface layer: review and analysis of laboratory and oceanic results. J. Geophys. Res. 84, 1683–1704.

Wulf, O. R., and L. S. Deming (1937) The distribution of atmospheric ozone in equilibrium with solar radiation and the rate of maintenance of the distribution. Terr. Magn. Atmos. Electr. 42, 195–202.

Wuosmaa, A. M., and L. P. Hager (1990) Methyl chloride transferase: a carbocation route for biosynthesis of halometabolites. Science 249, 160–162.

Wyers, G. P., R. P. Otjes, and J. Slanina (1993) A continuous-flow denuder for the measurement of ambient concentrations and surface exchange fluxes of ammonia. Atmos. Environ. 27A, 2085–2090.

Wyslouzil, B. E., J. H. Seinfeld, R. C. Flagan, and K. Okuyama (1991) Binary nucleation in acid-water systems. II. Sulfuric acid-water and a comparison with methanesulfonic acid-water. J. Chem. Phys. 94, 6842–6850.

Xue, H., M. D. L. S. Goncalves, M. Reutlinger, L. Sigg, and W. Stumm (1991) Copper(I) in fog water: determination and interaction with sulfite. Environ. Sci. Technol. 25, 1716–1722.

Yang, H., E. Olaguer, and K. K. Tung (1991) Simulation of the present-day atmospheric ozone, odd nitrogen, chlorine and other species using a coupled 2-D model in isentropic coordinates. J. Atmos. Sci. 48, 442–471.

Ye, R. W., B. A. Averill, and J. M. Tiedje (1994) Denitrification—production and consumption of nitric oxide. Appl. Environ. Microbiol. 60, 1053–1058.

Yienger, J. J., and H. Levy (1995) Empirical model of global soil-biogenic NO_x emissions. J. Geophys. Res. 100, 11,447–11,464.

Yin, F., D. Grosjean, and J. H. Seinfeld (1990a) Photooxidation of dimethyl sulfide and dimethyl disulfide. I. Mechanism development. J. Atmos. Chem. 11, 309–364.

Yin, F., D. Grosjean, R. C. Flagan, J. H. Seinfeld (1990b) Photooxidation of dimethyl sulfide and dimethyl disulfide. II. Mechanism evaluation. *J. Atmos. Chem.* 11, 365–399.

Yokouchi, Y., and Y. Ambe (1985) Aerosol formed from the chemical reaction of monoterpenes and ozone. *Atmos. Environ.* 19, 1271–1276.

Yokouchi, Y., and Y. Ambe (1986) Characterization of polar organics in airborne particulate matter. *Atmos. Environ.* 20, 1727–1734.

Yokouchi, Y., M. Okaniwa, Y. Ambe, and K. Fuwa (1983) Seasonal variation of monoterpenes in the atmosphere of a pine forest. *Atmos. Environ.* 17, 743–750.

Yokoyama, T., S. Nishinomiya, and H. Matsuda (1991) N_2O emissions from fossil fuel fired power plants. *Environ. Sci. Technol.* 25, 347–348.

Yoshida, N., H. Morimoto, M. Hirano, I. Koike, S. Matsuo, E. Wada, T. Saino, and A. Hattori (1989) Nitrification rates and ^{15}N abundances of N_2O and NO_3^- in the western North Pacific. *Nature* 342, 895–897.

Yoshinari, T. (1976) Nitrous oxide in the sea. *Marine Chem.* 4, 189–202.

Yoshino, K., D. E. Freeman, J. R. Esmond, and W. H. Parkinson (1987) High resolution absorption cross sections and band oscillator strength of the Schumann-Runge bands of oxygen at 79 K. *Planet. Space Sci.* 35, 1067–1075.

Yoshino, K., S.-C. Cheung, J. R. Esmond, W. H. Parkinson, D. E. Freeman, and S. L. Guberman (1988) Improved absorption cross sections of oxygen in the wavelength region 205–240 nm of the Herzberg continuum. *Planet. Space Sci.* 36, 1469–1475.

Yoshino, K., J. R. Esmond, A. S.-C. Cheung, D. E. Freeman, and W. H. Parkinson (1992) High resolution absorption cross sections in the transmission window region of the Schumann-Runge bands and Herzberg continuum of O_2. *Planet. Space Sci.* 40, 185–192.

Young, A. J., and A. W. Fairhall (1968) Radiocarbon from nuclear weapons tests. *J. Geophys. Res.* 73, 1185–1200.

Yue, G. K., and P. Hamill (1979) The homogeneous nucleation rates of $H_2SO_4 \cdot H_2O$ particles in the air. *J. Aerosol. Sci.* 10, 609–614.

Yvon, S. A., and J. H. Butler (1996) An improved estimate of the oceanic life time of atmospheric CH_3Br. *Geophys. Res. Lett.* 23, 53–56.

Yvon, S. A., and E. S. Saltzman (1996) Atmospheric sulfur cycling in the tropical Pacific marine boundary layer (12°S, 135°W): a comparison of field data and model results. 2. Sulfur dioxide. *J. Geophys. Res.* 101, 6911–6918.

Yvon, S. A., E. S. Saltzman, D. J. Cooper, T. S. Bates, and A. M. Thompson (1996) Atmospheric sulfur cycling in the tropical Pacific marine boundary layer (12°S, 135°W): a comparison of field data and model results. 1. Dimethylsulfide. *J. Geophys. Res.* 101, 6899–6909.

Zabel, F., A. Reimer, K. H. Becker, and E. H. Fink (1989) Thermal decomposition of alkyl peroxy nitrates. *J. Phys. Chem.* 93, 5500–5507.

Zafiriou, O. C. (1974) Photochemistry of halogens in the marine atmosphere. *J. Geophys. Res.* 79, 2730–2732.

Zafiriou, O. C. (1975) Reaction of methyl halides with sea water and marine aerosols. *J. Marine Res.* 33, 75–81.

Zafiriou, O. C., and M. McFarland (1981) Nitric oxide from nitrite photolysis from the central equatorial Pacific. *J. Geophys. Res.* 86, 3173–3182.

Zafiriou, O. C., J. Alford, M. Herrera, E. T. Peltzer, and R. B. Gagosian (1980) Formaldehyde in remote marine air and rain: flux measurements and estimates. *Geophys. Res. Lett.* 7, 341–344.

Zander, R. (1981) Recent observations of HCl and HF in the upper stratosphere. *Geophys. Res. Lett.* 8, 413–416.

Zander, R., M. R. Gunson, C. B. Farmer, C. P. Rinsland, F. W. Irion, and E. Mahieu (1992) The 1985 chlorine and fluorine inventories in the stratosphere based on ATMOS observations at 30° north latitude. *J. Atmos. Chem.* 15, 171–186.

Zartman, R. E., G. J. Wasserburg, and J. H. Reynolds (1961) Helium, argon and carbon in some natural gases. *J. Geophys. Res.* 66, 277–306.

Zeldovich, Y. B. (1946) Oxidation of nitrogen in combustion and explosion. *C. R. Acad. Sci. USSR* 51, 217–220.

Zellner, R. (1978) Recombination reactions in atmospheric chemistry. *Ber. Bunsenges. Phys. Chem.* 82, 1172–1179.

Zellner, R., A. Hoffmann, W. Malms, and V. Mörs (1997) Time-resolved studies of the NO_2 and OH formation in the integrated oxidation chain of hydrocarbons. In "Chemical Processes in Atmospheric Oxidation" (G. Le Bras, ed.), Transport and Chemical Transformation of Pollutants in the Troposphere, Vol. 3 (P. Borrell, P. M. Borrell, T. Cvitas, K. Kelly, and W. Seiler, series eds.), pp. 241–216. Springer-Verlag, Berlin.

Zepp, R. G., and M. O. Andreae (1994) Factors affecting the photochemical production of carbonyl sulfide in sea water. *Geophys. Res. Lett.* 21, 2813–2816.

Zettlemoyer, A. C. (ed.) (1969) "Nucleation." Marcel Dekker, New York.

Zettwoog, P., and R. Haulet (1978) Experimental results on the SO_2 transfer in the Mediterranean obtained with remote sensing devices. *Atmos. Environ.* 12, 795–796.

Zetzsch, C. (1994) Atmospheric oxidation processes of aromatics studied within LACTOZ and STEP. In "Physico-Chemical Behaviour of Atmospheric Pollutants" (G. Angeletti and G. Restelli, eds.), Proceedings of the Sixth European Symposium Varese, 1993, Vol. 1, pp. 118–136. European Commission, Brussels-Luxembourg.

Zetzsch, C., R. Koch, B. Bohn, R. Knispel, M. Siese, and F. Witte (1997) Adduct formation of OH with aromatics and unsaturated hydrocarbons and consecutive reactions with O_2 and NO_x to regenerate OH. In "Chemical Processes in Atmospheric Oxidation" (G. Le Bras, ed.), Transport and Chemical Transformation of Pollutants in the Troposphere, Vol. 3 (P. Borrell, P. M. Borrell, T. Cvitas, K. Kelly, and W. Seiler, series eds.), pp. 247–256. Springer-Verlag, Berlin.

Zhang, R., P. J. Wooldridge, J. P. D. Abbatt, and M. J. Molina (1993a) Physical chemistry of the H_2SO_4/H_2O binary system at low temperatures: stratospheric implications. *J. Phys. Chem.* 97, 7351–7358.

Zhang, R., P. J. Wooldridge, and M. J. Molina (1993b) Vapor pressure measurements for the $H_2SO_4/HNO_3/H_2O$ and $H_2SO_4/HCl/H_2O$ systems. Incorporation of stratospheric acids into background sulfate aerosols. *J. Phys. Chem.* 97, 8541–8548.

Zhang, W. Q., P. H. McMurry, S. V. Hering, and G. S. Casuccio (1993c) Mixing characteristics and water content of submicron aerosols measured in Los Angeles and at the Grand Canyon. *Atmos. Environ.* 27A, 1593–1608.

Zhuang, G., Z. Yi, R. A. Duce, and P. R. Brown (1992) Link between iron and sulfur suggested by detection of Fe(II) in remote marine aerosols. *Nature* 355, 537–539.

Ziajka, J., F. Beer, and P. Warneck (1994) Iron-catalysed oxidation of bisulphite in aqueous solution: evidence for a free radical chain mechanism. *Atmos. Environ.* 28, 2549–2552.

Ziereis, H., and F. Arnold (1986) Gaseous ammonia and ammonium ions in the free troposphere. *Nature* 321, 503–505.

Zimen, K. E., and F. K. Altenhein (1973) The future burden of industrial CO_2 on the atmosphere and the oceans. *Z. Naturforsch.* 28a, 1747–1752.

Zimen, K. E., P. Offermann, and G. Hartmann (1977) Source functions of CO_2 and future CO_2 burden in the atmosphere. *Z. Naturforsch.* 32a, 1544–1554.

Zimin, A. G. (1964) Mechanism of capture and precipitation of atmospheric contaminants by clouds and precipitation. *In* "Problems of Nuclear Meterology" (I. L. Karol and S. G. Malakhov, eds.), Rep. AE-tr-6128, pp. 139–182, U.S. Atomic Energy Commission, New York.

Zimmerman, P. R., R. B. Chatfield, J. Fishman, P. J. Crutzen, and P. L. Hanst (1978) Estimates of the production of CO and H_2 from the oxidation of hydrocarbon emissions from vegetation. *Geophys. Res. Lett.* **5**, 679–682.

Zimmerman, P. R., J. P. Greenberg, S. O. Wandiga, and P. J. Crutzen (1982) Termites: a potentially large source of atmospheric methane, carbon dioxide and molecular hydrogen. *Science* **218**, 563–565.

Zoller, W. H., E. S. Gladney, and R. A. Duce (1974) Atmospheric concentrations and sources of trace elements at the South Pole. *Science* **183**, 198–200.

Zumft, W. G. (1993) The biological role of nitric oxide in bacteria. *Arch. Microbiol.* **160**, 253–264.

Zuo, Y., and J. Hoigné (1992) Formation of hydrogen peroxide and depletion of oxalic acid in atmospheric water by photolysis of iron(III)-oxalato complexes. *Environ. Sci. Technol.* **26**, 1014–1022.

Zuo, Y., and J. Hoigné (1994) Photochemical decomposition of oxalic, glyoxylic and pyruvic acid catalyzed by iron in atmospheric waters. *Atmos. Environ.* **28**, 1231–1239.

INDEX

International Geophysics Series

EDITED BY

RENATA DMOWSKA
Division of Applied Sciences
Harvard University
Cambridge, Massachusetts

JAMES R. HOLTON
Department of Atmospheric Sciences
University of Washington
Seattle, Washington

H. THOMAS ROSSBY
Graduate School of Oceanography
University of Rhode Island
Narragansett, Rhode Island

* Out of print.

* Out of print.